CALCULUS
with Analytic Geometry

ARTHUR B. SIMON
California State University, Hayward

SCOTT, FORESMAN AND COMPANY
GLENVIEW, ILLINOIS
Dallas, Tex.   Oakland, N.J.   Palo Alto, Cal.   Tucker, Ga.   London, England

Library of Congress Cataloging in Publication Data

SIMON, ARTHUR B.
    Calculus with analytic geometry.

    Includes index.
    1. Calculus.   2. Geometry, Analytic.   I. Title.
QA303.S55493   515'.15   81-14552
ISBN   0-673-16044-0   AACR2

ISBN: 0-673-16044-0

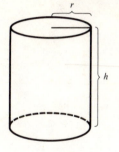

Right circular cylinder

Volume $= \pi r^2 h$

Lateral surface area $= 2\pi rh$

Total surface area $= 2\pi r(r + h)$

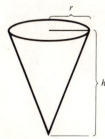

Right circular cone

Volume $= \dfrac{1}{3}\pi r^2 h$

Lateral surface area $= \pi r \sqrt{r^2 + h^2}$

Sphere   Volume $= \dfrac{4}{3}\pi r^3$

Surface area $= 4\pi r^2$

Equations of Lines

General form $ax + by + c = 0$, a and b not both zero

Slope-intercept form $y = mx + b$

Point-slope form $y - y_1 = m(x - x_1)$

Two-point form $y - y_1 = \dfrac{y_2 - y_1}{x_2 - x_1}(x - x_1)$

Dec 20
8:00

Equations of Conic Sections (centers or vertex at the origin)

Circle of radius r $x^2 + y^2 = r^2$

Parabola $x^2 = 4py$

Ellipse $\dfrac{x^2}{a^2} + \dfrac{y^2}{b^2} = 1$

Hyperbola $\dfrac{x^2}{a^2} - \dfrac{y^2}{b^2} = 1$

To my wife and partner
DOLORES

CONTENTS

*†The important properties of conic sections are discussed in Chapter 12 and also in Section 13–1. You have the option of studying the detailed presentation in Chapter 12 or the brief account in Section 13–1.

PREFACE

Now that calculus is a required course in such a wide range of disciplines, calculus classes are populated by students with diverse interests and levels of preparation. The modern text, therefore, must be accessible to a large majority of students and at the same time demonstrate the wide applicability of the subject. With this in mind, I have tried to write a calculus book that minimizes mathematical formalism and maximizes the variety of applications.

Each chapter is preceded by a short introduction that describes what new topics are introduced and explains why they are important. The introductions also contain advice to the student about studying the material. Most sections begin with an informal introduction followed by an intuitive discussion of the concepts. But elimination of all formality is neither possible nor desirable. Therefore, the final definitions and results are stated with as much precision as practical. In this way, correct statements can be learned from material that is readable and understandable. The results are usually accompanied by informal justifications; the formal proofs are presented as optional material either at the end of the section or at the end of the chapter. These proofs can be covered or not as time permits.

In addition to a complete coverage of the traditional topics of calculus, I would like to point out some unusual features of this text:

1. The function $f(x) = e^x$ and the trigonometric functions are introduced informally in Chapter 3 (Derivatives), after a thorough review of exponents and trigonometry in Chapter 1 (Introduction). Not only do these functions make the chain rule more meaningful, but they provide a wealth of interesting examples and motivating applications early in the course. The traditional (usual) development of these transcendental functions along with the natural logarithm is contained in Chapters 7 and 8.

2. Differential equations are introduced in Section 4–7 and are woven into the material throughout the rest of the book. They are used in Section 4–8 to enrich the discussion of velocity and acceleration, in Section 5–7 to compute the escape velocity of a rocket, in Chapter 7 to motivate the definition of the natural logarithm, in Section 8–3 to discuss simple harmonic motion, and so on. Finally, Chapter 18 is devoted entirely to differential equations and their applications.

3. There are six case studies marked *Optional Reading;* they can be read and discussed as time and desire permit. Each one presents a complete discussion and solution of a practical problem using the material just studied.

Section 8–5 Mercator and $\int \sec x \, dx$ (an application of the integral of the secant to map making)

Section 9–7 The Logistic Model of Population Growth (an application of partial fraction integration)

Section 10–4 The Predator-Prey Model (an application of L'Hôpital's rule and improper integrals)

Section 11–9 Infinite Series and Music Synthesizers (an application of series to producing musical sounds)

Section 13–6 Brachistochrones and Tautochrones (an application of parametric equations)

Section 15–10 An Application (of Lagrange Multipliers) to Economics

4. A section on Probability is included in Chapter 6 (Applications of the Definite Integral). Only the simplest concepts of probability are necessary to provide many interesting examples and applications. In later chapters, these same concepts give rise to practical applications of improper integrals and infinite series.

5. Calculator use is encouraged. After repeated warnings and an explicit example demonstrating how round-off procedures can produce gross errors, calculators are used to advantage in the following ways: (*a*) to visualize the function concept (function machine), (*b*) to illustrate limits such as $(1 + h)^{1/h}$ as h approaches 0, and (*c*) to ease the burden of tedious computations, for example, in numerical integration (Simpson's rule) and in estimating the sum of an infinite series.

6. Chapter 1 contains a thorough review of precalculus material. Though the trend in recent years is to present only a brief review in order to introduce the derivative as soon as possible, experience shows that students have no difficulty at all in learning how to differentiate and integrate. The real difficulties in calculus are more basic: (*a*) translating a problem into functional notation, (*b*) simplifying algebraic expressions, and (*c*) solving equations and inequalities. Therefore, Chapter 1 addresses these trouble spots in some detail. The topics reviewed, along with examples of each, are prominently displayed at the beginning of each section. Thus, at a glance, the instructor or individual student can decide to cover as much or as little of the material as seems appropriate.

7. Wherever it is appropriate, the exercises are grouped into four clearly marked categories. The first group contains the drill exercises, the second contains applications in the form of word problems, and exercises in the third group require the student to verify or prove something. Each category is graded in difficulty. The fourth group contains optional exercises that are apt to be more of a challenge than the others. Answers to underlined exercises appear in the back of the book.

There is ample material for the usual three-semester or four-quarter calculus course. The instructor will find plenty of room for any convenient approach. The formal definition of a limit and the formal proofs of most statements are presented as optional material, which can be covered at any desirable time or omitted altogether. There is also plenty of opportunity to rearrange the material. For example, Section 1–9, on Inverse Functions, can be delayed until Chapter 7 where its use is essential. The material on the exponential e^x and trigonometry need not be taken up until Chapters 7 and 8. Chapter 11, on Infinite Series, can be studied any time after Chapter 10. Another point of flexibility involves the conic sections. Some students have already studied conic sections and others have not. To accommodate both groups, Chapter 12 contains a detailed account of the conics whereas Section 13–1 contains only a brief review of their important properties. Either presentation is sufficient preparation for later work.

SUPPLEMENTAL MATERIALS

A student manual, containing worked-out solutions to many of the underlined exercises, is available through your book store. This manual was prepared by Professor Edward L. Keller, *California State University, Hayward.*

A teacher's manual, containing the answers not provided in the book, is available on request.

ACKNOWLEDGEMENTS

Reviewers I would like especially to thank Professors Anthony L. Peressini, *University of Illinois,* and David A. Smith, *Duke University,* for their careful reviews of each and every page of my manuscript and their many helpful suggestions for a better presentation. I am also grateful to the following professors who reviewed portions of the manuscript: John G. Bergman, *University of Delaware;* Garret J. Etgen, *University of Houston,* William Fox, *State University of New York-Stony Brook,* Frank Gilfeather, *University of Nebraska-Lincoln,* Joseph F. Krebs, *Boston College,* Russell J. Rowlett, *University of Tennessee,* William L. Siegmann, *Rensselaer Polytechnic Institute,* and Alberto Torchinsky, *Indiana University.*
Computer graphics The computer graphics were prepared by Professor Christopher L. Morgan, *California State University, Hayward.*
Exercises The answers to the exercises have been checked (and rechecked) by the following students: Chris Delp, Fai Ho, Octavius Lam, Wayne McClish, and Brad Swanson.
Typing I would like to thank my wife, Dolores, who not only did all of the typing, but also made many suggestions for improvements in the visual illustrations.
Editing I am indebted to everyone at Scott, Foresman who helped put this book together, but especially to my editor, Jack Pritchard. I have never seen an editor work harder or be of more assistance to a struggling author. I am extremely grateful and hope to have the opportunity to work with him again.

Arthur B. Simon
Moss Beach, California

INTRODUCTION TO CALCULUS

It was the seventeenth century, the era that gave birth to exciting new discoveries about molecular and gravitational forces, heat, energy, motion, electricity, and optics. But it was soon apparent that the language of classical mathematics—geometry, algebra, and trigonometry—was inadequate to describe these emerging theories about the laws of nature. There was a need for a new and richer methodology; one that combined the elementary notions of classical mathematics with the powerful concept of a limiting process. The result was *calculus*.

The compelling need for calculus was underscored by its parallel, but independent, development in England by Sir Isaac Newton (1642–1727) and in Germany by Gottfried Leibniz (1646–1716).

The student coming from geometry, algebra, and trigonometry to calculus must make the same leap as Newton and Leibniz. But it is a good idea, first, to be sure of the ground from which the leap is made. Therefore, Chapter 1 reviews the basic subjects of precalculus mathematics: the real numbers, the coordinate plane, functions, graphs, exponents, logarithms, and trigonometry. But each of these familiar topics is treated with an eye toward its use in calculus. We discuss, for example, how to express information from such fields as biology, business, and engineering in functional form (Section 1–4), and how to interpret the slope of a line as the average rate of change of a function (Section 1–5). This material will prepare you for our later discussions of derivatives and integrals, the main tools of calculus. The pace is brisk, but each section contains numerous examples and exercises to help you make the leap from geometry, algebra, and trigonometry to calculus.

1–1 THE REAL LINE

In this section, we review the following topics:

TOPIC	EXAMPLE
The real line	![real line from -2 to 2] $-2\ -1\ \ 0\ \ 1\ \ 2$
Order	$a < b < c$ / $a\ \ \ b\ \ \ c$
Intervals	(a,b) ⟨———⟩ $a\ \ b$: $[a, b]$ [———] $a\ \ b$
Solving inequalities	Find all x with $6 - 3x > 0$
Absolute value	$\|x\| = \begin{cases} x & \text{if } x \geq 0 \\ -x & \text{if } x < 0, \end{cases}$ $\ \|3\| = 3, \|-4\| = 4$
Distance	$\|x - y\|$ is the distance between x and y

The Real Line

The numbers that we use in the study of calculus are the real numbers. They consist of the *integers* (or whole numbers), the *rational numbers* (or fractions), and the *irrational numbers* (numbers, such as $\sqrt{2}$ and π, that cannot be written as fractions).

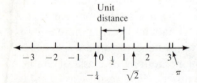

FIGURE 1–1. A portion of the real line.

Real numbers can be represented as points on a line. We draw a horizontal line, label some point 0, and to the right of 0, label another point 1. This establishes a unit distance on the line. Once this is done, then each point on the line corresponds to a unique real number; positive numbers to the right of 0 and negative numbers to the left (Figure 1–1). Conversely, each real number corresponds to a unique point; the number is called the **coordinate** of the point and the line itself is called a **coordinate line** or **real line.** Because of this correspondence, we often use the terms *number* and *point* interchangeably as a sort of shorthand. For example, we say the "point 5" instead of "the point corresponding to the number 5," or we say "the distance between -3 and 4" instead of "the distance between the points corresponding to the numbers -3 and 4."

The size of any two real numbers can be compared. Given real numbers a and b, then $a < b$ or $a = b$ or $a > b$. If a is positive we write $a > 0$, and if a is negative we write $a < 0$. The inequalities $a \leq b$ or $b \geq a$ denote that a is less than or equal to b. A string of inequalities such as $a < b < c$ means that a is less than b *and* b is less than c. Similar meanings may be attached to inequalities such as $a < b \leq c$ and $a \leq b < c$. On the coordinate line, the interpretation of $a < b$ is that b lies to the right of a. The inequalities $a < b < c$ mean that, on the line, b lies between a and c.

Now we define certain sets of real numbers called **intervals;** they are referred to frequently throughout this book. Intervals are divided into nine categories. In the definitions below, we assume that $a < b$.

INTERVAL	SYMBOL	DEFINITION	PICTURE
Open	(a, b)	all x with $a < x < b$	$\underset{a \quad b}{(\!\!-\!\!-\!\!)}$
Closed	$[a, b]$	all x with $a \le x \le b$	$\underset{a \quad b}{[\!\!-\!\!-\!\!]}$
Half-open	$(a, b]$	all x with $a < x \le b$	$\underset{a \quad b}{(\!\!-\!\!-\!\!]}$
Half-open	$[a, b)$	all x with $a \le x < b$	$\underset{a \quad b}{[\!\!-\!\!-\!\!)}$
Open	(a, ∞)	all x with $a < x$	$\underset{a}{(\!\!-\!\!\!\longrightarrow}$
Closed	$[a, \infty)$	all x with $a \le x$	$\underset{a}{[\!\!-\!\!\!\longrightarrow}$
Open	$(-\infty, b)$	all x with $x < b$	$\underset{b}{\longleftarrow\!\!-\!\!)}$
Closed	$(-\infty, b]$	all x with $x \le b$	$\underset{b}{\longleftarrow\!\!-\!\!]}$
Open	$(-\infty, \infty)$	all x	$\longleftarrow\!\!\!\longrightarrow$

The real numbers a and b are called **endpoints** of the interval (NOTE: ∞ is *not* a real number). The points between a and b are called **interior points.** Observe that open intervals do not include their endpoints, but closed intervals do. Intervals, such as (a, b) or $[a, b)$, with finite endpoints are called **bounded intervals;** intervals, such as (a, ∞) or $(-\infty, b]$, that involve ∞ are called **unbounded intervals.**

EXAMPLE 1   The following bounded intervals are sketched in Figure 1–2A. (a) The open interval $(-1, 2)$. (b) The closed interval $[1, 3]$. (c) The half-open interval $(-2, 2]$.   ∎

EXAMPLE 2   The following unbounded intervals are sketched in Figure 1–2B. (a) The open interval $(1, \infty)$. (b) The closed interval $(-\infty, 2]$.   ∎

Inequalities

In previous courses, you dealt with expressions involving a variable x. In calculus, it is important to know what values of x make the expression positive,

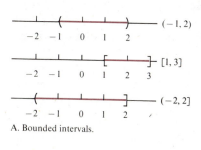

A. Bounded intervals.

B. Unbounded intervals.

FIGURE 1–2. Examples 1 and 2.

negative, or zero. To solve problems like this, we use the following ordering properties of real numbers:

ORDERING PROPERTIES
For any three real numbers a, b, and c.
(1.1) If $a < b$, then $a + c < b + c$ and $a - c < b - c$
(1.2) If $a < b$ and $c > 0$, then $ac < bc$ and $\dfrac{a}{c} < \dfrac{b}{c}$
(1.3) If $a < b$ and $c < 0$, then $ac > bc$ and $\dfrac{a}{c} > \dfrac{b}{c}$

The first property says that any number can be added to or subtracted from both sides of an inequality just as it can for an equality. The second says that the same is true when both sides are multiplied or divided by a positive number c. But the third property says that if c is *negative,* then multiplying or dividing by c *reverses* the inequality.

EXAMPLE 3   Determine the values of x for which the expression $6 - 3x$ is positive, negative, and zero.

Solution   The expression $6 - 3x$ is zero when $x = 2$. To determine where it is positive, we write $6 - 3x > 0$ and solve for x just as though we were solving an equation. The only thing to watch out for is the third ordering property (1.3).

$$6 - 3x > 0$$
$$-3x > -6 \qquad \text{(Subtract 6)}$$
$$x < 2 \qquad \text{(Divide by } -3)$$

Notice that the inequality is *reversed* if both sides are divided by the same *negative* number. To determine where the expression is negative, we set $6 - 3x < 0$, proceed as above, and find that $x > 2$. Therefore, the complete solution, written in terms of intervals, is as follows:

$$6 - 3x \begin{cases} > 0 \text{ for } x \text{ in } (-\infty, 2) \\ = 0 \text{ for } x = 2 \\ < 0 \text{ for } x \text{ in } (2, \infty) \end{cases}$$

This solution is sketched in Figure 1–3 where the sign of $6 - 3x$ is indicated by the symbols $+$ and $-$ ■

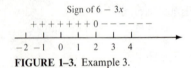

Sign of $6 - 3x$

FIGURE 1–3. Example 3.

If the expression is a quadratic, say $x^2 + x - 12$, then we factor it first

$$x^2 + x - 12 = (x - 3)(x + 4)$$

and treat each factor separately. Using the method of Example 3, we determine where $x - 3$ and $x + 4$ are positive, negative, and zero. An excellent way to keep

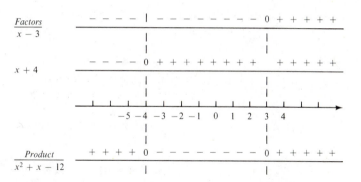

FIGURE 1–4. Example 4.

track of each factor is to draw a schematic diagram like the one in Figure 1–4. Then the product is positive where both factors have the same sign and negative where they have opposite signs.

EXAMPLE 4   Determine where $x^2 + x - 12$ is positive, negative, and zero.

Solution   The expression factors into $(x - 3)(x + 4)$. Working with each factor separately, we have

$$x - 3 \begin{cases} > 0 \text{ for } x \text{ in } (3, \infty) \\ = 0 \text{ for } x = 3 \\ < 0 \text{ for } x \text{ in } (-\infty, 3) \end{cases} \qquad x + 4 \begin{cases} > 0 \text{ for } x \text{ in } (-4, \infty) \\ = 0 \text{ for } x = -4 \\ < 0 \text{ for } x \text{ in } (-\infty, -4) \end{cases}$$

Now we enter this information in a diagram (Figure 1–4) and read off the answer:

$$x^2 + x - 12 \begin{cases} > 0 \text{ for } x \text{ in } (3, \infty) \text{ or } (-\infty, -4) \\ = 0 \text{ for } x = 3 \text{ or } -4 \\ < 0 \text{ for } x \text{ in } (-4, 3) \end{cases} \quad \blacksquare$$

Remarks:

(1) If the expression has three (or more) factors, then draw a diagram with one line for each factor.

(2) If the expression is a quotient, then treat the numerator and denominator separately. Draw a diagram showing the signs of each, and then determine the sign of the quotient. A word of caution: quotients are not defined at values of x that make the denominator zero.

Absolute Value and Distance

The absolute value of a number x is denoted by the symbol $|x|$.

(1.4)

> **DEFINITION**
> The **absolute value** of a number x is
> $$|x| = \begin{cases} x \text{ if } x \geq 0 \\ -x \text{ if } x < 0 \end{cases}$$

EXAMPLE 5

(a) $|3| = 3$ because $3 \geq 0$

(b) $|-4| = -(-4) = 4$ because $-4 < 0$

(c) If $x = 2$, then $|-3x + 10| = |-3 \cdot 2 + 10| = |4| = 4$

(d) If $x = 5$, then $|-3x + 10| = |-3 \cdot 5 + 10| = |-5| = 5$ ■

Several useful properties of absolute value are listed below.

For all real numbers x and y,

(1.5) $|x| \geq 0$ (1.6) $|-x| = |x|$

(1.7) $|xy| = |x||y|$ (1.8) $|x/y| = |x|/|y|$ $y \neq 0$

(1.9)   Triangle Inequality: $|x + y| \leq |x| + |y|$

(1.10) $\sqrt{x^2} = |x|$ (1.11) $|x| < a$ if and only if
                                    $-a < x < a$*

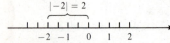

FIGURE 1–5. $|x| =$ distance between x and 0.

Geometrically, the absolute value of a number is the distance between the number and zero. For example, $|-2|$ is the distance between -2 and 0 (Figure 1–5). Moreover, absolute value is used to define the distance between *any* two numbers.

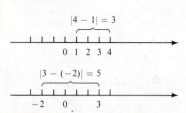

FIGURE 1–6. $|x - y| =$ distance between x and y.

(1.12)   **DEFINITION OF DISTANCE**

If x and y are real numbers, then the distance between x and y is $|x - y|$ (Figure 1–6)

The order in which x and y appear is not important because, by (1.6),

$$|x - y| = |-(x - y)| = |y - x|$$

The *minus sign*, however, is important.

EXAMPLE 6   Express the given information in terms of absolute value.

(a) The distance between x and 5 is 3. *Answer:*

$$|x - 5| = 3$$

---

*The phrase "if and only if" is used extensively in mathematics. Property (1.11) above says that $|x| < a$ if and only if $-a < x < a$. What this means is that the statement

if $|x| < a$, then $-a < x < a$

and its converse

if $-a < x < a$, then $|x| < a$

are *both* true.

(b) The distance between t and -4 is less than 2. *Answer:*

$$|t - (-4)| < 2 \quad \text{or} \quad |t + 4| < 2$$

Notice that $|t - 4|$ is the distance between t and 4 whereas $|t + 4| = |t - (-4)|$ is the distance between t and -4. ∎

EXAMPLE 7 Express the given information in terms of distance.

(a) $|x + 2| \geq 4$. *Answer:* the distance between x and -2 is greater than or equal to 4.

(b) $|5 - y| > 1$. *Answer:* the distance between y and 5 is greater than 1. ∎

EXAMPLE 8 Sketch the set of all x that satisfy

$$|3x + 1| \leq 2$$

Solution We use property (1.11) to write

$$|3x + 1| \leq 2 \quad \text{if and only if} \quad -2 \leq 3x + 1 \leq 2$$

It follows from properties (1.1)–(1.3) that

$$-2 \leq 3x + 1 \leq 2$$

$$-3 \leq 3x \leq 1 \qquad \text{(Subtract 1)}$$

$$-1 \leq x \leq \frac{1}{3} \qquad \text{(Divide by 3)}$$

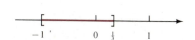

FIGURE 1–7. Example 8.

Therefore $|3x + 1| \leq 2$ if and only if x is in $[-1, 1/3]$ (Figure 1–7). ∎

EXERCISES

In Exercises 1–10, sketch the indicated intervals.

1. $(0, 3]$

2. $[-2, 0]$

3. $(-\infty, 2)$

4. $(-1, \infty)$

5. $(-\infty, -1)$

6. $(-\infty, 5)$

7. $[-1, 2)$

8. $(4, 5)$

9. $[-2, 3)$

10. $(-4, 1]$

In Exercises 11–32, determine where the given expression is positive, negative, and zero. Express the answer in terms of intervals.

11. $2x - 4$

12. $4x + 16$

13. $5 - 6x$

14. $3 - 7x$

15. $x(3 - x)$

16. $x(x + 1)$

17. $x^2 + 3x + 2$

18. $x^2 - 3x - 4$

19. $-3x^2 - 5x + 2$

20. $2x^2 + x - 1$

21. $x(x + 1)(2 - x)$

22. $x(x - 1)(x + 5)$

23. $x^3 + 2x^2 + x$

24. $x^3 - 6x^2 + 5x$

25. $x^4 - 16$

26. $x^4 - 1$

27. $\dfrac{x}{x + 2}$

28. $\dfrac{2x}{x - 1}$

29. $\dfrac{4 - x}{2x - 6}$

30. $\dfrac{2x - 1}{3x + 2}$

31. $x + \dfrac{1}{x - 2}$

32. $x + \dfrac{1}{x}$

(*Hint:* Combine fractions.)

Write the following numbers without absolute value signs.

33. $|5|$

34. $|-3|$

35. $|4 - 8|$

36. $|-2| - |6|$

37. $|-2| + |-6|$

38. $|4^2|$

39. $|x|$

40. $|-y|$

Express the given information in terms of absolute value.

41. The distance between x and 4 is $1/10$.
42. The distance between t and -1 is greater than 2.
43. The distance between $3u$ and -2 is less than $.01$.
44. The distance between $-2y$ and 7 is less than $.001$.

Express the given inequalities in terms of distance.

45. $|x + 3| \le 1$
46. $\left|y - \dfrac{1}{2}\right| > 2$
47. $|2t - 1| < .01$
48. $|-4x + 3| < .001$

In Exercises 49–56, find all x that satisfy the given inequality.

49. $|x + 3| < .5$
50. $|x - 2| < 1$
51. $|3x - 5| < 4$
52. $|2x + 1| \le 2$
53. $|2 - 4x| \le 1$
54. $|8 - 5x| < 1$
55. $\left|\dfrac{x - 3}{-2}\right| < .25$
56. $\left|\dfrac{x + 6}{3}\right| < 1/3$

Optional Exercises

57. Determine where $x^2 + 3x - 1$ is positive, negative, and zero.
58. Same as Exercise 57 for $x^2 + 3x + 3$.
59. Find all x that satisfy the inequality (a) $|x^2 - 4| < 2$ (b) $|x^2 - 9| < 1$.
60. Find all x that satisfy $|x^2 + 4| < 2$.
61. If x is in the closed interval $[-3, 6]$, what are the possible values of (a) $3x$ (b) $-2x$ (c) x^2?
62. If x is in the open interval $(0,1)$, what are the possible values of $1/x$?
63. If $|x + 1| \le 1$, what is the maximum possible distance between (a) $4x + 7$ and 3 (b) $2 - 3x$ and 5?
64. If $|x - 3| \le 1$, what is the maximum possible distance between x^2 and 9?

1–2   THE COORDINATE PLANE

In this section, we review the following topics:

TOPIC	EXAMPLE
The coordinate plane	
Distance formula	$\sqrt{(x_2 - x_1)^2 + (y_2 - y_1)^2}$
Graphs: lines, parabolas, and circles	$y = mx + b$ $y = x^2$ $x^2 + y^2 = 1$
Quadratic formula	$x = \dfrac{-b \pm \sqrt{b^2 - 4ac}}{2a}$

A plane in which each point is identified with an ordered pair (x, y) of real numbers* is called a **coordinate plane.** We construct a coordinate plane by drawing two perpendicular coordinate lines that meet at the 0 points of each line (Figure 1–8). These lines are called the **coordinate axes** and where they meet is called the **origin.** Usually, one axis is horizontal with its positive direction to the right; the other is vertical with its positive direction upward. We label the axes with letters; usually, the horizontal axis is called the *x*-**axis** and the vertical axis is called the *y*-**axis;** then the plane is called the *xy*-**plane** (if other letters are used, the names change accordingly). If P is any point in the plane, then vertical and horizontal lines through P determine an ordered pair of numbers (a, b) with a indicating the distance along the horizontal axis and b indicating the distance along the vertical axis (Figure 1–8). These numbers are called the **coordinates** of the point P. Conversely, given an ordered pair of numbers (a, b), they determine a unique point P with those coordinates. To *plot* a point (a, b) means to locate and mark that point in the coordinate plane as is illustrated in Figure 1–8B. The coordinate axes divide the plane into four **quadrants,** labeled counterclockwise I, II, III, and IV in Figure 1–8A.

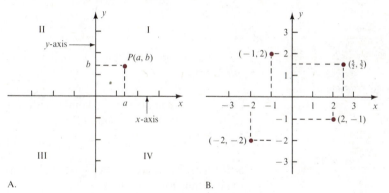

FIGURE 1–8. The coordinate plane. A. B.

Distance

Given any two points $P(x_1, y_1)$ and $Q(x_2, y_2)$ in a plane (we mean, of course, a coordinate plane, but most of the time we just say *plane*), we plot the points and construct the right triangle shown in Figure 1–9. If we use the symbol $d(P, Q)$ to denote the distance between P and Q, then the Pythogorean Theorem yields

$$d(P,Q) = \sqrt{|x_2 - x_1|^2 + |y_2 - y_1|^2}$$

---

*A pair of numbers x and y is an **ordered pair** if the order in which they appear is significant. The symbol (x, y) is used to indicate that x is the first number and y is the second. Although we have previously used the symbol (,) to denote an open interval, there is no danger of confusion because the context will always make clear which interpretation is intended.

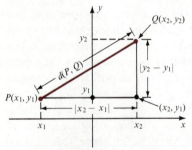

FIGURE 1–9. The distance between points $P(x_1, y_1)$ and $Q(x_2, y_2)$ is $d(P,Q) = \sqrt{(x_2 - x_1)^2 + (y_2 - y_1)^2}$.

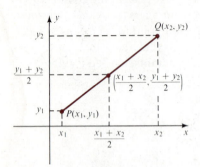

FIGURE 1–10. The midpoint formula.

But it follows from property (1.7) in Section 1–1 that $|x_2 - x_1|^2 = (x_2 - x_1)^2$ and the same is true for the y's. Therefore, we dispense with the absolute value signs and have

(1.13)

> **DISTANCE FORMULA**
> The distance between $P(x_1, y_1)$ and $Q(x_2, y_2)$ is
> $$d(P,Q) = \sqrt{(x_2 - x_1)^2 + (y_2 - y_1)^2}$$

Here is another useful formula:

(1.14)

> **MIDPOINT FORMULA**
> The midpoint of the line segment joining (x_1, y_1) and (x_2, y_2) has coordinates
> $$\left(\frac{x_1 + x_2}{2}, \frac{y_1 + y_2}{2} \right)$$

This is illustrated in Figure 1–10. The next example is an application of these two formulas.

EXAMPLE 1 Sketch the triangle with vertices $A(1, 1)$, $B(2, 4)$, and $C(4, 0)$ where the units are in inches. Show that it is an isosceles triangle (two sides of equal length) and find its area.

Solution The sketch is in Figure 1–11A. We find the lengths of the sides using the distance formula (1.13).

$$d(A, B) = \sqrt{(2 - 1)^2 + (4 - 1)^2} = \sqrt{10} \text{ inches}$$
$$d(B, C) = \sqrt{(4 - 2)^2 + (0 - 4)^2} = \sqrt{20} \text{ inches}$$
$$d(C, A) = \sqrt{(1 - 4)^2 + (1 - 0)^2} = \sqrt{10} \text{ inches}$$

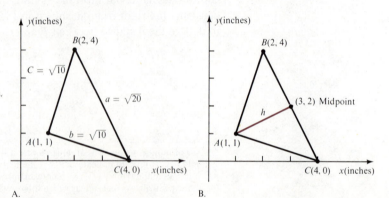

FIGURE 1–11. Example 1.

The sides labeled c and b in the figure are equal so the triangle is isosceles. To find the area, we use side $a = \sqrt{20}$ as the base and, because the triangle is isosceles, the height h will be the length of the line segment from A to the midpoint of side a (see Figure 1–11B). It follows from (1.14) that the midpoint is $((2 + 4)/2, (4 + 0)/2) = (3, 2)$. Now we use the distance formula again to find h.

$$h = \sqrt{(3 - 1)^2 + (2 - 1)^2} = \sqrt{5}$$

and the area is $(1/2)\sqrt{20}\sqrt{5} = (1/2)\sqrt{100} = 5$ square inches. ∎

Graphs: Lines, Parabolas, and Circles

One reason the coordinate plane is so important is that we can use it to draw pictures of equations with two variables. The pictures are called *graphs.*

(1.15)
> **DEFINITION**
> The **graph** of an equation in two variables x and y is the set of all points (x, y) in the plane whose coordinates satisfy the equation.

To *graph* an equation means to draw its graph in a coordinate plane. The procedure is as follows:

(1.16)
    (1) Mentally picture the general shape
    (2) If necessary, plot a few points
    (3) Sketch the graph

The examples below illustrate these three steps.

Lines: Any equation

(1.17) $y = mx + b$

where m and b are constants is called a **linear equation** in two variables. From your precalculus courses, you know that its graph is a *straight line.* Therefore, to graph a linear equation, simply locate any two points on the graph and draw the straight line through them. Two convenient points are the **x-intercept** and the **y-intercept;** that is, the points where the graph crosses the x- and y-axes. To find the x-intercept, we set $y = 0$ and solve for x; to find the y-intercept, set $x = 0$ and solve for y. Finding the intercepts is an aid in graphing *any* equation.

EXAMPLE 2   Graph the linear equation $y = 2x - 1$.

Solution   We follow the steps outlined above

(1) Mental picture: straight line.
(2) Plot two points: we will find the intercepts and keep track of them by making a *table of values.*

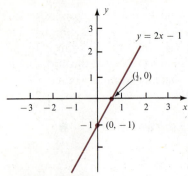

FIGURE 1–12. Example 2.

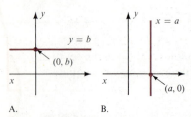

A. B.

FIGURE 1–13. Graphs of $y = b$ and $x = a$.

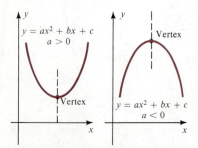

FIGURE 1–14. Parabolas have this general shape.

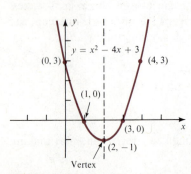

FIGURE 1–15. Example 3.

x	0	1/2
$y = 2x - 1$	-1	0

(3) Draw the line through these points (Figure 1–12). ∎

If $m = 0$ in equation (1.17), then $y = b$. The graph is a horizontal line through the point $(0, b)$; see Figure 1–13A. If a is some constant, then the graph of $x = a$ is a vertical line through the point $(a, 0)$; see Figure 1–13B.

Parabolas: An equation

$$y = ax^2 + bx + c \quad a \neq 0$$

where a, b, and c are constants is called a **quadratic equation** in two variables. The graph is always a *parabola* (Figure 1–14). The parabola opens upward if $a > 0$, but opens downward if $a < 0$. The *vertex* is the lowest or highest point on the graph depending on whether the parabola opens upward or downward. Parabolas are *symmetric* (that is, a mirror image of themselves) about the vertical line through the vertex. Most precalculus books derive the following formula for the x-coordinate of the vertex:

(1.18) | x-coordinate of the vertex is $-\dfrac{b}{2a}$

The easiest way to graph a parabola is to notice if it opens upward or downward, locate the vertex, and plot a point or two on either side of the vertex.

EXAMPLE 3 Graph the equation $y = x^2 - 4x + 3$.

Solution

(1) Mental picture: quadratic equation with $a = 1$, $b = -4$, and $c = 3$; the graph is a parabola that opens upward.

(2) Plot some points: by (1.18), the x-coordinate of the vertex is

$$-\frac{b}{2a} = -\frac{-4}{2(1)} = 2$$

When $x = 2$, $y = 2^2 - 4(2) + 3 = -1$; so $(2, -1)$ is the vertex. We now make a table of values to locate additional points.

x	0	1	2	3	4
$y = x^2 - 4x + 3$	3	0	-1	0	3

(3) Draw the parabola through these points (Figure 1–15). ∎

The x-intercepts of the graph of $y = ax^2 + bx + c$ are solutions of the equation

(1.19) $0 = ax^2 + bx + c$

The solutions (if there are any) can always be found using the **quadratic formula**

(1.20) $$x = \frac{-b \pm \sqrt{b^2 - 4ac}}{2a}$$

The expression $b^2 - 4ac$ is called the *discriminant;* there are two, one, or no solutions to (1.19), depending on whether the discriminant is positive, zero, or negative.

EXAMPLE 4   Graph the equation $y = -x^2 + 2x + 1$.

Solution   This is a quadratic equation with $a = -1, b = 2$, and $c = 1$. Its graph is a parabola that opens downward and its vertex is $(1, 2)$; check this out. The y-intercept is at $(0, 1)$ and the x-intercepts are the solutions of

$$0 = -x^2 + 2x + 1$$

By the quadratic formula,

$$x = \frac{-2 \pm \sqrt{2^2 - 4(-1)(1)}}{2(-1)} = \frac{-2 \pm \sqrt{8}}{-2}$$

$$= \frac{-2 \pm 2\sqrt{2}}{-2}$$

$$= 1 \pm \sqrt{2} \approx 2.4 \quad \text{or} \quad -.4$$

The graph is sketched in Figure 1–16.   ∎

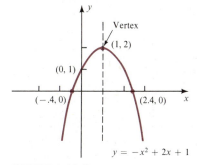

FIGURE 1–16. Example 4.

Circles: An equation

(1.21) $x^2 + y^2 + ax + by + c = 0$

is also a quadratic equation and, except for certain cases described below, its graph is a *circle.*

A circle with center $C(h, k)$ and radius r is the set of all points in the plane that are at a distance r from C. It follows that a point $P(x, y)$ is on the circle if and only if $d(P, C) = r$; that is,

$$\sqrt{(x - h)^2 + (y - k)^2} = r \tag{1.13}$$

Since both sides are positive, this equation is equivalent to

(1.22) $$(x - h)^2 + (y - k)^2 = r^2$$

Equation (1.22) is called the *standard equation* of a circle with center at (h, k) and radius r.

EXAMPLE 5 Find the standard equation of the circle (a) with center $(-1, 2)$ and radius 3 (b) with center $(2, -3)$ and passing through the point $(1, 1)$. Graph the circles.

Solution

(a) The center is $(-1, 2)$ so $h = -1$ and $k = 2$; the radius is $r = 3$. The standard equation is

$$(x + 1)^2 + (y - 2)^2 = 9$$

The circle is sketched in Figure 1–17A.

(b) The center is $(2, -3)$; the radius is the distance between the center and the point $(1, 1)$. Thus, $r = \sqrt{(2 - 1)^2 + (-3 - 1)^2} = \sqrt{17}$ and the standard equation is

$$(x - 2)^2 + (y + 3)^2 = 17$$

The circle is sketched in Figure 1–17B. ■

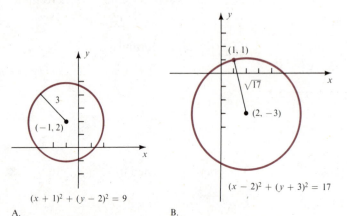

$(x + 1)^2 + (y - 2)^2 = 9$

$(x - 2)^2 + (y + 3)^2 = 17$

FIGURE 1-17. The circles in Example 5.   A.   B.

Let us return now to the quadratic equation (1.21)

$$x^2 + y^2 + ax + by + c = 0$$

By the method of *completing the squares* (we remind you that $x^2 + ax$ can be made into a square by adding $(a/2)^2 = a^2/4$; see Example 6 below), equation (1.21) can be rewritten as

(1.23) $(x - h)^2 + (y - k)^2 = d$

If $d < 0$, then there is no graph (why?). If $d = 0$, then the graph is the single point (h, k). But if $d > 0$, then (1.23) is the standard equation of a circle with center (h, k) and radius $\sqrt{d}$.

EXAMPLE 6 Complete the squares and write the equations (a) $x^2 + y^2 + 2x - 4y + 4 = 0$ and (b) $x^2 - 3x + y^2 + y + 3 = 0$ in the form (1.23). Then describe their graphs.

Solution

(a) To complete the squares, we write

$$(x^2 + 2x \quad) + (y^2 - 4y \quad) = -4$$
$$(x^2 + 2x + 1) + (y^2 - 4y + 4) = -4 + 1 + 4$$
$$(x + 1)^2 + (y - 2)^2 = 1$$

This is the standard equation with $h = -1$, $k = 2$, and $r = 1$. Therefore, the graph is a circle with center $(-1, 2)$ and radius 1.

(b)
$$(x^2 - 3x \quad) + (y^2 + y \quad) = -3$$
$$\left(x^2 - 3x + \frac{9}{4}\right) + \left(y^2 + y + \frac{1}{4}\right) = -3 + \frac{9}{4} + \frac{1}{4}$$
$$\left(x - \frac{3}{2}\right)^2 + \left(y + \frac{1}{2}\right)^2 = -\frac{1}{2}$$

There is no graph for this equation because the sum of two squares cannot be negative. ∎

If the origin $(0, 0)$ is the center of a circle, then $h = 0$, $k = 0$, and the standard equation reduces to

(1.24) $\boxed{x^2 + y^2 = r^2}$

Summary for Graphing

Lines: $y = mx + b$, $y = b$, or $x = a$. Plot two points and draw the line through them.

Parabolas: $y = ax^2 + bx + c$. Opens upward if $a > 0$ or downward if $a < 0$. Locate the vertex, using (1.18), plot points on either side, and draw the parabola. The quadratic formula (1.20) locates the x-intercepts.

Circles: $x^2 + y^2 + ax + by + c = 0$. Complete the squares to rewrite the equation as $(x - h)^2 + (y - k)^2 = d$. If $d > 0$, the graph is a circle with center (h, k) and radius $\sqrt{d}$. If $d < 0$, there is no graph.

EXERCISES

In Exercises 1–6, plot the points, find the distance between them, and find the midpoint of the line segment joining them.

1. $(1, 3)$ and $(2, 5)$

2. $(-3, 1)$ and $(1, -2)$

3. $(2, -3)$ and $(-3, 1)$

4. $(1, -1)$ and $(-2, 2)$

5. $(0, 0)$ and $(3, 4)$

6. $(-2, 3)$ and $(0, 0)$

In Exercises 7–14, draw the lines and indicate their intercept(s).

7. $y = -1$

8. $y = 4$

9. $x = 2$

10. $x = -1$

11. $y = x$

12. $y = 2x$

13. $y = 1 - 2x$

14. $y = -x + 2$

In Exercises 15–22, draw the parabolas; indicate the vertex and the intercept(s).

15. $y = x^2$

16. $y = -x^2$

17. $y = -x^2 + 1$

18. $y = x^2 - 1$

19. $y = 2x^2 - 3x + 1$

20. $y = 1 + x - x^2$

21. $y = x^2 + 2x + 2$

22. $y = -1 + x - 3x^2$

In Exercises 23–28, draw the circles; indicate the center and radius.

23. $x^2 + y^2 = 4$

24. $x^2 + 2x + y^2 = 0$

25. $x^2 + y^2 + 2x - 2y = 0$

26. $x^2 + y^2 - 4y = 0$

27. $x^2 + y^2 - x + y = 1/2$

28. $x^2 + y^2 - 3x - y = 1/2$

In Exercises 29–44, graph the equations (if there is a graph).

29. $y = -3x$

30. $x^2 + y^2 + 4 = 0$

31. $y = 1 - x - x^2$

33. $y = 3$

33. $x^2 + 6x + y^2 + 10 = 0$

34. $y = x^2 + 1$

35. $x = -2$

36. $y = 2x - 1$

37. $x^2 - 4x + y^2 + 2y = 4$

38. $x = 2$

39. $y = 3x^2 - x$

40. $x^2 + 2x + y^2 - 4y = 0$

41. $y = 0$

42. $y = 2x^2 - 3x + 1$

43. $y = -x^2$

44. $x = 0$

In Exercises 45–52, find the standard equation of the described circle.

45. Center $(0, 0)$, radius 1

46. Center $(0, 0)$, radius 4

47. Center $(-1, 3)$, radius $\sqrt{2}$

48. Center $(2, -4)$, radius $\sqrt{3}$

49. Center $(1, -2)$ passing through $(2, 4)$

50. Center $(0, 4)$ passing through $(1, 1)$

51. A diameter is the line segment joining $(2, -2)$ and $(8, 6)$

52. A diameter is the line segment joining $(1, -3)$ and $(4, 1)$

53. Suppose that $(3, a)$, $(-4, b)$, and $(x + h, c)$ are on the graph of $y = x^2 - 2x + 3$. Find $a, b,$ and c.

54. Suppose that $(-2, a)$, $(2, b)$, and $(x + h, c)$ are on the graph of $y = -2x^2 + x - 1$. Find $a, b,$ and c.

55. The vertices of a triangle are $A(1, 1)$, $B(-2, 0)$, and $C((-1 + \sqrt{3})/2, (1 - 3\sqrt{3})/2)$. Is it an equilateral triangle (all sides of equal length)?

56. Sketch the triangle with vertices $A(-4, 4)$, $B(5, -5)$, and $C(7, 6)$. Is it equilateral? Compute its area. (Units are in feet.)

57. A **median** of a triangle is a line segment joining a vertex to the midpoint of the opposite side. Are all three medians of the equilateral triangle in Exercise 55 of equal length? What about the medians of the triangle in Exercise 56?

58. Sketch the parallelogram with vertices $A(-1, -1)$, $B(2, 5)$, $C(3, 2)$, and $D(0, -4)$. Show that the diagonals intersect at their midpoints. Is the same thing true for *any* parallelogram?

59. Prove the Midpoint Formula (1.14): Take arbitrary points $P(x_1, y_1)$ and $Q(x_2, y_2)$ and show that the formula locates a point S such that $d(P, S) = d(S, Q) = \frac{1}{2}d(P, Q)$.

1–3 FUNCTIONS

Functions play an essential role in calculus; practically every page of this book contains some reference to them. In this, the first of two sections devoted to functions, we cover the following topics:

TOPIC	EXAMPLE
Function; domain and range	$f(-1) = -2$ ... $f(0) = 2$ Range -2 -1 0 1 2 Domain -2 -1 0 1 2
Graphs of functions	y, $f(x) = x^3$, x
Equations as functions	$y = x^2$, $y = \sqrt{1 - x^2}$
Evaluating functions	If $f(x) = x^2 + 3x$, then $f(\sqrt{2}) = (\sqrt{2})^2 + 3(\sqrt{2})$

In the next section, we describe how functions are applied to problems in business, science, and engineering.

Let us begin with the definition of a function.

DEFINITION

(1.25) Let X be any set of real numbers. A **function** f **on** X is a rule that assigns to each x in X *exactly one* number $f(x)$, read "f of x", called the **value** of f at x. The set X is called the **domain** of f and the set of all values is called the **range** of f.

FIGURE 1–18. Picturing a function in terms of arrows. The value of f at -2 is $f(-2) = 1$; the value of f at 0 is $f(0) = -1$; the value of f at any x is $f(x)$.

You can visualize a function in terms of arrows that indicate the value of f at a point x (Figure 1–18). Or you can think of a function as a machine; a number x is put into the machine and $f(x)$ comes out. A hand calculator, in fact, can be such a machine. If your calculator has an x^2 key, then enter any number, say 2.32, push $\boxed{x^2}$, and out comes its square 5.3824 (Figure 1–19). This is the squaring function $f(x) = x^2$. If the domain is not specified, then it is *automatically* taken to be the largest set of numbers x for which x^2 is defined. Since every real number can be squared, we take the domain to be all real numbers; then its range is the set of nonnegative numbers $[0, \infty)$. If we desire to restrict the domain, say to all numbers greater than 2, then we write $f(x) = x^2$, $x > 2$ or $f(x) = x^2$, x in $(2, \infty)$; the range now is the interval $(4, \infty)$.

The $\sqrt{x}$ key is another example of a function machine. Enter a number, say 6.25, push $\boxed{\sqrt{x}}$, and out comes the square root 2.5. This is the square root function $f(x) = \sqrt{x}$. If the domain is not specified, then it is *automatically* taken to be the largest set of numbers x for which $\sqrt{x}$ is defined. Since only nonnegative numbers have square roots that are real numbers, we take the domain to be all nonnegative real numbers. *Your calculator is programmed for this domain.*

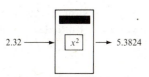

FIGURE 1–19. A calculator with an x^2 key is a function machine; this is the squaring function $f(x) = x^2$.

If you enter -4, for example, and push $\boxed{\sqrt{x}}$, you will get an error signal. The range, of course, is also the set of nonnegative numbers $[0, \infty)$.

Remark: The last digit of a number displayed by a calculator may not be correct because calculators are programmed to *round off*. Rounding off can produce gross errors, and we will give an example of this in the next chapter. A calculator with a 6, 8, or 10 digit display is correct only to 5, 7, or 9 digits, respectively. (Some calculators are programmed to carry extra digits in their memory, but others are not.) If you evaluate the irrational number $\sqrt{2}$ on an 8-digit calculator, the display reads 1.4142136. If that number is squared by hand, the result is not *exactly* 2. But the first seven digits are correct; we write $\sqrt{2} \approx 1.4142136$ to indicate that

$$1.41421355 \le \sqrt{2} < 1.41421365$$

On the other hand, the answer $\sqrt{6.25} = 2.5$ computed above is exact; if 2.5 is squared by hand, the result is exactly 6.25. ■

Most functions are not keys on a calculator. For instance, the function $f(x) = x$. This is called the *identity function;* its value at any x is x, itself. Also, the function $f(x) = 4$. This is an example of a *constant function;* its value at any x is the number 4. Here is another example:

EXAMPLE 1 Let $f(x) = 3x + 2$ for x in the interval $(1, 5)$. Find its domain, its range, and its value at $x = 2.5$.

Solution The domain is specified as $(1, 5)$; each x in $(1, 5)$ is assigned the number $3x + 2$. The range is easily seen to be the open interval from 5 (the value when $x = 1$) to 17 (the value when $x = 5$); thus, the range is $(5, 17)$. The value at 2.5 is

$$f(2.5) = 3(2.5) + 2 = 9.5 \quad ■$$

Functions are often expressed as equations. For instance, the squaring function can be written as $y = x^2$, the square root function can be written as $y = \sqrt{x}$, and the function in Example 1 can be written as $y = 3x + 2$, x in $(1, 5)$. When functions are written in this way, we say that *the equation defines y as a function of x*. The letter x denotes any number in the domain, and is called the **independent variable**. The letter y is called the **dependent variable**.

EXAMPLE 2 The equation $y = 1/(x^2 - 9)$ defines y as a function of x; x is the independent variable and y is the dependent variable. Find the domain, the range, and the value of y at $x = -2$.

Solution The domain is not specified so it is taken to be the largest set for which $1/(x^2 - 9)$ is defined; in this case, all real numbers $x \ne \pm 3$ (why?). The range is all real numbers $x \neq 0$ (why?). The value of y at $x = -2$ is

$$y = \frac{1}{(-2)^2 - 9} = -\frac{1}{5} \qquad \blacksquare$$

CAUTION: Not all equations define y as a function of x. For example, let $x^2 + y^2 = 4$. For some values of x, there are *two* possible values of y. If $x = 0$, then $y + \pm 2$, and if $x = 1$, then $y = \pm \sqrt{3}$. Thus, some numbers x are assigned two values, contrary to the definition of a function.

We now turn our attention to the graph of a function.

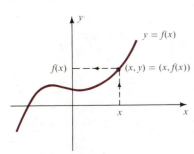

FIGURE 1–20. The graph of a function f; (x, y) is on the graph if and only if $y = f(x)$.

(1.26)

> **DEFINITION**
> Let f be a function on X. The *graph of f* is the set of all points (x, y) in the plane such that
>
> $$x \text{ is in } X \quad \text{and} \quad y = f(x)$$

To *graph* a function means to draw its graph in a coordinate plane. According to the definition,

(1.27)

$$(x, y) \text{ is on the graph if and only if } y = f(x)$$

It follows that the graph of f consists of all points

(1.28)

$$(x, f(x)) \quad \text{with } x \text{ in } X \qquad \text{(Figure 1–20)}$$

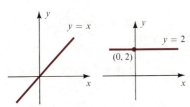

A. Graph of $f(x) = x$. B. Graph of $f(x) = 2$.
FIGURE 1–21. Graphs of functions.

In practice, we usually set $y = f(x)$ and draw the graph of that equation. For example, to graph the squaring function $f(x) = x^2$, we draw the graph of $y = x^2$ (a parabola); the graph of the identity function $f(x) = x$ is the (straight line) graph of $y = x$ (Figure 1–21A); the graph of the constant function $f(x) = 2$ is the (horizontal line) graph of $y = 2$ (Figure 1–21B).

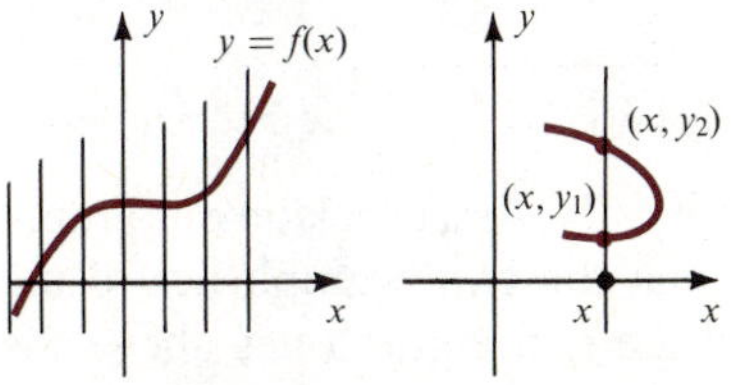

Remarks:

(1) If (x, y) is on the graph of f, then $y = f(x)$. If (x, y_1) and (x, y_2) are both on the graph of f, then $y_1 = f(x)$ and $y_2 = f(x)$. Since each x in the domain of f is assigned *exactly one* value, the conclusion is that $y_1 = y_2$. This observation has the following geometric interpretation (Figure 1–22):

A. Graph of a function. B. Not the graph of a function.

FIGURE 1–22. A vertical line can meet the graph of a function in *at most* one point.

(1.29)

$$A \text{ vertical line can meet the graph of a function in } at \ most \text{ one point.}$$

(2) The graphs of some functions are impossible to draw. For instance, the graph of

$$f(x) = \begin{cases} 1 & \text{if } x \text{ is rational} \\ 0 & \text{if } x \text{ is irrational} \end{cases}$$

is the collection of points one unit above the x-axis at each rational number and on the x-axis at each irrational number. ■

The next examples introduce some new functions and their graphs.

EXAMPLE 3 *(Absolute value function)* Let $f(x) = |x|$. Find its domain, its range, its value at -2 and 3, and sketch its graph.

Solution The absolute value is defined for all x, so its domain is $(-\infty, \infty)$. The absolute value of any number is nonnegative, so the range is $[0, \infty)$. Its values at -2 and 3 are

$$f(-2) = |-2| = 2 \quad \text{and} \quad f(3) = |3| = 3$$

To sketch its graph, we set

$$y = |x| = \begin{cases} x & \text{if } x \geq 0 \\ -x & \text{if } x < 0 \end{cases}$$

For $x \geq 0$, $y = x$; for $x < 0$, $y = -x$. This indicates that the graph of $y = |x|$ consists of two half-lines that meet at the origin (Figure 1–23). ■

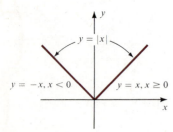

FIGURE 1–23. Example 3. Absolute value function.

EXAMPLE 4 The equation $y = \sqrt{x - 1}$ defines y as a function of x. Find its domain, its range, its values at 4 and 5, and sketch its graph.

Solution The domain is $[1, \infty)$ and the range is $[0, \infty)$ (why?). The values of y at 4 and 5 are

$$y(4) = \sqrt{4 - 1} = \sqrt{3} \quad \text{and} \quad y(5) = \sqrt{5 - 1} = 2$$

To sketch the graph, we make a table of values

x	1	2	3	4	5
$y = \sqrt{x - 1}$	0	1	$\sqrt{2} \approx 1.4$	$\sqrt{3} \approx 1.7$	2

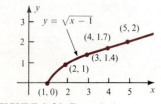

FIGURE 1–24. Example 4.

plot these points, and draw the graph as in Figure 1–24. As a check on our graph, we observe that the domain $[1, \infty)$ indicates that the entire graph lies to the right of $x = 1$; the range $[0, \infty)$ indicates that the entire graph lies above the x-axis. ■

EXAMPLE 5 The equation $y = \sqrt{1 - x^2}$ defines y as a function of x. Find the domain, the range, the values at 0 and 1, and sketch the graph.

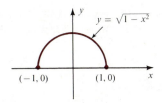

FIGURE 1–25. Example 5.

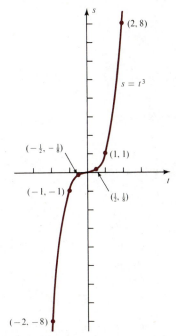

FIGURE 1–26. Example 6.

Solution   Since $1 - x^2$ cannot be negative (why?), it follows that $-1 \le x \le 1$; therefore, the domain is the closed interval $[-1, 1]$. The range is the closed interval $[0, 1]$ (check). The values at 0 and 1 are

$$y(0) = \sqrt{1 - 0^2} = 1 \quad \text{and} \quad y(1) = \sqrt{1 - 1^2} = 0$$

The domain and range tell us that the entire graph lies between the lines $x = -1$, $x = 1, y = 0$, and $y = 1$. We could make a table of values and plot the points, but it is easier to observe that $y = \sqrt{1 - x^2}$ if and only if $y \ge 0$ and

$$x^2 + y^2 = 1 \qquad \text{(square both sides of } y = \sqrt{1 - x^2}\text{)}$$

This is a circle with center at the origin and radius 1. It follows that the graph of $y = \sqrt{1 - x^2}$ is the semicircle shown in Figure 1–25.   ∎

We often use letters such as g, G, and F to denote functions and letters such as s, t, u, and v to denote variables. Here is an example:

EXAMPLE 6   Let $g(t) = t^3$. Find its domain, its range, its values at -1 and 2, and sketch its graph in a ts-plane.

Solution   The domain and range are both $(-\infty, \infty)$. The values of g at -1 and 2 are

$$g(-1) = (-1)^3 = -1 \quad \text{and} \quad g(2) = 2^3 = 8$$

The instructions are to graph g in a ts-plane. Therefore, we set $s = t^3$ and make a table of values:

t	-2	-1	$-1/2$	0	1/2	1	2
$s = t^3$	-8	-1	$-1/8$	0	1/8	1	8

The graph is sketched in Figure 1–26.   ∎

Evaluating a Function

A function is defined as a *rule* that assigns to each element in the domain exactly one value. When evaluating a function, it is important to think of the rule as a set of instructions that must be followed *no matter what the given domain element looks like*.

EXAMPLE 7   Let f be the function $f(x) = 3x + 5$. The instructions are: the value of f at *any real number* is three times the number plus five. Therefore,

$$f(-1) = 3(-1) + 5 = 2$$
$$f(\pi) = 3\pi + 5$$
$$f\left(\frac{1}{t}\right) = 3\left(\frac{1}{t}\right) + 5 = \frac{3}{t} + 5 \quad (t \ne 0)$$
$$f(x + h) = 3(x + h) + 5 = 3x + 3h + 5 \quad ∎$$

EXAMPLE 8   Let $g(t) = 2t^2 - 3t + 7$. The instructions are: the value of g at *any real number* is two times the square of the number minus three times the number plus seven. Therefore,

$$g(4) = 2(4)^2 - 3(4) + 7 = 27$$
$$g(y) = 2y^2 - 3y + 7$$
$$g(t + h) = 2(t + h)^2 - 3(t + h) + 7$$
$$= 2(t^2 + 2th + h^2) - (3t + 3h) + 7$$
$$= 2t^2 + 4th + 2h^2 - 3t - 3h + 7 \quad \blacksquare$$

EXERCISES

In Exercises 1–12, evaluate the functions at the indicated points.

1. $f(x) = 5$; $x = -3, 4.7, 1 + h$
2. $f(x) = -2$; $x = 1, 0, y^2 + 4$
3. $f(x) = x$; $x = -4, 1.2, 2 + h$
4. $f(x) = 3x + 1$; $x = 2, 1, t + 2$
5. $g(t) = t^2 - 3t + 1$; $t = \sqrt{2}, -3.7, -1 + h$
6. $g(t) = t^3$; $t = -3, 0, 1 + h$
7. $g(t) = \sqrt{2 - t^2}$; $t = \sqrt{2}, 1.3, 1$
8. $g(t) = \sqrt{t^2 + 1}$; $t = \sqrt{3}, 1, 8$
9. $F(x) = \dfrac{x^2 - 1}{x - 1}$; $x = .9, .99, .999$
10. $F(x) = \dfrac{x^2 - 1}{x + 1}$; $x = -.9, -.99, -.999$
11. $F(x) = \dfrac{x^2 - 4x + 3}{x - 3}$; $x = 3.1, 3.01, 3.001$
12. $F(x) = \dfrac{\sqrt{x + 1} - 1}{x}$; $x = .1, .01, .001$

In Exercises 13–20, find the domain and range of the function, and then compute the indicated value.

13. $g(t) = \sqrt{t - 3}$; $g(4)$
14. $g(t) = \sqrt{3 - t}$; $g(1)$
15. $F(u) = \sqrt{3 - u^2}$; $F(1)$
16. $F(u) = \sqrt{u^2 - 3}$; $F(2)$
17. $G(x) = \dfrac{1}{x + 1}$; $G(0)$
18. $G(x) = \dfrac{1}{\sqrt{x + 1}}$; $G(3)$

19. $f(x) = \dfrac{\sqrt{4 - x^2}}{x^2}$; $f(\sqrt{2})$
20. $f(x) = \dfrac{\sqrt{x^2 - 4}}{x^2 - 1}$; $G(\sqrt{8})$

In Exercises 21–40, find the domain, the range, and sketch the graph of the given function.

21. $f(x) = -1$
22. $f(x) = 3x + 1$
23. $g(x) = -2x$
24. $g(x) = 2 - x$
25. $f(x) = x^2 + 2x - 3$
26. $f(x) = -x^2 + x + 1$
27. $y = 3x^2 - 5x - 2$
28. $y = -3x^2 - 4x - 4$
29. $y = \sqrt{x}$
30. $y = -\sqrt{x}$
31. $y = -\sqrt{4 - x}$
32. $y = \sqrt{x + 1}$
33. $y = -\sqrt{4 - x^2}$
34. $y = \sqrt{9 - x^2}$
35. $f(x) = \dfrac{1}{x}$
36. $f(x) = \dfrac{1}{x + 1}$
37. $g(x) = \dfrac{1}{x^2}$
38. $g(x) = \dfrac{1}{\sqrt{x}}$
39. $F(x) = x^3 + 1$
40. $F(x) = \sqrt[3]{x}$

In Exercises 41–50, determine where $f(x)$ is positive, negative, and zero.

41. $f(x) = 3x - 1$
42. $f(x) = 4 - 3x$
43. $f(x) = x(8 + 4x)$
44. $f(x) = x(x - 1)$
45. $f(x) = -x^2 + 5x - 6$
46. $f(x) = x^2 + 2x - 3$
47. $f(x) = x + \dfrac{2}{x - 3}$
48. $f(x) = \dfrac{2x + 1}{3x - 2}$
   (*Hint:* Add fractions)

49. $f(x) = \sqrt{x} + \dfrac{2}{\sqrt{x}}$

50. $f(x) = \sqrt{x-1} - \dfrac{5}{\sqrt{x-1}}$

Optional Exercises

51. Let $f(x) = 3x + 7$. If $|x - 2| < .05$, how large can the distance be between $f(x)$ and 13?

52. Let $f(x) = 2 - 5x$. If $|x + 1| < .01$, how large can the distance be between $f(x)$ and 7?

53. Let $f(x) = 4x - 3$. If the distance between $f(x)$ and 5 is less than $\frac{1}{10}$, how large can the distance be between x and 2?

54. Let $f(x) = 9 - 2x$. If the distance between $f(x)$ and -1 is less than .001, how large can the distance be between x and 5?

1–4 APPLICATIONS OF FUNCTIONS

We continue our study of functions, concentrating now on the following two topics:

Topic	Example
Applications of functions	*Biology:* blood flow in an artery *Business:* straight-line depreciation *Engineering:* water entering a conical tank
Change of value	$\Delta f = f(x + \Delta x) - f(x)$

Functions are used in a wide variety of disciplines to describe how variable quantities are related. In these applications, we do not always use the letters x and y to denote the variables. We often use more suggestive letters such as t for time, A for area, and so on.

EXAMPLE 1 *(Depreciation)* A piece of equipment costs $1000 and has a scrap value of $200 at the end of 8 years. Using the straight-line method of depreciation, express the depreciation D as a function of time t and find the depreciation after 4.3 years.

Solution The straight-line method means to depreciate the equipment by equal amounts in equal time periods. Since the total depreciation is $800 over an 8-year period, the depreciation after one year is $100, after two years it is $200, and so on. In general, after t years

$$D = 100t \quad \text{dollars, } t \text{ in } [0, 8]$$

This equation expresses D as a function of t. Here, t is the independent variable, and D is the dependent variable. The domain is $[0, 8]$, the range is $[0, 800]$, and the value of D after 4.3 years is $100(4.3) = \$430$ (Figure 1–27). ■

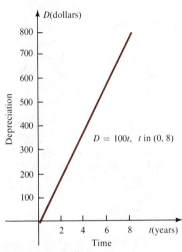

FIGURE 1–27. Example 1. Depreciation versus time.

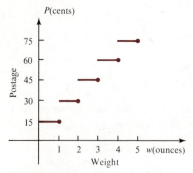

FIGURE 1–28. Example 2. Postage versus weight.

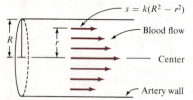

FIGURE 1–29. Example 3. Particles of blood flowing through an artery.

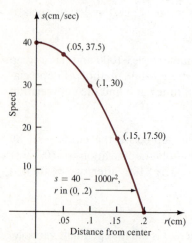

FIGURE 1–30. Example 3. Speed versus distance from the center.

Not all functions can be described by simple equations.

EXAMPLE 2 *(Postage)* Let each *w* in (0, 5] represent a possible weight (in ounces) of a letter. If it costs 15¢ per ounce to mail a letter, express the postage cost *P* as a function of *w*, and sketch the graph.

Solution If $0 < w \le 1$, then $P = 15$¢; if $1 < w \le 2$, then $P = 30$¢, and so on. We can express *P* as a function of *w* by writing

$$P = \begin{cases} 15 & \text{if } 0 < w \le 1 \\ 30 & \text{if } 1 < w \le 2 \\ 45 & \text{if } 2 < w \le 3 \\ 60 & \text{if } 3 < w \le 4 \\ 75 & \text{if } 4 < w \le 5 \end{cases}$$

The domain in this example is (0, 5], and the range is the set {15, 30, 45, 60, 75}. The graph is sketched in Figure 1–28. Notice that *P* is constant on each of the half open intervals (0, 1], (1, 2], and so on. ∎

EXAMPLE 3 *(Blood flow)* Assume that an artery is a cylindrical tube with a circular cross-section of radius R. Particles of blood flowing through the artery move at different speeds; the closer to the center, the faster they move (Figure 1–29). The relation between the speed *s* and the distance *r*, in centimeters (cm), from the center is described by the equation

$$s = k(R^2 - r^2) \quad \text{cm/sec, } r \text{ in } [0, R]^*$$

where *k* is a constant depending on artery length and blood pressure. This equation defines *s* as a function of *r*. A typical value for R in the human body is .2 cm, and a realistic value for *k* is 1000. For these values, $s = 1000(.2^2 - r^2)$ or

$$s = 40 - 1000r^2 \quad \text{cm/sec, } r \text{ in } [0, .2]$$

To find the speed of a blood particle .1 cm from the center, you compute the value $s = 40 - 1000(.1)^2 = 30$ cm/sec. The graph of $s = 40 - 1000r^2$, *r* in [0, .2], is part of a parabola (Figure 1–30). The graph shows how quickly the speed of the blood particle approaches zero as its position approaches the wall of the artery. ∎

EXAMPLE 4 A box manufacturer wants to make a box without a top by cutting squares from the corners of a rectangular piece of cardboard and bending up

---

*The function $s = k(R^2 - r^2)$ was derived experimentally by the French physician Jean Poiseuille. He was studying the energy loss due to friction as blood flows through an artery. His energy loss function is

$$L = \frac{kd}{R^4}$$

where *k* is a constant, *d* is the length, and *R* is the radius of the artery. It is reassuring to know that functions like $s = k(R^2 - r^2)$ and $L = kd/R^4$, which look so "unrealistic," actually have practical applications.

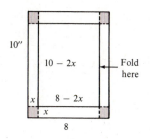

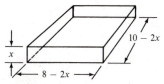

FIGURE 1–31. Example 4. Making a cardboard box.

the sides (Figure 1–31). If the cardboard is 8 by 10 inches, express the volume of the box as a function of the width of the squares that are cut out. What is the domain of this function?

Solution The width (in inches) of each square to be cut out is labeled x in Figure 1–31. The dimensions of the resulting box are x, $10 - 2x$, and $8 - 2x$. Therefore, the equation

$$V = x(10 - 2x)(8 - 2x)$$
$$= 80x - 36x^2 + 4x^3 \quad \text{cubic inches}$$

expresses the volume as a function of x. The side of the square that is cut out ranges from 0 to 4 inches (why?); therefore, the domain of this function is the interval $[0, 4]$, but notice that $V = 0$ if $x = 0$ or 4. ∎

Some quantities are functions of more than one variable. For example, the area of a rectangle is a function of *two* variables, length and width. If you are asked to express the quantity as a function of only one variable, then you must find a way to eliminate all other variables. This can be accomplished in many ways depending on the given problem. Here are three examples.

EXAMPLE 5 A farmer has 200 meters (m) of fencing to enclose a rectangular garden. Express the area of the garden as a function of the width. What is the domain of this function?

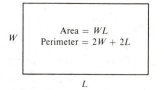

FIGURE 1–32. Example 5.

Solution Draw a picture (Figure 1–32) of the garden labeling the width W and the length L; the area, $A = WL$, is a function of two variables. One of the variables can be eliminated by establishing some relation between them. The given information indicates that the perimeter of the garden is 200; that is, $2W + 2L = 200$. Now it is possible to solve for either W or L in terms of the other. Since the problem calls for the area in terms of the width, we solve for L and obtain $L = 100 - W$. Therefore,

$$A = WL = W(100 - W)$$
$$= 100W - W^2 \quad \text{square meters}$$

expresses A as a function of W. The width in this problem can be any number from 0 to 100 (why?), so the domain of the area function is the interval $[0, 100]$, but notice that $A = 0$ if $W = 0$ or 100. ∎

EXAMPLE 6 Water is pouring into the conical tank shown in Figure 1–33A. Express the volume of the water as a function of its height. What is the domain?

Solution The volume of a cone is $V = \frac{1}{3}\pi r^2 h$, where r is the radius of the base and h is the height.* Because the problem calls for the volume in terms of h,

_____

*There is a list of such formulas on the inside cover.

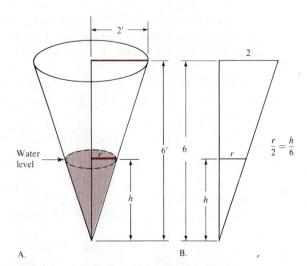

FIGURE 1–33. Example 6.

A.   B.

we must find a way of relating r and h in order to eliminate r. The diagram in Figure 1–33B shows that similar triangles can be used to write

$$\frac{r}{2} = \frac{h}{6} \quad \text{or} \quad r = \frac{1}{3}h$$

Thus, $V = \frac{1}{3}\pi r^2 h$ becomes

$$V = \frac{1}{3}\pi\left(\frac{1}{3}h\right)^2 h = \frac{\pi h^3}{27} \quad \text{cubic feet}$$

which expresses V as a function of h. The domain is the interval $[0, 6]$. ∎

EXAMPLE 7 Figure 1–34A shows a rectangle inscribed in a circle of radius 5 inches. Express the area of the rectangle as a function of x. What is the domain?

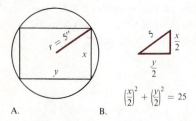

A.   B.
FIGURE 1–34. Example 7.

Solution The area is $A = xy$. The problem calls for the area in terms of x, so we must eliminate y. Figure 1–34B suggests using the right triangle to write $(x/2)^2 + (y/2)^2 = 25$ and solving for y

$$\frac{y^2}{4} = 25 - \frac{x^2}{4} = \frac{100 - x^2}{4}$$

$$y = \pm\sqrt{100 - x^2}$$

Since negative values for y do not make sense in this problem, we disregard the negative root. Then

$$A = xy = x\sqrt{100 - x^2} \quad \text{square inches}$$

expresses the area as a function of x; the domain is $[0, 10]$. ∎

Remark: In Chapter 4 we will return to some of the functions mentioned above. Using the methods of calculus, we will be able to advise the farmer what dimensions will yield the largest garden and the box manufacturer what size squares to cut out to make the box with the largest possible volume. We will also be able to predict how fast the water is rising in the conical tank.

Change in Value

Let f be a function and let x denote the independent variable. As you will see in later chapters, a large part of calculus is devoted to the study of the *rate of change* in the values of f with respect to changes in x. We can begin this study right now (on a modest scale) by computing a *change in value* of f. That is, if x changes from, say, x_1 to x_2, then we can compute the change in value $f(x_2) - f(x_1)$.

EXAMPLE 8 Let $f(x) = x^2 - 1$. Compute the change in value as the variable changes from 3 to 3.1; that is, compute $f(3.1) - f(3)$.

Solution

$$
\begin{aligned}
f(3.1) - f(3) &= [(3.1)^2 - 1] - [3^2 - 1] \\
&= 9.61 - 1 - 9 + 1 \\
&= .61 \quad \blacksquare
\end{aligned}
$$

EXAMPLE 9 *(Continuation of Example 8)* Compute the change in value as the variable changes from 3 to $3 + h$ where h is any nonzero number; that is, compute $f(3 + h) - f(3)$.

Solution

$$
\begin{aligned}
f(3 + h) - f(3) &= [(3 + h)^2 - 1] - [3^2 - 1] \\
&= 9 + 6h + h^2 - 1 - 9 + 1 \\
&= 6h + h^2
\end{aligned}
$$

Remark: The change in f from 3 to $3 + h$ is $6h + h^2$. If we let $h = .1$, then $3 + h = 3.1$ and $6h + h^2 = .61$. This agrees with the result of Example 8. The change $6h + h^2$ is valid for *all h*. $\quad \blacksquare$

Using the letter h to denote a change in the variable is a widely held practice. But it is also common practice to use the Greek letter Δ, capital *delta*. If f is a function and x is the independent variable, then Δx, read "delta x," represents a change in x and Δf, read "delta f," represents the change in value as the variable changes from x to $x + \Delta x$; that is,

(1.30) $\quad \boxed{\Delta f = f(x + \Delta x) - f(x)}$

In this notation, the result of Example 9 is

$$
\Delta f = f(3 + \Delta x) - f(3) = 6\Delta x + \Delta x^2
$$

The advantage of the Δ notation is that it indicates what is changing; Δx, Δf, and so on. If the function is expressed as an equation, say

$$y = x^2 - 1$$

then the letter y not only stands for the dependent variable, but it also replaces f in (1.30). That is, we write

$$\Delta y = y(x + \Delta x) - y(x)$$

to compute the change in y. For $y = x^2 - 1$, we have

$$\Delta y = y(x + \Delta x) - y(x) = [(x + \Delta x)^2 + 1] - [x^2 + 1]$$
$$= 6\Delta x + \Delta x^2$$

EXAMPLE 10   The radius of a circle is 2 inches. If the radius is changed by a small amount, what is the corresponding change in the area of the circle? What is the change in area if the new radius is 2.1 inches? If the new radius is 1.98 inches?

Solution   The function $A = \pi r^2$ expresses the area in terms of the radius. Let Δr represent a small change in r; that is, Δr is a small positive or negative number and the new radius is $2 + \Delta r$. If we let ΔA represent the corresponding change in the area, then

$$\Delta A = A(2 + \Delta r) - A(2) = \pi(2 + \Delta r)^2 - \pi(2)^2$$
$$= \pi(4 + 4\Delta r + \Delta r^2) - 4\pi$$
$$= (4\Delta r + \Delta r^2)\pi \quad \text{square inches}$$

This is the change in the area due to a change of Δr in the radius. If the radius changes from 2 to 2.1, then $\Delta r = .1$ and the formula above yields

$$\Delta A = [4(.1) + (.1)^2]\pi = .41\pi \quad \text{square inches}$$

If the radius changes from 2 to 1.98, then $\Delta r = -.02$ and

$$\Delta A = [4(-.02) + (-.02)^2]\pi = -.0796\pi \quad \text{square inches} \quad \blacksquare$$

Suppose that a particle* moves along a straight line and that its position s, to the left or right of the origin, is a function of time; this function is called the **position function** of the particle.

EXAMPLE 11   A particle moves along a straight line and its position function is $s = 3t - 5$ where s is measured in meters and t in seconds. What is the initial position? Where is the particle after one second? After two seconds? What is the change in position corresponding to change Δt in time?

---

*In this context, "particle" means any object whose position can be described by specifying a single point. The "particle" could be an electron, an automobile, or even our earth moving through space.

Solution The initial position is found by setting $t = 0; s = 3 \cdot 0 - 5 = -5$. Thus, the particle starts from a position 5 m to the *left* of the origin. After 1 second, $s = 3 \cdot 1 - 5 = -2$, so it is 2 m to the left of the origin. After 2 seconds, $s = 3 \cdot 2 - 5 = 1$, so it is 1 m to the *right* of the origin. Let Δt represent a change in t and Δs be the corresponding change in s. Then

$$\Delta s = s(t + \Delta t) - s(t) = [3(t + \Delta t) - 5] - [3t - 5]$$
$$= 3t + 3\Delta t - 5 - 3t + 5$$
$$= 3\Delta t$$

This means that the change in position is 3 times the change in time. ◼

Of particular interest to us is the graph of a position function. The graph of position versus time is called a **position curve.** The position function indicates where the object is at time t. The position curve provides additional information; it tells us the *direction of motion* at time t.

EXAMPLE 12 A particle moves on a straight line and its position function is $s = -t^2 + 4t - 3$ for $0 \le t \le 4$; s is measured in feet, t in seconds. Draw its position curve and describe its motion by means of a motion diagram.

Solution The graph is a part of a parabola with the t-coordinate of the vertex at $-4/(-2) = 2$. We make a table of values,

t	0	1	2	3	4
$s = -t^2 + 4t - 3$	-3	0	1	0	-3

plot the points, and sketch the curve as in Figure 1–35A. The shape of the position curve indicates the direction of motion. From $t = 0$ to $t = 2$, the curve is rising; this means that s is increasing, and the particle is moving to the *right*. From $t = 2$ to $t = 4$, the curve is falling; now s is decreasing, so motion is to the *left*. The diagram in Figure 1–35B describes the motion. (*Note:* The curve above

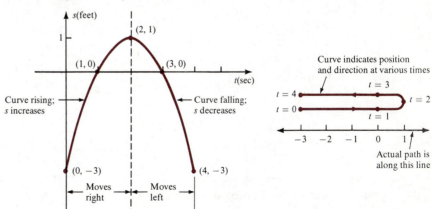

FIGURE 1–35. Example 12.   A. Position curve.   B. Motion diagram.

the line is *not* the path of motion. The particle travels along the coordinate line; the curve indicates the *position and direction* of motion at various times.) The diagram indicates that the particle starts at $s = -3$ and moves to the right until it arrives at $s = 1$ when $t = 2$. Then it moves to the left and is back at its starting point when $t = 4$. ∎

EXERCISES

In Exercises 1–12, compute $f(x + \Delta x) - f(x)$, the change in value of f corresponding to a change of Δx in x.

1. $f(x) = 3x + 7$

2. $f(x) = 4x - 2$

3. $f(x) = 8 - 5x$

4. $f(x) = 3 - 2x$

5. $f(x) = x^2 - 3x + 1$

6. $f(x) = 2x^2 + x - 4$

7. $f(x) = 2 + x - 4x^2$

8. $f(x) = x - 3x^2$

9. $f(x) = x^3$

10. $f(x) = x^3 + x - 1$

11. $f(x) = \dfrac{1}{x}$

12. $f(x) = \dfrac{1}{\sqrt{x}}$

In Exercises 13–24, sketch the graphs.

13. $y = |x - 1|$

14. $y = |x + 2|$

15. $y = x$

16. $y = 3x + 1$

17. $s = t^2 - 3t - 4$, t in $[-1, 4]$

18. $s = t - 2t^2$, t in $[0, 1]$

19. $V = 80x - 36x^2 + 4x^3$, x in $[0, 4]$

20. $V = \dfrac{\pi}{27} h^3$, h in $[0, 6]$

21. $y = \begin{cases} -3x & \text{if } x \le 0 \\ x^2 & \text{if } x > 0 \end{cases}$

22. $y = \begin{cases} x + 1 & \text{if } x \le 1 \\ 2 & \text{if } x > 1 \end{cases}$

23. $y = \begin{cases} x - 2 & \text{if } x \le -1 \\ -3 & \text{if } -1 < x < 0 \\ x^2 - 3 & \text{if } x \ge 0 \end{cases}$

24. $y = \begin{cases} -x^2 & \text{if } x \le 0 \\ 2x & \text{if } 0 < x < 1 \\ x + 1 & \text{if } x \ge 1 \end{cases}$

25. The period P (in seconds) of a pendulum of length L feet is given by the function $P = 2\pi\sqrt{L/32.2}$. Use $\pi \approx 3.14$ and approximate the period of a pendulum 2.30 feet long. What is the change in the period corresponding to a change of $\Delta L = .2$ feet?

26. If P dollars are deposited in a bank paying $r\%$ interest compounded n times per year, the amount accrued (principal and interest) at the end of a year is $A = P(1 + r/100n)^n$. How much is accrued in one year if \$1000 is deposited in a bank paying 8% compounded quarterly? What is the change in A if the interest rate is changed by $\Delta r = 1\%$?

27. A projectile is fired into the air. Its height h feet above the ground at time t seconds after firing is given by the function $h = -16t^2 + 160t$. How high is it when $t = 3$? When $t = 8$? What is the domain of this function?

28. A particle moves in a straight line and its position function is $s = t^2 - 6t + 8$ for $0 \le t \le 4$; s is measured in feet and t in seconds. What is the initial position? Where is the particle when $t = 1$? When $t = 3$? At what times will the particle be at the origin?

29. The volume V of a sphere of radius r is $V = \frac{4}{3}\pi r^3$. What is the volume when $r = 3$? Compute the change ΔV in volume corresponding to a change Δr in the radius; that is, when $r = 3 + \Delta r$.

30. The volume of the box constructed in Example 4 is given by the function $V = 80x - 36x^2 + 4x^3$. Compute ΔV, the change in volume corresponding to a change in x by an amount Δx.

31. The height of the projectile in Exercise 27 is given as a function of time. If the time is changed from t to $t + \Delta t$, what is the corresponding change in height? Use your answer to calculate the change in height if $\Delta t = 1$ and (a) $t = 2$ (b) $t = 4$. Explain why the two changes are different.

32. In Example 3, the speed of a particle of blood is given as $s = 40 - 1000r^2$. Compute the change in speed as r changes from .1 to .098.

33. The height of a triangle is three times its base. Express the area as a function of the base. What is the domain of this function?

34. A particle moves along a line and its position function is $s = 2t - 3$ for $-1 \leq t \leq 3$. Sketch the position curve and describe its motion by means of a motion diagram (see Figure 1–35B). What is the change in position corresponding to a change in time from t to $t + \Delta t$?

35. A particle moves along a line and its position function is $s = t^2 - 2t - 1$ for $-2 \leq t \leq 3$. Sketch the position curve and the motion diagram. What is the change in position corresponding to a change in time from t to $t + \Delta t$?

36. A particle moves on a line and its position function is $s = -t^2 + 4t + 3$ for $0 \leq t \leq 5$. Sketch the position curve and the motion diagram. If the time is changed from t to $t + \Delta t$, what is the corresponding change in position?

37. If $t = 0$ represents the beginning of the year 1979, the sales S of a certain company, in units of $100,000, are approximately

$$S = \begin{cases} t + 3 & \text{if } -2 \leq t \leq 0 \\ -\dfrac{1}{2}t + 3 & \text{if } 0 \leq t \leq 2 \\ t^2 - 2 & \text{if } 2 \leq t \leq 3 \end{cases}$$

where t is in years. Sketch the sales verses time graph. When would you buy or sell stock in this company?

38. The perimeter of a rectangle is 10 ft. Express the area as a function of its length. What is the domain of this function?

39. The area of a rectangle is 100 square centimeters. Express the perimeter as a function of the width. What is the domain of this function?

40. Two planes leave an airport at the same time. One flies north at 300 miles per hour and the other flies east at 400 miles per hour. Express the distance between them as a function of time (in hours).

41. The volume of a cone is $V = \frac{1}{3}\pi r^2 h$, where r is the radius of the base and h is the height. The tank in Figure 1–36 is being filled with water. Express the volume of water in the tank as a function of its height h.

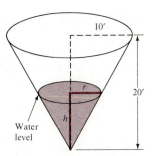

FIGURE 1–36. Exercise 41.

42. In Figure 1–37, each point P on the circle determines a right triangle OPQ. If the diameter is 10 inches as shown, express the area of the triangle in terms of one of its sides q. What is the domain of this function?

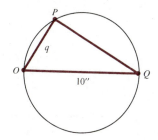

FIGURE 1–37. Exercise 42.

43. A window has the shape of a rectangle surmounted by a semicircle (Figure 1–38). If the perimeter is 12 feet, express the area of the window as a function of the radius r of the semicircle.

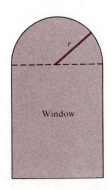

FIGURE 1–38. Exercise 43.

44. Let T be an isosceles triangle with base 4 and equal sides of length x. Express the area of T as a function of x. What is the domain?

45. Express the area of an equilateral triangle as a function of one of its sides. What is the domain?

46. A window has the shape of a rectangle surmounted by an equilateral triangle. If the perimeter of the window is 12 feet, express the area as a function of one side of the equilateral triangle. What is the domain?

47. The upper semicircle of the graph of $x^2 + 2x + y^2 - 4y = 4$ is the graph of a function. What is the function? What is the domain?

Optional Exercises

48. A piece of wire 20 inches long is cut at a point x inches from the left end. The left piece is bent into a circle and the right piece is bent into a square. Express the combined area of the circle and the square as a function of x. What is the domain of this function?

49. A metal can is to be constructed so as to have a volume of V cubic feet. The top and bottom are made from aluminum that costs a cents per square foot but the sides can be made of tin that costs $a/4$ cents per square foot. Express the cost of the can as a function of the radius r. ($V = \pi r^2 h$ and the surface area is $S = 2\pi rh + 2\pi r^2$.)

1–5   LINES

Lines are important in calculus. In particular, the discussion of derivatives in Chapter 3 requires that you know about the *slope* of a line; what it is, how to find it, and how to interpret it. You also have to know how to find an equation of a line. Therefore, in this section, we review the following topics:

TOPIC	EXAMPLE
Slope	$\quad$ slope $= \dfrac{\Delta y}{\Delta x}$
Sign of the slope	positive slope $\quad$ zero slope $\quad$ negative slope
Size of the slope	large slope $\quad$ small slope
Parallel lines	equal slopes
Perpendicular lines	slopes are negative reciprocals
Equations of lines	Point-slope form: $y - y_1 = m(x - x_1)$ Slope-intercept form: $y = mx + b$

After the review, we introduce the notion of the *average rate of change* of a function

$$\text{average rate of change:} \quad \frac{\Delta f}{\Delta x} = \frac{f(x + \Delta x) - f(x)}{\Delta x}$$

and relate it to the slope of a line.

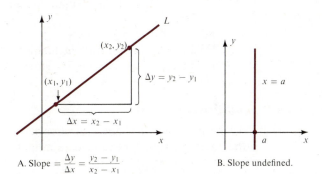

FIGURE 1–39. Slope of a line.

A. Slope $= \dfrac{\Delta y}{\Delta x} = \dfrac{y_2 - y_1}{x_2 - x_1}$

B. Slope undefined.

Figure 1–39 illustrates the slope concept. Letting $\Delta y = y_2 - y_1$ and $\Delta x = x_2 - x_1$ represent the change in y and x is consistent with our earlier use of the Δ notation.

DEFINITION
Let L be a nonvertical line and let (x_1, y_1) and (x_2, y_2) be any two distinct points on L. The **slope** of L, denoted by m, is the ratio

(1.31)
$$m = \frac{\Delta y}{\Delta x} = \frac{y_2 - y_1}{x_2 - x_1}$$

The slope of a vertical line is undefined.

Remarks:

(1) The slope of vertical line $x = a$ is undefined because x never changes, so $\Delta x = 0$.

(2) Figure 1–40 illustrates that

The slope of L can be computed using *any* two distinct points on L.

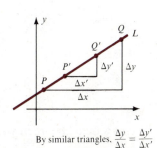

By similar triangles, $\dfrac{\Delta y}{\Delta x} = \dfrac{\Delta y'}{\Delta x'}$

FIGURE 1–40. Any two distinct points on L yield the same slope.

EXAMPLE 1 Let $(3, 2)$ and $(-4, 1)$ be two points on a line L. Find the slope of L.

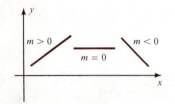

FIGURE 1–41. The sign of m indicates inclination.

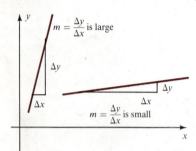

FIGURE 1–42. The size of $|m|$ indicates the steepness of inclination.

Solution To use (1.31), we let $x_1 = 3, y_1 = 2, x_2 = -4$, and $y_2 = 1$. Then the slope m is

$$m = \frac{1 - 2}{-4 - 3} = \frac{1}{7} \qquad \blacksquare$$

Figures 1–41 and 1–42 illustrate important interpretations of the slope of a line L. If $m > 0$, then L is inclined upward; as x increases, y also increases. If $m < 0$, then L is inclined downward; as x increases, y decreases. If $m = 0$, then L is horizontal. Moreover, if $|m|$ is large, then small changes in x produce large changes in y. If $|m|$ is small, then the opposite is true.

(1.32)

> The *sign* of m indicates inclination
> The *size* of $|m|$ indicates the steepness of inclination

Equation of a Line

Let L be any nonvertical line with slope m. If (x_1, y_1) is a point on L, then (x, y) is also on L if and only if the slope determined by the two points equals m. That is,

$$m = \frac{y - y_1}{x - x_1} \quad \text{or} \quad y - y_1 = m(x - x_2)$$

This observation leads to the following statement:

(1.33)

> Let L be a nonvertical line with slope m. If (x_1, y_1) is *any* point on L, then
> $$y - y_1 = m(x - x_1)$$
> is an equation of L; it is called the **point-slope form.**

Statement (1.33) says that if you know *one point and the slope,* then you know an equation of the line.

EXAMPLE 2 Find an equation of the line through $(-1, 2)$ with slope -4.

Solution We use (1.33) with $x_1 = -1, y_1 = 2$, and $m = -4$. Thus,

$$y - 2 = -4(x - (-1)) = -4x - 4$$

is an equation of the line (Figure 1–43). $\blacksquare$

The next example illustrates that if you know *two points* on a line, then you can find an equation of the line.

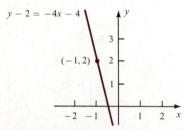

FIGURE 1–43. Example 2.

EXAMPLE 3 Find an equation of the line through the points $(0, -2)$ and $(1, 0)$.

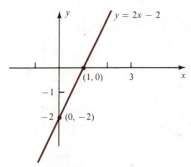

FIGURE 1–44. Example 3.

Solution   *First,* we find the slope using the two given points

$$m = \frac{\Delta y}{\Delta x} = \frac{0 - (-2)}{1 - 0} = 2$$

Second, we pick either point and use (1.33). With the point $(0, -2)$,

$$y - (-2) = 2(x - 0) \quad \text{or} \quad y = 2x - 2$$

You can easily verify that the other point $(1, 0)$ yields the same equation (Figure 1–44).   ■

If the point-slope form (1.33) is solved for y, we have

$$y = m(x - x_1) + y_1$$
$$= mx + (-mx_1 + y_1)$$

In this form, the coefficient of x is the slope of the line and, setting $x = 0$, we see that $-mx_1 + y_1$ (which we usually denote by the letter b) is the y-intercept. Therefore,

(1.34)

> A linear equation
>
> $$y = mx + b$$
>
> of a nonvertical line L is called the **slope-intercept form.** The coefficient of x is the slope of L and $(0, b)$ is the y-intercept.

EXAMPLE 4   Find the slope and y-intercept of the line $3y - 6x + 2 = 0$.*

Solution   To find the slope, we could determine two distinct points on the line, but it is easier to put the equation into slope-intercept form by solving for y. Thus,

$$3y = 6x - 2$$
$$y = 2x - \frac{2}{3}$$

In this form, we know by (1.34) that the slope is 2 and the y-intercept is $(0, -\frac{2}{3})$.   ■

Parallel and Perpendicular Lines

In later work, you will often be asked to find an equation of a line that is parallel or perpendicular to a given line. Knowing the slope of the given line, we use the following result to find the slope of the desired line:

---

*We often identify an equation with its graph; that is, we say "the line $3y - 6x + 2 = 0$" rather than "the line that is the graph of $3y - 6x + 2 = 0$." Similarly, we might say "the parabola $y = x^2$" or "the circle $x^2 + y^2 = 1$."

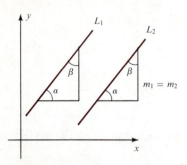

FIGURE 1–45. Slopes of parallel lines are equal.

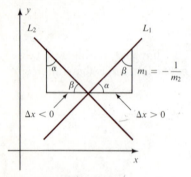

FIGURE 1–46. Slopes of perpendicular lines are negative reciprocals.

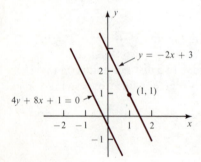

FIGURE 1–47. Example 5.

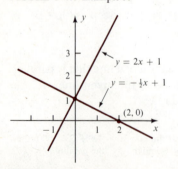

FIGURE 1–48. Example 6.

If L_1 and L_2 are nonvertical lines with slopes m_1 and m_2, then

(1.35)
(1) L_1 and L_2 are parallel if and only if $m_1 = m_2$
(2) L_1 and L_2 are perpendicular if and only if $m_1 = -1/m_2$

Figures 1–45 and 1–46 illustrate these statements. Both are proved using similar triangle arguments (see the figures), and we leave the details as an exercise.

EXAMPLE 5 Find an equation of the line through (1, 1) that is parallel to $4y + 8x + 1 = 0$.

Solution Using (1.35–1), we find the slope of the desired line by finding the slope of $4y + 8x + 1 = 0$. Solving for y yields

$$y = -2x - \frac{1}{4}$$

so the slope is -2. Now we know the slope -2 and a point (1, 1) on the line. The point-slope form gives

$$y - 1 = -2(x - 1) \quad \text{or} \quad y = -2x + 3$$

as an equation of the line (Figure 1–47). ∎

EXAMPLE 6 Find an equation of the line through (2, 0) that is perpendicular to $y = 2x + 1$.

Solution The given line has slope 2, so the desired line has slope $-\frac{1}{2}$ (why?). It follows that

$$y - 0 = -\frac{1}{2}(x - 2) \quad \text{or} \quad y = -\frac{1}{2}x + 1$$

is an equation of the line (Figure 1–48). ∎

Average Rate of Change

We close this section with an interesting interpretation of the slope of a line joining two points on the graph of a function. This interpretation will be useful in later work.

Suppose that f is a function and that x is in the domain of f; then $(x, f(x))$ is a point on the graph of f. Suppose further that there is a change Δx in x and that $x + \Delta x$ is also in the domain of f; then $(x + \Delta x, f(x + \Delta x))$ is another point on the graph of f. Figure 1–49 shows that the slope of the line joining these two points is

(1.36)
$$\frac{\Delta f}{\Delta x} = \frac{f(x + \Delta x) - f(x)}{\Delta x}$$

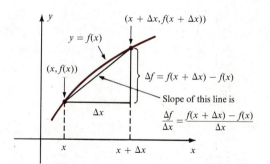

FIGURE 1–49. The slope of the line joining $(x, f(x))$ and $(x + \Delta x, f(x + \Delta x))$ is called the *average rate of change.*

The numerator, $f(x + \Delta x) - f(x)$, is the change in value of f; therefore, dividing by Δx produces the *average* change in value. The ratio (1.36) is called the **average rate of change** of f as the variable changes from x to $x + \Delta x$.

EXAMPLE 7   Find the average rate of change of $f(x) = x^2 + x + 1$ as the variable changes from $x = 1$ to $x = .7$.

Solution   Since x changes from 1 to .7, $\Delta x = -.3$. Thus,

$$\frac{\Delta f}{\Delta x} = \frac{f(.7) - f(1)}{-.3} = \frac{[.7^2 + .7 + 1] - [1^2 + 1 + 1]}{-.3}$$

$$= 2.7$$

is the average rate of change of f as the variable changes from $x = 1$ to $x = .7$ ∎

EXAMPLE 8   The average rate of change of $f(x) = 2 - 3x - x^2$ as the variable changes from x to $x + \Delta x$ is

$$\frac{\Delta f}{\Delta x} = \frac{f(x + \Delta x) - f(x)}{\Delta x}$$

$$= \frac{[2 - 3(x + \Delta x) - (x + \Delta x)^2] - [2 - 3x - x^2]}{\Delta x}$$

$$= \frac{2 - 3x - 3\Delta x - x^2 - 2x\Delta x - \Delta x^2 - 2 + 3x + x^2}{\Delta x}$$

$$= \frac{-3\Delta x - 2x\Delta x - \Delta x^2}{\Delta x}$$

$$= -3 - 2x - \Delta x   \quad ∎$$

EXERCISES

In Exercises 1–6, sketch the graph of the given equation.

1. $y = -2$ **2.** $y = x + 2$
3. $x = 3$ **4.** $y = -4x$
5. $y = 1 - 2x$ **6.** $y = 3x - 2$

In Exercises 7–16, find the slope of each line. Then sketch the graph and verify that   (a) the sign of the slope indicates how the line is inclined and   (b) the size of the slope indicates how steeply the line is inclined.

7. $y = x$

8. $y = x - 2$

9. $y + \frac{1}{2}x - 1 = 0$

10. $y - \frac{1}{3}x + 1 = 0$

11. $2y - 6x + 2 = 0$

12. $2x + 8y + 4 = 0$

13. $4y + 3x = 5$

14. $5y - 2x = 4$

15. $10x - 2y = 2$

16. $3x - y = 1$

In Exercises 17–22, find an equation of the line with the given properties.

17. Passing through $(-1, 2)$ and $(4, 1)$

18. Passing through $(4, -7)$ and $(-2, 6)$

19. Passing through $(-3, 2)$ with slope $\frac{1}{5}$

20. Passing through $(4, 0)$ with slope 3

21. Passing through $(\frac{1}{2}, -\frac{1}{3})$ and parallel to the graph of $3x - 4y + 2 = 0$

22. Passing through $(6, 1)$ and perpendicular to the graph of $x - 2y + 1 = 0$

In Exercises 23–30, compute the average rate of change $\Delta f/\Delta x$ of f as the variable changes from x_1 to x_2.

23. $f(x) = x^2; x_1 = 1, x_2 = 3$

24. $f(x) = -x^2; x_1 = 0, x_2 = 2$

25. $f(x) = 2x^2 - 3x + 1; x_1 = 4, x_2 = 3$

26. $f(x) = x^2 - 3x - 2; x_1 = -2, x_2 = -3$

27. $f(x) = x^3 + 1; x_1 = -1, x_2 = -.9$

28. $f(x) = -x^3; x_1 = 3, x_2 = 1$

29. $f(x) = 1/x; x_1 = .5, x_2 = .75$

30. $f(x) = \sqrt{x}; x_1 = 1, x_2 = 4$

In Exercises 31–40, compute the average rate of change $\Delta f/\Delta x$ of f as the variable changes from x to $x + \Delta x$. (Assume that both x and $x + \Delta x$ are in the domain of f.)

31. $f(x) = -2$

32. $f(x) = 7$

33. $f(x) = 3x - 2$

34. $f(x) = 5 - 4x$

35. $f(x) = x^2 - 2x + 3$

36. $f(x) = 2x^2 - 5$

37. $f(x) = \sqrt{x}$

38. $f(x) = 1/x$

39. $f(x) = x^3$

40. $f(x) = 1 - x^3$

41. Sketch the triangle with vertices $A(1, 0)$, $B(0, 2)$, and $C(-1, -1)$. (a) Does it look like a right triangle? (b) Verify that it is, using slopes.

42. Same Exercise 41 for $A(0, 4)$, $B(1, 0)$, and $C(-1, -\frac{1}{2})$.

43. Find an equation of the line passing through the center of the circle $x^2 - 2x + y^2 + 4y - 4 = 0$ and the point $(1, 1)$.

In Exercises 44–46, use the fact from geometry that a line tangent to a circle is perpendicular to the radius at the point of tangency.

44. Find an equation of the line that is tangent to the circle $x^2 + 6x + y^2 - 7 = 0$ at the point $(-3, 4)$.

45. Find an equation of line that is tangent to the circle $x^2 + y^2 = 1$ at each of points   (a) $(0, 1)$ (b) $(1, 0)$ (c) $(0, -1)$ and   (d) $(-1, 0)$.

46. Find an equation of the line tangent to the circle $x^2 + y^2 = 1$ at the point $(\frac{1}{2}, \sqrt{3}/2)$.

47. The sales of a certain retail store are found to be a linear function of time. At the end of two years, the sales were \$200,000; at the end of four years, the sales were \$150,000. Express sales as a function of time.

48. The Celsius (or centigrade) and Fahrenheit temperature scales are linearly related. Water boils at $100°C$ and $212°F$; water freezes at $0°C$ and $32°F$. Write F as a linear function of C. Normal body temperature is $98.6°F$; what is normal body temperature on the Celsius scale?

49. For a moving object, the average rate of change of distance is what we ordinarily call the *average speed* of the object. For example, suppose a drag strip racer leaves the starting position and after t seconds she is s feet down the track. Then the average rate of change of s as the variable changes from t to $t + \Delta t$ is

$$\frac{\Delta s}{\Delta t} = \frac{s(t + \Delta t) - s(t)}{\Delta t} \quad \text{ft/sec} = \text{average speed}$$

If $s = 2t^3 + 25t$, find the average speed of the racer during the third second (that is, from $t = 2$ to $t = 3$). What is her average speed during the sixth second?

50. A shell is fired straight up and its distance above the earth t seconds later is $s = -16t^2 + 192t$ feet. Find the average speed during the first three seconds and then during the next three seconds.

51. A ball is dropped from a building 256 ft high. Its height above the ground when $t = 1, 2, 3$ and 4 seconds is $240, 192, 112$, and 0 feet, respectively.

Sketch its position curve. What is its average speed during the first, second, third, and fourth second of its fall? What is the significance of the sign of the average speed?

52. (Continuation of Exercise 51) The ball is dropped from a height of 256 feet and it hits the ground when $t = 4$ seconds. Now it bounces up to a height of 64 feet and hits the ground again when $t = 8$ seconds. It bounces up to a height of 16 feet and hits the ground for the third time when $t = 10$ seconds. Sketch its position curve over the time interval $[0, 10]$.

53. Find the average rate of increase in the area of a circle as the radius increases from 3 inches to 3.2 inches.

54. Find the average rate of decrease in the volume of a sphere as the radius decreases from 3 inches to 2.8 inches.

55. Use Figures 1–45 and 1–46 to prove the statements in (1.35) about parallel and perpendicular lines.

1–6 EXPONENTS AND LOGARITHMS

Precalculus texts usually contain an elementary discussion of exponential and logarithmic functions. They cannot be too precise because careful definitions of these concepts require the methods of calculus. In this book, for example, the definitions do not appear until Chapter 7. Nevertheless, it is useful to have a working knowledge of both exponents and logarithms at an early stage, because they provide such a wealth of examples and applications. Therefore, this section reviews the following basic topics

TOPIC	EXAMPLE
Exponents (or powers)	$2^3 = 8$, $4^{1/2} = \sqrt{4} = 2$
Logarithms	$\log_{10} 100 = 2$, $\log_2 8 = 3$
Graphs	

In addition, a glimpse at a few applications (population growth, radioactive decay, and learning models) is included to whet your appetite.

Exponents

Let a be a positive real number and m be a positive integer. Then a^m represents the product of a with itself m times. The number a is called the **base** and m is called the **exponent** or **power**. We include the zero exponent by defining $a^0 = 1$; we include negative exponents by defining $a^{-m} = 1/a^m$.

The definitions above can be extended to include rational exponents as follows: If a is positive and m/n is any rational number with $n > 0$, then

$$a^{m/n} = (\sqrt[n]{a^m}) = (\sqrt[n]{a})^m$$

The denominator n must be positive because we are taking an nth root. The base must be positive because, if n is even (say $n = 2$), then $\sqrt[n]{a}$ is not defined for $a < 0$. The following properties of exponents hold for all positive real numbers a and b and all rational numbers x and y:

(1.37)
$$
\begin{aligned}
a^x a^y &= a^{x+y} & (ab)^x &= a^x b^x \\
\frac{a^x}{a^y} &= a^{x-y} & a^0 &= 1 \\
(a^x)^y &= a^{xy} & a^x &> 0
\end{aligned}
$$

The definition of a^x can be extended once more to include irrational exponents, for example, $a^{\sqrt{2}}$ and a^{π}, but this is where calculus is needed. As we stated earlier, a careful discussion of this extension does not occur until Chapter 7. In the meantime, let us *assume* that the extension has been made and that properties (1.37) remain valid for all real x and y.

(1.38)

Let $a > 0$. The function

$$f(x) = a^x$$

is called the **exponential function with base a.** Its domain is the set of all real numbers and its range is the set of all positive real numbers.

An exponential function such as 2^x is not to be confused with a function like x^2, which we call a *power function*. For a power function, the base is a variable and the power is fixed. For an exponential function, the opposite is true; the base is fixed and the exponent is allowed to vary.

The functions x^2 and 2^x are vastly different. For example, x^2 evaluated at $x = 100$ is $(100)^2 = 10,000$. However, 2^x evaluated at $x = 100$ is 2^{100}; this number is greater than a 1 followed by 30 zeros! The proof is as follows:

$$2^{10} = 1,024 > 10^3 \text{ and, therefore, } 2^{100} = (2^{10})^{10} > (10^3)^{10} = 10^{30}$$

To appreciate the size of the number 10^{30}, compare it with the fact that most of you reading this sentence have lived less than 10^9 *seconds,* and the fact that the most distant star ever observed is less than 10^{28} *inches* from our earth.

The behavior of $y = a^x$ depends on whether $a > 1$ or $a < 1$.

EXAMPLE 1   Sketch and compare the graphs of $y = 2^x$ and $y = (\frac{1}{2})^x$.

Solution   We make a table of values, carefully plot the points, and draw the curves as in Figure 1–50.

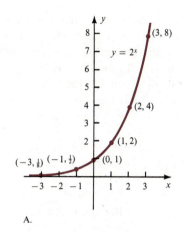

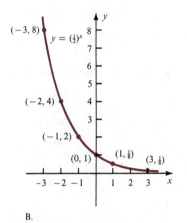

A.

B.

FIGURE 1–50. Example 1.

x	-3	-2	-1	0	1	2	3
$y = 2^x$	1/8	1/4	1/2	1	2	4	8
$y = (1/2)^x$	8	4	2	1	1/2	1/4	1/8

As you scan the graph of $y = 2^x$ from left to right, you can observe that the curve is rising. This means that the value of 2^x increases as x increases; that is, if $x_1 < x_2$, then $2^{x_1} < 2^{x_2}$. This will be true for any exponential function $y = a^x$ so long as $a > 1$. Contrast this with the graph of $y = (\frac{1}{2})^x$. Here, the values decrease as x increases, that is, if $x_1 < x_2$, then $(\frac{1}{2})^{x_1} > (\frac{1}{2})^{x_2}$. This is true for any exponential function $y = a^x$ so long as $a < 1$. ∎

(1.39)
> If $a > 1$ and $x_1 < x_2$, then $a^{x_1} < a^{x_2}$
> If $a < 1$ and $x_1 < x_2$, then $a^{x_1} > a^{x_2}$

In Figure 1–51A, we have drawn the graphs of $y = 2^x$ and $y = 3^x$ on the same set of axes. Notice that they have the same general shape, but $y = 3^x$ is steeper than $y = 2^x$ for positive x and flatter for negative x. This is true in general as a gets larger. A similar comparison is made in Figure 1–51B.

Remarks:

(1) Most calculators have a *power key* y^x that can represent an exponential function. For example, to evaluate $f(x) = 4^x$ at $x = 3.5$, you enter the base 4, push $\boxed{y^x}$, enter the exponent 3.5, and then push $\boxed{=}$; the display reads 128.* Thus,

$$f(3.5) = 4^{3.5} = 128$$

The base must be positive. If you try to evaluate $(-4)^{3.5}$, you will get an error signal; check this on your calculator.

(2) We remind you about the accuracy of calculator answers. The answer $4^{3.5} = 128$ is exact because $4^{3.5} = 4^{7/2} = (\sqrt{4})^7 = 128$. But if you compute $4^{1.25}$, the display reads 5.6568543. Thus $4^{1.25} \approx 5.6568543$ or, equivalently,

$$5.65685425 \le 4^{1.25} < 5.65685435$$

because we do not know if the last digit is correct.

Here are two applications of exponential functions. The first has a base $a > 1$; the second has a base $a < 1$.

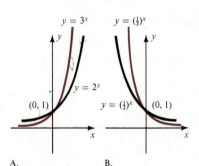

FIGURE 1–51. Comparison of the graphs of exponential functions.

*The sequence of entries

$\boxed{4}\ \boxed{y^x}\ \boxed{3}\ \boxed{.}\ \boxed{5}\ \boxed{=}$

used to evaluate $4^{3.5}$ is correct for most calculators in current use. Some calculators, however, require a slightly different sequence. Whenever a calculator computation is displayed in this book, students should consult their own calculator manual for the proper sequence of entries.

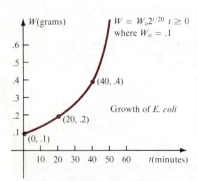

FIGURE 1–52. Example 2.
Exponential growth of bacteria.

EXAMPLE 2 *(Population growth)* Bacteria are typically one-celled animals. They reproduce by growing to a certain size, or weight, and then splitting into two cells. Each of these cells will grow until the critical weight is reached, and they will split; now there are four cells. Each of these will split, thereby producing eight cells, and so on. Growth patterns are usually studied in the laboratory by weighing the bacteria rather than counting them. A culture is prepared containing a known amount of bacteria and then the culture is weighed at regular intervals.

Suppose that the initial colony weighs W_o grams and that the weight is found to *double* every k minutes. Express the weight of the bacteria as a function of time.

Solution   At the start ($t = 0$), the bacteria weigh W_o. At the end of k minutes, the weight doubles, so they now weigh $2W_o$. At the end of $2k$ minutes, the weight has doubled again, so they weight $2 \cdot 2 \cdot W_o = W_o 2^2$. After $3k$ minutes, they weigh $W_o 2^3$, and so on. In general,

$$(1.40) \qquad W = W_o 2^{t/k} \quad t \geq 0$$

expresses the weight W(grams) as an exponential function of time t(minutes). See Figure 1–52.   ■

The next example uses the notion of average rate of change to illustrate the basis of concern about overpopulation.

EXAMPLE 3   *(Continuation of Example 2)* Laboratory tests show that

$$W = W_o 2^{t/20} \quad t \geq 0$$

approximates the growth behavior of the bacteria *Escherichia coli;* the population doubles about every 20 minutes (Figure 1–52). When $t = 40$ minutes, then

$$W = W_o 2^{40/20} = W_o 2^2$$

When $t = 60$ minutes, $W = W_o 2^3$. The average rate of change during that time period is

$$\frac{\Delta W}{\Delta t} = W_o \frac{2^3 - 2^2}{20} = \frac{W_o}{5} \quad \text{grams/min}$$

During the next 20 minutes, from $t = 60$ to $t = 80$, the average rate of change is

$$\frac{\Delta W}{\Delta t} = W_o \frac{2^4 - 2^3}{20} = \frac{2W_o}{5} \quad \text{grams/min}$$

Therefore, not only does the population double every 20 minutes, but its *average rate of growth also doubles.*   ■

EXAMPLE 4   *(Radioactive decay)* The half-life of carbon 14 is about 5740 years; that is, one-half of the present amount will be left at the end of that time. If you start with Q_o grams of carbon 14, then

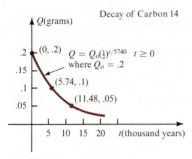

FIGURE 1–53. Example 4.
Radioactive decay of carbon 14.

$$Q = Q_0(\tfrac{1}{2})^{t/5740} \quad t \geq 0$$

is the amount present at time t (years). See Figure 1–53. ■

Logarithms

The range of the function $y = 2^x$ is the set of all positive numbers. Therefore, if x is any positive number, there must be some number b such that $2^b = x$. There can be only one such number b. (If $b \neq c$, then either $2^b < 2^c$ or $2^c < 2^b$ depending on whether $b < c$ or $c < b$). The number b is called the *logarithm base* 2 of x; it is denoted by $\log_2 x$. Thus,

$$\log_2 x = b \quad \text{if and only if } 2^b = x$$

Similar definitions can be made for any positive base $a \neq 1$.

(1.41)

Let $a > 0$ and $a \neq 1$. The **logarithm base** a (written $\log_a$) is a function on the set of positive numbers and

$$\log_a x = b \quad \text{if and only if } a^b = x$$

The range of $\log_a$ is the set of all real numbers.

EXAMPLE 5

    (a) $\log_2 16 = 4$ because $2^4 = 16$
    (b) $\log_{\frac{1}{3}} 9 = -2$ because $(\tfrac{1}{3})^{-2} = 9$
    (c) $\log_a 1 = 0$ because $a^0 = 1$ ■

Here are some basic properties of logarithms.

(1.42)

$$\log_a xy = \log_a x + \log_a y \qquad \log_a \frac{x}{y} = \log_a x - \log_a y$$

$$\log_a x^y = y \log_a x \qquad\qquad \log_a 1 = 0; \ \log_a a = 1$$

$$\log_a a^x = x \qquad\qquad\qquad a^{\log_a x} = x$$

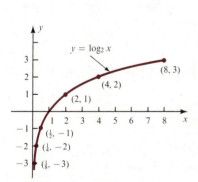

FIGURE 1–54. Example 6.

EXAMPLE 6   A table of values for $y = \log_2 x$ is

x	1/8	1/4	1/2	1	2	4	8
$y = \log_2 x$	-3	-2	-1	0	1	2	3

The graph is sketched in Figure 1–54. ■

Remark: Most calculators have a logarithm key $\boxed{\log x}$. This is the logarithm base 10. If you enter 3 and push $\boxed{\log x}$, the display reads 0.4771213. According to (1.41), it should be true that $10^{.4771213} = 3$; try it on your calculator (this is one way to check the accuracy of a calculator). Notice that the domain of any logarithm function is the set of *positive* numbers. If you enter -3 and push $\boxed{\log x}$, you will get an error signal. Some calculators have a second logarithm key $\boxed{\ln x}$; it uses the base $e \approx 2.7182818$. This number e is very important in the study of calculus, and we will have much more to say about it in later chapters.

Here is an application of logarithms.

EXAMPLE 7 *(Learning model)* Psychologists interested in learning behavior have established that the measure of retention of knowledge involves a logarithm. Experimenting with groups of students, they found that the average test score S on a topic covered in class t days before is approximated by

$$(1.43) \quad S = S_o - c \log_{10}(t + 1) \quad t \geq 0$$

where S_o is the average score if the test is given the same day the topic is covered and c is a constant that depends on the particular class. For example, if $S_o = 75$ and $c = 5$, then

$$S = 75 - 5 \log_{10}(t + 1)$$

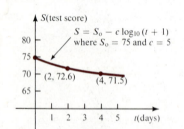

FIGURE 1–55. Example 7. Learning model.

Figure 1–55 indicates how the test scores decrease as t gets larger. After 2 days, the average score is 72.6; after 4 days, the average is only 71.5. ■

We close this section with the observation that exponential and logarithmic functions are different from functions such as $y = 2x + 3$ and $y = x^2 - 3x + 7$ in an essential way. For future use, we make the following classification of functions. A **polynomial** is any function of the form

$$y = a_n x^n + a_{n-1} x^{n-1} + \cdots + a_1 x + a_0 \quad a_n \neq 0$$

where the *coefficients* $a_n, a_{n-1}, \ldots, a_0$ are real numbers and the exponents are nonnegative integers. The number a_n is called the *leading coefficient* and n is called the *degree* of the polynomial. The functions

$$y = 2x^2 + 3x + 5 \quad \text{and} \quad y = -5x^3 + 2x - 7$$

are examples of polynomials of degree 2 and 3, respectively. Other types of functions are

$$y = \sqrt[3]{x^2 + 1} \quad \text{and} \quad y = \frac{x + 3}{x^2 + 2x + 1}$$

The first is the cube root of a polynomial; the second is a quotient of polynomials. All of these functions are examples of algebraic functions.

(1.44)

> An **algebraic function** is any function that can be expressed as a finite number of sums, differences, products, quotients, or roots of polynomials. Functions that are not algebraic are called **transcendental.**

Exponential and logarithmic functions are transcendental. The trigonometric functions, taken up in the next section, are also transcendental.

EXERCISES

In Exercises 1–15, compute the values (*without* the aid of a calculator).

1. $4^{3/2}$ **2.** $8^{1/3}$

3. $32^{-2/5}$ **4.** $27^{-4/3}$

5. 100^0 **6.** $\left(\dfrac{1}{10}\right)^2$

7. $\left(\dfrac{1}{10}\right)^{-3}$ **8.** $\log_4 16$

9. $\log_2 8$ **10.** $\log_{\frac{1}{2}}\dfrac{1}{8}$

11. $\log_{\frac{1}{3}} 81$ **12.** $\log_3 27$

13. $\log_2 \dfrac{1}{32}$ **14.** $\log_{10} 10$

15. $\log_2 1$

In Exercises 16–20, find x if the graph of $y = 3^x$ contains the given point.

16. $(x, \frac{1}{9})$ **17.** $(x, 27)$
18. $(x, \sqrt{3})$ **19.** $(x, 1)$
20. $(x, 9 \cdot 3^h)$

In Exercises 21–25, find x if the graph of $y = \log_4 x$ contains the given point.

21. $(x, 2)$ **22.** $(x, 4)$
23. $(x, 0)$ **24.** $(x, \frac{1}{16})$
25. $(x, \frac{1}{2})$

In Exercises 26–30, find a if the graph of $y = a^x$ contains the given point.

26. $(-2, \frac{1}{100})$ **27.** $(3, 8)$
28. $(1, \pi)$ **29.** $(1, 2)$
30. $(2 + h, 49 \cdot 7^h)$

In Exercises 31–35, find a if the graph of $y = \log_a x$ contains the given point.

31. $(4, 2)$ **32.** $(81, 2)$
33. $(81, -2)$ **34.** $(8, -3)$
35. $(\frac{1}{27}, 3)$

In Exercises 36–42, sketch the graphs of the given functions.

36. $y = 3^x$ **37.** $y = (\frac{1}{3})^x$
38. $y = 2^{t/3}$ **39.** $y = (\frac{1}{2})^{t-1}$
40. $y = \log_3 x$ **41.** $y = \log_{\frac{1}{3}} x$
42. $y = \log_2 |x|$

43. If the half-life of a certain radioactive isotope is 2 years and you start with 10 grams, express the amount present Q as a function of time t in years. How much will be left after 8 years? What is the average rate of change as the variable changes from $t = 2$ to $t = 4$? What does the negative sign mean?

44. Suppose the function in Example 7 relating test scores and elapsed time is $S = 80 - c \log_{10}(t + 1)$. If the average score was 75 after 9 days, what will be the average score after 20 days?

(Compound interest) If P dollars are invested at an annual interest rate r, then Pr is the interest earned in one year, and the investment will increase to $P + Pr = P(1 + r)$ dollars at the end of the first year. The interest earned in the second year is $P(1 + r) r$ and the accumulated amount at the end of the second year is $P(1 + r) + P(1 + r) r = P(1 + r)^2$. At the end of the third year, the investment will have grown to $P(1 + r)^3$, and so on. In general, at the end of n years, the accumulated amount A is

$$A = P(1 + r)^n$$

45. If Columbus had deposited $1 five hundred years ago in a bank that paid 5% $(r = .05)$ interest compounded annually, how much would have accumulated by now?

Most banks quote yearly rates, but they compound interest quarterly, monthly, or even daily. If the yearly rate r is compounded quarterly, the formula above must be adjusted. The amount accumulated after n years is now

$$A = P\left(1 + \frac{r}{4}\right)^{4n}$$

because the rate per quarter is $r/4$ and it is compounded 4 times a year. Similar adjustments must be made for monthly and daily compounding.

46. If $1,000 is invested at an annual rate of 8% compounded quarterly, how much is accumulated after 10 years?

47. Mr. Pettett wishes to accumulate $40,000 in 15 years to pay for his daughter's education. If he can invest his money at 8% compounded quarterly, how much should he invest now in order to accomplish his goal?

48. If $1,000 is invested at 6% compounded monthly, how long will it take to double?

49. Suppose $1 is invested for one year at 100%. If interest is compounded n times a year, the accumulated amount after one year is

$$A = \left(1 + \frac{1}{n}\right)^n$$

It may seem that, if interest is compounded often enough (that is, if n is very large), then A would grow without bound. You may be surprised to learn that this is false; on the contrary, A can be no larger than the irrational number $2.7182818\ldots$, which is always denoted by the letter e.* Check this out by computing the value of $(1 + 1/n)^n$ for $n = 1$, 10, 100, and 1,000.

---

*This is the same number that we referred to earlier as the base used for the $\boxed{\ln x}$ key of a calculator.

1–7 TRIGONOMETRY

In this section, we review the following basic topics from trigonometry:

TOPIC	EXAMPLE
Angle measurement	2π radians = 360 degrees
Trigonometric functions	$\sin t = \dfrac{y}{r}$ $\cos t = \dfrac{x}{r}$
Graphs	$s = \sin t$
Trigonometric identities	$\sin^2 t + \cos^2 t = 1$ $\sin(s + t) = \sin s \cos t + \cos s \sin t$
Triangle trigonometry	$\sin A = \dfrac{\text{side opposite}}{\text{hypotenuse}}$ Law of Cosines: $c^2 = a^2 + b^2 - 2ab \cos C$

After the review, we give four examples of applications.

Radian Measure

There are 360 *degrees* (360°) in a circle, or in one complete revolution. For calculus, however, it is more convenient to measure angles in *radians,* and there are 2π radians (rad) in one complete revolution. Therefore, the fundamental relationships between degrees and radians are

(1.45)
$$2\pi \text{ rad} = 360°$$
$$1 \text{ rad} = \frac{180°}{\pi} \approx 57.3°$$
$$1° = \frac{\pi}{180} \text{ rad} \approx .017 \text{ rad}$$

express 70° in rad

mag is v = v₀ + ½ at² t or something

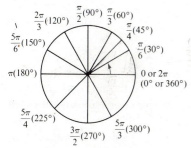

FIGURE 1–56. Some angles measured in degrees and radians.

EXAMPLE 1

(a) To convert degrees into radians, multiply by $\pi/180$.

$$30° = 30\left(\frac{\pi}{180}\right) = \frac{\pi}{6} \text{ rad} \quad 150° = 150\left(\frac{\pi}{180}\right) = \frac{5\pi}{6} \text{ rad}$$

(b) To convert radians into degrees, multiply by $180/\pi$.

$$\frac{\pi}{4} \text{ rad} = \frac{\pi}{4}\left(\frac{180}{\pi}\right) = 45° \quad \frac{5\pi}{3} \text{ rad} = \frac{5\pi}{3}\left(\frac{180}{\pi}\right) = 300°$$

Figure 1–56 shows some other conversions.   ∎

An angle of t radians represents a part, or fraction, of a circle; the fraction is $t/2\pi$. If the circle has radius r, then its circumference is $2\pi r$ and its area is πr^2. It follows that a central angle (an angle at the center of the circle) of t radians determines an arc of length rt on the circumference, and a sector of area $r^2t/2$ (Figure 1–57).

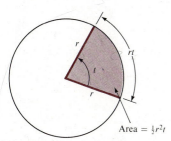

Area $= \frac{1}{2}r^2t$

FIGURE 1–57. The angle t is measured in radians.

(1.46)

In a circle of radius r, a central angle of t radians determines

(1) an arc of length rt on the circumference, and

(2) a sector of area $\frac{1}{2}r^2t$

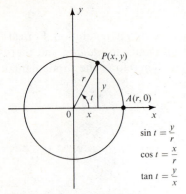

$$\sin t = \frac{y}{r}$$

$$\cos t = \frac{x}{r}$$

$$\tan t = \frac{y}{x}$$

FIGURE 1–58. Definition of the trigonometric functions.

Trigonometric Functions

Let C be a circle in the plane with radius r and center at the origin, and let t be any number. We now draw a central angle AOP of t radians (Figure 1–58) measured from the positive x-axis in the counterclockwise direction (this is the positive direction for angles) if $t > 0$ and in the clockwise (negative) direction if $t < 0$. The coordinates of the point P on the circle are labeled (x, y). The trigonometric functions *sine, cosine, tangent, cotangent, secant,* and *cosecant* are defined and abbreviated as follows

(1.47)

$$\sin t = \frac{y}{r} \qquad\qquad \cos t = \frac{x}{r}$$

$$\tan t = \frac{\sin t}{\cos t} = \frac{y}{x}(x \neq 0) \quad \cot t = \frac{1}{\tan t} = \frac{x}{y}(y \neq 0)$$

$$\sec t = \frac{1}{\cos t} = \frac{r}{x}(x \neq 0) \quad \csc t = \frac{1}{\sin t} = \frac{r}{y}(y \neq 0)$$

Using the similar triangles shown in Figure 1–59A, it is easy to see that the ratios in (1.47) depend only on t and not on r. Figure 1–59B illustrates that when $r = 1$, the coordinates of P are $(\cos t, \sin t)$. Observe that the *sine and cosine are defined for all real numbers,* whereas the other four functions are undefined at some points.

The following table shows some values of the sine, cosine, and tangent that occur frequently.

	Radians	0	$\pi/6$	$\pi/4$	$\pi/3$	$\pi/2$	π	$3\pi/2$	2π
	Degrees	0	30°	45°	60°	90°	180°	270°	360°
(1.48)	Sine	0	$1/2$	$\sqrt{2}/2$	$\sqrt{3}/2$	1	0	−1	0
	Cosine	1	$\sqrt{3}/2$	$\sqrt{2}/2$	$1/2$	0	−1	0	1
	Tangent	0	$\sqrt{3}/3$	1	$\sqrt{3}$	undef	0	undef	0

Remark: In this book, angles are measured in radians unless otherwise specified. For example,

sin 30 means the sine of 30 *radians*

The sine of 30 degrees is written sin 30°. Most calculators have both modes of angle measurement, and it is a good idea to check the mode before you begin computing. ∎

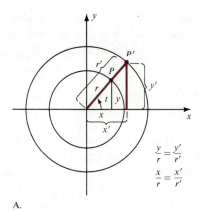

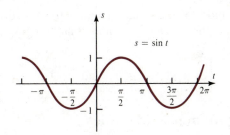

FIGURE 1–59. The ratios depend only on t.

A.

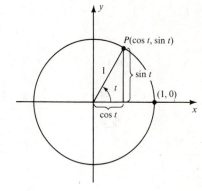

B.

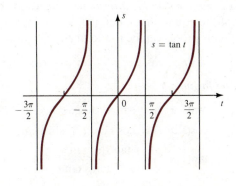

$s = \sin t$

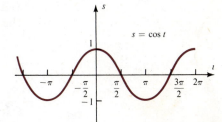

$s = \cos t$

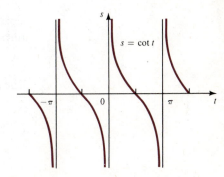

$s = \tan t$

$s = \cot t$

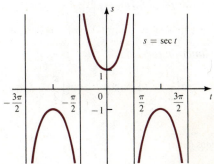

$s = \sec t$

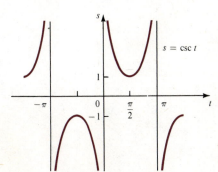

$s = \csc t$

FIGURE 1–60. Graphs of the trigonometric functions.

The definitions (1.47) of sine and cosine are in terms of the coordinates of a point P determined by a central angle of t radians. Since central angles of t and $t + 2n\pi$, where n is an integer, determine the same point P, we have that

$$(1.49) \qquad \sin(t + 2n\pi) = \sin t \quad \text{and} \quad \cos(t + 2n\pi) = \cos t$$

Thus, the values of sine and cosine are repeated endlessly over intervals of length 2π. This is described by saying that the sine and cosine are **periodic** with **period** 2π. It can be shown that the other trigonometric functions are also periodic; the secant and cosecant have period 2π, but the tangent and cotangent have period π.

Plotting a few points obtained from table (1.48), we can use the periodicity to sketch complete graphs of the sine, cosine, and tangent. These graphs along with those of the other trigonometric functions are sketched in Figure 1–60. The graphs indicate the domains and ranges; in particular, we note that

$$(1.50) \qquad -1 \leq \sin t \leq 1 \quad \text{and} \quad -1 \leq \cos t \leq 1$$

Trigonometric Identities

Equations that describe relationships between the trigonometric functions are called *trigonometric identities;* they are valid for all numbers t at which both sides of the equation are defined. We list below some of the most useful identities; their proofs can be found in any book on trigonometry.

$$(1.51) \qquad \sin^2 t + \cos^2 t = 1^* \quad \tan^2 t + 1 = \sec^2 t \quad 1 + \cot^2 t = \csc^2 t$$

The *sum formulas*

$$(1.52) \qquad
\begin{aligned}
\sin(s \pm t) &= \sin s \cos t \pm \cos s \sin t \\
\cos(s \pm t) &= \cos s \cos t \mp \sin s \sin t \\
\tan(s \pm t) &= \frac{\tan s \pm \tan t}{1 \mp \tan s \tan t}
\end{aligned}$$

The *negative angle formulas*

$$(1.53) \qquad \sin(-t) = -\sin t \quad \cos(-t) = \cos t$$

The *co-function formulas*

$$(1.54) \qquad \sin\left(\frac{\pi}{2} - t\right) = \cos t \quad \cos\left(\frac{\pi}{2} - t\right) = \sin t$$

---

*$\sin^2 t$ means $(\sin t)^2$

The *double angle formulas*

(1.55)
$$\sin 2t = 2 \sin t \cos t$$
$$\cos 2t = \cos^2 t - \sin^2 t = 2 \cos^2 t - 1 = 1 - 2 \sin^2 t$$

The *half angle formulas*

(1.56) $$\sin^2 t = \frac{1 - \cos 2t}{2} \quad \cos^2 t = \frac{1 + \cos 2t}{2}$$

Triangle Trigonometry

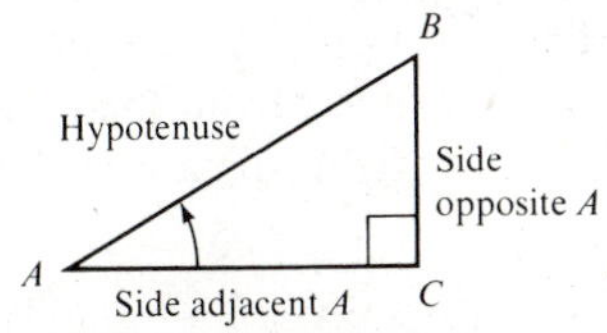

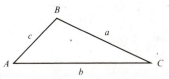

FIGURE 1–61. Right triangle.

Let A be an acute angle of a right triangle with the sides labeled as shown in Figure 1–61. If this triangle is thought of as corresponding to the right triangle in Figure 1–58, then

(1.57)

$$\sin A = \frac{\text{side opposite}}{\text{hypotenuse}} \quad \cos A = \frac{\text{side adjacent}}{\text{hypotenuse}}$$

$$\tan A = \frac{\text{side opposite}}{\text{side adjacent}} \quad \cot A = \frac{\text{side adjacent}}{\text{side opposite}}$$

$$\sec A = \frac{\text{hypotenuse}}{\text{side adjacent}} \quad \csc A = \frac{\text{hypotenuse}}{\text{side opposite}}$$

These formulas are useful in solving problems involving right triangles. For problems involving oblique (nonright) triangles, there are two other formulas relating sides and angles. Their proofs can be found in any trigonometry book.

FIGURE 1–62. Oblique triangle.

(1.58)

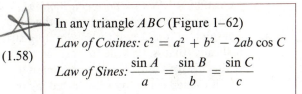

In any triangle ABC (Figure 1–62)

Law of Cosines: $c^2 = a^2 + b^2 - 2ab \cos C$

Law of Sines: $\dfrac{\sin A}{a} = \dfrac{\sin B}{b} = \dfrac{\sin C}{c}$

Some Applications

Applications of trigonometry usually fall into one of two categories; those involving triangles and those involving rhythmic, or periodic, motion. We begin with triangles.

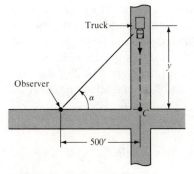

FIGURE 1–63. Example 2.

EXAMPLE 2 Figure 1–63 shows a truck heading south at a distance of y feet from the point C. An observer is 500 feet west of point C. Express the distance y as a function of the angle α (the Greek letter *alpha*). How far is the truck from C when $\alpha = 30°$?

Solution The tangent of α is the side opposite over the side adjacent (1.57); in this case, $y/500$. Therefore,

$$y = 500 \tan \alpha \quad \text{feet}$$

describes the distance y as a function of the angle α. When $\alpha = 30°$, then

$$y = 500 \tan 30°$$
$$\approx 500(.577) \approx 289 \text{ feet}$$

The value $\tan 30° \approx .577$ can be obtained from a table or a calculator. ∎

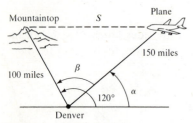

FIGURE 1–64. Example 3.

EXAMPLE 3 A radar screen at the Denver airport shows a plane and a mountaintop situated as shown in Figure 1–64. Express the distance s from the plane to the mountaintop as a function of the angle α. What is the distance when $\alpha = 65°$?

Solution In the figure, angle β (the Greek letter *beta*) is $120° - \alpha$. This angle is included between sides of known length. Therefore, the Law of Cosines (1.58) yields

$$s^2 = (100)^2 + (150)^2 - 2(100)(150) \cos (120 - \alpha)°$$

Since s is never negative (and, we hope, never zero), the positive square root of the right side expresses s as a function of α. If $\alpha = 65°$, then

$$s = \sqrt{(100)^2 + (150)^2 - 2(100)(150) \cos (120 - 65)°}$$
$$\approx 124 \text{ miles} \quad ∎$$

The next two examples concern periodic motion.

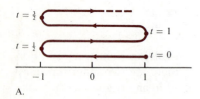

A.

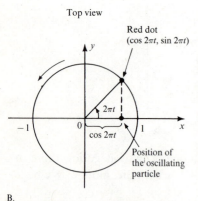

B.

FIGURE 1–65. Example 4. (A) The motion diagram of simple harmonic motion between 1 and -1. (B) Simulating this motion with a dot on the rim of a rotating disc.

EXAMPLE 4 *(Simple harmonic motion)* A particle oscillates between -1 and 1 on a coordinate line according to the motion diagram in Figure 1–65A. You can think of it as a point on a vibrating string. To describe this motion mathematically (that is, to write its position s as a function of time t), imagine a clear plastic disc of radius 1 with its center at 0 and a red dot painted on its rim. The disc is rotated counterclockwise making one revolution (2π radians) per second. From a top view, you would see the red dot start at 1, move left to -1, and then back to 1, simulating the motion of the particle (Figure 1–65B). As the disc rotates at 2π radians per second, a line joining the red dot and 0 sweeps through an angle of $2\pi t$ radians in t seconds. Since the red dot is on the unit circle, it follows (see Figure 1–59B) that its coordinates at time t are ($\cos 2\pi t$, $\sin 2\pi t$). But the x-coordinate of the red dot is the position of the particle at time t (Figure 1–65B). Therefore,

$$s = \cos 2\pi t \quad t \geq 0$$

describes the motion of the oscillating particle; this is called *simple harmonic motion.* ∎

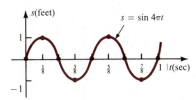

A. Position curve.

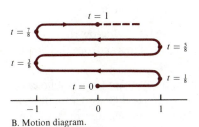

B. Motion diagram.

FIGURE 1–66. Example 5.

EXAMPLE 5   A particle moves along a line and its position at time t seconds is

$$s = \sin 4\pi t \quad \text{feet} \quad t \geq 0$$

Sketch the position curve, draw a motion diagram, and compute the average rate of change in position as t changes from 3 to 3.01 seconds.

Solution   To sketch the position curve, we make a table of values.

t	0	1/8	1/4	3/8	1/2	5/8	3/4	7/8	1
$s = \sin 4\pi t$	0	1	0	−1	0	1	0	−1	0

The position curve is a sine curve that makes two complete cycles in one second (Figure 1–66A). This is reflected in the motion diagram of Figure 1–66B. Again, this is *simple harmonic motion*. The average rate of change in position as t changes from 3 to 3.01 is

$$\frac{\Delta s}{\Delta t} = \frac{\sin[4\pi(3.01)] - \sin[4\pi(3)]}{3.01 - 3}$$

$$\approx \frac{.1253 - 0}{.01} = 12.53 \text{ ft/sec} \quad \blacksquare$$

EXERCISES

In Exercises 1–10, convert degrees into radians and radians into degrees. Evaluate the sine, cosine, and tangent at each angle.

1. 45°   **2.** −450°

3. −3π/2   **4.** π

5. −225°   **6.** 60°

7. 7π/4   **8.** −5π/6

9. 90°   **10.** 1080°

Let t and P be the angle and point, respectively, shown in Figure 1–59B. What are the coordinates of P if t has the following values?

11. π/2   **12.** π/3

13. π/4   **14.** π

15. 2π/3

Answer the same question if P now lies on a circle with center at the origin and radius 3 (instead of 1), and t has the following values:

16. 0   **17.** π/2

18. π/4   **19.** π/6

20. 5π/6

Answer the same question if the circle has radius r.

21. π/2   **22.** π/3

23. 5π/4   **24.** 7π/4

25. α

In Exercises 26–35, the given point P lies on the terminal side of an angle t as shown in Figure 1–67. Find the sine, cosine, and tangent of t.

26. (3, 4)   **27.** (1, 2)

28. (−3, 4)   **29.** (−1, 2)

30. (−√2, −1)   **31.** (−1, √3)

32. (2, −√3)   **33.** (4, −3)

34. (−√2/2, √2/2)   **35.** (6, 8)

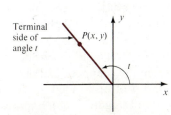

FIGURE 1–67. Exercises 26–35.

In Exercises 36–40, find the value(s) of t, $0 < t \le 2\pi$, that satisfy the equation.

36. $\sin t = \cos t$ **37.** $\sin^2 t + \cos^2 t = 1$

38. $\sin 2t = 1$ **39.** $\cos 3t = 0$

40. $\sin(t/2) = \sqrt{2}/2$

Simple harmonic motion can usually be described by either of the *sinusoidal* functions

$$y = a\sin(bx - c) \quad \text{or} \quad y = a\cos(bx - c)$$

The graph of a sinusoidal function is a sine or cosine graph that has been distorted by the effect of the constants a, b, and c. The distortions are analyzed as follows:

The values of y will range between $-|a|$ and $|a|$
The number $|a|$ is called the *amplitude*

41. Graph the functions $y = 2\sin x$, $y = \sin x$, and $y = \frac{1}{2}\sin x$ on the same set of axes and compare amplitudes.

If $b \ne 0$, the *period* is $2\pi/|b|$

42. Graph the functions $y = \cos 2x$, $y = \cos x$, and $y = \cos(x/2)$ on the same set of axes and compare periods.

If $b \ne 0$, the graph is shifted $|c/b|$ units to the right or left; c/b is called the *phase angle*.

43. Graph the functions $y = \sin(x + \pi/6)$, $y = \sin x$, and $y = \sin(x - \pi/6)$ on the same set of axes and compare phase angles.

44. Sketch the graph of $y = 2\cos(3x - 3\pi/2)$. What is the amplitude, period, and phase angle?

45. A lamppost is 12 feet high. A boy 5 feet tall is walking away from the lamppost, and the line from the lamp to the tip of his shadow makes an angle α with the ground (Figure 1–68). Express the length L of his shadow as a function of α. How long is his shadow when $\alpha = \pi/3$? When $\alpha = \pi/6$?

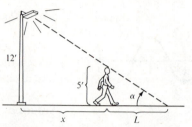

FIGURE 1–68. Exercises 45 and 46.

46. Express the length of the boy's shadow (in Exercise 45) as a function of the distance x from the boy to the base of the lamppost.

47. A rocket is fired straight up from Cape Canaveral. Express the height of the rocket as a function of its angle of elevation from a tracking unit 5.1 miles from the launch pad (Figure 1–69). If the angle of elevation changes from 45° to 60° in 3 seconds, what is the average speed of the rocket?

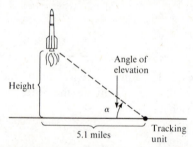

FIGURE 1–69. Exercise 47.

48. A forest ranger spots a fire in the direction North 38° East from his lookout post. A second ranger 10 miles due east of the first one spots the same fire; he gives the direction as North $\alpha°$ East (Figure 1–70). Express the distance s from the fire to the first ranger as a function of α. How far away is the fire if $\alpha = 13°$?

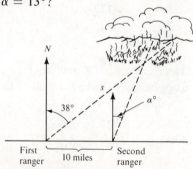

FIGURE 1–70. Exercise 48.

49. A particle moves along a line and its position at time t seconds is $s = 2 \cos(3t - 3\pi/2)$ feet $0 \le t \le 2\pi$. Draw its position curve (see Exercise 44) and a motion diagram. Find the average rate of change in position as t changes from 0 to .02 seconds.

50. *(Damped harmonic motion)* The motion of a point on a vibrating string is not actually simple harmonic motion; friction will cause the amplitude to decrease. This is called *damped harmonic motion*. An illustration is the position function

$$s = (1/2)^t \cos 2\pi t \quad 0 \le t \le 2$$

Sketch the position curve and the motion diagram.

51. Sketch the graph of $y = \sin(1/x)$ for x in the interval $(0, 2/\pi]$. *Hint:* As x ranges over the interval $[2/5\pi, 2/\pi]$, $1/x$ ranges over the interval $[\pi/2, 5\pi/2]$ so that $\sin(1/x)$ completes one cycle. As x ranges over the interval $[2/9\pi, 2/5\pi]$, $1/x$ ranges over the $[5\pi/2, 9\pi/2]$ and $\sin(1/x)$ completes another cycle. As x gets closer and closer to zero, the graph of $y = \sin(1/x)$ will oscillate faster and faster.

52. Use the addition formulas (1.52) to show that (a) $\sin(\pi/2 - x) = \cos x$ and (b) $\cos(\pi/2 - x) = \sin x$.

53. Let C be a circle of radius r. Each central angle determines a pie-shaped region. Show that the area A of the region determined by a central angle of x radians is $A = \frac{1}{2}r^2 x$.

Optional Exercises

54. Show that the average rate of change of $f(x) = \sin x$ as the variable changes from x to $x + \Delta x$ is

$$\frac{\Delta f}{\Delta x} = \sin x \frac{\cos \Delta x - 1}{\Delta x} + \cos x \frac{\sin \Delta x}{\Delta x}$$

55. Show that the average rate of change of $f(x) = 2^x$ as the variable changes from x to $x + \Delta x$ is

$$\frac{\Delta f}{\Delta x} = 2^x \frac{2^{\Delta x} - 1}{\Delta x}$$

1-8 COMBINATIONS OF FUNCTIONS

Many theorems on limits, derivatives, and integrals are stated in terms of combinations of functions. Your skill at applying the theorems depends, in part, on your ability to recognize the given combination. Therefore, we will review the following topics:

TOPIC	EXAMPLE
Arithmetic combinations	For $f(x) = x^2$ and $g(x) = 2^x$, Sum: $(f + g)(x) = f(x) + g(x) = x^2 + 2^x$ Product: $(fg)(x) = f(x)g(x) = x^2 2^x$
Composition	For $f(x) = \sin x$ and $g(x) = \sqrt{x}$, $f \circ g(x) = f(g(x)) = \sin \sqrt{x}$

Most functions are combinations of other, simpler, functions. For example, $F(x) = 2^x + x^2$ is a combination of two functions; the exponential $f(x) = 2^x$ and the power $g(x) = x^2$. We say that F is the *sum* of f and g and write

$$F = f + g$$

Another example, $G(x) = \sqrt{x}\sin x$ is a combination of $f(x) = \sqrt{x}$ and $g(x) = \sin x$. We say that G is the *product* of f and g and write

$$G = fg$$

Recognizing that a given function is some combination of simpler functions is an important step in computing limits (Chapter 2), derivatives (Chapter 3), and integrals (Chapter 5).

Suppose f and g are functions. The **sum** $f + g$ is a new function whose value at x is

(1.59) $\boxed{(f + g)(x) = f(x) + g(x)}$

The domain of $f + g$ is the set of all x such that both $f(x)$ and $g(x)$ are defined. The **difference** $f - g$, the **product** fg, and the **quotient** f/g are defined similarly:

(1.60) $\boxed{\begin{aligned} (f - g)(x) &= f(x) - g(x) \\ (fg)(x) &= f(x)g(x) \\ (f/g)(x) &= f(x)/g(x) \quad \text{provided } g(x) \neq 0 \end{aligned}}$

Another combination is a *constant c times a function f*

(1.61) $\boxed{(cf)(x) = cf(x)}$

EXAMPLE 1 Write $F(x) = 2^x\sqrt{x}$ as the product of two functions.

Solution We define $f(x) = 2^x$ and $g(x) = \sqrt{x}$. Then $F = fg$; that is

$$F(x) = 2^x\sqrt{x} = f(x)g(x) \quad \blacksquare$$

EXAMPLE 2 When we write $G(x) = \tan x = (\sin x)/(\cos x)$, we are expressing the tangent function as a quotient. If we define $f(x) = \sin x$ and $g(x) = \cos x$, then $G = f/g$; that is

$$G(x) = \tan x = \frac{\sin x}{\cos x} = \frac{f(x)}{g(x)}$$

Notice that $g(x) = \cos x$ is 0 whenever $x = \pm n\pi/2$ (n an integer), so the quotient is not defined at these points. $\quad \blacksquare$

EXAMPLE 3 The function

$$H(x) = \frac{2^x + 4\sqrt{x}}{x^2\sin x}$$

is a *quotient*. The numerator is the *sum* of $f(x) = 2^x$ and 4 times $g(x) = \sqrt{x}$; the denominator is the *product* of $h(x) = x^2$ and $k(x) = \sin x$. ∎

Composition

The combinations defined above are arithmetic: addition, subtraction, and so on. There is another, totally different, way to combine functions, called *composition*. Let f and g be functions and let x be in the domain of g. First evaluate g at x to obtain the number $g(x)$. If $g(x)$ is in the domain of f, then evaluate f at $g(x)$ to obtain the number $f(g(x))$.

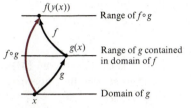

Range of $f \circ g$

$f \circ g$

Range of g contained in domain of f

Domain of g

FIGURE 1–71. The composition of functions. First evaluate g at x, then evaluate f at $g(x)$; $f \circ g(x) = f(g(x))$.

DEFINITION OF COMPOSITION

(1.62)

Let g be a function on X and let the range of g be contained in the domain of a function f. Then the **composite** $f \circ g$ (read "f circle g") is a function on X defined by

$$f \circ g(x) = f(g(x))$$

The composition of functions is illustrated in Figure 1–71 in terms of arrows. The arrow representing g takes x to $g(x)$; the arrow representing f takes $g(x)$ to $f(g(x))$; the arrow representing $f \circ g$ takes x directly to $f(g(x))$. Figure 1–72 shows the function machine version of a composite function. Enter any non-negative number x, and push $\boxed{\sqrt{x}}$ to obtain $\sqrt{x}$. If you now push $\boxed{\sin x}$ the result is $\sin \sqrt{x}$.

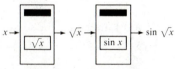

FIGURE 1–72. The function machine version of a composite function.

EXAMPLE 4 Let $f(x) = \sin x$ and $g(x) = \sqrt{x}$. Write the composite function $f \circ g$ and use a calculator to evaluate it at $x = 4$ and $x = 16$.

Solution The composite is

$$f \circ g(x) = f(g(x)) = f(\sqrt{x}) = \sin \sqrt{x}$$

To evaluate it, enter x, push $\boxed{\sqrt{x}}$, and then push $\boxed{\sin x}$.

$$\sin \sqrt{4} = \sin 2 \approx 0.9092974$$
$$\sin \sqrt{16} = \sin 4 \approx -0.7568025 \quad ∎$$

In any composition, the *order* is important. For $f \circ g$, first apply g, then apply f. For $g \circ f$, first apply f, then apply g. The composites $f \circ g$ and $g \circ f$ may not be the same.

EXAMPLE 5 Let $f(x) = x^2$ and $g(x) = \cos x$. Write the composite functions $f \circ g$ and $g \circ f$, and use a calculator to evaluate each at $x = 4$.

Solution

$$f \circ g(x) = f(g(x)) = f(\cos x) = \cos^2 x, \text{ but}$$
$$g \circ f(x) = g(f(x)) = g(x^2) = \cos x^2$$

The sequence for $f \circ g(4)$ is $\boxed{4}$, $\boxed{\cos x}$, $\boxed{x^2}$ ≈ 0.42725. The sequence for $g \circ f(4)$ is $\boxed{4}$, $\boxed{x^2}$, $\boxed{\cos x}$ ≈ -0.9576595. Thus, in this case, $f \circ g \neq g \circ f$. ■

Examples 4 and 5 illustrate how to compose two given functions. The next two examples illustrate the reverse process.

EXAMPLE 6   Write $F(x) = 3^{x^2}$ as a composition of two functions.

Solution   The instructions for F are to first square x, and then raise 3 to that power. If we let $g(x) = x^2$ and $f(x) = 3^x$, then

$$F(x) = 3^{x^2} = f(x^2) = f(g(x)) = f \circ g(x)$$

and we have written F as the composite $f \circ g$. (*Note:* 3^{x^2} is different from $(3^x)^2$; the latter means to first raise 3 to the power x, and then square that quantity. In terms of compositions, $3^{x^2} = f \circ g(x)$, but $(3^x)^2 = g \circ f(x)$.) ■

The next example shows that more than two functions can be composed.

EXAMPLE 7   Write $F(x) = \log_2 \sqrt{x^2 + 1}$ as a composition of a composition.

Solution   The instructions for F are to square x and add 1, then take the square root, and then take $\log_2$. Therefore, we let $h(x) = x^2 + 1$, $g(x) = \sqrt{x}$, and $f(x) = \log_2 x$. Then

$$F(x) = \log_2 \sqrt{x^2 + 1} = \log_2 \sqrt{h(x)}$$
$$= \log_2(g(h(x))) = f(g(h(x))) = f \circ g \circ h(x)$$

Thus, F is the composition $f \circ g \circ h$ of three functions. ■

Combinations of functions include sums, differences, products, quotients, constant times a function, and compositions. It will help in later work if you know how to describe a given function as one or more combinations.

EXAMPLE 8
(a) $\sqrt{x^2 + 1}$ is a *composition*. First evaluate $x^2 + 1$, then take the square root.
(b) $4x^3 / \sqrt{x^2 + 1}$ is a *quotient*. The numerator is 4 times the power function x^3, and the denominator is the *composition* in part (a).
(c) $2^x + \sin^2 x$ is a *sum*. The first term is the exponential 2^x, and the second is a *composition;* first compute $\sin x$, then square it.
(d) $(x^2 + x)(\sin x^2)$ is a *product*. The first factor is the *sum* $x^2 + x$, and the second factor is a *composition;* first find x^2, then take the sine.
(e) $\cos^2(x + 1)$ is *composition of a composition*. First find $x + 1$, then take the cosine, then square the result.
(f) $\cos^2(x + 1)^3$ is a *composition of a composition of a composition*. First find $x + 1$, then cube it, then take cosine, then square the result. ■

EXERCISES

In Exercises 1–10, compute the indicated values.

1. $f(x) = x^2, g(x) = x + 1; (4f + g)(3)$

2. $f(x) = 3^x, g(x) = x^2 - 1; (f/g)(2)$

3. $f(x) = \sin x, g(x) = x^3; (fg)(\pi)$

4. $f(x) = \sqrt{x}, g(x) = x^2 + 5; (f - 3g)(4)$

5. $f(x) = 1/x, g(x) = 2x - 1; f \circ g(5)$

6. $f(x) = \sqrt[3]{x + 1}, g(x) = x^3; f \circ g(0)$

7. $f(x) = x + 1, g(x) = 2^x; g \circ f(-1)$

8. $f(x) = 1/x, g(x) = \cos x; g \circ f(2/\pi)$

9. $f(x) = x^3 + 1, g(x) = 1/x, h(x) = \tan x; f \circ g \circ h(\pi/4)$

10. $f(x) = x - 1, g(x) = x + 1, h(x) = 13; f \circ g \circ h(-200)$

In Exercises 11–20, find $f \circ g(x)$ and $g \circ f(x)$, and note if they are the same.

11. $f(x) = \sin x, g(x) = x^2$

12. $f(x) = 2^x, g(x) = x^2$

13. $f(x) = 2x^2 - 1, g(x) = 7x$

14. $f(x) = x^3, g(x) = x^2 - 2$

15. $f(x) = x^3, g(x) = \sqrt{x}$

16. $f(x) = 3x, g(x) = 4x$

17. $f(x) = 4x + 1, g(x) = (x - 1)/4$

18. $f(x) = 3 - 2x, g(x) = (3 - x)/2$

19. $f(x) = x^3 - 1, g(x) = \sqrt[3]{x + 1}$

20. $f(x) = 1/x, g(x) = 1/x$

In Exercises 21–40, describe the given function as one or more combinations (see Example 8).

21. $(x^2 + 1)^2$

22. $\sqrt[3]{x} + \sin x$

23. $x^2/\sqrt{x^2 + 1}$

24. $x^2 \sqrt[3]{x} + \sin x$

25. $\sqrt{x^2/(x^2 + 1)}$

26. $\sqrt[3]{x^2(x + \sin x)}$

27. $2^{\sin x}$

28. 3^{x^2}

29. $\sin 2^x$

30. $(3^x)^2$

31. $\sin \sqrt[3]{x^2 + 1}$

32. $\cos(x^2 + 1)^3$

33. $[2x + (x + 1)^3]^4$

34. $[x^2(x + 3)^3]^4$

35. $\cos^3(x + 1)^2$

36. $\sin^2 \sqrt{x^3 + 1}$

37. 2^{2^x}

38. 4^{x^4}

39. $(x^2 + 4)^3(x^3 - 3x)^5$

40. $(x^2 + 4)^3/(x^3 - 3x)^5$

41. Show that "the average rate of change of a constant times a function is the constant times the average rate of change of the function." That is, as the variable changes from x to $x + \Delta x$, show that

$$\frac{\Delta(cf)}{\Delta x} = c\frac{\Delta f}{\Delta x}$$

42. Show that "the average rate of change of a sum is the sum of the average rates of change." That is,

$$\frac{\Delta(f + g)}{\Delta x} = \frac{\Delta f}{\Delta x} + \frac{\Delta g}{\Delta x}$$

Is a similar statement for the average rate of change of a product true?

A function f is *odd* if $f(-x) = -f(x)$ for each x in its domain; f is *even* if $f(-x) = f(x)$. For example, $f(x) = x^3$ is odd because $f(-x) = (-x)^3 = -x^3 = -f(x)$ and $g(x) = x^2$ is even because $g(-x) = (-x)^2 = x^2 = g(x)$. Determine if the functions below are odd or even or neither.

43. $f(x) = x^4 - 3x^2 + 2$

44. $f(x) = \sqrt{x^2 + 1}$

45. $g(x) = \sqrt[3]{x^3 + x}$

46. $g(x) = 4x^3 - 3x$

47. $f(x) = x^2 - 3x + 1$

48. $f(x) = 2x^3 - x^2$

49. $g(x) = \sin x$

50. $g(x) = \cos x$

51. $f(x) = 3^x$

52. $f(x) = \tan x$

53. If f and g are even functions, is fg even, odd, or, possibly neither? What about $f + g$?

54. Answer the same questions as in Exercise 53 if f and g are both odd.

55. Answer the same questions as in Exercise 53 if f is odd and g is even.

56. Let f be any function defined for all real numbers. (a) Show that $g(x) = f(x) + f(-x)$ is an even function. (b) Show that $h(x) = f(x) - f(-x)$ is an odd function. (c) Use parts (a) and (b) to show that f can be written as the sum of an even and an odd function.

57. (a) If f is an even function, show that (x, y) is on the graph of f if and only if $(-x, y)$ is also on the graph. What does this mean geometrically? (b) If f is odd, show that (x, y) is on the graph if and only if $(-x, -y)$ is also on the graph. What does this mean geometrically?

Optional Exercises

(Inverses) Two functions f and g are said to be **inverses** of each other if

$f \circ g(x) = x$ for all x in the domain of g,   and
$g \circ f(x) = x$ for all x in the domain of f

For example, $f(x) = 3x + 4$ and $g(x) = (x - 4)/3$ are inverses because

$$f \circ g(x) = f\left(\frac{x-4}{3}\right) = 3\left(\frac{x-4}{3}\right) + 4 = x, \quad \text{and}$$

$$g \circ f(x) = g(3x + 4) = \frac{(3x + 4) - 4}{3} = x$$

Verify that each of the following pairs of functions are inverses (show that $f \circ g(x) = x$ and $g \circ f(x) = x$).

58. $f(x) = 2x - 7, g(x) = (x + 7)/2$
59. $f(x) = 1/x, g(x) = 1/x$
60. $f(x) = x^3 + 1, g(x) = \sqrt[3]{x - 1}$
61. $f(x) = x^5, g(x) = \sqrt[5]{x}$
62. $f(x) = 2^x, g(x) = \log_2 x$
63. $f(x) = \log_4 x, g(x) = 4^x$
64. Make up two functions that are inverses of each other.
65. Make up two functions f and g for which $f \circ g(x) \neq g \circ f(x)$.
66. Make up two functions f and g that are not inverses but $f \circ g(x) = g \circ f(x)$.

1–9   INVERSE FUNCTIONS

The discussions in the preceding sections, covering familiar material, were presented in outline form. We suspect, however, that the subject of inverse functions represents new material for many of you. Therefore, our discussion of the following topics is presented in detail.

TOPIC	EXAMPLE
Inverse functions	If $f(x) = 2^x$, then $f^{-1}(x) = \log_2 x$
One-to-one functions	 one-to-one not one-to-one
Graphs of inverses	

It often happens that the composition of two functions f and g, in either order, is the identity function

$$f \circ g(x) = x \quad \text{and} \quad g \circ f(x) = x$$

Such functions are called *inverses*. For instance, if $f(x) = 5x$ and $g(x) = x/5$, then

$$f \circ g(x) = f\left(\frac{x}{5}\right) = 5\left(\frac{x}{5}\right) = x \quad \text{and} \quad g \circ f(x) = g(5x) = \frac{5x}{5} = x$$

Thus, f and g are inverses. Another important example is $f(x) = a^x$ and $g(x) = \log_a x$ for $a > 0$ and $a \neq 1$. In this case

$$f \circ g(x) = a^{\log_a x} = x \quad \text{and} \quad g \circ f(x) = \log_a a^x = x$$

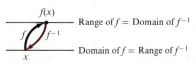

$f(x)$

Range of f = Domain of f^{-1}

Domain of f = Range of f^{-1}

x

FIGURE 1–73. Arrow representation of inverses.

DEFINITION

Let f and g be functions such that the range of f is the domain of g and the range of g is the domain of f. If

$$f \circ g(x) = f(g(x)) = x \quad \text{for all } x \text{ in the domain of } g \qquad (1.63)$$

and

$$g \circ f(x) = g(f(x)) = x \quad \text{for all } x \text{ in the domain of } f$$

then g is called the **inverse** of f and is denoted by the symbol f^{-1}, read "f inverse."*

The range of f is the domain of f^{-1} and the range of f^{-1} is the domain of f. The function f^{-1} reverses what f does and vice versa. This is illustrated in Figures 1–73 and 1–74. Most important are the relationships

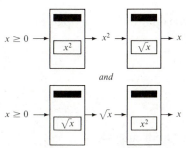

$x \geq 0 \longrightarrow \boxed{x^2} \longrightarrow x^2 \longrightarrow \boxed{\sqrt{x}} \longrightarrow x$

and

$x \geq 0 \longrightarrow \boxed{\sqrt{x}} \longrightarrow \sqrt{x} \longrightarrow \boxed{x^2} \longrightarrow x$

FIGURE 1–74. Machine representation of inverses.

(1.64) $\boxed{f^{-1}(f(x)) = x \text{ and } f(f^{-1}(x)) = x \text{ for all } x \text{ in the appropriate domain.}}$

EXAMPLE 1 Suppose f and f^{-1} are defined for all real numbers. Then

(a) $f^{-1}(f(17)) = 17$ and $f(f^{-1}(-\sqrt{2})) = -\sqrt{2}$

(b) If $f(3) = 5$, then $f^{-1}(5) = 3$. If $f^{-1}(\pi) = 12$, then $f(12) = \pi$.

In general,

(1.65) $\boxed{f(x) = y \quad \text{if and only if} \quad f^{-1}(y) = x}$ ∎

Remark: Pocket calculators cannot be used to find inverse functions, but they can be used to check statement (1.65) if the inverse is known. For example, if $f(x) = x^2$ for $x \geq 0$, then $f^{-1}(x) = \sqrt{x}$. Test this by entering a positive number like 36.834 in your calculator. Push $\boxed{x^2}$ and then $\boxed{\sqrt{x}}$. Also try it in reverse order. If there are no round-off errors, the final display will be 36.834. You can also test that $f(x) = a^x$ and $g(x) = \log_a x$ are inverses for base $a = 10$. To check that $\log_{10} 10^x = x$, the sequence is $\boxed{10}$, $\boxed{y^x}$, enter any number

---

*For functions, f^{-1} means "f inverse" and *not* $1/f$.

x, $\boxed{=}$, $\boxed{\log}$, and you get x back again. To check that $10^{\log_{10} x} = x$, the sequence is $\boxed{10}$, $\boxed{y^x}$, enter a positive number x (why positive?), $\boxed{\log}$, $\boxed{=}$ and you get x back.* Use your $\boxed{\ln}$ key to test that the same is true for the base $e \approx 2.7182818$. ∎

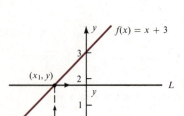

A. One-to-one.

Not every function has an inverse. The function $g(x) = x^2$ is an example. The reason is that $g(2)$ and $g(-2)$ both have the same value 4. If g had an inverse, then $g^{-1}(4)$ would have *two* values, 2 and -2, so g^{-1} is not a function. Whenever a function has the same value at two distinct points it will not have an inverse. Therefore, only those functions that assume distinct values at distinct points have inverses; such functions are said to be *one-to-one*.

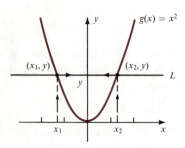

B. Not one-to-one.

FIGURE 1–75. A function is one-to-one if and only if each horizontal line meets its graph in at most one point.

> **DEFINITION**
> (1.66)
> A function f defined on X is **one-to-one on** X if, for every pair of distinct points x_1 and x_2 in X, we have
> $$f(x_1) \neq f(x_2)$$

A simple example is $f(x) = x + 3$. If $x_1 \neq x_2$, then $x_1 + 3 \neq x_2 + 3$, so $f(x_1) \neq f(x_2)$. An example of a function that is *not* one-to-one is $g(x) = x^2$. If $x_1 = 2$ and $x_2 = -2$, then $x_1 \neq x_2$, but $g(x_1)$ and $g(x_2)$ both equal 4.

The graph of a function f quickly reveals whether or not f is one-to-one (Figure 1–75). Suppose that L is a horizontal line; every point on L has the same second coordinate, say y. If L meets the graph of f at two distinct points (x_1, y) and (x_2, y), then $x_1 \neq x_2$, but $f(x_1)$ and $f(x_2)$ both equal y, so f is not one-to-one. The converse is also true and therefore,

> (1.67)
> A function f is one-to-one if and only if each horizontal line meets its graph in at most one point.

If f is not one-to-one, then it does not have an inverse. But if f is one-to-one and y is any element in its range, then there is exactly *one* point x such that $y = f(x)$. The rule that assigns to y the value x is a function on the range of f; this function is f^{-1}. It follows that

> (1.68)
> A function has an inverse if and only if it is one-to-one.

EXAMPLE 2   The graph of $f(x) = \log_2 x$ drawn in Figure 1–76A. Each horizontal line meets the graph in at most one point, so f is one-to-one. Therefore,

---

*The sequence may be slightly different for your calculator. See the footnote on page 41.

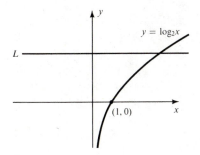

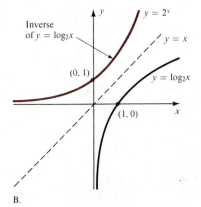

FIGURE 1–76. Example 2. The graphs of inverses are mirror images of each other through the line $y = x$.

by (1.68), it has an inverse. We already know what that inverse is; $f^{-1}(x) = 2^x$. Figure 1–76B shows both graphs drawn on the same set of axes. Notice that they are reflections of each other through the (dashed) line $y = x$. We will have more to say about this in a moment. ■

We know from (1.68), that only certain types of functions have inverses. Now suppose that a function is known to have an inverse; the question is, how do we find it? One way is to write

$$(1.69) \quad \boxed{f(f^{-1}(x)) = x}$$

and solve for $f^{-1}(x)$. For instance, if $f(x) = 2x + 7$, then

$$f(f^{-1}(x)) = 2(f^{-1}(x)) + 7$$

By (1.69), we set this equal to x and solve for $f^{-1}(x)$.

$$2(f^{-1}(x)) + 7 = x$$
$$2(f^{-1}(x)) = x - 7$$
$$f^{-1}(x) = \frac{x - 7}{2}$$

This is the inverse of f. Here are some other examples.

EXAMPLE 3 *(Temperature)* In the United States, temperature is reported in degrees *Fahrenheit,* whereas countries that use the metric system report the temperature in degrees *centigrade* (or *Celsius*). The function that converts Fahrenheit to centigrade is

$$f(x) = \frac{5}{9}(x - 32)$$

Find the inverse and sketch its graph.

Solution $f(f^{-1}(x)) = \frac{5}{9}(f^{-1}(x) - 32)$

by the definition of f. Using (1.69), we have

$$\frac{5}{9}(f^{-1}(x) - 32) = x$$

$$f^{-1}(x) - 32 = \frac{9}{5}x$$

$$f^{-1}(x) = \frac{9}{5}x + 32$$

The inverse of f is the formula that converts centigrade to Fahrenheit. The functions f and f^{-1} have been graphed on the same coordinate axes in Figure 1–77.

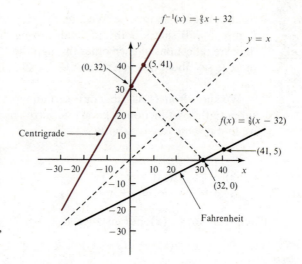

FIGURE 1–77. Example 3. Observe the reversal of coordinates; for example, (32, 0) is on the graph of f and (0, 32) is on the graph of f^{-1}.

Notice that *the graph of f^{-1} is the mirror image of the graph of f through the line $y = x$.* ∎

EXAMPLE 4   Find the inverse of $f(x) = x^3 + 4$.

Solution   *First,* use the definition of f to write out $f(f^{-1}(x))$. Since $f(x) = x^3 + 4$, we have

$$f(f^{-1}(x)) = [f^{-1}(x)]^3 + 4$$

Next, use (1.69) and solve for $f^{-1}(x)$.

$$[f^{-1}(x)]^3 + 4 = x$$
$$[f^{-1}(x)]^3 = x - 4$$
$$f^{-1}(x) = \sqrt[3]{x - 4}$$

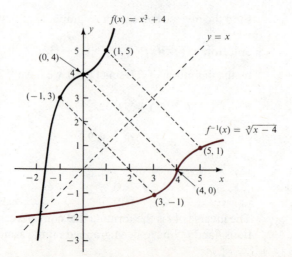

FIGURE 1–78. Example 4. Observe the reversal of coordinates; (x, y) is on the graph of f if and only if (y, x) is on the graph of f^{-1}.

The functions f and f^{-1} have been graphed on the same set of axes in Figure 1–78. Again observe that they are reflections of each other about the line $y = x$. ∎

The next example illustrates a procedure for finding the domain and range of an inverse.

EXAMPLE 5   Find the inverse of $f(x) = \sqrt{x + 2}$.

Solution   *First,* find the formula for f^{-1}. Since $f(f^{-1}(x)) = \sqrt{f^{-1}(x) + 2}$, you know that

$$\sqrt{f^{-1}(x) + 2} = x \tag{1.69}$$
$$f^{-1}(x) + 2 = x^2$$
$$f^{-1}(x) = x^2 - 2$$

Second, find the domain and range of f^{-1}. The domain of $f(x) = \sqrt{x + 2}$ is the set of all $x \geq -2$, and the range of f is the set of all $x \geq 0$. It follows that

$$\text{domain of } f^{-1} = \text{range of } f = \text{all} \quad x \geq 0$$
$$\text{range of } f^{-1} = \text{domain of } f = \text{all} \quad x \geq -2$$

Conclusion: The inverse of $f(x) = \sqrt{x + 2}$ is

$$f^{-1}(x) = x^2 - 2 \quad x \geq 0$$

The graphs of f and f^{-1} are sketched in Figure 1–79. ∎

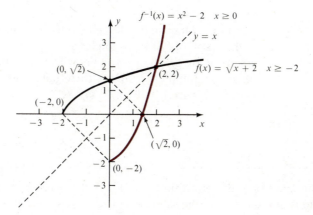

FIGURE 1–79. Example 5. The graphs of f and f^{-1}. The domain of f^{-1} is *automatically* taken to be the range of f; in this case, all $x \geq 0$.

When f and f^{-1} are graphed on the same set of axes, they are mirror images of each other. Let us see why this is so. A point (x, y) is on the graph of f if and only if $y = f(x)$. By (1.65), this means that $x = f^{-1}(y)$ which is equivalent to saying that (y, x) is on the graph of f^{-1}. Also observe that the point (y, x) is the reflection of the point (x, y) through the line $y = x$. Check this out in Figures 1–77, 78, and 79.

$$(1.70)$$

(1) (x, y) is on the graph of f if and only if (y, x) is on the graph of f^{-1}.

(2) The graphs of f and f^{-1} are reflections of each other through the line $y = x$.

This information can be used to sketch graphs of inverse functions.

EXAMPLE 6   Figure 1–80A shows the graph of a one-to-one function. Sketch the graph of its inverse.

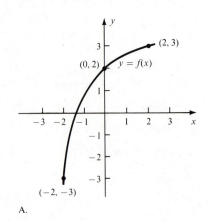

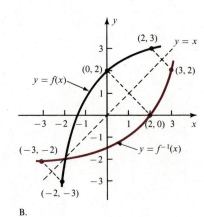

FIGURE 1–80. Example 6. Using the graph of $y = f(x)$ to sketch the graph of $y = f^{-1}(x)$.

Solution   Draw in the dashed line $y = x$; the graph of f^{-1} is the mirror image of the graph of f. It helps to plot a few points. Since $(-2, -3)$ $(0, 2)$, and $(2, 3)$ are all on the graph of f, you know from (1.70) that $(-3, -2)$, $(2, 0)$, and $(3, 2)$ lie on the graph of f^{-1}. Once these points are located, the graph of f^{-1} is easy to draw (Figure 1–80B).   ■

EXERCISES

In Exercises 1–20, find the inverses of the given functions, then sketch the graphs of f and f^{-1} on the same set of axes and indicate the domain of each.

1. $f(x) = 3x$

2. $f(x) = 5x$

3. $f(x) = x$

4. $f(x) = -x$

5. $f(x) = \frac{1}{2}x + 2$

6. $f(x) = 2x - 1$

7. $f(x) = -\frac{2}{3}x + 1$

8. $f(x) = -\frac{1}{4}x + 2$

9. $f(x) = -\sqrt{x}$

10. $f(x) = \sqrt{x}$

11. $f(x) = \frac{1}{2}\sqrt{-x}$

12. $f(x) = \sqrt{3x}$

13. $f(x) = \sqrt{x + 4}$

14. $f(x) = \sqrt{4 - x}$

15. $f(x) = \sqrt{x} + 2$

16. $f(x) = \sqrt{x} + 4$

17. $f(x) = (x + 1)^3 - 2$

18. $f(x) = 1 - x^3$

19. $f(x) = \sqrt[3]{x - 2} + 1$

20. $f(x) = \sqrt[3]{x + 1} - 2$

In Exercises 21–23, use the graphs to sketch the inverse of each function shown. What is the domain and range of the inverse?

21.

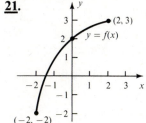

22.

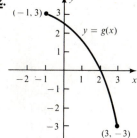

23.

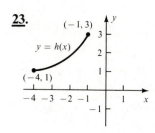

27. Let $f(x) = 4x - 7$. For what x is $f(x) = 9$? Without finding an explicit formula for f^{-1}, can you evaluate $f^{-1}(9)$? What is $f^{-1}(-3)$?

28. Let $f(x) = \sqrt[3]{x + 3} - 2$. Without finding an explicit formula for f^{-1}, evaluate $f^{-1}(-3)$.

29. Let $f(x) = x^2 - 4x + 3$ for $x \geq 2$. Without finding an explicit formula for f^{-1}, evaluate $f^{-1}(0)$.

30. On Monday, the topic of inverses is discussed in a calculus class; on the following Monday a test is given on inverses. From past experience, the professor knows that the learning model equation (Example 7 in Section 1–6) is $S = 80 - 7.5 \log_{10}(t + 1)$. What should the average score be on the test?

31. (Continuation of Exercise 30) How long (in days) after discussing inverses in class would it take for the average score to drop down to 70?

In Exercises 24–26, the figures show portions of the graphs of the sine, cosine, and tangent functions. Sketch the inverse functions and find the domain and range.

24.

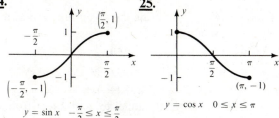

25.

26.

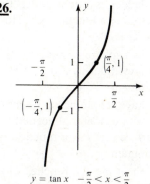

REVIEW EXERCISES

In Exercises 1–4, find the set of all x that satisfy the given inequality.

1. $|x + 2| < .01$

2. $|3x - 2| < 1/2$

3. $|4 - 2x| < .1$

4. $\left|\dfrac{x - 2}{4}\right| < .001$

In Exercises 5–12, determine where $f(x)$ is positive, negative, and zero.

5. $f(x) = x(3 - 2x)$

6. $f(x) = x^2 + 7x + 12$

7. $f(x) = x^3 - 4x^2 + 3x$

8. $f(x) = x^4 - 16$

9. $f(x) = \sqrt{x} - \dfrac{5}{\sqrt{x}}$

10. $f(x) = \dfrac{x - 3}{x + 4}$

11. $f(x) = x - \dfrac{6}{x + 5}$

12. $f(x) = \dfrac{x^2 - 3x - 4}{x^2 + 2x - 3}$

In Exercises 13–24, graph the function. Determine whether or not the function has an inverse. Find the inverse if there is one, and graph it.

13. $y = 1$

14. $y = -3$

15. $y = 2x - 1$

16. $y = 2 - x$

17. $y = x^2 - 5x + 6, x \geq 5/2$

18. $y = -x^2 + 2x + 3$

19. $y = x^3 + 1$

20. $y = (x + 1)^3$

21. $y = 1/x$

22. $y = \sqrt{x + 4}$

23. $y = 2 \sin 2x, -2\pi \leq x \leq 2\pi$

24. $y = (1/3)^x$

In Exercises 25–28, describe the graph of the given equation.

25. $x^2 - 3x + y^2 + 4y = -3$

26. $x^2 + 2x + y^2 - y = -3$

27. $x^2 + 2x + y^2 - 4y = -5$

28. $x^2 - x + y^2 - 3y = 0$

In Exercises 29–34, the indicated point is on the graph of the given function. Find x or y, as the case may be.

29. $f(x) = 3x + 4; (x, 10)$

30. $f(x) = x^3 - 2; (x, 6)$

31. $f(x) = x^2 - 3x + 1; (2 + h, y)$

32. $f(x) = \cos x; (\pi/4, y)$

33. $f(x) = \log_3 x; (x, 2)$

34. $f(x) = 3^x; (x, 27)$

In Exercises 35–38, find an equation of the line.

35. Passing through $(1, 4)$ with slope $-1/2$.

36. Passing through $(-3, 1)$ and $(0, 4)$.

37. Passing through $(-1/4, \sqrt{15}/4)$ and tangent to the unit circle $x^2 + y^2 = 1$.

38. Passing through $(4, 1)$ and parallel to the graph of $2x + y = 6$.

In Exercises 39–44, compute the average rate of change of f as the variable changes from x to $x + \Delta x$.

39. $f(x) = 5$

40. $f(x) = 4 - 3x$

41. $f(x) = x^2 + x - 3$

42. $f(x) = x^3$

43. $f(x) = 1/x$

44. $f(x) = \sqrt{x}$

In Exercises 45–48, the given point lies on the terminal side of an angle α in standard position. Find the sine, cosine, and tangent of α.

45. $(-3, 4)$

46. $(0, 5)$

47. $(-1, -\sqrt{3})$

48. $(1, -2)$

In Exercises 49–52, describe the given function as one or more combinations.

49. $\sqrt{\sin x^2}$

50. $2^{\sqrt{x^2 + 1}}$

51. $[x^2 + \sin^2 x]^4$

52. $2^x(x + \cos^2 x)^3$

53. A 6-foot fence runs parallel to, and 2 feet from, a tall building. A ladder extends from the ground, touches the top of the fence, and rests on the building. If the top of the ladder is y feet above the ground, express its length L as a function of y. How long is the ladder if $y = 12$ feet?

54. A cylinder is inscribed in a sphere of radius 4. Express the volume of the cylinder as a function of its radius. What is the domain?

55. A cylindrical can is to have a volume of 30 cubic inches. Express the amount of material needed for the can (that is, the surface area) as a function of its radius. What is the domain?

56. A rectangular piece of sheet metal 15 inches wide is made into a rain gutter by bending up the sides to make angles of 120° with the base. Express the cross-sectional area of the gutter as a function of the width of the base. If the gutter is 10 feet long and the base is 5 inches, how much water will it hold?

57. A point (x, y) is on the graph of $f(x) = x^2 - 3x + 1$. Express its distance to $(-1, 5)$ as a function of x.

58. The sum of two numbers x and y is 43. Express their product as a function of x.

59. The *outside* dimensions of a picture frame are to be 25 by 30 inches. How much wood (in square inches) is needed for a frame x inches wide? What is the domain?

60. An automobile is traveling so that its distance s miles from town at t hours is $s = 10t^2 - 3t$. Find the average speed during the third hour.

61. A large spherical balloon is being blown up. What is the average rate of increase in volume as the radius increases from 6 to 9 inches?

62. The x-axis, y-axis, and any line with negative slope passing through (3, 4) determine a right triangle. If the line crosses the x-axis at $(x, 0)$, express the area of the triangle as a function of x.

63. If the selling price of apples is 15¢ each, a merchant can sell 400 of them a day. For each penny increase in price, he sells 15 fewer apples. Express his revenue as a function of x, the *increase* in price. What is his revenue if he increases the price by 4¢? By 6¢? By 8¢?

64. Two planes leave an airport at the same time. The first travels due east at 300 mph and the second travels in the direction North 30° East at 400 mph. Express the distance between them as a function of time.

65. The base of an isosceles triangle is 4 inches. Express the area as a function of the angle α that the equal sides make with the base. What is the domain?

66. A projectile is fired straight up and its distance s feet above the ground at time t seconds is $s = -16t^2 + 160t$. Sketch its position curve.

67. A particle moves along a line and its position s in feet at time t seconds is $s = t^2 - 4t + 3$, $-1 \leq t \leq 4$. Sketch its position curve and the motion diagram.

68. The half-life of a radioactive substance is 1,000 years. If you start with 5 grams, express the amount present Q as a function of time t in years. How much will be left after 250 years?

LIMITS AND CONTINUITY

In calculus it is often necessary to examine the values that a function f takes *near* a point p rather than *at* a point p. That is, we shall study the numbers $f(x)$ when x is close, but not equal, to p. Indeed, f may not even be defined at p. Nevertheless, if $f(x)$ gets closer and closer to a number L as x gets closer and closer (but not equal) to p, then we say that the limit of $f(x)$ equals L as x approaches p. This is expressed in symbols by writing either

$$\lim_{x \to p} f(x) = L \quad \text{or} \quad f(x) \to L \text{ as } x \to p$$

The origin of the limit concept dates back to Isaac Newton (1642–1727) and Gottfried Leibniz (1646–1716), generally considered to be the "fathers" of calculus. They were looking for a mathematical way to describe motion; more specifically, a way to define the velocity of a moving particle at a given instant. Some contemporaries of Newton and Leibniz believed that instantaneous velocity was impossible to define. They argued as follows:

At a given instant, say t, the particle is at a single point, say s. But if the particle is at a single point, then it is not moving. If it is not moving, then it has no instantaneous velocity.

The argument is reasonable, but leaves unanswered the following question:

If the particle has no velocity at time t, shouldn't it stay at the point s indefinitely?

The solution to this puzzle depends on how you define instantaneous velocity.

There is no problem defining the average velocity; if Δs represents the change in position of the particle over a time interval from t to $t + \Delta t$, then the average velocity v_{av} is

$$v_{av} = \frac{\Delta s}{\Delta t}$$

over that time interval. It is very tempting to define the instantaneous velocity as the value of v_{av} when $\Delta t = 0$. But, of course, division by zero is not allowed. However, division by a number close (but not equal) to zero is allowed. Although there is no change in position over a time interval of length zero, there is a change in position over time intervals of any *nonzero* length Δt, *no matter how small.* Newton and Leibniz used different terminology and different symbols, but their ideas were similar. They defined the velocity at time t to be the ratio $\Delta s/\Delta t$ where Δt is the length of what they called an "infinitely small" or "infinitesimal" time interval. In modern language, we say that for each nonzero value of Δt, compute the average velocity $\Delta s/\Delta t$, and if these average velocities get closer and closer to some number v as Δt gets closer and closer (but not equal) to zero, then v is the instantaneous velocity of the particle at time t. In other words,

$$v = \lim_{\Delta t \to 0} \frac{\Delta s}{\Delta t}$$

Thus, the notion of a limit evolved from the study of motion. However, it soon became evident that limits could be applied to a wide variety of problems. Today, the limit is an indispensable tool in the social, life, and physical sciences, as well as in mathematics.

The first three sections of this chapter treat limits in an informal setting. Intuition and common sense are the order of the day. There are many illustrations of techniques for finding limits and a large number of practice exercises. Section 2–1 introduces the limit of a function, and several limit theorems are discussed in Section 2–2. In Section 2–3, limits are used to define a continuous function; such functions are particularly important in calculus.

There are two exceptions to the informal treatment. At the end of Section 2–2, a precise definition of limit and two examples of proofs are given. Also, Section 2–4 is devoted entirely to a rigorous discussion of limits. Although both exceptions are marked *Optional,* every student should at least read through them; those wanting a deeper understanding should study them.*

As you study the text and work the exercises, here are some points to emphasize. (1) Master the informal techniques for finding limits; use your intuition and imagination. (2) Learn the statements of the limit theorems; they will be referred to often. (3) Develop your intuition about continuous functions.

---

*Much of the early work in calculus was based on intuition. In fact, many of the results in this book concerning limits, derivatives, and integrals were already known by the year 1700. But the deeper the researchers probed, the more they realized the need for a solid foundation. That foundation, in the form of precise definitions and rigorous proofs, appeared about 1820 in the work of Augustin-Louis Cauchy (1789–1857).

2-1 LIMITS

The notion of a limit evolved from the study of motion. Let us follow history and begin with an example about velocity.*

EXAMPLE 1 A particle moves along a horizontal line and its position function is $s = t^2 - 7t$; s is measured in feet, t in seconds. Find its velocity when $t = 2$ seconds.

Solution As we mentioned in the introduction to this chapter, the average velocity v_{av} over a time interval from t to $t + \Delta t$ is

$$v_{av} = \frac{\Delta s}{\Delta t} = \frac{s(t + \Delta t) - s(t)}{\Delta t}$$

If the average velocities get closer and closer to some number v as Δt gets closer and closer (but not equal) to 0, then v is the instantaneous **velocity** at time t.

To find the velocity at time $t = 2$, let Δt represent a change in time from 2 to $2 + \Delta t$, and compute the *average* velocity for that time interval. That is, compute

$$v_{av} = \frac{\Delta s}{\Delta t} = \frac{s(2 + \Delta t) - s(2)}{\Delta t}$$

The change in time, Δt, can be positive or negative, but *never* zero. For example, if $\Delta t = 1$, then

$$v_{av} = \frac{s(2 + 1) - s(2)}{1} = \frac{[3^2 - 7(3)] - [2^2 - 7(2)]}{1} = -2 \text{ ft/sec}$$

This means that during the third second of motion (from $t = 2$ to $t = 3$) the average velocity is -2 ft/sec; the minus sign indicates that the motion is to the *left*. Now let $\Delta t = -.5$; then

$$v_{av} = \frac{s(1.5) - s(2)}{-.5} = \frac{-8.25 + 10}{-.5} = -3.5 \text{ ft/sec}$$

This means that during the time interval from $t = 1.5$ to $t = 2$, the average velocity is -3.5 ft/sec. The following table shows average velocities for various values of Δt.

$\dfrac{s(2 + \Delta t) - s(2)}{\Delta t}$	Δt	$\pm .1$	$\pm .01$	$\pm .001$	$\pm .0001$
	$\Delta t > 0$	-2.9	-2.99	-2.999	-2.9999
	$\Delta t < 0$	-3.1	-3.01	-3.001	-3.0001

---

*In everyday conversation, the words *velocity* and *speed* usually mean the same thing. But in scientific work, there is a distinction. Speed indicates only how fast an object is moving. Velocity indicates both speed *and* direction. (See Example 1 above).

You know that Δt can never be 0, but the table suggests that as Δt gets closer to 0, the average velocity gets closer and closer to -3 ft/sec. This fact is expressed in words, *the limit of $\Delta s/\Delta t$ as Δt approaches zero is -3*, or in symbols,

$$\lim_{\Delta t \to 0} \frac{\Delta s}{\Delta t} = -3$$

The limit of the average velocity is the (instantaneous) velocity. Thus, the velocity at $t = 2$ is -3 ft/sec, which means that the particle is traveling at a speed of 3 ft/sec in the negative direction (to the left). ∎

Although our primary interest lies in finding limits of quotients like $\Delta s/\Delta t$ as Δt approaches zero, the notion of a limit can be expanded to include general functions. Suppose that p is a point in an open interval (a, b) and that f is a function defined at all points of (a, b) except, perhaps, at p. If the numbers $f(x)$ get closer and closer to a fixed number L as x gets closer and closer (but not equal) to p, then we say that the **limit of f at p is L** (or the limit of $f(x)$ is L as x approaches p) and write

$$\lim_{x \to p} f(x) = L \quad \text{or} \quad f(x) \to L \text{ as } x \to p$$

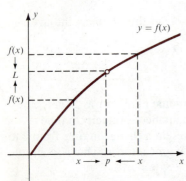

FIGURE 2–1. An illustration of $\lim_{x \to p} f(x) = L$. Even if f is defined at p, its value there does not matter; this is indicated by the small open circle above the point p. What does matter is that as x approaches p from *either* side, $f(x)$ approaches L.

Figure 2–1 illustrates the idea. The curve is part of the graph of a function f. As x approaches p *from either side* along the x-axis, $f(x)$ approaches L along the y-axis. The function f may not actually be defined at p, and $f(x)$ may never actually be equal to L; that does not matter. What does matter is that

> $f(x)$ can be made *arbitrarily close* to L by taking x *sufficiently close* to p.

Figure 2–2 illustrates a totally different situation. As x approaches p from the *right*, $f(x)$ approaches L. We say that L is the **right-hand limit of f at p**. This is expressed in symbols by writing

$$\lim_{x \to p^+} f(x) = L$$

The small plus sign indicates that x is approaching p from the right; x is larger than p. In the figure, as x approaches p from the *left*, $f(x)$ approaches the number M. We say that M is the **left-hand limit of f at p** and write

$$\lim_{x \to p^-} f(x) = M$$

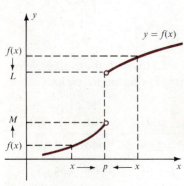

FIGURE 2–2. $\lim_{x \to p^+} f(x) = L$ and $\lim_{x \to p^-} f(x) = M$. Since $L \neq M$, $\lim_{x \to p} f(x)$ does not exist.

The small minus sign indicates that x is approaching p from the left; x is smaller than p. Since $L \neq M$, there is *no one* number that $f(x)$ gets close to as x approaches p from *either* side and, therefore,

$$\lim_{x \to p} f(x)$$

does not exist. The informal discussion along with the figures makes it reasonable to assert

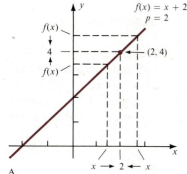

A.

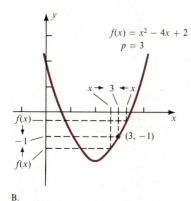

B.

FIGURE 2–3. (A) Example 2. (B) Example 3. In each case there is no break in the graph at the point p, and $f(x) \to f(p)$ as $x \to p$.

(2.1) $\boxed{\lim_{x \to p} f(x) = L \quad \text{if and only if} \quad \lim_{x \to p^+} f(x) = L \quad \text{and} \quad \lim_{x \to p^-} f(x) = L}$

This result is proved in Section 2–4.

EXAMPLE 2   Find $\lim_{x \to 2} x + 2$.

Solution   The function is $f(x) = x + 2$ and the number p is 2 (Figure 2–3A). To find the limit, ask yourself the question, "As x gets closer and closer (but not equal) to 2, does $x + 2$ get closer and closer to some one number?" If x is close to 2 (on either side), then $x + 2$ is close to 4. Therefore,

$$\lim_{x \to 2} x + 2 = 4$$

Here, $f(x) = x + 2$ is defined at $p = 2$ and the limit is $f(2) = 4$.   ∎

EXAMPLE 3   Find $\lim_{x \to 3} x^2 - 4x + 2$.

Solution   Here, $f(x) = x^2 - 4x + 2$ and $p = 3$ (Figure 2–3B). To find the limit, take each term separately and ask the question, "If x is close (but not equal) to 3, is x^2 close to some one number? Is $-4x$? Is 2?" If x is close to 3, then x^2 is close to 9, $-4x$ is close to -12, and 2 is 2 no matter what x is. Therefore, when x is close to 3 (on either side), then $x^2 - 4x + 2$ is close to $9 - 12 + 2 = -1$. Thus,

$$\lim_{x \to 3} x^2 - 4x + 2 = -1$$

Again, f is defined at $p = 3$ and the limit is $f(3) = -1$.   ∎

EXAMPLE 4   Find $\lim_{x \to 2} (x^2 - 4)/(x - 2)$.

Solution   The function is $f(x) = (x^2 - 4)/(x - 2)$ and p is 2. Notice that f is *not* defined at 2; the result of replacing x by 2 is $0/0$. Nevertheless, you can still ask, "Does $f(x)$ get close to some number L as x gets close (but not equal) to 2?" Indeed, if $x \neq 2$, then $f(x) = x + 2$ because

$$f(x) = \frac{x^2 - 4}{x - 2} = \frac{(x + 2)(x - 2)}{x - 2} = x + 2$$

Therefore, if x is close to 2 (on either side), then $f(x) = x + 2$ is close to 4, and

$$\lim_{x \to 2} \frac{x^2 - 4}{x - 2} = 4$$

Calculator Check Check this limit on your calculator. Compute the values of

$$f(x) = \frac{x^2 - 4}{x - 2}$$

for $x = 1.9, 2.1, 1.99, 2.01, 1.999,$ and 2.001. Are the values approaching 4? ∎

Remark: In Examples 2 and 3, the functions are defined at p and, in both cases, $f(x) \to f(p)$ as $x \to p$. Many functions have this property and they are discussed in Section 2–3. However, in Example 4, the function $f(x) = (x^2 - 4)/(x - 2)$ has the limit 4 at $p = 2$ even though it is not defined there. Moreover, if we decided to give this function a value at $p = 2$, *the limit would still be 4 no matter what value is assigned to $f(2)$.* For example, if

$$f(x) = \begin{cases} \dfrac{x^2 - 4}{x - 2} & \text{if } x \neq 2 \\ 5 & \text{if } x = 2 \end{cases}$$

then $f(x) \to 4$ as $x \to 2$ (Figure 2–4). Notice that f is defined at p, but the limit is *not* $f(p)$.

> The value, if any, of f at p has no effect on the *limit* of f at p; f may not be defined at p, and if it is, the value of $f(p)$ does not matter.

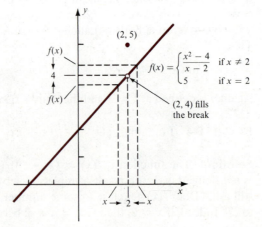

FIGURE 2–4. The function $(x^2 - 4)/(x - 2) \to 4$ as $x \to 2$ *no matter what value is assigned to it at the point* 2. If a break in the graph of f can be filled in with a *single* point (p, L), then $f(x) \to L$ as $x \to p$. In this case, $(2, 4)$ fills the break, so $f(x) \to 4$ as $x \to 2$.

The next two examples illustrate functions for which the limit does not exist.

EXAMPLE 5 Show that $\lim_{x \to 0} |x|/x$ does not exist.

Solution If $x > 0$, then $|x| = x$, so $|x|/x = x/x = 1$. If $x < 0$, then $|x| = -x$, so $|x|/x = -x/x = -1$. Thus,

Show that $\lim_{x \to 2} 1/(x - 2)$ does not exist.

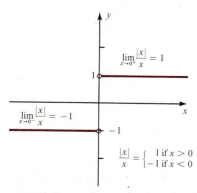

$$\frac{|x|}{x} = \begin{cases} 1 & \text{if } x > 0 \\ -1 & \text{if } x < 0 \end{cases}$$

The graph (Figure 2–5) suggests that

$$\lim_{x \to 0^+} \frac{|x|}{x} = 1 \quad \text{and} \quad \lim_{x \to 0^-} \frac{|x|}{x} = -1$$

and, by (2.1), the limit does not exist.

Question. Does $\lim_{x \to p} |x|/x$ exist at points p other than 0? What is the limit if $p = 2$? If $p = -3$? *Answer:* The limit does exist at points $p \neq 0$. The limit is 1 if $p > 0$ and the limit is -1 if $p < 0$. ■

FIGURE 2–5 Example 5. The limit does not exist at 0. If a break in the graph cannot be filled in with a *single* point, then the limit does not exist.

EXAMPLE 6 Show that $\lim_{x \to 2} 1/(x - 2)$ does not exist.

Solution Figure 2–6 shows the graph of this function. As x approaches 2 from the right, the denominator is very small but positive. The smaller the denominator, the larger the quotient; it does not approach one particular number. As x approaches 2 from the left, the denominator is very small but negative. The quotient gets larger and larger in the negative direction; again it does not approach one particular number. Therefore, in this example, even the right- and left-hand limits do not exist.

Question. Does $\lim_{x \to p} 1/(x - 2)$ exist at points p other than 2? What is the limit if $p = 3$? If $p = 0$? *Answer:* The limit exists at all other points p. The limit at $p = 3$ is 1; the limit at $p = 0$ is $-\frac{1}{2}$. ■

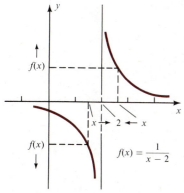

FIGURE 2–6. Example 6. The limit at $p = 2$ does not exist.

In Chapter 3, our primary interest is in limits of the form

$$\lim_{\Delta x \to 0} \frac{f(x + \Delta x) - f(x)}{\Delta x}$$

Zero cannot be substituted for Δx because the result is $0/0$. However, if f is an algebraic function, there is usually some algebraic manipulation that helps to reduce the quotient to manageable form. Here are three different examples.

EXAMPLE 7 Let $f(x) = 2x^2 - 3x + 4$. For any x, compute

$$\lim_{\Delta x \to 0} \frac{f(x + \Delta x) - f(x)}{\Delta x}$$

Solution Note that x is a fixed number; *it is Δx that gets close* (but not equal) *to zero.*

Step 1 Write out the quotient.

$$\frac{f(x + \Delta x) - f(x)}{\Delta x}$$

$$= \frac{[2(x + \Delta x)^2 - 3(x + \Delta x) + 4] - [2x^2 - 3x + 4]}{\Delta x}$$

Step 2 Perform the indicated algebra (write $2(x + \Delta x)^2 = 2x^2 + 4x\Delta x + 2\Delta x^2$, $-3(x + \Delta x) = -3x - 3\Delta x$, and so on); then combine like terms, and divide by Δx.

$$= \frac{2x^2 + 4x\Delta x + 2\Delta x^2 - 3x - 3\Delta x + 4 - 2x^2 + 3x - 4}{\Delta x}$$

$$= \frac{4x\Delta x + 2\Delta x^2 - 3\Delta x}{\Delta x}$$

$$= 4x + 2\Delta x - 3$$

Step 3 Only now, after all the algebra is done, do you think about the limit. If $\Delta x \neq 0$, then the quotient equals $4x + 2\Delta x - 3$; if Δx is close to zero, this sum is close to $4x - 3$. Therefore,

$$\lim_{\Delta x \to 0} \frac{f(x + \Delta x) - f(x)}{\Delta x} = 4x - 3$$

This limit holds for *any* x; for instance, if $x = 2$, then

$$\lim_{\Delta x \to 0} \frac{f(2 + \Delta x) - f(2)}{\Delta x} = 4(2) - 3 = 5 \qquad \blacksquare$$

EXAMPLE 8 Let $g(t) = 1/t$. For any $t \neq 0$, compute

$$\lim_{\Delta t \to 0} \frac{g(t + \Delta t) - g(t)}{\Delta t}$$

Solution Since $t \neq 0$ and Δt is close to 0, you can assume that $t + \Delta t \neq 0$.

Step 1 Write out the quotient.

$$\frac{g(t + \Delta t) - g(t)}{\Delta t} = \frac{\dfrac{1}{t + \Delta t} - \dfrac{1}{t}}{\Delta t}$$

Step 2 Perform the indicated algebra (find a common denominator and subtract the fractions); then divide by Δt.

$$= \frac{\dfrac{t - (t + \Delta t)}{(t + \Delta t)t}}{\Delta t}$$

$$= \frac{-\Delta t}{\Delta t(t + \Delta t)t}$$

$$= \frac{-1}{(t + \Delta t)t}$$

Step 3   Only now, after all the algebra is done, do you think about the limit. If $\Delta t \neq 0$, the quotient is $-1/(t + \Delta t)t$. If Δt is close to zero, then

$$t + \Delta t \quad \text{is close to } t$$

$$(t + \Delta t)t \quad \text{is close to } t^2$$

$$\frac{-1}{(t + \Delta t)t} \quad \text{is close to } \frac{-1}{t^2}$$

Therefore,

$$\lim_{\Delta t \to 0} \frac{\dfrac{1}{t + \Delta t} - \dfrac{1}{t}}{\Delta t} = -\frac{1}{t^2}$$

 Calculator Check   This limit holds for all $t \neq 0$; if $t = 2$, the limit is $-\dfrac{1}{2^2} = -.25$. Check this out on your calculator. Compute

$$\frac{\dfrac{1}{2 + \Delta t} - \dfrac{1}{2}}{\Delta t}$$

for $\Delta t = \pm.1, \pm.01,$ and $\pm.001.$ Do the values approach $-.25?$ ■

EXAMPLE 9   Let $f(x) = \sqrt{x}$. For any $x > 0$, compute

$$\lim_{\Delta x \to 0} \frac{f(x + \Delta x) - f(x)}{\Delta x}$$

Solution   Since $x > 0$ and Δx is close to 0, you can assume that $x + \Delta x > 0$.

Step 1   Write out the quotient.

$$\frac{f(x + \Delta x) - f(x)}{\Delta x} = \frac{\sqrt{x + \Delta x} - \sqrt{x}}{\Delta x}$$

Step 2   The algebraic manipulation that gets rid of the troublesome quotient is called *rationalizing the numerator.*

$$= \frac{\sqrt{x + \Delta x} - \sqrt{x}}{\Delta x} \cdot \frac{\sqrt{x + \Delta x} + \sqrt{x}}{\sqrt{x + \Delta x} + \sqrt{x}}$$

$$= \frac{(x + \Delta x) - x}{\Delta x(\sqrt{x + \Delta x} + \sqrt{x})}$$

$$= \frac{1}{\sqrt{x + \Delta x} + \sqrt{x}}$$

Step 3   Now take the limit. If Δx is close to 0, then $\sqrt{x + \Delta x} + \sqrt{x}$ is close to $2\sqrt{x}$; therefore,

$$\lim_{\Delta x \to 0} \frac{\sqrt{x + \Delta x} - \sqrt{x}}{\Delta x} = \frac{1}{2\sqrt{x}}$$

Calculator Check   This limit holds for all $x > 0$. If $x = 2.25$, then the limit is $1/2\sqrt{2.25} = 0.333\ldots$. Check this out on your calculator. Compute the values of

$$\frac{\sqrt{2.25 + \Delta x} - \sqrt{2.25}}{\Delta x}$$

for $\Delta x = \pm .1$, $\Delta x = \pm .01$, and $\Delta x = \pm .001$. Do the values approach $0.333\ldots$? ■

WARNING   Roundoff errors in calculators can result in wrong answers to limit problems. In the last example, when Δx is very small, the numbers $\sqrt{2.25 + \Delta x}$ and $\sqrt{2.25}$ are so close to each other that their difference $\sqrt{2.25 + \Delta x} - \sqrt{2.25}$ may be rounded off to 0 in a calculator! The values obtained for the quotient

$$\frac{\sqrt{2.25 + \Delta x} - \sqrt{2.25}}{\Delta x}$$

when $\Delta x = \pm .1$, $\pm .01$, $\pm .001$ do get closer and closer to $0.333\ldots$. But, if we continued with smaller and smaller Δx, the calculator readouts would actually move *away* from $0.333\ldots$. For example, computing with an 8-digit calculator (with no additional digits in the memory), we find that when $\Delta x = 10^{-6}$

$$\frac{\sqrt{2.25 + .000001} - \sqrt{2.25}}{.000001} = 0.300\ 000\ 0$$

and when $\Delta x = 10^{-7}$

$$\frac{\sqrt{2.25 + .0000001} - \sqrt{2.25}}{.0000001} = 0.000\ 000\ 0$$

This is absolutely false! Moreover, the same pattern will emerge using any calculator, but the more digits available, the smaller Δx has to be.

NEVERTHELESS, calculators can be used effectively to check limits or even to *guess the value* of a limit, if you know the limit exists. Just remember the limitations.

If f is an algebraic function, then algebraic manipulation helps find the limit. If f is transcendental (for example, exponential or trigonometric), additional methods are necessary. There really is an essential difference between algebraic and transcendental functions. We are not prepared, at this moment, to discuss the limits of transcendental functions in detail. Rather, we are content to use a calculator (subject to the limitations discussed above) to obtain reasonably close approximations of these limits.

EXAMPLE 10   Let $f(x) = 2^x$. For any x compute

$$\lim_{\Delta x \to 0} \frac{f(x + \Delta x) - f(x)}{\Delta x}$$

Solution

Step 1   As before, write out the quotient.

$$\frac{f(x + \Delta x) - f(x)}{\Delta x} = \frac{2^{x+\Delta x} - 2^x}{\Delta x}$$

Step 2   Use the properties of exponents to write $2^{x+\Delta x} = 2^x 2^{\Delta x}$; then factor out 2^x.

$$= \frac{2^x 2^{\Delta x} - 2^x}{\Delta x}$$

$$= 2^x \left(\frac{2^{\Delta x} - 1}{\Delta x} \right)$$

Step 3 2^x is some fixed number because x is; but, what about the quotient? You cannot substitute 0 for Δx because the result is $(2^0 - 1)/0 = 0/0$; moreover, no amount of algebraic manipulation will make this troublesome quotient go away. The best we can do, at this point, is to use a calculator. The entries in the table below were found with an 8-digit pocket calculator.

$\dfrac{2^{\Delta x} - 1}{\Delta x}$	Δx	$\pm.1$	$\pm.01$	$\pm.001$	$\pm.0001$
	$\Delta x > 0$	0.7177346	0.6955550	0.6933865	0.6931620
	$\Delta x < 0$	0.6696701	0.6907504	0.6929061	0.6931140

A reasonable guess is that the limit lies half way between the last two entries; that is, 0.6931380. Therefore,

$$\lim_{\Delta x \to 0} \frac{2^{x+\Delta x} - 2^x}{\Delta x} \approx (.6931380)2^x$$

Calculator Check   This limit holds for all x; if $x = -3$, the limit is $(.6931380)2^{-3}$ $= .6931380/8 \approx .0866423$. Check this out on your calculator. Compute

$$\frac{2^{-3+\Delta x} - 2^{-3}}{\Delta x}$$

for $\Delta x = \pm.1, \pm.01,$ and $\pm.001$. Do the values approach $.0866423$?   ∎

EXERCISES

In Exercises 1–4, the functions are quotients of polynomials; they are not defined at the given point p, but the limits exist. Find the limit at p.

1. $f(x) = \dfrac{x^2 - 16}{x - 4}$; $p = 4$

2. $f(x) = \dfrac{x^2 - 9}{x + 3}$; $p = -3$

3. $f(x) = \dfrac{2x^2 + 3x + 1}{x^2 - 2x - 3}$; $p = -1$

4. $f(x) = \dfrac{x^2 - 2x + 1}{x^2 + 6x - 7}$; $p = 1$

In Exercises 5–8, combine the fractions in the numerator and find the limit at p.

5. $g(t) = \dfrac{\dfrac{1}{t} + \dfrac{1}{3}}{t + 3}$; $p = -3$

6. $g(t) = \dfrac{\dfrac{1}{t} - \dfrac{1}{2}}{t - 2}$; $p = 2$

7. $G(w) = \dfrac{\dfrac{3}{w + 1} - \dfrac{1}{2}}{w - 5}$; $p = 5$

8. $G(w) = \dfrac{\dfrac{2}{w + 2} + 1}{w + 4}$; $p = -4$

In Exercises 9–12, rationalize the numerator and find the limit at p.

9. $f(x) = \dfrac{\sqrt{x} - 2}{x - 4}$; $p = 4$

10. $f(x) = \dfrac{3 - \sqrt{x}}{x - 9}$; $p = 9$

11. $g(t) = \dfrac{\sqrt{t + 1} - \sqrt{7}}{t - 6}$; $p = 6$

12. $g(t) = \dfrac{\sqrt{t - 3} - \sqrt{2}}{t - 5}$; $p = 5$

In Exercises 13–14, explain why the limit at p does not exist.

13. $f(x) = \dfrac{x}{x - 1}$; $p = 1$

14. $f(x) = \begin{cases} x^2 - 3x, & 0 \le x < 1 \\ x^3, & 1 < x \le 2 \end{cases}$; $p = 1$

▦ In Exercises 15–18, assume the limit exists, and use a calculator to guess a reasonable approximation to the limit.

15. $\lim\limits_{h \to 0} \dfrac{3^h - 1}{h}$

16. $\lim\limits_{t \to 0} \dfrac{\sin t}{t}$ (t in radians)

17. $\lim\limits_{h \to 0} \dfrac{\log_{10}(1 + h)}{h}$

18. $\lim\limits_{t \to 0} \dfrac{\sin t°}{t°}$ (t in degrees)*

In Exercises 19–36, find the limit, if it exists.

19. $\lim\limits_{x \to 2} 3$

20. $\lim\limits_{x \to -2} x^3 + 1$

21. $\lim\limits_{x \to -3} x^2 - 3x + 7$

22. $\lim\limits_{x \to 1} x - 3x^2$

23. $\lim\limits_{x \to 4} \dfrac{\dfrac{1}{x} - \dfrac{1}{4}}{x - 4}$

24. $\lim\limits_{x \to 6} \dfrac{\dfrac{1}{x} - \dfrac{1}{6}}{x - 6}$

25. $\lim\limits_{x \to 0} \dfrac{x^2 - 3x}{x}$

26. $\lim\limits_{x \to -2} \dfrac{x^2 + 4x + 4}{x + 2}$

27. $\lim\limits_{x \to 1} \dfrac{x^2 - 3x + 2}{x^2 - 2x + 1}$

28. $\lim\limits_{x \to 1} \dfrac{x - 1}{x^2 - 1}$

29. $\lim\limits_{x \to 3} \dfrac{\sqrt{x - 1} - \sqrt{2}}{x - 3}$

30. $\lim\limits_{x \to 2} \dfrac{\sqrt{x + 5} - \sqrt{7}}{x - 2}$

31. $\lim\limits_{x \to 2} \dfrac{x^3 - 8}{x - 2}$

32. $\lim\limits_{x \to 2} \dfrac{x^3 - 8}{x^2 - 4}$

33. $\lim\limits_{x \to 0} \sin x$

34. $\lim\limits_{x \to 0} \cos x$

35. $\lim\limits_{x \to 4} \dfrac{x - 4}{\sqrt{x} - 2}$

36. $\lim\limits_{x \to 0} \dfrac{1}{x} + 2$

In Exercises 37–54, assume that x and $x + \Delta x$ are in the domain of f, and find

$$\lim_{\Delta x \to 0} \frac{f(x + \Delta x) - f(x)}{\Delta x}$$

37. $f(x) = 4$

38. $f(x) = x + 2$

39. $f(x) = 3 - 8x$

40. $f(x) = 4x - 3$

41. $f(x) = 2x^2 - x$

42. $f(x) = x^2 - 4x$

43. $f(x) = x^3 - 4x$

44. $f(x) = 2x^3 + 1$

45. $f(x) = 1/(x + 1)$

46. $f(x) = 1/(x - 3)$

47. $f(x) = 1/2x$

48. $f(x) = -1/3x$

49. $f(x) = \sqrt{x + 4}$

50. $f(x) = \sqrt{x - 2}$

51. $f(x) = 4/\sqrt{x - 1}$

52. $f(x) = 3/\sqrt{x + 6}$

53. $f(x) = 3^x$

54. $f(x) = 10^x$

55. A particle moves along a line, and its position function is $s = t^2 - 4t + 1$ feet at t seconds. What is its velocity when $t = 1$ and $t = 2$? In what direction is it moving at these times?

---

*Compare the answers to Exercises 16 and 18. This is another reason we use radians rather than degrees.

56. Same as Exercise 55 but the position function is
$s = 1/t, t > 0$.

Test Your Intuition about Limits. For each of the following statements, ask yourself what happens as the variable gets closer and closer (but not equal) to p. Then decide whether the statement is true or false.

57. If $\lim\limits_{x \to p} f(x) = L$ and $\lim\limits_{x \to p} g(x) = M$, then $\lim\limits_{x \to p} f(x) + g(x) = L + M$.

58. If $\lim\limits_{x \to p} f(x) = L$ and $\lim\limits_{x \to p} g(x) = M$, then $\lim\limits_{x \to p} f(x) g(x) = LM$.

59. If $\lim\limits_{x \to p} f(x) = L$ and $\lim\limits_{x \to p} g(x) = M$, then $\lim\limits_{x \to p} f(x)/g(x) = L/M$.

60. If $\lim\limits_{x \to p} f(x)/g(x) = L$ and $\lim\limits_{x \to p} g(x) = 0$, then $\lim\limits_{x \to p} f(x) = 0$.

61. *If* $\lim\limits_{x \to p} f(x) = L$, then $\lim\limits_{h \to 0} f(p + h) = L$.

62. If $\lim\limits_{x \to p} f(x) = L$, then $f(p) = L$.

63. If $f(p) = L$, then $\lim\limits_{x \to p} f(x) = L$.

64. If $\lim\limits_{x \to p} f(x) = L$, then $\lim\limits_{x \to p} (f(x) - L) = 0$.

65. $\lim\limits_{x \to 0} \sin(1/x) = 1$

2–2 LIMIT THEOREMS

This section contains statements of several limit theorems. Actually, we have already used these theorems to find limits, but now it is important to state them formally so they can be referred to later. The intuitive approach is carried forward from the preceding section. Our arguments are intended to establish the *plausibility* of the statements; the precise proofs are presented in Section 2–4. At the end of this section, however, we give an example of a correct mathematical definition of limit and a precise proof.

Throughout this section *we assume that the functions discussed are defined at every point on an open interval except, possibly, at the point p.*

A function f may not have a limit at p. But if it does, there can be only one such limit because as x gets closer and closer to p, $f(x)$ cannot get closer and closer to two distinct numbers at the same time. Therefore,

(2.2) | **UNIQUENESS OF LIMITS**
A function can have at most one limit at p.

If f is a constant function, say $f(x) = c$, then for any x (close to p or not), $f(x) = c$. Therefore,

(2.3) | **LIMIT OF A CONSTANT**
The limit of a constant function is that constant. In symbols.

$$\lim_{x \to p} c = c$$

If f is the identity function $f(x) = x$ and x is very close to p, then $f(x)$ is very close to p (Figure 2–7). That is,

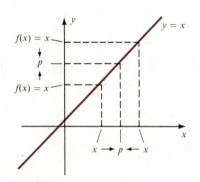

FIGURE 2–7. $\lim\limits_{x \to p} x = p$

(2.4)

> **LIMIT OF THE IDENTITY**
> The limit of the identity function $f(x) = x$ at any number p is p. In symbols,
>
> $$\lim_{x \to p} x = p$$

The next result is about combinations of functions. Suppose $f(x) \to L$ and $g(x) \to M$ both happen as $x \to p$. That is, if x is close to p, then $f(x)$ is close to L *and* $g(x)$ is close to M. It is natural to expect that $f(x) + g(x)$ is close to $L + M$, and that $f(x)g(x)$ is close to LM. Moreover, if $M \neq 0$, then $f(x)/g(x)$ is close to L/M.

(2.5)

> Suppose $\lim_{x \to p} f(x) = L$ and $\lim_{x \to p} g(x) = M$. Then
>
> (i) $\lim_{x \to p} (f + g)(x) = L + M$
>
> (ii) $\lim_{x \to p} (fg)(x) \quad = LM$
>
> (iii) $\lim_{x \to p} (f/g)(x) \quad = L/M$, provided $M \neq 0$

Theorem (2.5) can be stated in words: if both f and g have limits at p, then

(i) *The limit of the sum is the sum of the limits.*

(ii) *The limit of the product is the product of the limits.*

(iii) *The limit of the quotient is the quotient of the limits, provided the limit of the denominator is not zero.*

It is important to note that (2.5) can be extended to any finite number of functions. For example, the limit of a sum of three functions is the sum of the three limits; the limit of a product of five functions is the product of the five limits, and so on. Thus, if you think of ax^n as a product $a \cdot x \cdot x \cdots x$ of a constant function and n identity functions, then

$$\lim_{x \to p} ax^n = \left[\lim_{x \to p} a\right]\left[\lim_{x \to p} x\right] \cdots \left[\lim_{x \to p} x\right] \qquad \text{(2.5 ii)}$$

$$= ap^n \qquad \text{(2.3), (2.4)}$$

Since a polynomial is a sum of such products and is defined for all real numbers, it follows from (2.5i) that

(2.6)

> **LIMIT OF A POLYNOMIAL**
> The limit of any polynomial at any number p is the value of the polynomial at p. In symbols,
>
> $$\lim_{x \to p} a_n x^n + \cdots + a_1 x + a_0 = a_n p^n + \cdots + a_1 p + a_0$$

EXAMPLE 1

(a) $\lim_{x \to 3} 2x^2 - 3x + 1 = 2(3)^2 - 3(3) + 1 = 10$ (2.6)

(b) $\lim_{x \to -4} (x^2 - 3)(x + 2)^2 = \left[\lim_{x \to -4} x^2 - 3\right]\left[\lim_{x \to -4} x + 2\right]^2$ (2.5ii)

$$= [(-4)^2 - 3][-4 + 2]^2 \qquad (2.6)$$

$$= 52 \qquad \blacksquare$$

Before using the quotient theorem, be sure to check that the limit of the denominator is not zero.

EXAMPLE 2   The $\lim_{x \to 4} x^2 - 5x + 1 = -3 \neq 0$ and, therefore,

$$\lim_{x \to 4} \frac{x^2 - 2x - 6}{x^2 - 5x + 1} = \frac{\lim_{x \to 4} x^2 - 2x - 6}{\lim_{x \to 4} x^2 - 5x + 1} \qquad (2.5\text{iii})$$

$$= \frac{4^2 - 2(4) - 6}{4^2 - 5(4) + 1} \qquad (2.6)$$

$$= -\frac{2}{3} \qquad \blacksquare$$

EXAMPLE 3   Show $\lim_{x \to 4} (x^2 - 2x - 6)/(x^2 - 5x + 4)$ does not exist. In this case, the limit of the denominator is 0, so the quotient theorem cannot be used. But we can reason as follows. When x is close to 4, the numerator is close to 2 (check), but the denominator is close to 0. This is similar to the situation described in Example 6 of Section 2–1. The closer the denominator is to zero, the larger the quotient (either positive or negative). Therefore, the limit does not exist. $\blacksquare$

EXAMPLE 4   Find $\lim_{x \to 4} (x^2 - 2x - 8)/(x^2 - 5x + 4)$. In this case, the limit of the denominator is 0, but so is the limit of the numerator; both have factors of $x - 4$ which cancel out. First factor and cancel, then use (2.5iii)

$$\lim_{x \to 4} \frac{x^2 - 2x - 8}{x^2 - 5x + 4} = \lim_{x \to 4} \frac{(x - 4)(x + 2)}{(x - 4)(x - 1)} = \frac{6}{3} = 2 \qquad \blacksquare$$

Here are two more useful limit theorems. In the first one, we must be careful because if n is an even integer, then the nth root is not defined for negative numbers.

(2.7)

> **ROOT THEOREM**
>
> If $f(x) \to L$ as $x \to p$, then
>
> $$\lim_{x \to p} \sqrt[n]{f(x)} = \sqrt[n]{\lim_{x \to p} f(x)} = \sqrt[n]{L}$$
>
> provided n is odd *or* n is even and $L > 0$.

$$(2.8) \quad \boxed{\begin{array}{l} \textbf{ABSOLUTE VALUE THEOREM} \\ \text{If } f(x) \to L \text{ as } x \to p, \text{ then} \\[8pt] \displaystyle\lim_{x \to p} |f(x)| = \left|\lim_{x \to p} f(x)\right| \end{array}}$$

EXAMPLE 5

(a)
$$\lim_{x \to 1} \sqrt[3]{x^2 + 3x - 7} = \sqrt[3]{\lim_{x \to 1} x^2 + 3x - 7} \qquad (2.7)$$
$$= \sqrt[3]{-3} \qquad (2.6)$$

(b) $\displaystyle\lim_{x \to 4} x^{2/5} = \lim_{x \to 4} \sqrt[5]{x^2} = \sqrt[5]{16}$ ∎

EXAMPLE 6   Find $\lim_{x \to 4} (\sqrt{x} - 2)/(x - 4)$.

Solution   The limit of the denominator is zero so the quotient theorem cannot be used; evaluation at $x = 4$ yields $0/0$. In this case, you must somehow change the form of the function so that the theorems do apply. For this particular limit, rationalize the numerator. If $x \neq 4$, then

$$\frac{\sqrt{x} - 2}{x - 4} = \frac{\sqrt{x} - 2}{x - 4} \cdot \frac{\sqrt{x} + 2}{\sqrt{x} + 2} = \frac{x - 4}{(x - 4)(\sqrt{x} + 2)} = \frac{1}{\sqrt{x} + 2}$$

Now, the quotient and root theorems will apply. Therefore, by (2.5iii) and (2.7),

$$\lim_{x \to 4} \frac{\sqrt{x} - 2}{x - 4} = \lim_{x \to 4} \frac{1}{\sqrt{x} + 2} = \frac{1}{\sqrt{4} + 2} = \frac{1}{4}$$ ∎

We call the next result the *Squeeze Theorem*. A glance at Figure 2–8 shows why this name is appropriate. Essentially, the theorem says that if $f(x)$ and $g(x)$ are both close to L, then any number between them is also close to L.

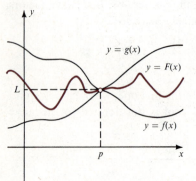

FIGURE 2–8. The Squeeze Theorem.

$$(2.9) \quad \boxed{\begin{array}{l} \textbf{SQUEEZE THEOREM} \\[6pt] \text{If} \quad \displaystyle\lim_{x \to p} f(x) = L \quad \text{and} \quad \lim_{x \to p} g(x) = L \\[6pt] \text{and } f(x) \le F(x) \le g(x) \text{ except, perhaps, at } x = p, \text{ then} \\[6pt] \displaystyle\lim_{x \to p} F(x) = L \end{array}}$$

The Squeeze Theorem is often used as a strategic tool to find an otherwise elusive limit.

EXAMPLE 7   Show that $\lim_{x \to 0} x \sin(1/x) = 0$.

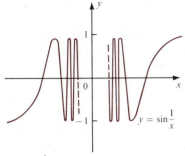

A. $\lim\limits_{x\to 0}\sin\dfrac{1}{x}$ does not exist.

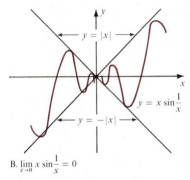

B. $\lim\limits_{x\to 0} x\sin\dfrac{1}{x} = 0$

FIGURE 2–9. Example 7.

Solution If the limit of $\sin(1/x)$ exists at $p = 0$, then we can use the theorem about products (2.5ii) to find the limit of $x\sin(1/x)$. Unfortunately, this is not the case. Figure 2–9A shows that as x gets closer and closer to 0, the values of $\sin(1/x)$ oscillate more and more rapidly between -1 and 1. They do not get close to any single number, so the limit does not exist. Therefore, to prove the limit of the product is 0, we try to squeeze it between functions whose limits are known to be 0. Observe that

$$|x\sin(1/x)| \le |x| \qquad\qquad \text{(because } |\sin(1/x)| \le 1)$$

and, therefore,

$$-|x| \le x\sin(1/x) \le |x| \qquad\qquad (|a| \le b \text{ means } -b \le a \le b)$$

See Figure 2–9B. Thus, $x\sin(1/x)$ is squeezed between $-|x|$ and $|x|$ both of which approach 0 as x approaches 0, by (2.8). It follows from the Squeeze Theorem that

$$\lim_{x\to 0} x\sin\left(\frac{1}{x}\right) = 0 \qquad \blacksquare$$

Definition of a Limit (*Optional*)

One number is "close to" another number means that the distance between them is small. The distance between numbers can be expressed in terms of absolute value. For example, the statement "$f(x)$ is close to L" can be written as $|f(x) - L| < E$ where E is a small positive number. Similarly, "x is close (but not equal) to p" can be written as $0 < |x - p| < D$ where D is a small positive number. Therefore, the statement "$f(x)$ is close to L when x is close (but not equal) to p" can be expressed in symbols

$$|f(x) - L| < E \quad \text{when} \quad 0 < |x - p| < D$$

If we think of E as *error* and D as *discrepancy*, the statement above suggests the following type of problem. A cylindrical shaft with a cross-sectional area of $L = \pi p^2$ square inches is to be milled. At various stages in the milling process, the lathe operator measures the diameter of the shaft, divides by 2 to obtain the radius x, and computes the cross-sectional area, $f(x) = \pi x^2$. If the radius is exactly p inches, then $f(p) = L$; that is *perfection*. But we cannot expect perfection; x will be slightly different from p, and $f(x)$ will be slightly different from L. What we can expect is that, given an acceptable error E (that is, $|f(x) - L| < E$ is acceptable), then it is possible to find a suitable discrepancy D in measuring the radius x such that

$$|f(x) - L| < E \quad \text{whenever} \quad 0 < |x - p| < D$$

Thus, the milling process is finished when the measured radius x is within D units of the number p.

In milling a cylindrical shaft, a single acceptable error is specified. However, in the theory of limits, *all* positive numbers are acceptable errors. For *each* error E, there must be *some* discrepancy D that will ensure an error less than E.

$$(2.10)$$

> **DEFINITION OF LIMIT**
> Let I be an open interval containing p, and let f be defined at all points of I except, perhaps, at p. Then $\lim_{x \to p} f(x) = L$ if, for every $\epsilon > 0$, there is a $\delta > 0$ such that
>
> $$|f(x) - L| < \epsilon \quad \text{whenever} \quad 0 < |x - p| < \delta^*$$

Remark: According to the definition, f may or may not be defined at p. Notice, however, that f *is defined at all x close to p*. That is, there are open intervals of the form (a, p) and (p, b) on which f is defined.

This definition is a precise way of saying that $f(x)$ approaches L as x approaches p. *Every* positive error is allowed. Once an error is specified, there must be *some* positive discrepancy with the property that *if x is within that discrepancy of p, then f(x) will be within the specified error of L* (Figure 2–10).

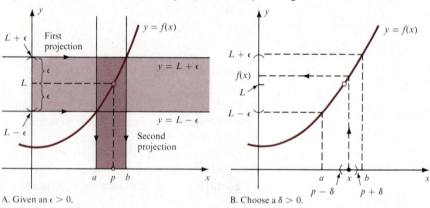

FIGURE 2–10. $f(x) \to L$ as $x \to p$. (A) For any given $\epsilon > 0$, first project the interval $(L - \epsilon, L + \epsilon)$ horizontally to the graph of f and then vertically to the x-axis. This determines an interval (a, b) which contains p. (B) Choose $\delta > 0$ so that $(p - \delta, p + \delta)$ is contained in (a, b). Now if any $x \neq p$ is chosen in this δ-interval, then the distance between $f(x)$ and L is less than ϵ.

A. Given an $\epsilon > 0$. B. Choose a $\delta > 0$.

Given an error ϵ, how do you find a discrepancy δ that "works?" For most functions, the easiest thing is to work backwards.

EXAMPLE 6 Find $\lim_{x \to 2} 3x + 1$ and prove your answer.

Solution Here, $f(x) = 3x + 1$ and $p = 2$. The intuitive approach suggests that the limit is 7. To *prove* this is correct, let ϵ represent *any* positive number. You must find a positive δ with the property that if $0 < |x - 2| < \delta$, then $|(3x + 1) - 7| < \epsilon$. Start with $|(3x + 1) - 7| < \epsilon$ and work backwards

$$|(3x + 1) - 7| = |3x - 6| < \epsilon$$

$$-\epsilon < 3x - 6 < \epsilon \qquad (|a| < b \text{ means } -b < a < b)$$

$$-\frac{\epsilon}{3} < x - 2 < \frac{\epsilon}{3} \qquad \text{(Divide by 3)}$$

$$|x - 2| < \frac{\epsilon}{3}$$

---

*It is customary to let the Greek letters ϵ (epsilon) and δ (delta) stand for error and discrepancy.

Notice that these steps are reversible; each step implies the preceding one. You are looking for a δ that "works" and the last line above provides it; let δ = ε/3. Now, if $0 < |x - 2| < δ$, then the last line is true and, therefore, each of the preceding lines, including the first one, $|(3x + 1) - 7| < ε$ is true. Thus, definition (2.10) is satisfied and this proves that $3x + 1 \to 7$ as $x \to 2$. *Note:* Discrepancies are not unique; clearly, any positive number less than ε/3 would also serve as a discrepancy in this example. ■

We conclude with an example of a proof using the definition of a limit.

Theorem If $\lim_{x \to p} f(x) = L$ and $\lim_{x \to p} g(x) = M$, then $\lim_{x \to p} (f + g)(x) = L + M$.

First, we do some reasoning to sort out the basic idea of the proof.

Basic Idea We want to show that if x is close (but not equal) to p, then $f(x) + g(x)$ is close to $L + M$; that is, $|(f(x) + g(x)) - (L + M)|$ is small. The Triangle Inequality (1.9) may be used to write

$$|(f(x) + g(x)) - (L + M)|$$
$$= |(f(x) - L) + (g(x) - M)| \quad \text{(Rearrange)}$$
$$\leq |f(x) - L| + |g(x) - M| \quad \text{(1.9)}$$

Now, if each of the numbers $|f(x) - L|$ and $|g(x) - M|$ is very small, then so is their sum.

Proof Let ε > 0. Because $\lim_{x \to p} f(x) = L$, there is a $δ_1 > 0$ such that

$$(2.11) \quad |f(x) - L| < ε/2 \quad \text{whenever} \quad 0 < |x - p| < δ_1$$

Because $\lim_{x \to p} g(x) = M$, there is a $δ_2 > 0$ such that

$$(2.12) \quad |g(x) - M| < ε/2 \quad \text{whenever} \quad 0 < |x - p| < δ_2$$

Now let δ be the smaller of $δ_1$ and $δ_2$. Then (2.11) and (2.12) hold with $δ_1$ and $δ_2$ replaced by δ. Thus, if $0 < |x - p| < δ$, then

$$|(f(x) + g(x)) - (L + M)| = |(f(x) - L) + (g(x) - M)|$$
$$\leq |f(x) - L| + |g(x) - M|$$
$$< \frac{ε}{2} + \frac{ε}{2}$$
$$= ε \quad ■$$

EXERCISES

In Exercises 1–20, find the limits. Justify each step.

1. $\lim_{x \to 0} 5$

2. $\lim_{x \to -2} \pi$

3. $\lim_{x \to -2} 4x + 1$

4. $\lim_{x \to 16} -2x + 32$

5. $\lim_{x \to 3} x^2 - 2x + 1$

6. $\lim_{x \to -3} 2x^2 + 4$

7. $\lim_{x \to 1} \dfrac{x^3 + 3}{x^2 - 2x + 4}$

8. $\lim_{x \to -2} \dfrac{x^4 - 7}{x + 7}$

9. $\lim_{t \to 1/2} \dfrac{\sqrt{t + 1}}{\sqrt{t}}$

10. $\lim_{t \to 4} \dfrac{t^2 + \sqrt{t}}{\sqrt[3]{t} - 5}$

11. $\lim\limits_{u \to 0} \dfrac{\sqrt{u+1}+1}{\sqrt{u+1}}$

12. $\lim\limits_{u \to 0} \dfrac{\sqrt{4-u}}{u-3}$

13. $\lim\limits_{x \to 9} \dfrac{\sqrt{x}-3}{x-9}$

14. $\lim\limits_{x \to 16} \dfrac{\sqrt{x}-4}{x-16}$

15. $\lim\limits_{x \to 2} \dfrac{\dfrac{1}{\sqrt{x}}-\dfrac{1}{\sqrt{2}}}{x-2}$

16. $\lim\limits_{x \to 49} \dfrac{\dfrac{1}{\sqrt{x}}-\dfrac{1}{7}}{x-49}$

17. $\lim\limits_{x \to -3} (x^2 - 7)^{14}(x + 7)$

18. $\lim\limits_{x \to 0}(x^4 - 1)^7(x + 1)^{10}$

19. $\lim\limits_{x \to 2} \dfrac{x^2 - 4x + 4}{x^2 + 3x - 10}$

20. $\lim\limits_{x \to 7} \dfrac{x^2 + x - 56}{x^2 - 8x + 7}$

In Exercises 21–26, x is measured in radians. Assume the limits exist, and use a calculator to guess an approximation to the limits.

21. $\lim\limits_{x \to 0} \dfrac{\sin x}{x}$

22. $\lim\limits_{x \to 0} \dfrac{\sin^2 x}{x}$

23. $\lim\limits_{x \to 0} \dfrac{\sin 3x}{5x}$

24. $\lim\limits_{x \to 0} \dfrac{\cos x - 1}{x}$

25. $\lim\limits_{x \to 0} \dfrac{\cos x - 1}{x^2}$

26. $\lim\limits_{x \to 0} \dfrac{[\cos x - 1]^2}{x^3}$

In Exercises 27–40, assume that x and $x + \Delta x$ are in the domain of f, and find

$$\lim\limits_{\Delta x \to 0} \dfrac{f(x + \Delta x) - f(x)}{\Delta x}$$

27. $f(x) = -37$

28. $f(x) = \sqrt{2}$

29. $f(x) = x/\pi$

30. $f(x) = -x/\sqrt{2}$

31. $f(x) = \pi/x$

32. $f(x) = -\sqrt{2}/x$

33. $f(x) = -3.6x^2$

34. $f(x) = 1.9x^2$

35. $f(x) = \sqrt{x}/9.31$

36. $f(x) = -\sqrt{x}/49$

37. $f(x) = 9.31/\sqrt{x}$

38. $f(x) = -49/\sqrt{x}$

39. $f(x) = x^2$

40. $f(x) = 2^x$
   (use a calculator)

If a particle moves along a line, and its position function is $s = s(t)$, then its *velocity function* is a function v whose value at any time t is

$$v(t) = \lim\limits_{\Delta t \to 0} \dfrac{s(t + \Delta t) - s(t)}{\Delta t}$$

Recall the exercises in Chapter 1 that dealt with determining the intervals on which a function is positive, negative, or zero. Determining those intervals for a velocity function tells you whether the particle is moving to the right, to the left, or standing still. For the position functions in Exercises 41–46, (a) find the velocity function, (b) find those intervals on which it is positive, negative, or zero, and (c) use that information to sketch a motion diagram.

41. $s(t) = 2t - 3$

42. $s(t) = 5 - 4t$

43. $s(t) = -4t^2 + 6t + 2$

44. $s(t) = 3t^2 - 4t + 1$

45. $s(t) = \dfrac{1}{3}t^3 - t + 1$

46. $s(t) = \dfrac{1}{3}t^3 - 2t^2 + 3t$

---

47. (This exercise is continued in Exercise 43 of the next section.)

(a) Find

$$\lim\limits_{\Delta x \to 0} \dfrac{f(x + \Delta x) - f(x)}{\Delta x}$$

for $f(x) = x^2$, $f(x) = x^3$, and $f(x) = x^4$.

(b) Write the answers obtained in part (a) in a column. Do you see a pattern being established? Can you guess what the limit would be if $f(x) = x^5$? Verify your guess by actually computing the limit.

(c) Guess a formula for the limit that holds for $f(x) = x^n$ whenever $n = 2, 3, 4, \ldots$.

(d) Does your formula also hold for $n = 1$; that is, for $f(x) = x$?

48. Let f be the function

$$f(x) = \begin{cases} 1 & \text{if } x \text{ is rational} \\ 0 & \text{if } x \text{ is irrational} \end{cases}$$

Its graph, which is impossible to draw, is the collection of points one unit above the x-axis at each rational number, and on the x-axis at each irrational number. *Its limit does not exist at any point p.* Nevertheless,

$$\lim\limits_{x \to 0} xf(x) = 0$$

Use the Squeeze Theorem to show this is true.

Optional Exercises

In Exercises 49–54, find the limits intuitively. Then, for any acceptable error ϵ, find a discrepancy δ that "works."

49. $\lim\limits_{x \to 3} x$

50. $\lim\limits_{x \to -2} 3x + 13$

51. $\lim\limits_{x \to 0} -2x + 5$

52. $\lim\limits_{x \to .5} \dfrac{2}{3}x - 1$

53. $\lim\limits_{x \to 3} \dfrac{x^2 - 9}{x - 3}$

54. $\lim\limits_{x \to -8} \dfrac{x^2 + 16x + 64}{x + 8}$

2–3   CONTINUOUS FUNCTIONS

A function f can have a limit at p even though it is not defined at p. For example,

$$f(x) = \frac{x^2 - 4}{x - 2}$$

is undefined at $x = 2$, yet the limit of f at 2 is 4 (Figure 2–11A). A function f can be defined at p, and have a limit at p, but the limit is not equal to $f(p)$. For example, if

$$f(x) = \begin{cases} \dfrac{x^2 - 4}{x - 2} & \text{if } x \neq 2 \\ 3 & \text{if } x = 2 \end{cases}$$

then the limit of f at $x = 2$ is 4 but $f(2) = 3$ (Figure 2–11B). But sometimes a function f is defined at p, has a limit at p, and the limit *equals* $f(p)$. For example, if

$$f(x) = \begin{cases} \dfrac{x^2 - 4}{x - 2} & \text{if } x \neq 2 \\ 4 & \text{if } x = 2 \end{cases}$$

then the limit of f at $x = 2$ is 4 and $f(2) = 4$ (Figure 2–11C). Such a function is said to be *continuous at p*.

FIGURE 2–11. In parts (A) and (B), there is a break in the graph at (2, 4); these functions are not continuous at $p = 2$. In part (C) the break is filled in; this function is continuous at $p = 2$.

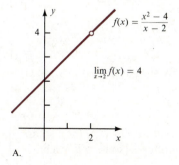

A.

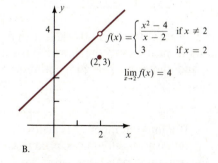

B.

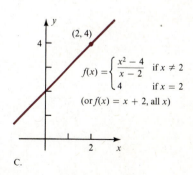

C.

Intuitively, a function is continuous at p if $f(x)$ is close to $f(p)$ whenever x is close to p. There are no breaks in the graph of a continuous function; it is a "continuous" curve. Most of the functions encountered in applications are continuous. For example, the velocity of a rocket as a function of time, the pressure exerted by a confined gas as a function of temperature, and the volume of an expanding sphere as a function of its radius are all continuous functions. But some important functions are not continuous. Populations, for instance, do not vary continuously with time; whenever someone is born or someone dies, the population "jumps" up or down by 1. Biologists, however, *assume* that populations are continuous functions of time. Then they can apply the methods of calculus to predict population growth patterns with great accuracy.

In this section, we define and examine some of the properties of continuous functions. Once more, the treatment is intuitive and informal.

(2.13)

Let f be defined on an open interval containing p. Then f is **continuous at** p if

$$\lim_{x \to p} f(x) = f(p)$$

If f is continuous at each point of its domain, we say simply that f is **continuous.**

In order for f to be continuous at p, three things must be true. Here is a check list:

(2.14)
(1) f must be defined on an open interval containing p
(2) the limit of f at p exists
(3) the limit equals $f(p)$

Let us use the check list to show that a polynomial f is continuous: (1) f is defined at every point, (2) the limit exists, by (2.6), and (3) the limit equals $f(p)$, also by (2.6). Therefore,

(2.15) | Polynomials are continuous.

If f and g are both continuous at p, then $f(x) \to f(p)$ *and* $g(x) \to g(p)$ as $x \to p$. It follows from the limit theorem about sums (2.5i), that $f(x) + g(x) \to f(p) + g(p)$ and, therefore, $f + g$ is continuous at p. By similar arguments, we conclude that

(2.16)

If a finite number of functions are all continuous at p, then so are their sums, differences, products, and quotients (so long as each denominator is nonzero).

EXAMPLE 1   Find all points of continuity for the given functions.

(a) $f(x) = x^3 - 2x + 7$. This is a polynomial, so it is continuous everywhere.

(b) $g(t) = (t^2 + 2t - 4)/(t^2 - 4)$. This is a quotient of polynomials.

By (2.16), it is continuous everywhere the denominator is nonzero; that is, everywhere except at $t = \pm 2$.   ■

Restricted Domain

The *blood flow* function $s = 40 - 1{,}000r^2$ cm/sec, introduced in Section 1–4, is a polynomial and is continuous. However, its domain is restricted, by the size of the artery, to the interval $[0, .2]$. Therefore, we cannot discuss the limit at either 0 or .2 (why not?). We can, however, discuss the right- and left-hand limits.

CONTINUITY AT ENDPOINTS
Suppose f is defined on the interval $[a, b]$. Then f is **continuous at the left endpoint** a if

(2.17)
$$\lim_{x \to a^+} f(x) = f(a)$$

It is **continuous at the right endpoint** b if

$$\lim_{x \to b^-} f(x) = f(b)$$

From now on, we will not specify whether p is an endpoint or not. The phrase "f is continuous at p" means that (2.17) holds if p is an endpoint of the domain of f, and (2.13) holds if p is an interior point of the domain of f.

With that hurdle out of the way, let us look at some examples.

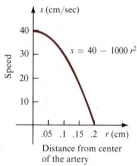

A. Blood flow.

FIGURE 2–12. Example 2. These functions are continuous.

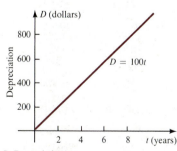

B. Depreciation.

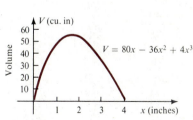

C. Volume of a cardboard box.

EXAMPLE 2

(a) The *blood flow* function $s = 40 - 1{,}000r^2$, for $0 \le r \le .2$, is continuous on its domain (Figure 2–12A).

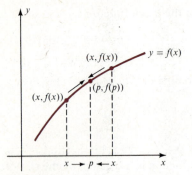

FIGURE 2–13. There are no breaks in the graph of a continuous function.

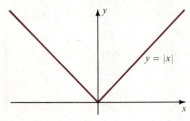

A. Absolute value.

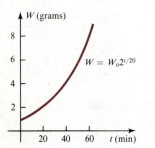

B. Population growth of *E. coli.*

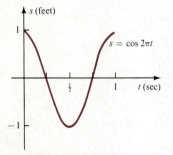

C. Simple harmonic motion.

FIGURE 2–14. Example 3. There are no breaks in the graphs; these functions are continuous on their domains.

(b) The *depreciation* function $D = 100t$ dollars, for $0 \le t \le 8$ (Example 1, Section 1–4), is a polynomial; it is continuous on its domain (Figure 2–12B).

(c) The *volume of a cardboard box* $V(x) = 80x - 36x^2 + 4x^3$, for $0 \le x \le 4$ (Example 4, Section 1–4) is also a polynomial; it is continuous on its domain (Figure 2–12C). ■

Continuity and Graphs

Sometimes it is possible to discuss continuity in terms of *breaks* in a graph. If f is continuous at p, then $(p, f(p))$ is on the graph of f. As x approaches p, then $f(x)$ approaches $f(p)$ so that, on the graph, $(x, f(x))$ approaches $(p, f(p))$; there is no break in the graph at $(p, f(p))$. See Figure 2–13. The argument also works in reverse. If there is no break in the graph at $(p, f(p))$, and x is very close to p, then $(x, f(x))$ is very close to $(p, f(p))$. This suggests that $f(x)$ is very close to $f(p)$, and that f must be continuous at p.

EXAMPLE 3 Use the graphs of the given functions to find all points of continuity.

(a) $f(x) = |x|$. The graph (Figure 2–14A) has no breaks, and the absolute value is continuous everywhere.

(b) The *population growth* function, $W = W_0 2^{t/20}$, $t \ge 0$, for the bacteria *E. coli* (Example 3 of Section 1–6) is graphed in Figure 2–14B. It has no breaks, and the exponential is continuous on its domain.*

(c) The *simple harmonic motion* function $s = \cos 2\pi t$, for $0 \le t \le 1$ (Example 4 of Section 1–7) is graphed in Figure 2–14C. It has no breaks, and the cosine is continuous on its domain. ■

The discussion so far suggests that there is, indeed, a large collection of continuous functions. There will be exceptions from time to time (see Example 4), but the functions that interest us most are continuous.

(2.18)

> The following functions are continuous wherever they are defined:
>
> (i) Algebraic (watch out for even roots and quotients)
> (ii) Absolute value
> (iii) Exponential
> (iv) Logarithmic (defined only for positive numbers)
> (v) Sine and Cosine
> (vi) Tangent (not defined at $x = \pm n\pi/2$; $n = 1, 3, 5, \ldots$)

Of course, not all functions are continuous. Here are some examples.

---

*As we mentioned earlier, population size is not continuous. The function $W = W_0 2^{t/20}$ is a continuous approximation to the true population function.

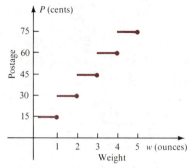

A. Postage function.

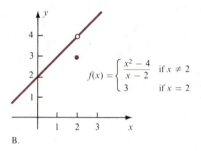

B.

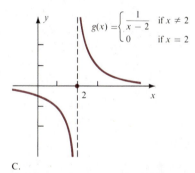

C.

FIGURE 2–15. Example 4. There are breaks in the graphs, and these functions are not continuous where the breaks occur.

EXAMPLE 4 Use the graphs of the given functions to find all points of continuity.

(a) The *postage function*

$$P(w) = \begin{cases} 15 & \text{if } 0 < w \le 1 \\ 30 & \text{if } 1 < w \le 2 \\ 45 & \text{if } 2 < w \le 3 \\ 60 & \text{if } 3 < w \le 4 \\ 75 & \text{if } 4 < w \le 5 \end{cases} \text{ cents}$$

was introduced in Example 2 of Section 1–4. There are breaks in the graph (Figure 2–15A) at $p = 1, 2, 3,$ and 4. At these points, the limit of P does not exist, so P is *not* continuous. However, it is continuous elsewhere.

(b) $$f(x) = \begin{cases} \dfrac{x^2 - 4}{x - 2} & \text{if } x \ne 2 \\ 3 & \text{if } x = 2 \end{cases}$$

There is a break in the graph (Figure 2–15B) at $p = 2$. The limit of f at $p = 2$ *does* exist, and equals 4. However, the limit does not equal $f(2) = 3$, so f is *not* continuous at $p = 2$. It is continuous elsewhere.

(c) $$g(x) = \begin{cases} \dfrac{1}{x - 2} & \text{if } x \ne 2 \\ 0 & \text{if } x = 2 \end{cases}$$

There is a break in the graph (Figure 2–15C) at $p = 2$. The limit of f at $p = 2$ does not exist, so p is not continuous there. It is continuous elsewhere.

Note: In part (b), the limit of f at 2 is 4. If $f(2)$ were 4 (instead of 3), then $f(x) \to f(2)$ as $x \to p$, so f would be continuous at $p = 2$. In part (a), the limit of P does not exist at $p = 1, 2, 3,$ and 4, so there is no way of defining P at these points to make it continuous. The same holds for the function g in part (c) at $p = 2$. ■

EXAMPLE 5 It is not possible to graph the function

$$f(x) = \begin{cases} 1 & \text{if } x \text{ is rational} \\ 0 & \text{if } x \text{ is irrational} \end{cases}$$

Nevertheless, because its values jump back and forth from 0 to 1 so frequently, we know that the limit does not exist at any point. Therefore, this is an example of a function that is defined everywhere, but not continuous anywhere! (See Exercise 42 for an example of a function that is defined everywhere, but is continuous at only *one* point.) ■

We conclude this section with two continuity theorems that will be referred to later.

$$
(2.19)
\boxed{
\begin{array}{l}
\textbf{COMPOSITION THEOREM} \\[4pt]
\text{Let } f \text{ and } g \text{ be functions such that } \lim_{x \to p} f(x) = L, \text{ and } g \text{ is} \\
\text{continuous at } L. \text{ Then} \\[6pt]
(1) \ \lim_{x \to p} g \circ f(x) = g\!\left(\lim_{x \to p} f(x)\right) = g(L) \\[6pt]
(2) \ \text{if } f \text{ is continuous at } p, \text{ so is } g \circ f
\end{array}
}
$$

A proof of (2.19) is given in Section 2–4.

$$
(2.20)
\boxed{
\begin{array}{c}
\text{A function } f \text{ is continuous at } p \text{ if and only if} \\[6pt]
\lim_{\Delta x \to 0} f(p + \Delta x) = f(p)
\end{array}
}
$$

A justification of (2.20) is given in Exercise 42.

EXERCISES

In Exercises 1–6, use the checklist (2.14) to show that the function is continuous at p.

1. $\sqrt{x^2 - 1} + 2x; \ p = 2$

2. $x^2 - 3x + 7; \ p = -1$

3. $\dfrac{x^2 + 2x - 5}{x^2 - 3x + 7}; \ p = -1$ **4.** $\sqrt[3]{x^4 - 8}; \ p = 0$

5. $\dfrac{1}{x - 2}; \ p = 3$ **6.** $\dfrac{x}{\sqrt{x + 1}}; \ p = 7$

In Exercises 7–26, find all points of continuity.

7. $3x^3 + 2x^2 - x + 1$ **8.** $\dfrac{1}{x^2 + 4}$

9. $\dfrac{\sqrt{x}}{\sqrt{1 - x^2}}$ **10.** $\dfrac{x^2 + 3x + 1}{x^2 + 4x + 3}$

11. $\dfrac{\sin x}{2^x}$ **12.** $3^x \sin x$

13. $4^x + \tan x$ **14.** $\tan x + \sin x$

15. $\dfrac{1}{\sqrt{1 - \cos x}}$ **16.** $\dfrac{1}{1 - 2^x}$

17. $\dfrac{1}{\sqrt{1 - 2^x}}$ **18.** $\dfrac{2^x}{\tan x}$

19. $\begin{cases} \dfrac{x^2 - 9}{x - 3} & \text{if } x \neq 3 \\ 6 & \text{if } x = 3 \end{cases}$

20. $\begin{cases} \dfrac{x}{x^2 + x} & \text{if } x \neq 0 \\ 1 & \text{if } x = 0 \end{cases}$

21. $\begin{cases} \dfrac{\dfrac{1}{x} - 1}{x - 1} & \text{if } x \neq 0 \text{ or } 1 \\ -1 & \text{if } x = 0 \text{ or } 1 \end{cases}$

22. $\begin{cases} \dfrac{\dfrac{1}{\sqrt{x}} - 1}{x - 1} & \text{if } x \neq 0 \text{ or } 1 \\ -\dfrac{1}{2} & \text{if } x = 0 \text{ or } 1 \end{cases}$

23. $\begin{cases} \dfrac{\dfrac{1}{x^2} - \dfrac{1}{4}}{x - 2} & \text{if } x > 0, x \neq 2 \\ -2 & \text{if } x = 2 \end{cases}$

24. $\begin{cases} \dfrac{\dfrac{1}{x} - \dfrac{1}{2}}{x - 2} & \text{if } x > 0, x \neq 2 \\ -\dfrac{1}{3} & \text{if } x = 2 \end{cases}$

25. $\begin{cases} 0 & \text{if } .9999 < x < 1.0001 \\ 1 & \text{for all other } x \end{cases}$

26. $\begin{cases} 0 & \text{if } x \neq 7 \\ \dfrac{1}{10^{30}} & \text{if } x = 7 \end{cases}$

If f is continuous at p, then $f(x) \to f(p)$ as $x \to p$. In Exercises 27–34, use this fact and the list of continuous functions (2.18) to find the limit, if it exists.

27. $\lim\limits_{x \to 0} \sin x$

28. $\lim\limits_{x \to 0} \tan x$

29. $\lim\limits_{x \to 0} \cos x$

30. $\lim\limits_{x \to \pi/2} \tan x$

31. $\lim\limits_{x \to -1} 2^x \sin x$

32. $\lim\limits_{x \to 2} (\log_2 x)(\cos x)$

33. $\lim\limits_{x \to 0} \log_3 x$

34. $\lim\limits_{x \to 0} 3^x \log_3(x + 1)$

In Exercises 35–36, find the number a that makes the function continuous at $p = -4$.

35. $\begin{cases} \dfrac{x^2 - 16}{x + 4} & \text{if } x \neq -4 \\ a & \text{if } x = -4 \end{cases}$

36. $\begin{cases} \dfrac{-\dfrac{1}{4} - \dfrac{1}{x}}{4 + x} & \text{if } x \neq -4 \\ a & \text{if } x = -4 \end{cases}$

In Exercises 37–38, find the numbers a and b that make the function continuous everywhere.

37. $\begin{cases} x^2 & \text{if } x \leq -1 \\ ax + b & \text{if } -1 < x < 0 \\ 4 - x^2 & \text{if } 0 \leq x \end{cases}$

38. $\begin{cases} x + 6 & \text{if } x \leq -1 \\ ax^2 + bx & \text{if } -1 < x < 2 \\ 2 & \text{if } 2 \leq x \end{cases}$

39. Compute the following pairs of limits:
(a) $\lim\limits_{x \to 2} x - 5$ and $\lim\limits_{h \to 0}(2 + h) - 5$
(b) $\lim\limits_{x \to -4} x^2$ and $\lim\limits_{h \to 0}(-4 + h)^2$
(c) $\lim\limits_{x \to 3} 2^x$ and $\lim\limits_{h \to 0} 2^{(3 + h)}$
(d) $\lim\limits_{x \to \pi/2} \sin x$ and $\lim\limits_{h \to 0} \sin\left(\dfrac{\pi}{2} + h\right)$

40. (Continuation of Exercise 39.) The limits in each part of Exercise 39 are of the form
$$\lim_{x \to p} f(x) \quad \text{and} \quad \lim_{h \to 0} f(p + h)$$
Check this out. Then give an intuitive argument that for any f, if either of these limits exists, then they both exist, and are equal.

41. (Continuation of Exercise 40.) Argue that f is continuous at p if and only if $f(p + h) \to f(p)$ as $h \to 0$.

42. Let f be defined by
$$f(x) = \begin{cases} x & \text{if } x \text{ is rational} \\ 0 & \text{if } x \text{ is irrational} \end{cases}$$
(a) Use the Squeeze Theorem to show that $f(x) \to 0$ as $x \to 0$.
(b) Show that f is continuous *only* at $p = 0$.*

43. (Continuation of Exercise 47 in Section 2–2.) In that exercise, we suggested that if $f(x) = x^n$, $n = 1, 2, 3, \ldots$, then
$$\lim_{\Delta x \to 0} \frac{f(x + \Delta x) - f(x)}{\Delta x} = nx^{n-1}$$
We now suggest that the same formula holds for $n = -1, -2, -3, \ldots$. Test it by computing the limit for $f(x) = 1/x, f(x) = 1/x^2$, and $f(x) = 1/x^3$.

TEST YOUR INTUITION ABOUT CONTINUITY. Decide whether the following statements are true or false:

44. If f is defined at p, then f is continuous at p.
45. If f is continuous at p, then f is defined at p.
46. If the limit of f at p exists, then f is continuous at p.
47. If f is continuous at p, then the limit of f at p exists.
48. Every function is continuous at some point of its domain.

*The behavior of a function can be even more bizarre. Exercise 20 in Section 2–4 contains an example of a function that is continuous at 0 and every irrational number, but not continuous at nonzero rational numbers!

49. If f is continuous at p, then it is also continuous at points close to p.

50. f is continuous at p if and only if $\lim_{h \to 0} f(p + h) - f(p) = 0$ (see Exercise 41 above).

51. If

$$\lim_{h \to 0} \frac{f(p + h) - f(p)}{h}$$

exists, then f is continuous at p.

52. Suppose that f is continuous on the interval $[0, 1]$, $f(0) = -1$, and $f(1) = 1$. Then there is some point

c between 0 and 1 such that $f(c) = 0$. *Hint:* Draw a picture, and remember that the graph of f has no breaks.

53. The statement in Exercise 52 remains valid even if there is some point p between 0 and 1 at which f is *not* continuous. (Draw a picture.)

54. There is a function f that is continuous on the half-open interval $(0, 1]$, and the values $f(x)$ get arbitrarily large as $x \to 0$.

55. Same as Exercise 54 only now f is continuous on the closed interval $[0, 1]$.

Optional Exercises

56. Write a definition for "f is continuous at p" in terms of ϵ's and δ's.

57. Use the definition in Exercise 56 to prove that the following functions are continuous at p:

(a) $f(x) = 3; p = 7$ (b) $f(x) = 2x + 9; p = -1$
(c) $f(x) = 3 - 4x; p = 0$

58. Make up a function that is defined everywhere, but is continuous only at π.

2–4 PROOFS OF THE LIMIT THEOREMS (*Optional*)

Saying that "as x gets closer and closer to p, $f(x)$ gets closer and closer to L" is a perfectly legitimate way to think and talk about limits on an informal basis. It is also the way to obtain a limit in many situations. However, it is quite useless when you want to *prove* something about limits. The reason is that the language is too vague; it doesn't say how close x and $f(x)$ are to p and L. It also does not convey the idea that $f(x)$ can be made *arbitrarily* close to L by making x *sufficiently close to p*. In this section, we present a precise definition of a limit, and then prove the limit theorems. (The numbers identifying each statement correspond to the numbers used in the first three sections and, therefore, may appear out of order here.)

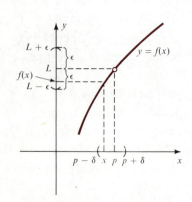

FIGURE 2–16. Given $\epsilon > 0$, there is a $\delta > 0$ such that $L - \epsilon < f(x) < L + \epsilon$ whenever $p - \delta < x < p + \delta$.

DEFINITION OF A LIMIT

(2.10)

Let I be an open interval containing p, and let f be defined at all points of I except, perhaps, at p. Then the **limit of f at p is L** ($\lim_{x \to p} f(x) = L$) if, for every $\epsilon > 0$, there is a $\delta > 0$ such that

$$|f(x) - L| < \epsilon \quad \text{whenever} \quad 0 < |x - p| < \delta$$

This definition is illustrated in Figure 2–16. The material at the end of Section 2–2 describes the motivation behind this definition and gives some examples. We suggest that you reread that part of the text before continuing on here.

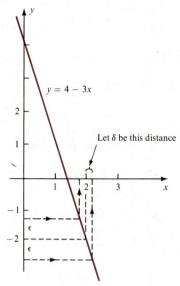

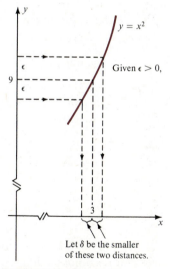

Given $\epsilon > 0$, find a $\delta > 0$ that "works."

FIGURE 2–17. Example 1.

FIGURE 2–18. Example 2.

Let δ be the smaller
of these two distances.

EXAMPLE 1 Prove that $\lim_{x \to 2} 4 - 3x = -2$.

Solution For any given $\epsilon > 0$, we must find a δ that "works." As was pointed out in Section 2–2, the easiest way to find a δ is to work backwards. We want $|(4 - 3x) - (-2)| < \epsilon$; therefore,

$$-\epsilon < (4 - 3x) - (-2) < \epsilon$$
$$-\epsilon < 6 - 3x < \epsilon$$
$$\frac{\epsilon}{3} > -2 + x > -\frac{\epsilon}{3} \qquad \text{(Divide by } -3)$$
$$|x - 2| < \frac{\epsilon}{3}$$

Because these steps are reversible, $\delta = \epsilon/3$ "works"; that is, if $0 < |x - 2| < \epsilon/3$, then $|(4 - 3x) - (-2)| < \epsilon$ (Figure 2–17). ■

The next example is somewhat more complicated.

EXAMPLE 2 Prove $\lim_{x \to 3} x^2 = 9$.

Solution We want $|x^2 - 9| < \epsilon$, so

$$-\epsilon < x^2 - 9 < \epsilon$$
$$9 - \epsilon < x^2 < 9 + \epsilon$$
$$\sqrt{9 - \epsilon} < x < \sqrt{9 + \epsilon} \qquad \text{(Take square roots*)}$$
$$\sqrt{9 - \epsilon} - 3 < x - 3 < \sqrt{9 + \epsilon} - 3 \qquad \text{(Subtract 3)}$$

Now let δ be the minimum of $|\sqrt{9 - \epsilon} - 3|$ and $\sqrt{9 + \epsilon} - 3$ (Figure 2–18); this is a δ that "works" (check). ■

Here is a precise definition for one-sided limits (Figure 2–19):

DEFINITION OF ONE-SIDED LIMITS

Let f be a function and p be a real number. The **right-hand limit of** f **at** p **is** L ($\lim_{x \to p+} f(x) = L$) if, for every $\epsilon > 0$, there is a $\delta > 0$, such that

$$|f(x) - L| < \epsilon \quad \text{whenever} \quad 0 < x - p < \delta$$

The **left-hand limit of** f **at** p **is** L ($\lim_{x \to p-} f(x) = L$) if, for every $\epsilon > 0$, there is a $\delta > 0$, such that

$$|f(x) - L| < \epsilon \quad \text{whenever} \quad 0 < p - x < \delta$$

---

*We know that $x > 0$ because $x \to 3$; therefore, $\sqrt{x^2} = x$. Also, if $0 < a < b$, then $\sqrt{a} < \sqrt{b}$ because $\sqrt{b} - \sqrt{a} = (b - a)/(\sqrt{b} + \sqrt{a}) > 0$. Therefore, taking square roots preserves the inequalities.

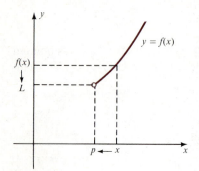

FIGURE 2–19. Illustration of right-hand limit.

With this definition, we can prove the following result:

(2.1) One-Sided Limit Theorem

$$\lim_{x \to p} f(x) = L \quad \text{if and only if} \quad \lim_{x \to p^+} f(x) = L \quad \text{and} \quad \lim_{x \to p^-} f(x) = L$$

Proof . First, suppose $f(x) \to L$ as $x \to p$, and prove the right- and left-hand limits equal L. If $\epsilon > 0$ is given, there is some $\delta > 0$, such that $|f(x) - L| < \epsilon$ whenever $0 < |x - p| < \delta$. This δ will "work" for the one-sided limits because

$$0 < x - p < \delta \text{ implies } 0 < |x - p| < \delta, \text{ so } |f(x) - L| < \epsilon \quad \text{and}$$
$$0 < p - x < \delta \text{ implies } 0 < |x - p| < \delta, \text{ so } |f(x) - L| < \epsilon$$

Now assume the one-sided limits both equal L and prove the limit is L. For any $\epsilon > 0$, there are two positive numbers, δ_1 and δ_2, that "work" for the left- and right-hand limits. Then $\delta = \min \{\delta_1, \delta_2\}$ "works" for the limit because if

$$0 < |x - p| < \delta, \quad \text{then either} \quad 0 < p - x < \delta \le \delta_1 \quad \text{or}$$
$$0 < x - p < \delta \le \delta_2$$

and, in either case, $|f(x) - L| < \epsilon$. ∎

Next, we prove the Uniqueness Theorem. The basic idea is that $f(x)$ cannot approach two distinct numbers at the same time. While this is intuitively obvious, the proof requires some expertise with ϵ's and δ's.

(2.2) Uniqueness Theorem A function can have at most one limit at p.

Proof Suppose that $f(x) \to L$ and $f(x) \to M$ as $x \to p$. If $L \ne M$, let $\epsilon = |L - M|/2$. There must be a $\delta > 0$, such that $|f(x) - L| < \epsilon$ *and* $|f(x) - M| < \epsilon$ whenever $0 < |x - p| < \delta$. But then

$$
\begin{aligned}
|L - M| &= |(L - f(x)) + (f(x) - M| && \text{(Add and subtract } f(x))^* \\
&\le |L - f(x)| + |f(x) - M| && \text{(Triangle Inequality)} \\
&= |f(x) - L| + |f(x) - M| && (|a| = |-a|) \\
&< \epsilon + \epsilon \\
&= 2\epsilon
\end{aligned}
$$

This shows that $2\epsilon > |L - M|$, or, equivalently, $\epsilon > |L - M|/2$. But ϵ was supposed to *equal* $|L - M|/2$. This contradiction proves that L cannot be different from M. ∎

The proofs of the next two theorems are left as an exercise.

(2.3) Limit of a Constant The limit of a constant function is that constant.

(2.4) Limit of the Identity The limit of the identity $f(x) = x$ at p is p.

---

*Adding and subtracting the same thing is a trick that is often used to advantage.

(2.5) Limit of a Combination   Suppose $\lim_{x \to p} f(x) = L$ and $\lim_{x \to p} g(x) = M$. Then

(i) $\lim_{x \to p} (f + g)(x) = L + M$

(ii) $\lim_{x \to p} (fg)(x) \quad = LM$

(iii) $\lim_{x \to p} (f/g)(x) \quad = L/M$ provided $M \neq 0$

Proof

(i) The proof was presented in Section 2–2

(ii) The idea is to show that $|f(x)g(x) - LM|$ is small whenever $|x - p|$ is small (but not zero). We can use the "add-subtract" trick, the triangle inequality, and factoring to write

(1)
$$|f(x)g(x) - LM| = |f(x)g(x) - f(x)M + f(x)M - LM|$$
$$\leq |f(x)| \, |g(x) - M| + |f(x) - L| \, |M|$$

Given an $\epsilon > 0$, our job is to find a $\delta > 0$ that makes each product in the last line smaller than $\epsilon/2$; then their sum is less than ϵ.

Proof: Let $\epsilon > 0$ be given. First, note that since $f(x) \to L$ as $x \to p$, there must be a $\delta_1 > 0$, such that

(2)   if $0 < |x - p| < \delta_1$, then $|f(x) - L| < \epsilon/2|M|$

This will make the second product in (1) less than $\epsilon/2$. The first product is more troublesome; $|g(x) - M|$ can be made small, but $f(x)$ can be extremely large, or *can it?* NO, it can't because it is supposed to be close to L. In fact, we can find a $\delta_2 > 0$, such that

(3)   if $0 < |x - p| < \delta_2$, then $|f(x)| < 2|L|$

Finally, since $g(x) \to M$ as $x \to p$, there is a $\delta_3 > 0$, such that

(4)   if $0 < |x - p| < \delta_3$, then $|g(x) - M| < \epsilon/4|L|$

The combination of (3) and (4) makes the first product in (1) less than $\epsilon/2$. Now, let $\delta = \min \{\delta_1, \delta_2, \delta_3\}$. If $0 < |x - p| < \delta$, then

$$|f(x) - L| < \epsilon/2|M|, |f(x)| < 2|L|, \text{ and } |g(x) - M| < \epsilon/4|L|$$

and it follows from (1) that $|f(x)g(x) - LM| < \epsilon$.

(iii) The idea here is to first prove $1/g(x) \to 1/M$ and then use part (ii), writing $f(x)/g(x)$ as a product $f(x)[1/g(x)]$. The proof that $1/g(x) \to 1/M$ is left as an exercise.   ■

The combination of (2.3), (2.4), and (2.5) establishes

(2.6) Limit of a Polynomial   The limit of any polynomial at any point is the value of the polynomial at the point.

A function f is continuous at p if $f(x) \to f(p)$ as $x \to p$. In terms of ϵ's and δ's, we have

---

DEFINITION OF CONTINUITY
A function f is **continuous at** p if for every $\epsilon > 0$, there is a $\delta > 0$, such that

$$|f(x) - f(p)| < \epsilon \quad \text{whenever} \quad |x - p| < \delta$$

(2.19) Composition Theorem   Let f and g be functions such that

$$\lim_{x \to p} f(x) = L \text{ and } g \text{ is continuous at } L$$

(i) Then $\lim_{x \to p} g \circ f(x) = g\left(\lim_{x \to p} f(x)\right) = g(L)$

(ii) If f is also continuous at p, so is $g \circ f$.

Proof

(i) Given $\epsilon > 0$, we must find a $\delta > 0$ that makes $|g(f(x)) - g(L)| < \epsilon$ whenever $0 < |x - p| < \delta$. Since g is continuous at L, there is a $\delta_1 > 0$ such that $|g(y) - g(L)| < \epsilon$ whenever $|y - L| < \delta_1$. Because $\delta_1 > 0$, we can use it as an ϵ. Since $f(x) \to L$ as $x \to p$, there must be a $\delta_2 > 0$, such that $|f(x) - L| < \delta_1$ whenever $0 < |x - p| < \delta_2$. This δ_2 is the δ that "works." If $0 < |x - p| < \delta_2$, then $|f(x) - L| < \delta_1$, so $|g(f(x)) - g(L)| < \epsilon$ (Figure 2–20).

(ii) The proof of this part is similar; only now, since f is continuous at p, we know that $L = f(p)$, and if $|x - p| < \delta_2$, then $|g(f(x)) - g(f(p))| < \epsilon$. ∎

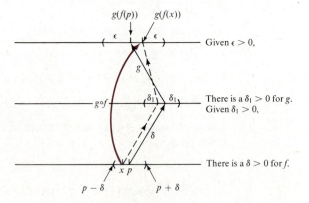

FIGURE 2–20. The composition theorem.

(2.7) Root Theorem   If $f(x) \to L$ as $x \to p$, then

$$\lim_{x \to p} \sqrt[n]{f(x)} = \sqrt[n]{\lim_{x \to p} f(x)} = \sqrt[n]{L}$$

provided n is odd *or* n is even and $L > 0$.

Proof   Let $g(x) = \sqrt[n]{x}$; then $g \circ f(x) = \sqrt[n]{f(x)}$. Therefore, this theorem follows from (2.19) above, if g is continuous at L. We will prove that $g(x) = \sqrt[n]{x}$ is continuous at *every* number p in its domain. Suppose $p > 0$ and n is any positive integer. Given $\epsilon > 0$, we must find a $\delta > 0$, such that $|\sqrt[n]{x} - \sqrt[n]{p}| < \epsilon$ whenever $|x - p| < \delta$. Working backwards,

$$-\epsilon < \sqrt[n]{x} - \sqrt[n]{p} < \epsilon$$
$$\sqrt[n]{p} - \epsilon < \sqrt[n]{x} < \sqrt[n]{p} + \epsilon$$

We can assume here that $\epsilon < \sqrt[n]{p}$ because, if we find a δ that "works" for an $\epsilon < \sqrt[n]{p}$, that same δ will "work" for any larger epsilon. If $\epsilon < \sqrt[n]{p}$, then $\sqrt[n]{p} - \epsilon$ is positive; therefore, taking n^{th} powers in the last line above preserves the inequalities.

$$(\sqrt[n]{p} - \epsilon)^n < x < (\sqrt[n]{p} + \epsilon)^n$$
$$(\sqrt[n]{p} - \epsilon)^n - p < x - p < (\sqrt[n]{p} + \epsilon)^n - p$$

These steps are reversible, so $\delta = \min \{|(\sqrt[n]{p} - \epsilon)^n - p|, (\sqrt[n]{p} + \epsilon)^n - p\}$ will "work" (check). If p happens to be negative and n is odd, then $-p > 0$ and the same technique applied above will show that $\sqrt[n]{x} \to \sqrt[n]{p}$. The case of $p = 0$ is simple; we take the right-hand limit if n is even, or the limit if n is odd. In either case, $\delta = \epsilon^n$ "works" (check). This completes the proof. ∎

(2.8) Absolute Value Theorem   If $f(x) \to L$ as $x \to p$, then

$$\lim_{x \to p} |f(x)| = \left| \lim_{x \to p} f(x) \right| = |L|$$

Proof   Let $g(x) = |x|$; then $g \circ f(x) = |f(x)|$. Therefore, this theorem follows from (2.19) above, if g is continuous at L. We claim that for any two numbers a and b,

(1) $\left| |a| - |b| \right| \leq |a - b|$*

To show g is continuous at L, let $\epsilon > 0$; the δ that "works" is ϵ itself. If $|x - L| < \epsilon$, then, by (1),

$$|g(x) - g(L)| = \left| |x| - |L| \right| \leq |x - L| < \epsilon \qquad ∎$$

(2.9) Squeeze Theorem   If $f(x) \to L$ and $g(x) \to L$ as $x \to p$, and $f(x) \leq F(x) \leq g(x)$ except, perhaps, at p, then

$$\lim_{x \to p} F(x) = L$$

Proof   Suppose $\epsilon > 0$. There are positive numbers δ_1 and δ_2 that "work" for f and g; let $\delta = \min \{\delta_1, \delta_2\}$. If $0 < |x - p| < \delta$, it follows that

$$L - \epsilon < f(x) \quad and \quad g(x) < L + \epsilon \qquad \text{(Why?)}$$

Since F is squeezed between f and g, we have

$$L - \epsilon < F(x) < L + \epsilon \quad \text{or} \quad |F(x) - L| < \epsilon \qquad ∎$$

---

*$|a| = |(a - b) + b| \leq |a - b| + |b|$; therefore, $|a| - |b| \leq |a - b|$
$|b| = |(b - a) + a| \leq |a - b| + |a|$; therefore, $|b| - |a| \leq |a - b|$
It follows that $\left| |a| - |b| \right| \leq |a - b|$.

EXERCISES

For each of the following limits, given an $\epsilon > 0$, find a $\delta > 0$ that "works."

1. $\lim\limits_{x \to 7} 8 = 8$ **2.** $\lim\limits_{x \to -2} x = -2$

3. $\lim\limits_{x \to 0} 3x + 17 = 17$ **4.** $\lim\limits_{x \to -3} 2 - 5x = 17$

5. $\lim\limits_{x \to 1} \dfrac{x}{5} - 2 = -\dfrac{9}{5}$ **6.** $\lim\limits_{x \to 2} x^2 = 4$

7. $\lim\limits_{x \to -4} x^2 = 16$ **8.** $\lim\limits_{x \to 7} \sqrt{x} = \sqrt{7}$

9. $\lim\limits_{x \to 27} \sqrt[3]{x} = 3$ **10.** $\lim\limits_{x \to 3} 2x^2 - 3x = 9$

11. Prove (2.3); the limit of a constant is that constant.

12. Prove (2.4); the limit of the identity function at p is p.

13. Use the theorems of this section to prove that an algebraic function is continuous wherever it is defined.

14. Prove that $f(x) \to L$ as $x \to p$ if and only if $f(p + h) \to L$ as $h \to 0$.

15. (Continuation of Exercise 14) Prove that f is continuous at p if and only if $f(p + h) \to f(p)$ as $h \to 0$.

16. Prove that if f is continuous at p and $f(p) > 0$, then $f(x) > 0$ for every x in some open interval containing p.

17. Fill in the blanks below to prove that if $p \neq 0$, then $1/x \to 1/p$ as $x \to p$.

(1) Given $\epsilon > 0$, we must find a $\delta > 0$ such that _____ whenever _____.

(2) We can assume $x \neq 0$ because _____.

(3) $\left| \dfrac{1}{x} - \dfrac{1}{p} \right| = \dfrac{|x - p|}{|x|\,|p|}$ because _____.

(4) Let δ be any positive number less than the minimum of $|p|/2$ and $|p|^2\epsilon/2$.

(5) If $0 < |x - p| < \delta$, then $|x| > |p|/2$ because _____, and it follows that

$$\frac{1}{|x|} < \frac{2}{|p|}$$

(6) If $0 < |x - p| < \delta$, then $|x - p| < |p|^2\epsilon/2$ because _____.

(7) Therefore, if $0 < |x - p| < \delta$, then

$$\left| \frac{1}{x} - \frac{1}{p} \right| = \frac{|x - p|}{|x|\,|p|} < \frac{2|p|^2\epsilon}{|p|\,|p|2} = \epsilon$$

18. Use the result in Exercise 17 and the Composition Theorem (2.19) to prove that if $g(x) \to M$ as $x \to p$, and $M \neq 0$, then $1/g(x) \to 1/M$.

19. Prove (2.5iii); if $f(x) \to L$ and $g(x) \to M$ as $x \to p$, and $M \neq 0$, then $f(x)/g(x) \to L/M$ as $x \to p$.

20. Every rational number is of the form m/n where m and n are integers. Every nonzero rational number can be reduced to *lowest terms;* that is, written so that the numerator and denominator have no common factors. This reduction is unique. Now define f on the closed interval $[0, 1]$ by

$$f(x) = \begin{cases} \dfrac{1}{n} & \text{if } x = \dfrac{m}{n} \text{ in lowest terms, } x \neq 0 \\ 0 & \text{if } x \text{ is irrational or } x = 0 \end{cases}$$

Fill in the blanks below to prove that f is continuous only at 0 and every irrational number.

(1) We claim that if $p \neq 0$ is rational, then the limit of f at p does not exist. If $p = m/n$, then $f(p) = 1/n$. Let $\epsilon = 1/2n^*$; then no $\delta > 0$ will "work". For any $\delta > 0$, there are irrational numbers x with $0 < |x - p| < \delta$, and for such x, $|f(x) - 1/n| > \epsilon$ because _____.

(2) Thus, f is not continuous at nonzero rational numbers because _____.

(3) Now let p be irrational; then $f(p) = 0$, and we claim that $f(x) \to 0$ as $x \to p$. Given an $\epsilon > 0$, we must find a $\delta > 0$ such that _____ whenever _____.

(4) Let N be a positive integer so large that $1/N < \epsilon$.

(5) Let $Q = \{q : q \text{ is in } [0, 1], q = m/n \text{ in lowest terms, and } n \leq N\}$. Q has only a *finite* number of elements because _____.

(6) Let $R = \{|q - p| : q \text{ is in } Q\}$. Then R has a *smallest* element because _____. Call this smallest number r_o.

(7) Let δ be any positive number less than r_o. Then δ "works" because _____.

(8) If $p = 0$, then $f(x) \to 0$ as $x \to p^+$ because _____.

(9) Therefore, f is continuous at 0 and at each irrational number because _____.

*Because p is in $[0, 1]$, we can assume that both m and n are positive.

REVIEW EXERCISES

In Exercises 1–30, find the limit at p, if it exists. Use the limit theorems, continuity, or a calculator, if necessary.

1. $x^2 + 4x - 8$; $p = -2$

2. $\dfrac{x^3 - 3x}{x + 1}$; $p = 4$

3. $\dfrac{\dfrac{1}{t} - \dfrac{1}{3}}{t - 3}$; $p = 3$

4. $\dfrac{\dfrac{1}{t} - \dfrac{1}{3}}{t + 3}$; $p = 3$

5. $\sin x$; $p = \pi/4$

6. $\dfrac{\sqrt{x + 1} - \sqrt{8}}{x - 7}$; $p = 7$

7. $\dfrac{1}{x^2 - 1}$; $p = -1$

8. $\tan x$; $p = \pi/4$

9. $\dfrac{\sqrt{x - 4} - \sqrt{10}}{x - 14}$; $p = 14$

10. $\cos x$; $p = 0$

11. $\dfrac{x^3 + 27}{x + 3}$; $p = -3$

12. $\dfrac{1}{x} - 37$; $p = 0$

13. $\dfrac{x^2 - 14x + 33}{x - 11}$; $p = 11$

14. $\dfrac{x - 9}{\sqrt{x} - 3}$; $p = 9$

15. $2^x \cos x$; $p = 0$

16. $\dfrac{\dfrac{1}{\sqrt{x}} - 1}{x - 1}$; $p = 1$

17. $\dfrac{x^2 - 3x + 8}{2^x}$; $p = 4$

18. 10^x; $p = 2$

19. $\dfrac{(2.7)^h - 1}{h}$; $p = 0$

20. $\dfrac{u^3 - 8}{u^2 - 4}$; $p = 2$

21. $\dfrac{t^3 - 64}{t^2 - 16}$; $p = 4$

22. $\dfrac{\sqrt{u + 1} + 1}{\sqrt{u + 1}}$; $p = 0$

23. $\dfrac{x - 16}{\sqrt{x} - 4}$; $p = 16$

24. $(x^2 + 2)^5$; $p = -1$

25. $\dfrac{\dfrac{1}{\sqrt{x}} - \dfrac{8}{x}}{x - 64}$; $p = 64$

26. $\dfrac{t - 2}{\dfrac{1}{\sqrt{t}} - \dfrac{1}{\sqrt{2}}}$; $p = 2$

27. $\dfrac{x^2 - 7}{x - 2}$; $p = 2$

28. $\dfrac{x^2 + \sqrt{x}}{8x}$; $p = 3$

29. $\dfrac{\sin 8x}{5x}$; $p = 0$

30. $\dfrac{\sin x}{x^2}$; $p = 0$

In Exercises 31–42, assume that x and $x + \Delta x$ are in the domain of f, and find

$$\lim_{\Delta x \to 0} \frac{f(x + \Delta x) - f(x)}{\Delta x}$$

31. $f(x) = 1.5$

32. $f(x) = 4x + 9$

33. $f(x) = 8 - 5x^2$

34. $f(x) = \sqrt{x}$

35. $f(x) = 9/x$

36. $f(x) = 1/(x + 1)$

37. $f(x) = 4^x$

38. $f(x) = \sqrt{2}$

39. $f(x) = 8/\sqrt{x}$

40. $f(x) = x^3$

41. $f(x) = (2.718)^x$

42. $f(x) = (2.71828)^x$

In Exercises 43–54, find all points of continuity for the given function.

43. $2^x/\sin x$

44. $\log_2 x$

45. $\begin{cases} \dfrac{x^2 - 3x - 4}{x^2 - 2x - 8} & \text{if } x \neq 4 \text{ or } -2 \\ 5/6 & \text{if } x = 4 \\ -2/3 & \text{if } x = -2 \end{cases}$

46. $\begin{cases} \dfrac{\sqrt{x} - 3}{x - 3} & \text{if } x \neq 3 \\ 2\sqrt{3} & \text{if } x = 3 \end{cases}$

47. $\dfrac{2^t}{\tan t}$

48. $\dfrac{1}{\sqrt{1 - x^2}}$

49. $\dfrac{\sqrt{x}}{\sqrt{1 - x^2}}$

50. $\dfrac{1}{3^x - 1}$

51. $\begin{cases} x \sin (1/x) & \text{if } x \neq 0 \\ 0 & \text{if } x = 0 \end{cases}$

52. $\begin{cases} x & \text{if } x \text{ is rational} \\ 0 & \text{if } x \text{ is irrational} \end{cases}$

53. $x^3 \cos x$

54. $\dfrac{1}{x^2 + 1}$

In Exercises 55–56, find the number a that makes the function continuous everywhere.

55. $\begin{cases} \dfrac{\dfrac{1}{x^2} - \dfrac{1}{6}}{x - \sqrt{6}} & \text{if } x \neq \sqrt{6} \\ a & \text{if } x = \sqrt{6} \end{cases}$

56. $\begin{cases} \dfrac{x^2 - 9}{x - 3} & \text{if } x \neq 3 \\ a & \text{if } x = 3 \end{cases}$

In Exercises 57–58, find the numbers a and b that make the function continuous everywhere.

57. $\begin{cases} \dfrac{x^4 + 2x^3 - 8x - 16}{x^2 - 4} & \text{if } x \neq \pm 2 \\ a & \text{if } x = 2 \\ b & \text{if } x = -2 \end{cases}$

58. $\begin{cases} x + 3 & \text{if } x < 0 \\ ax + b & \text{if } 0 \leq x \leq 1 \\ x^2 + 4 & \text{if } 1 < x \end{cases}$

In Exercises 59–62, a particle is moving along a horizontal line with the given position function. (a) Find the velocity function. (b) Find the intervals on which the velocity is positive, negative, and zero. (c) Sketch a motion diagram.

59. $s(t) = 4 - 3t$

60. $s(t) = t - 1$

61. $s(t) = t^2 - 4t$

62. $s(t) = 3t - t^3$

In Exercises 63–72, mark the statements True or False. In some cases, drawing a picture will help you decide.

63. If $f(2) < 0$ and f is continuous at $p = 2$, then there is an open interval containing 2 on which f takes only negative values.

64. If f is continuous everywhere except at $p = -1$, then $f(-1)$ is not defined.

65. If $f(p)$ is not defined, them $\lim_{x \to p} f(x)$ does not exist.

66. If $f(p)$ is not defined, then f is not continuous at p.

67. If f is continuous everywhere, then, for any x, $f(x + h) \to f(x)$ as $h \to 0$.

68. If f and g are continuous at p, so is f/g.

69. If
$$\lim_{h \to 0} \frac{f(x + h) - f(x)}{h}$$
exists, then f is continuous at x.

70. If f is continuous on the interval $[0, 1]$, then the values of f cannot become arbitrarily large (positive or negative).

71. If f is continuous on the interval $[a, b]$ and $f(a) < r < f(b)$, then there is some point c in (a, b) such that $f(c) = r$.

72. If the right- and left-hand limits of f at p exist and are equal, then f is continuous at p.

Optional Exercises

73. Write out the $\epsilon - \delta$ definitions of limit and continuity.

74. Evaluate the limits at p and prove your answer.

(a) $4x - 17$; $p = 3$ (b) $8 - 3x$; $p = -2$

(c) $\dfrac{x^2 - 16}{x - 4}$; $p = 4$ (d) x^2; $p = 1$

75. Without referring to the text, prove

(a) The Sum Theorem for limits

(b) The Product Theorem for limits

(c) The Absolute Value Theorem for limits

(d) The Squeeze Theorem

3

DERIVATIVES

We begin by posing two questions concerning a function f and a point x in its domain.

(1) What is meant by the *tangent* to the graph of f at the point $(x, f(x))$?

(2) What is the *instantaneous rate of change* in the values of f at the point x?

Both questions have historical significance. The early Greeks talked about tangents to circles, but many centuries were to pass before Descartes, in his *La géométrie* (published in 1637), satisfactorily described tangents to more general curves. The second question was central to the discussion of motion by Newton and Leibniz.

Apparently, the two questions are unrelated; the first is purely geometric, the second purely analytic. And yet, as our investigations will show, there is a single key that unlocks the answers to both questions. That key is a limit of the form

$$\lim_{\Delta x \to 0} \frac{f(x + \Delta x) - f(x)}{\Delta x}$$

Therefore, it is appropriate to spotlight this limit by giving it a name; we call it the derivative of f at x. As you will see, the derivative is a key that fits many locks.

Section 3–1 presents the definition and stresses the connection between the derivative, the rate of change, and the slope of a tangent line. In the next four sections we develop formulas that can be used to compute derivatives. The final section, 3–6, contains the important Mean Value Theorem and one of its consequences, the test for increasing and decreasing functions.

The major points to concentrate on in this chapter are: (1) The fact that the derivative, the rate of change, and the slope of a tangent line are synonymous. (2) The Chain Rule (Section 3–4). It has many applications and is referred to often in later sections. (3) The derivative formulas. For easy reference, there is a table of these formulas at the end of the chapter. (4) The test for increasing and decreasing functions. It is the basis for many of the applications presented in Chapter 4.

3–1 DERIVATIVE = SLOPE OF TANGENT = RATE OF CHANGE

Things change. Populations, the cost of living, and the position of a moving object all change with time. The volume of a balloon changes with its radius, the pressure of the gas inside the balloon changes with temperature, and the pull of the earth on a rocket changes with its distance from the earth. Derivatives describe the *rate* at which things change. Derivatives also describe tangent lines of graphs, and tangent lines tell us the general shape of a graph.

In this section, we define the derivative and look at three examples. After that, we show how the derivative describes tangent lines and rates of change.

(3.1)

> **DEFINITION OF DERIVATIVE**
> Suppose f is defined on an open interval I. Then the **derivative** of f is a *function* f' (read "f prime") whose value at any x in I is
>
> $$f'(x) = \lim_{\Delta x \to 0} \frac{f(x + \Delta x) - f(x)}{\Delta x}$$
>
> provided this limit exists.

If $f'(x)$ exists, we say that f is **differentiable at** x, and the number $f'(x)$ is called the **derivative of f at** x. Since f can be differentiable only at points where it is defined, it follows that the domain of f' is contained in the domain of f. If f is differentiable at each point of its domain, we say simply that f is **differentiable.** To **differentiate** a function f means to find its derivative f'.

Finding the derivative of f requires evaluation of a limit (3.1). Such limits were discussed at length in Section 2–1. Recall the three step procedure for an algebraic function:

(1) Write out the quotient.

(2) Perform the indicated algebra (combine fractions or rationalize, if necessary), and simplify.

(3) Evaluate the limit.

EXAMPLE 1   Differentiate $f(x) = x^2 + 2x - 4$. What is the domain of f'?

Solution

$$\frac{f(x + \Delta x) - f(x)}{\Delta x}$$

$$= \frac{[(x + \Delta x)^2 + 2(x + \Delta x) - 4] - [x^2 + 2x - 4]}{\Delta x}$$

Perform the indicated algebra, and simplify

$$= \frac{x^2 + 2x\Delta x + \Delta x^2 + 2x + 2\Delta x - 4 - x^2 - 2x + 4}{\Delta x}$$

$$= 2x + \Delta x + 2$$

Now take the limit

$$f'(x) = \lim_{\Delta x \to 0} \frac{f(x + \Delta x) - f(x)}{\Delta x} = \lim_{\Delta x \to 0} 2x + \Delta x + 2$$

$$= 2x + 2$$

Therefore, the derivative of $f(x) = x^2 + 2x - 4$ is the *function* $f'(x) = 2x + 2$; f and f' are both defined for all real numbers.   ■

The three steps need not be taken separately; sometimes it is convenient to run them all together.

EXAMPLE 2   Differentiate $f(x) = 1/x$. What is the domain of f'?

Solution f is not defined at 0, but if $x \neq 0$, and $|\Delta x|$ is small enough, then $x + \Delta x \neq 0$. Therefore,

$$f'(x) = \lim_{\Delta x \to 0} \frac{f(x + \Delta x) - f(x)}{\Delta x} = \lim_{\Delta x \to 0} \frac{\dfrac{1}{x + \Delta x} - \dfrac{1}{x}}{\Delta x}$$

$$= \lim_{\Delta x \to 0} \frac{\dfrac{x - (x + \Delta x)}{x(x + \Delta x)}}{\Delta x} \qquad \text{(Combine fractions)}$$

$$= \lim_{\Delta x \to 0} \frac{-1}{x(x + \Delta x)} = \frac{-1}{x^2} \qquad (2.5)$$

Thus, the derivative of $f(x) = 1/x$ is the *function* $f'(x) = -1/x^2$; it is defined everywhere except $x = 0$.   ■

The domains of the functions in Examples 1 and 2 did not have endpoints. If f is defined on an interval with one or two endpoints, we define the derivative at the endpoints by means of right- and/or left-hand limits.

Let f be defined on $[a, b]$. Then

(3.2)
$$f'(a) = \lim_{\Delta x \to 0^+} \frac{f(a + \Delta x) - f(a)}{\Delta x} \quad \text{and}$$

$$f'(b) = \lim_{\Delta x \to 0^-} \frac{f(b + \Delta x) - f(b)}{\Delta x}$$

provided these limits exist.

This is what we did to extend the definition of continuity to endpoints. Now, as then, we make the following convention. The phrase "f is differentiable at x" means that (3.2) holds if x is an endpoint of the domain of f, and (3.1) holds if x is an interior point of the domain of f.

EXAMPLE 3   Differentiate $f(x) = \sqrt{x}$. The domain of f is $[0, \infty)$; what is the domain of f'?

Solution   First, let us examine the derivative at points $x > 0$. If $x > 0$ and $|\Delta x|$ is small enough, then $x + \Delta x > 0$. Therefore,

$$f'(x) = \lim_{\Delta x \to 0} \frac{f(x + \Delta x) - f(x)}{\Delta x}$$

$$= \lim_{\Delta x \to 0} \frac{\sqrt{x + \Delta x} - \sqrt{x}}{\Delta x} \cdot \frac{\sqrt{x + \Delta x} + \sqrt{x}}{\sqrt{x + \Delta x} + \sqrt{x}} \quad \text{(Rationalize)}$$

$$= \lim_{\Delta x \to 0} \frac{(x + \Delta x) - x}{\Delta x(\sqrt{x + \Delta x} + \sqrt{x})}$$

$$= \lim_{\Delta x \to 0} \frac{1}{\sqrt{x + \Delta x} + \sqrt{x}} = \frac{1}{2\sqrt{x}} \quad \text{(2.7), (2.5)}$$

Thus, if $x > 0$, then $f'(x) = 1/2\sqrt{x}$. Now we look at the endpoint $x = 0$. By (3.2),

$$f'(0) = \lim_{\Delta x \to 0^+} \frac{f(0 + \Delta x) - f(0)}{\Delta x} \quad (x = 0)$$

$$= \lim_{\Delta x \to 0^+} \frac{\sqrt{0 + \Delta x} - \sqrt{0}}{\Delta x}$$

$$= \lim_{\Delta x \to 0^+} \frac{\sqrt{\Delta x}}{\Delta x}$$

$$= \lim_{\Delta x \to 0^+} \frac{1}{\sqrt{\Delta x}}$$

As Δx gets closer to 0, the quotient gets arbitrarily large; the limit does not exist and, therefore, f *is not differentiable* at $x = 0$. The domain of f' is the open interval $(0, \infty)$. ■

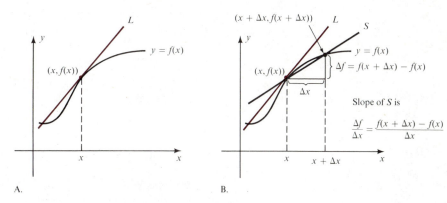

FIGURE 3–1. (A)The line L is a reasonable candidate for the tangent to the graph of f at $(x, f(x))$. (B) The slope of S is an approximation to the slope of L. This approximation becomes more accurate as $\Delta x \to 0$, so that

$$\text{slope of } L = \lim_{\Delta x \to 0} \frac{f(x + \Delta x) - f(x)}{\Delta x}$$

$$= f'(x)$$

A.   B.

Derivative = Slope of Tangent

Figure 3–1A shows the graph of a function f and a line L that fits our intuitive notion of a line tangent to the graph at the point $(x, f(x))$. The question is: *What is the slope of L?* To find the answer, we proceed as follows. Choose another point $x + \Delta x$ in the domain of f and draw the line S through the points $(x, f(x))$ and $(x + \Delta x, f(x + \Delta x))$. The slope of S is

$$\frac{\Delta f}{\Delta x} = \frac{f(x + \Delta x) - f(x)}{\Delta x}$$

as shown in Figure 3–1B. For each nonzero Δx, we obtain a line (like S) joining the points $(x, f(x))$ and $(x + \Delta x, f(x + \Delta x))$ whose slope is $\Delta f/\Delta x$. Now let $\Delta x \to 0$. We reason that the corresponding lines would approach the line L and, therefore, their slopes would approach the slope of L. That is, the **tangent line** of the graph of f at $(x, f(x))$ has slope

$$(3.3) \qquad m = \lim_{\Delta x \to 0} \frac{f(x + \Delta x) - f(x)}{\Delta x}$$

But (3.3) is exactly the definition of the derivative of f at x. Therefore,

(3.4)

> If f is differentiable at x, then
>
> (1) the graph of f has a tangent line L at $(x, f(x))$, and
> (2) the slope of L is $f'(x)$.

We can check, with an example, whether or not (3.4) agrees with our intuitive notion of a tangent line.

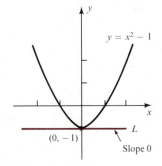

FIGURE 3–2. Example 4.

EXAMPLE 4   The graph of $f(x) = x^2 - 1$ is drawn in Figure 3–2. Intuitively, the tangent line at $(0, -1)$ is a horizontal line; and horizontal lines have slope 0. Verify that, at least in this case, statement (3.4) agrees with our intuition; that is, verify that $f'(0) = 0$.

Solution

$$f'(0) = \lim_{\Delta x \to 0} \frac{f(0 + \Delta x) - f(0)}{\Delta x} \qquad \text{((3.1) with } x = 0\text{)}$$

$$= \lim_{\Delta x \to 0} \frac{[(0 + \Delta x)^2 - 1] - [0^2 - 1]}{\Delta x}$$

$$= \lim_{\Delta x \to 0} \frac{\Delta x^2 - 1 + 1}{\Delta x}$$

$$= \lim_{\Delta x \to 0} \Delta x = 0 \qquad \blacksquare$$

Here is another check on (3.4). We know from elementary geometry that a line tangent to a circle is perpendicular to the radius at the point of tangency. The next example shows that (3.4) yields the correct slope for a line tangent to a circle.

EXAMPLE 5 Find the slope of the tangent to the graph of $f(x) = \sqrt{1 - x^2}$ at the point $(1/2, \sqrt{3}/2)$.

Solution Figure 3–3 shows the graph of f, which is the upper half of the unit circle $x^2 + y^2 = 1$, and the tangent line L. The radius from $(0, 0)$ to $(1/2, \sqrt{3}/2)$ has slope

$$\frac{\dfrac{\sqrt{3}}{2} - 0}{\dfrac{1}{2} - 0} = \sqrt{3}$$

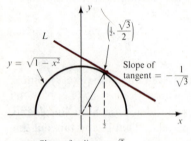

$y = \sqrt{1 - x^2}$

L

$\left(\frac{1}{2}, \frac{\sqrt{3}}{2}\right)$

Slope of tangent $= -\dfrac{1}{\sqrt{3}}$

Slope of radius $= \sqrt{3}$

FIGURE 3–3. Example 5.

Since a tangent to a circle is perpendicular to the radius, the slope of L is $-1/\sqrt{3}$. Let us show that the same result is obtained by computing $f'(\frac{1}{2})$.

$$f'(\tfrac{1}{2}) = \lim_{\Delta x \to 0} \frac{f(\tfrac{1}{2} + \Delta x) - f(\tfrac{1}{2})}{\Delta x}$$

$$= \lim_{\Delta x \to 0} \frac{\sqrt{1 - (\tfrac{1}{2} + \Delta x)^2} - \sqrt{1 - (\tfrac{1}{2})^2}}{\Delta x}$$

$$= \lim_{\Delta x \to 0} \frac{\sqrt{\tfrac{3}{4} - \Delta x - \Delta x^2} - \sqrt{\tfrac{3}{4}}}{\Delta x} \cdot \frac{\sqrt{\tfrac{3}{4} - \Delta x - \Delta x^2} + \sqrt{\tfrac{3}{4}}}{\sqrt{\tfrac{3}{4} - \Delta x - \Delta x^2} + \sqrt{\tfrac{3}{4}}}$$

$$= \lim_{\Delta x \to 0} \frac{(\tfrac{3}{4} - \Delta x - \Delta x^2) - \tfrac{3}{4}}{\Delta x(\sqrt{\tfrac{3}{4} - \Delta x - \Delta x^2} + \sqrt{\tfrac{3}{4}})}$$

$$= \lim_{\Delta x \to 0} \frac{-1 - \Delta x}{\sqrt{\tfrac{3}{4} - \Delta x - \Delta x^2} + \sqrt{\tfrac{3}{4}}}$$

$$= \frac{-1}{2\sqrt{\tfrac{3}{4}}} = -\frac{1}{\sqrt{3}} \qquad \blacksquare$$

Once you know that a tangent line passes through $(x, f(x))$ and that its slope is $f'(x)$, you can use the point-slope form (1.33) in Section 1–5 to write an equation for it.

$$(3.5)$$

If f is differentiable at x_0, then an equation of the tangent line at $(x_0, f(x_0))$ is

$$y - f(x_0) = f'(x_0)(x - x_0)$$

EXAMPLE 6 Find an equation of the line tangent to the graph of $f(x) = x^2 - 1$ at the point $(1, 0)$.

Solution Here $x_0 = 1$, so the slope of the tangent line is $f'(1)$.

$$f'(1) = \lim_{\Delta x \to 0} \frac{f(1 + \Delta x) - f(1)}{\Delta x}$$

$$= \lim_{\Delta x \to 0} \frac{[(1 + \Delta x)^2 - 1] - [1^2 - 1]}{\Delta x}$$

$$= \lim_{\Delta x \to 0} \frac{1 + 2\Delta x + \Delta x^2 - 1 - 0}{\Delta x}$$

$$= \lim_{\Delta x \to 0} 2 + \Delta x = 2$$

So the tangent line passes through $(1, 0)$ and its slope is 2. Using (3.5), we have

$$y - 0 = 2(x - 1) \quad \text{or} \quad y = 2x - 2$$

is an equation of the tangent line (Figure 3–4). ∎

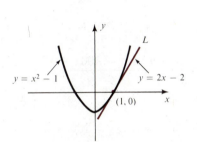

FIGURE 3–4. Example 6.

Derivative = Instantaneous Rate of Change

If s is the position of a particle at time t, then

$$(3.6) \quad \frac{s(t + \Delta t) - s(t)}{\Delta t}$$

represents the average velocity over the time interval from t to $t + \Delta t$. The limit as $\Delta t \to 0$ is the instantaneous velocity at t. In terms of rates of change, the quotient (3.6) is the average rate of change of position, and its limit is the instantaneous rate of change of position.

The example of velocity motivates us to apply the same terms to any function. We say that the quotient

$$\frac{f(x + \Delta x) - f(x)}{\Delta x}$$

is the average rate of change of f over the interval from x to $x + \Delta x$ and the limit

$$(3.7) \quad \lim_{\Delta x \to 0} \frac{f(x + \Delta x) - f(x)}{\Delta x}$$

is the **instantaneous rate of change of f at** x. But (3.7) is exactly the definition of the derivative of f at x. Therefore,

(3.8)
> If f is differentiable at x, then $f'(x)$ is the instantaneous rate of change of f at x.

We usually drop the *instantaneous* part and refer to $f'(x)$ simply as the *rate of change of f at x.*

EXAMPLE 7 Differentiate $f(x) = 100x - x^2$ and find the rate of change at $x = 40$ and $x = 70$.

Solution

$$f'(x) = \lim_{\Delta x \to 0} \frac{f(x + \Delta x) - f(x)}{\Delta x}$$

$$= \lim_{\Delta x \to 0} \frac{[100(x + \Delta x) - (x + \Delta x)^2] - [100x - x^2]}{\Delta x}$$

$$= \lim_{\Delta x \to 0} \frac{100x + 100\Delta x - x^2 - 2x\Delta x - \Delta x^2 - 100x + x^2}{\Delta x}$$

$$= \lim_{\Delta x \to 0} 100 - 2x - \Delta x = 100 - 2x$$

Thus, the derivative of $f(x) = 100x - x^2$ is the *function* $f'(x) = 100 - 2x$. According to (3.8),

$$f'(40) = 100 - 2(40) = 20 \quad \text{is the rate of change of } f \text{ at } x = 40$$
$$f'(70) = 100 - 2(70) = -40 \quad \text{is the rate of change of } f \text{ at } x = 70 \quad \blacksquare$$

Let us now interpret the results of Example 7. What significance is there in the fact that $f'(40) = 20$ and $f'(70) = -40$? Because the derivative is the rate of change, $f'(40) = 20$ means that at $x = 40$, a small change Δx in x will produce a change of about 20 *times* Δx in the value of f.* If Δx is positive, the value of f *increases;* if Δx is negative, the value of f *decreases.* The fact that $f'(70) = -40$ means that at $x = 70$, a small change Δx in x will produce a change of about $-40\Delta x$ in the value of f. In this case, a positive Δx produces a *decrease* in the value of f, whereas a negative Δx produces an *increase.* Here is an application.

EXAMPLE 8 A farmer has 200 feet of fencing to enclose a rectangular garden. What dimensions will yield the garden with the largest area?

Solution Figure 3–5 shows the rectangular garden. If one dimension is x, the other must be $100 - x$ because the perimeter is 200. Therefore, the area is

$$A(x) = x(100 - x) = 100x - x^2 \quad \text{sq ft, } 0 \le x \le 100$$

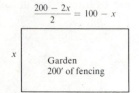

$$\frac{200 - 2x}{2} = 100 - x$$

x | Garden 200′ of fencing

FIGURE 3–5. Example 8.

---

*We say "about" 20 times Δx because as x changes, so may the rate of change of f.

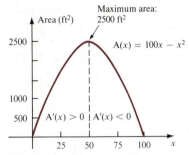

FIGURE 3–6. Example 8. The maximum area occurs when $x = 50$.

This is the function in Example 7, so $A'(x) = 100 - 2x$ (even at the endpoints).

If $A'(x) > 0$, then an increase in x will increase the area.

If $A'(x) < 0$, then a decrease in x will increase the area.

Since $A'(x) = 100 - 2x$ is positive for $x < 50$ and negative for $x > 50$, it follows that $x = 50$ yields the largest area. See Figure 3–6. ■

Other Notations for Derivatives

If a function is defined by an equation and y is the dependent variable, then the derivative is denoted by y'. For instance, if $y = 100x - x^2$, we write $y' = 100 - 2x$ (Example 7). Velocity is another example. When s represents the position of a particle at time t, then s' denotes the derivative of s with respect to t. If v represents the velocity at time t, we write

(3.9) $v = s'$

The velocity is defined as the rate of change of position s *with respect to time t.* The prime notation s' does not indicate the variable with respect to which the values of s are changing. In applications, it is often important to identify that variable. Therefore, we will sometimes write

$$\frac{ds}{dt} \quad \text{(read "dee s dee t")}$$

instead of s' to indicate that the rate of change of s is being computed with respect to the variable t. This notation was introduced by Leibniz. He let Δt represent a small change in t and he let $\Delta s = s(t + \Delta t) - s(t)$ represent the corresponding change in s. The quotient is $\Delta s/\Delta t$ and, as Δt approaches zero, it is as if the deltas "turn into" dees so that the derivative is written as ds/dt. Similarly, if y is a function of x, then

$$\frac{dy}{dx} \quad \text{(read "dee y dee x")}$$

is used to indicate that the rate of change of y is being computed with respect to the variable x. Sometimes we will want to display the function represented by s or y and write, for example,

$$\frac{d}{dt}(t^3 - 4t) \quad \text{or} \quad \frac{d}{dx}(100x - x^2)$$

The prime notation is easier to write, but the "dee" notation provides more information. We shall use the one that best fits the given circumstances.

EXAMPLE 9 (Volume of a sphere). The volume V of a sphere of radius r inches is

$$V = \tfrac{4}{3}\pi r^3 \quad \text{cu in}$$

If a spherical balloon is being inflated, what is the rate of change of the volume with respect to the radius when the radius is 2 inches?

Solution The rate of change of volume with respect to the radius is dV/dr. Thus,

$$
\begin{aligned}
\frac{dV}{dr} &= \lim_{\Delta r \to 0} \frac{\Delta V}{\Delta r} = \lim_{\Delta r \to 0} \frac{V(r + \Delta r) - V(r)}{\Delta r} \\
&= \lim_{\Delta r \to 0} \frac{\frac{4}{3}\pi(r + \Delta r)^3 - \frac{4}{3}\pi r^3}{\Delta r} \\
&= \lim_{\Delta r \to 0} \frac{\frac{4}{3}\pi(r^3 + 3r^2\Delta r + 3r\Delta r^2 + \Delta r^3) - \frac{4}{3}\pi r^3}{\Delta r} \\
&= \lim_{\Delta r \to 0} (4\pi r^2 + 4\pi r\Delta r + \frac{4}{3}\pi\Delta r^2) \\
&= 4\pi r^2 \quad \text{cu in/in}
\end{aligned}
$$

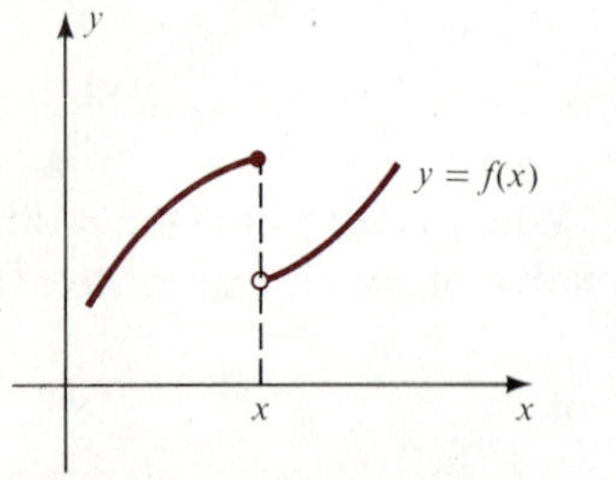

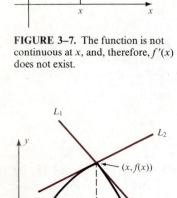

FIGURE 3–7. The function is not continuous at x, and, therefore, $f'(x)$ does not exist.

Thus, when $r = 2$, the volume is changing at the rate of 16π cubic inches per inch change in r. ■

We conclude this section with two important observations about the connection between continuity and differentiability. Suppose that f is differentiable at x. Then

$$
\begin{aligned}
\lim_{\Delta x \to 0}[f(x + \Delta x) - f(x)] &= \lim_{\Delta x \to 0}\left[\frac{f(x + \Delta x) - f(x)}{\Delta x} \cdot \Delta x\right] \\
&= \left[\lim_{\Delta x \to 0} \frac{f(x + \Delta x) - f(x)}{\Delta x}\right]\left[\lim_{\Delta x \to 0} \Delta x\right] \quad (2.5) \\
&= f'(x) \cdot 0
\end{aligned}
$$

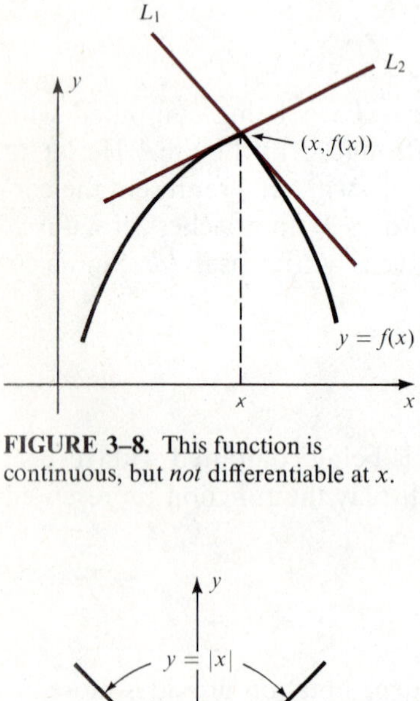

Thus, $\lim_{\Delta x \to 0}[f(x + \Delta x) - f(x)] = 0$ and, by (2.20), we have

FIGURE 3–8. This function is continuous, but *not* differentiable at x.

(3.10)

> **CONTINUITY THEOREM**
> If f is differentiable at x, then f is continuous at x.

It follows from the theorem that if f is not continuous at x, then f is not differentiable at x (Figure 3–7). However, it does *not* follow that continuity implies differentiability. Even if a function is continuous at x, *it may not be differentiable at x*. Figure 3–8 shows such a case. The sharp corner in the graph at $(x, f(x))$ suggests that the one-sided limits

$$
\lim_{\Delta x \to 0^+} \frac{f(x + \Delta x) - f(x)}{\Delta x} \quad \text{and} \quad \lim_{\Delta x \to 0^-} \frac{f(x + \Delta x) - f(x)}{\Delta x}
$$

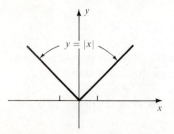

FIGURE 3–9. The absolute value is continuous at 0, but not differentiable at 0 (Exercise 47).

do not agree, and, therefore, $f'(x)$ does not exist. In particular, the absolute value function is continuous at 0, but it is *not* differentiable at 0 (Figure 3–9).

EXERCISES

Find $f'(x)$:

1. $f(x) = 3$ **2.** $f(x) = -7$

3. The values of a constant function do not change; therefore, its rate of change should be zero (Exercises 1 and 2). Let $f(x) = c$, and show that $f'(x) = 0$.

Find $f'(x)$:

4. $f(x) = x$ **5.** $f(x) = 5x$

6. $f(x) = 3 - 2x$ **7.** $f(x) = 4x - 1$

8. The graph of a linear function is a straight line; its rate of change should be the slope of that line (Exercises 4–7). Let $y = mx + b$ and show that $y' = m$.

Find dy/dx:

9. $y = -x^2$ **10.** $y = -3x^2$

11. $y = 4x^2 - 3x$ **12.** $y = x^2 + 2x$

13. $y = -2x^2 + 4x - 5$ **14.** $y = 5x^2 - 2x + 1$

15. Let $f(x) = ax^2 + bx + c$. Show that $f'(x) = 2ax + b$.

Find dy/dx:

16. $y = 3\sqrt{2}$ **17.** $y = 4x - 2$

18. $y = 6x^2 - 5x + 1$ **19.** $y = -7x^2 + 9x + 2$

20. $y = x^3$ **21.** $y = 3x^3$

22. $y = 4x^3 - 2x^2$ **23.** $y = -2x^3 + 4x^2$

24. $y = 1/(9x)$ **25.** $y = 1/(x + 1)$

26. $y = 7/x^2$ **27.** $y = 8\sqrt{x}$

28. $y = 3\sqrt{1 - x}$ **29.** $y = 2\sqrt{x - 4}$

30. $y = 2/\sqrt{6x}$ **31.** $y = 4/\sqrt{x - 2}$

In Exercises 32–34, carefully sketch the graph and draw tangent lines at the points $(x, f(x))$ for the given values of x. Guess the slopes of these lines, and then compare your guess with $f'(x)$.

32. $f(x) = x^2 + 2x + 1$; $x = -2, -1, 0$

33. $f(x) = x^3$; $x = -1, 0, 1$

34. $f(x) = \sqrt{x}$; $x = 1, 1/4, 1/16$. What happens to $f'(x)$ as $x \to 0^+$?

In Exercises 35–36, find an equation of the tangent line at $(x, f(x))$ for the given x.

35. $f(x) = \sqrt{4 - x^2}$; $x = 1$

36. $f(x) = \sqrt{x}$; $x = 4$

Exercises 37–40 refer to Figure 3–10.

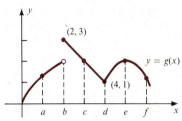

FIGURE 3–10. See Exercises 37–40.

37. Is g differentiable at a? If so, is $g'(a)$ positive, negative, or zero?

38. Same as Exercise 37 for points b, d, and f.

39. Compute the value of $g'(c)$.

40. Guess the value of $g'(e)$. Justify your guess.

41. At what rate is the area of a circle changing with respect to the radius when the radius is 3 centimeters?

42. At what rate is the area of a square changing with respect to one of its sides when the side is 4 meters?

43. A box without a top is to be made by cutting out equal squares and folding up the sides as shown in Figure 3–11A. If squares of x inches are cut, express the volume of the box as a function of x. At what rate is the volume changing with respect to x?

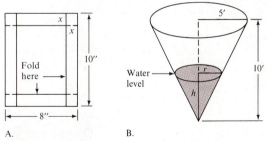

A.   B.

FIGURE 3–11. (A) Making a box in Exercise 43. (B) Filling a conical tank with water in Exercise 46.

44. (Continuation of Exercise 43). At what rate is the volume changing when $x = 1$ inch? To obtain the box with the largest volume, would you cut out squares with $x > 1$ or $x < 1$? Answer the same questions when $x = 2$ inches.

45. A projectile is fired straight up. Its height s feet above the ground t seconds later is $s = -16t^2 + 160t$. How fast and in which direction (up is positive) is it moving when $t = 3$ seconds? When $t = 6$ seconds? When will it be at its highest point?

46. The volume of a cone is $V = \frac{1}{3}\pi r^2 h$ where r is the radius of the base and h is the height. The conical tank in Figure 3–11B is being filled with water. Express the volume of water as a function of its height. Now compute the rate of change of the volume of water with respect to its height when $h = 3$ feet.

47. $y = |x|$ is continuous at 0 (why?). Nevertheless, show that it is *not* differentiable at 0.

48. Let $f(x) = 1$ if x is rational and 0 if x is irrational. Is f differentiable at any point? Why?

49. Let $f(x) = 3$, $g(x) = 2x$, and $F(x) = x^2$.
(a) $(f + g)(x) = 3 + 2x$. Verify that $(f + g)'(x) = f'(x) + g'(x)$
(b) $(f + F)(x) = 3 + x^2$. Verify that $(f + F)'(x) = f'(x) + F'(x)$
(c) Verify that $(f + g + F)'(x) = f'(x) + g'(x) + F'(x)$.

50. Let $f(x) = 7$, $g(x) = 4x$, and $F(x) = x^2$.
(a) Verify that $(fg)'(x)$ does *not* equal $f'(x)g'(x)$
(b) Verify that $(gF)'(x)$ does *not* equal $g'(x)F'(x)$.

Optional Exercises

51. Let $y = \sqrt[3]{x}$ and find dy/dx.

52. Let $y = 1/\sqrt[3]{x}$ and find dy/dx.

53. Prove that if f and g are differentiable at x, and c is a constant, then $(cf)'(x) = cf'(x)$ and $(f + g)'(x) = f'(x) + g'(x)$.

54. Prove that if f and g are differentiable at x, then $(fg)'(x) = f(x)g'(x) + g(x)f'(x)$.

3–2 DERIVATIVE FORMULAS

The derivative is defined by a limit

$$f'(x) = \lim_{h \to 0} \frac{f(x + h) - f(x)^*}{h}$$

Computing this limit for complicated functions like

$$f(x) = (x^2 + 3x - 2)(4x^3 + x^2 - 3x + 1) \quad \text{or}$$

$$f(x) = \frac{3x^2 - 5x + 7}{x^3 - 3x^2}$$

can be tedious. Fortunately, such computations are usually unnecessary because there are derivative formulas that enable us to compute the derivatives of nearly all functions in terms of just a few basic functions. Once the formulas are verified, most derivatives can be found without computing limits.

In this section, we state and prove several of these formulas. The proofs all follow the same pattern. We write out the quotient

---

*This is definition (3.1) with Δx replaced by h. There is no difference in meaning. Both notations, the one with Δx and the one above with h, are in common use.

$$\frac{f(x + h) - f(x)}{h}$$

which is called a **difference quotient** of f, and then evaluate the limit as $h \to 0$. Go over the examples carefully, and work as many exercises as you can. In that way the formulas will become second nature to you.

The difference quotient of a constant function $f(x) = c$ is

$$\frac{f(x + h) - f(x)}{h} = \frac{c - c}{h} = 0$$

and therefore,

(3.11)

CONSTANT

The derivative of a constant function is 0, or $\dfrac{dc}{dx} = 0.$

The difference quotient of the identity function $f(x) = x$ is

$$\frac{f(x + h) - f(x)}{h} = \frac{(x + h) - x}{h} = \frac{h}{h} = 1$$

and therefore,

(3.12)

IDENTITY

The derivative of the identity function is 1, or $\dfrac{dx}{dx} = 1.$

This formula holds no matter what letter is used to denote the variable; that is,

$$\frac{dy}{dy} = 1, \frac{dt}{dt} = 1, \frac{du}{du} = 1, \text{ and so on}$$

If c is a constant and f is a function, then $(cf)(x) = cf(x)$; the difference quotient is

$$\frac{(cf)(x + h) - (cf)(x)}{h} = \frac{cf(x + h) - cf(x)}{h} = c\left[\frac{f(x + h) - f(x)}{h}\right]$$

It follows that

(3.13)

CONSTANT TIMES A FUNCTION

If f is differentiable, then so is cf, and

$$(cf)' = cf', \text{ or } \frac{d}{dx}cf = c\frac{df}{dx}$$

Let us now examine a sum, $(f + g)(x) = f(x) + g(x)$. The difference quotient is

$$\frac{(f + g)(x + h) - (f + g)(x)}{h}$$

$$= \frac{[f(x + h) + g(x + h)] - [f(x) + g(x)]}{h}$$

$$= \left[\frac{f(x + h) - f(x)}{h}\right] + \left[\frac{g(x + h) - g(x)}{h}\right] \quad \text{(Rearrange)}$$

Now let $h \to 0$. If both f and g are differentiable at x, then the limit of the sum on the right exists, and equals the sum of the limits; namely, $f'(x) + g'(x)$. Thus, the limit of the quotient on the left exists, and it is $(f + g)'(x)$. It follows that

(3.14)

> **SUM**
> If f and g are both differentiable, then so is $f + g$, and
>
> $$(f + g)' = f' + g' \quad \text{or} \quad \frac{d}{dx}(f + g) = \frac{df}{dx} + \frac{dg}{dx}$$

A similar argument shows that (3.14) is also valid for the sum of three, four, or any finite number of functions. Furthermore, the analogous statement for differences is true; $(f - g)' = f' - g'$.

EXAMPLE 1   Let $y = 4 - 3x$ and find dy/dx.

Solution   Finding the limit of the difference quotient in this case is not difficult, but let us use the theorems to illustrate the method.

$$\frac{dy}{dx} = \frac{d}{dx}(4 - 3x) = \frac{d4}{dx} - \frac{d}{dx}(3x) \qquad \text{(Difference)}$$

$$= \frac{d4}{dx} - 3\frac{dx}{dx} \qquad \text{(cf)}$$

$$= 0 - 3(1) = -3 \qquad \text{(Constant; Identity)}$$

Thus,

$$\frac{d}{dx}(4 - 3x) = -3 \qquad \blacksquare$$

Power Functions

The next step is to develop a formula for the derivative of a power function. We claim that if n is a positive integer, then

(3.15)
$$\frac{d}{dx}(x^n) = nx^{n-1}$$

This is certainly true for $n = 1$ because

$$\frac{d}{dx}(x^1) = \frac{dx}{dx} = 1 = 1x^0$$

For $n = 2, 3$, and 4, formula (3.15) yields

$$\frac{d}{dx}(x^2) = 2x$$

$$\frac{d}{dx}(x^3) = 3x^2$$

$$\frac{d}{dx}(x^4) = 4x^3$$

To verify these equations, we examine the difference quotients.

For $n = 2$: $\dfrac{(x + h)^2 - x^2}{h} = \dfrac{x^2 + 2xh + h^2 - x^2}{h} = 2x + h$

For $n = 3$: $\dfrac{(x + h)^3 - x^3}{h} = \dfrac{x^3 + 3x^2h + 3xh^2 + h^3 - x^3}{h}$

$$= 3x^2 + 3xh + h^2$$

For $n = 4$: $\dfrac{(x + h)^4 - x^4}{h} = \dfrac{x^4 + 4x^3h + 6x^2h^2 + 4xh^3 + h^4 - x^4}{h}$

$$= 4x^3 + 6x^2h + 4xh^2 + h^3$$

In each case, the first term on the right is nx^{n-1} (that is, $2x = 2x^{2-1}$, $3x^2 = 3x^{3-1}$, and $4x^3 = 4x^{4-1}$), and every other term contains h as a factor. As $h \to 0$, the term nx^{n-1} remains fixed, but all the others approach zero. It can be shown that the same situation occurs for any positive integer n; therefore, (3.15) holds for all such n.

A pleasant consequence of the formulas proved so far is that polynomials are easy to differentiate.

(3.16)

POLYNOMIAL

If $f(x) = a_n x^n + a_{n-1} x^{n-1} + \cdots + a_1 x + a_0$, then

$$f'(x) = a_n n x^{n-1} + a_{n-1}(n - 1)x^{n-2} + \cdots + a_1$$

EXAMPLE 2   Let $y = 3x^2 + 2x - 4$. Then

$$\frac{dy}{dx} = \frac{d}{dx}(3x^2) + \frac{d}{dx}(2x) - \frac{d4}{dx} \qquad \text{(Sum)}$$

$$= 3\frac{d}{dx}(x^2) + 2\frac{dx}{dx} - 0 \qquad (cf; \text{Constant})$$

$$= 3(2x) + 2(1) \qquad (3.15); \text{(Identity)}$$

$$= 6x + 2 \qquad \blacksquare$$

The steps in Example 2 can be performed mentally.

EXAMPLE 3

(a) If $g(t) = -2t^3 + t^2 - 3t$, then

$$g'(t) = -6t^2 + 2t - 3$$

(b) $\dfrac{d}{du}(u^4 - 3u^3 + u + 2)$

$$= 4u^3 - 9u^2 + 1 \qquad \blacksquare$$

The fact that (3.15) holds for positive integers n led to a simple formula for differentiating polynomials, but that is only the tip of the iceberg. The formula

$$\frac{d}{dx}(x^n) = nx^{n-1}$$

also holds for *negative* integers. For instance, Example 2 of Section 3–1 shows that

(3.17) $\quad \dfrac{d}{dx}\left(\dfrac{1}{x}\right) = -\dfrac{1}{x^2}$

Substituting x^{-1} for $1/x$ and $-x^{-2}$ for $-1/x^2$ in (3.17) yields

$$\frac{d}{dx}(x^{-1}) = -x^{-2}$$

which is formula (3.15) with $n = -1$. It can be shown that the same is true for any negative integer. Thus,

$$\frac{d}{dx}(x^{-2}) = -2x^{-3}, \frac{d}{dx}(x^{-3}) = -3x^{-4}, \text{and so on}$$

Formula (3.15) also holds for *rational* powers. Example 3 in Section 3–1 shows that

(3.18) $\quad \dfrac{d}{dx}(\sqrt{x}) = \dfrac{1}{2\sqrt{x}}$

Substituting $x^{1/2}$ for $\sqrt{x}$ and $\frac{1}{2}x^{-1/2}$ for $1/2\sqrt{x}$ in (3.18) yields

$$\frac{d}{dx}(x^{1/2}) = \frac{1}{2}x^{-1/2}$$

which is formula (3.15) with $n = \frac{1}{2}$. It can be shown that the same is true for any rational power.

EXAMPLE 4

(a) If $y = x^2 + x^{-5}$, then

$$y' = 2x - 5x^{-6}$$

(b) If $s = t^{3/4} - t^{-3}$, then

$$s' = \frac{3}{4}t^{-1/4} + 3t^{-4}$$

(c) $\dfrac{d}{du}(3u^{1/3} - u^{-1/3})$

$$= u^{-2/3} + \frac{1}{3}u^{-4/3} \quad ■$$

Sometimes the form of a function must be changed in order to apply the power formula.

EXAMPLE 5

(a) Rewrite $1/x^5$ as the power x^{-5}; then

$$\frac{d}{dx}\left(\frac{1}{x^5}\right) = \frac{d}{dx}(x^{-5}) = -5x^{-6}$$

(b) Rewrite $\sqrt[3]{t^2}$ as the power $t^{2/3}$; then

$$\frac{d}{dt}(\sqrt[3]{t^2}) = \frac{d}{dt}(t^{2/3}) = \frac{2}{3}t^{-1/3} \quad ■$$

So far, we have actually verified the power formula for only a few values of n; namely, $n = 1, 2, 3, 4, -1$, and $\frac{1}{2}$. Obviously, this does not *prove* the formula, but does suggest its validity for many values of n. The fact is, *the power formula holds for every real number r.* Although a proof must wait until Chapter 7, we will state and use the result, in all its generality, from now on.

(3.19)

> **POWER**
>
> For any real number r, $\dfrac{d}{dx}(x^r) = rx^{r-1}$

Products and Quotients

Formula (3.14) says that the derivative of a sum is the sum of the derivatives. It would be nice if a similar statement were true for products, but a simple example shows that is not the case. If $f(x) = x^2$ and $g(x) = x^3$, then $(fg)(x) = f(x)g(x) = x^2x^3 = x^5$. Therefore,

$$(fg)'(x) = (x^5)' = 5x^4$$

whereas

$$f'(x)g'(x) = (x^2)'(x^3)' = (2x)(3x^2) = 6x^3$$

The example shows that the derivative of a product is *not* the product of the derivatives. This is unfortunate. There is, however, a formula for the derivative of a product; it is not as neat as we would like, but it has the undeniable advantage of being true.

(3.20)

> **PRODUCT**
>
> If f and g are differentiable, then so is fg, and
>
> $$(fg)' = fg' + gf', \text{ or } \frac{d}{dx}(fg) = f\frac{dg}{dx} + g\frac{df}{dx}$$

Before proving the product formula, let us look at an example.

EXAMPLE 6 If $y = (x^2 + 4)(x^2 - 5x + 8)$, then

$$
\begin{aligned}
y' &= (x^2 + 4)(x^2 - 5x + 8)' + (x^2 - 5x + 8)(x^2 + 4)' && \text{(Product)} \\
&= (x^2 + 4)(2x - 5) + (x^2 - 5x + 8)(2x) && \text{(Polynomial)} \\
&= 4x^3 - 15x^2 + 24x - 20 && \text{(Multiply; add like terms)}
\end{aligned}
$$

You can check this by multiplying first

$$y = (x^2 + 4)(x^2 - 5x + 8) = x^4 - 5x^3 + 12x^2 - 20x + 32$$

and then taking the derivative

$$y' = (x^4 - 5x^3 + 12x^2 - 20x + 32)' = 4x^3 - 15x^2 + 24x - 20$$

which agrees with the answer above. ∎

The proof of the product formula is enlightening; it shows why the derivative of a product is not the product of the derivatives. We write out the difference quotient, add and subtract $f(x + h)g(x)$ to obtain an expression we can work with, and then take the limit.

$$
\frac{(fg)(x + h) - (fg)(x)}{h}
$$

$$
= \frac{f(x + h)g(x + h) - f(x)g(x)}{h}
$$

$$= \frac{f(x + h)g(x + h) - f(x + h)g(x) + f(x + h)g(x) - f(x)g(x)}{h}$$

$$= f(x + h)\left[\frac{g(x + h) - g(x)}{h}\right] + g(x)\left[\frac{f(x + h) - f(x)}{h}\right]$$

If f is differentiable at x, then, by (3.10), it is continuous at x. From (2.20), it follows that $f(x + h) \to f(x)$ as $h \to 0$. The values of the other limits are unmistakable. Thus, taking the limit of both sides, we have

$$(fg)'(x) = f(x)g'(x) + g(x)f'(x)$$

which completes the proof of the product formula.

QUOTIENT

If f and g are differentiable, then f/g is differentiable whenever g is not zero, and

(3.21)

$$\left(\frac{f}{g}\right)' = \frac{gf' - fg'}{g^2} \quad \text{or} \quad \frac{d}{dx}\left(\frac{f}{g}\right) = \frac{g\dfrac{df}{dx} - f\dfrac{dg}{dx}}{g^2}$$

The proof of the quotient formula is obtained by first finding the derivative of the reciprocal $1/g$, and then using the product with $f/g = f[1/g]$. The proof is left as an exercise.

EXAMPLE 7   Let $y = (2x + 1)/(3x + 4)$. Then

$$y' = \frac{(3x + 4)(2x + 1)' - (2x + 1)(3x + 4)'}{(3x + 4)^2} \qquad \text{(Quotient)}$$

$$= \frac{(3x + 4)(2) - (2x + 1)(3)}{(3x + 4)^2}$$

$$= \frac{5}{(3x + 4)^2} \qquad \text{(Simplify)} \qquad \blacksquare$$

EXAMPLE 8

$$\left(\frac{3\sqrt[3]{u^2}}{4u + 1}\right)' = \frac{(4u + 1)(3u^{2/3})' - (3u^{2/3})(4u + 1)'}{(4u + 1)^2}$$

$$= \frac{(4u + 1)(2u^{-1/3}) - (3u^{2/3})(4)}{(4u + 1)^2}$$

To simplify the numerator, we write

$$\frac{2(4u + 1)}{u^{1/3}} - 12u^{2/3} = \frac{2(4u + 1)}{u^{1/3}} - \frac{12u}{u^{1/3}}$$

$$= \frac{-4u + 2}{u^{1/3}}$$

Therefore, the final form of the derivative is

$$\left(\frac{3\sqrt[3]{u^2}}{4u + 1}\right)' = \frac{-4u + 2}{u^{1/3}(4u + 1)^2} \qquad \blacksquare$$

Higher Order Derivatives

If f is differentiable, then f' is a function whose domain is contained in the domain of f. It is often useful to know the derivative of f'; that is, $(f')'$. This is called the **second derivative of f**. We interpret it as the *rate of change of f'*. But there is no reason to stop here. The derivative of the second derivative is called the **third derivative of f**, and there are fourth derivatives, fifth derivatives, and so on. The notations used are

The second derivative is denoted by f'' or $\dfrac{d^2f}{dx^2}$

The third derivative is denoted by f''' or $\dfrac{d^3f}{dx^3}$

For the nth derivative, with $n > 3$, we use the symbols

$$f^{(n)} \quad \text{or} \quad \frac{d^nf}{dx^n}$$

EXAMPLE 9   Find the second derivative of $y = 3x^4 - 5x^3 + x^2$.

Solution

$$y' = 12x^3 - 15x^2 + 2x$$
$$y'' = 36x^2 - 30x + 2 \qquad \blacksquare$$

EXAMPLE 10   Find the third derivative of $s = t^2 + \sqrt{t}$.

Solution   Write $s = t^2 + t^{1/2}$. Then

$$\frac{ds}{dt} = 2t + \frac{1}{2}t^{-1/2}$$

$$\frac{d^2s}{dt^2} = 2 - \frac{1}{4}t^{-3/2}$$

$$\frac{d^3s}{dt^3} = \frac{3}{8}t^{-5/2} \qquad \blacksquare$$

EXERCISES

In Exercises 1–24, differentiate the given functions (with respect to the obvious variable).

1. $y = 9x^3 - 3x^2 + 6x + 4$

2. $y = 6x^4 - 5x^2 + 17$

3. $s = 8 - \dfrac{t^2}{2} - \dfrac{t^4}{5}$

4. $s = t + \dfrac{t^3}{6} - \dfrac{t^5}{8}$

5. $v = 4u^{-1/2} + u^{4.5}$

6. $v = -2u^{2/3} + u^{-4/5}$

7. $y = -3\sqrt{x} - (1/\sqrt[3]{x})$

8. $y = 5\sqrt[3]{x^2} + (1/x^4)$

9. $s = (3t^2 - 4t)(t^3 + 7)$

10. $s = (4 - t^2)(9 - t^3)$

11. $v = \dfrac{1 - u - u^2}{2 - u}$

12. $v = \dfrac{3u^2 + u - 7}{u + 1}$

13. $y = (\sqrt{x} + \sqrt[3]{x})(\sqrt[4]{x} + \sqrt[5]{x})$

14. $y = (x^{-1} + x^{-2})(x^{-3} + x^{-4})$

15. $s = \dfrac{t^{-1} - t^{-2}}{t^{-3} - t^{-4}}$

16. $s = \dfrac{\sqrt{t} + \sqrt[3]{t}}{\sqrt[4]{t} + \sqrt[5]{t}}$

17. $v = \dfrac{1}{1 + u + u^2}$

18. $v = 1 + \dfrac{1}{u} + \dfrac{1}{u^2}$

19. $y = (x^2 - 2x + 1)^2$

20. $y = (3 - x - x^2)^2$

21. $s = (t + 1)(t^3 - 2t^2 + 3t)$

22. $s = (t - 2)(t^4 - t^2 + t)$

23. $v = \dfrac{3u^4 - 5}{(u + 1)^2}$

24. $v = \dfrac{(u - 5)^2}{u^5 + 4}$

In Exercises 25–36, find the second derivatives.

25. $y = \dfrac{x}{5} - 9$

26. $y = 4 - 7x$

27. $u = 3.6x^2 + 1.7x - 13$

28. $u = 8 - 2.6x - 5.9x^2$

29. $y = 2 - x + 13x^2 - 9x^3$

30. $y = x^5 - 3x^4$

31. $s = 3t^{1.7}$

32. $s = \dfrac{1}{4}t^{-4}$

33. $y = (3x^2 + 2)(x^3 - 7)$

34. $y = (x^2 - 5)(2x^2 + 7)$

35. $u = \dfrac{x^2 + 1}{x + 1}$

36. $u = \dfrac{1 - x^2}{x + 1}$

In Exercises 37–40, find the indicated derivative.

37. $y = 8x^4 - 3x^3 + x^2 - 1$; y'''

38. $y = 9x^2 - 3x + 5$; y'''

39. $y = \sqrt{x}$; $\dfrac{d^4y}{dx^4}$

40. $y = \dfrac{1}{x + 1}$; $\dfrac{d^4y}{dx^4}$

41. Newton's law of gravitation states that the magnitude of the force F exerted by a body of mass M on a body of mass m is

$$F = \frac{GmM}{r^2}$$

where G is the universal gravitation constant and r is the distance between the bodies. If the bodies are moving apart, write a formula that expresses the rate of change of F with respect to r.

42. Find an equation for each of the lines tangent to the curve $y = x^4 + 4x$ at the points $(2, 24)$ and $(-2, 8)$. For what values of x will the slope of the tangent be positive? Negative? Zero?

43. Find an equation for each of the lines tangent to the curve $y = x^3 - 2x^2 - 4x$ at the points $(3, -3)$, $(0, 0)$, and $(-2, -8)$. For what values of x will the slope of the tangent be positive? Negative? Zero?

44. A piece of cylindrical pipe is 10 inches long. What is the rate of change of the surface area with respect to the radius?

45. *(Productivity)* Economists define the total productivity T of a population to be $T = LP$ where L is the size of the labor force and P is the productivity of the average worker.* If the labor force L grows at the rate of about 4.5% per year (that is, $dL/dt = .045L$), and the productivity P of the average worker decreases at a rate of about .5% per year (that is, $dP/dt = -.005P$), what is the rate of growth (or decline) of the total productivity $T = LP$?

46. Data collected from a large number of samples indicate that the surface area (in square meters) of a human being is closely related to weight (in kilograms) by the equation

$$A = .11w^{2/3}$$

*The value of L at any time t is an integer; it is the number of people in the labor force. Thus, there are many "breaks" in the graph, and L is not differentiable. However, economists have found that treating L as though it were differentiable produces reliable results.

128   DERIVATIVES

What is the rate of change of surface area with respect to weight of a baby weighing 8 kg? (1 kg ≈ 2.2 lbs.)

47. The oxygen content C in the bloodstream is usually expressed as a percentage of saturation. In human beings, experimental data indicate that

$$C = 100 \frac{(.00013)P^{2.7}}{1 + (.00013)P^{2.7}} \text{ percent}$$

where P is the partial pressure of oxygen in torr (760 torr is about equal to air pressure at sea level). The rate of change of C with respect to P is a measure of how much oxygen the blood releases to the body in various situations, say during strenuous exercise. Find dC/dP when the pressure is 50 torr.

48. The position function of a particle moving along a horizontal line is $s = t^2 - 4t + 1$. In what time intervals will the particle be moving to the right? To the left? When will it be at rest?

49. Answer the same questions as in Exercise 48 if $s = \frac{1}{3}t^3 - \frac{5}{2}t^2 + 4t$.

50. Answer the same questions as in Exercise 48 if $s = t + (1/t), t > \frac{1}{2}$.

51. Air is pumped into a spherical weather balloon. The radius, in feet, is a function of time; $r = 2t + 1$. How fast is the radius changing when $t = 3$ seconds? How fast is the volume of the balloon changing with respect to the radius when $t = 3$ seconds?

Optional Exercises

55. The base and height of a rectangle are functions of time; $b = b(t)$ and $h = h(t)$. Suppose that they are changing in such a way that the area of the rectangle remains constant at 4 square feet. When $b = 3$, it is decreasing at the rate of .4 ft/sec. How fast is h changing at this moment?

56. Let

$$f(x) = \begin{cases} x & \text{if } x \text{ is rational} \\ 0 & \text{if } x \text{ is irrational} \end{cases}$$

$$g(x) = \begin{cases} x^2 & \text{if } x \text{ is rational} \\ 0 & \text{if } x \text{ is irrational} \end{cases}$$

Both functions are continuous *only* at 0 (see Exercise 42, Section 2–3). Of course, neither f nor g is differentiable at any nonzero point (why not?), and it seems reasonable to expect that they are not dif-

52. Show that if F, G, and H are all differentiable, then $(FGH)' = F'GH + FG'H + FGH'$. (*Hint:* Think of FGH as the product of *two* functions F and GH, and use the product formula twice). Use this formula to differentiate $y = (x + 1)(x^2 + 1)(x^3 + 1)$.

53. Fill in the blanks below to prove

(3.22) | If g is differentiable at x, and $g(x) \neq 0$, then $(1/g)'(x) = -g'(x)/g^2(x)$

(1) The difference quotient for $1/g$ is _____.
(2) Combine the fractions in the numerator to obtain _____.
(3) This can be rewritten as a product

$$-\frac{g(x + h) - g(x)}{h} \cdot \frac{1}{g(x + h)g(x)}$$

(4) $g(x + h) \to g(x)$ as $h \to 0$ because _____.
(5) Therefore, $(1/g)'(x) = -g'(x)/g^2(x)$ because _____.

54. Write f/g as a product $f[1/g]$. Now use the product formula and (3.22) above to prove the quotient formula: If f and g are differentiable, then $(f/g)' = (gf' - fg')/g^2$, wherever f/g is defined.

ferentiable at zero as well. But what seems reasonable is not always true. In this case, $f'(0)$ does not exist, but $g'(0)$ does! Verify this by examining the difference quotients of f and g at $x = 0$. What is the value of $g'(0)$?

57. Make up and prove a formula for the derivative of a product of four functions, $(FGHK)'$.

58. Find a function f such that $f'(x) = 0$ and $f(1) = 4$.
59. Find a function f such that $f'(x) = 7$ and $f(0) = 3$.
60. Find a function f such that $f'(x) = 2x$ and $f(-2) = 4$.
61. Find a function f such that $f'(x) = x^2 + x$ and $f(0) = 9$.
62. Find a function f such that $f''(x) = 3x^2 - 2x + 1$, $f'(1) = 8$, and $f(1) = 6$.

3-3   THE DERIVATIVE OF e^x

A complete description of exponential functions must be delayed until Chapter 7 because it involves material that we have not as yet discussed. But there is one exponential function, $f(x) = e^x$, that is so useful in examples and applications (population growth, radioactive decay, and so on) that we will introduce it right now in an informal way. The letter e denotes the irrational number 2.718281.... The motivation for choosing such an unlikely base for our exponential function is that this function is its own derivative! In a moment, you will see why this is so; in later sections, you will see why this is the key property that makes $f(x) = e^x$ so important. First we define e in terms of a limit, then use a calculator to approximate the value of the limit, and then work on the derivative.

The number e is defined by

(3.23)
$$e = \lim_{h \to 0}(1 + h)^{1/h}$$

under the assumption that this limit exists. As h approaches 0, the base $(1 + h)$ approaches 1 and the exponent $1/h$ gets larger and larger (positive if $h > 0$ and negative if $h < 0$). What is the value of a power if the base is very close to 1 and the exponent is very large? We can compute some values using a calculator. For example, if $h = \frac{1}{10}$, then $1/h = 10$ and

$$\left(1 + \frac{1}{10}\right)^{10} \approx 2.5937425$$

If $h = -\frac{1}{100}$, then $1/h = -100$ and

$$\left(1 - \frac{1}{100}\right)^{-100} \approx 2.7319990$$

The table below shows the results for several values of h.

	h	$\pm.1$	$\pm.01$	$\pm.001$
$(1 + h)^{1/h}$	$h > 0$	2.5937425	2.7048138	2.7169236
	$h < 0$	2.8679720	2.7319990	2.7196419

The entries in the table do appear to be squeezing down to a single number. This number is e; it is irrational and its value, correct to seven digits, is

(3.24)
$$e = 2.718281\ldots$$

Let us now find a formula for the derivative of $f(x) = e^x$. The difference quotient is

$$\frac{f(x + h) - f(x)}{h} = \frac{e^{x+h} - e^x}{h}$$

$$= \frac{e^x e^h - e^x}{h} \tag{1.37}$$

(3.25)
$$= e^x \frac{e^h - 1}{h} \tag{Factor}$$

The derivative is the limit as $h \to 0$. Now e^x is fixed because x is fixed; so it remains to evaluate

$$\lim_{h \to 0} \frac{e^h - 1}{h}$$

and we argue as follows. If h is very close to 0, then by (3.23),

$$e \approx (1 + h)^{1/h}$$

Now raise both sides to the power h,

$$e^h \approx [(1 + h)^{1/h}]^h = 1 + h \tag{1.37}$$

and continue with

$$e^h - 1 \approx h$$

$$\frac{e^h - 1}{h} \approx 1 \tag{Divide by $h \neq 0$}$$

This suggests that

(3.26)
$$\lim_{h \to 0} \frac{e^h - 1}{h} = 1$$

Combining (3.25) and (3.26), we have

EXPONENTIAL FORMULA

(3.27)
$$\frac{d}{dx}(e^x) = e^x$$

Here is a function that is its own derivative!

The derivative formulas of the preceding section can be applied to functions involving e^x.

EXAMPLE 1 $\dfrac{d}{dx}(4e^x) = 4\dfrac{d}{dx}e^x = 4e^x$ $\qquad$ (cf), (3.27) ∎

EXAMPLE 2   *(Product)* If $y = x^2 e^x$, then

$$y' = x^2(e^x)' + e^x(x^2)' \qquad \text{(Product)}$$
$$= x^2 e^x + 2xe^x \qquad \text{(3.27), (Power)}$$
$$= xe^x(x + 2) \qquad \text{(Simplify)} \quad \blacksquare$$

EXAMPLE 3   *(Quotient)* If $y = e^x/3x^4$, then

$$y' = \frac{3x^4(e^x)' - e^x(3x^4)'}{(3x^4)^2} \qquad \text{(Quotient)}$$

$$= \frac{3x^4 e^x - 12x^3 e^x}{9x^8} \qquad \text{(3.27), (Power)}$$

$$= \frac{(x - 4)e^x}{3x^5} \qquad \text{(Factor out } 3x^3\text{; simplify)} \quad \blacksquare$$

EXAMPLE 4   Find the second derivative of $y = xe^x$.

Solution   The first derivative is

$$y' = x(e^x)' + (x)'e^x \qquad \text{(Product)}$$
$$= xe^x + e^x \qquad \text{(3.27), (Identity)}$$

The second derivative is

$$y'' = (xe^x)' + (e^x)' \qquad \text{(Sum)}$$
$$= [xe^x + e^x] + e^x \qquad \text{(From the first part)}$$
$$= e^x(x + 2) \quad \blacksquare$$

If (x, y) is on the graph of $y = e^x$, it follows that the slope of the tangent line at (x, y) is $y' = e^x = y$ (Figure 3–12).

EXAMPLE 5   Find an equation of the line tangent to the graph of $y = e^x$ at the point $(-3, e^{-3})$.

Solution   The slope of the tangent line is e^{-3}. Therefore, an equation of the line is

$$y - e^{-3} = e^{-3}(x - (-3)), \text{ or}$$
$$y = e^{-3}x + 4e^{-3} \quad \blacksquare$$

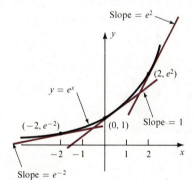

FIGURE 3–12. The curve $y = e^x$. The slope of the tangent line at (x, y) is y because $y' = e^x = y$.

Converting Exponentials to Base e

The number e is the base for the **natural logarithm,** which is abbreviated *ln* (many calculators have a $\boxed{\ln x}$ key). Thus, according to our discussion of inverses in Section 1–9, the functions $f(x) = e^x$ and $g(x) = \ln x$ are inverses. It follows from (1.63) that

(3.28)

$$
\begin{array}{l}
(1)\ e^{\ln x} = x, \text{ for } x > 0 \\
(2)\ \ln e^x = x, \text{ for all } x
\end{array}
$$

These identities can be used to convert any exponential a^x with $a > 0$ and $a \neq 1$ to an exponential with base e. We set

(3.29) $y = a^x$

This makes $y > 0$, and we can take the natural logarithm of both sides

$$\ln y = \ln a^x$$

The properties of logarithms, (1.42) in Section 1–6, yield

$$\ln y = x \ln a$$

It follows from this equality that

$$e^{\ln y} = e^{x \ln a}$$

Since $e^{\ln y} = y$ (3.28) and $y = a^x$ (3.29), the last equation becomes

(3.30) $a^x = e^{x \ln a}$, for all x

EXAMPLE 6 *(Population growth)* The population of bacteria in a culture is found to double in size every k minutes. Write the population P as an exponential function (with base e) of time t.

Solution   We have seen before (Example 2 in Section 1–6) that since the population *doubles* every k minutes, it can be expressed as the function

$$P = P_0 2^{t/k}$$

where P_0 is the initial population and t is time in minutes. Using (3.30) with $a = 2$ and $x = t/k$, we have that

(3.31) $P = P_0 e^{\frac{t}{k} \ln 2} = P_0 e^{\left(\frac{\ln 2}{k}\right)t}$

is the desired function.   ■

EXAMPLE 7 *(Radioactive decay)* If the half-life of a radioactive substance is k years, write the amount present Q as an exponential function (with base e) of time.

Solution   The amount present can be expressed as

$$Q = Q_0(1/2)^{t/k} \hspace{3cm} \text{(Example 4, Section 1–6)}$$

where Q_0 is the initial amount and t is measured in years. Using (3.30) with $a = \frac{1}{2}$ and $x = t/k$, we have

$$(3.32) \qquad Q = Q_o e^{\frac{t}{k}\ln(1/2)} = Q_o e^{-\left(\frac{\ln 2}{k}\right)t}$$

The minus sign comes from writing $\ln(1/2) = \ln 2^{-1} = -\ln 2$.   ∎

In the next section, we will learn how to differentiate the function $f(x) = e^{cx}$ for any constant c (this function is *not* its own derivative). Then we can compute the rates of change of population growth (3.31) and radioactive decay (3.32).

EXERCISES

1. Evaluate $(1 + h)^{1/h}$ for $h = \pm.0001$ and $h = \pm.00001$.

In Exercises 2, 3, and 4, assume that the following limit exists

$$\lim_{h \to 0} \frac{a^h - 1}{h}$$

2. Approximate the limit for $a = 2$. Then compare your approximation with the value of $\ln 2$.

3. Same as Exercise 2 with $a = 10$.

4. Approximate the limit for $a = 2.7$, $a = 2.718$, and $a = 2.718281$.

Differentiate the functions in Exercises 5–20.

5. $y = 3x^2 + e^x$ **6.** $y = e^x - x^3$

7. $y = x^4 e^x$ **8.** $y = 2xe^x$

9. $y = e^x/7$ **10.** $y = \dfrac{4}{x} + 8e^x$

11. $s = e^t(t^2 + t + 1)$ **12.** $s = t^2 e^t$

13. $u = e^x/(x^2 + 1)$ **14.** $u = e^x/3x^5$

15. $u = (x^2 + 1)/e^x$ **16.** $u = 3x^5/e^x$

17. $y = \sqrt{x}e^x$ **18.** $y = \sqrt[3]{x}e^x$

19. $y = \sqrt{x}/e^x$ **20.** $y = \sqrt[3]{x}/e^x$

In Exercises 21–26, find the second derivatives.

21. $y = x^2 e^x$ **22.** $y = x^4 e^x$

23. $y = e^x/x$ **24.** $y = \sqrt{x} + e^x$

25. $y = x^2/e^x$ **26.** $y = x^3/e^x$

In Exercises 27–30, find an equation of the line tangent to the curve at the indicated point.

27. $y = (x^2 - 3x + 2)(x + 1)$; $(0, 2)$

28. $y = (x + 1)/(x^2 + 1)$; $(1, 1)$

29. $y = x^2 e^x$; $(1, e)$

30. $y = (x - 11)(x + 9)$; $(1, -100)$

In Exercises 31–33, write the population or amount present as an exponential function (with base e) of time.

31. The population of a strain of bacteria doubles every 10 minutes. The initial population is P_o.

32. The population at time t (years) after 1970 of a certain country is given by formula $P = (3 \times 10^7) 2^{(.0427)t}$.

33. The half-life of radium is about 1600 years and the initial amount is 3 grams.

Optional Exercises

In Exercises 34–35, find a function $y = f(x)$ with the given derivative. Check your answers.

EXAMPLE   Given that $y' = x + 4e^x$, *guess* the answer $y = \frac{1}{2}x^2 + 4e^x$, and *check:* $(\frac{1}{2}x^2 + 4e^x)' = x + 4e^x$. If the answer doesn't check, then guess again.

34. $y' = 1 + e^x$ **35.** $y' = 3e^x - 4$

36. $y' = x - 2e^x$ **37.** $y' = x^2 - 5e^x$

38. $y' = x^3 - (e^x/2)$ **39.** $y' = x^4 + 2e^x$

40. $y' = xe^x + e^x$ (Think of y as a product.)

41. $y' = x^3 e^x + 3x^2 e^x$ **42.** $y' = x^2 e^x + 2xe^x$

43. $y' = \dfrac{xe^x - e^x}{x^2}$ (Think of y as a quotient.)

44. $y' = \dfrac{x^2 e^x - 2xe^x}{x^4}$ **45.** $y' = \dfrac{2x - x^2}{e^x}$

3–4 THE CHAIN RULE

Two basic derivative formulas are

(3.19) Power Formula: $(x^r)' = rx^{r-1}$

(3.27) Exponential Formula: $(e^x)' = e^x$

The symbol x represents the identity function $f(x) = x$, and, indeed, these formulas are valid for only the identity function. But many of the functions we work with do not fit into this category. For instance, $f(x) = \sqrt{1 - x^2} = (1 - x^2)^{1/2}$ is a power, but *not* a power of the identity. Therefore, you cannot use the power formula (3.19) to differentiate f. Another example is $g(x) = e^{-x^2}$; it is an exponential, but the exponent is *not* the identity function. Therefore, you cannot use the exponential formula (3.27) to differentiate g. In this section, we describe a method for differentiating such functions. In the process, formulas (3.19) and (3.27) will undergo a slight, but important, metamorphosis.

EXAMPLE 1 Let $f(x) = x^2 + 1$, and differentiate $[f(x)]^2 = (x^2 + 1)^2$.

Solution One approach is to first multiply out and then differentiate, but that will not help to illustrate our point. Rather, we will write $(x^2 + 1)^2$ as $(x^2 + 1)(x^2 + 1)$ and use the product formula.

$$
\begin{aligned}
[(x^2 + 1)^2]' &= [(x^2 + 1)(x^2 + 1)]' \\
&= (x^2 + 1)(x^2 + 1)' + (x^2 + 1)(x^2 + 1)' \\
&= (x^2 + 1)(2x) + (x^2 + 1)(2x) \\
&= 2(x^2 + 1)(2x)
\end{aligned}
$$

Or, substituting f for $x^2 + 1$ and f' for $2x$, we have

$$(f^2)' = 2ff' \qquad \blacksquare$$

EXAMPLE 2 Let $f(x) = x^2 + 1$, and differentiate $[f(x)]^3 = (x^2 + 1)^3$.

Solution This time write $(x^2 + 1)^3$ as $(x^2 + 1)(x^2 + 1)^2$. Then

$$
\begin{aligned}
[(x^2 + 1)^3]' &= [(x^2 + 1)(x^2 + 1)^2]' \\
&= (x^2 + 1)[(x^2 + 1)^2]' + (x^2 + 1)^2(x^2 + 1)' &&\text{(Product)} \\
&= (x^2 + 1)[2(x^2 + 1)(2x)] + (x^2 + 1)^2(2x) &&\text{(Example 1)} \\
&= 2[(x^2 + 1)^2(2x)] + (x^2 + 1)^2(2x) &&\text{(Rearrange)} \\
&= 3(x^2 + 1)^2(2x)
\end{aligned}
$$

Or, substituting f for $x^2 + 1$ and f' for $2x$, we have

$$(f^3)' = 3f^2f' \qquad \blacksquare$$

In both examples, the derivatives

$$(f^2)' = 2ff' \quad \text{and} \quad (f^3)' = 3f^2f'$$

can be obtained by substituting f for x in the power formula (3.19) and multiplying the result by f'. *This is no accident;* it is true for every power of every differentiable function. Not only that, a similar type of statement holds for any derivative formula.

(3.33)
> Any derivative formula valid for the identity function x, remains valid when x is replaced by a differentiable function f *provided* you remember to multiply by f'.

This is a rough interpretation of the important *Chain Rule* for the derivative of a composition of functions. A formal version is given later.

Formulas (3.19) and (3.27) can now be written as

(3.34) (New) **Power Formula:** $(f^r)' = rf^{r-1}f'$

(3.35) (New) **Exponential Formula:** $(e^f)' = e^f f'$

The formulas hold for any differentiable function f.

EXAMPLE 3   *(Powers)*

(a) $y = (7x^3 - 4x^2)^6$ is a power ($r = 6$) of a polynomial $f(x) = 7x^3 - 4x^2$. Therefore, by (3.34),

$$y' = [(7x^3 - 4x^2)^6]' = 6(7x^3 - 4x^2)^5(7x^3 - 4x^2)'$$
$$= 6(7x^3 - 4x^2)^5(21x^2 - 8x)$$

Without the chain rule, we would have had to actually multiply $7x^3 - 4x^2$ by itself six times, and then differentiate the resulting eighteenth-degree polynomial!

(b) $y = \sqrt[3]{x^2 + e^x} = (x^2 + e^x)^{1/3}$ is a power of a sum. By (3.34),

$$y' = [(x^2 + e^x)^{1/3}]' = \tfrac{1}{3}(x^2 + e^x)^{-2/3}(x^2 + e^x)'$$
$$= \frac{2x + e^x}{3\sqrt[3]{(x^2 + e^x)^2}} \quad \blacksquare$$

EXAMPLE 4   *(Exponentials)*

(a) Let $y = e^{3x}$; then, by (3.35),

$$y' = (e^{3x})' = e^{3x}(3x)' = 3e^{3x}$$

(b) Let $y = e^{-x^2}$; then, by (3.35),

$$y' = (e^{-x^2})' = e^{-x^2}(-x^2)'$$
$$= -2xe^{-x^2} \quad \blacksquare$$

Sometimes you have to use the chain rule more than once.

EXAMPLE 5 $y = e^{\sqrt{x^2+1}}$ is an exponential of a power of a polynomial. Therefore,

(1) $$y' = [e^{\sqrt{x^2+1}}]' = e^{\sqrt{x^2+1}}[\sqrt{x^2+1}]' \qquad (3.35)$$

Use the chain rule again to find

$$[\sqrt{x^2+1}]' = [(x^2+1)^{1/2}]' = \frac{1}{2}(x^2+1)^{-1/2}(x^2+1)' \qquad (3.34)$$

(2) $$= \frac{1}{2\sqrt{x^2+1}}(2x)$$

$$= \frac{x}{\sqrt{x^2+1}}$$

Therefore, combining (1) and (2),

$$[e^{\sqrt{x^2+1}}]' = \frac{xe^{\sqrt{x^2+1}}}{\sqrt{x^2+1}} \qquad \blacksquare$$

The chain rule may be combined with the product or any of the previously established derivative formulas.

EXAMPLE 6 *(Product)* If $y = e^{x^2}(3x^2 - 5x + 1)^3$, then

$$\begin{aligned}
y' &= e^{x^2}[(3x^2 - 5x + 1)^3]' + (3x^2 - 5x + 1)^3(e^{x^2})' &\text{(Product)}\\
&= e^{x^2}[3(3x^2 - 5x + 1)^2(6x - 5)] &\text{(3.34),}\\
&\quad + 2xe^{x^2}(3x^2 - 5x + 1)^3 &\text{(3.35)}\\
&= e^{x^2}(3x^2 - 5x + 1)^2(6x^3 - 10x^2 + 20x - 15) &\text{(Simplify)} \quad \blacksquare
\end{aligned}$$

Now let us turn to a more formal statement of the chain rule. The chain rule concerns the derivative of a composition of functions. We recall that a composition $f \circ g$ (read "f circle g") is defined as

$$f \circ g(x) = f(g(x))$$

(3.36)

CHAIN RULE

Let g be differentiable at x, and let f be differentiable at $g(x)$. Then $f \circ g$ is differentiable at x, and

$$(f \circ g)'(x) = f'(g(x)) \cdot g'(x)$$

The idea behind the proof is to rewrite the difference quotient for $f \circ g$ as a product. If $g(x + h) \neq g(x)$, then

$$\frac{f \circ g(x + h) - f \circ g(x)}{h} = \frac{f(g(x + h)) - f(g(x))}{h}$$

$$= \frac{f(g(x + h)) - f(g(x))}{g(x + h) - g(x)} \cdot \frac{g(x + h) - g(x)}{h}$$

Once it is shown that the limit, as $h \to 0$, of this product is $f'(g(x)) \cdot g'(x)$, then (3.36) is true under the assumption that $g(x + h) \neq g(x)$. The same proof can then be modified to eliminate this condition. The complete proof, in a slightly different form, is presented at the end of this section.

The chain rule can also be written in the "dee" notation. Let $y = f(u)$ and $u = g(x)$; then $y = f(u) = f(g(x)) = f \circ g(x)$, and

$$\frac{dy}{dx} = (f \circ g)'(x) = f'(g(x))g'(x) \tag{3.36}$$

$$= f'(u)\frac{du}{dx} \qquad (u = g(x))$$

$$= \frac{dy}{du}\frac{du}{dx} \qquad (y = f(u))$$

Therefore,

$$(3.37) \qquad \boxed{\frac{dy}{dx} = \frac{dy}{du}\frac{du}{dx}}$$

is another form of the chain rule.

EXAMPLE 7 Use (3.36) to differentiate $(x^2 - x + 1)^4$.

Solution Let $f(u) = u^4$ and $g(x) = x^2 - x + 1$; then

$$f \circ g(x) = f(x^2 - x + 1) = (x^2 - x + 1)^4$$

Furthermore, $f'(u) = 4u^3$ and $g'(x) = 2x - 1$. By (3.36),

$$(f \circ g)'(x) = f'(g(x))g'(x)$$
$$= 4(g(x))^3(2x - 1)$$
$$= 4(x^2 - x + 1)^3(2x - 1) \qquad \blacksquare$$

EXAMPLE 8 Use (3.37) to differentiate e^{x^2}.

Solution Let $y = e^u$ and $u = x^2$; then

$$y = e^u = e^{x^2}$$

Furthermore,

$$\frac{dy}{du} = e^u \quad \text{and} \quad \frac{du}{dx} = 2x$$

and, by (3.37),

$$\frac{dy}{dx} = \frac{dy}{du}\frac{du}{dx} = e^u(2x)$$

$$= 2xe^{x^2} \qquad\qquad\qquad (u = x^2)$$

Note: dy/du is a function of u (e^u in this case). To make dy/dx a function of x, you must substitute for u in terms of x ($u = x^2$ in this case). ■

Examples 7 and 8 are illustrations of how to use the chain rule within the framework of its technical language. But, in practice, it is clearly easier (and equally correct) to use the rough statement (3.33), and formulas (3.34) and (3.35) to find derivatives.

Observe that the chain rule in the "dee" notation

$$\frac{dy}{dx} = \frac{dy}{du}\frac{du}{dx}$$

says that the rate of change of y with respect to x equals the rate of change of y with respect to u *times* the rate of change of u with respect to x. This idea is important in applications.

EXAMPLE 9   A spherical weather balloon is being inflated with helium. The radius r after t minutes is $r = \sqrt{t}$ feet. How fast is the helium being pumped into the balloon after 4 minutes?

Solution   As helium is pumped in, the volume of the balloon changes. The question of how fast the helium is being pumped in can be rephrased as "What is the rate of change of the volume V with respect to time (that is, dV/dt) when $t = 4$?" You know that V is a function of r and r is given as a function of t, so the chain rule can be used to find dV/dt.

$$V = \frac{4}{3}\pi r^3 \quad \text{cu ft, so} \quad \frac{dV}{dr} = 4\pi r^2 \quad \text{cu ft/ft}$$

$$r = \sqrt{t} \quad \text{feet, so} \quad \frac{dr}{dt} = \frac{1}{2\sqrt{t}} \quad \text{ft/min}$$

Therefore,

$$(3.38) \qquad \frac{dV}{dt} = \frac{dV}{dr}\cdot\frac{dr}{dt}$$

$$= (4\pi r^2)\left(\frac{1}{2\sqrt{t}}\right) \quad \text{cu ft/min}$$

when $t = 4$, $r = \sqrt{4} = 2$. Thus,

$$\frac{dV}{dt} = (4\pi 2^2)\left(\frac{1}{2\sqrt{4}}\right) = 4\pi \text{ cu ft/min}$$

That is how fast the helium is being pumped in after 4 minutes. ■

EXAMPLE 10 If the helium in Example 9 is pumped in at a constant rate of 36π cu ft/min, how fast is the radius increasing when the radius is 1 foot? When it is 4 feet?

Solution In this example, the radius is not explicitly defined as a function of time. Even so, you can use the Chain Rule to find its rate of change by solving (3.38) for dr/dt.

$$\frac{dr}{dt} = \frac{dV/dt}{dV/dr}$$

You are told that the volume is increasing at a constant rate; $dV/dt = 36\pi$ cu ft/min. From Example 9, you know that $dV/dr = 4\pi r^2$ cu ft/ft. Therefore,

$$\frac{dr}{dt} = \frac{36\pi \text{ cu ft/min}}{4\pi r^2 \text{ cu ft/ft}} = \frac{9}{r^2} \text{ ft/min}$$

The radius is increasing at the rate of 9 ft/min when $r = 1$, but only $\frac{9}{16}$ ft/min when $r = 4$. ∎

Remark: One feature of the chain rule is that any new derivative formulas need to be established for only the identity function x. For example, in the next section, we prove that $(\sin x)' = \cos x$. Then, by the chain rule, we *automatically* know that

$$(\sin f)' = (\cos f)f'$$

for any differentiable function f.

Proof of the Chain Rule (*Optional*)

The derivative of a function f at x is defined to be

(1) $f'(x) = \lim_{h \to 0} \dfrac{f(x + h) - f(x)}{h}$

To prove the chain rule, it is convenient to use an alternate form of this definition. Set $x + h = t$; then $h = t - x$, and $t \to x$ as $h \to 0$. Therefore, (1) can be rewritten as

(2) $f'(x) = \lim_{t \to x} \dfrac{f(t) - f(x)}{t - x}$

(3.36) Chain Rule. If g is differentiable at x and f is differentiable at $g(x)$, then $f \circ g$ is differentiable at x and $(f \circ g)'(x) = f'(g(x))g'(x)$.

Proof Using the alternate form (2) above,

(3) $(f \circ g)'(x) = \lim_{t \to x} \dfrac{f(g(t)) - f(g(x))}{t - x}$

Therefore, it is sufficient to show that this limit is $f'(g(x))g'(x)$.

First, we define a new function F by

$$(4) \qquad F(u) = \begin{cases} \dfrac{f(u) - f(g(x))}{u - g(x)} & \text{if } u \neq g(x) \\ f'(g(x)) & \text{if } u = g(x) \end{cases}$$

This function is continuous at $g(x)$ because

$$\lim_{u \to g(x)} F(u) = \lim_{u \to g(x)} \frac{f(u) - f(g(x))}{u - g(x)} \qquad (4)$$

$$= f'(g(x)) \qquad (2)$$

$$= F(g(x)) \qquad (4)$$

Now we show that the limit in (3) can be rewritten in terms of F. For $t \neq x$, we claim that

$$(5) \qquad \frac{f(g(t)) - f(g(x))}{t - x} = F(g(t)) \left[\frac{g(t) - g(x)}{t - x} \right]$$

There are two cases. If $g(t) = g(x)$, then both sides of (5) equal 0. If $g(t) \neq g(x)$, then by (4),

$$F(g(t)) = \frac{f(g(t)) - f(g(x))}{g(t) - g(x)}$$

and again both sides of (5) are equal. Furthermore, since g is continuous at x (why?) and F is continuous at $g(x)$, it follows that $F(g(t)) \to F(g(x)) = f'(g(x))$ as $t \to x$. Therefore,

$$(f \circ g)'(x) = \lim_{t \to x} \frac{f(g(t)) - f(g(x))}{t - x} \qquad (3)$$

$$= \lim_{t \to x} F(g(t)) \left[\frac{g(t) - g(x)}{t - x} \right] \qquad (5)$$

$$= f'(g(x))g'(x)$$

which completes the proof.

EXERCISES

The functions in Exercises 1–6 are *powers*. Use (3.34) to find their derivatives.

1. $y = (x^2 - 2x + 1)^3$ **2.** $y = (x^3 - 5)^4$

3. $u = \sqrt{3x^2 + e^x}$ **4.** $u = \sqrt[3]{x^5 + 1}$

5. $y = \left(\dfrac{1}{x} + \sqrt{x} \right)^4$ **6.** $y = \sqrt{(e^x + 1)^3}$

The functions in Exercises 7–12 are *exponentials*. Use (3.35) to find their derivatives.

7. $y = e^{(x^2 + x + 1)}$ **8.** $y = e^{(x^3 - 3x^2)}$

9. $u = e^{1/x}$ **10.** $u = e^{1/x^2}$

11. $y = e^{\sqrt{x}}$ **12.** $y = e^{1/\sqrt{x}}$

In Exercises 13–16, use the chain rule as many times as necessary to differentiate the functions.

13. $f(x) = e^{\sqrt{x^4 + 1}}$ **14.** $f(x) = e^{(x^3 + x)^2}$

15. $g(x) = e^{e^{x^2}}$ **16.** $g(x) = (e^{\sqrt{x^2 + 1}} + 1)^4$

In Exercises 17–20, combine the chain rule with the product or quotient formulas, and differentiate.

17. $y = (x^2 - 3x + 1)^2(2x^3 + x)^3$

18. $y = (e^{1/x})\sqrt{4x + 1}$

19. $f(x) = \dfrac{e^{-5x}}{(x^2 - 3)^2}$

20. $g(x) = \dfrac{(2x + 1)^3}{(1 - 3x)^4}$

Differentiate the functions in Exercises 21–32.

21. $y = \left(\dfrac{1}{x} + 2 + x\right)^4$

22. $y = \sqrt{x^3 - 3x^2}$

23. $s = \sqrt[3]{(t^2 + 1)^4 + 3t}$

24. $s = [(t^3 - 2t)^2 + 1]^5$

25. $u = \left(\dfrac{3x - 1}{4x + 7}\right)^2$

26. $u = e^{e^x}$

27. $y = e^{x^e}$ $e^{(x^e)}$

28. $y = e^{-(x^2 - x)}$

29. $s = (e^{4t})\sqrt{t + 4}$

30. $s = (4t - 7)^2(2t + 1)^3$

31. $u = e^{2x}/(x^2 - 5)^2$

32. $u = \sqrt{x^2 + 1}/e^{3x}$

In Exercises 33–40, find the second derivative of the given function.

33. $y = (2 - 4x)^5$

34. $y = (1 - 2x - 3x^2)^4$

35. $u = (x^2 - 3)^3$

36. $u = \sqrt{e^x + 1}$

37. $y = e^{x^2}$

38. $y = e^{1/x}$

39. $u = e^{\sqrt{x}}$

40. $u = (x^2 + 1)^{-2}$

---

41. If $f(0) = 9$ and $f'(0) = 2$, evaluate the derivative of $y = \sqrt{f(x)}$ at $x = 0$.

42. If $g(1) = 0$ and $g'(1) = 7$, evaluate the derivative of $y = e^{g(x)}$ at $x = 1$.

43. If $F'(2) = -3$, evaluate the derivative of $s = F(t^2 - 3t + 2)$ at $t = 0$.

44. If $G'(1) = 14$, evaluate the derivative of $s = G(e^t)$ at $t = 0$.

45. Find an equation of the line tangent to the curve $y = (2x^2 + x - 3)^3$ at $(-2, 27)$.

46. A sandbag is dropped from a balloon 1000 feet above the earth. The pull of gravity will cause its velocity to increase at the constant rate of approximately 32 feet per second per second (in symbols, $dv/dt \approx 32$ ft/sec²). The air resistance, or *drag*, D is a function of the velocity. For this sandbag, $D = v^{3/2}$ pounds. How fast is the drag increasing with respect to time when the velocity is 64 ft/sec?

47. A point is moving along the curve $y = e^{2x}$. When it is at $(0, 1)$, x is changing at the rate of 4 cm/sec. How fast is y changing with respect to time at that point?

48. At what rate must helium be pumped into a weather balloon so that the radius at any time t (minutes) is $r = 3\sqrt[3]{t}$ feet?

49. A tank in the shape of an inverted cone with height 20 feet and top radius 10 feet is being filled with water at the rate of 3 cu ft/min. Express the volume of water in the tank as a function of its depth h and determine how fast the water is rising when $h = 2$ feet.

In Exercises 50–52, first write the population or present amount as an exponential function (with base e) in order to use (3.35).

50. If the population of a strain of bacteria doubles every 10 minutes, how fast is it growing after 2 hours? After 3 hours?

51. A study shows that the population of a certain country is given by $P = P_0 2^{(.0427)t}$ where P_0 is the population ten years ago and t is measured in years. What is the rate of increase (in percent) of the population per year? How long will it take for the population to double (that is, $P = 2P_0$)?

52. The half-life of radium is about 1600 years. If you start with 10 ounces, how much will be left after 200 years? At what rate (ounces per year) is it decaying at that time?

Optional Exercises

In Exercises 53–58, find a function $y = f(x)$ with the given derivative. Check your answers.

EXAMPLES

(a) Given that $y' = 2e^{2x} - x$, *guess* the answer $y = e^{2x} - (x^2/2)$, and *check*: $[e^{2x} - (x^2/2)]' = 2e^{2x} - x$.

(b) Given that $y' = 3(x^2 + 1)^2(2x)$, *guess* the answer $y = (x^2 + 1)^3$, and *check*: $[(x^2 + 1)^3]' = 3(x^2 + 1)^2(2x)$.

If the answer doesn't check, then guess again.

53. $y' = 2xe^{x^2}$ **54.** $y' = 3x^2e^{x^3}$ **57.** $y' = (x^2 + 1)^4(2x)$

55. $y' = x + x^2 + 2e^{2x}$ **56.** $y' = 5e^{5x} - (1/x^2)$ **58.** $y' = (x^3 - 2x + 1)^3(3x^2 - 2)$

3–5   DERIVATIVES OF SINE, COSINE, AND TANGENT

A point on a vibrating string oscillates, but as time goes on the amplitudes of the oscillations get smaller and smaller. This motion is called *damped harmonic motion*. A typical position function is

$$s = e^{-t}\sin t$$

The position curve is sketched in Figure 3–13. If we knew the derivative of the sine function, we could compute the velocity ds/dt and analyze the motion. Therefore, let us develop a formula for the derivative of the sine. As a bonus, the derivative formulas for cosine and tangent are obtained with very little effort. Derivatives of the cotangent, secant, and cosecant, which are not needed until later, are discussed in Chapter 8.

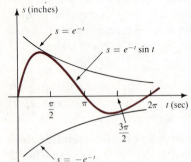

FIGURE 3–13. Damped Harmonic Motion.

To find the derivative of $f(x) = \sin x$, we first write out the difference quotient

$$\frac{f(x + h) - f(x)}{h} = \frac{\sin(x + h) - \sin x}{h}$$

Using an addition formula (1.52) in Section 1–7, $\sin(x + h) = \sin x \cos h + \cos x \sin h$, we continue with

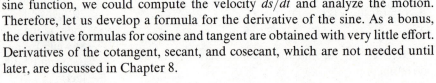

$$\frac{\sin(x + h) - \sin x}{h} = \frac{[\sin x \cos h + \cos x \sin h] - \sin x}{h} \quad (1.52)$$

$$= \cos x\left[\frac{\sin h}{h}\right] + \sin x\left[\frac{\cos h - 1}{h}\right] \quad \text{(Rearrange)}$$

Now let $h \to 0$. The $\sin x$ and $\cos x$ are fixed numbers, and we claim that

$$(3.39) \quad \lim_{h \to 0}\frac{\sin h}{h} = 1 \quad \text{and} \quad \lim_{h \to 0}\frac{\cos h - 1}{h} = 0$$

If our claim is correct, then

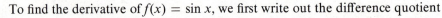

$$\frac{d}{dx}(\sin x) = \lim_{h \to 0}\frac{\sin(x + h) - \sin x}{h}$$

$$= \cos x\left[\lim_{h \to 0}\frac{\sin h}{h}\right] + \sin x\left[\lim_{h \to 0}\frac{\cos h - 1}{h}\right]$$

$$= (\cos x)[1] + (\sin x)[0] \qquad (3.39)$$
$$= \cos x$$

Therefore,

$$(3.40) \quad \boxed{\frac{d}{dx}(\sin x) = \cos x}$$

Now let us go back and justify our claim (3.39). The limit of $(\sin h)/h$ cannot be found by algebraic manipulation, or by substituting 0 for h. We could use a calculator (as we did in the case of exponentials), but there is a more interesting geometric argument. Figure 3–14 shows part of the unit circle C and a central angle $h > 0$, *measured in radians*. The indicated coordinates of A, B, and D are easily verified. Now,

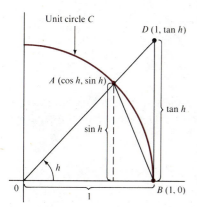

Unit circle C

$D\,(1, \tan h)$

$A\,(\cos h, \sin h)$

$\tan h$

$\sin h$

h

0

1

$B\,(1, 0)$

FIGURE 3–14. Area $\triangle 0AB$
$<$ area $\measuredangle 0AB <$ area $\triangle 0DB$.

$$\text{area of } \triangle 0AB = \frac{1}{2}(1)\sin h = \frac{1}{2}\sin h$$

$$\text{area of } \triangle 0DB = \frac{1}{2}(1)\tan h = \frac{1}{2}\frac{\sin h}{\cos h}$$

$$\text{area of } \measuredangle 0AB = \frac{1}{2}h(1)^2 = \frac{1}{2}h$$

The last assertion follows from the fact that in a circle of radius r, a central angle of 2π radians encloses the whole circle with area πr^2. Therefore, 1 radian encloses an area of $\pi r^2/2\pi = r^2/2$, and h radians enclose an area of $hr^2/2$.

The figure indicates that the areas above are related by the inequalities

$$\text{area of } \triangle 0AB < \text{area of } \measuredangle 0AB < \text{area of } \triangle 0DB$$

Therefore,

$$\frac{1}{2}\sin h < \frac{1}{2}h < \frac{1}{2}\frac{\sin h}{\cos h}$$

Since $h > 0$, all quantities are positive, and it follows that

$$1 < \frac{h}{\sin h} < \frac{1}{\cos h} \qquad \left(\text{Divide by } \frac{1}{2}\sin h\right)$$

$$\cos h < \frac{\sin h}{h} < 1 \qquad \text{(Positive reciprocals reverse inequalities)}$$

The last string of inequalities holds even if $h < 0$ because

$$\cos(-h) = \cos h \quad \text{and} \quad \frac{\sin(-h)}{-h} = \frac{-\sin h}{-h} = \frac{\sin h}{h} \qquad (1.53)$$

Thus, we have *squeezed* $(\sin h)/h$ between $\cos h$ and the constant function 1.

Since the cosine is continuous, $\cos h \to \cos 0 = 1$ as $h \to 0$, and it follows from the Squeeze Theorem that

$$\lim_{h \to 0} \frac{\sin h}{h} = 1$$

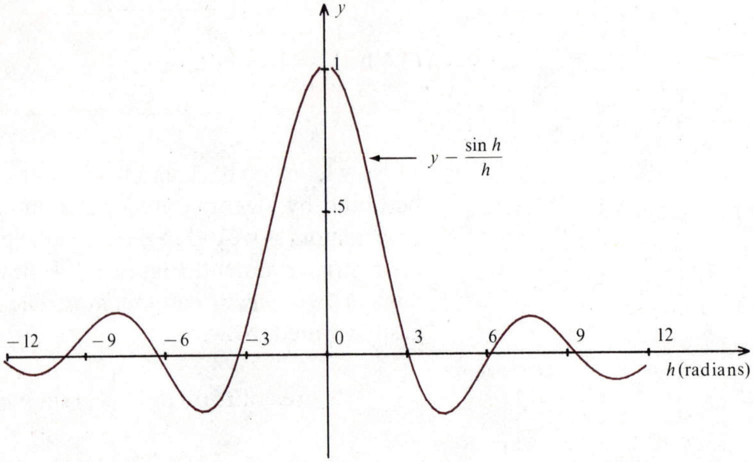

FIGURE 3–15. The computer-generated graph shows how $(\sin h)/h \to 1$ as $h \to 0$.

Figure 3–15 is a computer-generated graph that shows how $(\sin h)/h$ approaches 1 as h approaches 0. That is one part of our claim. The other part now follows because

$$\frac{\cos h - 1}{h} = \frac{\cos h - 1}{h} \cdot \frac{\cos h + 1}{\cos h + 1}$$

$$= \frac{\cos^2 h - 1}{h(\cos h + 1)}$$

$$= \frac{-\sin^2 h}{h(\cos h + 1)} \qquad (\sin^2 h + \cos^2 h = 1)$$

$$= -\sin h \left[\frac{\sin h}{h}\right]\left[\frac{1}{\cos h + 1}\right] \qquad \text{(Rearrange)}$$

The limits of all three factors exist as $h \to 0$, and $\sin h \to 0$ because the sine is continuous. Therefore,

$$\lim_{h \to 0} \frac{\cos h - 1}{h} = (0)(1)\left(\frac{1}{2}\right) = 0$$

Thus, (3.39) is verified, and now you know that *the derivative of the sine is the cosine*. With the help of the chain rule, formula (3.40) becomes

(3.41)
> **SINE FORMULA**
> If f is differentiable, then
> $$(\sin f)' = (\cos f)f'$$

EXAMPLE 1

$$\frac{d}{dx}(\sin x^3) = (\cos x^3)\frac{d}{dx}(x^3)$$ (3.41)

$$= 3x^2\cos x^3 \quad \blacksquare$$

EXAMPLE 2   *(Powers)*

$y = \sin^3 x$ is a power (think of $(\sin x)^3$) of a sine.

Therefore,

$$y' = 3\sin^2 x\,(\sin x)' \qquad\qquad \text{(Power)}$$

$$= 3\sin^2 x\cos x \qquad\qquad (3.41) \quad \blacksquare$$

EXAMPLE 3 $y = x^2\sin x$ is a product. Therefore,

$$y' = x^2(\sin x)' + (\sin x)(x^2)' \qquad\qquad \text{(Product)}$$

$$= x^2\cos x + 2x\sin x \quad \blacksquare$$

Let us return now to damped harmonic motion.

EXAMPLE 4   *(Damped harmonic motion)* A particle moves along a line, and its position at t seconds is $s = e^{-t}\sin t$ for $0 \le t \le 2\pi$ (Figure 3–13). Find the velocity function.

Solution

$$v = \frac{ds}{dt} = \frac{d}{dt}(e^{-t}\sin t)$$

$$= e^{-t}(\sin t)' + (\sin t)(e^{-t})'$$

$$= e^{-t}\cos t - e^{-t}\sin t$$

$$= e^{-t}(\cos t - \sin t) \quad \blacksquare$$

Derivatives of Cosine and Tangent

The acute angles in any right triangle are complementary; that is, their sum is 90° or $\pi/2$ radians. Figure 3–16 shows that

(3.42) $\sin x = \cos\left(\dfrac{\pi}{2} - x\right)$ and $\cos x = \sin\left(\dfrac{\pi}{2} - x\right)$

so the sine and cosine are said to be complementary functions.* Now the formula for the derivative of the cosine comes easily.

$$(\cos x)' = \left[\sin\left(\frac{\pi}{2} - x\right)\right]' \qquad\qquad (3.42)$$

$$\sin x = \frac{a}{c} = \cos\left(\frac{\pi}{2} - x\right)$$

$$\cos x = \frac{b}{c} = \sin\left(\frac{\pi}{2} - x\right)$$

FIGURE 3–16. x and $\pi/2 - x$ are complementary angles. Sine and cosine are complementary functions.

*The "co" in cosine stands for "complementary."

$$= \left[\cos\left(\frac{\pi}{2} - x\right)\right]\left(\frac{\pi}{2} - x\right)' \tag{3.41}$$

$$= -\cos\left(\frac{\pi}{2} - x\right)$$

$$= -\sin x \tag{3.42}$$

The derivative of the cosine is the negative of the sine. With the chain rule, we have

(3.43)

> **COSINE FORMULA**
> If f is differentiable, then
> $$(\cos f)' = -(\sin f)f'$$

The formula for the derivative of the tangent comes just as easily.

$$(\tan x)' = \left[\frac{\sin x}{\cos x}\right]' = \frac{(\cos x)(\sin x)' - (\sin x)(\cos x)'}{\cos^2 x} \qquad \text{(Quotient)}$$

$$= \frac{\cos^2 x + \sin^2 x}{\cos^2 x} \tag{3.41), (3.43}$$

$$= \frac{1}{\cos^2 x}$$

$$= \sec^2 x$$

Therefore, *the derivative of the tangent is the secant squared.* With the chain rule, we have

(3.44)

> **TANGENT FORMULA**
> If f is differentiable, then
> $$(\tan f)' = (\sec^2 f)f'$$

EXAMPLE 5

(a) $$\frac{d}{dx}\left(6\cos\frac{x}{2}\right) = 6\frac{d}{dx}\cos\frac{x}{2} \qquad (cf)$$

$$= 6\left(-\sin\frac{x}{2}\right)\left(\frac{x}{2}\right)' \tag{3.43}$$

$$= -3\sin\frac{x}{2}$$

(b) $$\frac{d}{dx}(4\tan 5x) = 4\frac{d}{dx}\tan 5x \qquad (cf)$$

$$= 4(\sec^2 5x)(5x)' \qquad (3.44)$$
$$= 20 \sec^2 5x \quad \blacksquare$$

EXAMPLE 6 *(Power)* $y = \cos^5 \sqrt{x}$ is a power of a cosine of a power. Therefore,

$$y' = [5 \cos^4 \sqrt{x}](\cos \sqrt{x})' \qquad \text{(Power)}$$
$$= [5 \cos^4 \sqrt{x}](-\sin \sqrt{x})(\sqrt{x})' \qquad (3.43)$$
$$= -\frac{5}{2\sqrt{x}}(\cos^4 \sqrt{x})(\sin \sqrt{x}) \quad \blacksquare$$

EXERCISES

In Exercises 1–20, differentiate the given functions.

1. $y = \sin 7x$

2. $y = \cos 3x$

3. $y = x^2 + 3 \tan 2x$

4. $y = e^x - 4 \sin 3x$

5. $s = e^{-t}\cos 2\pi t$

6. $s = x^2\tan x$

7. $u = 8 \sin(3x + 2)$

8. $u = 5 \cos(1 - 2x)$

9. $s = \tan(t^2 + t + 1)$

10. $s = \sin e^t$

11. $s = e^{\cos t}$

12. $s = e^{\tan t}$

13. $y = \sin^3 2x$

14. $y = 2 \cos^4 5x$

15. $y = \dfrac{\tan x}{3x + 1}$

16. $y = \dfrac{x^2 + 2}{\sin x}$

17. $y = \cos^3 \sqrt{3x + 1}$

18. $y = \sqrt{\tan(2 - x)^2}$

19. $s = e^{\sin^2 t}$

20. $s = \cos e^{e^t}$

Find the second derivatives of the functions in Exercises 21–30.

21. $y = \sin x$

22. $y = \cos x$

23. $y = \tan x$

24. $y = x \sin x$

25. $s = e^{-t}\cos t$

26. $s = e^t\sin t$

27. $u = \sin^2 x$

28. $u = \tan^2 x$

29. $y = \cos(1/x)$

30. $y = \cos \sqrt{x}$

31. If $w = 3u^2 + 2u - 4$ and $u = \sin x$, find dw/dx as a function of x.

32. If $v = e^{\sqrt{s}}$ and $s = \cos t$, find dv/dt as a function of t.

33. If $y = \sqrt{u^2 + 1}$ and $u = e^{3x}$, find dy/dx as a function of x.

34. Find an equation of the line tangent to the curve $y = \sin x$ at the point $(\pi/6, 1/2)$.

35. For $f(x) = \sin^2 x$, $0 \le x \le 2\pi$, on what intervals is the slope of the tangent to the graph positive? Negative? Zero? Sketch the graph of f.

36. A ladder 10 feet long leans against a house and makes an angle of x radians with the ground. If the top of the ladder is y feet above the ground, express y as a function of x. If the bottom of the ladder slips and starts to move away from the house, what is the rate of change of y with respect to x when $x = \pi/3$?

37. (Continuation of Exercise 36) If the angle x is a function of time t, $x = \pi/\sqrt{t}$, what is the rate of change of y with respect to t when $t = 9$ seconds?

38. A rocket is fired straight up from Cape Canaveral. At a point 3 miles from the launch pad, a man observes that the angle of elevation of the rocket is increasing at the rate of $10°$ per second. How fast is the rocket rising when the angle of elevation is $45°$?

39. An isosceles triangle T has equal sides of c inches, and the angle between them is x radians. Find the rate of change of the area of T with respect to x.

40. A lighthouse is on an island 1 mile from a straight shoreline. Its light makes 2 revolutions per minute. How fast is the light beam moving along the shore at the instant it makes an angle of $30°$ with the shoreline? An angle of $90°$?

Optional Exercises

41. A pie-shaped wedge with central angle x, is cut from a circular piece of paper of radius 6 inches.

The straight edges are joined to form a cone (Figure 3–17). Express the volume V of the cone as a function of x, and find dV/dx.

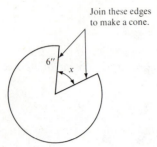

Join these edges to make a cone.

6″

FIGURE 3–17. Exercise 41.

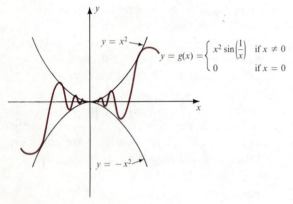

$y = x^2$

$y = g(x) = \begin{cases} x^2 \sin\left(\dfrac{1}{x}\right) & \text{if } x \neq 0 \\ 0 & \text{if } x = 0 \end{cases}$

$y = -x^2$

FIGURE 3–18. Exercise 42. g is differentiable at $x = 0$.

42. Let $f(x) = \begin{cases} x \sin(1/x) & \text{if } x \neq 0 \\ 0 & \text{if } x = 0 \end{cases}$

$g(x) = \begin{cases} x^2 \sin(1/x) & \text{if } x \neq 0 \\ 0 & \text{if } x = 0 \end{cases}$

The graph of f is in Figure 2–9B; the graph of g is in Figure 3–18. Show that $f'(0)$ does not exist, but that $g'(0)$ does exist! (Use the definition of derivative.) What is the value of $g'(0)$?

43. (Continuation of Exercise 42). Both f and g are differentiable at nonzero points. Find $f'(2/\pi)$ and $g'(2/\pi)$.

44. For any number t, $\cos t = \cos(-t)$. Therefore, $d/dt \cos t$ should equal $d/dt \cos(-t)$. Does it?

In Exercises 45–52, find a function $y = f(x)$ with the given derivative. Check your answers.

EXAMPLE. Given that $y' = x \cos x + \sin x$, *guess* the answer $y = x \sin x$, and *check:* $[x \sin x]' = x(\sin x)' + (\sin x)(x)' = x \cos x + \sin x$. (If your answer doesn't check, then guess again.)

45. $y' = e^x - \sin x$ **46.** $y' = e^x(\cos x + \sin x)$

47. $y' = -\sin 3x$ **48.** $y' = \sec^2 x + \dfrac{1}{2\sqrt{x}}$

49. $y' = e^x \cos e^x$ **50.** $y' = 3/\cos^2 3x$

51. $y' = 2 \sin x \cos x$ **52.** $y' = -3 \cos^2 x \sin x$

3–6 MEAN VALUE THEOREM; INCREASING AND DECREASING FUNCTIONS

The Mean Value Theorem penetrates deeply into much of the calculus, and we refer to it often. In this section, it will be used to formulate a test for increasing and decreasing functions that is basic to many of the applications presented in the next chapter.

(3.45)

> **MEAN VALUE THEOREM***
>
> If f is continuous on $[a, b]$ and differentiable on (a, b), then there is at least one point c in (a, b) such that
>
> $$f'(c) = \frac{f(b) - f(a)}{b - a}$$

*In mathematical language, "mean" means *average*.

We can interpret the theorem in terms of tangent lines. The line joining the points $(a, f(a))$ and $(b, f(b))$ has slope

$$\frac{f(b) - f(a)}{b - a}$$ (Figure 3–19)

The slope of the tangent line at $(c, f(c))$ has slope $f'(c)$. Since lines with the same slope are parallel, the Mean Value Theorem says that at least one tangent line of the graph of f is parallel to the line joining $(a, f(a))$ and $(b, f(b))$. This is illustrated in Figure 3–19.

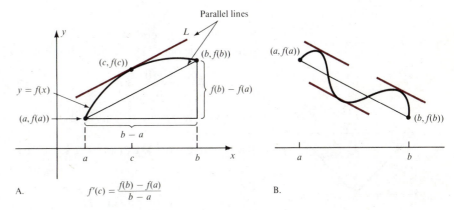

FIGURE 3–19 (A) *Mean Value Theorem.* (B) There may be many points between a and b for which the theorem holds.

We can also interpret the theorem in terms of rates of change. We think of a as x and b as $x + \Delta x$. Then

$$\frac{f(b) - f(a)}{b - a} = \frac{f(x + \Delta x) - f(x)}{\Delta x}$$

is the average rate of change of f over the interval from a to b. The instantaneous rate of change of f at any point c is $f'(c)$. Therefore, the Mean Value Theorem says that the instantaneous rate of change of f equals the average rate of change of f at least once in the interval (a, b). This can be illustrated with a familiar situation. If your average speed on an automobile trip is 50 mph, then the speedometer must read exactly 50 mph at least once during the trip.

The proof of the theorem is presented at the end of this section.

EXAMPLE 1 Let $f(x) = x^2 + 2x - 4$ on the interval $[-2, 2]$; then f is continuous on $[-2, 2]$ and differentiable on $(-2, 2)$. Find the point(s) c referred to in the Mean Value Theorem.

Solution Here, $a = -2, b = 2$, and

$$\frac{f(2) - f(-2)}{2 - (-2)} = \frac{4 - (-4)}{4} = 2$$

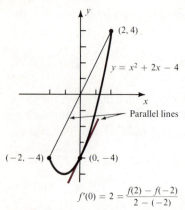

$$f'(0) = 2 = \frac{f(2) - f(-2)}{2 - (-2)}$$

FIGURE 3–20. Example 1.

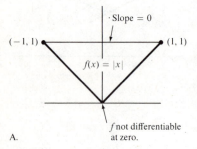

A.

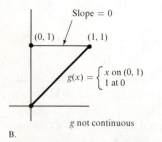

B.

FIGURE 3–21. If the hypotheses of the Mean Value Theorem are not met, the conclusion may not be true. (A) $f'(x)$ is never zero. (B) $g'(x)$ is never zero.

Therefore, we must find all points c in $(-2, 2)$ where $f'(c) = 2$. Since $f'(x) = 2x + 2$, we set

$$f'(c) = 2c + 2 = 2$$

and solve; $c = 0$ (Figure 3–20). ∎

Observe that if either of the hypotheses of the Mean Value Theorem is not met, then the conclusion may fail to be true. For example, $f(x) = |x|$ for $-1 \le x \le 1$ is continuous on $[-1, 1]$, but not differentiable at 0. In this case,

$$\frac{f(1) - f(-1)}{1 - (-1)} = 0$$

but $f'(x)$ is never zero. Now let $g(x) = x$ for $0 < x \le 1$ and $g(0) = 1$. Then g is differentiable on $(0, 1)$, but not continuous on $[0, 1]$. Once again,

$$\frac{g(1) - g(0)}{1 - 0} = 0$$

but $g'(x)$ is never zero (Figure 3–21).

Increasing and Decreasing Functions

A function $y = f(x)$ is said to be *increasing* if its values increase as x increases. It is *decreasing* if its values decrease as x increases. A geometric interpretation is illustrated in Figure 3–22.

These notions play an important role in our everyday lives. For example, personal income, the cost of living, and the amount of pollutants in the air are functions of time. Whether they are increasing or decreasing concerns us all. Here is the formal definition.

DEFINITION

Suppose f is defined on an interval I.

(i) f is **increasing on** I if, for any two numbers x_1 and x_2 in I,

(3.46)
$$x_1 < x_2 \quad \text{implies} \quad f(x_1) < f(x_2)$$

(ii) f is **decreasing on** I if, for any two numbers x_1 and x_2 in I,

$$x_1 < x_2 \quad \text{implies} \quad f(x_1) > f(x_2)$$

A linear function $y = mx + b$ is increasing or decreasing depending on whether the slope m of its graph is positive or negative (Figure 3–23). For an arbitrary function f, the slope of a tangent line is the value of f'. Could it be that a function f is increasing or decreasing depending on whether $f'(x)$ is positive or negative? Let's find out.

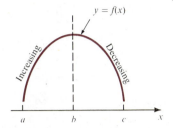

FIGURE 3–22. f is increasing on the interval $[a, b]$, but decreasing on $[b, c]$.

FIGURE 3–23. A line is the graph of an increasing or decreasing function depending on whether the slope is positive or negative.

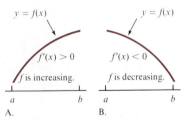

A. B.

FIGURE 3–24. Test for increasing and decreasing functions.

Suppose that f is continuous on $[a, b]$, differentiable on (a, b), and

(3.47) $f'(x) > 0$ for all x in (a, b)

Now let x_1 and x_2 be *any* two points in $[a, b]$ with $x_1 < x_2$. We can apply the Mean Value Theorem to the interval $[x_1, x_2]$, and assert that there must be some point c in the open interval (x_1, x_2) such that

$$f'(c) = \frac{f(x_2) - f(x_1)}{x_2 - x_1}$$

By (3.47), we know that $f'(c) > 0$; we also know that $x_2 - x_1 > 0$. It follows that $f(x_2) - f(x_1) > 0$, or

$$f(x_1) < f(x_2)$$

Thus, with the assumption (3.47) that $f'(x) > 0$, we proved that for any two numbers x_1 and x_2 in $[a, b]$,

$$x_1 < x_2 \quad \text{implies} \quad f(x_1) < f(x_2)$$

Thus, f is increasing. This is the first part of the next theorem; the proof of the second part is similar (Figure 3–24).

TEST FOR INCREASING AND DECREASING

Suppose that f is continuous on $[a, b]$ and differentiable on (a, b).

(3.48)

  (i) If $f'(x) > 0$ for all x in (a, b), then f is increasing on $[a, b]$.

  (ii) If $f'(x) < 0$ for all x in (a, b), then f is decreasing on $[a, b]$.

Finding the intervals on which a function f is increasing and decreasing is a two step process:

(1) Differentiate f.

(2) Locate the intervals on which $f'(x) > 0$ and $f'(x) < 0$.

EXAMPLE 2 A publisher estimates that if x thousand copies of a book are printed, then the cost per book is given by the function $C = x^2 - 16x + 76$ dollars, $0 \le x \le 16$. For what values of x is the cost increasing and decreasing?

Solution

$$C' = 2x - 16 \begin{cases} > 0 & \text{on } (8, 16) \\ = 0 & \text{at } x = 8 \\ < 0 & \text{on } (0, 8) \end{cases}$$

Thus, the cost per book is decreasing for $x < 8$, but starts increasing for $x > 8$ thousand copies. ∎

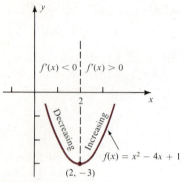

$f'(x) < 0$ $f'(x) > 0$

Decreasing   Increasing

$f(x) = x^2 - 4x + 1$

$(2, -3)$

FIGURE 3–25. Example 3.

The test for increasing and decreasing can also be used as an aid in graphing.

EXAMPLE 3   Find the intervals on which $f(x) = x^2 - 4x + 1$ is increasing and decreasing, and sketch the graph.

Solution   The derivative $f'(x) = 2x - 4$. It follows that

$$f'(x) = 2x - 4 \begin{cases} > 0 & \text{on } (2, \infty) \\ = 0 & \text{at } x = 2 \\ < 0 & \text{on } (-\infty, 2) \end{cases}$$

To sketch the graph, we locate the point $(2, f(2)) = (2, -3)$; f is increasing to the right of that point and decreasing to the left. This is enough information to make a rough sketch (Figure 3–25).   ■

EXAMPLE 4   Find the intervals on which $f(x) = x^3 + 2x^2 - 3x + 1$ is increasing and decreasing, and sketch the graph.

Solution   We must find the intervals on which $f'(x) = 3x^2 + 4x - 3$ is positive and negative. This quadratic function is not easy to factor, so we use the quadratic formula to find where it is zero.

$$x = \frac{-4 \pm \sqrt{4^2 - 4(3)(-3)}}{2(3)} = \frac{-4 \pm \sqrt{52}}{6}$$

$$= \frac{-2 \pm \sqrt{13}}{3} \qquad\qquad (\sqrt{52} = 2\sqrt{13})$$

Let $a = (-2 - \sqrt{13})/3 \approx -1.9$ and $b = (-2 + \sqrt{13})/3 \approx .5$; f' is 0 at these points. Testing with points to the right and left of a and b, we find that

$$f'(x) = 3x^2 + 4x - 3 \begin{cases} > 0 & \text{on } (-\infty, a) \text{ or } (b, \infty) \\ = 0 & \text{at } a \text{ and } b \qquad \text{(check)} \\ < 0 & \text{on } (a, b) \end{cases}$$

Thus, f is increasing on $(-\infty, a)$ and on (b, ∞), but it is decreasing on (a, b). The graph is sketched in Figure 3–26.   ■

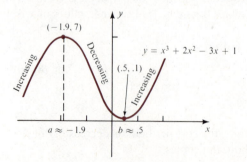

$(-1.9, 7)$

Increasing   Decreasing

$y = x^3 + 2x^2 - 3x + 1$

$(.5, .1)$ Increasing

$a \approx -1.9$ $b \approx .5$

FIGURE 3–26. Example 4.

EXAMPLE 5   Find the intervals on which $f(x) = \sin^2 x, 0 \le x \le 2\pi$ is increasing and decreasing, and sketch the graph.

Solution $f'(x) = 2 \sin x \cos x$

Since the sine and cosine have the same sign in the first and third quadrants but opposite signs otherwise,

$$f'(x) \begin{cases} > 0 & \text{on } (0, \pi/2) \text{ or } (\pi, 3\pi/2) \\ = 0 & \text{at } 0, \pi/2, \pi, 3\pi/2, 2\pi \\ < 0 & \text{at } (\pi/2, \pi) \text{ or } (3\pi/2, 2\pi) \end{cases}$$

This indicates where f is increasing and decreasing. The values of f are non-negative because the sine is squared. Therefore, the graph lies on or above the x-axis. See Figure 3–27.   ■

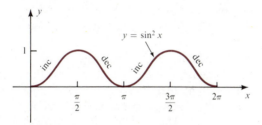

FIGURE 3–27. Example 5.

Note: Graph sketching is merely touched on here. We have a lot more to say about it in Chapter 4.

Finally, it should be pointed out that some functions are neither increasing nor decreasing on any interval. A constant function is a simple example. A more complicated one is

$$f(x) = \begin{cases} 1 & \text{if } x \text{ is rational} \\ 0 & \text{if } x \text{ is irrational} \end{cases}$$

Proof of the Mean Value Theorem (*Optional*)

Part of the proof of the theorem depends on the fact that a continuous function defined on a closed and bounded interval has a largest or *maximum value* and a smallest or *minimum value* on that interval. This is known as the *Extreme Value Theorem;* it is illustrated in Figure 3–28. Although a proof of it is beyond the scope of this book, the Extreme Value Theorem plays a central role here and in much of our later work. In the first section of the next chapter, it is stated formally, assigned a number (4.7), and boxed in to give it the attention it deserves.

We begin the proof of the Mean Value Theorem with a special case. If $f(a) = f(b) = 0$, then the points $(a, 0)$ and $(b, 0)$ are on the graph of f. The line joining them is horizontal, so its slope is zero. Rolle's Theorem, proved below, says that $f'(c) = 0$ for some point c in (a, b).

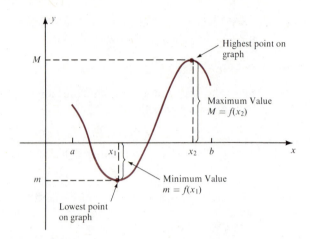

FIGURE 3–28. Illustration of the
Extreme Value Theorem.

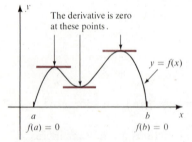

FIGURE 3–29. Illustration of Rolle's
Theorem.

Rolle's Theorem. If f is continuous on $[a, b]$, differentiable on (a, b), and $f(a) = f(b) = 0$, then there is at least one point c in (a, b) such that $f'(c) = 0$.

Proof. Figure 3–29 illustrates the theorem. First of all, if $f(x) = 0$ for all x in $[a, b]$, then f' is the constant function 0, and c can be any point in (a, b). If f is not constant, then it takes a nonzero value at some point in (a, b), say, a positive value. (The proof for a negative value is similar). It now follows from the Extreme Value Theorem (mentioned above) that f assumes its maximum value at some point c in (a, b); that is,

$$(3.49) \qquad f(x) \leq f(c) \quad \text{for all } x \text{ in } [a, b]$$

Since c is in (a, b), $f'(c)$ exists, and *we claim that $f'(c) = 0$.* To establish this claim, write

$$(3.50) \qquad \lim_{h \to 0^-} \frac{f(c + h) - f(c)}{h} = f'(c) = \lim_{h \to 0^+} \frac{f(c + h) - f(c)}{h}$$

If $|h|$ is small enough, then $c + h$ is in $[a, b]$, and it follows from (3.49) that $f(c + h) - f(c) \leq 0$. Therefore, the sign of the quotient in each limit above depends only on the sign of h, and we conclude that

$$\lim_{h \to 0^-} \frac{f(c + h) - f(c)}{h} \geq 0 \quad \text{and} \quad \lim_{h \to 0^+} \frac{f(c + h) - f(c)}{h} \leq 0$$

This, along with (3.50), implies that

$$f'(c) = 0$$

which concludes the proof of Rolle's Theorem.

Proof of the Mean Value Theorem (3.45). The idea is to introduce an auxiliary function g that reduces the hypotheses of the Mean Value Theorem to the hypotheses of Rolle's Theorem. That function is

$$g(x) = f(x) - f(a) - \frac{f(b) - f(a)}{b - a}(x - a)$$

defined for all x in $[a, b]$. There are three facts about g that are easily verified.

(1) g is continuous on $[a, b]$

(2) g is differentiable on (a, b), and

$$g'(x) = f'(x) - \frac{f(b) - f(a)}{b - a}$$

(3) $g(a) = g(b) = 0$

Thus, g satisfies the hypotheses of Rolle's Theorem, so there is at least one point c in (a, b) with $g'(c) = 0$. By (2) above, it follows that

$$0 = g'(c) = f'(c) - \frac{f(b) - f(a)}{b - a}$$

from which we obtain the conclusion of the Mean Value Theorem.

EXERCISES

In Exercises 1–10, determine whether or not the function satisfies the hypotheses of the Mean Value Theorem. If not, why not? If so, find the point(s) c referred to in the theorem.

1. $f(x) = x^2 + 3x + 2$ on $[-1, 0]$

2. $f(x) = \sqrt{x}$ on $[0, 1]$

3. $g(t) = \dfrac{t}{t - 1}$ on $[0, 2]$

4. $g(t) = \dfrac{t}{t - 1}$ on $[2, 3]$

5. $u = |x|$ on $[-1, 1]$

6. $u = x + \sqrt[3]{x}$ on $[-1, 1]$

7. $s = \sin t$ on $[0, \pi]$

8. $s = e^t$ on $[0, 1]$

9. $y = 2x^3 - x^2 - 2x$ on $[0, 1]$

10. $y = x^5 - 5x$ on $[-1, 1]$

In Exercises 11–24, determine the intervals on which the function is increasing and decreasing, and sketch the graph.

11. $y = x^2 + 2x + 2$

12. $y = 2x^2 - x + 1$

13. $s = -2t^2 + 4t - 1$

14. $s = -t^2 - t + 1$

15. $z = 2w^3 + 3w^2 - 12w$

16. $z = w^3 - 12w$

17. $y = x^3$

18. $y = x^4 + 1$

19. $y = 1/x$

20. $y = 1/x^2$

21. $z = w^{2/3}$

22. $z = w^{1/3}$

23. $s = \tan^2 t$

24. $s = \cos^2 t$

In Exercises 25–28, sketch the graph of an everywhere continuous function f that satisfies the given conditions. (There are many such functions, but each should have approximately the same shape.)

25. $f'(-1) = f'(1) = 0$, $f'(x) > 0$ on $(-1, 1)$, and $f'(x) < 0$ on $(-\infty, -1)$ and on $(1, \infty)$.

26. $f'(-1) = f'(0) = f'(1) = 0$, $f'(x) > 0$ on $(-1, 0)$ and on $(0, 1)$, and $f'(x) < 0$ on $(-\infty, -1)$ and on $(1, \infty)$.

27. $f'(-1) = f'(1) = 0$, $f'(x) > 0$ on $(-1, 0)$ and on $(0, 1)$, $f'(x) < 0$ on $(-\infty, -1)$ and on $(1, \infty)$, and $f'(0)$ does not exist.

28. $f'(-1) = f'(1) = 0$, $f'(0)$ does not exist, $f'(x) \leq 0$ on $(-\infty, 0)$, and $f'(x) \geq 0$ on $(0, \infty)$.

29. The number of people P (in thousands) infected t days after an epidemic begins is approximated by the function $P(t) = 1 + 60t - 3t^2$. When will the number of people infected start decreasing?

30. The sum of two numbers x and y is 52. For what values of x is the product xy increasing?

31. Given a piece of wire 20 inches long, pick a point x inches from the left end. Bend the wire at x into an L-shape. For what values of x is the distance from one end of the wire to the other decreasing?

32. A particle moves along a line with position function $s = t^3 - \frac{7}{2}t^2 + 4t$. For what values of t will it be moving to the right? To the left?

33. A rectangular beam is cut from a cylindrical log of radius 6 inches. The strength S of a rectangular beam varies jointly with the width w and the square of the height h. (That is, $S = kwh^2$ where k is a positive constant depending on the material.) For what values of h will S be increasing?

34. An isosceles triangle T has a perimeter of 2 inches and equal sides of x inches. For what values of x will the area of T be increasing?

35. A steel company can sell x tons of rolled steel depending on p, the price in dollars per ton; $x = p^2 - 50p + 100$, $10 \le p \le 50$. For what values of p will the *total revenue* be increasing?

36. (Continuation of Exercise 35). If the cost c of producing x tons of rolled steel is $c = 1,000 + 20x$ for what values of p will the *profit* be increasing?

37. *(Interpretation of the second derivative)* Suppose f is twice differentiable; that is, both f' and f'' exist for all x. The function f is increasing or decreasing depending on the sign of f'. In the same way, the function f' is increasing or decreasing depending on the sign of f''. Discover the effect of f'' by sketching a portion of the graph for each of the following situations which hold for all x.
    (a) $f'(x) > 0, f''(x) > 0$
    (b) $f'(x) > 0, f''(x) < 0$
    (c) $f'(x) < 0, f''(x) > 0$
    (d) $f'(x) < 0, f''(x) < 0$

38. Prove part (ii) of the test for increasing and decreasing.

Optional Exercises

39. Show that $\sin x \le x$ on $[0, \pi/2]$. *Hint:* Let $g(x) = \sin x - x$, and show that $g(0) = 0$ and g is decreasing. Carry on from there.

40. (Continuation of Exercise 39) Show that $x - x^3/6 \le \sin x$ on $[0, \pi/2]$.

41. Show that $1 + x + x^2/2 \le e^x$ for $x \ge 0$.

Table of Derivatives

The following table summarizes the derivative formulas developed so far. The formulas are written in abbreviated functional notation; f and g are assumed to be differentiable functions.

Derivative Formulas

1. $(c)' = 0$; c is a constant
2. $(x)' = 1$; x represents the identity function
3. $(cf)' = cf'$; c is a constant
4. $(f + g)' = f' + g'$
5. $(fg)' = fg' + gf'$
6. $(f/g)' = \dfrac{gf' - fg'}{g^2}$
7. $(f^r)' = rf^{r-1}f'$; r is any real number
8. $(e^f)' = (e^f)f'$
9. $(\sin f)' = (\cos f)f'$
10. $(\cos f)' = (-\sin f)f'$
11. $(\tan f)' = (\sec^2 f)f'$

REVIEW EXERCISES

In Exercises 1–8, use the definition
$$f'(x) = \lim_{h \to 0} \frac{f(x + h) - f(x)}{h}$$
to find the derivatives of the given functions.

1. $f(x) = 4$ **2.** $f(x) = 4 - 3x$

3. $f(x) = 3x^2 - 2x + 1$ **4.** $f(x) = x^3$

5. $f(x) = 9/x$ **6.** $f(x) = 1/x^2$

7. $f(x) = \sqrt{x + 1}$ **8.** $f(x) = 3/\sqrt{1 - x}$

In Exercises 9–30, differentiate the given functions.

9. $y = 8x^4 - 6x^3 + 5x^2$ **10.** $y = x - \dfrac{x^2}{8} - \dfrac{x^3}{9}$

11. $s = 7\sqrt[3]{(t + 1)^2}$ **12.** $s = 1 + \dfrac{2}{t} + \dfrac{3}{t^2}$

13. $u = (x + 1)(x^2 + 2x)$ **14.** $u = \dfrac{3x^2 - 2x}{1 - x}$

15. $y = \dfrac{\sqrt{3x - 4}}{e^x}$ **16.** $y = e^x(2x^3 + x)$

17. $y = (4x^3 - 3x^2)^5$ **18.** $y = \sqrt[3]{3 - 5x}$

19. $y = e^{\sqrt{2x+1}}$ **20.** $y = \sin^4 2x$

21. $s = e^{-t}\cos 2\pi t$ **22.** $s = e^{t^e}$

23. $y = \dfrac{(3x - 4)^3}{(5 - 4x)^2}$

24. $y = (x^2 + 1)^2(3x + 1)^3$

25. $y = \{3 - [6 - (1 - x^2)^3]^4\}^5$

26. $u = e^{\sin x}$

27. $y = \sin^2 \sqrt{1 - 6x}$ **28.** $u = \dfrac{\tan x}{1 - 3x}$

29. $y = e^{1.728}$ **30.** $y = 2x^{3.7} - 4x^{-2.1}$

In Exercises 31–36, find the second derivatives of the given functions.

Derivatives

31. $y = \sin(1/x)$ **32.** $y = \cos^2 x$

33. $y = (x^3 + 7)^3$ **34.** $y = e^{x^2}$

35. $y = \sqrt{e^x + 1}$

36. $y = (x^2 - 3)(1 - 4x)$

(*Optional*) In Exercises 37–50, find a function $y = f(x)$ with the given derivative. Check your answers.

37. $y' = 0$ **38.** $y' = -2$

39. $y' = 4x$ **40.** $y' = 2x - 3$

41. $y' = 5 - 3x$ **42.** $y' = 3x^2 - x + 1$

43. $y' = x^3 - 3x^2 + 4x - 1$

44. $y' = e^x$

45. $y' = 2xe^{x^2}$ **46.** $y' = 2 \sin x \cos x$

47. $y' = \sin 3x$ **48.** $y' = 2x(x^2 - 3)^4$

49. $y' = e^x(\sin x + \cos x)$ **50.** $y' = e^x(x^2 + 2x)$

In Exercises 51–56, determine whether or not the function satisfies the hypotheses of the Mean Value Theorem. If not, why not? If so, find the point(s) referred to in the theorem.

51. $y = x^{2/3}$ on $[0, 1]$

52. $y = x^2 - 5x + 7$ on $[1, 2]$

53. $y = 2x^3 - 3x^2 + 4x$ on $[-1, 0]$

54. $y = \cos x$ on $[-1, 1]$

55. $y = \dfrac{x}{x^2 - 2}$ on $[0, 2]$

56. $y = |x + 1|$ on $[-3, 0]$

In Exercises 57–64, determine the intervals on which the function is increasing and decreasing, and sketch the graph.

57. $y = 2 - 3x - x^2$ **58.** $y = 4x^2 - 8x + 1$

59. $y = x^{2/3}$ **60.** $y = \sqrt{1 - x^2}$

61. $y = x^3 - 3x^2 + 3$ **62.** $y = x + \dfrac{1}{x}$

63. $y = 2x - \dfrac{1}{x^2}$ **64.** $y = \sin^3 x$

In Exercises 65–66, sketch the graph of an everywhere continuous function f that satisfies the given conditions.

65. $f'(-1) = f'(0) = f'(1) = 0, f'(2)$ does not exist, $f'(x) > 0$ on $(-1, 0)$, on $(1, 2)$, and on $(2, \infty)$, and $f'(x) < 0$ on $(-\infty, -1)$ and on $(0, 1)$.

$f'(-1) = f'(1) = 0, f'(0)$ does not exist, $f'(x) \geq 0$ for all $x \neq 0$.

67. Find an equation of the line tangent to the curve $y = x^2 + 7x - 6$ at the point $(2, 12)$.

68. Find an equation of the line tangent to the curve $y = \sin x$ at the point $(\pi/4, \sqrt{2}/2)$.

69. Find an equation of the line with slope 2 that is tangent to the curve $y = 2 + 8x - 3x^2$.

70. If $f(1) = 3$ and $f'(1) = -2$, evaluate the derivative of $y = f^3(x)$ at $x = 1$.

71. If $f(6) = \pi/3$ and $f'(6) = 14$, evaluate the derivative of $y = \sin f(x)$ at $x = 6$.

72. If $f'(0) = -4$, evaluate the derivative of $y = f(\sin x)$ at $x = 0$.

73. A roller coaster has the shape of the curve $y = \sin^2 x$. When a rider is at the point $(5\pi/6, 1/4)$, he is moving 30 ft/sec horizontally. How fast is he falling at that moment?

74. A projectile is fired straight up, and its height at t seconds is $h = -16t^2 + 160t$ feet above the ground. For what values of t is h increasing and decreasing? What is the maximum height reached?

75. A rectangular plot with an area of 100 square meters is to be fenced in. One of the dimensions is x meters. For what values of x will the required amount of fencing be decreasing?

76. An isosceles triangle T has equal sides x inches and a fixed perimeter p inches. What is the rate of change of the area of T with respect to x?

77. The surface area of a sphere of radius r is $S = 4\pi r^2$. A spherical balloon is being inflated so that the radius at time t seconds is $r = \sqrt[3]{t}$ feet. At what rate is S increasing when $t = 8$ seconds?

78. The population of a strain of bacteria doubles every 30 minutes. If you start with 10 bacteria, how fast is the population increasing after 3 hours? After 5 hours?

79. The half-life of polonium is 140 days. If you start with Q_o grams, at what rate (grams/day) is it decaying after 7 days?

80. A small company estimates that its profit (in thousands) is given by $P = 100 + 2x - x^2$, where x represents the money (in thousands) spent on advertising. How much would you advise them to spend on advertising?

4

APPLICATIONS
OF DERIVATIVES

The preceding chapter describes what a derivative is, and how to find it. This chapter describes how to put that knowledge to work.

For example, in the first section, we develop a technique for finding the maximum and minimum values of a function. In the next section, this technique is applied to solve problems in a variety of disciplines. Here are just four examples.

Business: A one-quart can is to be manufactered. What radius and height will minimize the cost?

Engineering: A rectangular beam is to be cut from a circular log of radius *r*. What dimensions will maximize the strength of the beam?

Life Sciences: Do bees construct the cells of their hive so as to minimize the amount of wax needed to store a given amount of honey?

Mathematics: Among all triangles with a given perimeter, which one has the largest area?

Subsequent sections develop techniques for approximating functions with differentials, finding rates of change, describing the concavity of a graph, and solving a certain type of differential equation. Each technique is followed by applications.

The point to concentrate on in this chapter is to learn the techniques; once they are mastered, the ability to apply them requires only some practice.

4–1 MAXIMA AND MINIMA

Management can use derivatives to formulate policies that will maximize the profits of their company. Automotive engineers can use derivatives to compute the cylinder size that will minimize the fuel consumption of an engine. In fact, whenever a problem in business, engineering, or science can be put into functional form, derivatives can be used to find maximum and minimum values of the function. In this section, we describe a method for doing this; in the next section, we use the method to solve some practical problems.

We begin by considering functions that are defined on one or more open intervals. Functions whose domains have endpoints are treated afterwards.

At first, we define *local extrema*. Suppose f is a function and c is a point in its domain. The number $f(c)$ may not be the largest value that f assumes on its entire domain, but if it is the largest value that f assumes at all points near c, then $f(c)$ is called a *local maximum*. If $f(c)$ is the smallest value that f assumes at all points near c, then $f(c)$ is called a *local minimum* (Figure 4–1).

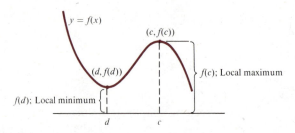

FIGURE 4–1. Local extrema. $f(c)$ is the largest value that f assumes at points near c; $f(d)$ is the smallest value that f assumes at points near d.

DEFINITION

Let c be a point in the domain of a function f.

(1) $f(c)$ is a **local maximum** of f if there is some open interval I about c such that

(4.1)
$$f(x) \leq f(c) \quad \text{for all } x \text{ in I}$$

(2) $f(c)$ is a **local minimum** of f if there is some open interval I about c such that

$$f(x) \geq f(c) \quad \text{for all } x \text{ in I}$$

Local maxima and minima are also called **local extreme values** or **local extrema.** Figure 4–2 suggests that local extreme values occur only at points where $f'(c) = 0$ or where $f'(c)$ does not exist. Let us explore this possibility.

Suppose that $f'(c) > 0$. Then

(4.2)
$$\frac{f(c + h) - f(c)}{h} > 0$$

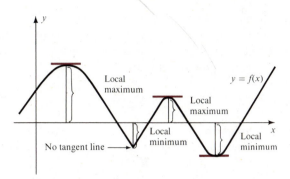

FIGURE 4–2. An illustration of local extrema. They occur only at points c where $f'(c) = 0$ or $f'(c)$ does not exist.

for all h close (but not equal) to 0. If $h > 0$, then the numerator $f(c + h) - f(c) > 0$ (why?), so $f(c + h) > f(c)$. This means that just to the right of c, f assumes values that are larger than $f(c)$. On the other hand, if $h < 0$, then $f(c + h) - f(c) < 0$, so $f(c + h) < f(c)$. This means that just to the left of c, f assumes values that are less than $f(c)$. It follows that *if $f'(c) > 0$, then $f(c)$ is not a local extreme value.* A similar argument can be applied if $f'(c) < 0$. Therefore, if $f(c)$ *is* a local extreme value, then there are only two possibilities.

(4.3)

> If $f(c)$ is a local extreme value of f, then either
>
> $$f'(c) = 0 \quad \text{or} \quad f'(c) \text{ does not exist}$$

A point c in the domain of f is called a **critical point** of f if $f'(c) = 0$ or $f'(c)$ does not exist. Therefore, the first step in finding local maxima and minima is to find the critical points. Extrema can occur *only* at critical points.

EXAMPLE 1   Find the local extreme values of $f(x) = 1 + 2x - x^2$.

Solution $f'(x) = 2 - 2x$ exists everywhere and is 0 only at $x = 1$. Therefore, $x = 1$ is the only critical point. By (4.3), $f(1) = 2$ is the *only possible* local extreme value. To determine if 2 is a maximum or minimum, write

$$f'(x) = 2 - 2x \begin{cases} > 0 & \text{on } (-\infty, 1) \\ < 0 & \text{on } (1, \infty) \end{cases}$$

To the left of $x = 1$, f is increasing (positive derivative); to the right of $x = 1$, f is decreasing (negative derivative). It follows that $f(1) = 2$ must be a local maximum (Figure 4–3).   ■

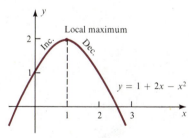

FIGURE 4–3. Example 1. $x = 1$ is the only critical point.

EXAMPLE 2   Find the local extreme values of $f(x) = |x|$.

Solution

$$f'(x) = \begin{cases} 1 & \text{if } x > 0 \\ -1 & \text{if } x < 0 \end{cases}$$

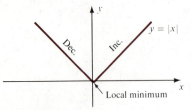

FIGURE 4–4. Example 2. $x = 0$ is the only critical point.

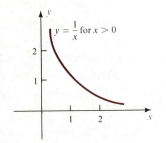

FIGURE 4–5. Example 3. No critical points; no extrema.

and does not exist at $x = 0$. Thus, 0 is the only critical point. The derivative is negative to the left of 0 and positive to the right. Therefore, f is decreasing to the left of 0 and increasing to the right of 0. It follows that $f(0) = 0$ must be a local minimum (Figure 4–4). ■

EXAMPLE 3 Find the local extreme values of $f(x) = 1/x$ for $x > 0$.

Solution $f'(x) = -1/x^2$; it exists, but is never zero, on the domain of f. Therefore, there are no critical points, so there are no local extreme values (Figure 4–5). ■

WARNING: As always, you should not read into a statement any more than is actually there. A case in point is statement (4.3). If $f'(c) = 0$ or $f'(c)$ does not exist, then (4.3) does *not* guarantee that $f(c)$ is a local extreme value. Here are two examples.

EXAMPLE 4 Let $f(x) = x^3$. Then

$$f'(x) = 3x^2 \begin{cases} > 0 & \text{if } x \neq 0 \\ = 0 & \text{if } x = 0 \end{cases}$$

Thus, f is increasing on *both* sides of the critical point $x = 0$. It follows that $f(0) = 0$ is *not* an extreme value (Figure 4–6A). ■

EXAMPLE 5 Let $g(x) = -\sqrt[3]{x}$. Then $g'(x) = -1/3x^{2/3}$ is negative everywhere except at $x = 0$ where it does not exist. Thus, g is decreasing on *both* sides of the critical point $x = 0$. It follows that $g(0) = 0$ is *not* an extreme value (Figure 4–6B). ■

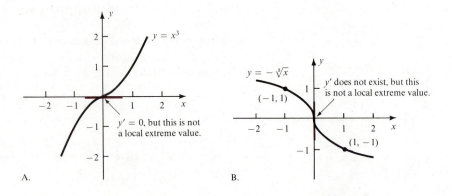

FIGURE 4–6. Examples 4 and 5.

Examples 1, 2, 4, and 5 above illustrate how the sign of the derivative to the right and left of a critical point is used to test for extreme values. This test is called the *First Derivative Test* (Figure 4–7).

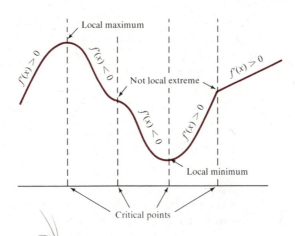

FIGURE 4–7. Illustration of the first derivative test.

FIRST DERIVATIVE TEST

Let c be a critical point of a function f and suppose that f is continuous at c. Suppose further that (a, b) is some interval containing c.

(4.4)

(1) If $f'(x) > 0$ for all x in the interval (a, c) and $f'(x) < 0$ for all x in the interval (c, b), then $f(c)$ is a local maximum.

(2) If the inequalities in (1) are reversed, then $f(c)$ is a local minimum.

(3) If $f'(x) > 0$ or $f'(x) < 0$ on both (a, c) and (c, b), then $f(c)$ is not a local extreme value.

There is another test for local extrema that in some cases is easier to apply than (4.4); it involves the second derivative of f. We suppose that f' is defined on an open interval about c, that $f'(c) = 0$, and that $f''(c) > 0$. If we replace f by f' in inequality (4.2) and also in the argument that follows it, we can conclude that just to the right of c, $f'(x) > f'(c)$ and just to the left of c, $f'(x) < f'(c)$. Since we are assuming that $f'(c) = 0$, this means that f is increasing to the right of c and decreasing to the left of c. Therefore, $f(c)$ is a local minimum of f (Figure 4–8A). If $f''(c) < 0$ the opposite occurs (Figure 4–8B). This is the *Second Derivative Test*.

SECOND DERIVATIVE TEST

Suppose that f is differentiable on an open interval about c and $f'(c) = 0$.

(4.5)

(1) If $f''(c) < 0$, then $f(c)$ is a local maximum.

(2) If $f''(c) > 0$, then $f(c)$ is a local minimum.

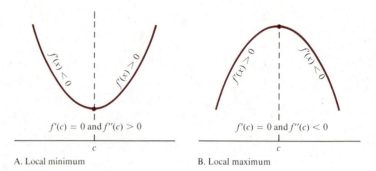

$f'(c) = 0$ and $f''(c) > 0$

$f'(c) = 0$ and $f''(c) < 0$

A. Local minimum

B. Local maximum

FIGURE 4–8. Illustration of the second derivative test.

Whether to use the first or second derivative test is largely a matter of convenience. The next two examples illustrate this.

EXAMPLE 6   Find the extreme values of $f(x) = x/(x^2 + 1)$.

Solution   *First,* find the critical points.

$$f'(x) = \frac{(x^2 + 1)(1) - (2x)x}{(x^2 + 1)^2} = \frac{1 - x^2}{(x^2 + 1)^2}$$

The only critical points are $x = \pm 1$.

Second, test for maxima and minima. It is more convenient to use the first derivative test than to differentiate again. To apply (4.4), we find an interval, say $(-2, 0)$, about -1 and observe that $f'(x) < 0$ on $(-2, -1)$ and $f'(x) > 0$ on $(-1, 0)$. It follows that $f(-1) = -1/2$ is a local minimum. A similar analysis shows that $f(1) = 1/2$ is a local maximum. ∎

EXAMPLE 7   Find and describe the extreme values of $f(x) = 2x^3 + 3x^2 - 12x$, and sketch the graph.

Solution   *First,* find the critical points.

$$f'(x) = 6x^2 + 6x - 12 = 6(x + 2)(x - 1)$$

So $x = -2$ and $x = 1$ are the only critical points.

Second, test for maxima and minima. We can use either the first or second derivative test, but for this problem, the latter seems easier. Therefore, we find the second derivative.

$$f''(x) = 12x + 6$$

Now, $f''(-2) = -18$, so $f(-2) = 20$ is a local maximum; $f''(1) = 18$, so $f(1) = -7$ is a local minimum. See Figure 4–9. ∎

Local maximum
$(-2, 20)$

$y = 2x^3 + 3x^2 - 12x$

$(1, -7)$ Local minimum

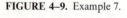

FIGURE 4–9. Example 7.

Extreme Values at End Points

Functions encountered in applications are usually defined on intervals with one or two endpoints. For instance, the area of a rectangular garden that can be

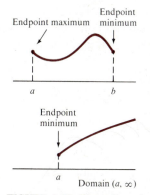

FIGURE 4–10. Illustrations of endpoint extrema.

bounded by 200 feet of fencing is a function of the width W, but W is restricted to the interval [0, 100] (Example 5, Section 1–4).

Suppose that the domain of f has one or more endpoints. If a is a left endpoint and $f(a) \geq f(x)$ for all x in some interval $[a, b)$, then $f(a)$ is called an **endpoint maximum;** if $f(a) \leq f(x)$ for all x in some interval $[a, b)$, then $f(a)$ is called an **endpoint minimum.** Similar definitions are made for right endpoints. The terms *extreme values* and *extrema* also apply to endpoint maxima and minima (Figure 4–10). Since extreme values can occur at endpoints, we will refer to them as critical points also. That is, from now on, a *critical point* is a point for which either (4.3) holds or else it is an endpoint. The first derivative test can be altered slightly to test endpoint extrema, as shown in the next example.

EXAMPLE 8   Find and describe the extreme points of $f(x) = 2x^3 + 3x^2 - 12x$ for $-3 \leq x \leq 2$, and sketch the graph.

Solution   This is the same function as in Example 7. We already know that $f(-2) = 20$ and $f(1) = -7$ are local extrema; but now we have the endpoints -3 and 2 to consider. A modification of the first derivative test can be used on these critical points. The idea is to determine the sign of f' just to the right of -3 and just to the left of 2. Since $f'(x) = 6x^2 + 6x - 12 = 6(x + 2)(x - 1)$, we see that

$$f'(x) > 0 \quad \text{on } (-3, -2), \text{ so } f \text{ is increasing on } [-3, -2]$$
$$f'(x) > 0 \quad \text{on } (1, 2), \text{ so } f \text{ is increasing on } [1, 2]$$

It follows that $f(-3) = 9$ is an endpoint minimum and $f(2) = 4$ is an endpoint maximum. See Figure 4–11.   ∎

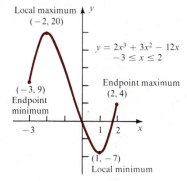

FIGURE 4–11. Example 8.

Absolute Extrema

Local or endpoint extrema are the largest and smallest values that f assumes *near* a point. But there may be extreme values with regard to the entire domain. For instance, Figure 4–11 shows that $f(-2) = 20$ is the largest and $f(1) = -7$ is the smallest value that f assumes on its whole doman $[-3, 2]$. These values are called the *absolute extrema*.

(4.6)

> **DEFINITION**
> Let f be defined on a set X.
>
> (1) $f(c)$ is an **absolute maximum** of f on X if
> $$f(c) \geq f(x) \quad \text{for all } x \text{ in } X.$$
> (2) $f(c)$ is an **absolute minimum** of f on X if
> $$f(c) \leq f(x) \quad \text{for all } x \text{ in } X.$$

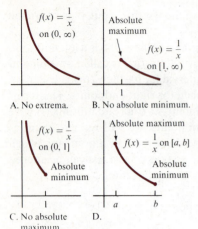

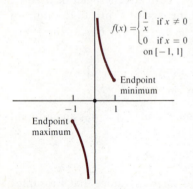

FIGURE 4–12. Various possibilities for absolute extrema.

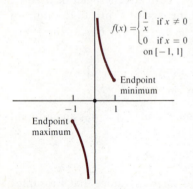

FIGURE 4–13. No absolute extrema.

A function may or may not have absolute extreme values. Figure 4–12 shows the graph of $f(x) = 1/x$ with several different domains. In only one case, when the domain is a closed and bounded interval $[a, b]$, are both absolute extrema present. Figure 4–13 shows the graph of

$$f(x) = \begin{cases} 1/x & \text{if } x \neq 0 \\ 0 & \text{if } x = 0 \end{cases} \quad x \text{ in } [-1, 1]$$

Here, the domain is a closed and bounded interval, but the function is not continuous and, again, there are no absolute extrema. It seems that absolute extrema have to do with both continuity and closed, bounded intervals. All of this leads us to the *Extreme Value Theorem*. We have already used it in proving the Mean Value Theorem (Section 3–6), and we will refer to it many times in the future.

(4.7)
> **EXTREME VALUE THEOREM**
> If f is continuous on a closed and bounded interval, then f has absolute maximum and absolute minimum values.

A proof is beyond the scope of this book, but can be found in any advanced calculus text.

(4.8)
> The absolute extreme values (if there are any) are the largest and smallest of the local and endpoint extrema.

EXAMPLE 9   Find the absolute maximum of

$$f(x) = (p - 2x)\sqrt{4px - p^2} \quad \text{for } p/4 \leq x \leq p/2$$

where p is a positive constant.

Solution   *First*, find the critical points (including endpoints).

$$f'(x) = (p - 2x)[\sqrt{4px - p^2}]' + [\sqrt{4px - p^2}](p - 2x)' \quad \text{(Product)}$$

$$= \frac{(p - 2x)(4p)}{2\sqrt{4px - p^2}} - 2\sqrt{4px - p^2}$$

$$= \frac{4(p^2 - 3px)}{\sqrt{4px - p^2}} \quad \text{(Simplify)}$$

A quotient is 0 only when the numerator is 0; in this case, when $x = p/3$. A quotient is undefined only when the denominator is 0; in this case, when $x = p/4$. Therefore, the critical points are $p/3$ and the endpoints $p/4$ and $p/2$.

Second, find the absolute maximum. Since f is continuous on a closed and bounded interval, we know the absolute maximum exists. Therefore, by (4.8), all we have to do is compare the values attained at the critical points and choose the largest one. Since

$$f(p/4) = 0, \quad f(p/3) = \frac{p^2}{3\sqrt{3}}, \quad \text{and} f(p/2) = 0$$

the choice is clear. ■

EXERCISES

In Exercises 1–14, locate and describe all local, end-point, and absolute extreme values, and sketch the graph.

1. $y = x^2 + 3x + 2$

2. $y = x^2 - 2x + 2$

3. $y = 1 + x - x^2$

4. $y = 3x - 2x^2$

5. $s = t^3 + 2t^2; [-2, 1]$

6. $s = t^3 - 12t; [-3, 3]$

7. $y = x^4 - 2x^3 + 1; [-1, 2]$

8. $y = \dfrac{1}{x^2 + 1}; [-5, 5]$

9. $y = x^2 + \dfrac{2}{x}; x > 0$

10. $y = 4x - \dfrac{2}{x^2}; x < 0$

11. $y = x + \sin x; x \geq 0$

12. $y = x + \cos x; x \geq 0$

13. $y = x - \tan x; [0, \pi/2]$ **14.** $y = e^{-x^2}$

In Exercises 15–29, find the absolute maximum and minimum values (if there are any) on the given domain.

15. $y = 36x + x^2$; all x

16. $y = 36x - x^2$; all x

17. $y = 14 + 6x - 2x^2; [0, 10]$

18. $y = x^3 - 2x^2; [-1, 2]$

19. $y = \left(x - \dfrac{3}{2}\right)\left(\dfrac{80}{x} - 2\right); [3/2, 40]$

20. $y = (x - 1)^2(x + 1); [-1, 1]$

21. $y = x\sqrt{p - x^2}; [-\sqrt{p}, \sqrt{p}]$

22. $y = \dfrac{x}{\sqrt{p - x^2}}; [0, \sqrt{p})$

23. $y = \dfrac{x^2}{x - 2}; x > 2$

24. $y = \dfrac{x - 2}{x^2}; [-2, -1]$

25. $y = \sqrt{x^2 - x + 1}$; all x

26. $y = \sqrt{4 - x^2}; [-2, 2]$

27. $y = px - \dfrac{(\pi - 4)x^2}{2}; [0, p]$

28. $y = \dfrac{\sqrt{x^2 + 4}}{3} + \dfrac{6 - x}{5}; [0, 6]$

29. $y = \dfrac{x^2}{4\pi} + \left(\dfrac{\sqrt{3}}{36}\right)(p - x)^2; [0, p]$

4-2   MAXIMUM-MINIMUM PROBLEMS

We are now prepared to put the material of the preceding section to practical use.

EXAMPLE 1   What is the largest rectangular area that can be enclosed with 200 feet of fencing?

Solution

(1) *Analyze the problem.* The area of a rectangle is to be maximized under the condition that the perimeter is 200 feet.

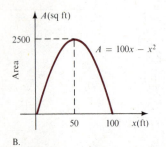

$$y = \frac{200 - 2x}{2} = 100 - x$$

x

Perimeter = 200′

A.

A(sq ft)

2500

Area

$A = 100x - x^2$

50   100 x(ft)

B.

FIGURE 4–14. Example 1. Maximizing area.

(2) *Draw a diagram.* Include all given information and label the variable parts with letters (Figure 4–14A).

(3) *Formulate a function.* Express the quantity to be maximized as a function of *one* of the variables, and determine the domain. In this case,

$$A = x(100 - x) = 100x - x^2 \quad 0 \le x \le 100$$

(4) *Locate the critical points (including endpoints).*

$$A' = 100 - 2x$$

Thus, 0, 50, and 100 are the critical points.

(5) *Test the critical points.* We want to maximize A. Because A is continuous on a closed and bounded interval, it must assume an absolute maximum value, and that value can occur only at a critical point.

$$A(0) = 0, \quad A(50) = 2,500, \quad \text{and } A(100) = 0$$

Conclusion. The largest rectangular area that can be enclosed with 200 feet of fencing is 2,500 square feet.

Remarks: *Testing the critical points.* In this example, the function A is continuous on a closed and bounded interval. To find its maximum value, simply compare the values at the critical points. In some problems, however, this is not the case, and other tests must be applied (see Example 2). ■

Step-by-Step Method

Solving a max-min problem consists of several steps as illustrated in Example 1.

(1) Analyze the problem (determine what is to be maximized or minimized and the conditions).

(2) Draw a diagram (if appropriate).

(3) Formulate a function.

(4) Locate the critical points.

(5) Test the critical points.

Taken separately, *each step is a familiar operation,* one that you already know how to perform. The only new skill required here is to put all the steps together to solve one problem.

EXAMPLE 2 A cylindrical can is to have a volume of V cubic inches. What dimensions will require the least amount of material to make such a can?

Solution

(1) *Analyze.* The surface area of a can is to be minimized under the condition that the volume is V.

(2) *Diagram.* Figure 4–15.

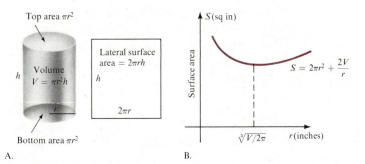

FIGURE 4–15. Example 2. Making a can with a minimum amount of material.

A.

B.

(3) *Formulate a function.* The surface area of the cylinder is

$$S = 2\pi r^2 + 2\pi rh*$$

The height h can be eliminated using the volume equation

$$V = \pi r^2 h, \quad \text{so } h = \frac{V}{\pi r^2}$$

It follows that

$$S = 2\pi r^2 + \frac{2V}{r} \quad r > 0$$

(4) *Locate critical points.*

$$S' = 4\pi r - \frac{2V}{r^2}$$

There are no endpoints in this problem. Set $S' = 0$ and solve.

$$r = \sqrt[3]{V/2\pi}$$

is the only critical point.

(5) *Test.* Unlike Example 1, we cannot simply compare the values at critical points because this function is not defined on a bounded closed interval. Some other test must be used. Let us try the second derivative test.

$$S'' = 4\pi + \frac{4V}{r^3} \quad r > 0$$

is always positive, so $S(\sqrt[3]{V/2\pi})$ is a minimum.

Conclusion. The least material is required when

$$r = \sqrt[3]{V/2\pi} \quad \text{and} \quad h = \frac{V}{\pi r^2} = \frac{2\pi(V/2\pi)}{\pi(V/2\pi)^{2/3}} = 2r \quad \left(\text{Write } V = 2\pi\frac{V}{2\pi}\right)$$

that is, when the height equals the diameter.   ∎

---

*Think of the lateral surface as a rectangle with height h and width $2\pi r$, the circumference of the cylinder.

In the remaining examples, the steps will not be set out in italics.

EXAMPLE 3   A rectangular beam is to be cut from a cylindrical log with a radius of r inches. What dimensions produce the beam with the largest cross-sectional area?

Solution   Cross-sectional area is to be maximized. Figure 4–16 shows the relationship between the variable quantities h and w and the fixed quantity r. To formulate the area function, we write

$$\frac{h}{2} = \sqrt{r^2 - (w/2)^2}, \quad \text{so } h = \sqrt{4r^2 - w^2}$$

Therefore,

$$A = wh = w\sqrt{4r^2 - w^2} \quad 0 \le w \le 2r$$

Now find the critical points (including endpoints).

$$A' = w(\sqrt{4r^2 - w^2})' + \sqrt{4r^2 - w^2}(w)' \qquad \text{(Product)}$$

$$= -\frac{w^2}{\sqrt{4r^2 - w^2}} + \sqrt{4r^2 - w^2}$$

$$= \frac{4r^2 - 2w^2}{\sqrt{4r^2 - w^2}} \qquad \text{(Simplify)}$$

Since $A' = 0$ only at $w = \sqrt{2}r$, the critical points are 0, $\sqrt{2}r$, and $2r$. The values at the endpoints are zero and $A(\sqrt{2}r)$ is positive; therefore, $A(\sqrt{2}r)$ is the absolute maximum value.

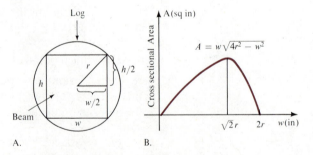

FIGURE 4–16. Example 3. Cutting the largest beam from a log.

Conclusion. The largest cross-sectional area is achieved when

$$w = \sqrt{2}r \quad \text{and} \quad h = \sqrt{4r^2 - (\sqrt{2}r)^2} = \sqrt{2}r$$

Since h and w are clearly interchangeable in this problem, the conclusion that $h = w$ is not surprising.   ■

EXAMPLE 4   *(Snell's Law)* According to Fermat's principle of optics, light traveling from point A to point B will follow the path that takes the least time.

Suppose that light has velocity v_1 in air and v_2 in water. *Snell's Law* says that light traveling from point A in air to point B in water will follow the path with

$$\frac{\sin \alpha_1}{v_1} = \frac{\sin \alpha_2}{v_2}$$

where α_1 and α_2 are shown in Figure 4–17A. Prove Snell's Law.

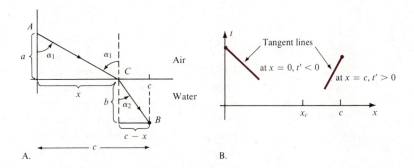

FIGURE 4–17. Example 4. Snell's Law.

Solution We want to find the path that takes the least time under the restrictions imposed by Fermat's principle. Since the velocity v_1 is constant in air, light will travel in a straight line from A to a point C on the water. The point C can be anywhere from $x = 0$ to $x = c$. The distance from A to C is $\sqrt{a^2 + x^2}$, so the time that it takes is

$$t_A = \frac{\sqrt{a^2 + x^2}}{v_1} \quad 0 \le x \le c$$

It will then travel in a straight line from C to B, and the time it takes is

$$t_W = \frac{\sqrt{b^2 + (c - x)^2}}{v_2} \quad 0 \le x \le c$$

Therefore, we want to minimize

$$t = t_A + t_W = \frac{\sqrt{a^2 + x^2}}{v_1} + \frac{\sqrt{b^2 + (c - x)^2}}{v_2} \quad 0 \le x \le c$$

Now we differentiate with respect to x

(4.9)
$$t' = \frac{x}{v_1 \sqrt{a^2 + x^2}} - \frac{(c - x)}{v_2 \sqrt{b^2 + (c - x)^2}}$$
$$= \frac{\sin \alpha_1}{v_1} - \frac{\sin \alpha_2}{v_2}$$

where α_1 and α_2 are the angles in the figure. Let x_C be the point where t' is zero; that is, the point where

$$\frac{\sin \alpha_1}{v_1} = \frac{\sin \alpha_2}{v_2}$$

The only critical points are 0, x_C, and c. To find the absolute minimum, we could compare the values of t at the critical points. But evaluation at x_C is too cumbersome. Another way to proceed is to argue that substituting the endpoints into (4.9) shows that t' is negative at $x = 0$ and positive when $x = c$. This *suggests* that t is a minimum at x_C (Figure 4–17B). But the surest method is to use the second derivative test. Since

$$t'' = \frac{a^2}{v_1(a^2 + x^2)^{3/2}} + \frac{b^2}{v_2(b^2 + (c - x)^2)^{3/2}} \qquad \text{(check this out)}$$

is always positive, $t(x_C)$ must be a minimum value. ∎

EXAMPLE 5 *(Inventory problem)* A car radio company plans to manufacture a total of T radios in the coming year. The radios are produced in *batches* or *runs,* and it costs S dollars to "set up" each run no matter how many radios are produced. After the radios are produced, they are stored as *inventory* at an annual cost of I dollars per radio. It is assumed that they are sold at a constant rate throughout the year. A large run means a low set up cost per radio, but high storage costs; a small run produces the opposite result. How many radios should be produced in each run to minimize the inventory costs?

Solution   Let x be the number of radios produced in each run. There will be T/x runs per year, so

$$S \cdot \frac{T}{x} = \text{setup cost per year}$$

The *average* number of unsold radios (that is, inventory) over the year is $x/2$, so

$$I \cdot \frac{x}{2} = \text{storage cost per year}$$

Therefore, the inventory cost function is

$$C = S \cdot \frac{T}{x} + I \cdot \frac{x}{2} \qquad 0 < x \le T$$

Now find the value of x that minimizes C.

$$C' = -\frac{ST}{x^2} + \frac{I}{2}$$

So $C' = 0$ when $x = \sqrt{2ST/I}$. Since $C'' = 2ST/x^3 > 0$, *this value of x minimizes C* (Figure 4–18).

Note: The answer is supposed to indicate the number of radios per run. In case $x = \sqrt{2ST/I}$ is not an integer, then the cost function should be tested at the two integers on either side of x to determine the correct run size. ∎

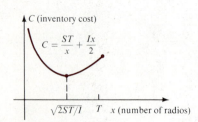

FIGURE 4–18. Example 5.

EXERCISES

<u>1</u>. Among all pairs of real numbers whose difference is 36, find a pair whose product is minimum.

2. Among all pairs of positive real numbers whose sum is 36, find the pair whose product is maximum.

<u>3</u>. What is the largest rectangular area that can be fenced in with 240 ft of fencing?

4. A rectangular area of 100 sq ft is to be fenced in but no fencing is needed along the building (see Figure 4–19). What is the minimum amount of fencing required?

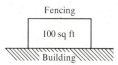

FIGURE 4–19. Exercise 4.

<u>5</u>. A farmer wants to enclose 3,600 sq ft of land which is subdivided into two pieces as shown in Figure 4–20. The outside fencing (solid line) cost \$2/ft and the inside fencing (dashed line) costs only \$1/ft. What dimensions will minimize the cost?

FIGURE 4–20. Exercise 5.

6. Given a piece of wire 20 inches long, pick a point x inches from the left end. Bend the wire at point x into an L-shape. What value of x will make the distance from one end of the wire to the other end a minimum? A maximum?

<u>7</u>. A square piece of cardboard 10 inches on a side is to be made into a box with no top by cutting out squares from the corners and folding up the sides. What size squares should be cut out to obtain the largest volume?

<u>8</u>. The accountant of a shirt manufacturing firm runs a cost analysis of producing a single shirt. After six months of careful data collecting, he finds that within the range of producing 1,000 to 2,000 shirts per day, the costs are as follows: The material and labor cost \$5 per shirt and the fixed costs (rent, insurance, etc.) are \$500 per day. But this production schedule causes an added cost of 2 cents times the square root of the number of shirts produced. This is due to increased employee error and machine breakdowns. How many shirts should be manufactured each day to minimize the cost per shirt? How much will each shirt cost?

9. The manufacturer in Exercise 8 can sell 2,000 shirts at \$7 each. For each 10¢ increase in price, he sells 20 less shirts. What price will maximize his revenue?

10. The same manufacturer wants to build a warehouse and divide it up as shown in Figure 4–21. The outside walls cost \$5 and the inside walls cost \$2 per running foot. If he needs a total of 10,000 sq ft of floor space, what dimensions should the warehouse be?

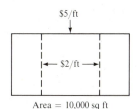

FIGURE 4–21. Exercise 10.

<u>11</u>. A piece of wire of length L is cut in two. One of the pieces is bent into the shape of a circle and the other into the shape of a square. Where should the wire be cut so that the total enclosed area of the two pieces is a minimum? A maximum?

<u>12</u>. If the angle of elevation of a cannon is x and a projectile is fired with a muzzle velocity v ft/sec, then the range is given by

$$R \approx \frac{v^2 \sin 2x}{32} \text{ feet}$$

What elevation produces the largest range?

<u>13</u>. A pie-shaped wedge with central angle x is cut from a circular piece of paper of radius r (Figure 4–22). If the straight edges are joined to form a cone, what value of x produces the cone with the largest volume?

Join these edges to form a cone.

FIGURE 4–22. Exercise 13.

14. At a certain moment, motorist *A* is 100 miles due west of motorist *B*. Motorist *A* is traveling north at 40 mi/hr and *B* is driving west at 60 mi/hr. What will be the minimum distance between them? *Hint*: Express the distance as a function of time.

15. A triangle is inscribed in a semicircle of radius 5 ft so that one side of the triangle is the diameter (see Figure 4–23). Of all such triangles, what is the maximum area attained?

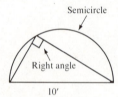

Semicircle

Right angle

10'

FIGURE 4–23. Exercise 15.

16. Same as Exercise 15 only this time inscribe a rectangle so that one side sits on the diameter.

17. The strength of a rectangular beam varies jointly with the width *w* and the square of the height *h*. (That is, the strength is kwh^2, where *k* is a constant depending on the material.) Find the dimensions of the strongest beam that can be cut from a cylindrical log of radius 6 inches.

18. The yield of fruit trees is reduced if they are planted too close together. If there are 30 orange trees per acre, then each tree produces 400 oranges. For each additional tree in that acre, the yield is reduced by 7 oranges. How many trees per acre yield the largest crop?

19. A page of a book is to have an area of 80 sq in. with 1 in. margins at the top and sides and a $\frac{1}{2}$ in. margin at the bottom. What dimensions will allow the largest area for printed matter?

20. A long rectangular piece of galvanized metal 12 in. wide is to be made into a rain gutter by bending up two sides at right angles to form a three-sided rectangular trough. How much should be turned up to maximize the capacity?

21. The *x*-axis, *y*-axis, and any line with negative slope passing through (1, 2) determine a right triangle. Find the line for which this triangle has minimum area.

22. (Continuation of Exercise 21) Find the line for which the area is a maximum.

23. A fence 8 ft tall stands 1 foot from a building. What is the shortest ladder that will extend from the ground outside the fence to the wall of the building?

24. Using Figure 4–24, run a power line from *A* to *C* as cheaply as possible. Below ground, you can run in a straight line for $26/ft. Above ground, you must run along the sides of the rectangle at a cost, including poles, of $24/ft.

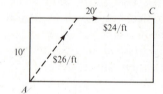

20'

$24/ft

10'

$26/ft

A

C

FIGURE 4–24. Exercise 24.

25. A man is in a desert at point *A* and wants to go to point *B* (see Figure 4–25). He can drive over the sand at only 2 mi/hr. Once he reaches the road, he can drive 20 mi/hr. What route will minimize his time of travel?

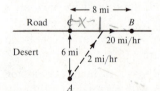

Road

8 mi

B

20 mi/hr

Desert

6 mi

2 mi/hr

A

FIGURE 4–25. Exercise 25.

26. A metal can is to be constructed so as to have volume *V*. The top and bottom are made from aluminum, which costs *a* cents/sq ft, but the sides can be made of tin, which costs *a*/4 cents/sq ft. Find the radius *r* and height *h* that yield the least cost for material. ($V = \pi r^2 h$ and the surface area $S = 2\pi rh + 2\pi r^2$.)

27. An aerosol can is to be made so that its volume is V. The top is in the shape of a cone and the bottom is a hemisphere (see Figure 4–26). The material for the top and bottom costs a cents and that of the sides costs $a/2$ cents/sq ft. What dimensions minimize the cost? (Surface area of a cone $= \pi rs$, surface of a hemisphere $= 2\pi r^2$.)

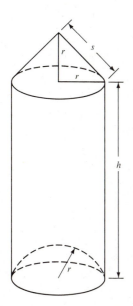

FIGURE 4–26. Exercise 27.

28. Find the point(s) on the graph of $y = x^2$ such that the distance to the point $(0, 3)$ is a minimum.

29. Find the point on the graph of $y = \sqrt{x}$ such that the distance to the point $(1, 0)$ is a minimum.

30. Show that if a function is differentiable for all x, then the shortest distance from (x_0, y_0) to points on the graph of $y = f(x)$ is along a line that is perpendicular to the graph (that is, perpendicular to the line tangent to the graph).

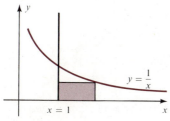

FIGURE 4–27. Exercise 31.

31. A rectangle is to have its base lying on the x-axis and be contained in the region between the x-axis, the line $x = 1$, and the graph of $y = 1/x$ (see Figure 4–27). Find the rectangle with the largest area. (There is no right end point.)

32. (Continuation of Exercise 31) If the rectangles are revolved about the x-axis, then cylinders are formed. Find the cylinder with the largest volume.

33. Find the cylinder of largest volume that can be inscribed in a sphere of radius 10.

34. A window is in the form of a rectangle surmounted by a semicircle. If the total perimeter is 14 feet, what dimensions yield the window that admits the most light?

35. Two hallways 3 and 4 feet wide meet at right angles (Figure 4–28A). What is the longest ladder that can be moved (parallel to the floor) around the corner?

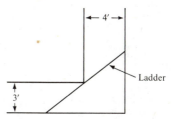

A. The hallway in Exercise 35.

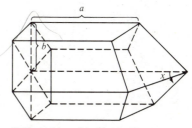

B. The beehive cell in Exercise 38.

FIGURE 4–28.

36. The stiffness of a rectangular beam varies jointly with the width w and the cube of the height (see Exercise 17). Find the dimensions of the stiffest beam that can be cut from a cylindrical log of radius 6 inches.

37. Among all triangles with a given base b and perimeter p, find the one with the greatest area.

38. In a bee hive, each cell is a regular hexagonal prism open at the front with a trihedral apex at the back. Examination of actual cells show that the measure of the apex angle is amazingly consistent; 54.7° ± .033°. It can be shown that the surface area S of the cell is a function of the apex angle x

$$S = bab + \frac{3}{2}b^2\left[\frac{\sqrt{3}}{\sin x} - \frac{1}{\tan x}\right], 0 < x < \pi/2$$

where a and b are as shown in Figure 4–28B. Do bees construct their cells so as to minimize the surface area?

Optional Exercises

39. Among all triangles with given area A, find the one with the least perimeter.

40. Among all triangles with a given perimeter p, find the one with the greatest area.

41. If h and b are positive, show that
$$2\sqrt{h^2 + (b/2)^2} < h + \sqrt{h^2 + b^2}$$

4–3 DIFFERENTIALS AND APPROXIMATIONS

Let $y = f(x)$ be differentiable at the point x. For a small change Δx in x, let Δy be the corresponding change in y,

$$\Delta y = f(x + \Delta x) - f(x)$$

Then the difference quotient is $\Delta y/\Delta x$ and

$$\lim_{\Delta x \to 0} \frac{\Delta y}{\Delta x} = f'(x)$$

It follows that if Δx is close to 0, then $\Delta y/\Delta x$ is close to $f'(x)$, or, put another way, Δy is close to $f'(x)\Delta x$. In symbols,

(4.10) $\Delta y \approx f'(x)\, \Delta x$

EXAMPLE 1 Compare Δy with $f'(x)\Delta x$ if $f(x) = x^2 + 2x - 1$, $x = 2$, and $\Delta x = .1$. Which quantity is easier to compute, Δy or $f'(x)\, \Delta x$?

Solution First find Δy; $f(x) = x^2 + 2x - 1$ so that

$$\Delta y = f(2.1) - f(2) = [2.1^2 + 2(2.1) - 1] - [2^2 + 2(2) - 1]$$
$$= 4.41 + 4.2 - 1 - 4 - 4 + 1$$
$$= .61$$

Now find $f'(x)\Delta x$; $f'(x) = 2x + 2$ and, therefore,

$$f'(2)\Delta x = (2 \cdot 2 + 2)(.1) = .6$$

Thus, Δy is very close to $f'(x)\Delta x$. Clearly, $f'(x)\Delta x$ is easier to compute than Δy. ■

The quantity $f'(x)\Delta x$ is called the **differential of** y and is denoted by the symbol dy. The number Δx, representing a small change in x, is called the **differential of** x and is denoted by dx. Thus,

(4.11) $\quad\boxed{dy = f'(x)\, dx}$

Choosing the symbols dx and dy for the differentials of x and y is no accident. It follows from (4.11) that the symbol dy/dx, which has been used to represent the function f', can now also be considered as a quotient of differentials.

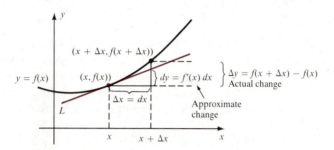

FIGURE 4–29. $\Delta y \approx dy$

Figure 4–29 illustrates a geometric comparison of Δy and dy. The figure suggests that Δy, the *actual* change in f, is approximated by dy, the change *along the tangent line L.*

Suppose f is differentiable at x, and $\Delta x = dx$ is a small change in x. Then

(4.12)

(i) $\Delta y = f(x + \Delta x) - f(x)$ is the actual change in f.

(ii) $dy = f'(x)\, dx$ is the change along the tangent line L.

(iii) $\Delta y \approx dy$

These observations are important because it is often easier to compute dy than Δy. (In Chapter 11, we develop a method for determining how good the approximation $\Delta y \approx dy$ really is.)

Differential approximations are especially useful in computing the consequences of errors in measurement. If the measurement of a quantity is M units with a possible error of E units, then the **relative error** is the quotient E/M. The relative error indicates the *average error per unit.* It can be expressed as a decimal or as a percentage.

EXAMPLE 2 ¹The diameter of a ball bearing is measured and found to be 4 cm. If the measurement is subject to a 1% error, estimate the possible error in the volume. What is the relative error in volume?

Solution   The radius r is 2 cm with a possible error of 1% or $\pm.02$ cm. The possible error ΔV in the volume can be approximated by the differential dV. Let $r = 2, dr = \pm.02$, and then

$$dV = V'(2)\, dr$$
$$= 4\pi 2^2(\pm .02)$$
$$= \pm .32\pi \text{ cu cm}$$

$$\left(V = \frac{4}{3}\pi r^3 \right)$$

The possible error in volume is approximately $.32\pi$ cu cm. Since $V(2) = 32\pi/3$ cu cm, the relative error in volume is

$$\frac{.32\pi}{32\pi/3} = .03 \text{ or } 3\% \qquad \blacksquare$$

The size of an object usually increases as its temperature rises. (One exception is water between 0 and 4°C; it actually contracts as the temperature increases.) Let L denote the length of an object, say, a long metal rod. Then L is a function of the temperature T. If L_0 is the length at temperature T_0, then the **linear thermal expansion coefficient at** T_0 is denoted by α and is defined by

$$(4.13) \qquad \alpha = \frac{1}{L_0} L'(T_0) \quad \text{or} \quad \alpha L_0 = L'(T_0)$$

If the coefficient is known for a certain substance, then the change in length brought about by a change in temperature can be estimated by differentials. Since $dL = L'(T_0)\, dT = \alpha L_0\, dT$, we have

---

(4.14)

The change in length ΔL corresponding to a change in temperature $\Delta T = dT$ is approximately

$$dL = \alpha L_0\, dT$$

where α is the coefficient of thermal expansion and L_0 is the length at T_0.

This is useful in engineering problems.

EXAMPLE 3 A 100 meter (m) span of a bridge is built of steel. The coefficient of linear expansion for steel is 11×10^{-6} per degree centigrade. How much will its length change from the coldest winter days of $-10°C$ to the warmest summer days of 40°C?

Note: The coefficients for various substances are found experimentally using temperatures near 20°C. However, for most materials, the error involved in using these same coefficients even at very high or low temperatures is negligible. That is why the coefficient for steel is given as 11×10^{-6} per degree.

Solution Use (4.14) with $\alpha = 11 \times 10^{-6}/°C$, $L_0 = 100$ m, and $dT = 40 - (-10) = 50°C$. Then,

$$dL = (11 \times 10^{-6}/°C)(100 \text{ m})(50°C)$$
$$= .055 \text{ m}$$

This is (approximately) the change in length of the bridge span. ∎

EXAMPLE 4 A pendulum clock keeps correct time at temperature T_0. How much time will it gain or lose per day if the temperature is changed by an amount dT?

Solution The period of a pendulum of length L feet is about

$$P = 2\pi\sqrt{L/g} \text{ seconds}$$

where g is a constant (about 32) due to the force of gravity. This is the time it takes for one oscillation of the pendulum. Thus, ΔP is the change in time per oscillation and

$$\frac{\Delta P}{P} = \text{change in time per second}$$

This is what we want to know.

Let L_0 be the length of the pendulum at temperature T_0, and use differentials to write

$$\frac{\Delta P}{P} \approx \frac{dP}{P} = \frac{P'(L_0)\,dL}{P(L_0)}$$

$$= \frac{\pi dL}{P(L_0)\sqrt{gL_0}} \qquad (P' = \pi/\sqrt{gL})$$

$$= \frac{dL}{2L_0} \qquad (P = 2\pi\sqrt{L/g})$$

$$= \frac{\alpha dT}{2} \quad \text{seconds/second} \qquad (4.14)$$

where α is the expansion coefficient of the pendulum.

Conclusion. If the temperature rises ($dT > 0$), then the pendulum becomes longer and the period increases; that is, it takes more time to complete one oscillation. Thus, time is *lost*. On the other hand, if the temperature decreases ($dT < 0$), then the opposite happens, and time is *gained*. In either case, since there are 86,400 seconds in a day, the accuracy of the clock will change by about 43,200 $\alpha\,dT$ sec/day. ∎

We conclude this section with an example of a problem that is slightly different from the others.

EXAMPLE 5 Approximate the value of $\sqrt[5]{31}$.

Solution A calculator read-out of $\sqrt[5]{31}$ is 1.98734, correct to six figures. But even without a calculator, it is easy to obtain a reasonably accurate estimate of $\sqrt[5]{31}$ using differentials. Notice that $\sqrt[5]{32} = 2$ is known. Therefore, if you let $f(x) = \sqrt[5]{x}$, let $x = 32$, and let $dx = -1$, then Δy, the actual difference between $\sqrt[5]{32} = 2$ and $\sqrt[5]{31}$, can be approximated by dy. Thus,

$$\Delta y \approx dy = f'(32)(-1)$$

$$= \frac{1}{5}(32)^{-4/5}(-1) \qquad\qquad (f(x) = x^{1/5})$$

$$= -\frac{1}{80} = -.0125$$

It follows that $\sqrt[5]{31} \approx 2 - .0125 = 1.9875$ which compares favorably with the calculator figure. ∎

EXERCISES

In Exercises 1–8, compute $\Delta y = f(x + \Delta x) - f(x)$, and compare it with $dy = f'(x)\,dx$.

1. $y = 2x^2 - x$; $x = 1, dx = .01$
2. $y = x^3 + 2x$; $x = -1, dx = .02$
3. $y = x^3 + 2x^2 + 1$; $x = 1, dx = .01$
4. $y = x^4$; $x = -3, dx = .02$
5. $y = 1/x$; $x = 3, dx = -.01$
6. $y = 1/x^2$; $x = -1, dx = .01$
7. $y = \sqrt{x}$; $x = 1.69, dx = -.25$
8. $y = 1/\sqrt{x}$; $x = 1.69, dx = -.25$

In Exercises 9–16, use differentials to approximate the given numbers, and check the results with a calculator.

9. $\sqrt[5]{33}$
10. $\sqrt[4]{80}$
11. $\sqrt{65} + \sqrt[6]{65}$
12. $(2.03)^4 - 4(2.03)^3$
13. $\sin 31°$ (convert to radians)
14. $\cos 59°$
15. $e^{.03}$
16. $e^{-.01}$

In Exercises 17–19, show that the stated approximations are valid. These approximations are important in physics, chemistry, and engineering.

17. $(1 + x)^n \approx 1 + nx$, where $|x|$ is small, and n is any real number.
18. $\sqrt{a^2 + x} \approx a + x/2a$, where $|x|$ is small and a is positive.
19. $\sqrt[3]{a^3 + x} \approx a + x/3a^2$, where $|x|$ is small and a is any nonzero number.

20. If the diameter of a sphere is found to be 30 cm with a possible error of 2%, estimate the possible error in computing the volume. What is the relative error in computing the volume?

21. According to Stefan's Law, the rate of emission R of radiant energy from the surface of an object is $R = kT^4$, where T is the absolute temperature and k is a constant. Estimate the change in R if there is a 1% change in T.

22. A metal rod is 24 inches long and 4 inches in diameter. The surface (except the ends) is to be covered with a .01 inch coating of nickel. Estimate how much nickel is needed.

23. A metal sphere has a radius of 2 inches. How much nickel is needed to plate the sphere with a .02-inch thickness of nickel? (Surface area of a sphere is $4\pi r^2$.)

24. Approximately how much paint is needed to paint a cube 2 ft on a side with a .02-inch thickness of paint?

25. The **volume thermal expansion cofficient** at T_0 is denoted by β and defined by

$$\beta = \frac{1}{V_0} V'(T_0)$$

where V_0 is the volume at T_0. One liter ($= 1000$ cu cm) of benzene is heated from $0°C$ to $30°C$. Estimate its new volume if the coefficient β for benzene is $1240 \times 10^{-6}/°C$. (*Hint:* $dV = \beta V_0 dT$, which is similar to (4.14) for length.)

26. An aluminum sheet has a hole in it that is 2 cm in diameter. As the aluminum is heated from $20°C$ to $100°C$, will the hole increase or decrease in diameter? By how much? (The linear expansion coefficient for aluminum is $24 \times 10^{-6}/°C$.)

27. A plate glass window 6 ft by 6 ft (measured at $-15°C$) is to be used as part of a store front. The temperature during the year ranges from $-15°C$

to 40°C. In making the window frame, how much room for expansion should be allowed on either side and at the top? (The coefficient α for plate glass is $9 \times 10^{-6}/°C$.)

28. One problem with approximations by differentials is that (4.12) does not provide information as to how small dx should be for a given tolerance of error in dy. As an example, let $f(x) = x^{100}$ and show that $f(1.08) \approx 9$ using differentials. Now, using a logarithm table or a calculator, verify that $f(1.08)$ is actually over 2,000.

29. Notwithstanding Exercise 28, it is true that $|\Delta y - dy|$ is small in comparison to Δx. What is meant by that statement is that

$$\lim_{\Delta x \to 0} \frac{\Delta y - dy}{\Delta x} = 0$$

Can you prove that this equation is true? *Hint:* Write Δy and dy in terms of $f(x)$ and dx.

4–4 IMPLICIT DIFFERENTIATION

The equation

$$y = x^2 - 3x + 4$$

defines y as a function of x. Given any value of x, the instructions for finding the corresponding value of y are explicitly stated; square x, subtract three times x, and add four. The equation

$$y\sqrt{x} - 9x = 7 - x^2 y$$

also defines y as a function of x. But this time the function is not altogether obvious; the instructions are not stated explicitly. In this case, we say that the equation *implicitly* defines y as a function of x. The equation can be solved for y

$$y = \frac{9x + 7}{\sqrt{x} + x^2}$$

to yield an explicitly defined function with domain $x > 0$.

An example from chemistry is the ideal gas law

(4.15) $PV = nRT$

where P is the pressure, V the volume, n the number of moles of gas, R a constant (≈ 8.31), and T the absolute temperature. Equation (4.15) *implicitly* defines each of the variables as a function of the others. Suppose that n and T are constant, and we want to know the rate of change of pressure P with respect to volume V. We can easily solve this equation for P and find dP/dV.

But it is often the case that an equation cannot easily be solved for one variable in terms of the other(s). For example, it is not easy to solve

(4.16) $y^5 - 4y^2 = x^4 + x^2$

for y in terms of x.

Suppose that x and y are related by (4.16) and that we need to know the rate of change of y with respect to x. If we assume that (4.16) implicitly defines y as a *differentiable* function of x, we can differentiate both sides of the equation with respect to x. BUT, we must always use the chain rule on terms that involve y. For instance, if y is assumed to be a differentiable function of x, then

$$\frac{d}{dx} y^5 = 5y^4 \frac{dy}{dx}$$

or, in the prime notation,

$$(y^5)' = 5y^4 y'$$

EXAMPLE 1 Assume that (4.16) implicitly defines y as a differentiable function of x. Find dy/dx.

Solution We differentiate both sides of (4.16) with respect to x, remembering to use the chain rule on terms involving y.

$$\frac{d}{dx}(y^5 - 4y^2) = \frac{d}{dx}(x^4 + x^2)$$

$$5y^4 \frac{dy}{dx} - 8y \frac{dy}{dx} = 4x^3 + 2x$$

Now, we simply solve the last equation for dy/dx

$$\frac{dy}{dx} = \frac{4x^3 + 2x}{5y^4 - 8y} \qquad \blacksquare$$

The method used in Example 1 to find dy/dx is called **implicit differentiation.** It can be used whenever an equation implicitly defines a differentiable function. For the remainder of this section, we will assume that this is the case; that is, all equations discussed below are assumed to implicitly define each of the variables as a differentiable function of the other variables.

EXAMPLE 2 Find an equation of the line tangent to the graph of $x^2 y^2 - 3x + 2y^3 = 0$ at the point $(1, 1)$.

Solution The slope of the tangent line is y'; we will use implicit differentiation to find y' when $x = y = 1$.

CAUTION: The term $x^2 y^2$ must be treated as a product of functions.

$$(x^2 y^2)' = x^2(2yy') + y^2(2x) \qquad \text{(Product)}$$

Thus, beginning with the original equation, we have

$$(x^2 y^2 - 3x + 2y^3)' = (0)'$$

$$(2x^2 yy' + 2xy^2) - 3 + 6y^2 y' = 0$$

Solving for y' yields

$$y' = \frac{3 - 2xy^2}{2x^2 y + 6y^2}$$

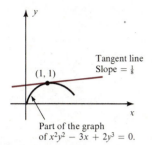

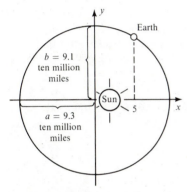

FIGURE 4–30. Example 2.

FIGURE 4–31. Example 3. The orbit of the earth.

Substitute $x = y = 1$ to obtain $y' = \frac{1}{8}$; this is the slope of the tangent line at $(1, 1)$. Therefore, an equation of the tangent line is

$$y - 1 = \frac{1}{8}(x - 1) \quad \text{or} \quad y = \frac{1}{8}x + \frac{7}{8} \quad \text{(Figure 4–30)} \quad \blacksquare$$

Kepler's first law of planetary motion asserts that planets move in elliptical orbits about the sun. An equation for an ellipse is

$$(4.17) \qquad \frac{x^2}{a^2} + \frac{y^2}{b^2} = 1 \quad a \text{ and } b \text{ positive constants}$$

We can apply this to the motion of our earth.

EXAMPLE 3 In a coordinate plane with units in tens of millions of miles, locate the sun at $(1.9, 0)$. With $a = 9.3$ and $b = 9.1$, equation (4.17) describes the orbit of the earth (Figure 4–31). How fast is y changing with respect to x when $x = 5$ and $y > 0$?

Solution To find dy/dx, you can differentiate both sides of (4.17) with respect to x. BUT, once again, be sure to use the chain rule when you differentiate y^2/b^2.

$$\frac{d}{dx}\left(\frac{y^2}{b^2}\right) = \frac{2y}{b^2}\frac{dy}{dx}$$

With that in mind,

$$\frac{d}{dx}\left[\frac{x^2}{a^2} + \frac{y^2}{b^2}\right] = \frac{d}{dx}(1)$$

$$\frac{d}{dx}\left(\frac{x^2}{a^2}\right) + \frac{d}{dx}\left(\frac{y^2}{b^2}\right) = 0$$

$$\frac{2x}{a^2} + \frac{2y}{b^2}\frac{dy}{dx} = 0$$

$$\frac{dy}{dx} = -\frac{b^2 x}{a^2 y} \quad y \neq 0$$

To find the value of dy/dx when $x = 5$ and $y > 0$, set

$$\frac{5^2}{a^2} + \frac{y^2}{b^2} = 1$$

and solve for y.

$$y = b\sqrt{1 - (5/a)^2}$$

Therefore,

$$\frac{dy}{dx} = -\frac{b^2 x}{a^2 y}$$

$$= -\frac{b^2(5)}{a^2 b \sqrt{1 - (5/a)^2}}$$

$$= -\frac{(9.1)(5)}{(9.3)^2 \sqrt{1 - (5/9.3)^2}} \approx -.6329 \qquad \blacksquare$$

Implicit differentiation can be used even when a variable is missing from the equation. For instance, as the earth moves along its orbit

$$\frac{x^2}{a^2} + \frac{y^2}{b^2} = 1$$

its coordinates, x and y, are both (differentiable) functions of time t. Therefore, it is permissible to differentiate with respect to t *even though t does not appear in the equation!*

EXAMPLE 4 Suppose it is known that when the coordinates (in Figure 4–31) of the earth are (x, y) with $y \neq 0$, then x is changing at the rate of S mph. How fast is y changing at that moment?

Solution Since x and y are assumed to be functions of t, the chain rule must be used on both.

$$\frac{d}{dt}\left[\frac{x^2}{a^2} + \frac{y^2}{b^2}\right] = \frac{d}{dt}(1)$$

$$\frac{2x}{a^2}\frac{dx}{dt} + \frac{2y}{b^2}\frac{dy}{dt} = 0$$

$$\frac{dy}{dt} = -\frac{b^2 x}{a^2 y}\frac{dx}{dt}$$

$$= -\frac{b^2 S x}{a^2 y} \quad \text{mph} \qquad \left(\frac{dx}{dt} = S\right) \qquad \blacksquare$$

The question might arise as to whether a derivative arrived at by implicit differentiation would agree with one derived in the usual way. The answer is *yes*. Here is an example.

EXAMPLE 5 Suppose $x^2 + 3x + y = 4 + 2y$. Find y' by (a) implicit differentiation and (b) solving for y.

Solution

(a) By implicit differentiation:

$$(x^2 + 3x + y)' = (4 + 2y)'$$

$$2x + 3 + y' = 2y'$$

Therefore,

$$y' = 2x + 3$$

(b) Solving for y yields $y = -4 + x^2 + 3x$ and

$$y' = 2x + 3 \qquad \blacksquare$$

Here is another question that might arise. Suppose that an equation defines y as more than one function of x. For example, the equation $y^2 = x$ defines at least two functions of x

$$y = \sqrt{x} \quad \text{and} \quad y = -\sqrt{x} \quad x > 0*$$

The question is: if dy/dx is found by implicit differentiation, for which of these functions does it hold? The surprising answer is: *all of them.*

EXAMPLE 6 The equation $x^2 + y^2 = 1$ defines (at least) two functions

$$y = \sqrt{1 - x^2} \quad \text{and} \quad y = -\sqrt{1 - x^2} \quad -1 \le x \le 1$$

Show that the formula for y' obtained by implicit differentiation holds for both of these functions in the open interval $(-1, 1)$.

Solution We find y' by implicit differentiation first and then by using the equations above.

$$(x^2 + y^2)' = (1)'$$
$$2x + 2yy' = 0$$
$$y' = -\frac{x}{y}$$

If $y = \sqrt{1 - x^2}$, then

$$y' = \frac{-2x}{2\sqrt{1 - x^2}} = -\frac{x}{y}$$

If $y = -\sqrt{1 - x^2}$, then

$$y' = \frac{2x}{2\sqrt{1 - x^2}} = \frac{x}{-y} = -\frac{x}{y}$$

These formulas hold for all x with $-1 < x < 1$ (Figure 4–32). $\blacksquare$

The next example illustrates how to find second derivatives by implicit differentiation.

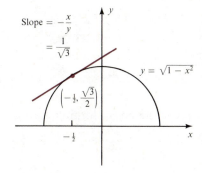

Slope $= -\dfrac{x}{y}$

$= \dfrac{1}{\sqrt{3}}$

$y = \sqrt{1 - x^2}$

$\left(-\frac{1}{2}, \frac{\sqrt{3}}{2}\right)$

$-\frac{1}{2}$

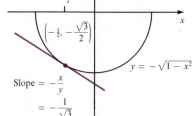

$-\frac{1}{2}$

$\left(-\frac{1}{2}, -\frac{\sqrt{3}}{2}\right)$

$y = -\sqrt{1 - x^2}$

Slope $= -\dfrac{x}{y}$

$= -\dfrac{1}{\sqrt{3}}$

FIGURE 4–32. Example 6. The same formula $-x/y$ yields the derivative of both functions.

_____

*The equation $y^2 = x$ actually defines infinitely many functions. The function

$$y = \begin{cases} \sqrt{x} & \text{for } 0 \le x \le p \\ -\sqrt{x} & \text{for } p < x \end{cases}$$

is different for each value of $p > 0$.

EXAMPLE 7 Let $s^2 - s + e^t = \sin t$. Find the second derivative s'' with respect to t.

Solution

$$(s^2 - s + e^t)' = (\sin t)'$$
$$2\,ss' - s' + e^t = \cos t$$
$$s' = \frac{\cos t - e^t}{2s - 1}$$

Now differentiate implicitly again, using the quotient rule.

$$s'' = \left[\frac{\cos t - e^t}{2s - 1}\right]'$$
$$= \frac{(2s - 1)(-\sin t - e^t) - (\cos t - e^t)(2s')}{(2s - 1)^2}$$

Substituting $(\cos t - e^t)/(2s - 1)$ for s' yields

$$s'' = \frac{(2s - 1)^2(-\sin t - e^t) - 2(\cos t - e^t)^2}{(2s - 1)^3} \qquad \blacksquare$$

Implicit differentiation can also be used to prove the power formula for rational powers

$$(x^{m/n})' = \frac{m}{n}x^{m/n - 1}$$

We set $y = x^{m/n}$; then $y^n = x^m$. Assuming that this equation defines y as a differentiable function of x, we have

$$(y^n)' = (x^m)'$$
$$ny^{n-1}y' = mx^{m-1}$$
$$y' = \frac{m}{n}x^{m-1}y^{1-n}$$
$$= \frac{m}{n}x^{m-1}(x^{m/n})^{1-n}$$
$$= \frac{m}{n}x^{m/n - 1}$$

which completes the proof.

EXERCISES

In Exercises 1–14, find dy/dx.

1. $y = 3x^2 + \sqrt{y}$

2. $y + x^3 = x - \sqrt[3]{y}$

3. $x = \sqrt{y + \sqrt{y}}$

4. $y = \sqrt{x + \sqrt{x}}$

5. $y^2 + xy + x^2 = 5$

6. $x^3 + y^3 = xy$

7. $\dfrac{xy}{x + y} = 1$

8. $\dfrac{x}{y} + \dfrac{y}{x} + x^2y^2 = 1$

9. $y = \sin xy$ **10.** $x = e^{xy}$

11. $y = \cos(x + y)$ **12.** $x = \cos xy$

13. $\sqrt{x} + \sqrt{y} = \sqrt{a}$ **14.** $x^{2/3} + y^{2/3} = a^{2/3}$

In Exercises 15–18, find d^2y/dx^2.

15. $x^3 + y^3 = 1$ **16.** $x^4 + y^4 = 1$

17. $\dfrac{xy}{x + y} = 1$ **18.** $x^2y^2 = 1$

In Exercises 19–24, find an equation of the line tangent to the graph of the given equation at the given point. Sketch a portion of the graph near the point.

19. $\dfrac{x^2}{16} + \dfrac{y^2}{9} = 1; \left(2, \dfrac{3\sqrt{3}}{2}\right)$ **20.** $xy = 5; (-1, -5)$

21. $x^3 + y^3 - 9xy = 0; (4, 2)$

22. $\dfrac{1}{x} + \dfrac{1}{y} = 2; (1, 1)$

23. $\left(\dfrac{x}{a}\right)^n + \left(\dfrac{y}{b}\right)^n = 2; (a, b)$ **24.** $y^2 = 20x; (5, 10)$

25. Find all points on the graph of $x^2 + y^2 = 1$ where the tangent line is parallel to the line $2x - y + 1 = 0$.

26. Let $P(x, y)$ be any point on the circle $x^2 + y^2 = 1$. Prove that the radius and the tangent line through P are perpendicular.

Optional Exercises

Exercises 32–34 refer to any triangle with angles labeled A, B, and C and the sides opposite those angles labeled a, b, and c.

32. Show that the area of a triangle is $\frac{1}{2}$ of the product of any two sides and the sine of the included angle; say, Area $= \frac{1}{2}ac \sin B$.

27. The equation $x^2 + y^2 = 1$ defines the function

$$y = \begin{cases} \sqrt{1 - x^2} & \text{if } -1 \le x < -\dfrac{1}{2} \\ -\sqrt{1 - x^2} & \text{if } -\dfrac{1}{2} \le x \le \dfrac{1}{2} \\ \sqrt{1 - x^2} & \text{if } \dfrac{1}{2} < x \le 1 \end{cases}$$

Find the value of y' at the points $(-1/\sqrt{2}, 1/\sqrt{2})$, $(0, -1)$, and $(1/\sqrt{2}, 1/\sqrt{2})$ first by implicit differentiation, and then using the function defined above.

28. Find all points on the graph of $4x^2 - 9y^2 = 36$ where the tangent line is perpendicular to the line $2y + 5x = 10$.

29. A particle moves on the circle $x^2 + y^2 = 1$ and its coordinates are functions of time. When it is at $(-1/2, -\sqrt{3}/2)$, the x-coordinate is changing at the rate of 3 inches/sec. How fast is y changing?

30. (Continuation of Exercise 29) At what rate is the distance between the particle and the point $(1, 0)$ changing?

31. A particle is moving along the curve $4x^2 - 9y^2 = 36$. When it is at $(3, 0)$, the y coordinate is changing at the rate of -4 ft/sec. How fast is x changing?

33. If a, c, and B are all functions of time t, what is the rate of change of the area with respect to time when $a = 2$, $c = 3$, $B = \pi/4$, $da/dt = -1$, $dc/dt = 2$, and $dB/dt = -2$?

34. Let b be the base and h_b be the corresponding height of the triangle. Show that $db/dB = h_b$.

4–5   RELATED RATES

The equation

$$V = \frac{4}{3}\pi r^3$$

defines the volume of a sphere as a function of its radius. The rate of change of V with respect to r is *related* to r by the equation

$$\frac{dV}{dr} = 4\pi r^2$$

Suppose that r, in turn, is a function of another variable, t. Now the rate of change of V with respect to t is *related* to the rate of change of r by the chain rule

$$\frac{dV}{dt} = \frac{dV}{dr} \cdot \frac{dr}{dt}$$

Problems involving relationships between rates of change are called **related rate** problems.

EXAMPLE 1   One way to make coffee is to pour boiling water over coffee grounds that are at the bottom of a cone made of filter paper (Figure 4–33). The cone is 6 inches deep and has a radius of 3 inches. If coffee filters through at the constant rate of $\frac{1}{8}$ cup (≈ 1.8 cubic inches) per second, how fast is the water level in the filter falling when its depth is 4 inches?

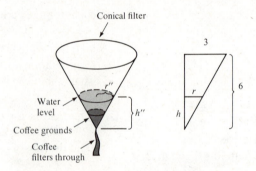

FIGURE 4–33. Example 1.

Solution   Referring to Figure 4–33, the question can be rephrased in terms of derivatives:

What is $\dfrac{dh}{dt}$ when $h = 4$ inches?

By the chain rule,

$$(4.18) \qquad \frac{dV}{dt} = \frac{dV}{dh} \cdot \frac{dh}{dt}$$

where V is the volume of water in the filter. It is given that this volume is decreasing at the rate of 1.8 cu in/sec; that is, $dV/dt = -1.8$. If dV/dh is known at $h = 4$, then (4.18) can be solved for dh/dt. Therefore, the problem reduces to expressing $V = \frac{1}{3}\pi r^2 h$ as a function of h alone and computing dV/dh.

To eliminate r from the volume formula, use similar triangles (Figure 4–33) to write $r = h/2$. Thus,

$$V = \frac{1}{3}\pi \left(\frac{h}{2}\right)^2 h = \frac{1}{12}\pi h^3$$

$$\frac{dV}{dh} = \frac{1}{4}\pi h^2$$

and $dV/dh = 4\pi$ when $h = 4$. Now, by (4.18),

$$-1.8 = (4\pi)\frac{dh}{dt}$$

$$\frac{dh}{dt} = -\frac{1.8}{4\pi} \approx -.14 \text{ in/sec}$$

That is how fast the water level is falling. ∎

EXAMPLE 2 Figure 4–34 shows a ladder 13 ft long leaning against a wall. The bottom of the ladder slips and is moving away from the wall at a rate of 2 ft/sec. How fast is the top of the ladder falling when the bottom is 5 ft from the wall?

Solution (similar to the method used in Example 1). Using the variables in the figure, it is given that $dx/dt = 2$ ft/sec and you are asked to find dy/dt when $x = 5$ ft. The chain rule formula to use is

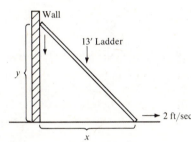

(4.19) $\dfrac{dy}{dt} = \dfrac{dy}{dx} \cdot \dfrac{dx}{dt}$

FIGURE 4–34. Example 2.

The relationship between x and y is

$$x^2 + y^2 = 13^2 \quad \text{or} \quad y = \pm\sqrt{169 - x^2}$$

but only the positive square root is meaningful in this problem. Therefore,

$$\frac{dy}{dx} = \frac{-x}{\sqrt{169 - x^2}} \quad \text{and} \quad \frac{dy}{dx}(5) = \frac{-5}{12}$$

It follows from (4.19) that $dy/dt = -\frac{5}{12} \cdot 2 = -\frac{5}{6}$ ft/sec.

Alternate solution Use the relationship

(4.20) $x^2 + y^2 = 13^2 = 169$

and the fact that x and y are each functions of t to differentiate both sides of (4.20) with respect to t. This is implicit differentiation and you must remember to use the chain rule. Thus,

(4.21) $2x\dfrac{dx}{dt} + 2y\dfrac{dy}{dt} = \dfrac{d}{dt}(169) = 0$

You are given that $dx/dt = 2$ and $x = 5$; when $x = 5$, equation (4.20) yields $y = 12$. Therefore, (4.21) becomes

$$2(5)(2) + 2(12)\frac{dy}{dt} = 0$$

from which it follows that $dy/dt = -\frac{5}{6}$ ft/sec. ∎

In Example 2, the alternate method of differentiating with respect to t seems easier than the first solution because you do not have to solve for y in terms of x and then find dy/dx. The alternate method could also have been used in Example 1: If $V = \frac{1}{3}\pi r^2 h$ is differentiated with respect to t, then

$$\frac{dV}{dt} = \frac{1}{3}\pi\left[r^2\frac{dh}{dt} + h\left(2r\frac{dr}{dt}\right)\right]$$ (Product)

Now you would have to evaluate r and dr/dt when $h = 4$ and solve for dh/dt. This could be accomplished using the relationship $r = h/2$, but in this case, the first method seems easier than the alternate method.

Step-by-Step Method

Solving a related rate problem consists of several steps as illustrated in Examples 1 and 2 above.

(1) Analyze the problem (identify the variable whose rate you are asked to find; identify the variables whose rates are given).

(2) Draw a diagram (if appropriate).

(3) Write an equation relating the variables. If possible, use the given information to eliminate unwanted variables.

(4) Write the desired quantity as a function of the other variable(s) and differentiate (Example 1) *or* use implicit differentiation (alternate solution in Example 2).

EXAMPLE 3   Two lines meet at point A and form an angle of $30°$ (Figure 4–35). A particle is moving along one line towards A (negative direction) with a speed of 16 ft/sec. A second particle is moving along the other line away from A (positive direction) with a speed of 12 ft/sec. At a certain moment their distances from A are 10 ft and 3 ft, respectively. How fast is the distance between them changing at that moment?

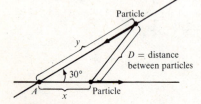

FIGURE 4–35. Example 3.

Solution   Referring to Figure 4–35, use the Law of Cosines to write

(4.22) $D^2 = x^2 + y^2 - 2xy\cos 30°$
$$= x^2 + y^2 - \sqrt{3}xy$$ $(\cos 30° = \sqrt{3}/2)$

and differentiate both sides with respect to t (this is easier than solving for x or y).

$$2D\frac{dD}{dt} = 2x\frac{dx}{dt} + 2y\frac{dy}{dt} - \sqrt{3}\left[x\frac{dy}{dt} + y\frac{dx}{dt}\right]$$

Now simply substitute the given information $x = 3$, $y = 10$, $dx/dt = 12$, and $dy/dt = -16$ to obtain

$$2D\frac{dD}{dt} = 2(3)(12) + 2(10)(-16) - \sqrt{3}[3(-16) + 10(12)]$$

$$\approx -373$$

Now use (4.22) to find the value of D at that moment.

$$D = \sqrt{x^2 + y^2 - \sqrt{3}xy}$$
$$= \sqrt{3^2 + 10^2 - \sqrt{3}(3)(10)} \approx 7.55$$

Therefore,

$$\frac{dD}{dt} \approx \frac{-373}{2(7.55)} \approx -24.7 \text{ ft/sec}$$

Conclusion. The distance between the particles is decreasing at the rate of about 25 ft/sec. ■

EXERCISES

1. You are fishing and, if your luck is anything like mine, hook into a floating log. If the tip of your rod is 6 ft above the water and you reel in the line at the rate of 2 ft/sec, how fast is the log moving towards you when there is still 10 ft of line out?

2. A balloon is to be blown up so that its radius increases at the rate of 10 in/sec. When the radius is 2 inches, at what rate must air be pumped in?

3. Motorist A is 20 mi east of an intersection and heading towards it at 50 mi/hr. Motorist B is 15 mi north of the same intersection and heading away from it at 60 mi/hr. Are the cars approaching each other or moving apart? How fast?

4. Two particles start from the origin. The first moves along the x-axis and its position at time t is $x(t) = 3t - t^2$. The second moves along the y-axis and its position at time t is $y(t) = \frac{1}{3}t^3 - 2t^2 + 3t$. How fast are they moving apart or together after 1 second? 2 seconds?

5. A balloon is being filled with air in such a way that the radius r at any time t (minutes) is $r(t) = 3\sqrt[3]{t}$ ft. How fast is air being pumped in?

6. A girl flying a kite lets out 3 ft of string per second. The kite is moving horizontally at an altitude of 100 ft. Assuming there is no sag in the string, how fast is the kite moving when there is 150 ft of string out?

7. A thin circular metal disc is being heated and its diameter is increasing at the rate of .0001 inch/sec. How fast is the surface area (of one side) changing when its area is 4π square inches?

8. Sand is pouring from a steam shovel at the rate of 10 cu ft/min. It forms a conical pile whose vertical cross section is always an equilateral triangle. How fast is the height of the pile increasing when the height is 3 ft?

9. If the sand of Exercise 8 forms a conical pile whose height is always the same as its radius and the height is increasing at the rate of 3 ft/min, how much sand is being poured when the height is 2 ft?

10. The ideal gas law $PV = nRT$ relates the pressure P, the volume V, the number of moles of gas n, the absolute temperature T, and the constant R (≈ 8.31). Suppose a fixed amount of a certain gas (that is, n is constant) is heated so that its temperature increases at the rate of $5°$ per minute. If the pressure decreases at the rate of 2 atmospheres per minute, how fast is the volume increasing (or decreasing) when P = 10 atmospheres and V is 3 cubic centimeters?

11. If the amount of gas and the temperature are held constant, then the ideal gas law (Exercise 10) takes the form $PV = c$, where c is a constant. This is known as Boyle's Law. If the volume is 50 cubic inches and the pressure is 2 atmospheres (≈ 30 pounds per square inch) and the volume is decreasing at the rate of 3 cubic inches per minute, is the pressure increasing or decreasing? How fast?

12. A car traveling east speeds past an intersection. An officer in a patrol car is 150 ft south of the intersection. The officer takes a radar reading and finds that the speeder is 250 ft from his car and that this distance is increasing at the rate of 80 ft/sec. How fast is the speeder going?

13. Part of a roller coaster has the shape of the curve $y = 25x \sin (\pi/2)x$ for $0 \le x \le 4$. Sketch the curve. When the coaster car is at $(2, 0)$, it is moving to the right at a speed of 2 ft/sec. How rapidly is it falling?

14. A particle moves along the graph of $y = x^3 - 2x + 1$. If the x-coordinate is increasing at the rate of 3 units per second, is y increasing or decreasing at $(2, 5)$? How fast?

15. The ends of a 10 ft long water trough are in the shape of an isosceles triangle (with its vertex at the bottom). The trough is 4 ft across at the top and its deepest part is 2 ft. If the trough is being filled with water at the rate of 5 cu ft/min, how fast is the water rising after 2 minutes?

16. A man 6 ft tall walks 5 ft/sec away from the base of a lamp post 12 feet high. What does your intuition say about the rate at which the tip of his shadow is moving? Does it move faster as the man gets farther away from the light? Check.

17. Referring to the man in Exercise 16, at what rate is the length of his shadow increasing?

18. A light is at the top of a 48 ft pole. A rock is dropped from the same height but 10 ft away from the pole. Its height at time t is $h = 48 - 16t^2$. How fast is the shadow of the rock moving along the ground after 1 second? After 2 seconds?

19. The illumination I from a light source varies directly with the strength S of the source and inversely with the square of the distance s from the source (that is, $I = kS/s^2$ for some constant k). If, for a certain light, I is 100 units at a distance of 4 ft and the light is moving away at 5 ft/sec, how fast is I changing?

20. The force (measured in dynes) of repulsion of two like charges q_1 and q_2 (measured in statcoulombs) is q_1q_2/r^2 where r is the distance (measured in centimeters) between them. If charges of 4 and 7 statcoulombs are moving apart at the rate of 8 cm/sec, how fast is the force changing when $r = 2$ centimeters?

21. A spherical piece of ice is melting so that its radius is decreasing at the rate of .1 millimeters/sec. How fast is the volume decreasing when the radius is 1 centimeter? (1 millimeter = .1 centimeter = .001 meter.)

22. Suppose the ice in Exercise 21 is suspended above a conical paper cup measuring 4 cm across the top and 3 cm deep and as the ice melts, the water drips into the cup. When the radius of the ice is 1 cm, the height of the water in the cup is 2 cm. If you assume a continual flow of water, how fast is the water in the cup rising?

23. If the ice in Exercises 21 and 22 was floating in the water in the cup (instead of being suspended above it), how fast is the water rising when the radius of the ice is 1 cm and the height of the water is 2 cm?

24. The flight paths of a small plane and a jetliner are lines that intersect at A forming an angle of 60°. Both are flying towards A at constant speeds of 100 mph and 500 mph, respectively. At a given moment, both planes are 10 miles from A. How fast is the distance between them decreasing at that moment?

25. A lighthouse is situated 2 mi from a straight beach. The light makes 3 revolutions per minute counterclockwise. How fast is the light beam moving along the beach when it is 4 miles from the point A on the beach directly opposite the lighthouse?

26. A conical tank (vertex down) is 20 ft deep and 20 ft across the top. If water is pouring in at the rate of 3 cu ft/min, how fast is the water rising when it reaches a height of 2 ft?

27. Turn the conical tank of Exercise 26 upside down (vertex up). Given the same data, answer the same question.

Optional Exercise

28. An east-west road intersects a north-south road at point A. Motorist B is 10 mi east of A and traveling east at 40 mi/hr. Motorist C is 8 mi north of A and traveling south at 60 mi/hr. If they continue to drive at the same speeds, how far will each be from A when the distance between them is a minimum?

4–6   CONCAVITY AND GRAPHING

Let $C = C(t)$ be the cost-of-living index at time t. *Inflation* means that C is increasing, and the *rate of inflation* is the rate at which C is increasing. If the government announces that the rate of inflation is decreasing, does that mean that the cost of living is going down? Let us investigate this and similar questions in terms of functions and derivatives.

If f is increasing on an interval I, then $f(x_1) < f(x_2)$ whenever $x_1 < x_2$ are points in I, so the graph of f is rising. A similar type of statement holds if f is decreasing on I, in this case, the graph of f is falling. If f' is increasing on I, then

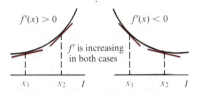

$f'(x) > 0$ $f'(x) < 0$

f' is increasing in both cases

x_1 x_2 I x_1 x_2 I

FIGURE 4–36. Concave up. In the figure on the left, $f'(x_2)$ is more positive than $f'(x_1)$; in the one on the right, $f'(x_2)$ is less negative than $f'(x_1)$. In either case, the curve bends upward.

$$(4.23) \qquad f'(x_1) < f'(x_2)$$

whenever $x_1 < x_2$ are points in I. Whether f itself is increasing or decreasing on I, inequality (4.23) forces the curve to bend upwards (Figure 4–36). Such a curve is said to be *concave up*. If f' is decreasing on I, the opposite is true; the curve bends downward and is said to be *concave down* (Figure 4–37).

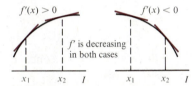

$f'(x) > 0$ $f'(x) < 0$

f' is decreasing in both cases

x_1 x_2 I x_1 x_2 I

FIGURE 4–37. Concave down. In the figure on the left, $f'(x_2)$ is less positive than $f'(x_1)$; in the one on the right, $f'(x_2)$ is more negative than $f'(x_1)$. In either case, the curve bends downward.

(4.24)

> **DEFINITION**
> The graph of f on an interval I is
> (1) **concave up** if f' is increasing on I.
> (2) **concave down** if f' is decreasing on I.

Of particular importance is the point where a curve changes from concave up to concave down or vice versa (Figure 4–38).

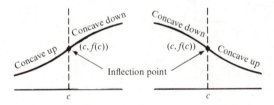

Concave down   Concave down

Concave up $(c, f(c))$ $(c, f(c))$ Concave up

Inflection point

c c

FIGURE 4–38. Inflection points $(c, f(c))$.

(4.23)

> **DEFINITION**
> A point $(c, f(c))$ on the graph of f is called an **inflection point** if the graph of f is concave one way on an interval (a, c) and concave the opposite way on an interval (c, b).

If f is differentiable on an interval I, then f is increasing or decreasing on I depending on whether $f'(x)$ is positive or negative. If f' is differentiable on I, then f' is increasing or decreasing on I depending on whether $f''(x)$ is positive or negative. It follows that the sign of f'' determines the concavity of the graph of f.

> **CONCAVITY TEST**
> If f is twice differentiable on an interval I, then the graph of f
>
> (4.26)
> (i) is concave up if $f''(x) > 0$ on I.
> (ii) is concave down if $f''(x) < 0$ on I.
> (iii) has an inflection point $(c, f(c))$ if $f''(x)$ has opposite signs on either side of c.

EXAMPLE 1 Sketch the graph of $y = x^2 - 4x + 2$ and indicate all important information.

Solution

$$y' = 2x - 4 \begin{cases} > 0 & \text{on } (2, \infty) \\ = 0 & \text{at } 2 \\ < 0 & \text{on } (-\infty, 2) \end{cases}$$

Thus, y is decreasing on $(-\infty, 2)$ and increasing on $(2, \infty)$; it follows that $(2, -2)$ is an absolute minimum. Since $y'' = 2$ is always positive, the curve is concave up, and there are no inflection points. See Figure 4–39. ∎

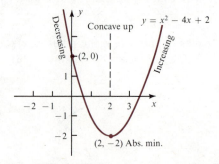

FIGURE 4–39. Example 1.

EXAMPLE 2 Sketch the graph of $y = -x^3 + 2x^2 + x - 1$ and indicate all important information.

Solution $y' = -3x^2 + 4x + 1$ and $y'' = -6x + 4$. The zeros of y' are

$$\frac{-2 \pm \sqrt{7}}{-3} \approx -.2 \text{ and } 1.5 \qquad \text{(Quadratic formula)}$$

We test the values of y' on either side of $-.2$ and 1.5 to determine where y' is positive and negative, and find that

$$y'(x) \begin{cases} > 0 & \text{on } (-.2, 1.5) \\ = 0 & \text{at } -.2 \text{ and } 1.5 \\ < 0 & \text{on } (-\infty, -.2) \text{ or } (1.5, \infty) \end{cases}$$

This indicates where y is increasing and decreasing, and the possible location of local extrema.

$$y'' = -6x + 4 \begin{cases} > 0 & \text{on } (-\infty, 2/3) \\ = 0 & \text{at } 2/3 \\ < 0 & \text{on } (2/3, \infty) \end{cases}$$

This indicates the concavity of the graph and that $(2/3, y(2/3)) \approx (.67, .26)$ is an inflection point. See Figure 4–40. ■

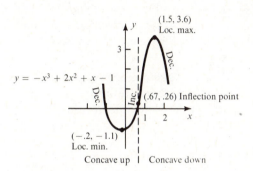

FIGURE 4–40. Example 2.

EXAMPLE 3 *(Inflation)* A newly elected president makes the following statement: "For the past four years, the rate of inflation has been increasing. But now there is a turnaround, and the rate of inflation is decreasing." Let us analyze this statement. Let $C = C(t)$ be the cost-of-living index at time t. Inflation means that $C'(t) > 0$; that is, C is increasing. The rate of inflation is increasing or decreasing depending on the sign of $C''(t)$. The president was careful to talk about the *rate* of inflation. To say that the rate of inflation was increasing before means that the curve was concave up. It is possible (though unlikely) that C itself was decreasing during that time! To say that the rate of inflation is now decreasing simply means that the curve is concave down. It is possible (and even likely) that C is rising and will continue to do so. The "turnaround" he spoke of is an inflection point.

Conclusion. We, as individuals, are primarily concerned with the current value of C. Although the president's remark promises that the rise in the cost of living may level off, it does not mean that each of us will be financially better off during his term of office. We may have actually been better off under the previous administration! ■

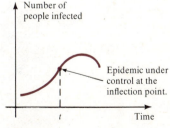

FIGURE 4–41. Example 4.

EXAMPLE 4 *(Epidemics)* When an epidemic first occurs, the number of people infected increases at an increasing rate (the sick people vs. time curve is concave up). After a while, the increase in the number of people infected decreases (the curve is concave down), and the epidemic subsides. Doctors consider the epidemic to be *under control* at the inflection point (Figure 4–41). ■

Examples 1 and 2 above illustrate how to use the first and second derivatives as aids in sketching the graph of a function. Here is a summary:

Step-by-Step Method of Graphing

(1) Find f' and f''.

(2) Determine the intervals on which f' is positive or negative and the points where it is zero. This information indicates where f is increasing and decreasing and locates the possible extrema.

(3) Determine the intervals on which f'' is positive or negative. This information indicates where the curve is concave up and down and locates the possible inflection points.

(4) Plot the possible extrema and inflection points. Plot any other points (the intercepts, for example) that you think may be helpful. Plot extra points on either side of any point where f is not continuous or where f' does not exist.

(5) Sketch the graph near any points where f is not continuous or where f' does not exist (if there are any). For the rest of the graph, draw a smooth curve through the plotted points using the information gathered in steps (2) and (3) above.

EXAMPLE 5   Sketch the graph of

$$s = t^{1/3}(t + 4) \quad -4 \le t \le 2$$

and indicate all important information.

Solution

$$s' = t^{1/3}(1) + (t + 4)\left[\frac{1}{3}t^{-2/3}\right] \qquad \text{(Product)}$$

$$= \frac{4}{3} \cdot \frac{t + 1}{t^{2/3}} \qquad \text{(Simplify)}$$

Therefore,

$$s' \begin{cases} > 0 & \text{on } (-1, 0) \text{ or } (0, 2] \\ < 0 & \text{on } [-4, -1) \\ = 0 & \text{at } -1 \\ & \text{is undefined at } 0 \end{cases}$$

Using the quotient rule on s', we find that

$$s'' = \frac{4}{9} \cdot \frac{t - 2}{t^{5/3}}$$

and it follows that

$$s'' \begin{cases} > 0 & \text{on } [-4, 0) \\ < 0 & \text{on } (0, 2) \end{cases}$$

Now we plot some points; $(-1, -3)$ is a local minimum, $(0, 0)$ is an inflection point, and $(-4, 0)$ and $(2, 7.6)$ are endpoint extrema. Since s' does not exist at 0, we plot points on either side of 0; say $(.5, 3.6)$ and $(-.5, -2.8)$. The graph is sketched in Figure 4-42. ∎

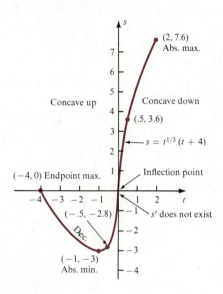

FIGURE 4-42. Example 5.

EXAMPLE 6 Sketch the graph of $y = x + (1/x)$ and indicate all important information.

Solution

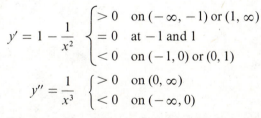

$$y' = 1 - \frac{1}{x^2} \begin{cases} > 0 & \text{on } (-\infty, -1) \text{ or } (1, \infty) \\ = 0 & \text{at } -1 \text{ and } 1 \\ < 0 & \text{on } (-1, 0) \text{ or } (0, 1) \end{cases}$$

$$y'' = \frac{1}{x^3} \begin{cases} > 0 & \text{on } (0, \infty) \\ < 0 & \text{on } (-\infty, 0) \end{cases}$$

y, y', and y'' do not exist at 0. From the information on y', it follows that $(-1, -2)$ is a local maximum and $(1, 2)$ is a local minimum. From the information on y'', it follows that the graph is concave up on $(0, \infty)$ and concave down on $(-\infty, 0)$, but there is no inflection point, because $y(0)$ does not exist.

It remains to examine the behavior near $x = 0$. As x approaches 0, $x + (1/x)$ gets arbitrarily large; positive if $x > 0$, but negative if $x < 0$. Putting all this information together, we sketch the graph as in Figure 4-43. ∎

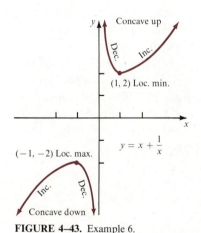

FIGURE 4-43. Example 6.

EXERCISES

In Exercises 1-20, sketch the graphs and indicate all important information.

1. $y = 2x^3 + x^2 - x$ **2.** $y = x^3 - x^2$

3. $s = t^4 - 4t^2 + 1$ **4.** $s = t^4 + t$

5. $v = 3u^5 - 5u^3$ **6.** $v = 2u^6 - 6u^4$

7. $z = \dfrac{1}{w - 4}$ **8.** $z = \dfrac{1}{w + 3}$

9. $y = e^{-x^2}$ **10.** $y = x + \sin x$

11. $y = (x - 4)^{2/3}$ **12.** $y = x^{2/3}(x^2 - 8)$

13. $s = (t - 1)(t - 2)(t - 3)$

14. $s = (t + 2)(1 - t^2)$

15. $y = |x^2 + 3x + 2|$ **16.** $y = \sqrt{|x^2 - 1|}$

17. $y = \dfrac{2x^2 - 1}{x^2 - 4}$ **18.** $y = \dfrac{x^2 + 2}{x - 3}$

19. $s = \dfrac{t + 1}{t^2 + 1}$ **20.** $s = \sqrt{t^4 - t^2 + 5}$

---

21. Find a function f such that $f'(x) = x^2 - 2x$ and $(1, 0)$ is an inflection point.

22. Find a function $s = s(t)$ such that $s' = 3t^2 - 2t + 1$, and $s(1) = -2$.

23. Find numbers a and b so that $f(x) = ax^3 + bx^2$ passes through $(1, 1)$ and has an inflection point when $x = 1/6$.

24. Find a function $s = s(t)$ such that $s'' = -\pi^2 \sin \pi t$, $s'(1/2) = 1$, and $s(0) = 2$.

In Exercises 25 and 26, sketch a continuous function that satisfies all the given conditions.

25. $f(-3) = f(-1/2) = f(1/2) = 0$; $f'(x) > 0$ on $(-\infty, -2)$ or $(0, \infty)$ and $f'(x) < 0$ on $(-2, 0)$; $f''(x) > 0$ on $(-1, 1)$ or $(2, \infty)$ and $f''(x) < 0$ on $(-\infty, -1)$ or $(1, 2)$.

26. $f'(x) > 0$ on $(-\infty, -2)$ or $(2, \infty)$ and $f'(x) < 0$ on $(-2, 2)$; $f'(x)$ does not exist at -2 and 2; $f''(x) > 0$ on $(-\infty, -2)$ or $(-2, -1)$ and $f''(x) < 0$ on $(1, 2)$ or $(2, \infty)$.

27. The oxygen content C in the bloodstream is usually expressed as a percentage of saturation. In human beings,

$$C = 100 \frac{aP^{2.7}}{1 + aP^{2.7}} \quad \text{percent}$$

where $a = .00013$ and P is the partial pressure in torr (760 torr $\approx$ air pressure at sea level). The derivative dC/dP is a measure of how much oxygen the blood releases to the body in various situations, say during strenuous exercise. For what value of P will dC/dP be a maximum? Sketch the graph of C for $0 \le P$.

28. If the cost of producing x units of a commodity is denoted by $C(x)$, then C is called the *cost function*. The derivative dC/dx is called the **marginal cost function**.* Economists interpret the value of the marginal cost at any integer x to be the cost of producing *one additional unit*. Thus, the minimum marginal cost is important. If

$$C(x) = 300 + .5x - .05x^2 + .005x^3 \quad \text{dollars}$$

find the minimum marginal cost. Sketch the graph of the marginal cost function.

29. (Continuation of Exercise 28) The **average cost** of producing x units is $A(x) = C(x)/x$. For the cost function in Exercise 28, sketch the graphs of A and marginal cost on the same axes. Find the value of x where the curves intersect. Show that for this value of x, $A(x)$ is a minimum value of A.

---

30. (Continuation of Exercise 29) Show that for any cost function, the average cost is a minimum when the marginal cost equals the average cost.

31. Show that if $(c, f(c))$ is an inflection point of the graph of f, then $f''(c) = 0$ or $f''(c)$ does not exist.

32. (Continuation of Exercise 31) If $(c, f(c))$ is an inflection point, then c is a critical point of f'. At which of the inflection points in Figure 4–38 does f' have a local maximum? A local minimum?

4–7   ANTIDERIVATIVES; THE DIFFERENTIAL EQUATION $y' = f(x)$

If we know the velocity of an object at any time t, can we find its position at any time t? In general, if we know the derivative of a function, can we work backwards and recapture the function? This is an important question because many

---

*C is actually defined only for integers x. However, assuming that C is a differentiable function defined on an interval leads to useful results.

quantities, such as marginal cost, velocity, radioactive decay, and electric current, are defined in terms of rates of change. If the answer is *yes*, then we can find the cost function from the marginal cost, we can find the position function from the velocity, and so on. This section discusses the possibility of recapturing a function from its derivative.

EXAMPLE 1   A particle moves along a horizontal line. When $t = 2$ seconds, the particle is 10 feet to the left of 0. If its velocity at any time t is $v = 2t + 5$ ft/sec, where is the particle when $t = 4$ seconds?

Solution   Let s be the position function. To answer the question above, we must evaluate s at 4. But first, we must *find* the position function s. The given information can be written as

(1) $s' = 2t + 5$ $(s' = v)$

(2) $s(2) = -10$ (Position at $t = 2$ is -10)

The idea is to find a function s that satisfies both (1) and (2). The function $s(t) = t^2 + 5t$ satisfies part (1); its derivative is $2t + 5$. But then, $s(2) = 2^2 + 5(2) = 14$, and we must have $s(2) = -10$. Therefore, let us add the constant -24 to $s(t)$. Now it is easy to verify that

$$s(t) = t^2 + 5t - 24$$

satisfies both (1) and (2), so it is the position function we seek. When $t = 4$, $s(4) = 4^2 + 5(4) - 24 = 12$. Therefore, the particle is 12 feet to the right of 0 when $t = 4$ seconds.   ∎

There is a serious flaw in the solution of Example 1. We found a function s that satisfied the given information, and boldly declared it to be *the* position function. But how do we know this solution is unique? Perhaps there are other functions satisfying (1) and (2), and they might have different values at $t = 4$. Let us investigate that possibility.

In Example 1, we have to find a function s whose derivative is $2t + 5$. Finding a function with a given derivative is called *antidifferentiation*. In general, if f is a given function, then any function F with $F' = f$ is called an *antiderivative* of f.

(4.27)

DEFINITION
Let f be a function on $[a, b]$. A function F is an **antiderivative of f on** $[a, b]$ if

  (i) F is continuous on $[a, b]$ and

  (ii) $F' = f$ on (a, b).

In practice, the domain of F and f are not usually specified; we simply say that "F is an antiderivative of f," and mean that $F' = f$ on some suitable interval.

Antiderivatives are never unique. Indeed, if $F' = f$, then $F + C$, where C is any constant, is also an antiderivative of f because $(F + C)' = F' = f$.

EXAMPLE 2   Let $f(x) = 6x$. Then $F(x) = 3x^2$ is an antiderivative of f because $F'(x) = (3x^2)' = 6x$. But you can easily check that

$$G(x) = 3x^2 + 9 \quad \text{and} \quad H(x) = 3x^2 - \sqrt{2}$$

are also antiderivatives of f.   ■

Clearly, any two functions that differ by a constant have the same derivative. Moreover, the converse is also true. See Figure 4–44.

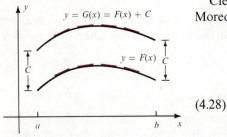

FIGURE 4–44. *Equal Derivatives Theorem.*

EQUAL DERIVATIVES THEOREM

(4.28)   If F and G are both continuous on $[a, b]$ and $F'(x) = G'(x)$ on (a, b) then

$$G(x) = F(x) + C \quad \text{on } [a, b]$$

where C is some constant.

To prove this, we set $H = G - F$, and show that H is a constant function. Clearly, H is continuous on $[a, b]$ and $H'(x) = G'(x) - F'(x) = 0$ on (a, b). If x_1 and x_2 are *any* two points in $[a, b]$ with $x_1 < x_2$, we can apply the Mean Value Theorem (3.45) to the interval $[x_1, x_2]$ and write

$$H'(c) = \frac{H(x_2) - H(x_1)}{x_2 - x_1}$$

for some point c in (x_1, x_2). But $H'(c) = 0$, and it follows that $H(x_1) = H(x_2)$. This means that H is constant and concludes the proof.

An immediate corollary to the Equal Derivatives Theorem is that any continuous function whose derivative is zero on an interval must be constant.

(4.29)   If H is continuous on $[a, b]$ and $H'(x) = 0$ on (a, b), then H is constant on $[a, b]$.

The Equal Derivatives Theorem says that if F and G are both antiderivatives of f, then $G = F + C$. Put another way, if F is *some* antiderivative of f, then *every* antiderivative of f is of the form $F + C$. The symbol $F + C$ denotes the set of all possible antiderivatives, and is called the **general antiderivative** of f.* For

---

*It is also called the **indefinite integral** of f, which is denoted by the symbol $\int f(x)\, dx$. We will discuss indefinite integrals in the next chapter.

your convenience, here is a short list of general antiderivatives. The entries are easily verified.

(4.30)

FUNCTION	GENERAL ANTIDERIVATIVE
(1) ax^n; $\begin{array}{c} n \neq -1 \\ a \text{ constant} \end{array}$	$a\left[\dfrac{x^{n+1}}{n+1}\right] + C$
(2) e^x	$e^x + C$
(3) $\sin x$	$-\cos x + C$
(4) $\cos x$	$\sin x + C$
(5) $\sec^2 x$	$\tan x + C$

At first, finding antiderivatives may be a bit confusing because you are used to going in the other direction, that is, finding derivatives. But you will soon learn to keep the two operations separate. In the meantime, it is a good idea to quickly check your answers every time you find an antiderivative.

EXAMPLE 3   Find the general antiderivative of $f(x) = x^4 - 3x^3 + x^2 - 2x + 1$.

Solution   Using formula (1) in (4.30), we have

$$F(x) = \frac{x^5}{5} - 3\left(\frac{x^4}{4}\right) + \frac{x^3}{3} - 2\left(\frac{x^2}{2}\right) + x + C$$

$$= \frac{1}{5}x^5 - \frac{3}{4}x^4 + \frac{1}{3}x^3 - x^2 + x + C$$

Check: $F' = x^4 - 3x^3 + x^2 - 2x + 1$ ■

EXAMPLE 4   Find the general antiderivative of $y = e^x - 4 \sin x$.

Solution   By formulas (2) and (3) in the table, we have that

$$F(x) = e^x + 4 \cos x + C$$

is the answer. Check this out.   ■

The Differential Equation $y' = f(x)$

Any equation that involves the derivative of an unknown function is called a **differential equation.** Any function that satisfies the equation is a **(particular) solution.** To **solve** a differential equation means to find all solutions, and the set of all solutions is called the **general solution.** Some differential equations are easy to solve; others are very difficult. In this book we discuss only the simpler ones.

Let us consider a differential equation of the form $y' = f(x)$. A function $y = F(x)$, is a particular solution if and only if F is an antiderivative of f. It follows from the Equal Derivatives Theorem that every solution must be of the form $F + C$ (Figure 4–45). That is,

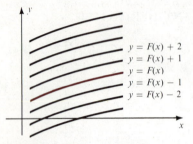

$$y = F(x) + 2$$
$$y = F(x) + 1$$
$$y = F(x)$$
$$y = F(x) - 1$$
$$y = F(x) - 2$$

(4.31)

FIGURE 4–45. Graphs of solutions to a differential equation $y' = f(x)$.

The general solution to a differential equation of the form $y' = f(x)$ is

$$y = F(x) + C$$

where F is any antiderivative of f.

EXAMPLE 5   Solve the equation $y' = 1/x^2 + \sqrt{x}$.

Solution   The general solution is the general antiderivative of the function $f(x) = 1/x^2 + \sqrt{x}$. Rewrite f in terms of powers; $f(x) = x^{-2} + x^{1/2}$, and then

$$F(x) = \frac{x^{-1}}{-1} + \frac{x^{3/2}}{3/2} + C \qquad (4.30)$$

$$= -\frac{1}{x} + \frac{2}{3}\sqrt{x^3} + C$$

is the answer. *Check this out;* let $y = F(x)$, and make sure that $y' = 1/x^2 + \sqrt{x}$. ∎

It often happens in practice that we seek a particular solution that satisfies the condition of having a given value at a specified point. This is called a **boundary** (or **initial**) **condition** (Figure 4–46).

EXAMPLE 6   Find the solution of $y' = 4 + \sec^2 x$ with boundary condition $y(\pi/4) = \pi$.

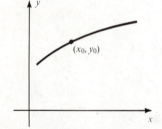

(x_0, y_0)

FIGURE 4–46. The graph of the particular solution of $y' = f(x)$ that satisfies the initial condition $f(x_0) = y_0$; it passes through the point (x_0, y_0).

Solution   *First,* find the general solution

$$y = 4x + \tan x + C. \qquad \text{(check)}$$

Every solution is of this form.
    Second, find the constant C, using the condition $y(\pi/4) = \pi$.

$$\pi = y(\pi/4) = 4(\pi/4) + \tan(\pi/4) + C$$
$$\pi = \pi + 1 + C$$

Therefore, $C = -1$, and $y = 4x + \tan x - 1$ is the *only* solution that satisfies the equation *and* the boundary condition. ∎

EXAMPLE 7   Returning now to Example 1, we see that

$$s = t^2 + 5t + C$$

is the general solution of $s' = 2t + 5$. Every solution is of this form. To satisfy the initial condition $s(2) = -10$, we write

$$-10 = s(2) = 2^2 + 5(2) + C$$
$$-10 = 4 + 10 + C$$

and $C = -24$. Thus, $s = t^2 + 5t - 24$ is, indeed, *the* position function.   ■

Differential equations that involve only first derivatives are called **first order** equations. Those that also involve second derivatives are **second order** equations. The differential equation $y'' = f(x)$ is thought of as $(y')' = f(x)$. Solving it requires *two* antidifferentiations, and the general solution contains *two* constants. To find a particular solution, there must be *two* initial (or boundary) conditions.

EXAMPLE 8   Find the particular solution of $y'' = 2x + 3$ that satisfies the initial conditions $y'(0) = 3$ and $y(0) = -4$.

Solution   *First*, find the general solution:

$$(y')' = 2x + 3$$

(1) $$y' = x^2 + 3x + C_1$$ (check)

(2) $$y = \frac{1}{3}x^3 + \frac{3}{2}x^2 + C_1 x + C_2$$ (check)

Second, find the constants C_1 and C_2:

Because $y'(0) = 3$ and $y(0) = -4$, equations (1) and (2) yield

$$3 = y'(0) = 0^2 + 3(0) + C_1$$

and

$$-4 = y(0) = \frac{1}{3}0^3 + \frac{3}{2}0^2 + C_1 \cdot 0 + C_2$$

It follows that $C_1 = 3$ and $C_2 = -4$. Thus,

$$y = \frac{1}{3}x^3 + \frac{3}{2}x^2 + 3x - 4$$

is the solution.   ■

EXAMPLE 9   If the cost of producing x units of a commodity is denoted by $C(x)$, then C is called the **cost function,** and the derivative C' is called the **marginal cost function.** Economists interpret the value of the marginal cost at any integer x to be the cost of producing *one additional unit.* A lamp shade company estimates that its marginal cost is $C'(x) = 10 - .02x$ dollars. If its fixed costs (rent, insurance, depreciation, and so on) are $250 per day, what is the cost of producing x lamp shades per day?

Solution   The general solution of $C'(x) = 10 - .02x$ is

$$C(x) = 10x - .01x^2 + K$$

where K is used to denote the arbitrary constant because C is already used to represent the cost function. The initial condition is obtained from the fixed costs; we are told that $250 must be spent per day even if no lamp shades are produced, so $C(0) = 250$. It follows that $K = 250$ and

$$C(x) = 10x - .01x^2 + 250 \text{ dollars}$$

is the desired cost function.   ■

EXERCISES

In Exercises 1–10, find the general solution of the first order equation.

1. $y' = 5$

2. $y' = -3$

3. $y' = 6x - 4$

4. $y' = 2 - 5x$

5. $s' = 6t^2 - 3t + 1$

6. $s' = 3t^2 + 4t - 5$

7. $s' = \sqrt{t} - \dfrac{1}{\sqrt{t}}$

8. $s' = t^{-3} + \sin t$

9. $y' = x^{1/3} - x^{-4}$

10. $y' = \sqrt[3]{x} - \dfrac{1}{x^2}$

In Exercises 11–20, find the general solution of the second order equation.

11. $y'' = 18x$

12. $y'' = 12x^2$

13. $y'' = \sin x$

14. $y'' = \cos x$

15. $s'' = e^t + 1$

16. $s'' = e^t$

17. $s'' = \sin t + \cos t$

18. $s'' = \sin t - \cos t$

19. $y'' = x^{-3} + x^{-1/2}$

20. $y'' = \sqrt[3]{x} + 1$

In Exercises 21–26, find the particular solution that satisfies the given boundary condition.

21. $y' = 3 - 2x; y(1) = 3$

22. $y' = 4x - 5; y(-1) = 2$

23. $y' = \sqrt{x} - 2; y(0) = -4$

24. $y' = x^{-4} + x^{-3}; y(1) = 0$

25. $s' = e^t + \sec^2 t; s(0) = 1$

26. $s' = \cos t + \sin t; s(0) = 0$

In Exercises 27–30, find the particular solution that satisfies both boundary conditions.

27. $y'' = x + 1; y'(1) = -2$ and $y(1) = 0$

28. $y'' = \sqrt{x}; y'(4) = 0$ and $y(4) = -3$

29. $s'' = \cos t; s'(\pi/6) = 0$ and $s(0) = 0$

30. $s'' = 1 + e^t; s'(0) = 3$ and $s(0) = 1$

In Exercises 31–36, the solution involves a product or quotient. The procedure is to *guess* a general solution and *check*. If the solution does not check, then guess again.

31. $y' = e^x(\cos x + \sin x)$

32. $y' = e^x(2x + x^2)$

33. $y' = 3x^2\cos x - x^3\sin x$

34. $y' = \dfrac{1}{2\sqrt{x}}\tan x + \sqrt{x}\sec^2 x$

35. $y' = \dfrac{e^x(x + 1) - e^x}{(x + 1)^2}$

36. $y' = \dfrac{x^2\cos x - 2x\sin x}{x^4}$

In Exercises 37–44, finding the general solution requires the chain rule. *Guess* a solution and *check*.

37. $s' = e^{3t}$

38. $s' = \sin 5x$

39. $y' = x(x^2 + 1)^4$

40. $y' = (x^2 + 1)(x^3 + 3x)^2$

41. $y' = \dfrac{3x^2 + 1}{\sqrt{x^3 + x}}$

42. $y' = \dfrac{4x}{(2x^2 + 3)^3}$

43. $s' = \sin t \cos t$

44. $s' = -\sin t \cos t$

45. The velocity function of a particle moving on a horizontal line is $v = 6t^2 - 4t - 3$ ft/sec for $-2 \le t \le 2$. If it starts (that is, when $t = -2$) from a point 4 feet to the right of 0, where is it when $t = 1$? In which direction is it moving at that moment?

46. A weight is attached to a spring that is hanging from a support. The weight is pulled down (negative direction) 6 inches and then released. It oscillates up and down with a velocity of $v = 6 \cos(t - (\pi/2))$

inches/sec. Where is it when $t = \pi/2$? In which direction is it moving?

47. Prove statement (4.29); if H is continuous and $H' = 0$, then H is constant.

Optional Exercises

The Differential Equation $y' = ry$

It often happens that the unknown function is part of the differential equation; for example $y' = ry$. Such equations are not as simple to solve as those of the form $y' = f(x)$. The exercises below will establish that

(4.32)

> The general solution to a differential equation of the form $y' = ry$ is
>
> $$y = Ce^{rx}$$
>
> when C is any constant.

48. Show that $y = e^{rx}$ satisfies the equation $y' = ry$.

49. You are probably tempted to say that $y = e^{ry} + C$ is the general solution, but that is *false*. In fact, $y = e^{rx} + C$ is not even a solution unless $C = 0$. For example, show that $y = e^{rx} + 1$ is not a solution.

50. On the other hand, $y = Ce^{rx}$ *is* a solution. Show that $y = 5e^{rx}$ satisfies $y' = ry$.

51. Now suppose that $y = F(x)$ is *any* solution; that is, $F' = rF$. Show that $(F/e^{rx})' = 0$.

52. Therefore, $F/e^{rx} = C$ because _____, and $F = Ce^{rx}$. This means that the general solution of $y' = ry$ is $y = Ce^{rx}$.

53. Find the particular solution that satisfies $y' = -\frac{1}{10}y$ with initial condition $y(0) = 100$.

54. A particle starts ($t = 0$) from a point 5 units to the left of 0 and moves so that its velocity at time t is twice its position at time t (that is, $v(t) = 2s(t)$). Where is the particle when $t = 3$?

55. It is reasonable to assume that population P grows at a rate proportional to its size; that is, $P' = rP$. In a study of a colony of rabbits, r is found to be 0.3. If the initial population is 30, how many rabbits will there be when $t = 12$ months? (t is measured in months.)

4–8   MOTION; VELOCITY AND ACCELERATION

In this section, all motion is assumed to be rectilinear; that is, along a straight line.

Let s be the **position function** of a moving particle. The particle is on one side or the other of the origin depending on the sign of $s(t)$, and $|s(t)|$ represents its *distance* to the origin. The **velocity** v is defined to be the rate of change of position with respect to time; that is $v = s'(t)$. The sign of $v(t)$ indicates the *direction* of motion. The **speed** is defined to be $|v(t)|$; it is measured in meters per second, miles per hour, and so on. Now, just as the position changes with time, so may the velocity. The rate of change of velocity with respect to time is called the **acceleration;** it is also a function of t and usually denoted by a.* Thus, $a = v'(t)$

---

*The gas pedal of an automobile is sometimes called an *accelerator* because it is used to change the velocity.

$= s''(t)$. The sign of $a(t)$ indicates whether v is an increasing or decreasing function. Because acceleration measures the rate of change of velocity, it is discussed in units of distance per time per time. For instance, feet per second per second (ft/sec²), miles per hour per second (mi/hr/sec), and so on.

POSITION – VELOCITY – ACCELERATION

s is the position function;
sign of $s(t)$ indicates on which side of the origin the particle lies.
$|s(t)|$ is the distance to the origin.

(4.33) $\quad v = s'(t)$ is the velocity function;
sign of $v(t)$ indicates the direction of motion.
$|v(t)|$ is the speed.

$a = v'(t) = s''(t)$ is the acceleration function;
sign of $a(t)$ indicates increasing or decreasing velocity.

EXAMPLE 1 Describe the motion of the particle whose position function has the graph in Figure 4–47A.

Solution Unless otherwise stated, the motion will always be along a horizontal line with the positive direction to the right of zero.

The graph in Figure 4–47A is a smooth curve so you can assume that s is twice differentiable; that is, $v = s'(t)$ and $a = s''(t)$ both exist. The graph is concave up which indicates that the second derivative $a(t) > 0$; this means that the velocity v is always increasing. The fact that the curve is falling for $0 < t < 2$ and rising for $2 < t < 4$ indicates that the first derivative $v(t) < 0$ on the first interval and $v(t) > 0$ on the second; furthermore, $v(2) = 0$ because the tangent line at $(2, -2)$ is horizontal.

The next observation is that a positive acceleration does *not* necessarily mean that the object is speeding up. For the case in point, the particle starts moving at $t = 0$ and comes to rest at $t = 2$ ($v(2) = 0$). Therefore, it must be slowing down even though the acceleration is always positive! The reason is that $v(t) < 0$ for $0 < t < 2$. The fact that $a(t) > 0$ means that v is increasing, and when a negative velocity increases, its absolute value (the speed) becomes smaller. (For example, if -3 is increased to -2, then there is a decrease in the absolute value from 3 to 2.) However, in the interval $2 < t < 4$, both $a(t)$ and $v(t)$ are positive so that the particle is, indeed, speeding up.

These observations, along with the positional information obtained from the graph, will allow us to draw the motion diagram shown in Figure 4–47B.

Note: The curve above the line is *not* the path of the particle; it is intended to show only the manner in which the particle moves along the line.

The diagram shows that the particle starts ($t = 0$) at a point 1 unit to the right of 0 and moves to the left ($v(t) < 0$). It is slowing down ($|v|\downarrow$) because $v(t) < 0$ and $a(t) > 0$. It comes to rest at $t = 2$ ($v(2) = 0$), then reverses direction ($v(t) > 0$)

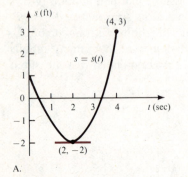

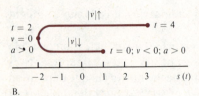

FIGURE 4–47. (A) The position function referred to in Example 1. (B) The motion diagram.

and is speeding up ($|v|\uparrow$) because $v(t) > 0$ and $a(t) > 0$. It ends up at a point 3 units to the right when $t = 4$. ∎

EXAMPLE 2 Describe the motion of a particle whose position function is $s = \sin(\pi/2)t$ for $0 \le t \le 4$.

Solution This is the motion of alternating current or of an object oscillating on a spring; it is called **simple harmonic motion.**

$$s = \sin\frac{\pi}{2}t$$

$$v = s'(t) = \frac{\pi}{2}\cos\frac{\pi}{2}t$$

$$a = s''(t) = -\left(\frac{\pi}{2}\right)^2 \sin\frac{\pi}{2}t$$

Using the graph of the position function (Figure 4–48A) and the equations above for v and a, the motion can be described by the motion diagram in Figure 4–48B. It starts ($t = 0$) at the origin ($s(0) = 0$) and moves to the right for 1 second. It is slowing down because $v(t) > 0$ and $a(t) < 0$ and comes to rest when $t = 1$ ($v(1) = \pi/2 \cos \pi/2 = 0$) at the point 1 ($s(1) = \sin \pi/2 = 1$). It then reverses direction ($v(t) < 0$ for $1 < t < 3$). For $1 < t < 2$, it is speeding up because both $v(t)$ and $a(t)$ are negative but for $2 < t < 3$, it is slowing down because $v(t) < 0$ and $a(t) > 0$. It comes to rest at $t = 3$ (why?) and reverses direction again; for $3 < t < 4$ its speed increases (why?) and at $t = 4$ it is back at the origin. ∎

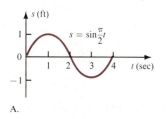

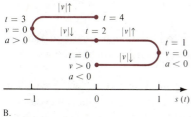

FIGURE 4–48. (A) The position curve of the particle in Example 2. (B) The motion diagram.

Finding $s(t)$ and $v(t)$ from $a(t)$

Given the position, you can find the velocity and acceleration by differentiation. Now, thanks to your experience in solving differential equations, you can also work backwards; given the acceleration, solve $a = s''(t)$ for $v = s'(t)$ and $s = s(t)$.

EXAMPLE 3 A particle starts from rest at a point 3 units to the left of the origin. Its acceleration at any time $t \ge 0$ is $a = t + 2$ ft/sec^2. Find its velocity and position functions.

Solution The statement of the problem contains two initial conditions. The particle starts from rest means $v(0) = 0$; it starts from a point 3 units to the left of the origin means $s(0) = -3$. Now then, solve the differential equations as follows:

$$a = s''(t) = t + 2 \quad \text{ft/sec}^2 \qquad \text{(Given)}$$

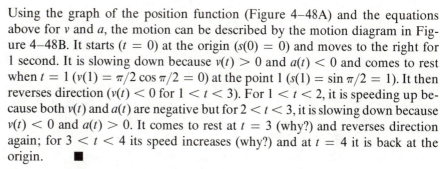

$$(1) \qquad v = s'(t) = \frac{1}{2}t^2 + 2t + C_1 \quad \text{ft/sec} \qquad \text{(Check)}$$

$$(2) \qquad s = \frac{1}{6}t^3 + t^2 + C_1t + C_2 \quad \text{ft} \qquad \text{(Check)}$$

where C_1 and C_2 are constants that are determined by the initial conditions $v(0) = 0$ and $s(0) = -3$. Equations (1) and (2) yield

$$0 = v(0) = \frac{1}{2} \cdot 0^2 + 2 \cdot 0 + C_1$$

$$-3 = s(0) = \frac{1}{6} \cdot 0^3 + 0^2 + C_1 \cdot 0 + C_2$$

It follows that $C_1 = 0$ and $C_2 = -3$. Therefore,

$$v = \frac{1}{2}t^2 + 2t \quad \text{and} \quad s = \frac{1}{6}t^3 + t^2 - 3$$

are the velocity and position functions. ■

Near the surface of the earth all objects are subject to a constant acceleration due to the pull of gravity. This acceleration is usually denoted by the letter g and

$$g \approx 32 \text{ ft/sec}^2 \approx 9.8 \text{ meters/sec}^2$$

This fact is used in discussing the free-fall motion of objects near the Earth's surface. Unless otherwise stated, the effect of air resistance is neglected.

EXAMPLE 4   A ball is thrown straight down from a height of 192 ft with an initial velocity of 64 ft/sec. When will it strike the ground, and what is its velocity when it hits? See Figure 4–49.

Solution   The motion is along a vertical line. For this problem it is convenient to let $s(t)$ be the distance below the starting point at time t. Therefore, $s(0) = 0$ and the positive direction is *down*. The initial conditons are $v(0) = 64$ and $s(0) = 0$. You start with the equation $a = 32$.

$$a = s''(t) = 32 \text{ ft/sec}^2$$
$$v = s'(t) = 32t + C_1 \quad \text{ft/sec}$$
$$s = 16t^2 + C_1 t + C_2 \quad \text{ft}$$

The initial conditions yield

$$64 = v(0) = 32 \cdot 0 + C_1$$
$$0 = s(0) = 16 \cdot 0^2 + C_1 \cdot 0 + C_2$$

so $C_1 = 64$ and $C_2 = 0$; therefore,

$$v = 32t + 64 \quad \text{and} \quad s = 16t^2 + 64t$$

When the ball hits the ground, it will be 192 ft from the original position, so set $s = 192$ and solve for t:

$$192 = 16t^2 + 64t$$
$$t^2 + 4t - 12 = 0$$

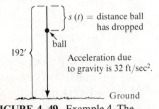

$s(t) =$ distance ball has dropped

ball

192′

Acceleration due to gravity is 32 ft/sec².

Ground

FIGURE 4–49. Example 4. The positive direction is *down;* acceleration is positive.

$$(t + 6)(t - 2) = 0$$
$$t = -6 \text{ or } 2$$

Only the solution $t = 2$ makes sense, so the ball will strike the ground in 2 seconds. Its velocity at that time is $v(2) = 32 \cdot 2 + 64 = 128$ ft/sec. ∎

EXAMPLE 5 A projectile is fired straight up with a muzzle velocity of 256 ft/sec. How high will it rise and when will it strike the ground? See Figure 4–50.

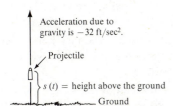

Acceleration due to gravity is -32 ft/sec².

Projectile

 $s(t)$ = height above the ground

Ground

FIGURE 4–50. Example 5. The positive direction is *up;* acceleration is negative.

Solution This time, it is convenient to let $s(t)$ represent the height of the projectile at time t. Therefore, $s(0) = 0$ and the positive direction is *up*. Thus, gravity works *against* the motion, that is, it slows down the projectile. In this case, we take the acceleration to be -32.

$$a = s''(t) = -32 \text{ ft/sec}^2$$
$$v = s'(t) = -32t + C_1 \quad \text{ft/sec}$$
$$s = -16t^2 + C_1 t + C_2 \quad \text{ft}$$

The boundary conditions $v(0) = 256$ and $s(0) = 0$ yield $C_1 = 256$ and $C_2 = 0$. Thus,

$$v = -32t + 256 \quad \text{and} \quad s = -16t^2 + 256t$$

The projectile rises, comes to rest momentarily, and falls back to earth. When the projectile reaches its highest point, the velocity will be zero. Therefore, set $v = 0$ and solve for t:

$$0 = -32t + 256 \quad \text{or} \quad t = 8 \text{ sec}$$

Since $s(t)$ is the height at time t and

$$s(8) = -16 \cdot 8^2 + 256 \cdot 8 = 1{,}024 \text{ ft}$$

the projectile will rise to a height of 1,024 feet. It takes 8 seconds to rise, so it will take another 8 seconds, or 16 seconds in all, to hit the ground. ∎

EXERCISES

In Exercises 1 and 2, draw a motion diagram (as in Examples 1 and 2) to describe the motion of a particle whose position function is graphed in the indicated figure.

1.

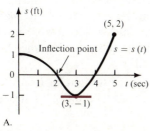

A.

2.

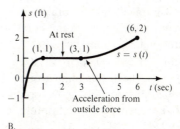

B.

FIGURE 4–51. Position curves for Exercises 1 and 2.

In Exercises 3–8, draw a motion diagram to describe the motion of a particle with the given position function.

3. $s(t) = t + 3$ **4.** $s(t) = 4 - t$

5. $s(t) = t^3 - 6t^2 + 9t$ **6.** $s(t) = t^3 - 3t - 1$

7. $s(t) = e^{-t}$

8. $s(t) = \cos \frac{\pi}{2} t, 0 \le t \le 4$

In Exercises 9–12, find $s(1)$, $v(1)$, and $a(1)$ from the given information.

9. $a(t) = -2; v(0) = 0, s(2) = 4$

10. $a(t) = t^3; v(0) = -1, s(0) = 3$

11. $v(t) = e^{-3t}; s(0) = -\frac{4}{3}$

12. $v(t) = t^2 + t - 1; s(0) = 2$

13. The position function of a particle in a linear accelerator at time t seconds is $s(t) = t^2$ feet. If the accelerator is 2 miles long, with what velocity will the particle hit the target? (1 mile = 5,280 feet).

14. Answer the same question as in Exercise 13 if the accelerator is 10,000 feet long and the position function is $s(t) = t^2 - 150t + 10,000$.

15. A ball rolls down an incline with a velocity $v(t) = t^2 + 2t + 6$ ft/sec. How far will it travel in 2 seconds? What is its acceleration after 1 second?

16. A ball is rolled up an incline so that its velocity is $v(t) = -3t^2 - 2t + 16$ ft/sec. How far will it roll up the incline before it starts rolling down?

17. A woman is driving an automobile at 60 mi/hr and applies the brakes. If the brakes cause a deceleration of 10 mi/hr² (that is, an acceleration of -10 mi/hr²), *guess* how long will it take to stop and how far she will travel after the brakes are applied. Now check your estimate. Work the same problem if the deceleration is 7,200 mi/hr².

18. A landing jet has a velocity of 300 mi/hr (= 440 ft/sec) as its wheels touch the ground. If the brakes are applied immediately and supply a deceleration of 40 ft/sec², what is the minimum length of runway needed?

19. A bullet is shot straight up with a muzzle velocity of 320 ft/sec. How high will it rise and when will it strike the ground?

20. Answer the same questions as in Exercise 19 if the muzzle velocity is 150 mi/hr.

21. A man, standing on a cliff 100 feet above the ocean, throws a rock directly upward with an initial velocity of 128 ft/sec. How high above the ocean will the rock rise?

22. A ball falls out of a window 160 ft above the sidewalk. When will it strike the sidewalk?

23. Answer the same question as in Exercise 22 if the ball is thrown down with a velocity of 48 ft/sec.

24. A marble is dropped from a stationary helicopter 1,600 feet above the ground. What is its velocity when it hits the ground?

25. How long does it take for a mounted rocket starting from rest to travel 100 miles on a level track if it accelerates at the constant rate of 20 mi/hr/sec.?

26. An automobile is traveling at 40 mi/hr and suddenly accelerates at the constant rate of 10 mi/hr/min. How fast is it moving after 5 minutes of acceleration? How far will it have traveled in those 5 minutes?

27. A car is traveling 50 mi/hr in a 25 mi/hr speed zone. A police officer in a patrol car starts from rest as the speeder passes him and accelerates at the rate of 5 mi/hr/sec. How long does it take the officer to catch the speeder?

28. In Exercise 27, how fast is the police officer traveling when he catches up with the speeder? How far has he traveled?

29. If the speeder in Exercise 27 was going 70 mi/hr (instead of 50) and the police car had a maximum speed of 100 mi/hr, how long would it take to catch the speeder?

30. A particle starts from the origin with a velocity of 4 ft/sec. If it moves with a constant acceleration of -2 ft/sec², what is its maximum distance from the origin in the first 3 seconds? 5 seconds?

31. As a speeder, going 70 mi/hr, passes a police car parked on the roadside, the police car gives chase with a constant acceleration of 5 mi/hr/sec. At what time will the maximum distance between the cars be attained?

32. A rocket is fired straight up. Its first engine provides a constant acceleration of 52 ft/sec² for 1 min at which time a second engine ignites, and the first engine falls off. When will the first engine strike the ground?

33. To determine the height of a building, a calculus student drops a small stone from the roof. Three seconds later the student hears the sound of the

impact. How high is the building? (Sound travels about 1,100 ft/sec in air.)

34. If the positive direction is up in free-fall motion, show that the height above the ground at time t is

$$s(t) = -16t^2 + v_0 t + s_0$$

where v_0 and s_0 are the initial velocity and height.

Optional Exercises

35. The acceleration of a certain particle at any time t is -3 times its velocity at time t. If it starts with a velocity of 5 ft/sec from a point 10 feet to the right of the origin, where is it when $t = 3$? Is it speeding up or slowing down? (See (4.32) in Section 4–7.)

36. *(Free-fall with air resistance)* An object falling under the influence of gravity and air resistance has an acceleration at time t of

(4.34) $$a = g - Dv(t)$$

where $g = 32$ ft/sec², $v(t)$ is the velocity, and D (for *drag*) is a constant depending on the mass and shape of the object. Suppose an object starts at a height of s_0 feet above the earth with an initial

velocity of v_0 ft/sec and drag constant D. Then (4.34) becomes the differential function

(4.35) $$v' = g - Dv$$

with initial conditions $v(0) = v_0$ and $s(0) = 0$ (the positive direction is down). This equation looks simple enough but its solution is actually rather complicated. Show that

(4.36) $$v = -\frac{1}{D}[(g - Dv_0)e^{-Dt} - g]$$

is the velocity function; that is, show that it satisfies (4.35) above and that $v(0) = v_0$.* Use (4.36) and $s(0) = 0$ to find the position function.

4–9   NEWTON'S METHOD

Linear equations can be solved using the simple operations of arithmetic. Quadratic equations can always be solved with the quadratic formula. But, for equations involving polynomials of higher degree or transcendental functions, it is usually difficult, and often impossible, to obtain an exact solution. Examples of such equations are

$$x^4 = 3 - x \quad \text{and} \quad \sin x - e^{-x} = 0$$

There are, however, techniques for approximating solutions. One of them, *Newton's method*, is described in this section.

   Our first observation is that finding the solutions of an equation can be reduced to finding the *zeros* of a function.

(4.37)
> **DEFINITION**
> Given a function f, the number r is a **zero** of f if $f(r) = 0$.

*If you have a candidate for a solution to (4.35) that is much simpler than (4.36), try it out; that is, see if it satisfies (4.35) and $v(0) = v_0$.

For instance, if the first equation above is rewritten as $x^4 + x - 3 = 0$, then any zero of the function $f(x) = x^4 + x - 3$ is a solution to the original equation $x^4 = 3 - x$. Newton's method is used to approximate the zeros of a function. First we will give an example, and then summarize the procedure.

EXAMPLE 1 Approximate a solution of $x^4 = 3 - x$.

Solution

(1) Rewrite the equation as $x^4 + x - 3 = 0$ and set

$$f(x) = x^4 + x - 3$$

Then any zero of f is a solution of $x^4 = 3 - x$.

(2) The zeros of f occur where the graph of f crosses the x-axis. Therefore, we sketch the graph of f and *estimate* where it crosses the axis. Figure 4–52 indicates that f has two zeros; let us approximate the one between 1 and 2. Choose a number near the zero as a first approximation; we have chosen the number 2.

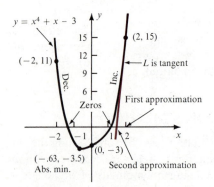

FIGURE 4–52. Example 1.

(3) Now draw the tangent line L at $(2, f(2)) = (2, 15)$. This line closely resembles the graph of f near the point $x = 2$, and *we can easily compute where L crosses the axis.* An equation of L is

$$y - f(2) = f'(2)(x - 2)$$
$$y - 15 = 33(x - 2) \qquad (f'(x) = 4x^3 + 1)$$
$$y = 33x - 51$$

Setting $y = 0$ yields $x = \frac{51}{33} \approx 1.5454545$. Our first approximation was $x = 2$, but Figure 4–52 suggests that 1.5454545 is better.

(4) Repeat step (3) with 1.5454545 in place of $x = 2$. Find an equation of the tangent line at $(1.5454545, f(1.5454545))$, set $y = 0$, and solve. This time the answer is 1.275864.

Continue the process until two consecutive answers are the same. The next five approximations are

1.1764081

1.1642032

1.1640352

1.1640351

1.1640351

Thus, the positive solution of $x^4 = 3 - x$ is approximately 1.1640351. ■

Newton's Method for approximating the zeros of a function is based on two facts.

(i) If f is differentiable at x, then the tangent line at $(x, f(x))$ can be used to approximate the values of f near x.

(ii) The point where a line crosses the x-axis is easy to determine.

Let x_1 be the first approximation. An equation of the tangent line at $(x_1, f(x_1))$ is

$$y - f(x_1) = f'(x_1)(x - x_1)$$

Let x_2 be the point where this line crosses the x-axis; that is, when $x = x_2$, then $y = 0$. Thus, $0 - f(x_1) = f'(x_1)(x_2 - x_1)$, and if $f'(x_1) \neq 0$, then

$$x_2 = x_1 - \frac{f(x_1)}{f'(x_1)}$$

Now let x_2 be the second approximation, and repeat the process to obtain

$$x_3 = x_2 - \frac{f(x_2)}{f'(x_2)} \quad \text{provided } f'(x_2) \neq 0$$

Continue with x_3 as the third approximation, and so on (Figure 4–53). At each step, if x_n is the nth approximation, then

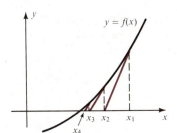

FIGURE 4–53. Geometric interpretation of Newton's method.

(4.38) $\quad \boxed{x_{n+1} = x_n - \frac{f(x_n)}{f'(x_n)} \quad \text{provided } f'(x_n) \neq 0}$

Remarks:

(1) For the function $f(x) = x^4 + x - 3$ in Example 1, formula (4.38) takes the form

$$x_{n+1} = x_n - \frac{x_n^4 + x_n - 3}{4x_n^3 + 1}$$

$$= \frac{3(x_n^4 + 1)}{4x_n^3 + 1} \qquad \text{(Simplify)}$$

This *iteration* formula (x_{n+1} is defined in terms of x_n) is particularly easy to use on a programmable calculator; store x_n and then recall it to make the calculation above that produces x_{n+1}; then the process repeats to produce x_{n+2}, and so on.

(2) You can stop the process when two successive approximations are equal because it follows from (4.38) that when $x_{n+1} = x_n$, then $f(x_n) = 0$. That is, x_n is a zero of f (at least to within the accuracy limits of your calculator).

EXAMPLE 2   Find the point of intersection of the curves $y = \sin x$ and $y = e^{-x}$, $0 \le x \le \pi/2$ (Figure 4–54).

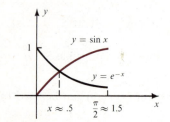

FIGURE 4–54. Example 2.

Solution   The curves will intersect when $\sin x = e^{-x}$; that is, when $\sin x - e^{-x} = 0$. Therefore, the x-coordinate of the point of intersection is a zero of the function

$$f(x) = \sin x - e^{-x}$$

that lies between 0 and $\pi/2$. Since $f'(x) = \cos x + e^{-x}$, it follows from (4.38) that

$$(4.39) \qquad x_{n+1} = x_n - \frac{\sin x_n - e^{-x_n}}{\cos x_n + e^{-x_n}}$$

Now sketch the curves to make a reasonable first approximation. From Figure 4–54, it appears that x is about one-third of the way to $\pi/2 \approx 1.5$, so let $x_1 = .5$ be the first approximation. The entries below were computed with an 8-digit calculator using equation (4.39).

$$x_1 = .5$$
$$x_2 = .5856438$$
$$x_3 = .5885294$$
$$x_4 = .5885327$$
$$x_5 = .5885327$$

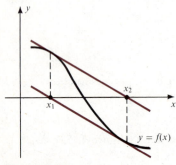

FIGURE 4–55. Pitfall (1).

Observe that $x_4 = x_5$. When $x_{n+1} = x_n$, it follows from (4.38) that $f(x_n) = 0$ (at least to 8 digits). Therefore, .5885327 is the x-coordinate of the intersection point. The y-coordinate is $\sin .5885327$ (or $e^{-.5885327}$). Thus,

$$(.5885327, .5551412)$$

is the intersection of the two curves.   ∎

Remarks: Some care must be taken in choosing the first approximation. Be alert for the following two pitfalls:

(1) It could happen that the first approximation x_1 yields a second approximation x_2 which, in turn, yields the same x_1 as a third approximation (see Figure 4–55). The approximations are alternately x_1 and x_2; nothing is accomplished. Let $f(x) = \sqrt{1 - x}$ for $x \le 1$ and $f(x) = \sqrt{x - 1}$ for $x > 1$; try $x_1 = 0.5$.

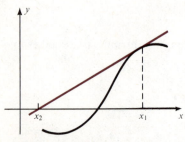

FIGURE 4–56. Pitfall (2).

(2) If the first approximation x_1 is chosen too close to a critical point, the second approximation could be much worse than the first (see Figure 4–56). The positive root of $f(x) = x^2 - 2x - 3$ is $x = 3$; try 1.1 as a first approximation.

EXERCISES

For each equation in Exercises 1–7, approximate the indicated solution.

1. $x^3 = x + 1$ (there is only one solution)

2. $x^3 = 3 - 5x$ (positive solution)

3. $x^4 + 2x^3 = 5x^2 - 1$ (solution between 1 and 2)

4. $x^4 + 4x = 1$ (solution between 0 and 1)

5. $x^4 + 4x = 1$ (solution between -2 and -1)

6. $x^5 = 1 - x$ (only one solution)

7. $\cos x = e^{-x}$ (solution between 0 and $\pi/2$)

In Exercises 8–13, approximate the intersection point for each pair of curves.

8. $y = x^4$ and $y = x + 2, 0 \le x \le 2$

9. $y = x^4$ and $y = x + 2, -2 \le x \le 0$

10. $y = x^3$ and $y = 1 - x, 0 \le x \le 1$

11. $y = 2x$ and $y = e^{-x}$

12. $y = e^x$ and $y = -2x^3$

13. $y = \sin x$ and $y = 2x - 5$

14. The function $f(x) = x^4 + 10x^2 - 12x$ has a local minimum value when x is between 0 and 1. Approximate this value of f.

15. The graph of $f(x) = x^6 + 20x^3 - 15x^2$ has an inflection point whose x-coordinate is between 0 and 1. Approximate the coordinates of this inflection point.

REVIEW EXERCISES

In Exercises 1–10, sketch the graphs and indicate the extrema, intervals of increasing or decreasing, and concavity.

1. $y = 3 - x - 2x^2$

2. $y = 3x^2 + 2x - 1$

3. $y = 2x^3 + 3x^2$

4. $y = 2x^4 - 4x^2$

5. $y = x^2 + \dfrac{2}{x}$

6. $y = x + \sin x$

7. $y = (x + 1)^{2/3}$

8. $y = e^{-x^2}$

9. $y = x^{1/3}(x + 4)$

10. $y = \dfrac{x^2 + 1}{x + 1}$

In Exercises 11–16, use differentials to approximate the given numbers, and check your results with a calculator.

11. $\sqrt{15.9}$

12. $\sqrt{82} - \sqrt[3]{82}$

13. $2(1.01)^5 - 3(1.01)^4$

14. $e^{.01}$

15. $\tan 44.8°$

16. $\cos 121°$

In Exercises 17–22, find dy/dx.

17. $y + \sqrt{y} = x^3$

18. $y = \sqrt{x + y}$

19. $x^2 - 3xy + y^2 = 15$

20. $\dfrac{xy}{x + y} = 1$

21. $x = \cos(x/y)$

22. $y = e^{xy}$

In Exercises 23–26, find d^2y/dx^2.

23. $xy = x + y$

24. $x^2 + y^2 = 4$

25. $x = \sin y$

26. $x = e^y$

In Exercises 27–38, find the general solution to the given differential equation.

27. $y' = 3x^2 + 4x - 2$

28. $y' = x^4 + x^3 + x^2 + x + 1$

29. $s' = \sqrt{t} - (1/t^2)$

30. $s' = t^{-2/3} + t^{-4}$

31. $s' = \sin 2\pi t$

32. $s' = e^{-4t}$

33. $y'' = e^{2x} - \cos 3x$

34. $y'' = x^{-3} + x$

35. $y' = e^x(\cos x + \sin x)$

36. $y' = x(x^2 - 1)^3$

37. $s'' = t^3 - 3t^2 + t$

38. $s'' = 3\cos(4t - \pi)$

In Exercises 39–42, find an equation of the line tangent to the graph of the given equation at the given point.

39. $9x^2 - 16y^2 = 144$; $(5, 9/4)$
40. $(x/3)^4 + (y/6)^4 = 2$; $(3, 6)$
41. $x^{2/3} + y^{2/3} = 2$; $(1, 1)$ **42.** $xy = 3$; $(-1, -3)$

In Exercises 43–46, the symbols s, v, and a represent the position, velocity, and accleration of a particle moving along a horizontal line. In each case, find $s(1)$, $v(1)$, and $a(1)$ from the given information.

43. $a(t) = t^2 - 2t + 1$; $v(0) = -2$, $s(0) = 1$
44. $a(t) = 4\pi^2 \sin 2\pi t$; $v(\frac{1}{4}) = 2$, $s(0) = -2$
45. $v(t) = t(t^2 + 1)$; $s(2) = 4$
46. $v(t) = 2e^{3t}$; $s(0) = 2$

47. A piece of wire 10 inches long is cut into two pieces. One piece is bent into the shape of a circle and the other into an equilateral triangle. Where should the wire be cut so that the total enclosed area of the two pieces is a maximum? A minimum?

48. Two planes leave New York at the same time. One flies in the direction N30°E at 500 mph. The other flies due east at 550 mph. How fast is the distance between them changing after 30 minutes?

49. One plane is flying 500 mph at a constant altitude. At a particular moment another plane is flying 550 mph, and is 1 mile directly above the first plane. They are both heading along straight lines that meet at a point A, 10 miles from the first plane. What is the minimum distance between them as they approach point A?

50. The top of a silo is in the shape of a hemisphere with a radius of 20 feet. If the outside is coated with a sheet of ice 2 inches thick, approximately how many cubic inches of ice are there?

51. (Continuation of Exercise 50). The ice is melting (that is, the volume is changing) at a rate proportional to its surface area. How fast is the thickness of the ice changing?

52. Find all points on the graph of $x^{2/3} + y^{2/3} = 1$ where the tangent line is parallel to $y = x$ or $y = -x$.

53. A department store sells approximately 3,000 boxes of a particular manufacturer's shirt (packed 6 to a box) at about a constant rate throughout the year. It costs $2 per year to store each box, and it costs $25 in office expenses to send in an order. How many boxes of shirts should be ordered each time to minimize the inventory costs?

54. A particle is moving along the curve $5x^2 - 6xy + 5y^2 = 32$. When it is at $(2\sqrt{2}, 2\sqrt{2})$ the x-coordinate is changing at the rate of -3 ft/sec. How fast is y changing?

55. (Continuation of Exercise 54) How fast is the distance between the particle and origin changing at that moment?

56. What point on the curve $y = \sqrt{2x + 3}$ is nearest to the origin?

57. The period P of a pendulum is $P = 2\pi\sqrt{L/32}$ sec, where L is the length in feet. A brass pendulum is 6 ft long at 20°C. Aproximate the change in the period if the temperature rises to 30°C? (The coefficient of linear expansion for brass is 19×10^{-6}.)

58. Towns A and B want to draw water from the same pumping station; their positions are shown in Figure 4–57. At what point along the river should the pumping station be located so as to minimize the length of pipeline?

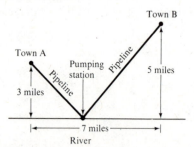

FIGURE 4–57. Exercise 58.

59. Rework Exercise 58 *without* using any calculus.

60. Find a constant c such that the line joining $(1, 4)$ and $(6, -1)$ is tangent to $y = c/(x + 1)$.

61. A motorist drives from town A to town B at 60 mph. He has more time on the return trip, and drives the same route at a leisurely 30 mph. What is his average speed for the round trip? *Hint:* 45 mph is *not* the answer.*

*This problem appears in an article by M. R. Spiegel in *American Mathematical Monthly*, vol. 59 (1952): 99–100.

In Exercises 62–63, sketch a continuous function that satisfies all the given conditions.

62. $f'(-1) = f'(1) = 0$; $f(0) = 1$, but $f'(0)$ does not exist; $f'(2)$ does not exist; $f'(x) > 0$ on $(-1, 0)$ or $(0, 1)$ or $(2, \infty)$; $f'(x) < 0$ on $(-\infty, -1)$ or $(1, 2)$; $f''(x) > 0$ on $(-\infty, 0)$; $f''(x) < 0$ on $(0, 2)$ or $(2, \infty)$.

63. $f(x) > 0$ for all x; $f(0) = 1$; $f'(x) < 0$ on $(0, \infty)$; $f'(x) > 0$ on $(-\infty, 0)$; $f''(x) \geq 0$ for all $x \neq 0$.

64. From a diving board 20 ft above the water, a diver jumps up with a velocity of 16 ft/sec. With what speed does the diver hit the water below?

65. Approximate the coordinates of the point of intersection of $y = x^4$ and $y = x + 2$, $-2 \leq x \leq 0$.

5

INTEGRATION

The study of black and white relations in Mathematics

Integration predates differentiation by almost 2,000 years! The early Greeks, Eudoxus (ca. 370 B.C.), Archimedes (287–212 B.C.), and others, used a crude form of integration to compute areas and volumes of geometric figures. Kepler (1571–1630) also used integration to formulate his second law of planetary motion: *A line joining a planet to the sun sweeps over equal areas in equal intervals of time.* However, these early forms of integration were unwieldy and limited in scope. It was not until the latter part of the 17th century, when Newton and Leibniz established a workable system relating the integral to the derivative, that integration became a viable operation. From that day to this, it remains one of the most powerful mathematical tools at our disposal.

In this chapter we define the integral, and describe various methods of evaluating it. The next chapter is devoted entirely to applications.

In Section 5–1 we present a technique for computing areas and in Section 5–2 this technique is used as a model for defining the definite integral. Section 5–3 contains the statement that expresses the connection between integration and differentiation; it is called the Fundamental Theorem of Calculus. This is the summit; the place from which you get a panoramic view of all of calculus. Section 5–4 explains how to interpret the definite integral so you can use it to solve problems. The next four sections explore some of the other properties and uses of integration. Section 5–9 (Optional) contains the proofs that are omitted in earlier sections.

Virtually every section of this chapter contains important material, but a few items do stand out. (1) Learn the definition of a Riemann sum (Section 5–2) and the definition of upper and lower sums (Section 5–4). They will help you understand the integral and its applications. (2) The Fundamental Theorem of Calculus is easily the most important result in this book; make sure you understand it thoroughly and know how to use it. (3) The method of substitution (Section 5–6) is one of the basic tools for evaluating integrals. Be sure to work enough of the exercises to feel comfortable using it.

5–1 A TECHNIQUE FOR COMPUTING AREAS

Suppose that f is a continuous, nonnegative function defined on an interval $[a, b]$. Let S be the region bounded by the graph of f, the x-axis, and the lines $x = a$ and $x = b$ (Figure 5–1). Then S is called the **region determined by f from a to b**. In this section, we develop a special technique for defining and computing the area of S. The technique is important on two counts.

(1) It is used as a model for defining the integral of f.

(2) It is the basis for most applications of the integral. Some illustrations are given at the end of this section.

The technique is best described by means of examples. But first, an important remark.

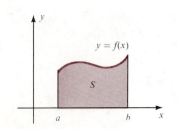

FIGURE 5–1. S is the region determined by f from a to b.

Remark: In Examples 1 and 2, the regions are familiar, and the areas are already known. The purpose of the examples is to *illustrate the special technique* in a simple situation. Therefore, please forget that you already know the area, and concentrate on the technique itself. It will be applied to more complicated situations shortly.

EXAMPLE 1 Let $f(x) = x + 1$ on the interval $[1, 3]$. The region determined by f from 1 to 3 is the trapezoid shown in Figure 5–2. It is easy to compute the area (6 square units), but that is not the purpose of this example. The purpose is to describe a method of approximating areas with rectangles in a special way.

Partition the interval $[1, 3]$ into two subintervals $[1, 2]$ and $[2, 3]$. Each of these subintervals is to be the base of a rectangle whose height is the value of f at some point in the subinterval. In Figure 5–3A we have drawn rectangles of height $f(1) = 2$ and $f(2) = 3$ respectively. The sum of the areas of the two rectangles is

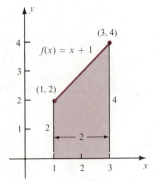

FIGURE 5–2. Example 1.

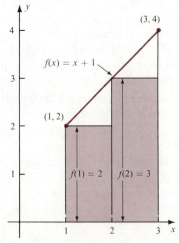

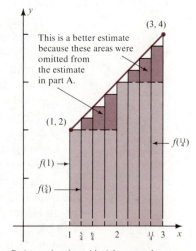

FIGURE 5–3. Example 1.

A. Approximation with two rectangles.

B. Approximation with eight rectangles.

(5.1) $f(1)(2-1) + f(2)(3-2) = 2(1) + 3(1) = 5$ square units

This is a first approximation. Now subdivide the interval [1, 3] into eight subintervals [1, 5/4], [5/4, 6/4], . . . , [11/4, 3]. Over each of these subintervals construct a rectangle whose height is the value of f at the left endpoint. The result is the eight rectangles shown in Figure 5–3B. The sum of their areas is

(5.2) $f(1)\left(\dfrac{1}{4}\right) + f\left(\dfrac{5}{4}\right)\left(\dfrac{1}{4}\right) + \cdots + f\left(\dfrac{11}{4}\right)\left(\dfrac{1}{4}\right) = 2\left(\dfrac{1}{4}\right) + \dfrac{9}{4}\left(\dfrac{1}{4}\right) + \cdots + \dfrac{15}{4}\left(\dfrac{1}{4}\right)$

$$= \dfrac{92}{16} = 5.75 \text{ square units}$$

The figure indicates why this is a better approximation than the previous one. If we continue to make more subdivisions, the estimates will become even more accurate. The table below illustrates this.

(5.3)

NUMBER OF SUBINTERVALS	SUM OF AREAS OF RECTANGLES $f(x) = x + 1$ **FROM** $x = 1$ **TO** $x = 3$
16	$f(1)\left(\dfrac{1}{8}\right) + f\left(\dfrac{9}{8}\right)\left(\dfrac{1}{8}\right) + \cdots + f\left(\dfrac{23}{8}\right)\left(\dfrac{1}{8}\right) = 5.875$
32	$f(1)\left(\dfrac{1}{16}\right) + f\left(\dfrac{17}{16}\right)\left(\dfrac{1}{16}\right) + \cdots + f\left(\dfrac{47}{16}\right)\left(\dfrac{1}{16}\right) = 5.9375$
64	$f(1)\left(\dfrac{1}{32}\right) + f\left(\dfrac{33}{32}\right)\left(\dfrac{1}{32}\right) + \cdots + f\left(\dfrac{95}{32}\right)\left(\dfrac{1}{32}\right) = 5.9687$

It can be shown that the exact area, 6 square units, is obtained by taking the limit of these sums as the length of the subintervals approach zero. ■

Let us pause now to write down the steps of the technique used in Example 1.

> (1) Partition the interval [a, b] into subintervals.
>
> (2) Over each subinterval construct a rectangle whose height is the value of f at some point in the subinterval. (In the example, we always used the left endpoint, but actually any point will do.)
>
> (3) Compute the sum of the areas of the rectangles; this is an approximation of the area of the region determined by f from a to b.
>
> (4) Repeat steps (1), (2), and (3) using more and more subintervals of shorter and shorter lengths. The limit of the approximations obtained in (3) as the lengths of the subintervals approach zero is the exact area.

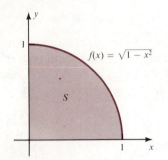

FIGURE 5–4. The region determined by f from 0 to 1.

EXAMPLE 2   Let $f(x) = \sqrt{1 - x^2}$ on the interval $[0, 1]$. The region determined by f from 0 to 1 is the quarter circle shown in Figure 5–4. Let us see how the technique outlined above works in this case.

(1) Partition the interval $[0, 1]$ into two subintervals $[0, 1/2]$ and $[1/2, 1]$ each with length $1/2$.

(2) Choose a point from each subinterval; this time, for a change, we choose the midpoints $1/4$ and $3/4$. Construct rectangles with heights $f(1/4)$ $= \sqrt{1 - (1/4)^2} \approx .9682$ and $f(3/4) = \sqrt{1 - (3/4)^2} \approx .6614$ (Figure 5–5A).

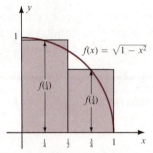

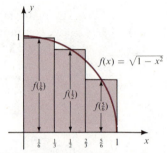

A. Approximation with two rectangles.   B. Approximation with three rectangles.

FIGURE 5–5. Example 2.

(3) The sum of the areas of the two rectangles is

(5.4) $f\left(\dfrac{1}{4}\right)\left(\dfrac{1}{2}\right) + f\left(\dfrac{3}{4}\right)\left(\dfrac{1}{2}\right) \approx (.9682)\left(\dfrac{1}{2}\right) + (.6614)\left(\dfrac{1}{2}\right) = .8148$ square units

(4) Repeat steps (1), (2), and (3). The table below shows the calculations using three, four, and five subintervals of equal length. In each case, the height of the rectangle is the value of f at the midpoint of the subinterval. Figure 5–5B illustrates the case of three subintervals.

(5.5)

NUMBER OF SUBINTERVALS	SUM OF AREAS OF RECTANGLES $f(x) = \sqrt{1 - x^2}$ FROM $x = 0$ TO $x = 1$
3	$f\left(\dfrac{1}{6}\right)\left(\dfrac{1}{3}\right) + f\left(\dfrac{1}{2}\right)\left(\dfrac{1}{3}\right) + f\left(\dfrac{5}{6}\right)\left(\dfrac{1}{3}\right) \approx .8016$
4	$f\left(\dfrac{1}{8}\right)\left(\dfrac{1}{4}\right) + f\left(\dfrac{3}{8}\right)\left(\dfrac{1}{4}\right) + f\left(\dfrac{5}{8}\right)\left(\dfrac{1}{4}\right) + f\left(\dfrac{7}{8}\right)\left(\dfrac{1}{4}\right) \approx .7960$
5	$f\left(\dfrac{1}{10}\right)\left(\dfrac{1}{5}\right) + f\left(\dfrac{3}{10}\right)\left(\dfrac{1}{5}\right) + \cdots + f\left(\dfrac{9}{10}\right)\left(\dfrac{1}{5}\right) \approx .7929$

It can be shown that the limit of these sums as the lengths of the subintervals approach zero is $\pi/4 \approx .7854$, which is the area of the quarter circle.   ■

In the first example, with $f(x) = x + 1$, computing the combined areas of the rectangles was not difficult, and the calculations were carried out up to 64 subintervals. In the second example, with $f(x) = \sqrt{1 - x^2}$, the computations were tedious (even with a calculator), so we stopped at 5 subintervals. But in both cases we illustrated our point; *as the lengths of the subintervals get smaller, the sum of the areas of the rectangles approaches the exact area of the region.*

Some Applications of the Technique

Suppose f is a positive constant function defined on an interval $[x_1, x_2]$. The region determined by f from x_1 to x_2 is a rectangle (Figure 5–6). If w is any point in $[x_1, x_2]$, then $f(w)$ is the height of the rectangle, and its area is the product

$$(5.6) \qquad \text{area} = f(w)(x_2 - x_1)$$

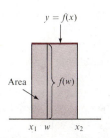

FIGURE 5–6. f is a positive constant on $[x_1, x_2]$. The region determined by f is a rectangle; its area is $f(w)(x_2 - x_1)$, where w is any point in $[x_1, x_2]$.

The simple product (5.6) is subject to a large number of interpretations. For example, if an object moves at a *constant* positive rate, say $v(w)$ ft/sec, over the time interval from $t = x_1$ to $t = x_2$, then

$$\text{distance} = (\text{rate})(\text{time})$$
$$= v(w)(x_2 - x_1) \text{ feet}$$

If a *constant* force, say $F(w)$ pounds, is applied to move an object in a straight line from point x_1 to point x_2, then, by definition,

$$\text{work} = (\text{force})(\text{distance})$$
$$= F(w)(x_2 - x_1) \text{ foot-pounds}$$

Finally, suppose that electricity is consumed at a *constant* rate, say $E(w)$ kilowatts/hour, over a time period from $t = x_1$ to $t = x_2$. Then

$$\text{total consumption} = (\text{consumption rate})(\text{time})$$
$$= E(w)(x_2 - x_1) \text{ kilowatts}$$

(5.7) | The product used to compute the area of a rectangle is subject to many interpretations; for example, *distance, work,* or *total consumption.*

Figure 5–7 illustrates these interpretations.

In Example 1, equation (5.1)

$$f(1)(2 - 1) + f(2)(3 - 2) = 2(1) + 3(1) = 5 \text{ square units}$$

is the sum of the areas of two rectangles; it is a first approximation to the area of the region determined by $f(x) = x + 1$ on $[1, 3]$. If an object is moving at the rate $v(t) = t + 1$ over the time interval $[1, 3]$, it follows from (5.7) that *the sum above can also be interpreted as a first approximation to the distance the object*

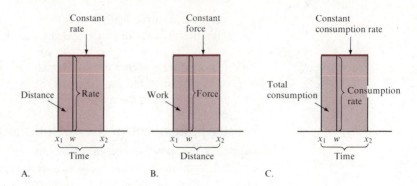

FIGURE 5–7. Three different interpretations of the area of a rectangle. (A) Distance = (rate)(time) (B) Work = (force)(distance) (C) Total consumption = (consumption rate)(time).

moves from time $t = 1$ *to* $t = 3$. Moreover, the approximations become increasingly accurate as the subintervals get smaller, because the rate v is continuous and, therefore, comes closer and closer to being a constant function on each subinterval. Therefore, it is reasonable to define the distance the particle moves to be the limit of these approximations. Similar remarks are valid for work and consumption.

EXAMPLE 3 *(Consumption)* A household is using electricity at the rate of $E(t) = t + 1$ kw/hr. How much electricity is consumed from $t = 1$ to $t = 3$ hours?

Solution This is the same function and interval encountered in Example 1 ($f(x) = x + 1$ on $[1, 3]$). Let us follow the steps in that example. Partition the interval $[1, 3]$ into two subintervals $[1, 2]$ and $[2, 3]$. Next, choose the left endpoints 1 and 2. If we assume *constant* consumption rates of $E(1) = 2$ over the first subinterval and $E(2) = 3$ over the second, then

(5.8) $E(1)(2 - 1) + E(2)(3 - 2) = 2(1) + 3(1) = 5$ kilowatts

represents a first approximation to the amount of electricity consumed. Observe that (5.8) is numerically the same as (5.1), but the *interpretation* is different. By similar arguments, the sums (5.2) and (5.3) in Example 1 can be interpreted as representing better and better approximations to the exact amount consumed. Therefore, we define the amount of electricity consumed to be the limit of these approximations; that is, 6 kilowatts (numerically, the same answer as in Example 1). ■

EXAMPLE 4 A particle moves along a line at the rate of $v(t) = \sqrt{1 - t^2}$ ft/sec. How far does it travel from time $t = 0$ to $t = 1$?

Solution This is the same function and interval encountered in Example 2 ($f(x) = \sqrt{1 - x^2}$ on $[0, 1]$). Let us follow the steps in that example. Partition the interval $[0, 1]$ into two subintervals $[0, 1/2]$ and $[1/2, 1]$. Next, choose the midpoints $1/4$ and $3/4$. If we assume *constant* rates of $v(1/4) \approx .9682$ over the first subinterval and $v(3/4) \approx .6614$ over the second, then

(5.9) $v\left(\dfrac{1}{4}\right)\left(\dfrac{1}{2} - 0\right) + v\left(\dfrac{3}{4}\right)\left(1 - \dfrac{1}{2}\right) \approx .8148$ feet

represents a first approximation to the distance traveled. Observe that (5.9) is numerically the same as (5.4) in Example 2, but the *interpretation* is different. By similar arguments, the sums in table (5.5) can be interpreted as representing better and better approximations to the total distance traveled.

Conclusion. The particle travels $\pi/4 \approx .7854$ feet (numerically, the same answer as in Example 2).   ■

There is an alternate method for computing the distance a particle moves along a horizontal line. If its velocity v is nonnegative over a time interval $[a, b]$, then it is always moving to the right. In this case, the distance it moves is its position at time b minus its position at time a; that is, $s(b) - s(a)$ where s is the position function. Therefore, if you can solve the differential equation $s' = v(t)$ for s, you can find the distance.

EXAMPLE 5   A particle starts from the origin ($s(0) = 0$), and its velocity is $v(t) = t + 1$ ft/sec. How far does it move from time $t = 2$ to $t = 4$ seconds?

Solution   Set $s' = t + 1$ and solve for s.

$$s = \frac{t^2}{2} + t + C$$

and the side condition $s(0) = 0$ means that $C = 0$. Therefore, the distance moved from $t = 2$ to $t = 4$ is

$$s(4) - s(2) = \left[\frac{4^2}{2} + 4\right] - \left[\frac{2^2}{2} + 2\right] = 8 \text{ ft}   ■$$

EXERCISES

In Exercises 1–5 sketch the region determined by the given function, compute its exact area, and then approximate the area with rectangles according to the instructions. Observe that the approximations become more accurate as the lengths of the subintervals get smaller.

1. $f(x) = x - 1$ defined on $[1, 4]$. Partition the interval into three, six, and twelve subintervals of equal length, and each time use the left endpoint of each subinterval to determine the height of the rectangles.

2. $f(x) = 2x$ defined on $[1, 3]$. Partition into two, four, and eight subintervals of equal length, and use the right endpoint of each subinterval.

3. $f(x) = \sqrt{4 - x^2}$ defined on $[0, 2]$. Partition into two, three, and four subintervals of equal length, and use the midpoint.

4. $f(x) = -3x$ defined on $[-1, 0]$. Partition into three, six, and twelve subintervals of equal length, and use the left endpoint.

5. $f(x) = |x|$ defined on $[-1, 1]$. Partition into two, four, and eight subintervals of equal length, and use the midpoint.

In Exercises 6–8, sketch the region determined by the given function. You may not know the areas of these regions, but you can approximate them using our special technique. In each case, partition into two, three, and four subintervals of equal length, and use the midpoints. The exact answers are in the back of the book.

6. $f(x) = 1/x$ defined on $[1, 2]$.

7. $f(x) = x^2$ defined on $[0, 1]$.

8. $f(x) = -x^3$ defined on $[-1, 0]$.

9. As a container moves along an assembly line, it is being filled with merchandise. When the container is at point x, the force needed to keep it moving is $F(x) = x + 4$ pounds. How much work is done in moving the carton from $x = 0$ to $x = 10$ feet? (Compute the area determined by F from 0 to 10. See Examples 3 or 4.)

10. If a particle moves along a line at the rate of $v(t) = 2t - 6$ ft/sec, how far does it move from time $t = 3$ to $t = 5$ seconds?

In Exercises 11–17, use the method of Example 5 to find the distance. You can assume that the particle starts from the origin; $s(0) = 0$.

11. $v(t) = t - 1$ from $t = 1$ to $t = 4$ (compare with Exercise 1).

12. $v(t) = 2t$ from $t = 1$ to $t = 3$ (compare with Exercise 2).

13. $v(t) = -3t$ from $t = -1$ to $t = 0$ (compare with Exercise 4).

14. $v(t) = t^2$ from $t = 0$ to $t = 1$ (compare with Exercise 7).

15. $v(t) = -t^3$ from $t = -1$ to $t = 0$ (compare with Exercise 8).

16. $v(t) = t + 4$ from $t = 0$ to $t = 10$ (compare with Exercise 9).

17. $v(t) = 2t - 6$ from $t = 3$ to $t = 5$ (compare with Exercise 10).

TEST YOUR INTUITION ABOUT AREAS. Exercises 18–25 refer to two continuous, nonnegative functions f and g both defined on the same interval $[a, b]$. Let A be the area of the region determined by f and B be the area of the region determined by g from a to b. It will help to answer the questions below if you sketch the graph of a function (or functions) which meets the stated requirements.

18. If $f(x) = c$ for every x in $[a, b]$, is it true that $A = c(b - a)$?

19. If $f(x) \le M$ for every x in $[a, b]$, is it true that $A \le M(b - a)$?

20. If $m \le f(x)$ for every x in $[a, b]$, what is the comparison between the numbers A and $m(b - a)$?

21. If $f(x) \le g(x)$ for every x in $[a, b]$, what is the comparison between the numbers A and B?

22. Is it true that the area of the region determined by $2f$ from a to b equals $2A$?

23. Is it true that the area of the region determined by $f + g$ from a to b will equal $A + B$?

24. If $f(x) \le g(x)$ for every x in $[a, b]$, what is the area of the region determined by $g - f$ from a to b?

25. Suppose $a < c < b$. Let C be the area of the region determined by f from a to c and D be the area of the region determined by f from c to b. What is the relationship between the numbers A, C, and D?

5–2   THE DEFINITE INTEGRAL

The technique developed in the preceding section will now be used as a model for defining the definite integral. But first, we introduce a time and labor saving notation known as the *sigma notation*.

The Sigma Notation

The examples and exercises of the preceding section involved long sums. To reduce the work of writing out such sums, we use the Greek letter Σ, capital *sigma*. For example, the symbol

$$\sum_{i=1}^{4} 2i$$

is read "the sum from $i = 1$ to 4 of $2i$;" that means to first let $i = 1$, then 2, then 3, and then 4. In each case, multiply i by 2 and add the results. Thus,

$$\sum_{i=1}^{4} 2i = 2 \cdot 1 + 2 \cdot 2 + 2 \cdot 3 + 2 \cdot 4 = 20$$

Letters other than i can be used and the numbers need not begin with 1. For instance, $\sum_{n=3}^{5}(2n + 1)$ is read "the sum from $n = 3$ to 5 of $2n + 1$," and

$$\sum_{n=3}^{5}(2n + 1) = (2 \cdot 3 + 1) + (2 \cdot 4 + 1) + (2 \cdot 5 + 1) = 27$$

Another example is

$$\sum_{k=0}^{3} 2^k = 2^0 + 2^1 + 2^2 + 2^3 = 15$$

The sigma notation can be combined with functional notation. Thus, if f is a function, then

$$\sum_{i=1}^{4} f\left(\frac{i}{i+1}\right) = f\left(\frac{1}{2}\right) + f\left(\frac{2}{3}\right) + f\left(\frac{3}{4}\right) + f\left(\frac{4}{5}\right)$$

In table (5.3) of the previous section, the last entry is a sum of the areas of 64 rectangles; it reads

$$f(1)\left(\frac{1}{32}\right) + f\left(\frac{33}{32}\right)\left(\frac{1}{32}\right) + \cdots + f\left(\frac{95}{32}\right)\left(\frac{1}{32}\right)$$

In the sigma notation, this can be neatly rewritten as

$$\sum_{n=32}^{95} f\left(\frac{n}{32}\right)\left(\frac{1}{32}\right)$$

The Definite Integral

We return now to the task of defining the integral. Let f be a function defined on an interval $[a, b]$, and partition $[a, b]$ into subintervals $[x_0, x_1], [x_1, x_2], \ldots,$ $[x_{n-1}, x_n]$. Choose any point w_1 from the first subinterval $[x_0, x_1]$ and multiply $f(w_1)$ by $x_1 - x_0$. Choose any point w_2 from the second subinterval $[x_1, x_2]$ and multiply $f(w_2)$ by $x_2 - x_1$. Continue taking such products, one for each subinterval, and then compute the sum

$$\sum_{i=1}^{n} f(w_i)(x_i - x_{i-1}) = f(w_1)(x_1 - x_0)$$

$$+ f(w_2)(x_2 - x_1) + \cdots + f(w_n)(x_n - x_{n-1})$$

If the sums obtained in this way get closer and closer to one fixed number as the lengths of the subintervals get closer and closer to zero, then that number is called the *definite integral of f from a to b*, and it is denoted by the symbol

$$\int_a^b f(x)\, dx$$

which is read "the integral from a to b of f of x dee x."

The paragraph above is only a rough outline of the definition. Let us go over it again, step by step, and this time include the details. It is assumed throughout that f is *any* function defined on the interval $[a, b]$; f *is not necessarily continuous or nonnegative*.

Step 1.   *Partition the interval $[a, b]$.*

A **partition** P of an interval $[a, b]$ is any finite set of points $\{x_0, x_1, \ldots, x_n\}$ such that

$$a = x_0 < x_1 < \cdots < x_{n-1} < x_n = b$$

The partition P divides the interval $[a, b]$ into **subintervals** $[x_0, x_1], [x_1, x_2], \ldots, [x_{n-1}, x_n]$. The ith subinterval is $[x_{i-1}, x_i]$; its length is denoted by Δx_i. That is,

$$\Delta x_1 = x_1 - x_0, \Delta x_2 = x_2 - x_1, \ldots, \Delta x_i = x_i - x_{i-1}, \ldots$$

The largest of the lengths $\Delta x_1, \Delta x_2, \ldots, \Delta x_n$ is called the **norm** of P, and is denoted by the symbol $\|P\|$. Of particular interest to us are partitions in which all subintervals have the same length Δx; such partitions are called **regular.**

EXAMPLE 1   The regular partition of $[0, 1]$ with n subintervals is

$$P_n = \left\{0, \frac{1}{n}, \frac{2}{n}, \ldots, \frac{n}{n}\right\}$$

The length of each subinterval is $\Delta x = 1/n$ (Figure 5–8).   ∎

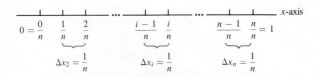

FIGURE 5–8. A regular partition of $[0, 1]$.

Step 2.   *Form the Riemann sums.*

Each partition P divides $[a, b]$ into subintervals $[x_0, x_1], [x_1, x_2]$, and so on. From each subinterval choose a single point; w_1 in $[x_0, x_1]$, w_2 in $[x_1, x_2]$, and, in general,

$$w_i \text{ in } [x_{i-1}, x_i]$$

Now form the sum

$$\sum_{i=1}^n f(w_i)\, \Delta x_i = f(w_1)\, \Delta x_1 + f(w_2)\, \Delta x_2 + \cdots + f(w_n)\, \Delta x_n$$

This is called a *Riemann sum.**

DEFINITION

Let f be a function defined on $[a, b]$, and let $P = \{x_0, x_1, \ldots, x_n\}$ be a partition of $[a, b]$. A **Riemann sum** of f corresponding to the partition P is any sum

$$R_P = \sum_{i=1}^{n} f(w_i) \, \Delta x_i$$

where w_i is an element of $[x_{i-1}, x_i]$.

EXAMPLE 2   Let $f(x) = x$ on $[0, 1]$. Find the Riemann sum corresponding to the regular partition P_n in Example 1 choosing w_i as the right endpoint i/n of each subinterval.

Solution P_n is a regular partition with n subintervals, so each $\Delta x_i = 1/n$.

$$R_{P_n} = \sum_{i=1}^{n} f(w_i) \, \Delta x_i = \sum_{i=1}^{n} f\left(\frac{i}{n}\right)\left(\frac{1}{n}\right) \qquad (w_i = i/n)$$

(5.10)
$$= \sum_{i=1}^{n} \left(\frac{i}{n}\right)\left(\frac{1}{n}\right) \qquad (f(x) = x)$$

$$= \left(\frac{1}{n}\right)\left(\frac{1}{n}\right) + \left(\frac{2}{n}\right)\left(\frac{1}{n}\right) + \cdots + \left(\frac{n}{n}\right)\left(\frac{1}{n}\right)$$

$$= \frac{1}{n^2}[1 + 2 + \cdots + n] \qquad \text{(Simplify)} \qquad ■$$

Observe that if f is nonnegative, then each term $f(w_i) \, \Delta x_i$ of a Riemann sum can be thought of as the area of a rectangle with base Δx_i and height $f(w_i)$. See Figure 5–9A. Thus, a Riemann sum can be thought of as a sum of areas of rectangles. If f is continuous and nonnegative, the Riemann sum is an approximation to the *area under the curve.*† This is illustrated in Figure 5–9B.

Step 3.   *Define the integral.*

Each partition P of $[a, b]$ has a norm $\|P\|$ equal to the length of the longest subinterval; corresponding to each partition are many Riemann sums R_P. Now consider all possible partitions and all corresponding Riemann sums of f. Suppose that as the norms of the partitions get closer and closer to zero, the corresponding Riemann sums get closer and closer to some particular number.

---

*In honor of the famous mathematician G. F. B. Riemann, (1826–1866).

†The area of the region determined by f from a to b is often referred to simply as "the area under the curve."

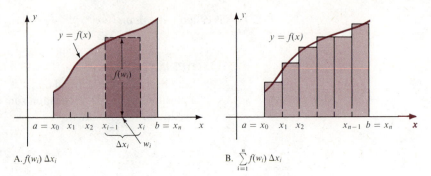

FIGURE 5–9. Riemann sums and areas.

A. $f(w_i)\,\Delta x_i$

B. $\sum_{i=1}^{n} f(w_i)\,\Delta x_i$

Then we call that number the **definite integral of f from a to b**. This is expressed in symbols by writing

$$\int_a^b f(x)\,dx = \lim_{\|P\| \to 0} \sum_{i=1}^{n} f(w_i)\,\Delta x_i{}^*$$

If $\int_a^b f(x)\,dx$ exists, that is, if the limit of the Riemann sums exists, we say that f is **integrable** on $[a, b]$. The value of the integral is a *number;* finding that number is called **evaluating** the integral. The integral sign $\int$ is an elongated S which stands for *sum.* The numbers a and b are called the **limits of integration** and f is called the **integrand.** The symbol dx indicates that x is the independent variable; that is, x varies over the interval $[a, b]$. Other letters are often used in place of x. Thus, as long as f, a, and b remain fixed,

$$\int_a^b f(x)\,dx, \quad \int_a^b f(t)\,dt, \quad \text{and} \quad \int_a^b f(u)\,du$$

all represent the same number. For this reason, the independent variable is often referred to as the **dummy variable.** The integral defined above is sometimes called the **Riemann integral.**

---

*In terms of ϵ's and δ's, the definition is as follows:

$$\int_a^b f(x)\,dx = \lim_{\|P\| \to 0} \sum_{i=1}^{n} f(w_i)\,\Delta x_i$$

means that for every $\epsilon > 0$, there is a $\delta > 0$ such that, if $P = \{x_0, x_1, \ldots, x_n\}$ is any partition with $\|P\| < \delta$, and w_i is any point in $[x_{i-1}, x_i]$, then

$$\left| \int_a^b f(x)\,dx - \sum_{i=1}^{n} f(w_i)\,\Delta x_i \right| < \epsilon$$

Although this is not exactly the same as the limit discussed in Chapter 2, the idea is so similar that the meaning here should be clear.

EXAMPLE 3   Write $\int_0^1 x\,dx$ as a limit of Riemann sums.

Solution   The integrand is $f(x) = x$, and the limits of integration 0 and 1 indicate that the interval is [0, 1]. Thus,

$$(5.11) \qquad \int_a^b x\,dx = \lim_{\|P\| \to 0} \sum_{i=1}^{n} f(w_i)\,\Delta x_i$$

$$= \lim_{\|P\| \to 0} [w_1(x_1 - x_0) + w_2(x_2 - x_1) + \cdots + w_n(x_n - x_{n-1})]$$

where w_i is *any* point in the ith subinterval determined by the partition P.   ■

Not all functions are integrable; there are two examples of nonintegrable functions in the exercises. But in most applications, the functions encountered may be taken as continuous, and *the definite integral of a continuous function always exists*. A proof of this important statement is beyond the scope of this book, but much of our work with integrals depends on it. Therefore, it deserves to be given a special name and highlighted.

(5.12)
> **CONTINUOUS FUNCTION THEOREM**
> If f is continuous on $[a, b]$, then f is integrable on $[a, b]$.

The next example illustrates the power of this theorem.

EXAMPLE 4   Evaluate $\int_0^1 x\,dx$.

Solution   According to the definition, the value of this integral is the limit (5.11) of Example 3. Because the partitions P and the points w_i are arbitrary, evaluating this limit is an arduous, if not impossible, task. However, the integrand $f(x) = x$ is continuous, so by (5.12), we know *the integral exists*. Therefore, if we can find a particular set of Riemann sums R_P that have a limit as $\|P\| \to 0$, then *that limit is the value of the integral*. The Riemann sums we are thinking of are the ones like (5.10) in Example 2 corresponding to regular partitions. For each $n = 1, 2, 3, \ldots$, let P_n be the regular partition of [0, 1] with n subintervals, each of length $1/n$. As n increases, $\|P\| = 1/n \to 0$. Now, from (5.10),

$$R_{P_n} = \frac{1}{n^2}[1 + 2 + \cdots + n]$$

$$= \frac{1}{n^2}\left[\frac{n(n+1)}{2}\right] \qquad \qquad \text{(See footnote)}$$

---

*The formula $1 + 2 + \cdots + n = n(n+1)/2$ and its proof can be found in any algebra book. See, for example, the author's book *Algebra and Trigonometry with Analytic Geometry* (San Francisco: W. H. Freeman and Co., 1979), pp. 440–444.

$$= \frac{1}{2}\left[\frac{n^2 + n}{n^2}\right] \qquad \text{(Rearrange)}$$

$$= \frac{1}{2}\left[1 + \frac{1}{n}\right]$$

As n gets large, the last line approaches $\frac{1}{2}$. Thus,

$$\int_0^1 x \, dx = 1/2 \qquad \blacksquare$$

Some Applications of the Integral

The definition of the definite integral presented above is valid for any function f defined on $[a, b]$; it does not depend on f being continuous or nonnegative. However, for the applications we are about to discuss, let us once again assume that f is *continuous and nonnegative*.

The integral of f is defined as the limit of Riemann sums as the norms of the partitions approach zero. Let P be any partition of $[a, b]$. It follows that

(1) If $\|P\|$ is very close to zero, then

$$R_P \text{ is very close to } \int_a^b f(x) \, dx$$

But every Riemann sum is just the sum of areas of rectangles. Figure 5–9 suggests that each sum approximates the area under the curve. Moreover, the smaller the norm, the better the approximation. Thus,

(2) If $\|P\|$ is very close to zero, then every corresponding Riemann sum

$$R_P \text{ is very close to the area under the curve}$$

Statements (1) and (2) suggest that we make the following definition:

DEFINITION

(5.13) If f is continuous and nonnegative on $[a, b]$, then $\int_a^b f(x) \, dx =$ the area under the curve.

EXAMPLE 5 In Example 4, we found that $\int_0^1 x \, dx = 1/2$. The integrand is $f(x) = x$, and the limits of integration 0 and 1 indicate that the interval is $[0, 1]$. Figure 5–10 shows that the region determined by f from 0 to 1 is a triangle whose area is, indeed, equal to $1/2$ square units. $\blacksquare$

EXAMPLE 6 Compute the value of $\int_1^2 \sqrt{4 - u^2} \, du$.

Solution The integrand is $f(u) = \sqrt{4 - u^2}$, and the limits of integration 1 and 2 indicate that f is defined on the interval $[1, 2]$. To evaluate this integral as a limit of Riemann sums, even with regular partitions, is a monumental task.

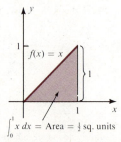

FIGURE 5–10. Example 5.

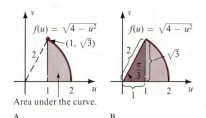

A.   Area under the curve.

B.

FIGURE 5–11. Example 6.

Fortunately, in view of (5.13), we can evaluate the integral by computing the area under the curve, which is part of a circle of radius 2 (Figure 5–11A). To find this area, think of it as the area of a pie-shaped wedge minus the area of a triangle (Figure 5–11B). The area of part of a circle of radius r subtended by a central angle of t radians is $t(r^2/2)$. In this example, $r = 2$ and $t = \pi/3$ (because its cosine is $\frac{1}{2}$). The triangle has base 1 and height $\sqrt{3}$. Thus, the area we seek is

$$A = \frac{\pi}{3}\left(\frac{2^2}{2}\right) - \frac{1}{2}(1)(\sqrt{3}) \approx 1.228 \text{ square units}$$

Therefore, by (5.13),

$$\int_1^2 \sqrt{4 - u^2}\, du = \frac{2\pi}{3} - \frac{\sqrt{3}}{2} \approx 1.228 \qquad \blacksquare$$

In Section 5–1, we observed that the area of a rectangle has many interpretations; for example, distance, work, and total consumption. Since each Riemann sum is simply a sum of areas of rectangles, it follows that such sums can be interpreted as approximations to distance, work, and total consumption. For instance, suppose that an object is moving at the rate of $v(t)$ ft/sec over the time interval $[a, b]$. If $P = \{t_0, t_1, \ldots, t_n\}$ is any partition of $[a, b]$, if w_i is chosen from $[t_{i-1}, t_i]$, and we assume that the object moves at the *constant* rate of $v(w_i)$ ft/sec on $[t_{i-1}, t_i]$, then

$$R_P = \sum_{i=1}^n v(w_i)\, \Delta t_i$$

is an approximation to the total distance traveled. If $\|P\|$ is close to zero, then R_P is close to both the distance traveled and $\int_a^b v(t)\, dt$. This suggests that we define the value of the integral to be the distance traveled. Similar reasoning holds for work and total consumption.

DEFINITIONS

Let v, F, and E be continuous and nonnegative functions on $[a, b]$.

(1) If v represents the rate at which an object moves, then

$$\int_a^b v(t)\, dt = \text{distance traveled}$$

(5.14)

(2) If F represents the force applied to move an object, then

$$\int_a^b F(x)\, dx = \text{work}$$

(3) If E represents the consumption rate of a commodity, then

$$\int_a^b E(t)\, dt = \text{total consumption}$$

EXAMPLE 7 Gasoline is consumed by an engine at the rate of $E(t) = \sqrt{4 - t^2}$ gal/hr. How much is used in two hours of operation, from $t = 0$ to $t = 2$?

Solution The answer, according to (5.14), is

$$\text{total consumption} = \int_0^2 \sqrt{4 - t^2}\, dt \quad \text{gallons}$$

The graph of the integrand $E(t) = \sqrt{4 - t^2}$ on $[0, 2]$ is a quarter of a circle of radius 2. The value of the integral is the area under this curve. Therefore, the engine consumes

$$\int_0^2 \sqrt{4 - t^2}\, dt = \frac{1}{4}(\pi 2^2) \approx 3.14 \text{ gallons of gasoline} \quad \blacksquare$$

EXERCISES

In Exercises 1–14, evaluate the integrals by interpreting the value of the definite integral as the area under a curve. (Pictures will help.)

1. $\int_0^3 \frac{1}{2} x\, dx$

2. $\int_3^5 2x\, dx$

3. $\int_{-5}^{-2} (1 - t)\, dt$

4. $\int_{-4}^0 (-t)\, dt$

5. $\int_3^6 (x - 1)\, dx$

6. $\int_2^4 (x - 1)\, dx$

7. $\int_0^2 (6z + 3)\, dz$

8. $\int_{-1}^1 (2z + 3)\, dz$

9. $\int_0^3 \sqrt{9 - t^2}\, dt$

10. $\int_{-4}^0 \sqrt{16 - t^2}\, dt$

11. $\int_{-1}^1 \sqrt{1 - x^2}\, dx$

12. $\int_{-2}^2 \sqrt{4 - x^2}\, dx$

13. $\int_{-1}^2 \sqrt{4 - u^2}\, du$

14. $\int_{-4}^{-2} \sqrt{16 - u^2}\, du$

In Exercises 15–20, sketch the regions determined by the given functions. Approximate the areas with Riemann sums using *regular* partitions of two, three, and four subintervals, and choosing w_i to be the midpoint. The exact area is given in the back of the book.

15. $f(x) = x^2$ on $[0, 2]$

16. $f(x) = x^3$ on $[0, 1]$

17. $f(x) = \sqrt{x}$ on $[0, 1]$

18. $f(x) = \sqrt[3]{x}$ on $[0, 1]$

19. $f(x) = \sin x$ on $[0, \pi]$

20. $f(x) = \cos x$ on $\left[-\frac{\pi}{2}, \frac{\pi}{2}\right]$

21. A bucket of sand is lifted off the ground to a height of 6 feet. As it rises, sand is pouring out of a hole in the bottom. The force required to lift the bucket when it is x feet above the ground is $F(x) = 20 - 2x$ pounds. How much work is done?

22. A particle moves along a line at the rate of $f(t) = 3t - 2$ ft/sec. How far does it move from time $t = 1$ to $t = 3$ seconds?

In Exercises 23–28, an object starts from the origin $(s(0) = 0)$ and moves at the indicated rate. Use the *alternate method* to find the distance the object moves (see Example 5 in Section 5–1).

23. $v(t) = \frac{1}{2}t$ from $t = 0$ to $t = 3$ (compare with Exercise 1).

24. $v(t) = 1 - t$ from $t = -5$ to $t = -2$ (compare with Exercise 3).

25. $v(t) = t - 1$ from $t = 3$ to $t = 6$ (compare with Exercise 5).

26. $v(t) = 6t + 3$ from $t = 0$ to $t = 2$ (compare with Exercise 7).

27. $v(t) = 20 - 2t$ from $t = 0$ to $t = 6$ (compare with Exercise 21).

28. $v(t) = 3t - 2$ from $t = 1$ to $t = 3$ (compare with Exercise 22).

TEST YOUR INTUITION ABOUT INTEGRALS. Answer the following questions by interpreting the value of the definite integral as the area under a curve (draw pictures). You may assume the functions are continuous so that the integrals do exist.

29. Does $\int_a^b c \, dx = c(b - a)$?

30. If $f(x) \geq 0$ for all x in $[a, b]$, is $\int_a^b f(x) \, dx \geq 0$?

31. If $m \leq f(x) \leq M$ for all x in $[a, b]$, is it true that $m(b - a) \leq \int_a^b f(x) \, dx \leq M(b - a)$?

32. If $f(x) \leq g(x)$ for all x in $[a, b]$, what is the comparison between the numbers $\int_a^b f(x) \, dx$ and $\int_a^b g(x) \, dx$?

33. Does $\int_a^b (f(x) + g(x)) \, dx = \int_a^b f(x) \, dx + \int_a^b g(x) \, dx$?

34. Suppose $a < c < b$. Is it true that $\int_a^b f(x) \, dx = \int_a^c f(x) \, dx + \int_c^b f(x) \, dx$?

35. If $\int_a^b f(x) \, dx = A$ and $c \geq 0$, what is the value of $\int_a^b (f(x) + c) \, dx$?

Optional Exercises

36. Using regular partitions, the formula $1 + 2^2 + \cdots + n^2 = n(n + 1)(2n + 1)/6$, and Example 4 as a guide, show that $\int_0^1 x^2 \, dx = 1/3$.

37. Using regular partitions, the formula $1 + 2^3 + \cdots + n^3 = n^2(n + 1)^2/4$, and Example 4 as a guide, show that $\int_0^1 x^3 \, dx = 1/4$.

38. Given that $\int_0^\pi \sin x \, dx = 2$, evaluate

(a) $\displaystyle\int_0^{\pi/2} \sin x \, dx$ (b) $\displaystyle\int_{-\pi/2}^{\pi/2} \cos x \, dx$

(c) $\displaystyle\int_0^\pi (1 + \sin x) \, dx$ (d) $\displaystyle\int_{\pi/2}^{3\pi/2} \sin\left(x - \frac{\pi}{2}\right) dx$

39. Make up two integral problems similar to those in Exercises 1–14 and evaluate them.

Examples of Nonintegrable Functions

40. Define f on the interval $[0, 1]$ by

$$f(x) = \begin{cases} 1 & \text{if } x \text{ is rational} \\ 0 & \text{if } x \text{ is irrational} \end{cases}$$

(a) Let $P = \{x_0, x_1, \ldots, x_n\}$ be any partition of $[0, 1]$.

(b) In each subinterval pick w_i to be rational and show that the corresponding Riemann sum is 1.

(c) For that same partition, now pick each w_i to be irrational and show that the corresponding Riemann sum is 0.

(d) Steps (b) and (c) show that for each partition there correspond two Riemann sums, one of which equals 1 and the other 0. Explain why this implies that the function is not integrable on $[0, 1]$.

41. Define f on $[0, 1]$ by

$$f(x) = \begin{cases} 1/x & \text{if } x \neq 0 \\ 0 & \text{if } x = 0 \end{cases}$$

(a) Let $P = \{x_0, x_1, \ldots, x_n\}$ be any partition of $[0, 1]$.

(b) Let M be any number greater than 1 (think of M as very large, say ten billion). Then

$$0 < \frac{x_1}{M} < x_1 \quad \text{(why?)}$$

so that x_1/M is a point in the first subinterval of the partition. Pick $w_1 = x_1/M$ and pick any points $w_2, w_3, \ldots, w_n$ in the other subintervals.

(c) Show that the corresponding Riemann sum

$$\sum_{i=1}^n f(w_i) \Delta x_i = f\left(\frac{x_1}{M}\right) \Delta x_1$$
$$+ f(w_2) \Delta x_2 + \cdots + f(w_n) \Delta x_n$$

is larger than M.

(d) Explain why part (c) implies that this function is not integrable on $[0, 1]$.

5-3   THE FUNDAMENTAL THEOREM OF CALCULUS

The definite integral is useful because it can be interpreted in many ways. Be that as it may, the definite integral is use*less* if the only way to evaluate it is to know the area under a curve or to compute a limit of Riemann sums. The first way is too limited, the latter too difficult. True, modern computers could be programmed to approximate such limits, but the point is that back in the 17th century the subject would never have gotten off the ground. But that is precisely when the Fundamental Theorem of Calculus was discovered. It provided an easy method of evaluating integrals using antiderivatives, and *Calculus* was born.

We begin our discussion with an observation about a particle moving along a horizontal line. Let $s = s(t)$ and $v = v(t)$ be the position and velocity functions defined on the time interval $[a, b]$; both are assumed to be continuous. The discussion in the previous two sections suggests that if v is nonnegative, then

$$\int_a^b v(t)\, dt = \text{distance traveled}$$

But if $v \geq 0$, then the particle is always moving to the right. In that case, the distance traveled is the position at time b minus the position at time a; that is, $s(b) - s(a)$. Therefore,

(5.15) $$\int_a^b v(t)\, dt = s(b) - s(a)$$

We know that the position function s is an antiderivative of the velocity function v. Therefore, (5.15) says that

> *The value of the integral of a nonnegative velocity function is found by evaluating an antiderivative at the limits of integration.*

But if this is true for a velocity function, shouldn't it also be true for any continuous, nonnegative function? If so, is it also true for any continuous function, nonnegative or not? The answer to both questions is, *Yes*. That is part of the Fundamental Theorem of Calculus.

The first step leading to the Fundamental Theorem is to use the integral to define a new function. Suppose f is continuous on $[a, b]$; then f is continuous on $[a, x]$ for each x between a and b. By the Continuous Function Theorem (5.12), the integral of f from a to x exists. Therefore,

$$G(x) = \int_a^x f(u)\, du*$$

defines a function G on $[a, b]$ whose value at any x is the area under the curve from a to x (Figure 5–12).

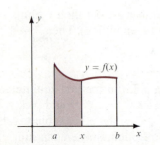

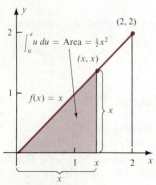

FIGURE 5–12. For each x in $[a, b]$, set $G(x) = \int_a^x f(u)\, du$. Then $G(x)$ is the area under the curve from a to x.

EXAMPLE 1 Let $f(x) = x$ on $[0, 2]$. Find an explicit formula for $G(x) = \int_0^x f(u)\, du = \int_0^x u\, du$.

Solution Sketch the graph of f (Figure 5–13), pick any point x in $[0, 2]$, and compute the area under the curve from 0 to x. The figure shows that the region is a triangle with area $\frac{1}{2}x^2$. Therefore,

$$G(x) = \frac{1}{2}x^2 \qquad \blacksquare$$

FIGURE 5–13. Example 1.

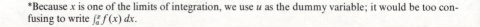

*Because x is one of the limits of integration, we use u as the dummy variable; it would be too confusing to write $\int_a^x f(x)\, dx$.

EXAMPLE 2   Let $f(x) = x - 1$ on $[2, 4]$. Find an explicit formula for $G(x)$ $= \int_2^x f(u)\,du = \int_2^x (u - 1)\,du$.

Solution   Sketch the graph of f (Figure 5–14), pick any x in $[2, 4]$, and compute the area under the curve from 2 to x. The region is a trapezoid; its parallel sides have length 1 and $x - 1$, and the distance between them is $x - 2$. The area is $\frac{1}{2}(x - 2)[1 + (x - 1)] = \frac{1}{2}x^2 - x$. Therefore,

$$G(x) = \frac{1}{2}x^2 - x \qquad \blacksquare$$

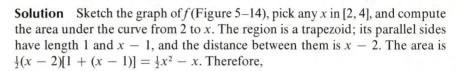

The important thing to notice is that in both examples, *G is an antiderivative of f on* $[a, b]$; that is, G is continuous on $[a, b]$ and $G'(x) = f(x)$ on (a, b).

In Example 1: $G(x) = \frac{1}{2}x^2$ is continuous on $[0, 2]$ and $G'(x) = x = f(x)$ on $(0, 2)$.

In Example 2: $G(x) = \frac{1}{2}x^2 - x$ is continuous on $[2, 4]$ and $G'(x) = x - 1 = f(x)$ on $(2, 4)$.

The fact is that this is *always* the case!

(5.16) | If f is continuous on $[a, b]$ and $G(x) = \int_a^x f(u)\,du$, then G is an antiderivative of f on $[a, b]$.

This result is the heart of the Fundamental Theorem of Calculus. It tells us about the connection between differentiation and integration. A geometric argument that $G'(x) = f(x)$ is as follows (refer to Figure 5–15):

$$G(x + h) = \int_a^{x+h} f(u)\,du \quad \text{is the area from } a \text{ to } x + h$$

$$G(x) = \int_a^x f(u)\,du \qquad \text{is the area from } a \text{ to } x$$

$$G(x + h) - G(x) \qquad \text{is the area from } x \text{ to } x + h$$

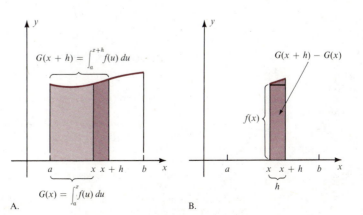

FIGURE 5–15. A geometric argument that $G' = f$.

If h is very small, then f is almost constant on the interval $[x, x + h]$ because f is continuous. Therefore, the area from x to $x + h$ is very close to the area of the rectangle with base h and height $f(x)$. That is,

(5.17) $G(x + h) - G(x) \approx f(x) \cdot h$ or $\dfrac{G(x + h) - G(x)}{h} \approx f(x)$

The closer h is to zero, the better this approximation will be. Therefore,

(5.18) $G'(x) = \lim\limits_{h \to 0} \dfrac{G(x + h) - G(x)}{h} = f(x)$

Remark: The above argument is not a proof; it leaves too many questions unanswered. For example, what happens if $f(x)$ is negative? What happens if h is negative? Does (5.18) really follow from (5.17)? And so on. Nevertheless, it does make the result seem plausible. A precise proof is presented in Section 5–9.*

In the next paragraph we will encounter an integral of the form $\int_a^a f(x)\, dx$. Since it can be thought of as the area under the curve from a to a, it seems reasonable to assign it the value zero.

$$\int_a^a f(x)\, dx = 0$$

Now we are ready to state the theorem.

FUNDAMENTAL THEOREM OF CALCULUS
If f is continuous on $[a, b]$, then

(5.19)

(1) $G(x) = \int_a^x f(u)\, du$ exists for each x in $[a, b]$ and, furthermore, G is an antiderivative of f on $[a, b]$.

(2) If F is *any* antiderivative of f on $[a, b]$, then

$$\int_a^b f(x)\, dx = F(b) - F(a)$$

Part (1) is just the statement (5.16) which we have already discussed. To prove part (2), let F be any antiderivative of f on $[a, b]$. This means that

F is continuous on $[a, b]$ and $F'(x) = f(x)$ on (a, b)

If $G(x) = \int_a^x f(u)\, du$, then G is also an antiderivative of f, and it follows from (4.26), the Equal Derivative Theorem, that

---

*The proof depends on some properties of integrals that are not presented until Section 5–5. There is no harm in this reversal of order because we are careful not to use the fact that $G' = f$ to prove those properties in Section 5–5. The advantage of presenting the material in this order is to gain earlier use of the Fundamental Theorem.

(5.20) $G(x) = F(x) + C$ for all x in $[a, b]$

Since $G(a) = \int_a^a f(u)\, du = 0$, we know from (5.20) that

$$0 = G(a) = F(a) + C$$

and it follows that $C = -F(a)$. Now (5.20) becomes

$$G(x) = F(x) - F(a) \text{ for all } x \text{ in } [a, b]$$

Since $G(b) = \int_a^b f(u)\, du = \int_a^b f(x)\, dx$, we conclude that

$$\int_a^b f(x)\, dx = G(b) = F(b) - F(a)$$

This is what we wanted to prove.

You now have at your disposal a mighty mechanism for evaluating integrals.

(5.21)   If f is continuous and F is *any* antiderivative of f on $[a, b]$, then $\int_a^b f(x)\, dx$ is the number $F(b) - F(a)$.

The difference, $F(b) - F(a)$, is often denoted by one the symbols

$$F(x)\big|_a^b \quad \text{or} \quad [F(x)]_a^b$$

EXAMPLE 3   Evaluate $\int_{-2}^4 (x^3 + 3x^2)\, dx$.

Solution   The first step is to find an antiderivative of the integrand $f(x) = x^3 + 3x^2$. One such function is

$$F(x) = \frac{x^4}{4} + x^3$$

Now, according to (5.21),

$$\int_{-2}^4 (x^3 + 3x^2)\, dx = F(x)\bigg|_{-2}^4 = F(4) - F(-2)$$

$$= \left[\frac{4^4}{4} + 4^3\right] - \left[\frac{(-2)^4}{4} + (-2)^3\right]$$

$$= [64 + 64] - [4 + (-8)]$$

$$= 132 \quad \blacksquare$$

Remark: *Any* antiderivative may be used to evaluate an integral. For instance, $G(x) = x^4/4 + x^3 + 7$ is also an antiderivative of $f(x) = x^3 + 3x^2$. Thus,

$$\int_{-2}^4 (x^3 + 3x^2)\, dx = \left[\frac{x^4}{4} + x^3 + 7\right]_{-2}^4$$

$$= \left[\frac{4^4}{4} + 4^3 + 7\right] - \left[\frac{(-2)^4}{4} + (-2)^3 + 7\right] = 132$$

which agrees with the answer obtained in Example 3. The constant 7 cancels out. Clearly, the simplest antiderivative should be used to evaluate an integral.

EXAMPLE 4

(a) $\displaystyle\int_1^2 x^2\, dx = \left.\frac{x^3}{3}\right|_1^2 = \frac{2^3}{3} - \frac{1^3}{3} = \frac{7}{3}$

(b) $\displaystyle\int_0^4 (3x^2 - 2x - 1)\, dx = \left[x^3 - x^2 - x\right]_0^4 = (4^3 - 4^2 - 4) - (0) = 44$

(c) $\displaystyle\int_0^{\pi/2} \sin t\, dt = \left[-\cos t\right]_0^{\pi/2} = \left(-\cos\frac{\pi}{2}\right) - (-\cos 0) = 1$

(d) $\displaystyle\int_{-1}^1 e^u\, du = \left.e^u\right|_{-1}^1 = e^1 - e^{-1} = e - \frac{1}{e}$ ■

The following example illustrates a few simple techniques, such as multiplying out the integrand and the use of fractional and negative exponents.

EXAMPLE 5

(a) $\displaystyle\int_{-2}^0 (x^2 - 1)^2\, dx = \int_{-2}^0 (x^4 - 2x^2 + 1)\, dx = \left[\frac{x^5}{5} - \frac{2x^3}{3} + x\right]_{-2}^0$ (Check)

$$= 0 - \left(-\frac{32}{5} + \frac{16}{3} - 2\right) = \frac{46}{15}$$

(b) $\displaystyle\int_0^4 (3x + 1)(x - 1)\, dx = \int_0^4 (3x^2 - 2x - 1) = 44$ (Example 4b)

(c) $\displaystyle\int_1^3 \sqrt{t - 1}\, dt = \int_1^3 (t - 1)^{1/2}\, dt = \left[\frac{2}{3}(t - 1)^{3/2}\right]_1^3$ (Check)

$$= \frac{2}{3}[(3 - 1)^{3/2} - (1 - 1)^{3/2}] = \frac{4}{3}\sqrt{2}$$

(d) $\displaystyle\int_{-4}^{-1} \frac{1}{(u - 3)^2}\, du = \int_{-4}^{-1} (u - 3)^{-2}\, du = \left[-(u - 3)^{-1}\right]_{-4}^{-1}$ (Check)

$$= -[(-1 - 3)^{-1} - (-4 - 3)^{-1}]$$

$$= -\left(-\frac{1}{4} + \frac{1}{7}\right) = \frac{3}{28}$$ ■

Remark: Integrands can now be grouped in three categories.

(1) An explicit antiderivative is known (examples are x^n, e^x, $\sin x$, and so on).

(2) An explicit antiderivative is not known, but the area under the curve is known (for example, $\sqrt{1 - x^2}$).

(3) Neither an explicit antiderivative nor the area is known (for example, $1/x$).

Integrals of integrands in (1) and (2) can be evaluated; those in (3) cannot, *as yet!*

EXERCISES

In Exercises 1–8, find an explicit formula for $G(x)$ $= \int_a^x f(u)\, du$. Check to see if $G'(x) = f(x)$. (See Examples 1 and 2.)

1. $f(x) = 3$ on $[1, 3]$ **2.** $f(x) = 2$ on $[0, 3]$

3. $f(x) = 2x$ on $[0, 3]$ **4.** $f(x) = 5x$ on $[0, 1]$

5. $f(x) = 3x - 2$ on $[1, 2]$

6. $f(x) = -2x + 1$ on $[-1, 0]$

7. $f(x) = -x + 1$ on $[-3, -1]$

8. $f(x) = x + 3$ on $[0, 3]$

In Exercises 9–16, evaluate the integrals in two ways. First, by computing areas; second, by using the Fundamental Theorem. Then compare answers.

9. $\displaystyle\int_0^3 4\, dx$ **10.** $\displaystyle\int_{-2}^4 3\, dx$

11. $\displaystyle\int_0^2 2x\, dx$ **12.** $\displaystyle\int_{-2}^0 -2x\, dx$

13. $\displaystyle\int_{-1}^0 -x\, dx$ **14.** $\displaystyle\int_{-2}^0 -3x\, dx$

15. $\displaystyle\int_3^4 (2x + 4)\, dx$ **16.** $\displaystyle\int_1^2 (-2x + 4)\, dx$

In Exercises 17–36, evaluate the integrals using the Fundamental Theorem.

17. $\displaystyle\int_{-2}^2 3x^2\, dx$ **18.** $\displaystyle\int_1^3 6x^2\, dx$

19. $\displaystyle\int_{-2}^2 4x^3\, dx$ **20.** $\displaystyle\int_1^3 4x^3\, dx$

21. $\displaystyle\int_{-2}^0 (3x^2 - 2x + 1)\, dx$

22. $\displaystyle\int_0^2 (-3x^2 + 2x - 4)\, dx$

23. $\displaystyle\int_1^4 \sqrt{x}\, dx$ **24.** $\displaystyle\int_0^8 \frac{1}{\sqrt[3]{x}}\, dx$

25. $\displaystyle\int_0^\pi \sin t\, dt$ **26.** $\displaystyle\int_{-\pi}^0 \cos t\, dt$

27. $\displaystyle\int_1^4 \frac{3}{x^2}\, dx$ **28.** $\displaystyle\int_1^2 \left(2x - \frac{1}{x^3}\right) dx$

29. $\displaystyle\int_{-3}^0 2e^x\, dx$ **30.** $\displaystyle\int_0^4 -2e^x\, dx$

31. $\displaystyle\int_{-2}^0 (t + 2)(t - 3)\, dt$ **32.** $\displaystyle\int_{-1}^1 (t^2 + 1)(t - 1)\, dt$

33. $\displaystyle\int_0^1 (u + 1)^3\, du$ **34.** $\displaystyle\int_0^1 (u - 1)^3\, du$

35. $\displaystyle\int_2^3 \frac{4}{\sqrt{x - 1}}\, dx$ **36.** $\displaystyle\int_0^2 \frac{5}{\sqrt[4]{x + 1}}\, dx$

In Exercises 37–44, evaluate the integrals using the Fundamental Theorem or by computing areas. If neither method is possible at this time, say so.

37. $\displaystyle\int_0^1 \sqrt{1 - x^2}\, dx$ **38.** $\displaystyle\int_{-2}^0 (4 - x^2)\, dx$

39. $\displaystyle\int_1^2 \frac{1}{x}\, dx$ **40.** $\displaystyle\int_0^1 \frac{1}{1 + x^2}\, dx$

41. $\displaystyle\int_0^1 \sqrt{4 - x^2}\, dx$ **42.** $\displaystyle\int_0^1 \frac{1}{\sqrt{4 - x^2}}\, dx$

43. $\displaystyle\int_0^1 e^{3x}\, dx$ **44.** $\displaystyle\int_0^\pi \cos 2x\, dx$

45. If f is continuous for all x and a is any fixed number, what is the value of

$$\frac{d}{dx}\left[\int_a^x f(u)\, du\right]$$

at any x?

46. Find a continuous function f such that $\int_0^x f(u)\, du = x^2 + 3x$ for all x. *Hint:* See Exercise 45.

47. (Continuation of Exercise 46) Find a continuous function f and a constant a such that $\int_a^x f(u)\, du = x^2 + 3x - 4$ for all x.

Optional Exercises

48. (a) Show that $\int_0^1 x(1 - x)^2\, dx = \int_0^1 x^2(1 - x)\, dx$.

    (b) Show that $\int_0^1 x^2(1 - x)^3\, dx = \int_0^1 x^3(1 - x)^2\, dx$.

    (c) Make a conjecture about $\int_0^1 x^n(1 - x)^m\, dx$. Can you prove it?

49. Suppose that f is continuous for all x. If $F(x) = \int_0^{2x} f(u)\,du$, find $F'(x)$. *Hint:* Set $G(x) = \int_0^x f(u)\,du$ and $g(x) = 2x$. Then $F(x) = G(g(x))$; now use the chain rule.

50. (Continuation of Exercise 49) Suppose f is continuous for all x. Find $F'(x)$ if

(a) $F(x) = \displaystyle\int_0^{x^2} f(u)\,du$

(b) $F(x) = \displaystyle\int_0^{\sin x} f(u)\,du$

5–4 INTERPRETATIONS OF THE DEFINITE INTEGRAL

Riemann sums are defined for all functions, and the Fundamental Theorem is true for all continuous functions. But until now, we have interpreted the integral (as area, distance, and so on) only for continuous, *nonnegative* functions. In this section, the restriction that a continuous function be nonnegative is removed. In so doing, we expand old interpretations, and introduce a new one, that of *net profit*.

We begin by defining two special Riemann sums that are important in applications.

Upper and Lower Riemann Sums

Suppose f is continuous on $[a, b]$. Let $P = \{x_0, x_1, \ldots, x_n\}$ be any partition of $[a, b]$. Among the many Riemann sums of f corresponding to P, there are two special ones that we want to single out. Since f is continuous, we know by (4.7), the Extreme Value Theorem, that f will assume an absolute maximum value M_i and an absolute minimum value m_i on each subinterval $[x_{i-1}, x_i]$. These special values of f will be used to define the special Riemann sums.

DEFINITION OF UPPER AND LOWER SUMS

Suppose f is continuous on $[a, b]$ and $P = \{x_0, x_1, \ldots, x_n\}$ is any partition of $[a, b]$. In each subinterval $[x_{i-1}, x_i]$, pick a point u_i such that $f(u_i) = M_i$ is the maximum value of f on $[x_{i-1}, x_i]$. Then the **upper sum U_P associated with P** is the Riemann sum

$$U_P = \sum_{i=1}^{n} f(u_i)\,\Delta x_i = \sum_{i=1}^{n} M_i\,\Delta x_i$$

Similarly, if we pick v_i in $[x_{i-1}, x_i]$ such that $f(v_i) = m_i$ is the minimum value of f on $[x_{i-1}, x_i]$, then the **lower sum L_P associated with P** is the Riemann sum

$$L_P = \sum_{i=1}^{n} f(v_i)\,\Delta x_i = \sum_{i=1}^{n} m_i\,\Delta x$$

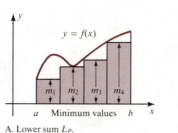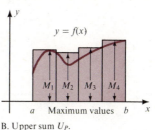

A. Lower sum L_P.   B. Upper sum U_P.

FIGURE 5–16. Lower and upper Riemann sums.

Observe that all other Riemann sums associated with a given partition must lie between the lower and upper sums for that partition. Figure 5–16 shows the rectangles formed for these special Riemann sums. The significance of lower and upper sums will become apparent as you read through the examples below.

Area

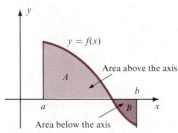

FIGURE 5–17. $\int_a^b f(x)\, dx = A - B$

If f is continuous and nonnegative, the integral of f is the area under the curve. If f is allowed to take negative values, then the integral of f is the area above the x-axis *minus* the area below the x-axis. Thus, referring to Figure 5–17,

$$(5.22) \qquad \int_a^b f(x)\, dx = A - B$$

EXAMPLE 1 Evaluate $\int_0^3 (x - 1)\, dx$ by (a) the Fundamental Theorem, and (b) computing the area above the x-axis minus the area below.

Solution

(a) $\displaystyle \int_0^3 (x - 1)\, dx = \left(\frac{x^2}{2} - x \right) \Big|_0^3 = \frac{3}{2}$

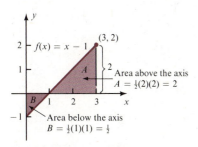

FIGURE 5–18. Example 1. $\int_0^3 (x - 1)\, dx = A - B = \frac{3}{2}$

(b) The graph of f is sketched in Figure 5–18. The region above the axis is a triangle whose area is $A = \frac{1}{2}(2)(2) = 2$ square units. The region below the axis is a triangle whose area is $B = \frac{1}{2}(1)(1) = \frac{1}{2}$ square units. The difference

$$A - B = 2 - \frac{1}{2} = \frac{3}{2} \text{ square units}$$

agrees (numerically) with the value of the integral in (a).   ■

Now let us verify (5.22) for any continuous function by squeezing the number $A - B$ between lower and upper sums.

A typical Riemann sum of f is

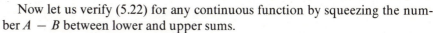

$$R_P = \sum_{i=1}^{n} f(w_i)\, \Delta x_i = f(w_1)\, \Delta x_1 + f(w_2)\, \Delta x_2 + \cdots + f(w_n)\, \Delta x_n$$

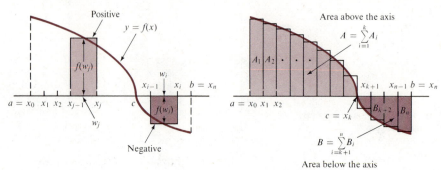

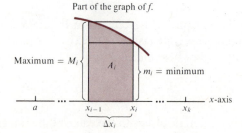

FIGURE 5–19. Approximating areas above and below the *x*-axis.

A.

B.

If $f(w_i) \geq 0$, the rectangle formed lies above the *x*-axis and its area is $f(w_i)\,\Delta x_i$. But what happens if $f(w_i) < 0$? In that case the rectangle formed lies *below* the *x*-axis, and the product $f(w_i)\,\Delta x_i$ is the *negative* of the area (Figure 5–19A). Thus, the Riemann sum R_P approximates the area above the axis *minus* the area below (Figure 5–19B).

Let f be the function in Figure 5–19 and let $P = \{x_0, x_1, \ldots, x_n\}$ be any partition of $[a, b]$. We may assume that P contains the point c where f is zero; that is, assume that $c = x_k$ as shown in the figure.* For each $i = 1, 2, \ldots, k$, let A_i represent the exact area of the region determined by f from x_{i-1} to x_i; these regions all lie above the *x*-axis. For each $i = k + 1, k + 2, \ldots, n$, let B_i represent the area of the region determined by f from x_{i-1} to x_i; these regions all lie below the *x*-axis. Finally, let m_i and M_i represent the minimum and maximum values assumed by f on $[x_{i-1}, x_i]$. For the areas A_i above the axis,

$$m_i\,\Delta x_i \leq A_i \leq M_i\,\Delta x_i \qquad \text{(Figure 5–20)}$$

FIGURE 5–20. A_i is the exact area of the shaded region; m_i and M_i are positive, and $m_i\,\Delta x_i \leq A_i \leq M_i\,\Delta x_i$.

Summing from $i = 1$ to $i = k$,

$$\text{(5.23)} \qquad \sum_{i=1}^{k} m_i\,\Delta x_i \leq \sum_{i=1}^{k} A_i \leq \sum_{i=1}^{k} M_i\,\Delta x_i$$

For the areas B_i below the axis,

$$m_i\,\Delta x_i \leq -B_i \leq M_i\,\Delta x_i \qquad \text{(Figure 5–21)}$$

*If c is not in P, form a new partition including the points of P and c. Adding points to a partition can do no harm, and it may actually improve the Riemann sum approximations to the area.

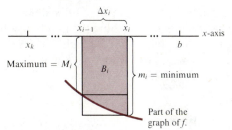

FIGURE 5–21. B_i is the exact area of the shaded region; this time m_i and M_i are negative, and $m_i \Delta x_i \le -B_i \le M_i \Delta x_i$.

Summing from $i = k + 1$ to $i = n$,

$$(5.24) \qquad \sum_{i=k+1}^{n} m_i \Delta x_i \le \sum_{i=k+1}^{n} (-B_i) \le \sum_{i=k+1}^{n} M_i \Delta x_i$$

Now add the inequalities (5.23) and (5.24). The sums on the left, when added together, yield the lower sum L_P associated with P. The sums on the right yield the upper sum U_P. The sums in the middle yield

$$\sum_{i=1}^{k} A_i + \sum_{i=k+1}^{n} (-B_i) = A - B$$

where A is the exact area above, and B is the exact area below the axis. Thus, for any partition P, we have that

$$L_P \le A - B \le U_P$$

Since L_P and U_P are Riemann sums, they *both* approach the number $\int_a^b f(x)\, dx$ as $\|P\|$ approaches zero. Since $A - B$ is always between them, it follows from the Squeeze Theorem (2.9) that

$$\int_a^b f(x)\, dx = A - B$$

This is what we wanted to verify.

Now suppose that we integrate $|f|$ instead of f. The absolute value has the effect of flipping the region below the axis up above the axis (Figure 5–22B). Therefore,

$$\int_a^b |f(x)|\, dx = A + B$$

These notions are summarized below.

(5.25)

Suppose f is continuous on $[a, b]$, and the region determined by f has area A above the x-axis and area B below. Then

(1) $\int_a^b f(x)\, dx = A - B$, called the **net area.**

(2) $\int_a^b |f(x)|\, dx = A + B$, called the **total area.**

To evaluate the integral of $|f|$, first determine where f is positive and negative. You can do that by sketching the graph or solving the inequality $f(x) \geq 0$. Suppose there is a point c in $[a, b]$ such that $f(x) \geq 0$ on $[a, c]$ and $f(x) \leq 0$ on $[c, b]$. Then the integral of f from c to b is the *negative* of the area B below the axis. Therefore, by (5.25),

(5.26)

$$\text{If } f(x) \geq 0 \text{ on } [a, c] \text{ and } f(x) \leq 0 \text{ on } [c, b], \text{ then}$$
$$\int_a^b |f(x)| \, dx = \int_a^c f(x) \, dx - \int_c^b f(x) \, dx$$

EXAMPLE 2   Find the total area bounded by the graph of $f(x) = x^2 - 2$ from 0 to 2.

Solution   The graph of f (Figure 5–22) indicates that

$$f(x) \leq 0 \text{ on } [0, \sqrt{2}] \quad \text{and} \quad f(x) \geq 0 \text{ on } [\sqrt{2}, 2]$$

Therefore, by (5.26), the total area is

$$\int_0^2 |x^2 - 2| \, dx = -\int_0^{\sqrt{2}} (x^2 - 2) \, dx + \int_{\sqrt{2}}^2 (x^2 - 2) \, dx$$
$$= -\left[\frac{x^3}{3} - 2x\right]_0^{\sqrt{2}} + \left[\frac{x^3}{3} - 2x\right]_{\sqrt{2}}^2 \approx 2.44 \text{ square units}$$

■

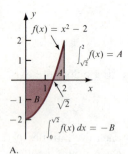

A.

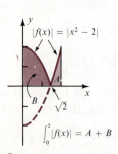

B.

FIGURE 5–22. Example 2.

Motion

Let $v = v(t)$ be the velocity over a time interval $[a, b]$ of a particle moving on a line. If v is nonnegative, then we have interpreted the integral of v as the distance traveled. That is, if $v \geq 0$ on $[a, b]$, then

(5.27) $\int_a^b v(t) \, dt = \text{distance traveled}$

Let us verify (5.27) with upper and lower sums. If $P = \{t_0, t_1, \ldots, t_n\}$ is any partition of $[a, b]$, then

$$\sum_{i=1}^n v(w_i) \, \Delta t_i = v(w_1) \, \Delta t_1 + \cdots + v(w_n) \, \Delta t_n$$

Each term $v(w_i) \, \Delta t_i$ represents the distance the particle would move if v were *constant* at $v(w_i)$ over a time interval of length Δt_i. Let D_i be the exact distance it moves in the time interval $[t_{i-1}, t_i]$, and let M_i and m_i be the maximum and minimum values of v in that interval. Then

(5.28) $m_i \, \Delta t_i \leq D_i \leq M_i \, \Delta t_i \quad \text{for each } i = 1, 2, \ldots, n$

If D is the exact distance the particle moves in the time interval $[a, b]$, then $D = \sum_{i=1}^{n} D_i$, and it follows from (5.28) that

$$L_P \le D \le U_P$$

Thus, if P is *any* partition of $[a, b]$, then the exact distance lies somewhere between the lower and upper sums of v associated with P. As $\|P\|$ approaches zero, both L_P and U_P approach the number $\int_a^b v(t)\, dt$. Since D is always between them, it follows that

$$D = \int_a^b v(t)\, dt = \text{distance traveled}$$

which is what we wanted to verify.

Now suppose that v can also have negative values. On those intervals where $v > 0$, the particle moves in the positive direction; where $v < 0$, it moves in the negative direction. Just as in the case of area, the integral will *automatically* compute the distance moved in the positive direction and *subtract* the distance moved in the negative direction. Also similar to the case of area, the integral of $|v| = speed$ is the sum of the distances moved in both directions.

(5.29)

Suppose $v = v(t)$ is the velocity over the time interval $[a, b]$ of a particle moving on a line. If D^+ and D^- represent the distances traveled in the positive and negative directions, then

(1) $\int_a^b v(t)\, dt = D^+ - D^-$, called the **net distance.**

(2) $\int_a^b |v(t)|\, dt = D^+ + D^-$, called the **total distance.**

EXAMPLE 3   A particle moves on a horizontal line with velocity $v = \sin t$ ft/sec, $t \ge 0$. (a) Where is it after 5 seconds? (b) How far has it traveled (total distance) in the time interval $[0, 5]$?

Solution

(a) To answer the first part, we compute the *net* distance traveled.

$$\int_0^5 \sin t\, dt = \left[-\cos t \right]_0^5 = -\cos 5 + \cos 0$$

$$\approx .7163 \text{ feet} \qquad \text{(Use radians)}$$

Since the net distance is positive, the particle is about .7163 feet to the *right* of its starting point.

(b) To compute the total distance, we must integrate $|\sin t|$. The graph of v (Figure 5–23) indicates that

$$\sin t \ge 0 \text{ on } [0, \pi] \quad \text{and} \quad \sin t \le 0 \text{ on } [\pi, 5]$$

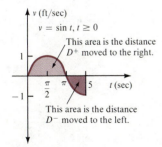

FIGURE 5–23. Example 3.

Since the integral of $\sin t$ from π to 5 is the *negative* of the distance D^-, it follows from (5.29) that the total distance is

$$\int_0^5 |\sin t|\, dt = \int_0^\pi \sin t\, dt - \int_\pi^5 \sin t\, dt$$

$$= \Big[-\cos t\Big]_0^\pi - \Big[-\cos t\Big]_\pi^5$$

$$\approx 2 + 1.2837$$

$$= 3.2837 \text{ feet} \quad \blacksquare$$

Net Profit

Suppose that the rate of profit of a steel company is given by a continuous function $p = p(t)$ over a time interval $[a, b]$. It is important to remark that a *negative* profit is interpreted as a *loss*. Now let P be any partition and form the Riemann sum

$$\sum_{i=1}^n p(w_i)\, \Delta t_i = p(w_1)\, \Delta t_1 + \cdots + p(w_n)\, \Delta t_n$$

Each term $p(w_i)\, \Delta t_i$ represents the profit (or loss) the company would make if the rate of profit were *constant* at $p(w_i)$ over the time interval Δt_i. For instance, if they make (or lose) $p(w_i)$ dollars/year for Δt_i years, then their profit (or loss) would be $p(w_i)\, \Delta t_i$ dollars. Let $\$_i$ be the actual profit or loss in the time interval $[t_{i-1}, t_i]$ and let M_i and m_i be the maximum and minimum values of the profit function p in that interval. Then

(5.30) $\quad m_i\, \Delta t_i \le \$_i \le M_i\, \Delta t_i \quad$ for each $i = 1, 2, \ldots, n$

If $\$$ is the *net profit* (that is, losses subtracted from gains) the steel company makes in the time interval $[a, b]$, then $\$ = \sum_{i=1}^n \$_i$ and it follows from (5.30) that

$$L_P \le \$ \le U_P$$

Thus, if P is any partition of $[a, b]$, the exact net profit of the company lies somewhere between the lower and upper sums of the profit function p. It follows, as in the two previous examples, that the net profit is the integral of p.

(5.31)

> If the rate of profit of a company is a continuous function $p = p(t)$ over a time interval $[a, b]$, then
>
> $$\int_a^b p(t)\, dt$$
>
> is the **net profit** of the company over that interval.

EXAMPLE 4 A company that manufactures silicon chips for computers began operations in 1970. Over the first several years, its rate of profit p in millions of

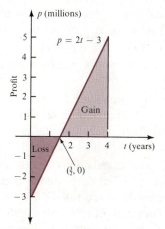

FIGURE 5–24. Example 4.

dollars per year is $p(t) = 2t - 3$. What is its net profit over the first four years of operation?

Solution   The graph of p (Figure 5–24) shows that the company lost money at first but then began to make a profit. The net profit is the integral $\int_0^4 (2t - 3)\, dt$ which *automatically* subtracts losses from gains.

$$\int_0^4 (2t - 3)\, dt = \left[t^2 - 3t\right]_0^4 = 4 \text{ million dollars net profit} \quad \blacksquare$$

EXAMPLE 5   An oil company predicts that the profit on a barrel of oil t years from now will be $t^2 + 3$ dollars. If it can produce B barrels a year, what will be its profit over the next two years?

Solution   The company will make a profit at the rate of $B \cdot (t^2 + 3)$ dollars per year. Therefore, over the next two years, its profit will be

$$\int_0^2 B(t^2 + 3)\, dt = B\left(\frac{t^3}{3} + 3t\right)\Big|_0^2$$

$$= \frac{26}{3} B \text{ dollars} \quad \blacksquare$$

EXERCISES

In Exercises 1–4 evaluate the integrals by (a) the Fundamental Theorem and (b) computing the net area.

1. $\int_{-1}^{1} (2 - 3x)\, dx$ **2.** $\int_0^2 (2x - 1)\, dx$

3. $\int_0^{2\pi} \sin x\, dx$ **4.** $\int_{-1}^{1} x\, dx$

Evaluating integrals is like learning to play a musical instrument; you should practice every day. In Exercises 5–22, evaluate the integrals by the Fundamental Theorem or by computing areas. If neither method is possible at this time, say so.

5. $\int_0^{\pi/4} \sec^2 x\, dx$ **6.** $\int_0^1 \cos 3x\, dx$

7. $\int_0^1 e^{4x}\, dx$ **8.** $\int_1^2 \frac{1}{x}\, dx$

9. $\int_0^1 \frac{1}{1 + x^2}\, dx$ **10.** $\int_0^1 \frac{1}{(1 + x)^2}\, dx$

11. $\int_0^1 \sqrt{4 - t^2}\, dt$ **12.** $\int_{-1}^0 \sqrt{4 - t^2}\, dt$

13. $\int_{-2}^{-1} \sqrt{4 - u^2}\, du$ **14.** $\int_0^4 \sqrt{16 - u^2}\, du$

15. $\int_{-1}^2 (x^4 - x^2)\, dx$ **16.** $\int_0^2 (x^3 + x)\, dx$

17. $\int_0^2 (x^2 - 1)^2\, dx$ **18.** $\int_{-1}^1 (x + 1)^3\, dx$

19. $\int_3^8 \sqrt{x + 1}\, dx$ **20.** $\int_0^5 \sqrt{x + 4}\, dx$

21. $\int_0^7 \frac{1}{\sqrt[3]{t + 1}}\, dt$ **22.** $\int_1^{16} \frac{1}{\sqrt[4]{t}}\, dt$

Evaluate the integrals in Exercises 23–25.

23. $\int_0^2 (x^2 - 1)\, dx$ and $\int_0^2 |x^2 - 1|\, dx$

24. $\int_0^\pi \cos t\, dt$ and $\int_0^\pi |\cos t|\, dt$

25. $\int_0^3 (x^2 - 4x + 3)\, dx$ and $\int_0^3 |x^2 - 4x + 3|\, dx$

Exercises 26–29 illustrate that the same integral can be interpreted many ways depending on the meaning of the integrand.

26. (Refer to Exercise 23) (a) What is the total area under the curve $y = x^2 - 1$ from 0 to 2? How much of the area lies above the x-axis? How much below the axis? (b) If the velocity of a particle moving along a line is $v(t) = t^2 - 1$ ft/sec, where is the particle 2 seconds after it starts? How far has it moved in those 2 seconds? (c) If a company's rate of profit is $p(t) = t^2 - 1$ millions of dollars per year, what is its net profit after 2 years of operation?

27. (Refer to Exercise 25) (a) What is the total area under the curve $y = x^2 - 4x + 3$ from 0 to 3? How much of the area lies above the x-axis? How much below the axis? (b) If the velocity of a particle moving along a line is $v(t) = t^2 - 4t + 3$ mph, where is it after 3 hours? How far has it moved in those hours? (c) If a company makes a profit at the rate of $p(t) = t^2 - 4t + 3$ millions of dollars per year, what is its net profit in the first three years?

28. Suppose $\int_a^b f(x)\, dx = -3$ and $\int_a^b |f(x)|\, dx = 8$. (a) What is the total area under the curve $y = f(x)$ from a to b? (b) How much of the area is above the x-axis? How much is below the axis? (c) If $f(t)$ is the velocity in feet per second of a particle moving along a line, what is the significance of the number -3? What is the significance of the number 8? (d) If $f(t)$ is the rate of profit of a company in millions of dollars per year, how would you interpret the number -3?

29. Answer the same type of questions asked in Exercise 28 if $\int_a^b f(x)\, dx = 6$ and $\int_a^b |f(x)|\, dx = 10$.

30. It is estimated that the rate of profit p on a barrel of oil t years from now will be $p = 2t + 6$ dollars per barrel. If an oil well can produce 50,000 barrels of oil per year, what is the total profit in the first three years of operation?

31. If the oil well in Exercise 30 runs dry in four years, what is the total profit?

32. With our current energy situation, it is probably more realistic to assume that profits on oil will rise exponentially. Rework Exercise 30 with the rate of profit function $p(t) = e^t + 5$ dollars per barrel.

33. A company manufactures calculators that sell for $15. The present cost is $8 but they predict their costs will rise continuously at the rate of 75¢ per calculator per year. They cannot raise their selling price because of the fierce competition. If they sell 25,000 calculators each year, what will be their total profit over the next four years?

34. A particle starts from the origin $s(0) = 0$ and moves along a line with a velocity $v = 4t - 3$ ft/sec. Where is the particle when $t = 2$? What is the total distance it has moved?

35. A particle starts from the point $(2, 0)$ with a velocity of -5 ft/sec. It moves along the t-axis with a constant acceleration of 2 ft/sec^2. Where is the particle after 4 seconds? What is the total distance it has moved?

36. A printing company starts operations in 1980. The owner has researched some of the established printing companies in the area and concludes that he can expect a rate of profit of $p(t) = 10,000\,(t^2 - 1)$ dollars per year for the first few years. Aside from the money for equipment, he has $10,000 working capital. Will he run out of money before he starts making a profit? How long will it be before his net profit is positive?

Optional Exercises

You probably do not know the antiderivatives of the integrands in Exercises 37 and 38, but perhaps you can think of another way to evaluate the integrals (draw pictures).

37. $\int_1^3 \sqrt{-x^2 + 4x - 3}\, dx$

38. $\int_2^3 (1 + \sqrt{-x^2 + 6x - 8})\, dx$

39. If f is continuous and positive on $[a, b]$, show that $F(x) = \int_a^x f(u)\, du$ is an increasing function.

5-5 ·SOME PROPERTIES OF THE DEFINITE INTEGRAL

This section presents several useful properties of integrals. Some are proved here in an informal way; proofs of a more technical nature are presented in Section 5–9.*

If $f(x) = c$ on $[a, b]$, then each product $f(w_i)\, \Delta x_i$ equals $c\, \Delta x_i$, and every Riemann sum of f is simply c times $\Delta x_1 + \cdots + \Delta x_n = b - a$. Therefore,

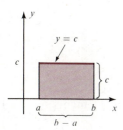

FIGURE 5–25. $\int_a^b c\,dx = c(b - a)$

(5.32)

> **INTEGRAL OF A CONSTANT**
> If c is any constant, then
> $$\int_a^b c\, dx = c(b - a)$$

(Figure 5–25)

If c is a constant, then every Riemann sum of the function cf is just c times a Riemann sum of f. It follows that

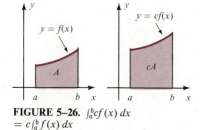

FIGURE 5–26. $\int_a^b cf(x)\, dx$ $= c\int_a^b f(x)\, dx$

(5.33)

> **INTEGRAL OF A CONSTANT TIMES A FUNCTION**
> If f is continuous on $[a, b]$, and c is any constant, then
> $$\int_a^b cf(x)\, dx = c\int_a^b f(x)\, dx$$

(Figure 5–26)

Remark: Statement (5.33) says that a constant can be factored into or out of an integral. This is sometimes referred to as *taking the constant inside (or outside) the integral.* It is an extremely useful tool in evaluating integrals. You will see how it works in the next section.

(5.34)

> **INTEGRAL OF A SUM OR DIFFERENCE**
> If f and g are continuous on $[a, b]$, then
> $$\int_a^b (f + g)(x)\, dx = \int_a^b f(x)\, dx + \int_a^b g(x)\, dx$$
> and
> $$\int_a^b (f - g)(x)\, dx = \int_a^b f(x)\, dx - \int_a^b g(x)\, dx$$

---

*We must resist the temptation of using the Fundamental Theorem to prove these properties because, as the footnote on page 238 points out, the proof of that theorem depends on some of the properties presented here.

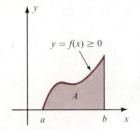

FIGURE 5–27. $\int_a^b f(x)\, dx = A \geq 0$

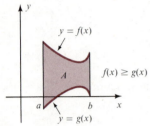

FIGURE 5–28. $\int_a^b f(x)\, dx - \int_a^b g(x)\, dx$
$= \int_a^b (f - g)(x)\, dx \geq 0$

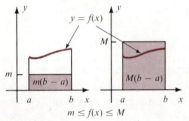

FIGURE 5–29. $m(b - a) \leq \int_a^b f(x)\, dx$
$\leq M(b - a)$

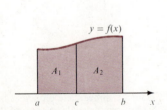

FIGURE 5–30. $\int_a^c f(x)\, dx + \int_c^b f(x)\, dx$
$= \int_a^b f(x)\, dx$

The formal proof of (5.34) is presented in Section 5–9. The basic idea of the proof is that every Riemann sum of $f + g$ can be separated into a Riemann sum of f plus a Riemann sum of g.

Now suppose that $f(x)$ is nonnegative on $[a, b]$. Then every Riemann sum of f is also nonnegative, and therefore the integral of f is nonnegative (Figure 5–27). Moreover, if $f(x) \geq g(x)$ on $[a, b]$ then $(f - g)$ is nonnegative on $[a, b]$. Thus, by (5.34), the integral of f is greater than or equal to the integral of g (Figure 5–28).

(5.35)

> **INTEGRAL OF A NONNEGATIVE FUNCTION**
> If f is continuous and $f(x) \geq 0$ for all x in $[a, b]$, then
> $$\int_a^b f(x)\, dx \geq 0$$
> Furthermore, if f and g are any two continuous functions and $f(x) \geq g(x)$ for all x in $[a, b]$, then
> $$\int_a^b f(x)\, dx \geq \int_a^b g(x)\, dx$$

(5.36)

> **COROLLARY**
> If f is continuous and $m \leq f(x) \leq M$ for all x in $[a, b]$, then
> $$m(b - a) \leq \int_a^b f(x)\, dx \leq M(b - a)$$

(Figure 5–29)

The next property can be quite useful in evaluating the integral of certain types of functions.

(5.37)

> If f is continuous on $[a, b]$ and $a < c < b$, then
> $$\int_a^b f(x)\, dx = \int_a^c f(x)\, dx + \int_c^b f(x)\, dx$$

Statement (5.37) is easily understood in terms of area. It says that the area under the curve from a to b equals the area from a to c plus the area from c to b (Figure 5–30). However, the analytical proof is not so simple; it is presented in Section 5–9.

EXAMPLE 1   Suppose $f(x) = \begin{cases} x^2 & \text{for } 0 \le x \le 1 \\ -x + 2 & \text{for } 1 \le x \le 2 \end{cases}$

Evaluate $\int_0^2 f(x)\,dx$.

Solution   The graph (Figure 5–31) indicates that you should break up the domain into two parts and compute the integral on each part separately. Thus,

$$\int_0^2 f(x)\,dx = \int_0^1 f(x)\,dx + \int_1^2 f(x)\,dx \tag{5.37}$$

$$= \int_0^1 x^2\,dx + \int_1^2 (-x + 2)\,dx$$

$$= \frac{1}{3} + \frac{1}{2} = \frac{5}{6} \quad \blacksquare$$

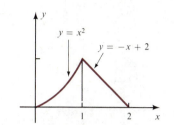

FIGURE 5–31. Example 1.

Statement (5.37) also provides a convenient method for evaluating the integral of an absolute value.

EXAMPLE 2   Evaluate $\int_0^2 |x^2 - 2|\,dx$.

Solution   The integrand is the function of Example 2 in Section 5–4 (see Figure 5–22). Here we can write

$$|x^2 - 2| = \begin{cases} -x^2 + 2 & \text{for } 0 \le x \le \sqrt{2} \\ x^2 - 2 & \text{for } \sqrt{2} \le x \le 2 \end{cases}$$

Therefore, by (5.37),

$$\int_0^2 |x^2 - 2|\,dx = \int_0^{\sqrt{2}} (-x^2 + 2)\,dx + \int_{\sqrt{2}}^2 (x^2 - 2)\,dx \approx 2.44 \quad \blacksquare$$

In writing the symbol $\int_a^b$, we have always assumed that a is less than b. In order to free ourselves of this assumption, let us agree that reversing the limits of integration has the effect of reversing the sign of the integral. In other words

(5.38)

> **DEFINITION**
> If f is continuous on $[a, b]$ then
> $$\int_b^a f(x)\,dx = -\int_a^b f(x)\,dx$$

EXAMPLE 3

(a) $\int_3^1 f(x)\,dx = -\int_1^3 f(x)\,dx$

(b) $\displaystyle\int_0^{-2} f(x)\,dx = -\int_{-2}^0 f(x)\,dx$ ∎

Combining (5.37) and (5.38), we conclude that

(5.39)

> If f is continuous on an interval I, then
>
> $$\int_a^b f(x)\,dx = \int_a^c f(x)\,dx + \int_c^b f(x)\,dx$$
>
> for *any* three points a, b, and c in I regardless of their order.

EXAMPLE 4   If f is continuous on $[a, b]$, and x and $x + h$ are in $[a, b]$, it follows from (5.39) that $\int_a^{x+h} f(u)\,du = \int_a^x f(u)\,du + \int_x^{x+h} f(u)\,du$. Therefore, whether h is positive or negative, we have

(5.40) $$\int_a^{x+h} f(u)\,du - \int_a^x f(u)\,du = \int_x^{x+h} f(u)\,du$$

This equation will be referred to in Section 5–9 when we prove the first part of the Fundamental Theorem.   ∎

The Mean Value Theorem for derivatives (Section 3–6) has an analog for integrals.

(5.41)

> **MEAN VALUE THEOREM FOR INTEGRALS**
> If f is continuous on $[a, b]$, then there is some point c in (a, b) such that
>
> $$f(c) = \frac{1}{b-a}\int_a^b f(x)\,dx$$

A proof of the Mean Value Theorem is outlined in an exercise.

If both sides of (5.41) are multiplied by $b - a$, then the equation becomes

$$f(c)(b - a) = \int_a^b f(x)\,dx$$

In geometric terms, this says that if $f(x) \geq 0$, then the area of the rectangle with base $b - a$ and height $f(c)$ equals the area under the curve (Figure 5–32A). It is as if the number $f(c)$ were the average of all the values of f. For that reason we make the following definition:

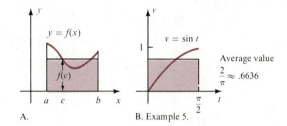

FIGURE 5–32. Illustration of the Mean Value Theorem for Integrals; the rectangle has the same area as the area under the curve.

A.

B. Example 5.

DEFINITION OF AVERAGE VALUE

If f is continuous on $[a, b]$, then the **average value** of f on $[a, b]$

(5.42)   is the number

$$\frac{1}{b - a} \int_a^b f(x)\, dx$$

The Mean Value Theorem guarantees that a continuous function actually assumes its average value at some point c in the open interval (a, b). Here is an application of the average value of a function.

EXAMPLE 5   A particle moves along a line with velocity $v(t) = \sin t$ m/sec from $t = 0$ to $t = \pi/2$. What is its average velocity over this time period?

Solution   According to definition (5.42), the average velocity is

$$\frac{1}{(\pi/2) - 0} \int_0^{\pi/2} \sin t\, dt = \frac{2}{\pi} \left[-\cos t \right]_0^{\pi/2}$$

$$= \frac{2}{\pi} \approx .6636 \text{ m/sec}$$

Since $v(t) \geq 0$ on $[0, \pi/2]$, we interpret the average velocity as follows: If the particle moved at this *constant* velocity, then the distance it would travel (the area of the rectangle in Figure 5–32B) equals the distance it actually travels (the area under the sine curve).   ■

Remark: The properties and definitions above are all stated in terms of continuous functions. That is because the functions encountered in most applications can be taken as continuous. However, you should be aware that there are integrable functions that are not continuous; examples will be presented later. For now, it suffices to say that all but one of the properties remain valid if the word "continuous" is replaced by "integrable;" the only exception is The Mean Value Theorem (5.41).

EXERCISES

Evaluate the integrals in Exercises 1–12.

1. $\int_{-1}^{3} dx$

2. $\int_{-4}^{-2} 5\, dx$

3. $\int_{4}^{6} |-2x + 9|\, dx$

4. $\int_{0}^{2} |3x - 1|\, dx$

5. $\int_{1}^{3} |x^2 - 4|\, dx$

6. $\int_{0}^{2} |1 - x^2|\, dx$

7. $\int_{0}^{3} |x^2 + 2x - 3|\, dx$

8. $\int_{0}^{5} |x^2 - 3x - 4|\, dx$

9. $\int_{-1}^{2} (x - 2)(3x + 8)\, dx$

10. $\int_{-1}^{0} (x + 3)(x - 1)\, dx$

11. $\int_{0}^{1} \frac{1}{(x + 1)^3}\, dx$

12. $\int_{0}^{1} \frac{1}{\sqrt{x + 1}}\, dx$

Exercises 13 and 14 illustrate that the definition $\int_{b}^{a} f(x)\, dx = -\int_{a}^{b} f(x)\, dx$ is consistent with the assumption that the Fundamental Theorem applies to both integrals.

13. Evaluate $\int_{-2}^{1} 3x^2\, dx$ and $\int_{1}^{-2} 3x^2\, dx$ using the Fundamental Theorem.

14. Evaluate $\int_{-3}^{0} 1/\sqrt{x + 5}\, dx$ and $\int_{0}^{-3} 1/\sqrt{x + 5}\, dx$ using the Fundamental Theorem.

Exercises 15–17 illustrate that $\int_{a}^{b} f(x)\, dx = \int_{a}^{c} f(x)\, dx + \int_{c}^{b} f(x)\, dx$.

15. Evaluate $\int_{-3}^{4} (2x - 3)\, dx$ and $\int_{-3}^{0} (2x - 3)\, dx + \int_{0}^{4} (2x - 3)\, dx$.

16. Evaluate $\int_{-2}^{1} 3x^2\, dx$ and $\int_{-2}^{3} 3x^2\, dx + \int_{3}^{1} 3x^2\, dx$.

17. Evaluate $\int_{9}^{4} \sqrt{x}\, dx$ and $\int_{9}^{16} \sqrt{x}\, dx + \int_{16}^{4} \sqrt{x}\, dx$.

In Exercises 18–19, evaluate $\int_{a}^{b} f(x)\, dx$.

18. $f(x) = \begin{cases} x & \text{for } 0 \le x \le 2 \\ -x + 4 & \text{for } 2 \le x \le 4 \end{cases}$ $a = 0, b = 4$

19. $f(x) = \begin{cases} x^2 & \text{for } -1 \le x \le 2 \\ 4 & \text{for } 2 \le x \le 3 \\ x + 1 & \text{for } 3 \le x \le 5 \end{cases}$ $a = -1, b = 5$

20. Find the average value of the functions $y = x$, $y = x^2$, and $y = x^3$ over the interval $[-1, 1]$. What is the limit of the average values of $y = x^n$ over the interval $[-1, 1]$ as n gets larger and larger?

21. Same as Exercise 20 only this time find the average values over the interval $[0, 2]$.

22. The output of a power station t hours after midnight is $1{,}000 + 4{,}800t - 200t^2$ kilowatts. What is the maximum power output of the station? The minimum power? What is the average power output in a 24 hour period?

23. Consider the x-axis from -1 to 1 as a diameter of the unit circle $x^2 + y^2 = 1$. What is the average length of the chords (line segments extending from one part of the circle to another) that are perpendicular to the diameter?

24. Use (5.32) and (5.35) to show that if f is continuous and $m \le f(x) \le M$ on $[a, b]$, then $m(b - a) \le \int_{a}^{b} f(x)\, dx \le M(b - a)$.

25. Show that if f is continuous on $[a, b]$, then $-\int_{a}^{b} |f(x)|\, dx \le \int_{a}^{b} f(x)\, dx \le \int_{a}^{b} |f(x)|\, dx$ from which you can conclude that $\left| \int_{a}^{b} f(x)\, dx \right| \le \int_{a}^{b} |f(x)|\, dx$.

Optional Exercises

26. Fill in the blanks to prove (5.41) the Mean Value Theorem for integrals: If f is continuous on $[a, b]$, then there is some point c in (a, b), such that

$$f(c) = \frac{1}{b - a} \int_{a}^{b} f(x)\, dx$$

Proof

(1) Let $F(x) = \int_{a}^{x} f(u)\, du$; then F is an antiderivative of f on $[a, b]$ by the _____.

(2) So F is continuous on $[a, b]$ and differentiable on (a, b). It follows from the _____

that there is some point c in (a, b) such that

$$F'(c) = \frac{F(b) - F(a)}{b - a}$$

(3) $F'(c) = f(c)$ because _____, $F(b)$ = _____, and $F(a)$ = _____.

(4) It follows from steps (2) and (3) that

$$f(c) = \frac{1}{b - a}\int_a^b f(x)\, dx$$

which completes the proof.

The usual definition of the *average* of n numbers is the *sum of the numbers divided by n.* The technical name for this average is the **arithmetic average** (or arithmetic mean). The next exercise shows the connection between the arithmetic average and the average value of a continuous function.

27. Let f be continuous on $[a, b]$. Show that the average value $1/(b - a)\int_a^b f(x)\, dx$ is the limit of arithmetic averages of values of f. *Hint:* Let P be a partition that divides $[a, b]$ into n subintervals of equal length $(b - a)/n$. If R_P is any Riemann sum corresponding to P, show that $1/(b - a)\, R_P$ is an arithmetic average of values of f. Then take the limit as $\|P\| \to 0$.

28. Suppose f and its derivative f' are continuous on $[a, b]$. The rate of change of f at any x is $f'(x)$. The *average* rate of change of f on $[a, b]$ is

$$\frac{f(b) - f(a)}{b - a}$$

Is it true that the average value of f' equals the average rate of change of f on $[a, b]$? Explain.

29. Let f be defined on $[0, 1]$ by

$$f(x) = \begin{cases} 1 & \text{if } x \text{ is rational} \\ -1 & \text{if } x \text{ is irrational} \end{cases}$$

Is f integrable? Is $|f|$ integrable? Explain your answers.

5–6 THE INDEFINITE INTEGRAL AND THE METHOD OF SUBSTITUTION

The first step in evaluating a definite integral is to find an antiderivative of the integrand. In this section we will take a closer look at finding antiderivatives.

The integral $\int f(x)\, dx$ written without limits of integration is called the **indefinite integral of** f. It stands for the set of all antiderivatives of f. Given a particular antiderivative F of f, we know that if G is also an antiderivative of f, then $G = F + C$ where C is some constant. Therefore, we make the following formal definition.

(5.43)

> **DEFINITION OF THE INDEFINITE INTEGRAL**
> $\int f(x)\, dx = F(x) + C$ if and only if F is an antiderivative of f.

The arbitrary constant C is called the **constant of integration** and the process of finding all antiderivatives is called **evaluating the** (indefinite) **integral.** The domains of f and F are not usually specified; it is always assumed that F is an antiderivative of f on *some* interval.

EXAMPLE 1   Evaluate $\int(3x^2 - 2x)\, dx$.

Solution Find any antiderivative of $f(x) = 3x^2 - 2x$; the function $F(x) = x^3 - x^2$ will do. Then, by definition,

$$\int (3x^2 - 2x)\, dx = x^3 - x^2 + C \quad \blacksquare$$

EXAMPLE 2

(a) $\int dx = x + C$ because $x' = 1$

(b) $\int (x + 3)\, dx = \frac{1}{2}x^2 + 3x + C$ because $\left(\frac{1}{2}x^2 + 3x\right)' = x + 3$

(c) $\int \dfrac{dx^*}{\sqrt{x-4}} = 2\sqrt{x-4} + C$ because $(2\sqrt{x-4})' = \dfrac{1}{\sqrt{x-4}}$ $\quad \blacksquare$

The entries in the table below are easy to verify.

(5.44)

INDEFINITE INTEGRALS

1. $\int cf(x)\, dx = c\int f(x)\, dx$; c any constant

2. $\int (f + g)(x)\, dx = \int f(x)\, dx + \int g(x)\, dx$

3. If $r \neq -1$, then $\int x^r dx = \dfrac{1}{r+1}x^{r+1} + C$

4. $\int e^x dx = e^x + C$

5. $\int \sin x\, dx = -\cos x + C$

6. $\int \cos x\, dx = \sin x + C$

7. $\int \sec^2(x)\, dx = \tan x + C$

The u-Substitution

Formula 3 in (5.44) is sufficient to handle simple integrals such as

$$\int x^3 dx, \quad \int \frac{dx}{x^4}, \quad \text{and} \quad \int \sqrt[3]{x^5}\, dx$$

*The symbol dx is written in the numerator to save space. The integrand is $1/\sqrt{x-4}$.

In each case, the integrand can be rewritten as a power of the identity function x. But it is often necessary to integrate the power of a function other than the identity.

EXAMPLE 3   Evaluate $\int 2x(x^2 + 1)^3 dx$.

Solution   Since $F(x) = \frac{1}{4}(x^2 + 1)^4$ is an antiderivative of the integrand (check this out), we have that

$$\int 2x(x^2 + 1)^3 dx = \frac{1}{4}(x^2 + 1)^4 + C \qquad \blacksquare$$

It was not mere chance that led us to the antiderivative in Example 3; it was a systematic method called **u-substitution.** When a function, say g, other than the identity is raised to a power, we try to arrange the integral in the form $\int [g(x)]^r g'(x)\, dx$. Then we let

$$u = g(x) \quad \text{and} \quad du = g'(x)\, dx^*$$

so that

$$\int \underbrace{[g(x)]^r}_{u} \underbrace{g'(x)\, dx}_{du} = \int u^r du = \frac{1}{r+1} u^{r+1} + C \qquad (5.44\text{–}3)$$

$$= \frac{1}{r+1} [g(x)]^{r+1} + C \qquad (u = g(x))$$

This is called a *u-substitution* because we substitute u for $g(x)$ and du for $g'(x)\, dx$. Notice, however, that the final answer is in terms of the *original variable,* and not left in terms of u.

Here are several examples of the *u*-substitution method, beginning with a reworking of Example 3.

EXAMPLE 3 *(Reworked)*   Evaluate $\int 2x(x^2 + 1)^3 dx$ using *u*-substitution.

Solution   The function raised to a power is $g(x) = x^2 + 1$ and $g'(x) = 2x$. Let $u = x^2 + 1$; then $du = (x^2 + 1)'\, dx = 2x\, dx$. Therefore,

$$\int 2x(x^2 + 1)^3 dx = \int \underbrace{(x^2 + 1)^3}_{u} \underbrace{2x\, dx}_{du} = \int u^3 du = \frac{1}{4} u^4 + C$$

$$= \frac{1}{4}(x^2 + 1)^4 + C$$

Check. $\left[\frac{1}{4}(x^2 + 1)^4\right]' = \frac{1}{4} \cdot 4(x^2 + 1)^3(x^2 + 1)' = 2x\,(x^2 + 1)^3 \qquad \blacksquare$

*This is the **differential** of the function $u = g(x)$; see Section 4–3. This use of the differential is a reason for including the symbol dx inside the integral sign.

It often happens that a constant factor must be introduced inside the integral to obtain *exactly du = g'(x) dx*. But constants move freely through an integral sign (see the remark following (5.33) in Section 5–5). Therefore, a nonzero constant factor can be introduced inside the integral *provided* we compensate by multiplying outside the integral by the reciprocal of the constant. That is, if $c \neq 0$, then

(5.45) $$\int g(x) \, dx = \frac{1}{c} \int cg(x) \, dx$$

EXAMPLE 4 Evaluate $\displaystyle\int \frac{x}{(x^2 - 3)^5} \, dx$.

Solution Think of $x/(x^2 - 3)^5$ as $x(x^2 - 3)^{-5}$. If $u = x^2 - 3$, then $du = 2x \, dx$ and $u^{-5} \, du = (x^2 - 3)^{-5} \, 2x \, dx$. This differs from the given integrand by a factor of 2. Applying (5.45) with $c = 2$, we multiply the integrand by 2 and compensate by multiplying the integral by $\frac{1}{2}$. Thus,

$$\int x(x^2 - 3)^{-5} dx = \frac{1}{2} \int \underbrace{(x^2 - 3)^{-5}}_{u} \underbrace{2x \, dx}_{du} = \frac{1}{2} \int u^{-5} du$$

$$= \frac{1}{2}\left(-\frac{1}{4}u^{-4}\right) + C$$

$$= -\frac{1}{8}(x^2 - 3)^{-4} + C$$

Check. $\left[-\frac{1}{8}(x^2 - 3)^{-4}\right]' = \left(-\frac{1}{8}\right)(-4)(x^2 - 3)^{-5}(2x) = x(x^2 - 3)^{-5}$ ∎

EXAMPLE 5 Evaluate $\int 5x^2 \sqrt[3]{x^3 + 1} \, dx$.

Solution Let $u = x^3 + 1$; then $du = 3x^2 dx$. In order to obtain *exactly du = 3x^2dx*, you must multiply the integrand by $\frac{3}{5}$ and compensate with $\frac{5}{3}$ outside the integral.

$$\int 5x^2 \sqrt[3]{x^3 + 1} \, dx = \frac{5}{3} \int \underbrace{(x^3 + 1)^{1/3}}_{u} \underbrace{3x^2 dx}_{du} = \frac{5}{3} \int u^{1/3} du$$

$$= \frac{5}{3}\left(\frac{3}{4}u^{4/3}\right) + C$$

$$= \frac{5}{4}(x^3 + 1)^{4/3} + C$$

Check: $\left[\frac{5}{4}(x^3 + 1)^{4/3}\right]' = \frac{5}{4} \cdot \frac{4}{3}(x^3 + 1)^{1/3}(3x^2) = 5x^2(x^3 + 1)^{1/3}$ ∎

WARNING Only *constants* can be taken in and out of integrals. If the integral cannot be put into the form $\int [g(x)]^r g'(x)\, dx$ by introducing a constant factor, then the present method of substitution cannot be used with the power rule.

The *u*-substitution is valid not only for powers but for a much larger class of functions. Suppose an integrand has the form $f(g(x))g'(x)$ and that F is any antiderivative of f. The *u*-substitution yields

(5.46)
$$\int f(g(x))g'(x)\, dx = \int f(u)\, du = F(u) + C$$
$$= F(g(x)) + C$$

Check.
$$[F(g(x))]' = F'(g(x))g'(x) \qquad \text{(Chain rule)}$$
$$= f(g(x))g'(x) \qquad (F' = f)$$

In Chapters 7 and 8 the method of substitution will be applied to integrands involving exponential, logarithmic, and trigonometric functions (also see Exercises 42–49).

Evaluating a Definite Integral

Substitution can be used to evaluate a definite integral in either of two ways. The first way does not involve a change of limits.

EXAMPLE 6 Evaluate $\int_0^1 x\sqrt{x^2 + 1}\, dx$.

Solution First find an antiderivative; if $u = x^2 + 1$, then $du = 2x\, dx$. Therefore,

$$\int x\sqrt{x^2 + 1}\, dx = \frac{1}{2}\int (x^2 + 1)^{1/2} 2x\, dx = \frac{1}{2}\int u^{1/2}\, du$$
$$= \frac{1}{3} u^{3/2} + C$$
$$= \frac{1}{3}(x^2 + 1)^{3/2} + C \qquad \text{(Check)}$$

It follows now from the Fundamental Theorem that

$$\int_0^1 x\sqrt{x^2 + 1} = \left[\frac{1}{3}(x^2 + 1)^{3/2}\right]_0^1$$
$$= \frac{1}{3}(\sqrt{8} - 1) \qquad \blacksquare$$

The second way does involve a change of limits and is called a **change of variable**.

CHANGE OF VARIABLE THEOREM
If g' is continuous on $[a, b]$ and f is continuous on the range of g, then

(5.47)
$$\int_a^b f(g(x))g'(x)\,dx = \int_{g(a)}^{g(b)} f(u)\,du$$

where $u = g(x)$.

Proof   If F is any antiderivative of f on the range of g, then

$$\int_{g(a)}^{g(b)} f(u)\,du = F(g(b)) - F(g(a))$$

$$= [F(g(x))]_a^b$$

$$= \int_a^b f(g(x))g'(x)\,dx \qquad (5.46)$$

EXAMPLE 7   Use a change of variable to evaluate the integral $\int_0^1 x\sqrt{x^2 + 1}\,dx$ of Example 6.

Solution   Let $u = g(x) = x^2 + 1$; then $du = g'(x)\,dx = 2x\,dx$. Now

$$g(0) = 0^2 + 1 = 1 \quad \text{and} \quad g(1) = 1^2 + 1 = 2$$

Therefore, by (5.47),

$$\int_0^1 x\sqrt{x^2 + 1}\,dx = \frac{1}{2}\int_0^1 (x^2 + 1)^{1/2}\,2x\,dx \qquad \text{(Set up the integrand for } u\text{-substitution)}$$

$$= \frac{1}{2}\int_1^2 u^{1/2}\,du \qquad \text{(Make } u\text{-substitution and change limits)}$$

$$= \frac{1}{3}u^{3/2}\Big|_1^2 = \frac{1}{3}(\sqrt{8} - 1)$$

which agrees with the answer in Example 6.   ■

Examples 6 and 7 illustrate two methods for evaluating a definite integral. Neither is preferred over the other. In any given problem, choose the one that seems easier.

EXERCISES

In Exercises 1–34, evaluate the indefinite integrals. Check your answers *before* looking in the back of the book.

1. $\int 3x^2(x^3 + 1)^4\,dx$

2. $\int \frac{3x^2}{(x^3 + 1)^4}\,dx$

3. $\int \frac{2x}{\sqrt{x^2 + 2}}\,dx$

4. $\int 2x\sqrt[3]{x^2 + 2}\,dx$

5. $\int (4x + 1)^9\,dx$

6. $\int \sqrt[9]{4x + 1}\,dx$

7. $\int x^2(x^3 - 3)^5 \, dx$

8. $\int \frac{x^2}{\sqrt{x^3 - 3}} \, dx$

9. $\int (2x + 2)(x^2 + 2x - 3)^2 \, dx$

10. $\int (3x^2 - 2)(x^3 - 2x + 1)^4 \, dx$

11. $\int \frac{(2x + 3)}{\sqrt{x^2 + 3x - 2}} \, dx$

12. $\int \frac{(-3x^2 + 4x)}{(-x^3 + 2x^2)^3} \, dx$

13. $\int (x + 1)(x^2 + 2x - 3)^2 \, dx$

14. $\int (4x - 6)(x^2 - 3x + 1)^4 \, dx$

15. $\int \frac{(6x^2 - 16x + 2)}{\sqrt{x^3 - 4x^2 + x}} \, dx$

16. $\int \frac{(x^2 + 2x)}{\sqrt[3]{x^3 + 3x^2}} \, dx$

17. $\int \frac{1}{x^2}\left(1 + \frac{1}{x}\right)^4 \, dx$

18. $\int \frac{4}{x^2}\left(2 - \frac{1}{x}\right)^2 \, dx$

19. $\int \frac{(1 + \sqrt{x})^3}{\sqrt{x}} \, dx$

20. $\int \frac{(1 - \sqrt[3]{x})^4}{\sqrt[3]{x^2}} \, dx$

21. $\int \frac{\sqrt{1 + \sqrt[3]{x}}}{\sqrt[3]{x^2}} \, dx$

22. $\int \frac{\sqrt[3]{1 + \sqrt{x}}}{\sqrt{x}} \, dx$

23. $\int \frac{dx}{\sqrt{x}(1 + \sqrt{x})^2}$

24. $\int \frac{dx}{\sqrt[3]{x^2}(1 + \sqrt[3]{x})^2}$

25. $\int (x^2 + 4)^2 \, dx$

26. $\int \frac{x^3 + 2x^2 - 1}{x^2} \, dx$

27. $\int (3x + 1)^3 \, dx$

28. $\int \frac{dx}{(3x + 1)^3}$

29. $\int (x^2 + 1)^3 \, dx$

30. $\int \frac{\sqrt[3]{x} + \sqrt{x}}{\sqrt[4]{x}} \, dx$

In Exercises 31–34, evaluate the integrals assuming that $r \neq \pm 1$ and $a \neq 0$.

31. $\int (ax + b)^r \, dx$

32. $\int \frac{dx}{(ax + b)^r}$

33. $\int x^{n-1}(ax^n + b)^r \, dx$

34. $\int \frac{x^{n-1} \, dx}{(ax^n + b)^r}$

In Exercises 35–38, evaluate the definite integrals. Just for practice, use *both* methods discussed in the text and compare the answers.

35. $\int_0^1 \frac{x}{(x^2 + 3)^3} \, dx$

36. $\int_1^4 \frac{(\sqrt{x} + 2)^3}{\sqrt{x}} \, dx$

37. $\int_0^2 \frac{(x^2 + x)}{(2x^3 + 3x^2 + 4)^2} \, dx$

38. $\int_0^{\sqrt{5}} x\sqrt{x^2 + 4} \, dx$

39. To use the method of substitution, you are allowed to introduce constant factors inside and outside the integral. Only *constants* can be used. To see what happens if you introduce factors that involve the independent variable, evaluate $\int (x^2 - 3)^2 \, dx$ as follows: (1) Let $u = x^2 - 3$; then $du = 2x \, dx$ (2) Multiply inside the integral by $2x$ and outside by $1/(2x)$ (3) Evaluate; what is your answer? (4) Check; is the answer correct?

40. (Continuation of Exercise 39) The method of substitution cannot be used to evaluate $\int (x^2 - 3)^2 \, dx$. Can you think of a valid method to evaluate this integral?

41. Which of the integrals below cannot be evaluated by any method discussed so far?

(a) $\int \sqrt{x + 6} \, dx$

(b) $\int \sqrt{x^2 + 6} \, dx$

(c) $\int (x^2 + 4)^2 \, dx$

(d) $\int (x^2 + 4)^{-2} \, dx$

(e) $\int \frac{x}{(1 + x^2)^3} \, dx$

(f) $\int \frac{x}{1 + x^2} \, dx$

Optional Exercises

In Exercises 42–49, use the substitution equation (5.46) to evaluate the indefinite integrals. Check your answers.

42. $\int \cos 3x \, dx$

43. $\int \sin \frac{x}{4} \, dx$

44. $\int x \sin x^2 \, dx$

45. $\int x^2 \cos x^3 \, dx$

46. $\int e^{-x} \, dx$

47. $\int x e^{x^2} \, dx$

48. $\int e^x \cos e^x \, dx$

49. $\int e^{\sin x} \cos x \, dx$

5–7 APPLICATIONS OF THE INDEFINITE INTEGRAL

Two major applications of the indefinite integral are evaluating definite integrals and solving differential equations. The first application was discussed in the preceding section; a small part of the second is discussed here.

Many laws of nature are described in terms of differential equations. An example from physics is Newton's second law of motion

(5.48) $F = ma$

where a is the acceleration of an object, m is its mass, and F is the force acting on it. Since $a = dv/dt$, (5.48) is a differential equation

(5.49) $F = m\dfrac{dv}{dt}$

Later in this section we will use (5.49) to compute the *escape velocity* of a rocket; that is, the velocity it must attain to escape the pull of the earth's gravity. Because differential equations describe relationships between rates of change, they arise naturally in applied sciences such as astronomy, biology, chemistry, economics, engineering, mathematics, and physics.

We began our study of differential equations in Section 4–7 with the simplest of all such equations, $y' = f(x)$. In this section, we take one small step up the ladder of difficulty to equations of the form

(5.50) $y' = \dfrac{f(x)}{g(y)}$

If we multiply both sides by dx and observe that the left side is $y'\, dx = dy$, then (5.50) becomes

$$dy = \frac{f(x)}{g(y)} dx \quad \text{or} \quad g(y)\, dy = f(x)\, dx$$

This manipulation is referred to as *separating the variables*. An equivalent method of separating the variables is to think of y' as a quotient dx/dy. Then

$$\frac{dy}{dx} = \frac{f(x)}{g(y)}$$

$$g(y)\, dy = f(x)\, dx$$

The final "separated" equation is the same in either case. Integrating both sides of this equation yields

$$\int g(y)\, dy = \int f(x)\, dx + C$$

The result is an equation in x and y that represents the general solution to (5.50). If this equation can be solved for y in terms of x to obtain an explicit general solution, then we will do so. This is the most desirable situation. But it often happens that the resulting equation is difficult, or even impossible, to solve for y. In this case, the general solution is defined *implicitly;* that is, any function $y = F(x)$ implicitly defined by the equation is a solution. Differential equations like (5.50), in which the variables can be separated, are called **separable.**

EXAMPLE 1   Solve the equation $y' = 2x/y^2$.

Solution   First we replace y' by dy/dx, and then separate the variables

$$\frac{dy}{dx} = \frac{2x}{y^2}$$

$$y^2 dy = 2x\ dx$$

Integrating both sides of this equation yields

$$\int y^2 dy = \int 2x\ dx$$

$$\frac{y^3}{3} = x^2 + C$$

$$y^3 = 3(x^2 + C)$$

Solving for y in terms of x yields

$$y = \sqrt[3]{3(x^2 + C)}$$

as an explicit general solution.

Check.   To check a solution means to verify that it satisfies the original differential equation. In this case, using the solution above,

$$y' = (\sqrt[3]{3(x^2 + C)})' = \frac{2x}{[\sqrt[3]{3(x^2 + C)}]^2}$$

Since $y = \sqrt[3]{3(x^2 + C)}$, we have

$$y' = \frac{2x}{y^2}$$

which is the original equation.   ∎

EXAMPLE 2   Solve the equation $y' = y^2(2x + 1)$.

Solution   First, we replace y' by dy/dx, and then separate the variables

$$\frac{dy}{dx} = y^2(2x + 1)$$

$$y^{-2} dy = (2x + 1)\ dx$$

Integrating both sides of this equation yields

$$\int y^{-2} dy = \int (2x + 1)\, dx$$

$$-y^{-1} = x^2 + x + C$$

Solving for y, we have

(5.51) $\quad y = -\dfrac{1}{x^2 + x + C} = -(x^2 + x + C)^{-1}$

is the (explicit) general solution.

Check. $\quad y' = (x^2 + x + C)^{-2}(2x + 1).$

To make sure that this agrees with the original equation, we continue with

$$y' = [-(x^2 + x + C)^{-1}]^2(2x + 1)$$

and it follows from (5.51) that

$$y' = y^2(2x + 1)$$

which is the original equation. ■

EXAMPLE 3 Solve $(3y^2 + 2y + 4)y' = 1$.

Solution Replace y' by dy/dx, separate the variables, and integrate

$$(3y^2 + 2y + 4)\frac{dy}{dx} = 1$$

$$(3y^2 + 2y + 4)dy = dx$$

$$y^3 + y^2 + 4y = x + C$$

This equation is too difficult to solve for y; as it stands, it is the *implicit* general solution.

Check. We use implicit differentiation to check this solution.

$$3y^2y' + 2yy' + 4y' = 1 \quad \text{or} \quad (3y^2 + 2y + 4)y' = 1$$

which is the original equation. ■

We will now demonstrate how the indefinite integral is used to compute *escape velocity.* A rocket is fired straight up (positive direction) with an initial velocity v_0. When the engines shut down (say the fuel has been exhausted), then the rocket is in a free-fall situation. On the one hand, it is subject to the force of motion

$$F = m\frac{dv}{dt} \tag{5.49}$$

and on the other hand, it is subject to the force of the earth's gravity (we neglect all other forces). According to Newton's law of gravitation, this force is

$$F = -\frac{mgR^2}{s^2}$$

where g is the gravitational constant 32 ft/sec², R is the radius of the earth (about 21×10^6 ft), and s is the distance from the rocket to the center of the earth. The minus sign indicates that g is working against the motion (Figure 5–33). Setting the two forces equal to each other results in the differential equation

$$m\frac{dv}{dt} = -\frac{mgR^2}{s^2}$$

If we cancel the m's and write

$$\frac{dv}{dt} = \frac{ds}{dt}\frac{dv}{ds} = v\frac{dv}{ds}$$

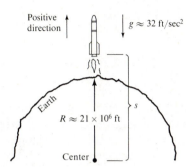

Positive direction

$g \approx 32$ ft/sec²

Earth

$R \approx 21 \times 10^6$ ft

s

Center

FIGURE 5–33. A rocket is fired straight up.

the equation above becomes the separable equation

(5.52) $vdv = -gR^2s^{-2}ds$

EXAMPLE 4   Solve equation (5.52). If the initial velocity is 10,000 mph, how high above the earth's surface will the rocket go?

Solution   Integrating both sides of (5.52) yields

(5.53) $\dfrac{v^2}{2} = \dfrac{gR^2}{s} + C$

This is the general solution to equation (5.52). Now we will find how high the rocket goes.

Let H be the maximum height attained by the rocket. At the moment the rocket is at that height, $v = 0$ and $s = R + H$. Substituting these values in (5.53), we have

$$0 = \frac{gR^2}{R + H} + C \quad \text{or} \quad C = -\frac{gR^2}{R + H}$$

so that (5.53) becomes

(5.54) $\dfrac{v^2}{2} = \dfrac{gR^2}{s} - \dfrac{gR^2}{R + H}$

The initial velocity v_0 is attained on the ground,* when $s = R$, so

(5.55) $\dfrac{v_0{}^2}{2} = gR - \dfrac{gR^2}{R + H}$

---

*The velocity v_0 is actually achieved shortly after takeoff at a height that is negligible compared to the radius of the earth.

Solving for H, we have

$$H = \frac{2gR^2}{2gR - v_0^2} - R$$

In this example, $v_0 = 10{,}000$ mph $\approx 14{,}667$ ft/sec. Therefore,

$$H = \frac{2(32)(21 \times 10^6)^2}{2(32)(21 \times 10^6) - (14{,}667)^2} - (21 \times 10^6)$$

$$\approx 4 \times 10^6 \text{ ft (or about 758 mi)} \quad \blacksquare$$

EXAMPLE 5 *(Escape velocity)* If the rocket escapes the earth's pull of gravity, then there is no maximum height attained. That is, its height H *increases without bound.* Refer now to equation (5.55) in Example 4. Each height H determines an initial velocity. If we allow H to get larger and larger, the term

$$\frac{gR^2}{R + H}$$

approaches 0 and, therefore, v_0^2 approaches $2gR$. Thus,

$$(5.56) \quad v_0 = \sqrt{2gR} \approx 36{,}600 \text{ ft/sec} \approx 25{,}000 \text{ mph}$$

is the escape velocity. $\blacksquare$

Our final example is from economics.

EXAMPLE 6 *(Marginal cost)* If the total cost to produce x units of a certain item is given by a *cost function* $S = S(x)$,* then the derivative dS/dx is called the **marginal cost function.** Economists interpret the value of dS/dx at an integer x to be the cost of producing one more unit. If the marginal cost function is known, as is often the case, the cost function can be found by solving a differential equation.

Suppose a company manufactures locks. They have fixed costs of $10,000 per month (rent, insurance, and so on) and their marginal cost function is

$$\frac{dS}{dx} = 3 - .2x - .03x^2$$

Find the cost of making x locks per month.

Solution Separate the variables and integrate

$$S = \int dS = \int (3 - .2x - .03x^2)\, dx$$

$$= 3x - .1x^2 - .01x^3 + C$$

---

*The letter C is not used for "cost" because of the possible confusion with the constant of integration.

The fixed costs are \$10,000 means that $S(0) = 10,000$ (why?); it follows that $C = 10,000$, and the cost of making x locks per month is

$$S = 3x - .1x^2 - .01x^3 + 10,000 \qquad \blacksquare$$

We close this section with the observation that not all differential equations are separable. For instance, the variables in the simple equation

$$y' = \frac{1}{x + y}$$

cannot be separated.

EXERCISES

In Exercises 1–10, solve the separable equations. Check your answers *before* looking in the back of the book.

1. $xyy' + x^2 = 0$

2. $y' = 2x^2 - 3x + 4$

3. $y' = x\sqrt{y}$

4. $x^2y' = x^2 + 1$

5. $y' = e^{x-y}$

6. $y' = \sqrt{xy}$

7. $\dfrac{\sin y}{\cos x} y' = 1$

8. $\dfrac{y'}{x^2 + 1} = 5$

9. $2x(y^2 + 1)^3 + yy' = 0$

10. $y' = x\sqrt{1 + x^2}$

In Exercises 11–20, solve only those equations that are separable.

11. $(x + y)y' = x$

12. $(x^2 + y^2)y' = y^2$

13. $yy' = x\sqrt{1 + x^2}\sqrt{1 + y^2}$

14. $x^2y' - y^2 = 0$

15. $y' = 4x^2y + x$

16. $xy' + y = \sin x$

17. $\sqrt{x}yy' = 3$

18. $yy' = 3x^2 - 2x + 1$

19. $yy' = \sqrt{(1 + y^2)^3}$

20. $y' = y^2\sec^2 x$

21. A particle moves along a straight line with velocity $v(t) = t\sqrt{1 + t^2}$ ft/sec, $t \geq 0$. If it starts at the origin, what is the position function?

22. A particle moves along a line, and its acceleration is the square root of its velocity at any time t. If it starts ($t = 0$) with a velocity of 1 ft/sec., what is the velocity function?

23. (Continuation of Exercise 22) How far does the particle travel in the first 5 seconds?

24. *(Marginal revenue)* If the total revenue from selling x units of an item is given by a *revenue function* $R = R(x)$, then the derivative dR/dx is called the **marginal revenue function.** Economists interpret its value at an integer x as the revenue derived from selling one more item. If the revenue from selling the first 100 refrigerators is \$39,600, and the marginal revenue function is

$$\frac{dR}{dx} = 400 + 2,000x^{-3/2} \quad \text{for } 1 \leq x \leq 250$$

what is the revenue function?

25. (Continuation of Exercise 24) Because of high fixed costs, the total cost of making the first 100 refrigerators is \$44,700. If the marginal cost function is

$$\frac{dS}{dx} = 300 + 30,000x^{-2} \quad \text{for } 1 \leq x \leq 250$$

what is the net profit on selling 250 refrigerators?

26. A rocket is fired straight up with an initial velocity of 20,000 mph. How high will it go?

27. A rocket is fired straight up with escape velocity $v_0 = \sqrt{2gR}$. Show that in this case the constant C in equation (5.53) of Example 4 is 0. Then $v = \sqrt{2g}Rs^{-1/2}$; find the position function of the rocket. Does s increase without bound (no maximum) as t increases?

5–8   NUMERICAL INTEGRATION

To evaluate a definite integral using the Fundamental Theorem you must first find an antiderivative of the integrand. As you know, this is not always easy to do. In fact, there are some continuous functions for which simple antiderivatives cannot be found; the innocent looking $y = e^{-x^2}$ is an example. Nevertheless, the definite integral of a continuous function *does* exist and in this section we discuss some methods of approximating its value.

Suppose f is continuous on $[a, b]$. If you do not know an antiderivative of f, how can you approximate the value of $\int_a^b f(x)\,dx$? The most familiar approximations are Riemann sums. If P is a partition of $[a, b]$ with a very small norm, then

$$\int_a^b f(x)\,dx \approx \sum_{i=1}^{n} f(w_i)\,\Delta x_i$$

The trouble is that if $\|P\|$ is very small and if the partition P and the points w_i are not chosen conveniently, the corresponding Riemann sum is quite tedious to calculate. In general, it is simpler to use one of two other approximation methods, the **Trapezoidal Rule** or **Simpson's Rule.** Both require only simple computations and are ideally suited to the use of hand-held calculators.

Trapezoidal Rule

Let f be continuous on $[a, b]$ and, for convenience of illustration, suppose that f is nonnegative. Let P be a regular partition of $[a, b]$. If there are n subintervals, then each one has length $(b - a)/n$. Now x_{i-1} and x_i are, respectively, the left and right endpoints of the ith subinterval. Write the two Riemann sums

$$\sum_{i=1}^{n} f(x_{i-1})\,\Delta x_i \quad \text{and} \quad \sum_{i=1}^{n} f(x_i)\,\Delta x_i$$

and take their average

$$\frac{1}{2}\left[\sum_{i=1}^{n} f(x_{i-1})\,\Delta x_i + \sum_{i=1}^{n} f(x_i)\,\Delta x_i\right] = \sum_{i=1}^{n} \frac{1}{2}[f(x_{i-1}) + f(x_i)]\,\Delta x_i$$

While neither of the Riemann sums above may be good approximations to $\int_a^b f(x)\,dx$, it turns out that their average is. Each term $\frac{1}{2}[f(x_{i-1}) + f(x_i)]\,\Delta x_i$ of the average is the area of a trapezoid which closely approximates the area under the curve from x_{i-1} to x_i (Figure 5–34).

Let us examine this average more closely. Each $\Delta x_i = (b - a)/n$, so this constant along with $\frac{1}{2}$ can be factored out of the sum. Furthermore, with the exception of $f(x_0) = f(a)$ and $f(x_n) = f(b)$, each term $f(x_i)$ appears twice.* Therefore,

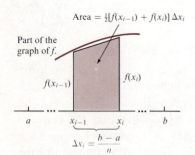

Area $= \frac{1}{2}[f(x_{i-1}) + f(x_i)]\,\Delta x_i$

Part of the graph of f.

$f(x_{i-1})$ $f(x_i)$

a x_{i-1} x_i b

$\Delta x_i = \dfrac{b-a}{n}$

FIGURE 5–34. Trapezoid approximation.

$*\displaystyle\sum_{i=1}^{n}[f(x_{i-1}) + f(x_i)] =$

$$[f(x_0) + f(x_1)] + [f(x_1) + f(x_2)] + [f(x_2) + f(x_3)] + \cdots + [f(x_{n-1}) + f(x_n)]$$

$$\sum_{i=1}^{n} \frac{1}{2}[f(x_{i-1}) + f(x_i)] \Delta x_i = \frac{b-a}{2n} \sum_{i=1}^{n} [fx_{i-1}) + f(x_i)]$$

$$= \frac{b-a}{2n}[f(a) + 2f(x_1) + \cdots + 2f(x_{n-1}) + f(b)]$$

This is the trapezoidal rule

> **TRAPEZOIDAL RULE**
> If f is continuous on $[a, b]$ and $P = \{x_0, x_1, \ldots, x_n\}$ is a *regular*
> (5.57) partition, then
>
> $$\int_a^b f(x)\, dx \approx \frac{b-a}{2n}[f(a) + 2f(x_1) + \cdots + 2f(x_{n-1}) + f(b)]$$

EXAMPLE 1   Use the trapezoidal rule with $n = 5$ to approximate $\int_1^2 dx/x$.

Solution   To use the trapezoidal rule you must be systematic.

(1) *Find the points of the partition.* Since $n = 5$, each subinterval has length
$(b-a)/n = (2-1)/5 = \frac{1}{5}$. Thus,

$$x_0 = a = 1, x_1 = \tfrac{6}{5}, x_2 = \tfrac{7}{5}, \ldots x_5 = b = 2$$

(2) *Compute $f(a), f(b),$ and $2f(x_i)$ for $1 \leq i \leq n - 1$* (using a calculator, if necessary). In this example, $f(x) = 1/x$.

$$
\begin{aligned}
f(a) = f(1) &= 1.0000 \\
2f(x_1) = 2f(\tfrac{6}{5}) &= 1.6667 \\
2f(x_2) = 2f(\tfrac{7}{5}) &= 1.4286 \\
2f(x_3) = 2f(\tfrac{8}{5}) &= 1.2500 \\
2f(x_4) = 2f(\tfrac{9}{5}) &= 1.1111 \\
f(b) = f(2) &= .5000
\end{aligned}
$$

(3) *Add and multiply by $(b-a)/2n$.* The sum of the last column above is 6.9564
and $(b-a)/2n = \frac{1}{10}$. Thus,

$$\int_1^2 \frac{dx}{x} \approx \frac{6.9564}{10} = .69564$$

It will be shown in Chapter 7 that the value of $\int_1^2 dx/x$ is .6931 correct to four
decimal places. The relative error in the answer obtained above is

$$\frac{.6956 - .6931}{.6931} \approx .003 = .3\%$$

Thus, with just five subdivisions, the trapezoidal rule produces an error of only
.3%. Surprising, isn't it?   ■

Simpson's Rule

Instead of using trapezoids, Simpson's rule uses parabolas to approximate the area under a curve; it is usually more accurate than the trapezoidal rule.

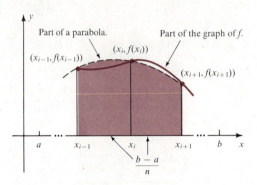

FIGURE 5–35. Parabola approximation.

Again let f be continuous and nonnegative on $[a, b]$. Let $P = \{x_0, x_1, \ldots, x_n\}$ be a regular partition with n an *even* integer and let x_{i-1}, x_i, and x_{i+1} be any three successive points in the partition. Figure 5–35 shows part of a parabola drawn through the points $(x_{i-1}, f(x_{i-1}))$, $(x_i, f(x_i))$, and $(x_{i+1}, f(x_{i+1}))$. It can be shown (see Exercise 22) that the area under this parabola from x_{i-1} to x_{i+1} is

$$\frac{b - a}{3n}[f(x_{i-1}) + 4f(x_i) + f(x_{i+1})]$$

For the first three points a, x_1, x_2 in the partition, we have

$$\frac{b - a}{3n}[f(a) + 4f(x_1) + f(x_2)]$$

For the next three points x_2, x_3, x_4, we have

$$\frac{b - a}{3n}[f(x_2) + 4f(x_3) + f(x_4)]$$

Continue in this manner, taking three points each time. If the areas so obtained are added up, the sum approximates the area under the curve $y = f(x)$ from a to b; that is, they approximate $\int_a^b f(x)\, dx$.

The

(5.58)

> **SIMPSON'S RULE**
>
> If f is continuous on $[a, b]$ and $P = \{x_0, x_1, \ldots, x_n\}$ is a *regular* partition with n an even integer, then
>
> $$\int_a^b f(x)\, dx \approx \frac{b - a}{3n}[f(a) + 4f(x_1) + 2f(x_2)$$
> $$+ 4f(x_3) + \cdots + 2f(x_{n-2}) + 4f(x_{n-1}) + f(b)]$$

EXAMPLE 2   Estimate $\int_0^1 \sqrt{1 + x^2}\, dx$ using Simpson's rule with $n = 6$.

Solution   The steps are similar to those for the trapezoidal rule.

(1) *Find the points of the partition.*

$$x_0 = a = 0, x_1 = \tfrac{1}{6}, x_2 = \tfrac{2}{6}, \ldots, x_6 = b = 1$$

(2) *Compute $f(a), f(b), 4f(x_i)$ for $i = 1, 3, \ldots$, and $2f(x_i)$ for $i = 2, 4, \ldots$. In this example, $f(x) = \sqrt{1 + x^2}$.*

$$
\begin{aligned}
f(a) &= f(0) &&= 1.0000 \\
4f(x_1) &= 4f(\tfrac{1}{6}) &&= 4.0552 \\
2f(x_2) &= 2f(\tfrac{2}{6}) &&= 2.1082 \\
4f(x_3) &= 4f(\tfrac{3}{6}) &&= 4.4721 \\
2f(x_4) &= 2f(\tfrac{4}{6}) &&= 2.4037 \\
4f(x_5) &= 4f(\tfrac{5}{6}) &&= 5.2068 \\
f(b) &= f(1) &&= 1.4142
\end{aligned}
$$

(3) *Add and multiply by $(b - a)/3n$.* In this example, the sum is 20.6602 and $(b - a)/3n = \tfrac{1}{18}$. Thus,

$$\int_0^1 \sqrt{1 + x^2}\, dx \approx \frac{20.6602}{18} \approx 1.1478$$

As you will see in Chapter 9, this answer approximates the value of the integral correct to four decimal places.   ■

If f is continuous on $[a, b]$, then both of the methods described above give good estimates of $\int_a^b f(x)\, dx$.* Furthermore, as you may have noticed, only a few values of f are needed. In other words, if you know that a function is continuous and you know its values at a few equally spaced points in an interval, then you can estimate its integral even though you do not know an explicit formula for the function itself. This observation has many useful applications.

EXAMPLE 3   An experiment is being performed which produces oxygen at a continuous rate. The rate of oxygen produced is measured each minute and the results tabulated.

Minutes	0	1	2	3	4	5	6	7	8	9	10
Oxygen cu ft/min.	0	1.4	1.8	2.2	3.0	4.2	4.1	3.6	2.9	2.0	1.2

Estimate the total amount of oxygen produced in 10 minutes.

---

*Exercise 23 contains an error estimate for Simpson's rule.

Solution Suppose that f is the continuous function whose value at time t is the amount (cu ft/min) of oxygen being produced. The total amount of oxygen produced in ten minutes is the value of $\int_0^{10} f(t)\, dt$. Let us use the trapezoidal rule to estimate this integral.

$$\frac{b-a}{2n}[f(0) + 2f(1) + \cdots + 2f(9) + f(10)]$$

$$= \frac{1}{2}[0 + 2.8 + \cdots + 4.0 + 1.2] = 25.8 \text{ cu. ft.} \qquad \blacksquare$$

1-6

(18 problems) Sngg

EXERCISES

In Exercises 1–6, use both the trapezoidal rule and Simpson's rule with $n = 4$ to estimate the integrals. Then compare the estimates with the value obtained using the Fundamental Theorem.

1. $\displaystyle\int_0^1 x^2\, dx$ **2.** $\displaystyle\int_0^1 x^3\, dx$

3. $\displaystyle\int_0^1 x\sqrt{1 + x^2}\, dx$ **4.** $\displaystyle\int_0^1 \frac{x}{(1 + x^2)^2}\, dx$

5. $\displaystyle\int_0^\pi \sin x\, dx$ **6.** $\displaystyle\int_0^1 e^x\, dx$

The Midpoint Rule. A Riemann sum with the function f evaluated at the midpoint of each subinterval is another way to approximate a definite integral.

MIDPOINT RULE

If f is continuous on $[a, b]$ and $P = \{x_0, x_1, \ldots, x_n\}$ is a regular partition, then

(5.59) $\displaystyle\int_a^b f(x)\, dx \approx \frac{b - a}{n}\left[f\left(\frac{x_1 + x_0}{2}\right)\right.$

$+ f\left(\frac{x_2 + x_1}{2}\right)$

$\left. + \cdots + f\left(\frac{x_n + x_{n-1}}{2}\right)\right]$

The midpoint rule is sometimes as accurate as Simpson's rule and often more accurate than the trapezoidal rule.

In Exercises 7–12, rework Exercises 1–6 using the midpoint rule with $n = 4$ and compare answers.

7. Exercise 1   **8.** Exercise 2
9. Exercise 3   **10.** Exercise 4
11. Exercise 5   **12.** Exercise 6

13. Use the trapezoidal rule with $n = 12$ to estimate $\int_0^1 \sqrt{1 + x^2}\, dx$. Compare with the estimate derived in Example 2. (For this integral, Simpson's rule with $n = 6$ is *more accurate* than the trapezoidal rule with $n = 12$.)

14. Use Simpson's rule with $n = 4$ to estimate $\int_1^2 dx/x$. Compare with the estimate derived in Example 1. (For this integral, Simpson's rule with $n = 4$ is more accurate than the trapezoidal rule with $n = 5$.)

15. In Chapter 9 you will learn that $\int_0^1 dx/(1 + x^2)$ $= .7854$ correct to four decimal places. Estimate this integral using Simpson's rule with $n = 4$.

16. In Chapter 9 you will learn that $\int_0^1 dx/\sqrt{1 + x^2}$ $= .8814$ correct to four decimal places. Estimate this integral using Simpson's rule with $n = 4$.

---

17. The velocity in miles per minute of a test car is recorded every fifteen seconds during the first two minutes of a run at the Bonneville salt flats. The results are recorded in the table below. Estimate the distance the car moves using Simpson's Rule.

Minutes	0	1/4	1/2	3/4	1	5/4	3/2	7/4	2
Velocity mi/min.	0	1.2	1.8	2.3	3.2	3.6	4.0	4.5	5.0

18. (Continuation of Exercise 17) What is the average velocity in miles per minute of the test car over the first two minutes?

19. The temperature in San Francisco on September 9, 1979 was recorded hourly from 8 am to 8 pm. The results are tabulated below. Find the average temperature during that period by (a) adding up the temperatures and dividing by 13 (b) considering the temperature as a continuous function of time and using the trapezoidal rule (c) same as (b) but use Simpson's rule. Which answer do you think is most accurate?

Hour	8	9	10	11	12	1	2	3	4	5	6	7	8
Temperature	57	57	58	60	61	63	66	70	73	71	67	64	61

20. Doctors can measure the rate at which a heart pumps blood through the body in the following way: A known amount of dye is injected into a vein near the heart and the concentration of dye passing through the aorta is measured at regular intervals. If the heart pumps blood at the *constant* rate of R liters per second and the *concentration* of dye at time t is $C(t)$ milligrams per liter, then the *amount* of dye flowing through the aorta is $R \cdot C(t)$ milligrams per second. It takes about 20 seconds for blood to circulate through the body so that

$$D = \int_0^{20} R \cdot C(t)\, dt = R \int_0^{20} C(t)\, dt$$

is the total amount of dye injected. Since D is known, this equation can be solved for R once an estimate for $\int_0^{20} C(t)\, dt$ is computed. The data in the following table shows the concentration at one second intervals after injecting $D = 5$ milligrams of dye. The table is taken from *Measuring Cardiac Output*, by B. Horelick and S. Koont, UMAP, Unit 71.

Seconds	0	1	2	3	4	5	6	7	8	9	10
Concentration	0	0	0	.1	.6	.9	1.4	1.9	2.7	3.0	3.7

Seconds	11	12	13	14	15	16	17	18	19	20
Concentration	4.0	4.1	4.0	3.8	3.7	2.9	2.2	1.5	1.1	.9

Use Simpson's rule to determine R, the rate at which the heart is pumping blood.

21. The blood pressure of a patient is recorded every $\frac{1}{3}$ second during a cardiac cycle, and tabulated as follows:

Time	0	1/3	2/3	1	4/3
Pressure	80	112	90	72	80

What is the average pressure during that cycle?

Optional Exercises

22. An important part of Simpson's rule is the formula
$$\frac{b-a}{3n}\,[f(x_{i-1}) + 4f(x_i) + f(x_{i+1})] \quad \text{for the}$$
area under the parabola shown in Figure 5–35. The following steps verify this formula.

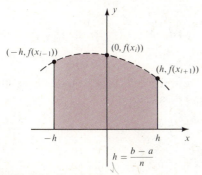

FIGURE 5–36. Exercise 22.

(1) Translate the shaded region of Figure 5–35 so that x_i coincides with the origin. Set $h = (b - a)/n$ so that x_{i-1} becomes $-h$ and x_{i+1} becomes h. The new parabola passes through $(-h, f(x_{i-1}))$, $(0, f(x_i))$, and $(h, f(x_{i+1}))$; see Figure 5–36. The area remains the same but the computations are made simpler by this translation.

(2) Let $y = Ax^2 + Bx + C$ be an equation of the parabola passing through the three points indicated above. Evaluate $\int_{-h}^{h}(Ax^2 + Bx + C)\, dx$ to find the area under the curve in terms of A, B, C, and h.

(3) The coordinates of the three points must satisfy the equation of the curve $y = Ax^2 + Bx + C$. Use this fact to write down three equations and then show that $f(x_i) = C$ and $2Ah^2 = f(x_{i-1}) + f(x_{i+1}) - 2f(x_i)$.

(4) Substitute these values into the area formula obtained in Step (2); the formula you want is the result.

23. The following error estimate can be proved using advanced methods

> If M is a positive number and $|f^{(4)}(x)| \le M$ for all x in $[a, b]$, then the error involved in using Simpson's rule (5.58) is no greater than
>
> $$\frac{M(b-a)^5}{180n^4}$$

Differentiate the integrand four times, determine M, and estimate the error in the answers to (a) Exercise 5 and (b) Exercise 14 above.

5–9 PROOFS OF INTEGRAL THEOREMS (OPTIONAL)

The proofs are based on the ϵ-δ definition of the integral given in the footnote on page 230.

If f and g are continuous on $[a, b]$, then

$$(5.34) \qquad \int_a^b (f + g)(x)\, dx = \int_a^b f(x)\, dx + \int_a^b g(x)\, dx$$

Proof Notice that any Riemann sum of $f + g$ can be separated into Riemann sums of f and g as follows

$$\sum_{i=1}^{n} (f + g)(w_i)\, \Delta x_i = \sum_{i=1}^{n} f(w_i)\, \Delta x_i + \sum_{i=1}^{n} g(w_i)\, \Delta x_i$$

We now introduce the following notation. If P is any partition of $[a, b]$, we write $R_{P_{f+g}}$, R_{P_f}, and R_{P_g} to represent Riemann sums of $f + g$, f, and g corresponding to P. Therefore for any given $R_{P_{f+g}}$, the remark above implies that

$$(5.59) \qquad R_{P_{f+g}} = R_{P_f} + R_{P_g} \qquad\qquad (R_{P_f} \text{ and } R_{P_g} \text{ depend on } R_{P_{f+g}})$$

Since f and g are continuous, we know that all three integrals in (5.34) exist. To simplify the writing, let

$$I = \int_a^b (f + g)(x)\, dx, \quad J = \int_a^b f(x)\, dx, \text{ and } K = \int_a^b g(x)\, dx$$

We want to show that $I = J + K$.

If ϵ is any positive number, it follows from the definition of the integral that there are three positive numbers δ_1, δ_2, and δ_3 such that

$$(5.60) \qquad \begin{aligned} |R_{P_{f+g}} - I| &< \epsilon \quad \text{whenever} \quad \|P\| < \delta_1 \\ |R_{P_f} - J| &< \epsilon \quad \text{whenever} \quad \|P\| < \delta_2 \\ |R_{P_g} - K| &< \epsilon \quad \text{whenever} \quad \|P\| < \delta_3 \end{aligned}$$

Let δ be the smallest of the numbers δ_1, δ_2, and δ_3. If P is any partition of $[a, b]$ with $\|P\| < \delta$, it follows from (5.59) and (5.60) that

$$\begin{aligned}
|I - (J + K)| &= |(I - R_{P_{f+g}}) + (R_{P_{f+g}} - J - K)| \\
&= |(I - R_{P_{f+g}}) + (R_{P_f} - J) + (R_{P_g} - K)| \\
&\le |I - R_{P_{f+g}}| + |R_{P_f} - J| + |R_{P_g} - K| \\
&< \epsilon + \epsilon + \epsilon = 3\epsilon
\end{aligned}$$

This shows that the distance between I and $J + K$ is less than three times any positive number. It follows that $I = J + K$ and completes the proof. ∎

To prove the first part of the Fundamental Theorem we need to know that $\int_a^b f(x)\,dx = \int_a^c f(x)\,dx + \int_c^b f(x)\,dx$ regardless of the order of a, b, and c. This is property (5.39) in Section 5–5. We begin by proving the special case of $a < c < b$.

If f is continuous on $[a, b]$ and $a < c < b$, then

(5.37)
$$\int_a^b f(x)\,dx = \int_a^c f(x)\,dx + \int_c^b f(x)\,dx$$

Proof   We shall use the symbols P, P_1, and P_2 to represent any partitions of the intervals $[a, b]$, $[a, c]$, and $[c, b]$, respectively. Notice that partitions P_1 and P_2 taken together become a partition P of $[a, b]$ and that $\|P\|$ is no larger than the maximum of $\|P_1\|$ and $\|P_2\|$ (Figure 5–37). Furthermore, if R_{P_1} is a Riemann sum of f corresponding to P_1 and R_{P_2} is a Riemann sum of f corresponding to P_2, then their sum $R_{P_1} + R_{P_2}$ is a Riemann sum R_P of f corresponding to P. Finally, since f is continuous on $[a, b]$, it follows that all three integrals exist. To simplify the writing, let $I = \int_a^b f(x)\,dx$, $J = \int_a^c f(x)\,dx$, and $K = \int_c^b f(x)\,dx$. We want to show that $I = J + K$.

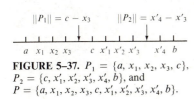

$\|P_1\| = c - x_3$ $\|P_2\| = x'_4 - x'_3$

FIGURE 5–37. $P_1 = \{a, x_1, x_2, x_3, c\}$, $P_2 = \{c, x'_1, x'_2, x'_3, x'_4, b\}$, and $P = \{a, x_1, x_2, x_3, c, x'_1, x'_2, x'_3, x'_4, b\}$.

Let ϵ be any positive number. Since I, J, and K exist, it follows from the definition of the integral that there are three positive numbers δ_1, δ_2, and δ_3 such that

(5.61)
$$\begin{aligned}
|R_{P_1} - J| &< \epsilon \quad \text{whenever} \quad \|P_1\| < \delta_1 \\
|R_{P_2} - K| &< \epsilon \quad \text{whenever} \quad \|P_2\| < \delta_2 \\
|R_P - I| &< \epsilon \quad \text{whenever} \quad \|P\| < \delta_3
\end{aligned}$$

Let δ be the smallest of the numbers δ_1, δ_2, and δ_3. Pick any partition P_1 of $[a, c]$ with $\|P_1\| < \delta$ and any partition P_2 of $[c, b]$ with $\|P_2\| < \delta$. Taken together, P_1 and P_2 form a partition P of $[a, b]$, and it follows that $\|P\|$ is also less than δ. For these particular partitions, all three inequalities in (5.61) hold and, furthermore, $R_P = R_{P_1} + R_{P_2}$. Therefore,

$$\begin{aligned}
|I - (J + K)| &= |(I - R_P) + (R_P - J - K)| \\
&= |(I - R_P) + (R_{P_1} - J) + (R_{P_2} - K)| \\
&\le |I - R_P| + |R_{P_1} - J| + |R_{P_2} - K| \\
&< \epsilon + \epsilon + \epsilon = 3\epsilon
\end{aligned}$$

This shows that the distance between I and $J + K$ is less than three times any positive number. It follows that $I = J + K$ and completes the proof. ∎

The proof of (5.39), that the order of a, b, and c is irrelevant, involves six cases because there are six ways of ordering three numbers. For example, suppose that $b < a < c$. Then

$$\int_b^c f(x)\, dx = \int_b^a f(x)\, dx + \int_a^c f(x)\, dx \qquad (5.37)$$

$$-\int_c^b f(x)\, dx = -\int_a^b f(x)\, dx + \int_a^c f(x)\, dx \qquad \text{(Reverse limits; (5.38))}$$

from which it follows that

$$\int_a^b f(x)\, dx = \int_a^c f(x)\, dx + \int_c^b f(x)\, dx$$

The proofs of the other cases are similar.

Now we can prove part one of the Fundamental Theorem.

If f is continuous on $[a, b]$, then $F(x) = \int_a^x f(u)\, du$ exists for each x in $[a, b]$ and, furthermore, F is an antiderivative of f on $[a, b]$.

Proof Since f is continuous on $[a, b]$, it is certainly continuous on $[a, x]$ and, therefore, $\int_a^x f(u)\, du$ exists. Next we will show that $F'(x) = f(x)$ on (a, b) and then that F is continuous on $[a, b]$.

Proof that $F'(x) = f(x)$ on (a, b): Let x be any element of (a, b) and let h be any nonzero number such that $x + h$ is also in (a, b). Then, by equation (5.40) in Section 5–5, we can write

$$(5.62) \qquad \frac{F(x + h) - F(x)}{h} = \frac{\displaystyle\int_a^{x+h} f(u)\, du - \int_a^x f(u)\, du}{h}$$

$$= \frac{1}{h} \int_x^{x+h} f(u)\, du$$

Let m_h and M_h be the minimum and maximum values of f on the interval $[x, x + h]$ (or $[x + h, x]$ if $h < 0$). Then $m_h \leq f(u) \leq M_h$ and it follows from (5.36) that

$$m_h \cdot h \leq \int_x^{x+h} f(u)\, du \leq M_h \cdot h \quad \text{if } h > 0$$

$$m_h \cdot h \geq \int_x^{x+h} f(u)\, du \geq M_h \cdot h \quad \text{if } h < 0$$

Dividing through by h, we have, in *either case,* that

$$(5.63) \qquad m_h \leq \frac{1}{h} \int_x^{x+h} f(u)\, du \leq M_h$$

Because f is continuous at x, it follows that both of the numbers m_h and M_h approach $f(x)$ as h approaches zero. Thus, it follows from (5.63) and the Squeeze Theorem (2.9) that

$$\lim_{h \to 0} \frac{1}{h} \int_x^{x+h} f(u)\, du = f(x)$$

Combining this with (5.62) above, we have

$$F'(x) = \lim_{h \to 0} \frac{F(x + h) - F(x)}{h} = \lim_{h \to 0} \frac{1}{h} \int_x^{x+h} f(u)\, du = f(x)$$

and that completes the proof.

Proof that F is continuous on $[a, b]$: Since F is differentiable on (a, b), it is continuous there. To prove F is continuous at a, we must show that

$$\lim_{x \to a^+} F(x) = F(a)$$

Now $F(x) = \int_a^x f(u)\, du$ and if m_x and M_x are the extreme values of f on the interval $[a, x]$, it follows as before that

$$m_x(x - a) \le \int_a^x f(u)\, du \le M_x(x - a)$$

As $x \to a^+$, both $m_x(x - a)$ and $M_x(x - a)$ approach zero and, therefore, $F(x) = \int_a^x f(u)\, du \to 0$ (why?). But $0 = F(a)$ so that, indeed, $F(x) \to F(a)$ as $x \to a^+$. The case of the other endpoint b is similar. Here, $F(b) - F(x) = \int_a^b f(u)\, du - \int_a^x f(u)\, du = \int_x^b f(u)\, du$ can be squeezed between $m_x(b - x)$ and $M_x(b - x)$. Thus, $F(b) - F(x) \to 0$ as $x \to b^-$, which is just another way of saying that $F(x) \to F(b)$ as $x \to b^-$. ∎

REVIEW EXERCISES

In Exercises 1–24, evaluate the indefinite integrals. If it is not possible using the methods discussed so far, say so. Check your answers *before* looking in the back of the book.

1. $\displaystyle\int \sqrt{x + 1}\, dx$

2. $\displaystyle\int (x - 1)(x + 2)\, dx$

3. $\displaystyle\int \sqrt{x^2 + 1}\, dx$

4. $\displaystyle\int \frac{1}{(x - 1)(x + 2)}\, dx$

5. $\displaystyle\int x\sqrt{x^2 + 1}\, dx$

6. $\displaystyle\int \frac{2x + 1}{(x - 1)(x + 2)}\, dx$

7. $\displaystyle\int 4 \sec^2 x\, dx$

8. $\displaystyle\int (x - \sqrt{x})\, dx$

9. $\displaystyle\int \frac{1}{x}\, dx$

10. $\displaystyle\int (\sin x + \cos x)\, dx$

11. $\displaystyle\int \frac{1}{x^2}\left(1 + \frac{1}{x}\right)^4 dx$

12. $\displaystyle\int (5x - 3)^6\, dx$

13. $\displaystyle\int \frac{(1 - \sqrt[3]{x})^3}{\sqrt[3]{x^2}}\, dx$

14. $\displaystyle\int 5e^x\, dx$

15. $\displaystyle\int x(x^2 - 1)^4\, dx$

16. $\displaystyle\int (x^2 + 2)^2\, dx$

17. $\displaystyle\int \frac{x^3 - 3x^2 + 1}{x^2}\, dx$

18. $\displaystyle\int \frac{1}{1 + x^2}\, dx$

19. $\displaystyle\int \frac{1}{\sqrt{1 + x^2}}\, dx$

20. $\displaystyle\int \frac{x}{\sqrt{1 + x^2}}\, dx$

21. $\int \dfrac{\sqrt[4]{x} - \sqrt[5]{x}}{\sqrt[20]{x}} \, dx$

22. $\int \sqrt[3]{1 - 8x} \, dx$

23. $\int \dfrac{1}{\sqrt{x}(1 + \sqrt{x})^2} \, dx$

24. $\int \dfrac{\sqrt[3]{1 + \sqrt{x}}}{\sqrt{x}} \, dx$

In Exercises 25–40, evaluate the definite integrals with the Fundamental Theorem (using substitution or a change of variable, if necessary) or by computing the area under the curve. If neither method is feasible, then estimate the integral using Simpson's rule with $n = 6$.

25. $\int_{-\pi}^{\pi} \sin x \, dx$

26. $\int_0^2 (x^2 - 3x + 1) \, dx$

27. $\int_0^1 x\sqrt{1 - x^2} \, dx$

28. $\int_1^0 \sqrt{1 - x^2} \, dx$

29. $\int_{-4}^{-2} \sqrt{16 - x^2} \, dx$

30. $\int_{-2}^{-4} x\sqrt{16 - x^2} \, dx$

31. $\int_0^1 \dfrac{1}{1 + x^2} \, dx$

32. $\int_4^6 \dfrac{1}{x} \, dx$

33. $\int_{-3}^3 (x^5 - 3x^3 + 4x) \, dx$

34. $\int_0^{\pi/4} \sec^2 x \, dx$

35. $\int_4^1 \dfrac{(\sqrt{x} + 1)^4}{\sqrt{x}} \, dx$

36. $\int_0^1 \sqrt{4 - x^2} \, dx$

37. $\int_0^1 \dfrac{1}{\sqrt{1 + x^2}} \, dx$

38. $\int_4^9 \dfrac{\sqrt{1 + \sqrt{x}}}{\sqrt{x}} \, dx$

39. $\int_2^3 \dfrac{x^2}{\sqrt[3]{x^3 - 3}} \, dx$

40. $\int_3^2 \dfrac{x}{\sqrt{1 + x^2}} \, dx$

In Exercises 41–50, solve only the separable differential equations. If it is not separable, say so. Check your answers *before* looking in the back of the book.

41. $y' = x\sqrt{y}$

42. $(x + y)y' = x\sqrt{y}$

43. $yy' = \sqrt{1 + y^2}$

44. $yy' = x\sqrt{1 + x^2}\sqrt{1 + y^2}$

45. $(\cos^2 x)y' = y^2$

46. $\sqrt{xy}\, y' = 7$

47. $yy' = e^x\sqrt{y}$

48. $y' = \sqrt{y} \cos x$

49. $xy' - y = e^x$

50. $y' = (x^2 + 1)^2$

In Exercises 51–54, evaluate the integrals.

51. $\int_1^3 |-3x + 6| \, dx$

52. $\int_4^2 |8 - x^2| \, dx$

53. $\int_0^3 |(x - 1)(x - 2)| \, dx$

54. $\int_{-\pi}^{\pi} |\cos x| \, dx$

---

55. To stretch a spring x inches from its natural length requires a force $F = kx$ pounds where k is a constant depending on the spring; this is known as *Hooke's Law*. If $k = 10$, how much work is done in stretching the spring 6 inches from its natural length?

56. Find a continuous function f and a constant a such that $\int_a^x f(u) \, du = 2x^2 - 3x + 1$ for all x.

57. Make up and solve a problem similar to the one in Exercise 56.

58. What is the net area bounded by the graph of $f(x) = x^2 + 3x - 4$ from $x = -5$ to $x = 2$? What is the total area?

59. A particle moves along a horizontal line with acceleration $a = 2t - 8$ ft/sec^2 at any time $t \geq 0$. If it starts with a velocity of 12 ft/sec, what is the velocity function? For what values of t is the particle slowing down? Speeding up?

60. (Continuation of Exercise 59) If the particle starts from 0, what is the position function?

61. (Continuation of Exercise 60) Where is the particle after 6 seconds? What is the total distance it has traveled in those 6 seconds?

62. At time t hours, $0 \leq t \leq 24$, past midnight a power station generates $E = -t^2 + 24t + 1$ thousands of kilowatts of power. What is the peak production, and when does it occur? What is the total production in a 24-hour period? What is the average hourly output of power?

APPLICATIONS OF THE DEFINITE INTEGRAL

The preceding chapter describes what an integral is, and how to evaluate it. This chapter describes how the definite integral is used to solve problems in a wide variety of disciplines. Here are just a few examples.

Business: For what length of time should a company guarantee its product if it wants no more than 10% returns?

Engineering: What is the length of the cable between the towers of the Golden Gate Bridge?

Life Sciences: At what rate does blood flow through an artery?

Mathematics: What is the volume of a torus (a donut-shaped solid)?

Physical Sciences: What horsepower must an engine produce to accelerate a 2,000 pound automobile from 0 to 60 mph in 10 seconds?

The definite integral is used to compute area, volume, work, arc length, center of mass, probability, and the force of a liquid. In each case, an appropriate integrand must be found. Sometimes we use the method of squeezing a quantity between upper and lower sums (as we did in Section 5–4 to compute area, distance, and profit). However, when it is more convenient to do so, we use the method of approximating the quantity with ordinary Riemann sums.

The main point of this chapter is for you to observe how the single idea of integration can be employed in so many different ways. This is what makes integration so important.

6-1 AREAS OF PLANE REGIONS

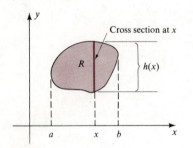

FIGURE 6–1. Vertical cross section.

The same considerations that lead to the computation of areas under a curve can also be used to find areas of more general regions. Figure 6–1 shows a region R in the plane; its projection onto the x-axis determines the interval $[a, b]$. For each x in $[a, b]$, the vertical line through x will intersect R. The part of that line contained in R is called the (vertical) **cross section of R at** x. These cross sections can be used to define a function on $[a, b]$.

DEFINITION
Let R be a plane region whose projection onto the x-axis determines the interval $[a, b]$. The (vertical) **cross section function** is a function h whose value at each x in $[a, b]$ is the length of the cross section at x (Figure 6–1).

Notice that $h(x) \geq 0$ for all x in $[a, b]$. We claim that if h is continuous on $[a, b]$ then $\int_a^b h(x)\, dx$ is the area of the region.

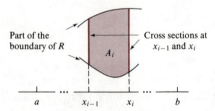

FIGURE 6–2. A_i is the area between the cross sections.

To establish this claim, let $P = \{x_0, x_1, \ldots, x_n\}$ be any partition of $[a, b]$. Each point of P will determine a cross section of R, and let A_i be the area of that part of R between the cross sections at x_{i-1} and x_i (Figure 6–2). Since h is continuous, it will have a minimum value m_i and a maximum value M_i on $[x_{i-1}, x_i]$. The area of a region with a constant cross section m_i and length Δx_i is $m_i \Delta x_i$. Similarly, the area of a region is $M_i \Delta x_i$ if its cross section is a constant M_i. This suggests that

(6.1) $m_i \Delta x_i \leq A_i \leq M_i \Delta x_i$

Now $L_P = \sum_{i=1}^n m_i \Delta x_i$ and $U_P = \sum_{i=1}^n M_i \Delta x_i$ are lower and upper sums of h and the area of R is $A = \sum_{i=1}^n A_i$. Summing each of the three quantities in (6.1) from $i = 1$ to $i = n$ yields

$$L_P \leq A \leq U_P$$

for any partition P. As you have seen before, this implies that A is the integral of h.

Let R be a plane region that determines an interval $[a, b]$ on the x-axis. If the (vertical) cross section function h is continuous on $[a, b]$, then

(6.2)

$$A = \int_a^b h(x)\, dx$$

is the area of R.

The regions of interest to us are usually bounded by graphs of functions.

EXAMPLE 1   Compute the area of the region bounded by the curves $y = x + 1$ and $y = x^2$ from $x = 0$ to $x = 1$.

Solution   Sketch the region (Figure 6–3). For each x in $[0, 1]$, the cross section extends from the curve $y = x^2$ up to the curve $y = x + 1$. Thus, $h(x) = (x + 1) - x^2$ and the area is

$$A = \int_0^1 (x + 1 - x^2)\, dx = \left[\frac{x^2}{2} + x - \frac{x^3}{3} \right]_0^1$$

$$= \frac{7}{6} \text{ square units} \qquad \blacksquare$$

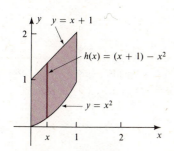

FIGURE 6–3. Example 1.

If the values of the limits a and b are not given explicitly, they can usually be found algebraically.

EXAMPLE 2   Find the area of the region bounded by $y = x + 1$ and $y = x^2$.

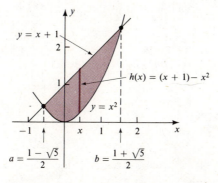

FIGURE 6–4. Example 2. Solve the equation $x + 1 = x^2$ to find a and b.

$$a = \frac{1 - \sqrt{5}}{2} \qquad b = \frac{1 + \sqrt{5}}{2}$$

Solution   Sketch the region (Figure 6–4). This region determines the interval $[a, b]$ whose endpoints are the points of intersection of the two graphs; that is, they are the solutions to the equation

$$x + 1 = x^2$$

Write $-x^2 + x + 1 = 0$ and use the quadratic formula to find $x = (-1 \pm \sqrt{5})/(-2)$. Therefore,

$$a = \frac{1 - \sqrt{5}}{2} \approx -.62 \quad \text{and} \quad b = \frac{1 + \sqrt{5}}{2} \approx 1.6$$

For each x between a and b, the cross section at x extends from the curve $y = x^2$ up to the curve $y = x + 1$. Thus, $h(x) = (x + 1) - x^2$ and the area is

$$A = \int_{(1-\sqrt{5})/2}^{(1+\sqrt{5})/2} (x + 1 - x^2)\, dx$$

$$\approx \left[\frac{x^2}{2} + x - \frac{x^3}{3} \right]_{-.62}^{1.6} \approx 1.9 \text{ square units} \qquad \blacksquare$$

Examples 1 and 2 illustrate a general principle.

(6.3)

> Let f and g be continuous with $f(x) \geq g(x)$ for all x in $[a, b]$. Let R be the region bounded by the curves $y = f(x)$ and $y = g(x)$ from a to b. Then the cross section function is $h(x) = f(x) - g(x)$, so the area of R is
>
> $$A = \int_a^b [f(x) - g(x)]\, dx$$

Computing the area of a region R bounded by curves $y = f(x)$ and $y = g(x)$ that satisfy the hypotheses of (6.3) is relatively straightforward. In the remainder of this section, we describe some variations that can occur and how to treat them.

Sometimes, $f(x) \geq g(x)$ on only part of the interval $[a, b]$ and $g(x) \geq f(x)$ on the rest. In that case, the cross section function becomes $h(x) = |f(x) - g(x)|$ and the area bounded by the graphs of f and g is

(6.3a)

> $$A = \int_a^b |f(x) - g(x)|\, dx$$

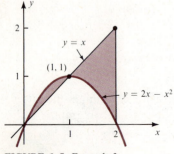

FIGURE 6–5. Example 3.

EXAMPLE 3 Find the area bounded by the curves $y = x$ and $y = 2x - x^2$ from $a = 0$ to $b = 2$.

Solution Sketch the region (Figure 6–5). On the interval $[0, 1]$, $2x - x^2 \geq x$; on the interval $[1, 2]$, $x \geq 2x - x^2$. Therefore, by (6.3a),

$$A = \int_0^2 |(2x - x^2) - x|\, dx$$

$$= \int_0^1 (2x - x^2 - x)\, dx + \int_1^2 - (2x - x^2 - x)\, dx$$

$$= \left[\frac{x^2}{2} - \frac{x^3}{3}\right]_0^1 + \left[\frac{x^3}{3} - \frac{x^2}{2}\right]_1^2$$

$$= \frac{1}{6} + \frac{5}{6} = 1 \text{ square unit} \quad \blacksquare$$

Sometimes the cross section function is determined by totally different functions on different parts of the region. Figure 6–6 is an illustration; it shows that on $[a, c]$, $h(x)$ is determined by the functions f_1 and g whereas on $[c, b]$, $h(x)$ is determined by f_2 and g. For the region in the figure, the area is

$$A = \int_a^c [f_1(x) - g(x)]\, dx + \int_c^b [f_2(x) - g(x)]\, dx$$

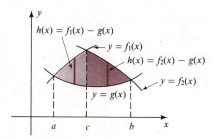

FIGURE 6–6. The cross section functions are different on different parts of $[a, b]$.

EXAMPLE 4 Set up the integral (or integrals) whose value is the area of the region bounded by the graphs of $y = \sqrt{x}$, $y = -\sqrt{x}$, and $y = x - 2$.

Solution Sketch the region (Figure 6–7). After finding the points of intersection, we see that

on the interval $[0, 1]$, $h(x) = \sqrt{x} - (-\sqrt{x}) = 2\sqrt{x}$

on the interval $[1, 4]$, $h(x) = \sqrt{x} - (x - 2) = \sqrt{x} - x + 2$

It follows that the area is

$$A = \int_0^1 2\sqrt{x}\, dx + \int_1^4 (\sqrt{x} - x + 2)\, dx \quad \blacksquare$$

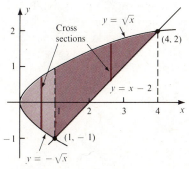

FIGURE 6–7. Example 4.

Most area problems involving regions of the type illustrated in Example 4 require the evaluation of two (or more) integrals. But, under certain conditions, there is an alternate approach that may save some work. The conditions are that the boundary curves of the region R can be considered as the *graphs of functions of y*.

Suppose that a region is determined by two or more functions of y from $y = c$ to $y = d$. In this situation, we use *horizontal cross sections*. For each y in the interval $[c, d]$ on the y-axis, $h(y)$ is the length of the (horizontal) cross section. If h is continuous, then the area of the region is

$$(6.4) \qquad A = \int_c^d h(y)\,dy$$

Example 4 is a case in point. The object is to write the boundary curves $y = -\sqrt{x}$, $y = \sqrt{x}$, and $y = x - 2$ as functions of y. The equation $y = x - 2$ can be solved for x and rewritten as $x = y + 2$. Solving $y = -\sqrt{x}$ and $y = \sqrt{x}$ for x both yield the equation $x = y^2$. Therefore, we can think of the region in Example 4 as bounded by the graphs of $x = y + 2$ and $x = y^2$, and apply (6.4).

EXAMPLE 5   Compute the area of the region shown in Figure 6–7 using horizontal cross sections.

Solution   Redraw the region (Figure 6–8) and rewrite the boundary curves as functions of y; $x = y^2$ and $x = y + 2$. Project the region onto the y-axis to determine the interval $[-1, 2]$. For *each y* in this interval, the (horizontal) cross section extends from the curve $x = y^2$ on the left to the curve $x = y + 2$ on the right. Thus, $h(y) = (y + 2) - y^2$ and it follows from (6.4) that the area of the region is

$$A = \int_{-1}^2 (y + 2 - y^2)\,dy = \left[\frac{y^2}{2} + 2y - \frac{y^3}{3}\right]_{-1}^2$$

$$= \frac{27}{6} \quad \text{square units}$$

So, integrating with respect to x required the evaluation of two integrals (Example 4) whereas integrating with respect to y required only one. We leave it as an exercise to verify that the answers obtained in either case agree.   ∎

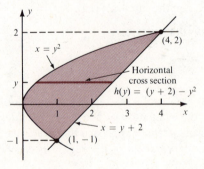

FIGURE 6–8. Example 5.

EXERCISES

In Exercises 1–4, sketch the regions bounded by the given curves and compute the area using vertical cross sections.

1. $y = x$, $y = x^2$ **2.** $y = x$, $y = x^3$

3. $y = x^2$, $y = \dfrac{3}{2}x + 1$, $y = 1$

4. $y = \sin x, y = \cos x, x = 0$

In Exercises 5–6, compute the area using horizontal cross sections.

5. $x = y^3, y = -x + 2, y = 0$

6. $x = y^3, y = -x + 2, x = 0$

In Exercises 7–24, compute the area using any method. (If any require Simpson's rule, then estimate the area using $n = 6$.)

7. $y = x, y = -x, x = 1$

8. $y = |x| - 1, y = 1 - |x|$

9. $y = x^2, y = 18 - x^2$

10. $y = -2x, y = 3 - x^2$

11. $y^2 = x + 1, y = 2x - 1$

12. $y^2 = x + 4, y^2 = 2x$

13. $y^2 = 4x, y = 2x - 4$

14. $y = x\sqrt{4 - x^2}, y = 0, 0 \le x \le 2$

15. $y^2 = x - 1, y = x - 3$

16. $y = \sqrt{1 - x^2}, y = x^2 - 1$

17. $y = x^2, x = y^2$ **18.** $y = x^3, x = y^2$

19. $y^2 = 1 - x, y^2 = x$

20. $y = \sqrt{x} + 1, y = x + 1$

21. $y^3 = x, y = x$ **22.** $y = x^3, y = x$

23. $y = \dfrac{1}{x}, y = \dfrac{5}{2} - x$

24. $y = x^3 - 2x^2 + x, y = x^2 + 2x$

25. Evaluate the integrals in Example 4 and compare the sum with the answer obtained in Example 5.

26. Two functions f and g are continuous and $f(x) \ge g(x)$ for all x in [0, 1]. Their values at certain points are recorded in the table below. Use Simpson's rule to estimate the area of the region bounded by their graphs from $x = 0$ to $x = 1$.

x	0	1/6	1/3	1/2	2/3	5/6	1
$f(x)$	2	1.8	1.6	1.4	1.2	1	1.2
$g(x)$	0	1.3	1.4	1.2	-.5	-.8	-1

27. Use Simpson's rule with $n = 4$ to estimate the area of the region bounded by the curves $y = 1/x, y = 2, x = 1$.

Optional Exercises

28. The line $x + y = 2$ divides the region bounded by the circle $(x - 2)^2 + (y - 2)^2 = 4$ into two parts. Set up an integral(s) whose value is the area of the smaller part.

29. Set up an integral(s) to find the area of the intersection of the two regions bounded by $(x - 2)^2 + (y - 2)^2 = 4$ and $(x - 3)^2 + (y - 2)^2 = 4$.

30. Set up an integral(s) to find the area of the region bounded by the curves $y = |4 - x^2|$ and $y = 4$.

31. Find the area of the region bounded by the curves $y = \sin x$ and $y = \sin 2x, 0 \le x \le \pi$.

6–2 VOLUMES BY SLICING

In this section and the next, we are concerned with the computation of volume. There are several techniques available and we begin with the method called "slicing." You will notice a similarity between this method of computing volumes and the method used in the last section to compute areas. However, we are now working in three dimnensions instead of two, so it may require a little more geometric imagination on your part.

Figure 6–9 shows a solid S with the property that for each x in $[a, b]$, the plane through x that is perpendicular to the x-axis meets the solid in a plane region.

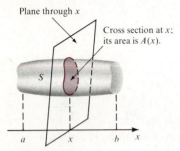

FIGURE 6–9. Cross sections of solids.

That region is called the **cross section of S at** x. These cross sections can be used to define a function on $[a, b]$.

DEFINITION

Let S be a solid with the property that for each x in $[a, b]$, there is a cross section of S at x. The **cross section function** is a function A whose value at each x in $[a, b]$ is the area of the cross section at x (Figure 6–9).

We claim that if each cross section of S is determined by some point in $[a, b]$, and A is continuous on $[a, b]$, then $\int_a^b A(x)\,dx$ is the volume of S.

To establish this claim, let $P = \{x_0, x_1, \ldots, x_n\}$ be any partition of $[a, b]$. Each point of P determines a cross section of S, and let V_i be the volume of that part of S between the cross sections at x_{i-1} and x_i (Figure 6–10). Since A is continuous, it will have a minimum value m_i and a maximum value M_i on $[x_{i-1}, x_i]$. If a solid has a constant cross sectional area m_i over an interval of length Δx_i, then its volume is $m_i \Delta x_i$. Similarly, if the cross sectional area of a solid is a constant M_i, then its volume is $M_i \Delta x_i$. This suggests that

$$m_i \Delta x_i \leq V_i \leq M_i \Delta x$$

As in previous cases, this means that the volume V of S is squeezed between the lower and upper sums of A and implies that V is the integral of A.

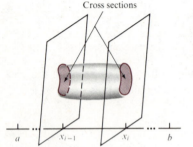

FIGURE 6–10. V_i is the volume between the cross sections at x_{i-1} and x_i.

Let S be a solid with the property that each cross section is determined by some point x in $[a, b]$. If the cross section function A is continuous on $[a, b]$, then

(6.5)

$$V = \int_a^b A(x)\,dx$$

is the volume of S.

The technique described above is called the method of *slicing*. The procedure is as follows:

To compute the volume of a solid S by the method of slicing:

(1) Sketch the solid. Be careful to position it so that you can obtain cross sections.

(2) Compute the area of an arbitrary cross section.

(3) Integrate the cross section function.

EXAMPLE 1   Find the volume of a right circular cone with a base of radius 1 and height 4.

Solution   Sketch the cone (Figure 6–11). A typical cross section is a disc of radius r; its area is πr^2. But the cross section function should be a function of the variable that determines the cross section; in this case, x. Therefore, you must express r in terms of x. To find r, use the similar triangles in the figure to write

$$\frac{r}{1} = \frac{x}{4} \quad \text{or} \quad r = \frac{x}{4}$$

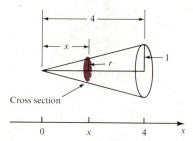

FIGURE 6–11. Example 1.

Thus, $A(x) = \pi x^2 / 16$, and the volume of the cone is

$$V = \frac{\pi}{16} \int_0^4 x^2 \, dx = \frac{\pi}{16} \left[\frac{x^3}{3} \right]_0^4 = \frac{4\pi}{3} \quad \text{cubic units}$$

Remark: It may be easier for you to visualize a cone in the position shown in Figure 6–12. If so, then by all means, draw it that way and use cross sections at y. In this case r must be expressed in terms of y. Again use similar triangles; the figure indicates that the ratios this time are $r/1 = (4 - y)/4$. Now $A(y) = \pi(4 - y)^2/16$ and

$$V = \frac{\pi}{16} \int_0^4 (4 - y)^2 \, dy = \frac{\pi}{16} \left[-\frac{(4 - y)^3}{3} \right]_0^4 \qquad \text{(Check)}$$

$$= \frac{4\pi}{3} \quad \text{cubic units} \quad \blacksquare$$

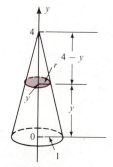

FIGURE 6–12. See the remark following Example 1.

EXAMPLE 2   Find the volume of a pyramid with a square base 4 units on a side and height 6 units.

Solution   Sketch the pyramid in a way that is easy for you to visualize. We have chosen to draw it as shown in Figure 6–13 because it simplifies the computations. For any x in $[0, 6]$, the cross section is a *square* $2y$ units on a side. To express y in terms of x, use similar triangles again

$$\frac{y}{2} = \frac{x}{6} \quad \text{or} \quad y = \frac{x}{3}$$

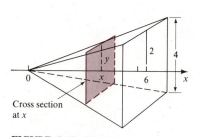

FIGURE 6–13. Example 2.

Thus, $2y = 2x/3$ and $A(x) = 4x^2/9$. The volume of the pyramid is then

$$V = \frac{4}{9} \int_0^6 x^2 \, dx = \frac{4}{9} \left[\frac{x^3}{3} \right]_0^6 = 32 \text{ cubic units} \quad \blacksquare$$

The next example is a little different in that the solid is described in terms of its cross sections.

EXAMPLE 3   The base of a solid S is the plane region bounded by the curve $y = \sqrt{1 - x^2}$ from $x = 0$ to $x = 1$. Each cross section perpendicular to the x-axis is a square with one side lying in the base. Compute the volume of S.

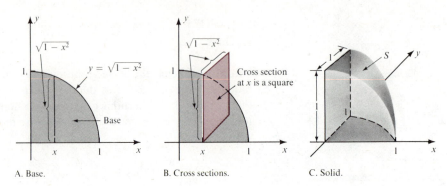

FIGURE 6–14. Example 3.   A. Base.   B. Cross sections.   C. Solid.

Solution   We have drawn the solid in three stages (Figure 6–14); this is a good approach when the solid is unfamiliar. The cross section at x is a square whose area is $(\sqrt{1-x^2})^2 = 1 - x^2$, so the volume of S is

$$V = \int_0^1 (1 - x^2)\, dx = \frac{2}{3} \text{ cubic units} \qquad \blacksquare$$

EXAMPLE 4   A wedge S is cut from a tree by making one cut parallel to the ground and another at an angle of 30°. Both cuts meet at the center of the tree (Figure 6–15). If the radius of the tree is 6 inches, compute the volume of the wedge.

Solution   The figure indicates that for each x in $[-6, 6]$, the cross section of the wedge at x is a right triangle whose area is $(\frac{1}{2})(y)(y \tan 30°) = (1/2\sqrt{3})y^2$. Now y is the length of a line segment extending from the point x to the outer edge of the tree. It is one leg of a right triangle whose other leg has length $|x|$ and whose hypotenuse is 6 (see the figure); therefore, $y = \sqrt{36 - x^2}$. It follows that the area of the cross section is $(1/2\sqrt{3})(36 - x^2)$ and that the volume of the wedge is

$$V = \frac{1}{2\sqrt{3}} \int_{-6}^{6} (36 - x^2)\, dx = \frac{144}{\sqrt{3}} \text{ cubic inches} \qquad \blacksquare$$

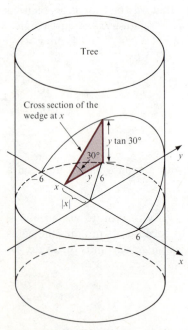

FIGURE 6–15. Example 4.

EXERCISES

1-13

To keep in practice with the technique of u-substitution, evaluate the following indefinite integrals. Check your answers *before* looking in the back of the book.

1. $\int (5x + 3)^4\, dx$

2. $\int \frac{x^2}{\sqrt{x^3 + 1}}\, dx$

3. $\int x^2 \sqrt{x^3 + 5}\, dx$

4. $\int (x + 1)(x^2 + 2x - 3)^4\, dx$

5. $\int \frac{1}{x^2}\left(1 - \frac{1}{x}\right)^4\, dx$

6. $\int \frac{(1 + \sqrt{x})^4}{\sqrt{x}}\, dx$

In Exercises 7–10, evaluate the definite integrals by *u*-substitution or by a change of variable.

7. $\displaystyle\int_0^1 \frac{x}{(x^2+1)^2}\,dx$ **8.** $\displaystyle\int_1^9 \frac{(1+\sqrt{x})^2}{\sqrt{x}}\,dx$

9. $\displaystyle\int_{-\sqrt{3}}^{\sqrt{3}} x\sqrt{x^2+1}\,dx$

10. $\displaystyle\int_0^1 (x+1)(x^2+2x)^3\,dx$

---

11. Find the volume of a right circular cone with a base of radius *r* and height *h* (see Example 1).

12. Find the volume of a pyramid with a square base *k* units on a side and height *h* (see Example 2).

13. Find the volume of a wedge cut from a right circular cylinder of radius *r*. The wedge is determined by two planes; the first is perpendicular to the axis of the cylinder and the second is at an angle of α to the first. The planes intersect at the center of the cylinder (see Example 4).

In Exercises 14–16, the solids are described in terms of cross sections (see Example 3). Find the volumes.

14. The base is the plane region bounded by the curves $y = \sqrt{x}$, $y = 0$, and $x = 2$. Each cross section perpendicular to the *x*-axis is a square with one side lying in the base.

15. Same base as in Exercise 14 but now the cross sections are semicircles with the diameter lying in the base.

16. The base is the plane region bounded by the curves $y = -x^2 + 2x$ and $y = 0$. Each cross section perpendicular to the *x*-axis is an isosceles right triangle with one leg in the base.

In Exercises 17–22, find the volumes of the indicated solids.

17. A sphere of radius *r*.

18. Same base as Exercise 16 but now the cross sections are equilateral triangles with one side in the base.

19. The base is the semicircular region between the curve $y = \sqrt{1-x^2}$ and the *x*-axis. Each cross section perpendicular to the *x*-axis is an equilateral triangle with one side in the base.

20. A horn-shaped solid. For each *x* in [0, 1] the cross section at *x* is a circle whose diameter is the line segment extending from the curve $y = x^{1/3}$ up to the curve $y = 2x^{1/3}$.

21. Let *R* be the plane region bounded by $y = 1/x$, $y = 0$, $x = 1$, and $x = 2$. The solid *S* is formed by revolving *R* about the *x*-axis.

22. Let *R* be the plane region bounded by the graphs of $y = x^2$ and $y = 3$. The solid *S* is formed by revolving *R* about the *y*-axis. *Hint:* Use horizontal cross sections and integrate with respect to *y*.

In Exercises 23–27 *set up* a definite integral whose value is the volume of the indicated solid.

23. The base of the solid is the plane region determined by $y = \sin x$ for $0 \le x \le \pi$. Each cross section perpendicular to the *x*-axis is a square with one side in the base.

24. Same base as in Exercise 23 but each cross section is an equilateral triangle with one side in the base.

25. The solid obtained by revolving the base described in Exercise 23 about the *x*-axis.

26. The solid obtained by revolving the region shown in Figure 6–16 about the *x*-axis.

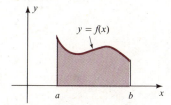

FIGURE 6–16. Exercise 26.

27. The solid obtained by revolving the region shown in Figure 6–17 about the *x*-axis.

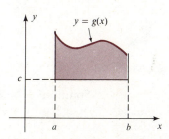

FIGURE 6–17. Exercise 27.

28. A rectangular swimming pool is 20 ft wide and 36 ft long. The bottom of the pool slopes gradually with no sudden drops. The depth of the water x ft from one end is measured at 6 ft intervals and the results are tabulated below:

x	0	6	12	18	24	30	36
depth (ft)	3	5	6.5	7.5	9.5	10	12

Use Simpson's rule to estimate the volume of water in the pool.

29. Estimate the volume of water if the pool in Exercise 28 is circular (rather than rectangular) with a diameter of 36 ft.

Optional Exercises

30. Given a sphere of radius r, what is the volume remaining after a hole of radius $r/2$ is bored through its center?

31. Two right circular cylinders of radius r intersect at right angles. What is the volume of their intersection?

32. Two planes are perpendicular to each other. Each contains a circle of radius r and the two circles have a common diameter which lies on the intersection of the two planes. A solid S is described as follows: Each cross section of S perpendicular to the common diameter of the circles is a square whose corners lie on the circles. Find the volume of S.

33. The base of a solid is the semicircular region bounded by $y = \sqrt{1 - x^2}$ and $y = 0$. Each cross section perpendicular to the x-axis is a regular hexagon (a polygon with six equal sides) with one side in the base. Find the volume of the solid.

6–3   SOLIDS OF REVOLUTION

Many familiar solids are obtained by revolving a plane region about a line. They are called **solids of revolution** and we say that the solid is **generated** by the region. For example (see Figure 6–18), a right circular cone can be generated by a right triangular region, a sphere can be generated by a semicircular region, and a torus, or donut, can be generated by a circular region. In this section, we present two methods for computing the volume of a solid of revolution.

Disc or Washer Method

This is a slicing method like the one used in the last section. A solid of revolution is generated by revolving a region about a line; the line is called the **axis of revolution.** Each cross section perpendicular to the axis of revolution is either a *disc* or a *washer* (a washer is a disc with a hole in the middle). Figure 6–19 illustrates the two possibilities. The area of each cross section defines the cross section function A and the integral of A is the volume of the solid generated.

EXAMPLE 1   Show that the volume of a sphere of radius r is $\frac{4}{3}\pi r^3$.

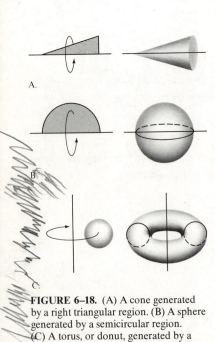

FIGURE 6–18. (A) A cone generated by a right triangular region. (B) A sphere generated by a semicircular region. (C) A torus, or donut, generated by a circular region.

Solution   Think of the sphere as being generated by the semicircular region bounded by the graph of $y = \sqrt{r^2 - x^2}$ from $x = -r$ to $x = r$. Figure 6–20 shows that for each x in $[-r, r]$, the cross section is a disc of radius $\sqrt{r^2 - x^2}$. Therefore, $A(x) = \pi(\sqrt{r^2 - x^2})^2 = \pi(r^2 - x^2)$ is the cross section function and the volume of the sphere is

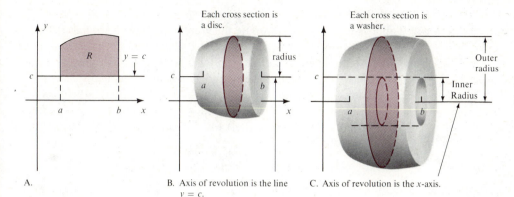

A.

B. Axis of revolution is the line
   $y = c.$

C. Axis of revolution is the x-axis.

FIGURE 6–19. (A) The region R.
(B) If R is revolved about the line
$y = c$, then each cross section is a disc;
its area is $\pi(\text{radius})^2$. (C) If R is revolved
about the x-axis, then each cross section
is a washer. A washer has an outer
radius and an inner radius; its area is
$\pi(\text{outer radius})^2 - \pi(\text{inner radius})^2$.

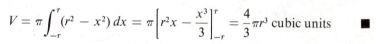

$$V = \pi \int_{-r}^{r} (r^2 - x^2)\, dx = \pi \left[r^2 x - \frac{x^3}{3} \right]_{-r}^{r} = \frac{4}{3}\pi r^3 \text{ cubic units} \qquad \blacksquare$$

EXAMPLE 2 Given a sphere of radius r, find the volume remaining after a hole of radius $r/2$ is bored through its center.

Solution Consider the remaining solid as being generated by the region shown in Figure 6–21. The limits a and b are found by solving the equation

$$\frac{r}{2} = \sqrt{r^2 - x^2}$$

Squaring both sides and solving for x yields

$$a = -\frac{\sqrt{3}}{2}r \quad \text{and} \quad b = \frac{\sqrt{3}}{2}r$$

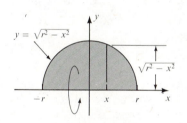

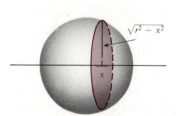

FIGURE 6–20. Example 1.

For each x in this interval, the cross section of the solid is a washer with an outer radius of $\sqrt{r^2 - x^2}$ and an inner radius of $r/2$. It follows that

$$A(x) = \pi (\sqrt{r^2 - x^2})^2 - \pi \left(\frac{r}{2}\right)^2$$

$$= \pi \left(\frac{3}{4}r^2 - x^2\right)$$

Therefore, the volume is

$$V = \pi \int_{-\sqrt{3}r/2}^{\sqrt{3}r/2} \left(\frac{3}{4}r^2 - x^2\right) dx = \pi \left[\frac{3}{4}r^2 x - \frac{x^3}{3}\right]_{-\sqrt{3}r/2}^{\sqrt{3}r/2}$$

$$= \frac{\sqrt{3}}{2}\pi r^3 \text{ cubic units} \qquad \blacksquare$$

If a region is revolved about the y-axis or a line parallel to the y-axis, you can use horizontal cross sections. Just as in the last section, the boundaries of the region must be converted to graphs of functions of y.

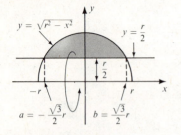

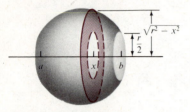

FIGURE 6–21. Example 2.

EXAMPLE 3 Revolve the region bounded by the graphs of $y = x^2 - 1$, $x = 2$, and $y = 0$ about the y-axis and compute the volume of the solid generated.

Solution Sketch the region and the solid generated by revolving it about the y-axis (Figure 6–22). Convert the boundaries to graphs of functions of y; $y = x^2 - 1$ becomes $x = \sqrt{y + 1}$ (x is positive) and the other is the constant $x = 2$. The limits are $a = 0$ and $b = 3$. For each y in $[0, 3]$ the figure illustrates that the cross section at y is a (horizontal) washer with outer radius 2 and inner radius $\sqrt{y + 1}$. Therefore,

$$A(y) = \pi 2^2 - \pi(\sqrt{y + 1})^2 = \pi(3 - y)$$

Thus, the volume of the solid generated is

$$V = \pi \int_0^3 (3 - y)\, dy = \frac{9\pi}{2} \text{ cubic units} \qquad \blacksquare$$

FIGURE 6–22. Example 3.

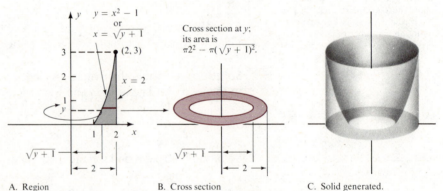

A. Region   B. Cross section   C. Solid generated.

To compute the volume of a solid of revolution with the *disc* or *washer* method:

(1) Sketch the generating region along with the axis of revolution.

(2) If the axis of revolution is parallel to the x-axis, then the boundaries of the region must be graphs of functions of x.

(3) Express the outer radius and inner radius of a typical cross section as functions of x (if the cross sections are discs, then the inner radii are 0).

(4) Find the limits of integration a and b, and then the volume generated is

$$V = \int_a^b \pi[(\text{outer radius})^2 - (\text{inner radius})^2]\, dx$$

If the axis of revolution is parallel to the y-axis, then each of the steps above must be carried out in terms of y rather than x.

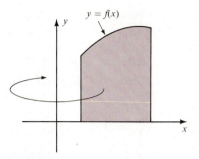

FIGURE 6–23. The shaded region is revolved about the *y*-axis to generate a solid *S*.

Shell Method

A solid S is generated by revolving the region in Figure 6–23 about the *y*-axis. The volume of S could be computed using horizontal washers but there is an alternate technique that, in some situations, is simpler to apply. It is called the **shell method.**

The justification for the shell method, which is quite different from slicing, uses a clever technique that is worth describing in detail.

Let $P = \{x_0, x_1, \ldots, x_n\}$ be any partition of $[a, b]$, and let S_i be that part of the plane region between the vertical lines through x_{i-1} and x_i (Figure 6–24A). If V_i is the volume of the solid generated by revolving S_i about the *y*-axis, then the total volume generated is $V = \sum_{i=1}^{n} V_i$.

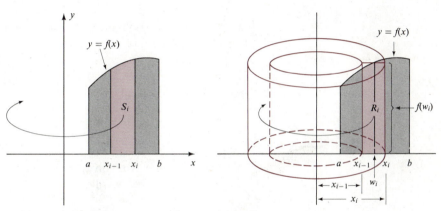

FIGURE 6–24. The shell method of computing volumes.

A. Region generates a solid.   B. Cylindrical shell.

Let w_i be any point in $[x_{i-1}, x_i]$, and let R_i be the rectangle with base Δx_i and height $f(w_i)$. If R_i is revolved about the *y*-axis, a *cylindrical shell* is formed (Figure 6–24B). The radius of the outer wall is x_i, the radius of the inner wall is x_{i-1}, and both have height $f(w_i)$. It follows that the volume of the shell is

$$\pi x_i^2 f(w_i) - \pi x_{i-1}^2 f(w_i) = \pi f(w_i)(x_i^2 - x_{i-1}^2)$$
$$= \pi f(w_i)(x_i + x_{i-1})(x_i - x_{i-1}) \quad \text{(Factor)}$$
$$= \pi(x_i + x_{i-1})f(w_i)\Delta x_i$$

Since the area of R_i approximates the area of S_i, it follows that $V_i \approx \pi(x_i + x_{i-1})f(w_i)\Delta x_i$ so that

$$(1) \qquad V = \sum_{i=1}^{n} V_i \approx \sum_{i=1}^{n} \pi(x_i + x_{i-1})f(w_i)\,\Delta x_i$$

As it stands, the sum on the right is *not* a Riemann sum because $(x_i + x_{i-1})f(w_i)$ is not the value of a function at a point of $[x_{i-1}, x_i]$. But, *if* the sum on the right *were* the Riemann sum of some function g, then equation (1) would justify saying that the volume V is the limit of those sums; that is, the integral of g. Here is where the clever part comes in. Since w_i can be any point in $[x_{i-1}, x_i]$, choose it to be the *midpoint*; that is $w_i = (x_i + x_{i-1})/2$. Now

$$(x_i + x_{i-1})f(w_i) = 2\left[\frac{x_i + x_{i-1}}{2}\right]f(w_i) = 2w_i f(w_i)$$

and equation (1) becomes

(2) $$V \approx \sum_{i=1}^{n} 2\pi w_i f(w_i)\, \Delta x_i$$

The sum on the right *is* a Riemann sum of the continuous function $g(x) = 2\pi x f(x)$. Therefore,

(3) $$\boxed{V = \int_a^b 2\pi x f(x)\, dx}$$

The shell method can also be used if the boundaries are graphs of functions of y and the region is revolved about the x-axis or a line parallel to it. The procedure is summarized as follows:

To compute the volume of a solid of revolution with the *shell method:*

(1) Sketch the generating region along with the axis of revolution.

(2) If the axis of revolution is parallel to the y-axis, then the boundaries must be graphs of functions of x; if it is parallel to the x-axis, then the boundaries must be graphs of functions of y.

(3) Find the limits of integration.

(4) Determine the (average) radius and height of a typical shell, and set $g = 2\pi(\text{radius})(\text{height})$.

(5) The volume generated is the integral of g.

EXAMPLE 4 Use the shell method to compute the volume of the solid generated in Example 3.

Solution   The generating region is bounded by the graph of $f(x) = x^2 - 1$ from $x = 1$ to $x = 2$ and the y-axis is the axis of revolution. Draw a picture and verify that each shell will have radius x and height $x^2 - 1$. Therefore the volume of the solid generated is

$$V = 2\pi \int_1^2 x(x^2 - 1)\, dx = 2\pi \int_1^2 (x^3 - x)\, dx$$

$$= 2\pi \left[\frac{x^4}{4} - \frac{x^2}{2} \right]_1^2 = \frac{9\pi}{2} \text{ cubic units}$$

which agrees with the answer in Example 3.   ■

   The volume of the solid in Examples 3 and 4 can be computed by either method, washer or shell. The choice is yours. However, the next two examples illustrate cases where only one of the methods is *practical*.

EXAMPLE 5   Revolve the region bounded by the graph of $y = (x - 1)^{1/3}$ from $x = 1$ to $x = 9$ about the y-axis. Set up integrals using (a) the washer method and (b) the shell method. Which one is not practical?

Solution

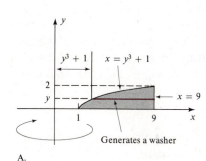

$y^3 + 1$ $x = y^3 + 1$

$x = 9$

Generates a washer

A.

(a) *Washer method.* Sketch the region with the boundaries converted to graphs of functions of y; $y = (x - 1)^{1/3}$ becomes $x = y^3 + 1$ (Figure 6–25A). For each y in $[0, 2]$, the cross section is a washer with an outer radius of 9 and an inner radius of $y^3 + 1$. Thus,

$$V = \pi \int_0^2 [81 - (y^3 + 1)^2]\, dy$$

which can be evaluated by rewriting the integrand as a polynomial.

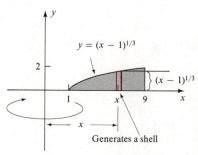

$y = (x - 1)^{1/3}$

$(x - 1)^{1/3}$

Generates a shell

B.

FIGURE 6–25. Example 5.

(b) *Shell method.* Sketch the region again as in Figure 6–25B. For each x in $[1, 9]$, the shell generated will have height $(x - 1)^{1/3}$ and radius x. Therefore, the volume of the solid is

$$V = 2\pi \int_1^9 x(x - 1)^{1/3}\, dx$$

Conclusion. The second integral looks simpler but, at this point, you probably do not know an antiderivative of the integrand (and will not until Chapter 9). Therefore, use the washer method; the shell method is not practical (at least not for now).   ■

EXAMPLE 6   Rotate the region bounded by the graph of $x = y^3 - 3y + 2$ from $y = 0$ to $y = 1$ about the line $y = -1$, and set up an integral that yields the volume of the solid generated.

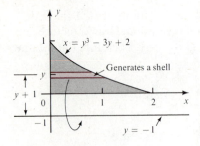

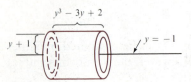

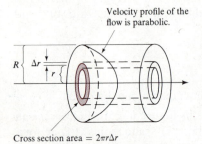

FIGURE 6–26. Example 6.

Velocity profile of the flow is parabolic.

Cross section area $= 2\pi r\Delta r$

FIGURE 6–27. Example 7. Blood flow in an artery.

Solution Sketch the region (Figure 6–26). To use the disc or washer method in this example, you must convert the boundaries to graphs of functions of x. In this case, solving the equation $x = y^3 - 3y + 2$ for y in terms of x is a formidable task; not impossible, but difficult. Therefore, the washer method is not practical. On the other hand, the shell method is quite simple. For each y in [0, 1], the figure shows that the generated shell has a radius $y + 1$ and height $y^3 - 3y + 2$. Thus,

$$V = 2\pi \int_0^1 (y + 1)(y^3 - 3y + 2)\, dy$$

This integral can be evaluated by expanding the integrand. ∎

Our final example illustrates the shell method in a practical application.

EXAMPLE 7 *(Rate of blood flow)* In Example 3 of Section 1–4, the velocity v of a blood particle flowing through an artery of radius R is related to its distance r from the central axis by the equation

$$v(r) = k(R^2 - r^2) \quad \text{cm/sec}, 0 \le r \le R$$

where k is a positive constant. Let us compute the volume of blood per second that flows past a given cross section of the artery.

Imagine that the artery is divided into very thin cylindrical shells of width Δr_i, one inside the other, and that all blood particles within a particular shell move with the same velocity (Figure 6–27). If the midpoint between the walls of a shell S_i is at a distance r_i from the center of the artery, then the volume of blood flowing through that shell is approximately

$$V_i \approx 2\pi r_i v(r_i)\, \Delta r_i \quad \text{cu cm/sec}$$

The sum from $i = 1$ to n is an approximation to the total volume V of blood per second flowing through the cross section. It follows that

$$V = \int_0^R 2\pi r v(r)\, dr \quad \text{cu cm/sec}$$

In Exercise 32, you are asked to evaluate this integral. ∎

Summary

Washer method:

$$V = \int_a^b \pi[(\text{outer radius})^2 - (\text{inner radius})^2]\, dx$$

Shell method:

$$V = \int_a^b 2\pi(\text{radius})(\text{height})\, dx$$

If you can answer the questions below, you should have no trouble with solids of revolution (the answers are on page 300).

1. The region R in Figure 6–28A is revolved about the line indicated below. In each case, answer the following two questions: (i) What is the most practical method, washer or shell, for computing the volume? (ii) Depending on your answer to (i), what function is integrated to find the volume?

   (a) the x-axis   (b) the line $y = C$ (c) the line $y = D$
   (d) the y-axis   (e) the line $x = c$ (f) the line $x = d$

2. Repeat the questions for the region R in Figure 6–28B.

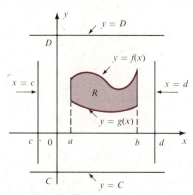

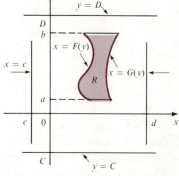

FIGURE 6–28. Summary.   A. Review Question 1.   B. Review Question 2.

EXERCISES

In Exercises 1–10, sketch the region bounded by the graphs of the given equations and use the disc or washer method to compute the volume of the solid generated by revolving the region about the indicated axis of revolution.

1. $y = x^2, x = 2, y = 0$; x-axis
2. $x = \sqrt{y}, y = 4, x = 0$; y-axis
3. Same as Exercise 1; $y = -1$
4. Same as Exercise 2; $x = -1$
5. $y = \sqrt{\sin x}, x = \pi/2, y = 0$; x-axis
6. $y = \sqrt{\cos x}, x = -\pi/2, x = \pi/2, y = 0$; x-axis
7. $x = y^2, x = y$; x-axis

8. Same as Exercise 7; $y = -2$
9. Same as Exercise 7; $y = 2$
10. $y = x^3, y = x$; y-axis

In Exercises 11–20, follow the instructions above but now use the shell method.

11. $y = \sqrt{x}, x = 2, y = 0$; y-axis
12. $x = y^2, y = 4, x = 0$; x-axis
13. Same as Exercise 11; $x = -1$
14. Same as Exercise 12; $y = -1$
15. $x = \sqrt{4 - y^2}, y = 0, y = 1, x = 0$; x-axis
16. $y = \sqrt{1 - x^2}, x = 0, x = 1, y = 0$; y-axis

17. $y = -x^2 + 2x, y = 0$; y-axis
18. $y = x^2 - 2x + 1, x = 0, x = 2, y = 0$; y-axis
19. $y = \sqrt{x}, y = x^2$; y-axis
20. Same as Exercise 19; x-axis

In Exercises 21–23, compute the volume with both methods and verify that the answers agree.

21. $y = x^3, x = 2, y = 0$; y-axis
22. $x = -y^2 + 1, y = 0, y = 1, x = 0$; x-axis
23. $x = -(y - 1)^2 + 1, x = 0$; x-axis

In Exercises 24–28, compute the volume with the method you think is most practical.

24. $x = y^2 - 1, x = 0, x = 3, y = 0$; x-axis
25. $y = x^3 - 3x + 2, y = 0, x = 0$; y-axis
26. $x = 4 - y^2, x = 3y, y = 0$; x-axis
27. $y = 1/x, x = 1, x = 2, y = 0$; x-axis
28. $y = \sqrt{x^2 - 1}, x = 1, x = 3, y = 0$; y-axis

---

29. The volume of a sphere of radius r is $4\pi r^3/3$. How much of this volume is contained between two parallel planes that are each at distance $d(0 < d < r)$ from the center?

30. A hemispherical punch bowl has a radius of 1 foot. At the beginning of a party it was filled to within 1 inch of the top. After 30 minutes there was only 2 inches of punch left at the bottom. How many gallons of punch have been consumed? (There are 231 cubic inches in a gallon.)

31. The region shown in Figure 6–29 is revolved about the y-axis to form a container. The units are in feet. If water is poured into the container at the rate of

3 cu ft/sec, how fast is the water rising when its depth (in the container) is 1 foot?

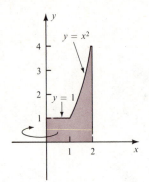

FIGURE 6–29. Exercise 31.

32. Evaluate the integral in Example 7, and show that the volume of blood flow in an artery is proportional to the 4th power of its radius. Compute the volume if the radius is .2 cm and $v = 40 - 1{,}000r^2$.

Optional Exercises

Find the volume generated by revolving the regions described about the indicated axes of revolution.

33. $y = e^x, y = 1, x = 1$; x-axis
34. $y = e^{x^2}, y = 1, x = 1$; y-axis
35. Triangle with vertices (2, 2), (4,4) and (2, 6); x-axis
36. $y = \sin x, y = \cos x, x = 0, x = \pi/4$; x-axis (*Hint:* $\cos^2 x - \sin^2 x = \cos 2x$)
37. Same region as Exercise 36; $y = -1$
38. Donut generated by revolving $(x - 3)^2 + (y - 4)^2 = 4$ about the x-axis. Set up the integral, and use Simpson's rule with $n = 4$ to estimate it.

6–4 WORK, ENERGY, AND POWER

When an applied force moves an object, work is done. Work involves energy, and energy involves power. In this section you will see the role that calculus can play in a discussion of work, energy, and power.

Answers to questions on page 299. 1(a) washer, $\pi[f(x)]^2 - \pi[g(x)]^2$, (b) washer, $\pi[f(x) - C]^2 - \pi[g(x) - C]^2$, (c) washer, $\pi[D - g(x)]^2 - \pi[D - f(x)]^2$, (d) shell, $2\pi x[f(x) - g(x)]$, (e) shell, $2\pi(x - c)[f(x) - g(x)]$, (f) shell, $2\pi(d - x)[f(x) - g(x)]$. 2(a) shell, $2\pi y[G(y) - F(y)]$, (b) shell, $2\pi(y - C)[G(y) - F(y)]$, (c) shell, $2\pi(D - y)[G(y) - F(y)]$, (d) washer, $\pi[G(y)]^2 - \pi[F(y)]^2$, (e) washer, $\pi[G(y) - c]^2 - \pi[F(y) - c]^2$, (f) washer, $\pi[d - F(y)]^2 - \pi[d - G(y)]^2$.

In the British system of measurement, the unit of force is the pound (lb). If a constant force of F lbs. moves an object d feet along a line, then the work W done by the force is defined as the product

$$W = F \cdot d \quad \text{ft-lbs*}$$

Work can be positive or negative depending on whether the force is applied in the direction of motion or opposite to it. For example, if you apply a constant force of 10 lbs to push a carton along the floor a distance of 3 ft, then you have done 30 ft-lbs of work. At the same time, friction exerts a force of 10 lbs in the opposite direction, so the work done by the friction is -30 ft-lbs.

In most applications the force is not constant and the work done is computed by evaluating a definite integral. The problem, of course, is to find the correct integrand. The examples below illustrate that there are two types of integrands corresponding to two different types of work problems.

Work on a Single Particle System

Suppose that a force F is applied to move a single object, or particle along a (horizontal) line from point a to point b. The sign of the work done does not depend on whether a is to the right or left of b; it depends only on whether the force is applied in the direction of motion or opposite to it. Therefore, let us assume that $a < b$ and that F is a continuous function on $[a, b]$. Thus, $F(s)$ is the force applied when the particle is at position $s^\dagger$.

Let $P = \{s_0, s_1, \ldots, s_n\}$ be any partition of $[a, b]$, let W_i be the actual work done by F in moving the particle from s_{i-1} to s_i, and let M_i and m_i be the maximum and minimum values of F on $[s_{i-1}, s_i]$. This suggests that $m_i \Delta s_i \leq W_i \leq M_i \Delta s_i$, and summing from $i = 1$ to n yields

$$L_P \leq W \leq U_P$$

where $W = \sum_{i=1}^{n} W_i$ is the total work done. Since L_P and U_P are lower and upper sums of F, we conclude that

(6.6)

> An applied force $F = F(s)$ moves a particle along a line from position a to position b. If F is continuous, then
>
> $$W = \int_a^b F(s) \, ds$$
>
> is the work done by the force.

---

*In the metric system, the unit of force is the Newton (N) and $1N \approx .225$ lbs. Distance is measured in meters (m) and the measurements of work are in N-m or joules(J); $1J = 1$N-m $\approx .738$ ft-lbs.

†We use s here instead of x. This is consistent with our earlier use of s to denote the position of a particle.

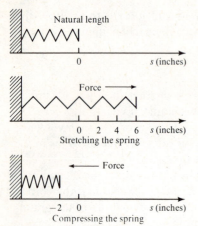

FIGURE 6–30. Example 1.

The first example is about a spring that is anchored at one end (Figure 6–30). We consider the free end as the single particle to be moved. The force $F(s)$ required to move the free end s units from the spring's natural length is proportional to s; that is,

$$(6.7) \qquad \boxed{F(s) = ks}$$

where k is a constant called the **spring constant.** This is known as **Hooke's Law** (Robert Hooke, 1635–1703). The value of k depends on the material used, the tightness of the coil, and so forth.

EXAMPLE 1 It takes a force of 20 lbs to stretch a spring 4 inches beyond its natural length. Find the work done in (a) stretching the spring 6 inches and (b) compressing it 2 inches.

Solution The work done is the integral of the force function; this you already know by (6.6). The problem, then, is to find the force function.

First find the spring constant k. According to Hooke's Law (6.7), $F(s) = ks$. When $s = 4$, the force is 20 lbs. Thus, $20 = k \cdot 4$ so

$$k = 5 \quad \text{and the force function is} \quad F(s) = 5s$$

In Figure 6–30, the spring is drawn in its natural state and the free end is at the point labeled 0.

(a) By (6.6), the work done in stretching the spring 6 inches (that is, from position 0 to position 6) is

$$W = \int_0^6 5s \, ds = 90 \text{ in-lbs.}$$

(b) The work done in compressing the spring 2 inches (that is, from position 0 to position −2) is

$$W = \int_0^{-2} 5s \, ds = 10 \text{ in-lbs.}$$

Notice that in both cases, the force is applied in the direction of motion, so the work is positive. ∎

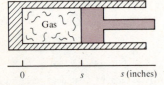

FIGURE 6–31. Expanding gas exerts a fore against the piston and does work.

Here is another example of the same type. A gas in a cylinder is expanding adiabatically against a piston (Figure 6–31). The word *adiabatic* means that the system is insulated from its surroundings so that no heat enters or leaves the gas. Under these conditions, the pressure P and the volume V of the gas are related by the equation

$$(6.8) \qquad PV^k = C \quad \text{or} \quad P = CV^{-k}$$

where k and C are constants that depend on the gas. As the gas expands, the piston moves and work is done. Computing this work is important in determining the efficiency of an engine.

Suppose the cylinder in Figure 6–31 has a constant cross-sectional area of A sq in. and that the piston is s inches from the end. Then the volume of the gas is

(6.9) $V = As$ cu in.

If the pressure of the gas is P lbs/sq in., then the total force on the piston is $F = PA$ lbs. Therefore, the force function is

(6.10)

$$F(s) = PA = CV^{-k}A \qquad (6.8)$$
$$= C(As)^{-k}A \qquad (6.9)$$
$$= CA^{1-k}s^{-k} \quad \text{lbs}$$

The integral of F is the work done by the gas.

EXAMPLE 2   Suppose the cylinder in Figure 6–31 has a radius of 3 inches. If the gas in it satisfies the equation $PV^{1.6} = 100$, find the work done in moving the piston from $s = 6$ to $s = 10$ inches.

Solution   In this example the constants k and C are 1.6 and 100, and the cross-sectional area is $\pi(3)^2 = 9\pi$ sq in. Therefore, by (6.10),

$$F(s) = 100(9\pi)^{-.6}s^{-1.6} \text{ lbs}$$

It now follows from (6.6) that the work done is

$$W = \frac{100}{(9\pi)^{.6}} \int_6^{10} s^{-1.6} \, ds$$

$$= \frac{100}{(9\pi)^{.6}} \left[\frac{s^{-.6}}{-.6} \right]_6^{10} \approx 2.0 \text{ in-lbs} \quad \blacksquare$$

Work on a Multiple Particle System

A second type of work problem occurs when the object being moved consists of many parts and each part is moved a different distance. Removing the water from a conical tank is an example (Figure 6–32). The water near the top is moved a shorter distance than the water near the bottom.

EXAMPLE 3   The conical tank in Figure 6–32 is filled with water. An apparatus is rigged to pump water from the surface over the top of the tank. Find the work done by the pump in removing all of the water.

Solution   The work is given by an integral; that much you may suspect. The question is, how do you find the correct integrand? In problems of this sort it

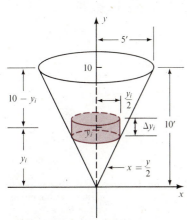

FIGURE 6–32. Example 3. Moving a "slice" of water.

is more convenient to use ordinary Riemann sums rather than the specialized upper and lower sums.

Refer to Figure 6–32. Let $P = \{y_0, y_1, \ldots, y_n\}$ be any partition of $[0, 10]$. Horizontal planes through the points of P will slice the water into n parts. The figure shows that the volume V_i of the ith slice is approximated by a disc of radius $y_i/2$ and thickness Δy_i; that is,

$$V_i \approx \pi \left(\frac{y_i}{2}\right)^2 \Delta y_i \quad \text{cu ft}$$

Since the weight of water is about 62.5 lbs/cu ft, the force needed to lift this slice is 62.5 times the volume. The figure also shows that the ith slice must be lifted a distance of about $10 - y_i$ feet. It follows that the work W done in removing all of the water is approximated by

$$W \approx \sum_{i=1}^{n} [10 - y_i] \left[62.5\pi \left(\frac{y_i}{2}\right)^2 \Delta y_i \right]$$

The sum on the right is a Riemann sum of the continuous function $62.5\pi(10 - y)(y^2)/4$. Since these approximations improve as the norm of the partition approaches zero, we have that

$$W = \frac{62.5\pi}{4} \int_0^{10} (10y^2 - y^3)\, dy$$

$$= \frac{62.5\pi}{4} \left(\frac{10^4}{12}\right) \approx 40{,}906 \text{ ft-lbs} \quad \blacksquare$$

Energy

An interesting theorem from mechanics is that the work done on a particle of mass m^* is related to the kinetic energy (K.E.) of the particle. The K.E. of the particle is defined to be $\frac{1}{2}mv^2$ where v is the velocity. If v_a and v_b represent the velocity when the particle is at positions a and b, respectively, then the work done in moving from a to b is

(6.11)
$$W = \frac{1}{2}mv_b{}^2 - \frac{1}{2}mv_a{}^2$$

This is the **work-energy equation;** it says that the work done is the change in the kinetic energy.

---

*The terms **mass** and **weight** are often confused. The mass of an object is constant (except at speeds approaching the speed of light). A man has the same mass whether he is on earth or on the moon. His weight, however, varies with his location. *Weight is the force that gravity exerts on an object.* According to Newton's second law of motion, $F = ma$ where a is acceleration. On earth, gravity causes an acceleration of $g \approx 32$ ft/sec^2. Therefore, on or near the surface of the earth, the weight of an object of mass m is $mg \approx 32\,m$ lbs.

EXAMPLE 4 An object weighing w lbs starts from rest at point a and slides down a smooth (frictionless) plane inclined at an angle θ. Find the work done by the force of gravity and the velocity of the object when it gets to the bottom.

Solution Sketch the incline (Figure 6–33A). The forces acting on the object are the weight $w = mg$ and the force N exerted by the plane. Because N is perpendicular to the motion, it does no work. The part of the force mg that points in the same direction as the motion is labeled F; that force is doing the work. The force diagram (Figure 6–33B) indicates that $F = mg \sin \theta$. This force is constant; therefore, if the particle moves s feet along the incline, then the work done is

$$(6.12) \qquad W = Fs = (mg \sin \theta)s \quad \text{ft-lbs}$$

To find the velocity at the end of the slide, we use the work-energy equation (6.11). The object starts from rest, so $v_a = 0$ and it follows that $W = \frac{1}{2}mv_b^2$. Solving for v_b, we have

$$v_b = \sqrt{\frac{2W}{m}} = \sqrt{\frac{2(mg \sin \theta)s}{m}} \qquad (6.12)$$
$$= \sqrt{2gs(\sin \theta)} \quad \text{ft/sec}$$

Remark: The product $s(\sin \theta)$ is the vertical height h of the object at the starting position a. Thus, the final velocity is $v_b = \sqrt{2gh}$; it depends *only* on the vertical height and not on the mass or the shape of the incline (Figure 6–33C). Since $W = \frac{1}{2}mv_b^2$, the work done by gravity is also independent of the shape of the incline. Also observe the similarity between the formula $v_b = \sqrt{2gh}$ and the escape velocity formula $v_e = \sqrt{2gR}$ discussed in Example 5 of Section 5–7. ∎

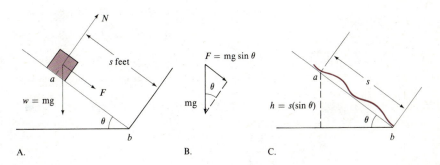

FIGURE 6–33. Example 4.

A.   B.   C.

Power

Power measures the *rate* at which work is done. That is, if W can be considered as a function of time t, then dW/dt measures the power consumed.

Suppose a force F moves an object starting at position a. When the object is at position s, the work done is

$$W = \int_a^s F(u) \, du$$

The dummy variable is u because s is one of the limits of integration. It follows that

$$\frac{dW}{ds} = F(s) \qquad \text{(Why?)}$$

If the position is a function of time, that is, $s = s(t)$, then, by the chain rule, we have

(6.13)
$$p = \frac{dW}{dt} = \frac{dW}{ds}\frac{ds}{dt} = F(s(t))v(t)$$

where $p = dW/dt$ is called the **power input** of the force F. The unit of power is ft-lbs/sec and

$$1 \text{ horsepower} = 550 \text{ ft-lbs/sec*}$$

EXAMPLE 5   A man is driving a 3,000 lb automobile on a level road at 60 mph ($= 88$ ft/sec). He comes to a hill that rises 4 ft vertically for every 100 ft of road. How much additional horsepower must the engine produce to maintain the speed of 60 mph?

Solution   The automobile is traveling up a plane inclined at an angle θ whose sine is $4/100 = .04$ (Figure 6–34). It follows, as in Example 4, that the additional force necessary to counteract the weight is

$$F = mg \sin \theta = (3,000)(.04) = 120 \text{ lbs}$$

Therefore, by (6.13), the extra power needed is

$$p = Fv = (120)(88) = 10,560 \text{ ft-lbs/sec}$$

or, $10,560/550 = 19.2$ horsepower.   ∎

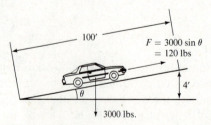

FIGURE 6–34. Example 5.

---

*In the metric system the unit of power is joules/sec commonly called a **watt**; 1 horsepower ≈ 746 watts.

EXAMPLE 6   If a motor that produces $\frac{1}{5}$ horsepower is used to pump the water out of the conical tank in Example 3 (Figure 6–32), how long will it take?

Solution   A $\frac{1}{5}$ horsepower motor will do work at the rate of $(\frac{1}{5})(550) = 110$ ft-lbs/sec. Since the total work done in removing the water is 40,906 ft-lbs (see Example 3), it will take

$$\frac{40{,}906}{110} \approx 372 \text{ sec} \quad \text{or about 6.2 min}$$

to empty the tank.   ■

EXERCISES

Evaluate the integrals below using a u-substitution.

1. $\displaystyle\int \frac{\sqrt{1 + x^{2/3}}}{x^{1/3}}\, dx$

2. $\displaystyle\int xe^{x^2}\, dx$ (Let $u = x^2$)

3. $\displaystyle\int e^{2x}\, dx$

4. $\displaystyle\int \frac{\sin \sqrt{x}}{\sqrt{x}}\, dx$ (Let $u = \sqrt{x}$)

Evaluate the following integrals using a change of variable.

5. $\displaystyle\int_0^{\sqrt{3}} x\sqrt{1 + x^2}\, dx$ **6.** $\displaystyle\int_4^9 \frac{(1 + \sqrt{x})^2}{\sqrt{x}}\, dx$

7. A spring with spring constant $k = 3$ is compressed from its natural length of 10 inches to a length of 8 inches. How much work is done?

8. The natural length of a spring is 4 inches. A force of 10 lbs will stretch it $\frac{1}{2}$ inch. How much work is done in stretching this spring (a) to a total length of 7 inches? (b) From a length of 7 inches to a length of 10 inches?

9. The natural length of a spring is 4 inches and its spring constant is 2. A force compresses the spring to a length of 3 inches. How much work is done by the *spring?*

10. The natural length of a spring is 8 inches. If it takes 24 in-lbs of work to stretch it from a length of 10 inches to a length of 12 inches, what is the spring constant?

11. A cylindrical tank of radius 25 ft and height 20 ft is filled with water. How much work is required to pump the water out (a) over the top of the tank and (b) through a pipe that rises 10 ft above the top of the tank?

12. (Continuation of Exercise 11). If a 3 horsepower motor is used to pump out the water, how many hours will it take?

13. One compartment of the hold of an oil tanker is 100 ft long and 20 ft deep; its end panels are trapezoids measuring 50 ft across at the top and 30 ft across at the bottom. Suppose the compartment is full of crude oil that weighs 50 lbs/cu ft. A 5 horsepower pump is rigged to remove the oil from the surface and pump it over the top of a storage tank 30 ft above the boat. How long will it take?

14. (Continuation of Exercise 13). How long would it take if the pump were attached to a pipe that remains on the bottom of the compartment during the whole operation? In this case, is the *shape* of the compartment important?

15. The tank in Example 3 has to be emptied in at most 2 hrs. What size pump (that is, what horsepower) is needed?

16. According to Coulomb's Law (Charles Augustin de Coulomb, 1736–1806) the force, in Newtons, with which two electrons repel each other is inversely proportional to the square of the distance (in meters) between them. That is, $F = k/s^2$ where k is a constant. If one electron is fixed, find the work done (in terms of k) in moving another electron from 5 meters away to 2 meters away.

17. Air is contained in a cylinder with a radius of 4 inches. When air expands adiabatically, the pressure and volume are related by the equation $PV^{1.4} = 100$. How much work is done as the air expands from a volume of 30 cu in. to a volume of 50 cu in?

18. (Continuation of Exercise 17). In an adiabatic expansion, no heat is added to the system. In an *isothermal* expansion, heat is added to keep the gas at a constant temperature. In this case, the equation relating the pressure and volume of air is $PV = 100$. Set up an integral whose value is the work done if the expansion in Exercise 17 is isothermal.

Exercises 19–23 do not require integration.

19. A bucket of sand weighing 100 lbs is hoisted 10 ft by a cable of negligible weight. As the bucket rises, sand pours out of a hole in the bottom at a constant rate and only 80 lbs of sand are left when it reaches the top. How much work is done?

20. (Continuation of Exercise 19). The same bucket with the same hole is used to lift the sand but now take into account that the cable weighs 5 lbs/ft. How much work is done? (Assume 10 ft of cable.)

21. A 500 lb weight is pulled up a smooth incline that makes an angle of 30° with the horizontal. The weight is moved a distance of 10 ft along the incline. (a) How much work is done? (b) If the weight is released, what is its velocity at the bottom?

22. What horsepower must an engine produce to accelerate a 2,000 lb automobile from 0 to 60 mph ($= 88$ ft/sec) in 10 seconds on a level road?

Optional Exercises

26. An object weighing 1 lb is moved by a force along the x-axis so that its position s in feet at any time t seconds is $s = (t + 1)^3$. Find the work done by the force as the object moves from $s = 1$ to $s = 4$. (*Hint:* Use Newton's second law of motion $F = ma$ to determine the force; m is the mass and a is the acceleration.)

23. (Continuation of Exercise 22). What horsepower must the engine produce if the road rises 3 ft for every 100 ft of road? (Assuming constant acceleration, the auto will travel 440 ft along the road in 10 seconds.)

24. If the mass M of the earth is regarded as concentrated at its center and a particle of mass m is r miles from the earth's center, then Newton's law of gravitation says that the force of attraction is

$$F = \frac{GMm}{r^2} \quad \text{lbs} \quad (G \text{ is a gravitation constant})$$

How much work is done in lifting a rocket of mass m from the earth's surface to a height of 10,000 miles? (The radius of the earth is about 4,000 miles.)

25. In an adiabatic expansion, pressure and volume are related by an equation $PV^k = C$. According to (6.10) the work done in moving a piston from position s_1 to position s_2 is

$$W = \int_{s_1}^{s_2} CA^{1-k}s^{-k}\, ds$$

When the piston is at position s the volume of the gas is $V = As$. Use this equation to perform a change of variable in the integral above and show that

$$W = \int_{V_1}^{V_2} P\, dV$$

where V_1 and V_2 are the volumes corresponding to s_1 and s_2.

27. It requires 24 in-lbs of work to stretch a spring from a length of 5 inches to a length of 7 inches. It requires an additional 66 in-lbs of work to stretch it from 7 inches to 10 inches. Find the spring constant and the spring's natural length.

6–5   ARC LENGTH

In this section, we describe a method for computing the *arc length* of a curve; more specifically, the arc length of the graph of a function. Applications include finding the length of wire needed to string a power line and computing the distance a particle moves along a curved path.

As you may suspect, the length s of a graph is defined in terms of an integral. The integrand, however, may come as a surprise.

DEFINITION

Suppose that f is differentiable on an interval I and that f' is continuous on I. If I contains the interval $[a, b]$, then the **arc length** s of the graph of f from $x = a$ to $x = b$ is

(6.14)

$$s = \int_a^b \sqrt{1 + [f'(x)]^2}\, dx$$

Let us see how this unexpected integrand comes about.

Suppose f is continuous on an interval I containing $[a, b]$ and that $P = \{x_0, x_1, \ldots, x_n\}$ is a partition of $[a, b]$. Locate the points $Q_0(x_0, f(x_0))$, $Q_1(x_1, f(x_1)), \ldots, Q_n(x_n, f(x_n))$ on the graph of $y = f(x)$ and join them, in succession, by straight lines (Figure 6–35A). The length of the line segment from Q_{i-1} to Q_i is $\sqrt{\Delta x_i^2 + [f(x_i) - f(x_{i-1})]^2}$ (Figure 6–35B). Clearly, the length of this line approximates the length of the arc from Q_{i-1} to Q_i, especially if these two points are very close to each other. Thus, the length s of the graph is approximated by the sum

(6.15) $$s \approx \sum_{i=1}^{n} \sqrt{\Delta x_i^2 + [f(x_i) - f(x_{i-1})]^2}$$

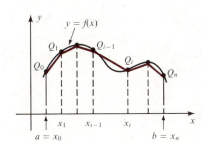

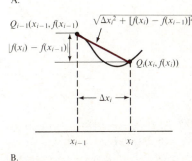

A.

B.

FIGURE 6–35. Estimating arc length.

But the sum on the right is not a Riemann sum of f nor of any other function. However, if f is assumed to be *differentiable* on I, then we can use the Mean Value Theorem* to write

$$f(x_i) - f(x_{i-1}) = f'(w_i)\Delta x_i \quad i = 1, 2, \ldots, n$$

for some point w_i between x_{i-1} and x_i. Now (6.15) becomes

$$s \approx \sum_{i=1}^{n} \sqrt{\Delta x_i^2 + [f'(w_i)\Delta x_i]^2} \qquad (w_i \text{ in } (x_{i-1}, x_i))$$

$$= \sum_{i=1}^{n} \sqrt{1 + [f'(w_i)]^2}\, \Delta x_i \qquad (\text{Factor out } \Delta x_i^2)$$

---

*Recall that the Mean Value Theorem states that if f is continuous on $[a, b]$ and differentiable on (a, b), then

$$f(b) - f(a) = f'(c)(b - a)$$

for some point c between a and b.

The last sum *is* a Riemann sum of the function $\sqrt{1 + (f')^2}$. If we now assume that f' is continuous on I, then s is approximated by Riemann sums of a continuous function. These approximations are more and more accurate as the norms of the partitions approach zero. Therefore, s is given by the integral in (6.14).

EXAMPLE 1   Find the length of the graph of $y = x^{3/2}$ from $x = 0$ to $x = 4$ (Figure 6–36).

Solution   First find y' and then $(y')^2$.

$$y' = \frac{3}{2}x^{1/2} \quad \text{and} \quad (y')^2 = \frac{9}{4}x$$

By (6.14), the arc length is

$$s = \int_0^4 \sqrt{1 + \frac{9}{4}x}\, dx$$

$$= \frac{4}{9} \int_0^4 \left(1 + \frac{9}{4}x\right)^{1/2} \frac{9}{4}\, dx$$

$$= \frac{4}{9} \cdot \frac{2}{3}\left[\left(1 + \frac{9}{4}x\right)^{3/2}\right]_0^4$$

$$= \frac{8}{27}(\sqrt{1{,}000} - 1) \approx 9.1 \text{ units} \quad \blacksquare$$

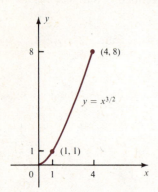

FIGURE 6–36. Example 1.

It may happen that you cannot evaluate the integral of $\sqrt{1 + (f')^2}$ by any of the methods discussed so far. In that case you can use Simpson's rule to estimate the integral.

EXAMPLE 2   Use Simpson's rule with $n = 4$ to estimate the length of the sine curve from $x = 0$ to $\pi/2$ (Figure 6–37).

Solution $y = \sin x$; $y' = \cos x$; $(y')^2 = \cos^2 x$. Thus,

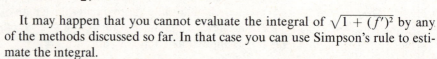

$$s = \int_0^{\pi/2} \sqrt{1 + \cos^2 x}\, dx$$

Recall our systematic method for using Simpson's rule:

(1) **Find the points of the partition.**
   $x_0 = a = 0, x_1 = \pi/8, x_2 = \pi/4, x_3 = 3\pi/8, x_4 = b = \pi/2$

(2) **Compute** $f(a), f(b), 4f(x_i)$ **for** $i = 1, 3, \ldots$, **and** $2f(x_i)$ **for** $i = 2, 4, \ldots$. In this example, $f(x) = \sqrt{1 + \cos^2 x}$.

$$
\begin{aligned}
f(a) &= f(0) &&= 1.4142 \\
4f(x_1) &= 4f(\pi/8) &&= 5.4458 \\
2f(x_2) &= 2f(\pi/4) &&= 2.4494
\end{aligned}
$$

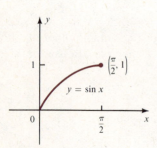

FIGURE 6–37. Example 2.

$$4f(x_3) = 4f(3\pi/8) = 4.2829$$
$$f(b) = f(\pi/2) = 1.0000$$

(3) **Add and multiply by** $(b - a)/3n$. In this example, the sum is 14.5923 and $(b - a)/3n = \pi/24 \approx .1309$. Therefore,

$$s = \int_0^{\pi/2} \sqrt{1 + \cos^2 x}\, dx \approx (14.5923)(.1309) \approx 1.9101 \text{ units} \quad\blacksquare$$

Sometimes it is necessary to think of a curve as being the graph of $x = g(y)$ instead of $y = f(x)$. In that case, the length formula takes the form

(6.16) $$s = \int_a^b \sqrt{1 + [g'(y)]^2}\, dy$$

EXAMPLE 3 Find the length of $y = x^{2/3}$ from $x = -1$ to $x = 8$.

Solution The derivative

$$y' = \frac{2}{3} x^{-1/3}$$

is undefined at $x = 0$. Because $f(x) = x^{2/3}$ is not differentiable on an interval containing $[-1, 8]$, formula (6.14) cannot be used for this problem.*

An alternate approach is to solve $y = x^{2/3}$ for x. First write $y^3 = x^2$ and then

$$x = \pm y^{3/2}$$

The $\pm$ indicates that the curve must be broken up into the graphs of two functions, $x = y^{3/2}$ and $x = -y^{3/2}$. The graphs are sketched in Figure 6–38, which clearly indicates that

$$x = y^{3/2} \text{ for } 0 \le y \le 4 \quad \text{and} \quad x = -y^{3/2} \text{ for } 0 \le y \le 1$$

Now we find the lengths of each curve separately. The derivatives are $x' = \frac{3}{2} y^{1/2}$ and $x' = -\frac{3}{2} y^{1/2}$, and it follows that the total length of the curve is

$$\int_0^1 \sqrt{1 + \frac{9}{4} y}\, dy + \int_0^4 \sqrt{1 + \frac{9}{4} y}\, dy$$

$$= \frac{4}{9} \left[\frac{2}{3} \left(1 + \frac{9}{4} y \right)^{3/2} \right]_0^1 + \frac{4}{9} \left[\frac{2}{3} \left(1 + \frac{9}{4} y \right)^{3/2} \right]_0^4$$

$$\approx 10.5 \text{ units} \quad\blacksquare$$

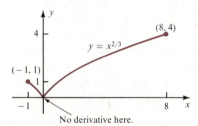

A. As a function of x.

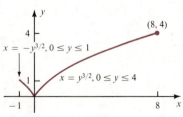

B. Rewrite as two functions of y.

FIGURE 6–38. Example 3.

---

*However, formula (6.14) can be used to find the arc length of $y = x^{2/3}$ from $x = a$ to $x = b$ whenever 0 is not in the interval $[a, b]$.

EXERCISES

In Exercises 1–8, find the lengths of the curves.

1. $y = x^{3/2}$ from $x = 4$ to $x = 8$

2. $x = -y^{3/2}$ from $y = 8$ to $y = 12$

3. $x = \dfrac{y^4}{4} + \dfrac{1}{8y^2}$ from $y = 1$ to $y = 2$

4. $y = \dfrac{1}{4x} + \dfrac{x^3}{3}$ from $x = 2$ to $x = 3$

5. $y^3 = x^2$ from $(-2, \sqrt[3]{4})$ to $(3, \sqrt[3]{9})$

6. $(y - 2)^2 = (x + 1)^3$ from $(-1, 2)$ to $(0, 3)$

7. $y = \dfrac{x^4 + 12}{12x}$ from $x = 1$ to $x = 2$

8. $(y + 3)^3 = (x - 1)^2$ from $(1, -3)$ to $(0, -2)$

In Exercises 9–12, set up a definite integral whose value is the length of the curve.

9. $y = e^x$ from $x = 0$ to $x = 1$

10. $y = \cos x$ from $x = -\pi/2$ to $x = 0$

11. $y = \sqrt{x}$ from $(0, 0)$ to $(4, 2)$

12. $3x^2 + x - 2y = 0$ from $(0, 0)$ to $(1, 2)$

In Exercises 13–14, use Simpson's rule with $n = 4$ to estimate the lengths of the curves.

13. $y = e^{x/2}$ from $x = 2$ to $x = 4$

14. $y = x^2$ from $x = 0$ to $x = 1$

---

15. A particle starts at $(0, 0)$ and travels along the path $y = x^2$. Let s be the distance it moves along the path. If its x-coordinate is increasing at the constant rate of 3 ft/sec, what is the rate of change of s with respect to time when $x = 5$?

16. (Continuation of Exercise 15). Answer the same question if the x-coordinate is a function of time: $x = t^2 - 4t$.

17. The parabola $y = ax^2 + bx + c$ passes through $(1, 1)$, $(2, 0)$, and $(3, 4)$. Find a, b, and c and set up a definite integral whose value is the length of the parabola from $x = 1$ to $x = 3$.

18. In a suspension bridge the cable hangs from fixed supports and the load is hung from the cable. If coordinate axes are drawn as indicated in Figure 6–39, then the cable takes the shape of a parabola $y = Cx^2$ where C is a constant and depends on the load. For the Golden Gate Bridge, the towers are about 300 ft high and the distance between them is about 4,000 ft. Use Simpson's rule (you decide the value of n) to estimate the amount of cable between the towers (both sides of the bridge).

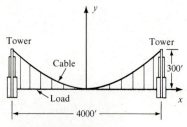

FIGURE 6–39. Exercise 18.
Golden Gate Bridge.

19. A cable carrying electric power is supported by towers 300 ft apart. The cable sags due to its own weight and forms what is called a **catenary curve**. If coordinate axes are introduced as shown in Figure 6–40, the equation of the curve is $y = \frac{25}{2}(e^{x/150} + e^{-x/150})$. Set up a definite integral whose value is the length of cable between the towers.

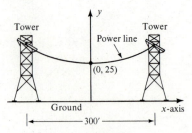

FIGURE 6–40. Exercise 19.

Optional Exercise

20. The graph of $x^{2/3} + y^{2/3} = 1$ is shown in Figure 6–41. Find its length.

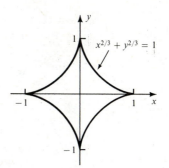

FIGURE 6–41. Exercise 20.

6–6   CENTER OF MASS

A girl and boy are playing on a seesaw. Even if they have never been on a seesaw before, they quickly learn that the heavier one must sit closer to the center to achieve a balance. They realize, somehow, that their *weight* and *position* affect the apparatus (Figure 6–42). These simple observations about a seesaw can be generalized and put to good use.

If an object of mass m is located at x on a coordinate line, then the **moment about the origin** of the object is defined as the product mx. The moment is positive or negative; its sign depends on the sign of x. More generally, if a system of n objects of mass $m_1, m_2, \ldots, m_n$ are located on the line at points $x_1, x_2, \ldots, x_n$, the moment M_0 about the origin of the system is the sum of the individual moments; that is,

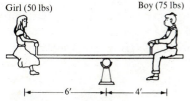

Girl (50 lbs)   Boy (75 lbs)

FIGURE 6–42. Children balance a seesaw.

$$(6.17) \qquad M_0 = \sum_{i=1}^{n} m_i x_i = m_1 x_1 + m_2 x_2 + \cdots + m_n x_n$$

If the total mass $M = \sum_{i=1}^{n} m_i$ of the system were concentrated at a single point $\bar{x}$, then the moment would be $M\bar{x}$. The point $\bar{x}$ for which this moment is the same as the moment M_0 is called the **center of mass** (or **center of gravity**) of the system. Thus, $M\bar{x} = M_0$, or

$$(6.18) \qquad \bar{x} = \frac{M_0}{M} = \frac{\displaystyle\sum_{i=1}^{n} m_i x_i}{\displaystyle\sum_{i=1}^{n} m_i} \qquad \text{is the center of mass}$$

EXAMPLE 1   Particles of mass 10, 20, and 30 are located at points -3, -1, and 4. Find the moment about the origin and the center of mass of the system.

Solution   Draw a diagram (Figure 6–43A). By (6.17), the moment of the system is

$$M_0 = 10(-3) + 20(-1) + 30(4) = 70$$

The effect of this positive moment is that the line tends to rotate clockwise about the origin. To find the center of mass, compute the total mass of the system; $M = 10 + 20 + 30 = 60$. Therefore, by (6.18), the center of mass is

$$\bar{x} = \frac{M_0}{M} = \frac{70}{60} = \frac{7}{6}$$

The system behaves as if the entire mass were concentrated at the single point $\bar{x} = \frac{7}{6}$ (Figure 6–43B). In other words, if the system were supported by a wire attached to the point $\bar{x}$ (like a mobile), it would be perfectly balanced (Figure 6–43C).   ∎

The notions of moment and center of mass can be extended to points in the plane. Let masses m_1, m_2, ..., m_n be located at points (x_1, y_1), (x_2, y_2), ..., (x_n, y_n). Now we define moments with respect to the coordinate axes rather than the origin. In Figure 6–44, the moment of $m_1 = 10$ about the x-axis is the product $m_1 y_1 = 10(3) = 30$ and its moment about the y-axis is the product $m_1 x_1 = 10(-1) = -10$. The sums

$$(6.19) \qquad M_x = \sum_{i=1}^{n} m_i y_i \quad \text{and} \quad M_y = \sum_{i=1}^{n} m_i x_i$$

define the **moments of the system about the x-axis and y-axis,** respectively. The coordinates $(\bar{x}, \bar{y})$ of the center of mass of the system are defined by the equations

$$(6.20) \qquad \bar{x} = \frac{M_y}{M} \quad \text{and} \quad \bar{y} = \frac{M_x}{M}$$

EXAMPLE 2   Suppose masses of 10, 20, and 30 are located at $(-1, 3)$, $(2, -1)$, and $(3, 1)$. Find the moments M_x and M_y and the center of mass.

Solution   Refer to Figure 6–44A. By (6.19), the moments are

$$M_x = 10(3) + 20(-1) + 30(1) = 40$$
$$M_y = 10(-1) + 20(2) + 30(3) = 120$$

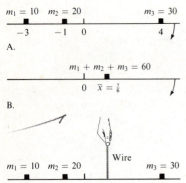

FIGURE 6–43. Example 1.

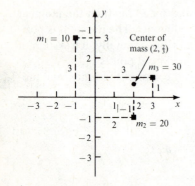

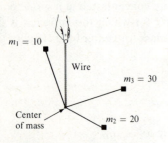

FIGURE 6–44. Example 2. The system is in perfect balance (like a mobile) when supported at the center of mass.

The center of mass is given by equations (6.20)

$$\bar{x} = \frac{120}{60} = 2 \quad \text{and} \quad \bar{y} = \frac{40}{60} = \frac{2}{3}$$

If the masses are attached to the center of mass (2, 2/3) by thin weightless rods and the entire system is supported by a wire attached to (2, 2/3), it would be in perfect balance (Figure 6-44B). ■

The notions of moment and center of mass can also be extended to regions in the plane. It helps to think of a region as a very thin plate or **lamina.** We assume that the lamina has constant density (mass per unit of area); such laminas are called **homogeneous.*** If the density is δ and the area of the region is A, then the mass of the lamina is

(6.21) $M = \delta A$

For the purpose of computing moments, this mass is considered to be concentrated at the center of mass. Intuitively, the center of mass of a lamina is the point about which the lamina is perfectly balanced (Figure 6-45). For a homogeneous lamina in the shape of a regular polygon, rectangle or circle, the center of mass is its geometric center.

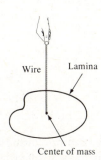

Wire   Lamina

Center of mass

FIGURE 6-45 The lamina is balanced about its center of mass.

EXAMPLE 3   Find the center of mass of the region in Figure 6-46A.

Solution   Divide the region into two rectangles. Rectangle I has area 4 so its mass is 4δ; its center of mass is its geometric center (2, 5/2). Rectangle II has area 2 so its mass is 2δ and its center of mass is at (7/2, 1). If we assume that the mass of each rectangle is concentrated at its center of mass, then what we have is a system of two masses, $m_1 = 4\delta$ and $m_2 = 2\delta$ located at (2, 5/2) and (7/2, 1), respectively. Therefore, by (6.20), the coordinates of the center of mass are

$$\bar{x} = \frac{m_1 x_1 + m_2 x_2}{m_1 + m_2} = \frac{4\delta(2) + 2\delta(7/2)}{4\delta + 2\delta} = \frac{15\delta}{6\delta} = \frac{5}{2}$$

and

$$\bar{y} = \frac{m_1 y_1 + m_2 y_2}{m_1 + m_2} = \frac{4\delta(5/2) + 2\delta(1)}{4\delta + 2\delta} = \frac{12\delta}{6\delta} = 2$$

The center of mass is (5/2, 2). ■

In Example 3, we were fortunate that the region could be divided into rectangles. If that is not the case, perhaps the region can be *approximated* by rectangles. Then the moments and the mass are found by evaluating definite integrals.

EXAMPLE 4   Find the center of mass of a triangular lamina with density δ.

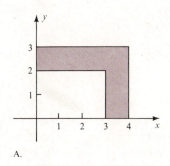

A.

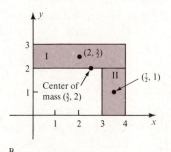

B.

FIGURE 6-46. Example 3.

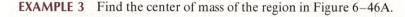

*Moments and centers of mass of laminas with variable densities are included in the discussion of multiple integrals in Chapter 16.

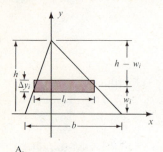

A.

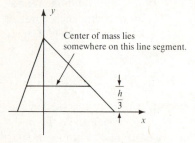

B

FIGURE 6–47. Example 4.

Solution Choose any side of the triangle as a base b. Then place the triangle so that this base is on the x-axis and the height h is on the y-axis (Figure 6–47A). Now let us find M_x, the moment about the x-axis.

Partition $[0, h]$ and construct (horizontal) approximating rectangles. The area of the ith rectangle is $l_i \Delta y_i$ so its mass is $\delta l_i \Delta y_i$. The center of mass is at the center of the rectangle and it follows that if w_i is the midpoint of the ith interval, then

$$(6.22) \qquad (M_x)_i = (\delta l_i \Delta y_i) w_i$$

is the moment of the rectangle about the x-axis. The length l_i can be found by considering similar triangles (see the figure).

$$\frac{l_i}{b} = \frac{h - w_i}{h} \quad \text{or} \quad l_i = \frac{b}{h}(h - w_i)$$

Then (6.22) becomes

$$(M_x)_i = \delta \frac{b}{h} w_i (h - w_i) \Delta y_i$$

The sum of the moments $(M_x)_i$ from $i = 1$ to n is a Riemann sum of the continuous function $\delta b y(h - y)/h$ and it follows that the moment M_x of the triangle is

$$(6.23) \qquad M_x = \delta \frac{b}{h} \int_0^h y(h - y) \, dy$$

$$= \delta \frac{b}{h} \left[\frac{hy^2}{2} - \frac{y^3}{3} \right]_0^h = \frac{\delta bh^2}{6}$$

The mass of the triangle is $M = \delta A = \delta bh/2$ so the y coordinate of the center of mass is

$$\bar{y} = \frac{M_x}{M} = \frac{h}{3}$$

Therefore, the center of mass lies $\frac{1}{3}$ of the way from the base to the opposite vertex (Figure 6–47B). The same is true if any other side is chosen as the base. It follows that the center of mass of a triangular lamina is at the intersection of the medians (lines joining the vertices to the midpoints of the opposite sides).* ∎

EXAMPLE 5 Find the center of mass of a lamina with density δ and the shape of the region bounded by the curves $y = x^2$ and $y = x^3$.

Solution Sketch the region along with a typical approximating rectangle (Figure 6–48). The figure shows that if w_i is the midpoint of the ith subinterval, then $(w_i, (w_i^2 + w_i^3)/2)$ is the center of mass of this rectangle. Its mass is $\delta A = \delta(w_i^2 - w_i^3)\Delta x_i$ and, therefore, its moments are

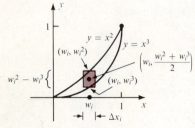

FIGURE 6–48. Example 5.

---

*A theorem from geometry says that the medians of a triangle intersect at a point which is $\frac{1}{3}$ of the distance from any side to the opposite vertex.

$$(M_x)_i = [\delta(w_i^2 - w_i^3)\Delta x_i][(w_i^2 + w_i^3)/2]$$

$$= \frac{\delta}{2}(w_i^4 - w_i^6)\Delta x_i$$

$$(M_y)_i = [\delta(w_i^2 - w_i^3)\Delta x_i]w_i$$

$$= \delta(w_i^3 - w_i^4)\Delta x_i$$

The sums of these moments are Riemann sums and approximate the moments M_x and M_y of the lamina. It follows that

(6.23)
$$M_x = \frac{\delta}{2}\int_0^1 (x^4 - x^6)\,dx = \frac{\delta}{35}$$

$$M_y = \delta\int_0^1 (x^3 - x^4)\,dx = \frac{\delta}{20}$$

The total mass of the lamina is

(6.24)
$$M = \delta A = \delta\int_0^1 (x^2 - x^3)\,dx = \frac{\delta}{12}$$

Dividing the results of (6.23) by (6.24) we have

$$\bar{x} = \frac{M_y}{M} = \frac{3}{5} \quad \text{and} \quad \bar{y} = \frac{M_x}{M} = \frac{12}{35} \quad \blacksquare$$

The procedure illustrated in Examples 4 and 5 can be summarized as follows:

To compute the center of mass of a homogeneous lamina:

(1) Sketch the lamina along with a typical approximating rectangle.

(2) The center of mass of the rectangle is its geometric center. Find the coordinates of that point.

(3) Calculate the mass of the rectangle and the moments with respect to the *x*- and *y*-axes. Summing over all such rectangles yields Riemann sums of continuous functions.

(4) Now use integration to compute M_x, M_y, and the total mass M. Then

$$\bar{x} = \frac{M_y}{M} \quad \text{and} \quad \bar{y} = \frac{M_x}{M}$$

Momentum and the Center of Mass

The **momentum** P of a particle of mass m moving with a velocity v is defined as the product $P = mv$. If masses $m_1, m_2, \ldots, m_n$ are moving with velocities $v_1, v_2, \ldots, v_n$ respectively, then the momentum of the system is defined to be the sum

(6.25) $$P = m_1 v_1 + m_2 v_2 + \cdots + m_n v_n$$

Now suppose that $M = m_1 + m_2 + \cdots + m_n$ is the total mass of the system and that the particles are at positions $x_1, x_2, \ldots, x_n$. If $\bar{x}$ is the center of mass, then

(6.26) $$M\bar{x} = m_1 x_1 + m_2 x_2 + \cdots + m x_n$$

If the positions are functions of time

$$x_1 = x_1(t), x_2 = x_2(t), \ldots, x_n = x_n(t)$$

then $x_i' = v_i$, the velocity of the ith particle. Differentiating both sides of (6.26) with respect to time, we have

(6.27) $$M\bar{v} = m_1 v_1 + m_2 v_2 + \cdots + m_n v_n$$

where $\bar{v} = \bar{x}'$ is the velocity of the center of mass. By (6.25), the right side of (6.27) is the momentum of the system and the conclusion is that

(6.28)

> The momentum of a system is the total mass times the velocity of the center of mass. That is
> $$P = M\bar{v}$$

EXAMPLE 6 A bullet weighing .1 lbs moving horizontally with a velocity of 1,000 ft/sec strikes and embeds itself in a wooden block weighing 2 lbs. (Figure 6–49). The block is initially at rest on a smooth (frictionless) table. Find the velocity of the bullet and block after impact.

Solution The system consists of two particles; the bullet with mass .1/32* and the block with mass 2/32. The total mass is 2.1/32. Just before the bullet strikes, the velocities of the bullet and block are 1,000 ft/sec and 0, respectively. It follows from (6.27) that

$$\frac{2.1}{32}\bar{v} = \frac{.1}{32}(1{,}000) + \frac{2}{32}(0)$$

$$\bar{v} = \frac{100}{2.1} \approx 47.6 \text{ ft/sec}$$

Since there is no friction to slow the system down, we conclude that the center of mass (hence the block and bullet) will continue to move at about 47.6 ft/sec after impact. ∎

EXAMPLE 7 A projectile weighing 30 lbs explodes at the top of its flight into two pieces of 10 and 20 lbs. Describe how the pieces will fall.

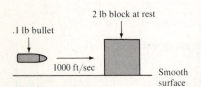

2 lb block at rest

.1 lb bullet

1000 ft/sec

Smooth surface

FIGURE 6–49. Example 6.

---

*Recall the relationship between **mass** m and **weight** w; on or near the surface of the earth, $w = mg \approx 32\,m$.

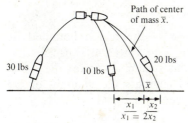

FIGURE 6–50. Example 7.

Solution We assume that gravity is the only external force acting on the projectile; in that case, the path is a parabola. After the explosion, the center of mass will continue along the path just as though there were no explosion because the momentum of the system is preserved. The 10 lb piece will land x_1 units on one side of the center of mass and the 20 lb piece will land x_2 units on the other side (see Figure 6–50). It follows that $x_1 = 2x_2$ (why?). ∎

EXERCISES

1. Particles of mass 5, 10, and 20 are located at $(-1, 4)$, $(1, 3)$, and $(2, -1)$. Find the center of mass of the system.

2. Same as Exercise 1 for particles of mass 3, 7, and 8 located at $(5, -2)$, $(1, 3)$, and $(-2, -4)$.

3. Particles of mass 5 and 8 are located on the x-axis at -6 and 3. (a) Find the center of mass. (b) If the particles are moving with velocities of 3 and -4 ft/sec, respectively, what is the velocity of the center of mass?

4. Same as Exercise 3 for particles of mass 4 and 7 located at points -4 and 2. Their velocities are -2 and 3 ft/sec.

In Exercises 5–15, sketch the region described and find the center of mass of a homogeneous lamina with that shape.

5. Region bounded by $y = x^2$, $y = 0$, and $x = 1$.

6. Region bounded by $y = x^2$, $y = 0$, $x = 1$, and $x = -1$.

7. Region bounded by $y = x^2$ and $y = x$.

8. Region bounded by $y = x^3$ and $y = x$.

9. Region bounded by $y = \sqrt{x}$ and $y = x^2$.

10. Region bounded by $y = \sqrt{x}$, $y = 0$, and $x = 4$.

11. The semicircular region bounded by $y = \sqrt{1 - x^2}$ and $y = 0$. $0, 4/3\pi$

12. The quarter-circle bounded by $y = \sqrt{1 - x^2}$ and the positive coordinate axes.

13. Region bounded by $y = \sqrt{1 - x^2}$, $y = x$, and $y = 0$.

14. Region bounded by $y = \sqrt{1 - x^2}$, $y = x$, and $x = 0$.

15. The triangular region with vertices $(0, 0)$, $(1, 2)$, and $(3, 0)$. Verify that the center of mass is the intersection of the medians.

In Exercises 16–19, find the center of mass of the regions shown (see Example 3).

16.

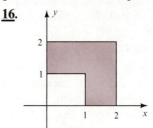

FIGURE 6–51.

17.

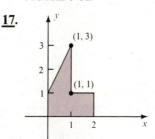

FIGURE 6–52.

18.

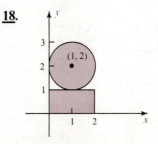

FIGURE 6–53.

19.

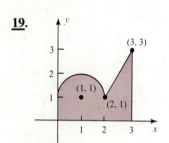

FIGURE 6–54.

20. A 2,000 lb automobile traveling north at 60 mph collides head-on with a 6,000 lb truck traveling south at 40 mph. The car and the truck stick together. What is the direction and velocity of the wreckage after the collision?

21. An open freight car weighing 10 tons (1 ton = 2,000 lbs) is rolling along a (frictionless) track at 5 mph. Rain is falling vertically into the car. Will the rain slow the car down? What is the speed of the car after it accumulates 500 lbs of water?

22. A rocket is fired so that its path is the parabola $y = \sqrt{3}x - (x/500)^2$; x and y are measured in feet. The engine is $\frac{1}{3}$ of the mass of the entire rocket. At the top of the flight, the engine drops off and lands at a point approximately 333,000 ft from the launch pad. How far from the launch pad does the rest of the rocket land?

Volume and the Center of Mass

23. The center of mass of the region bounded by $y = x^2$, $y = 0$, and $x = 1$ is $(3/4, 3/10)$; see Exercise 5. Rotate this region about (a) the y-axis (b) the x-axis and compute the volumes of the solids generated. Then verify that in each case

(6.29)

> The volume generated equals the area of the region times the distance the center of mass travels in making one revolution.

24. Repeat Exercise 23 for the region bounded by $y = x^2$ and $y = x$ (the center of mass is $(1/2, 2/5)$; see Exercise 7).

Remark: Statement (6.29) is called **Pappus' Theorem.** The fact that the statement is true is not *too* surprising; what is surprising is that Pappus lived almost 1,700 years ago! (Pappus of Alexandria, ca. 300.)

Work and the Center of Mass

25. In Example 3 of Section 6–4, the work done in removing the water from an inverted conical tank was computed to be about 40,906 ft-lbs. (a) Show that the center of mass of the tank of water is the point $(0, 15/2)$ by partitioning the interval $[0, 10]$ and approximating with horizontal discs. (b) Show that the work done in removing the water can be computed by assuming that all of the water is concentrated at the center of mass.

6–7 PROBABILITY DENSITY AND EXPECTED VALUE

The notion of the center of mass was originally introduced in the theory of mechanics to study systems of particles. In this section you will see that this notion has useful applications in other fields as well.

If particles of mass 50 and 950 are located at points $-.75$ and 1.50 on the x-axis, then the center of mass is

$$\bar{x} = \frac{50(-.75) + 950(1.50)}{1,000} \approx 1.39$$

In any mechanical analysis of the system (motion, momentum, and so on) the above calculation is interpreted to mean that the system can be treated as though the entire mass were concentrated at the point 1.39. Now let us see how this same calculation can be interpreted in an entirely different way.

EXAMPLE 1   A lock company manufactures 1,000 locks on a certain day. From past experience it is known that about 5% of all locks produced are defective and have to be redone. Thus, of the 1,000 locks, about 50 of them are defective and 950 are nondefective. Suppose that on each defective lock the company loses 75¢ and that on each nondefective lock the company makes $1.50. The total profit that day is $50(-.75) + 950(1.50)$ dollars, and the *average profit per lock* is

$$(6.30) \qquad \frac{50(-.75) + 950(1.50)}{1,000} \approx 1.39 \text{ dollars}$$

This is the same calculation used above to find the center of mass; this time it yields the average profit. The idea is the same but the interpretation is different. ∎

Remark: It is important to observe that (6.30) can be rewritten as

$$(-.75)\frac{50}{1,000} + (1.50)\frac{950}{1,000} = (-.75)(.05) + (1.50)(.95) \approx 1.39$$

Notice that the number 1,000 does not appear on the right. This means that the information that the company produced 1,000 locks is not essential. What is essential is that for *any* amount of production, $5\% = .05$ are defective locks $95\% = .95$ are nondefective. This allows the following interpretation: *on average,* the company can *except* to make about $1.39 for every lock produced.

To discuss expected profit and similar topics, we introduce the following terminology. Let X be a function whose range is a finite set of numbers x_1, $x_2, \ldots, x_n$; the domain of X depends on the particular problem involved. In probability theory such functions are called **discrete random variables.** Now define a function p on the numbers $x_1, x_2, \ldots, x_n$ so that

$$(6.31) \qquad \boxed{p(x_i) \geq 0 \quad \text{and} \quad \sum_{i=1}^{n} p(x_i) = 1}$$

You can think of $p(x_i)$ as the *probability* that the value of X is x_i. The function p is called the **probability density function** (abbreviated pdf) of X. Finally, the **expected value** of X, denoted by $E(X)$, is defined to be the sum

$$(6.32) \qquad \boxed{E(X) = x_1 p(x_1) + x_2 p(x_2) + \cdots + x_n p(x_n)}$$

To solidfy these concepts, let us return to Example 1 about the lock company. Let X be the function whose value is the profit per lock; its domain is the set of locks produced. Its value at any particular lock is either $x_1 = -.75$ (if the lock is defective) or $x_2 = 1.50$ (if the lock is nondefective). So X is a discrete random variable with two values. The probability that it takes the value $-.75$ is $5\% = .05$; the probability that it takes the value 1.50 is $95\% = .95$. Therefore, the pdf of X is the function p with $p(-.75) = .05$ and $p(1.50) = .95$ (notice that $.05 + .95 = 1$). By (6.32), the *expected* (or *average*) value of the profit per lock is

$$E(X) = (-.75)(.05) + (1.50)(.95) \approx 1.39 \text{ dollars}$$

The general procedure for solving problems of this kind is as follows:

> (1) Formulate a random variable X whose expected value provides the answer to the question asked.
>
> (2) Find the pdf of X.
>
> (3) Use (6.32) to find $E(X)$.

EXAMPLE 2   An opaque jar contains three balls, one white and two black. The jar is turned over and one ball falls out. If the ball is white the "house" pays you $9.00; if the ball is black you pay the house $5.00. What is the expected profit for the house on each play?

Solution   Let X be the house's profit on each play; the value of X can be -9 or $+5$. The probabilities are not given but they are easy to figure out. Since there is one white ball and two black ones, the probability that X is -9 or $+5$ is $\frac{1}{3}$ and $\frac{2}{3}$, respectively. Therefore, the expected profit for the house per play is

$$E(X) = -9(\tfrac{1}{3}) + 5(\tfrac{2}{3}) = \tfrac{1}{3} \approx 33 \text{ cents}$$

This means that, in the *long run,* the house can expect to make about 33¢ every time the jar is turned over.   ■

The notion of expected value can be extended by allowing the function X to take all values in an interval $[a, b]$; then X is called a **continuous random variable.** In this case, the probability density function (pdf) is assumed to be a continuous function $p = p(x)$ defined on $[a, b]$ such that

(6.33) $\qquad p(x) \geq 0 \quad \text{and} \quad \displaystyle\int_a^b p(x)\, dx = 1$

These properties are the counterparts to the properties expressed in (6.31). The **expected value** of a continuous random variable is defined to be

(6.34) $\qquad E(X) = \displaystyle\int_a^b xp(x)\, dx *$

---

*This is analogous to the case of a solid rod whose length extends from $x = a$ to $x = b$. If the density (mass per unit of length) of the rod is given by the continuous function $\delta = \delta(x)$, then the total mass of the rod is $M = \int_a^b \delta(x)\, dx$ and its center of mass is $[\int_a^b x\delta(x)\, dx]/M$. In (6.34) above, $M = 1$.

EXAMPLE 3 *(Expected life of a product)* A company wants to compute the expected lifetime of its product. Samples of the product are selected randomly and subjected to continuous use for a maximum of 100 hours. If a particular sample wears out before then, its lifetime t, in hours, is recorded. After a large number of samples are tested, the pdf is found to be $p(t) = \frac{3}{1300}(11 - t^{1/2})$. What is the expected lifetime of the product?

Solution Let X be the function whose value at any particular sample is its lifetime or 100, whichever is smaller. Then X is a continuous random variable with values in the interval [0, 100]. The pdf of X is given as $p(t) = \frac{3}{1300}(11 - t^{1/2})$. Thus, by (6.34), the expected lifetime is

$$E(X) = \int_0^{100} tp(t)\, dt$$

$$= \frac{3}{1,300} \int_0^{100} (11t - t^{3/2})\, dt \approx 34.6 \text{ hours} \quad \blacksquare$$

Suppose X is a continuous random variable with values in $[a, b]$ and pdf p. An interesting observation is that if c and d are points in the interval $[a, b]$, then $\int_c^d p(x)\, dx$ represents the proportion of times that the values of X fall between c and d. That is,

(6.35)

> Let X be a continuous random variable with values in $[a, b]$ and pdf p. If c and d are points in $[a, b]$, then
>
> $$\int_c^d p(x)\, dx$$
>
> is the probability that the values of X lie in the interval $[c, d]$.

For instance, in Example 3, the probability that the product will last for 12 hours or less is

(6.36) $\displaystyle \int_0^{12} p(t)\, dt = \frac{3}{1,300} \int_0^{12} (11 - t^{1/2})\, dt \approx .24$

The probability that the product will last between 12 and 24 hours is

(6.37) $\displaystyle \int_{12}^{24} p(t)\, dt = \frac{3}{1,300} \int_{12}^{24} (11 - t^{1/2})\, dt \approx .19$

EXAMPLE 4 Suppose the company in Example 3 estimates that one hour of continuous use in testing is equivalent to one month of actual use by a customer. With this assumption, they guarantee their product as follows: a full refund will be given if the product fails within one year, and a one-half refund if the product fails in the period from one to two years. If the product costs $10 to produce and sells for $18, what is the expected profit per item?

Solution   Let Y be the profit per item (we have already used the letter X to denote the lifetime of the product). The values of Y are -10 (full refund), -1 (one-half refund) or $+8$ (no refund). If we assume that every customer takes advantage of the guarantee, then the probability that Y takes the values of -10, -1, and 8 are, respectively, .24, .19, and .57. This follows from (6.36) and (6.37). Therefore

$$E(Y) = (-10)(.24) + (-1)(.19) + 8(.57) = 1.97 \text{ dollars}$$

is the expected profit per item.   ■

EXAMPLE 5 *(Expected life of bacteria)*   The lifetime of a certain strain of bacteria, measured in hours, is a continuous random variable X with values in the time interval $[0, 27]$; its pdf is $p(t) = \frac{4}{81}(3 - \sqrt[3]{t})$. (a) What is the probability that a bacterium will live at most 8 hours? (b) What is the expected life of a bacterium?

Solution

(a) The probability that a single bacterium will live at most 8 hours is

$$\int_0^8 p(t)\, dt = \frac{4}{81} \int_0^8 (3 - t^{1/3})\, dt$$

$$= \frac{4}{81}\left[3t - \frac{3}{4}t^{4/3}\right]_0^8 \approx .59$$

(b) The expected lifetime is

$$\int_0^{27} tp(t)\, dt = \frac{4}{81} \int_0^{27} (3t - t^{4/3})\, dt$$

$$= \frac{4}{81}\left[\frac{3}{2}t^2 - \frac{3}{7}t^{7/3}\right]_0^{27} \approx 7.7 \text{ hours}$$ ■

EXERCISES

In Exercises 1–10, find the expected value of the random variable X if its pdf is the indicated function.

1. $p(-3) = .31, p(0) = .41, p(2) = .28$

2. $p(1) = .14, p(5) = .83, p(7) = .03$

3. $p(1) = p(2) = p(3) = p(4) = p(5) = p(6) = 1/6$*

4. $p(n) = n/10$ for $n = 1, 2, 3, 4$

5. $p(x) = 1/4$ on $[0, 4]$ **6.** $p(x) = 1/3$ on $[-2, 1]$

7. $p(x) = x/6$ on $[2, 4]$ **8.** $p(x) = x^2/9$ on $[0, 3]$

9. $p(t) = .6t^{-3/2}$ on $[4, 25]$

10. $p(t) = 3(1 - t^{1/2})$ on $[0, 1]$

11. Gordon's Bakery puts out a fruit cake that costs about \$2 to make and sells for \$3.50. In the baking process, the cakes pass through the oven on a conveyor belt. About 5% of the cakes are overdone;

---

*Here the values of X are the numbers that turn up when one die is cast. Each value $1, 2, \ldots, 6$ is equally likely with a probability of $1/6$. When a random variable has a constant pdf, then it is said to be **uniformly distributed.** Other examples are in Exercises 5 and 6. In this case, the expected value is the arithmetic average of the values of X.

these are sold at an outlet store for 75¢. About 10% of the cakes are underdone and have to be rebaked individually at an extra cost of 40¢. What is the expected profit per fruit cake?

12. A professor has given a five question exam to a large number of students over the years and has computed the following probabilities:

Number of wrong answers	0	1	2	3	4	5
Probability	.10	.35	.25	.15	.10	.05

What is the expected number of wrong answers?

13. In a random sample of 2,000 adult American males, the height of each man is recorded to the nearest one-half foot. The results are:

Height (feet)	5	5.5	6	6.5	7
Number of men	400	750	615	200	35

Assuming this is a representative sample, formulate a discrete pdf using the table and then compute the expected height of an adult American male.

Exercises 14–16 are about the game of *craps* as played in a gambling casino. You make a (line) bet of $1 and roll out two dice. Each die has 6 numbers, so there are 36 possible (ordered) pairs of numbers that can turn up on each roll. The sum of the two numbers is called the *number rolled.* For example, if any of the pairs

$$(1, 3), (2, 2), \text{ or } (3, 1)$$

turn up, the number rolled is 4. Because these are the *only* pairs that yield the number 4, we say that *there are three ways to roll a 4.* It follows that the probability of rolling a 4 is $3/36 = 1/12$. For another example, there are four ways to roll a nine

$$(3, 6), (4, 5), (5, 4), \text{ and } (6, 3)$$

so the probability of rolling a 9 is $4/36 = 1/9$. The table below shows the number of ways to make each number rolled.

Number rolled	2	3	4	5	6	7	8	9	10	11	12
Number of ways	1	2	3	4	5	6	5	4	3	2	1

The game is played as follows:

(a) **On the first roll.** If a 7 or 11 is rolled, you win $1; if a 2, 3, or 12 is rolled you lose your $1 bet. If any

other number is rolled, it becomes your "point" and you continue to roll the dice until your point or a 7 is rolled.

(b) **Continue to roll the dice.** If your point is rolled again *before* a 7, you win $1; if not, you lose $1.

14. Given that a 2, 3, 7, 11, or 12 comes up on the first roll, what is your expected profit or loss?

15. Given that your point is 6, what is your expected profit or loss?

16. For the game of craps, what is your expected profit or loss?

17. A piece of lab equipment is used continuously. If it is used for 6 hrs in a row at full capacity, it is likely to break down. Suppose the time span, in hours, during which it is used at full capacity has the probability density

$$p(t) = \begin{cases} \dfrac{t}{24} & \text{if } 0 \le t \le 6 \\[2mm] -\dfrac{t}{8} + 1 & \text{if } 6 \le t \le 8 \end{cases}$$

How long, on average, is the equipment used at full capacity? Is it likely to break down?

18. The air pressure P on an airplane wing depends on the velocity of the wind. Within a certain range of wind velocity, the pressure is distributed according to the function $p(x) = \sqrt{x/.012}$ for $0 \le x \le .3$. What is the average (or expected) pressure on the wing?

Sometimes it is necessary to convert a discrete pdf, obtained by a statistical survey, into a continuous pdf. There are several methods of doing this, but most of them are not easily described. However, there is one method that is readily available to us. That is to assume that the pdf, whose values at certain points are obtained from the survey, is continuous. Then estimate the integral using Simpson's rule and adjust the values of the pdf in order to make the integral equal 1. The adjusted pdf is then used to compute the expected value. Exercises 19 and 20 illustrate this method.

19. The data on heights in Exercise 13 yields a discrete pdf p. Assume now that p is a continuous function. Use the information in Exercise 13 and Simpson's rule to estimate $\int_5^7 p(x)\, dx$.

20. (Continuation of Exercise 19) Define a continuous function $\bar{p}$ by

$$\bar{p}(x) = \frac{p(x)}{\displaystyle\int_5^7 p(x)\,dx}$$

Then $\int_5^7 \bar{p}(x)\,dx = 1$ so $\bar{p}$ is a continuous pdf. Use Simpson's rule and $\bar{p}$ to calculate the expected height of an adult American male. Compare with the answer obtained in Exercise 13.

21. The probability density for the lifetime of a light bulb, in units of 1,000 hrs, is $p(t) = .105e^{-.01t}$ for $0 \le t \le 10$. Use Simpson's rule with $n = 4$ to estimate the average life of a light bulb.

22. (Continuation of Exercise 21) What is the probability that a light bulb will burn between 2,000 and 10,000 hrs?

23. (Continuation of Exercise 22) If the company making the light bulb wants no more than 10% returns, what is the longest lifetime they can guarantee for their light bulbs?

24. *(Lifetime of a plant)* The lifetime of a certain plant, measured in days, is a continuous random variable X with values in the interval [0, 100]; its pdf is $p(t) = \frac{3}{1000}(10 - \sqrt{t})$. (a) What is the probability that a plant will live at most 25 days? (b) What is the expected life of a plant?

6–8 FORCE OF A LIQUID

An object submerged in a liquid is subjected to a force exerted by the liquid. The deeper the object is submerged, the greater the force. You may have experienced this force while scuba diving or even swimming near the bottom of a deep pool.

If a horizontal thin plate, or lamina, is submerged in a liquid, the force on the top of it is the weight of liquid directly above it (see Figure 6–55). If the plate has a surface area of A sq ft, if it is submerged to a depth of h ft, and if the weight of the liquid is δ lbs/cu ft, then the force is

$$F = \delta h A \quad \text{lbs}$$

Now let us rearrange the lamina so that it is vertical. A fact from physics is that at any point below the surface, *the liquid exerts its force equally in all directions*. This means that the force on the vertical plate increases as the depth increases (Figure 6–56).

Figure 6–57 shows the front view of a vertical lamina submerged in a liquid. To determine the force on one side of the plate, draw the coordinate axes as indicated. For convenience, the positive direction of the y-axis is *downward*. Now partition the interval $[a, b]$ and approximate the force on a typical strip of the plate. Pick any point w_i in $[y_{i-1}, y_i]$ and let $W(w_i)$ be the width of the plate at a depth of w_i ft. The area of the shaded rectangle is $W(w_i)\Delta y_i$ and its approximate depth is w_i. It follows that the force F_i on the rectangle is

$$F_i \approx \delta w_i W(w_i)\Delta y_i$$

The sum of these forces from $i = 1$ to $i = n$ is a Riemann sum of the function $\delta y W(y)$ and it approximates the total force F on the lamina. As the norms of the partitions approach zero, the approximations become more accurate. Thus, if the width function W is continuous, then we define

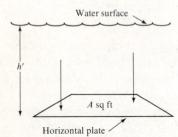

FIGURE 6–55. Force on a horizontal plate.

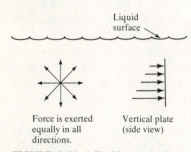

FIGURE 6–56. A liquid exerts its force equally in all directions. Naturally, the force on a vertical plate increases as the depth increases (indicated by longer arrows).

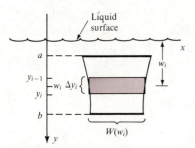

FIGURE 6–57. The force on the shaded rectangle is $\delta w_i W(w_i)\Delta y_i$.

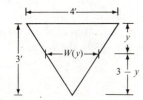

FIGURE 6–58. Example 1.

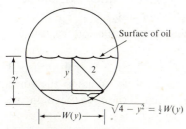

FIGURE 6–59. Example 2.

$$(6.38) \qquad F = \int_a^b \delta y W(y)\, dy$$

as the force of the liquid on the lamina.

EXAMPLE 1   A trough is filled with water. Each end is an isosceles triangle as shown in Figure 6–58. If water weighs 62.5 lbs/cu ft, find the force on one end of the trough.

Solution   In order to use formula (6.38), you must find the width function W. Figure 6–58 shows that if $0 \le y \le 3$, you can use similar triangles to write

$$\frac{W(y)}{4} = \frac{3-y}{3} \quad \text{or} \quad W(y) = \frac{4(3-y)}{3}$$

Thus, by (6.38), the force of the water is

$$F = \int_0^3 \delta y W(y)\, dy$$

$$= 62.5 \left(\frac{4}{3}\right) \int_0^3 y(3-y)\, dy = 375 \text{ lbs} \qquad \blacksquare$$

EXAMPLE 2   An oil drum 4 ft in diameter is lying on its side. It is half full of oil that weighs 50 lbs/cu ft. Find the force on one end of the drum.

Solution   Sketch the circular end of the drum as in Figure 6–59. For any depth y with $0 \le y \le 2$, the width is $W(y) = 2\sqrt{4-y^2}$. Therefore, by (6.38), the force is

$$F = \int_0^2 \delta y W(y)\, dy$$

$$= 50 \int_0^2 2y\sqrt{4-y^2}\, dy$$

$$= -50 \int_0^2 (4-y^2)^{1/2}(-2y)\, dy \qquad \text{(Set up for } u\text{-substitution)}$$

$$= -50 \int_4^0 u^{1/2}\, du \qquad \text{(Change of variable)}$$

$$= \frac{800}{3} \text{ lbs} \qquad \blacksquare$$

Our final observation is that the concept of *center of mass* can be used to advantage to compute the force of a liquid.

(6.39)

> Suppose a vertical plate of constant density and area A sq ft is entirely submerged in a liquid weighing δ lbs/cu ft. If the center of mass of the plate is at a depth of $\bar{h}$ ft below the surface, then the force on one face of the plate is
>
> $$F = \delta \bar{h} A \quad \text{lbs}$$

It is as though all of the liquid near the surface of the plate is concentrated at the center of mass. The proof of (6.39) is left as an exercise, but let us at least test its validity for Example 1 above.

The center of mass of a triangle is $\frac{1}{3}$ of the way from any base to the opposite vertex (Example 4, Section 6–6). Thus, the center of mass of the triangular end of the trough in Example 1 is 1 foot below the surface. The area of the triangle is $\frac{1}{2}(4)(3) = 6$ sq ft and $\delta = 62.5$ lbs/cu ft. Therefore, by (6.39), the force is

$$F = \delta \bar{h} A = (62.5)(1)(6) = 375 \text{ lbs}$$

which agrees with our answer in Example 1.

If you happen to know the center of mass of the submerged plate, then (6.39) provides an easy way to compute the force.

EXERCISES

1. A dam in the shape of an isosceles trapezoid is 100 ft across at the surface of the river and 50 ft across at the bottom. If the river is 50 ft deep find the force of the water on the dam.

2. A square gate at the bottom of the dam in Exercise 1 is 4 ft on a side. What is the force on the gate?

3. Some of the water from the trough in Example 1 is pumped out. If the water remaining has a depth of 2 ft, what is the force on one end?

4. Set up the definite integral whose value is the force on one end of the drum in Example 2 when the oil has a depth of 1 foot.

5. The ends of a water trough have the shape of the region bounded by $-y = x^2 - 4$ and $y = 0$. (Positive direction of the y-axis is downward.) If the trough is full of water, set up an integral whose value is the force on one end.

6. Follow the same instructions as in Exercise 5 if the ends of the trough have the shape of the region bounded by $-y = x^2 - 9$, $y = 5$, and $y = 0$.

7. What is the force on the (horizontal) bottom of trough in Exercise 6 if the trough is 6 ft long?

8. Apply formula (6.39) to the oil drum in Example 2. (See Exercise 11 in Section 6–6.)

9. If the drum in Example 2 is full of oil, what is the force on one end?

10. The oil drum in Example 2 is 6 ft long. It is turned upright and filled with oil. (a) What is the force on the bottom? (b) What is the total force on the curved surface?

11. Prove formula (6.39) by filling in the blanks below.

(1) Sketch a front view of the plate with the x-axis at the surface of the liquid and the positive direction of the y-axis downward. Suppose that the plate extends from $y = a$ to $y = b$ and that for any point y in $[a, b]$, the width of the plate is $W(y)$. Suppose further that the plate has an area of A sq ft and a constant density, say D.

(2) The moment of the plate about the x-axis is the value of the integral _____.

(3) The mass of the plate is _____.

(4) It follows from steps (2) and (3) that if $\bar{y}$ is the y-coordinate of the center of mass of the plate, then $\bar{y}A =$ _____.

(5) Formula (6.39) follows from step (4) because _____.

REVIEW EXERCISES

In Exercises 1–10, evaluate the definite integrals with the Fundamental Theorem (using substitution or a change of variable, if necessary) or by computing the area under the curve. If neither method is feasible, then estimate the integral using Simpson's rule with $n = 6$.

1. $\int_0^1 x\sqrt{1 - x^2}\, dx$

2. $\int_1^2 \frac{1}{x}\, dx$

3. $\int_1^2 \sqrt{9 - x^2}\, dx$

4. $\int_0^{\pi/4} \sin x\, dx$

5. $\int_0^{\pi/4} \sec^2 x\, dx$

6. $\int_1^2 \frac{x^2}{\sqrt[3]{3x^3 + 1}}\, dx$

7. $\int_2^4 \frac{\sqrt{1 + (1/x)}}{x^2}\, dx$

8. $\int_{-1}^1 \sqrt{1 - x^2}\, dx$

9. $\int_{-1}^1 x(4x^2 - 3)\, dx$

10. $\int_0^1 e^x\, dx$

If Simpson's rule is needed for the exercises below, let $n = 6$.

11. Find the area bounded by the curves $y = 2 - x^2$ and $y = -x$.

12. A wedge is cut from a right circular cylinder of radius r by two planes that intersect on the diameter. One is perpendicular to the axis of the cylinder and the other makes an angle α with the first. Find the volume of the wedge.

13. The base of a solid is an equilateral triangle with sides 5 inches. Each cross section is a square with one side lying in the base. Find the volume of the solid.

14. A hemispherical bowl of radius r is filled with water to a depth of $r/2$. How much water is in the bowl?

15. The region determined by $y = \sin x$ from $x = 0$ to $x = \pi$ is revolved about the x-axis. Find the volume generated.

16. Rotate the region in Exercise 15 about the y-axis, and compute the volume generated.

17. Find the length of the curve $y = \sqrt{1 - x^2}$, $0 \le x \le 1$.

18. Find the length of the curve $y = x^{2/3}$, $-8 \le x \le -1$.

19. Find the length of the curve $y = x^{2/3}$, $-8 \le x \le 1$.

20. The ends of a water trough 8 feet long and 4 feet deep are regular trapezoids with parallel sides of 4 and 6 feet. If the trough is filled with water, what is the force on one end?

21. A force of 9 lbs is needed to stretch a spring 2 inches beyond its natural length. Find the work done in stretching the spring 4 inches beyond its natural length.

22. If the spring in Exercise 21 has a natural length of 3 inches, how much work is done in stretching it from 4 to 6 inches?

23. A cylindrical tank 6 feet in diameter is lying on its side. If it is half filled with oil weighing 50 lbs/cu ft, what is the force on one end?

24. Find the area bounded by the curves $2y^2 = x + 4$ and $y^2 = x$.

25. The region determined by $f(x) = 2x - x^2$ from $x = 0$ to $x = 2$ is revolved about the y-axis. Find the volume generated.

26. A cube of steel 1 foot on a side is completely submerged in a tank of oil weighing 50 lbs/cu ft. If the steel is suspended so that it is not touching the bottom of the tank, what is the difference in the force on the top and the bottom of the cube?

27. Find the center of gravity of the semicircular region bounded by the curve $y = \sqrt{r^2 - x^2}$, $-r \le x \le r$.

28. Find the centers of gravity of the shaded regions shown in Figure 6–60A and B.

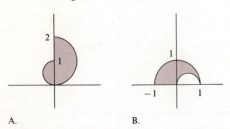

A. B.

FIGURE 6–60. Exercise 28. The curved boundaries are semicircles.

29. A hemispherical tank 20 feet in diameter is filled with water. Find the work done in pumping all of the water to a point 6 feet above the top of the tank.

30. (Continuation of Exercise 29) If a motor delivering 3 horsepower is used to pump the water, how long will it take to empty the tank?

31. The life of a certain species of plant, measured in days, is a continuous random variable X with values in the time interval [0, 100]; its pdf is $p(t) = \frac{3}{5000}(10 + \sqrt{100 - t})$. What is the probability that a plant will live at least 25 days?

32. In the game of craps, given that your point is 8, what is your expected profit or loss? What if your point is 9?

33. The lifetime of a light bulb, measured in units of 1,000 hours, is a continuous random variable X with values in the time interval [0, 10]; its pdf is $p(t) = .105e^{-.01t}$. What is the probability that a light bulb will burn out in the first 2,000 hours?

34. A bullet weighing w lbs moving horizontally with a velocity of v ft/sec strikes and imbeds itself in a wooden block weighing W lbs. The block is initially at rest on a smooth (frictionless) table. Find the velocity of the bullet and block after impact.

35. (a) Find the center of mass of the region bounded by a triangle with vertices (0, 0), (1, 1), and (2, 0). (b) Find the volume of the solid generated by revolving this region about the x-axis. (c) About the y-axis.

36. A woman is driving an automobile weighing W lbs at v mph. She comes to a hill that rises h ft for every H ft of road. How much additional horsepower is needed to maintain the original speed?

7

LOGARITHMIC AND EXPONENTIAL FUNCTIONS

Two major applications of exponential functions, population growth and radioactive decay, were discussed in earlier chapters. The first application was based on the observation that populations of bacteria tend to double in size during fixed periods of time. The second was based on the observation that radioactive substances tend to halve in size during fixed periods of time. A single observation that encompasses both situations, and is more in keeping with the language of calculus, is that *the rate of change is proportional to the amount present.* That is, if y represents a function whose value at time t is the amount present (of population or of a radioactive substance), then

(7.1) $y' = ky$

Expressed in this way, we shall see that population growth and radioactive decay are just two examples of a much broader class of phenomena whose behavior can be described by equation (7.1). Other examples discussed in this chapter are pollutants in the air or a body of water, Newton's law of cooling, and chemicals dissolving in a liquid.

To solve equation (7.1), we can use the definition of the differential $dy = y' \, dx$ and divide both sides by y (assuming $y \neq 0$) to obtain

(7.2) $\dfrac{1}{y} dy = k \, dx$

If L is any antiderivative of the function $1/y$, then $\int 1/y \, dy = L(y) + C$. Therefore, integrating both sides of (7.2) yields

(7.3) $L(y) = kx + C$

This equation defines y implicitly as a function of x; that is, it is an implicit solution of (7.1). If L has an inverse, then, by (1.64) in Section 1–9, $L^{-1}(L(y)) = y$, and it follows from (7.3) that

(7.4) $y = L^{-1}(kx + C)$

is an explicit solution of (7.1).

The wide applicability of equation (7.1) is motivation enough to seek its solution, but there is also an immediate bonus. The function L mentioned above turns out to satisfy all of the usual properties of a logarithm. In Section 7–1, we carefully define the function L, call it the natural logarithm and denote it by the symbol ln. The implicit solution (7.3) then reads

$$\ln y = kx + C$$

To find an explicit solution, we take a short detour in Section 7–2 to review inverses and discuss their derivatives. Section 7–3 picks up the main idea again and shows that the inverse of the natural logarithm is the natural exponential e^x. Therefore, the explicit solution (7.4) can be written as

$$y = e^{(kx + C)}$$

Section 7–4 applies this solution to problems about pollution, population, and so on. The remaining sections describe various other applications of logarithms and exponentials including a new differentiation technique and a definition of a^x that holds for any positive number a and all real x.

7–1 THE NATURAL LOGARITHM

Motivated by the discussion in the introduction to this chapter, we define a function L in terms of an integral, and call it the natural logarithm. Then we develop the derivative and logarithmic properties, sketch the graph, and determine the range of this function. These are the first steps on our way to solving the equation $y' = ky$.

The function $f(x) = 1/x$ defined on $(0, \infty)$ is continuous on its domain. Therefore, the equation

$$L(x) = \int_1^x \frac{1}{u}\, du \quad x > 0$$

defines a function L on $(0, \infty)$. This function is called the **natural logarithm,** and is denoted by ln. Thus,

(7.5) $$\boxed{\ln x = \int_1^x \frac{1}{u}\, du \quad x > 0}$$

In a moment, you will see why this function is called a logarithm.

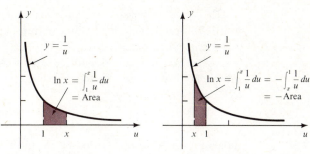

FIGURE 7–1. The value of ln in terms of area.

A. $\ln x > 0$ if $x > 1$

B. $\ln x < 0$ if $x < 1$

Since ln is defined in terms of the integral of a nonnegative function, its value at x can be interpreted in terms of area (Figure 7–1). We see that if $x > 1$, then $\ln x > 0$, but if $0 < x < 1$, then $\ln x < 0$. If $x = 1$, then

(7.6)
$$\ln 1 = 0$$

because $\int_1^1 (1/u)\, du = 0$.

The derivative of ln follows directly from the Fundamental Theorem of Calculus (5.19):

$$\frac{d}{dx}\ln x = \frac{1}{x}$$

By the chain rule, we have a new derivative formula

(7.7)

LOGARITHM FORMULA
If f is differentiable and $f(x) > 0$, then

$$\frac{d}{dx}\ln f(x) = \frac{1}{f(x)}f'(x) = \frac{f'(x)}{f(x)}$$

EXAMPLE 1

(a) $\dfrac{d}{dx}\ln 3x = \dfrac{(3x)'}{3x} = \dfrac{3}{3x} = \dfrac{1}{x}$

(b) $\dfrac{d}{dt}\ln(2t^2 - 3t + 7) = \dfrac{4t - 3}{2t^2 - 3t + 7}$

(c) $\dfrac{d}{dx}\ln\cos x = \dfrac{-\sin x}{\cos x} = -\tan x$ ∎

Remark: The natural logarithm is defined only for positive numbers. Therefore, it is understood that the derivatives obtained in Example 1 hold only where

$3x$, $2t^2 - 3t + 7$, and $\cos x$ are positive. Later, in Section 7–5, we will extend the definition to include all nonzero numbers.

The result of Example 1(a) is especially interesting; it shows that the derivative of $\ln 3x$ equals the derivative of $\ln x$. Actually, this is true not only for the number 3, but for any positive constant. If $a > 0$, then

$$\frac{d}{dx}\ln ax = \frac{(ax)'}{ax} = \frac{a}{ax} = \frac{1}{x} = \frac{d}{dx}\ln x$$

holds for all positive x. It follows from the Equal Derivatives Theorem (4.28), in Section 4–7, that $\ln ax$ and $\ln x$ can differ only by a constant,

$$\ln ax = \ln x + C \quad x > 0$$

Set $x = 1$; then $\ln a = \ln 1 + C$. Since $\ln 1 = 0$, it follows that $C = \ln a$, and the equation above becomes

(7.8) $\boxed{\ln ax = \ln x + \ln a \quad a \text{ and } x > 0}$

Equation (7.8) is the logarithm property: *the logarithm of a product is the sum of the logarithms.* All of the usual properties of logarithms follow from it. For example,

$$\ln x^2 = \ln xx = \ln x + \ln x = 2 \ln x$$
$$\ln x^3 = \ln x^2 x = \ln x^2 + \ln x = 2 \ln x + \ln x = 3 \ln x$$

and, in general, for any positive n,

(7.9) $\ln x^n = n \ln x$

This formula holds for negative n as well because

$$0 = \ln 1 = \ln (x^n x^{-n}) = \ln x^n + \ln x^{-n} \tag{7.8}$$

Therefore,

$$\ln x^{-n} = -\ln x^n = -n \ln x$$

Other logarithmic properties can be proved in similar fashion (see Exercises 44–46).

(7.10) $\boxed{\begin{array}{l} \text{(a) } \ln xy = \ln x + \ln y; \; x \text{ and } y \text{ positive} \\[4pt] \text{(b) } \ln \dfrac{x}{y} = \ln x - \ln y; \; x \text{ and } y \text{ positive} \\[4pt] \text{(c) } \ln x^y = y \ln x; \; x \text{ positive, } y \text{ rational} \end{array}}$

Now you see why ln is called a *logarithm*.

EXAMPLE 2   Properties (7.10) can be used to simplify differentiation computations.

(a) $\dfrac{d}{dx} \ln{[(x^2 - 3)(2x^3 - 5x)]} = \dfrac{d}{dx}[\ln{(x^2 - 3)} + \ln{(2x^3 - 5x)}]$ (7.10a)

$$= \frac{2x}{x^2 - 3} + \frac{6x^2 - 5}{2x^3 - 5x}$$ (7.7)

$$= \frac{10x^4 - 33x^2 + 15}{(x^2 - 3)(2x^3 - 5x)}$$ (Add fractions)

(b) $\dfrac{d}{dx} \ln{\dfrac{x - 1}{\sqrt{x - 1}}} = \dfrac{d}{dx}[\ln{(x - 1)} - \ln{\sqrt{x + 1}}]$ (7.10b)

$$= \frac{d}{dx}\left[\ln{(x - 1)} - \frac{1}{2}\ln{(x + 1)}\right]$$ (7.10c)

$$= \frac{1}{x - 1} - \frac{1}{2(x + 1)}$$ (7.7)

$$= \frac{x + 3}{2(x^2 - 1)} \quad \blacksquare$$

EXAMPLE 3   Find y' if $y = \ln{\sqrt[3]{(x^2 + 1)/(x - 1)}}$

Solution

$$y = \frac{1}{3}[\ln{(x^2 + 1)} - \ln{(x - 1)}]$$ (7.10)

$$y' = \frac{1}{3}\left[\frac{2x}{x^2 + 1} - \frac{1}{x - 1}\right]$$ (7.7)

$$= \frac{x^2 - 2x - 1}{3(x^2 + 1)(x - 1)} \quad \blacksquare$$

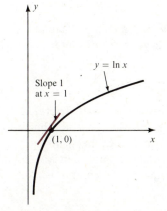

FIGURE 7-2. The function $y = \ln x$ is increasing and its graph is concave downward.

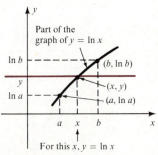

FIGURE 7-3. There is no way to connect the points $(a, \ln a)$ and $(b, \ln b)$ by a solid curve (no breaks) without crossing every horizontal line between them.

Let us now sketch the graph of $y = \ln x$. The first derivative $1/x$ is positive because x is. Therefore, *ln is an increasing function.* (This means that ln is one-to-one, and has an inverse; tuck this information away for later use.) The second derivative $-1/x^2$ is always negative, so the graph is concave downward (Figure 7-2).

Because we will soon want to discuss the inverse, it is important to know the domain and range of the natural logarithm.

(7.11)
> The domain of ln is $(0, \infty)$.
> The range of ln is $(-\infty, \infty)$.

According to definition (7.5), the domain is $(0, \infty)$. To prove that the range is $(-\infty, \infty)$, we note that ln is continuous because it is differentiable. Since there are no breaks in the graph of a continuous function, this suggests that if $\ln a < y < \ln b$, then $y = \ln x$ for some x between a and b (Figure 7-3). Actually, this is just a special case of a theorem that is usually proved in more advanced courses.

INTERMEDIATE VALUE THEOREM

(7.12)   If f is continuous on $[a, b]$ and $f(a) < y < f(b)$, then $y = f(x)$ for some x in $(a\ b)$.

In light of this theorem, we can prove that the range of ln is $(-\infty, \infty)$ by showing that every real number y lies between some two values of ln.

The function $y = \ln x$ is increasing, but it is not clear from the graph that its values get arbitrarily large. Is it possible that there is some number $N > 0$ such that $\ln x \le N$ for all $x > 0$? The answer is *NO* because for any integer $n = 1, 2, \ldots$, we know that

$$\ln 2^n = n \ln 2$$

Since $\ln 2 > 0$, the numbers $n \ln 2$ will eventually become larger than any given number N. Therefore, we see that the values of ln do get arbitrarily large. Similarly,

$$\ln\left(\frac{1}{2}\right)^n = n \ln\left(\frac{1}{2}\right)$$

and, since $\ln\left(\frac{1}{2}\right) < 0$, the values of ln also get arbitrarily small. Therefore, if y is any real number, there are values of ln that are larger than y and other values that are smaller than y. That is, there are numbers a and b such that $\ln a < y < \ln b$. It follows from (7.12) that $y = \ln x$ for some x and that the range of ln is $(-\infty, \infty)$.

Properties of logarithms can be used to advantage to sketch graphs.

EXAMPLE 4   Sketch the graph of $y = \ln 3x^2$.

Solution   Since $y = \ln 3x^2 = \ln 3 + 2 \ln x$, the graph is simply two times the graph of $y = \ln x$ translated up ln 3 units. Look up ln 3 in a table or find it with a calculator; $\ln 3 \approx 1.1$. The graph of

$$y = \ln 3x^2 \approx 1.1 + 2 \ln x$$

is shown in Figure 7–4.   ∎

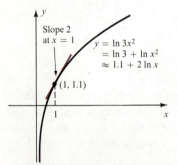

Slope 2 at $x = 1$

$y = \ln 3x^2$
$= \ln 3 + \ln x^2$
$\approx 1.1 + 2 \ln x$

$(1, 1.1)$

FIGURE 7–4. Using the properties of logarithms to sketch a graph.

EXERCISES

In Exercises 1–20, find y' if y is the given function.

1. $\ln 10x$

2. $\ln(-5x)$

3. $\ln 4x^3$

4. $\ln(2x)^2$

5. $\ln\sqrt{4 - 9x}$

6. $\ln\sqrt[3]{x + 1}$

7. $\ln\dfrac{x^2 - 1}{x^2 + 1}$

8. $\ln(x^2 + 1)(x + 3)$

9. $\ln\sqrt[3]{(x^2 + 1)(x + 3)}$

10. $\ln\sqrt{\dfrac{x^2 - 1}{x^2 + 1}}$

11. $\ln\tan x$

12. $\ln\sin x$

13. $[\ln(x + 1)]^2$

14. $[\ln(x^2 + 1)]^4$

15. $x \ln x$

16. $(x + 1)\ln(x + 1)$

17. $\dfrac{\ln(1/x)}{x^2}$

18. $\ln \sqrt[3]{\dfrac{x}{x+1}}$

19. $\ln(\ln x)$

20. $\dfrac{1}{\ln x}$

In Exercises 21–26, use implicit differentiation to find y'.

21. $\ln y = 3x$

22. $\ln xy = 3x$

23. $x \ln y + y \ln x = 5$

24. $\ln y = x \ln x$

25. $\ln(y/x) = 3x^2$

26. $[\ln y]^2 = 5x + 1$

In Exercises 27–34, sketch the graphs.

27. $y = \ln x^3$

28. $y = \ln 3x$

29. $y = \ln \dfrac{x^2}{2}$

30. $y = \ln 2x^3$

31. $y = \ln(x + 1)$

32. $y = \ln(x - 1)$

33. $y = \ln|x|$

34. $y = \ln|x - 1|$

---

35. Find an equation of the line tangent to the curve $y = x^2 + x \ln x$ at the point $(1, 1)$.

36. Find the slope of the curve $y = \ln x$ for $x = .001$ and $x = 1{,}000$. What does this say about the shape of curve for x close to 0 and for large x?

37. Use Simpson's rule with $n = 4$ to estimate $\ln 2$ ($\ln 2 = \int_1^2 1/u \, du$).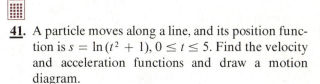

38. Use Simpson's rule with $n = 4$ to estimate $\ln(\tfrac{1}{2})$.

39. For each $n = 1, 2, 3, 4$, and 5, use a calculator to compute first $\ln 2^n$ and then $n \ln 2$. Do the answers agree?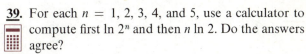

40. Same as Exercise 39 for $n = -5, -4, \ldots, -1$.

41. A particle moves along a line, and its position function is $s = \ln(t^2 + 1), 0 \le t \le 5$. Find the velocity and acceleration functions and draw a motion diagram.

42. The area determined by the curve $y = 1/\sqrt{x}$ from $x = 1$ to $x = 2$ is revolved about the x-axis. Find the volume of the generated solid using the disc method.

43. The area determined by the curve $y = 1/x^2$ from $x = 1$ to $x = 2$ is revolved about the y-axis. Find the volume of the generated solid using the shell method.

---

44. Use (7.9) and the fact that $\ln(x/y) = \ln xy^{-1}$ to prove property (7.10b): $\ln(x/y) = \ln x - \ln y$.

45. Let q be any nonzero integer. Use (7.9) and the fact that $\ln x = \ln(x^{1/q})^q$ to prove that $\ln x^{1/q} = (1/q) \ln x$.

46. Use the result in Exercise 45 and the fact that $x^{p/q} = (x^{1/q})^p$ to prove property (7.10c); $\ln x^y = y \ln x$ for rational y.

Optional Exercises

47. Since ln is increasing (why?), it is one-to-one and, therefore, has an inverse. Denote this inverse by the letter E. What are the domain and range of E? Sketch the graph of $y = E(x)$.

48. (Continuation of Exercise 47) Assume that E is differentiable. Use the chain rule and the equation $\ln(E(x)) = x$ to show that $E'(x) = E(x)$.

7–2 THE DERIVATIVE OF AN INVERSE

In this section, we review inverses and discuss the derivative of an inverse. Although the immediate application concerns logarithms and exponentials, the subject is presented in a more general setting because it is used again in the next chapter on trigonometric functions.

We begin with a brief list of properties of inverses. This material was covered in detail in Section 1–9.

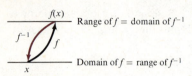

FIGURE 7–5. Arrow representation of inverses.

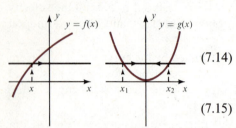

A. One-to-one. B. Not one-to-one

FIGURE 7–6. A function has an inverse if and only if it is one-to-one.

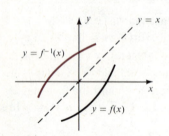

FIGURE 7–7. The graphs are mirror images.

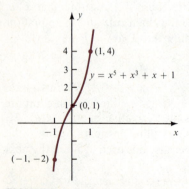

FIGURE 7–8. Example 1.

DEFINITION

Functions f and g with the property that

(7.13)
$$f \circ g(x) = x \text{ for all } x \text{ in the domain of } g, \text{ and}$$
$$g \circ f(x) = x \text{ for all } x \text{ in the domain of } f$$

are **inverses** of each other. The inverse of a function f is sometimes denoted by the symbol f^{-1} (read "f inverse").

(7.14)
The range of f is the domain of f^{-1}.
The domain of f is the range of f^{-1}.

(7.15)
$$f^{-1}(f(x)) = x \quad \text{and} \quad f(f^{-1}(x)) = x$$
for all x in the appropriate domain (Figure 7–5).

(7.16) A function has an inverse if and only if it is one-to-one (Figure 7–6).

(7.17) The graphs of $y = f(x)$ and $y = f^{-1}(x)$ are mirror images of each other through the line $y = x$ (Figure 7–7).

Now suppose that f is differentiable. If $f'(x) > 0$ for all x in the domain, then f is an increasing function; that is,

$$\text{if } x_1 < x_2 \quad \text{then} \quad f(x_1) < f(x_2)$$

Surely such functions are one-to-one. If $f'(x) < 0$, then f is decreasing and, again, it is one-to-one. These observations together with (7.16) above yield

(7.18) If f is differentiable on an interval I and $f'(x) > 0$ or $f'(x) < 0$ for all x in I, then f has an inverse on I.

Statement (7.18) is extremely useful, as the next examples illustrate.

EXAMPLE 1 Let $f(x) = x^5 + x^3 + x + 1$ for all x. Does f have an inverse?

Solution The answer is YES because

$$f'(x) = 5x^4 + 3x^2 + 1$$

is positive for all x. This means that f is increasing (Figure 7–8), which implies it is one-to-one, which implies it has an inverse. ∎

EXAMPLE 2 Show that $f(x) = x^3 + x^2 - x$ defined on $[-1, 1/3]$ has an inverse.

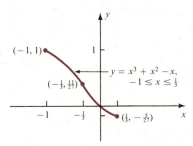

FIGURE 7–9. Example 2.

Solution

$$f'(x) = 3x^2 + 2x - 1$$
$$= (3x - 1)(x + 1)$$

For each x in $(-1, 1/3)$, the first factor is negative and the second factor is positive. It follows that f' is decreasing on $[-1, 1/3]$, so it has an inverse (Figure 7–9). ∎

EXAMPLE 3   Show that $y = \ln x$ has an inverse.

Solution   The natural logarithm is defined for $x > 0$, and

$$\frac{d}{dx} \ln x = \frac{1}{x} > 0$$

Therefore, ln is increasing (Figure 7–10) and it has an inverse. ∎

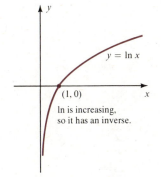

FIGURE 7–10. Example 3.

Many properties of a function are inherited by its inverse. For example, suppose f is increasing. If $f(x_1) < f(x_2)$, then $x_1 < x_2$ (why?) and it follows that

$$f^{-1}(f(x_1)) = x_1 < x_2 = f^{-1}(f(x_2))$$

Therefore, f^{-1} is increasing. A similar argument holds if f is decreasing.

(7.19)   If f is an increasing function, so is f^{-1}.
   If f is a decreasing function, so is f^{-1}.

If f is continuous, then there are no breaks in its graph. The graph of f^{-1}, which is a mirror image of the graph of f, will also have no breaks; so f^{-1} is continuous. This geometric argument suggests that

(7.20)   If f is continuous, so is f^{-1}.

Most important is the relationship between the derivatives of a function and its inverse. Let us suppose that f is differentiable at x. The graph of f has a tangent line L_1 at $(x, f(x))$ with slope $f'(x)$. Figure 7–11 indicates that the graph of f^{-1} has a tangent line L_2 at $(f(x), x)$ that is the mirror image of L_1. The two lines L_1 and L_2 intersect at (a, a). Knowing two points on each line, we can compute their slopes.

$$\text{Slope of } L_1 = \frac{f(x) - a}{x - a}$$

$$\text{Slope of } L_2 = \frac{x - a}{f(x) - a} = \frac{1}{\text{slope of } L_1}$$

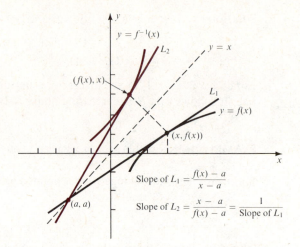

Slope of $L_1 = \dfrac{f(x) - a}{x - a}$

Slope of $L_2 = \dfrac{x - a}{f(x) - a} = \dfrac{1}{\text{Slope of } L_1}$

FIGURE 7–11. Derivatives of inverses.

Since the slope of L_1 also equals $f'(x)$, we have

$$\text{Slope of } L_2 = \frac{1}{f'(x)}$$

If $f'(x) \neq 0$, then the slope of L_2 is defined. This means that f^{-1} is differentiable at $f(x)$, and

$$[f^{-1}]'(f(x)) = \frac{1}{f'(x)}$$

This geometric argument justifies the following statement:

INVERSE DERIVATIVE THEOREM
If f has an inverse g, if f is differentiable at x, and if $f'(x) \neq 0$, then g is differentiable at $f(x)$ and

(7.21)

$$g'(f(x)) = \frac{1}{f'(x)}$$

As you can see, we have used the symbol g to denote the inverse of f; the reason is that the notation $[f^{-1}]'$ is too cumbersome. In the Leibniz notation, the last line of (7.21) reads very simply

$$\frac{dx}{dy} = \frac{1}{dy/dx}$$

Remarks: Let f be a function and let g be its inverse.

(1) To use (7.21), you must know the values of x, $f(x)$, and $f'(x)$.

(2) Formula (7.21) requires that f' be evaluated at x and that g' be evaluated at $f(x)$. Note the difference.

(3) If $f'(x) = 0$, then g is not differentiable at $f(x)$.

An interesting feature of (7.21) is that you can compute the values of g' without having an explicit definition of g.

EXAMPLE 4   Let f be the function in Example 1, $f(x) = x^5 + x^3 + x + 1$. Let $g = f^{-1}$, and evaluate g' at $f(1) = 4$ and $f(-2) = -41$.

Solution   *First,* find the derivative of f; $f'(x) = 5x^4 + 3x^2 + 1$. *Second,* use the formula in (7.21) with $x = 1$ and then with $x = -2$.

$$g'(4) = \frac{1}{f'(1)} = \frac{1}{9}$$

$$g'(-41) = \frac{1}{f'(-2)} = \frac{1}{93} \quad \blacksquare$$

In Example 4, the value of x and $f(x)$ are given. If only the value of $f(x)$ is given, then you must solve an equation of the form $f(x) = c$ to find x. If f is a linear or quadratic function, then such an equation is easy to solve. If f is more complicated, it may be possible to *guess* a solution.

EXAMPLE 5   Let f be the function in Example 2, $f(x) = x^3 + x^2 - x$ on $[-1, 1/3]$. Let $g = f^{-1}$, and evaluate $g'(0)$.

Solution   To use (7.21), you must know the value of x in $[-1, 1/3]$ for which $f(x) = 0$. It is easy to see that $x^3 + x^2 - x = 0$ when $x = 0$. Therefore, since $f'(x) = 3x^2 + 2x - 1$, we have

$$g'(0) = \frac{1}{f'(0)}$$

$$= \frac{1}{3(0)^2 + 2(0) - 1} = -1 \quad \blacksquare$$

EXAMPLE 6   By Example 3, we know that the natural logarithm has an inverse; let E be the inverse. Find $E'(0)$.

Solution   We have to find the number x such that $\ln x = 0$. From (7.6) in the previous section, we know that $x = 1$. Since the derivative of $\ln$ evaluated at 1 equals 1, it follows that

(7.22) $E'(0) = \dfrac{1}{1} = 1 \quad \blacksquare$

EXERCISES

In Exercises 1–10, determine whether or not the given function has an inverse. Give reasons.

1. $f(x) = x^3 + x - 4$
2. $f(x) = x^2 + 3x - 4$
3. $f(x) = x^2 - x + 2$
4. $f(x) = x^5 + 3x^3$

5. $f(x) = x^2 - 4x + 2; x \geq 2$

6. $f(x) = x^2 - 6x + 1; x \leq 3$

7. $f(x) = \sin x; -\dfrac{\pi}{2} \leq x \leq \dfrac{\pi}{2}$

8. $f(x) = \cos x; -\dfrac{\pi}{2} \leq x \leq \dfrac{\pi}{2}$

9. $f(x) = \displaystyle\int_0^x \dfrac{1}{1 + t^2} \, dt$

10. $f(x) = \displaystyle\int_0^x \dfrac{1}{1 - t^2} \, dt; -1 < x < 1$

In Exercises 11–24, c is a point in the *domain* of f. Let $g = f^{-1}$ and evaluate $g'(f(c))$.

11. $f(x) = \sqrt{x}; c = 4$

12. $f(x) = \sqrt{5x}; c = 5$

13. $f(x) = (x - 2)^3 + 1; c = 0$

14. $f(x) = 1 - x^3; c = -2$

15. $f(x) = \sqrt[3]{x + 4} - 1; c = -5$

16. $f(x) = \sqrt[3]{x + 1} + 2; c = 7$

17. $f(x) = \displaystyle\int_1^x 3t^2 \, dt; c = 1$

18. $f(x) = \displaystyle\int_{-1}^x t^4 \, dt; c = 1$

19. $f(x) = x^5 + x^3 + x + 2; c = -1$

20. $f(x) = x^7 + x + 1; c = 0$

21. $f(x) = \sqrt{4x + 1}; c = 0$

22. $f(x) = \sqrt{5 - 3x}; c = 1$

23. $f(x) = \displaystyle\int_0^x \dfrac{1}{1 + t^2} \, dt; c = 0$

24. $f(x) = \displaystyle\int_1^x (3t^2 + 2) \, dt; c = 1$

In Exercises 25–32, a point c in the *range* of f is given. Let $g = f^{-1}$ and evaluate $g'(c)$.

25. $f(x) = x^5 + x^3 + x + 2; c = 5$

26. $f(x) = x^7 + x + 1; c = 1$

27. $f(x) = \sqrt{9 - 2x}; c = 4$ **28.** $f(x) = \sqrt{x}; c = 5$

29. $f(x) = \displaystyle\int_1^x \dfrac{1}{u} \, du, x > 0; c = 0$

30. $f(x) = \displaystyle\int_0^x \dfrac{1}{1 + t^2} \, dt; c = 0$

31. $f(x) = \sin x, -\dfrac{\pi}{2} \leq x \leq \dfrac{\pi}{2}; c = \dfrac{1}{2}$

32. $f(x) = \cos x, 0 \leq x \leq \pi; c = 0$

Optional Exercise

33. Suppose that E is a function that takes only positive values and has the property that $E'(x) = E(x)$ for all x. Show that E has an inverse L and that $L'(x) = 1/x$.

7–3   THE NATURAL EXPONENTIAL

Our search for the solution of $y' = ky$ will be completed when we find the inverse of the natural logarithm. The inverse of any logarithm is an exponential function $F(x) = a^x$ where a is the *base* of the logarithm. In precalculus courses, the most common base is 10; the logarithm with base 10 is denoted by the symbol log_{10}, and

(7.23) $\log_{10} x = y$ means that $10^y = x$

It follows from this definition that if $L(x) = \log_{10} x$ and $E(x) = 10^x$, then

$$L(E(x)) = \log_{10} 10^x = x \quad \text{and}$$
$$E(L(x)) = 10^{\log_{10} x} = x$$

so L and E are inverses.

Remark: If your calculator has a *log* key, you can test the statement above. Enter any positive number, say 1.849. Compute log $10^{1.849}$ and then $10^{\log 1.849}$. Try it with other positive numbers.

Another base usually mentioned in algebra courses is 2; the logarithm with base 2 is denoted by the symbol log_2, and

(7.24) $\log_2 x = y$ means that $2^y = x$

Again it follows that if $L(x) = \log_2 x$ and $E(x) = 2^x$, then

$$L(E(x)) = \log_2 2^x = x \quad \text{and}$$
$$E(L(x)) = 2^{\log_2 x} = x$$

so L and E are inverses.

Observe that in both examples above

    (1) The base is that number whose logarithm is 1.

    (2) The inverse of the logarithm is the exponential function with that base.

Motivated by these observations, we will determine the base of the natural logarithm and then formulate the exponential function with that base.

It is already established that the range of ln includes all real numbers, so ln $x = 1$ for some positive x. Since ln is also one-to-one, there is only one such number; call it e. (Our choice of the letter e is no accident. Later we will prove that the base of the natural logarithm really is the number $e = \lim_{h \to 0}(1 + h)^{1/h} \approx 2.71828\ldots$ introduced in Chapter 3.) Thus,

(7.25) $\boxed{\ln e = 1}$

Since e is positive, the meaning of the symbol $e^{p/q}$ is firmly established for any rational number p/q; it is the pth power of the qth root of e. But what about irrational exponents; for example $e^{\sqrt{2}}$ or e^{π}? Observe that for rational exponents, ln $e^{p/q} = (p/q) \ln e$ (7.10c), and that ln $e = 1$ (7.25). Therefore,

$$\ln e^{p/q} = p/q$$

This relation makes the following definition reasonable.

(7.26)
> **DEFINITION of** e^x
> For *any* number x, e^x is that number whose natural logarithm is x. That is,
>
> $$e^x = y \quad \text{if and only if} \quad \ln y = x$$

Thus, $e^{\sqrt{2}}$ is that number whose natural logarithm is $\sqrt{2}$, and e^{π} is that number whose natural logarithm is π. *Now e^x is defined for real numbers x.* Moroever, this definition says that

(7.27)

$$\begin{aligned} \ln e^x &= x \quad \text{for all } x \\ e^{\ln x} &= x \quad \text{for } x > 0 \end{aligned}$$

The function $y = e^x$ is called the **natural exponential.** Statement (7.27) shows that *the natural logarithm and the natural exponential are inverse functions.*

Let us now examine some of the properties of e^x; most of them follow from the fact that it is the inverse of $\ln x$.

(7.28)   The domain of e^x is $(-\infty, \infty)$ and the range is $(0, \infty)$

$$e^x > 0 \quad \text{for all } x$$

(7.29)   The graph of $y = e^x$ is the reflection of the graph of $y = \ln x$ through the line $y = x$ (Figure 7–12).

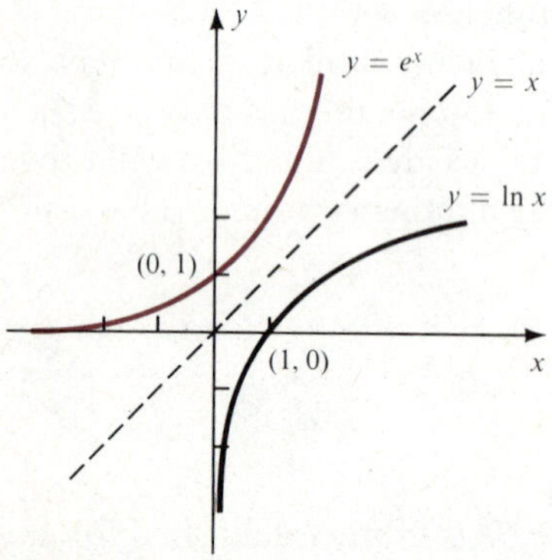

Furthermore, the usual rules of exponents hold

(7.30)

$$e^x e^y = e^{x+y}, e^x / e^y = e^{x-y}, \text{ and if } y \text{ is rational,* then}$$
$$(e^x)^y = e^{xy}$$

FIGURE 7–12. The graph of $y = e^x$ is the reflection of the graph of $y = \ln x$ through the line $y = x$.

The proofs of these properties are straightforward; for example, to prove the first one, write

$$e^x e^y = e^{\ln e^x e^y} = e^{(\ln e^x + \ln e^y)} = e^{x+y}$$

The other proofs are similar.

Now for the derivative. Because the derivative of $\ln x$ is never zero, it follows from (7.21) that its inverse e^x is differentiable at every x. We can find the derivative of e^x by differentiating both sides of the equation

$$\ln e^x = x$$

with respect to x. Thus, by (7.7),

$$\frac{1}{e^x}(e^x)' = 1$$

and it follows that $(e^x)' = e^x$. With the chain rule, we have

---

*So far, the only exponential defined for all real numbers is the one with base e. The definition will be extended in Section 7–6 to any positive base a. But until then, if $a \neq e$, then a^y is defined only for rational y.

$$(7.31) \quad \boxed{\begin{array}{c} \text{If } f \text{ is differentiable, then} \\ (e^f)' = e^f f' \end{array}}$$

EXAMPLE 1

(a) $\dfrac{d}{dx} e^{3x} = e^{3x}(3x)' = 3e^{3x}$

(b) $\dfrac{d}{dx} e^{x^2} = e^{x^2}(x^2)' = 2xe^{x^2}$ ∎

EXAMPLE 2

(a) $\dfrac{d}{dx}(x^2 e^{3x}) = x^2(e^{3x})' + e^{3x}(x^2)'$ (Product)

$\qquad\qquad = 3x^2 e^{3x} + 2xe^{3x}$

(b) $\dfrac{d}{dx}\left(\dfrac{e^{3x}}{x^2}\right) = \dfrac{x^2(e^{3x})' - e^{3x}(x^2)'}{(x^2)^2}$ (Quotient)

$\qquad\qquad = \dfrac{3x^2 e^{3x} - 2xe^{3x}}{x^4}$ ∎

EXAMPLE 3

$$\frac{d}{dx}\sqrt{x + e^{4x}} = \frac{d}{dx}(x + e^{4x})^{1/2}$$

$$= \frac{1}{2}(x + e^{4x})^{-1/2}(1 + 4e^{4x})$$ ∎

EXAMPLE 4   Use implicit differentiation to find y' if $\ln y = xe^y$.

Solution   Differentiate both sides with respect to x. Thus,

$$\frac{1}{y} y' = xy'e^y + e^y$$

Solving for y' yields

$$y' = \frac{e^y}{\dfrac{1}{y} - xe^y}$$

$$= \frac{ye^y}{1 - xye^y}$$ ∎

EXAMPLE 5 Determine where $y = e^{-x^2}$ is increasing, decreasing, concave up, concave down, and sketch the graph.

Solution The first derivative is

$$y' = -2xe^{-x^2} \begin{cases} >0 & \text{for } x < 0 \\ =0 & \text{for } x = 0 \\ <0 & \text{for } x > 0 \end{cases}$$

Therefore, $y = e^{-x^2}$ is increasing on $(-\infty, 0)$ and decreasing on $(0, \infty)$; its absolute maximum occurs at $x = 0$.

The second derivative is

$$y'' = -2x(e^{-x^2})' + e^{-x^2}(-2x)' = 4x^2 e^{-x^2} - 2e^{-x^2}$$
$$= 2e^{-x^2}(2x^2 - 1)$$

Thus,

$$y'' \begin{cases} >0 & \text{for } |x| > 1/\sqrt{2} \\ =0 & \text{for } |x| = 1/\sqrt{2} \\ <0 & \text{for } |x| < 1/\sqrt{2} \end{cases}$$

Therefore, the graph is concave up for x in $(1/\sqrt{2}, \infty)$ or $(-\infty, -1/\sqrt{2})$ and concave down for x in $(-1/\sqrt{2}, 1/\sqrt{2})$. The graph is sketched in Figure 7–13. ∎

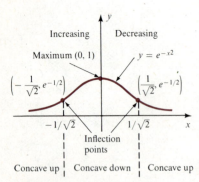

FIGURE 7–13. Example 5.

The Differential Equation $y' = ky$

We now have the necessary tools to solve the equation $y' = ky$. Rewrite it in terms of differentials $dy/dx = ky$, separate the variables,

$$\frac{1}{y} dy = k \, dx$$

and integrate both sides to obtain

$$\ln y = kx + C \quad \text{for } y > 0$$

Therefore,

$$e^{\ln y} = e^{(kx+C)} = e^{kx}e^C$$

Since e^C is a positive constant (call it K) and $y = e^{\ln y}$, we have

(7.32) | $y = Ke^{kx}$ is the general solution of $y' = ky$

The next two examples illustrate the type of problems encountered in applications.

EXAMPLE 6

(a) Find the general solution of $y' = 3y$.

(b) Find the particular solution that has the value 5 when $x = 0$; that is, $y(0) = 5$.

Solution

(a) According to (7.32), the general solution is

$$y = Ke^{3x}$$

where K is a constant.

(b) To find the particular solution with $y(0) = 5$, set $x = 0$ and solve for K. Thus,

$$5 = y(0) = Ke^{3\cdot 0} = K \qquad\qquad (e^0 = 1)$$

So $K = 5$ and $y = 5e^{3x}$ is the solution.

Check.   If $y = 5e^{3x}$, then $y(0) = 5$ and

$$
\begin{aligned}
y' &= 5(e^{3x})' = 5(3e^{3x}) \\
&= 3(5e^{3x}) \\
&= 3y
\end{aligned}
$$

Thus, $y' = 3y$ which is the given differential equation.   ∎

EXAMPLE 7

(a) Find the particular solution of $y' = -2y$ with $y(0) = 8$.

(b) For what value of x will $y(x) = 5$?

Solution

(a) The general solution is $y = Ke^{-2x}$. The value of K is determined by the condition $y(0) = 8$. Thus

$$8 = y(0) = Ke^{-2\cdot 0} = K$$

Therefore, $y = 8e^{-2x}$ is the solution.

Check.

$$
\begin{aligned}
y(0) = 8 \quad\text{and}\quad y' &= 8(e^{-2x})' = 8(-2e^{-2x}) \\
&= -2(8e^{-2x}) \\
&= -2y
\end{aligned}
$$

(b) To find the value of x with $y(x) = 5$, write

$$5 = y(x) = 8e^{-2x}$$

$$\frac{5}{8} = e^{-2x}$$

Now take ln of both sides and solve for x,

$$\ln \frac{5}{8} = \ln e^{-2x} = -2x$$

$$x = -\frac{1}{2}\ln \frac{5}{8}$$

$$\approx .235$$

Check. $y(.235) = 8e^{-2(.235)} \approx 5$ ∎

EXERCISES

In Exercises 1–20, find y' if y is the given function.

1. e^{-2x}
2. e^{4x}

3. e^{5x^2}
4. e^{-x^3}

5. $e^{\sqrt{x}}$
6. $\sqrt{e^x}$

7. $x^2 e^x$
8. xe^{x^2}

9. $e^{\sin x}$
10. $e^{\tan x}$

11. $\dfrac{\cos x}{e^x}$
12. $e^x \sin x$

13. $e^{1/x}$
14. $\sqrt{1 + e^x}$

15. $e^{\ln x}$
16. $\ln e^x$

17. $\ln \dfrac{e^x}{e^x + 1}$
18. $\ln \sqrt{1 + e^x}$

19. $e^{x \ln x}$
20. $\ln \cos e^x$

In Exercises 21–24, find y' using implicit differentiation.

21. $y = e^{xy}$
22. $ye^x + xe^y = 0$

23. $\ln y = x^2 + e^y$
24. $\ln y = e^{xy}$

In Exercises 25–32, determine where the function is increasing and decreasing, where the graph is concave up and down, and sketch the graph.

25. $y = e^{-x}$
26. $y = e^{-4x}$

27. $y = \dfrac{1}{2}(e^x + e^{-x})$
28. $y = \dfrac{1}{2}(e^x - e^{-x})$

29. $y = xe^{-x}$
30. $y = xe^x$

31. $y = e^{|x|}$
32. $y = e^{|x+1|}$

In Exercises 33–36, find the particular solution that satisfies the given condition.

33. $y' = -7y$; $y(0) = 2$
34. $y' = \sqrt{2}y$; $y(0) = \pi$

35. $y' = 5y$; $y(1) = e^3$
36. $y' = -6y$; $y(1/6) = e$

37. Find an equation of the line tangent to the curve $y = e^x$ at the point $(0, 1)$.

38. Find an equation of the line tangent to the curve $y = (x - 2)e^x$ at the point $(2, 0)$.

39. Find a point on the curve $y = e^{-3x}$ where the tangent line passes through the origin.

40. A particle moves along a line with position function $s = 2e^{4t} + 3e^{-4t}$. Show that its acceleration is proportional to its position.

41. Find the particular solution of $y' = 5y$ for which $y(0) = 20$. For what value of x will $y(x) = 40$?

42. Find the particular solution of $y' = -4y$ for which $y(2) = 1$. For what value of x will $y(x) = 1/2$?

43. Suppose y satisfies the equation $y' = ky$ and $y > 0$. If $k > 0$, is y an increasing or decreasing function? What if $k < 0$?

44. For what value(s) of c will $y = e^{cx}$ satisfy the differential equation $y'' + 4y' - 5y = 0$?

45. Prove that $e^x/e^y = e^{x-y}$ and $(e^x)^y = e^{xy}$ for rational y.

Optional Exercises

46. Let $f(x) = e^x - 1 - x$ for $x > 0$. Show (a) f is increasing (b) $f(0) = 0$, and conclude that (c) $e^x > 1 + x$ for all $x > 0$.

47. (Continuation of Exercise 46) Let $g(x) = e^x - 1 - x - x^2/2$ for $x > 0$. Show (a) g is increasing, (b) $g(0) = 0$, and conclude that (c) $e^x > 1 + x + x^2/2$ for all $x > 0$.

48. (Continuation of Exercises 46 and 47) Show that

$$e^x > 1 + x + \frac{x^2}{2} + \frac{x^3}{6} \qquad \text{for all } x > 0.$$

7-4 APPLICATIONS

Finding the solution of $y' = ky$ was hard work; we had to discuss the natural logarithm, inverses, and the natural exponential. But the applications make the effort worthwhile.

In the examples that follow, y is a function of time t, and the rate of change of y is proportional to the value of y. That is, $y' = ky$; the general solution is $y = Ke^{kt}$. If the initial value $y(0)$ is known, then $y(0) = Ke^{k \cdot 0} = K$. Therefore,

(7.33)

> If y is a function of time t and $y' = ky$, then,
> $$y = y(0)e^{kt}$$

Pollution

EXAMPLE 1 *(A polluted lake)*   A chemical company dumps its waste into a lake, and the concentration of pollutants is becoming dangerously high. If the local government orders the company to dump its waste elsewhere, is it possible to predict when the concentration of pollutants in the lake will be reduced to a safe level?

Solution   In order to solve the problem, we make the following (reasonable) assumptions and definitions of notation:

(1) No new pollutants are added to the lake.

(2) The volume of the lake is constant, V cu ft.

(3) There is an inlet admitting a constant flow of I cu ft/day of unpolluted water. If V is to be constant, then there must be an outlet through which I cu ft/day of polluted water flows (Figure 7-14).

(4) At time t days, the lake contains $P = P(t)$ cu ft of pollutants.

(5) At time t days, let $y(t) = P(t)/V$ be the *concentration* of pollutants. We assume the concentration at any time t is uniform throughout the lake.

With these assumptions, we proceed as follows. I cu ft of polluted water is expelled from the lake each day and in each cubic foot, there is a concentration of $y = y(t)$. Thus, Iy is the rate at which pollutants flow out of the lake. But $P = P(t)$ is the amount of pollutant at time t, so $-dP/dt = -P'$ is the rate at which the amount of pollutants is decreasing. It follows that

$$-P' = Iy$$

Since $y = P/V$, or $Vy = P$, we can substitute Vy' for P' in the equation above to obtain $-Vy' = Iy$, or

$$y' = -\frac{I}{V}y$$

It follows from (7.33) that the solution of this equation is

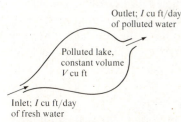

Outlet; I cu ft/day of polluted water

Polluted lake, constant volume V cu ft

Inlet; I cu ft/day of fresh water

FIGURE 7-14. Example 1.

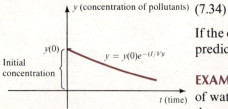

FIGURE 7–15. Example 1. The concentration of pollutants decreases exponentially.

$$(7.34) \qquad y = y(0)e^{-(I/V)t}$$

If the concentration is known on some particular day, let that be $y(0)$; then (7.34) predicts the concentration on any future day (Figure 7–15). ∎

EXAMPLE 2   Suppose the lake in Example 1 has a constant volume of 10^9 cu ft of water, that water flows out of the lake at the rate of 10^5 cu ft/day, and that the concentration of pollutants in the lake on a certain day is 10^{-6}. How long will it take for the concentration to decrease to 10^{-7}?

Solution   Use equation (7.34) with $y(0) = 10^{-6}$, $I = 10^5$ and $V = 10^9$:

$$y = 10^{-6}e^{-10^{-4}t}$$

The question is, for what value of t will $y = 10^{-7}$? In other words, solve the following equation for t:

$$10^{-7} = 10^{-6}e^{-10^{-4}t}$$

$$\frac{1}{10} = e^{-10^{-4}t}$$

Now take the natural logarithm of both sides; remember that $\ln e^x = x$. Thus,

$$\ln \frac{1}{10} = -10^{-4}t$$

$$t = -10^4 \ln \frac{1}{10} \quad \text{days}$$

This is the answer. If you happen to have a natural logarithm table or a calculator with an ln key, then you can compute that $t \approx 23{,}000$ days or about 63 years!

Remark: This is a hypothetical problem constructed for its striking conclusion. In real problems about polluted lakes, the nature of the pollutants, the dangerous and safe levels of concentration, the volume of the lake, and other pertinent data would have to be known before anyone could decide what to do about reducing the pollution. ∎

Population Growth

Under ideal conditions, populations grow in proportion to the population present at any given time. For example, if bacteria divide (that is, reproduce) every hour and if 10 bacteria are present at time $t = 0$, then 10 new bacteria are produced during the first hour. But now each of the 20 bacteria divide so there are 20 new bacteria produced during the second hour, 40 new bacteria produced during the third hour, and so on.*

Let $y = y(t)$ be the number of bacteria present at time t. Then, the rate of increase of y is proportional to y; that is, $y' = ky$ where k depends on the strain

---

*This is not quite accurate because some of the bacteria will die, some will fail to divide, and so on. A more realistic model (The Logistics Model) is presented in Chapter 9.

of bacteria, the culture in which they are grown, the temperature, and so forth. If $y(0)$ is the population at time $t = 0$, then

(7.35) $y = y(0)e^{kt}$

is the population at time t.

EXAMPLE 3   A certain strain of bacteria is being grown in a culture. If there are 100 bacteria at the start and 1,000 two hours later, how many will there be at the end of five hours?

Solution   Use equation (7.35) with $y(0) = 100$; then

$$y = 100e^{kt}$$

In the pollution problem, $k = -I/V$ is known. In population problems, k will be different in each case. However, the side condition $y(2) = 1,000$ can be used to solve for k. If $t = 2$, then $y = 1,000$ so the equation above becomes

$$1,000 = 100e^{k\cdot 2}$$
$$10 = e^{2k}$$

Now use the natural logarithm,

$$\ln 10 = 2k$$
$$k = \frac{1}{2}\ln 10$$

Therefore, the population function for this strain of bacteria is

$$y = 100e^{(1/2)(\ln 10)t}$$

When $t = 5$, there will be $100e^{(5\ln 10)/2}$ bacteria. If you have a logarithm table or a calculator, you can compute that $y(5) \approx 31,600$ bacteria!   ■

Radioactive Decay

Radioactive decay works very much like population growth. The rate of decay (negative rate of change) is proportional to the amount present at any given time. The constant of proportionality k depends on the radioactive substance. Therefore, if $y = y(t)$ is the amount present at time t and $y(0)$ is the amount at $t = 0$, then

(7.36) $y = y(0)e^{kt}$

EXAMPLE 4   Cobalt 60 has a half-life of 5.3 years. Starting with 10 grams of cobalt 60, how much will be left after 2 years?

Note: The **half-life** of a radioactive substance is the time it takes for one-half of the amount present to radiate away.

Solution   Use equation (7.36) with $y(0) = 10$; then

$$y = 10e^{kt}$$

The half-life of 5.3 years provides the side condition $y(5.3) = 5$ (which is one-half of $y(0) = 10$). This can be used to solve for k as in the population example.

$$5 = 10e^{k(5.3)}$$

$$\frac{1}{2} = e^{5.3k}$$

$$\ln \frac{1}{2} = 5.3k$$

$$k = \frac{1}{5.3} \ln \frac{1}{2}$$

Thus, the amount of cobalt 60 at any time t is

$$y = 10e^{(1/5.3)(\ln 1/2)t}$$

When $t = 2$, there will be

$$10e^{(2/5.3)(\ln 1/2)} \approx 7.7 \text{ grams} \quad \blacksquare$$

no continuous interest
no predator/prey
no carbon dating).

Newton's Law of Cooling

Newton's law of cooling states that the rate of change of temperature of an (inanimate) body is proportional to the *difference* between its temperature and that of the surrounding medium. If $y = y(t)$ is the temperature of the body at time t, A is the constant temperature of the surrounding medium, and $y > A$, then

$$y' = k(y - A)$$

We solve this equation as usual:

$$\frac{dy}{y - A} = k \, dt$$

$$\ln(y - A) = kt + C \qquad \text{(Integrate both sides)}$$

$$y - A = e^{kt}e^{C}$$

$$y - A = Ke^{kt} \qquad (K = e^{C})$$

If the initial temperature $y(0)$ is known, then setting $t = 0$, we see that $K = y(0) - A$, and the equation above becomes

$$(7.37) \quad y - A = [y(0) - A]e^{kt}$$

EXAMPLE 5   A turkey at 325° is removed from the oven and placed in a room of constant temperature 70°. Two minutes later, the temperature of the turkey is 300°. How long must you wait for the turkey to cool down to eating temperature of 110°?

Solution Use equation (7.37) with $y(0) = 325$ and $A = 70$; then

$$y - 70 = 255e^{kt}$$

The side condition $y(2) = 300$ can be used to find k. Setting $t = 2$, we have

$$230 = 255e^{2k}$$

$$\frac{230}{255} = e^{2k}$$

$$k = \frac{1}{2}\ln\frac{230}{255} \approx -\frac{1}{20}$$

Now the temperature y at any time t is (approximately)

$$y - 70 = 255e^{-t/20}$$

To find out how long you must wait for the turkey to cool down, take ln of both sides and solve for t:

$$\ln(y - 70) = \ln 225 + \ln e^{-t/20} = \ln 225 - \frac{t}{20}$$

$$t = -20\ln\left(\frac{y - 70}{225}\right)$$

For $y = 110°$, $t \approx 35$ minutes. Bon appétit! ∎

EXERCISES

Find the value of k in Exercises 1–4.

1. $10 = 3e^{2k}$ 2. $4 = 2e^{-k}$
3. $y = Ke^{kt}$; $y(0) = 4$; $y(2) = 3$
4. $y = Ke^{kt}$; $y(1) = 5$; $y(6) = 10$
5. If $y = 7e^{1.3t}$, find t such that $y(t) = 8$.
6. If $y = 6e^{-3.6t}$, find t such that $y(t) = 4$.
7. In Example 2, it took about 63 years to reduce the concentration of pollutants to 1/10 of the original amount. How long would it take to reduce the original amount by half?
8. How long would it take to reduce the concentration of pollutants of the lake in Example 1 by 10%?
9. Cotton mills are plagued by the fact that in the milling process, cotton lint fills the air and can cause serious lung damage to the workers. Suppose there are 200,000 cu ft of air in the mill and at a certain time a measurement shows there is 10^{-4} cu ft of lint per cu ft of air (that is, the concentration is 10^{-4}).

That is considered dangerous and the milling machines are shut down. An exhaust fan at one end of the mill can move 1,000 cu ft of air per minute and fresh air is sucked in at the other end. Assuming that the lint concentration is uniform throughout the mill at all times, how long will it take to reduce the concentration down to the "safe" level of 10^{-5}?

10. A strain of bacteria is being grown in a culture. If the amount of bacteria rises from 1 million to 2 million in 3 hours, how many bacteria will there be after 30 hours?
11. How many hours will it take the bacteria in Exercise 10 to reach a population of 10 million?
12. In 1960 the population of the Unite'd States was 180 million and in 1970 it was 203 million. Assuming conditions remain the same as during those 10 years, predict what the population will be in the year 2000.

#9 reduce to 10^{-5} in 30 min find flow rate

13. Use the information in Exercise 12 to predict how long it will take (after 1960) for the population of the United States to double. (Assume conditions remain the same).

14. The world population in 1960 was 3 billion and in 1970 it was about 3.5 billion. Assuming all conditions remain the same, how long will it take (after 1960) for the world population to double?

15. Uranium 238 has half-life of 4.51×10^9 years. Assuming the age of the earth to be about 4 billion years, how much of the original amount of uranium 238 is still in existence?

16. The amount of a certain radioactive substance is reduced from 10 grams to 7 grams in 20 days. What is its half-life?

17. *(Carbon dating)* Carbon 14 is radioactive with a half-life of approximately 5,740 years. Living plants and animals reach a balance between the amount of intake of carbon 14 and the amount lost by its continual decay. When life ceases, so does the intake of carbon 14. By carefully measuring the amount of carbon 14 remaining in an artifact (say a piece of wood or the bones of an animal), it is possible to estimate when it died. A fallen tree from the eruption that formed Crater Lake in Oregon was found to contain only 44% of its original carbon 14. How old is Crater Lake?

18. The Dead Sea Scrolls are about 2,000 years old. How much of the original carbon 14 is still left in them?

19. If the Gross National Product (GNP) of a country increases at an annual rate of 4%, how long (years) will it take to double the present GNP?

20. Suppose a metal rod is immersed in a large body of water whose temperature is held constant at 20°C. If the rod cools from 50° to 40°C in 2 minutes, what is its temperature after 6 minutes?

21. A thermometer reading 70°C is plunged into a tub of water whose temperature is kept constant at

10°C. If the thermometer reads 60° after 30 seconds, what will it read after 2 minutes?

22. Suppose that 10 pounds of sugar is dumped into a large body of water and that it dissolves at a rate equal to the undissolved amount at any time t. How long (minutes) will it take for 80 percent of the sugar to dissolve?

23. A particle moves on a line and its acceleration at any time t is 1/5 of its velocity. If its initial velocity is 1 ft/sec, how far will it travel in the first 5 seconds?

24. *(Predator-prey model)* Suppose that a population consists of two classes, the *predators* and the *preys* (for example, foxes and rabbits). Let $x = x(t)$ and $y = y(t)$ be the number of predators and preys at time t. A biologist-mathematician A. J. Lotka (1880–1944) and a mathematician V. Volterra (1860–1940) developed a pair of equations that describe the relationship between x, y, and t

$$\frac{dy}{dt} = Ay - Bxy$$

$$\frac{dx}{dt} = -Cx + Dxy$$

where A, B, C, and D are positive constants. The first equation says that the prey population increases exponentially if there are no predators (if $x = 0$, then $dy/dt = Ay$), and it decreases in proportion to the product of x and y. The second equation says that the predator population decreases exponentially if there are no prey, and it increases in proportion to xy.

If the first equation is divided by the second, the result is the separable equation

$$\frac{dy/dt}{dx/dt} = \frac{dy}{dx} = \frac{(A - Bx)y}{(-C + Dy)x}$$

Find the general solution that expresses y implicitly as a function of x (do not attempt to solve for y). *There is a detailed discussion of the predator-prey model at the end of Chapter 10.*

7–5 DIFFERENTIATION AND INTEGRATION

To each derivative formula, there corresponds an integral formula. In this section, we use the derivative formula for the natural logarithm to obtain a new and useful integral formula. But first, to make the integral formula as general as possible, we extend the domain of the natural logarithm to include all nonzero numbers. At the end of this section, we also introduce the technique of logarithmic differentiation.

The natural logarithm is defined only for positive numbers. But the domain can be extended by defining a new function

$$f(x) = \ln |x| \quad x \neq 0$$

The graph of $y = \ln |x|$ is drawn in Figure 7–16.

If $x > 0$, then $f(x) = \ln x$ and $f'(x) = 1/x$. If $x < 0$, then $f(x) = \ln (-x)$ and

$$f'(x) = \frac{d}{dx} \ln (-x) = \frac{1}{-x} \frac{d}{dx}(-x) = \frac{1}{-x}(-1) = \frac{1}{x}$$

Thus, in either case

$$\frac{d}{dx} \ln |x| = \frac{1}{x} \quad x \neq 0$$

It follows from the chain rule that

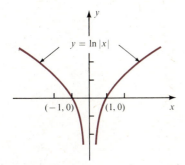

FIGURE 7–16. $y = \ln |x|$ is defined for all $x \neq 0$.

(7.38)

> If f is differentiable, then
>
> $$\frac{d}{dx} \ln |f(x)| = \frac{f'(x)}{f(x)} \quad f(x) \neq 0$$

EXAMPLE 1

(a) $\dfrac{d}{dx} \ln |x^2 - 3x - 4| = \dfrac{(x^2 - 3x - 4)'}{x^2 - 3x - 4}$

$$= \frac{2x - 3}{x^2 - 3x - 4}$$

(b) $\dfrac{d}{dx} \ln |\cos x| = \dfrac{-\sin x}{\cos x} = -\tan x$ ■

EXAMPLE 2

$$\frac{d}{dx} \ln \left| \frac{x - 1}{x + 1} \right| = \frac{d}{dx} \ln |x - 1| - \frac{d}{dx} \ln |x + 1|$$

$$= \frac{1}{x - 1} - \frac{1}{x + 1}$$

$$= \frac{2}{x^2 - 1} \quad ■$$

The corresponding integration formulas are

(7.39)
$$\int \frac{1}{x}\,dx = |x| + C \quad \text{and} \quad \int \frac{f'(x)}{f(x)}\,dx = |f(x)| + C$$

EXAMPLE 3

(a) $\displaystyle\int \frac{3}{x}\,dx = 3\int \frac{1}{x}\,dx = 3\ln|x| + C$

Check.

$$\frac{d}{dx}\,3\ln|x| = 3\frac{d}{dx}\ln|x| = \frac{3}{x}$$

(b) $\displaystyle\frac{2x}{x^2 - 1}$ is of the form $\displaystyle\frac{f'(x)}{f(x)}$ where $f(x) = x^2 - 1$.

Therefore,

$$\int \frac{2x}{x^2 - 1}\,dx = \ln|x^2 - 1| + C$$

Check.

$$\frac{d}{dx}\ln|x^2 - 1| = \frac{1}{x^2 - 1}(x^2 - 1)' = \frac{2x}{x^2 - 1} \qquad \blacksquare$$

If the integrand differs from those in (7.39) by a constant factor, then the *u-substitution* method can be used.

EXAMPLE 4   Find $\int x^2/(1 - x^3)\,dx$.

Solution   Set $u = 1 - x^3$; then $du = -3x^2\,dx$. Therefore,

$$\int \frac{x^2}{1 - x^3}\,dx = -\frac{1}{3}\int \underbrace{\frac{1}{1 - x^3}}_{u}\underbrace{(-3x^2)\,dx}_{du} = -\frac{1}{3}\int \frac{1}{u}\,du$$

$$= -\frac{1}{3}\ln|u| + C$$

$$= -\frac{1}{3}\ln|1 - x^3| + C$$

Check.

$$\frac{d}{dx}\left(-\frac{1}{3}\ln|1 - x^3|\right) = -\frac{1}{3}\left(\frac{1}{1 - x^3}\right)(1 - x^3)' = \frac{x^2}{1 - x^3}$$

Remark: If you recognize straight off that the numerator of the integrand above needs a factor of -3, then you can dispense with the *u*-substitution, and simply write

$$\int \frac{x^2}{1-x^3}\,dx = -\frac{1}{3}\int \frac{-3x^2}{1-x^3}\,dx = -\frac{1}{3}\ln|1-x^3| + C$$

But if you feel more confident using substitution, then, by all means, continue to do so. ■

EXAMPLE 5 Evaluate $\int_1^2 e^{3x}/(e^{3x}-1)\,dx$.

Solution Set $u = e^{3x} - 1$; then $du = 3e^{3x}\,dx$. Therefore,

$$\int \frac{e^{3x}}{e^{3x}-1}\,dx = \frac{1}{3}\int \frac{3e^{3x}}{e^{3x}-1}\,dx = \frac{1}{3}\int \frac{1}{u}\,du$$

$$= \frac{1}{3}\ln|u| + C$$

$$= \frac{1}{3}\ln|e^{3x}-1| + C$$

Check.

$$\frac{d}{dx}\left(\frac{1}{3}\ln|e^{3x}-1|\right) = \frac{1}{3}\left(\frac{1}{e^{3x}-1}\right)(e^{3x}-1)' = \frac{e^{3x}}{e^{3x}-1}$$

Now evaluate at the limits of integration.

$$\int_1^2 \frac{e^{3x}}{e^{3x}-1}\,dx = \left[\frac{1}{3}\ln|e^{3x}-1|\right]_1^2$$

$$= \frac{1}{3}[\ln|e^6-1| - \ln|e^3-1|]$$

$$\approx 1.016$$

Remark: The evaluation above could have been carried out using a change of variable; that is, changing the limits. When $x = 1$, then $u = e^{3x} - 1 = e^3 - 1 \approx 19.08$ and when $x = 2$, then $u = e^6 - 1 \approx 402.4$. Thus,

$$\int_1^2 \frac{e^{3x}}{e^{3x}-1}\,dx = \frac{1}{3}\int_{19.08}^{402.4} \frac{1}{u}\,du = \frac{1}{3}[\ln 402.4 - \ln 19.08] \approx 1.016 \quad ■$$

The use of formula (7.39) is not automatic every time the integrand is a quotient. The next example illustrates that the form of an integrand is often deceiving, and you must be alert to recognize the correct procedure for evaluating the integral.

EXAMPLE 6 Evaluate (a) $\int (\ln x)/x\,dx$ and (b) $\int 1/(x \ln x)\,dx$.

Solution

(a) Rewrite the integrand as $(\ln x)(1/x)$. Then let $u = \ln x$ and $du = (1/x)\,dx$. Thus,

$$\int \frac{\ln x}{x}\, dx = \int (\ln x)\left(\frac{1}{x}\right) dx = \int u\, du = \frac{u^2}{2} + C \qquad \text{(Power)}$$

$$= \frac{1}{2}[\ln x]^2 + C$$

Check.

$$\frac{d}{dx}\left[\frac{1}{2}(\ln x)^2\right] = \left(\frac{1}{2}\right)(2)(\ln x)(\ln x)' = \frac{\ln x}{x}$$

(b) Rewrite the integrand as $(1/x)/\ln x$. Again let $u = \ln x$ and $du = (1/x)\, dx$. Then

$$\int \frac{1}{x \ln x}\, dx = \int \frac{(1/x)}{\ln x}\, dx = \int \frac{1}{u}\, du = \ln |u| + C$$

$$= \ln |\ln x| + C$$

Check.

$$\frac{d}{dx} \ln |\ln x| = \frac{1}{\ln x}(\ln x)' = \frac{1}{x \ln x} \qquad \blacksquare$$

Corresponding to the derivative formula $[e^f]' = f'e^f$ is the integral formula

(7.40) $$\boxed{\int f'(x)e^{f(x)}\, dx = e^{f(x)} + C}$$

EXAMPLE 7

(a) $\displaystyle \int 5e^{5x}\, dx = e^{5x} + C$

(b) $\displaystyle \int 2xe^{x^2}\, dx = e^{x^2} + C \qquad \blacksquare$

EXAMPLE 8 Evaluate $\int e^{\sqrt{x}}/\sqrt{x}\, dx$.

Solution Let $u = \sqrt{x}$ and $du = 1/2\sqrt{x}\, dx$. Then

$$\int \frac{e^{\sqrt{x}}}{\sqrt{x}}\, dx = 2 \int \underbrace{e^{\sqrt{x}}}_{e^u} \underbrace{\left(\frac{1}{2\sqrt{x}}\right) dx}_{du} = 2 \int e^u\, du = 2e^u + C$$

$$= 2e^{\sqrt{x}} + C$$

Check.

$$\frac{d}{dx} 2e^{\sqrt{x}} = 2e^{\sqrt{x}}(\sqrt{x})' = \frac{e^{\sqrt{x}}}{\sqrt{x}} \qquad \blacksquare$$

EXAMPLE 9   Evaluate $\int_0^{\ln 2} e^x \sqrt{e^x - 1}\, dx$.

Solution   Use a u-substitution keeping the same limits of integration or use a change of variable. Let $u = e^x - 1$ and $du = e^x\, dx$. To change variables, observe that when $x = 0$, $u = e^0 - 1 = 0$; when $x = \ln 2$, $u = e^{\ln 2} - 1 = 2 - 1 = 1$. Thus,

$$\int_0^{\ln 2} e^x \sqrt{e^x - 1}\, dx = \int_0^1 u^{1/2}\, du = \frac{2}{3} u^{3/2} \Big|_0^1 = \frac{2}{3} \qquad \blacksquare$$

Logarithmic Differentiation

To differentiate a complicated product

$$y = f_1(x) f_2(x) \ldots f_n(x)$$

it sometimes helps to first take the logarithm of both sides,

$$\ln |y| = \ln |f_1(x)| + \ln |f_2(x)| + \cdots + \ln |f_n(x)|$$

and then differentiate

$$\frac{y'}{y} = \frac{f_1'(x)}{f_1(x)} + \frac{f_2'(x)}{f_2(x)} + \cdots + \frac{f_n'(x)}{f_n(x)}$$

This method is called **logarithmic differentiation.**

EXAMPLE 10   Use logarithmic differentiation to find y' if

$$y = x(x + 1)^2 (x^2 + 1)^3$$

Solution   First write

$$\ln |y| = \ln |x| + 2\ln |x + 1| + 3\ln |x^2 + 1|$$

Then differentiate

$$\frac{y'}{y} = \frac{1}{x} + \frac{2}{x + 1} + \frac{6x}{x^2 + 1}$$

Now multiply both sides by y to obtain the derivative

$$y' = [x(x + 1)^2 (x^2 + 1)^3] \left[\frac{1}{x} + \frac{2}{x + 1} + \frac{6x}{x^2 + 1} \right] \qquad \blacksquare$$

The same method can also be used for quotients.

EXAMPLE 11   Find y' if

$$y = \frac{\sqrt{x^2 + 1}}{(x - 1)^3}$$

Solution   First write

$$\ln|y| = \frac{1}{2}\ln|x^2 + 1| - 3\ln|x - 1|$$

Then differentiate

$$\frac{y'}{y} = \frac{x}{x^2 + 1} - \frac{3}{x - 1}$$

and conclude with

$$y' = \frac{\sqrt{x^2 + 1}}{(x - 1)^3}\left[\frac{x}{x^2 + 1} - \frac{3}{x - 1}\right] \quad \blacksquare$$

Remark: Logarithmic differentiation not only simplifies finding the derivative in certain cases, but it also provides the answer in a "nearly factored" form that is easy to work with.

EXERCISES

In Exercises 1–20, find y' if y is the given function. In some cases, logarithmic differentiation will make the computations easier.

1. $\ln\left|\frac{2}{x}\right|$

2. $\ln\left|\frac{x}{4}\right|$

3. $\ln(x^2 - 5x + 2)$

4. $\ln|4x^3 - 3x|$

5. $x(x^2 - 3)^2\sqrt{x^2 + 1}$

6. $(x - 1)(x - 2)(x - 3)$

7. $\ln|x(x^2 - 3)^2\sqrt{x^2 + 1}|$

8. $\ln|(x - 1)(x - 2)(x - 3)|$

9. $\ln\left[\frac{x\sqrt{x^2 + 4}}{(x + 2)^3}\right]$

10. $\ln\left[\frac{e^x(x + 1)}{x + 5}\right]$

11. $\frac{x\sqrt{x^2 + 4}}{(x + 2)^3}$

12. $\frac{e^x(x + 1)}{x + 5}$

13. $\frac{xe^x\cos x}{x + 1}$

14. $\frac{(x + 1)\sqrt{x^2 - 9}}{x - 1}$

15. $[\ln(x^2 + 1)]^4$

16. $[\ln(x^2 - 1)]^3$

17. $e^{7x}\ln x$

18. $(\sin x)(\ln|x|)$

19. $\ln\sqrt{\left|\frac{x + 1}{x - 2}\right|}$

20. $\frac{\ln(x + 1)}{x + 1}$

Evaluate the integrals in Exercises 21–40. Check your answers.

21. $\int \frac{x}{4 - x^2}\, dx$

22. $\int \frac{1}{1 + 7x}\, dx$

23. $\int \frac{x}{(4 - x^2)^2}\, dx$

24. $\int \frac{1}{\sqrt{1 + 7x}}\, dx$

25. $\int_2^3 \frac{x^2}{x^3 + 1}\, dx$

26. $\int_1^4 \frac{1}{x + 2}\, dx$

27. $\int \frac{x + 4}{x^2 + 8x + 9}\, dx$

28. $\int \frac{(x - 1)^2}{x^3 - 3x^2 + 3x}\, dx$

29. $\int \frac{(\ln x)^3}{x}\, dx$

30. $\int \frac{\ln(x + 8)}{x + 8}\, dx$

31. $\int \frac{\sqrt{x}}{1 + x\sqrt{x}}\, dx$

32. $\int \frac{1}{x(\ln x)^2}\, dx$

33. $\int_0^5 e^{-2x}\, dx$

34. $\int_{-2}^0 xe^{x^2}\, dx$

35. $\int \frac{e^x + 4}{e^x}\, dx$

36. $\int \frac{x^2 + 1}{x}\, dx$

37. $\int \frac{e^x}{e^x + 4}\, dx$

38. $\int \frac{x}{x^2 + 1}\, dx$

39. $\int \frac{e^x}{(e^x + 4)^2}\, dx$

40. $\int \frac{(e^x + 4)^2}{e^x}\, dx$

41. Find the area of the region between the curves $y = e^{-x}$ and $y = 1/x$ from $x = 1$ to $x = 2$.

42. The region bounded by $y = e^{-x^2}$, $y = 0$, $x = 0$, and $x = 1$ is revolved about the y-axis. Find the volume generated.

43. The base of a solid is the region bounded by $y = 1/\sqrt{x}$, $y = 0$, $x = 1$, and $x = 2$. If each cross-section perpendicular to the x-axis is a square, what is the volume?

44. A particle moves along a line with acceleration $a(t) = 1/(t + 2)^2$ ft/sec². If its initial velocity is 3 ft/sec, how far will it travel in the first 2 seconds?

45. Neither of the integrals

$$\int_0^{\ln 2} 2\pi x (2 - e^x)\, dx \quad \text{and} \quad \int_1^2 \pi (\ln y)^2 \, dy$$

can be evaluated with the methods discussed so far. But they both represent the volume of a certain solid, and, therefore, have equal values. Find the solid.

7-6   THE FUNCTIONS a^x AND $\log_a x$

In elementary mathematics, exponents have a precise meaning only if they are rational. Although numbers such as

$$10^\pi \quad \text{and} \quad 2^{\sqrt{2}}$$

are *assumed* to exist, their values are never discussed. But now our knowledge of natural logarithms and exponentials can be used to assign exact values to them.

Exponentials with Base a

Suppose that $a > 0$ and r is rational. Then

$$a^r = e^{\ln a^r} = e^{r \ln a}$$

Motivated by this equality, we make the following definition:

DEFINITION OF a^x

If $a > 0$ and x is *any* real number, then

(7.41) $$a^x = e^{x \ln a}$$

The function $y = a^x$ is called the **exponential with base** a.

Thus, $10^\pi = e^{\pi \ln 10}$ and $2^{\sqrt{2}} = e^{\sqrt{2} \ln 2}$.

The first thing to notice is that the third property of logarithms (7.10c) is now valid for any real exponent. For x positive and y *any* number, $x^y = e^{y \ln x}$; therefore,

(7.42) $$\ln x^y = \ln(e^{y \ln x}) = y \ln x$$

Now we can show that the usual properties of exponents hold for any $a > 0$ and all x and y.

(7.43)

(a) $a^x a^y = a^{x+y}$

(b) $a^x / a^y = a^{x-y}$

(c) $(a^x)^y = a^{xy}$

For example, to prove part (c), write

$$(a^x)^y = e^{y \ln a^x} \qquad (7.41)$$

$$= e^{xy \ln a} \qquad (7.42)$$

$$= a^{xy} \qquad (7.41)$$

The derivative of a^x is obtained by rewriting it as $e^{x \ln a}$.

$$\frac{d}{dx} a^x = \frac{d}{dx} e^{x \ln a} = e^{x \ln a}(x \ln a)'$$

$$= (\ln a)(e^{x \ln a})$$

$$= (\ln a)a^x$$

Invoking the chain rule, we have

$$(7.44) \quad \boxed{\begin{array}{l} \text{(a)} \ \dfrac{d}{dx} a^x = (\ln a)a^x \\[2mm] \text{(b)} \ \text{If } f \text{ is differentiable, then} \\[2mm] \qquad \dfrac{d}{dx} a^{f(x)} = (\ln a)a^{f(x)}f'(x) \end{array}}$$

EXAMPLE 1

(a) $\dfrac{d}{dx} 3^x = (\ln 3)3^x$

(b) $\dfrac{d}{dx} 2^{\sin x} = (\ln 2)2^{\sin x}(\cos x)$ ∎

WARNING: It is important to keep in mind the distinction between functions like a^x and x^a. The first is an *exponential* with derivative $(\ln a)a^x$; the second is a *power* with derivative ax^{a-1}.

EXAMPLE 2 If $y = 2^{(x^2 + x)} + (x^2 + x)^2$, then

$$y' = (\ln 2)2^{(x^2 + x)}(2x + 1) + 2(x^2 + x)(2x + 1) \quad ∎$$

An example of a third type of exponential function is x^x with $x > 0$; it is different from a^x and x^a in that the base and the exponent *both* vary.

EXAMPLE 3 To differentiate $y = x^x$, write it as $y = e^{x \ln x}$. Then

$$y' = [e^{x \ln x}]' = e^{x \ln x}[x \ln x]'$$

$$= e^{x \ln x}\left[x\left(\frac{1}{x}\right) + (1) \ln x\right]$$

$$= x^x(1 + \ln x)$$

Remark: The derivative can also be found with logarithmic differentiation.

$$y = x^x$$

$$\ln y = \ln x^x = x \ln x \tag{7.42}$$

$$\frac{y'}{y} = x\left(\frac{1}{x}\right) + (1)\ln x$$

$$y' = y(1 + \ln x)$$

$$= x^x(1 + \ln x) \qquad \blacksquare$$

The corresponding integral formulas are

$$(7.45) \qquad \boxed{\begin{aligned} &\text{(a)} \int a^x \, dx = \frac{1}{\ln a} a^x + C \\[2mm] &\text{(b)} \int a^{f(x)} f'(x) \, dx = \frac{1}{\ln a} a^{f(x)} + C \end{aligned}}$$

EXAMPLE 4

(a) $\displaystyle \int 4^x \, dx = \frac{1}{\ln 4} 4^x + C$

(b) $\displaystyle \int \frac{5^{\sqrt{x}}}{2\sqrt{x}} \, dx = \frac{1}{\ln 5} 5^{\sqrt{x}} + C \qquad \blacksquare$

EXAMPLE 5   Evaluate $\int 10^{x^2} x \, dx$.

Solution   Use a u-substitution; let $u = x^2$ and $du = 2x \, dx$. Or, if you immediately recognize that a factor of 2 is required in the integrand, simply write

$$\int 10^{x^2} x \, dx = \frac{1}{2} \int 10^{x^2} (2x) \, dx = \frac{1}{2 \ln 10} 10^{x^2} + C$$

Check.

$$\frac{d}{dx}\left[\frac{1}{2 \ln 10} 10^{x^2} \right] = \frac{1}{2 \ln 10} \frac{d}{dx} 10^{x^2} = \frac{1}{2 \ln 10}[(\ln 10) \, 10^{x^2}(2x)]$$

$$= 10^{x^2} x \qquad \blacksquare$$

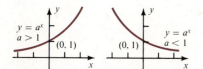

FIGURE 7-17. The graphs of $y = a^x$ for $a > 1$ and $a < 1$. Notice that, in either case, $a^x > 0$ for all x.

The graphs of $y = a^x$ for $a > 1$ and $a < 1$ are shown in Figure 7-17.

Logarithms with Base a

Just as $\ln x$ and e^x are inverses, so are $\log_a x$ and a^x.

(7.46)

DEFINITION OF $\log_a x$
If $a > 0$, then the **logarithm with base** a is denoted by $\log_a$, and for all positive x,

$$\log_a x = y \quad \text{means that} \quad a^y = x$$

It follows that

(7.47)

$$a^{\log_a x} = x \text{ for } x > 0, \text{ and}$$
$$\log_a a^x = x \text{ for all } x$$

Moreover, the usual properties of logarithms hold for all positive x and y.

(7.48)

(a) $\log_a xy = \log_a x + \log_a y$

(b) $\log_a \dfrac{x}{y} = \log_a x - \log_a y$

(c) $\log_a x^y = y \log_a x$

There is an interesting relationship between $\ln$ and $\log_a$. Since $x = a^{\log_a x}$, it follows that

$$\ln x = \ln a^{\log_a x} = (\log_a x)(\ln a)$$

Solving this equation for $\log_a x$ yields

(7.49)

$$\log_a x = \frac{\ln x}{\ln a}$$

The derivative of $\log_a x$ can be found by differentiating both sides of (7.49). Because $\ln a$ is a constant, we have

$$\frac{d}{dx} \log_a x = \frac{1}{x \ln a}$$

The same considerations that led to enlarging the domain of $\ln x$ to include all nonzero x also apply to $\log_a x$. Thus,

(a) $\dfrac{d}{dx} \log_a |x| = \dfrac{1}{x \ln a}$

(7.50)

(b) If f is differentiable and $f(x) \neq 0$, then

$$\frac{d}{dx} \log_a |f(x)| = \frac{f'(x)}{f(x) \ln a}$$

EXAMPLE 6   Find the derivative of $y = \log_2 \sqrt{|x^2 - 4|}$.

Solution

$$y = \frac{1}{2} \log_2 |x^2 - 4| \tag{7.48}$$

$$y' = \frac{1}{2} \left[\frac{2x}{(x^2 - 4) \ln 2} \right]$$

$$= \frac{x}{(x^2 - 4) \ln 2} \quad \blacksquare$$

EXAMPLE 7   Find y' if $y = \log_{10} |(x + 1)/(x - 1)|^3$.

Solution

$$y = 3[\log_{10} |x + 1| - \log_{10} |x - 1|] \tag{7.48}$$

$$y' = 3 \left[\frac{1}{(x + 1) \ln 10} - \frac{1}{(x - 1) \ln 10} \right] \tag{7.50}$$

$$= \frac{-6}{(x^2 - 1) \ln 10} \quad \blacksquare$$

We close this section with a proof that the base of the natural logarithm really is the number e introduced in Chapter 3. That is,

(7.51)

$$e = \lim_{h \to 0} (1 + h)^{1/h}$$

First, observe that

$$\lim_{h \to 0} \ln (1 + h)^{1/h} = \lim_{h \to 0} \frac{1}{h} \ln (1 + h)$$

(7.52)
$$= \lim_{h \to 0} \frac{\ln (1 + h) - \ln 1}{h} \qquad (\ln 1 = 0)$$

$$= 1 \qquad (\ln'(1) = 1)$$

Then write,

$$(1 + h)^{1/h} = e^{\ln(1 + h)^{1/h}} \tag{7.41}$$

and take the limit of each side

$$\lim_{h \to 0}(1 + h)^{1/h} = \lim_{h \to 0} e^{\ln(1+h)^{1/h}}$$
$$= e^{\lim_{h \to 0} \ln(1+h)^{1/h}}$$
$$= e \qquad\qquad (7.52)$$

Switching limits in the second to last line is legitimate because of the Composition Theorem (2.19) in Section 2–3 and the fact that e^x is differentiable, hence continuous, at each x. This proves (7.51). The number e, correct to 15 decimal places is

$$e = \lim_{h \to 0}(1 + h)^{1/h} \approx 2.718281828459045$$

EXERCISES

In Exercises 1–22, find y' if y is the given function.

1. 6^x
2. π^x
3. $4^{\sqrt{x}}$
4. 5^{x^2}
5. $3^{\sin x}$
6. $9^{\cos x}$
7. $\log_5(x^2 - 3x + 1)$
8. $\log_8|\tan x|$
9. $\log_7\sqrt{\dfrac{x+1}{x-1}}$
10. $\log_3\sqrt[3]{x^2 - 5}$
11. $10^{\sqrt{x^2+1}}$
12. $9^{(x^2+2)}$
13. $\log_3|\ln x|$
14. $\ln(\log_3 x)$
15. $x^2 2^x$
16. $x^{\sqrt{2}}\sqrt{2^x}$
17. $(x+1)^{(x+1)}$
18. $x^{\sin x}$
19. $(\sin x)^x$
20. $(2^x + x^2)^x$
21. x^{x^x}
22. $(x^2+1)^{\ln x}$

Evaluate the integrals in Exercises 23–34.

23. $\displaystyle\int 2^x\,dx$
24. $\displaystyle\int 10^x\,dx$
25. $\displaystyle\int 3^{x^2}x\,dx$
26. $\displaystyle\int 4^{-x^2}x\,dx$
27. $\displaystyle\int_{-2}^0 6^{-2x}\,dx$
28. $\displaystyle\int_0^4 2^{3x}\,dx$

29. $\displaystyle\int \frac{10^x}{\sqrt{1+10^x}}\,dx$
30. $\displaystyle\int 10^x\sqrt{1+10^x}\,dx$
31. $\displaystyle\int_0^1 \frac{10^x}{1+10^x}\,dx$
32. $\displaystyle\int_0^1 \frac{10^x}{(1+10^x)^2}\,dx$
33. $\displaystyle\int \frac{1}{x\log_2 x}\,dx$
34. $\displaystyle\int \frac{e^x + e^{-x}}{e^x - e^{-x}}\,dx$

35. Find x if $3^x = 2^{x+1}$.
36. Find x if $\log_2 x = 1 + \log_3 x$.
37. Find the area of the region bounded by the graphs of $y = 2^x$, $y = 3^x$, and $x = 1$.
38. Compute the volume generated by revolving the region bounded by the graphs of $y = 2^x - 1$, $y = 0$, and $x = 1$ about the x-axis.

39. Show that $a^x a^y = a^{x+y}$ and $a^x/a^y = a^{x-y}$ for $a > 0$ and all x and y.
40. Show that the properties of logarithms (7.48) hold for $a > 0$ and all x and y.

7–7 ADDITIONAL APPLICATIONS *(Optional)*

In the polluted lake example of Section 7–4, it is assumed that no new pollutants are added. The resulting differential equation

$$y' = -\frac{I}{V}y$$

is separable and has the solution $y = y(0)e^{-(I/V)t}$, where $y(t)$ is the pollution concentration at time t. But a more realistic model is obtained if new pollutants are allowed to enter the lake through the inlet. If $g(t)$ represents the concentration of pollutants added at time t, the resulting differential equation turns out to be

$$y' = -\frac{I}{V}y + g(t)$$

Equations of this type may or may not be separable. If it is not separable, then solving it requires a new technique.

An equation

(7.53) $y' + f(x)y = g(x)$

is called a (first order) **linear differential equation.** We assume that f and g are continuous functions of x. To solve (7.53), we multiply both sides of the equation by $e^{\int f(x)\,dx}$.

The expression $e^{\int f(x)\,dx}$ is called an *integrating factor.* Multiplying both sides of (7.53) by this integrating factor *always* transforms the left side into the derivative of a product

$$y'e^{\int f(x)\,dx} + yf(x)e^{\int f(x)\,dx} = \frac{d}{dx}[ye^{\int f(x)\,dx}] \qquad\qquad (Check!)$$

Therefore, (7.53) is equivalent to

(7.54) $d[ye^{\int f(x)\,dx}] = [g(x)e^{\int f(x)\,dx}]\,dx$

The general solution of (7.53) is obtained by integrating both sides of (7.54).*

EXAMPLE 1   Solve the equation $y' + (1/x)y = 1$.

Solution   This is a linear equation (7.53) with $f(x) = 1/x$ and $g(x) = 1$; it is not separable. Multiply both sides of the equation by

$$e^{\int f(x)\,dx} = e^{\int 1/x\,dx} = e^{\ln x} = x \qquad\qquad \text{(Do not add a constant of integration here)}$$

to obtain

(7.55) $xy' + y = x$

Notice that the left side of (7.55) is the derivative of the product xy; $(xy)' = xy' + y$. Therefore, (7.55) can be written as

$$\frac{d}{dx}(xy) = x$$

---

*Multiplying both sides of a linear differential equation by $e^{\int f(x)\,dx}$ to make the left side the derivative of a product is an ingenious device. Although it was used extensively by Leonard Euler (1707–1783), it is not known (at least to this author) if he actually discovered the method. But whoever did discover it deserves our greatest admiration.

or

$$d(xy) = x\,dx$$

Integrating both sides of this equation yields

$$xy = \frac{x^2}{2} + C \qquad\qquad \text{(Add the constant here)}$$

and it follows that

$$y = \frac{x}{2} + \frac{C}{x}$$

is the general solution of $y' + (1/x)y = 1$.

Check.

$$\left(\frac{x}{2} + \frac{C}{x}\right)' + \frac{1}{x}\left(\frac{x}{2} + \frac{C}{x}\right) = \frac{1}{2} - \frac{C}{x^2} + \frac{1}{2} + \frac{C}{x^2} = 1 \qquad \blacksquare$$

EXAMPLE 2   The equation $y' + 2xy = 2x$ is both linear and separable. Find its solution two ways and compare the answers.

Solution   *Linear.* In this linear equation, $f(x) = 2x$ and $g(x) = 2x$. The integrating factor is

$$e^{\int 2x\,dx} = e^{x^2} \qquad\qquad \text{(No constant here)}$$

Multiplying through, we have

$$y'e^{x^2} + 2xye^{x^2} = 2xe^{x^2}$$
$$d[\,ye^{x^2}] = 2xe^{x^2}\,dx \qquad\qquad (7.54)$$

Integrating both sides yields

$$ye^{x^2} = e^{x^2} + C \qquad\qquad \text{(Add constant here)}$$
$$(7.56) \qquad y = 1 + Ce^{-x^2}$$

This is the general solution obtained by the linear method.

Separable. The equation $y' + 2xy = 2x$ is separable because $y' = 2x - 2xy = 2x(1 - y)$; so

$$\frac{y'}{1-y} = 2x \quad\text{or}\quad \frac{1}{1-y}\,dy = 2x\,dx$$

Integration yields $-\ln|1 - y| = x^2 + C$, which can be solved for y by writing

$$\ln|1 - y| = -x^2 - C$$
$$|1 - y| = e^{(-x^2 - C)} = e^{-C}e^{-x^2}$$
$$(7.57) \qquad y = 1 \pm e^{-C}e^{-x^2}$$

If we set $K = \pm e^{-C}$, a constant, then the solution here, $y = 1 + Ke^{-x^2}$, agrees with the solution (7.56) obtained above.* ■

Let us now see how linear differential equations can be used to solve pollution and similar problems.

EXAMPLE 3 *(Polluted lake — revisited)* A lake has a constant volume V with I cu ft of water per day flowing in and out. The waste from several factories enters the lake with a pollution concentration of p lbs per cu ft. On a certain day $(t = 0)$, the lake has a concentration of $3p$ lbs per cu ft. If the authorities order the factories to dispose of 50% of their waste elsewhere, what will the concentration be after t days?

Solution Let $y = y(t)$ be the concentration of pollutants in the lake at time t days. Pollutants are flowing out of the lake at the rate of $(I/V)y$ lbs per cu ft per day, and into the lake at the rate of $0.5pI/V$ lbs per cu ft per day. Therefore, the rate of change of y can be described by the differential equation

$$y' = -\frac{I}{V}y + \frac{0.5pI}{V}$$

Set $K = I/V$ and rewrite the equation as

$$y' + Ky = 0.5pK$$

This is a linear equation with $f(t) = K$ and $g(t) = 0.5pK$. (It is also separable, but we will use an integrating factor to solve it.) The integrating factor is

$$e^{\int K dt} = e^{Kt}$$

Thus,

$$y'e^{Kt} + e^{Kt}Ky = 0.5pKe^{Kt}$$
$$d(e^{Kt}y) = 0.5pKe^{Kt}dt$$
$$e^{Kt}y = 0.5pe^{Kt} + C \qquad \text{(Integrate)}$$
$$y = 0.5p + Ce^{-Kt}$$

This is the general solution (*check*). It is given that $y(0) = 3p$, so

(7.58) $\quad y = 0.5p + 2.5pe^{-It/V}$ $\hfill (K = I/V)$

is the concentration at time t. ■

EXAMPLE 4 How long will it take for the pollution level of the lake in Example 3 to drop to a concentration of $2p$?

Solution Using (7.58) you want to solve the equation

---

*Technically, there is a difference. In the linear solution (7.56), C can be 0, in which case, $y = 1$. In the separable solution, y can never take the value 1 because, in separating the variables, we divided by $1 - y$. This is reflected in the solution (7.57) by the fact that $e^{-C}e^{-x^2}$ is never 0.

$$2p = 0.5p + 2.5pe^{-It/V}$$

for t. Rewrite the equation as

$$\frac{2p - 0.5p}{2.5p} = e^{-It/V}$$

and take the natural logarithm of both sides

$$\ln\frac{1.5}{2.5} = -It/V$$

$$t = -\frac{V(\ln 1.5 - \ln 2.5)}{I}$$

$$\approx \frac{0.5V}{I} \text{ days} \quad \blacksquare$$

A similar type of problem is encountered in free-falling bodies when air resistance (*drag*) is taken into consideration. Suppose that the drag D is proportional to the velocity; that is, $D = kv, k > 0$. The acceleration due to gravity (32 ft/sec^2) is impeded by D. The resulting differential equation is

(7.59)
$$v' = 32 - kv \qquad\qquad (\text{Acceleration} = v')$$
$$v' + kv = 32$$

which is linear with $f(x) = k$ and $g(x) = 32$.

EXAMPLE 5 An object is dropped from a stationary balloon. If the drag is $0.1v$, find the velocity at any time t.

Solution The velocity function satisfies equation (7.59) with $k = 0.1$. The integrating factor is $e^{0.1t}$ and

$$v'e^{0.1t} + e^{0.1t}(0.1)v = 32e^{0.1t}$$
$$d(e^{0.1t}v) = 32e^{0.1t}\, dt$$
$$e^{0.1t}v = 320e^{0.1t} + C$$
$$v = 320 + Ce^{-0.1t}$$

Because the object is dropped, you know that $v(0) = 0$, so $C = -320$. Therefore,

$$v = 320(1 - e^{-0.1t})$$

at any time t. $\blacksquare$

EXERCISES

In Exercises 1–20, indicate whether the given equation is separable and/or linear, or neither. Then solve the ones that are separable or linear. Check your answers.

1. $2y' - 8y = 6$

2. $y' - 5y = 0$

3. $3y' - xy = 0$

4. $3y' + 6y = -2$

5. $y' = \dfrac{1}{x + y}$

6. $y' = \dfrac{3x}{1 + y}$

7. $s' = t(1 - s)$

8. $\sqrt{st} = 3s'$

9. $y' + \dfrac{y}{x} = x$

10. $y' + 2y = e^x$

11. $y' + y = e^{-x}$

12. $xy' = 2(1 - y)$

13. $(1 + e^x)y' + e^x y = 1$

14. $y' + y = \dfrac{1}{1 + e^x}$

15. $s' = \dfrac{t}{1 + s}$

16. $s' = \dfrac{s + t}{s - t}$

17. $(1 + x)y' + 2y = 4(1 + x)$

18. $t \, ds = (s + 1) \, dt$

19. $y' - xy = x$

20. $y' + \dfrac{y}{x} = \dfrac{1}{x^2}$

---

21. The concentration of pollutants in a 5×10^6 cu ft reservoir is 0.01 lbs per cu ft. The polluted water is drained off at the rate of 10^5 cu ft per day and replaced with water containing 0.001 lbs per cu ft of pollutants. When will the pollution level reach 0.005 lbs per cu ft?

22. Suppose that a pollutant-free lake has a constant volume of 10^9 cu ft with a constant intake (and outflow) of 10^5 cu ft/day. A factory is built nearby with the result that the intake will now contain a concentration of pollutants of 10^{-6} lbs per cu ft. Assuming the pollutants are uniformly distributed, how long will it take for the lake to have a concentration of 10^{-8} of pollutants?

23. A 500-gallon tank contains a brine with 1/2 lbs of salt per gallon. If the brine is drawn off at the rate of 50 gallons per minute and replaced with a brine containing 1/10 lbs of salt per gallon, what is the concentration of salt at time t?

24. *(Aerodynamics)* Show that the maximum velocity (called *terminal velocity*) of a freely falling object, with drag $D = kv$, cannot be greater than $32/k$ (take positive direction as down).

25. (RL *Circuits*) In the electric circuit pictured in Figure 7–18, E is the voltage, R the resistance, and L the inductance. If $I = I(t)$ is the current in the cir-

cuit at time t, it follows from Kirchoff's first rule that

$$L\frac{dI}{dt} + RI = E \quad L, R, \text{ and } E \text{ constant}$$

(a) Suppose that $I(0) = E/R$. Show that the inductance L does not affect the circuit. What is $I(t)$ in this case?

(b) What is the behavior of $I(t)$ if $I(0) > E/R$? If $I(0) < E/R$?

(c) If $I(0) = 0$, what is the largest possible value of $I(t)$? Show that $I(t)$ is (about) 63% of its maximal value when $t = L/R$, called the *time constant* of the circuit.

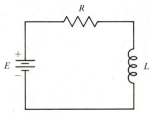

FIGURE 7–18. Exercise 25. A typical *RL* circuit.

26. A 10-gallon tank is full of a solution containing 5 lbs of sugar. A solution containing 3 lbs of sugar per gallon is added at the rate of 2 gallons per minute. The solutions mix instantly and the new solution is drained off at a rate of 3 gallons per minute. (a) When will the tank be emptied? (b) What is the concentration of sugar in the tank at time t?

27. *(Cotton mill — revisited)* Suppose the cotton mill of Exercise 9 in Section 7–4 is lint-free on a certain morning. When the milling machines are started up, they produce 0.2 cu ft of lint per minute. How many exhaust fans, each moving 1,000 cu ft of air per minute are needed to keep the concentration at about the level of 10^{-5}?

REVIEW EXERCISES

In Exercises 1–4, find the inverses of the given functions. Then sketch the graphs of f and f^{-1} on the same set of axes.

1. $f(x) = 3x - 4$

2. $f(x) = \sqrt{x + 1}$

3. $f(x) = \sqrt{x} + 2$

4. $f(x) = 1 - x^3$

In Exercises 5–8, let $g = f^{-1}$ and evaluate $g'(f(x))$ for the specified value of x in the *domain* of f.

5. $f(x) = x^3 + 3x + 1; x = 0$

6. $f(x) = \sqrt{3x + 1}; x = 1$

7. $f(x) = \int_1^x \frac{1}{3 + t^2} dt; x = 2$

8. $f(x) = \int_0^x t^2 dt; x = 4$

In Exercises 9–12, let $g = f^{-1}$ and evaluate $g'(y)$ for the specified value of y in the *range* of f.

9. $f(x) = x^3 + 3x + 1; y = 5$

10. $f(x) = \sqrt{x}; y = 3$

11. $f(x) = \int_0^x \frac{1}{1 + t^2} dt; y = 0$

12. $f(x) = \cos x, 0 \le x \le \pi; y = 1/2$

In Exercises 13–30, find y' if y is the given function.

13. $\ln \frac{x^2 - 1}{x^2 + 1}$

14. $[\ln(3x + 4)]^4$

15. $\ln|\sin x|$

16. $e^x \ln|x|$

17. $e^{\sin t}$

18. $\ln \sqrt[3]{t + 1}$

19. $\ln \sqrt[3]{(x^2 + 1)(x^2 + 2)}$

20. $\ln \sqrt{1 + e^x}$

21. xe^{x^2}

22. $e^x \sin x$

23. 3^t

24. $\ln|\cos e^t|$

25. x^x

26. $(x + 1)^x$

27. $9^{\sqrt{x}}$

28. $10^{\sqrt{x^2 + 1}}$

29. $x^4 4^x$

30. $(2^x)^x$

In Exercises 31–44, evaluate the integrals. *Check your answers.*

31. $\int \frac{1}{1 + 8x} dx$

32. $\int_1^2 \frac{x^2}{x^3 + 1} dx$

33. $\int \frac{1}{x(\ln x)^2} dx$

34. $\int \frac{\sqrt{x}}{1 + x\sqrt{x}} dx$

35. $\int \frac{x^4 + 1}{x^3} dx$

36. $\int \frac{e^x}{e^x + 1} dx$

37. $\int \frac{x^3}{x^4 + 1} dx$

38. $\int \frac{e^x + 1}{e^x} dx$

39. $\int_1^2 \frac{\ln x}{x} dx$

40. $\int \frac{x + 3}{x^2 + 6x + 1} dx$

41. $\int 3^{-2x} dx$

42. $\int x4^{x^2} dx$

43. $\int \frac{9^x}{1 + 9^x} dx$

44. $\int \frac{1}{x \log_3 x} dx$

In Exercises 45–48, sketch the graphs.

45. $y = \ln|x|$

46. $y = \ln x^2$

47. $y = xe^x$

48. $y = e^{-x^2}$

In Exercises 49–56, find the general solutions, or the particular solutions if a side condition is given. *Check your answers. Exercises 53–56 are optional.*

49. $y' = -3y; y(0) = 1$

50. $y' = 2y$

51. $y' - y = 1$

52. $y' = y(1 + x); y(0) = 2$

53. $s' = 2s + e^t$

54. $s' + s = e^{-t}$

55. $y' + \frac{y}{10 - x} = 10 - x$

56. $y' - 2xy = x$

57. *(Atmospheric pressure)* At constant temperature, the rate of change of atmospheric pressure p with respect to altitude h is proportional to p. If the pressure at sea level is 15 lbs/sq in and the pressure at 10,000 ft is 10 lbs/sq in, what is the pressure at 20,000 ft?

58. An outdoor thermometer reads 20°F. It is brought into a room whose constant temperature is 70°. After 10 minutes it reads 40°. When will it read 60°?

59. A particle moves along a line so that its acceleration at any time t equals $v(t) + 6$ ft/sec^2, where v is its velocity. If it starts from rest, how far will it move in 3 seconds?

60. Find the area of the region bounded by the curves $y = 1/x, y = 1/x^2$, and the line $x = 2$.

61. The region determined by $y = 2^x$ from $x = 0$ to $x = 1$ is revolved about the x-axis. Find the volume generated.

62. Find the minimum value of $y = e^{kx} + e^{-kx}$; k is some constant.

63. *(Continuous compounding)* If P dollars are invested at $r\% (= r/100)$ and compounded m times a year, then the accumulated amount A after t years is

$$A = P\left(1 + \frac{r}{100m}\right)^{mt}$$

If m is allowed to increase without bound, then the interest is said to **compound continuously.** Set $h = r/100m$ and show that the accumulated amount after t years is given by the formula

$$A = Pe^{rt/100}$$

64. *(Cost of real estate)* At the time of writing, the cost of housing was reported to be increasing at the rate of 7% per year (continuous compounding). If this rate continues and the cost of a house is now $50,000, what will it be 5 years later? (See Exercise 63 above.)

65. *(Optional)* A ruptured sewer pipe is dumping water containing 1/2 lbs/cu ft of pollutants into the inlet of a reservoir which holds 50,000 cu ft of pollutant-free water. If water enters and leaves the reservoir at the rate of 1,000 cu ft/day, how long will it take for the reservoir to contain 1/10 lbs/cu ft of pollutants?

66. *(Optional)* An object is dropped from a very high altitude. If the drag is $D = 0.2v$, what is the velocity after 3 seconds? What is its terminal velocity? (Positive direction is down.)

67. Find an equation of the line tangent to the curve $2y = y^x$ at the point $(2, 2)$.

68. For what values of c will $y = e^{cx}$ satisfy the equation $y'' + 6y' - 7y = 0$?

69. *(Inflation)* At the time of writing, inflation in the United States was at the rate of 18% per year (compounded continuously—see Exercise 63). Suppose this rate continues, and you have a fixed retirement income of $25,000 per year. In 4 years what will this income be worth in terms of current dollars?

TRIGONOMETRIC AND HYPERBOLIC FUNCTIONS

Rhythmic, or periodic, phenomena are an essential part of life. Our pulse beats; the earth revolves; the sun rises; the pendulum swings. Electrons oscillating in a wire transmit electricity, and vibrating air molecules bring us sound. One way that scientists describe these phenomena is with trigonometric functions.

In Section 8–1, we develop derivative formulas for the cotangent, secant, and cosecant, and discuss the integrals of all six trigonometric functions. The next section shows how to invert the trigonometric functions. The inverses lead to new and useful integral formulas. In Section 8–3, there is a discussion of simple harmonic motion; it illustrates how trigonometric functions are used to describe some of the rhythmic phenomena mentioned above.

Section 8–4 introduces the hyperbolic functions which have a curious relationship to the trigonometric functions. Curious because these decidedly nonperiodic functions share so many of the other properties enjoyed by the trigonometric functions. Here, too, several new integral formulas are developed.

There are three focal points in this chapter. (1) The inverses of the trigonometric functions, (2) the applications to simple harmonic motion, and (3) the new integral formulas; they will be referred to often in the next chapter.

Section 8–5, *Mercator and the ∫ sec x dx*, is optional reading material. It contains an interesting application of trigonometry and calculus to the art of map making.

8–1 TRIGONOMETRIC FUNCTIONS

The basic definitions and properties of the trigonometric functions are outlined in Section 1–7. The derivatives of the sine, cosine, and tangent are discussed in Section 3–5. In this section, we derive formulas for the derivatives of the cotangent, secant, and cosecant, and the integrals of all six functions.

Derivatives

The derivative formulas for the trigonometric functions are

(8.1)

> If f is a differentiable function, then
>
> (1) $(\sin f)' = (\cos f)f'$
> (2) $(\cos f)' = (-\sin f)f'$
> (3) $(\tan f)' = (\sec^2 f)f'$
> (4) $(\cot f)' = (-\csc^2 f)f'$
> (5) $(\sec f)' = (\sec f)(\tan f)f'$
> (6) $(\csc f)' = -(\csc f)(\cot f)f'$

As we mentioned above, the first three formulas are derived in Section 3–5. The last three are derived by converting to sines and cosines as follows:

$$(\cot x)' = \left[\frac{\cos x}{\sin x}\right]' = \frac{(\sin x)(\cos x)' - (\cos x)(\sin x)'}{\sin^2 x} \qquad \text{(Quotient)}$$

$$= \frac{-\sin^2 x - \cos^2 x}{\sin^2 x}$$

$$= \frac{-1}{\sin^2 x} \qquad (\sin^2 x + \cos^2 x = 1)$$

$$= -\csc^2 x$$

This, along with the chain rule, verifies formula (4) above. For the secant, we have

$$(\sec x)' = \left[\frac{1}{\cos x}\right]' = \frac{(\cos x)(1)' - (1)(\cos x)'}{\cos^2 x}$$

$$= \frac{\sin x}{\cos^2 x}$$

$$= \frac{1}{\cos x} \cdot \frac{\sin x}{\cos x}$$

$$= (\sec x)(\tan x)$$

This, along with the chain rule, verifies formula (5) above. We leave the verification of formula (6) for you.

EXAMPLE 1 Find $\dfrac{d}{dx} \cot x^2$.

Solution By formula (4),

$$\frac{d}{dx} \cot x^2 = (-\csc^2 x^2)(x^2)'$$

$$= -2x \csc^2 x^2 \qquad \blacksquare$$

EXAMPLE 2 If $\cos y = \csc x$, use implicit differentiation to find y'.

Solution Differentiate both sides with respect to x using formulas (2) and (6).

$$(-\sin y)y' = -(\csc x)(\cot x)$$

$$y' = \frac{\csc x \cot x}{\sin y} \quad \blacksquare$$

EXAMPLE 3 Find an equation of the line tangent to the curve $y = \sec x$ at the point $(\pi/3, 2)$.

Solution The derivative is

$$y' = \sec x \tan x$$

Therefore, the slope of the tangent line when $x = \pi/3$ is

$$\text{slope} = \sec\frac{\pi}{3}\tan\frac{\pi}{3}$$

$$= 2\sqrt{3}$$

It follows that

$$y - 2 = 2\sqrt{3}\left(x - \frac{\pi}{3}\right)$$

is an equation of the line (Figure 8–1). $\blacksquare$

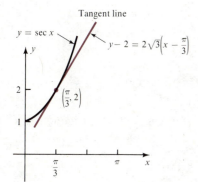

Tangent line

$y = \sec x$

$y - 2 = 2\sqrt{3}\left(x - \frac{\pi}{3}\right)$

$\left(\frac{\pi}{3}, 2\right)$

FIGURE 8–1. Example 3.

EXAMPLE 4 Sketch the graph of $y = \cos^2 x$ for $0 \le x \le 2\pi$.

Solution We find the first and second deirvatives

$$y' = -2\cos x \sin x$$
$$y'' = -2[\cos^2 x - \sin^2 x] \qquad \text{(Product rule)}$$

Since the sine and cosine agree in sign only in the first and third quadrants, it follows that

$$y' \begin{cases} <0 & \text{on } (0, \pi/2) \text{ or } (\pi, 3\pi/2) \\ =0 & \text{at } 0, \pi/2, \pi, 3\pi/2, 2\pi \\ >0 & \text{on } (\pi/2, \pi) \text{ or } (3\pi/2, 2\pi) \end{cases}$$

Locating the intervals on which $|\cos x| > |\sin x|$, we see that

$$y'' \begin{cases} <0 & \text{on } [0, \pi/4) \text{ or } (3\pi/4, 5\pi/4) \text{ or } (7\pi/4, 2\pi] \\ =0 & \text{at } \pi/4, 3\pi/4, 5\pi/4, 7\pi/4 \\ >0 & \text{on } (\pi/4, 3\pi/4) \text{ or } (5\pi/4, 7\pi/4) \end{cases}$$

Now we know where the curve is increasing, concave up, and so on. The sketch is in Figure 8–2. $\blacksquare$

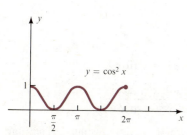

$y = \cos^2 x$

FIGURE 8–2. Example 4.

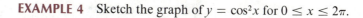

Integrals

Each of the derivative formulas (8.1) has a corresponding integral formula.

EXAMPLE 5   Find $\int x \sec x^2 \tan x^2 \, dx$.

Solution   Let $u = x^2$; then $du = 2x \, dx$ and

$$\int x \sec x^2 \tan x^2 \, dx = \frac{1}{2} \int \sec u \tan u \, du$$

$$= \frac{1}{2} \sec u + C \qquad\qquad (8.1\text{--}5)$$

$$= \frac{1}{2} \sec x^2 + C$$

Check.

$$\frac{d}{dx}\left[\frac{1}{2} \sec x^2\right] = \frac{1}{2}(\sec x^2 \tan x^2)(2x) = x \sec^2 \tan x^2 \qquad \blacksquare$$

In addition, there are integral formulas for each of the trigonometric functions:

(8.2)

> If $u = f(x)$ is differentiable, then
>
> (1) $\int \sin u \, du = -\cos u + C$
> (2) $\int \cos u \, du = \sin u + C$
> (3) $\int \tan u \, du = -\ln |\cos u| + C$
> (4) $\int \cot u \, du = \ln |\sin u| + C$
> (5) $\int \sec u \, du = \ln |\sec u + \tan u| + C$
> (6) $\int \csc u \, du = \ln |\csc u - \cot u| + C$

Formulas (1) and (2) are obvious, and formulas (3) and (4) are easily verified.

$$\int \tan u \, du = \int \frac{\sin u}{\cos u} \, du = -\ln |\cos u| + C$$

$$\int \cot u \, du = \int \frac{\cos u}{\sin u} \, du = \ln |\sin u| + C$$

The integral of the secant, formula (5), is obtained by writing

$$\sec x = \frac{\sec x \, (\tan x + \sec x)}{\sec x + \tan x} = \frac{\sec x \tan x + \sec^2 x}{\sec x + \tan x}$$

and observing that the numerator is the derivative of the denominator. A similar manipulation can be used to verify formula (6).

EXAMPLE 6   Find $\int e^x \tan e^x \, dx$.

Solution   Let $u = e^x$; then $du = e^x \, dx$ and

$$\int e^x \tan e^x \, dx = \int \tan u \, du$$

$$= -\ln |\cos u| + C \qquad\qquad (8.2\text{–}3)$$

$$= -\ln |\cos e^x| + C$$

Check.

$$\frac{d}{dx}[-\ln |\cos e^x|] = -\frac{1}{\cos e^x}[\cos e^x]'$$

$$= -\frac{1}{\cos e^x}(-\sin e^x)e^x$$

$$= e^x \tan e^x \qquad\blacksquare$$

EXAMPLE 7   Find $\int \sec 4x \, dx$.

Solution   Let $u = 4x$; then $du = 4dx$ and

$$\int \sec 4x \, dx = \frac{1}{4}\int \sec u \, du$$

$$= \frac{1}{4}\ln |\sec u + \tan u| + C \qquad\qquad (8.2\text{–}5)$$

$$= \frac{1}{4}\ln |\sec 4x + \tan 4x| + C$$

Just for practice, check this answer yourself. $\blacksquare$

If you want to see an interesting application of $\int \sec x \, dx$ to map making, take a few moments to read through Section 8–5 (pp 403–406).

EXERCISES

In Exercises 1–12, sketch the graphs.

1. $y = \tan x$ **2.** $y = \cot x$

3. $y = \sec x$ **4.** $y = \csc x$

5. $y = \sin^2 x$ **6.** $y = \tan^2 x$

7. $y = \cot^2 x$ **8.** $y = \sec^2 x$

9. $y = \sin 2x$ **10.** $y = 2 \cos x$

11. $y = 2 \cos(x - \pi)$ **12.** $y = 3 \sin(x/2)$

In Exercises 13–18, find the value(s) of x, $0 \le x \le \pi$, that satisfy the given equation.

13. $\sec x = -2$ **14.** $\sec x = 2/\sqrt{3}$

15. $\csc x = 2$ **16.** $\csc x = \sqrt{2}$

17. $\cot x = 0$ **18.** $\cot x = 1$

19. For what values of x, $0 \le x \le 2\pi$, are $\sec x$, $\csc x$, and $\cot x$ equal to 0?

20. For what values of x, $0 \le x \le 2\pi$, are $\sec x$, $\csc x$, and $\cot x$ undefined?

In Exercises 21–30, find y' if y is the given function.

21. $\tan 3x^2$ **22.** $\cos(x^4 + x)$

23. $\sin^2 2x$

24. $\tan^2 x$

25. $\sec x^2$

26. $\csc \ln x$

27. $\cot^2 x$

28. $\dfrac{1 + \sin x}{1 + \cos x}$

29. $\csc(x^2 + 1)^2$

30. $|\sin x|, 0 \le x \le 2\pi$

In Exercises 31–36, find dy/dx.

31. $y = \tan xy$

32. $y = \cot(x^2 + y^2)$

33. $x = \csc xy$

34. $x = \sin(x + y)$

35. $x = \ln(\cot y)$

36. $y = x \sec y$

In Exercises 37–50, evaluate the integrals. Check your answers.

37. $\displaystyle\int x \csc^2 x^2 \, dx$

38. $\displaystyle\int \sec 3x \tan 3x \, dx$

39. $\displaystyle\int_{\pi/6}^{\pi/4} \csc 2x \cot 2x \, dx$

40. $\displaystyle\int_0^{\pi/4} \sec^2 5x \, dx$

41. $\displaystyle\int e^x \tan e^x \, dx$

42. $\displaystyle\int x \tan x^2 \, dx$

43. $\displaystyle\int \dfrac{\sec \sqrt{x}}{\sqrt{x}} \, dx$

44. $\displaystyle\int \csc 9x \, dx$

45. $\displaystyle\int \csc x \cot x \, dx$

46. $\displaystyle\int \sec x \tan x \, dx$

47. $\displaystyle\int \csc^2 x \cot^2 x \, dx$

48. $\displaystyle\int \sec^2 x \tan^2 x \, dx$

49. $\displaystyle\int \dfrac{\sec^2 x}{\tan x} \, dx$

50. $\displaystyle\int \dfrac{\csc^2 x}{\cot x} \, dx$

51. *(Branching angle of blood vessels)* In his work concerning the energy loss due to friction as blood flows through a blood vessel, Jean Poiseuille discovered the loss to be kd/r^4 where k is a constant, d is the length, and r is the radius of the blood vessel

(see the footnote on page 24). Suppose that a blood vessel of length d_1 and radius r branches off from one of length d_2 and radius R. Then the energy loss L is the sum

$$L = \frac{kd_1}{r^4} + \frac{kd_2}{R^4}$$

Show that L can be rewritten as a function of the angle α at which they branch:

$$L = k \left[\frac{b \csc \alpha}{r^4} + \frac{a - b \cot \alpha}{R^4} \right]$$

where a, b, and α are as shown in Figure 8–3.

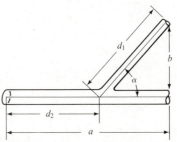

FIGURE 8–3. Exercise 51. The branching of blood vessels.

52. Find an equation of the line normal (that is, perpendicular to the tangent) to the curve $y = \cot x$ at $(\pi/4, 1)$.

53. Find the area of the triangle bounded by the coordinate axes and the line tangent to $y = \csc x$ at $(\pi/6, 2)$.

54. Find the area of the region determined by $y = \cot x$ from $x = \pi/6$ to $x = \pi/2$.

55. The region determined by $y = \sec x$ from $x = 0$ to $x = \pi/4$ is revolved about the x-axis. Find the volume generated.

8–2   INVERSE TRIGONOMETRIC FUNCTIONS

The inverses of the trigonometric functions play an important role in calculus because they lead to new and useful integration formulas. But only one-to-one functions can be inverted, and since the trigonometric functions are periodic, they assume each value infinitely many times. How, then, can these functions have inverses? The idea is to restrict their domains to intervals on which they are one-to-one.

Inverse Sine

There are many intervals on which the sine is one-to-one, but it is standard to use the interval $[-\pi/2, \pi/2]$. The graph of

$$f(x) = \sin x \quad x \text{ in } [-\pi/2, \pi/2]$$

is drawn in Figure 8–4A. The derivative $f'(x) = \cos x$ is positive on the open interval $(-\pi/2, \pi/2)$, so f is increasing, and it follows that f has an inverse. There are two symbols commonly used for the inverse sine, *sin⁻¹* and *arc sin;* most of the time we will use sin⁻¹.*

(8.3)

$$\sin^{-1}x = y \text{ means that } \sin y = x$$
Domain of $\sin^{-1}$ is $[-1, 1]$.
Range of $\sin^{-1}$ is $[-\pi/2, \pi/2]$.

The graph of $y = \sin^{-1}x$ is the reflection of the graph of $y = \sin x$ through the line $y = x$ (Figure 8–4B).

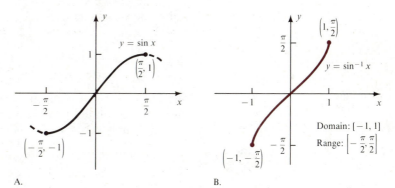

FIGURE 8–4. Inverting the sine.   A.   B.

A good way to think about the inverse sine is

$$\sin^{-1}x \text{ is that number in } \left[-\frac{\pi}{2}, \frac{\pi}{2}\right] \text{ whose sine is } x$$

EXAMPLE 1

(a) $\sin^{-1}\dfrac{1}{2} = \dfrac{\pi}{6}$ because $-\dfrac{\pi}{2} \leq \dfrac{\pi}{6} \leq \dfrac{\pi}{2}$ and $\sin\dfrac{\pi}{6} = \dfrac{1}{2}$

*The symbol sin⁻¹ will *always* stand for the inverse sine. If we want to express csc $x = 1/\sin x$ in exponential form, we will write $(\sin x)^{-1}$.

(b) $\sin^{-1}\left(-\dfrac{\sqrt{2}}{2}\right) = -\dfrac{\pi}{4}$ because $-\dfrac{\pi}{2} \le -\dfrac{\pi}{4} \le \dfrac{\pi}{2}$ and

$$\sin\left(-\dfrac{\pi}{4}\right) = -\dfrac{\sqrt{2}}{2} \qquad \blacksquare$$

It is often necessary to evaluate combinations such as $\tan(\sin^{-1}\frac{3}{4})$ or $\cos(\sin^{-1}x)$. An easy way to do this is to draw right triangle representations of the inverse function.

EXAMPLE 2   Evaluate (a) $\tan(\sin^{-1}\frac{3}{4})$ and (b) $\cos(\sin^{-1}x)$.

Solution

(a) Draw a right triangle with an acute angle labeled $\alpha = \sin^{-1}(3/4)$. Then the lengths of the opposite side and the hypotenuse are 3 and 4 because α is the angle with

$$\sin \alpha = \dfrac{\text{side opposite}}{\text{hypotenuse}} = \dfrac{3}{4} \qquad \text{(Figure 8–5A)}$$

It follows that the side adjacent is $\sqrt{4^2 - 3^2} = \sqrt{7}$ and now the tangent can be read off

$$\tan\left(\sin^{-1}\dfrac{3}{4}\right) = \dfrac{\text{side opposite}}{\text{side adjacent}} = \dfrac{3}{\sqrt{7}}$$

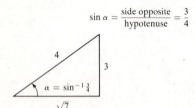

$\sin \alpha = \dfrac{\text{side opposite}}{\text{hypotenuse}} = \dfrac{3}{4}$

A. $\tan\left(\sin^{-1}\frac{3}{4}\right) = \dfrac{3}{\sqrt{7}}$

$\sin \alpha = \dfrac{\text{side opposite}}{\text{hypotenuse}} = \dfrac{x}{1} = x$

B. $\cos(\sin^{-1}x) = \sqrt{1 - x^2}$

FIGURE 8–5. Example 2.

(b) Draw a right triangle. The acute angle $\alpha = \sin^{-1}x$ has opposite side x and hypotenuse 1 because α is the angle with $\sin \alpha = x$ (Figure 8–5B). It follows that the side adjacent is $\sqrt{1 - x^2}$, and now the cosine can be read off

$$\cos(\sin^{-1}x) = \dfrac{\text{side adjacent}}{\text{hypotenuse}} = \sqrt{1 - x^2}$$

Remark: With a slight adjustment, the procedure above also works for negative angles. Simply disregard the negative sign for a moment, set up the right triangle, compute the desired value, and *then affix the proper sign.* For instance, $\alpha = \sin^{-1}(-3/4)$ is in the fourth quadrant where the tangent is negative. Therefore,

$$\tan\left[\sin^{-1}\left(-\dfrac{3}{4}\right)\right] = -\dfrac{3}{\sqrt{7}}$$

The formula

$$\cos(\sin^{-1}x) = \sqrt{1 - x^2}$$

holds for all x in $[-1, 1]$ because the cosine is nonnegative on $[-\pi/2, \pi/2]$. $\blacksquare$

Being inverses, the sine and inverse sine satisfy the following identities:

(8.4)

$$\sin(\sin^{-1}x) = x \text{ for all } x \text{ in } [-1, 1]$$

$$\sin^{-1}(\sin x) = x \text{ for all } x \text{ in } \left[-\frac{\pi}{2}, \frac{\pi}{2}\right]$$

Remark: Calculators are programmed with the proper domain and range of the inverse sine. For instance, the domain of $\sin^{-1}$ is the interval $[-1, 1]$. If you enter the number 2 and push $\boxed{\text{INV}}\;\boxed{\text{SIN}}$ (*Note:* some calculators have an $\boxed{\text{ARCSIN}}$ key), you will get an error signal. Try it (but make sure you are in the *radian mode*). Moreover, the range of $\sin^{-1}$ is $[-\pi/2, \pi/2]$. Therefore, $\sin^{-1}(\sin 2)$ *cannot* equal 2; it must be the number between $-\pi/2$ and $\pi/2$ whose sine equals sin 2. Thus,

$$\boxed{2}\;\boxed{\text{SIN}}\;\boxed{\text{INV}}\;\boxed{\text{SIN}} = 1.1415927$$

But whenever x is in $[-\pi/2, \pi/2]$, then $\sin^{-1}(\sin x) = x$. For example,

$$\boxed{-.321}\;\boxed{\text{SIN}}\;\boxed{\text{INV}}\;\boxed{\text{SIN}} = -.321$$

Try several values on your calculator.

Since the derivative of the sine is never 0 in $(-\pi/2, \pi/2)$, it follows from the Inverse Derivative Theorem (7.21) that $\sin^{-1}$ is differentiable on $(-1, 1)$. Moreover, differentiating both sides of the first equation in (8.4) yields

$$[\sin(\sin^{-1}x)]' = 1$$
$$\cos(\sin^{-1}x)[\sin^{-1}x]' = 1 \qquad \text{(Chain rule)}$$
$$\sqrt{1 - x^2}[\sin^{-1}x]' = 1 \qquad \text{(Figure 8–5B)}$$
$$[\sin^{-1}x]' = \frac{1}{\sqrt{1 - x^2}}$$

The last line above and the chain rule imply that

(8.5)

If $u = f(x)$ is differentiable and $|u| < 1$, then

$$\frac{d}{dx}\sin^{-1}u = \frac{1}{\sqrt{1 - u^2}}\frac{du}{dx}$$

Now we have a new integral formula as well:

(8.6)

$$\int \frac{du}{\sqrt{1 - u^2}} = \sin^{-1}u + C$$

EXAMPLE 3 $\dfrac{d}{dx}\sin^{-1}3x = \dfrac{1}{\sqrt{1-(3x)^2}}(3x)' = \dfrac{3}{\sqrt{1-9x^2}}$ ■

EXAMPLE 4   Evaluate $\displaystyle\int \dfrac{1}{\sqrt{1-9x^2}}\,dx$

Solution   The form of the integrand suggests that we use formula (8.6). Therefore, we write

$$\int \dfrac{1}{\sqrt{1-9x^2}}\,dx = \int \dfrac{1}{\sqrt{1-(3x)^2}}\,dx$$

With $u = 3x$, a factor of 3 is needed in the numerator, so we continue with

$$= \dfrac{1}{3}\int \dfrac{3}{\sqrt{1-(3x)^2}}\,dx$$

to obtain the exact form of (8.6) with $u = 3x$. It follows that

$$\int \dfrac{1}{\sqrt{1-9x^2}}\,dx = \dfrac{1}{3}\sin^{-1}3x + C$$

Check.

$$\dfrac{d}{dx}\left[\dfrac{1}{3}\sin^{-1}3x\right] = \dfrac{1}{3}\dfrac{1}{\sqrt{1-(3x)^2}}(3x)' = \dfrac{1}{\sqrt{1-9x^2}}$$

Remark: If the coefficient of x^2 is not a perfect square (as is the number 9), we still follow the same pattern as above using square roots. For example,

$$\int \dfrac{1}{\sqrt{1-2x^2}}\,dx = \int \dfrac{1}{\sqrt{1-(\sqrt{2}x)^2}}\,dx = \dfrac{1}{\sqrt{2}}\int \dfrac{\sqrt{2}}{\sqrt{1-(\sqrt{2}x)^2}}\,dx$$

$$= \dfrac{1}{\sqrt{2}}\sin^{-1}\sqrt{2}x + C$$

Make a quick check to verify this answer.   ■

Inverse Tangent

The interval on which to invert the tangent is $(-\pi/2, \pi/2)$. Again, either of the symbols tan^{-1} or $arc\ tan$ is used. Since the range of the tangent is $(-\infty, \infty)$, we have

(8.7)

$\tan^{-1}x = y$ means that $\tan y = x$.
Domain of $\tan^{-1}$ is $(-\infty, \infty)$.
Range of $\tan^{-1}$ is $(-\pi/2, \pi/2)$.

The graphs are drawn in Figure 8–6. A good way to think of $\tan^{-1}$ is

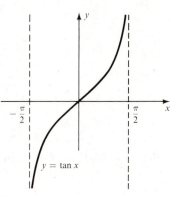
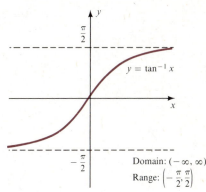

FIGURE 8–6. Inverting the tangent.   A.   B.

$$\tan^{-1}x \text{ is that number in } \left(-\frac{\pi}{2}, \frac{\pi}{2}\right) \text{ whose tangent is } x$$

The derivative of the tangent, $\sec^2 x$, is never zero, so the inverse, $\tan^{-1}$, is differentiable everywhere. Again, we differentiate both sides of the identity $\tan(\tan^{-1}x) = x$ and obtain

$$[\tan(\tan^{-1}x)]' = 1$$

$$\sec^2(\tan^{-1}x)[\tan^{-1}x]' = 1 \qquad \text{(Chain rule)}$$

$$(1 + x^2)[\tan^{-1}x]' = 1 \qquad \text{(Figure 8–7)}$$

$$[\tan^{-1}x]' = \frac{1}{1 + x^2}$$

The last line above and the chain rule imply that

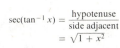

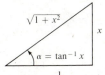

FIGURE 8–7. $\sec^2(\tan^{-1}x) = 1 + x^2$

(8.8)

$$\begin{array}{l} \text{If } u = f(x) \text{ is differentiable, then} \\[2mm] \dfrac{d}{dx}\tan^{-1}u = \dfrac{1}{1 + u^2}\dfrac{du}{dx} \end{array}$$

Once more there is a new and useful integral formula:

(8.9)

$$\int \frac{du}{1 + u^2} = \tan^{-1}u + C$$

The next example illustrates a method of rewriting integrands in order to use formulas (8.6) and (8.9).

EXAMPLE 5

(a) $\displaystyle\int \frac{1}{4+x^2}\,dx = \int \frac{1}{4\left(1+\dfrac{x^2}{4}\right)}\,dx$ (4 is factored out)

$\displaystyle = \frac{1}{4}\int \frac{1}{1+\left(\dfrac{x}{2}\right)^2}\,dx$

Continuing with $u = x/2$, we need a $1/2$ in the numerator, so

$\displaystyle = \frac{2}{4}\int \frac{1/2}{1+\left(\dfrac{x}{2}\right)^2}\,dx$

$\displaystyle = \frac{1}{2}\tan^{-1}\frac{x}{2} + C$ (8.9)

Check.

$$\frac{d}{dx}\left[\frac{1}{2}\tan^{-1}\frac{x}{2}\right] = \frac{1}{2}\left[\frac{1}{1+\left(\dfrac{x}{2}\right)^2}\right]\left(\frac{x}{2}\right)' = \frac{1}{4}\left(\frac{1}{1+\left(\dfrac{x}{2}\right)^2}\right) = \frac{1}{4+x^2}$$

(b) $\displaystyle\int \frac{1}{\sqrt{9-x^2}}\,dx = \int \frac{1}{3\sqrt{1-(x/3)^2}}\,dx$ (Factor out 9)

$\displaystyle = \int \frac{1/3}{\sqrt{1-(x/3)^2}}\,dx$ $(u = x/3)$

$\displaystyle = \sin^{-1}\frac{x}{3} + C$ (8.6)

Check.

$$\frac{d}{dx}\sin^{-1}\frac{x}{3} = \frac{1}{\sqrt{1-(x/3)^2}}\left(\frac{x}{3}\right)' = \frac{1}{3\sqrt{1-(x/3)^2}} = \frac{1}{\sqrt{9-x^2}} \qquad \blacksquare$$

Inverse Secant

The secant is inverted on all points in $[0, \pi]$ except $x = \pi/2$. The number $\pi/2$ is eliminated because $\sec(\pi/2)$ is undefined. The graphs of the secant and its inverse are sketched in Figure 8–8.

(8.10)

$\sec^{-1}x = y$ means that $\sec y = x$

Domain of $\sec^{-1}$ is all points x with $|x| \ge 1$.
Range of $\sec^{-1}$ is all points x in $[0, \pi]$, $x \ne \pi/2$.

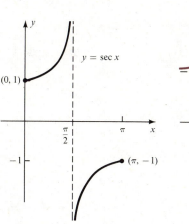

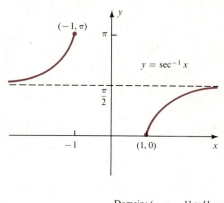

Domain: $(-\infty, -1]$ or $[1, \infty)$

Range: $\left[0, \dfrac{\pi}{2}\right)$ or $\left(\dfrac{\pi}{2}, \pi\right]$

FIGURE 8–8. Inverting the secant. A. B.

The derivative of the secant is never 0 on $(0, \pi/2)$ or $(\pi/2, \pi)$, so its inverse is differentiable on $(-\infty, -1)$ or $(1, \infty)$. If both sides of the identity $x = \sec(\sec^{-1}x)$ are differentiated, then

$$1 = \sec(\sec^{-1}x)\tan(\sec^{-1}x)\frac{d}{dx}(\sec^{-1}x)$$

$$= |x|\sqrt{x^2 - 1}\,\frac{d}{dx}(\sec^{-1}x)$$

The last line above is true because $\sec(\sec^{-1}x) = x$ and $\tan(\sec^{-1}x) = \pm\sqrt{x^2 - 1}$ (Figure 8–9). The absolute value sign is needed because if $x > 1$, then $0 < \sec^{-1}x < \pi/2$ and $\tan(\sec^{-1}x) > 0$; if $x < -1$, then $\pi/2 < \sec^{-1}x < \pi$ and $\tan(\sec^{-1}x) < 0$. In either case, the product above is *positive;* hence, the absolute value sign.

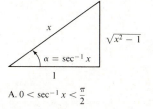

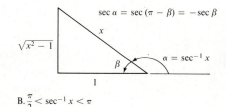

$\sec\alpha = \sec(\pi - \beta) = -\sec\beta$

FIGURE 8–9. $\tan(\sec^{-1}x)$

$= \begin{cases} \sqrt{x^2 - 1} & \text{if } 0 < \sec^{-1}x < \pi/2 \\ -\sqrt{x^2 - 1} & \text{if } \pi/2 < \sec^{-1}x < \pi \end{cases}$

A. $0 < \sec^{-1}x < \dfrac{\pi}{2}$

B. $\dfrac{\pi}{2} < \sec^{-1}x < \pi$

With this established, we have

(8.11)

> If $u = f(x)$ is differentiable and $|u| > 1$, then
>
> $$\frac{d}{dx}\sec^{-1}u = \frac{1}{|u|\sqrt{u^2 - 1}}\frac{du}{dx}$$

and

$$(8.12) \qquad \boxed{\int \frac{du}{|u|\sqrt{u^2 - 1}} = \sec^{-1}u + C}$$

EXAMPLE 6 Evaluate $\displaystyle\int \frac{1}{|x|\sqrt{9x^2 - 1}}\,dx$

Solution The form of the integrand suggests that we use formula (8.12). Therefore, we write

$$\int \frac{1}{|x|\sqrt{9x^2 - 1}}\,dx = \int \frac{1}{|x|\sqrt{(3x)^2 - 1}}\,dx$$

With $u = 3x$, a factor of 3 is needed in the numerator. But instead of compensating with a factor of $1/3$ outside the integral, we write

$$= \int \frac{3}{|3x|\sqrt{(3x)^2 - 1}}\,dx$$

to obtain the exact form of (8.12) with $u = 3x$. It follows now that

$$\int \frac{1}{|x|\sqrt{9x^2 - 1}}\,dx = \sec^{-1}3x + C \qquad \blacksquare$$

EXAMPLE 7 Find the area of the region determined by $y = 1/x\sqrt{x^2 - 1}$ from $x = -2$ to $x = -2/\sqrt{3}$.

Solution Because x is negative, the curve lies below the x-axis. Therefore, the area is

$$
\begin{aligned}
-\int_{-2}^{-2/\sqrt{3}} \frac{1}{x\sqrt{x^2 - 1}}\,dx &= \int_{-2}^{-2/\sqrt{3}} \frac{1}{(-x)\sqrt{x^2 - 1}}\,dx \\
&= \int_{-2}^{-2/\sqrt{3}} \frac{1}{|x|\sqrt{x^2 - 1}}\,dx \\
&= \sec^{-1}x \,\big|_{-2}^{-2/\sqrt{3}} \qquad\qquad (8.12) \\
&= \sec^{-1}(-2/\sqrt{3}) - \sec^{-1}(-2) \\
&= \frac{5\pi}{6} - \frac{2\pi}{3} = \frac{\pi}{6} \text{ sq units} \qquad \blacksquare
\end{aligned}
$$

Inverses of the Co-functions

The cosine is inverted on the interval $[0, \pi]$, the cotangent is inverted on the intervals $(-\pi/2, 0)$ or $(0, \pi/2)$, and the cosecant is inverted on the intervals $[-\pi/2, 0)$ or $(0, \pi/2]$. It turns out that the derivatives are just the negatives of those we have already derived; that is,

$$\frac{d}{dx}\cos^{-1}x = -\frac{d}{dx}\sin^{-1}x$$

(8.13) $\quad \dfrac{d}{dx}\cot^{-1}x = -\dfrac{d}{dx}\tan^{-1}x$

$$\frac{d}{dx}\csc^{-1}x = -\frac{d}{dx}\sec^{-1}x$$

The verifications are left as an exercise.

Summary

Here is a table of the inverses discussed in this section. To save space, the entries are in terms of x rather than an arbitrary differentiable function $f(x)$, and the constants of integration are omitted in the last column.

(8.14)

Function	Domain	Range	Derivative	Integral
$\sin^{-1}$	$[-1, 1]$	$\left[-\dfrac{\pi}{2}, \dfrac{\pi}{2}\right]$	$\dfrac{1}{\sqrt{1-x^2}}$	$\displaystyle\int \dfrac{dx}{\sqrt{1-x^2}} = \sin^{-1}x$
$\cos^{-1}$	$[-1, 1]$	$[0, \pi]$	$-\dfrac{1}{\sqrt{1-x^2}}$	$\displaystyle\int \dfrac{-dx}{\sqrt{1-x^2}} = \cos^{-1}x$
$\tan^{-1}$	all x	$\left(-\dfrac{\pi}{2}, \dfrac{\pi}{2}\right)$	$\dfrac{1}{1+x^2}$	$\displaystyle\int \dfrac{dx}{1+x^2} = \tan^{-1}x$
$\cot^{-1}$	all $x \neq 0$	$\left(-\dfrac{\pi}{2}, \dfrac{\pi}{2}\right) x \neq 0$	$\dfrac{-1}{1+x^2}$	$\displaystyle\int \dfrac{-dx}{1+x^2} = \cot^{-1}x$
$\sec^{-1}$	$|x| \geq 1$	$[0, \pi]x \neq \dfrac{\pi}{2}$	$\dfrac{1}{|x|\sqrt{x^2-1}}$	$\displaystyle\int \dfrac{dx}{|x|\sqrt{x^2-1}} = \sec^{-1}x$
$\csc^{-1}$	$|x| \geq 1$	$\left[-\dfrac{\pi}{2}, \dfrac{\pi}{2}\right] x \neq 0$	$\dfrac{-1}{|x|\sqrt{x^2-1}}$	$\displaystyle\int \dfrac{-dx}{|x|\sqrt{x^2-1}} = \csc^{-1}x$

EXERCISES

Find the *exact* values of the expressions in Exercises 1–12. (You may use a calculator to check your answers, but not to find them.)

1. $\sin^{-1}\left(-\dfrac{1}{2}\right)$

2. $\tan^{-1}(-1)$

3. arc sec$\left(-\dfrac{2}{\sqrt{3}}\right)$

4. arc csc $\sqrt{2}$

5. arc cos 1

6. $\tan\left(\sin^{-1}\dfrac{2}{3}\right)$

7. $\cos\left(\tan^{-1}\dfrac{1}{4}\right)$

8. $\sin\left(\sin^{-1}\dfrac{1}{4}\right)$

9. $\sin^{-1}(\sin \pi)$

10. $\tan^{-1}(\tan 3\pi)$

11. $\cos^{-1}(\cos 3\pi)$

12. $\cos^{-1}(\cos 2\pi)$

In Exercises 13–23, find y' if y is the given function.

13. $\sin^{-1}2x^3$

14. $\tan^{-1}e^x$

15. $\sec^{-1}\sqrt{x}$

16. $\cos^{-1}(x^2 + x)$

17. $e^{\sin^{-1}x}$

18. $e^{\tan^{-1}x}$

19. $\ln \sec^{-1} 5x$

20. $\ln|\sin^{-1} 2x|$

21. $\cos(\sin^{-1} x)$

22. $\tan(\sec^{-1} x), x > 1$

23. $\sin^{-1}(\cos x), 0 \le x \le \pi/2$

24. (a) The answer to Exercise 21 is $-x/\sqrt{1-x^2}$ which is the same as the derivative of $y = \sqrt{1-x^2}$. Explain why this is so. (b) The answer to Exercise 23 is -1. Can you explain this unexpected result?

Evaluate the integrals in Exercises 25–38.

25. $\displaystyle\int \frac{1}{\sqrt{1-4x^2}}\,dx$

26. $\displaystyle\int \frac{1}{\sqrt{1-3x^2}}\,dx$

27. $\displaystyle\int \frac{1}{1+5x^2}\,dx$

30. $\displaystyle\int \frac{4}{1+9x^2}\,dx$

29. $\displaystyle\int \frac{1}{|x|\sqrt{3x^2-1}}\,dx$

30. $\displaystyle\int \frac{1}{|x|\sqrt{4x^2-1}}\,dx$

31. $\displaystyle\int_0^1 \frac{1}{\sqrt{4-x^2}}\,dx$

32. $\displaystyle\int_0^2 \frac{1}{4+x^2}\,dx$

33. $\displaystyle\int \frac{x}{\sqrt{4-x^2}}\,dx$

34. $\displaystyle\int \frac{x}{4+x^2}\,dx$

35. $\displaystyle\int \frac{1}{|x|\sqrt{x^2-9}}\,dx$

36. $\displaystyle\int \frac{1}{|4x|\sqrt{x^2-1}}\,dx$

37. $\displaystyle\int \frac{\tan^{-1}\sqrt{x}}{\sqrt{x}(1+x)}\,dx$

38. $\displaystyle\int \frac{x \sin^{-1} 3x^2}{\sqrt{1-9x^4}}\,dx$

39. *(Minimal energy loss in blood vessels)* In Exercise 51 of Section 8–1, the equation

$$L = k\left[\frac{b \csc \alpha}{r^4} + \frac{a - b \cot \alpha}{R^4}\right]$$

expresses the energy loss as a function of the branching angle α of two blood vessels. If $R = .2$ cm and $r = .1$ cm, find the angle α that minimizes L.

40. Let S be the region determined by the graph of $y = 1/\sqrt{1-x^2}$ from $x = 0$ to $x = 1/2$. (a) Find the area of S. (b) Find the volume generated by revolving S about the y-axis.

·oximation of π) Show that

$$\int_0^1 \frac{1}{1+x^2}\,dx = \frac{\pi}{4}$$

·mate the integral with Simpson's rule to obtain an approximation of π (carry ·s as your calculator displays).

42. *(Refraction and reflection)* When a ray of light travels in one medium at a velocity v_1 and passes into another medium where it travels at a velocity v_2, then the angles of incidence α_1 and α_2 are related by Snell's law

$$\frac{\sin \alpha_1}{v_1} = \frac{\sin \alpha_2}{v_2} \qquad \text{(Figure 8–10)}$$

Express α_2 as a function of α_1 and determine the largest value of α_1 for which α_2 is defined. For angles greater than this, the light is reflected off the surface of the second medium. Why?

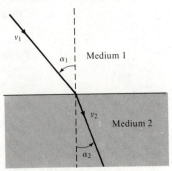

FIGURE 8–10. Exercise 42.

43. A picture is 3 feet high. It is hung on a wall with its base 4 feet above a viewer's eyes. How far from the wall should the viewer stand so as to maximize the angle subtended by the picture (see Figure 8–11)?

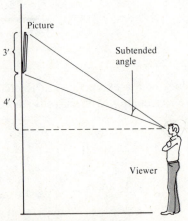

FIGURE 8–11. Exercise 43.

45. Show that equations (8.13) are true.

44. Suppose that a is a positive constant. Use differentiation to show these formulas are correct:

(8.15)

$$\int \frac{1}{\sqrt{a^2 - u^2}} \, du = \sin^{-1} \frac{u}{a} + C$$

$$\int \frac{1}{a^2 + u^2} \, du = \frac{1}{a} \tan^{-1} \frac{u}{a} + C$$

$$\int \frac{1}{|u| \sqrt{u^2 - a^2}} \, du = \frac{1}{a} \sec^{-1} \frac{u}{a} + C$$

Optional Exercise

46. Sometimes an integrand can be put into one of the forms (8.15) by completing a square. Evaluate

(a) $\displaystyle\int \frac{1}{2 + 2x + x^2} \, dx$

(b) $\displaystyle\int \frac{1}{\sqrt{-3 - 4x - x^2}} \, dx$

8-3   SIMPLE HARMONIC MOTION

Fasten one end of a spring to an overhead support, and attach a weight to the other end. The spring will stretch, and the weight will come to rest in an equilibrium position. Now pull the weight down beyond this position. You know, intuitively, that when the weight is released, it will oscillate up and down. If the spring friction is neglected, the motion of the weight is called *simple harmonic motion.*

Simple harmonic motion plays a central role in the analysis of any type of oscillation; for example, in the study of sound waves or the motion of a bird's wing. Therefore, it is worth developing a mathematical model for it. The form of the model is a differential equation.

Let us return to the example of the spring. The weight $W = mg$ (m = mass, g = gravity constant) stretches the spring p units beyond its natural length to reach the equilibrium position. According to Hooke's Law, (6.7) in Section 6–4, this happens when

$$mg = kp \quad (k = \text{spring constant})$$

Now the weight is displaced an additional c units downward (Figure 8–12), and released. If $s = s(t)$ denotes its position relative to the equilibrium point and friction is neglected, then there are two forces acting on the weight: mg due to gravity and $-k(p + s)$ due to the spring (the minus sign indicates that the spring works against the motion). The sum of these forces is

(8.16) $\quad mg - k(p + s) = mg - kp - ks$

$$= -ks \qquad\qquad (mg = kp)$$

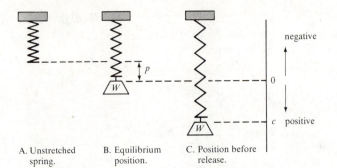

FIGURE 8–12. A weight W is attached to a spring and pulled down c units beyond the equilibrium position. When the weight is released it will oscillate up and down.

A. Unstretched spring. B. Equilibrium position. C. Position before release.

Moreover, by Newton's second law of motion, the force acting on a moving object is ma where $a = s''$ is its acceleration. Therefore, the force ms'' must equal the force (8.16); that is,

$$ms'' = -ks \quad \text{or} \quad s'' = -\frac{k}{m}s$$

To emphasize that the constant k/m is positive, we set it equal to a square, say b^2. Thus,

$$(8.17) \quad b^2 = \frac{k}{m}$$

and the differential equation

$$(8.18) \quad \boxed{s'' = -b^2 s}$$

describes the simple harmonic motion of the weight. In simple harmonic motion, *the acceleration is proportional to the position and opposite in sign.*

Equation (8.18) is a (second order) *linear differential equation.* Because it involves the second derivative of the unknown function, the general solution will contain *two* arbitrary constants, and it takes *two* initial conditions to find a particular solution.

It is not difficult to see that

$$s = A \cos(bt + C)$$

where A and C are arbitrary constants, satisfies (8.18).

$$s' = -bA \sin(bt + C)$$
$$s'' = -b^2 A \cos(bt + C)$$
$$= -b^2 s$$

Although we will not prove it here, the fact is that this is the general solution. That is, *every solution of $s'' = -b^2 s$ can be written as*

$$(8.19) \qquad \boxed{s = A \cos(bt + C)*}$$

If t is replaced by $t + 2\pi/b$, then

$$s\left(t + \frac{2\pi}{b}\right) = A \cos\left(b\left(t + \frac{2\pi}{b}\right) + C\right)$$
$$= A \cos(bt + 2\pi + C)$$
$$= A \cos(bt + C) \qquad (\cos(x + 2\pi) = \cos x)$$
$$= s(t)$$

This shows that simple harmonic motion is periodic; the **period** T is

$$(8.20) \qquad \boxed{T = \frac{2\pi}{b}}$$

If t is measured in seconds, then the motion repeats itself every $2\pi/b$ seconds. The reciprocal $b/2\pi$ is the number of cycles per second; this is called the **frequency** f.

$$(8.21) \qquad \boxed{f = \frac{b}{2\pi}}$$

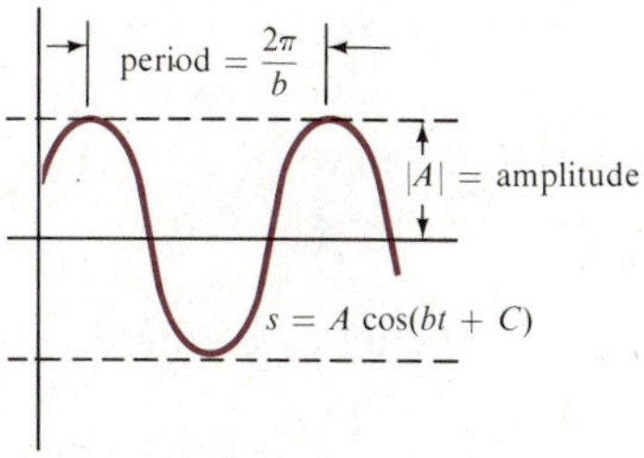

Because $|\cos x| \le 1$, the function $A \cos(bt + C)$ varies between $|A|$ and $-|A|$. The number $|A|$ is called the **amplitude**. The quantity $bt + C$ is called the **phase** of the motion; the constant C is the phase when $t = 0$. See Figure 8–13.

period $= \dfrac{2\pi}{b}$

$|A| = $ amplitude

$s = A \cos(bt + C)$

FIGURE 8–13. Simple harmonic motion.

EXAMPLE 1   The motion described by the position function $s = 3 \cos(2t + \pi/2)$ is simple harmonic motion. If s is measured in feet and t in seconds, then

$$\text{Amplitude} = 3 \text{ feet}$$

$$\text{Period} = \frac{2\pi}{b} = \pi \text{ sec/cycle}$$

$$\text{Frequency} = \frac{1}{\pi} \text{ cycles/sec}$$

$$\text{Velocity: } v = s' = -6 \sin\left(2t + \frac{\pi}{2}\right) \text{ ft/sec}$$

$$\text{Acceleration: } a = s'' = -12 \cos\left(2t + \frac{\pi}{2}\right) \text{ ft/sec}^2$$

*Other, equivalent, forms are $s = a \sin(bt + c)$ and $s = a \cos bt + c \sin bt$. See Exercise 19.

$$\text{Initial Position: } s(0) = 3 \cos\left(2 \cdot 0 + \frac{\pi}{2}\right) = 0$$

$$\text{Initial Velocity: } v(0) = -6 \sin\left(2 \cdot 0 + \frac{\pi}{2}\right) = -6 \text{ ft/sec} \qquad \blacksquare$$

In most problems about harmonic motion, you have to determine a position function. In that case, begin with

$$s = A \cos(bt + C)$$

and use the given information to find the values of A, b, and C.

EXAMPLE 2 Find an equation that describes the simple harmonic motion with period $\pi/2$, $s(0) = 2$ feet, and $v(0) = 4$ ft/sec.

Solution The period is always $2\pi/b$, so in this example,

$$\frac{\pi}{2} = \frac{2\pi}{b}, \text{ so } b = 4$$

So far then,

$$s = A \cos(4t + C)$$
$$v = s' = -4A \sin(4t + C)$$

Because the initial conditions are $s(0) = 2$ and $v(0) = 4$, we have

(8.22) $2 = A \cos C$ and $-1 = A \sin C$

The simplest procedure to find A and C is to add the squares of these two equations:

$$2^2 + (-1)^2 = A^2 cos^2 C + A^2 sin^2 C$$
$$= A^2(\cos^2 C + \sin^2 C)$$
$$= A^2$$

So $A = \pm\sqrt{5}$; if we choose $A = \sqrt{5}$, then

$$\frac{2}{\sqrt{5}} = \cos C \quad \text{and} \quad -\frac{1}{\sqrt{5}} = \sin C \qquad\qquad (8.22)$$

It follows that C is in the fourth quadrant, and $C = \sin^{-1}(-1/\sqrt{5}) \approx -.4636$. The final equation is

$$s = \sqrt{5} \cos(4t - .4636)$$

Note: If A is chosen to be $-\sqrt{5}$, then (8.22) becomes

$$-\frac{2}{\sqrt{5}} = \cos C \quad \text{and} \quad \frac{1}{\sqrt{5}} = \sin C$$

In this case, C is in the second quadrant, and $C = \cos^{-1}(-2/\sqrt{5}) \approx 2.678$. Then $s = -\sqrt{5} \cos(4t + 2.678)$. Both position functions describe the same motion. $\blacksquare$

EXAMPLE 3   A weight of 10 lbs stretches a certain spring 2 inches. The weight is displaced 6 inches more, and released. Find a position function and draw the position curve.

Solution   Any position function is of the form

$$s = A \cos(bt + C)$$

All you have to do is evaluate the constants. Recall that in the case of a spring, $b^2 = k/m$ (8.17). The 10 pound weight stretches the spring 2 inches so

$$10 = k2, \text{ so } k = 5$$

Also, $10 = mg$, so $m = 10/g$. Therefore,

$$b^2 = \frac{5}{10/g} = \frac{g}{2} \approx 16, \text{ so } b \approx 4 \qquad (g \approx 32 \text{ ft/sec}^2)$$

To find A and C, use the initial conditions $s(0) = 6$ and $v(0) = 0$ to write

$$6 = A \cos C \quad \text{and} \quad 0 = -4A \sin C$$

The first equation implies that $A \neq 0$, so the second equation implies that

$$\sin C = 0, \text{ or } C = 0^*$$

Now, from the first equation,

$$6 = A \cos 0 = A$$

Therefore, a position curve is

$$s = 6 \cos 4t$$

whose graph is shown in Figure 8–14.   ■

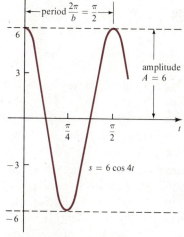

FIGURE 8–14. Example 3. The motion of a weight on a spring.

EXAMPLE 4   *(Bird in flight).* Although the tip of a bird's wing moves along the arc of a circle rather than straight up and down (Figure 8–15A), the motion

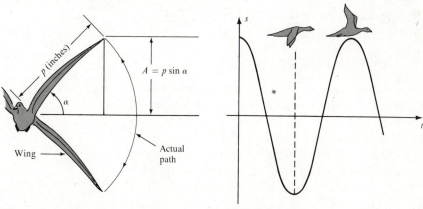

FIGURE 8–15. Example 4. The motion of the tip of a bird's wing.

A. Front view.

B. Side view.

*C could equal any multiple of π. We choose $C = 0$ for simplicity.

is closely approximated by a simple harmonic motion curve (Figure 8–15B). If the wing is p inches long and moves through an angle α above and below the horizontal, then the amplitude is

$$A = p \sin \alpha \qquad \text{(Figure 8–15A)}$$

If the wings make f complete cycles/sec, then

$$f = \frac{b}{2\pi}, \text{ so } b = 2\pi f \qquad (8.21)$$

For simplicity, let the phase be 0 when $t = 0$; this means that $C = 0$. Thus,

$$s = (p \sin \alpha) \cos (2\pi f)t$$

describes the motion of the tip of the bird's wing. ■

EXERCISES

In Exercises 1–7, find particular solutions. As a review, differential equations other than second order linear have been included.

1. $y'' = -4y; y(0) = 0, y'(0) = -1$
2. $y'' = x^2 + x - 1; y(0) = 1, y'(0) = 5$
3. $y' = -2y; y(0) = 8$
4. $y' + xy = x; y(0) = 3$
5. $y'' + 9y = 0; y(0) = 1, y'(0) = 3$
6. $y' = \dfrac{1 + y}{1 + x^2}; y(0) = 2$
7. $\sqrt{1 - x^2}y' = e^{-y}; y(0) = \ln 2$

In Exercises 8–10, find a position function for simple harmonic motion.

8. Period $2\pi/5$, $s(0) = 1$, and $v(0) = -5$.
9. Frequency 4, amplitude 2, and $s(0) = 1$.
10. Amplitude 5, phase at $t = 0$ is $\dfrac{\pi}{2}$, and $v(0) = -10$.

---

11. If $s = 2 \cos(\pi t + \pi)$, find the amplitude, period, and frequency of the motion.
12. (Continuation of Exercise 11) What is the starting position and initial velocity? When is the velocity a maximum? A minimum? When is the acceleration a maximum? A minimum?
13. A 24-lb weight stretches a spring 4 inches. Describe the motion (that is, find a position function) if the weight is released from a point 5 inches below the equilibrium position.
14. A 10-lb weight attached to the end of a spring is oscillating with a frequency of 3 cycles/sec. What is the spring constant?
15. (LC *Circuits*) Figure 8–16 shows a capacitor connected to an inductor with an open switch. The capacitor has a capacity of C μF (microfarads $= C \times 10^{-6}$ farads) and the inductance is L μH (microhenries $= L \times 10^{-6}$ henries). At $t = 0$, the switch is closed and a charge $Q = Q(t)$ μC (microcoulombs) flows through the inductor. By definition, the current I is

$$(8.23) \qquad I = \frac{dQ}{dt} \quad \text{amps}$$

According to Kirchoff's rule

$$(8.24) \qquad L\frac{dI}{dt} + \frac{Q}{C} = 0$$

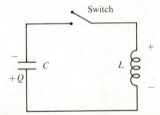

FIGURE 8–16. Exercise 15. An *LC* circuit.

Show that (8.23) and (8.24) lead to a differential equation of the form $y'' = -b^2 y$ where $b = 1/\sqrt{LC}$. Use this equation to find a function for the charge Q at time t ($I(0) = 0$ and let Q_0 be the initial charge).

16. (Continuation of Exercise 15) A 3 μF capacitor is charged with $Q_0 = 20\mu C$ and then shorted across a $5\mu H$ inductor. (a) What is the frequency ($f = b/2\pi$) of oscillations in the circuit? (b) At what time after the short occurs will the current be a maximum?

17. *(Jogging)* The arm of a jogger swings rhythmically back and forth in simple harmonic motion about the shoulder joint. If the distance from the shoulder to the elbow is 14 inches and the arm swings through an angle of 30° on each side of the body, completing 2/3 of a cycle per sec, what is the position of the elbow at time t? (Assume the elbow is at position 0 when $t = 0$.)

18. *(Hummingbird flight)* The wings of a ruby-throated hummingbird each measure about 1.7 inches. While hovering, each wing makes about 8 complete oscillations per second and moves through an angle of about 80° above and below the horizontal. Derive a function that approximates the simple harmonic motion of a point on the tip of the bird's wing. (Assume $s(0) = 0$.)

19. Suppose $s = A \cos(bt + C)$. Show that there are constants a and c such that $s = a \sin(bt + c)$. (*Hint:* $\cos x = \sin(\pi/2 - x)$.) What is the relationship between A, C, a, and c? Show that there are constants p and q such that $s = p \cos bt + q \sin bt$. What is the relationship between A, C, p, and q? (*Hint:* Use a sum formula.)

Start here for test up to Ch 10

8-4 THE HYPERBOLIC FUNCTIONS

The graph of the equation $x^2 + y^2 = 1$ is the unit circle. If the $+$ is changed to a $-$, the equation becomes

(8.25) $x^2 - y^2 = 1$

and its graph is the unit hyperbola (Figure 8–17). The two functions $x = \cos t$ and $y = \sin t$ satisfy the equation $x^2 + y^2 = 1$. We now introduce two functions

$$x = \frac{1}{2}(e^t + e^{-t}) \quad \text{and} \quad y = \frac{1}{2}(e^t - e^{-t})$$

that satisfy (8.25).

$$x^2 = \frac{1}{4}(e^{2t} + 2 + e^{-2t})$$

$$y^2 = \frac{1}{4}(e^{2t} - 2 + e^{-2t})$$

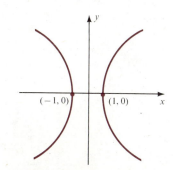

FIGURE 8–17. The unit hyperbola.

so $x^2 - y^2 = 1$. The cosine and sine are called *circular* functions. The two new functions are called *hyperbolic* functions, and are defined by

$$\cosh x = \frac{1}{2}(e^x + e^{-x}) \quad \text{and} \quad \sinh x = \frac{1}{2}(e^x - e^{-x})$$

The first is the **hyperbolic cosine** and is pronounced to rhyme with "gosh"; the second is the **hyperbolic sine** and is pronounced "cinch". These functions arise frequently in engineering problems (see, for example, Exercise 19 of Section 6–5), and also lead to some new integration formulas. The identity

$$(8.26) \qquad \boxed{\cosh^2 x - \sinh^2 x = 1}$$

established above is reminiscent of the fundamental identity $\cos^2 x + \sin^2 x = 1$; only the sign is different. This is typical of the curious relationship between these so differently defined functions. For example,

$$\frac{d}{dx}\cosh x = \frac{d}{dx}\left[\frac{1}{2}(e^x + e^{-x})\right]$$

$$= \frac{1}{2}(e^x - e^{-x}) = \sinh x$$

whereas $(\cos x)' = -\sin x$; again, a difference in sign. But sometimes the formulas are exactly analogous.

$$\frac{d}{dx}\sinh x = \frac{d}{dx}\left[\frac{1}{2}(e^x - e^{-x})\right]$$

$$= \frac{1}{2}(e^x + e^{-x}) = \cosh x$$

which copies the formula $(\sin x)' = \cos x$. The following list of properties of cosh and sinh are easily verified. Observe that each one copies the corresponding one for cosine and sine or differs only in sign.

$$(8.27)$$

> (1) $\cosh^2 x - \sinh^2 x = 1$
>
> (2) $\cosh(-x) = \cosh x$; $\sinh(-x) = -\sinh x$ [even / odd handwritten]
>
> (3) $\cosh(x + y) = \cosh x \cosh y + \sinh x \sinh y$
> $\sinh(x + y) = \sinh x \cosh y + \cosh x \sinh y$
>
> (4) $\cosh 2x = \cosh^2 x + \sinh^2 x$
> $\sinh 2x = 2 \sinh x \cosh x$
>
> (5) $\dfrac{d}{dx}\cosh x = \sinh x$; $\dfrac{d}{dx}\sinh x = \cosh x$
>
> (6) $y = A \cosh(bx + C)$ is the solution to $y'' = b^2 y$

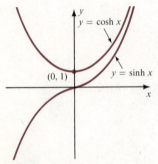

FIGURE 8–18. The graphs of the hyperbolic functions $y = \cosh x$ and $y = \sinh x$.

To graph $y = \cosh x$ and $y = \sinh x$, you can easily find the intervals of increasing and decreasing, the concavity, and so on. Both graphs are drawn in Figure 8–18. Observe that $\cosh x \geq 1$ and that $\cosh x > \sinh x$. But the difference

$$\cosh x - \sinh x = \frac{1}{2}(e^x + e^{-x}) - \frac{1}{2}(e^x - e^{-x}) = e^{-x}$$

approaches 0 as positive x gets large.

Because of the striking similarity (except for the graphs) to the trigonometric functions, it is natural to introduce the hyperbolic tangent, cotangent, secant, and cosecant as follows

$$\tanh x = \frac{\sinh x}{\cosh x} \quad \coth x = \frac{1}{\tanh x}$$

$$\operatorname{sech} x = \frac{1}{\cosh x} \quad \operatorname{csch} x = \frac{1}{\sinh x}$$

Unlike the tangent, the hyperbolic tangent is defined for all x because $\cosh x$ is never 0. The graph of $y = \tanh x$ is shown in Figure 8–19.

The derivative formulas are

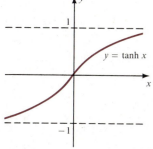

FIGURE 8–19. The graph of $y = \tanh x$.

(8.28)
$$\frac{d}{dx}\tanh x = \operatorname{sech}^2 x$$

$$\frac{d}{dx}\coth x = -\operatorname{csch}^2 x$$

$$\frac{d}{dx}\operatorname{sech} x = -\operatorname{sech} x \tanh x$$

$$\frac{d}{dx}\operatorname{csch} x = -\operatorname{csch} x \coth x$$

Inverse Hyperbolic Functions

Of the six hyperbolic functions, only cosh and sech are not one-to-one (see the graphs of cosh, sinh, and tanh). The cosh and sech are inverted for $x \geq 0$; the rest are inverted on their entire domains.

The derivatives of the inverse hyperbolic functions can be found in the usual way. For example,

$$\frac{d}{dx}\sinh(\sinh^{-1}x) = \frac{d}{dx}x$$

$$[\cosh(\sinh^{-1}x)]\frac{d}{dx}\sinh^{-1}x = 1 \qquad \text{(Chain rule)}$$

Now we use (8.26) and the fact that $\cosh x > 0$ to write

(8.29) $$\frac{d}{dx}\sinh^{-1}x = \frac{1}{\cosh(\sinh^{-1}x)} = \frac{1}{\sqrt{1 + \sinh^2(\sinh^{-1}x)}}$$

$$= \frac{1}{\sqrt{1 + x^2}}$$

But there is another way to find this derivative that is much more informative.

Unlike the trigonometric functions, it is possible to express the inverse hyperbolic functions explicitly. Since the hyperbolic functions are defined in terms of exponentials, you might expect that their inverses can be expressed in terms of logarithms.

$$y = \sinh^{-1}x \quad \text{means} \quad x = \sinh y$$

Thus,

$$x = \sinh y = \frac{1}{2}(e^y - e^{-y})$$

$$2x = e^y - e^{-y}$$

$$e^y - 2x - e^{-y} = 0 \qquad \text{(Rearrange)}$$

$$e^{2y} - 2xe^y - 1 = 0 \qquad \text{(Multiply by } e^y)$$

The last equation is a quadratic $t^2 + bt - 1 = 0$, where $t = e^y$ and $b = -2x$. It follows from the quadratic formula that

$$e^y = \frac{2x \pm \sqrt{4x^2 + 4}}{2}$$

$$= x \pm \sqrt{x^2 + 1}$$

Because $e^y > 0$, the minus sign must be discarded. Taking the natural logarithm of both sides yields

$$y = \ln(x + \sqrt{x^2 + 1})$$

In other words,

$$y = \sinh^{-1}x = \ln(x + \sqrt{x^2 + 1})$$

Using this formula, the derivative of $\sinh^{-1}x$ is

$$\frac{d}{dx}\sinh^{-1}x = \frac{d}{dx}[\ln(x + \sqrt{x^2 + 1})]$$

$$= \frac{1}{x + \sqrt{x^2 + 1}}\left[1 + \frac{x}{\sqrt{x^2 + 1}}\right]$$

$$= \frac{1}{\sqrt{x^2 + 1}}$$

which agrees with (8.29).

Each of the other inverse hyperbolic functions can also be expressed as logarithms. The $\coth^{-1}$ and csch^{-1} are omitted from the following list because they are seldom used.

The derivative formulas can be found with the usual method for inverses, or by differentiating the expressions in (8.30)

$$\cosh^{-1}x = \ln(x + \sqrt{x^2 - 1}) \quad x \geq 1$$
$$\sinh^{-1}x = \ln(x + \sqrt{x^2 + 1}) \quad \text{all } x$$
(8.30)
$$\tanh^{-1}x = \frac{1}{2}\ln\left(\frac{1 + x}{1 - x}\right) \quad |x| < 1$$
$$\text{sech}^{-1}x = \ln\left(\frac{1 + \sqrt{1 - x^2}}{x}\right) \quad 0 < x \leq 1$$

If $u = f(x)$ is differentiable, then

$$\frac{d}{dx}\cosh^{-1}u = \frac{1}{\sqrt{u^2 - 1}}\frac{du}{dx} \quad |u| > 1$$

$$\frac{d}{dx}\sinh^{-1}u = \frac{1}{\sqrt{1 + u^2}}\frac{du}{dx}$$
(8.31)
$$\frac{d}{dx}\tanh^{-1}u = \frac{1}{1 + u^2}\frac{du}{dx} \quad |u| < 1$$

$$\frac{d}{dx}\text{sech}^{-1}u = \frac{-1}{u\sqrt{1 - u^2}}\frac{du}{dx} \quad 0 < u < 1$$

Each of these derivative formulas has a corresponding integral formula:

If $u = f(x)$ is differentiable, then

$$\int \frac{du}{\sqrt{u^2 - 1}} = \cosh^{-1}u + C = \ln|u + \sqrt{u^2 - 1}| + C$$

$$\int \frac{du}{\sqrt{1 + u^2}} = \sinh^{-1}u + C = \ln|u + \sqrt{1 + u^2}| + C$$
(8.32)
$$\int \frac{du}{1 - u^2} = \begin{cases} \tanh^{-1}u + C & \text{if } |u| < 1 \\ \frac{1}{2}\ln\left|\frac{1 + u}{1 - u}\right| + C & \text{if } |u| \neq 1 \end{cases}$$

$$\int \frac{du}{u\sqrt{1 - u^2}} = -\text{sech}^{-1}|u| + C = -\ln\left|\frac{1 + \sqrt{1 - u^2}}{u}\right| + C$$

EXAMPLE 1

(a) $\displaystyle \int \frac{dx}{\sqrt{4x^2 - 1}} = \int \frac{dx}{\sqrt{(2x)^2 - 1}} = \frac{1}{2}\int \frac{2\,dx}{\sqrt{(2x)^2 - 1}} \qquad (u = 2x)$

$$= \frac{1}{2}\cosh^{-1}2x + C$$

(b) $\int \dfrac{dx}{\sqrt{4+x^2}} = \int \dfrac{dx}{2\sqrt{1+(x/2)^2}} = \int \dfrac{(1/2)\,dx}{\sqrt{1+(x/2)^2}}$ $(u=x/2)$

$$= \sinh^{-1}\dfrac{x}{2} + C \quad \blacksquare$$

To evaluate a definite integral, we use the logarithm forms of the inverse functions.

EXAMPLE 2

$$\int_0^{0.1} \dfrac{dx}{1-9x^2} = \dfrac{1}{3}\int_0^{0.1} \dfrac{3\,dx}{1-(3x)^2} \qquad (u=3x)$$

$$= \dfrac{1}{3}\tanh^{-1} 3x \Big|_0^{0.1}$$

$$= \dfrac{1}{6}\left[\ln\left|\dfrac{1+3x}{1-3x}\right|\right]_0^{0.1}$$

$$= \dfrac{1}{6}(\ln 1.3 - \ln 0.7) \quad \blacksquare$$

EXERCISES

Differentiate the functions in Exercise 1–10.

1. $y = \sinh 4x$
2. $y = \tanh 3x$
3. $y = \cosh(\cos x)$
4. $y = \sinh(x^2/5)$
5. $y = \tanh^{-1} x^2$
6. $y = \cosh^{-1}(1+x^2)$
7. $y = \operatorname{sech}^{-1} e^x$
8. $y = \tanh^{-1}\ln x$
9. $y = \cosh^{-1}\sqrt{x}$
10. $y = \sinh^{-1}(\cosh x)$

Evaluate the integrals in Exercises 11–20.

11. $\int \dfrac{dx}{\sqrt{x^2-4}}$
12. $\int \dfrac{dx}{\sqrt{x^2+4}}$
13. $\int \dfrac{dx}{\sqrt{1+9x^2}}$
14. $\int \dfrac{dx}{\sqrt{9x^2-1}}$
15. $\int \dfrac{dx}{x\sqrt{1-4x^2}}$
16. $\int \dfrac{dx}{x\sqrt{4-x^2}}$
17. $\int_0^3 \dfrac{dx}{\sqrt{1+x^2}}$
18. $\int_2^4 \dfrac{dx}{\sqrt{x^2-1}}$
19. $\int_5^6 \dfrac{dx}{1-x^2}$
20. $\int_5^6 \dfrac{dx}{1-4x^2}$

21. Find the arc length of the curve $y=\cosh x$ from $x=0$ to $x=1$.

22. Find the area of the region determined by $y=\sinh x$ from $x=0$ to $x=1$.

23. *(Air resistance)* Suppose that a body of mass m falling towards the earth encounters a drag proportional to the square of its velocity; $D=kv^2$. If it starts from rest, find its velocity at time t. What is its terminal velocity? (See Equation (7.59) in Section 7–7.)

24. *(Hanging cable)* When a cable sags of its own weight (for instance, a power line), the shape it assumes is called a *catenary curve*. It can be shown that, with the proper choice of coordinate axes, the function representing a catenary curve satisfies the differential equation

$$y'' = c\sqrt{1+(y')^2} \quad c \text{ a constant.}$$

Show that $y=(1/c)\cosh cx$ is a solution.

25. Verify the formulas (8.30) expressing the inverses as logarithms.

26. Verify the derivative formulas (8.31).

27. Verify the integral formulas (8.32).

Optional Exercise

28. *(Hyperbolic radian)* The connection between the trigonometric and hyperbolic functions is even deeper than we suggested in this section.

For any number t, the point P (cos t, sin t) lies on the unit circle. If t represents radians, then the area of the shaded sector determined by P (Figure 8–20A) is

$$\text{Area} = \frac{1}{2}t \quad \text{sq units} \qquad \text{((1.46) in Section 1–7)}$$

There is an analogous statement for the hyperbolic functions.

For any number u, the point P (cosh u, sinh u) lies on the unit hyperbola. Show that the area of the shaded sector determined by P (Figure 8–20B) is

$$\text{Area} = \frac{1}{2}u \quad \text{sq units}$$

In this context, u is called a *hyperbolic radian*.

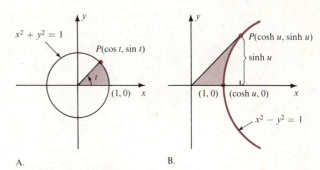

A.   B.

FIGURE 8–20. Exercise 28. The area of the shaded region in (A) is $\frac{1}{2}t$. The area in (B) is $\frac{1}{2}u$.

Hint:

(1) Verify that the area A of the sector is a function of u:

$$A(u) = \frac{1}{2}\cosh u \sinh u - \int_1^{\cosh u} \sqrt{x^2 - 1}\, dx$$

(2) Show that $A'(u) = \frac{1}{2}$, and conclude that $A = \frac{1}{2}u$.

8–5   MERCATOR AND THE ∫sec x dx *(Optional reading)**

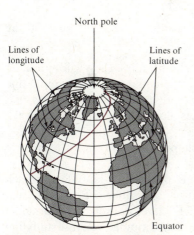

FIGURE 8–21. A flight with a constant bearing of 45° E of N from the Galapagos Islands in the Pacific to Franz Josef Land in the Arctic Ocean.

Imagine yourself as the navigator of a ship or, better yet, a plane and it is up to you to plot a course from point A to point B. The easiest thing to do is get out a map, join A to B with a straight line, and measure the angle from north (clockwise) to the line; say it is 45°. Then tell your captain to keep the plane headed in the direction north 45° east. The question is, would this constant heading get you to point B? The answer is YES, *provided* you use a Mercator map.

Figure 8–21 is a glope map. It shows the flight path with a constant heading of north 45° east from the Galapagos Islands in the Pacific to Franz Joseph Land in the Arctic. The spiral path makes an angle of 45° with each line of *longitude* (the north-south lines). Figure 8–22 is a plane chart; that is, a map on which the lines of longitude and *latitude* (the east-west lines) form congruent rectangles. On such a map, angles are distorted and the figure shows the curved flight path with constant bearing north 45° east. Figure 8–23 is a Mercator map. The vertical lines of longitude are evenly spaced but the horizontal lines of latitude are not; the space between them increases toward the poles. On this map, the path of constant bearing is a straight line!

Gerhardus Mercator first published his world map in 1569. It was hailed as a tremendous achievement and recognized as the first significant improvement in

---

*This discussion is adapted from Philip Tuchinsky, *Mercator's World Map and the Calculus*, UMAP (Unit 206), Newton, Mass., 1978.

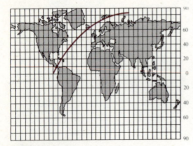

FIGURE 8–22. Plot of flight on "plane chart" such as was in use for charts of small areas in Mercator's time. Angles are not true and a straight line would not give a path of constant bearing.

FIGURE 8–23. Flight at constant bearing on a Mercator projection. Straight line path is easily constructed, measured, and followed.

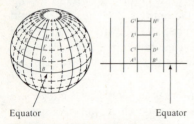

Equator   Equator

FIGURE 8–24. Corresponding points on meridians and map: EF on the globe stretches to $E'F'$ on the map.

map design in over 1,400 years. True, there are serious shortcomings: northern regions appear grossly exaggerated in area, the polar regions cannot be shown, and distances are distorted. But outweighing these disadvantages is the fact that here is a map on which constant heading paths appear as straight lines. Navigation is made easier because such paths are simple to construct, measure, and follow.

The Mercator Map and the $\int \sec x \, dx$

Let us now discuss the construction of the Mercator map, and discover why, almost a century before Newton and Leibniz launched the calculus, Mercator found himself in need of the $\int \sec x \, dx$.

We begin by introducing some terminology and setting forth our assumptions. On a globe (Figure 8–21), the east-west lines of *latitude* are circles parallel to the equator; the north-south lines of *longitude* are great circles (circles with their center at the center of the globe) passing through both poles. Lines of latitude and longitude are called *meridians*. A constant heading path will meet each meridian of longitude at the same angle; such paths are called *rhumb lines* on a map. Clearly, all lines of latitude and longitude are rhumb lines as well as the paths shown in Figures 8–21, 22, and 23. Only on Mercator's map do all rhumb lines appear as straight lines.

To construct this map, we make the following assumptions:

(1) distances along the equator are to scale

(2) distances off the equator will be distorted (stretched) in such a way to make rhumb lines appear as straight lines

(3) the earth is a sphere (we will not compensate for the slight bulge at the equator)

Since all lines of latitude and longitude are rhumb lines, they will appear as straight lines on our map. First, we represent the equator as a horizontal line across the middle. The lines of longitude are represented by vertical lines, and they must be evenly spaced because distances along the equator are to scale. The lines of latitude (the tropics of Cancer and Capricorn, the Arctic Circles, and so on) will be horizontal lines. The whole problem boils down to spacing these lines so as to make rhumb lines appear as straight lines.

Figure 8–24 shows that the distances along parallels of latitude approach zero as we move toward the poles, but those distances will be equal on our map because, there, the lines of longitude are parallel. Therefore, horizontal distances must be stretched. To see how to place the lines of latitude, we must study the horizontal stretching. Figure 8–25 shows a wedge of the earth and the corresponding part on the map. Arc $\widehat{AB}$ is part of the equator spanned by a central angle of θ (the Greek letter *theta*) radians. If R is the radius of the earth (approximately 4,000 miles), then $R\theta$ is the actual length of the arc $\widehat{AB}$ and of the corresponding line segment $\overline{A'B'}$ on the map. Now consider the arc $\widehat{PQ}$ at ϕ (the Greek letter *phi*) radians north latitude; $\widehat{PQ}$ is part of the circle centered

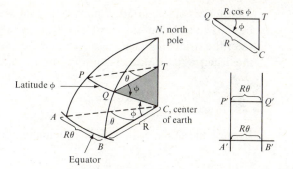

FIGURE 8–25. A wedge of the earth and its corresponding part of the map.

at T, directly north of the earth's center C. Using the shaded triangle in the figure, we see that the radius of that circle is

$$\overline{QT} = \overline{QC} \cos \phi = R \cos \phi \qquad (R = \overline{QC})$$

so that the actual length of the arc $\widehat{PQ}$ is

$$\widehat{PQ} = \overline{QT}\theta = [R \cos \phi]\theta$$

On the map, however, the corresponding line segment has length $\overline{P'Q'} = \overline{A'B'} = R\theta$. Since

$$\frac{\overline{P'Q'}}{\widehat{PQ}} = \frac{R\theta}{[R \cos \phi]\theta} = \frac{1}{\cos \phi} = \sec \phi$$

we see that

(8.33) $$\overline{P'Q'} = \widehat{PQ} \sec \phi$$

If rhumb lines are to appear as straight lines, then it is necessary for our map to preserve angles; that is, if two paths on earth meet at a certain angle, then their images on our map will also meet at that angle. Statement (8.33) indicates that east-west distances at latitude ϕ on earth are stretched by a factor of $\sec \phi$ on the map. Mercator's great insight was to realize that if angles are to be preserved, then *vertical distances at latitude ϕ will also have to be stretched by a factor of* $\sec \phi$.

This is where calculus comes in because as we move north-south, the latitude ϕ, and hence the stretching factor $\sec \phi$, is continuously changing. In Figure 8–25, the arc $\widehat{BQ}$ has actual length $R\phi$; we have to figure out the length of $\overline{B'Q'}$ on the map taking into account the stretching factor $\sec \alpha$ as α changes from 0 to ϕ. To do that, we let $\{\alpha_0, \alpha_1, \ldots, \alpha_n\}$ be a regular partition of $[0, \phi]$ into small

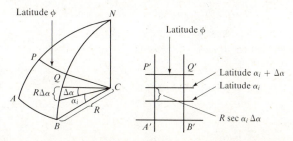

FIGURE 8–26. Stretching vertical distances requires the $\int_0^\phi \sec \alpha \, dx$.

subintervals of length $\Delta\alpha$; Figure 8–26 shows a typical subarc of $\overset{\frown}{BQ}$ determined by α_i. The length of this subarc is $R\Delta\alpha$, and if $\Delta\alpha$ is small enough, we can consider the stretching factor $\sec\alpha_i$ to remain constant and transfer to the map as the distance $R\sec\alpha_i\Delta\alpha$. Therefore,

$$\overline{B'Q'} \approx \sum_{i=1}^{n} R\sec\alpha_i\Delta\alpha$$

Because the sum on the right is a Riemann sum, we see that setting

$$\overline{B'Q'} = \int_{0}^{\phi} R\sec\alpha\,d\alpha$$

will produce the desired result. This shows that $\int \sec x\,dx$ is needed to construct a Mercator map.

Mercator, of course, could not know about integrals and antiderivatives. Although it is not known how he computed the spacings on his map, the method of "continual summing" was available (it was used as far back as the early Greeks to compute areas). Perhaps Mercator knew of this technique and used it. Because of the inaccuracies of geographical data in his time, and because of some minor errors in his calculations, his original world map (Figure 8–27) bears slight resemblance to a modern map. Nevertheless, the fact that he could understand the problem at all and devise a correct method of solution is a tribute to his genius.

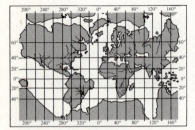

FIGURE 8–27. Sketch of Mercator's map of 1569.

REVIEW EXERCISES

In Exercises 1–20, find y'.

1. $y = \tan^3(x^3 + x)$

2. $y = \sin(\tan x)$

3. $y = \tan(x^2)$

4. $y = \ln|\sec x|$

5. $y = \dfrac{2\tan(2x)}{1 - \tan^2(2x)}$

6. $y = \csc x \tan x$

7. $y = \tan(x^2 y)$

8. $y = x^2\sin(x + y)$

9. $y = \sin^{-1}(x^3 + x^2)$

10. $y = \tan^{-1}(\ln|x|)$

11. $y = e^{x\sin^{-1}x}$

12. $y = \ln|\sin^{-1}(\ln|x|)|$

13. $y = \sin^{-1}(\sin^{-1}x)$

14. $y = \sec^{-1}(\tan\sqrt{x})$

15. $y = \sinh(5x^2 + 4)$

16. $y = \cosh^2(2x^2)$

17. $y = \tanh(\ln\operatorname{sech} x)$

18. $y = \operatorname{sech}^{-1}(x\sqrt{x})$

19. $y = \sinh^{-1}(\tanh^{-1}x)$

20. $y = \tanh^{-1}(\ln|\cosh x|)$

In Exercises 21–42, evaluate the integral.

21. $\displaystyle\int 3x^2\cos(x^3)\,dx$

22. $\displaystyle\int \cos^2 2x - \sin^2 2x\,dx$

23. $\displaystyle\int \frac{\sin\sqrt{x}\cos\sqrt{x}}{\sqrt{x}}\,dx$

24. $\displaystyle\int e^{2x}\sin(e^{2x})\,dx$

25. $\displaystyle\int \sec^5 x\tan x\,dx$

26. $\displaystyle\int \frac{\sec^2 x}{\sqrt{1 + \tan^2 x}}\,dx$

27. $\displaystyle\int [2x\sec^2(x^2 + x) + \sec^2(x^2 + x)]\,dx$

28. $\displaystyle\int \tan^5 x\sec^2 x\,dx$

29. $\displaystyle\int \frac{1}{\sqrt{9 - x^2}}\,dx$

30. $\displaystyle\int \frac{1}{\sqrt{1 - 9x^2}}\,dx$

31. $\displaystyle\int \frac{5}{1 + 25x^2}\,dx$

32. $\displaystyle\int \frac{dx}{|2x|\sqrt{4x^2 - 4}}$

33. $\displaystyle\int \frac{2x\,dx}{1 + x^4}$

34. $\displaystyle\int \frac{(\sec^{-1}x)^2}{|x|\sqrt{x^2 - 1}}\,dx$

35. $\displaystyle\int \frac{1}{x^2 + 4x}\,dx$ (Hint: complete the square)

36. $\displaystyle\int \frac{1}{\sqrt{24 - 2x - x^2}}\,dx$

37. $\int \dfrac{dx}{\sqrt{\dfrac{x^2}{4} - 1}}$

38. $\int \dfrac{dx}{\sqrt{\dfrac{x^2}{4} + 4}}$

39. $\int \dfrac{2\,dx}{x\sqrt{1 - x^4}}$

40. $\int \dfrac{dx}{-x^2 - 4x - 3}$

41. $\int \dfrac{dx}{\sqrt{4 + x^2}}$

42. $\int \dfrac{dx}{25 - 16x^2}$

---

43. Given $y'' + 16y = 0$, find the general solution.

44. Given $y'' + 16y = 0$, find the particular solution if $y(0) = 1$, and $y'(0) = 4$.

In Exercises 45–47, find a position function for each simple harmonic motion.

45. Period $\dfrac{\pi}{3}$, $s(0) = 1$, and $v(0) = -6$.

46. Frequency 2, amplitude 4, and $s(0) = 4$.

47. Amplitude 3, phase at $t = 0$ is $\dfrac{\pi}{6}$, and $v(0) = -6$.

48. A 30-lb weight stretches a spring 2 inches. Find the position function that describes the motion if the weight is released from a point 7 inches below the equilibrium point.

49. A $4\mu F$ capacitor is charged with $Q = 10\mu C$ and then shorted across a $25\mu H$ inductor. (a) What is the frequency of oscillation in the circuit? (b) At what time after the short occurs will the current be a maximum?

50. Find the area of the region determined by $y = \sin 2x$ from $x = 0$ to $x = \pi/4$.

51. Find the area of the region determined by $y = \cosh x$ from $x = 0$ to $x = 1$.

52. Find the line which is tangent to the curve $y = \sin^{-1}x$ at $(1, \pi/2)$.

53. Find an equation of the line which is normal to the curve $y = 2 \sin x$ at $(\pi/6, 1)$.

54. Use differentials to approximate $\sin^{-1}0.49$.

55. Determine the concavity of $\sin^{-1}(x^2)$ over the interval $(-1, 1)$.

56. Find the area of the region determined by $(4 + x^2)y^2 = 1$ ($y > 0$) from $x = 0$ to $x = 2$.

57. (Continuation of Exercise 56). If the region is revolved about the x-axis, find the volume generated.

9

METHODS OF INTEGRATION

i.e. make people sit on the bus together only provide 1 drinking fountain

All but one of our present integral formulas have been derived from known differentiation formulas. The lone exception is $\int 1/x\, dx$, which was used to define the natural logarithm. Such formulas are inadequate to handle many problems that arise in business, science, and engineering. Here are some examples.

Business/Economics: Finding the present value of future revenue may involve integrals of the form $\int e^{-ax} \sin bx\, dx$.

Chemistry/Physics: Predicting the location of the electron in a hydrogen atom involves an integral of the form $\int ax^2 e^{-bx}\, dx$.

Engineering: Finding the center of mass of a lamina with the shape of the region bounded by one hump of a sine curve and the x-axis involves integrals of the form $\int x \sin x\, dx$ and $\int \sin^2 x\, dx$.

Life Science: The logistic model of population growth involves an integral of the form $\int a/x(b-x)\, dx$.

Mathematics: Finding the length of a portion of the parabola $y = x^2/2$ involves an integral of the form $\int \sqrt{1+x^2}\, dx$.

Not one of the integrals above can be evaluated with the formulas developed so far. We obviously need to improve our methods of integration.

Each section in this chapter introduces a new method for evaluating certain types of integrals. But underlying all of these methods is a single principle: to somehow change the form of an integral that we cannot evaluate into one that we can evaluate. Your skill at evaluating integrals will depend on how good you are at choosing the right method and then applying it correctly.

409

Below, you will find a list of the integral formulas derived so far. We will refer to any formula in this list as a *basic formula*. Actually, the basic formulas are part of a Table of Integrals that we have put together and placed in Appendix A toward the end of the book.

Although it is a good idea for you to learn how to use a Table of Integrals, and our table is referred to often in this chapter, we would be remiss not to set out a word of caution about the table: unless you are instructed to do so, *do not use the Table of Integrals for your homework assignments.* There are at least two good reasons for this. (1) The exercises are designed to develop your understanding of the methods and skill at applying them. Once this is accomplished, you will be able to use the table more intelligently. (2) Working through the exercises will improve your skill in manipulating and simplifying integrals.

Section 9–7, *The Logistic Model of Population Growth,* is optional reading material. It describes how the model is set up, and how the method of integration called *partial fractions* is used to derive the logistic equation.

Basic Formulas

1. $\displaystyle\int x^n \, dx = \begin{cases} \dfrac{x^{n+1}}{n+1} + C & n \neq -1 \\ \ln|x| + C & n = -1 \end{cases}$

2. $\displaystyle\int e^x \, dx = e^x + C$

3. $\displaystyle\int a^x \, dx = \dfrac{a^x}{\ln a} + C$

4. $\displaystyle\int \sin x \, dx = -\cos x + C$

5. $\displaystyle\int \cos x \, dx = \sin x + C$

6. $\displaystyle\int \sec^2 x \, dx = \tan x + C$

7. $\displaystyle\int \csc^2 x \, dx = -\cot x + C$

8. $\displaystyle\int \sec x \tan x \, dx = \sec x + C$

9. $\displaystyle\int \csc x \cot x \, dx = -\csc x + C$

10. $\displaystyle\int \tan x \, dx = -\ln|\cos x| + C$

11. $\displaystyle\int \cot x \, dx = \ln|\sin x| + C$

12. $\displaystyle\int \sec x \, dx = \ln|\sec x + \tan x| + C$

13. $\displaystyle\int \csc x \, dx = \ln|\csc x - \cot x| + C$

14. $\displaystyle\int \frac{dx}{\sqrt{a^2 - x^2}} = \sin^{-1}\frac{x}{a} + C$

15. $\displaystyle\int \frac{dx}{a^2 + x^2} = \frac{1}{a}\tan^{-1}\frac{x}{a} + C$

16. $\displaystyle\int \frac{dx}{|x|\sqrt{x^2 - a^2}} = \frac{1}{a}\sec^{-1}\frac{x}{a} + C$

17. $\displaystyle\int \frac{dx}{\sqrt{a^2 + x^2}} = \ln|x + \sqrt{a^2 + x^2}| + C$

18. $\displaystyle\int \frac{dx}{\sqrt{x^2 - a^2}} = \ln|x + \sqrt{x^2 - a^2}| + C$

19. $\displaystyle\int \frac{dx}{a^2 - x^2} = \frac{1}{2a} \ln \left| \frac{x + a}{x - a} \right| + C$

20. $\displaystyle\int \frac{dx}{x\sqrt{a^2 - x^2}} = -\frac{1}{a} \ln \left| \frac{a + \sqrt{a^2 - x^2}}{x} \right| + C$

9–1   INTEGRATION BY PARTS

Because it can be applied to such a variety of integrals, integration by parts is a powerful tool. With it, we can evaluate integrals such as

$$\int x \cos x \, dx, \quad \int \sin^{-1}x \, dx, \quad \text{and} \quad \int e^x \sin x \, dx$$

In this section, we describe how the method works; in the next section, we illustrate how it is used to solve practical problems and to develop new integral formulas.

The method of integration by parts is derived from the product rule. The product rule, in differential form, reads $d(uv) = u \, dv + v \, du$ which can be rewritten as

$$u \, dv = d(uv) - v \, du$$

Now we integrate both sides of this equation and obtain

$$\int u \, dv = uv + C - \int v du \qquad \left(\int d(uv) = uv + C \right)$$

But the integral on the right already has a built-in arbitrary constant, so we will drop the C and write

(9.1) $$\boxed{\int u \, dv = uv - \int v \, du}$$

This is the *integration by parts* formula. As the name suggests, the integral is partitioned into parts; a function part u and a differential part dv. Then the value of the integral is uv minus the integral of $v \, du$. Here is how it works.

EXAMPLE 1   Evaluate $\int x \cos x \, dx$.

Solution   Without the factor x in the integrand, this would be an elementary problem. But as it stands, we need integration by parts. To integrate by parts, first decide what part is u. Then the remaining part is dv. One possibility for this integral is to let $u = x$ and $dv = \cos x \, dx$. According to (9.1), you also need to know v and du. A simple way to keep track of these functions is to write them as follows.

$$u = x \qquad dv = \cos x \, dx$$
$$du = dx \qquad v = \sin x$$

The function $v = \sin x$ is obtained by integrating $dv = \cos x \, dx$, but, for the reason stated earlier, no constant is added yet. This array, together with (9.1) yields

$$\underbrace{\int x}_{u} \underbrace{\cos x \, dx}_{dv} = \underbrace{x}_{u} \underbrace{\sin x}_{v} - \int \underbrace{\sin x}_{v} \underbrace{dx}_{du}$$

Observe that integration by parts converts the difficult problem of evaluating $\int x \cos x \, dx$ into the simple problem of evaluating $\int \sin x \, dx$. Thus,

$$\int x \cos x \, dx = x \sin x + \cos x + C \qquad \text{(Add constant now)}$$

Check.

$$\frac{d}{dx}(x \sin x + \cos x) = \underbrace{(x \cos x + \sin x)}_{\text{product rule}} - \sin x = x \cos x \qquad \blacksquare$$

The Role of Guesswork. In order to evaluate an integral $\int u \, dv$ by parts, you eventually have to evaluate the integral $\int v \, du$. If the second integral is more difficult to evaluate than the first, nothing is gained. But that does not necessarily mean that the original integral cannot be evaluated. Perhaps you chose the wrong u and dv. For instance, in Example 1, if we tried

$$u = \cos x \qquad dv = x \, dx$$
$$du = -\sin x \, dx \qquad v = x^2/2$$

Then

$$\underbrace{\int x \cos x \, dx}_{u \, dv} = \underbrace{\frac{1}{2}x^2}_{v} \underbrace{\cos x}_{u} + \int \underbrace{\frac{1}{2}x^2}_{v} \underbrace{\sin x \, dx}_{du}$$

Although this equation is correct, you are no closer to the desired answer because the last integral contains x^2 rather than x; it is more difficult to evaluate than the original integral. When this happens, ERASE and start over. Remember, evaluating an integral is a *trial* and *check* procedure.

For a definite integral, equation (9.1) becomes

(9.2)
$$\int_a^b u \, dv = uv \Big|_a^b - \int_a^b v \, du$$

EXAMPLE 2 Evaluate $\int_0^1 x\, e^x\, dx$.

Solution Again, the factor x forces us to integrate by parts. Prepare to use integration by parts by choosing u and dv; then compute du and v. *Try*

$$u = x \qquad dv = e^x\, dx$$
$$du = dx \qquad v = e^x$$

Eventually you will have to integrate $v\, du$; you can save some writing by mentally *testing $v\, du = e^x\, dx$*. This is easy to integrate, so proceed with

$$\int \underbrace{x}_{u}\ \underbrace{e^x\, dx}_{dv} = \underbrace{x}_{u}\ \underbrace{e^x}_{v} - \int \underbrace{e^x}_{v}\ \underbrace{dx}_{du}$$
$$= x\, e^x - e^x + C$$

Check.

$$\frac{d}{dx}(x\, e^x - e^x) = (x\, e^x + e^x) - e^x = x\, e^x$$

Now use (9.2):

$$\int_0^1 x\, e^x = \left[x\, e^x - e^x \right]_0^1 = (1e^1 - e^1) - (0e^0 - e^0) = 1 \qquad \blacksquare$$

Integration by parts is a "natural" for evaluating integrals such as $\int x \cos x\, dx$ and $\int xe^x\, dx$ because the integrands are products that fit easily into the pattern of $u\, dv$. It is not so apparent, however, that integrals such as $\int \ln x\, dx$ and $\int \sin^{-1} x\, dx$ can also be integrated by parts. But they can.

EXAMPLE 3 Evaluate $\int \ln ax\, dx$, $a > 0$.

Solution *Try*

$$u = \ln ax \qquad dv = dx$$
$$du = \frac{1}{x}dx \qquad v = x$$

Test (mentally): $v\, du = x(1/x)\, dx = a\, dx$, so go ahead with

$$\int \underbrace{\ln ax}_{u}\ \underbrace{dx}_{dv} = \underbrace{x}_{v}\ \underbrace{\ln ax}_{u} - \int \underbrace{x}_{v}\ \underbrace{\left(\frac{1}{x}\right)dx}_{du}$$
$$= x \ln ax - x + C$$

$$\boxed{\int \ln ax\, dx = x \ln ax - x + C}$$

This is Formula **74** in the Table of Integrals in Appendix A.

Check.

$$\frac{d}{dx}[\text{answer}] = \underbrace{\left[x\left(\frac{1}{x}\right) + \ln ax\right]}_{\text{Product rule}} - 1 = \ln ax \qquad \blacksquare$$

EXAMPLE 4   Integrate by parts and verify Formula **64** in the Table of Integrals.

$$\int \sin^{-1}ax \, dx = x \sin^{-1}ax + \frac{1}{a}\sqrt{1 - a^2x^2} + C$$

Solution   *Try*

$$u = \sin^{-1}ax \qquad dv = dx$$

$$du = \frac{a \, dx}{\sqrt{1 - a^2x^2}} \qquad v = x$$

Test: $v \, du = (ax/\sqrt{1 - a^2x^2}) \, dx = ax(1 - a^2x^2)^{-1/2} \, dx$ is a basic power, so proceed with

$$\int \underbrace{\sin^{-1}ax}_{u} \underbrace{dx}_{dv} = \underbrace{x}_{v} \underbrace{\sin^{-1}ax}_{u} - \int \underbrace{\frac{ax}{\sqrt{1 - a^2x^2}}dx}_{v \, du}$$

$$= x \sin^{-1}ax + \frac{1}{a}\sqrt{1 - a^2x^2} + C$$

and the formula is verified. $\blacksquare$

It often happens that in the process of evaluating $\int v \, du$, the original integral $\int u \, dv$, or something very similar to it, turns up. The next two examples illustrate what to do in this situation.

EXAMPLE 5   Use integration by parts to evaluate $\int \sec^3 x \, dx$.

Solution   *Try*

$$u = \sec^3 x \qquad\qquad dv = dx$$

$$du = 3 \sec^2 x (\sec x \tan x) \, dx \qquad v = x$$

$$= 3 \sec^3 x \tan x \, dx$$

Test: $v \, du$ does not look promising; we ERASE and try again. This time we write $\sec^3 x \, dx$ as $\sec x \sec^2 x \, dx$ and *try*

$$u = \sec x \qquad\qquad dv = \sec^2 x \, dx$$

$$du = \sec x \tan x \, dx \qquad v = \tan x$$

Test: $v\,du = \tan^2 x \sec x\,dx$ still does not look promising, but maybe something can be done using the identity $\tan^2 x = \sec^2 x - 1$. For instance,

(9.3) $v\,du = \tan^2 x \sec x\,dx = \sec^3 x\,dx - \sec x\,dx$

Let's try it.

$$\int \sec^3 x\,dx = \int \underbrace{\sec x}_{u}\ \underbrace{\sec^2 x\,dx}_{dv} = \underbrace{\sec x}_{u}\ \underbrace{\tan x}_{v} - \int \underbrace{\tan x}_{v}\ \underbrace{\sec x \tan x\,dx}_{du}$$

$$= \sec x \tan x - \int (\sec^3 x - \sec x)\,dx \qquad (9.3)$$

$$= \sec x \tan x - \int \sec^3 x + \int \sec x\,dx$$

Observe that the original integral $\int \sec^3 x\,dx$ appears on both sides. The result is an equation of the form $a = b - a + c$, and solving for a yields $a = \frac{1}{2}(b + c)$. Therefore,

$$\int \sec^3 x\,dx = \frac{1}{2}\left[\sec x \tan x + \int \sec x\,dx\right]$$

Our Basic Formula (12) is $\int \sec x\,dx = \ln|\sec x + \tan x|$, so

$$\int \sec^3 x\,dx = \frac{1}{2}\sec x \tan x + \frac{1}{2}\ln|\sec x + \tan x| + C$$

Check.

$$\frac{d}{dx}[\text{answer}] = \frac{1}{2}[(\sec x \sec^2 x + \tan x \sec x \tan x) + \sec x]$$

$$= \frac{1}{2}[\sec^3 x + \underbrace{(\sec^2 x - 1)}_{\tan^2 x}\sec x + \sec x] = \sec^3 x \qquad \blacksquare$$

EXAMPLE 6   Evaluate $\int e^x \sin x\,dx$.

Solution   *Try*

$$u = e^x \qquad dv = \sin x\,dx$$
$$du = e^x\,dx \qquad v = -\cos x$$

Test: $v\,du = -e^x \cos x\,dx$ looks very much like the original integral, but let us proceed and see what happens.

$$\int e^x \sin x\,dx = -e^x \cos x + \int e^x \cos x\,dx$$

Now integrate $\int e^x \cos x$ by parts. *Try*

$$u = e^x \qquad dv = \cos x\,dx$$
$$du = e^x\,dx \qquad v = \sin x$$

Test: $v \, du = e^x \sin x \, dx$ is our original integral. Thus,

$$\int e^x \sin x \, dx = -e^x \cos x + \left[e^x \sin x - \int e^x \sin x \, dx \right]$$

Remove the brackets, and proceed as in Example 5. Then

$$\int e^x \sin x \, dx = \frac{1}{2} e^x [\sin x - \cos x] + C$$

Check.

$$\frac{d}{dx}[\text{answer}] = \frac{1}{2} e^x (\cos x + \sin x) + \frac{1}{2} e^x (\sin x - \cos x) = e^x \sin x \quad \blacksquare$$

EXERCISES

Use integration by parts to evaluate the following integrals. Check your answers *before* looking in the back of the book.

1. $\int x \sin x \, dx$

2. $\int x \sin 2x \, dx$

3. $\int x \cos 3x \, dx$

4. $\int x e^{3x} \, dx$

5. $\int x e^{2x} \, dx$

6. $\int x e^{x/2} \, dx$

7. $\int \ln 4x \, dx$

8. $\int \ln \dfrac{x}{3} \, dx$

9. $\int \sin^{-1} \dfrac{x}{2} \, dx$

10. $\int \sin^{-1} 5x \, dx$

11. $\int e^{2x} \sin 3x \, dx$

12. $\int e^{3x} \sin 2x \, dx$

13. $\int e^x \cos 4x \, dx$

14. $\int e^{-x} \cos x \, dx$

15. $\int \sec^3 5x \, dx$

16. $\int \sec^3 2x \, dx$

17. $\int x e^{-x} \, dx$

18. $\int x e^{-2x} \, dx$

19. $\int x \ln x \, dx$

20. $\int x \ln \dfrac{x}{2} \, dx$

21. $\int x \ln(x + 1) \, dx$

22. $\int x \ln(x + 4) \, dx$

23. $\int x \sqrt{x + 1} \, dx$

24. $\int x \sqrt{x + 5} \, dx$

25. $\int_1^e x^2 \ln x \, dx$

26. $\int_1^e x^3 \ln x \, dx$

27. $\int_0^1 \tan^{-1} x \, dx$

28. $\int_0^1 \cos^{-1} x \, dx$

29. $\int_0^{\pi/2} e^x \cos 2x \, dx$

30. $\int_0^{\pi/2} e^x \sin 2x \, dx$

9-2   APPLICATIONS

In this section, we describe three applications of integration by parts. The first, from the fields of business and economics, discusses the meaning of the term *present value* and how to compute the present value of future revenue. The second application uses integration by parts to introduce a new kind of integral formula called a *reduction formula*. The third, from the fields of chemistry and physics, discusses the electron cloud of a hydrogen atom and uses a reduction formula to predict the location of the electron.

Present Value

Suppose that you will receive A dollars t years from now, say, from an inheritance, a maturing bond, or whatever. If you need to borrow some money right now and pledge your future income as collateral, how much will a bank lend you? That is, what is the *present value* of your future income? Under the assumption of continuous compounding at an interest rate r, economists compute the present value $P.V.$ of A dollars t years from now by the formula

$$P.V. = Ae^{-rt}$$

This equation is derived by noting that interest accumulates exponentially.*

Now suppose that instead of a lump sum at the end of so many years, you are to receive a revenue of $R(t)$ dollars at time t for the next n years. If R is a continuous function, we refer to it as a *continuous revenue*. What is the present value of this revenue? To calculate it, we partition the time interval $[0, n]$ into small subintervals of length Δt. The revenue over the ith subinterval is approximately $R(t_i)\Delta t$. Therefore, its present value $P.V._i$ is

$$P.V._i \approx R(t_i)e^{-rt_i}\Delta t$$

The sum of the present values is a Riemann sum of $R(t)e^{-rt}$. This suggests that the present value of a continuous revenue $R(t)$ over a period of n years be defined as

(9.4)
$$P.V. = \int_0^n R(t)e^{-rt}\, dt$$

EXAMPLE 1   What is the present value of a revenue $R(t) = 1,000t$ dollars (t in years) for the next 3 years if interest is 12% compounded continuously?

Solution   According to (9.4), the present value is

$$P.V. = \int_0^3 1,000te^{-0.12t}\, dt$$

Evaluation of this integral requires integration by parts. Let

$$u = t \quad dv = e^{-0.12t}\, dt$$

$$du = dt \quad v = -\frac{1}{0.12}e^{-0.12t}$$

---

*$(1 + r/n)^{nt}$ is the amount accumulated after t years if one dollar is invested at interest r compounded n times a year. For continuous compounding, we let n get larger and larger; then $(1 + r/n)^{nt} \to e^{rt}$. See Exercise 63 at the end of Chapter 7.

Then

$$P.V. = 1{,}000\left[-\frac{t}{0.12}e^{-0.12t}\Big|_0^3 + \frac{1}{0.12}\int_0^3 e^{-0.12t}\,dt\right]$$

$$= 1{,}000\left[-\frac{t}{0.12}e^{-0.12t} - \frac{1}{(0.12)^2}e^{-0.12t}\right]_0^3$$

$$= 1{,}000\left[-\frac{1}{0.04}e^{-0.36} - \frac{1}{0.0144}e^{-0.36} + \frac{1}{0.0144}\right]$$

$$\approx 3{,}553 \text{ dollars} \quad \blacksquare$$

Reduction Formulas

In the process of integration by parts, it often happens that the integrand $v\,du$ does not fit into one of the basic formulas, but it does reduce a power that appears in the original integrand $u\,dv$. In that case, we use a succession of integrations by parts that eventually lead to an answer.

EXAMPLE 2 Evaluate $\int x^3 \cos x\,dx$.

Solution *Try*

$$u = x^3 \qquad dv = \cos x\,dx$$
$$du = 3x^2\,dx \qquad v = \sin x$$

Test (mentally): $v\,du = 3x^2 \sin x\,dx$. This is not one of the basic formulas, but the power x^3 is reduced to x^2, so let's continue with

$$\int \underset{u}{x^3}\,\underset{dv}{\cos x\,dx} = \underset{uv}{x^3 \sin x} - \int \underset{v}{\sin x}\,\underset{du}{3x^2\,dx}$$

$$= x^3 \sin x - 3\int x^2 \sin x\,dx$$

Now we evaluate $\int x^2 \sin x\,dx$ with a second integration by parts. *Try*

$$u = x^2 \qquad dv = \sin x\,dx$$
$$du = 2x\,dx \qquad v = -\cos x$$

Test: $v\,du = -2x \cos x$; again the power of x is reduced, and

$$\int \underset{u}{x^2}\,\underset{dv}{\sin x\,dx} = \underset{uv}{-x^2 \cos x} - \int \underset{v}{-\cos x}\,\underset{du}{2x\,dx}$$

$$= -x^2 \cos x + 2\int x \cos x\,dx$$

Combining this with the result of the first integration above, we have

$$\int x^3 \cos x\,dx = x^3 \sin x + 3x^2 \cos x - 6\int x \cos x\,dx$$

We could perform a third integration by parts to evaluate $\int x \cos x \, dx$, but we already know its value, $x \sin x + \cos x$, from Example 1 in Section 9–1. Therefore,

$$\int x^3 \cos x \, dx = x^3 \sin x + 3x^2 \cos x - 6x \sin x - 6 \cos x + C$$

is the final answer.

Check.

$$\frac{d}{dx}[\text{answer}] = [x^3 \cos x + 3x^2 \sin x] + [3x^2(-\sin x) + 6x \cos x]$$

$$- [6x \cos x + 6 \sin x] - [-6 \sin x]$$

$$= x^3 \cos x \quad \blacksquare$$

In Example 2, each time we integrated by parts, the power of x was reduced. Formulas that reduce powers are called **reduction formulas.** There are several of them in the Table of Integrals in Appendix A. Here is an example of how to derive and use a reduction formula.

EXAMPLE 3 Find a reduction formula for $\int x^n e^{ax} \, dx$, and use it to evaluate $\int x^3 e^{2x} \, dx$.

Solution *Try*

$$u = x^n \qquad\qquad dv = e^{ax} \, dx$$

$$du = nx^{n-1} \, dx \qquad v = \frac{1}{a}e^{ax} \qquad\qquad \textit{(Check)}$$

Then $\int u \, dv = uv - \int v \, du$ yields

$$\boxed{\int x^n e^{ax} \, dx = \frac{1}{a}x^n e^{ax} - \frac{n}{a} \int x^{n-1} e^{ax} \, dx}$$

This is Formula **70** in the Table of Integrals. Each time it is applied, the power of x is reduced by 1. Repeated application will eventually lead to evaluating the integral $\int e^{ax} \, dx$.

Without the reduction formula above, we can evaluate $\int x^3 e^{2x} \, dx$ using three integrations by parts (as we did in Example 2). But, just as an illustration, let us use the formula. With $n = 3$ and $a = 2$, we have

$$\int x^3 e^{2x} \, dx = \frac{1}{2}x^3 e^{2x} - \frac{3}{2} \int x^2 e^{2x} \, dx$$

Apply **70** again to the last integral with $n = 2$ and $a = 2$.

$$= \frac{1}{2}x^3 e^{2x} - \frac{3}{2}\left[\frac{1}{2}x^2 e^{2x} - \int xe^{2x} \, dx\right]$$

Apply **70** once more with $n = 1$ and $a = 2$.

$$= \frac{1}{2}x^3e^{2x} - \frac{3}{4}x^2e^{2x} + \frac{3}{2}\left[\frac{1}{2}xe^{2x} - \frac{1}{2}\int e^{2x}\,dx\right]$$

$$= \frac{1}{2}x^3e^{2x} - \frac{3}{4}x^2e^{2x} + \frac{3}{4}xe^{2x} - \frac{3}{8}e^{2x} + C$$

Check.

$$\frac{d}{dx}[\text{answer}] = \left(x^3e^{2x} + \frac{3}{2}x^2e^{2x}\right) - \left(\frac{3}{2}x^2e^{2x} + \frac{3}{2}xe^{2x}\right)$$

$$+ \left(\frac{3}{2}xe^{2x} + \frac{3}{4}e^{2x}\right) - \frac{3}{4}e^{2x} = x^3e^{2x} \qquad\blacksquare$$

The Electron Cloud

The hydrogen atom is composed of one proton in the nucleus and one electron moving about the nucleus. But the electron does not move in a fixed orbit. In quantum mechanics, the nucleus is thought of as imbedded in a fog, or cloud, of negative charge. At the state of lowest energy, called the *ground state*, the shape of this electron cloud is a sphere with center at the nucleus (Figure 9–1). What we want to do is calculate the probability of finding the electron between spheres of radii r_1 and r_2.

The probability density function (*pdf*) is

(9.5) $$p(r) = \frac{1}{\pi a_0^3}e^{-2r/a_0}$$

where r is the distance (meters) from the nucleus to the electron and $a_0 = .529 \times 10^{-10}$ is called the Bohr radius. Equation (9.5) is a consequence of the famous Schrödinger wave equation (which we are not prepared to discuss here). Now we partition the interval $[r_1, r_2]$ into small subintervals of length Δr. The subintervals determine thin spherical shells, and the volume of the ith shell is approximately the surface area $4\pi r_i^2$ times the thickness Δr. It follows from (9.5) that the probability of finding the electron in this shell is approximately

$$4\pi r_i^2 \Delta r\, p(r_i) = \frac{4r_i^2}{a_0^3}e^{-2r_i/a_0}\Delta r$$

Adding up the probabilities and taking the limit as $\Delta r \to 0$, we have that

(9.6) $$\boxed{P = \int_{r_1}^{r_2} \frac{4}{a_0^3}r^2e^{-2r/a_0}\,dr}$$

is the probability of finding the electron between spheres of radius r_1 and r_2. Figure 9–2 shows the graph of the integrand.

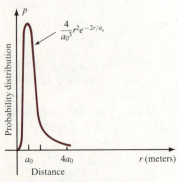

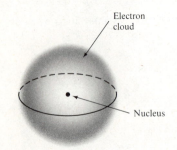

FIGURE 9–1. The electron cloud of a hydrogen atom.

FIGURE 9–2. The probability of finding the electron within a sphere of radius r is the area under the curve from 0 to r.

EXAMPLE 4   What is the probability P of finding the electron between spheres of radii $2a_0$ and $4a_0$?

Solution   According to (9.6),

$$P = \frac{4}{a_0{}^3} \int_{2a_0}^{4a_0} r^2 e^{-2r/a_0}\, dr$$

We can integrate by parts twice or use Formula **70** twice (setting $a = -2/a_0$). In either case (check this out),

$$P = -\left(\frac{2}{a_0{}^2}r^2 + \frac{2}{a_0}r + 1\right)e^{-2r/a_0}\Big|_{2a_0}^{4a_0}$$

$$= -41e^{-8} + 13e^{-4} \approx .22 \quad \blacksquare$$

EXERCISES

Use integration by parts to evaluate the integrals in Exercises 1–10. Check your answers.

1. $\int x \sin x\, dx$

2. $\int x \sin 3x\, dx$

3. $\int_0^{1/2} \cos^{-1} 2x\, dx$

4. $\int_0^1 \tan^{-1}x\, dx$

5. $\int xe^{-x}\, dx$

6. $\int e^x \cos x\, dx$

7. $\int \sin(\ln x)\, dx$

8. $\int x \ln x\, dx$

9. $\int \csc^3 x\, dx$

10. $\int x \sec^2 x\, dx$

In Exercises 11–30, evaluate the integrals by any method. (Some fit into one of the basic formulas, some require u-substitution, and some require integration by parts.)

11. $\int_0^1 \frac{x}{\sqrt{4 - x^2}}\, dx$

12. $\int x^2 \cos x\, dx$

13. $\int x^2 e^{-x}\, dx$

14. $\int_{-2}^0 (x^2 + 3x + 2)\, dx$

15. $\int xe^{x^2}\, dx$

16. $\int \frac{x}{4 + 9x^2}\, dx$

17. $\int \frac{2x}{\sqrt{1 - x^4}}\, dx$

18. $\int x \tan^{-1}x\, dx$

19. $\int x^2 \sin x\, dx$

20. $\int x^{-2} \sin(x^{-1})\, dx$

21. $\int x(1 - x)^{1/3}\, dx$

22. $\int \sec^2 x\, dx$

23. $\int \left(\frac{1}{x - 1} - \frac{1}{x - 2}\right)\, dx$

24. $\int e^x \sinh x\, dx$

25. $\int \sec^3 2x\, dx$

26. $\int x(2x - 1)^6\, dx$

27. $\int e^{2x} \sin 3x\, dx$

28. $\int x \cosh 2x\, dx$

29. $\int \tan^{-1} 2x\, dx$

30. $\int \frac{x}{\sqrt{1 - x^2}}\, dx$

31. Derive a reduction formula for $\int \sin^n ax\, dx$. Let $u = \sin^{n-1}ax$ and $dv = \sin ax\, dx$; then $v\, du$ contains $\cos^2 ax$, which can be replaced by $1 - \sin^2 ax$. The answer is Formula **53** in the Table of Integrals.

$$\int \sin^n ax\, dx = \frac{-\sin^{n-1}ax \cos ax}{na}$$
$$+ \frac{n - 1}{n}\int \sin^{n-2}ax\, dx$$

Use Formula **53** to compute $\int_0^{\pi/2}\sin^3x\, dx$.

32. Use Formulas **54** and **52** to evaluate $\int_0^{\pi/2}\cos^4x\, dx$.

33. Verify the reduction Formula **56**

$$\int \sec^n ax \, dx = \frac{\sec^{n-2} ax \tan ax}{a(n-1)}$$
$$+ \frac{n-2}{n-1} \int \sec^{n-2} ax \, dx$$

and use it to evaluate $\int \sec^4 x \, dx$.

34. Use Formula **57** to evaluate $\int \tan^3 x \, dx$.

35. Derive a formula for $\int x^n \ln ax \, dx, n \neq -1$. (Let $u = \ln ax$ and $dv = x^n \, dx$.) Use your answer to evaluate $\int_1^2 x^3 \ln 3x \, dx$.

36. Derive a reduction formula for $\int x^n \sin ax \, dx$ (check with Formula **62** in the table).

37. Derive a reduction formula for $\int x^n \cos ax \, dx$ (check with Formula **63** in the table).

38. Derive a reduction formula for $\int \tan^n ax \, dx$ (check with Formula **57** in the table).

---

39. The region determined by $y = \sin x$ from $x = 0$ to $x = \pi$ is revolved about the y-axis. Find the volume generated.

40. The region bounded by the graphs of $y = (1 + x)^{1/3}$, $y = 0, x = 1$, and $x = 2$ is revolved about the y-axis. Find the volume generated.

41. If X is a continuous random variable with values in $[0, \pi/2]$ and its *pdf* (probability density function) is $p(x) = \cos x$, what is the expected value of X?

42. (*Electron probability cloud*) The probability that the electron of a hydrogen atom is located within a sphere of radius r_1 meters from the nucleus is given by the integral

$$\int_0^{r_1} \frac{4}{a_0^3} r^2 e^{-2r/a_0} \, dr \qquad (9.6)$$

(a) Evaluate the integral.   (b) Show that the probability is always less than 1. Thus, the probability that the electron is outside of any fixed sphere is positive. This means that, theoretically, the electron can be arbitrarily far from the nucleus.   (c) Find the probability of finding the electron in a sphere of radius 2.22×10^{-10} meters.

43. (*Present value of revenue*) If the revenue from an investment is assumed to flow at a rate of $R = R(t)$ dollars at time t, and this revenue is to be discounted at a constant annual rate of interest r for n years, then the present value of the revenue is

$$P.V. = \int_0^n R(t) e^{-rt} \, dt \qquad (9.4)$$

For a "growth company," $R(t)$ is a steadily increasing function; for a "cyclical company," $R(t)$ fluctuates periodically. Find the present value of a 3-year investment in a cyclical company that is expected to bring in revenue at the rate of $R(t) = 1,000 \sin t$ dollars. Assume that money is discounted at an annual rate of 8%.

44. Find general solutions of these linear equations and check your answers.

(a) $y' + y = x$ (b) $y' + (y/x) = \sin x$

9–3   TRIGONOMETRIC INTEGRALS

Formulas **53–55** in the Table of Integrals are reduction formulas for powers of sine and cosine. But these formulas are seldom used because there are alternate methods that are more direct and easier to apply. The plan of attack depends on whether the powers are odd or even.

Odd Powers of Sine and Cosine

If n is odd, then $n - 1$ is even. Therefore,

$$\sin^n x = \sin x \sin^{n-1} x = (\sin x)(\sin^2 x)^{(n-1)/2}$$

When $1 - \cos^2 x$ is substituted for $\sin^2 x$, the result is the sine times a polynomial in cosines. Once this is written out, a basic power formula is used to evaluate the integral. A similar treatment works for an odd power of the cosine.

EXAMPLE 1

$$\int \sin^5 x \, dx = \int \sin x (\sin^2 x)^2 \, dx$$

$$= \int \sin x (1 - \cos^2 x)^2 \, dx$$

$$= \int \sin x (1 - 2 \cos^2 x + \cos^4 x) \, dx$$

$$= \int \sin x \, dx - 2 \int \sin x \cos^2 x \, dx + \int \sin x \cos^4 x \, dx$$

$$= -\cos x + \frac{2}{3} \cos^3 x - \frac{1}{5} \cos^5 x + C$$

Check.

$$\frac{d}{dx} [\text{answer}] = \sin x - 2 \sin x \cos^2 x + \sin x \cos^4 x$$

$$= \sin x (1 - 2 \cos^2 x + \cos^4 x)$$

$$= \sin x (1 - \cos^2 x)^2 = \sin^5 x \qquad \blacksquare$$

This method works equally well for products of powers of sine and cosine, provided that *one of the powers is odd.*

EXAMPLE 2

$$\int \sin^6 x \cos^3 x \, dx = \int \sin^6 x \cos^2 x \cos x \, dx$$

$$= \int \sin^6 x (1 - \sin^2 x) \cos x \, dx$$

$$= \int \sin^6 x \cos x \, dx - \int \sin^8 x \cos x \, dx$$

$$= \frac{1}{7} \sin^7 x - \frac{1}{9} \sin^9 x + C \qquad \textit{(Check)} \qquad \blacksquare$$

Even Powers of Sine and Cosine

For even powers, one or more of the following formulas can be used to advantage (see (1.55) and (1.56) in Section 1–7).

$$(1) \quad \frac{1}{2} \sin 2x = \sin x \cos x$$

(2) $\sin^2 x = \dfrac{1}{2} - \dfrac{1}{2}\cos 2x$

(3) $\cos^2 x = \dfrac{1}{2} + \dfrac{1}{2}\cos 2x$

For example, Formulas **51** and **52** in the table are

$$\int \sin^2 ax \, dx = \frac{x}{2} - \frac{1}{4a}\sin 2ax + C$$

$$\int \cos^2 ax \, dx = \frac{x}{2} + \frac{1}{4a}\sin 2ax + C$$

To verify **51**, use half angle formula (2); then

$$\int \sin^2 ax \, dx = \int \left(\frac{1}{2} - \frac{1}{2}\cos 2ax\right) dx$$

$$= \frac{1}{2}\int dx - \frac{1}{2}\left(\frac{1}{2a}\right)\int 2a \cos 2ax \, dx$$

$$= \frac{x}{2} - \frac{1}{4a}\sin 2ax + C$$

A similar treatment is valid for the cosine.

If n is even but greater than 2, then write

$$(\sin^2 x)^{n/2} = \left(\frac{1}{2} - \frac{1}{2}\cos 2x\right)^{n/2} \quad \text{or} \quad (\cos^2 x)^{n/2} = \left(\frac{1}{2} + \frac{1}{2}\cos 2x\right)^{n/2}$$

The result in either case is a polynomial in $\cos 2x$, which is not difficult to integrate.

EXAMPLE 3

$$\int \cos^4 x \, dx = \int (\cos^2 x)^2 \, dx$$

$$= \int \left(\frac{1}{2} + \frac{1}{2}\cos 2x\right)^2 dx$$

$$= \frac{1}{4}\int dx + \frac{1}{2}\int \cos 2x \, dx + \frac{1}{4}\int \cos^2 2x \, dx$$

The first two integrals are easy, and the last can be evaluated by writing $\cos^2 2x = \frac{1}{2} + \frac{1}{2}\cos 4x$ (or by using Formula **52** with $a = 2$). Thus,

$$\int \cos^4 x \, dx = \frac{1}{4} \int dx + \frac{1}{2} \int \cos 2x \, dx + \frac{1}{4} \int \left(\frac{1}{2} + \frac{1}{2}\cos 4x\right) dx$$

$$= \frac{3x}{8} + \frac{1}{4}\sin 2x + \frac{1}{32}\sin 4x + C \qquad \text{(Check)} \qquad ■$$

In a product of even powers of sine and cosine, double angle formula (1) comes into play.

EXAMPLE 4

$$\int \sin^4 x \, \cos^2 x \, dx = \int \sin^2 x (\sin x \cos x)^2 \, dx$$

$$= \int \left(\frac{1}{2} - \frac{1}{2}\cos 2x\right)\left(\frac{1}{2}\sin 2x\right)^2 dx$$

$$= \frac{1}{8} \int \sin^2 2x \, dx - \frac{1}{8} \int \cos 2x \sin^2 2x \, dx$$

The first integral can be evaluated using Formula **51** with $a = 2$, and the second is a basic power. Thus,

$$\int \sin^4 x \, \cos^2 x \, dx = \frac{1}{8}\left(\frac{x}{2} - \frac{1}{8}\sin 4x\right) - \frac{1}{48}\sin^3 2x + C$$

$$= \frac{x}{16} - \frac{1}{64}\sin 4x - \frac{1}{48}\sin^3 2x + C \quad \text{(Check)} \, . \qquad ■$$

Powers of Tangent

The reduction formula

57.
$$\int \tan^n ax \, dx = \frac{\tan^{n-1}ax}{a(n-1)} - \int \tan^{n-2}ax \, dx$$

can be derived using the identity $\tan^2 x = \sec^2 x - 1$. Thus,

$$\int \tan^n ax \, dx = \int \tan^{n-2}ax \, \tan^2 ax \, dx$$

$$= \int \tan^{n-2}ax(\sec^2 ax - 1) \, dx$$

$$= \int \tan^{n-2}ax \, \sec^2 ax \, dx - \int \tan^{n-2}ax \, dx$$

$$= \frac{\tan^{n-1}ax}{a(n-1)} - \int \tan^{n-2}ax \, dx$$

EXAMPLE 5 To evaluate $\int \tan^5 x \, dx$, you could use Formula **57** twice; once with $n = 5$ and then with $n = 3$. But if the table is not available, then write $\tan^5 x = \tan^3 x \tan^2 x$ and proceed as follows:

$$\int \tan^5 x \, dx = \int \tan^3 x (\sec^2 - 1) \, dx$$

$$= \int \tan^3 x \sec^2 x \, dx - \int \tan^3 x \, dx$$

$$= \int \tan^3 x \sec^2 x \, dx - \int \tan x (\sec^2 x - 1) \, dx$$

$$= \int \tan^3 x \sec^2 x \, dx - \int \tan x \sec^2 x \, dx + \int \tan x \, dx$$

$$= \frac{\tan^4 x}{4} - \frac{\tan^2 x}{2} - \ln|\cos x| + C \quad \blacksquare$$

Powers of Secant

For *even* powers of secant, we write

$$\sec^n x = (\sec^2 x)^{(n-2)/2} \sec^2 x = (1 + \tan^2 x)^{(n-2)/2} \sec^2 x$$

EXAMPLE 6

$$\int \sec^4 3x \, dx = \int \sec^2 3x \sec^2 3x \, dx$$

$$= \int (1 + \tan^2 3x) \sec^2 3x \, dx$$

$$= \int \sec^2 3x \, dx + \int \tan^2 3x \sec^2 3x \, dx$$

$$= \frac{1}{3}\tan 3x + \frac{1}{9}\tan^3 3x + C \quad \blacksquare$$

For *odd* powers of secant, we use integration by parts (see Example 5 of Section 9–1). Also, if a table is handy, there is a reduction formula

56. $\quad \boxed{\displaystyle\int \sec^n ax \, dx = \frac{\sec^{n-2} ax \tan ax}{a(n-1)} + \frac{n-2}{n-1}\int \sec^{n-2} ax \, dx}$

Products of Tangents and Secants; $\int \tan^m x \sec^n x \, dx$

(1) If $n = 2$, the integral is a basic power. If n is even but larger than 2, write

$$\tan^m x (\sec^2 x)^{(n-2)/2} \sec^2 x = \tan x (1 + \tan^2 x)^{(n-2)/2} \sec^2 x$$

Then expand and integrate the powers of $\tan x$ times $\sec^2 x$.

(2) If n and m are both odd, write

$$(\tan^2 x)^{(m-1)/2} \sec^{n-1} x \sec x \tan x = (\sec^2 x - 1)^{(m-1)/2} \sec^{n-1} x \sec x \tan x$$

Then expand and integrate the powers of $\sec x$ times $\sec x \tan x$.

(3) If n is odd and m is even, write the product entirely in terms of the secant and integrate by parts (or use Formula **56**).

EXAMPLE 7

$$\int \tan^3 x \sec^4 x \, dx = \int \tan^3 x (\sec^2 x)(\sec^2 x) \, dx$$

$$= \int \tan^3 x (1 + \tan^2 x)(\sec^2 x) \, dx$$

$$= \int \tan^3 x \sec^2 x \, dx + \int \tan^5 x \sec^2 x \, dx$$

$$= \frac{1}{4} \tan^4 x + \frac{1}{6} \tan^6 x + C \qquad \blacksquare$$

EXAMPLE 8

$$\int \tan^3 x \sec x \, dx = \int \tan^2 x \sec x \tan x \, dx$$

$$= \int (\sec^2 x - 1) \sec x \tan x \, dx$$

$$= \int \sec^2 x (\sec x \tan x) \, dx - \int \sec x \tan x \, dx$$

$$= \frac{1}{3} \sec^3 x - \sec x + C \qquad \blacksquare$$

EXAMPLE 9

$$\int \tan^2 x \sec x \, dx = \int (\sec^2 x - 1) \sec x \, dx$$

$$= \int \sec^3 x \, dx - \int \sec x \, dx$$

$$= \left[\frac{1}{2} \sec x \tan x + \frac{1}{2} \ln|\sec x + \tan x| \right]$$

$$- \ln|\sec x + \tan x| + C$$

$$= \frac{1}{2} \sec x \tan x - \frac{1}{2} \ln|\sec x + \tan x| + C \qquad \blacksquare$$

Product of Sine and Cosine

We want to evaluate $\int \sin ax \cos bx \, dx$. If $a = \pm b$, there is no difficulty; an answer is $(\sin^2 ax)/2a$. But, if $a \neq \pm b$, then

59.
$$\int \sin ax \cos bx \, dx = -\frac{\cos(a+b)x}{2(a+b)} - \frac{\cos(a-b)x}{2(a-b)} + C$$

To verify this, we use the sum formulas for the sine:

$$\sin(a+b)x = \sin(ax+bx) = \sin ax \cos bx + \cos ax \sin bx$$
$$\sin(a-b)x = \sin(ax-bx) = \sin ax \cos bx - \cos ax \sin bx$$

Adding these equations yields

$$\sin ax \cos bx = \frac{1}{2}\sin(a+b)x + \frac{1}{2}\sin(a-b)x$$

Integrating both sides proves Formula **59**.

EXAMPLE 10

$$\int \sin 2x \cos 5x \, dx = -\frac{\cos 7x}{2(7)} - \frac{\cos(-3)x}{2(-3)} + C$$

$$= -\frac{\cos 7x}{14} + \frac{\cos 3x}{6} + C \qquad \blacksquare$$

Similar formulas hold for integrals of $\sin ax \sin bx$ and $\cos ax \cos bx$ (see Exercise 31).

EXERCISES

Evaluate the integrals in Exercises 1–10 using the methods of this section.

1. $\int \cos^3 x \, dx$

2. $\int \tan^2 4x \, dx$

3. $\int \sin^3 x \cos^2 x \, dx$

4. $\int \sec^4 3x \, dx$

5. $\int \tan^5 x \sec^4 x \, dx$

6. $\int \sin^6 x \, dx$

7. $\int \sin^2 x \cos^4 x \, dx$

8. $\int \cos 2x \sin 3x \, dx$

9. $\int \tan^5 x \sec^3 x \, dx$

10. $\int \tan^5 3x \, dx$

Evaluate the integrals in Exercises 11–30 using any method.

11. $\int \frac{\sin x}{1 + 2\cos x} \, dx$

12. $\int \frac{1}{\cos^2 x} \, dx$

13. $\int \sec^2 3x \tan 3x \, dx$

14. $\int x\sqrt{4-x^2} \, dx$

15. $\int \frac{\cos^3 x}{\sin^2 x} \, dx$

16. $\int \frac{\sec^2 x}{2 + \tan x} \, dx$

17. $\int x^3 \ln 3x \, dx$

18. $\int \tan^3 x \sec x \, dx$

19. $\int \frac{1}{4 + x^2} \, dx$

20. $\int \sec^3 x \, dx$

21. $\int \sec^5 x \, dx$

22. $\int \sin x \sqrt{1 + \cos x} \, dx$

23. $\int \sin^{3/2} x \cos x \, dx$

24. $\int xe^x \, dx$

25. $\int e^{2x} \sin 4x \, dx$

26. $\int e^{4x} \cos 2x \, dx$

27. $\int \sin \sqrt[3]{x}\, dx$ (let $y = \sqrt[3]{x}$, so $x = y^3$, and $dx = 3y^2 dy$)

28. $\int \ln(x + 1)\, dx$ (let $y = x + 1$, so $x = y - 1$, and $dx = dy$)

29. $\int_0^{2\pi} \sin mx \cos nx\, dx$ **30.** $\int_0^{2\pi} \cos^2 nx\, dx$

31. Use the sum formulas for $\cos(a + b)x = \cos(ax + bx)$ and $\cos(a - b)x = \cos(ax - bx)$ to verify Formulas **60** and **61** in the Table of Integrals.

$$\int \sin ax \sin bx\, dx =$$

$$\int \cos ax \cos bx\, dx =$$

32. Find the length of the curve $y = \ln(\cos x)$ from $x = 0$ to $x = \pi/3$.

33. A particle moves along a line, and its velocity function is $v = \cos^2 t$ ft/sec. Find its acceleration and position functions.

34. *(Measuring current)* Ammeters are used to measure current I in an electrical system. Most *ac* ammeters, however, are constructed to show the *root mean square* I_{rms} of the current. By definition,

$$I_{rms} = \left[\frac{1}{T} \int_0^T I^2\, dt \right]^{1/2}$$

where T is the period of the alternating current. What will an ammeter read if $I = 3 \sin(100t)$?

35. Find the center of mass of a lamina with density δ and the shape of the region bounded by the graphs of (a) $y = e^x$, $y = 0$, $x = 0$, and $x = 1$ (b) $y = \sec^2 x$, $y = 0$, $x = 0$, and $x = 1$.

Optional Exercises

36. Evaluate $\int x \sin^2 x\, dx$.

37. Find the center of mass of a lamina with density δ and the shape of the region bounded by the graph of $y = \sin^2 x$ and the x-axis from $x = 0$ to $x = \pi$.

9-4   INTEGRALS INVOLVING $a^2 + x^2$, $a^2 - x^2$, AND $x^2 - a^2$

Formulas **14-41** in the Table of Integrals treat integrals involving the quadratic expressions $a^2 + x^2$, $a^2 - x^2$, and $x^2 - a^2$. All of them can be derived by a single method known as **trigonometric substitution**. In trigonometric substitution, x is replaced by one of the functions $a \tan u$, $a \sin u$, or $a \sec u$, depending on the quadratic. Of course, dx must also be replaced by the appropriate differential.

	QUADRATIC	REPLACE x BY	REPLACE dx BY
(9.7)	$a^2 + x^2$	$x = a \tan u$	$dx = a \sec^2 u\, du$
	$a^2 - x^2$	$x = a \sin u$	$dx = a \cos u\, du$
	$x^2 - a^2$	$x = a \sec u$	$dx = a \sec u \tan u\, du$

If x is replaced by $a \tan u$, then $u = \tan^{-1}(x/a)$. Thus, u is the angle whose tangent is x/a; this is illustrated in Figure 9-3A. Figures 9-3B and C illustrate the other two substitutions.

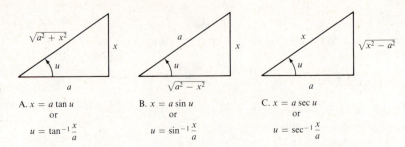

FIGURE 9–3. Diagrams illustrating the various trigonometric substitutions. Observe how the particular quadratic $a^2 + x^2$, $a^2 - x^2$, or $x^2 - a^2$ shows up according to the substitution used.

A. $x = a \tan u$
or
$u = \tan^{-1} \dfrac{x}{a}$

B. $x = a \sin u$
or
$u = \sin^{-1} \dfrac{x}{a}$

C. $x = a \sec u$
or
$u = \sec^{-1} \dfrac{x}{a}$

If an integral involves one of the quadratics $a^2 + x^2$, $a^2 - x^2$, or $x^2 - a^2$ and it cannot be evaluated by methods discussed so far, then the replacements displayed in table (9.7) will usually transform the integrand into a recognizable form.

The Quadratic $a^2 + x^2$

If x is replaced by $a \tan u$, then

(9.8)

$$a^2 + x^2 = a^2 + a^2 \tan^2 u = a^2(1 + \tan^2 u) = a^2 \sec^2 u$$

and

$$dx = a \sec^2 u \, du$$

EXAMPLE 1   Evaluate $\int \sqrt{9 + x^2} \, dx$.

Solution   The quadratic is $a^2 + x^2$ with $a = 3$. If $x = 3 \tan u$, then, by (9.8), $9 + x^2 = 9 \sec^2 u$ and $dx = 3 \sec^2 u \, du$. Therefore,

$$\int \sqrt{9 + x^2} \, dx = \int \sqrt{9 \sec^2 u}\,(3 \sec^2 u) \, du$$

$$= 9 \int \sec^3 u \, du \qquad \text{(See Remark 1 below)}$$

$$= 9 \left[\frac{\sec u \tan u}{2} + \frac{1}{2} \ln |\sec u + \tan u| \right] + C$$

To obtain the last line, integrate by parts or use Formula **56**. The final answer, of course, must be rewritten in terms of x, and this is where Figure 9–3 plays its part. Since we used the substitution $x = 3 \tan u$, the values of $\sec u$ and $\tan u$ can be read off from Figure 9–3A with $a = 3$.

$$\sec u = \frac{\sqrt{9 + x^2}}{3} \quad \text{and} \quad \tan u = \frac{x}{3}$$

Therefore,

$$\int \sqrt{9 + x^2}\, dx = \left(\frac{9}{2}\right) \frac{\sqrt{9 + x^2}}{3}\left(\frac{x}{3}\right) + \frac{9}{2}\ln\left|\frac{\sqrt{9 + x^2}}{3} + \frac{x}{3}\right| + C$$

$$= \frac{x}{2}\sqrt{9 + x^2} + \frac{9}{2}\ln\left|\frac{\sqrt{9 + x^2} + x}{3}\right| + C$$

Remark 1: Since $\sqrt{\sec^2 u} = |\sec u|$, we should write $|\sec u|\sec^2 u$ instead of $\sec^3 u$. But we are assuming that $x = 3 \tan u$ implies that $u = \tan^{-1}(x/3)$. Therefore, $-\pi/2 < u < \pi/2$ and $\sec u > 0$.

Remark 2: In the final answer, it is common practice to write

$$\ln\left|\frac{\sqrt{9 + x^2} + x}{3}\right| = \ln|\sqrt{9 + x^2} + x| - \ln 3$$

and absorb $-\ln 3$ into the arbitrary constant. Thus,

$$\int \sqrt{9 + x^2}\, dx = \frac{x}{2}\sqrt{9 + x^2} + \frac{9}{2}\ln|\sqrt{9 + x^2} + x| + C$$

is the answer given by Formula **21** with $a = 3$. We will follow this procedure whenever applicable. ∎

EXAMPLE 2   Evaluate $\int dx/x^2\sqrt{4 + x^2}$.

Solution   Let $x = 2 \tan u$. Then $4 + x^2 = 4 \sec^2 u$ and $dx = 2 \sec^2 u\, du$. Therefore,

$$\int \frac{dx}{x^2\sqrt{4 + x^2}} = \int \frac{2\sec^2 u}{4\tan^2 u\sqrt{4\sec^2 u}}\, du$$

$$= \frac{1}{4}\int \frac{\sec u}{\tan^2 u}\, du = \frac{1}{4}\int \frac{1}{\cos u}\cdot\frac{\cos^2 u}{\sin^2 u}\, du$$

$$= \frac{1}{4}\int \frac{\cos u}{\sin^2 u}\, du = \frac{1}{4}\int (\cos u)(\sin u)^{-2}\, du$$

$$= -\frac{1}{4}(\sin u)^{-1} + C$$

$$= -\frac{\sqrt{4 + x^2}}{4x} + C \qquad\qquad \text{(Figure 9–3A)} \qquad ∎$$

The Quadratic $a^2 - x^2$

If x is replaced by $a \sin u$, then

$$a^2 - x^2 = a^2 - a^2\sin^2 u = a^2(1 - \sin^2 u) = a^2\cos^2 u$$

(9.9)   and

$$dx = a \cos u\, du$$

EXAMPLE 3   Evaluate $\int \sqrt{5 - x^2}\, dx$.

Solution   The quadratic is $a^2 - x^2$ with $a = \sqrt{5}$. If $x = \sqrt{5} \sin u$, then, by (9.9), $5 - x^2 = 5 \cos^2 u$ and $dx = \sqrt{5} \cos u\, du$. Therefore,

$$\int \sqrt{5 - x^2}\, dx = \int \sqrt{5 \cos^2 u}(\sqrt{5} \cos u)\, du$$

$$= 5 \int \cos^2 u\, du \qquad\qquad (\cos u \geq 0)$$

$$= 5 \left(\frac{u}{2} + \frac{1}{4} \sin 2u \right) + C \qquad \left(\cos^2 u = \frac{1}{2} + \frac{1}{2} \cos 2u \right)$$

The answer contains $\sin 2u$, but Figure 9–3B provides information only about u. Therefore, use the double angle formula $\sin 2u = 2 \sin u \cos u$ and continue with

$$= \frac{5}{2} u + \frac{5}{4}(2 \sin u \cos u) + C$$

To convert to x's, use Figure 9–3B with $a = \sqrt{5}$; $u = \sin^{-1}(x/\sqrt{5})$ and

$$\sin u \cos u = \frac{x}{\sqrt{5}} \frac{\sqrt{5 - x^2}}{\sqrt{5}}$$

Thus,

$$\int \sqrt{5 - x^2}\, dx = \frac{5}{2} \sin^{-1} \frac{x}{\sqrt{5}} + \frac{1}{2} x \sqrt{5 - x^2} + C$$

Remark: The area of the circle $x^2 + y^2 = 5$ is 4 times the area under the graph of $y = \sqrt{5 - x^2}, 0 \leq x \leq \sqrt{5}$ (Figure 9–4). Using the formula above we compute the area of the circle as

$$A = 4 \int_0^{\sqrt{5}} \sqrt{5 - x^2}\, dx = 4 \left[\frac{5}{2} \sin^{-1} \frac{x}{\sqrt{5}} + \frac{1}{2} x \sqrt{5 - x^2} \right]_0^{\sqrt{5}}$$

$$= 4 \left[\frac{5}{2} \cdot \frac{\pi}{2} - 0 \right] = 5\pi$$

It is reassuring to note that the answer is, indeed, π times the radius squared. ∎

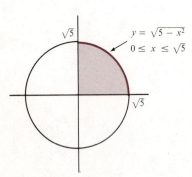

$y = \sqrt{5 - x^2}$
$0 \leq x \leq \sqrt{5}$

FIGURE 9–4. Example 3.

EXAMPLE 4   Evaluate $\int \sqrt{16 - x^2}/x\, dx$.

Solution   Let $x = 4 \sin u$. Then $16 - x^2 = 16 \cos^2 u$ and $dx = 4 \cos u\, du$.

$$\int \frac{\sqrt{16 - x^2}}{x}\, dx = \int \frac{\sqrt{16 \cos^2 u}}{4 \sin u}(4 \cos u)\, du$$

$$= 4 \int \frac{\cos^2 u}{\sin u}\, du$$

$$= 4 \int \frac{1 - \sin^2 u}{\sin u}\, du$$

$$= 4 \int \csc u \, du - 4 \int \sin u \, du$$

$$= 4 \ln |\csc u - \cot u| + 4 \cos u + C$$

$$= 4 \ln \left| \frac{4 - \sqrt{16 - x^2}}{x} \right| + \sqrt{16 - x^2} + C$$

The last line is obtained from Figure 9–3B with $a = 4$; this answer is also given by Formula **30** with $a = 4$. ∎

The Quadratic $x^2 - a^2$

If x is replaced by $a \sec u$, then

(9.10)

$$x^2 - a^2 = a^2 \sec^2 u - a^2 = a^2(\sec^2 u - 1) = a^2 \tan^2 u$$

and

$$dx = a \sec u \tan u \, du$$

This case is slightly different from the others. For the replacement $x = a \sec u$, we restrict u so that $0 \le u < \pi/2$. This ensures that $\tan u \ge 0$ and that $\sqrt{\tan^2 u} = \tan u$.

EXAMPLE 5   Evaluate $\int \sqrt{x^2 - 16}/x \, dx$.

Solution   The quadratic is $x^2 - a^2$ with $a = 4$. If $x = 4 \sec u$, then, by (9.10), $x^2 - 16 = 16 \tan^2 u$ and $dx = 4 \sec u \tan u \, du$. Therefore

$$\int \frac{\sqrt{x^2 - 16}}{x} \, dx = \int \frac{\sqrt{16 \tan^2 u}}{4 \sec u} (4 \sec u \tan u) \, du$$

$$= 4 \int \tan^2 u \, du \qquad\qquad (\tan u \ge 0)$$

$$= 4 \int (\sec^2 u - 1) \, du$$

$$= 4 \tan u - 4u + C$$

$$= \sqrt{x^2 - 16} - 4 \sec^{-1} \left(\frac{x}{4} \right) + C \qquad \text{(Figure 9–3C)}$$

This answer is also given by Formula **38** with $a = 4$. ∎

EXAMPLE 6   Evaluate $\int \sqrt{x^2 - 1} \, dx$.

Solution   Let $x = \sec u$. Then $x^2 - 1 = \tan^2 u$ and $dx = \sec u \tan u \, du$.

$$\int \sqrt{x^2 - 1} \, dx = \int \sqrt{\tan^2 u} \, (\sec u \tan u) \, du$$

$$= \int \tan^2 u \, \sec u \, du \qquad\qquad (\tan u \geq 0)$$

$$= \int \sec^3 u \, du - \int \sec u \, du \qquad (\tan^2 = \sec^2 - 1)$$

$$= \frac{1}{2} \sec u \, \tan u - \frac{1}{2} \ln |\sec u + \tan u| + C$$

$$= \frac{x}{2} \sqrt{x^2 - 1} - \frac{1}{2} \ln |x + \sqrt{x^2 - 1}| + C \quad \text{(Figure 9–3C)}$$

The answer above is also given by Formula **36** with $a = 1$.   ■

EXERCISES

Evaluate the integrals in Exercises 1–10 using trigonometric substitution.

1. $\displaystyle\int \frac{x^2}{\sqrt{x^2 - 4}} \, dx$

2. $\displaystyle\int \frac{x^2}{\sqrt{x^2 + 4}} \, dx$

3. $\displaystyle\int \frac{1}{(x^2 + 4)^2} \, dx$

4. $\displaystyle\int_0^2 \sqrt{x^2 + 1} \, dx$

5. $\displaystyle\int_0^1 \frac{1}{\sqrt{4 - x^2}} \, dx$

6. $\displaystyle\int \frac{\sqrt{x^2 - 3}}{x^2} \, dx$

7. $\displaystyle\int \frac{1}{(x^2 - 5)^{3/2}} \, dx$

8. $\displaystyle\int \frac{1}{x^2 \sqrt{2 + x^2}} \, dx$

9. $\displaystyle\int \frac{\sqrt{2 + x^2}}{x^2} \, dx$

10. $\displaystyle\int (5 - x^2)^{3/2} \, dx$

Evaluate the integrals in Exercises 11–20 using any method.

11. $\displaystyle\int \sqrt{1 - 9x^2} \, dx$ (Factor the 9 out first.)

12. $\displaystyle\int \sqrt{4 - (x + 1)^2} \, dx$ (Let $y = x + 1$.)

13. $\displaystyle\int \sec^{-1} \sqrt{x} \, dx$ (Let $y = \sqrt{x}$. Then $x = y^2$ and $dx = 2y \, dy$.)

14. $\displaystyle\int e^x \sin x \, dx$

15. $\displaystyle\int_0^3 \frac{1}{81 + 9x^2} \, dx$

16. $\displaystyle\int_0^1 \sqrt{1 - x^2} \, dx$

17. $\displaystyle\int \frac{1}{3 - x^2} \, dx$

18. $\displaystyle\int x^2 \sqrt{x^2 - 1} \, dx$

19. $\displaystyle\int \frac{1}{x^2 \sqrt{1 - x^2}} \, dx$

20. $\displaystyle\int \frac{\cos 2x}{1 + \sin 2x} \, dx$

21. Derive the indicated formulas and check your answers in the Table of Integrals.

(a) Formula **25** $\displaystyle\int \frac{x^2}{\sqrt{a^2 + x^2}} \, dx$

(b) Formula **27** $\displaystyle\int \frac{dx}{(a^2 + x^2)^{3/2}}$

(c) Formula **31** $\displaystyle\int \frac{\sqrt{a^2 - x^2}}{x^2} \, dx$

(d) Formula **32** $\displaystyle\int \frac{x^2}{\sqrt{a^2 - x^2}} \, dx$

(e) Formula **33** $\displaystyle\int \frac{1}{x^2 \sqrt{a^2 - x^2}} \, dx$

(f) Formula **34** $\displaystyle\int \frac{1}{(a^2 - x^2)^{3/2}} \, dx$

(g) Formula **39** $\displaystyle\int \frac{\sqrt{x^2 - a^2}}{x^2} \, dx$

(h) Formula **40** $\displaystyle\int \frac{x^2}{\sqrt{x^2 - a^2}} \, dx$

(i) Formula **41** $\displaystyle\int \frac{1}{x^2 \sqrt{x^2 - a^2}} \, dx$

(j) Formula **42** $\displaystyle\int \frac{1}{(x^2 - a^2)^{3/2}} \, dx$

22. Find the length of the graph of the parabola $y = x^2/2$ from $x = 0$ to $x = 2$.

23. The region determined by $y = x\sqrt{1 + x^2}$ from $x = 0$ to $x = 1$ is revolved about the y-axis. Find the volume generated.

24. A lamina has density δ and the shape of the region in Exercise 23. Find its moment about the y-axis.

25. Find the average value of $f(x) = x^2/\sqrt{4 + x^2}$ on the interval $[0, 2]$.

26. Find the center of mass of a lamina with density δ and the shape of the region bounded by the graphs of $y = 1/(x^2 + 4)$, $y = 0$, $x = 0$, and $x = 3$.

Optional Exercises

27. Verify the following formulas in the Table of Integrals:

(a) Formula **22** $\int x^2 \sqrt{a^2 + x^2}\, dx$

(b) Formula **23** $\int \dfrac{\sqrt{a^2 + x^2}}{x}\, dx$

(c) Formula **24** $\int \dfrac{\sqrt{a^2 + x^2}}{x^2}\, dx$

(d) Formula **29** $\int x^2 \sqrt{a^2 - x^2}\, dx$

(e) Formula **35** $\int (a^2 - x^2)^{3/2}\, dx$

(f) Formula **37** $\int x^2 \sqrt{x^2 - a^2}\, dx$

9-5   PARTIAL FRACTIONS

A quotient of two polynomials is called a *rational function*. Three examples are

$$\frac{1}{4 + x^2}, \quad \frac{2x + 2}{x^2 + 2x + 3}, \quad \text{and} \quad \frac{x - 1}{x^2 - 5x + 6}$$

You already know how to evaluate the integral of some rational functions. For the first two examples above,

$$\int \frac{1}{4 + x^2}\, dx = \frac{1}{2} \tan^{-1} \frac{x}{2} + C$$

$$\int \frac{2x + 2}{x^2 + 2x + 3}\, dx = \ln |x^2 + 2x + 3| + C$$

But the integral of the third is something new. One possibility is to rewrite this function as a sum of simpler rational functions that can be integrated. Theoretically, this can always be done, and the purpose of this section is to describe a systematic way to accomplish it.

If $f(x)$ and $g(x)$ are polynomials, then the rational function $f(x)/g(x)$ is said to be **proper** if the degree of the numerator $f(x)$ is less than the degree of the denominator $g(x)$. It is a theorem from algebra that *any proper rational function $f(x)/g(x)$ can be rewritten as a sum of rational functions of the form*

(9.11) $\dfrac{A}{(x - a)^m}$ or $\dfrac{Bx + C}{(x^2 + bx + c)^n}$

where m and n are nonnegative integers, and $x^2 + bx + c$ is **irreducible** (cannot be factored). The functions in (9.11) are called **partial fractions,** and the sum of the partial fractions is called the **partial fraction decomposition** of $f(x)/g(x)$.

Let us return to the third example of a rational function given above. It is easy to verify (simply add the two fractions on the right) that

(9.12) $\dfrac{x-1}{x^2-5x+6} = \dfrac{2}{x-3} - \dfrac{1}{x-2}$

The sum on the right is the partial fraction decomposition of the function on the left. Once this decomposition is obtained, we integrate both sides to obtain

$$\int \frac{x-1}{x^2-5x+6}\,dx = \int \frac{2}{x-3}\,dx - \int \frac{1}{x-2}\,dx$$

(9.13) $\qquad\qquad = 2\ln|x-3| - \ln|x-2| + C$

Check.

$$\frac{d}{dx}[\text{answer}] = \frac{2}{x-3} - \frac{1}{x-2} = \frac{2(x-2)-(x-3)}{(x-3)(x-2)} = \frac{x-1}{x^2-5x+6}$$

This method is called *integrating by partial fractions.*

Integrating a rational function by partial fractions is a four-step process.

(9.14)

> To find the integral of a rational function $f(x)/g(x)$ by partial fractions:
>
> (1) Make sure the function is proper; if not, divide.
> (2) Factor the denominator $g(x)$ into linear and irreducible quadratic factors.
> (3) Find the partial fraction decomposition using the factors obtained in step (2).
> (4) Integrate.

The five examples below illustrate the method. Each one describes a slightly different case.

EXAMPLE 1 *(Each linear factor appears only once)* Evaluate

$$\int \frac{-x^2-4x+2}{x^3-3x^2+2x}\,dx$$

Solution   Follow the steps in (9.14):

(1) The integrand is proper.

(2) Factor the denominator,

$$x^3-3x^2+2x = x(x^2-3x+2) = x(x-1)(x-2)$$

(3) Find the partial fractions. The denominators will be the factors x, $x-1$, and $x-2$. Therefore, you want to find constants A, B, and C such that

$$\frac{-x^2 - 4x + 2}{x^3 - 3x^2 + 2x} = \frac{A}{x} + \frac{B}{x - 1} + \frac{C}{x - 2}$$

Clearing fractions yields

(9.15) $-x^2 - 4x + 2 = A(x - 1)(x - 2) + Bx(x - 2) + Cx(x - 1)$

The functions on either side of (9.15) must agree at *every* value of x. Substituting any three different values will result in three equations in the three unknowns A, B, and C. The choice of values is up to you; any three will do. But a careful choice will cause some terms to drop out and simplify the computations. In the present case, substituting 0, 1, and 2 for x in (9.15) yields

$$\begin{aligned}
(x = 0) & \qquad 2 = 2A \\
(x = 1) & \qquad -3 = -B \\
(x = 2) & \qquad -10 = 2C
\end{aligned}$$

So $A = 1$, $B = 3$, and $C = -5$. Therefore,

$$\frac{-x^2 - 4x + 2}{x^3 - 3x^2 + 2x} = \frac{1}{x} + \frac{3}{x - 1} - \frac{5}{x - 2}$$

You can check this out by adding the fractions on the right.

(4) Integrate,

$$\int \frac{-x^2 - 4x + 2}{x^3 - 3x^2 + 2x} dx = \ln|x| + 3\ln|x - 1| - 5\ln|x - 2| + C \quad \textit{(Check)} \qquad \blacksquare$$

EXAMPLE 2 *(Some linear factors repeat)*   Evaluate

$$\int \frac{-x^2 - x + 16}{(x + 1)(x - 3)^2} dx$$

Solution

(1) The integral is proper (the denominator has degree 3).

(2) The denominator is already factored.

(3) Write

$$\frac{-x^2 - x + 16}{(x + 1)(x - 3)^2} = \frac{A}{(x + 1)} + \frac{B}{(x - 3)} + \frac{C}{(x - 3)^2}$$

Remark: You need *both* terms

$$\frac{B}{x - 3} \quad \text{and} \quad \frac{C}{(x - 3)^2}$$

The method will not work otherwise. If a factor is squared, you need two terms; if a factor is cubed, you need three terms, and so on. In general, each factor $(x - a)^m$ gives rise to an expression of the form

$$\frac{A_1}{x - a} + \frac{A_2}{(x - a)^2} + \cdots + \frac{A_m}{(x - a)^m}$$

in the partial fraction decomposition. We continue now by clearing fractions,

$$(9.16) \quad -x^2 - x + 16 = A(x - 3)^2 + B(x + 1)(x - 3) + C(x + 1)$$

and giving x three different values to obtain three equations in the three unknowns A, B, and C. Two good choices are 3 and -1 because some terms will drop out. Any convenient value, say $x = 0$, will do for the third. With these values, (9.16) yields

$$(x = 3) \qquad 4 = 4C$$
$$(x = -1) \qquad 16 = 16A$$
$$(x = 0) \qquad 16 = 9A - 3B + C$$

So, $C = 1$, $A = 1$, and $B = -2$. Therefore,

$$\frac{-x^2 - x + 16}{(x + 1)(x - 3)^2} = \frac{1}{x + 1} - \frac{2}{x - 3} + \frac{1}{(x - 3)^2}$$

(4) Integrate,

$$\int \frac{-x^2 - x + 16}{(x + 1)(x - 3)^2} dx = \ln|x + 1| - 2\ln|x - 3| - \frac{1}{x - 3} + C \quad \text{(Check)} \qquad \blacksquare$$

EXAMPLE 3 *(Each quadratic factor appears only once)*   Evaluate

$$\int \frac{5x^2 - 3x + 1}{x^3 - 2x^2 + x - 2} dx$$

Solution

(1) The integrand is proper.

(2) Factor the denominator,

$$x^3 - 2x^2 + x - 2 = (x^3 - 2x^2) + (x - 2)$$
$$= x^2(x - 2) + (x - 2)$$
$$= (x - 2)(x^2 + 1)$$

The quadratic $x^2 + 1$ is irreducible.

(3) Write

$$\frac{5x^2 - 3x + 1}{x^3 - 2x^2 + x - 2} = \frac{A}{x - 2} + \frac{Bx + C}{x^2 + 1}$$

Remark: The numerator of each quadratic factor must be of the form $Bx + C$. We continue by clearing fractions

$$5x^2 - 3x + 1 = A(x^2 + 1) + Bx(x - 2) + C(x - 2)$$

and letting x take the values 0, 1, and 2.

$$(x = 0) \qquad 1 = A \qquad\quad - 2C$$
$$(x = 1) \qquad 3 = 2A - B - C$$
$$(x = 2) \qquad 15 = 5A$$

Solving these three equations yields $A = 3$, $C = 1$, and $B = 2$. Therefore,

$$\frac{5x^2 - 3x + 1}{x^3 - 2x^2 + x - 2} = \frac{3}{x - 2} + \frac{2x + 1}{x^2 + 1}$$

(4) Integrate. To integrate the second partial fraction, split it into two pieces.

$$\frac{2x + 1}{x^2 + 1} = \frac{2x}{x^2 + 1} + \frac{1}{x^2 + 1}$$

Thus,

$$\int \frac{5x^2 - 3x + 1}{x^3 - 2x^2 + x - 2} dx = \int \frac{3}{x - 2} dx + \int \frac{2x}{x^2 + 1} dx + \int \frac{1}{x^2 + 1} dx$$

$$= 3 \ln|x - 2| + \ln(x^2 + 1) + \tan^{-1}x + C \qquad \blacksquare$$

EXAMPLE 4 *(Some quadratic factors repeat)* Evaluate

$$\int \frac{1}{x(1 + x^2)^2} dx$$

Solution

(1) The integrand is proper.

(2) The denominator is factored.

(3) Write

$$\frac{1}{x(1 + x^2)^2} = \frac{A}{x} + \frac{Bx + C}{1 + x^2} + \frac{Dx + E}{(1 + x^2)^2}$$

Remark: Here, also, you need *both* terms

$$\frac{Bx + C}{1 + x^2} \quad \text{and} \quad \frac{Dx + E}{(1 + x^2)^2}$$

In general, each irreducible factor $(x^2 + bx + c)^n$ gives rise to an expression of the form

$$\frac{B_1x + C_1}{x^2 + bx + c} + \frac{B_2x + C_2}{(x^2 + bx + c)^2} + \cdots + \frac{B_nx + C_n}{(x^2 + bx + c)^n}$$

We continue by clearing fractions,

$$1 = A(1 + x^2)^2 + Bx^2(1 + x^2) + Cx(1 + x^2) + Dx^2 + Ex$$

This time we need five values of x because there are five unknowns.

$$
\begin{array}{ll}
(x = 0) & 1 = A \\
(x = 1) & 1 = 4A + 2B + 2C + D + E \\
(x = -1) & 1 = 4A + 2B - 2C + D - E \\
(x = 2) & 1 = 25A + 20B + 10C + 4D + 2E \\
(x = -2) & 1 = 25A + 20B - 10C + 4D - 2E
\end{array}
$$

This is a system of five linear equations with five unknowns. Solving such systems, by *eliminating the unknowns,* is discussed in precalculus courses. The solution to the system above is $A = 1$, $B = D = -1$, and $C = E = 0$. Thus,

$$\frac{1}{x(1 + x^2)^2} = \frac{1}{x} - \frac{x}{1 + x^2} - \frac{x}{(1 + x^2)^2}$$

(4) Integrate,

$$\int \frac{1}{x(1 + x^2)^2}\, dx = \ln|x| - \frac{1}{2}\ln|1 + x^2| + \frac{1}{2(1 + x^2)} + C \quad \textit{(Check)} \qquad \blacksquare$$

EXAMPLE 5 *(The integrand is not proper)* Evaluate

$$\int \frac{x^3 - 3x^2 - 3x + 11}{x^2 - 5x + 6}\, dx$$

Solution

(1) The integrand is not proper, so divide using long division. The result is

$$\frac{x^3 - 3x^2 - 3x + 11}{x^2 - 5x + 6} = x + 2 + \frac{x - 1}{x^2 - 5x + 6}$$

Division *always* yields a polynomial plus a *proper* rational function for which you can find a partial fraction decomposition.

(2) and (3) It follows from (9.12) and step (1) above that

$$\frac{x^3 - 3x^2 - 3x + 11}{x^2 - 5x + 6} = x + 2 + \frac{2}{x - 3} - \frac{1}{x - 2}$$

(4) Integrate,

$$\int \frac{x^3 - 3x^2 - 3x + 11}{x^2 - 5x + 6}\, dx = \frac{x^2}{2} + 2x + 2\ln|x - 3| - \ln|x - 2| + C \qquad \blacksquare$$

EXERCISES

Evaluate the integrals in Exercises 1–10 by the method of partial fractions.

1. $\int \dfrac{5}{x^2 + 5x + 6}\, dx$

2. $\int \dfrac{-5x + 9}{(x - 1)(x - 2)(x - 3)}\, dx$

3. $\int \dfrac{x^3}{x^2 + 2x - 3}\, dx$

4. $\int \dfrac{x}{x^2 - 4x + 4}\, dx$

5. $\int \dfrac{x^5}{x^2 - 4x + 4}\, dx$

6. $\int \dfrac{x}{x^4 - 1}\, dx$

7. $\int \dfrac{2x^3 + 10x}{(x^2 + 1)^2}\, dx$

8. $\int \dfrac{5}{(x - 1)(x^2 + 4)}\, dx$

9. $\int \dfrac{x^4 + 9x^2 + 15}{(x - 1)(x^2 + 4)^2}\, dx$

10. $\int \dfrac{4x^3 + 2x^2 - 5x - 18}{(x - 4)(x + 1)^3}\, dx$

Evaluate the integrals in Exercises 11–22 by any method.

11. $\int_0^1 \dfrac{\tan^{-1}x}{1+x^2}\,dx$ **12.** $\int \dfrac{1}{\sqrt{4-x^2}}\,dx$

13. $\int \cot^{-1}x\,dx$

14. $\int e^{\sqrt{x}}\,dx$ (Let $y=\sqrt{x}$. Then $x=y^2$)

15. $\int_3^8 \dfrac{\cos\sqrt{x+1}}{\sqrt{x+1}}\,dx$ **16.** $\int e^x \sin e^x\,dx$

17. $\int \dfrac{1}{(1+x^2)^3}\,dx$ **18.** $\int \sin^3 x \cos^4 x\,dx$

19. $\int \dfrac{-2x}{(x+1)(x^2+1)}\,dx$ **20.** $\int \dfrac{2x^2+3}{x^2(x-1)}\,dx$

21. $\int_0^{1/2} \sin^{-1}x\,dx$ **22.** $\int_0^1 \tan^{-1}x\,dx$

In Exercises 23–30, find the general solutions.

23. $y' = y^2 - 4y + 3$ **24.** $y' - 3xy = 8$
25. $y'' = -2y$ **26.** $y' = y^2 - 1$
27. $y' = ky(S-y)$; k and S positive constants

28. $y' = \dfrac{ay - bxy}{-cx + dxy}$; a, b, c, and d positive constants (implicit solution)

29. $y' = x^3 + 2x^2 - 3x$ **30.** $y' = \sqrt{1+x^2}$

31. An initial population of 50 bacteria in a culture is found to grow according to the equation

$$\frac{dy}{dt} = .0015y(200 - y)$$

Find the population $y = y(t)$ at any time t.

32. *(Chemical reactions)* Chemicals often combine to form a compound at a rate proportional to the amount present at any time. If A_0 grams of chemical A is mixed with B_0 grams of chemical B, then the compound C is produced at the rate

$$\frac{dC}{dt} = k(A_0 - C)(B_0 - C) \quad k > 0$$

Find the amount $C = C(t)$ of the compound present at time t if (a) $A_0 = B_0$ and (b) if $A_0 \neq B_0$.

9–6 MISCELLANEOUS METHODS YOU CAN USE.

It often happens that an integrand is not in a form that can be directly handled by any of the methods discussed so far. Here are three examples.

$$\int \frac{dx}{\sqrt{x^2 - 4x + 5}} \qquad \int \frac{dx}{1 + \sin x} \qquad \int \frac{dx}{\sqrt{x} + \sqrt[3]{x}}$$

In such cases, the integrand must first be transformed, by substitution or some other method, into some recognizable form. Over the years, a long list of transformation techniques has evolved, and some of them are quite ingenious. In this section, we describe three of the most useful ones.

Completing the Square

When a quadratic $ax^2 + bx + c$ appears in the integrand, completing the square will often help you recognize the method needed to evaluate the integral.

EXAMPLE 1 Evaluate $\int dx/\sqrt{x^2 - 4x + 5}$.

Solution First complete the square: $x^2 - 4x + 5 = (x^2 - 4x + 4) + (5 - 4) = (x - 2)^2 + 1$. Now the substitutions $y = x - 2$ and $dy = dx$ will transform the integrand into a recognizable form.

$$\int \frac{dx}{\sqrt{x^2 - 4x + 5}} = \int \frac{dx}{\sqrt{(x-2)^2 + 1}}$$

$$= \int \frac{dy}{\sqrt{y^2 + 1}} \qquad (y = x - 2)$$

$$= \ln|y + \sqrt{1 + y^2}| + C$$

The last line is obtained by trigonometric substitution or by Formula **17**. Finally, replacing y by $x - 2$ yields

$$= \ln|(x-2) + \sqrt{1 + (x-2)^2}| + C$$

Check.

$$\frac{d}{dx}[\text{answer}] = \frac{1}{(x-2) + \sqrt{1 + (x-2)^2}}\left[1 + \frac{x-2}{\sqrt{1 + (x-2)^2}}\right]$$

$$= \frac{1}{\sqrt{1 + (x-2)^2}} = \frac{1}{\sqrt{x^2 - 4x + 5}} \quad \blacksquare$$

Completing the square is also used in conjunction with the method of partial fractions.

EXAMPLE 2   Evaluate

$$\int \frac{x^2 + 5x + 8}{(x-1)(x^2 + 4x + 9)} dx$$

Solution   The integrand is a proper rational function, so find its partial fraction decomposition (do not complete the square yet, it may not be necessary). The denomiantor is already factored and $x^2 + 4x + 9$ is irreducible ($b^2 - 4ac = 16 - 36 < 0$). Using the methods of the previous section, it follows that

$$\int \frac{x^2 + 5x + 8}{(x-1)(x^2 + 4x + 9)} dx = \int \frac{1}{x-1} dx + \int \frac{1}{x^2 + 4x + 9} dx$$

NOW we complete the square in the second integral; $x^2 + 4x + 9 = (x + 2)^2 + 5$. This calls for the substitutions $y = x + 2$ and $dy = dx$. Continuing,

$$= \ln|x - 1| + \int \frac{dx}{(x+2)^2 + 5}$$

$$= \ln|x - 1| + \int \frac{dy}{y^2 + 5}$$

$$= \ln|x - 1| + \frac{1}{\sqrt{5}} \tan^{-1} \frac{y}{\sqrt{5}} + C$$

This last line is obtained by trigonometric substitution or by Formula **15** with $a = \sqrt{5}$. Replacing y by $x + 2$ yields

$$\int \frac{x^2 + 5x + 8}{(x-1)(x^2 + 4x + 9)} dx = \ln|x - 1| + \frac{1}{\sqrt{5}} \tan^{-1} \frac{(x+2)}{\sqrt{5}} + C \quad \blacksquare$$

Rational Functions of Sine and Cosine

If the integrand is a rational function of sines and cosines, the substitution

$$(9.17) \quad \boxed{y = \tan \frac{x}{2} \quad -\pi < x < \pi}$$

will transform it into a rational function of polynomials in y. To prove this, we note that

$$\cos \frac{x}{2} = \frac{1}{\sec(x/2)} = \frac{1}{\sqrt{1 + \tan^2(x/2)}} = \frac{1}{\sqrt{1 + y^2}}$$

$$\sin \frac{x}{2} = \left(\tan \frac{x}{2}\right)\left(\cos \frac{x}{2}\right) = \frac{y}{\sqrt{1 + y^2}}$$

It follows from these identities and the double angle formulas that

$$\cos x = 1 - 2\sin^2 \frac{x}{2} = 1 - \frac{2y^2}{1 + y^2} = \frac{1 - y^2}{1 + y^2}$$

$$\sin x = 2\left(\sin \frac{x}{2}\right)\left(\cos \frac{x}{2}\right) = \frac{2y}{1 + y^2}$$

Furthermore, by (9.17) $x = 2\tan^{-1}y$, so

$$dx = \frac{2}{1 + y^2}\,dy$$

Putting all this information together we have

$$(9.18) \quad \boxed{\begin{array}{l} \text{If } y = \tan \dfrac{x}{2},\ -\pi < x < \pi, \text{ then} \\[2mm] \cos x = \dfrac{1 - y^2}{1 + y^2},\ \sin x = \dfrac{2y}{1 + y^2},\ \text{and } dx = \dfrac{2}{1 + y^2}\,dy \end{array}}$$

These replacements will produce a rational function in y.

EXAMPLE 3   Evaluate $\int dx/(1 + \sin x)$.

Solution   Set $y = \tan(x/2)$ and use (9.18) to write

$$\int \frac{dx}{1 + \sin x} = \int \frac{1}{1 + \dfrac{2y}{1 + y^2}} \left[\frac{2}{1 + y^2}\right] dy$$

$$= \int \frac{1 + y^2}{y^2 + 2y + 1} \left[\frac{2}{1 + y^2} \right] dy$$

$$= 2 \int \frac{1}{(y + 1)^2} \, dy$$

$$= \frac{-2}{y + 1} + C$$

$$= \frac{-2}{1 + \tan \dfrac{x}{2}} + C \qquad\qquad \left(y = \tan \frac{x}{2} \right)$$

Check. $d/dx[-2(1 + \tan x/2)^{-1}] = (\sec^2 x/2)/(1 + \tan x/2)^2$. After writing $\tan(x/2) = \sin(x/2)/\cos(x/2)$, it is not difficult to verify that the denominator equals $(1 + \sin x)\sec^2(x/2)$. ∎

Other Substitutions

A careful choice of substitution can clear radicals from the integrand.

EXAMPLE 4 Evaluate $\int dx/(\sqrt{x} + \sqrt[3]{x})$.

Solution Set y equal to some power of x that will eliminate both the square root and the cube root. Try

$$y = x^{1/6} \text{ or, equivalently, } x = y^6$$

Then $\sqrt{x} = \sqrt{y^6} = y^3$, $\sqrt[3]{x} = \sqrt[3]{y^6} = y^2$, and $dx = 6y^5 \, dy$.

Thus,

$$\int \frac{dx}{\sqrt{x} + \sqrt[3]{x}} = \int \frac{6y^5}{y^3 + y^2} \, dy = 6 \int \frac{y^3}{y + 1} \, dy \qquad \text{(Cancel out } y^2\text{)}$$

$$= 6 \int \left(y^2 - y + 1 - \frac{1}{y + 1} \right) dy \qquad \text{\textit{(Divide)}}$$

$$= 2y^3 - 3y^2 + 6y - 6 \ln|y + 1| + C$$

To put the answer in terms of x, we use $y = x^{1/6}$. Then $y^3 = x^{3/6} = \sqrt{x}$, $y^2 = \sqrt[3]{x}$, and so on. Therefore,

$$\int \frac{dx}{\sqrt{x} + \sqrt[3]{x}} = 2\sqrt{x} - 3\sqrt[3]{x} + 6\sqrt[6]{x} - 6 \ln|\sqrt[6]{x} + 1| + C \qquad ∎$$

EXAMPLE 5 Evaluate $\int x^3/\sqrt[3]{1 + x^2} \, dx$.

Solution Make a substitution that will clear the radical; try

$$y = \sqrt[3]{1 + x^2} \quad \text{or} \quad y^3 = 1 + x^2$$

Then $x^2 = y^3 - 1$. But rather than solve for x to find dx, it is easier in this case to write $2x \, dx = 3y^2 \, dy$, and

$$x \, dx = \frac{3}{2} y^2 \, dy$$

Therefore,

$$
\begin{aligned}
\int \frac{x^3}{\sqrt[3]{1 + x^2}} \, dx &= \int \frac{x^2}{\sqrt[3]{1 + x^2}} \, x \, dx \\
&= \int \frac{y^3 - 1}{y} \left(\frac{3}{2} y^2 \right) dy \\
&= \frac{3}{2} \int (y^4 - y) \, dy \\
&= \frac{3}{10} y^5 - \frac{3}{4} y^2 + C \\
&= \frac{3}{10} (1 + x^2)^{5/3} - \frac{3}{4} (1 + x^2)^{2/3} + C \qquad \blacksquare
\end{aligned}
$$

EXERCISES

Evaluate the integrals in Exercises 1–10 by the methods discussed in this section.

1. $\displaystyle\int \frac{dx}{x^2 + 2x + 5}$

2. $\displaystyle\int \frac{dx}{\sqrt{1 - 4x - x^2}}$

3. $\displaystyle\int \frac{dx}{1 + \cos x}$

4. $\displaystyle\int \frac{dx}{1 + 2 \cos x}$

5. $\displaystyle\int \frac{\sqrt{x}}{1 + x} \, dx$

7. $\displaystyle\int x \sqrt{2 - 3x} \, dx$

7. $\displaystyle\int \frac{x}{\sqrt{x + 4}} \, dx$

8. $\displaystyle\int \frac{x - 1}{(2x + 3)^3} \, dx$

9. $\displaystyle\int \sqrt{1 + e^x} \, dx$ (Try $y = \sqrt{1 + e^x}, x = \ln(y^2 - 1)$.)

10. $\displaystyle\int \frac{dx}{5 \sec x - 3}$ $\left(\text{Write } \sec x = \frac{1}{\cos x}.\right)$

Evaluate the integrals in Exercises 11–30 by any method.

11. $\displaystyle\int \sin^3 x \cos^2 x \, dx$

12. $\displaystyle\int_0^{2\pi} \sin 3x \sin 2x \, dx$

13. $\displaystyle\int \frac{1 - \cos x}{1 + \sin x} \, dx$

14. $\displaystyle\int \frac{x}{(x + 2)^{3/2}} \, dx$

15. $\displaystyle\int x^2 \ln x \, dx$

16. $\displaystyle\int x e^{-x} \, dx$

17. $\displaystyle\int e^{\sqrt{x}} \, dx$

18. $\displaystyle\int \sqrt{\frac{1 - \cos x}{2}} \, dx$

19. $\displaystyle\int \sin^2 x \cos^2 x \, dx$

20. $\displaystyle\int \tan 3x \, dx$

21. $\displaystyle\int \frac{x + 1}{\sqrt{x^2 + 2x + 2}} \, dx$

22. $\displaystyle\int \frac{dx}{\sqrt{2 - 5x^2}}$

23. $\displaystyle\int \frac{x}{\sqrt{x^2 + 2x + 2}} \, dx$

24. $\displaystyle\int \frac{x}{\sqrt{5 + 4x - x^2}} \, dx$

25. $\displaystyle\int \frac{x}{x^2 - 2x - 3} \, dx$

26. $\displaystyle\int \frac{x^3}{x^2 - 2x + 1} \, dx$

27. $\displaystyle\int x^2 \cos x \, dx$

28. $\displaystyle\int \frac{\sqrt{x}}{1 + \sqrt[4]{x}} \, dx$

29. $\displaystyle\int x^3 \sqrt{x^2 + 1} \, dx$

30. $\displaystyle\int \tan^{-1} x \, dx$

9–7   THE LOGISTIC MODEL OF POPULATION GROWTH
(Optional reading)

Under ideal conditions, populations tend to grow exponentially. But ideal conditions cannot last indefinitely in a limited environment. As the population grows, the food supply diminishes, and the environment becomes overcrowded and polluted. Extensive experiments with yeast cells, fruit flies, and so on, indicate that populations grow slowly at first, and then increase more and more rapidly up to a certain point. After that, the growth rate begins to decline and approaches zero. The differential equation that describes this behavior is called the **logistic equation.**

(9.19)
$$\frac{dy}{dt} = ky(S - y)$$

where $y = y(t)$ is the population at time t, and k and S are positive constants that depend on the given conditions. Let us see how this equation is derived.

We begin by observing that the rate of growth of population is the difference between the *birth rate* and the *mortality* (or *death*) *rate;* that is,

(9.20) $\dfrac{dy}{dt} =$ birth rate $-$ mortality rate

Two important auxiliary concepts are the *average* birth rate and the *average* mortality rate

(9.21) $\dfrac{\text{birth rate}}{y}$ and $\dfrac{\text{mortality rate}}{y}$

measured in births or deaths per unit of time per individual. We shall now make four assumptions.

(1) There is a maximum population size S that the given environment can support.

(2) $y(t) < S$ for all t and the initial population $y_0 = y(0)$ is small compared to S; say, $y_0 < S/2$.

(3) We are dealing with a closed society; that is, individuals enter the population only by birth and leave it only by dying.

(4) The average birth rate and average death rate (9.21) are *linear* functions of y. (This is supported by experimental evidence.)

Because of overcrowding, pollution, and so on, the average birth rate *decreases linearly* as the population increases; that is,

$$\frac{\text{birth rate}}{y} = B - k_B y \quad \text{or} \quad \text{birth rate} = (B - k_B y)y \quad B \text{ and } k_B \text{ positive}$$

Because of dwindling food supplies, lack of space, and so on, the average mortality rate *increases linearly* as the population increases; that is,

$$\frac{\text{mortality rate}}{y} = M + k_M y \text{ or mortality rate} = (M + k_M y)y \quad M \text{ and } k_M \text{ positive}$$

It follows from these equations and (9.20) that

$$\frac{dy}{dt} = \text{birth rate} - \text{mortality rate}$$

$$= (B - k_B y)y - (M + k_M y)y$$

(9.22) $$= [B - M - (k_B + k_M)y]y$$

$$= (k_B + k_M)\left[\frac{B - M}{k_B + k_M} - y\right]y$$

The constant $(B - M)/(k_B + k_M)$ has a special significance. The population ceases to grow and is, therefore, at a maximum when the average birth rate equals the average mortality rate (Figure 9–5). Equating these averages and solving for y yields

$$B - k_B y = M + k_M y$$

$$y = \frac{B - M}{k_B + k_M}$$

Thus, $S = (B - M)/(k_B + k_M)$ is the maximum size of the population. If we use S and set $k = k_B + k_M$, then (9.22) is the logistic equation (9.19)

$$\boxed{\frac{dy}{dt} = ky(S - y)}$$

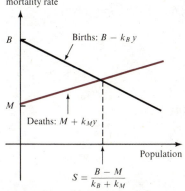

Average birth or mortality rate

B Births: $B - k_B y$

M

Deaths: $M + k_M y$

Population

$$S = \frac{B - M}{k_B + k_M}$$

FIGURE 9–5. Aveage birth rate decreases and average mortality rate increases as population increases. Where these lines cross $(S = (B - M)/(k_B + k_M))$ is the maximum population size.

and our derivation is complete. It has been verified experimentally that, under carefully controlled conditions, this equation accurately reflects population growth in a limited environment.

Now we will solve (9.19) and graph the solution. First, we separate the variables

$$\frac{1}{y(S - y)} dy = kdt$$

and, to simplify later calculations, we rewrite the equation as

$$\frac{1}{y(y - S)} = -kdt$$

Using the technique of partial fractions, we see that

$$\frac{1}{y(y - S)} = \frac{1}{S}\left[\frac{1}{y - S} - \frac{1}{y}\right] \qquad \text{(Partial fractions)}$$

Therefore,

$$\frac{1}{S} \int \left[\frac{1}{y-S} - \frac{1}{y} \right] dy = -k \int dt$$

The integral of the left side is

$$\frac{1}{S}[\ln|y-S| - \ln|y|] = \frac{1}{S} \ln \left| \frac{y-S}{y} \right|$$

Since $y > 0$ and $y - S < 0$ (by assumption 2), it follows that

$$\ln \frac{|y-S|}{y} = -Skt + C$$

(9.23) $\qquad \dfrac{|y-S|}{y} = e^C e^{-Skt}$ $\qquad\qquad$ (Exponential of each side)

$$\frac{y-S}{y} = -e^C e^{-Kst}$$

We multiply through by y, collect the y-terms on one side, and solve

(9.24) $\qquad y = \dfrac{S}{1 + e^C e^{-Skt}}$

This is the general solution of the logistic equation. If $y_0 = y(0)$ is the initial population, then setting $t = 0$ in (9.23) yields $e^C = (S - y_0)/y_0$. Let c be this constant; then (9.24) becomes the **logistic function**

(9.25) $\qquad \boxed{ y = \dfrac{S}{1 + ce^{-Skt}} \qquad c = \dfrac{S-y_0}{y_0} }$

To graph the logistic function, we note that $y < S$, but as t increases, the denominator approaches 1 and, therefore, y approaches S. Next, we determine the signs of the first and second derivatives. The logistic equation (9.19) shows that $dy/dt = ky(S - y)$ is always positive, so y is increasing. Moreover, implicit differentiation, along with the product rule, yields

$$\frac{d^2y}{dt^2} = ky\left(-\frac{dy}{dt}\right) + (S-y)\left(k\frac{dy}{dt}\right)$$

$$= k(S - 2y)\left(\frac{dy}{dt}\right)$$

Since k and dy/dt are positive, it follows that

$$\frac{d^2y}{dt^2} \begin{cases} > 0 & \text{if } S > 2y \\ = 0 & \text{if } S = 2y \\ < 0 & \text{if } S < 2y \end{cases}$$

Therefore, the curve is concave up if $S > 2y$, concave down if $S < 2y$, and there is an inflection point when $y = S/2$. The inflection point occurs when

$$\frac{S}{2} = \frac{S}{1 + ce^{-Skt}}$$ (9.25)

$$2 = 1 + ce^{-Skt}$$

$$c = e^{Skt}$$

$$t = \frac{\ln c}{Sk}$$

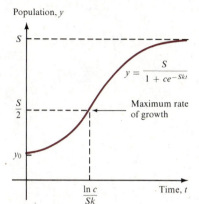

Population, y

$y = \dfrac{S}{1 + ce^{-Skt}}$

Maximum rate of growth

$\dfrac{\ln c}{Sk}$ Time, t

FIGURE 9–6. Graph of the logistic function.

Putting all this information together, the graph is sketched in Figure 9–6. It shows that from $t = 0$ to $t = \ln c/Sk$, population grows slowly at first but then at an ever increasing rate. This part of the curve resembles exponential growth. At $t = \ln c/Sk$, the population is $S/2$; this is the point of *maximum rate of growth*. After that the growth rate decreases and approaches zero.

Test problems

REVIEW EXERCISES

In Exercises 1–100, evaluate the integrals.

1. $\displaystyle\int \sec^2 x \ln(\tan x)\,dx$

2. $\displaystyle\int \frac{x^2\,dx}{\sqrt{1 - x^2}}$

3. $\displaystyle\int \ln x \sqrt{x}\,dx$

4. $\displaystyle\int x \ln(x - 1)\,dx$

5. $\displaystyle\int \cos^6 x\,dx$

6. $\displaystyle\int \tan^3(2x)\,dx$

7. $\displaystyle\int \sin 5x \cos 2x\,dx$

8. $\displaystyle\int \sin 4x \sin 2x\,dx$

got

9. $\displaystyle\int \sqrt{16 + 4x^2}\,dx$

not on test

10. $\displaystyle\int e^x \sqrt{4 + e^{2x}}\,dx$ $u = e^x$

11. $\displaystyle\int \sin^2 x \cos x \sqrt{1 + \sin^2 x}\,dx$ do by parts $u = \sin x$
$dv = $ rest

12. $\displaystyle\int \frac{\sqrt{25 + x^2}}{5x}\,dx$

13. $\displaystyle\int \frac{4\,dx}{x^2 - 2x - 3}$

14. $\displaystyle\int \frac{2x + 6}{x^3 + 5x^2 + 6x}\,dx$

15. $\displaystyle\int \frac{2x\,dx}{x^2 - 2x + 1}$

16. $\displaystyle\int \frac{x^4 - x^2}{x^2 + 3x - 4}\,dx$ got not on test (damn!)

17. $\displaystyle\int \frac{dx}{x^2 + 4x + 8}$

18. $\displaystyle\int \frac{dx}{\sqrt{-x^2 - 4x - 3}}$

got

19. $\displaystyle\int \frac{dx}{1 + 2 \sin x}$

20. $\displaystyle\int \frac{x}{1 + \sqrt{x}}\,dx$

21. $\displaystyle\int (25 - 4x^2)^{3/2}\,dx$

22. $\displaystyle\int \frac{dx}{(25 - 4x^2)^{3/2}}$

23. $\displaystyle\int \frac{x^2\,dx}{\sqrt{25 - 4x^2}}$

24. $\displaystyle\int \frac{\sqrt{36 - 16x^2}}{x^2}\,dx$

25. $\displaystyle\int x3^x\,dx$

26. $\displaystyle\int (\ln x)^2\,dx$

got

27. $\displaystyle\int x^2 \cos 2x\,dx$

28. $\displaystyle\int x^2 \tan^{-1} x\,dx$

29. $\displaystyle\int \cos 3x \cos 7x\,dx$

30. $\displaystyle\int \sin^6 x\,dx$

31. $\displaystyle\int \sec^4(2x)\,dx$

32. $\displaystyle\int e^{2x} \cos 3x\,dx$ not on test

33. $\displaystyle\int \frac{dx}{4 \csc x + 1}$ hyp opp

34. $\displaystyle\int \frac{dx}{\sqrt{x} - \sqrt[4]{x}}$

35. $\displaystyle\int \frac{2x - 7}{(4x - 3)^2}\,dx$

36. $\displaystyle\int x\sqrt{3 - 2x}\,dx$

37. $\displaystyle\int \frac{5x + 9}{(x + 1)(x + 3)(x + 4)}\,dx$

38. $\displaystyle\int \frac{x^3 - 5}{x^2(x - 1)}\,dx$

39. $\displaystyle\int \frac{5}{(x - 2)(x^2 + 1)}\,dx$

not on test

40. $\displaystyle\int \frac{3x^2}{x^6 - 1}\,dx$

41. $\displaystyle\int \frac{2x^4 + 8x^2}{x^2 + 2}\, dx$

42. $\displaystyle\int \frac{1 - \sin x}{1 + \cos x}\, dx$

79. $\displaystyle\int \tan^{-1} 5x\, dx$

80. $\displaystyle\int x \cos^{-1}x\, dx$

43. $\displaystyle\int \sqrt{4x^2 - 9}\, dx$

44. $\displaystyle\int \frac{dx}{x^2\sqrt{4x^2 - 25}}$

81. $\displaystyle\int e^{2x} \sin 3x\, dx$

82. $\displaystyle\int \tan^{-1}\sqrt{x}\, dx$

45. $\displaystyle\int x^2\sqrt{x^2 - 9}\, dx$

46. $\displaystyle\int \frac{dx}{(x^2 - 7)^{3/2}}$

83. $\displaystyle\int \cos\sqrt{x}\, dx$

84. $\displaystyle\int \frac{x}{x^2 - 9}\, dx$

47. $\displaystyle\int x(1 + x)^{1/4}\, dx$

48. $\displaystyle\int x \sec x \tan x\, dx$

85. $\displaystyle\int x^3 e^{x^2}\, dx$

86. $\displaystyle\int \frac{1}{1 - \cos^2 x}\, dx$

49. $\displaystyle\int \sin^7(2x)\, dx$

50. $\displaystyle\int e^{\sqrt[3]{x}}\, dx$

87. $\displaystyle\int \frac{1}{\sec^2 x + \tan^2 x}\, dx$

88. $\displaystyle\int \frac{x}{\sqrt{1 + x}}\, dx$

51. $\displaystyle\int x \sin^{-1}x\, dx$

52. $\displaystyle\int \sin x \cos 2x\, dx$

89. $\displaystyle\int \frac{x}{(x - 1)^2}\, dx$

90. $\displaystyle\int x \sin^{-1}x\, dx$

53. $\displaystyle\int \frac{1}{(x^2 + 25)^{3/2}}\, dx$

54. $\displaystyle\int \sin^2 x \cos x\, dx$

91. $\displaystyle\int \frac{1}{x^2\sqrt{4 - x^2}}\, dx$

92. $\displaystyle\int e^{\sqrt{x}}\, dx$

55. $\displaystyle\int \frac{x^3 + 1}{x(x - 1)^3}\, dx$

56. $\displaystyle\int \frac{1}{x\sqrt{x^2 + 1}}\, dx$

93. $\displaystyle\int \frac{\cos\sqrt{x}}{\sqrt{x}}\, dx$

94. $\displaystyle\int \frac{x}{1 + \sqrt{x}}\, dx$

57. $\displaystyle\int x^2 \sin 5x\, dx$

58. $\displaystyle\int \frac{x}{x^2 + 3x + 2}\, dx$

95. $\displaystyle\int \frac{\sin^{-1}x}{\sqrt{1 - x^2}}\, dx$

96. $\displaystyle\int \frac{1}{\sin x \cos x}\, dx$

59. $\displaystyle\int \sin^3 x \cos^3 x\, dx$

60. $\displaystyle\int \sin(\ln x)\, dx$

97. $\displaystyle\int e^x \cos 2x\, dx$

61. $\displaystyle\int e^x \sec e^x\, dx$

62. $\displaystyle\int \csc^2\!\left(\frac{x}{2}\right) dx$

98. $\displaystyle\int \frac{\cos x}{\sin^2 x + 2\sin x + 1}\, dx$

63. $\displaystyle\int e^x \sec^2 e^x\, dx$

64. $\displaystyle\int \frac{x}{\sec x}\, dx$

99. $\displaystyle\int \ln\sqrt{x - 1}\, dx$

100. $\displaystyle\int \frac{1}{x^3 + 1}\, dx$

65. $\displaystyle\int e^x\sqrt{1 + e^x}\, dx$

66. $\displaystyle\int \sin^2 8x\, dx$

101. The region bounded by the graphs $y = \sin^3 x$, $x = 0$, $x = \pi/2$, and $y = 0$ is revolved about the x-axis. Find the volume generated.

67. $\displaystyle\int \frac{1}{(\sqrt{x})^3 + \sqrt{x}}\, dx$

68. $\displaystyle\int x \tan x\, dx$

102. Find the area of the region bounded by $y = x^2 \tan^{-1}x$, $x = 0$, and $x = 1$.

69. $\displaystyle\int x^2 e^{-4x}\, dx$

70. $\displaystyle\int e^{\ln x}\, dx$

103. Given $y' + 2y = x^2$. Find the particular solution that satisfies $y(0) = 1$.

71. $\displaystyle\int \sec^2 x e^{\tan x}\, dx$

72. $\displaystyle\int \frac{1 - \sin x}{\cos x}\, dx$

104. Find the length of the curve $y = 4x^3/3$ from $x = 0$ to $x = 4$.

73. $\displaystyle\int \frac{\sqrt{9 - 4x^2}}{x^2}\, dx$

105. Find the center of mass of a lamina with density δ and the shape of the region bounded by the graphs $y = e^{\sqrt{x}}$, $x = 0$, $x = 2$, and $y = 0$.

74. $\displaystyle\int (x + 3)(x^2 - 4)\, dx$

106. Given that the velocity of a particle is $v(t) = t \tan^{-1}t$, find the acceleration and position functions.

75. $\displaystyle\int (x + 5)(x - 1)\, dx$

76. $\displaystyle\int x \cos x\, dx$

107. An initial population of 50 bacteria in a culture is found to grow according to the equation $dy/dt = 0.007y(100 - y)$. Find the population at $t = 1{,}000$.

77. $\displaystyle\int \frac{\sin x}{\sqrt{1 + \cos x}}\, dx$

78. $\displaystyle\int \frac{\sin x}{(1 - \cos x)^2}\, dx$

L'HÔPITAL'S RULE AND IMPROPER INTEGRALS

The material of this chapter extends the concepts of limit and integral to include the "infinite" case. In Section 1, we introduce and discuss limits of the form

$$\lim_{x \to \infty} f(x) = L \quad \text{and} \quad \lim_{x \to p} f(x) = \infty$$

In Section 3, we introduce and discuss integrals of the form

$$\int_a^\infty f(x)\, dx \quad \text{and} \quad \int_{-\infty}^\infty f(x)\, dx$$

These extensions provide an extra degree of freedom in working with and applying limits and integrals. The exercises include applications to graphing, to the present value of an asset, such as land, that brings in revenue indefinitely, and to probability when the density function is defined on an interval $[a, \infty)$. The extensions also produce some unexpected results. For instance, Section 3 contains an example of a region with finite area that generates a solid of revolution with infinite volume!

Section 2 describes a way of evaluating limits known as L'Hôpital's rule.* Instead of using limits to compute derivatives, L'Hôpital's rule used derivatives to compute limits. The rule is extremely useful and will be referred to often in later work.

The results of all three sections play an important role in Section 4, where we analyze the population patterns of a predator and its prey.

*Pronounced "Lo-pi-tal."

10–1   INFINITE LIMITS; LIMITS AT INFINITY

At times it is convenient to use the infinity symbol ∞ in connection with limits. For instance, the limit

$$\lim_{x \to p} f(x) = \infty$$

conveys the idea that $f(x)$ gets larger and larger as x approaches p. On the other hand, the limit

$$\lim_{x \to \infty} g(x) = L$$

conveys the idea that $g(x)$ gets closer and closer to the number L as x gets larger and larger. Examples of each type are

$$\lim_{x \to 0} \frac{1}{x^2} = \infty \quad \text{and} \quad \lim_{x \to \infty} \frac{1}{x} = 0$$

Our treatment of these limits is informal and relies on your intuition. At the end of the section, however, we do present some formal definitions and proofs.

Infinite Limits

We say that $f(x)$ *increases without bound* (or *gets arbitrarily large*) if, for any preassigned positive number N, no matter how large, there are numbers x such that $f(x) > N$. With this terminology,

(10.1)

> If $f(x)$ increases without bound as x gets closer and closer (but not equal) to p, then we write
>
> $$\lim_{x \to p} f(x) = \infty$$

Of course, if the values of f *decrease without bound* (the meaning should be clear), then we write

$$\lim_{x \to p} f(x) = -\infty$$

Similar meanings are applied to one-sided limits

$$\lim_{x \to p^+} f(x) = \infty, \; \lim_{x \to p^-} f(x) = -\infty, \text{ and so on}$$

Figure 10–1 shows the geometric interpretations attached to these limits.

Remark: It must be clearly understood that ∞ and $-\infty$ are *not* real numbers. If $f(x) \to \infty$ as $x \to p$, we still say the limit of f at p *does not exist.* The symbol $\lim_{x \to p} f(x) = \infty$ is used merely to indicate that $f(x)$ increases without bound when x is close (but not equal) to p.

EXAMPLE 1   Let $f(x) = (x^2 + 1)/(1 - x)$. Find the one-sided limits at $x = 1$ and sketch the graph near that point.

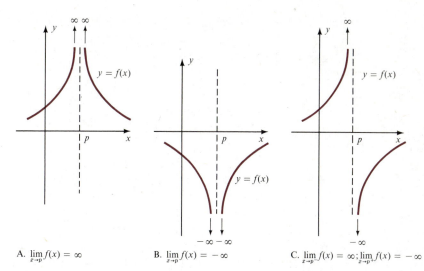

FIGURE 10–1. Geometric interpretations of infinite limits.

A. $\lim\limits_{x \to p} f(x) = \infty$ B. $\lim\limits_{x \to p} f(x) = -\infty$ C. $\lim\limits_{x \to p^-} f(x) = \infty; \lim\limits_{x \to p^+} f(x) = -\infty$

Solution   When x is close (but not equal) to 1, the numerator is close to 2 and the denominator is close to 0 so the quotient is a large positive or negative number. As $x \to 1^+$, the denominator $1 - x$ is negative; as $x \to 1^-$, the denominator is positive. Therefore,

$$\lim_{x \to 1^+} \frac{x^2 + 1}{1 - x} = -\infty \quad \text{and} \quad \lim_{x \to 1^-} \frac{x^2 + 1}{1 - x} = \infty$$

The part of the graph near $x = 1$ is shown in Figure 10–2.   ■

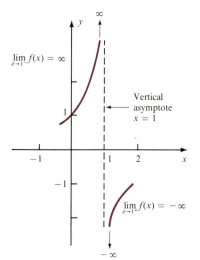

FIGURE 10–2. Example 1.

EXAMPLE 2   Sketch the graph of $g(x) = (3x + 2)/(x^2 - 4)$ near the points $x = \pm 2$.

Solution   The function g is not defined at $x = \pm 2$. To learn about the behavior of g near those points, we use one-sided limits. First, for $x = 2$. As $x \to 2^+$ or $x \to 2^-$, the numerator $3x + 2 \to 8$ and the denominator $x^2 - 4 \to 0$. Therefore, the quotient approaches $\pm \infty$. Since the numerator is positive, the sign of the quotient depends on the sign of the denominator $x^2 - 4 = (x + 2)(x - 2)$. When x is slightly larger than 2, both factors are positive; when x is slightly smaller than 2, the first factor is positive, but the second is negative (test with $x = 1.9$). It follows that

$$\lim_{x \to 2^+} g(x) = \infty \quad \text{and} \quad \lim_{x \to 2^-} g(x) = -\infty$$

Now, for $x = -2$. The analysis is similar to the one above. As $x \to -2^+$ or $x \to -2^-$, the quotient approaches $\pm \infty$. This time the numerator $3x + 2 \to -4$, so it is negative. The denominator $x^2 - 4 = (x + 2)(x - 2)$ is negative if x is slightly larger than -2 (test with $x = -1.9$), but it is positive if x is slightly smaller than -2 (test with $x = -2.1$). It follows that

$$\lim_{x \to -2^+} g(x) = \infty \quad \text{and} \quad \lim_{x \to -2^-} g(x) = -\infty$$

The graph of g near $x = \pm 2$ is shown in Figure 10–3. We will return to this function and complete its graph in Example 7.   ■

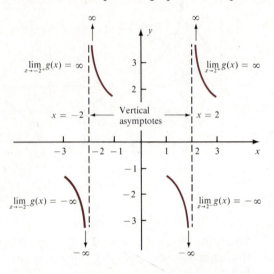

FIGURE 10–3. Example 2.

In Figures 10–2 and 10–3, we have labeled some lines as vertical asymptotes.*

(10.2)   If $f(x) \to \pm\infty$ as $x \to p$, then the line $x = p$ is called a **vertical asymptote** of the graph of f.

Notation: We use the symbol $f(x) \to \pm\infty$ as a shorthand notation for $f(x) \to \infty$ *or* $f(x) \to -\infty$. The same is true for the symbol $x \to \pm\infty$.

Limits at Infinity ? no such place   means x increase w/out bound

What happens to the values of f as x increases without bound?

(10.3)   If $f(x)$ can be made arbitrarily close to L by taking x sufficiently large, then we say that *the limit as x approaches infinity of $f(x)$ is L,* and write

$$\lim_{x \to \infty} f(x) = L$$

Other symbols, such as

$$\lim_{x \to -\infty} f(x) = L \quad \text{and} \quad \lim_{x \to \infty} f(x) = \infty$$

are also used. Their meanings should be clear. Figure 10–4 shows the geometric interpretations attached to these limits.

*From the Greek word *synpiptein*, meaning "to fall together."

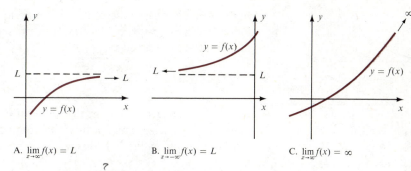

FIGURE 10–4. Geometric interpretations of limits at infinity.

A. $\lim_{x \to \infty} f(x) = L$ B. $\lim_{x \to -\infty} f(x) = L$ C. $\lim_{x \to \infty} f(x) = \infty$

When limits at infinity are real numbers, then the limit theorems hold. For example, if $\lim_{x \to \infty} f(x) = L$ and $\lim_{x \to \infty} g(x) = M$, then

$$\lim_{x \to \infty} (f + g)(x) = \lim_{x \to \infty} f(x) + \lim_{x \to \infty} g(x) = L + M$$

The product, quotient, and root theorems also hold. Even the squeeze theorem is valid.

EXAMPLE 3   Let $f(x) = 1 + (2/x)$. Find the limits as $x \to \pm \infty$ and sketch the graph for large x.

Solution   As $x \to \pm \infty$, the quotient $2/x \to 0$. Thus,

$$\lim_{x \to \pm \infty} 1 + \frac{2}{x} = \lim_{x \to \pm \infty} 1 + \lim_{x \to \pm \infty} \frac{2}{x} = 1$$

When x is large and positive, then $2/x$ is small and positive so $f(x)$ is slightly larger than 1. When x is large and negative, then $2/x$ is small and negative so $f(x)$ is slightly less than 1. This is pictured in Figure 10–5. ∎

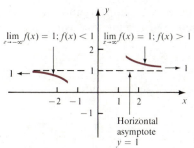

$\lim_{x \to -\infty} f(x) = 1; f(x) < 1$ $\lim_{x \to \infty} f(x) = 1; f(x) > 1$

Horizontal asymptote $y = 1$

FIGURE 10–5. Example 3.

EXAMPLE 4   Let $g(x) = (6x^2 - 10x - 57)/(3x^2 + 4)$. Find the limits as $x \to \pm \infty$ and sketch the graph for large x.

Solution   The limit theorem for quotients cannot be used because the limits of the numerator and the denominator are infinite. There is, however, a way to convert quotients of polynomials into a form so that the limit theorem does apply. Simply pick the largest power x^n that appears in the quotient (x^2 in this problem) and divide every term by this power.

$$\frac{6x^2 - 10x - 57}{3x^2 + 4} = \frac{\dfrac{6x^2}{x^2} - \dfrac{10x}{x^2} - \dfrac{57}{x^2}}{\dfrac{3x^2}{x^2} + \dfrac{4}{x^2}}$$

$$= \frac{6 - \dfrac{10}{x} - \dfrac{57}{x^2}}{3 + \dfrac{4}{x^2}}$$

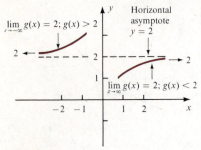

$\lim_{x \to -\infty} g(x) = 2; g(x) > 2$

Horizontal asymptote $y = 2$

$\lim_{x \to \infty} g(x) = 2; g(x) < 2$

FIGURE 10–6. Example 4.

In this form, it is easy to see that the numerator approaches 6 and the denominator approaches 3 as x approaches plus or minus infinity. Therefore, the limit theorem for quotients does apply, and

$$\lim_{x \to \pm\infty} \frac{6x^2 - 10x - 57}{3x^2 + 4} = \frac{6}{3} = 2$$

Because of the term $-10x$ in the numerator, the quotient will be slightly less than 2 when x is large and positive; the quotient is slightly greater than 2 when x is large and negative (test this with, say, $x = \pm 10$.) See Figure 10–6. ■

In Figures 10–5 and 10–6, we have labeled some lines as horizontal asymptotes.

(10.4) If $f(x) \to L$ as $x \to \pm\infty$, then the line $y = L$ is called a **horizontal asymptote** of the graph of f.

The technique of dividing by the largest power that appears in a quotient of two polynomials (see Example 4) can be used to justify the following general rule.

(10.5) If f and g are polynomials with leading coefficients a and b, respectively, then

$$\lim_{x \to \pm\infty} \frac{f(x)}{g(x)} = \begin{cases} 0 & \text{if degree of } f < \text{degree of } g \\ \dfrac{a}{b} & \text{if degree of } f = \text{degree of } g \\ \pm\infty & \text{if degree of } f > \text{degree of } g \end{cases}$$

The sign in the last statement is determined by the signs of $f(x)$ and and $g(x)$ as $x \to \pm\infty$.

EXAMPLE 5

(a) $\displaystyle\lim_{x \to \pm\infty} \frac{4x^2 + 5}{-x^4 - 3} = 0$ because degree of $f <$ degree of g.

(b) $\displaystyle\lim_{x \to \pm\infty} \frac{5x^3 + x^2}{3x^3 - 5x} = \frac{5}{3}$ because degree of $f =$ degree of g. ■

When the degree of $f >$ degree of g, the quotient *always* tends to $\pm\infty$ as $x \to \pm\infty$. The sign is determined by the signs of f and g. In general, the sign of a polynomial as $x \to \pm\infty$ agrees with the sign of the term of highest power.

EXAMPLE 6

(a) $\lim\limits_{x \to \infty} \dfrac{x^3 - 9x^2 - 8x - 4}{-2x^2 + 5x + 1} = -\infty$

As $x \to \infty$, the numerator is positive because the term of highest power x^3 is positive; but the denominator is negative because the term of highest power $-2x^2$ is negative.

(b) $\lim\limits_{x \to -\infty} \dfrac{-4x^3 - 20x^2 + 4x}{5x^2 - 30x - 8} = \infty$

As $x \to -\infty$, the numerator is positive because $-4x^3$ is positive, and the denominator is also positive because $5x^2$ is positive. ∎

Infinite limits and limits at infinity can be used as aids in graph sketching.

EXAMPLE 7 Sketch the graph of $g(x) = (3x + 2)/(x^2 - 4)$.

Solution Find the vertical and horizontal asymptotes, if any. This is the same function as in Example 2 above so we already know that it has two vertical asymptotes (see Figure 10–3). To find its horizontal asymptotes, if any exist, take the limit as $x \to \pm\infty$. In this case

$$\lim_{x \to \pm\infty} \frac{3x + 2}{x^2 - 4} = 0$$

so $y = 0$ (the x-axis) is a horizontal asymptote. As $x \to \infty$, both numerator and denominator are positive, so the graph is slightly above the x-axis. As $x \to -\infty$, the quotient is negative (why?), so the graph is slightly below the x-axis. The information gathered so far is shown in Figure 10–7A. We could almost complete

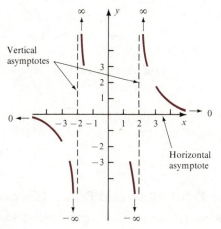

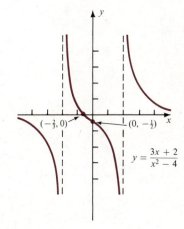

FIGURE 10–7. Example 7.
(A) Information obtained about asymptotes. (B) The complete graph.

A.

B.

the graph by just looking at this figure, especially the parts in the upper right and lower left corners. But let us go on and find the intercepts and derivatives.

When $x = 0$, then $y = -1/2$ and when $y = 0$, then $x = -2/3$; thus $(0, -1/2)$ and $(-2/3, 0)$ are the intercepts. The derivatives are

$$g'(x) = \frac{-(3x^2 + 4x + 12)}{(x^2 - 4)^2} \quad \text{and} \quad g''(x) = \frac{6x^3 + 12x^2 + 72x + 16}{(x^2 - 4)^3}$$

At all points $x \neq \pm 2$, $g'(x) < 0$; so g is always decreasing and there are no extreme values. Testing $g''(x)$ with several values of x, we find that g is concave down for $x < -2$, concave up for $x > 2$, and there is an inflection point somewhere between $x = -2/3$ and $x = 0$ ($g''(-2/3) > 0$ but $g''(0) < 0$). This is enough information to sketch the graph shown in Figure 10–7B. ■

The next example illustrates another technique for finding limits at infinity.

EXAMPLE 8 Find $\lim_{x \to \infty} \sqrt{x}(\sqrt{x + 1} - \sqrt{x})$. Before going on, guess what the limit is. Then compare with the solution below.

Solution The limit theorems do not apply (why?). There is, however, a way to change the form of this function so that the limit theorems do apply. You can *rationalize* as follows:

$$\sqrt{x}(\sqrt{x + 1} - \sqrt{x}) = \sqrt{x}(\sqrt{x + 1} - \sqrt{x})\frac{\sqrt{x + 1} + \sqrt{x}}{\sqrt{x + 1} + \sqrt{x}}$$

$$= \frac{\sqrt{x}(x + 1 - x)}{\sqrt{x + 1} + \sqrt{x}}$$

$$= \frac{\sqrt{x}}{\sqrt{x + 1} + \sqrt{x}}$$

$$= \frac{1}{\sqrt{1 + \dfrac{1}{x}} + 1} \qquad \text{(Divide by } \sqrt{x}\text{)}$$

Now the root, sum, and quotient theorems for limits do apply because all the limits are real numbers. Since $(\sqrt{1 + (1/x)} + 1) \to 2$ as $x \to \infty$, we have

$$\lim_{x \to \infty} \sqrt{x}(\sqrt{x + 1} - \sqrt{x}) = \lim_{x \to \infty} \frac{1}{\sqrt{1 + \dfrac{1}{x}} + 1} = \frac{1}{2} \qquad ■$$

Some Definitions and Proofs (Optional)

The precise definitions for infinite limits and limits at infinity are as follows:

(10.6)

If f is defined at every point in an open interval (a, b) except, perhaps, at p, then

$$\lim_{x \to p} f(x) = \infty$$

if, for every $N > 0$, there is a $\delta > 0$ such that

$$f(x) > N \quad \text{whenever} \quad 0 < |x - p| < \delta$$

(10.7)

If f is defined on an interval (a, ∞), then

$$\lim_{x \to \infty} f(x) = L$$

if, for every $\epsilon > 0$, there is an $N > a$ such that

$$|f(x) - L| < \epsilon \quad \text{whenever} \quad x > N$$

Similar definitions can be made for one-sided infinite limits and for limits as $x \to -\infty$. Here are two examples of proofs using these definitions.

Theorem.   For any positive number c, $\lim_{x \to 0} c/x^2 = \infty$.

Proof   Let N be any fixed positive number. We must find a $\delta > 0$ such that

$$\frac{c}{x^2} > N \quad \text{whenever} \quad 0 < |x - 0| < \delta$$

Working backwards from $c/x^2 > N$, we find that δ can be $\sqrt{c/N}$. If

$$0 < |x - 0| = |x| < \sqrt{c/N}$$

then

$$x^2 = |x|^2 < (\sqrt{c/N})^2 = \frac{c}{N}$$

and

$$\frac{c}{x^2} > N$$

By (10.6), this proves the theorem.

Theorem.   For any number c, $\lim_{x \to \infty} c/x = 0$.

Proof   Let ϵ be any fixed positive number. We must find an $N > 0$ such that

$$\left| \frac{c}{x} - 0 \right| < \epsilon \quad \text{whenever} \quad x > N$$

Working backwards from $|c/x| < \epsilon$, we find that N can be $|c|/\epsilon$. If

$$x > \frac{|c|}{\epsilon}$$

then x is positive, and

$$\epsilon > \frac{|c|}{x} = \left|\frac{c}{x}\right| = \left|\frac{c}{x} - 0\right|$$

By (10.7), this proves the theorem.

EXERCISES

In Exercises 1–10, find the indicated limits.

1. $\dfrac{x^2 - 4}{x - 1}$; $x \to 1^+$ and $x \to 1^-$

2. $\dfrac{3x^2 + 4x - 8}{x(x + 1)}$; $x \to 0^+$ and $x \to 0^-$

3. $\dfrac{x}{x^2 - x - 6}$; $x \to 3^+$ and $x \to 3^-$

4. $\dfrac{x^2}{x^2 + x - 2}$; $x \to 1^+$ and $x \to 1^-$

5. $\dfrac{-x}{x^2 - 2x + 1}$; $x \to 1^+$ and $x \to 1^-$

6. $\dfrac{x^3}{x^2 + 2x + 1}$; $x \to -1^+$ and $x \to -1^-$

7. $\dfrac{x^2}{(x + 2)^3}$; $x \to -2^+$ and $x \to -2^-$

8. $\dfrac{-x}{(x - 5)^3}$; $x \to 5^+$ and $x \to 5^-$

9. $\dfrac{2x^2 + 3}{x^3 + x^2 - 6x}$; $x \to -3^+$ and $x \to -3^-$

10. $\dfrac{2x^2 + 3}{x^3 + x^2 - 6x}$; $x \to 2^+$ and $x \to 2^-$

In Exercises 11–20, find the limits as $x \to \infty$ and as $x \to -\infty$.

11. $\dfrac{x^2 + 3x + 1}{2x^2 - 48}$

12. $\dfrac{x^3 + x + 1}{4x^3 - 3x}$

13. $\dfrac{1,000x + 1,000,000}{x^2}$

14. $\dfrac{-5,000x^2 - 5,000,000}{x^3}$

15. $\dfrac{x^2}{1,000x + 1,000,000}$

16. $\dfrac{x^3}{-5,000x^2 - 5,000,000}$

17. $\dfrac{50 - 3x^2 - x^3}{x^2}$

18. $\dfrac{-x^2}{3x + 983}$

19. $\sqrt{\dfrac{x}{4x + 1}}$

20. $\sqrt{\dfrac{9x - 10}{x}}$

Compute the following limits and compare them with the results of Example 8.

21. $\lim\limits_{x \to \infty} (\sqrt{x + 1} - \sqrt{x})$

22. $\lim\limits_{x \to \infty} x(\sqrt{x + 1} - \sqrt{x})$

In Exercises 23–30, find the vertical and horizontal asymptotes (if there are any) and sketch the graph.

23. $3 + \dfrac{1}{x}$

24. $\dfrac{2}{x} - 1$

25. $\dfrac{2x}{x^2 + x - 6}$

26. $\dfrac{-x}{x^2 - 2x + 1}$

27. $\dfrac{2x^2}{x^2 - 2x + 1}$

28. $\dfrac{5x}{3x - 2}$

29. $\dfrac{4x^2 + 4x + 1}{2x^2 - x - 1}$

30. $\dfrac{x^2 - 4x + 4}{2x^2 - 4x}$

Optional Exercises

31. (a) Use (10.6) to prove that $1/(x + 1)^2 \to \infty$ as $x \to -1$.

  (b) Use (10.7) to prove that $x^{-2} \to 0$ as $x \to \infty$.

32. If f and g are functions with $0 < f(x) < g(x)$ for all x and $g(x) \to 0$ as $x \to \infty$, prove that $f(x) \to 0$.

If $f(x) \to \infty$ as $x \to \infty$, prove that $g(x) \to \infty$ also.

33. Write out precise definitions similar to (10.6) and (10.7) for the following limits:

  (a) $\lim\limits_{x \to p^+} f(x) = -\infty$ (b) $\lim\limits_{x \to -\infty} f(x) = L$

  (c) $\lim\limits_{x \to \infty} f(x) = -\infty$

10-2 L'HÔPITAL'S RULE

A quotient $f(x)/g(x)$ is said to have the **indeterminate form 0/0 at** p if $f(x)$ and $g(x)$ both tend to 0 as x approaches p. The quotients

$$\frac{x}{x^2 - 3x} \quad \text{and} \quad \frac{\sin x}{x}$$

both have the indeterminate form 0/0 at $p = 0$, and yet their limits at 0 are quite different:

$$\lim_{x \to 0} \frac{x}{x^2 - 3x} = \lim_{x \to 0} \frac{1}{x - 3} = -\frac{1}{3} \quad \text{and} \quad \lim_{x \to 0} \frac{\sin x}{x} = 1$$

That is why the word *indeterminate* is used. In the same way, a quotient $f(x)/g(x)$ has the **indeterminate form** ∞/∞ **at** p if $f(x)$ and $g(x)$ both tend to $\pm\infty$ as x approaches p.

We already know how to find the limit of some quotients of indeterminate form; for instance, the ones above. Now we will discuss a method, known as *L'Hôpital's rule,* that will enable us to find many more.* At the end of this section there is a precise statement and partial proof of this rule. We begin, however, with an informal statement and several examples.

> **L'HÔPITAL'S RULE (Informal statement)**
>
> If $f(x)/g(x)$ has an indeterminate form 0/0 or ∞/∞ at p, then
>
> (10.8)
> $$\lim_{x \to p} \frac{f(x)}{g(x)} = \lim_{x \to p} \frac{f'(x)}{g'(x)}$$
>
> provided the last limit exists or is infinite. The same is true if p is replaced by $p^+, p^-,$ or $\pm\infty$.

Applying the rule to the two examples used earlier, we have

$$\lim_{x \to 0} \frac{x}{x^2 - 3x} = \lim_{x \to 0} \frac{(x)'}{(x^2 - 3x)'} = \lim_{x \to 0} \frac{1}{2x - 3} = -\frac{1}{3}$$

and

$$\lim_{x \to 0} \frac{\sin x}{x} = \lim_{x \to 0} \frac{(\sin x)'}{(x)'} = \lim_{x \to 0} \frac{\cos x}{1} = 1$$

Remark: To be completely correct in the two examples above, we should have determined whether or not the limits

---

*In 1696, the Marquis de L'Hôpital (1661–1704) published the first calculus textbook. It consisted of a series of lectures given by L'Hôpital's teacher Johann Bernoulli (1667–1748). It was actually Bernoulli who formulated the rule for finding the limit of an indeterminate form. But, because it appeared first in L'Hôpital's book, it is always (incorrectly) referred to as L'Hôpital's rule.

$$\lim_{x \to 0} \frac{(x)'}{(x^2 - 3x)'} \quad \text{and} \quad \lim_{x \to 0} \frac{(\sin x)'}{(x)'}$$

exist *before* equating them to the expressions on the left. But, in order to simplify the procedure, it is customary to write the equalities as indicated.

If you are wondering how it is that derivatives can be used to compute limits, consider the following special case. Suppose that f and g are differentiable in an open interval containing p. If $f(p) = g(p) = 0$, then f/g has the indeterminate form $0/0$ at p; moreover, for $x \neq p$, we have

$$\frac{f(x)}{g(x)} = \frac{f(x) - f(p)}{g(x) - g(p)} = \frac{\dfrac{f(x) - f(p)}{x - p}}{\dfrac{g(x) - g(p)}{x - p}}$$

If $g'(p) \neq 0$, the quotient theorem for limits applies to the last expression, so that

$$\lim_{x \to p} \frac{f(x)}{g(x)} = \frac{f'(p)}{g'(p)}$$

This special case is only remotely connected with the proof of L'Hôpital's rule (which is presented at the end of this section), but it does provide some indication of the link between derivatives and limits of indeterminate forms.

Using L'Hôpital's rule is a three-step process.

(10.9)

> (1) Make sure that $f(x)/g(x)$ has an indeterminte form; if it hasn't, *the rule cannot be used.*
> (2) Differentiate f and g *separately.*
> (3) Find the limit of $f'(x)/g'(x)$; this limit is also the limit of $f(x)/g(x)$.

EXAMPLE 1   Find $\lim_{x \to 0} (e^x - 1)/x$.

Solution   Follow the steps in (10.9):

(1) This quotient has the indeterminate form $0/0$

(2) $(e^x - 1)' = e^x$ and $(x)' = 1$

(3) $\lim_{x \to 0} (e^x/1) = 1$; therefore,

$$\lim_{x \to 0} \frac{e^x - 1}{x} = 1$$

Remark: To save space, we will combine steps (2) and (3). Thus, after checking that the quotient has an indeterminate form, we will write

$$\lim_{x \to 0} \frac{e^x - 1}{x} = \lim_{x \to 0} \frac{e^x}{1} = 1 \qquad \blacksquare$$

EXAMPLE 2   Find $\lim_{x \to 0^+} \sin \sqrt{x}/x$.

Solution   The quotient has indeterminate form $0/0$. Thus,

$$\lim_{x \to 0^+} \frac{\sin \sqrt{x}}{x} = \lim_{x \to 0^+} \frac{\frac{1}{2\sqrt{x}} \cos \sqrt{x}}{1} = \infty \qquad \blacksquare$$

It often happens that the quotient $f'(x)/g'(x)$ also has an indeterminate form, and L'Hôpital's rule can be applied a second time. In fact, the rule can be used as often as is necessary, *provided each new quotient has an indeterminate form.*

EXAMPLE 3   Find $\lim_{x \to 0} (1 - \cos x)/(x^3 + x^2)$.

Solution   The quotient has indeterminate form $0/0$. Therefore,

$$\lim_{x \to 0} \frac{1 - \cos x}{x^3 + x^2} = \lim_{x \to 0} \frac{\sin x}{3x^2 + 2x}$$

Now the last quotient also has indeterminate form $0/0$, so L'Hôpital's rule is applied again:

$$= \lim_{x \to 0} \frac{\cos x}{6x + 2}$$

This quotient does *not* have an indeterminate form, so the process *stops*. It follows that

$$\lim_{x \to 0} \frac{1 - \cos x}{x^3 + x^2} = \lim_{x \to 0} \frac{\sin x}{3x^2 + 2x} = \lim_{x \to 0} \frac{\cos x}{6x + 2} = \frac{1}{2} \qquad \blacksquare$$

EXAMPLE 4   Find $\lim_{x \to \infty} x^{5/2}/e^x$.

Solution   The quotient has indeterminate form ∞/∞; so does the quotient of its derivatives $(5/2)x^{3/2}$ and e^x. In fact, the quotients will continue to have an indeterminate form until the exponent of x becomes negative. Thus,

$$\lim_{x \to \infty} \frac{x^{5/2}}{e^x} = \lim_{x \to \infty} \frac{(5/2)x^{3/2}}{e^x} = \lim_{x \to \infty} \frac{(15/4)x^{1/2}}{e^x} = \lim_{x \to \infty} \frac{(15/8)x^{-1/2}}{e^x}$$

The last quotient is *not* an indeterminate form because $x^{-1/2} \to 0$ and $e^x \to \infty$. The quotient approaches 0, so

$$\lim_{x \to \infty} \frac{x^{5/2}}{e^x} = 0 \qquad \blacksquare$$

Example 4 is a special case of a more general statement that is important in later work.

(10.10)

> For *any* positive number *a*,
>
> $$\lim_{x \to \infty} \frac{x^a}{e^x} = 0$$

The proof is similar to that exhibited in Example 4. We simply continue to differentiate the numerator and denominator until the exponent of *x* becomes negative. The process stops and we see that the limit in (10.10) is valid. This result holds for any positive *a*, no matter how large; it shows that

> e^x increases faster than any positive power of *x*

The next example also contains an important limit.

EXAMPLE 5   Show that

(10.11)

> For *any* positive number *a*,
>
> $$\lim_{x \to \infty} \frac{\ln x}{x^a} = 0$$

Solution   The quotient has indeterminate form ∞/∞. Therefore,

$$\lim_{x \to \infty} \frac{\ln x}{x^a} = \lim_{x \to \infty} \frac{(1/x)}{ax^{a-1}} = \frac{1}{a} \lim_{x \to \infty} \frac{1}{x^a} = 0$$

This result holds for any positive *a*, no matter how small; it shows that

> $\ln x$ increases more slowly than any positive power of *x*   ■

Other Indeterminate Forms

If $f(x) \to 0$ and $g(x) \to \pm\infty$ as $x \to p$, then the product $f(x)g(x)$ is said to have the **indeterminate form $0 \cdot \infty$ at p.** We can convert this into one of the indeterminate forms 0/0 or ∞/∞ by writing

$$f(x)g(x) = \frac{f(x)}{1/g(x)} \quad \text{or} \quad f(x)g(x) = \frac{g(x)}{1/f(x)}.$$

EXAMPLE 6   Find $\lim_{x \to 0^+} x \ln x$.

Solution   This has indeterminate form $0 \cdot \infty$ and we convert it to ∞/∞ by writing

$$x \ln x = \frac{\ln x}{(1/x)}$$

Now L'Hôpital's rule applies and

$$\lim_{x \to 0^+} x \ln x = \lim_{x \to 0^+} \frac{\ln x}{(1/x)} = \lim_{x \to 0^+} \frac{(1/x)}{(-1/x^2)}$$

The last quotient simplifies to $-x$, so the limit is 0. Therefore,

$$\lim_{x \to 0^+} x \ln x = 0 \quad \blacksquare$$

Other indeterminates of the form $0°$, $\infty°$, and 1^∞ arise from functions such as $f(x)^{g(x)}$. We treat these forms by writing

$$(10.12) \quad f(x)^{g(x)} = e^{g(x) \ln f(x)}$$

Now the product $g(x) \ln f(x)$ will have the form $0 \cdot \infty$, which can be dealt with by the method just discussed. If

$$\lim_{x \to p} g(x) \ln f(x) = L$$

then it follows from (10.12), that

$$(10.13) \quad \lim_{x \to p} f(x)^{g(x)} = e^L$$

EXAMPLE 7 Find $\lim_{x \to 0^+} x^x$.

Solution This has indeterminate form $0°$. First we write

$$x^x = e^{x \ln x}$$

and note that $x \ln x$ has the form $0 \cdot \infty$ at 0. From Example 6, we know that $x \ln x \to 0$ as $x \to 0^+$, and it follows from (10.13) that

$$\lim_{x \to 0^+} x^x = e^0 = 1 \quad \blacksquare$$

EXAMPLE 8 Show that $\lim_{x \to \infty} (1 + r/x)^x = e^r$ for any real number r.

Solution This has indeterminate form 1^∞. First we write

$$\left(1 + \frac{r}{x}\right)^x = e^{x \ln(1 + r/x)}$$

The product $x \ln(1 + rx^{-1})$ has the form $\infty \cdot 0$ as $x \to \infty$. Thus,

$$\lim_{x \to \infty} x \ln(1 + rx^{-1}) = \lim_{x \to \infty} \frac{\ln(1 + rx^{-1})}{x^{-1}} = \lim_{x \to \infty} \frac{(-rx^{-2})/(1 + rx^{-1})}{-x^{-2}} = r$$

and it follows from (10.13) that

$$\lim_{x \to \infty} \left(1 + \frac{r}{x}\right)^x = e^r \quad \blacksquare$$

CAUTION: An error that students frequently make is to stop after computing the limit $g(x) \ln f(x)$. Remember that *the final answer is e raised to that power.*

Another indeterminate form that occurs frequently is $\infty - \infty$. Such forms can usually be converted into one of the forms already discussed.

EXAMPLE 9   Find $\lim_{x \to 0}(1/(\sin x) - 1/x)$.

Solution   Convert this $\infty - \infty$ form into $0/0$ form by combining fractions. Thus,

$$\lim_{x \to 0}\left(\frac{1}{\sin x} - \frac{1}{x}\right) = \lim_{x \to 0} \underbrace{\frac{x - \sin x}{x \sin x}}_{0/0} = \lim_{x \to 0}\left.\frac{1 - \cos x}{x \cos x + \sin x}\right\} 0/0 \text{ again}$$

$$= \lim_{x \to 0}\left.\frac{\sin x}{2 \cos x - x \sin x}\right\} \textit{not } 0/0$$

$$= 0 \quad \blacksquare$$

Statement and Proof of L'Hôpital's Rule *(Optional)*

(10.14)

> **L'HÔPITAL'S RULE**
>
> Suppose that $f(x)/g(x)$ has the indeterminate form $0/0$ or ∞/∞ at p. Suppose further that f and g are differentiable on open intervals (a, p) and (p, b) and that $g'(x)$ is never zero in either interval. Then
>
> $$\lim_{x \to p}\frac{f(x)}{g(x)} = \lim_{x \to p}\frac{f'(x)}{g'(x)}$$
>
> provided the latter limit exists or is infinite. This result remains valid if p is replaced by $p^+, p^-,$ or $\pm\infty$.

To prove this statement, we must first prove an extended version of the Mean Value Theorem (3.45). It is named for A. L. Cauchy (1789–1857).

(10.15)

> **CAUCHY MEAN VALUE THEOREM**
> Suppose that f and g are continuous on $[a, b]$ and differentiable on (a, b). If $g'(x)$ is never zero in (a, b), then there is a number c in (a, b) such that
>
> $$\frac{f(b) - f(a)}{g(b) - g(a)} = \frac{f'(c)}{g'(c)}$$

Proof   First, we observe that $g(a) \neq g(b)$; if it were, then by the Mean Value Theorem, $g'(c) = 0$ for some c in (a, b) contrary to the hypothesis. Next, we introduce a new function

$$F(x) = [f(b) - f(a)]g(x) - [g(b) - g(a)]f(x)$$

for x in $[a, b]$. It is easy to see that F is continuous on $[a, b]$, differentiable on (a, b), and $F(a) = F(b)$. Again by the Mean Value Theorem, there is a point c in (a, b) with $F'(c) = 0$; that is,

$$0 = [f(b) - f(a)]g'(c) - [g(b) - g(a)]f'(c)$$

This implies the conclusion of the theorem and completes the proof.

Now for the proof of L'Hôpital's rule.

Proof of (10.14)   Let $f(x)/g(x)$ have the indeterminate form $0/0$ and suppose that $\lim_{x \to p} f'(x)/g'(x) = L$. We must show that $\lim_{x \to p} f(x)/g(x)$ is also equal to L.

Since $f(x) \to 0$ and $g(x) \to 0$ as $x \to p$, we may assume that $f(p) = 0$ and $g(p) = 0$. (Remember that the value of f at p does not effect its limit at p.) This assumption makes f and g continuous on the whole interval (a, b). If $x \neq p$, we can apply (10.15) to either interval $[x, p]$ or $[p, x]$, as the case may be, and assert that there must be a number c *between* x and p such that

$$\frac{f(x) - f(p)}{g(x) - g(p)} = \frac{f'(c)}{g'(c)}$$

Or, since $f(p) = g(p) = 0$,

$$\frac{f(x)}{g(x)} = \frac{f'(c)}{g'(c)}$$

Because c is always between x and p, the limit of the left side as $x \to p$ equals L, the limit of the right side. This completes the proof. A similar argument would prove the statement if p is replaced by p^+ or p^- or the quotient $f(x)/g(x)$ tends to infinity.

To prove (10.14) in case p is replaced by ∞, we write

$$\lim_{x \to \infty} \frac{f'(x)}{g'(x)} = \lim_{t \to 0^+} \frac{f'(1/t)}{g'(1/t)} \qquad \text{(Replace } x \text{ by } 1/t)$$

$$= \lim_{t \to 0^+} \frac{f(1/t)}{g(1/t)} \qquad \text{(Already proved above)}$$

$$= \lim_{x \to \infty} \frac{f(x)}{g(x)}$$

The case of the indeterminate ∞/∞ is more complicated, and we omit its proof.

EXERCISES

Find the following limits; L'Hôpital's rule may or may not apply.

1. $\lim\limits_{x \to 0} \dfrac{\sin x}{3x}$

2. $\lim\limits_{x \to 0} \dfrac{\tan x}{4x}$

3. $\lim\limits_{x \to 3} \dfrac{x^2 - 4x + 3}{x^2 - 2x - 3}$

4. $\lim\limits_{x \to 0} \dfrac{x^2 - 4x + 3}{x^2 - 2x - 3}$

5. $\lim\limits_{x \to \infty} \dfrac{x^2 - 4x + 3}{x^2 - 2x - 3}$

6. $\lim\limits_{x \to \infty} \dfrac{\cos x}{x}$

7. $\lim\limits_{x \to 0} \dfrac{2^x - 3^x}{x}$

8. $\lim\limits_{x \to 0} \dfrac{\ln x}{\sqrt{x}}$

9. $\lim\limits_{x \to 0} \dfrac{e^x - x - 1}{x}$

10. $\lim\limits_{x \to \pi/2^-} \dfrac{2 + \sec x}{3 \tan x}$

11. $\lim\limits_{x \to 0} \dfrac{e^x - x - 1}{x^2}$

12. $\lim\limits_{x \to \infty} \dfrac{x^3 - 3x^2}{x^4}$

13. $\lim\limits_{x \to \infty} \left(\dfrac{1}{x}\right)^x$

14. $\lim\limits_{x \to \infty} \dfrac{x \ln x}{x + \ln x}$

15. $\lim\limits_{h \to 0} (1 + h)^{1/h}$

16. $\lim\limits_{h \to 0} (1 + h)^{4/h}$

17. $\lim\limits_{t \to \infty} \left(1 + \dfrac{3}{t}\right)^t$

18. $\lim\limits_{x \to \infty} \dfrac{\ln x}{e^x}$

19. $\lim\limits_{x \to 0} \dfrac{3x - 1}{4x}$

20. $\lim\limits_{x \to \pi/2} \dfrac{1 - \sin x}{\cos x}$

21. $\lim\limits_{x \to \infty} \dfrac{x^{10,000}}{e^{.00001x}}$

22. $\lim\limits_{x \to \infty} \dfrac{x^{.00001}}{\ln x}$

23. $\lim\limits_{x \to 0} \dfrac{\cos\left(\dfrac{\pi}{2} + x\right)}{x}$

24. $\lim\limits_{x \to -3} \dfrac{x + 3}{x}$

25. $\lim\limits_{x \to 0} \dfrac{\sin^{-1}x}{x}$

26. $\lim\limits_{x \to 0} \dfrac{\sin x}{1 - \cos x}$

27. $\lim\limits_{x \to 0^+} (\tan x)^{\sin x}$

28. $\lim\limits_{x \to \infty} (\sqrt{x^2 + x} - x)$

29. $\lim\limits_{x \to 0^+} x \ln x^2$

30. $\lim\limits_{x \to 0} \dfrac{1 - \cos 2x}{x^2}$

31. $\lim\limits_{x \to 1} \left(\dfrac{1}{\ln x} - \dfrac{1}{x - 1}\right)$

32. $\lim\limits_{x \to 0} \dfrac{e^x - 1}{x(x + 1)}$

33. $\lim\limits_{x \to 0^+} x^{1/\ln x}$

34. $\lim\limits_{x \to \infty} \dfrac{x + \sin 2x}{x - \sin 2x}$

35. $\lim\limits_{x \to 0} \dfrac{\sin x - x \cos x}{x^2 \sin x}$

36. $\lim\limits_{x \to 0} \dfrac{\tan \pi x}{e^x - 1}$

37. $\lim\limits_{x \to -3^+} \dfrac{\ln(x + 3)}{x + 3}$

38. $\lim\limits_{x \to 3^+} \dfrac{\ln(x - 2)}{x - 3}$

39. $\lim\limits_{x \to 3^+} \dfrac{\ln(x - 2)}{(x - 3)^2}$

40. $\lim\limits_{x \to 0} (e^x + x)^{1/x}$

41. $\lim\limits_{x \to 0} \dfrac{\sin x}{\sin 2x}$

42. $\lim\limits_{x \to 0^+} \dfrac{\ln \sin x}{\ln \sin 2x}$

43. $\lim\limits_{x \to 0} \dfrac{\ln \cos x}{\ln \cos 2x}$

44. $\lim\limits_{x \to 0^+} x^{\sin x}$

45. $\lim\limits_{x \to 0} \dfrac{\sqrt{9 + x} - \sqrt{9 - x}}{x}$

46. $\lim\limits_{x \to \infty} \dfrac{\sqrt{1 + x^2}}{x}$

47. $\lim\limits_{x \to 0} \left(\dfrac{1}{\sin^2 x} - \dfrac{1}{x^2}\right)$

48. $\lim\limits_{x \to 0} \dfrac{e^x - 2^x}{x}$

49. $\lim\limits_{x \to \infty} \left(1 - \dfrac{3}{x}\right)^{2x}$

50. $\lim\limits_{x \to 0} (1 - 3x)^{1/x}$

10–3   IMPROPER INTEGRALS

Suppose that f is continuous and nonnegative on the infinite interval $[a, \infty)$. The region between the graph of f and the x-axis is shown in Figure 10–8. This region goes on and on; it has no right-hand boundary. Our first inclination might be to declare that such a region must have infinite area. But that could prove to be a hasty judgement.

EXAMPLE 1   Discuss the area of the region under the graph of $y = 1/x^2$ for $x \geq 1$.

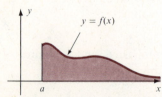

FIGURE 10–8. The function f is defined on $[a, \infty)$. The region under the graph has no right-hand boundary.

Solution   Figure 10–9 shows the region under the graph of $y = 1/x^2$ for $x \geq 1$. For each $t > 1$,

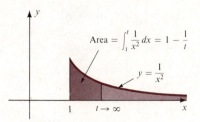

$$\text{Area} = \int_1^t \frac{1}{x^2}\,dx = 1 - \frac{1}{t}$$

$$y = \frac{1}{x^2}$$

$1 \qquad t \to \infty$

FIGURE 10–9. Example 1.

$$\int_1^t \frac{1}{x^2}\,dx = \left[-\frac{1}{x}\right]_1^t = 1 - \frac{1}{t}$$

is the area under the graph from $x = 1$ to $x = t$. *This area is less than 1 (square unit) for every t, no matter how large.* Moreover,

$$\lim_{t \to \infty} \int_1^t \frac{1}{x^2}\,dx = \lim_{t \to \infty}\left(1 - \frac{1}{t}\right) = 1$$

Therefore, it is reasonable to say that *the area under the graph of $y = 1/x^2$ for $x \geq 1$ is 1 square unit.* ■

As this example suggests, it is useful to extend the notion of a definite integral to include infinite intervals. To do that, we introduce some new symbols:

$$\int_a^\infty f(x)\,dx, \int_{-\infty}^b f(x)\,dx, \text{ and } \int_{-\infty}^\infty f(x)\,dx$$

These are called **improper integrals.**

Suppose f is continuous on an unbounded interval $[a, \infty)$. For each $t > a$, the integral

$$\int_a^t f(x)\,dx$$

exists, and we now take the limit as $t \to \infty$. If

$$\lim_{t \to \infty} \int_a^t f(x)\,dx = L$$

then we write

$$\int_a^\infty f(x)\,dx = L$$

and say that the improper integral **converges** (to L). If the limit does not exist, then we say that the improper integral **diverges.** The improper integral

$$\int_{-\infty}^b f(x)\,dx = \lim_{t \to -\infty} \int_t^b f(x)\,dx$$

is treated in a similar way. We'll come back to the third type after some examples.

EXAMPLE 2   Determine whether or not $\int_1^\infty 1/x^2\,dx$ converges.

Solution   This is the integrand in Example 1, and we have seen that

$$\lim_{t \to \infty} \int_1^t \frac{1}{x^2}\,dx = 1$$

Therefore, we say that the integral *converges* to 1 and write

$$\int_1^\infty \frac{1}{x^2}\,dx = 1 \qquad ■$$

EXAMPLE 3   Determine whether or not $\int_{-\infty}^{0} 1/(1 + x^2)\, dx$ converges.

Solution   To test for convergence, replace $-\infty$ by t and take the limit:

$$\lim_{t \to -\infty} \int_{t}^{0} \frac{1}{1 + x^2}\, dx = \lim_{t \to -\infty} \left[\tan^{-1}x \right]_{t}^{0}$$

$$= \lim_{t \to -\infty} (-\tan^{-1}t)$$

$$= \frac{\pi}{2}$$

Therefore, this integral converges and

$$\int_{-\infty}^{0} \frac{1}{1 + x^2}\, dx = \frac{\pi}{2} \quad ∎$$

EXAMPLE 4   Determine if $\int_{1}^{\infty} 1/x\, dx$ converges and relate the answer to the area under the curve $y = 1/x$ for $x \geq 1$.

Solution   To test for convergence, replace ∞ by t and take the limit:

$$\lim_{t \to \infty} \int_{1}^{t} \frac{1}{x}\, dx = \lim_{t \to \infty} \left[\ln x \right]_{1}^{t} = \lim_{t \to \infty} \ln t = \infty$$

Therefore, this integral *diverges*. In terms of area, you can see that for each $t > 1$, the area under the curve from $x = 1$ to $x = t$ is $\ln t$ square units. As $t \to \infty$, so does the area. In this case, it is reasonable to say that *the area under the curve $y = 1/x$ for $x \geq 1$ is infinite.*   ∎

EXAMPLE 5 *(Probability on unbounded intervals).*   In Section 6–7, we introduced the notion of continuous random variables and their probability density functions (*pdf*). These ideas can now be extended.

Most products are manufactured with built-in obsolescence. For example, refrigerators are built to last about 7 years. Of course, it is possible for a given refrigerator to last a lot longer. Let X be the continuous random variable with values in $[0, \infty)$ that represents the lifetime (in years) of a refrigerator. Its *pdf* is $p(x) = \frac{1}{7}e^{-x/7}$; $p(x) \geq 0$ and

$$\int_{0}^{\infty} \frac{1}{7}e^{-x/7}\, dx = \lim_{t \to \infty} \int_{0}^{t} \frac{1}{7}e^{-x/7}\, dx = \lim_{t \to \infty} \left[-e^{-x/7} \right]_{0}^{t}$$

$$= \lim_{t \to \infty} [1 - e^{-t/7}]$$

$$= 1$$

The expected value $E(X)$ is the integral of $xp(x)$. In this example,

$$E(X) = \int_{0}^{\infty} x \left[\frac{1}{7}e^{-x/7} \right] dx = \lim_{t \to \infty} \int_{0}^{t} x \left[\frac{1}{7}e^{-x/7} \right] dx$$

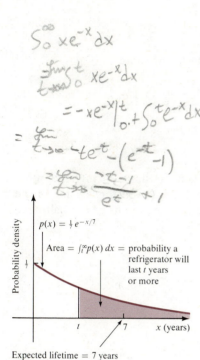

The last integral is evaluated by parts:

$$\int_0^t x\left[\frac{1}{7}e^{-x/7}\right] dx = \left[-xe^{-x/7}\right]_0^t + \int_0^t e^{-x/7}\, dx$$

$$= (-te^{-t/7} + 0) - 7(e^{-t/7} - 1)$$

$$= \frac{-t - 7}{e^{t/7}} + 7$$

Using L'Hôpital's rule, it is easy to see that as $t \to \infty$, the limit of the first term is 0, and it follows that

$$E(X) = 7 \text{ years}$$

What is the probability that any given refrigerator will last 10 years or more? The answer is obtained by integrating the *pdf* from 10 to ∞ (Figure 10–10).

$$\int_{10}^\infty \frac{1}{7} e^{-x/7}\, dx = \lim_{t \to \infty} \int_{10}^t \frac{1}{7} e^{-x/7}\, dx = \lim_{t \to \infty}\left[-e^{-x/7}\right]_{10}^t$$

$$= e^{-10/7} \approx 0.2 \quad \blacksquare$$

We return now to the third type of improper integral. If f is continuous on the whole real line, then

$$\int_{-\infty}^\infty f(x)\, dx$$

is said to *converge* if

$$\int_{-\infty}^a f(x)\, dx \quad \text{and} \quad \int_a^\infty f(x)\, dx$$

both converge for some number a. In that case, we set

$$\int_{-\infty}^\infty f(x)\, dx = \int_{-\infty}^a f(x)\, dx + \int_a^\infty f(x)\, dx$$

EXAMPLE 6 Find the area of the region bounded by the graph of $y = e^{-|x|}$ and the *x*-axis.

Solution The region is shown in Figure 10–11; its area is given by the improper integral

$$\int_{-\infty}^\infty e^{-|x|}\, dx$$

provided this integral converges. We will show that

$$\int_{-\infty}^0 e^{-|x|}\, dx \quad \text{and} \quad \int_0^\infty e^{-|x|}\, dx$$

both converge.

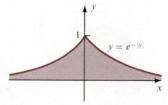

FIGURE 10–11. Example 6.

Below the figure 10-10 area:

Probability density

$p(x) = \frac{1}{7}e^{-x/7}$

Area $= \int_t^\infty p(x)\, dx =$ probability a refrigerator will last t years or more

t 7 x (years)

Expected lifetime $= 7$ years

FIGURE 10–10. Example 5.

$$\int_{-\infty}^{0} e^{-|x|}\, dx = \lim_{t \to -\infty} \int_{t}^{0} e^{-|x|}\, dx = \lim_{t \to -\infty} \int_{t}^{0} e^{x}\, dx \qquad (x < 0)$$

$$= \lim_{t \to -\infty} (1 - e^{t}) = 1$$

It is easy to see that $\int_{0}^{\infty} e^{-|x|}\, dx$ also converges to 1. Therefore,

$$\int_{-\infty}^{\infty} e^{-|x|}\, dx = 2$$

and the area is 2 square units. ■

Unbounded Integrands

Another kind of improper integral occurs when the limits of integration are finite, but the integrand is unbounded. Figure 10–12 illustrates some of the different possibilities. In general, suppose that f is continuous on the half-open interval $(a, b]$ and that $f(x) \to \pm\infty$ as $x \to a^{+}$. If

$$\lim_{t \to a^{+}} \int_{t}^{b} f(x)\, dx = L$$

then we write

$$\int_{a}^{b} f(x)\, dx = L$$

and say that the improper integral *converges* (to L). If the limit does not exist, then we say that the integral *diverges*. If f is continuous on $[a, b)$ and $f(x) \to \pm\infty$ as $x \to b^{-}$, then the improper integral

$$\int_{a}^{b} f(x)\, dx = \lim_{t \to b^{-}} \int_{a}^{t} f(x)\, dx$$

is treated in a similar way. If f is continuous on $[a, c)$ and $(c, b]$, but $f(x) \to \pm\infty$ as $x \to c$, then we define

$$\int_{a}^{b} f(x)\, dx = \int_{a}^{c} f(x)\, dx + \int_{c}^{b} f(x)\, dx$$

provided *both* integrals on the right converge.

EXAMPLE 7   Find the area between the graph of $y = 1/\sqrt{x}$ and the x-axis from $x = 0$ to $x = 1$ (Figure 10–13).

Solution   As in the case of a region over an unbounded interval, it is reasonable to define the area of the region in Figure 10–13 as the value of the improper integral

$$\int_{0}^{1} \frac{1}{\sqrt{x}}\, dx$$

provided it converges. Since the integrand becomes infinite as $x \to 0^{+}$, we have

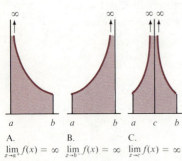

A.
$$\lim_{x \to a^{+}} f(x) = \infty$$
B.
$$\lim_{x \to b^{-}} f(x) = \infty$$
C.
$$\lim_{x \to c} f(x) = \infty$$

FIGURE 10–12. In each case, $\int_{a}^{b} f(x)\, dx$ is an *improper integral*.

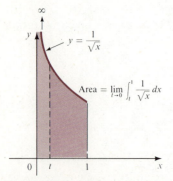

$$y = \frac{1}{\sqrt{x}}$$

$$\text{Area} = \lim_{t \to 0} \int_{t}^{1} \frac{1}{\sqrt{x}}\, dx$$

FIGURE 10–13. Example 7.

$$\int_0^1 \frac{1}{\sqrt{x}}\,dx = \lim_{t\to 0^+} \int_t^1 \frac{1}{\sqrt{x}}\,dx = \lim_{t\to 0^+}\left[2\sqrt{x}\right]_t^1 = \lim_{t\to 0^+}(2 - 2\sqrt{t}) = 2$$

Thus, the area is 2 square units. ■

EXAMPLE 8   Show that $\int_0^3 (x-1)^{-2/3}\,dx$ converges, and find its value.

Solution   The graph of $y = (x-1)^{-2/3}$ is sketched in Figure 10–14. As it indicates, we must break up the integral into two parts

$$\int_0^1 (x-1)^{-2/3}\,dx \quad \text{and} \quad \int_1^3 (x-1)^{-2/3}\,dx$$

and show that both converge. For the first part,

$$\int_0^1 (x-1)^{-2/3}\,dx = \lim_{t\to 1^-} \int_0^t (x-1)^{-2/3}\,dx$$

$$= \lim_{t\to 1^-}\left[3(x-1)^{1/3}\right]_0^t$$

$$= \lim_{t\to 1^-} 3(t-1)^{1/3} - (-3) = 3$$

For the second,

$$\int_1^3 (x-1)^{-2/3}\,dx = \lim_{t\to 1^+} \int_t^3 (x-1)^{-2/3}\,dx$$

$$= \lim_{t\to 1^+}\left[3(x-1)^{1/3}\right]_t^3$$

$$= \lim_{t\to 1^+} 3\cdot 2^{1/3} - 3(t-1)^{1/3} = 3\sqrt[3]{2}$$

Therefore, the integral $\int_0^3 (1-x)^{-2/3}\,dx$ converges, and its value is $3(1 + \sqrt[3]{2})$. ■

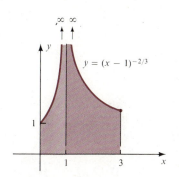

FIGURE 10–14. Example 8.

$y = (x-1)^{-2/3}$

The Comparison Test

It is often useful to know whether or not an improper interval converges even though we cannot determine its exact value. For that purpose, we have the following test:

COMPARISON TEST
Suppose f and g are continuous and $0 \le f(x) \le g(x)$ for all x in $[a, \infty)$.

(10.16)   (1) If $\displaystyle\int_a^\infty f(x)\,dx$ diverges, then $\displaystyle\int_a^\infty g(x)\,dx$ diverges.

(2) If $\displaystyle\int_a^\infty g(x)\,dx$ converges, then $\displaystyle\int_a^\infty f(x)\,dx$ converges.

Although a proof of the Comparison Test is beyond the scope of this book, the test itself is intuitively reasonable. Since $0 \le f(x) \le g(x)$, we have

$$0 \le \int_a^t f(x)\, dx \le \int_a^t g(x)\, dx \quad \text{all } t \ge a$$

If the smaller integral diverges, so does the larger one (part (1) of the test). If the larger integral tends to a finite limit, so does the smaller one (part (2) of the test). A similar test is also valid for the other improper integrals.

EXAMPLE 9   Determine whether or not $\int_1^\infty e^{-x^2}\, dx$ converges.

Solution   There is no elementary formula for evaluating

$$\int e^{-x^2}\, dx$$

However,

$$0 \le e^{-x^2} < e^{-x} \quad \text{for all } x \ge 1$$

and

$$\int_1^\infty e^{-x}\, dx = \lim_{t \to \infty} \int_1^t e^{-x}\, dx = \lim_{t \to \infty}[-e^{-t} + e^{-1}] = \frac{1}{e}$$

Therefore, by the comparison test (10.16),

$$\int_1^\infty e^{-x^2}\, dx \text{ converges} \quad \blacksquare$$

EXERCISES

In Exercises 1–30, determine whether or not the improper integrals converge. If convergent, find the value. (Not all of them are improper.)

1. $\displaystyle\int_2^\infty \frac{1}{x \ln x}\, dx$

2. $\displaystyle\int_0^1 \frac{1}{\sqrt{1 - x^2}}\, dx$

3. $\displaystyle\int_0^\infty \frac{1}{1 + x^2}\, dx$

4. $\displaystyle\int_0^\infty e^{-2x}\, dx$

5. $\displaystyle\int_0^1 \frac{1}{x^{3/2}}\, dx$

6. $\displaystyle\int_0^1 \frac{1}{x^2}\, dx$

7. $\displaystyle\int_0^1 \frac{1}{x^{2/3}}\, dx$

8. $\displaystyle\int_0^1 \frac{1}{\sqrt{1 - x}}\, dx$

9. $\displaystyle\int_0^1 x \ln x\, dx$

10. $\displaystyle\int_0^2 \frac{x}{\sqrt{4 - x^2}}\, dx$

11. $\displaystyle\int_1^\infty \frac{x}{(1 + x^2)^2}\, dx$

12. $\displaystyle\int_3^\infty \frac{x}{\sqrt{x^2 - 9}}\, dx$

13. $\displaystyle\int_0^\infty x e^{-x^2}\, dx$

14. $\displaystyle\int_2^\infty \frac{1}{x^2 - 1}\, dx$

15. $\displaystyle\int_1^\infty \frac{1}{x^2 - 1}\, dx$

16. $\displaystyle\int_0^{\pi/2} \sec^2 x\, dx$

17. $\displaystyle\int_0^{\pi/2} \tan x\, dx$

18. $\displaystyle\int_0^\infty \frac{1}{\sqrt[3]{x + 1}}\, dx$

19. $\displaystyle\int_0^\infty \sin x\, dx$

20. $\displaystyle\int_{-\infty}^{-1} \frac{1}{x^3}\, dx$

21. $\displaystyle\int_0^1 \frac{1}{x^2} \sin \frac{1}{x}\, dx$

22. $\displaystyle\int_0^\infty \frac{1}{4 + x^2}\, dx$

23. $\displaystyle\int_{-\infty}^\infty \cos^2 x\, dx$

24. $\displaystyle\int_{-\infty}^{-1} \frac{1}{x}\, dx$

25. $\displaystyle\int_0^\infty \frac{1}{x^2 + 3x + 2}\, dx$

26. $\displaystyle\int_0^\infty \sin^2 x \cos x\, dx$

27. $\displaystyle\int_0^\infty e^{-x} \cos x\, dx$

28. $\displaystyle\int_{-2}^{-1} \frac{1}{x\sqrt{x^2 - 1}}\, dx$

29. $\int_1^\infty \dfrac{1}{x^r} dx$ for $\begin{cases} r > 1 \\ r = 1 \\ r < 1 \end{cases}$ (Three separate cases.)

30. $\int_0^1 \dfrac{1}{x^r} dx$ for $\begin{cases} r > 1 \\ r = 1 \\ r < 1 \end{cases}$ (Three separate cases.)

In Exercises 31–34, use the Comparison Test to determine whether or not the integrals converge.

31. $\int_0^\infty \dfrac{1}{1 + x^3} dx$

32. $\int_0^\infty \dfrac{x^2}{1 + x^2 + x^4} dx$

33. $\int_2^\infty \dfrac{1}{\ln x} dx$

34. $\int_0^\infty \dfrac{1}{x \sqrt{|\ln x|}} dx$

35. By Example 4, the region between the x-axis and the graph of $y = 1/x$, $x \geq 1$ has *infinite* area. But the volume generated by revolving it about the x-axis is *finite!* Find the volume.

36. By Example 7, the region between the x-axis and the graph of $y = 1/\sqrt{x}$, $0 < x \leq 1$, has *finite* area. But the volume generated by revolving it about the x-axis is *infinite!* Show that this is so.

37. Sometimes you must evaluate an improper integral in order to evaluate a proper one. For example, evaluate $\int_0^1 \sin^{-1} x \, dx$. (This is proper, but the integration by parts formula yields an improper integral.)

38. The integral $\int_1^\infty 1/x \sqrt{x^2 - 1} \, dx$ is improper in both respects; an unbounded interval and an unbounded integrand. Evaluate it by evaluating the sum

$$\int_1^2 \frac{1}{x \sqrt{x^2 - 1}} dx + \int_2^\infty \frac{1}{x \sqrt{x^2 - 1}} dx$$

39. *(Present value of revenue).* If an indestructible capital asset, such as land, brings in revenue at the rate of $R(t)$ dollars per year, and this revenue is to be discounted at an annual rate of $r\%$, then the *present value* of the asset is

$$P.V. = \int_0^\infty R(t)e^{-.01rt} \, dt$$

(See Example 1 in Section 9–2.) Find the present value of the income from a piece of land that brings in (a) $R(t) = \$1,000$ per year (b) $R(t) = 1,000 + 100t$ dollars per year. Assume that money is discounted at an annual rate of 8%.

40. *(Circumference of a circle).* Compute the circumference of a circle of radius r by finding the arc length of the curve $y = \sqrt{r^2 - x^2}$ for $0 \leq x \leq r$. (This indicates that our definitions are consistent.)

41. *(Probability).* Suppose that X is a continuous random variable with values in $[0, \infty)$ and a probability density function $p(x) = e^{-x}$, $x \geq 0$. (a) What is the expected value of X? (b) What is the probability that the values of X are in the interval $[0.5, \infty)$?

42. (Continuation of Exercise 41). Suppose the values of X represent the lifetime (in years) of a piece of equipment that costs \$2 to make and sells for \$5. The manufacturer guarantees a full refund if the product fails within 6 months. What is the manufacturer's expected profit on each item? (*Hint:* Look at Example 4 in Section 6–7.)

Optional Exercises

43. Let

$$f(x) = \begin{cases} \cos x & \text{if } \cos x \geq 0 \\ 0 & \text{if } \cos x < 0 \end{cases} \quad \text{and}$$

$$g(x) = \begin{cases} 0 & \text{if } \cos x \geq 0 \\ -\cos x & \text{if } \cos x < 0 \end{cases}$$

Then $f(x)$ and $g(x)$ are nonnegative and $f(x) - g(x) = \cos x$ *(check).* Use f and g and the Comparison Test to show that

$$\int_1^\infty \frac{\cos x}{x^2} dx$$

converges.

44. (Continuation of Exercise 43). Use integration by parts and the results of Exercise 43 to show that

$$\int_1^\infty \frac{\sin x}{x} dx$$

converges.

10–4 THE PREDATOR-PREY MODEL *(Optional reading)*

Let us suppose that a population consists of two classes, the *predators* and the *prey*. Examples abound in nature; foxes as predators and rabbits as prey, pelicans as predators and herring as prey, and so on. Let us take as our example a greenhouse that contains aphids and ladybugs. Aphids are insects that suck the life juices from plants and ladybugs are beetles that suck the life juices from aphids.

Let $y = y(t)$ and $x = x(t)$ represent the population of ladybugs (the predators) and aphids (the prey) at time t. Our predator-prey model is based on three assumptions about x and y.

(1) The supply of vegetation will support an unrestricted growth of aphids. Therefore, in the absence of ladybugs, x grows exponentially and

(10.17) $\quad \dfrac{dx}{dt} = Ax \quad A > 0$

(2) In the absence of aphids, there is no food for the ladybugs so their population decreases exponentially.

(10.18) $\quad \dfrac{dy}{dt} = -Cy \quad C > 0$

(3) A certain proportion of encounters between aphids and ladybugs will result in the death of an aphid. The number of encounters is directly proportional to the product of populations x and y.

Figure 10–15 shows the cyclical nature of the two populations. As x increases, there is more food for the ladybugs, so y increases. When the ladybugs begin to kill aphids faster than the aphids can reproduce, then x begins to decrease. This means less food for the predators, so y decreases.

Assumption (3) suggests that equations (10.17) and (10.18) be adjusted as follows

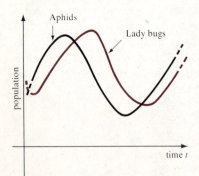

FIGURE 10–15. The cyclical nature of the populations of predators and prey.

(10.19) $\quad \boxed{\begin{aligned} \dfrac{dx}{dt} &= Ax - Bxy \\[6pt] \dfrac{dy}{dt} &= -Cy + Dxy \end{aligned}}$

where A, B, C, and D are positive constants. These are known as the *Lotka-Volterra equations.**

Our first observation is that the constant populations

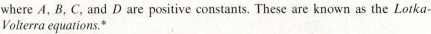

*A. J. Lotka (1880–1949) was an American biophysicist and V. Volterra (1860–1940) was an Italian mathematician.

(10.20) $x = \dfrac{C}{D}$ and $y = \dfrac{A}{B}$

are a particular solution of these equations. If x and y are these constants, then dx/dt and dy/dt are 0 and

$$Ax - Bxy = A\frac{C}{D} - B\frac{C}{D}\frac{A}{B} = 0$$

$$-Cy + Dxy = -C\frac{A}{B} + D\frac{C}{D}\frac{A}{B} = 0$$

We will return to this particular solution later.

Although we cannot find elementary general solutions for $x(t)$ and $y(t)$, we can still obtain a lot of information about their relationship. The first step is to divide the second equation in (10.19) by the first:

$$\frac{dy/dt}{dx/dt} = \frac{dy}{dx} = \frac{-Cy + Dxy}{Ax - Bxy} = \frac{(-C + Dx)y}{(A - By)x}$$

Separating the variables yields

$$\left(\frac{A}{y} - B\right)dy = \left(-\frac{C}{x} + D\right)dx$$

or

$$\left(\frac{A}{y} - B\right)dy - \left(-\frac{C}{x} + D\right)dx = 0$$

Integrating both sides, we have

$$A \ln y - By + C \ln x - Dx = K$$

where K is the constant of integration. This equation can be expressed in terms of exponentials as follows:

(10.21) $(y^A e^{-By})(x^C e^{-Dx}) = e^K$

This is the key relationship between x and y. For one thing, it shows that neither x nor y can increase without bound. To establish this, we note that each of the factors

$$\frac{y^A}{e^{By}} \quad \text{and} \quad \frac{x^C}{e^{Dx}}$$

in (10.21) is bounded (each is less than some positive number M) because the exponential increases faster than any power of x. Now, if $x \to \infty$, then

$$\frac{x^C}{e^{Dx}} \to 0$$

and the product on the left of (10.21) would equal 0 contrary to the fact that $e^K \neq 0$. Therefore, x cannot increase without bound. A similar argument applies to y.

(10.22) | The populations x and y are bounded.

Now let us return to the first Lotka-Volterra equation. Rewrite it as

$$\frac{dx}{x} = A\,dt - By\,dt$$

and integrate from $t = 0$ to $t = w$

$$\ln x(w) - \ln x(0) = Aw - B\int_0^w y(t)\,dt$$

Now divide both sides by w

$$\frac{\ln x(w)}{w} - \frac{\ln x(0)}{w} = A - \frac{B}{w}\int_0^w y(t)\,dt$$

and let $w \to \infty$. Since x is bounded (10.22), the left side approaches 0, and it follows that

$$(10.23) \quad \frac{1}{w}\int_0^w y(t)\,dt \to \frac{A}{B} \quad \text{as } w \to \infty$$

A similar treatment of the second Lotka-Volterra equation leads to the conclusion that

$$(10.24) \quad \frac{1}{w}\int_0^w x(t)\,dt \to \frac{C}{D} \quad \text{as } w \to \infty$$

The expressions

$$\frac{1}{w}\int_0^w y(t)\,dt \quad \text{and} \quad \frac{1}{w}\int_0^w x(t)\,dt$$

represent the average value of y and x over the time interval $[0, w]$. Thus, (10.23) and (10.24) indicate that *the average values of y and x tend to A/B and C/D, respectively, as $t \to \infty$.* Note that these are the constant solutions (10.20).

Figure 10–16 is a computer-generated graph of equation (10.21) for various values of K. Notice that the point $(C/D, A/D)$ is at the "center" of each loop. The figure indicates that the values of x and y recur in cycles about their averages. Many conclusions about the use of pesticides, the trapping of fur-bearing animals, and so on, can be drawn from this model.

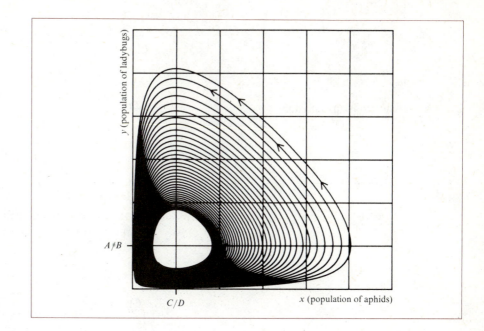

FIGURE 10–16. Graphs of ladybug vs. aphid populations.

REVIEW EXERCISES

In Exercises 1–5, find the indicated limits.

1. $\dfrac{x^2 - 16}{x - 2}$; $x \to 2^+$ and $x \to 2^-$

2. $\dfrac{2x}{x^2 - x - 2}$; $x \to 2^+$ and $x \to 2^-$

3. $\dfrac{-x}{x^2 + 4x + 4}$; $x \to -2^+$ and $x \to -2^-$

4. $\dfrac{x^2}{x^2 - 2x + 1}$; $x \to 1^+$ and $x \to 1^-$

5. $\dfrac{x^2 - 9}{x + 3}$; $x \to -3^+$ and $x \to -3^-$

In Exercises 6–11, find the limits as $x \to \infty$, and as $x \to -\infty$.

6. $\dfrac{2x^2 + 2x + 1}{3x^2 + 4}$

7. $\dfrac{2x + 100}{5x^2}$

8. $\dfrac{10x^2 + 5x + 7}{x\sqrt{|x|}}$

9. $\sqrt{\dfrac{4x}{9x + 2}}$

10. $\dfrac{7x^3 + 2x^2 + 4x + 1}{\frac{1}{2}x^3 + 4x + 3}$

11. $\sqrt{x^2 + 2} - \sqrt{x^2 + 1}$

In Exercises 12–16, find the vertical and horizontal asymptotes of the graphs (if there are any).

12. $y = \dfrac{x^2 - 9}{x - 3}$

13. $y = \dfrac{\sqrt{4x^2 + 2x + 1}}{2x + 7}$

14. $y = \dfrac{\sqrt{2x^2 + 5x + 2}}{x^2 - 1}$

15. $y = \dfrac{x + 3}{x^2 + x - 6}$

16. $\dfrac{x^3 - 1}{x + 1}$

In Exercises 17–37, find the indicated limits (L'Hôpital's rule may be necessary).

17. $\lim\limits_{x \to 0} \dfrac{2\sin(x/2)}{x}$

18. $\lim\limits_{x \to 0} \dfrac{\sqrt{-\cos^2 x + 1}}{x}$

19. $\lim\limits_{x \to 0} \dfrac{x^2 + 1}{x^2 - 1}$

20. $\lim\limits_{x \to \pi} \dfrac{1 - \sin(x/2)}{\cos(x/2)}$

21. $\lim\limits_{x \to 0^+} \dfrac{e^{-x} - 1}{x^2}$

22. $\lim\limits_{x \to 0} \dfrac{4^x - 2^x}{x^2}$

23. $\lim\limits_{x \to \infty} \sqrt[x]{x}$

24. $\lim\limits_{x \to 1} \dfrac{\ln x}{x - 1}$

25. $\lim\limits_{x \to \pi/2^-} \dfrac{2 \tan x}{3 + \sec x}$

26. $\lim\limits_{x \to \infty} \dfrac{x^3 - 3x^2}{4x^2 + 1}$

27. $\lim\limits_{x \to \infty} \dfrac{\ln x}{x \ln x + x^2 \ln x}$

28. $\lim\limits_{x \to \infty} \dfrac{x^a}{e^x}$ where a is any constant

29. $\lim\limits_{x \to \infty} \dfrac{e^x}{\ln(x^2)}$

30. $\lim\limits_{x \to 0} \dfrac{2 \sin x - 2 \sin(x/2)\cos(x/2)}{x}$

31. $\lim\limits_{x \to 3^+} \dfrac{\ln(x - 2)}{x^2 - 9}$

32. $\lim\limits_{x \to 0^+} x^{\tan^2 x}$

33. $\lim\limits_{x \to 0} \dfrac{e^x - 1}{2^x - 1}$

34. $\lim\limits_{x \to 0} \dfrac{\sqrt{9 + x} - 3}{x}$

37. $\lim\limits_{x \to \infty} \left(1 + \dfrac{1}{e^x}\right)^x$

In Exercises 38–49, determine whether or not the improper integrals converge. If one does converge, find the value.

38. $\displaystyle\int_1^\infty \dfrac{dx}{x(\ln x)^2}$

39. $\displaystyle\int_1^2 \dfrac{dx}{\sqrt{x - 1}}$

40. $\displaystyle\int_{-\infty}^\infty \dfrac{1}{4 + x^2}\, dx$

41. $\displaystyle\int_1^\infty x^{-(5/2)}\, dx$

42. $\displaystyle\int_0^\infty xe^{-x}\, dx$

43. $\displaystyle\int_0^\infty \cos 2x\, dx$

44. $\displaystyle\int_4^\infty \dfrac{x\, dx}{(x^2 - 4)^3}$

45. $\displaystyle\int_1^2 \dfrac{dx}{(x - 2)(x - 1)}$

46. $\displaystyle\int_0^\infty 2^{-x}\, dx$

47. $\displaystyle\int_0^{\pi/2} \dfrac{\sec^2 x}{\tan x}\, dx$

48. $\displaystyle\int_{-1}^0 \dfrac{\ln(x + 1)}{x + 1}\, dx$

49. $\displaystyle\int_1^\infty \dfrac{dx}{x\sqrt{x^2 + 1}}$

In Exercises 50–53, use the Comparison Test to determine whether or not the integrals converge.

50. $\displaystyle\int_0^\infty \dfrac{dx}{\sqrt[4]{1 + x^3}}$

51. $\displaystyle\int_0^\infty \dfrac{x^3}{1 + x^2 + x^4 + x^6}\, dx$

52. $\displaystyle\int_1^\infty \dfrac{1}{\sqrt{x} \ln x}\, dx$

53. $\displaystyle\int_0^\infty \dfrac{x^2}{e^x}\, dx$

54. The present value of a plot of land is $\int_0^\infty R(t)e^{-\alpha t}\, dt$.
   (a) Find the present value of the land if $R(t) = \$500$, and α is 0.7.
   (b) Find the present value of the land if $R(t) = \$500 + \alpha t$ and α is 0.7.

11

INFINITE SERIES

The sum of infinitely many numbers can be finite. For instance,

$$\frac{3}{10} + \frac{3}{100} + \frac{3}{1,000} + \cdots = 0.333 \cdots = \frac{1}{3}$$

Infinite sums often arise in practical situations: Economists estimate that a certain percentage of each dollar in circulation is recirculated over and over again. Let us suppose the percentage is 90%. A dollar is printed and circulated; 90% is spent, then 90% of the 90% is spent, and so on. The total value of one dollar to the economy is defined to be the infinite sum

$$.90 + (.90)(.90) + (.90)(.90)(.90) + \cdots$$

As you will see in Section 11–2, the sum is 9 dollars. Similar applications occur in probability, science, and engineering. The connection with calculus, our primary concern, is that certain functions can be expressed as infinite sums; then new and useful information about the functions is obtained by applying the methods of calculus to the sums.

Section 1 serves as an introduction to the subject. Section 2 contains the basic definitions and a lot of examples. These examples are important because subsequent material is based on them. Sections 3, 4, and 5 describe the tests that can be applied to determine whether or not an infinite sum has a finite value. These tests are preparation for Sections 6, 7, and 8 which contain the heart of the matter; an analysis of the infinite sums that represent functions. Such sums can be differentiated, integrated, and manipulated as though they were polynomials.

Section 9, *Infinite Series and Music Synthesizers,* is optional reading. It is an application of infinite sums to reproducing musical sounds.

11–1 SEQUENCES

In conversation, the word *sequence* means a succession or an order of succession. For example, we speak of a "sequence of events" meaning there is a first event, a second event, and so on. In this book, a "sequence of numbers," means an infinite list of numbers

(11.1) $a_1, a_2, a_3, \ldots, a_n, \ldots$

where a_1 is the first number (or *term*), a_2 is the second, a_n is the *n*th, and so on. We picture a sequence, geometrically, as points on a line (Figure 11–1).

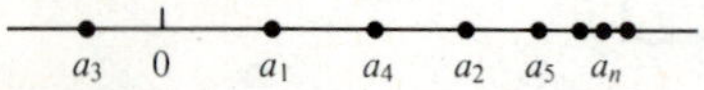

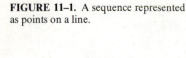

FIGURE 11–1. A sequence represented as points on a line.

Sometimes the sequence (11.1) is written more compactly as

$$\{a_n\}_{n=1}^{\infty} \quad \text{or, simply,} \quad \{a_n\}$$

This means to let $n = 1$, then let $n = 2$, then 3, and so on. Some examples are

$$\left\{\frac{1}{n}\right\}_{n=1}^{\infty} \quad \text{is the sequence } 1, \frac{1}{2}, \frac{1}{3}, \ldots$$

$$\left\{2^n\right\}_{n=1}^{\infty} \quad \text{is the sequence } 2, 4, 8, \ldots$$

It is not necessary to always begin a sequence with $n = 1$, nor is it necessary to always use the letter n.

$$\left\{\frac{1}{k(k+1)}\right\}_{k=3}^{\infty} \quad \text{is the sequence } \frac{1}{3 \cdot 4}, \frac{1}{4 \cdot 5}, \frac{1}{5 \cdot 6}, \ldots$$

$$\left\{\left(-\frac{1}{3}\right)^p\right\}_{p=0}^{\infty} \quad \text{is the sequence } 1, -\frac{1}{3}, \frac{1}{9}, -\frac{1}{27}, \ldots$$

Sequences have a lot in common with functions. Clearly, if f is any function defined on $[1, \infty)$, then $\{f(n)\}$ is a sequence. Conversely, if $\{a_n\}$ is any sequence, then we can define a function g on the set of positive integers $\{1, 2, 3, \ldots\}$ by setting $g(1) = a_1, g(2) = a_2$, and so on. In fact, the formal definition of a sequence is given in terms of functions:

(11.2)

> **DEFINITION**
> A **sequence** is a function defined on the set of positive integers.

Now sequences can be represented as graphs of functions (Figure 11–2). An advantage of defining a sequence in this way is that we can talk about limits of sequences, increasing or decreasing sequences, and so on, just as we did with functions.

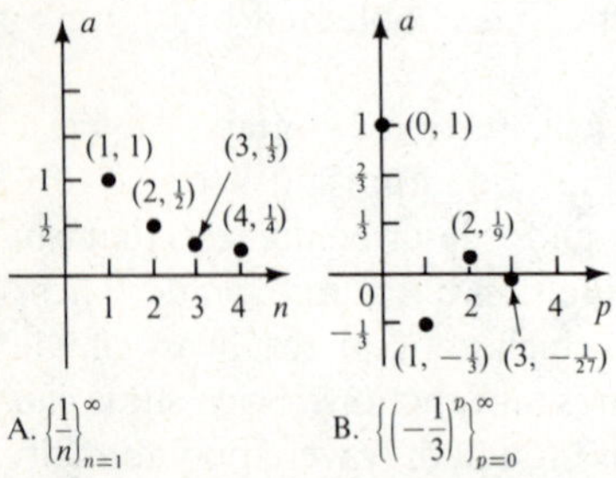

For example, if $\{a_n\}$ is a sequence, then

$$\lim_{n \to \infty} a_n = L$$

means that as n gets larger and larger, the numbers a_n get closer and closer to

A. $\left\{\frac{1}{n}\right\}_{n=1}^{\infty}$ B. $\left\{\left(-\frac{1}{3}\right)^p\right\}_{p=0}^{\infty}$

FIGURE 11–2. Sequences as graphs of functions.

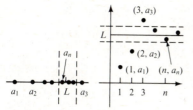

A. Points on a line   B. Graph of a function.

FIGURE 11–3. Two ways to illustrate $\lim_{n \to \infty} a_n = L$.

the number L (Figure 11–3). The precise definition is given at the end of this section, but the following statement conveys the general idea:

(11.3)

> If a_n can be made arbitrarily close to L by taking n sufficiently large, then we say that *the limit of the sequence* $\{a_n\}$ *is L,* and write
> $$\lim_{n \to \infty} a_n = L \quad \text{or} \quad a_n \to L \quad \text{as} \quad n \to \infty$$
> In this case, we sometimes say that the sequence **converges** to L.

This is analogous to the interpretation (10.3) of the symbol

$$\lim_{x \to \infty} f(x) = L$$

where f is defined on an interval $[a, \infty)$. Moreover, the following result is evident:

(11.4)

> Suppose that $\{a_n\}$ is a sequence. Suppose further, that f is a function defined on $[1, \infty)$ and that $f(n) = a_n$ for $n = 1, 2, 3, \ldots$
> $$\text{If } \lim_{x \to \infty} f(x) = L, \quad \text{then} \quad \lim_{n \to \infty} a_n = L$$

This statement saves us the trouble of repeating the limit results proved earlier.*

To find the limit of a sequence $\{a_n\}$ using (11.4), simply formulate a function f defined on $[1, \infty)$ for which (1) $f(n) = a_n$ and (2) you know the limit as $x \to \infty$.

EXAMPLE 1 Show that $\lim_{n \to \infty} n/(n + 1) = 1$.

Solution To use (11.4), we define the function $f(x) = x/(x + 1)$. Then (1) $f(n) = n/(n + 1)$ and (2) from previous work, we know that

$$\lim_{x \to \infty} \frac{x}{x + 1} = 1$$

Therefore, by (11.4), the limit of $\{n/n + 1\}$ is 1. ∎

We can also use L'Hôpital's rule to find limits of sequences.

EXAMPLE 2 Show that $\lim_{n \to \infty} n^{1/n} = 1$.

Solution We define $f(x) = x^{1/x} = e^{(1/x)\ln x}$. By L'Hôpital's rule, the limit of the exponent is

$$\lim_{x \to \infty} \frac{\ln x}{x} = \lim_{x \to \infty} \frac{1/x}{1} = 0$$

---

*This is a fine point, but one worth mentioning: *The converse of* (11.4) *is false.* For example, if $f(x) = \sin \pi x$, then $\lim_{x \to \infty} f(x)$ does not exist; but the sequence $\{f(n)\} = \{\sin n\pi\}$ is just a succession of 0's which certainly has a limit.

Therefore, $f(x) \to 1$ as $x \to \infty$, and it follows from (11.4) that $n^{1/n} \to 1$ as $n \to \infty$. ∎

We can even use the Squeeze Theorem.

EXAMPLE 3   Show that $\lim_{n \to \infty} (\sin n)/n = 0$.

Solution   Define $f(x) = (\sin x)/x$ for $x \geq 1$. Then

$$-\frac{1}{x} \leq \frac{\sin x}{x} \leq \frac{1}{x}$$

and it follows from the Squeeze Theorem that

$$\lim_{x \to \infty} f(x) = 0$$

Now apply (11.4).   ∎

The sum, difference, product, and quotient theorems carry over from functions:

(11.5)

> Let $\{a_n\}$ and $\{b_n\}$ be sequences with
>
> $$\lim_{n \to \infty} a_n = L \quad \text{and} \quad \lim_{n \to \infty} b_n = M$$
>
> Then
>
> $$\lim_{n \to \infty} (a_n + b_n) = L + M \quad \lim_{n \to \infty} (a_n - b_n) = L - M$$
>
> $$\lim_{n \to \infty} a_n b_n = LM \quad \lim_{n \to \infty} \frac{a_n}{b_n} = \frac{L}{M}, \quad \text{if } M \neq 0$$

Not every sequence has a limit; for example,

$$\{(-1)^n\} = -1, 1, -1, 1, \ldots$$

Another example of a sequence that does not have limit is

$$\{2^n\} = 2, 4, 8, \ldots$$

In this case, the terms 2^n increase without bound and we write

$$\lim_{n \to \infty} 2^n = \infty$$

> If the numbers a_n increase (or decrease) without bound, then we write
>
> $$\lim_{n \to \infty} a_n = \infty \quad \text{or} \quad \lim_{n \to \infty} a_n = -\infty$$
>
> as the case may be.

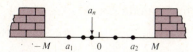

FIGURE 11–4. The sequence $\{a_n\}$ is bounded; every a_n is trapped between $-M$ and M.

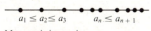

Monotonic increasing.

Monotonic decreasing.

FIGURE 11–5. Monotonic sequences.

What we want to do now is consider some conditions under which a sequence will always have a limit. A sequence $\{a_n\}$ is **bounded** if there is some positive number M such that $|a_n| \leq M$ for all n. That is,

$$-M \leq a_n \leq M \qquad \text{(Figure 11–4)}$$

A sequence $\{a_n\}$ is **monotonic** if

$$a_1 \leq a_2 \leq a_3 \leq \cdots \quad \text{(nondecreasing)}$$

or   (Figure 11.5)

$$a_1 \geq a_2 \geq a_3 \geq \cdots \quad \text{(nonincreasing)}$$

The fact that a sequence is bounded does not guarantee that it has a limit; for example, $\{(-1)^n\}$. The fact that a sequence is monotonic does not guarantee that it has a limit; for example, $\{2^n\}$. However, if a sequence is *both bounded and monotonic*, then it must have a limit.

(11.6)   **THE BOUNDED CONVERGENCE THEOREM**
A bounded monotonic sequence converges (has a limit).

A correct proof requires advanced methods, but we can argue intuitively as follows (refer to Figure 11–6). Suppose that $\{a_n\}$ is bounded and nondecreasing. Since it is bounded, $a_n \leq M$ for some M. Now let L be the *smallest* number with that property; that is,

$$L \text{ is the smallest number such that } a_n \leq L \text{ for all } n$$

(It seems intuitively clear that there *is* such a number L, but it is precisely this point that has to be established in a rigorous proof.) If x is a number very close to but less than L, then there must be some a_n between x and L (otherwise every $a_n \leq x$, so L isn't the smallest such number). Therefore, a_n is very close to L, and since the sequence is nondecreasing, we know that $a_{n+1}, a_{n+2}, \ldots$ are also close to L because

all a_n's $\leq L$

A. L is *smallest* number such that $a_n \leq L$.

At least one a_n between x and L.

B. $x < L$

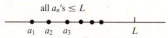

$a_n \leq a_{n+1} \leq a_{n+2} \leq \cdots \leq L.$

C.

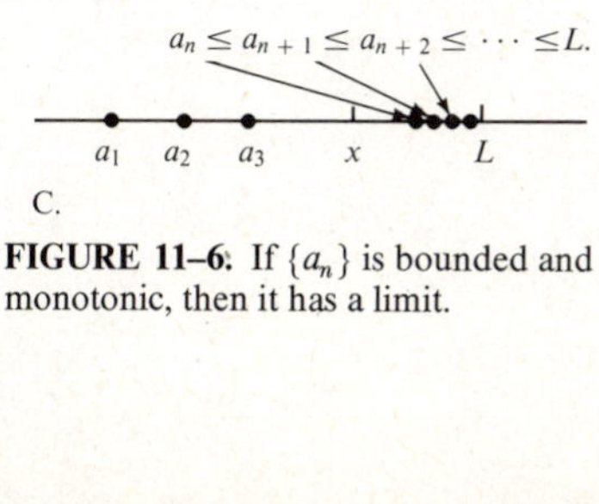

FIGURE 11–6. If $\{a_n\}$ is bounded and monotonic, then it has a limit.

$$a_n \leq a_{n+1} \leq a_{n+2} \leq \cdots \leq L \qquad \text{(Figure 11–6)}$$

This completes our argument because it shows that a_n can be made arbitrarily close to L by taking n sufficiently large. A similar argument holds for nonincreasing sequences.

Notice that (11.6) says nothing about the value of the limit; it merely asserts the existence of one.

EXAMPLE 4   The sequence

$$1, 1.1, 1.11, 1.111, \ldots$$

is nondecreasing and bounded by $M = 2$. Therefore, it must have a limit. ∎

We conclude the regular portion of this section with three special limits that are referred to in later work.

(11.7)   If $r > 0$, then $\lim\limits_{n \to \infty} \left(\dfrac{1}{n}\right)^r = 0$

Let $f(x) = (1/x)^r = 1/x^r$; then $x^r \to \infty$, so $1/x^r \to 0$ as $x \to \infty$.

(11.8) $\lim\limits_{n \to \infty} \left(1 + \dfrac{r}{n}\right)^n = e^r$ for any number r

This follows directly from Example 8 in Section 10–2.

(11.9) $\lim\limits_{n \to \infty} r^n \begin{cases} = 0 & \text{if } |r| < 1 \\ = 1 & \text{if } r = 1 \\ \text{does not exist if } |r| > 1 \text{ or } r = -1 \end{cases}$

If $r = 1$, the sequence is $1, 1, 1, \ldots$ and the limit is 1. If $r = -1$, the sequence is $-1, 1, -1, 1, \ldots$ and the limit does not exist. For the other cases, we write

$$|r^n| = |r|^n = e^{n \ln |r|}$$

If $|r| < 1$, then $\ln|r| < 0$, but if $|r| > 1$, then $\ln|r| > 0$. Thus, as $n \to \infty$, the exponent of e becomes a very large negative or positive number, depending on the size of $|r|$, and now it is easy to see that (11.9) is true.

Remark: Although we do not permit negative bases for general exponential functions, we do allow them in sequences because the exponents are integers and there are no problems with even roots of negative numbers. Thus, in (11.9), if $r = -2$, then $r^2 = (-2)(-2) = 4$, $r^3 = (-2)(-2)(-2) = -8$, and so on.

Definition of the Limit *(Optional)*

The precise definition of the limit of a sequence is as follows:

(11.10)

The **limit** of a sequence $\{a_n\}$ is L if, given any $\epsilon > 0$, there is a positive integer N such that

$$|a_n - L| < \epsilon \quad \text{whenever } n \geq N$$

In that case, we write

$$\lim\limits_{n \to \infty} a_n = L$$

EXAMPLE 5   Use (11.10) to prove that $\lim_{n \to \infty} 1/n = 0$.

Solution   Let $\epsilon > 0$. Working backwards to find N, we want

$$\frac{1}{n} = \left| \frac{1}{n} - 0 \right| < \epsilon$$

so take N to be any integer larger than $1/\epsilon$. If $N > 1/\epsilon$ and $n \geq N$, then

$$\left| \frac{1}{n} - 0 \right| = \frac{1}{n} \leq \frac{1}{N} < \epsilon$$

proving that $1/n \to 0$ as $n \to \infty$.   ∎

EXERCISES

For each sequence in Exercises 1–23, find the limit as $n \to \infty$, if it exists.

1. $\left\{ \dfrac{1}{2^n} \right\}$

2. $\{n\}$

3. $\left\{ \dfrac{\ln n}{n} \right\}$

4. $\left\{ \dfrac{1}{n^2 + 3n} \right\}$

5. $\left\{ \dfrac{k+1}{k+2} - \dfrac{k}{k+1} \right\}$ (*Hint:* Combine fractions.)

6. $\{ \sqrt{k+1} - \sqrt{k} \}$ (*Hint:* Rationalize.)

7. $\{\cos n\pi\}$

8. $\{\tan^{-1} n\pi\}$

9. $\left\{ \dfrac{(3n+1)(2n-3)}{n^2} \right\}$

10. $\left\{ \left(\dfrac{1}{n} \right)^3 \right\}$

11. $\{3^{1/n}\}$

12. $\{4^{n/2}\}$

13. $\left\{ \dfrac{n^5}{e^n} \right\}$

14. $\left\{ \dfrac{n^5}{\sqrt{e^n}} \right\}$

15. $\left\{ \left(1 + \dfrac{1}{n} \right)^n \right\}$

16. $\left\{ \left(1 + \dfrac{2}{n} \right)^n \right\}$

17. $\left\{ \displaystyle\int_0^n \dfrac{1}{1+x^2}\, dx \right\}$

18. $\left\{ \dfrac{\cos n\pi}{n} \right\}$

19. $\{n^{1/n}\}$

20. $\{(2n)^{1/n}\}$

21. $\left\{ \dfrac{\tan^{-1} n}{n} \right\}$

22. $\{ \sqrt{n^2+1} - n \}$

23. $\left\{ \left(1 - \dfrac{3}{n} \right)^n \right\}$

In the next section, we will be summing the numbers in a sequence; the sum of the first n numbers is called the *nth partial sum* of the sequence. For each sequence $\{a_n\}$ in Exercises 24–30, find the first four partial sums,

$$S_1 = a_1$$
$$S_2 = a_1 + a_2$$
$$S_3 = a_1 + a_2 + a_3$$
$$S_4 = a_1 + a_2 + a_3 + .a_4$$

24. $\{n\}$

25. $\{2^n\}$

26. $\left\{ \dfrac{1}{n} \right\}$

27. $\{\cos n\pi\}$

28. $\{(1/n) - 1/(n+1)\}$ (*Hint:* Write out the full sum before adding, and take advantage of cancellation.)

29. $\left\{ \ln \dfrac{n}{n+1} \right\}$ (There is cancellation here also.)

30. $\left\{ \left(\dfrac{1}{2} \right)^n \right\}$

In Exercises 31–34, use (11.6) to show that the given sequences have limits.

31. $\{2 + (.1)^n\}_{n=0}^{\infty}$

32. $\left\{ \displaystyle\int_0^n \dfrac{1}{1+x^3}\, dx \right\}$

33. $\left\{ \displaystyle\sum_{k=1}^{n} (.1)^k \right\}$

34. $\left\{ \displaystyle\sum_{k=1}^{n} \dfrac{3}{10^k} \right\}$

35. Find the limit

$$\lim_{n \to \infty} \sum_{k=1}^{n} \frac{3}{10^k}$$

and express it as a rational number.

11–2 INFINITE SERIES

In this section, we discuss the possibility of summing infinitely many numbers. As we mentioned earlier, the concept is not entirely new to you. For instance, the number 1/3 can be thought of as an infinite sum

$$\frac{1}{3} = 0.333\cdots = \frac{3}{10} + \frac{3}{100} + \frac{3}{1,000} + \cdots$$

More precisely, we start with the sequence

$$\left\{\frac{3}{10^n}\right\} = \frac{3}{10}, \frac{3}{100}, \frac{3}{1,000}, \cdots, \frac{3}{10^n}, \cdots$$

add the first n terms

$$\sum_{k=1}^{n} \frac{3}{10^k} = \frac{3}{10} + \frac{3}{100} + \frac{3}{1,000} + \cdots + \frac{3}{10^n}$$

and then take the limit as $n \to \infty$. The more terms we add, the closer the sum is to 1/3. Thus, it is natural to introduce the symbol $\sum_{n=1}^{\infty}$ and write

$$\frac{1}{3} = \sum_{n=1}^{\infty} \frac{3}{10^n}$$

This example illustrates the general idea of infinite sums. Begin with a sequence, sum the first n terms of the sequence, and then take the limit of the sums.

DEFINITION

Let $\{a_n\}$ be a sequence. For each n,

$$S_n = \sum_{k=1}^{n} a_k = a_1 + a_2 + \cdots + a_n$$

is called the **n-th partial sum** of $\{a_n\}$. The sequence $\{S_n\}$ is called the **sequence of partial sums** of $\{a_n\}$. The infinite sum

(11.11)
$$\sum_{n=1}^{\infty} a_n = a_1 + a_2 + \cdots + a_n + \cdots$$

is called an (infinite) **series**. If $S_n \to S$ as $n \to \infty$, then we say that the series **converges** (to S) and write

$$\sum_{n=1}^{\infty} a_n = \lim_{n \to \infty} \sum_{k=1}^{n} a_k = S$$

In this case, S is called the **sum** of the series. If $\{S_n\}$ does not have a limit, then the series is said to **diverge**. In this case, the series has no sum.

Series have a lot in common with improper integrals over intervals of the form $[a, \infty)$. You may have already noticed a similarity in the terminology and definitions. The connection is deepened by the observation that certain infinite series can also be thought of in terms of area. If each a_n is nonnegative, then a_n is the area of the rectangle with height a_n and base $[n, n + 1]$ (Figure 11–7).

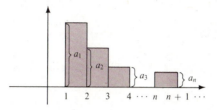

FIGURE 11–7. The series $\sum_{n=1}^{\infty} a_n$ can be interpreted as a sum of areas of rectangles.

Notation: When we are talking about series in general, or when we are not concerned with evaluating the sum explicitly, we will use the shorter notation

$$\sum a_n$$

to denote an infinite series.

Given a series $\sum a_n$, we will want to know if it converges and, if it does converge, what its sum is. Definition (11.11) says that a series converges or diverges according to whether or not its sequence $\{S_n\}$ of partial sums has a limit. If $S_n \to S$ as $n \to \infty$, then S is the sum of the series; that is,

$$\sum_{n=1}^{\infty} a_n = \lim_{n \to \infty} S_n = S$$

In the first two examples below, we are able to find an explicit formula for S_n and evaluate the limit.

EXAMPLE 1 *(Telescoping series).*   Find the sum

$$\sum_{n=1}^{\infty} \frac{1}{n(n + 1)} = \frac{1}{1 \cdot 2} + \frac{1}{2 \cdot 3} + \frac{1}{3 \cdot 4} + \cdots$$

Solution   Our first observation is that

$$\frac{1}{n(n + 1)} = \frac{1}{n} - \frac{1}{n + 1} \qquad \text{(Partial fractions)}$$

Therefore, the first few partial sums are

$$S_1 = 1 - \frac{1}{2}$$

$$S_2 = \left(1 - \frac{1}{2}\right) + \left(\frac{1}{2} - \frac{1}{3}\right) = 1 - \frac{1}{3}$$

$$S_3 = \left(1 - \frac{1}{2}\right) + \left(\frac{1}{2} - \frac{1}{3}\right) + \left(\frac{1}{3} - \frac{1}{4}\right) = 1 - \frac{1}{4}$$

Notice that all of the numbers cancel, except the first and last; this is true for each S_n,

$$S_n = \left(1 - \frac{1}{2}\right) + \left(\frac{1}{2} - \frac{1}{3}\right) + \cdots + \left(\frac{1}{n-1} - \frac{1}{n}\right) + \left(\frac{1}{n} - \frac{1}{n+1}\right) = 1 - \frac{1}{n+1}$$

Thus, each partial sum collapses (or *telescopes*) and

$$S_n = 1 - \frac{1}{n+1}$$

It is easy to see that $S_n \to 1$ as $n \to \infty$; therefore,

$$\sum_{n=1}^{\infty} \frac{1}{n(n+1)} = 1 \qquad \blacksquare$$

EXAMPLE 2 *(Geometric series).*   Any series of the form

$$\sum_{n=0}^{\infty} ar^n = a + ar + ar^2 + \cdots$$

is called a **geometric series.** The numbers a and r are fixed and nonzero; a is the first term and each term thereafter is obtained by multiplying the preceding one by r. (We start this series with $n = 0$ in order to make the first term ar^0 equal a.) Let us find a formula for S_n.

$$S_n = a + ar + ar^2 + \cdots + ar^{n-1}$$

Multiplying both sides by the number r yields

$$rS_n = ar + ar^2 + ar^3 + \cdots + ar^n$$

If we now subtract these two equations, there will be a lot of cancellation; only two terms, a and ar^n, remain.

$$S_n - rS_n = a - ar^n = a(1 - r^n)$$

The left side equals $(1 - r)S_n$, and if $r \neq 1$, then

$$S_n = \frac{a(1 - r^n)}{1 - r} = \frac{a}{1 - r}(1 - r^n)$$

Now let us find the limit of S_n. This limit depends only on the limit of r^n; everything else is fixed. From (11.9), we know that

$$\lim_{n \to \infty} r^n = 0 \text{ if } |r| < 1 \text{ and does not exist if } |r| > 1 \text{ or } r = -1$$

From this it follows that

$$\lim_{n \to \infty} S_n = \frac{a}{1 - r} \text{ if } |r| < 1 \text{ and does not exist if } |r| > 1 \text{ or } r = -1$$

Moreover, if $r = 1$, the series $\sum ar^n = a + a + a + \cdots$ surely diverges. Putting all this information together we have

---

GEOMETRIC SERIES

(11.12) $$\sum_{n=0}^{\infty} ar^n = \frac{a}{1 - r} \text{ if } |r| < 1 \text{ and diverges if } |r| \geq 1$$

■

EXAMPLE 3   Show that $\sum_{n=0}^{\infty} 1/2^n$ is a geometric series and find its sum.

Solution   If we rewrite $1/2^n$ as $(1/2)^n$, then

$$\sum_{n=0}^{\infty} \frac{1}{2^n} = \sum_{n=0}^{\infty} \left(\frac{1}{2}\right)^n = 1 + \frac{1}{2} + \frac{1}{4} + \cdots$$

is a geometric series with first term $a = 1$ and $r = 1/2$. Therefore, by (11.12),

$$\sum_{n=0}^{\infty} \frac{1}{2^n} = \frac{1}{1 - (1/2)} = 2 \quad ■$$

EXAMPLE 4

(a) $\sum_{n=0}^{\infty} 5(-\frac{3}{4})^n = 5 - 5(\frac{3}{4}) + 5(\frac{9}{16}) - \cdots$ is a geometric series with first term 5 and $r = -3/4$. Therefore,

$$\sum_{n=0}^{\infty} 5\left(-\frac{3}{4}\right)^n = \frac{5}{1 - (-3/4)} = \frac{20}{7}$$

(b) $\sum_{n=0}^{\infty} 4^n = 1 + 4 + 16 + \cdots$ is a geometric series with $r = 4$; it clearly diverges.   ■

EXAMPLE 5   Express the repeating decimal $0.222 \cdots$ as a geometric series and find its sum.

Solution   Write

$$0.222 \cdots = \frac{2}{10} + \frac{2}{100} + \frac{2}{1,000} + \cdots$$

This is a geometric series with $a = 2/10$ and $r = 1/10$. Therefore, by (11.12),

$$0.222 \cdots = \sum_{n=0}^{\infty} \frac{2}{10}\left(\frac{1}{10}\right)^n = \frac{2/10}{1 - (1/10)} = \frac{2}{9} \quad ■$$

A more general situation is obtained if we allow the series to start with any integer p rather than 0. In this case

$$\sum_{n=p}^{\infty} ar^n = ar^p + ar^{p+1} + ar^{p+2} + \cdots$$

is still called a geometric series, but now the first term is ar^p. The same computations as before with S_n and rS_n lead us to

---

GEOMETRIC SERIES

(11.12a)

$$\sum_{n=p}^{\infty} ar^n = \frac{ar^p}{1 - r} \text{ if } |r| < 1 \text{ and diverges if } |r| \geq 1$$

EXAMPLE 6   Find the sum of $\sum_{n=2}^{\infty} 4/3^n$.

Solution   Rewrite $4/3^n$ as $4(1/3)^n$; then

$$\sum_{n=2}^{\infty} \frac{4}{3^n} = \sum_{n=2}^{\infty} 4\left(\frac{1}{3}\right)^n = \frac{4}{9} + \frac{4}{27} + \cdots$$

is a geometric series with first term $4/9$ and $r = 1/3$. By (11.12a), we have

$$\sum_{n=2}^{\infty} \frac{4}{3^n} = \frac{4/9}{1 - (1/3)} = \frac{2}{3} \qquad \blacksquare$$

In the next example, we discuss the important *harmonic series*. Although it is not practical to find an explicit formula for the partial sums, we can show that they increase without bound, so the series diverges.

EXAMPLE 7 *(Harmonic series)*.   The series

$$\sum_{n=1}^{\infty} \frac{1}{n} = 1 + \frac{1}{2} + \frac{1}{3} + \cdots$$

is called the **harmonic series**. We claim that the harmonic series *diverges* (to infinity). To see this, we group the first few terms in the following way:

$$\text{1st term: } 1$$

$$\text{Next term: } \frac{1}{2}$$

$$\text{Sum the next 2 terms: } \frac{1}{3} + \frac{1}{4} > 2\left(\frac{1}{4}\right) = \frac{1}{2}$$

$$\text{Sum the next 4 terms: } \frac{1}{5} + \frac{1}{6} + \frac{1}{7} + \frac{1}{8} > 4\left(\frac{1}{8}\right) = \frac{1}{2}$$

Sum the next 8 terms: $\dfrac{1}{9} + \dfrac{1}{10} + \cdots + \dfrac{1}{16} > 8\left(\dfrac{1}{16}\right) = \dfrac{1}{2}$

It follows that

$$S_1 = 1, S_2 = \frac{3}{2}, S_4 > 2, S_8 > \frac{5}{2}, S_{16} > 3$$

Using this same process, it can be shown that

$$S_{2^n} \geq 1 + \frac{n}{2} \quad \text{for } n = 0, 1, 2, \ldots$$

This shows that the partial sums increase without bound; therefore,

(11.13)

> **HARMONIC SERIES**
>
> $\displaystyle\sum_{n=1}^{\infty} \frac{1}{n}$ diverges to infinity ■

Convergent series can be added or multiplied by constants.

(11.14)

> If Σa_n and Σb_n both converge, then $\Sigma(a_n + b_n)$ also converges, and
>
> $$\sum_{n=1}^{\infty} (a_n + b_n) = \sum_{n=1}^{\infty} a_n + \sum_{n=1}^{\infty} b_n$$
>
> Furthermore, if c is any number, then Σca_n converges, and
>
> $$\sum_{n=1}^{\infty} ca_n = c \sum_{n=1}^{\infty} a_n$$

Although the proof of (11.14) is straightforward, it is instructive to run through at least the first part. Suppose that $\Sigma a_n = a$ and $\Sigma b_n = b$. If S_n and T_n represent the partial sums of these two series, then $S_n \to a$ and $T_n \to b$ as $n \to \infty$. It follows from (11.5) that $S_n + T_n \to a + b$. But

$$S_n + T_n = \sum_{k=1}^{n} a_k + \sum_{k=1}^{n} b_k = \sum_{k=1}^{n} (a_k + b_k)$$

is the nth partial sum of the series $\Sigma(a_n + b_n)$. Therefore,

$$\sum_{n=1}^{\infty} (a_n + b_n) = \lim_{n \to \infty}(S_n + T_n) = a + b = \sum_{n=1}^{\infty} a_n + \sum_{n=1}^{\infty} b_n$$

A similar argument proves the part about multiplying by a constant.

EXAMPLE 8 Show that $\sum_{n=1}^{\infty}(1/n(n + 1) - 4(\frac{1}{2})^{n-1})$ converges and find its sum.

Solution By Example 1, $\sum_{n=1}^{\infty}1/n(n + 1)$ converges to 1. By Example 3, $\sum_{n=1}^{\infty}(\frac{1}{2})^{n-1}$ converges to 2. Therefore, by (11.14),

$$\sum_{n=1}^{\infty}\left(\frac{1}{n(n + 1)} - 4\left(\frac{1}{2}\right)^{n-1}\right) = \sum_{n=1}^{\infty}\frac{1}{n(n + 1)} - 4\sum_{n=1}^{\infty}\left(\frac{1}{2}\right)^{n-1}$$
$$= 1 - 8 = -7 \quad \blacksquare$$

EXAMPLE 9 Show that $\sum 1/2n$ diverges.

Solution We reason as follows: if this series converges, then, by (11.14), it can be multiplied by 2 and still converge. But

$$2\sum_{n=1}^{\infty}\frac{1}{2n} = \sum_{n=1}^{\infty}2\left(\frac{1}{2n}\right) = \sum_{n=1}^{\infty}\frac{1}{n}$$

and we know the harmonic series diverges (Example 7). Therefore, $\sum 1/2n$ must diverge. $\quad \blacksquare$

Our final observation in this section is that

(11.15) | The convergence of a series $\sum a_n$ is unaffected by changing a *finite* number of terms.

That is to say, if P is any positive integer, then

$$\sum_{n=P}^{\infty} a_n \quad \text{and} \quad \sum_{n=1}^{\infty} a_n$$

both converge or *both* diverge. The first finite number of terms does not affect the convergence. Of course, if they converge, the *sums* may be different.

EXERCISES

The series in Exercises 1–4 are telescoping series. Find the nth partial sum. Determine whether or not the series converge. If so, find their sums.

1. $\sum_{n=1}^{\infty}\frac{3}{n(n + 1)}$

2. $\sum_{k=2}^{\infty}\frac{2}{k^2 - 1}$ (Use partial fractions)

3. $\sum_{k=1}^{\infty}\ln\frac{k}{k + 1}$

4. $\sum_{n=1}^{\infty}\frac{1}{\sqrt{n + 1} + \sqrt{n}}$ (Rationalize)

The series in Exercises 5–8 are geometric series. Determine whether or not they converge. If they converge, find their sums.

5. $\displaystyle\sum_{n=0}^{\infty}\left(\frac{2}{3}\right)^n$

6. $\displaystyle\sum_{n=3}^{\infty}\left(\frac{1}{2}\right)^n$

7. $\displaystyle\sum_{m=0}^{\infty}\left(-\frac{1}{2}\right)^m$

8. $\displaystyle\sum_{n=0}^{\infty}2^n$

In Exercises 9–22, determine whether or not the series converge. If so, find the sums.

9. $\displaystyle\sum_{n=2}^{\infty}\left(-\frac{1}{3}\right)^n$

10. $\displaystyle\sum_{n=0}^{\infty}4\left(\frac{1}{4}\right)^n$

11. $\displaystyle\sum_{n=3}^{\infty}\left(\frac{4}{3}\right)^n$

12. $\displaystyle\sum_{n=0}^{\infty}\left(\frac{9}{10}\right)^n$

13. $\displaystyle\sum_{q=1}^{\infty}\frac{3}{q}$

14. $\displaystyle\sum_{q=1}^{\infty}\frac{1}{6q}$

15. $\displaystyle\sum_{n=1}^{\infty}\frac{2n+1}{n^2(n+1)^2}$ (Telescoping)

16. $\displaystyle\sum_{n=1}^{\infty}\frac{1}{(n+1)(n+2)}$

17. $\displaystyle\sum_{n=1}^{\infty}\left[\left(\frac{1}{3}\right)^{n-1}+\frac{1}{n(n+1)}\right]$

18. $\displaystyle\sum_{n=1}^{\infty}\left(\frac{1}{3^n}+\frac{1}{3n}\right)$

19. $\displaystyle\sum_{n=0}^{\infty}\left(\frac{5}{2}\right)^{-n}$

20. $\displaystyle\sum_{n=0}^{\infty}5^{-n}$

21. $\displaystyle\sum_{n=0}^{\infty}\frac{1+2^n}{6^n}$

22. $\displaystyle\sum_{n=0}^{\infty}\frac{3^n-4^n}{5^n}$

23. Express the repeating decimal 1.111 . . . as a geometric series and find the sum.

24. Express the repeating decimal 0.121212 . . . as a geometric series and find the sum.

25. Express the repeating decimal 0.123123 . . . as a geometric series and find the sum.

26. Does the repeating decimal 0.999 . . . equal 1?

27. (Bouncing ball). A ball is dropped from a height of 10 meters. Because of friction, each time it strikes the ground it rebounds only 2/3 of the previous height. How far will the ball travel (up and down) before coming to rest? (See Exercise 31 for a different aspect of this problem.)

28. (Swinging pendulum). A pendulum is 20 inches long. It is displaced by an angle $\pi/6$ from the vertical and released. Because of friction, the maximum angle on each successive swing is 9/10 of the previous one (Figure 11–8). How far does the bob on the end of the pendulum travel (along the arc) before coming to rest?

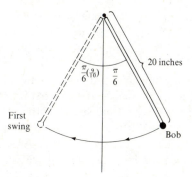

FIGURE 11–8. Exercise 28.

29. (Harmonic series).   Example 6 establishes, by algebraic means, that the harmonic series $\Sigma 1/n$ diverges. A geometric argument is made by thinking of the series in terms of area. Show that the harmonic series diverges by comparing the nth partial sum S_n with the integral $\int_1^{n+1}1/x\,dx$ (refer to Figure 11–9).

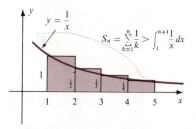

FIGURE 11–9. Exercise 29.

30. The telescoping series

$$\sum_{n=1}^{\infty}\left(\frac{1}{n}-\frac{1}{n+1}\right)$$

converges to 1 (Example 1). But

$$\sum_{n=1}^{\infty} \frac{1}{n} \quad \text{and} \quad \sum_{n=1}^{\infty} \frac{1}{n+1}$$

are harmonic series (the second one just starts with $1/2$ instead of 1) and they diverge. Therefore,

$$\sum_{n=1}^{\infty} \left(\frac{1}{n} - \frac{1}{n+1} \right) \neq \sum_{n=1}^{\infty} \frac{1}{n} - \sum_{n=1}^{\infty} \frac{1}{n+1}$$

Does this contradict (11.14)? Explain.

Optional Exercises

31. *(Bouncing ball riddle).* A rubber ball is thrown straight up and allowed to bounce until it comes to rest. Suppose that each time it strikes the ground with velocity v, a fixed percentage of the energy is absorbed by the ball, and it rebounds with velocity $v/2$. Thus, if v is the velocity on the first impact, then $v/2$ is the velocity on the second impact, $v/4$ is the velocity on the third impact, and, in general, $v/2^{n-1}$ is the velocity on the nth impact. At some point, the velocity is so small that there is not enough energy to push the ball upward, and the action stops. Let us disregard this physical analysis, and assume that the ball actually bounces infinitely many times. Show that, even in this case, the ball will come to rest in $4v/g$ seconds (g is the gravity constant). How do you reconcile this with the fact that the impact velocity $v/2^{n-1}$ is never zero?

32. Use a geometric approach (described in Exercise 29) to show that the series $\sum 1/n^2$ converges.

11–3   TESTS FOR CONVERGENCE AND DIVERGENCE

If an explicit formula for the partial sums is known, as in the case of telescoping or geometric series, then it is easy to compute the sum of the series. But for most series, finding the partial sums is either impractical or impossible. Nevertheless, if a series is known to converge, then there is some hope of estimating its sum. Of course, if the series diverges, then there is no point in searching for its sum. Therefore, it becomes important to know if a series converges or not. In this section (and the next), we develop various tests for convergence and divergence, and describe some methods of estimating sums.

Convergence and Divergence Tests

We begin with a *divergence* test.

(11.16)

> **nth TERM TEST**
>
> If $\lim_{n \to \infty} a_n \neq 0$, then $\sum a_n$ diverges*

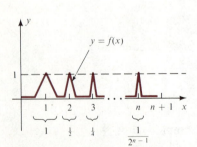

FIGURE 11–10. $\int_1^{\infty} f(x)\, dx$ is the sum of the areas of the triangular regions; $\sum_{n=1}^{\infty} (\frac{1}{2})^n = 1$. Thus, the integral converges, but $\lim_{x \to \infty} f(x) \neq 0$.

*Series and improper integrals part company here; that is, $\int_1^{\infty} f(x)\, dx$ can converge even though $\lim_{x \to \infty} f(x) \neq 0$. Figure 11–10 shows an example. The graph of f coincides with the x-axis except that near each integer n, it forms a triangle of height 1 and base $(1/2)^{n-1}$. The area of this triangle is $(1/2)^n$; therefore,

$$\int_1^{\infty} f(x)\, dx = \sum_{n=1}^{\infty} \left(\frac{1}{2} \right)^n = 1$$

The integral converges, but $\lim_{x \to \infty} f(x)$ does not exist.

We will prove (11.16) by verifying an equivalent statement; we will show that *if Σa_n converges, then $a_n \to 0$ as $n \to \infty$.* Let S_n and S_{n+1} be the n and $(n+1)$st partial sums of Σa_n. Since we are assuming that the series converges, it follows that $\lim_{n \to \infty} S_n$ exists and it is not hard to see that

$$\lim_{n \to \infty} S_n = \lim_{n \to \infty} S_{n+1}$$

We now observe that $S_{n+1} - S_n = a_{n+1}$ and it follows that

$$\lim_{n \to \infty} a_n = \lim_{n \to \infty} a_{n+1} = \lim_{n \to \infty}(S_{n+1} - S_n) = 0$$

This proves that if Σa_n converges, then $a_n \to 0$, which is equivalent to saying that if a_n does not approach 0, then Σa_n diverges.

CAUTION: Do not read into the nth term test anything more than is actually there. It says that if a_n does not approach 0, then Σa_n diverges. If a_n does approach 0, then *the nth term test is no test at all and must be discarded.*

EXAMPLE 1

(a) $\Sigma n/(n+1)$ diverges by the nth term test because $n/(n+1) \to 1$ as $n \to \infty$.

(b) The nth term test fails for $\Sigma (n+1)/n^2$ because $(n+1)/n^2 \to 0$ as $n \to \infty$.
The series may or may not converge (in Example 3, it is shown to diverge). ∎

COMPARISON TEST

Let Σa_n and Σb_n be series with

(11.17) $0 \le a_n \le b_n$ $n = 1, 2, 3, \ldots$

(1) If Σa_n diverges, then Σb_n diverges.
(2) If Σb_n converges, then Σa_n converges.

Notice the similarity to the comparison test (10.16) for improper integrals. If Σa_n and Σb_n are series with $0 \le a_n \le b_n$, let $\{S_n\}$ and $\{T_n\}$ be their sequences of partial sums. Then each are nondecreasing sequences (adding nonnegative terms) and

$$0 \le S_n \le T_n \quad \text{for all } n$$

By the Bounded Convergence Theorem (11.6), such sequences must converge if they are bounded. Therefore, if Σa_n diverges, it can only mean that $\{S_n\}$ increases without bound. This forces $\{T_n\}$ to also increase without bound, so Σb_n also diverges, proving part (1) of the test. If Σb_n converges, then $\{T_n\}$ converges, say to T. Since $\{T_n\}$ is nondecreasing, we know that $T_n \le T$ for all n. But then $S_n \le T$ for all n, so $\{S_n\}$ is bounded. Therefore, $\{S_n\}$ and, in turn, Σa_n also converge. This completes the proof of part (2) of the test.

Any series of nonnegative terms that is known to converge or diverge can be used in the comparison test. But the two used most frequently are the geometric and harmonic series.

EXAMPLE 2   Show that $\sum (\cos^2 k)/2^k$ converges.

Solution   This is a series of nonnegative terms; we will compare it with a geometric series. Since

$$\frac{\cos^2 k}{2^k} \le \frac{1}{2^k} = \left(\frac{1}{2}\right)^k$$

and $\sum(\frac{1}{2})^k$ converges, we know that $\sum(\cos^2 k)/2^k$ converges.   ■

EXAMPLE 3   In Example 1(b), the nth term test failed for $\sum(n + 1)/n^2$. We can, however, compare it with the harmonic series.

$$\frac{n + 1}{n^2} = \frac{1}{n} + \frac{1}{n^2} > \frac{1}{n}$$

Since $\sum(1/n)$ diverges, so does $\sum(n + 1)/n^2$.   ■

(11.18)

> **INTEGRAL TEST**
> A series $\sum a_n$ is given with $a_n \ge 0$. If f is a decreasing nonnegative function defined on $[1, \infty)$ with $f(n) = a_n$ for $n = 1, 2, 3, \ldots$, then
>
> $$\sum_{n=1}^{\infty} a_n \quad \text{and} \quad \int_1^{\infty} f(x)\, dx$$
>
> both converge or both diverge.

FIGURE 11–11. Illustration of the integral test. The figure indicates that $\sum_{k=2}^{\infty} a_n \le \int_1^{\infty} f(x)\, dx \le \sum_{k=1}^{\infty} a_k$.

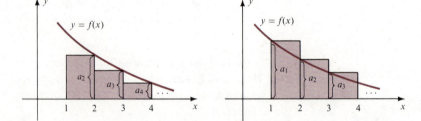

The hypotheses that f is decreasing and $f(n) = a_n$ insure that $a_n \ge f(x) \ge a_{n+1}$ for all x in $[n, n + 1]$. Therefore, Figure 11–11 shows a typical case, and the figure clearly indicates that

$$\sum_{k=2}^{n} a_n \le \int_1^n f(x)\, dx \le \sum_{k=1}^{n} a_n$$

From the left-hand inequality and the Bounded Convergence Theorem (11.6), we see that if the integral converges, so does the series. From the right-hand inequality, we see that if the integral diverges, so does the series. This proves (11.18).

To test the convergence of a series Σa_n with (11.18), simply formulate a nonnegative function f defined on $[1, \infty)$ for which (1) f is decreasing, (2) $f(n) = a_n$, and (3) you know whether $\int_1^\infty f(x)\, dx$ converges or diverges.

EXAMPLE 4   Show that Σne^{-n} converges.

Solution   Define the nonnegative function $f(x) = xe^{-x}$ on $[1, \infty)$. Then f is decreasing (its derivative $e^{-x}(1 - x)$ is negative for $x > 1$), $f(n) = ne^{-n}$, and we can show that $\int_1^\infty xe^{-x}\, dx$ converges (integrate by parts with $u = x$ and $dv = e^{-x}\, dx$),

$$\int_1^\infty xe^{-x}\, dx = \lim_{t \to \infty} [-xe^{-x} - e^{-x}]_1^t = \frac{2}{e}$$

Therefore, by (11.18), Σne^{-n} converges.   ∎

The integral test can also be used to discuss another important class of series, the *p*-series. Any series of the form

$$\sum_{n=1}^\infty \frac{1}{n^p} \quad p > 0$$

is called a **p-series**. If we define $f(x) = 1/x^p$, then f meets the criteria of the integral test, and

$$\int_1^t \frac{1}{x^p} = \begin{cases} \ln t & \text{if } p = 1 \\ \dfrac{t^{(-p+1)} - 1}{-p + 1} & \text{if } p \neq 1 \end{cases}$$

Now let $t \to \infty$. If $p = 1$, $\ln t \to \infty$; if $p < 1$, then $-p + 1 > 0$ and $t^{(-p+1)} \to \infty$. In either case, the integral diverges. But if $p > 1$, then $-p + 1 < 0$ and $t^{(-p+1)} \to 0$; in this case, the integral converges. Therefore,

---

THE p-SERIES

(11.19) $$\sum \frac{1}{n^p} \quad p > 0$$

converges if $p > 1$ and diverges if $p \leq 1$.

EXAMPLE 5

(a) $\Sigma 1/n^2$ is a *p*-series with $p = 2$; it converges.

(b) $\Sigma 1/\sqrt{n}$ is a *p*-series with $p = 1/2$; it diverges.   ∎

The *p*-series is also frequently used in the comparison test.

EXAMPLE 6   Show that $\Sigma 1/(n^2 + 3n + 1)$ converges.

Solution   We compare with the convergent p-series $\sum(1/n^2)$. Since

$$\frac{1}{n^2 + 3n + 1} < \frac{1}{n^2}$$

it follows that $\sum 1/(n^2 + 3n + 1)$ converges.   ■

Estimating a Sum

The comparison and integral tests do not say anything about the value of the sum of a convergent series. They can, however, be used to estimate the sum in certain cases. The general idea is to break up the sum into a **finite part** and a **remainder**:

$$(11.20) \quad \sum_{n=1}^{\infty} a_n = \underbrace{\sum_{n=1}^{N} a_n}_{\text{finite part}} + \underbrace{\sum_{n=N+1}^{\infty} a_n}_{\text{remainder}}$$

The sum of the remainder is called the **truncation error.** The finite part is computed by hand or by calculator, and the truncation error is estimated by comparison with another series or an integral of known value.

> To estimate a sum within an allowable error E, you have to determine N so that the truncation error is less than E. Then the finite part is the desired estimate.

EXAMPLE 7   Estimate $\sum_{n=1}^{\infty} 1/(2^n + 1)$ with an error less than .005, and then round off to two decimal places.

Solution   Since

$$\frac{1}{2^n + 1} \leq \frac{1}{2^n} = \left(\frac{1}{2}\right)^n$$

we see that for any N,

$$\sum_{n=N+1}^{\infty} \frac{1}{2^n + 1} \leq \sum_{n=N+1}^{\infty} \left(\frac{1}{2}\right)^n = \frac{(1/2)^{N+1}}{1 - (1/2)} = \frac{1}{2^N}$$

Since we want the error to be less than $.005 = 1/200$, it follows that $N = 8$ will do ($1/2^8 < 1/200$). Therefore,

$$\sum_{n=1}^{8} \frac{1}{2^n + 1} = \frac{1}{3} + \frac{1}{5} + \frac{1}{9} + \cdots + \frac{1}{257}$$

$$\approx 0.7606592 \qquad \text{(8-digit calculator)}$$

is within the specified error of .005 because

$$\sum_{n=9}^{\infty} \frac{1}{2^n + 1} \le \sum_{n=9}^{\infty} \left(\frac{1}{2}\right)^n = \frac{1}{2^8} < \frac{1}{200}$$

Rounding off to two decimal places, we have

$$\sum_{n=1}^{\infty} \frac{1}{2^n + 1} \approx 0.76$$

This answer is correct to two decimal places. ■

Remark: Each operation with an 8-digit calculator can involve a round-off error of 5×10^{-9}. In Example 7, we added 8 numbers, so the accumulated error is at most 40×10^{-9}, well within the limits of the specified error of .005. However, if a very large number of terms are added, the error can be significant.

EXAMPLE 8 Estimate $\sum_{n=1}^{\infty} 1/n^2$ correct to one decimal place (that is, with an error less than .05).

Solution Figure 11–12 indicates that

$$\sum_{n=N+1}^{\infty} \frac{1}{n^2} \le \int_{N}^{\infty} \frac{1}{x^2} \, dx = \lim_{t \to \infty} \left[-\frac{1}{x} \right]_{N}^{t} = \frac{1}{N}$$

Since the error is to be less than $.05 = 1/20$, we take $N = 21$. Therefore,

$$\sum_{n=1}^{21} \frac{1}{n^2} = 1 + \frac{1}{4} + \frac{1}{9} + \cdots + \frac{1}{441}$$

$$\approx 1.5984308$$

is within the specified error of .05 because $\sum_{n=22}^{\infty} 1/n^2 \le 1/21$. Rounding off to one decimal place yields,

$$\sum_{n=1}^{\infty} \frac{1}{n^2} \approx 1.6$$

Note: It can be shown that the exact sum is $\pi^2/6 = 1.64 \dots$ ■

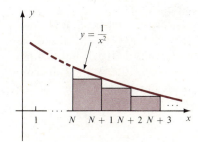

FIGURE 11–12. Example 8. $\sum_{n=N+1}^{\infty} 1/n^2 \le \int_{N}^{\infty} 1/x^2 \, dx$.

EXERCISES

In Exercises 1–4, test for divergence with the nth term test. If the test fails, say so.

1. $\sum \dfrac{\ln n}{n}$

2. $\sum \cos n\pi$

3. $\sum \dfrac{n}{\sqrt{n^2 + 1}}$

4. $\sum \dfrac{n}{2^n}$

In Exercises 5–8, test for convergence with the comparison test.

5. $\sum \dfrac{1}{\sqrt{n^2 - 1}}$

6. $\sum \dfrac{\sin^2 k}{2^k}$

7. $\sum \dfrac{1}{\ln k}$

8. $\sum \dfrac{1}{n^3 - 5}$

In Exercises 9–12, test for convergence with the integral test.

9. $\displaystyle\sum \frac{1}{k \ln k}$ **10.** $\displaystyle\sum \frac{1}{k(\ln k)^2}$

11. $\displaystyle\sum \frac{1}{1 + n^2}$ **12.** $\displaystyle\sum \frac{1}{n\sqrt{n^2 - 1}}$

In Exercises 13–30, test for convergence. In each case, state which test is used.

13. $\displaystyle\sum \frac{n^2}{(n + 1)(n + 2)}$ **14.** $\displaystyle\sum \frac{1}{\sqrt[3]{n^2 - 1}}$

15. $\displaystyle\sum n^2 e^{-n}$ **16.** $\displaystyle\sum n4^{-n}$

17. $\displaystyle\sum \frac{1}{n^4 + n^2 + 2}$ **18.** $\displaystyle\sum \frac{e^n}{3^{n+1}}$

19. $\displaystyle\sum \frac{k}{3^k}$ **20.** $\displaystyle\sum \frac{1}{\sqrt{k + 3}}$

21. $\displaystyle\sum \frac{1}{k\sqrt{\ln k}}$ **22.** $\displaystyle\sum \frac{1}{(n + 1)(n + 2)}$

23. $\displaystyle\sum \frac{1}{\sqrt{n^3 - 4}}$ **24.** $\displaystyle\sum \frac{\sqrt{n}}{n^2 + 1}$

25. $\displaystyle\sum \frac{2 + \sin n}{n^2}$ **26.** $\displaystyle\sum \frac{1}{k\sqrt{k + 1}}$

27. $\displaystyle\sum \frac{1}{k(\ln k)(\ln \ln k)}$ **28.** $\displaystyle\sum \frac{3^n}{2^n}$

29. $\displaystyle\sum \frac{n^2 + 2^n}{n + 3^n}$ **30.** $\displaystyle\sum \frac{\arctan n}{2^n}$

In Exercises 31–35, estimate the sum correct to two decimal places (that is, with an error less than .005).

31. $\displaystyle\sum_{n=1}^{\infty} \frac{1}{3^n + 1}$ **32.** $\displaystyle\sum_{n=1}^{\infty} ne^{-n}$

33. $\displaystyle\sum_{n=1}^{\infty} \frac{1}{n^4}$ **34.** $\displaystyle\sum_{n=1}^{\infty} \frac{n}{1 + n^5}$

35. *(Discrete probability).* A function p defined on 1, 2, 3, . . . is a *probability density function* (pdf) if

$$p(n) \geq 0 \quad \text{and} \quad \sum_{n=1}^{\infty} p(n) = 1$$

Verify that $p(n) = (1/10)(9/10)^{n-1}$ is a pdf.

36. *(Testing a product).* Light bulbs coming off an assembly line are tested. The probability that any given bulb lights up is 9/10. Let X be the number of tests required to find the first defective bulb. Then X is a discrete random variable with possible values 1, 2, 3, ... If $X = n$, it means that the first $n - 1$ bulbs lit up (probability $(9/10)^{n-1}$), and the nth bulb did not (probability $1/10$). It follows that pdf of X is $p(n) = (1/10)(9/10)^{n-1}$. Find the sum $\sum_{n=5}^{\infty} p(n)$ and interpret this sum in terms of the testing.

11–4   ADDITIONAL TESTS

We continue our discussion of tests for convergence and divergence.

The *root test* and the *ratio test* are two of the most useful tests because only the given series is considered; no comparisons or integrations are needed.

ROOT TEST

Let $\sum a_n$ be a series of nonnegative terms, and suppose that

$$\lim_{n \to \infty} \sqrt[n]{a_n} = L$$

(11.21)

(1) If $L < 1$, the series *converges*.

(2) If $L > 1$, (or $L = \infty$), the series *diverges*.

(3) If $L = 1$, the test *fails*.

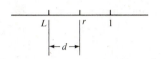

FIGURE 11–13. Root test; d is half the distance from L to 1 and $r = L + d$.

To justify the first part of the root test, suppose that $\sqrt[n]{a_n} \to L$ and $L < 1$. Let d equal one-half of the distance from L to 1, and set $r = L + d$; then $r < 1$ (Figure 11–13). We know that $\sqrt[n]{a_n}$ can be made arbitrarily close to L by taking n sufficiently large. But if $\sqrt[n]{a_n}$ is close enough to L, then it will be less than r; that is,

$$\sqrt[n]{a_n} < r \quad \text{for all large } n$$

It follows that

$$a_n < r^n$$

and, since $r < 1$, the series $\sum a_n$ converges by comparison with the convergent geometric series $\sum r^n$. This completes our argument. The justification of the second part is similar and is left as an exercise. To show that the test fails if $L = 1$, we observe that

$$\left(\frac{1}{n}\right)^{1/n} \to 1 \quad \text{and} \quad \left(\frac{1}{n^2}\right)^{1/n} \to 1 \qquad \text{(L'Hôpital's rule; see Example 2 in Section 11–1)}$$

Yet $\sum (1/n)$ diverges and $\sum (1/n^2)$ converges; therefore, *if $L = 1$, the root test must be discarded.*

The root test is used primarily on series that involve powers of n.

EXAMPLE 1   Test the convergence of $\sum n/2^n$.

Solution   We use the root test

$$\left(\frac{n}{2^n}\right)^{1/n} = \frac{n^{1/n}}{2} \to \frac{1}{2} < 1 \qquad \text{(Example 2 in Section 11–1)}$$

Therefore, the series converges.   ■

EXAMPLE 2   Test the convergence of $\sum n^n/3^{2n+1}$.

Solution   Use the root test

$$\left[\frac{n^n}{3^{2n+1}}\right]^{1/n} = \frac{n}{3^{2+(1/n)}} \to \infty$$

Therefore, the series diverges.   ■

EXAMPLE 3   Test the convergence of $\sum n/\sqrt{n^2 - 4}$.

Solution   Trying the root test yields

$$\left(\frac{n}{\sqrt{n^2 - 4}}\right)^{1/n} = \frac{n^{1/n}}{(n^2 - 4)^{1/2n}} \to \frac{1}{1} = 1 \qquad \text{(L'Hôpital's rule)}$$

The test fails and must be discarded (but the series diverges by the nth term test).   ■

RATIO TEST
Let $\sum a_n$ be a series of nonnegative terms, and suppose that

(11.22)
$$\lim_{n \to \infty} \frac{a_{n+1}}{a_n} = L$$

(1) If $L < 1$, the series *converges*.
(2) If $L > 1$ (or $L = \infty$), the series *diverges*.
(3) If $L = 1$, the test *fails*.

The justification for the ratio test is similar to the one for the root test. A proof of the first part is given at the end of this section.

The ratio test is especially effective on series that involve *factorials*. The symbol $n!$ is read "n factorial" and is defined as

$$n! = n(n - 1)(n - 2) \ldots 3 \cdot 2 \cdot 1$$

By convention, $0! = 1$, so the first few factorials are

$$0! = 1$$
$$1! = 1$$
$$2! = 2 \cdot 1 = 2$$
$$3! = 3 \cdot 2 \cdot 1 = 6$$
$$4! = 4 \cdot 3 \cdot 2 \cdot 1 = 24$$

In applying the ratio test, we often need the equality

(11.23) $$(n + 1)! = (n + 1)n!$$

(which follows directly from the definition) to compute ratios of factorials. For instance,

(11.24) $$\frac{n!}{(n + 1)!} = \frac{n!}{(n + 1)n!} = \frac{1}{n + 1}$$

EXAMPLE 4 Test the convergence of $\sum 2^n/n!$.

Solution We use the ratio test. Here,

$$a_n = \frac{2^n}{n!} \quad \text{and} \quad a_{n+1} = \frac{2^{n+1}}{(n + 1)!}$$

Therefore,

$$\lim_{n \to \infty} \frac{a_{n+1}}{a_n} = \lim_{n \to \infty} \frac{2^{n+1}/(n+1)!}{2^n/n!}$$

$$= \lim_{n \to \infty} \frac{2^{n+1}n!}{2^n(n+1)!} \qquad \text{(Algebra)}$$

$$= \lim_{n \to \infty} \frac{2}{n+1} \qquad \text{(11.24)}$$

$$= 0$$

By the ratio test, the series *converges.* ■

EXAMPLE 5   Test the convergence of $\sum n^n/n!$.

Solution   Again we use the ratio test. This time

$$a_n = \frac{n^n}{n!} \quad \text{and} \quad a_{n+1} = \frac{(n+1)^{n+1}}{(n+1)!}$$

Therefore,

$$\lim_{n \to \infty} \frac{a_{n+1}}{a_n} = \lim_{n \to \infty} \frac{(n+1)^{n+1}/(n+1)!}{n^n/n!}$$

$$= \lim_{n \to \infty} \frac{(n+1)(n+1)^n n!}{n^n(n+1)n!} \qquad \text{(Algebra and (11.23))}$$

$$= \lim_{n \to \infty} \left(\frac{n+1}{n}\right)^n \qquad \text{(More algebra)}$$

$$= \lim_{n \to \infty} \left(1 + \frac{1}{n}\right)^n = e \qquad \text{((11.8) in Section 11–1)}$$

Since $e > 1$, the series *diverges.* ■

Here is another useful test that is closely related to the comparison test (11.17).

(11.25)

> **LIMIT COMPARISON TEST**
> Let $\sum a_n$ and $\sum b_n$ be series with nonnegative terms. If
> $$\lim_{n \to \infty} \frac{a_n}{b_n} = L > 0$$
> then both series converge or both diverge.

To prove (11.25), we note that if n is large enough, then

$$\frac{L}{2} < \frac{a_n}{b_n} < \frac{3L}{2} \quad \text{or} \quad \frac{L}{2}b_n < a_n < \frac{3L}{2}b_n$$

Now we use the last string of inequalities and the comparison test. If $\sum a_n$ converges, then $(L/2)b_n < a_n$ implies that $\sum b_n$ converges. If $\sum a_n$ diverges, then $a_n < (3L/2)b_n$ implies that $\sum b_n$ diverges. This completes the proof.

EXAMPLE 6   Test the convergence of $\sum \sin(1/n)$.

Solution   The fact that $(\sin x)/x \to 1$ as $x \to 0$, is a clue to try the limit comparison test using the series $\sum(1/n)$. Thus, with

$$\sum \sin \frac{1}{n} \quad \text{and} \quad \sum \frac{1}{n}$$

we have

$$\frac{\sin(1/n)}{1/n} \to 1 \quad \text{as } n \to \infty$$

Since $\sum(1/n)$ diverges, so does $\sum \sin(1/n)$.   ∎

Proof of the Ratio Test (*Optional*)

Let $\sum a_n$ be a series of nonnegative terms with

$$\lim_{n \to \infty} \frac{a_{n+1}}{a_n} = L < 1$$

For any $\epsilon > 0$, it follows from definition (11.10) that there is some integer N such that

$$\left| \frac{a_{n+1}}{a_n} - L \right| < \epsilon \quad \text{whenever } n \geq N$$

If we choose $\epsilon = (1 - L)/2$, then

$$\left| \frac{a_{n+1}}{a_n} - L \right| < \frac{1 - L}{2} \quad n \geq N$$

which implies that

$$\frac{a_{n+1}}{a_n} < \frac{1 - L}{2} + L = \frac{1 + L}{2} \quad n \geq N$$

Now set $r = (1 + L)/2$; this is the midpoint between L and 1 (Figure 11–13). Therefore, $r < 1$ and

$$a_{n+1} < a_n r \quad \text{whenever } n \geq N$$

It follows that

$$a_{n+2} < a_{n+1}r < (a_n r)r = a_n r^2$$
$$a_{n+3} < a_{n+2}r < (a_n r^2)r = a_n r^3$$

and, in general, $a_{n+m} < a_n r^m$. Therefore, by comparison with the convergent geometric series

$$\sum_{m=1}^{\infty} a_n r^m$$

(a_n is fixed)

we see that Σa_n converges. This proves part (1) of the test; the proof of part (2) is similar.

EXERCISES

Stop here have a test ch 10 to here

In Exercises 1–4, test for convergence with the root test. If the test fails, say so.

1. $\displaystyle\sum \frac{n^2}{3^n}$

2. $\displaystyle\sum \frac{3^{n+4}}{n^n}$

3. $\displaystyle\sum \frac{n^2 - 5n + 2}{n^3 + 6n + 1}$

4. $\displaystyle\sum \frac{4^n}{9^{n/2}}$

In Exercises 5–8, test for convergence with the ratio test. If the test fails, say so.

5. $\displaystyle\sum \frac{n^4}{n!}$

6. $\displaystyle\sum \frac{4^n}{n!}$

7. $\displaystyle\sum \frac{n!}{n^n}$

8. $\displaystyle\sum \frac{1}{n^8}$

In Exercises 9–12, test for convergence with the limit comparison test.

9. $\displaystyle\sum \sin \frac{1}{n^2}$

10. $\displaystyle\sum \tan \frac{1}{n}$

11. $\displaystyle\sum \frac{1}{n^2 - 1}$

12. $\displaystyle\sum \frac{n}{(n + 1)(n + 2)}$

In Exercises 13–30, test for convergence with any test previously discussed. State what test is used (there may be more than one test possible).

13. $\displaystyle\sum \frac{\ln k}{k}$

14. $\displaystyle\sum \frac{2n + 1}{2^n}$

15. $\displaystyle\sum \frac{\ln k}{k^3}$

16. $\displaystyle\sum \frac{1}{\sqrt[3]{n^2 + 1}}$

17. $\displaystyle\sum \frac{1}{\sqrt{n^3 + 1}}$

18. $\displaystyle\sum \frac{k + \ln k}{k^3}$

19. $\displaystyle\sum \frac{n!}{(n + 2)!}$

20. $\displaystyle\sum \frac{100^n}{n!}$

21. $\displaystyle\sum \frac{\ln k}{(1.01)^k}$

22. $\displaystyle\sum \frac{n!}{n^{100}}$

23. $\displaystyle\sum \tan \frac{1}{n^2}$

24. $\displaystyle\sum \frac{(n!)^2}{(2n)!}$

25. $\displaystyle\sum \left(\frac{1}{2}\right)^{1/n}$

26. $\displaystyle\sum \frac{n!}{n^n}$

27. $\displaystyle\sum \frac{1}{(2n - 1)!}$

28. $\displaystyle\sum \frac{2n}{3n + 8}$

29. $\displaystyle\sum \frac{2n}{n^2 - 4}$

30. $\displaystyle\sum \frac{n + \sin n}{n^3 - n^2 + n + 1}$

31. Show that $\Sigma 2^n n!/n^n$ converges, but $\Sigma 3^n n!/n^n$ diverges.

32. Suppose $0 < q < 1$. Then consider the geometric series

$$q + q^2 + q^3 + q^4 + q^5 + \cdots = \frac{q}{1 - q}$$

$$q^2 + q^3 + q^4 + q^5 + \cdots = \frac{q^2}{1 - q}$$

$$q^3 + q^4 + q^5 + \cdots = \frac{q^3}{1 - q}$$

and so on. By adding the columns, show that

$$\sum_{n=1}^{\infty} n q^n = \frac{q}{(1 - q)^2}$$

and evaluate $\sum_{n=1}^{\infty} n(1/3)^n$.

33. *(Expected value).* If X is a discrete random variable with possible values $1, 2, 3, \ldots$ and a pdf $p = p(n)$, then the *expected value* of X is

$$\sum_{n=1}^{\infty} n p(n)$$

Use the formula developed in Exercise 32 to find the expected value of X if   (a) $p(n) = (1/3)(3/4)^n$

(b) if $p(n) = \begin{cases} \dfrac{1}{2^k} & \text{if } n = 2^k, k = 1, 2, \ldots \\ 0 & \text{otherwise} \end{cases}$

34. *(Testing a product).* In Exercise 36 of the last section, light bulbs were being tested. The probability that a given bulb lights up is $9/10$ and X is the number of tests required to find the first defective bulb. The pdf of X is $p(n) = (1/10)(9/10)^{n-1}$. What is the expected number of tests before the first defective bulb is found? Does the answer agree with your intuition?

35. *(The economic value of a dollar).* Let us assume that 95% of each dollar is recirculated into the economy. That is, when a dollar is put into circulation, 95% of it is spent. Then 95% of the 95% is spent, and so on. What is the total economic value of a dollar?

36. Suppose $\sqrt[n]{a_n} \to L > 1$. Give an informal argument that $\sum a_n$ diverges.

37. *(Log test*).* Let $\sum a_n$ be a series of nonnegative terms and suppose that

$$\lim_{n \to \infty} \log_n a_n = L$$

Then the series converges if $L < -1$ (or $L = -\infty$), diverges if $L > -1$, and the test fails if $L = -1$. Give an informal argument (similar to the one for the root test) for the case $L < -1$. Use the log test to show that $\sum(\ln k/k)^{\ln k}$ converges. *Hint:* $\log_k a_k = (\ln a_k)/(\ln k)$.

11–5   ALTERNATING SERIES; ABSOLUTE CONVERGENCE

A series whose terms alternate in sign is called an **alternating series** (Figure 11–14). If $a_n \geq 0$, an alternating series is written

$$\sum_{n=1}^{\infty}(-1)^{n-1}a_n = a_1 - a_2 + a_3 - a_4 + \cdots$$

(or $\sum_{n=1}^{\infty}(-1)^n a_n$ if the first term is to be negative).

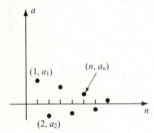

A. Sequence of terms.

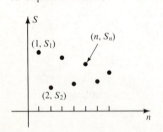

B. Sequence of partial sums.

FIGURE 11–14. Alternating series.

> **ALTERNATING SERIES TEST**
> If $\{a_n\}$ is a nonincreasing sequence of positive terms and
>
> (11.26) $$\lim_{n \to \infty} a_n = 0$$
>
> then the alternating series $\sum(-1)^{n-1}a_n$ converges and its sum is no greater than the first term a_1.

A striking consequence of this test is that if every other sign in the *divergent* harmonic series is changed from $+$ to $-$, the result is the *convergent* alternating series

$$1 - \frac{1}{2} + \frac{1}{3} - \frac{1}{4} + \cdots$$

Evidently, the negative terms subtract off enough from the partial sums of the harmonic series to give them a finite limit. (In Section 11–7, we show that the sum of this alternating series is $\ln 2 \approx 0.693$).

---

*The author is indebted to Professor Jack Zelver for pointing out this test.

To prove (11.26), we first consider the partial sums $S_2, S_4, \ldots, S_{2n}, \ldots$ that contain an even number of terms. Since

$$S_{2n} = (a_1 - a_2) + (a_3 - a_4) + \cdots + (a_{2n-1} - a_{2n})$$

and $a_k - a_{k+1} \geq 0$, we see that S_{2n} is a sum of positive terms and

$$0 \leq S_2 \leq S_4 \leq \cdots \leq S_{2n} \leq \cdots$$

That is, $\{S_{2n}\}$ is a monotonic sequence. Moreover, we can rewrite this sum as

$$S_{2n} = a_1 - (a_2 - a_3) - (a_4 - a_5) - \cdots - (a_{2n-2} - a_{2n-1}) - a_{2n}$$

which shows that $S_{2n} \leq a_1$ for all n. Therefore, by the Bounded Convergence Theorem (11.6), we know that this sequence has a limit

$$\lim_{n \to \infty} S_{2n} = S \quad \text{and} \quad S \leq a_1$$

Now we consider the partial sums S_{2n+1} that contain an odd number of terms. But $S_{2n+1} = S_{2n} + a_{2n+1}$, so

$$\lim_{n \to \infty} S_{2n+1} = \lim_{n \to \infty}(S_{2n} + a_{2n+1}) = S \qquad (a_{2n+1} \to 0)$$

Since both sets of partial sums have the limit S, it follows that

$$\lim_{n \to \infty} S_n = S \quad \text{and} \quad S \leq a_1$$

and the proof of the alternating test is complete.

The alternating test requires that the sequence of terms be nonincreasing. This can be verified algebraically or by using calculus.

EXAMPLE 1 Test the convergence of $\sum (-1)^{n-1} \sqrt{n}/(n + 1)$.

Solution This is an alternating series and the terms tend to 0 as n tends to infinity. We will verify that the sequence of terms is nonincreasing in two ways:

Algebraic Method. Show $a_n - a_{n+1} \geq 0$

$$\frac{\sqrt{n}}{n+1} - \frac{\sqrt{n+1}}{n+2} = \frac{(n+2)\sqrt{n} - (n+1)\sqrt{n+1}}{(n+1)(n+2)} \qquad \text{(Combine fractions)}$$

The denominator is positive; now rationalize the numerator

$$(n+2)\sqrt{n} - (n+1)\sqrt{n+1} = \frac{(n+2)^2 n - (n+1)^3}{(n+2)\sqrt{n} + (n+1)\sqrt{n+1}}$$

This denominator is also positive, and the numerator is

$$(n+2)^2 n - (n+1)^3 = n^3 + 4n^2 + 4n - n^3 - 3n^2 - 3n - 1$$

$$= n^2 + n - 1 \geq 0 \quad \text{for } n \geq 1$$

It follows that $a_n \geq a_{n+1} > 0$ and the series converges.

Calculus Method. Define $f(x)$ on $[1, \infty)$ with $f(n) = a_n$, and show $f'(x) < 0$.

Let $f(x) = \sqrt{x}/(x + 1)$; then, by the Quotient rule,

$$f'(x) = \frac{(x + 1)(1/2\sqrt{x}) - \sqrt{x}}{(x + 1)^2}$$

$$= \frac{-x + 1}{2\sqrt{x}(x + 1)^2} \qquad \text{(Combine fractions)}$$

Therefore, $f'(x) < 0$ for $x > 1$, and f is decreasing. This shows that $a_n \geq a_{n+1}$; so the series converges. ∎

The proof of the alternating test shows that the sum of a convergent alternating series is less than the first term. Since the remainder of an alternating series is itself an alternating series, it follows that

(11.27)

> The truncation error in an alternating series is less than a_{N+1}; that is,
>
> $$\sum_{n=N+1}^{\infty} (-1)^n a_n \leq a_{N+1}$$

WARNING: The truncation error estimate in (11.27) can be used only for alternating series. In other cases, the methods of Section 11–3 must be employed.

EXAMPLE 2 Estimate $\sum_{n=1}^{\infty} (-1)^{n-1} 1/n$ with an error less than .05. How many terms must be added to obtain accuracy to three decimal places?

Solution By (11.27), we know that

$$\sum_{n=N+1}^{\infty} (-1)^{n-1} \frac{1}{n} \leq \frac{1}{N + 1}$$

To obtain a truncation error less than $.05 = 1/20$, we let $N = 20$. Thus,

$$\sum_{n=1}^{20} (-1)^{n-1} \frac{1}{n} = 1 - \frac{1}{2} + \frac{1}{3} - \frac{1}{4} + \cdots - \frac{1}{20}$$

$$\approx 0.69019997$$

is the desired estimate correct to one decimal place. To obtain three decimal-place accuracy, we would have to sum at least the first 2,000 terms because 2,001 is the first number whose reciprocal is less than .0005. ∎

Absolute and Conditional Convergence

The alternating series $\sum (-1)^{n-1}(1/n)$ and $\sum (-1)^{n-1}(1/n^2)$ both converge. But there is a difference.

$$\sum \left| (-1)^{n-1} \frac{1}{n} \right| = \sum \frac{1}{n} \text{ diverges}$$

and

$$\sum\left|(-1)^{n-1}\frac{1}{n^2}\right| = \sum\frac{1}{n^2} \text{ converges}$$

Without absolute values, the first series converges because the negative terms subtract off enough to make the sum finite; this series is said to converge *conditionally*. The second series converges even if the minus signs are replaced by plus signs; this series is said to converge *absolutely*.

(11.28)

DEFINITION

(1) $\sum a_n$ is **absolutely convergent** if $\sum|a_n|$ converges.
(2) $\sum a_n$ is **conditionally convergent** if $\sum a_n$ converges but $\sum|a_n|$ diverges.

Since most of the convergence tests require that the terms of the series be non-negative, you can anticipate that absolute convergence will play an important role in our work with infinite series.

Note that if $\sum a_n$ is a series of nonnegative terms, then $a_n \doteq |a_n|$ so that absolute convergence and convergence are the same. Moreover, a finite number of negative terms would not effect convergence. But what if there are infinitely many positive and negative terms and $\sum|a_n|$ converges; does that imply that $\sum a_n$ converges? To find out, we set $b_n = a_n + |a_n|$; then $b_n \geq 0$ for all n and

$$0 \leq b_n \leq 2|a_n|$$

If $\sum|a_n|$ converges, then $\sum 2|a_n|$ converges, and then $\sum b_n$ converges by comparison. It follows from the sum theorem (11.14) that

$$\sum_{n=1}^{\infty} b_n - \sum_{n=1}^{\infty}|a_n| = \sum_{n=1}^{\infty}(b_n - |a_n|) = \sum_{n=1}^{\infty} a_n$$

and, therefore, $\sum a_n$ converges.

(11.29)

If $\sum|a_n|$ converges, then $\sum a_n$ converges.

Any series can be classified as (1) absolutely convergent, (2) conditionally convergent, or (3) divergent. To classify a given series $\sum a_n$, first determine if the series is absolutely convergent by testing $\sum|a_n|$. If this series diverges, then test $\sum a_n$ to see if it is conditionally convergent or divergent.

EXAMPLE 3   Classify the series $\sum(\sin n)/n^2$.

Solution   First test for absolute convergence. Since

$$\left|\frac{\sin n}{n^2}\right| \le \frac{1}{n^2}$$

we know, by comparison with a p-series, that

$$\sum \left|\frac{\sin n}{n^2}\right| \text{ converges}$$

Therefore, the series is absolutely convergent.   ■

EXAMPLE 4   Classify $\sum(-1)^{n-1}1/\sqrt{n}$.

Solution   First test for absolute convergence. Since

$$\left|(-1)^{n-1}\frac{1}{\sqrt{n}}\right| \ge \frac{1}{n}$$

we know, by comparison with the harmonic series, that

$$\sum\left|(-1)^{n-1}\frac{1}{\sqrt{n}}\right| \text{ diverges}$$

The series is *not* absolutely convergent. Now test for conditional convergence. By the alternating test,

$$\sum(-1)^{n-1}\frac{1}{\sqrt{n}} \text{ converges}$$

so the series is conditionally convergent.   ■

The Series of Positive and Negative Terms *(Optional)*

Suppose $\sum a_n$ is a series with infinitely many positive and negative terms. For each n, let

$$p_n = \begin{cases} a_n & \text{if } a_n \ge 0 \\ 0 & \text{if } a_n < 0 \end{cases} \quad \text{and} \quad q_n = \begin{cases} a_n & \text{if } a_n < 0 \\ 0 & \text{if } a_n \ge 0 \end{cases}$$

Thus, the p_n's are the positive terms and the q_n's are the negative terms. We call $\sum p_n$ and $\sum q_n$ the *series of positive and negative terms of* $\sum a_n$. It is not difficult to verify that

(11.30)
$$p_n = \frac{a_n + |a_n|}{2} \quad q_n = \frac{a_n - |a_n|}{2}$$
$$p_n + q_n = a_n \quad p_n - q_n = |a_n|$$

We can now prove the following result:

(11.31)
(1) $\sum a_n$ is absolutely convergent if and only if $\sum p_n$ and $\sum q_n$ both converge. In that case, $\sum a_n = \sum p_n + \sum q_n$.
(2) If $\sum a_n$ is conditionally convergent, then $\sum p_n$ and $\sum q_n$ both diverge.

If $\sum |a_n|$ converges, then so does $\sum a_n$, by (11.29). Therefore,

$$\sum \frac{a_n}{2} + \sum \frac{|a_n|}{2} = \sum \frac{a_n + |a_n|}{2} = \sum p_n \qquad (11.30)$$

So, $\sum p_n$ converges; similarly, $\sum q_n$ converges. On the other hand, if $\sum p_n$ and $\sum q_n$ both converge, then

$$\sum p_n - \sum q_n = \sum (p_n - q_n) = \sum |a_n| \qquad (11.30)$$

Therefore, $\sum a_n$ is absolutely convergent. Moreover, again by (11.30), we have

$$\sum p_n + \sum q_n = \sum (p_n + q_n) = \sum a_n$$

That completes the proof of part (1). To prove part (2), suppose that $\sum a_n$ converges conditionally *and* $\sum p_n$ converges; then

$$\sum 2p_n - \sum a_n = \sum (2p_n - a_n) = \sum |a_n| \qquad (11.30)$$

which implies that $\sum a_n$ is absolutely convergent, contrary to our assumption. It follows that $\sum p_n$ diverges and a similar argument holds for $\sum q_n$.

The results in (11.31) contain the essential difference between absolutely and conditionally convergent series. Exercise 36 contains a surprising fact about conditionally convergent series; *they can be rearranged to add up to any number whatsoever.*

EXERCISES

In Exercises 1–30, classify the series as absolutely convergent, conditionally convergent, or divergent.

1. $\sum \left(-\frac{1}{2}\right)^n$

2. $\sum (-1)^{n-1} \frac{n+1}{n}$

3. $\sum (-1)^{k-1} \frac{k^2+1}{k^3+1}$

4. $\sum (-1)^{n-1} \frac{n^3}{e^n}$

5. $\sum (-1)^{k-1} \frac{\ln k}{k}$

6. $\sum (-1)^{k-1} \frac{\ln k}{(1.01)^k}$

7. $\sum \frac{\cos n}{n^2}$

8. $\sum \frac{\sin(n + \frac{1}{2})\pi}{n}$

9. $\sum (-1)^k \frac{\sqrt{k}}{k+1}$

10. $\sum (-1)^k \frac{1}{2k-1}$

11. $\sum (-1)^{n-1} \frac{n}{n+1}$

12. $\sum \left(\frac{1}{2}\right)^{1/n}$

13. $\sum \frac{n!}{(-n)^n}$

14. $\sum n^{1/n}$

15. $\sum \left(-\frac{1}{n}\right)^n$

16. $\sum n \left(-\frac{4}{5}\right)^n$

17. $\sum (-1)^n \frac{n!}{n^n}$

18. $\sum (-1)^n \frac{n^n}{n!}$

19. $\sum (-1)^n \sin \frac{1}{n}$

20. $\sum (-1)^n \sin \frac{1}{n^2}$

21. $\sum (-1)^{k-1} \frac{1}{k(\ln k)^2}$

22. $\sum (-1)^{k-1} \frac{1}{(\ln k)^k}$

23. $\sum (-1)^n \frac{1 \cdot 3 \cdot 5 \cdots (2n-1)}{2 \cdot 5 \cdot 8 \cdots (3n-1)}$

24. $\sum \frac{(-10)^n}{n!}$

25. $\sum \frac{n(-3)^n}{5^n}$

26. $\sum (-1)^n \frac{\arctan n}{n^2}$

27. $\sum (-1)^{n-1} \frac{1 \cdot 3 \cdot 5 \cdots (2n-1)}{n!}$

28. $\sum \dfrac{(n!)^2}{(2n)!}$

29. $\sum \dfrac{\cos n\pi}{\sqrt[3]{n^2+1}}$ **30.** $\sum \dfrac{\cos n\pi}{\sqrt{n^3+1}}$

31. $\displaystyle\sum_{n=1}^{\infty} (-1)^{n-1}\dfrac{1}{n!}$ **32.** $\displaystyle\sum_{n=1}^{\infty} (-1)^{n-1}\dfrac{1}{n^2}$

33. $\displaystyle\sum_{n=1}^{\infty} (-1)^{n-1} n\left(\dfrac{1}{2}\right)^n$ **34.** $\displaystyle\sum_{n=1}^{\infty} \dfrac{(-1)^{n-1}}{(2n-1)!}$

In Exercises 31–34, estimate the sums correct to two decimal places.

Optional Exercises

35. Rearranging the order of the terms of a conditionally convergent series can *change the sum!* Let S be sum of the alternating harmonic series.

(1) $S = 1 - \dfrac{1}{2} + \dfrac{1}{3} - \dfrac{1}{4} + \dfrac{1}{5} - \dfrac{1}{6} + \cdots$

From Example 2, we know that $S \neq 0$. Since the series converges we can multiply it by $1/2$,

(2) $\dfrac{1}{2}S = \dfrac{1}{2} - \dfrac{1}{4} + \dfrac{1}{6} - \dfrac{1}{8} + \dfrac{1}{10} - \dfrac{1}{12} + \cdots$

Adding 0's does not change a sum, so (2) can be written as

(3) $\dfrac{1}{2}S = 0 + \dfrac{1}{2} + 0 - \dfrac{1}{4} + 0 + \dfrac{1}{6} + \cdots$

Since (1) and (3) both converge, we can add them term by term

(4) $\dfrac{3}{2}S = (1+0) + \left(-\dfrac{1}{2} + \dfrac{1}{2}\right) + \left(\dfrac{1}{3} + 0\right)$

$+ \left(-\dfrac{1}{4} - \dfrac{1}{4}\right) + \cdots$

$= 1 + \dfrac{1}{3} - \dfrac{1}{2} + \dfrac{1}{5} + \dfrac{1}{7} - \dfrac{1}{4} + \cdots$

Verify that (1) and (4) contain the same terms but in a different order (do only the first 20 terms). Yet one has the sum S and the other has the sum $3S/2$.

36. Show that *the terms of a conditionally convergent series can be rearranged to sum to any preassigned number whatsoever!* *Hint:* The series of positive terms diverges to infinity and the series of negative terms diverges to minus infinity (11.31). Given any positive number L, add enough positive terms to obtain a sum larger than L; then add enough negative terms to obtain a sum smaller than L; then start adding positive terms to obtain a sum larger than L; and so on.

11–6   POWER SERIES

A finite sum of the form

$$c_0 + c_1 x + c_2 x^2 + \cdots + c_n x^n$$

is a *polynomial in x*. An infinite sum of the form

(11.32) $\displaystyle\sum_{n=0}^{\infty} c_n x^n = c_0 + c_1 x + c_2 x^2 + \cdots + c_n x^n + \cdots$

is called a **power series in x**. A power series can be thought of as a polynomial of "infinite" degree. In this section, we discuss some remarkable results about the convergence of power series. Then, in the next two sections, we investigate the important relationship between power series and functions.

If x is replaced by 0, the series (11.32) converges to c_0.* For other values of x, the convergence can be determined in any of the usual ways.

EXAMPLE 1 Consider the power series $\sum_{n=1}^{\infty} x^n/n$. If x is replaced by $1/2$, the series becomes

$$\sum_{n=1}^{\infty} \frac{1}{n}\left(\frac{1}{2}\right)^n = \sum_{n=1}^{\infty} \frac{1}{n2^n}$$

which converges by the root test, the ratio test, or the comparison test. For $x = -1$, the series becomes

$$\sum_{n=1}^{\infty} \frac{(-1)^n}{n}$$

which converges by the alternating series test. For $x = 1$, the result is the divergent harmonic series. ∎

The following result is the basis of much of our later work with power series.

(11.33)

Let $\sum c_n x^n$ be a power series, and let a be any nonzero number.

(1) If the series converges for $x = a$, then it converges absolutely for all x with $|x| < |a|$.

(2) If the series diverges for $x = a$, then it diverges for all x with $|x| > |a|$.

For the first part, suppose that $a \neq 0$ and $\sum c_n a^n$ converges. It follows from the nth term test that $|c_n a^n| < 1$ for all large n. If $|x| < |a|$, then for all large n,

$$|c_n x^n| = |c_n a^n| \frac{|c_n x^n|}{|c_n a^n|} \qquad\qquad (a \neq 0)$$

$$< \frac{|x^n|}{|a_n|} \qquad\qquad (|c_n a^n| < 1)$$

$$= \left(\frac{|x|}{|a|}\right)^n$$

Therefore, since $|x|/|a| < 1$, $\sum |c_n x^n|$ converges by comparison with the convergent geometric series $\sum(|x|/|a|)^n$. This shows that $\sum c_n x^n$ converges absolutely if $|x| < |a|$. The proof of the second part is left as an exercise.

EXAMPLE 2 The series $\sum x^n/n$ in Example 1 converges for $x = -1$, but diverges for $x = 1$. It follows from (11.33) that the series converges absolutely for all x with $|x| < |-1| = 1$ and diverges for all x with $|x| > 1$. Thus, the power

---

*For power series that start with $n = 0$, we adopt the convention that $x^0 = 1$ even when $x = 0$.

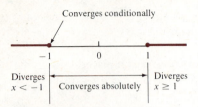

FIGURE 11–15. Example 2. The convergence of $\sum x^n/n$.

series $\sum x^n/n$ converges for all x in the interval $[-1, 1)$ and diverges elsewhere (Figure 11–15). ∎

The series in Example 2 converges only for x in a certain interval. This is typical for power series. It follows from (11.33) and properties of the real numbers that

(11.34)

> For any power series $\sum c_n x^n$, exactly one of the following statements is true:
>
> (1) It converges only for $x = 0$.
>
> (2) It converges for all real x.
>
> (3) There is some positive real number r such that it converges absolutely for all x in the interval $(-r, r)$ and diverges for x outside of the interval $[-r, r]$.

The number r above is called the **radius of convergence** of the series. If the series converges only for $x = 0$, we set $r = 0$; if it converges for all x, we set $r = \infty$. The series may or may not converge at the endpoints $-r$ and r; these points must be tested separately. The totality of points for which the series converges is called the **interval of convergence** (Figure 11–16). The radius of convergence is usually determined with the ratio test. Once the radius is known, the interval of convergence is found by testing the endpoints.

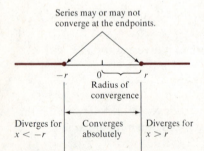

FIGURE 11–16. Typical convergence picture for a power series. The interval of convergence is one of the intervals $(-r, r), [-r, r], [-r, r),$ or $(-r, r]$.

EXAMPLE 3   Find the interval of convergence of $\sum(x/3)^n$.

Solution   The ratio test is for series with nonnegative terms so we must use absolute values. Thus,

$$a_{n+1} = \left|\left(\frac{x}{3}\right)^{n+1}\right|, \; a_n = \left|\left(\frac{x}{3}\right)^n\right|, \text{ and}$$

$$\lim_{n \to \infty} \frac{a_{n+1}}{a_n} = \lim_{n \to \infty} \frac{|x/3|^{n+1}}{|x/3|^n} = \left|\frac{x}{3}\right|$$

The series will converge (absolutely) whenever $|x/3| < 1$ and diverge whenever $|x/3| > 1$. It follows that $r = 3$ is the radius of convergence. At the endpoints -3 and 3, the resulting series are $\sum(-1)^n$ and $\sum 1^n$ both of which diverge. Therefore, the interval of convergence of $\sum(x/3)^n$ is $(-3, 3)$. ∎

EXAMPLE 4   Find the interval of convergence of $\sum x^n/n^2$.

Solution   Use the ratio test with $a_{n+1} = |x^{n+1}|/(n + 1)^2$ and $a_n = |x^n|/n^2$. Then

$$\lim_{n \to \infty} \frac{a_{n+1}}{a_n} = \lim_{n \to \infty} \frac{|x^{n+1}|}{(n + 1)^2} \cdot \frac{n^2}{|x^n|} = \lim_{n \to \infty} |x| \left[\frac{n}{n + 1}\right]^2 = |x|$$

The series converges when $|x| < 1$ and diverges when $|x| > 1$; thus, $r = 1$. At both endpoints -1 and 1, the series is a convergent p-series, so the interval of convergence is $[-1, 1]$. ∎

The radius of convergence can also be found using the root test.

EXAMPLE 5   Find the interval of convergence of $\sum n^n x^n$.

Solution   The root test for $\sum n^n x^n$ yields

$$\lim_{n \to \infty} |n^n x^n|^{1/n} = \lim_{n \to \infty} n|x|$$

This limit is infinite for every nonzero value of x. Therefore, the radius of convergence is $r = 0$, and the series converges only for $x = 0$. ∎

The next example is important for later work.

EXAMPLE 6   Show that $\sum x^n/n!$ converges for all x.

Solution   The ratio test for $\sum x^n/n!$ yields

$$\lim_{n \to \infty} \frac{|x^{n+1}|}{(n + 1)!} \cdot \frac{n!}{|x^n|} = \lim_{n \to \infty} \frac{|x|}{n + 1} = 0$$

for any value of x. Thus, $r = \infty$ and this series converges for all real x. ∎

Sometimes it is convenient to consider series of the form $\sum c_n(x - a)^n$. They are called **power series in $x - a$.** The theory is the same as for power series in x, but with a slight modification. Suppose r is the radius of convergence of $\sum c_n(x - a)^n$. If $r = 0$, then the series converges only for $x = a$; if $r = \infty$, the series converges for all real x; if r is some positive number, then the series converges absolutely for all x in the interval $(a - r, a + r)$ and diverges outside of $[a - r, a + r]$. Convergence at the endpoints is tested separately (Figure 11–17).

EXAMPLE 7   Find the interval of convergence of $\sum(-3)^n(x - 2)^n/n$.

Solution   This is a power series in $(x - a)$ with $a = 2$ and $c_n = (-3)^n/n$. Either test can be used; the root test yields

$$\lim_{n \to \infty} \left| \frac{(-3)^n(x - 2)^n}{n} \right|^{1/n} = \lim_{n \to \infty} \frac{|-3||x - 2|}{n^{1/n}} = 3|x - 2|$$

This series will converge if $3|x - 2| < 1$ and diverge if $3|x - 2| > 1$. It follows that $r = 1/3$ and the series converges absolutely on the interval $(2 - \frac{1}{3}, 2 + \frac{1}{3})$ $= (5/3, 7/3)$. Now replace x by the endpoints and test for convergence.

$$x = \frac{5}{3}; \sum \frac{(-3)^n(x - 2)^n}{n} = \sum \frac{(-3)^n(-1/3)^n}{n} = \sum \frac{1}{n} \qquad \text{(diverges)}$$

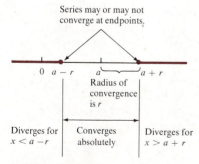

Series may or may not converge at endpoints.

$$0 \quad a - r \quad a \quad a + r$$

Radius of convergence is r

Diverges for $x < a - r$ | Converges absolutely | Diverges for $x > a + r$

FIGURE 11–17. Convergence of a power series $\sum c_n(x - a)^n$.

$$x = \frac{7}{3}; \quad \sum \frac{(-3)^n (x-2)^n}{n} = \sum \frac{(-3)^n (1/3)^n}{n} = \sum \frac{(-1)^n}{n} \qquad \text{(converges)}$$

Therefore, the interval of convergence is $(5/3, 7/3]$. ■

EXAMPLE 8   Find the interval of convergence of $\sum (x+4)^n / n$.

Solution   This is a power series in $(x - a)$, but this time $a = -4$ (note the minus sign). Either test can be used; the ratio test yields

$$\lim_{n \to \infty} \frac{|x+4|^{n+1}}{n+1} \cdot \frac{n}{|x+4|^n} = |x+4|$$

It follows that the radius of convergence is $r = 1$ and the series converges absolutely on the interval $(-4-1, -4+1) = (-5, -3)$. A test of the endpoints shows that the interval of convergence is $[-5, -3)$. ■

EXERCISES

In Exercises 1–28, find the interval of convergence.

1. $\sum \frac{n^2}{2^n} x^n$

2. $\sum \frac{n}{n^2 + 4} x^n$

3. $\sum \frac{\ln k}{k^2} x^k$

4. $\sum n! x^n$

5. $\sum \frac{3^{2n}}{n+1} (x-2)^n$

6. $\sum \frac{n}{e^n} (x-e)^n$

7. $\sum n^p x^n, p > 0$

8. $\sum (-1)^n (x + \pi)^n$

9. $\sum \frac{(5x)^n}{n!}$

10. $\sum \frac{(x-1)^n}{(2n)^n}$

11. $\sum \frac{(2n)!(x+1)^n}{n!}$

12. $\sum \frac{(n-1)x^n}{n^5}$

13. $\sum \frac{4^{\sqrt{n}} x^n}{\sqrt{n}}$

14. $\sum \frac{x^n}{n!}$

15. $\sum p^n (x+2)^n, p > 0$

16. $\sum \frac{n^5}{(n+1)!} x^n$

17. $\sum \frac{n+1}{10^n} (x-3)^n$

18. $\sum \frac{1}{(-4)^n} x^{2n+1}$

19. $\sum \frac{n^2}{2^{3n}} (x+1)^n$

20. $\sum \frac{(-x)^n}{n}$

21. $\sum \frac{n!(x-1)^n}{1 \cdot 3 \cdot 5 \cdots (2n-1)}$

22. $\sum \frac{1 \cdot 3 \cdot 5 \cdots (2n-1)}{2 \cdot 4 \cdot 6 \cdots (2n)} x^n$

23. $\sum (2x)^{2n}$

24. $\sum \frac{(3x-1)^{2n}}{n}$

25. $\sum \frac{x^{2n+1}}{(2n+1)!}$

26. $\sum \frac{x^{2n}}{(2n)!}$

27. $\sum \frac{n^n}{n!} x^n$

28. $\sum \frac{n!}{n^n} x^n$

29. *(Derivatives and integrals).* Suppose that $r \neq 0$ is the radius of convergence of $\sum c_n x^n$ and that

$$\lim_{n \to \infty} \left| \frac{c_{n+1}}{c_n} \right| = \frac{1}{r}$$

For each term $c_n x^n$, the derivative is $c_n n x^{n-1}$ and the integral is $c_n x^{n+1}/(n+1)$. Use the ratio test to find the radius of convergence of the power series $\sum c_n n x^{n-1}$ and $\sum c_n x^{n+1}/(n+1)$.

30. (Continuation of Exercise 29). For each x with $|x| < 1$, the geometric series $\sum_{n=0}^{\infty} x^n$ converges to the number $1/(1-x)$. The derivative of each term of the series is nx^{n-1}. For each x with $|x| < 1$, does $\sum_{n=1}^{\infty} nx^{n-1}$ converge to the derivative of $1/(1-x)$? (See Exercise 32 of Section 11–4.)

31. Show that the second part of (11.33) is true; if $\sum c_n x^n$ diverges for $x = a$, then it diverges for all x with $|x| > |a|$. *Hint:* Use the first part of (11.33) to show that if $|x| > |a|$, then $\sum c_n x^n$ cannot converge.

11-7 POWER SERIES AS FUNCTIONS

A power series $\sum c_n x^n$ with an interval of convergence $(-r, r)$ can be thought of as a function f. For each x in $(-r, r)$, $f(x)$ is the sum of the series for that value of x; that is,

$$f(x) = \sum_{n=0}^{\infty} c_n x^n \quad x \text{ in } (-r, r)$$

You might say that a power series has a dual personality; it represents an infinite series and it also represents a function. In this section, we examine several power series, and find the functions they represent. After that, we discuss the derivative and integral of a power series.

The Power Series as a Function

EXAMPLE 1 Find a function that is represented by the geometric power series

$$\sum_{n=0}^{\infty} x^n = 1 + x + x^2 + \cdots$$

and find the sum of the series for $x = 1/2$ and $-1/3$.

Solution The radius of convergence is obtained by the ratio test,

$$\lim_{n \to \infty} \frac{|x^{n+1}|}{|x^n|} = |x|$$

so $r = 1$. Since the series diverges at both endpoints -1 and 1, the interval of convergence is $(-1, 1)$. From previous work with geometric series, we know that for each x in $(-1, 1)$, the sum is $1/(1 - x)$. Therefore,

$$\sum_{n=0}^{\infty} x^n = \frac{1}{1 - x} \quad x \text{ in } (-1, 1)$$

Thus, the series represents the function $f(x) = 1/(1 - x)$. The sums for $x = 1/2$ and $-1/3$ are obtained by evaluating f at these points.

$$f\left(\frac{1}{2}\right) = \frac{1}{1 - (1/2)} = 2 \quad \text{and} \quad f\left(-\frac{1}{3}\right) = \frac{1}{1 - (-1/3)} = \frac{3}{4}$$

Figure 11-18 compares the graphs of the infinite sum $f(x) = 1/(1 - x)$ and the partial sum $S_{11}(x) = 1 + x + x^2 + \cdots + x^{10}$. There is very close agreement on the interval $(-.75, .75)$. There would be closer agreement over a larger interval if we used the graphs of S_{100}, S_{1000}, and so on, but no finite sum will agree with f over the entire interval $(-1, 1)$. We will continue this discussion in the next section. ∎

The next few examples illustrate how simple variations of the geometric power series lead to different functions.

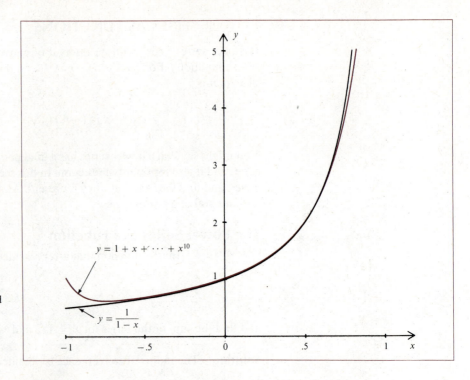

FIGURE 11–18. Example 1.
Comparison of the computer-generated
graphs of the infinite sum $\sum_{n=0}^{\infty} x^n$
$= 1/1 - x$ and the partial sum
$S_{11} = 1 + x + x^2 \ldots + x^{10}$. There
is close agreement on the interval
$(-.75, .75)$.

EXAMPLE 2   Find a function that is represented by the series

$$\sum_{n=0}^{\infty} (-x)^n = 1 - x + x^2 - x^3 + \cdots$$

and find the sum for $x = 1/4$ and $-1/8$.

Solution   This is the geometric power series with x replaced by $-x$. It follows
from Example 1 that the sum is $1/(1 - (-x)) = 1/(1 + x)$. That is,

$$\sum_{n=0}^{\infty} (-x)^n = \frac{1}{1 + x} \quad x \text{ in } (-1, 1)$$

This series represents the function $f(x) = 1/(1 + x)$. See Figure 11–19. The
sums for $x = 1/4$ and $-1/8$ are

$$f\left(\frac{1}{4}\right) = \frac{1}{1 + (1/4)} = \frac{4}{5} \quad \text{and} \quad f\left(-\frac{1}{8}\right) = \frac{1}{1 + (-1/8)} = \frac{8}{7} \quad \blacksquare$$

EXAMPLE 3   Find a function that is represented by the series

$$\sum_{n=0}^{\infty} (-x)^{n+1} = -x + x^2 - x^3 + \cdots$$

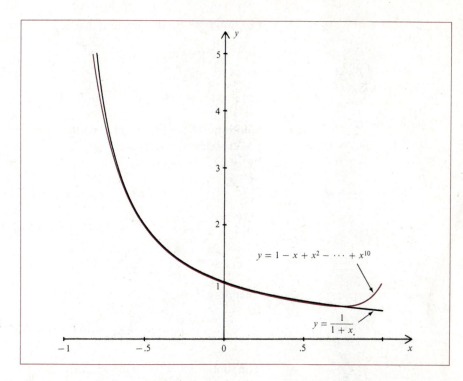

FIGURE 11–19. Example 2. Comparison of the computer-generated graphs of the infinite sum $\sum_{n=0}^{\infty} (-x)^n = 1/1 + x$ and the partial sum $S_{11} = 1 - x + x^2 - \ldots + x^{10}$.

Solution If you rewrite each term $(-x)^{n+1}$ as $(-x)(-x)^n$, then this series is simply $-x$ times the series in Example 2. Thus,

$$\sum_{n=0}^{\infty} (-x)^{n+1} = (-x) \sum_{n=0}^{\infty} (-x)^n = \frac{-x}{1+x} \quad x \text{ in } (-1, 1) \qquad \blacksquare$$
$$\text{by (11.14)}$$

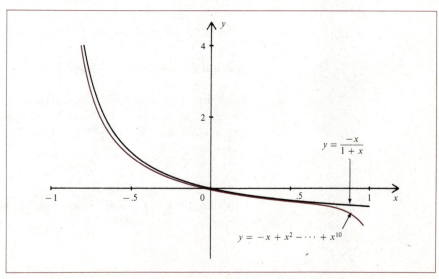

FIGURE 11–20. Example 3. Comparison of $f(x) = \sum_{n=0}^{\infty} (-x)^{n+1} = -x/1 + x$ and $S_{10} = -x + x^2 - \ldots + x^{10}$.

EXAMPLE 4   Find a function that is represented by the series

$$\sum_{n=0}^{\infty} \frac{x^{2n}}{3^n} = 1 + \frac{x^2}{3} + \frac{x^4}{9} + \cdots$$

and find the sum for $x = \sqrt{2}$ and $-1/2$.

Solution   If the terms are rewritten as $(x^2/3)^n$, we have the basic geometric power series with x replaced by $x^2/3$. The interval of convergence is computed by solving the inequality

$$\left| \frac{x^2}{3} \right| < 1 \quad \text{or} \quad x^2 < 3$$

Thus, $-\sqrt{3} < x < \sqrt{3}$, and for each x the sum is

$$\frac{1}{1 - (x^2/3)} = \frac{3}{3 - x^2}$$

Therefore,

$$\sum_{n=0}^{\infty} \frac{x^{2n}}{3^n} = \frac{3}{3 - x^2} \quad x \text{ in } (-\sqrt{3}, \sqrt{3}) \qquad \text{(Figure 11–21)}$$

The sums for $x = \sqrt{2}$ and $-1/2$ are

$$\frac{3}{3 - (\sqrt{2})^2} = 3 \quad \text{and} \quad \frac{3}{3 - (-1/2)^2} = \frac{12}{11} \qquad \blacksquare$$

The next example involves a power series in $x - a$.

EXAMPLE 5   Find a function that is represented by the series

$$\sum_{n=0}^{\infty} (1 - x)^n = 1 + (1 - x) + (1 - x)^2 + \cdots$$

and find the sum for $x = 1/5$ and $3/2$.

Solution   This is the basic geometric series with x replaced by $1 - x$. The series converges for all x with $|1 - x| < 1$, so the interval of convergence is $(0, 2)$. The sum at any x is $1/(1 - (1 - x)) = 1/x$. Therefore,

$$\sum_{n=0}^{\infty} (1 - x)^n = \frac{1}{x} \quad x \text{ in } (0, 2) \qquad \text{(Figure 11–22)}$$

The sums for $x = 1/5$ and $3/2$ are 5 and $2/3$. $\blacksquare$

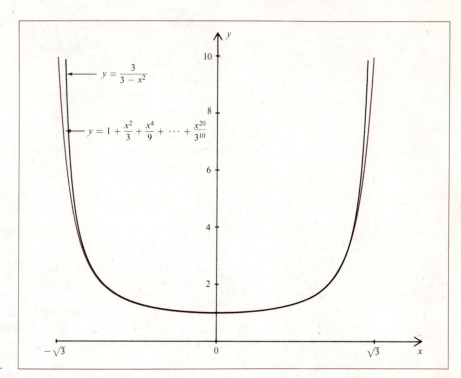

FIGURE 11–21. Example 4.
Comparison of $f(x) = \sum_{n=0}^{\infty} x^{2n}/3^n$
$= 3/3 - x^2$ on $(-\sqrt{3}, \sqrt{3})$ and
$S_{11} = 1 + x^2/3 + x^4/9 + \ldots + x^{20}/3^{10}$.

$$y = \frac{3}{3 - x^2}$$

$$y = 1 + \frac{x^2}{3} + \frac{x^4}{9} + \cdots + \frac{x^{20}}{3^{10}}$$

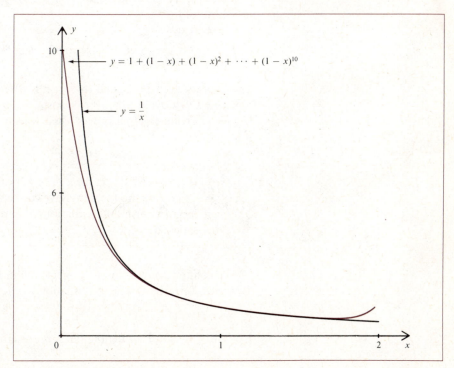

FIGURE 11–22. Example 5.
Comparison of $f(x) = \sum_{n=0}^{\infty} (1 - x)^n$
$= 1/x$ on $(0, 2)$ and $S_{11} = 1 + (1 - x)$
$+ (1 - x)^2 + \ldots + (1 - x)^{10}$.

$$y = 1 + (1 - x) + (1 - x)^2 + \cdots + (1 - x)^{10}$$

$$y = \frac{1}{x}$$

Term-by-term Differentiation and Integration

Each term of a power series is a function $c_n x^n$; its derivative is $nc_n x^{n-1}$. It can be shown that both series

$$\sum_{n=0}^{\infty} c_n x^n \quad \text{and} \quad \sum_{n=1}^{\infty} nc_n x^{n-1}$$

have the same radius of convergence. Moreover, these series and the functions they represent are related as stated below (the proof is omitted here).

(11.35)

If $f(x) = \sum_{n=0}^{\infty} c_n x^n$ for x in $(-r, r)$, then f is differentiable and

$$f'(x) = \sum_{n=1}^{\infty} nc_n x^{n-1} \text{ for } x \text{ in } (-r, r)$$

A similar statement is true for power series in $x - a$.

The new series $\sum nc_n x^{n-1}$ is obtained from the original series $\sum c_n x^n$ by differentiating each term. This is called **term-by-term differentiation.** It is valid only *inside* the interval of convergence; it may not hold at the endpoints (see Exercise 39).

Statement (11.35) can be applied with f' in place of f to obtain the second derivative of f:

$$f''(x) = \sum_{n=2}^{\infty} n(n-1)c_n x^{n-2} \quad x \text{ in } (-r, r)$$

This process can be repeated to obtain further derivatives. The symbol $f^{(k)}$ denotes the kth derivative of f. If $f^{(k)}$ exists, we say that f has derivatives of *order k.* If f has derivatives of all orders, then we say that f is **infinitely differentiable.** It follows that

(11.36)

If $f(x) = \sum_{n=0}^{\infty} c_n x^n$ for x in $(-r, r)$, then f is infinitely differentiable on $(-r, r)$ and the derivatives of f can be obtained using term-by-term differentiation. A similar statement is true for power series in $x - a$.

We continue now with our project of finding functions that are represented by series. Term-by-term differentiation is a new tool.

EXAMPLE 6   Find a function that is represented by the series

$$\sum_{n=1}^{\infty} nx^{n-1} = 1 + 2x + 3x^2 + \cdots$$

Solution   Each term nx^{n-1} of the series is the derivative of x^n. Therefore, this series is the result of term-by-term differentiation of the basic geometric power series $\sum x^n$. Since

$$\sum_{n=0}^{\infty} x^n = \frac{1}{1-x} \quad x \text{ in } (-1, 1)$$

it follows from (11.35) that

$$\sum_{n=1}^{\infty} nx^{n-1} = \left[\frac{1}{1-x}\right]' = \frac{1}{(1-x)^2} \quad x \text{ in } (-1, 1)$$

Compare this result with Exercise 32 of Section 11–4.   ■

EXAMPLE 7   Find a function that is represented by the series

$$\sum_{n=2}^{\infty} n(n-1)(x-2)^{n-2} = 2 + 6(x-2) + 12(x-2)^2 + \cdots$$

Solution   Each term is the second derivative of $(x-2)^n$. Since

$$\sum_{n=0}^{\infty} (x-2)^n = \frac{1}{1-(x-2)} = \frac{1}{3-x} \quad x \text{ in } (1, 3)$$

we have, by (11.35), that

$$\sum_{n=2}^{\infty} n(n-1)(x-2)^n = \left[\frac{1}{3-x}\right]'' = \frac{2}{(3-x)^3} \quad x \text{ in } (1, 3) \qquad ■$$

Term-by-term integration of a power series is also valid inside the interval of convergence, and the consequent relationships are stated below. Again, the proof is omitted,

---

If $f(x) = \sum_{n=0}^{\infty} c_n x^n$ for x in $(-r, r)$, then

$$\int f(x)\, dx = \sum_{n=0}^{\infty} \frac{c_n}{n+1} x^{n+1} + C \quad x \text{ in } (-r, r)$$

(11.37)   Furthermore, if a and b are in $(-r, r)$, then

$$\int_a^b f(x)\, dx = \sum_{n=0}^{\infty} \int_a^b c_n x^n\, dx$$

Similar statements are true for power series in $x - a$.

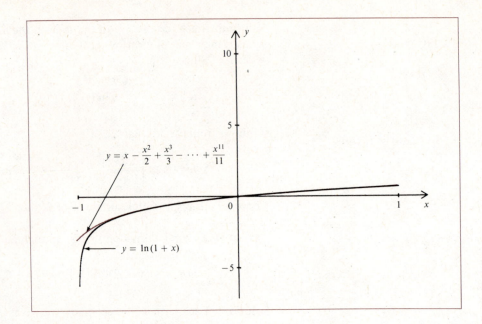

FIGURE 11–23. Example 8.
Comparison of $f(x) = \ln(1 + x)$ on
$(-1, 1)$ and $S_{11} = x - x^2/2 + x^3/3$
$- \ldots + x^{11}/11$.

EXAMPLE 8 Find a function that is represented by the series

$$\sum_{n=0}^{\infty} \frac{(-1)^n}{n + 1} x^{n+1} = x - \frac{x^2}{2} + \frac{x^3}{3} - \cdots$$

Solution Each term of this series is the integral of $(-1)^n x^n = (-x)^n$. Since

$$\sum_{n=0}^{\infty} (-x)^n = \frac{1}{1 + x} \text{ for } x \text{ in } (-1, 1) \qquad \text{(Example 2)}$$

it follows from (11.37) that

$$\sum_{n=0}^{\infty} \frac{(-1)^n}{n + 1} x^{n+1} = \int \frac{dx}{1 + x} = \ln(1 + x) + C \quad x \text{ in } (-1, 1)$$

Because the series and $\ln(1 + x)$ are both 0 for $x = 0$, we conclude that $C = 0$. Thus,

$$\sum_{n=0}^{\infty} \frac{(-1)^n}{n + 1} x^{n+1} = \ln(1 + x) \quad x \text{ in } (-1, 1)$$

Figure 11–23 compares the graphs of $y = \ln(1 + x)$ and $y = S_{11}(x)$. ∎

EXAMPLE 9 Use the results of Example 8 to approximate $\ln(1.1)$ correct to four decimal places.

Solution   By Example 8, we know that for x in $(-1, 1)$

$$\ln(1 + x) = \sum_{n=0}^{\infty} \frac{(-1)^n}{n + 1} x^{n+1} = x - \frac{x^2}{2} + \frac{x^3}{3} - \frac{x^4}{4} + \cdots$$

To compute $\ln(1.1)$, let $x = 0.1$ and then

$$\ln(1.1) = \ln(1 + 0.1) = 0.1 - \frac{(0.1)^2}{2} + \frac{(0.1)^3}{3} - \frac{(0.1)^4}{4} + \cdots$$

$$= 0.1 - .005 + .000333 - .000025 + \cdots$$

This is an alternating series, so by (11.26), the truncation error in adding the first three terms is less than the fourth term, .000025. Thus,

$$\ln(1.1) \approx 0.1 - .005 + .0003 = 0.0953$$

correct to four decimal places.   ■

   We conclude this section with an observation that will be used in the next section. Suppose that

$$f(x) = \sum_{n=0}^{\infty} c_n(x - a)^n \quad x \text{ in } (a - r, a + r)$$

The statement corresponding to (11.36) for power series in $x - a$ is

$$f^{(k)}(x) = \sum_{n=k}^{\infty} n(n - 1) \cdots (n - k + 1)c_n(x - a)^{n-k} \quad x \text{ in } (a - r, a + r)$$

If $x = a$, then the first term $(n = k)$ is $k(k - 1) \cdots 2 \cdot 1 \cdot c_k = k!c_k$, and all the remaining terms are zero. Therefore, for $x = a$, we have

$$f^{(k)}(a) = k!c_k \quad \text{or} \quad c_k = \frac{f^{(k)}(a)}{k!}$$

This is true for all $k = 0, 1, 2, 3, \ldots$, if we agree to let the symbol $f^{(0)}$ stand for the function f. The conclusion is that

(11.38)

If $f(x) = \sum_{n=0}^{\infty} c_n(x - a)^n$ for x in $(a - r, a + r)$, then

$$c_n = \frac{f^{(n)}(a)}{n!}$$

and

$$f(x) = \sum_{n=0}^{\infty} \frac{f^{(n)}(a)}{n!}(x - a)^n \quad x \text{ in } (a - r, a + r)$$

This statement is referred to in the next section.

EXERCISES

In Exercises 1–4, use the basic geometric power series to find a function that is represented by the given series.

1. $\sum_{n=0}^{\infty} x^{n+1}$

2. $\sum_{n=0}^{\infty} \frac{x^n}{3^{n+1}}$

3. $\sum_{n=0}^{\infty} (-1)^n x^{2n}$

4. $\sum_{n=0}^{\infty} (3 - x)^n$

In Exercises 5–8, use term-by-term differentiation to find a function that is represented by the given series.

5. $\sum_{n=1}^{\infty} (-n)(-x)^{n-1}$

6. $\sum_{n=0}^{\infty} (n + 1)x^n$

7. $\sum_{n=1}^{\infty} (-1)^n 2n x^{2n-1}$ (Use Exercise 3 above.)

8. $\sum_{n=1}^{\infty} \frac{2n}{3^n} x^{2n-1}$ (Use Example 4 in the text.)

In Exercises 9–12, use term-by-term integration to find a function that is represented by the given series.

9. $\sum_{n=0}^{\infty} \frac{x^{n+1}}{n+1}$

10. $\sum_{n=0}^{\infty} \frac{(-1)^n}{n+2} x^{n+2}$

11. $\sum_{n=0}^{\infty} \frac{(-1)^n}{2n+1} x^{2n+1}$ (Use Exercise 3 above.)

12. $\sum_{n=0}^{\infty} \frac{-1}{n+1}(1 - x)^{n+1}$ (Use Example 5 in the text.)

In Exercises 13–20, find a function that is represented by the given series.

13. $\sum_{n=0}^{\infty} \frac{(-1)^n}{2^{n+1}}(x - 2)^n$

14. $\sum_{n=0}^{\infty} (x^n + x^{n+1})$

15. $\sum_{n=0}^{\infty} \frac{x^n}{5^{n+1}}$

16. $\sum_{n=2}^{\infty} n(n - 1)x^{n-2}$

17. $\sum_{n=2}^{\infty} \frac{n(n - 1)}{5^{n+1}} x^{n-2}$

18. $\sum_{n=0}^{\infty} x^{3n}$

19. $\sum_{n=0}^{\infty} 4^{n+2} x^n$

20. $\sum_{n=0}^{\infty} \frac{4^{n+1}}{n+1} x^{n+1}$

In Exercises 21–30, find the sum at the indicated value of x.

21. $\sum_{n=1}^{\infty} n x^{n-1}, x = 1 - \sqrt{2}$

22. $\sum_{n=0}^{\infty} \frac{6^{n+1}}{n+1} x^{n+1}, x = 0.1$

23. $\sum_{n=0}^{\infty} 2(n + 1)x^{2n+1}, x = 1/2$

24. $\sum_{n=0}^{\infty} x^{4n+1}, x = -1/2$

25. $\sum_{n=0}^{\infty} \frac{(-1)^n}{2n + 1} x^{2n+1}, x = 0.3$

26. $\sum_{n=2}^{\infty} n(n - 1)x^{n-2}, x = 1/3$

27. $\sum_{n=0}^{\infty} \frac{3^n}{2^{n+1}} x^{n+1}, x = 0.1$

28. $\sum_{n=0}^{\infty} \frac{-1}{n+1}(1 - x)^{n+1}, x = 3/2$

29. $\sum_{n=2}^{\infty} \frac{n(n - 1)}{5^{n+1}} x^{n-2}, x = 4$

30. $\sum_{n=0}^{\infty} \frac{4^{n+1}}{n+1} x^{n+1}, x = 0$

In Exercises 31 and 32, estimate the value correct to four decimal places.

31. $\ln(1.2)$

32. $\ln(0.9)$

In Exercises 33 and 34, use Exercise 11 above to estimate the value correct to four decimal places.

33. $\tan^{-1}(0.1)$

34. $\tan^{-1}(-0.2)$

35. Use the results of Exercise 11 above to find a power series that represents $\tan^{-1} x^2$ in the interval $(-1, 1)$.

36. Find a power series that represents $\ln(1 + x^2)$ in the interval $(-1, 1)$.

37. Set

$$f(x) = \sum_{n=0}^{\infty} \frac{x^n}{n!}$$

(a) What is the radius of convergence? (b) Find $f'(x)$. (c) Show that $f'(x) = f(x)$ for all x. (d) What familiar function do you think f is?

38. Set

$$f(x) = \sum_{n=0}^{\infty} \frac{(-1)^n}{(2n+1)!} x^{2n+1}$$

and

$$g(x) = \sum_{n=0}^{\infty} \frac{(-1)^n}{(2n)!} x^{2n}$$

(a) What is the radius of convergence of each? (b) Find $f'(x)$ and $g'(x)$.   (c) Show that $f'(x) = g(x)$ and $g'(x) = -f(x)$.   (d) What familiar functions do you think f and g are?

39. *(Differentiating at an endpoint).* Show that

$$\sum_{n=1}^{\infty} \frac{x^n}{n}$$

converges on $[-1, 1)$ but that term-by-term differentiation is not valid at $x = -1$.

Optional Exercise

40. Show that $\sum_{n=2}^{\infty} n^2 x^{n-2} = (4 - 3x + x^2)/(1 - x)^3$ for x in $(-1, 1)$. *Hint:* Let $f(x) = \sum_{n=0}^{\infty} x^n$ and compute $f(x) + f'(x) + f''(x)$.

11–8   TAYLOR SERIES

Given a power series in x or $x - a$, you have seen that it is sometimes possible to find a function that it represents. Now let us reverse the process. Given a function, is it possible to find a power series that represents it? More precisely, if f is a function, is there a power series $\sum c_n(x - a)^n$ such that

$$f(x) = \sum_{n=0}^{\infty} c_n(x - a)^n$$

for all x in some interval $(a - r, a + r)$?

Any function that can be represented by a power series *must be infinitely differentiable.* In fact, it was demonstrated at the end of the last section that

(11.38)

If $f(x) = \sum_{n=0}^{\infty} c_n(x - a)^n$ for x in $(a - r, a + r)$, then

$$c_n = \frac{f^{(n)}(a)}{n!}$$

and

$$f(x) = \sum_{n=0}^{\infty} \frac{f^{(n)}(a)}{n!} (x - a)^n \quad x \text{ in } (a - r, a + r)$$

This does *not* mean that every infinitely differentiable function has a power series representation. (An example is given in Exercise 36.) But it does mean that any power series representation of f (if there is one) must look like the one in (11.38). For example, $f(x) = e^x$ is infinitely differentiable and $f^{(n)}(x) = e^x$. Thus, $f^{(n)}(a) = e^a$ and *if* e^x has a power series representation, then it must be

$$(11.39) \quad e^x = \sum_{n=0}^{\infty} \frac{e^a}{n!}(x - a)^n = e^a + e^a(x - a) + \frac{e^a}{2!}(x - a)^2$$
$$+ \frac{e^a}{3!}(x - a)^3 + \cdots$$

or, in case $a = 0$,

$$(11.40) \quad e^x = \sum_{n=0}^{\infty} \frac{1}{n!}x^n = 1 + x + \frac{x^2}{2!} + \frac{x^3}{3!} + \cdots$$

We will eventually show that both of these equations are valid for all real x.

Let us begin with the following result:

TAYLOR'S FORMULA WITH REMAINDER*
Let I be an interval containing a and let f be a function with derivatives of order $n + 1$ on I. Then, for each x in I,

$$(11.41) \quad f(x) = f(a) + f'(a)(x - a) + \frac{f''(a)}{2!}(x - a)^2$$
$$+ \cdots + \frac{f^{(n)}(a)}{n!}(x - a)^n + \frac{f^{(n+1)}(c)}{(n + 1)!}(x - a)^{n+1}$$

where c is some number between a and x

The proof of Taylor's formula is given at the end of this section. The sum of the first $n + 1$ terms on the right is called the *n*th **Taylor polynomial of f about a** and is denoted by $P_n(x)$. The last term is called the *n*th **remainder** and is denoted by $R_n(x)$. Thus, a shortened version of Taylor's formula is

$$(11.42) \quad f(x) = P_n(x) + R_n(x)$$

For example, when $n = 0$,

---

*In 1715, an English mathematician, Brook Taylor (1685–1731), published this result except that his last term was

$$\frac{1}{n!} \int_a^x (x - t)^n f^{(n+1)}(t)\, dt$$

The form above was published in 1797 by Joseph Lagrange (1736–1813), one of the greatest mathematicians of the eighteenth century.

$$P_0(x) = f(a) \quad \text{and} \quad R_0(x) = f'(c)(x - a)$$

When $n = 1$,

$$P_1(x) = f(a) + f'(a)(x - a) \quad \text{and} \quad R_1(x) = \frac{f''(c)}{2}(x - a)^2$$

and so on, where c is always some point between a and x.*

For the purpose of approximating values of functions, we restate (11.42) in the following way:

(11.43) $\boxed{\; f(x) \approx P_n(x) \text{ with an error no greater than } |R_n(x)| \;}$

EXAMPLE 1   Find the nth Taylor polynomial of $f(x) = e^x$ about $a = 0$ and estimate the value of e correct to three decimal places.

Solution   To find the nth Taylor polynomial about a, we have to evaluate f and the first n derivatives at a. In this example, $f^{(n)}(x) = e^x$ and $a = 0$ so $f^{(n)}(0) = 1$. Therefore,

$$P_n(x) = 1 + x + \frac{x^2}{2!} + \cdots + \frac{x^n}{n!}$$

The remainder is

$$R_n(x) = \frac{e^c}{(n + 1)!}x^{n+1} \quad \text{where } 0 < c < x$$

To estimate the value of $e = f(1)$ correct to three places, we write

$$e = f(1) = P_n(1) + R_n(1)$$

and determine the value of n such that $|R_n(1)| < .0005$. Since $e^c < 3$ for any value of c in $(0, 1)$, we compute

$$|R_1(1)| < \frac{3}{2!} = 1.5, \; |R_2(1)| < \frac{3}{3!} = 0.5, \; |R_3(1)| < \frac{3}{4!} = 0.125$$

and continue until the answer is less than .0005. In this particular case, that first occurs when $n = 7$.

$$|R_7(1)| < \frac{3}{8!} \approx .0000744$$

---

*When $n = 0$, (11.42) becomes

$$f(x) = f(a) + f'(c)(x - a) \quad \text{or} \quad f'(c) = \frac{f(x) - f(a)}{x - a}$$

which is simply a restatement of the Mean Value Theorem. For this reason, Taylor's formula is sometimes referred to as an extended mean value theorem.

and

$$e \approx P_7(1) = 1 + 1 + \frac{1}{2!} + \frac{1}{3!} + \cdots + \frac{1}{7!} = 2.718$$

correct to three decimal places. Figure 11–24 illustrates how the Taylor polynomials approximate $y = e^x$. ∎

As Example 1 and Figure 11–24 indicate, the smaller the remainder $|R_n(x)|$, the better $P_n(x)$ approximates $f(x)$. Moreover, *if f has derivatives of all orders and $R_n(x) \to 0$ as $n \to \infty$*, it follows that

$$f(x) = \lim_{n \to \infty} P_n(x) = \sum_{n=0}^{\infty} \frac{f^{(n)}(a)}{n!}(x - a)^n$$

This series is called the **Taylor series of f about a.** If $a = 0$, then the series above reduces to

$$f(x) = \sum_{n=0}^{\infty} \frac{f^{(n)}(0)}{n!}x^n$$

and is called the **Maclaurin series of f,** named for the brilliant Scottish mathematician Colin Maclaurin (1698–1746). The following theorem sums up the discussion so far.

TAYLOR SERIES THEOREM
Suppose f has derivatives of all orders on an interval I containing the point a. Suppose further, that for each x in I, $R_n(x) \to 0$ as

(11.44) $n \to \infty$. Then f is represented by its Taylor series on I, that is,

$$f(x) = \sum_{n=0}^{\infty} \frac{f^{(n)}(a)}{n!}(x - a)^n \quad x \text{ in } I$$

Verifying that $R_n(x) \to 0$ can be troublesome. But if you can show that there is some constant M such that $|f^{(n)}(x)| \leq M$ for all n and all x in I, then

$$|R_n(x)| = \left| \frac{f^{(n+1)}(c)}{(n + 1)!}(x - a)^{n+1} \right| \leq M \frac{|x - a|^{n+1}}{(n + 1)!} \to 0$$

The last expression approaches 0 because $\sum b^n/n!$ converges for every real number b (Example 6, Section 11–6) and therefore, by the nth term test, $b^n/n! \to 0$. It now follows from (11.44) that

---

(11.45) If $|f^{(n)}(x)| \leq M$ for all n and all x in I, then f is represented by its Taylor series on I.

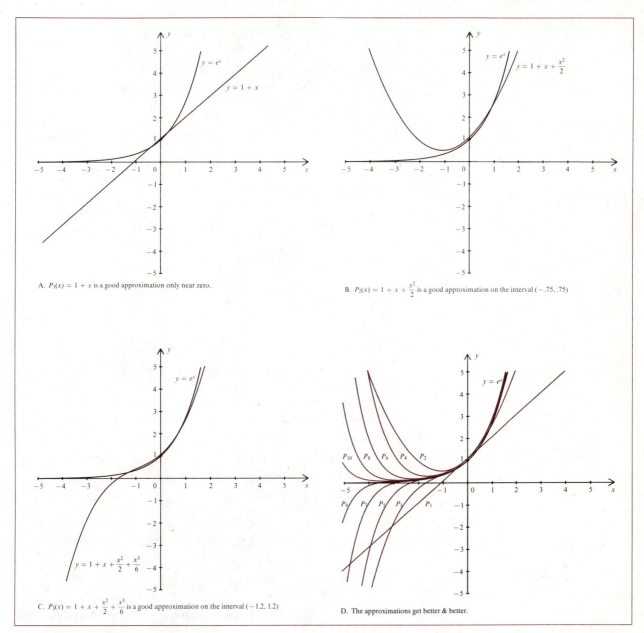

A. $P_1(x) = 1 + x$ is a good approximation only near zero.

B. $P_2(x) = 1 + x + \dfrac{x^2}{2}$ is a good approximation on the interval $(-.75, .75)$

C. $P_3(x) = 1 + x + \dfrac{x^2}{2} + \dfrac{x^3}{6}$ is a good approximation on the interval $(-1.2, 1.2)$

D. The approximations get better & better.

FIGURE 11–24. Example 1. Computer generated graphs that illustrate how the Taylor polynomials approximate $y = e^x$.

EXAMPLE 2 Show that the Taylor and Maclaurin series for $f(x) = e^x$ represents f for all real x.

Solution $f(x) = e^x$ is infinitely differentiable everywhere and $f^{(n)}(a) = e^a$. Therefore, the Taylor series of e^x about a is

$$\sum_{n=0}^{\infty} \frac{e^a}{n!}(x-a)^n = e^a + e^a(x-a) + \frac{e^a}{2!}(x-a)^2 + \cdots$$

According to (11.45), this series will represent f on an interval I containing a if we can find a constant M such that $|f^{(n)}(x)| \leq M$ for all n and all x in I. Let r be any positive number and set $I = (a - r, a + r)$ and $M = e^{a+r}$. Because e^x is an increasing function, it follows that $|f^{(n)}(x)| = e^x < e^{a+r} = M$ for all n and all x in I. Therefore, the series represents f on I. Since r is any positive number, we conclude that

(11.46)

> For any number a,
>
> $$e^x = e^a + e^a(x-a) + \frac{e^a}{2!}(x-a)^2 + \frac{e^a}{3!}(x-a)^3 + \cdots$$
>
> holds for all x. If $a = 0$, then
>
> $$e^x = 1 + x + \frac{x^2}{2!} + \frac{x^3}{3!} + \cdots$$
>
> for all x.   ■

EXAMPLE 3 Find the Maclaurin series for $f(x) = \sin x$ and show that it represents f for all real x.

Solution To write a Maclaurin series for f, you must find $f^{(n)}$ and evaluate it at 0. Thus,

$$
\begin{aligned}
f(x) &= \sin x & f(0) &= 0 \\
f'(x) &= \cos x & f'(0) &= 1 \\
f''(x) &= -\sin x & f''(0) &= 0 \\
f'''(x) &= -\cos x & f'''(0) &= -1
\end{aligned}
$$

and this cycle is repeated endlessly. It follows that the Maclaurin series for $\sin x$ is

$$\sum_{n=0}^{\infty} \frac{(-1)^n}{(2n+1)!} x^{2n+1} = x - \frac{x^3}{3!} + \frac{x^5}{5!} - \frac{x^7}{7!} + \cdots$$

Since $|f^{(n)}(x)| \leq 1$ for all n and all x, it follows from (11.45) that this series represents $\sin x$ for all x. That is,

(11.47)

$$\sin x = x - \frac{x^3}{3!} + \frac{x^5}{5!} - \frac{x^7}{7!} + \cdots \quad \text{for all } x$$

(Figure 11–25)   ■

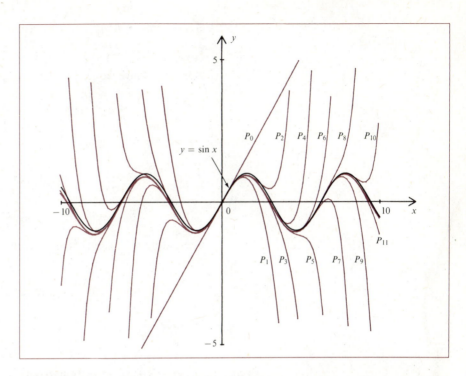

FIGURE 11–25. Example 3. Computer-generated graphs that illustrate how the Taylor polynomials approximate $y = \sin x$.

Generally, you must compute $f^{(n)}$ for all n in order to write the Taylor or Maclaurin series of f. But you can often save a lot of work by observing that *the methods of the preceding section may also be used to advantage here.*

EXAMPLE 4

(a) If x is replaced by $-x$ in the Maclaurin series for e^x, we have

$$e^{-x} = 1 - x + \frac{x^2}{2!} - \frac{x^3}{3!} + \cdots$$

(b) If the Maclaurin series for $\sin x$ (obtained in Example 3) is differentiated term-by-term, we obtain the Maclaurin series of $\cos x$,

$$\cos x = 1 - \frac{x^2}{2!} + \frac{x^4}{4!} - \frac{x^6}{6!} + \cdots$$

$$= \sum_{n=0}^{\infty} \frac{(-1)^n}{(2n)!} x^{2n} \quad \blacksquare$$

The same techniques can be used to approximate integrals; they are, in general, easier to apply than Simpson's rule.

Maclurian series for sin

$\sin x = x - \frac{x^3}{3!} + \frac{x^5}{5!} - \frac{x^7}{7!} + \dots$

EXAMPLE 5 Approximate $\int_0^{0.1} e^{-x^2}\, dx$ correct to four decimal places.

Solution Recall that e^{-x^2} has no elementary antiderivative. But if we replace x by $-x^2$ in the Maclaurin series for e^x, then

$$e^{-x^2} = 1 - x^2 + \frac{x^4}{2!} - \frac{x^6}{3!} + \cdots$$

Now we integrate term-by-term,

$$\int_0^{0.1} e^{-x^2}\, dx = x \Big|_0^{0.1} - \frac{x^3}{3}\Big|_0^{0.1} + \frac{x^5}{10}\Big|_0^{0.1} - \cdots$$

$$= \frac{1}{10} - \frac{1}{3,000} + \frac{1}{1,000,000} - \cdots$$

This is an alternating series and the third term is already less than our required error of .00005. Therefore,

$$\int_0^{0.1} e^{-x^2}\, dx \approx \frac{1}{10} - \frac{1}{3,000} \approx .0997$$

correct to four decimal places. ■

Proof of Taylor's Formula (*Optional*)

Suppose f has $n + 1$ derivatives on an interval I containing the points a and b. We want to show that

$$f(b) = P_n(b) + R_n(b) = \sum_{k=0}^{n} \frac{f^{(k)}(a)}{k!}(b - a)^k + \frac{f^{(n+1)}(c)}{(n + 1)!}(b - a)^{n+1}$$

where c is between a and b. Let g be the function defined on the interval I by

$$g(x) = f(b) - \sum_{k=0}^{n} \frac{f^{(k)}(x)}{k!}(b - x)^k - [f(b) - P_n(b)]\frac{(b - x)^{n+1}}{(b - a)^{n+1}}$$

When $x = a$, the sum in the definition of g becomes $P_n(b)$ and it follows that $g(a) = 0$. It is also clear that $g(b) = 0$. Moreover,

$$g'(x) = -\frac{f^{(n+1)}(x)}{n!}(b - x)^n + [f(b) - P_n(b)](n + 1)\frac{(b - x)^n}{(b - a)^{n+1}}$$

This can be verified using the product rule on the terms $f^{(k)}(x)(b - x)^k$ of the sum. Applying the Mean Value Theorem, we know that $g'(c) = 0$ for some c between a and b. Evaluating g' at c and equating the result to zero yields

$$f(b) - P_n(b) = \frac{f^{(n+1)}(c)}{(n + 1)!}(b - a)^{n+1}$$

If b is replaced by x, this is the same as Taylor's formula.

The table below contains the most basic Maclaurin series. The first few were obtained in the preceding section, and verification of the last one is left as an exercise.

(11.48)

Geometric:
$$\frac{1}{1-x} = \sum_{n=0}^{\infty} x^n = 1 + x + x^2 + \cdots \quad \text{on } (-1, 1)$$

Logarithmic:
$$\ln(1 + x) = \sum_{n=0}^{\infty} \frac{(-1)^n}{n+1} x^{n+1}$$
$$= x - \frac{x^2}{2} + \frac{x^3}{3} - \cdots \quad \text{on } (-1, 1)$$

Arctan:
$$\tan^{-1}x = \sum_{n=0}^{\infty} \frac{(-1)^n}{2n+1} x^{2n+1}$$
$$= x - \frac{x^3}{3} + \frac{x^5}{5} - \cdots \quad \text{on } (-1, 1)$$

Exponential:
$$e^x = \sum_{n=0}^{\infty} \frac{1}{n!} x^n = 1 + x + \frac{x^2}{2!} + \cdots \quad \text{all } x$$

Sine:
$$\sin x = \sum_{n=0}^{\infty} \frac{(-1)^n}{(2n+1)!} x^{2n+1}$$
$$= x - \frac{x^3}{3!} + \frac{x^5}{5!} - \cdots \quad \text{all } x$$

Cosine:
$$\cos x = \sum_{n=0}^{\infty} \frac{(-1)^n}{(2n)!} x^{2n} = 1 - \frac{x^2}{2!} + \frac{x^4}{4!} - \cdots \quad \text{all } x$$

Binomial:
$$(1 + x)^b$$
$$= 1 + \sum_{n=1}^{\infty} \frac{b(b-1)(b-2)\cdots(b-n+1)}{n!} x^n$$
$$= 1 + bx + \frac{b(b-1)}{2!} x^2 + \frac{b(b-1)(b-2)}{3!} x^3 + \cdots \quad \text{on } (-1, 1)$$

EXERCISES

In Exercises 1–12, find the Maclaurin series and the interval on which it represents the given function. (Whenever possible, use the methods illustrated in Example 4.)

1. $\dfrac{1}{1 + x^2}$

2. $\sin x^2$

3. $\ln(1 - x)$

4. e^{2x}

5. $\cosh x$

6. $\sinh x$

7. xe^{-x} **8.** $x^2 e^x$

9. $\sqrt{1 + x}$ (Use the binomial series in (11.48) with $b = 1/2$.)

10. $\cos^2 x$ (Use the half-angle formula $\cos^2 x = (1 + \cos 2x)/2$.)

11. $\cos 5x$ **12.** $x \sin x$

In Exercises 13–18, find the Taylor series of the given function about the indicated point a.

13. $\ln x; a = 1$ **14.** $e^x; a = 2$

15. $e^{-x}; a = -3$ **16.** $10^x; a = 1$

17. $\dfrac{1}{x}; a = 2$ **18.** $\cos x; a = \pi/4$

In Exercises 19–26, find the Taylor polynomial $P_3(x)$ for the given function about the indicated point a.

19. $\sin x; a = \pi/6$ **20.** $\sin^{-1} x; a = 1/2$

21. $\tan x; a = \pi/4$ **22.** $xe^x; a = -1$

23. $\ln \cos x; a = 0$ **24.** $e^x \tan x; a = 0$

25. $\sqrt[3]{1 + x}; a = 0$ **26.** $(1 + x)^{-3}; a = 0$

In Exercises 27–30, use power series to estimate the values.

27. $\sqrt{e}$ correct to four decimal places.

28. $\sqrt[3]{e}$ correct to five decimal places.

29. $\int_0^1 \sin x^2 \, dx$ correct to four decimal places.

30. $\int_0^1 e^{-x^2} \, dx$ correct to four decimal places.

31. Find the Taylor polynomial $P_3(x)$ of $f(x) = \sqrt{x}$ about (a) $a = 1$, (b) $a = 2$, and (c) $a = 4$.

32. Find the Taylor polynomial $P_3(x)$ of $f(x) = \sin x$ about (a) $a = 0$, (b) $a = \pi/6$, and (c) $a = \pi/3$.

33. The Maclaurin series for $\tan^{-1} x$ converges at $x = 1$. Use this series to estimate $\pi/4$ correct to 5 decimal places.

34. The function $f(x) = \tan^{-1} x$ is infinitely differentiable at every x and yet the radius of convergence of its Maclaurin series is $r = 1$. But it is possible to write a Taylor series for f about any point a. Choose a suitable point a and derive an estimate of $\tan^{-1}(3/2)$ correct to 5 decimal places.

35. (Binomial series). (a) Using the ratio test, show that

$$1 + \sum_{n=1}^{\infty} \frac{b(b - 1)(b - 2) \cdots (b - n + 1)}{n!} x^n$$

converges (absolutely) for $|x| < 1$. (b) Show that if b is a positive integer, then the series reduces to the binomial expansion $(1 + x)^b$. (c) Show that the series is the Maclaurin series of $(1 + x)^b$ for any value of b.

36. It can be shown (see Exercise 37 below) that the function

$$f(x) = \begin{cases} e^{-1/x^2} & \text{if } x \neq 0 \\ 0 & \text{if } x = 0 \end{cases}$$

has derivatives of all orders at every x and that $f^{(n)}(0) = 0$ for all n. Therefore, its Maclaurin series is just a sum of 0's. Since f is clearly not the zero function, does this contradict the results of this section? Explain.

Optional Exercise

37. For the function defined in Exercise 36, show that

$$f'(x) = \begin{cases} 2x^{-3} e^{-1/x^2} & \text{if } x \neq 0 \\ 0 & \text{if } x = 0 \end{cases}$$

(Hint: To find $f'(0)$, go back to the (limit) definition of derivative and then use L'Hôpital's rule.) With the same method, show that $f''(0) = 0$.

11–9 INFINITE SERIES AND MUSIC SYNTHESIZERS
(Optional Reading)*

In Sections 11–7 and 11–8, you learned that certain functions can be represented by power series. Another useful representation of functions is *trigonometric*,

---

*The data on music synthesizers was obtained from Herbert A. Deutsch, *Synthesis* (Port Washington, N.Y.: Alfred Publishing, 1976).

or *Fourier*, series.* If f and its first derivative have at most a finite number of finite discontinuities on the interval $[-\pi, \pi]$, then it can be shown that the series

(11.49) $\quad \dfrac{a_0}{2} + \displaystyle\sum_{n=1}^{\infty} a_n \cos nx + b_n \sin nx$

where

(11.50)
$$a_n = \frac{1}{\pi} \int_{-\pi}^{\pi} f(x) \cos nx \, dx \quad \text{for } n = 0, 1, 2, \ldots$$

$$b_n = \frac{1}{\pi} \int_{-\pi}^{\pi} f(x) \sin nx \, dx \quad \text{for } n = 1, 2, 3, \ldots$$

converges to $f(x)$ at points of continuity and to the average of the right- and left-hand limits at points of discontinuity. The series (11.49) is called the **Fourier series of** f, and the numbers a_n and b_n in (11.50) are called the **Fourier coefficients of** f.

It is an interesting exercise, which we leave for you, to verify that if $f(x)$ equals the series in (11.49), then the coefficients given in (11.50) are valid. For example, if

$$f(x) = \frac{a_0}{2} + \sum_{n=1}^{\infty} a_n \cos nx + b_n \sin nx$$

then

$$f(x) \cos mx = \frac{a_0}{2} \cos mx + \sum_{n=1}^{\infty} a_n \cos nx \cos mx + b_n \sin nx \cos mx$$

Now integrate both sides from $-\pi$ to π; and show that the right side equals πa_m (assuming that term-by-term integration is permissible).

EXAMPLE 1   Find the Fourier series for

$$f(x) = x \quad \text{for } -\pi < x < \pi$$

Solution   According to (11.50), we have

$$a_0 = \frac{1}{\pi} \int_{-\pi}^{\pi} x \, dx \quad a_n = \frac{1}{\pi} \int_{-\pi}^{\pi} x \cos nx \, dx \quad b_n = \frac{1}{\pi} \int_{-\pi}^{\pi} x \sin nx \, dx$$

Since x and $x \cos nx$ are odd functions (a function g is odd if $g(-t) = -g(t)$), it follows that all of the a's are 0. This leaves only the third integral, which we integrate by parts with $u = x$ and $dv = \sin nx \, dx$. Then

---

*Jean Fourier (1768–1830) was a brilliant French mathematician. It was in his work on heat conductivity that he showed that functions can be represented by series of sines and cosines.

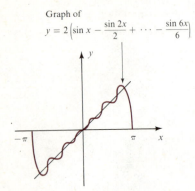

Graph of
$$y = 2\left(\sin x - \frac{\sin 2x}{2} + \cdots - \frac{\sin 6x}{6}\right)$$

FIGURE 11–26. Approximating $f(x) = x$ with the first six terms of its Fourier series.

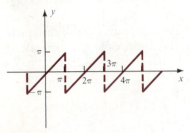

FIGURE 11–27. The function in Example 1, extended periodically, is called a *sawtooth waveform.*

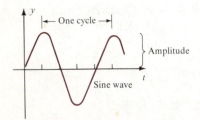

One cycle

Amplitude

Sine wave

FIGURE 11–28. Waveform produced by a tuning fork.

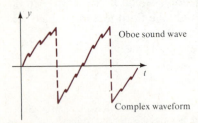

Oboe sound wave

Complex waveform

FIGURE 11–29. Waveform produced by an oboe; it is very similar to a sawtooth waveform.

$$b_n = \frac{1}{\pi} \int_{-\pi}^{\pi} x \sin nx \, dx = \frac{1}{\pi}\left[-\frac{x}{n}\cos nx \right]_{-\pi}^{\pi} - \frac{1}{\pi} \int_{-\pi}^{\pi} -\frac{1}{n}\cos nx \, dx$$

$$= \frac{1}{\pi}\left[-\frac{x}{n}\cos nx + \frac{1}{n^2}\sin nx \right]_{-\pi}^{\pi}$$

Since $\sin n\pi = 0$ and $-\cos n\pi = (-1)^{n+1}$, we have $b_n = (-1)^{n+1}2/n$. Therefore, by (11.49),

$$(11.51) \quad f(x) = 2\left(\sin x - \frac{\sin 2x}{2} + \frac{\sin 3x}{3} - \cdots \right)$$

is the Fourier series of f. Figure 11–26 shows the function f and the approximation to it using the first six terms of the series. ∎

It is important to note that sines and cosines are periodic. Therefore, a Fourier series represents a periodic function. It follows that if we extend the domain of the function in Example 1 in such a way that the extended function is periodic (Figure 11–27), then the Fourier series (11.51) will converge to the extended function at all points of continuity. At the points of discontinuity, $\pm n\pi$, the series converges to the average of the right- and left-hand limits; in this case, 0.

But why would we ever want to approximate such a simple function as $f(x) = x$ with a complicated Fourier series? And, besides, what has all of this to do with music? These are good questions; let's find the answers.

Sound waves are produced by a vibrating source that causes the air molecules near it to oscillate with simple harmonic motion. In traditional musical instruments, the vibrations are produced by some physical force such as drawing the bow across a violin string, blowing air over a reed, or plucking a guitar string. The sound produced by these vibrations is barely audible. They must be amplified either by the instrument itself (*acoustic* instruments) or electronically by an amplifier (*electric* instruments). For the first type, we actually hear the sound produced by the vibrating instrument itself. For the second type, we hear the sound produced by the vibrating cone of a loudspeaker. But in both cases, the original vibrations are caused by some physical force. In a music synthesizer, the original vibrations take the form of electronically produced sine waves.

Musical machines have a long history that can be traced back to the ancient Greeks whose *hydraulis,* a reed instrument operated by water pressure, was invented in about 300 B.C. Beethoven himself composed a work (the *Battle Symphony*) for a mechanical instrument. Our interest, however, is in synthesizers, which date from about 1965 when Robert Moog produced his now-famous *Moog Synthesizer.*

To discuss synthesizers, we need to know some basic facts about sound. Sound waves are periodic. The "purest" sound produced, say by a tuning fork, is a simple sine wave (Figure 11–28). The frequency determines the *pitch* and the amplitude determines the *loudness.* The term *complex waveform* refers to any waveform other than the sine wave; almost all sound waves are composed of complex waveforms (Figure 11–29). The complex waveform is the result of *mixing*, or combining, several waveforms of different frequencies and amplitudes. This determines the *timbre*, or quality, of the sound. Thus, the tuning fork and the oboe can produce waveforms with the same pitch but with totally different timbres.

An *oscillator* is a device that produces an alternating electric current whose frequency is determined by some controlling device. The frequency is measured in cycles per second or *hertz* (after Heinrich Hertz, a nineteenth-century German physicist). The range of an audio oscillator is usually between 20 and 20,000 Hz, which is the range of human hearing. A synthesizer consists of several oscillators that feed into a mixer. The mixer combines the various frequencies to the desired waveform. This wave is then amplified and fed to a loudspeaker which produces the sound (Figure 11–30).

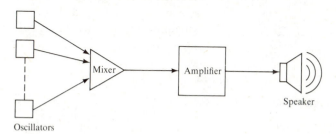

FIGURE 11–30. Simplified diagram of a synthesizer hookup.

For example, to *synthesize* an oboe sound, we set six oscillators to produce the waves

$$y = \sin t, y = -\frac{\sin 2t}{2}, y = \frac{\sin 3t}{3}, \ldots, y = -\frac{\sin 6t}{6}$$

respectively, and let the mixer add the waves to obtain the waveform

$$Y = \sin t - \frac{\sin 2t}{2} + \cdots - \frac{\sin 6t}{6}$$

Now we amplify this waveform by a factor of 2 and the result is the first six terms of the Fourier series obtained in Example 1. This wave approximates the sawtooth waveform which, in turn, is very close to the waveform of an oboe (Figure 11–29). Therefore, the speaker will produce an oboe sound.

Another interesting waveform is the *square waveform* (Figure 11–31A). It has a woody character, like a clarinet. To synthesize a square waveform, we define

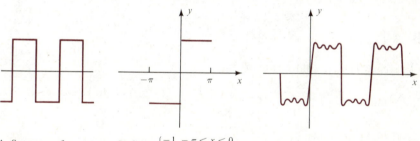

FIGURE 11–31. Synthesizing a clarinet sound.

A. Square waveform.

B. $f(x) = \begin{cases} -1 & -\pi < x < 0 \\ 1 & 0 < x < \pi \end{cases}$

C. Graph of the first four terms of the Fourier series.

$$f(x) = \begin{cases} -1 & -\pi < x < 0 \\ 1 & 0 < x < \pi \end{cases}$$

(Figure 11-31B)

Its Fourier series is (and you should check this out)

$$\frac{4}{\pi}\left(\sin x + \frac{\sin 3x}{3} + \frac{\sin 5x}{5} + \cdots\right)$$

Figure 11–31C shows an approximation using the first four terms. It follows that a clarinet sound can be synthesized using four or more oscillators.

Any given waveform, no matter how complicated, can be approximated with arbitrary accuracy by its Fourier series. Once the series is known, the sound can be *synthesized*. Modern synthesizers contain multi-functional oscillators that have as many as six basic waveforms available at the flick of a switch. But the underlying principle outlined above still applies.

REVIEW EXERCISES

For each sequence in Exercises 1–6, find the limit if it exists.

1. $\left\{\left(\frac{8}{9}\right)^n\right\}$

2. $\left\{\frac{n^2}{e^{n/2}}\right\}$

3. $\left\{\left(1 + \frac{1}{n}\right)^{3n}\right\}$

4. $\{\sec^{-1} n\pi\}$

5. $\left\{\left(\frac{1}{4}\right)^{-n/2}\right\}$

6. $\left\{\left(\frac{1}{3}\right)^{1/n}\right\}$

For each series in Exercises 7–12, determine whether it converges. If so, find the sum.

7. $\sum_{n=1}^{\infty}\left(\frac{7}{11}\right)^n$

8. $\sum_{n=0}^{\infty}\left(-\frac{5}{6}\right)^n$

9. $\sum_{k=2}^{\infty} \ln\left(\frac{k-1}{k+1}\right)$

10. $\sum_{n=1}^{\infty} \frac{2n-9}{(n+4)^2(n+5)^2}$

11. $\sum_{n=0}^{\infty}\left(\frac{7}{2}\right)^{-n}$

12. $\sum_{n=1}^{\infty} \frac{1}{10^n}$

In Exercises 13–24, test for convergence.

13. $\sum_{n=0}^{\infty} \tan(n\pi)$

14. $\sum_{n=0}^{\infty} \frac{1}{n^2+5}$

15. $\sum_{k=2}^{\infty} \frac{1}{k\ln(k^2)}$

16. $\sum_{n=1}^{\infty} n^3 e^{-2n}$

17. $\sum_{n=0}^{\infty} \frac{n+2^n}{n^2+3^n}$

18. $\sum_{n=1}^{\infty} \frac{2^n}{3^n}$

19. $\sum_{n=1}^{\infty} \frac{n^{2n}}{e^{n^2}}$

20. $\sum_{n=1}^{\infty} \frac{3^{n+4}}{e^n}$

21. $\sum_{n=1}^{\infty} \frac{2^n}{n!}$

22. $\sum_{n=1}^{\infty} \frac{(2n)!}{n^n}$

23. $\sum_{n=1}^{\infty} \frac{n^2}{(n^2+1)(n+2)}$

24. $\sum_{n=1}^{\infty} \tan\left(\frac{1}{n^2}\right)$

In Exercises 25–30, classify the series as absolutely convergent, conditionally convergent, or divergent.

25. $\sum_{n=1}^{\infty}\left(-\frac{1}{2^n}\right)^n$

26. $\sum_{n=1}^{\infty} n\left(-\frac{2}{3}\right)^n$

27. $\sum_{n=1}^{\infty}(-1)^n \frac{n^n}{(2n)!}$

28. $\sum_{n=1}^{\infty}(-1)^{n-1}\frac{e^n}{n^3}$

29. $\sum_{k=1}^{\infty}(-1)^{k-1}\frac{\sqrt{k}}{k\sqrt{k}+1}$

30. $\sum_{n=1}^{\infty} \frac{(-2)^n}{n!}$

In Exercises 31–36, find the interval of convergence.

31. $\sum_{n=1}^{\infty} \frac{n^2}{4^n}x^{2n}$

32. $\sum_{k=2}^{\infty} \frac{\ln k}{k!}x^k$

33. $\displaystyle\sum_{n=1}^{\infty} \frac{2^n}{2n}(x-1)^n$

34. $\displaystyle\sum_{n=1}^{\infty} \frac{(5x)^n}{(3n)!}$

35. $\displaystyle\sum_{n=1}^{\infty} \frac{10^n}{2n}(x+2)^n$

36. $\displaystyle\sum_{n=1}^{\infty} \frac{(2n)!}{n^n}$

In Exercises 37–42, find a function that is represented by the given series.

37. $\displaystyle\sum_{n=0}^{\infty} x^{2n+1}$

38. $\displaystyle\sum_{n=0}^{\infty} \frac{(2x)^n}{3^{2n}}$

39. $\displaystyle\sum_{n=1}^{\infty} 4nx^{4n-1}$

40. $\displaystyle\sum_{n=0}^{\infty} 3^{n+1}x^{n+1}$

41. $\displaystyle\sum_{n=1}^{\infty} n(n+1)x^{n-1}$

42. $\displaystyle\sum_{n=1}^{\infty} \frac{2^n}{n+1}x^{n+1}$

In Exercises 43–48, find the Taylor series that represents the given function.

43. $\sin^2 x$; $a = 0$

44. $\dfrac{1}{x^2}$; $a = 2$

45. 2^x; $a = 0$

46. $x \ln x$; $a = 1$

47. e^{-2x}; $a = \dfrac{1}{2}$

48. xe^x; $a = 0$

49. Express the repeating decimal $0.127127127\ldots$ as a geometric series, and find the sum.

50. A ball is dropped from a height of 20 meters. Because of friction, each time it strikes the ground it rebounds only $\frac{3}{4}$ of the previous height. Find the total distance the ball travels before it comes to rest.

51. Estimate $\sum_{n=1}^{\infty} 1/(n^4 + 1)$ correct to two decimal places.

52. Estimate $\ln(0.99)$ correct to 3 decimal places.

53. Estimate $\tan^{-1}(1.01)$ correct to 4 decimal places.

54. Approximate $\int_0^{0.5} xe^{-x^3}\, dx$ correct to 3 decimal places.

55. Approximate $\int_0^1 \cos x^2\, dx$ correct to 3 decimal places.

End O' First Year

12

CONIC SECTIONS

The conic sections are curves that can be represented as intersections of planes and cones. As Figure 12–0 shows, they are circles, parabolas, ellipses, and hyperbolas. These familiar curves have been known and used by artists, architects, and mathematicians since the time of the early Greeks. More recently, they have been applied to science and engineering. For example, the orbits of planets are ellipses, the supporting cable on a suspension bridge is in the shape of a parabola, and hyperbolas can be used to locate the source of a radio signal. Moreover, the surfaces generated by revolving conic sections about a line will be of special interest to us in our study of partial derivatives and multiple integrals.

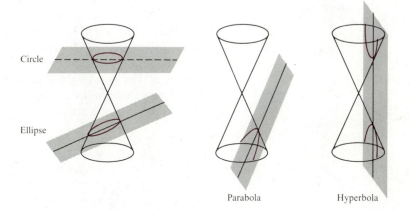

FIGURE 12–0. The intersection of various planes with a cone produces conic sections.

In Sections 2, 3, and 4, the format is as follows:

(1) The particular conic is defined in geometric terms.
(2) Using the origin as the vertex (for parabolas) or the center (for ellipses and hyperbolas), an equation of the conic is developed.
(3) The more general equation, with the vertex or center at an arbitrary point in the plane is derived using a translation of axes.

545

Since translations of axes are used throughout the chapter, this concept is discussed in Section 1. The last section treats the most general case in which the axes of the conics are not necessarily parallel to the coordinate axes.

If you are already familiar with the conic sections and need only a review of parabolas, ellipses, and hyperbolas, then you can omit this chapter and study Section 13–1 instead.

12–1 REMARKS ON TRANSLATION OF AXES

Translation of axes is an effective method of changing variables in order to find an equation of a curve. For example, it is easy to find an equation of a parabola if its vertex is at the origin, say $y = x^2$. To find an equation of a congruent parabola with its vertex at the point (h, k), we draw a new pair of coordinate axes (an x'-axis and a y'-axis) parallel to the original x- and y-axes with the new origin at (h, k). See Figure 12–1. Each point (x, y) in the original coordinate system also has coordinates (x', y') in the new system. Because the parabola is congruent to $y = x^2$ and has its vertex at the origin of the new system, it has an equation

$$y' = x'^2$$

As you will see in a moment, the change of variables by translating the axes is $y' = y - k$ and $x' = x - h$. Substitution of these values in the equation above yields

$$y - k = (x - h)^2$$

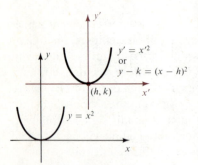

FIGURE 12–1. Using a translation of axes to find an equation of a parabola with its vertex at (h, k).

as an equation of the parabola in the original coordinate system. This method of finding equations of curves will be used throughout the chapter.

A parallel shift of one or both coordinate axes is called a **translation of axes.** Figure 12–2 shows a translation of axes; the x-axis is shifted k units upward to the new horizontal x'-axis, and the y-axis is shifted h units to the right to the new vertical y'-axis. The new origin $0'$ is located at the point whose xy-coordinates are (h, k). Every point in the plane now has two pairs of coordinates, (x, y) and (x', y'). These coordinates are related by the **equations of translation**

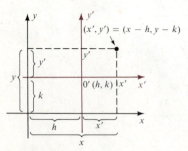

FIGURE 12–2. A translation of axes with the new origin (h, k).

$$(12.1) \quad \boxed{x' = x - h \quad \text{and} \quad y' = y - k}$$

In any translation, the x-axis is moved up or down depending on whether k is positive or negative, and the y-axis is moved right or left depending on whether h is positive or negative. Equations (12.1) hold in all cases.

EXAMPLE 1 Translate the x-axis 2 units upward and the y-axis 3 units to the left. Where is the new origin and what are the new coordinates of $(1, 2)$, $(-1, 4)$, and $(3, -1)$?

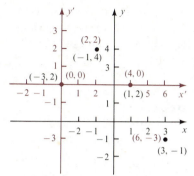

FIGURE 12–3. Example 1. The new origin is $(-3, 2)$. Each point has two sets of coordinates and $(x', y') = (x + 3, y - 2)$.

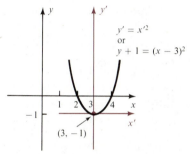

FIGURE 12–4. Example 3.

Solution   The old origin is translated 2 units up and 3 units left, so the new origin is at the point $(h, k) = (-3, 2)$. According to (12.1), with $h = -3$ and $k = 2$, $(x', y') = (x - (-3), y - 2), = (x + 3, y - 2)$. See Figure 12–3. Thus

$$(1, 2) \text{ becomes } (1 - (-3), 2 - 2) = (4, 0)$$
$$(-1, 4) \text{ becomes } (2, 2)$$
$$(3, -1) \text{ becomes } (6, -3) \quad ■$$

EXAMPLE 2   Find an equation of the circle of radius 2 and center at $(2, -2)$.

Solution   If the axes are translated so that the new origin is at $(h, k) = (2, -2)$, the equation is $x'^2 + y'^2 = 4$. The equations of this translation are $x' = x - 2$ and $y' = y - (-2) = y + 2$. Therefore,

$$(x - 2)^2 + (y + 2)^2 = 4$$

is an equation of the circle.   ■

EXAMPLE 3   Find an equation of the parabola that is congruent to $y = x^2$ and has its vertex at $(3, -1)$.

Solution   If the axes are translated so that the new origin is $(3, -1)$, the equation of the parabola is $y' = x'^2$. The equations of translations are $x' = x - 3$ and $y' = y - (-1) = y + 1$. Therefore,

$$y + 1 = (x - 3)^2$$

is an equation of the parabola (Figure 12–4).   ■

12–2   PARABOLAS

Given a line D and a point F not on D,

(12.2) | A **parabola** is the set of points in the plane that are equidistant from D and F.

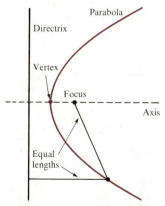

FIGURE 12–5. A parabola is the set of points equidistant from a fixed point (the focus) and a fixed line (the directrix).

The line D is called the **directrix** and the point F is called the **focus** (Figure 12–5). The line through the focus that is perpendicular to the directrix is the **axis;** notice that *a parabola is symmetric about its axis*. The point where the axis crosses the parabola is called the **vertex;** notice that the vertex is half way between the focus and directrix.

Parabolas occur in nature as the paths followed by some comets traveling through the universe. The path of a projectile fired from a gun is also a parabola if air friction is ignored. Moreover, if a light source is placed at the focus of a parabolic mirror, then the rays are reflected parallel to the axis to form a powerful beam (Figure 12–6). That is why the inside surface of automobile headlamps

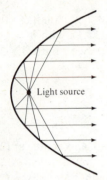

FIGURE 12–6. Rays from a light source at the focus of a parabolic mirror will be reflected as a beam of parallel rays.

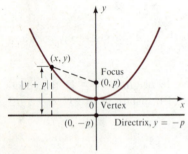

FIGURE 12–7. If the focus of a parabola is at $(0, p)$, $p > 0$, and the directrix is the horizontal line $y = -p$, then the vertex is at $(0, 0)$, and the coordinates of any point (x, y) on the parabola must satisfy the equation $x^2 = 4py$.

and searchlights are parabolic mirrors. Similarly, incoming rays parallel to the axis of a parabolic surface are reflected through the focus. That is why parabolic shapes are used in the construction of telescopes, radar scopes, and solar energy collecting devices.

Algebraic equations for parabolas can be derived from the geometric definition (12.2). A parabola has the simplest possible equation if its vertex is at the origin and its focus is on one of the coordinate axes. Suppose the focus is on the y-axis at the point $(0, p)$ with $p > 0$ and the directrix is the line $y = -p$ (Figure 12–7). If a point (x, y) is on the parabola, it means that its distance from the focus equals its distance to the directrix. Therefore,

$$\sqrt{(x - 0)^2 + (y - p)^2} = |y + p| \qquad \text{(Figure 12–7)}$$
$$(x - 0)^2 + (y - p)^2 = (y + p)^2 \qquad \text{(Square both sides)}$$
$$x^2 + y^2 - 2py + p^2 = y^2 + 2py + p^2$$
$$x^2 = 4py$$

Since these steps are reversible, it follows that $x^2 = 4py$ is an equation of the parabola in Figure 12–7. If the focus is on the y-axis below the origin or on the x-axis to the right or left of the origin, similar equations are derived. Thus, there are four possibilities for parabolas with their vertices at the origin.

(12.3)

> Any equation that can be put into one of the following forms (with $p > 0$) is an equation of a parabola with its vertex at the origin.
>
> (1) $x^2 = 4py$; focus $(0, p)$, directrix $y = -p$, opens upward.
> (2) $x^2 = -4py$; focus $(0, -p)$, directrix $y = p$, opens downward.
> (3) $y^2 = 4px$; focus $(p, 0)$, directrix $x = -p$, opens to the right.
> (4) $y^2 = -4px$; focus $(-p, 0)$, directrix $x = p$, opens to the left.

When x is the squared term, the parabola opens upward or downward.

EXAMPLE 1 Find the focus and directrix, and sketch the graph of (a) $x^2 = 4y$ and (b) $\frac{1}{2}x^2 + y = 0$.

Solution

(a) $x^2 = 4y$ is the form (1) above with $p = 1$. Therefore, the focus is at $(0, 1)$, $y = -1$ is the directrix, and the parabola opens upward. The graph can be drawn using the methods of calculus or by simply plotting a few points (Figure 12–8A).

(b) Rewrite $\frac{1}{2}x^2 + y = 0$ as $x^2 = -2y$; this is the form (2) above with $p = 1/2$. Therefore, the focus is at $(0, -1/2)$ and the directrix is $y = 1/2$. This parabola opens downward (Figure 12–8B). ∎

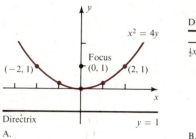

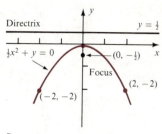

FIGURE 12–8. Example 1. If x is the squared term, then the parabola opens (A) upward or (B) downward.

When y is the squared term, the parabola will open to the right or left.

EXAMPLE 2   Find the focus and directrix, and sketch the graph of (a) $y^2 = 8x$ and (b) $y^2 + x = 0$.

Solution

(a) $y^2 = 8x$ is the form (3) above with $p = 2$. Therefore, the focus is at $(2, 0)$ and the directrix is $x = -2$. This parabola opens to the right (Figure 12–9A).

(b) Rewrite $y^2 + x = 0$ as $y^2 = -x$ which is the form (4) above with $p = 1/4$. The focus is at $(-1/4, 0)$, the directrix is $x = 1/4$, and the parabola opens to the left (Figure 12–9B).   ■

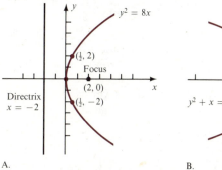

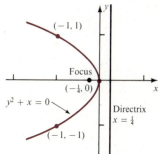

FIGURE 12–9. Example 2. If y is the squared term, then parabola opens (A) to the right or (B) to the left.

Now let us consider a more general situation. Suppose the vertex is at the point (h, k), and the directrix is parallel to the x-axis and p units below the vertex. Then, as Figure 12–10A indicates, the focus is at $(h, k + p)$, the directrix is the line $y = k - p$, and the parabola opens upward. If the axes are translated to make (h, k) the new origin (Figure 12–10B), the parabola has the equation

$$x'^2 = 4py'$$

With the equations of translation $x' = x - h$ and $y' = y - k$, we obtain

$$(x - h)^2 = 4p(y - k)$$

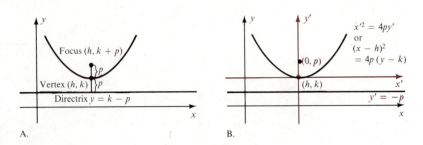

FIGURE 12–10. (A) Parabola with vertex at (h, k), directrix at distance p below the vertex and parallel to the x-axis. (B) The same parabola after the axes are translated so that (h, k) is the origin.

as an equation of the parabola in the xy-coordinates. This is called a **standard form** of an equation of a parabola. All together, there are four possible standard forms for parabolas.

(12.4)

Any equation that can be put into one of the following standard forms (with $p > 0$) is an equation of a parabola with its vertex at (h, k).

I. $(x - h)^2 = 4p(y - k)$; focus $(h, k + p)$, directrix $y = k - p$, opens upward.

II. $(x - h)^2 = -4p(y - k)$; focus $(h, k - p)$, directrix $y = k + p$, opens downward.

III. $(y - k)^2 = 4p(x - h)$; focus $(h + p, k)$, directrix $x = h - p$, opens to the right.

IV. $(y - k)^2 = -4p(x - h)$; focus $(h - p, k)$, directrix $x = h + p$, opens to the left.

EXAMPLE 3   Find the vertex, focus, and directrix, and sketch the graph of (a) $x^2 + 4x = -2y - 2$ and (b) $2y^2 - 4y = 6x - 5$.

Solution

(a) The squared term is x, so try to put the equation in standard form I or II by completing the square,

$$x^2 + 4x + 4 = -2y - 2 + 4$$
$$(x + 2)^2 = -2(y - 1)$$

This is form II with $p = 1/2$, $h = -2$, and $k = 1$. The parabola has vertex $(-2, 1)$, focus $(-2, 1/2)$, directrix $y = 3/2$, and it opens downward. Use calculus methods or plot a few points and sketch its graph (Figure 12–11A).

(b) The squared term is y, so try for type III or IV by dividing through by 2 and completing the square,

$$y^2 - 2y + 1 = 3x - \frac{5}{2} + 1$$

$$(y - 1)^2 = 3\left(x - \frac{1}{2}\right)$$

This is form III with $p = 3/4$, $h = 1/2$, and $k = 1$. Therefore, we have a parabola with vertex $(1/2, 1)$, focus $(5/4, 1)$, and directrix $x = -1/4$ (Figure 12–11B). ∎

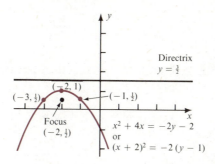

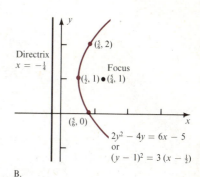

FIGURE 12–11. Example 3. Translated parabolas.

A.

B.

The methods of Examples 1, 2, and 3 can be summed up as follows:

(12.5)

If a quadratic equation

$$Ax^2 + Bx + Cy + D = 0 \quad \text{or} \quad Ax + By^2 + Cy + D = 0$$

can be put into one of the standard forms in (12.4) by completing the square, then it represents a parabola whose axis is either vertical or horizontal. The standard form indicates the vertex, focus, directrix, and how the parabola opens.*

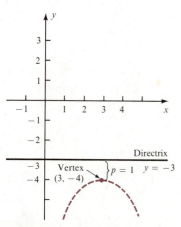

FIGURE 12–12. Example 4. Since the directrix is above the vertex, the parabola must open downward; the standard form is $(x - h)^2 = -4p(y - k)$.

Given the algebraic equation of a parabola, you have seen how to convert it into geometric information about the vertex, focus, and so on. The next two examples illustrate how to use geometric information to find an equation of a parabola; it is enough to know the coordinates of the vertex, the value of p, and how the parabola opens.

EXAMPLE 4   Find an equation of the parabola with vertex $(3, -4)$ and directrix $y = -3$.

Solution   First make a rough sketch indicating the given geometric information. Figure 12–12 shows that the directrix is parallel to the x-axis and is one

---

*Some quadratic equations in these forms do not represent parabolas. For example, $x^2 = 0$ represents a single point, $x^2 - 1 = 0$ represents a pair of vertical lines, and $x^2 + 1 = 0$ has no solutions. These are called *degenerate* cases.

unit above the vertex. Therefore, x is the squared term, $p = 1$, and the parabola opens downward. The standard form to use is

$$(x - h)^2 = -4p(y - k)$$

We know that $p = 1$, and because the coordinates of the vertex are h and k, we know that $h = 3$ and $k = -4$. Thus,

$$(x - 3)^2 = -4(y + 4)$$

is an equation of the parabola. ∎

EXAMPLE 5 Find an equation of the parabola that has a horizontal axis, vertex $(-1, 2)$, and passes through $(1, 1)$.

Solution Make a rough sketch. Figure 12–13 indicates that the parabola opens to the right. Thus,

$$(y - k)^2 = 4p(x - h)$$

is the standard form to use. We know that $h = -1$ and $k = 2$, so $(y - 2)^2 = 4p(x + 1)$. Since $(1, 1)$ is on the curve, we find $4p$ by solving the equation

$$(1 - 2)^2 = 4p(1 + 1)$$

$$4p = \frac{1}{2}$$

Therefore, $(y - 2)^2 = \frac{1}{2}(x - 1)$ is an equation of the parabola. ∎

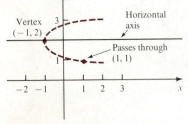

FIGURE 12–13. Example 5. The given information indicates that the parabola must open to the right; the standard form is $(y - k)^2 = 4p(x - h)$.

The methods of Examples 4 and 5 can be summed up as follows:

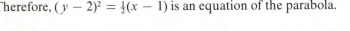

(12.6)　To find an equation of a parabola from given geometric information, draw a rough sketch and determine what standard form applies. Then use the information and the standard form to find h, k, and p.

EXERCISES

In Exercises 1–12, find the vertex, focus, and directrix, and sketch the graph.

1. $y^2 - 4x = 0$

2. $y^2 + 6x = 0$

3. $4x^2 = y$

4. $\frac{1}{2}x^2 = -3y$

5. $(x - 1)^2 = 2y$

6. $(y - 3)^2 = 2x + 4$

7. $4(y + 2)^2 + x = 0$

8. $(y - 3)^2 - 2x + 4 = 0$

9. $x^2 - 2x + 4y = 3$

10. $y^2 + 4y - x + 2 = 0$

11. $y^2 + 4y + x + 6 = 0$

12. $2x^2 + 5x - 2y = 3$

In Exercises 13–22, find an equation of the parabola with the given properties.

13. Vertex $(0, 0)$, focus $(0, 3)$.

14. Vertex $(0, 0)$, directrix $x = -2$.

15. Vertex $(-1, 4)$, directrix $y = 0$.

16. Vertex $(3, -2)$, focus $(5, -2)$.

17. Vertex $(-1, -3)$, vertical axis, passes through $(0, 4)$.

18. Vertex $(2, -1)$, horizontal axis, passes through $(5, 0)$.

19. Focus $(6, -1)$, directrix $y = 3$.

20. Focus $(-4, -2)$, directrix $x = 1$.

21. Vertical axis, parabola passes through $(0, 3), (-1, 9)$, and $(2, 3)$. *Hint:* Draw a picture, determine which standard form is appropriate, and derive three equations using the given points.

22. Horizontal axis, parabola passes through $(0, -2)$, $(1, 0)$, and $(4, 2)$.

23. Find the area of the region bounded by the parabola $y^2 = 4x - 4$ and a vertical line through its focus. What is the volume generated by revolving this region about the x-axis?

24. The region bounded by the parabola $y = 1 - x^2$ and the x-axis is revolved about the directrix of the parabola. What is the volume generated?

25. What is the arc length of the parabola $x^2 = 2y$ from $x = 0$ to $x = 2$?

26. Find an equation of the line tangent to the parabola $ay^2 + by + cx + d = 0$ at the point (x_1, y_1).

27. The reflector of a searchlight is in the shape of a parabolic mirror that is 4 ft in diameter at the opening and 2 ft deep. How far along the axis from the vertex should the light source be placed?

28. The cable between the towers of a suspension bridge forms a parabola (Figure 12–14). If the towers are 300 meters apart and 100 meters high, what is an equation of the parabola formed? (Use the axes indicated in the figure.) How much cable is needed on one side of the bridge?

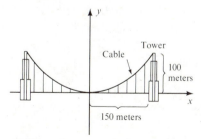

FIGURE 12–14. Exercise 28. The cable in a suspension bridge forms a parabola.

29. *(Flight path of a projectile).* A shell is fired from a gun positioned at the origin with an initial velocity v_0 ft/sec at an angle θ with the horizontal. If air friction is neglected, the position of the shell (that is, its x- and y-coordinates) at any time t is given by the equations

$$x = (v_0 \cos \theta)t \quad \text{and} \quad y = (v_0 \sin \theta)t - 16t^2$$

(We will prove this in Chapter 14.) Show that the path of the shell is a parabola by finding a single equation in x and y. (*Hint:* Solve the first equation for t and then substitute in the second equation.)

30. (Continuation of Exercise 29) At what angle should the shell be fired in order to maximize the range?

31. *(Parallel reflections).* It is a law of physics that when light is reflected, the angle of incidence equals the angle of reflection. Let $y^2 = 4px$ be the parabola cross-section of a parabolic mirror (Figure 12–15). The light from a light source at the focus $(p, 0)$ strikes the surface at point $P(x_1, y_1)$ making an angle α_1 with the tangent line L and is reflected at an angle α_2. By the laws of physics, $\alpha_1 = \alpha_2$. Find the angle α_1 (in terms of p, x_1, and y_1) and show that if $\alpha_1 = \alpha_2$, then the line of reflection L_1 is parallel to the x-axis.

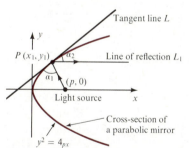

FIGURE 12–15. Exercise 31. Light from a light source at the focus is reflected parallel to the axis.

32. Use calculus to show that the x-coordinate of the vertex of the parabola $y = ax^2 + bx + c$ is $-b/2a$. Use calculus to show that the parabola opens upward or downward depending on whether $a > 0$ or $a < 0$.

33. (a) By completing the square, show that the graph of any quadratic equation of the form

$$Ax^2 + Bx + Cy + D = 0 \quad A \neq 0$$

is a parabola if $C \neq 0$. (b) What are the possibilities if $C = 0$?

Optional Exercises

34. A shell is fired with an initial velocity of 1,000 ft/sec. (see Exercise 29). At what angle should the gun be aimed to hit a target 5,000 feet away?

35. A line passing through the focus meets a parabola in the points P and Q. Show that the lines tangent to the parabola at P and Q are perpendicular to each other and intersect on the directrix.

36. Find equations of the two lines passing through $(1, -3)$ that are tangent to the parabola $y = x^2$.

12–3 ELLIPSES

Given two distinct points F_1 and F_2,

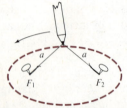

FIGURE 12–16. Construction of an ellipse. The pencil traces out the set of points, the sum of whose distances from F_1 and F_2 is $2a$.

(12.7) | An **ellipse** is the set of all points in the plane whose distances from F_1 and F_2 have a constant sum.

It is easy to construct an ellipse. On a piece of paper, insert two thumbtacks at points marked F_1 and F_2 and fasten the ends of a piece of string to the tacks. Pull the string taut with a pencil to form an isosceles triangle with equal sides of length a (Figure 12–16). Keeping the string taut, the pencil will trace out the set of points, the sum of whose distances from F_1 and F_2 is $2a$; that is, it traces out an ellipse.

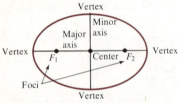

FIGURE 12–17. The parts of an ellipse. An ellipse is symmetric about its center and its axes.

The two points F_1 and F_2 are called **foci** (plural of *focus*); the point midway between the foci is the **center** (Figure 2–17). An ellipse has two perpendicular *axes;* the one through the foci is called the **major axis,** and the other passes through the center and is called the **minor axis.** There are four **vertices.** Observe that *an ellipse is symmetric about its center and its axes.*

Algebraic equations for ellipses can be derived from the geometric definition (12.7). An ellipse has the simplest equation if its center is at the origin and both foci lie on one coordinate axis. Figure 12–18 shows an ellipse with foci $(-c, 0)$ and $(c, 0)$. If a point (x, y) is on the ellipse, it means that the sum $d_1 + d_2$ is a constant greater than $2c$. Taking our cue from Figure 12–16, we set

$$d_1 + d_2 = 2a \quad \text{where } a > c$$

In terms of the distance formula,

$$\sqrt{(x + c)^2 + (y - 0)^2} + \sqrt{(x - c)^2 + (y - 0)^2} = 2a$$

or

$$\sqrt{(x + c)^2 + y^2} = 2a - \sqrt{(x - c)^2 + y^2}$$

Squaring both sides and simplifying yields

$$-a^2 + cx = -a\sqrt{(x - c)^2 + y^2}$$

Squaring both sides again and simplifying, we end up with

$$(a^2 - c^2)x^2 + a^2y^2 = a^2(a^2 - c^2)$$

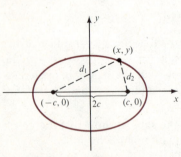

FIGURE 12–18. An ellipse with foci at $(-c, 0)$ and $(c, 0)$. The sum of the distances $d_1 + d_2 = 2a$, a constant greater than $2c$.

or

$$\frac{x^2}{a^2} + \frac{y^2}{a^2 - c^2} = 1 \qquad (a > c)$$

If we set $b^2 = a^2 - c^2$, then

$$(12.8) \qquad \frac{x^2}{a^2} + \frac{y^2}{b^2} = 1$$

The coordinates of any point on the ellipse in Figure 12–18 will satisfy this equation. It can also be shown that any point whose coordinates satisfy this equation will lie on the ellipse. Therefore, (12.8) is an equation of an ellipse with center at the origin, and foci at $(-c, 0)$ and $(c, 0)$ where $c = \sqrt{a^2 - b^2}$. If we start with the foci on the y-axis, say at $(0, -c)$ and $(0, c)$, then the same procedure yields the same equation (12.8), but the roles of a and b are reversed. In this case, the equation is

$$(12.8a) \qquad \frac{x^2}{b^2} + \frac{y^2}{a^2} = 1$$

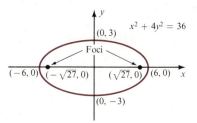

FIGURE 12–19. Example 1.

EXAMPLE 1 Sketch the graph of $x^2 + 4y^2 = 36$ and locate the foci and vertices.

Solution Divide through by 36 to rewrite the equation with a 1 on the right.

$$\frac{x^2}{36} + \frac{y^2}{9} = 1$$

This is equation (12.8) with $a = 6$ and $b = 3$. Therefore $c = \sqrt{36 - 9} = \sqrt{27}$ and the foci are at $(-\sqrt{27}, 0)$ and $(\sqrt{27}, 0)$. The vertices are where the ellipse crosses the axes (Figure 12–19). ∎

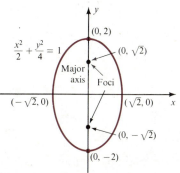

FIGURE 12–20. Example 2. Since the larger denominator is under y^2, the foci are on the y-axis.

EXAMPLE 2 The equation

$$\frac{x^2}{2} + \frac{y^2}{4} = 1$$

is equation (12.8a) with $b = \sqrt{2}$ and $a = 2$. As before, $c = \sqrt{4 - 2} = \sqrt{2}$, but this time the foci are on the y-axis at $(0, -\sqrt{2})$ and $(0, \sqrt{2})$. See Figure 12–20. ∎

Let us now consider a more general ellipse with its center at (h, k) and its major axis parallel to the x-axis. A translation of axes (Figure 12–21) to the new origin (h, k) results in the equation

$$\frac{x'^2}{a^2} + \frac{y'^2}{b^2} = 1$$

The equations of translation $x' = x - h$ and $y' = y - k$ yield

$$\frac{(x - h)^2}{a^2} + \frac{(y - k)^2}{b^2} = 1$$

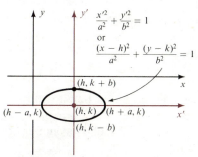

FIGURE 12–21. An ellipse with center (h, k) and major axis parallel to the x-axis.

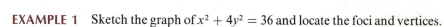

This is a **standard form** of an ellipse.

(12.9)

Any equation that can be written in one of the standard forms

$$\frac{(x-h)^2}{a^2} + \frac{(y-k)^2}{b^2} = 1 \quad \text{or} \quad \frac{(x-h)^2}{b^2} + \frac{(y-k)^2}{a^2} = 1$$

with $a > b$ is an equation of an ellipse with center (h, k) and axes parallel to the coordinate axes.

In the first case, the foci are on the horizontal axis at $(h \pm c, k)$ where $c^2 = a^2 - b^2$. The vertices are at $(h \pm a, k)$ and $(h, k \pm b)$

In the second case, the foci are on the vertical axis at $(h, k \pm c)$ where $c^2 = a^2 - b^2$. The vertices are at $(h \pm b, k)$ and $(h, k \pm a)$.

EXAMPLE 3 Show that $4x^2 + y^2 - 16x + 2y + 1 = 0$ is an ellipse and sketch its graph.

Solution To obtain the standard form, we complete the squares

$$4(x^2 - 4x \quad) + (y^2 + 2y \quad) = -1$$
$$4(x^2 - 4x + 4) + (y^2 + 2y + 1) = -1 + 16 + 1$$
$$4(x - 2)^2 + (y + 1)^2 = 16$$
$$\frac{(x - 2)^2}{4} + \frac{(y + 1)^2}{16} = 1$$

According to (12.9), this is an ellipse with center at $(2, -1)$, $b = 2$, and $a = 4$ (Figure 12–22). ■

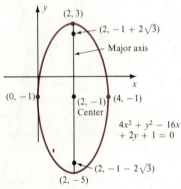

FIGURE 12–22. Example 3.

The methods of Examples 1, 2, and 3 can be summed up as follows:

(12.10)

Any quadratic equation

$$Ax^2 + By^2 + Cx + Dy + E = 0 \quad \text{with } AB > 0$$

that can be put into a standard form (12.9) represents an ellipse whose axes are parallel to the coordinate axes. The standard form indicates the center, foci, and vertices.*

Given the algebraic equation of an ellipse, you have seen how to convert it into geometric information. The next two examples illustrate how to use geometric information to find an equation of a particular ellipse; it is enough to know the coordinates of the center and the values of a and b.

---

*Some quadratic equations of this form are degenerate and do not represent ellipses. For example, $x^2 + y^2 = 0$ represents a single point and $x^2 + y^2 + 1 = 0$ has no solutions.

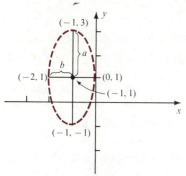

FIGURE 12-23. Example 4.

EXAMPLE 4 Find an equation of the ellipse with vertices $(0, 1)$, $(-2, 1)$, $(-1, 3)$, and $(-1, -1)$.

Solution You must find the center and the constants a and b. Make a rough sketch by plotting the vertices (Figure 12-23). This shows that the center is $(-1, 1)$ and that $b = 1$, $a = 2$. Therefore, an equation of the ellipse is

$$(x + 1)^2 + \frac{(y - 1)^2}{4} = 1 \quad \blacksquare$$

Another geometric aspect of ellipses is that of *eccentricity*.

DEFINITION
The **eccentricity** of the ellipse

(12.11) $\quad \dfrac{(x - h)^2}{a^2} + \dfrac{(y - k)^2}{b^2} = 1 \quad$ or $\quad \dfrac{(x - h)^2}{b^2} + \dfrac{(y - k)^2}{a^2} = 1$

is c/a, where $c^2 = a^2 - b^2$

The eccentricity is always between 0 and 1. Figure 12-24 shows the effect of eccentricity on the shape of the ellipse. Exercise 39 shows that the eccentricity can be used to give a focus-directrix definition of an ellipse similar to the one for parabolas.

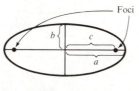

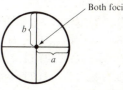

FIGURE 12-24. The eccentricity indicates the shape of the ellipse.

c/a close to 1;
b is small compared to a.

c/a close to 0;
b is almost equal to a.

$c/a = 0$; circle
$b = a$

EXAMPLE 5 Find an equation of the ellipse with foci $(1, -1)$ and $(3, -1)$, and eccentricity $3/4$.

Solution The center lies midway between the foci, so $(2, -1)$ is the center. We now use the given eccentricity to find a and b. In this example, the major axis is horizontal (why?) and the eccentricity is $c/a = 3/4$. Since c is the distance from the center to one focus, we know that $c = 1$. It follows that $a = 4/3$ and $b = \sqrt{a^2 - c^2} = \sqrt{7}/3$. Therefore,

$$\frac{9(x - 2)^2}{16} + \frac{9(y + 1)^2}{7} = 1$$

is an equation of the ellipse. $\quad \blacksquare$

EXERCISES

In Exercises 1–16, determine whether the equation represents a circle, parabola, or ellipse. In each case, find the pertinent information and sketch the graph.

1. $\dfrac{x^2}{4} + \dfrac{y^2}{4} = 1$ **2.** $4x^2 + 9y^2 = 36$

3. $6x^2 + y = 1$ **4.** $4x^2 + 4y^2 = 100$

5. $x^2 - 10x + 4y^2 = 0$ **6.** $3x - 4y^2 + 16y = 1$

7. $\dfrac{(x + 1)^2}{5} + \dfrac{(y - 4)^2}{7} = 1$

8. $\dfrac{x - 2}{5} + \dfrac{(y - 4)^2}{7} = 1$

9. $\dfrac{(x + 1)^2}{5} + \dfrac{(y - 4)^2}{5} = 1$

10. $3x^2 + y^2 - y = 4$

11. $4x^2 + 9y^2 - 16x - 18y = 0$

12. $2x^2 + y^2 + x + y = 3$

13. $x^2 + 5x + y^2 - 3y = 0$

14. $x^2 + 5x - 3y = 5$

15. $x^2 + 2y^2 + 4x + 4y + 2 = 0$

16. $5x^2 + 3y^2 + 6y + 2 = 0$

In Exercises 17–26, find an equation of each ellipse.

17. Vertices $(2, 4)$, $(2, 0)$, $(3, 2)$, and $(1, 2)$.

18. Vertices $(-6, -2)$, $(0, -2)$, $(-3, 2)$, and $(-3, -6)$.

19. Foci $(-3, 2)$, $(1, 2)$ and eccentricity $\tfrac{1}{3}$.

20. Foci $(1, -2)$, $(1, -4)$ and eccentricity $\tfrac{4}{5}$.

21. Foci $(0, 3)$, $(4, 3)$ and major axis of length 6.

22. Foci $(2, -1)$, $(2, -6)$ and major axis of length 9.

23. Center $(-3, -2)$ and three vertices $(-7, -2)$, $(-3, 0)$, and $(-3, -4)$.

24. Center $(4, 0)$ and three vertices $(4, 1)$, $(1, 0)$, and $(7, 0)$.

25. One focus $(2, 2)$ and two vertices $(2, 3)$ and $(2, -1)$.

26. One focus $(-1, 4)$ and two vertices $(-3, 4)$ and $(4, 4)$.

Exercises 27–31 refer to the ellipse

$$\frac{x^2}{a^2} + \frac{y^2}{b^2} = 1 \quad a > b$$

Note the similarity with corresponding results for circles.

27. Find the area of the region bounded by the ellipse.

28. Find the volume generated by revolving the region bounded by the ellipse about x-axis. The solid formed is shaped like a football and is called an *ellipsoid*.

29. Find the volume of a right elliptic "cone" with height h and base the ellipse above.

30. Find an equation of the line tangent to the ellipse at the point (x_1, y_1).

31. Find the dimensions of the rectangle with the largest area that can be inscribed in the ellipse.

32. Determine (approximately) the points of intersection of the graphs of the following pairs of equations.

$$\begin{cases} 4x^2 + 9y^2 = 36 \\ x - 2y = 1 \end{cases} \quad \begin{cases} 4x^2 + 9y^2 = 36 \\ 9x^2 + y^2 = 9 \end{cases}$$

33. A particle moves along a curve in such a way that its position (that is, its x- and y-coordinates) at any time t are given by the equations

$$x = a \cos t \quad y = b \sin t$$

Show that the path of the particle is the ellipse $b^2x^2 + a^2y^2 = a^2b^2$. Is the motion for $t \geq 0$ clockwise or counterclockwise?

34. (Continuation of Exercise 33). The position of one particle at time t is given by $x = 3 \cos t$, $y = 2 \sin t$; the position of a second particle at time t is given by $x = \cos t$, $y = 3 \sin t$. (a) Write equations in x and y for their paths and show that the paths intersect in four points. (b) If each particle starts its motion at $t = 0$, will they ever collide?

35. What point (x, y) on the ellipse $b^2x^2 + a^2y^2 = a^2b^2$ is closest to the focus $(c, 0)$? What point is farthest away? Justify both answers.

36. *(Earth's orbit)*. The orbit of the earth is approximately an ellipse with a major axis of about 18.52×10^7 miles and eccentricity .02. The sun is at one focus. How close does the earth get to the sun? How far away? (See Exercise 35.)

37. *(Haley's comet)*. Comets that appear periodically travel in elliptical paths. Haley's comet appears every 76 years. The minor and major axes of its path are, respectively, 85.2×10^7 and 335×10^7 miles in length. What is the eccentricity?

read

38. *(Reflections).* Show that the angles α_1 and α_2 in Figure 12–25 are equal. This means that light from one focus will be reflected through the other focus.

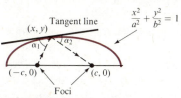

FIGURE 12–25. Exercise 38. Light from one focus is reflected through the other focus.

39. *(A focus-directrix definition of an ellipse).* Choose any numbers a and e* with $a > 0$ and $0 < e < 1$. Let F be the point $(ae, 0)$ and D be the vertical line $x = a/e$ (Figure 12–26). F is called a *focus* and D is called a *directrix.* (a) Show that the set of points $P(x, y)$ whose distances from F and D satisfy the equation

$$d(P, F) = ed(P, D)$$

that is,

$$\sqrt{(x - ae)^2 + y^2} = e\left|\frac{a}{e} - x\right|$$

is an ellipse.   (b) Show that e is the eccentricity of the ellipse.   (c) Find an equation for the set of points whose distance from $(2, 0)$ is $1/3$ of their distance from the line $x = 18$.

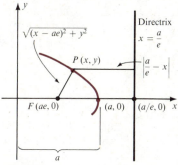

FIGURE 12–26. Exercise 39.

40. By completing the squares, show that the graph of any quadratic equation of the form

$$Ax^2 + By^2 + Cx + Dy + E = 0 \quad AB > 0$$

is an ellipse, a single point, or contains no points at all.

12–4 HYPERBOLAS

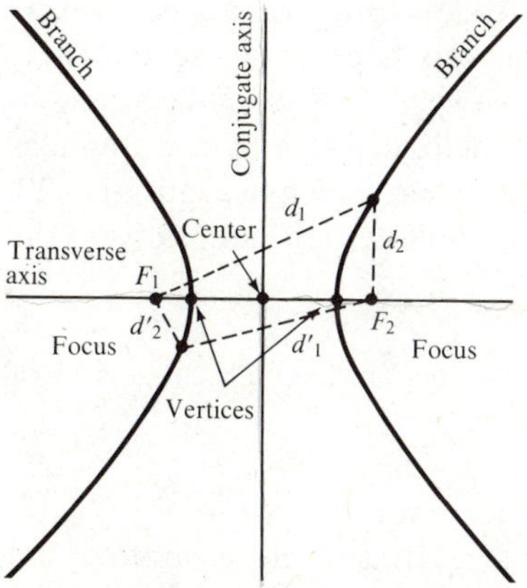

Given two distinct points F_1 and F_2,

(12.12) | A **hyperbola** is the set of all points in the plane whose distances from F_1 and F_2 have a constant positive difference.

The two points F_1 and F_2 are the **foci.** These, and the other parts of a hyperbola are labeled in Figure 12–27. It has two **branches** (unconnected parts), two **vertices,** and its **center** is midway between the foci. The line through the foci is called the **transverse axis,** and the line through the center perpendicular to the transverse axis is the **conjugate axis.** Notice that *a hyperbola is symmetric about its center and its axes.*

The paths of some comets are hyperbolic, with our sun as a focus. Hyperbolas can also be used to locate a signal source. Suppose two fixed stations report the

FIGURE 12–27. The parts of a hyperbola. Points F_1 and F_2 are the foci. The vertices are where the hyperbola crosses the transverse axis. For all points on either branch, $|d_1 - d_2|$ is a constant, which must be smaller than the distance between the foci.

---

*Not to be confused with the base of the natural logarithm.

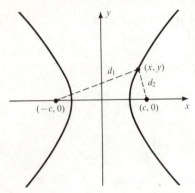

FIGURE 12–28. A hyperbola whose foci lie on the *x*-axis.

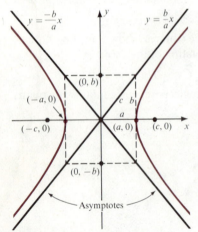

FIGURE 12–29. The hyperbola $x^2/a^2 - y^2/b^2 = 1$. The foci are at $(\pm c, 0)$, where $c^2 = a^2 + b^2$; the vertices are at $(\pm a, 0)$. The straight lines $y = \pm(b/a)x$ are asymptotes of the hyperbola.

exact time a signal is received. Then the difference in time multiplied by the speed of the signal is the difference of the distances from the source to the stations. Thus, the source lies on a hyperbola with the two stations as foci. If a third station also reports, the source can be pinpointed at the intersection of two hyperbolas.

The equation of a hyperbola is simplest if the foci are on a coordinate axis, say, at $(-c, 0)$ and $(c, 0)$; the origin is then the center of the hyperbola (Figure 12–28). For every point (x, y) on either branch, the number $|d_1 - d_2|$ is a constant which must be less than $2c$. Again, it is convenient to denote this constant by $2a$; notice that $a < c$. Thus, $|d_1 - d_2| = 2a$ and in terms of the distance formula,

$$\left| \sqrt{(x + c)^2 + (y - 0)^2} - \sqrt{(x - c)^2 + (y - 0)^2} \right| = 2a$$

After some algebraic manipulations similar to those we used to find the equation of an ellipse, the final equation is

$$\frac{x^2}{a^2} - \frac{y^2}{c^2 - a^2} = 1$$

Now set $b^2 = c^2 - a^2$, to get

$$(12.13) \qquad \frac{x^2}{a^2} - \frac{y^2}{b^2} = 1$$

The coordinates of any point on the hyperbola in Figure 12–28 will satisfy this equation. It can also be shown that any point whose coordinates satisfy this equation will lie on the hyperbola. Therefore, (12.13) is an equation of a hyperbola with center at the origin and foci $(-c, 0)$ and $(c, 0)$, where $c^2 = a^2 + b^2$. The vertices are where the branches cross the transverse axis (the *x*-axis in this case). Set $y = 0$ to find the vertices at $(\pm a, 0)$; notice that $|x| \geq a$. See Figure 12–29.

Figure 12–29 also indicates that the two lines $y = \pm(b/a)x$ are *asymptotes* of the hyperbola; in the first quadrant, when x is very large, y is close to $(b/a)x$; in the second quadrant, when $-x$ is very large, y is very close to $-(b/a)x$; and so on. (See Exercise 35.) The asymptotes are a big help in sketching the graph, as is the box from $x = -a$ to $x = a$ and $y = -b$ to $y = b$ shown in Figure 12–29. The hyperbola stays outside this box, and the asymptotes are extensions of the diagonals of the box.

EXAMPLE 1 Find the foci, vertices, and asymptotes, and sketch the graph of $x^2/4 - y^2/9 = 1$.

Solution Because $a = 2$ and $b = 3$, you know that $c = \sqrt{a^2 + b^2} = \sqrt{13}$. Thus, the foci are $(\pm \sqrt{13}, 0)$, the vertices are $(\pm 2, 0)$, and the asymptotes are the lines $y = \pm \frac{3}{2}x$. The graph is shown in Figure 12–30. ∎

A hyperbola also has a simple equation if the foci are on the *y*-axis, say at $(0, -c)$ and $(0, c)$. This interchanges the roles of *x* and *y*; and the resultant equation is

$$(12.13a) \qquad \frac{y^2}{a^2} - \frac{x^2}{b^2} = 1$$

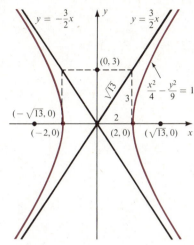

FIGURE 12–30. Example 1.

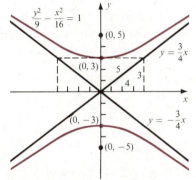

FIGURE 12–31. Example 2.

The relationship $b^2 = c^2 - a^2$ still holds; the vertices are now at $(0, \pm a)$; the asymptotes are the lines $y = \pm(a/b)x$.

EXAMPLE 2 Find the foci, vertices, and asymptotes, and sketch the graph of $16y^2 - 9x^2 = 144$.

Solution Rewrite the equation as

$$\frac{y^2}{9} - \frac{x^2}{16} = 1$$

Then $a = 3, b = 4$, and $c = \sqrt{a^2 + b^2} = 5$. Because the positive term contains y, the foci are on the y-axis at $(0, \pm c)$. Thus, the foci are at $(0, \pm 5)$, the vertices are at $(0, \pm 3)$, and the asymptotes are the lines $y = \pm\frac{3}{4}x$. (Figure 12–31). ∎

Let us now consider a more general hyperbola with its center at (h, k) and its axes parallel to the coordinate axes. Just as in the case of parabolas and ellipses, a translation of axes yields the equations

$$\frac{(x-h)^2}{a^2} - \frac{(y-k)^2}{b^2} = 1 \quad \text{or} \quad \frac{(y-k)^2}{a^2} - \frac{(x-h)^2}{b^2} = 1$$

These are the **standard equations** of hyperbolas.

(12.14)

> Any equation that can be put into one of the following standard forms is an equation of a hyperbola with a center at (h, k).
>
> (1) $(x - h)^2/a^2 - (y - k)^2/b^2 = 1$; transverse axis is horizontal, foci are at $(h \pm c, k)$ where $c^2 = a^2 + b^2$, and vertices are at $(h \pm a, k)$.
>
> (2) $(y - k)^2/a^2 - (x - h)^2/b^2 = 1$; transverse axis is vertical, foci are at $(h, k \pm c)$ where $c^2 = a^2 + b^2$, and vertices are at $(h, k \pm a)$.
>
> The asymptotes are $y = \pm(b/a)(x - h) + k$ or $y = \pm(a/b)(x - h) + k$

EXAMPLE 3 Find the center, foci, and asymptotes, and sketch the graph of $2x^2 - 4x - 4y^2 - 16y - 6 = 0$.

Solution Complete the squares:

$$2(x^2 - 2x \quad) - 4(y^2 + 4y \quad) = 6$$
$$2(x^2 - 2x + 1) - 4(y^2 + 4y + 4) = 6 + 2 - 16$$
$$2(x - 1)^2 - 4(y + 2)^2 = -8$$
$$-\frac{(x - 1)^2}{4} + \frac{(y + 2)^2}{2} = 1$$

The center is at $(1, -2)$; because the positive term contains y, the transverse axis is vertical. Now $a = \sqrt{2}, b = 2$, and $c = \sqrt{6}$. The foci are at $(1, -2 \pm \sqrt{6})$, the vertices are at $(1, -2 \pm \sqrt{2})$, and the asymptotes are $y = \pm(\sqrt{2}/2)(x - 1) - 2$. The graph is shown in Figure 12–32.

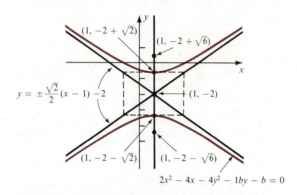

FIGURE 12–32. Example 3.

Notice that, in the original equation, the positive squared term contained x, which you might have thought would place the transverse axis in a horizontal position. However, in the end this was not so; first appearances can be deceiving. ■

The methods of Examples 1, 2, and 3 can be summed up as follows:

(12.15)

> Any quadratic equation
> $$Ax^2 + By^2 + Cx + Dy + E = 0 \text{ with } AB < 0$$
> that can be put into one of the standard forms (12.14) represents a hyperbola whose axes are parallel to the coordinate axes. The standard form indicates the center, foci, vertices, and asymptotes.*

Given the algebraic equation of a hyperbola, you have seen how to convert it into geometric information. The next two examples illustrate how to use geometric information to find an equation of a hyperbola; it is enough to know the coordinates of the center, the values of a and b, and whether the transverse axis is horizontal or vertical.

EXAMPLE 4 Find an equation of the hyperbola with a vertex at $(2, 0)$ and asymptotes $3y + 2x = 7$ and $3y - 2x = -1$.

Solution The asymptotes give two pieces of information: they meet at the center and their slopes are $\pm b/a$ or $\pm a/b$. Solve the two equations

---

*A degenerate case is $x^2 - y^2 = 0$; the graph is two lines $x = \pm y$.

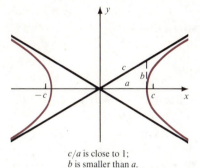

c/a is close to 1;
b is smaller than a.

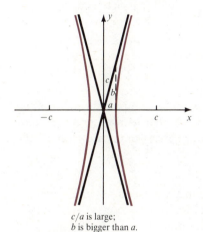

c/a is large;
b is bigger than a.

FIGURE 12–33. The effect of eccentricity on the shape of a hyperbola.

$$3y + 2x = 7$$
$$3y - 2x = -1$$

simultaneously to find the center at (2, 1). Because the center is at (2, 1) and a vertex is at (2, 0), it follows (draw a picture) that the transverse axis is vertical and that $a = 1$. Finally, notice that the slopes of the asymptotes are $\pm a/b = \pm \frac{2}{3}$ which implies that $b = 3/2$, because $a = 1$. Thus,

$$(y - 1)^2 - \frac{4(x - 2)^2}{9} = 1$$

is an equation of the hyperbola. ■

Analogous to an ellipse, a hyperbola also has an eccentricity.

(12.16) | **DEFINITION**
The **eccentricity** of a hyperbola is c/a.

The eccentricity of a hyperbola is greater than 1. Figure 12–33 shows the effect of eccentricity on the shape of a hyperbola. Exercise 32 shows that the eccentricity can be used to give a focus-directrix definition of a hyperbola similar to the ones for parabolas and ellipses.

EXAMPLE 5 Find an equation of the hyperbola with foci $(-1, 6)$ and $(5, 6)$ and eccentricity 2.

Solution The foci lie on a horizontal line and are 6 units apart. This tells you three things: the transverse axis is horizontal, the center is at (2, 6); and $c = 3$. Because the eccentricity $c/a = 2$, you know that $a = 3/2$. Finally, $b^2 = c^2 - a^2 = \frac{27}{4}$. It follows that an equation for this hyperbola is

$$\frac{4(x - 2)^2}{9} - \frac{4(y - 6)^2}{27} = 1 \qquad ■$$

EXAMPLE 6 *(Locating a signal source).*   Radio (or any electromagnetic) waves travel at the speed of light, 3×10^8 meters/sec. A radio signal is received by one station located at $A(-4, 0)$ and 5 microseconds (1 microsecond = 10^{-6} seconds) later by another station located at $B(4, 0)$. If the units of the coordinate system are in kilometers, describe the possible location of the sending station.

Solution The *difference* in the distances from the sending station to stations A and B is

$$(3 \times 10^8)(5 \times 10^{-6}) = 15 \times 10^2 \text{ meters}$$
$$= 1.5 \text{ kilometers}$$

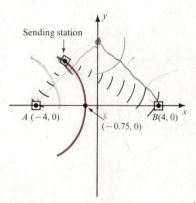

Sending station

$A(-4, 0)$ $(-0.75, 0)$ $B(4, 0)$

FIGURE 12–34. Example 6.

$\frac{x^2}{.5625} - \frac{y^2}{15.4325} - 1 = 0$

$c^2 = b^2 + a^2$
$16 = b^2 + .5625$
$b^2 = 15.4375$

Therefore, the sending station must lie somewhere on the branch nearest station A of the hyperbola with foci $(\pm 4, 0)$, center at the origin, and vertices at $(\pm a, 0)$ where $a = 1.5/2 = 0.75$ (Figure 12–34). ■

EXERCISES

In Exercises 1–16, determine whether the equation represents a circle, parabola, ellipse, or hyperbola. In each case, find the pertinent information and sketch the graph.

1. $\dfrac{x^2}{4} - \dfrac{y^2}{4} = 1$

2. $\dfrac{x^2}{4} + \dfrac{y^2}{4} = 1$

3. $\dfrac{y^2}{4} - \dfrac{x^2}{9} = 1$

4. $-\dfrac{y^2}{4} - \dfrac{x}{9} + 1 = 0$

5. $-3x^2 - 4y^2 + 12 = 0$

6. $3x^2 - 4y^2 = 1$

7. $\dfrac{(x + 1)^2}{16} - \dfrac{(y - 2)^2}{9} = 1$

8. $\dfrac{(x + 1)^2}{16} + \dfrac{(y - 2)^2}{9} = 1$

9. $9(x + 1) - 16(y - 2)^2 = 144$

10. $8(x + 1)^2 + 8(y - 2)^2 = 200$

11. $-2x^2 - y^2 - 4x + 2y = -3$

12. $2x^2 - y^2 + x + y = 3$

13. $y^2 - 2x^2 + 8x + 4y = 0$

14. $2x^2 - y + x = 3$

15. $x^2 - 4x - 3y^2 + 6y = -4$

16. $4x^2 + 8x - y^2 = 32$

In Exercises 17–24, find an equation of each hyperbola.

17. Foci $(\pm 4, 0)$ and vertices $(\pm 1, 0)$.

18. Foci $(0, \pm 3)$ and vertices $(0, \pm 2)$.

19. Foci $(1, \pm 3)$ and eccentricity $\frac{3}{2}$.

20. Foci $(2, 0)$ and $(2, 6)$ and eccentricity 4.

21. A vertex $(-3, 2)$ and asymptotes $y = 3x + 5$, $3x + y + 1 = 0$.

22. A vertex $(0, -1)$ and asymptotes $y + 4 = \pm \frac{1}{2}x$.

23. A focus $(0, -1)$ and asymptotes $y - 3 = \pm 2x$.

24. A focus $(0, -1)$ and asymptotes $y = 4x + 7$, $4x + y = -9$.

25. Find the area of the region bounded by the hyperbola $x^2 - y^2 = 4$ and the vertical line through its focus $(2\sqrt{2}, 0)$.

26. Find the volume generated by revolving the region in Exercise 25 about the x-axis.

27. *(Locating a sending source).* Station B receives a radio signal 100 microseconds after station A. If the stations are 50 kilometers apart, describe the possible location of the source. See Example 6.

28. (Continuation of Exercise 27). Station B is able to fix the direction of the incoming signal. It reports that the angle between the lines connecting B with the source and B with A is 45°. Find the location of the source.

29. *(Hyperbolic sine and cosine).* A particle moves along a curve in such a way that its position (that is, its x- and y-coordinates) at any time t are given by the equations

$$x = (a \cosh t) + h \quad \text{and} \quad y = (b \sinh t) + k$$

Find an equation in x and y that describes the path of the particle. (*Hint:* First write $(x - h)/a$ and $(y - k)/b$ in terms of t.)

30. Show that the line tangent to the hyperbola $bx^2 - ay^2 = a^2b^2$ at the point $P(x_1, y_1)$ has an equation $b^2xx_1 - a^2yy_1 = a^2b^2$.

31. *(Reflections).* Show that the tangent line to the hyperbola $b^2x^2 - a^2y^2 = a^2b^2$ at $P(x_1, y_1)$ bisects the angle formed by the lines drawn from P to the foci (Figure 12–35). Then justify the statement that

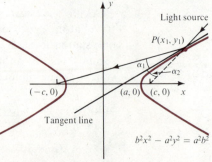

FIGURE 12–35. Exercise 31. Light directed toward one focus is reflected through the other focus.

light from outside of a hyperbola that is directed toward one focus is reflected through the other focus.

32. *(A focus-directrix definition of a hyperbola).* Choose any numbers a and e with $a > 0$ and $e > 1$. Let F be the point $(ae, 0)$ and D be the vertical line $x = a/e$ (Figure 12–36). F is called a *focus* and D is called a *directrix.*   (a) Show that the set of points $P(x, y)$ whose distances from F and D satisfy the equation

$$d(P, F) = ed(P, D)$$

that is,

$$\sqrt{(x - ae)^2 + y^2} = e\left|\frac{a}{e} - x\right| \quad \text{(Figure 12–36)}$$

is a hyperbola.   (b) Show that e is the eccentricity of the hyperbola.   (c) Find an equation for the set of points whose distance from $(6, 0)$ is 3 times their distance from the line $x = 2/3$.

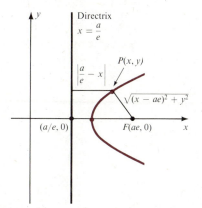

FIGURE 12–36. Exercise 32.

33. By completing the squares, show that the graph of any equation of the form

$$Ax^2 + By^2 + Cx + Dy + E = 0 \quad AB < 0$$

is a hyperbola or a pair of intersecting lines.

34. *(Conjugate hyperbolas).* Consider the equation

$$\frac{x^2}{a^2} - \frac{y^2}{b^2} = E$$

If $E \neq 0$, the graph is a hyperbola; its transverse axis is horizontal if $E > 0$ and vertical if $E < 0$. This pair of hyperbolas are called *conjugate hyperbolas.* (a) Sketch both hyperbolas for $E = \pm 1$ on the same set of coordinate axes.   (b) What is the graph if $E = 0$?

35. *(Asymptotes).* Show that $y = bx/a$ is an asymptote to the hyperbola $b^2x^2 - a^2y^2 = a^2b^2$ in the first quadrant by showing that if (x, y) is on the hyperbola with $x > 0, y > 0$, then

$$\lim_{x \to \infty}\left|\frac{bx}{a} - y\right| = 0$$

12–5   ROTATION OF AXES

You have seen how a translation of axes is used to simplify equations. Here we discuss another useful tool, **rotation of axes.**

When axes are rotated, the origin is kept fixed and the coordinate axes are rotated counterclockwise through some angle θ. This establishes new coordinate axes; the x'-axis and y'-axis (Figure 12–37). Each point P in the plane now has two sets of cordinates, (x, y) and (x', y'). To see the relationship between them, let r be the distance from P to the origin (the same in both coordinate systems). Figure 12–38A indicates that

$$x = r\cos(\alpha + \theta) \quad \text{and} \quad y = r\sin(\alpha + \theta)$$

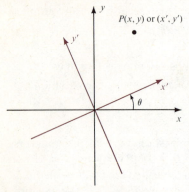

FIGURE 12–37. A rotation of axes through an angle θ.

Using the addition formulas for sine and cosine, we have

$$x = r \cos \alpha \cos \theta - r \sin \alpha \sin \theta$$
$$y = r \cos \alpha \sin \theta + r \sin \alpha \cos \theta$$

Figure 12–38B indicates that $r \cos \alpha = x'$ and $r \sin \alpha = y'$. Substituting these values in the equations above yields

(2.17) $$x = x' \cos \theta - y' \sin \theta \quad \text{and} \quad y = x' \sin \theta + y' \cos \theta$$

These equations produce the xy-coordinates if you know the $x'y'$-coordinates. To reverse the procedure, we solve (2.17) simultaneously for x' and y'. The result is

(2.18) $$x' = x \cos \theta + y \sin \theta \quad \text{and} \quad y' = -x \sin \theta + y \cos \theta$$

Formulas (2.17) and (2.18) are called the **equations of rotation.**

Given an equation in terms of x and y, we can apply (2.17) and *transform* it into an equation in x' and y'. You will see the advantages of doing this in a moment, but first, we present an example.

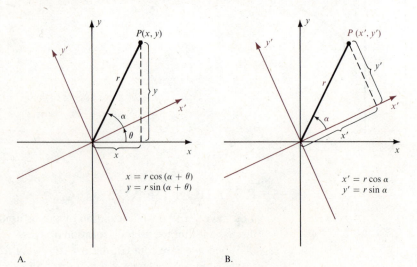

FIGURE 12–38. Establishing a relationship between the xy- and $x'y'$-coordinates.

A.

B.

EXAMPLE 1 Rotate the axes through an angle of $\pi/4$, transform the equation $xy = 1$ into an equation in x' and y', and sketch the graph.

Solution   We use (2.17) with $\theta = \pi/4$. Then $\sin \theta = \cos \theta = \sqrt{2}/2$, and

$$x = x'\left(\frac{\sqrt{2}}{2}\right) - y'\left(\frac{\sqrt{2}}{2}\right) = \frac{\sqrt{2}}{2}(x' - y')$$

$$y = x'\left(\frac{\sqrt{2}}{2}\right) + y'\left(\frac{\sqrt{2}}{2}\right) = \frac{\sqrt{2}}{2}(x' + y')$$

Since the equation we want to transform is $xy = 1$, we have

$$\left[\frac{\sqrt{2}}{2}(x' - y')\right]\left[\frac{\sqrt{2}}{2}(x' + y')\right] = 1$$

which reduces to

$$\frac{x'^2}{2} - \frac{y'^2}{2} = 1$$

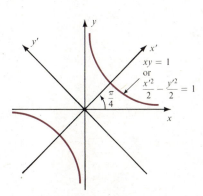

FIGURE 12–39. Example 1. The rotation of $\pi/4$ transforms $xy = 1$ into a standard form of a hyperbola.

This is the *standard form of a hyperbola* with center at the origin and vertices $(\pm \sqrt{2}, 0)$ on the x'-axis (Figure 12–39).   ■

Example 1 illustrates how a rotation can be used to transform an equation involving an xy-term into one that does not. Let us investigate this in more detail.

Eliminating the xy-Term

The most general quadratic equation in two variables is

$$(12.19) \quad Ax^2 + Bxy + Cy^2 + Dx + Ey + F = 0$$

If $B = 0$, the methods of the preceding sections can be used to graph and analyze the equation. If $B \neq 0$, then (12.19) can *always* be transformed by rotation into a quadratic equation in x' and y' that does not contain an $x'y'$-term.

To prove this, we suppose that the x and y axes are rotated through an angle θ. If the equations of rotation (12.17) are used to substitute for x and y, then the general quadratic (12.19) becomes

$$(12.20) \quad A'x'^2 + B'x'y' + C'y'^2 + D'x' + E' + F' = 0$$

where the coefficient B' of the $x'y'$-term is

$$(12.21) \quad \begin{aligned} B' &= 2(C - A)\sin \theta \cos \theta + B(\cos^2\theta - \sin^2\theta) \\ &= (C - A)\sin 2\theta + B \cos 2\theta \end{aligned}$$

Checking the first equality in (12.21) is left as an exercise; the second equality follows from the double-angle formulas (1.55). To eliminate the $x'y'$-term from (12.20), we must find an angle θ that makes $B' = 0$. It follows from (12.21) that $B' = 0$ when

$$(12.22) \quad \cot 2\theta = \frac{A - C}{B} \quad \text{or} \quad 2\theta = \cot^{-1}\frac{A - C}{B}$$

Since we are assuming that $B \neq 0$, there will always be such an angle. Notice that $0 < 2\theta < \pi$ (the range of $\cot^{-1}$), so $0 < \theta < \pi/2$.

We are now equipped to analyze even the most general quadratic equations. If there is an xy-term present, then there are several steps to follow.

(12.23)

To analyze and graph an equation

$$Ax^2 + Bxy + Cy^2 + Dx + Ey + F = 0 \quad \text{with } B \neq 0$$

(1) Determine the rotation that makes $B' = 0$ using

$$\cot 2\theta = \frac{A - C}{B}$$

(2) Find $\sin \theta$ and $\cos \theta$. If necessary, use (1) to find $\cos 2\theta$, and then apply the half-angle formulas

$$\sin \theta = \sqrt{\frac{1 - \cos 2\theta}{2}} \quad \text{and} \quad \cos \theta = \sqrt{\frac{1 + \cos 2\theta}{2}} \quad (0 < \theta < \pi/2)$$

(3) Make the substitutions (12.17), multiply out, and simplify. This equation will not contain an $x'y'$ term.

(4) Employ the methods of the preceding sections.

EXAMPLE 2 Analyze and sketch the graph of $8x^2 - 4xy + 5y^2 = 36$.

Solution We follow the steps in (12.23).

(1) In this equation, $A = 8$, $B = -4$, and $C = 5$. Therefore

$$\cot 2\theta = \frac{A - C}{B} = -\frac{3}{4}$$

(2) It follows (Figure 12–40A) that $\cos 2\theta = -3/5$; so

$$\sin \theta = \sqrt{\frac{1 - (-3/5)}{2}} = 2/\sqrt{5}$$

$$\cos \theta = \sqrt{\frac{1 + (-3/5)}{2}} = 1/\sqrt{5}$$

(3) From equations (12.17),

$$x = \frac{x' - 2y'}{\sqrt{5}} \quad \text{and} \quad y = \frac{2x' + y'}{\sqrt{5}}$$

Therefore, the given equation becomes

$$8\left(\frac{x' - 2y'}{\sqrt{5}}\right)^2 - 4\left(\frac{x' - 2y'}{\sqrt{5}}\right)\left(\frac{2x' + y'}{\sqrt{5}}\right) + 5\left(\frac{2x' + y'}{\sqrt{5}}\right)^2 = 36$$

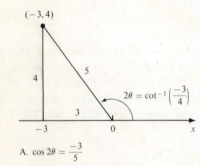

A. $\cos 2\theta = \dfrac{-3}{5}$

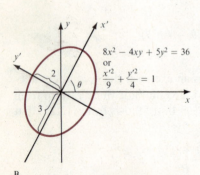

B.
FIGURE 12–40. Example 2. Rotation

$8x^2 - 4xy + 5y^2 = 36$
or
$\dfrac{x'^2}{9} + \dfrac{y'^2}{4} = 1$

When this is multiplied out and like terms are collected, we have

$$4x'^2 + 9y'^2 = 36$$

(4) Put in standard form,

$$\frac{x'^2}{9} + \frac{y'^2}{4} = 1$$

is the equation of an ellipse with center at the origin and major axis on the x'-axis; the x'-axis makes an angle of

$$\theta = \frac{1}{2}\cot^{-1}\left(-\frac{3}{4}\right) \approx 1.1 \text{ radians or } 63°$$

with the x-axis. The curve is sketched in Figure 12–40B.   ■

The next example combines rotation and translation.

EXAMPLE 3   Analyze and sketch the graph of $x^2 + 2xy + y^2 + 8\sqrt{2}x = 0$.

Solution

(1) In this equation, $A = C = 1$; so $(A - C)/B = 0$, $2\theta = \cot^{-1}0 = \pi/2$, and $\theta = \pi/4$.

(2) $\sin(\pi/4) = \cos(\pi/4) = \sqrt{2}/2$

(3) By (12.17),

$$x = \frac{\sqrt{2}}{2}(x' - y') \quad \text{and} \quad y = \frac{\sqrt{2}}{2}(x' + y')$$

Substitute these values into the given equation to obtain

$$\frac{1}{2}(x' - y')^2 + (x' - y')(x' + y') + \frac{1}{2}(x' + y')^2 + 8(x' - y') = 0$$

which simplifies to

$$2x'^2 + 8x' - 8y' = 0$$

(4) Completing the square yields

$$(x' + 2)^2 = 4(y' + 1)$$

This is the standard form of a parabola whose vertex is at $(-2, -1)$ and opens upward *in the $x'y'$-coordinate system*. The graph is sketched in Figure 12–41.   ■

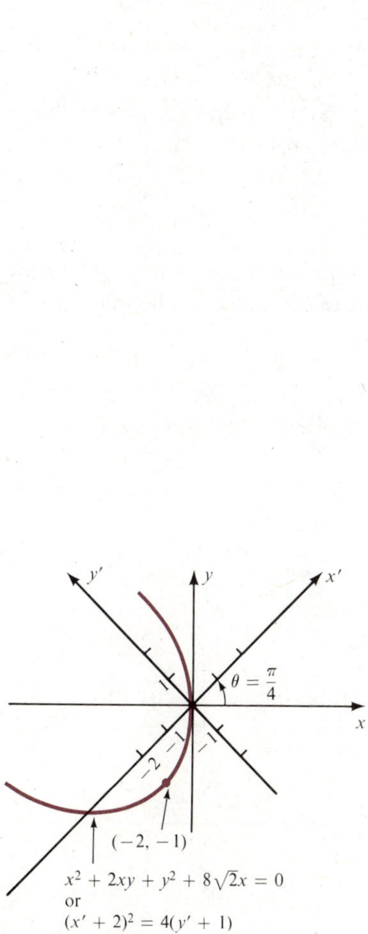

$x^2 + 2xy + y^2 + 8\sqrt{2}x = 0$
or
$(x' + 2)^2 = 4(y' + 1)$

FIGURE 12–41. Example 3. Rotation and translation.

We conclude with some observations that lead to a neat description of conic sections. The general quadratic

$$Ax^2 + Bxy + Cy^2 + Dx + Ey + F = 0$$

can be transformed by a rotation into a new quadratic

$$A'x'^2 + B'x'y' + C'y'^2 + D'x' + E'y' + F' = 0$$

where $B' = 0$ if the angle of rotation is properly chosen. Once $B' = 0$, it follows from our work in the preceding sections that (except for degenerate cases) the original equation represents

    (1) a parabola if $A' = 0$ or $C' = 0$; that is, $A'C' = 0$

    (2) an ellipse or circle if A' and C' agree in sign; that is, $A'C' > 0$

    (3) a hyperbola if A' and C' differ in sign; that is, $A'C' < 0$

In the general form of a quadratic, we call the quantity $B^2 - 4AC$ the *discriminant*. After rotation, the discriminant is $B'^2 - 4A'C'$ which reduces to $-4A'C'$ if $B' = 0$. Thus, it follows from (1), (2), and (3) above, that the shape of the curve is determined by the discriminant after a rotation that makes $B' = 0$. But it can be shown (Exercise 24) that

$$B^2 - 4AC = B'^2 - 4A'C'$$

after any rotation; that is, the discriminant is *invariant* under rotations. It now follows from the discussion above, that the shape of the curve can be determined by the *original* discriminant.

(12.24)

> Except for degenerate cases, the graph of any quadratic equation
>
> $$Ax^2 + Bxy + Cy^2 + Dx + Ey + F = 0$$
>
> is
>
> (1) a parabola if $B^2 - 4AC = 0$
> (2) an ellipse or circle if $B^2 - 4AC < 0$
> (3) a hyperbola if $B^2 - 4AC > 0$
>
> Moreover, every conic section is the graph of such an equation.

EXERCISES

In Exercises 1–22, analyze and sketch the graph (if there is one) of each equation. Use (12.24) as a check.

1. $5x^2 - 6xy + 5y^2 = 32$

2. $2x^2 + 3xy + 2y^2 = 9$

3. $3x^2 - 10xy + 3y^2 = -32$

4. $7x^2 + 12xy - 2y^2 = 7$

5. $x^2 - 2xy + y^2 + 1 = 0$

6. $16x^2 - 24xy + 9y^2 = 25$

7. $x^2 - 2xy + y^2 - 3x = 0$

8. $x^2 - 4xy + 4y^2 = 0$

9. $13x^2 - 10xy + 13y^2 = 72$

10. $5x^2 - 8xy + 5y^2 = 9$

11. $xy = 5$ **12.** $xy = -3$

13. $3x^2 + 10xy + 3y^2 - 2x - 14y - 5 = 0$

14. $4x^2 - 8xy - 2y^2 + 20x - 4y + 15 = 0$

15. $4x^2 + 4xy + y^2 - 24x + 38y - 139 = 0$

16. $16x^2 - 24xy + 9y^2 + 56x - 42y + 49 = 0$

17. $x^2 - 3xy + y^2 + 10x + 10y + 10 = 0$
18. $x^2 - 4xy + 9y^2 - 6y = 0$
19. $x^2 - xy + y^2 - x - y = 20$
20. $x^2 + xy + y^2 + x + y = 20$
21. $5x^2 + 12xy = 4$
22. $x^2 + 2xy + y^2 - x + y = 0$

---

23. Show that after a rotation of axes through an angle θ, the coefficients in (12.20) are related to the coefficients in (12.19) as follows

$\checkmark A' = A \cos^2\theta + B \cos\theta \sin\theta + C \sin^2\theta$
$B' = 2(C - A)\sin\theta \cos\theta + B(\cos^2\theta - \sin^2\theta)$
$\checkmark C' = A \sin^2\theta - B \cos\theta \sin\theta + C \cos^2\theta$
$\checkmark D' = D \cos\theta + E \sin\theta$
$E' = E \cos\theta - D \sin\theta$
$F' = F$

24. Use the results of Exercise 23 to show that the discriminant $B^2 - 4AC$ is invariant under rotations; that is, $B^2 - 4AC = B'^2 - 4A'C'$.

REVIEW EXERCISES

In Exercises 1–10, determine whether the equation represents a circle, parabola, ellipse, or hyperbola, and sketch the graph.

1. $\dfrac{x^2}{4} + \dfrac{y^2}{9} = 1$ **2.** $x^2 = 1 - \dfrac{y^2}{4}$

3. $y^2 - 4x = 4$ **4.** $x^2 = y - 4x$

5. $2x^2 + 2y^2 = 8$
6. $2x^2 + y^2 + 4x + 4y = 0$
7. $4x^2 - 9y^2 - 16x + 18y = 0$
8. $x^2 + 4x + 2y + y^2 - 1 = 0$

9. $x^2 = -1 + \dfrac{y^2}{4}$

10. $2x^2 + y^2 - 2y = 3$

In Exercises 11–18, determine whether the equation represents a circle, parabola, ellipse, or hyperbola, and sketch the graph.

11. $4x^2 - 6xy + 4y^2 - 32 = 0$
12. $3x^2 - 8xy + 3y^2 - 16 = 0$
13. $x^2 - 2xy + y^2 - 4y = 0$ **14.** $xy = 4$
15. $4x^2 + 4xy + y^2 - 2x + 4y = 0$
16. $x^2 - xy + y^2 - 3x - 3y = 10$
17. $5y^2 - 12xy - 4 = 0$
18. $25x^2 - 20xy + 4y^2 = 36$

In Exercises 19–22, find an equation of the parabola with the given properties.

19. Vertex $(1, 2)$, focus $(3, 2)$.
20. Vertex $(-2, 4)$, directrix $y = 2$.
21. Vertex $(-1, -5)$, vertical axis, passes through $(1, 3)$.
22. Focus $(5, -1)$, directrix $y = 2$.

In Exercises 23–26, find an equation of an ellipse with the given properties.

23. Vertices $(5, 4)$, $(-3, 4)$, $(1, 6)$, $(1, 2)$.
24. Foci $(-4, 3)$, $(2, 3)$, and eccentricity $\frac{1}{4}$.
25. Center $(-2, 3)$, and three vertices $(1, 3)$, $(-5, 3)$, and $(-2, 4)$.
26. One focus $(4, 1)$, and two vertices $(5, 1)$, $(-3, 1)$.

In Exercises 27–30, find an equation of a hyperbola with the given properties.

27. Foci $(\pm 5, 0)$, and vertices $(\pm 2, 0)$.
28. Foci $(1, 3)$ and $(1, 9)$, and eccentricity $\frac{3}{2}$.
29. A vertex $(-1, 2)$, and asymptotes $y = \pm 3x + 2$.
30. A focus $(0, -2)$, and asymptotes $y = \pm 2x$.
31. In a Cassegrain telescope, two reflective surfaces are positioned so that incoming rays of light are directed as shown in Figure 12–42. What shapes are the surfaces and how are they positioned relative to one another?

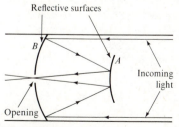

FIGURE 12–42. Exercise 31. A Cassegrain telescope.

13

POLAR COORDINATES AND PARAMETRIC EQUATIONS

In this chapter we extend our study of curves in the plane by introducing two new concepts: polar coordinates and parametric equations. An immediate consequence is that a large collection of new and interesting curves becomes available to us. Figure 13–0 shows three examples. A far-reaching consequence is that we are laying the foundation for our subsequent study of vector functions and the calculus of several variables.

In the optional section, entitled *Brachistochrones and Tautochrones,* we combine optics, physics, and parametric equations to find the solutions of two famous problems. It is worthwhile reading.

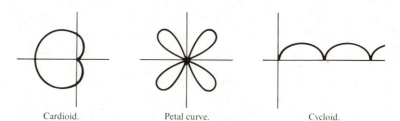

FIGURE 13–0. Examples of curves obtained using polar coordinates and parametric equations.

Cardioid.   Petal curve.   Cycloid.

573

13–1 PARABOLAS, ELLIPSES, AND HYPERBOLAS*

The curves in the title, along with circles, are called *conic sections* because they are the intersections of a plane and a cone (Figure 13–1). This section is a brief review of the properties of conic sections that are needed in later work. A full account of each conic section is contained in the preceding chapter.

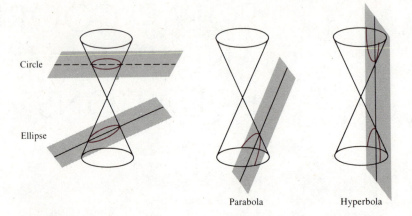

FIGURE 13–1. The intersection of various planes with a cone produces conic sections.

Circle

Ellipse

Parabola

Hyperbola

Definitions

Although parabolas, ellipses, and hyperbolas are the intersections of planes and cones, it is more convenient to study these curves from the point of view of the following definitions:

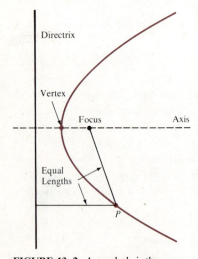

FIGURE 13–2. A parabola is the curve consisting of all points that are equidistant from a fixed point, called the focus, and a fixed line, called the directrix.

Directrix

Vertex

Focus   Axis

Equal Lengths

P

(13.1)   Given a line D and a point F not on D, a **parabola** is the set of all points in the plane that are equidistant from D and F.

Figure 13–2 shows a parabola along with the names of its parts.

(13.2)   Given two distinct points F_1 and F_2, an **ellipse** is the set of all points in the plane whose distances from F_1 and F_2 have a constant sum.

Figure 13–3 shows an ellipse along with the names of its parts.

(13.3)   Given two distinct points F_1 and F_2, a **hyperbola** is the set of all points in the plane whose distances from F_1 and F_2 have a constant positive difference.

*Those who have studied Chapter 12 can omit this section.

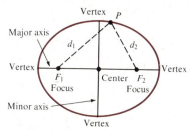

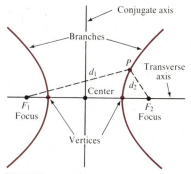

FIGURE 13–3. An ellipse with foci F_1 and F_2. The sum $d_1 + d_2$ is constant for all points on the ellipse.

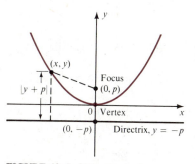

FIGURE 13–4. A hyperbola with foci F_1 and F_2. For all points on either branch, $|d_1 - d_2|$ is a constant.

Figure 13–4 shows a hyperbola along with the names of its parts.

It is sometimes convenient to consider the following definitions of the conic sections: Let F be a point, D be a line, and e be any positive number.* If $d(P, F)$ and $d(P, D)$ represent the distances from a point P to F and to D, respectively, then the set of all points P in the plane with

$$d(P, F) = e[d(P, D)]$$

is

a parabola	if $e = 1$
an ellipse	if $0 < e < 1$
a hyperbola	if $e > 1$

The number e is called the **eccentricity** of the conic section.

Equations

Equations of conic sections are derived directly from definitions (13.1)–(13.3) and the distance formula. For example, if the focus of a parabola is at $(0, p)$ and its directrix is the line $y = -p$, then a point (x, y) is on the parabola means that

$$\underbrace{\sqrt{(x - 0)^2 + (y - p)^2}}_{\substack{\text{distance from} \\ (x, y) \text{ to } (0, p)}} = \underbrace{|y + p|}_{\substack{\text{distance from} \\ (x, y) \text{ to line } y = -p}} \qquad \text{(see Figure 13–5)}$$

After squaring both sides, this equation reduces to

$$x^2 = 4py$$

Similar calculations result in equations such as

$$\frac{x^2}{a^2} + \frac{y^2}{b^2} = 1 \quad \text{and} \quad \frac{x^2}{a^2} - \frac{y^2}{b^2} = 1$$

for ellipses and hyperbolas, respectively.

Standard Forms

Each of the equations above is a quadratic equation in x and y. If we start with a quadratic equation

$$Ax^2 + By^2 + Cx + Dy + E = 0$$

then, by completing the squares, it is usually possible to write the quadratic in one of the *standard forms* below. For parabolas, we have

FIGURE 13–5. If the focus of a parabola is at $(0, p)$, $p > 0$, and the directrix is the horizontal line $y = -p$, then the vertex is at $(0, 0)$, and the coordinates of any point (x, y) on the parabola must satisfy the equation $x^2 = 4py$.

---

*This letter e is not to be confused with the base of the natural logarithm.

Any equation that can be put into one of the following standard forms (with $p > 0$) is an equation of a parabola with its vertex at (h, k):

(1) $(x - h)^2 = 4p(y - k)$; focus $(h, k + p)$, directrix $y = k - p$, opens upward.

(2) $(x - h)^2 = -4p(y - k)$; focus $(h, k - p)$, directrix $y = k + p$, opens downward.

(3) $(y - k)^2 = 4p(x - h)$; focus $(h + p, k)$, directrix $x = h - p$, opens to the right.

(13.4)

(4) $(y - k)^2 = -4p(x - h)$; focus $(h - p, k)$, directrix $x = h + p$, opens to the left.

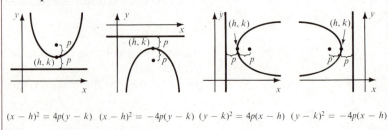

$(x - h)^2 = 4p(y - k)$ $(x - h)^2 = -4p(y - k)$ $(y - k)^2 = 4p(x - h)$ $(y - k)^2 = -4p(x - h)$

The corresponding data for ellipses is

Any equation that can be put into one of the standard forms

$$\frac{(x - h)^2}{a^2} + \frac{(y - k)^2}{b^2} = 1 \quad \text{or} \quad \frac{(x - h)^2}{b^2} + \frac{(y - k)^2}{a^2} = 1$$

with $a > b$ is an equation of an ellipse with center (h, k) and axes parallel to the coordinate axes.

In the first case, the foci are on the horizontal axis at $(h \pm c, k)$ where $c^2 = a^2 - b^2$. The vertices are at $(h \pm a, k)$ and $(h, k \pm b)$.

In the second case, the foci are on the vertical axis at $(h, k \pm c)$ where $c^2 = a^2 - b^2$. The vertices are at $(h \pm b, k)$ and $(h, k \pm a)$.

(13.5)

$$\frac{(x - h)^2}{a^2} + \frac{(y - k)^2}{b^2} = 1 \qquad \frac{(x - h)^2}{b^2} + \frac{(y - k)^2}{a^2} = 1$$

For hyperbolas, the situation is

Any equation that can be put into one of the following standard forms is an equation of a hyperbola with its center at (h, k):

(1) $\dfrac{(x - h)^2}{a^2} - \dfrac{(y - k)^2}{b^2} = 1$

Transverse axis is horizontal, foci are at $(h \pm c, k)$ where $c^2 = a^2 + b^2$, and vertices at $(h \pm a, k)$. The lines

$$y = \pm \frac{b}{a}(x - h) + k$$

are asymptotes.

(2) $\dfrac{(y - k)^2}{a^2} - \dfrac{(x - h)^2}{b^2} = 1$

Transverse axis is vertical, foci are at $(h, k \pm c)$ where $c^2 = a^2 + b^2$, and vertices at $(h, k \pm a)$. The lines

$$y = \pm \frac{a}{b}(x - h) + k$$

are asymptotes.

(13.6)

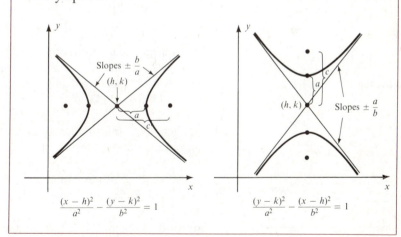

Some quadratic equations do not represent conic sections. For example, the graph of $x^2 + y^2 = 0$ is a single point $(0, 0)$, the graph of $x^2 = y^2$ is the intersecting lines $x = y$ and $x = -y$, and $x^2 + y^2 + 1 = 0$ has no graph at all. These are called *degenerate* cases. Aside from these, the graph of a quadratic equation is a conic section. To analyze and sketch the graph, complete the squares and match the result with one of the standard forms above.

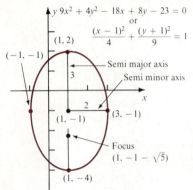

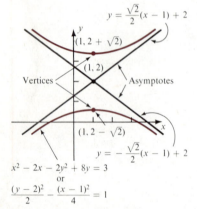

FIGURE 13–6. Example 1.

EXAMPLE 1 Analyze and sketch the graph of $9x^2 + 4y^2 - 18x + 8y - 23 = 0$.

Solution First, complete the squares

$$9(x^2 - 2x + \quad) + 4(y^2 + 2y + \quad) = 23$$
$$9(x^2 - 2x + 1) + 4(y^2 + 2y + 1) = 23 + 9 + 4$$
$$9(x - 1)^2 + 4(y + 1)^2 = 36$$

Now divide through by 36 to obtain a 1 on the right.

$$\frac{(x - 1)^2}{4} + \frac{(y + 1)^2}{9} = 1$$

This is the standard form of an ellipse (13.5.2) with $h = 1, k = -1, a = 3$, and $b = 2$. The major axis is vertical, the foci are at $(1, -1 \pm \sqrt{5})$, and the vertices are at $(1, 2), (1, -4), (-1, -1)$, and $(3, -1)$. See Figure 13–6. ■

EXAMPLE 2 Analyze and sketch the graph of $x^2 - 2x - 2y^2 + 8y = 3$.

Solution First complete the squares.

$$(x^2 - 2x + 1) - 2(y^2 - 4y + 4) = 3 + 1 - 8$$
$$(x - 1)^2 - 2(y - 2)^2 = -4$$

and then divide through by -4 to obtain a 1 on the right.

$$\frac{(y - 2)^2}{2} - \frac{(x - 1)^2}{4} = 1$$

This is the standard form of a hyperbola (13.6.2) with $h = 1, k = 2, a = \sqrt{2}$, and $b = 2$. The transverse axis is vertical, the vertices are at $(1, 2 \pm \sqrt{2})$, the foci are at $(1, 2 \pm \sqrt{6})$, and the asymptotes are $y = \pm(\sqrt{2}/2)(x - 1) + 2$ (Figure 13–7.) ■

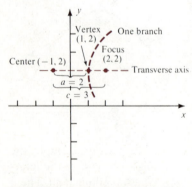

FIGURE 13–7. Example 2.

If geometric information is given, then an equation is found by finding the appropriate standard form and fitting the given information into it.

EXAMPLE 3 Find an equation of the hyperbola with center at $(-1, 2)$, one vertex at $(1, 2)$, and one foucs at $(2, 2)$.

Solution First decide what standard form to use; a picture usually helps. Figure 13–8 suggests that the transverse axis is horizontal so

$$\frac{(x - h)^2}{a^2} - \frac{(y - k)^2}{b^2} = 1 \tag{13.6.1}$$

is the standard form to use. We know $h = -1, k = 2$, and $a = 2$, so all that remains is to determine b. Since $b^2 = c^2 - a^2$ and $c = 3$, we have $b^2 = 9 - 4 = 5$.

FIGURE 13–8. Example 3. Given this data, find an equation of the hyperbola.

Thus

$$\frac{(x + 1)^2}{4} - \frac{(y - 2)^2}{5} = 1$$

is an equation of the hyperbola. ■

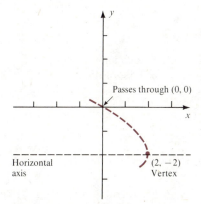

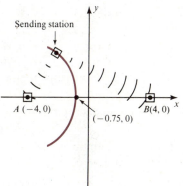

FIGURE 13–9. Example 4. Given this data, find an equation of the parabola.

EXAMPLE 4   Find an equation of the parabola with its vertex at $(2, -2)$, a horizontal axis, and passing through the origin.

Solution   Draw a picture indicating the given data (Figure 13–9). The figure indicates that the parabola opens to the left so that

$$(y - k)^2 = -4p(x - h) \tag{13.4.4}$$

is the appropriate standard form. We know that $h = 2$ and $k = -2$ and that $(0, 0)$ is on the parabola. It follows that

$$(0 + 2)^2 = -4p(0 - 2) \quad \text{or} \quad p = \frac{1}{2}$$

Therefore, $(y + 2)^2 = -2(x - 2)$ is an equation of the parabola. ■

EXAMPLE 5 *(Locating a signal source).*   Radio (or any electromagnetic) waves travel at the speed of light, 3×10^8 meters/sec. A radio signal is received by one station located at $A(-4, 0)$ and 5 microseconds (1 microsecond = 10^{-6} seconds) later by another station located at $B(4, 0)$. If the units of the coordinate system are in kilometers, describe the possible location of the sending station.

Solution   The *difference* in the distances from the sending station to stations A and B is

$$(3 \times 10^8)(5 \times 10^{-6}) = 15 \times 10^2 \text{ meters}$$
$$= 1.5 \text{ kilometers}$$

Therefore, the sending station must lie somewhere on the branch nearest station A of the hyperbola with foci $(\pm 4, 0)$, center at the origin, and vertices at $(\pm a, 0)$ where $a = 1.5/2 = .75$ (Figure 13–10). ■

FIGURE 13–10. Example 5.

General Quadratics

The most general quadratic equation in two variables is

$$Ax^2 + Bxy + Cy^2 + Dx + Ey + F = 0$$

If B, the coefficient of the xy-term is not zero, then a *rotation of axes* (see Section 12–5) can be used to produce an equivalent equation without an xy-term. Once this is done, the methods described above are sufficient to analyze and sketch the graph of the equation. Although we will not discuss rotation of axes here, it can be shown that

Except for degenerate cases, the graph of a quadratic equation

$$Ax^2 + Bxy + Cy^2 + Dx + Ey + F = 0$$

is

(13.7)

(1) a parabola if $B^2 - 4AC = 0$

(2) an ellipse if $B^2 - 4AC < 0$

(3) a hyperbola if $B^2 - 4AC > 0$

Moreover, every conic section is the graph of such an equation. (The quantity $B^2 - 4AC$ is called the *discriminant* of the equation.)

EXERCISES

In Exercises 1–4, graph the parabola, and label the important parts.

1. $y = x^2 + 4x$ **2.** $2x^2 + 5y - 3x + 4 = 0$

3. $y^2 + x + y = 0$ **4.** $x = y^2 - 2y + 1$

In Exercises 5–8, graph the ellipses and label the important parts.

5. $3x^2 + 4y^2 = 12$ **6.** $4x^2 + y^2 = 16$

7. $9x^2 + 5y^2 - 20y - 25 = 0$

8. $2x^2 + 3y^2 + 16x - 6y + 29 = 0$

In Exercises 9–12, graph the hyperbolas and label the important parts.

9. $9y^2 - 4x^2 = 36$ **10.** $x^2 - 4y^2 = 4$

11. $9(x + 1)^2 - 16(y - 2)^2 = 144$

12. $y^2 - 2x^2 + 8x + 4y = 0$

In Exercises 13–16, use (13.7) to name the conic section.

13. $5x^2 - 6xy + 5y^2 = 32$

14. $3x^2 - 10xy + 3y^2 + 32 = 0$

15. $x^2 - 2xy + y^2 - 3x = 0$

16. $2y^2 + \sqrt{3}xy - x^2 - 2 = 0$

In Exercises 17–22, find an equation of the described curve.

17. A parabola with its vertex at $(0, 0)$ and focus at $(0, 3)$.

18. A parabola with its vertex at $(-1, -3)$, a vertical axis, and passing through $(0, 4)$.

19. An ellipse with vertices $(2, 4), (2, 0), (3, 2)$, and $(1, 2)$.

20. An ellipse with one focus $(2, 2)$ and two vertices $(2, 3)$ and $(2, -1)$.

21. A hyperbola with foci $(\pm 4, 0)$ and vertices $(\pm 1, 0)$.

22. A hyperbola with vertices $(0, \pm 3)$ and passing through $(1, 4)$.

---

23. Find the arc length of the parabola $x^2 = 2y$ from $x = 0$ to $x = 2$.

24. A region is bounded by the hyperbola $x^2 - y^2 = 4$ and a vertical line through one focus $(2\sqrt{2}, 0)$. Find the volume generated by revolving this region about the x-axis.

25. Find the area of the region bounded by the ellipse $b^2x^2 + a^2y^2 = a^2b^2$.

26. Find the volume of the right elliptic "cone" with height h and base the ellipse in Exercise 25.

27. (*Locating a sending source*). Station B receives a radio signal 100 microseconds after Station A receives the same signal. If the stations are 50 kilometers apart, describe the possible location of the sending source.

28. A particle moves along a curve in such a way that its position (that is, its x- and y-coordinates) at any time t are given by the equations

$$x = [a \cos t] + h \quad y = [b \sin t] + k$$

Describe the path of the particle. (*Hint:* First solve for $(x - h)/a$ and $(y - k)/b$.) Is the motion clockwise or counterclockwise?

29. (Continuation of Exercise 28). What is the path if the position at time t is

$$x = [a \cosh t] + h \quad y = [b \sinh t] + k$$

30. What point (x, y) on the ellipse $b^2x^2 + a^2y^2 = a^2b^2$ is closest to the focus $(c, 0)$? What point is farthest away? Justify both answers.

31. (Earth's orbit). The orbit of the earth is approximately an ellipse with a major axis of about 18.52×10^7 miles and eccentricity .02. The sun is at one focus. How close does the earth get to the sun? How far away? (The eccentricity is the distance from the center to one focus divided by distance from the center to the corresponding vertex.)

32. (Flight path of a projectile). A shell is fired from a gun positioned at the origin with an initial velocity v_0 ft/sec at an angle θ with the horizontal. If air friction is neglected, the x- and y-coordinates of the shell at time t are given by

$$x = (v_0 \cos \theta)t \quad y = (v_0 \sin \theta)t - 16t^2$$

(We will prove this in the next chapter.) Describe the flight path of the shell.

---

33. Show that an equation of the line tangent to (a) the parabola $x^2 = 4py$ at the point (x_1, y_1) is $x_1x - 2p(y + y_1) = 0$, (b) the ellipse $b^2x^2 + a^2y^2 = a^2b^2$ at (x_1, y_1) is $b^2xx_1 + a^2yy_1 = a^2b^2$, and (c) the hyperbola $b^2x^2 - a^2y^2 = a^2b^2$ at (x_1, y_1) is $b^2xx_1 - a^2yy_1 = a^2b^2$.

Reflective properties of conic sections

34. Show that angles α_1 and α_2 in Figure 13-11 are equal. This means that light coming from the focus will be reflected parallel to the axis, making a powerful beam. That is why reflectors in automobile

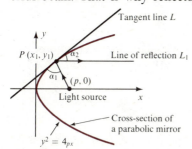

FIGURE 13-11. Exercise 34. Light from the focus is reflected parallel to the axis.

headlamps and searchlights are made in the shape of parabolic mirrors. (Also see Figure 12-6.)

35. Show that angles α_1 and α_2 in Figure 13-12 are equal. This means that light from one focus will be reflected through the other focus.

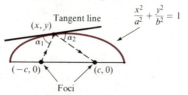

FIGURE 13-12. Exercise 35. Light from one focus is reflected through the other focus.

36. Show that angles α_1 and α_2 in Figure 13-13 are equal. This means that light from outside the hyperbola that is directed toward one focus is reflected through the other focus.

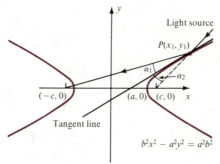

FIGURE 13-13. Exercise 36. Light directed to one focus is reflected through the other focus.

37. In a Cassegrain telescope, two reflective surfaces are positioned so that incoming rays of light are directed as shown in Figure 13-14. What shapes are the surfaces and how are they positioned relative to one another? (See Exercises 34-36 above.)

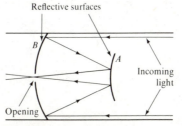

FIGURE 13-14. Exercise 37. A Cassegrain telescope.

13-2   POLAR COORDINATES

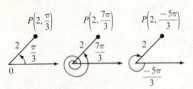

FIGURE 13-15. Each point has many different polar coordinates.

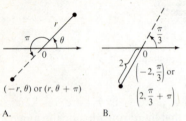

A.   B.

FIGURE 13-16. The point $(-r, \theta)$ is understood to coincide with the point $(r, \theta + \pi)$.

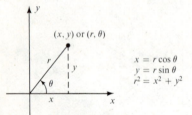

FIGURE 13-17. The relationships between polar and rectangular coordinates.

It is often convenient to locate points P in the plane with *polar coordinates*. A line segment OP is drawn from the origin to P. The length of OP is denoted by r, and the angle that OP makes with the positive x-axis is denoted by θ. The ordered pair (r, θ) represents the **polar coordinates** of P. The angle θ is measured in degrees or radians; it is positive when measured counterclockwise and negative otherwise. Observe that polar coordinates are not unique. For example, $(2, \pi/3)$, $(2, 7\pi/3)$, and $(2, -5\pi/3)$ all represent the same point (Figure 13-15). Moreover, we also allow r to be negative, where it is understood that $(-r, \theta)$ represents the same point as $(r, \theta + \pi)$ as shown in Figure 13-16A. For instance, to locate the point $(-2, \pi/3)$, you measure off $\pi/3$ radians and then go *backward* 2 units (Figure 13-16B).

The relationships between the polar and rectangular coordinates of a point are expressed by the equations

$$(13.8) \quad \boxed{\begin{array}{c} x = r \cos \theta \quad y = r \sin \theta \\ x^2 + y^2 = r^2 \end{array}}$$

See Figure 13-17. These equations hold for all values of r and θ.

Polar Equations and their Graphs

A polar equation relates r and θ. The graph of the equation is the set of all points (r, θ) whose coordinates, in some form, satisfy the equation.

EXAMPLE 1   The simplest polar equations are $r = $ a constant and $\theta = $ a constant. Graph the equations (a) $r = 2$ and (b) $\theta = \pi/6$.

Solution

(a) r is fixed at $r = 2$ and there are no restrictions on θ; that is, θ varies from $-\infty$ to ∞. The result is a circle of radius 2 with center at the origin (Figure 13-18A).

(b) θ is fixed at $\theta = \pi/6$ and r varies from $-\infty$ to ∞. The result is a straight line through the origin (Figure 13-18B).   ■

The technique of changing equations from polar coordinates to rectangular coordinates and vice versa is a useful aid in graphing.

EXAMPLE 2

(a) Change $r = 2a \cos \theta$ into rectangular coordinates.

(b) Change $x^2 + y^2 - 2ay = 0$ into polar coordinates. Then sketch the graphs.

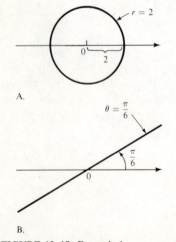

A.

B.

FIGURE 13-18. Example 1.

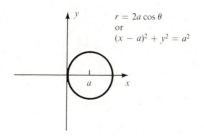

$r = 2a \cos \theta$
or
$(x - a)^2 + y^2 = a^2$

A.

$x^2 + (y - a)^2 = a^2$
or
$r = 2a \sin \theta$

B.

FIGURE 13–19. Example 2. Polar equations of circles.

Solution

(a) We multiply both sides of $r = 2a \cos \theta$ by r (this is a good trick to learn) to obtain

$$r^2 = 2ar \cos \theta$$
$$x^2 + y^2 = 2ax \qquad (13.8)$$
$$x^2 - 2ax + y^2 = 0 \qquad \text{(Rearrange)}$$
$$(x - a)^2 + y^2 = a^2 \qquad \text{(Complete the square)}$$

This is an equation of the circle shown in Figure 13–19A.

(b) We recognize $x^2 + y^2 - 2ay = 0$ as an equation of the circle sketched in Figure 13–19B. To find its polar equation, we write

$$x^2 + y^2 - 2ay = 0$$
$$r^2 - 2ar \sin \theta = 0 \qquad (13.8)$$
$$r(r - 2a \sin \theta) = 0$$

If $r = 0$, the graph is just the origin. Therefore, $r - 2a \sin \theta = 0$, or

$$r = 2a \sin \theta$$

is the desired polar equation. ∎

The methods in Example 2 can be used to verify that

(13.9)
> The graph of any polar equation of the form
> $$r = \pm 2a \cos \theta \quad \text{or} \quad r = \pm 2a \sin \theta \quad (a > 0)$$
> is a circle with center $(\pm a, 0)$ or $(0, \pm a)$ and radius a.

Aside from changing into rectangular coordinates, here are some other aids in graphing polar equations.

(1) *Symmetry.* The entries in the following table are easily verified using Figure 13–20.

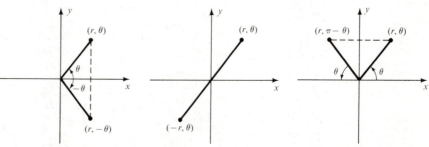

FIGURE 13–20. Symmetries of polar coordinates.

A. Symmetric about the x-axis. B. Symmetric about the origin. C. Symmetric about the y-axis.

	If the point below is on the graph whenever (r, θ) is on the graph,	then the graph is symmetric about the
(13.10)	$(r, -\theta)$	x-axis
	$(-r, \theta)$	origin
	$(r, \pi - \theta)$	y-axis

(2) Use the *derivative $dr/d\theta$* to determine where r is increasing and decreasing.

(3) Make a table of values and *plot some points*.

EXAMPLE 3   Sketch the graph of $r = a(1 - \cos \theta)$, $a > 0$.

Solution   In this example, it is not easy to change into rectangular coordinates, so other methods are used. Let us test for symmetry. Since $\cos \theta = \cos(-\theta)$, $(r, -\theta)$ is on the graph whenever (r, θ) is; therefore, by (13.10), the graph is symmetric about the x-axis. After testing the other two cases, we see that this is the only symmetry. Next, we find the derivative

$$\frac{dr}{d\theta} = a \sin \theta$$

which is positive for $0 < \theta < \pi$, so r is increasing for those values of θ. Finally, we make a table of values and plot the points (Figure 13–21A).

θ	0	$\pi/3$	$\pi/2$	$2\pi/3$	π
$r = a(1 - \cos \theta)$	0	$a/2$	a	$3a/2$	$2a$

Since $dr/d\theta > 0$, we connect the points by a smooth curve with r increasing. By symmetry, this part of the curve can be reflected across the x-axis to obtain the complete graph in Figure 13–21B. This heart-shaped curve is called a *cardioid*. ∎

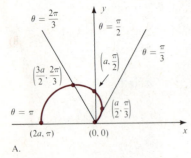

A.

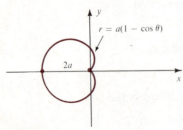

B.

FIGURE 13–21. Example 3. A cardioid.

The method in Example 3 can be used to verify that

	The graph of any polar equation of the form
(13.11)	$r = a(1 \pm \cos \theta)$ or $r = a(1 \pm \sin \theta)$ $(a > 0)$
	is a cardioid.

EXAMPLE 4   Sketch the graph of $r = a \sin 2\theta$, $a > 0$.

Solution   This curve has all three symmetries. For example, suppose that (r, θ) is on the graph. To show that $(r, -\theta)$ is also on the graph, we observe that $(r, -\theta)$

$= (-r, -\theta + \pi)$ and that the coordinates of the latter point satisfy the equation because

$$a \sin 2(-\theta + \pi) = a \sin(-2\theta + 2\pi) = -a \sin 2\theta = -r$$

The other symmetries can be shown in similar fashion. Thus, we need to analyze only that portion of the curve in the first quadrant.

The derivative is $dr/d\theta = 2a \cos 2\theta$. As θ varies from 0 to $\pi/4$, 2θ varies from 0 to $\pi/2$, the derivative is positive, and r is increasing. But as θ varies from $\pi/4$ to $\pi/2$, 2θ varies from $\pi/2$ to π, the derivative is negative, and r is decreasing. We now plot a few points

θ	0	$\pi/12$	$\pi/4$	$5\pi/12$	$\pi/2$
$r = a \sin 2\theta$	0	$a/2$	a	$a/2$	0

and connect them by a curve with increasing and decreasing r according to the information above (Figure 13–22A). After this, the symmetry gives us the *petal curve* in Figure 13–22B. ■

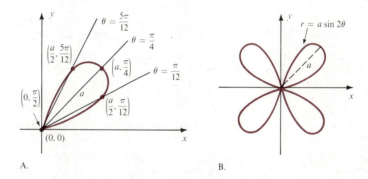

FIGURE 13–22. Example 4. A four-petal curve.

A.

B.

The method of Example 4 can be used to verify that

> The graph of any polar equation of the form
>
> $$r = a \sin n\theta \quad \text{or} \quad r = a \cos n\theta \quad (a > 0, n > 1)$$
>
> is a petal curve. If n is even, there are $2n$ petals and all three symmetries are present; if n is odd, there are n petals and there is symmetry with the y-axis for sine and the x-axis for cosine

(13.12)

Some of the other unusual curves obtained as graphs of polar equations are shown in Figure 13–23.

CAUTION: Because each point has many polar coordinates, it can happen that a point (r, θ) is on the graph of a polar equation *even though its coordinates, as they appear, do not satisfy the equation.* For example, the coordinates of $(0, 0)$

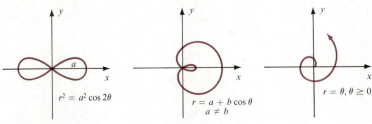

FIGURES 13–23. Graphs of some other polar equations.

A. A *lemniscate.*   B. A *limaçon.*   C. A *spiral.*

do not satisfy the equation $r = 1 + \cos \theta$. However, $(0, 0)$ and $(0, \pi)$ represent the same point, and the coordinates of $(0, \pi)$ do satisfy the equation. This situation is compounded when we are dealing with two curves.

EXAMPLE 5   Find all points of intersection of the two cardioids $r = 1 + \cos \theta$ and $r = 1 - \cos \theta, 0 \le \theta \le 2\pi$.

Solution   First we solve the two equations simultaneously. Since $r = 1 + \cos \theta$ and $r = 1 - \cos \theta$, we have

$$1 + \cos \theta = 1 - \cos \theta \quad \text{or} \quad \cos \theta = 0$$

It follows that $\theta = \pi/2$ or $3\pi/2$ and that

$$\left(1, \frac{\pi}{2}\right) \quad \text{and} \quad \left(1, \frac{3\pi}{2}\right)$$

are points of intersection. But Figure 13–24 shows that the origin is also on both curves; $(0, \pi)$ is on $r = 1 + \cos \theta$ and the same point $(0, 0)$ is on $r = 1 - \cos \theta$. This simply means that the curves do not pass through the origin for the same value of θ. You can think of the graphs as the paths of two particles and θ as time. The particles reach the origin at different times and will not collide there. They will collide, however, at the other two points of intersection. ■

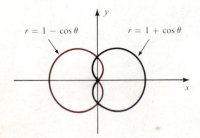

FIGURE 13–24. Example 5.

Conic Sections

The conic sections can be neatly described in polar coordinates. The idea is based on the eccentricity definition; given a point F (called the *focus*), a line D (called the *directrix*), and a positive number e (called the *eccentricity*), then the set of points P satisfying the equation

$$d(P, F) = e[d(P, D)]$$

is a parabola, ellipse, or hyperbola depending on the size of e.

EXAMPLE 6   Let F be a point, D be a vertical line, and e be any positive number. Find a polar equation whose graph is the set of points P with

$$d(P, F) = e[d(P, D)]$$

Change to rectangular coordinates and show that the graph is a conic section.

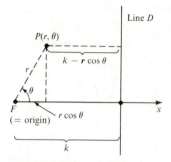

FIGURE 13–25. Example 6.
$d(P, F) = e[d(P, D)]$ means that
$r = e(k - r \cos \theta)$

Solution   We let the origin be the point F, and k units to the right is the vertical line D. It is easy to see from Figure 13–25 that the set of points $P(r, \theta)$ that meet the condition above also satisfy the equation

(13.13) $r = e(k - r \cos \theta)$

or, solving for r, we have

(13.14) $r = \dfrac{ke}{1 + e \cos \theta}$

This is the polar equation. To change into rectangular coordinates, it is easier to use (13.13). By (13.8), we replace r by $\pm \sqrt{x^2 + y^2}$ and $r \cos \theta$ by x; then (13.13) becomes

$$\pm \sqrt{x^2 + y^2} = e(k - x)$$
$$x^2 + y^2 = e^2 k^2 - 2ke^2 x + e^2 x^2 \quad \text{(Square both sides)}$$
(13.15) $(1 - e^2)x^2 + 2ke^2 x + y^2 = e^2 k^2$ (Rearrange)

Equation (13.15) is the rectangular version of (13.14). If $0 < e < 1$, then the coefficients of x^2 and y^2 are both positive so the graph is an ellipse. If $e = 1$, the x^2 term drops out and the graph is a parabola. If $e > 1$, the coefficient of x^2 is negative and the graph is a hyperbola. Thus, (13.14) represents a conic section for any value of e. ■

The same method illustrated in Example 6 can be used to show that

(13.16)

> A polar equation having one of the four forms
> $$r = \frac{ke}{1 \pm e \cos \theta} \quad \text{or} \quad r = \frac{ke}{1 \pm e \sin \theta}$$
> is an ellipse if $0 < e < 1$, a parabola if $e = 1$, and a hyperbola if $e > 1$.

EXERCISES

In Exercises 1–6, identify the curve and change to polar coordinates.

1. $x^2 + y^2 = 9$

2. $(x - 2)^2 + y^2 = 4$

3. $2xy = 1$

4. $y = 5x$

5. $ax + by = c$

6. $y^2 = 4x$

In Exercises 7–14, change to rectangular coordinates and identify the curve.

7. $r = \cos \theta$

8. $r = \sin \theta$

9. $\theta = c$

10. $r = 2 \csc \theta$

11. $r = \dfrac{1}{1 + \cos \theta}$ (Example 6.)

12. $r = \dfrac{3}{1 - 3\sin \theta}$

13. $r = \dfrac{2}{2 - \sin \theta}$

14. $r = \dfrac{8}{4 - \cos \theta}$

In Exercises 15–40, sketch the graphs.

15. $r = 2$

16. $r = 1$

17. $\theta = -\pi/3$

18. $\theta = \pi/2$

19. $r(2 \sin \theta + \cos \theta) = 1$

20. $r(\cos \theta - \sin \theta) = 2$

$\theta = \sin r$?

21. $r = -4 \sin \theta$

22. $r = -2 \cos \theta$

23. $r = 1 - \sin \theta$

24. $r = 2 + \cos \theta$

25. $r = \dfrac{3}{1 - 3 \sin \theta}$

26. $r = \dfrac{1}{1 + \cos \theta}$

27. $r = \dfrac{8}{4 - \cos \theta}$

28. $r = \dfrac{2}{2 - \sin \theta}$

29. $r = \cos 4\theta$

30. $r = \cos 3\theta$

31. $r = \sin 3\theta$

32. $r = \sin 4\theta$

33. $r^2 = \sin 2\theta$

34. $r^2 = 4 \cos 2\theta$

35. $r = 1 + 2 \cos \theta$

36. $r = 1 + 2 \sin \theta$

37. $r = \tan \theta$

38. $r = \cot \theta$

39. $r\theta = 1, r > 0$

40. $r = 2^\theta$

41. Show that $(1, \pi)$ is a point on the graph of $r^2 = \sec \theta$.

42. Show that $(-1, 3\pi/2)$ is a point on the graph of $r = 1 + 2 \cos \theta$.

43. Show that $(2, \pi)$ is on the graphs of both $r^2 = 4 \cos \theta$ and $r = 3 + \cos \theta$.

44. Show that $(-\sqrt{2}/2, -\pi/4)$ is on the graphs of both $r = \cos 3\theta$ and $r = \sin 3\theta$.

In Exercises 45–46, find the points at which the curves intersect.

45. $r = 2 \sin \theta, r = 2 \cos \theta$

46. $r = -\sin \theta, r = 1 + \sin \theta$

47. (a) Suppose that one particle is traveling along the curve $r = 2 \sin \theta$ and another along the curve $r = 2 \cos \theta$ so that their positions at time θ is (r, θ). Will they ever collide? If so, where? (b) Answer the same questions if the curves are $r = -\sin \theta$ and $r = 1 + \sin \theta$. (See Exercises 45 and 46.)

Test 1

13–3 TANGENT LINES AND AREA

In this section, we discuss tangent lines and area in terms of polar coordinates.

You will recall that tangent lines play a significant role in analyzing equations in rectangular coordinates. In this and later sections, you will see that the same is true for polar equations.

Let $r = f(\theta)$ be a polar equation that defines r as a differentiable function of θ on some interval and let L be a line tangent to the graph at $P(r, \theta)$ as shown in Figure 13–26. To find the slope of L, we reason as follows: If the curve were given in rectangular coordinates, then the slope of L would be the value of dy/dx. Since

$$(13.17) \quad x = r \cos \theta \quad \text{and} \quad y = r \sin \theta$$

it follows that x and y are *products* of differentiable functions of θ and, by the chain rule,

$$\frac{dy}{dx} = \frac{dy/d\theta}{dx/d\theta} = \frac{r \cos \theta + (\sin \theta)dr/d\theta}{-r \sin \theta + (\cos \theta)dr/d\theta} \tag{13.17}$$

If we assume that $y'(\theta)$ and $x'(\theta)$ are not both 0 simultaneously, then this result follows:

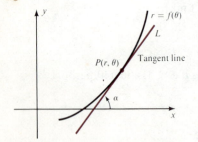

FIGURE 13–26. Finding the slope of a line tangent to the curve $r = f(\theta)$ at the point (r, θ).

(13.18)

Let $r = f(\theta)$ satisfy the assumptions stated above; then the slope of a line tangent to the graph at (r, θ) is

$$\text{Slope} = \frac{y'(\theta)}{x'(\theta)} = \frac{r \cos \theta + (\sin \theta)dr/d\theta}{-r \sin \theta + (\cos \theta)dr/d\theta} \quad (x'(\theta) \neq 0)$$

If $x'(\theta) = 0$, the tangent line is vertical at that point.

EXAMPLE 1   Find the slope of the line tangent to the spiral $r = \theta(\theta \geq 0)$ at $\theta = 0, \pi/2, \pi, 3\pi/2$, and 2π. Use this information to sketch the graph.

Solution   Here, $r = \theta$ and $dr/d\theta = 1$. By (13.18), the slope of the tangent line at $(r, \theta) = (\theta, \theta)$ is

$$\frac{\theta \cos \theta + \sin \theta}{-\theta \sin \theta + \cos \theta}$$

Therefore,

θ	0	$\pi/2$	π	$3\pi/2$	2π
Slope of tangent	0	$-\dfrac{2}{\pi} \approx -0.7$	π	$-\dfrac{2}{3\pi} \approx -0.2$	2π

This information is used to sketch the tangent lines and they, in turn, are used to sketch the graph (Figure 13–27). ∎

Graphs of functions in rectangular coordinates can have at most one tangent line at a point, but graphs of functions in polar coordinates can have more than one. This is illustrated in the next example.

EXAMPLE 2   Find the slopes of the lines tangent to the lemniscate $r^2 = \cos 2\theta$ at the origin.

Solution   To find the slope using (13.18), we need to know $dr/d\theta$. Differentiating implicitly, we have

$$2r\frac{dr}{d\theta} = -2 \sin 2\theta \quad \text{or} \quad \frac{dr}{d\theta} = \frac{-\sin 2\theta}{r}$$

Therefore, by (13.18), the slope at (r, θ) is

$$\frac{r \cos \theta + (\sin \theta)\left(\dfrac{-\sin 2\theta}{r}\right)}{-r \sin \theta + (\cos \theta)\left(\dfrac{-\sin 2\theta}{r}\right)}$$

which reduces to

$$(13.19) \qquad \frac{r^2 \cos \theta - \sin \theta \sin 2\theta}{-r^2 \sin \theta - \cos \theta \sin 2\theta}$$

We cannot use $(0, 0)$ to find the slopes at the origin because those coordinates do not satisfy the equation $r^2 = \cos 2\theta$. But $(0, \pi/4)$ and $(0, 3\pi/4)$ both represent the origin and their coordinates do satisfy the equation. Substitution of these values of r and θ into (13.19) yields slopes of ± 1. Also notice that at $\theta = 0$, the denominator $x'(0) = 0$ and, therefore, the tangent lines at $(\pm 1, 0)$ are vertical (Figure 13–28). ∎

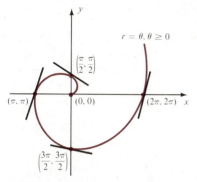

FIGURE 13–27. Example 1. Tangent lines are drawn at various points as an aid in graphing the spiral $r = \theta$, $\theta \geq 0$.

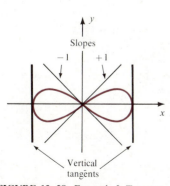

FIGURE 13–28. Example 2. Tangent lines (there are *two* at the origin) of the lemniscate $r^2 = \cos 2\theta$.

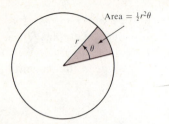

FIGURE 13–29. The area of a circular sector is $r^2\theta/2$ if θ is measured in *radians*.

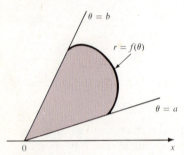

FIGURE 13–30. A region bounded by $r = f(\theta)$ and the lines $\theta = a$ and $\theta = b$.

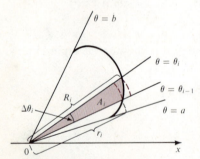

FIGURE 13–31. The area A_i is between the areas of the circular sectors determined by r_i and R_i.

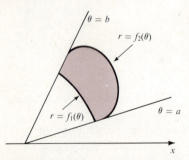

FIGURE 13–32. A region bounded by two graphs $r = f_1(\theta)$ and $r = f_2(\theta)$.

Areas

In rectangular coordinates, we approximate areas with rectangles. In polar coordinates, we approximate areas with circular sectors. Recall that the area of a circular sector (pictured in Figure 13–29) is $r^2\theta/2$ if θ is measured in *radians*.

Suppose a region is bounded by the graph of a polar equation $r = f(\theta)$ and the lines $\theta = a$ and $\theta = b$ with $0 < b - a \le 2\pi$ (Figure 13–30). Let $P = \{\theta_0, \theta_1, \ldots, \theta_n\}$ be any partition of $[a, b]$. If f is continuous, then it will have a minimum value r_i and maximum value R_i on each subinterval $[\theta_{i-1}, \theta_i]$. If A_i is the area between the lines $\theta = \theta_{i-1}$ and $\theta = \theta_i$, then

$$\frac{1}{2}r_i^2\Delta\theta_i \le A_i \le \frac{1}{2}R_i^2\Delta\theta_i$$

where $\Delta\theta_i = \theta_i - \theta_{i-1}$ (Figure 13–31). It follows that the total area A is squeezed between the sums

$$\sum_{i=1}^{n}\frac{1}{2}r_i^2\Delta\theta_i \le A \le \sum_{i=1}^{n}\frac{1}{2}R_i^2\Delta\theta_i$$

These sums are Riemann sums of the continuous function $r = f(\theta)$. Taking the limit as $\Delta\theta_i \to 0$, we define the area as follows:

(13.20) $\boxed{\text{Area} = \int_a^b \frac{1}{2}r^2 d\theta = \int_a^b \frac{1}{2}[f(\theta)]^2 d\theta}$

EXAMPLE 3   Find the area enclosed by the lemniscate $r^2 = \cos 2\theta$.

Solution   By symmetry (Figure 13–28), we can calculate the area between $\theta = 0$ and $\theta = \pi/4$ and multiply the result by 4. Thus, by (13.20),

$$\text{Area} = 4\int_0^{\pi/4} \frac{1}{2}r^2 d\theta = \int_0^{\pi/4} 2\cos 2\theta d\theta \qquad (r^2 = \cos 2\theta)$$

$$= [\sin 2\theta]_0^{\pi/4}$$

$$= 1 \text{ square unit} \qquad \blacksquare$$

A slightly more general case is a region bounded by the lines $\theta = a$ and $\theta = b$ and the graphs of two polar equations $r = f_1(\theta)$ and $r = f_2(\theta)$ as shown in Figure 13–32. If $f_2(\theta) \ge f_1(\theta)$ for all θ in $[a, b]$, then it is easy to see that the area of the region is

(13.21) $\text{Area} = \int_a^b \frac{1}{2}([f_2(\theta)]^2 - [f_1(\theta)]^2)d\theta$

EXAMPLE 4   Find the area inside the petal curve $r = 2\cos 2\theta$, but outside the circle $r = 1$.

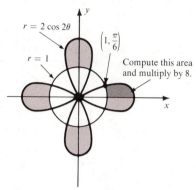

$r = 2 \cos 2\theta$

$\left(1, \frac{\pi}{6}\right)$

$r = 1$

Compute this area and multiply by 8.

FIGURE 13-33. Example 4.

Solution First sketch the curves (Figure 13-33). The points of intersection can be found by setting $1 = 2 \cos 2\theta$. But, because of symmetry, we need to compute only the area of one-half of one of the shaded regions and then multiply by 8. Thus, by (13.21),

$$\text{Area} = 8 \int_0^{\pi/6} \frac{1}{2}[4 \cos^2 2\theta - 1^2]d\theta$$

$$= 8 \int_0^{\pi/6} 2 \cos^2 2\theta d\theta - 4 \int_0^{\pi/6} 1^2 d\theta$$

$$= 8 \int_0^{\pi/6} (\cos 4\theta + 1)d\theta - 4 \int_0^{\pi/6} d\theta$$

half-angle formula

$$= 8 \left[\frac{1}{4} \sin 4\theta + \theta \right]_0^{\pi/6} - 4 \Big[\theta \Big]_0^{\pi/6}$$

$$= 8 \left(\frac{\sqrt{3}}{8} + \frac{\pi}{6} \right) - \frac{2\pi}{3} = \sqrt{3} + \frac{2\pi}{3} \text{ sq units} \quad \blacksquare$$

EXERCISES

In Exercises 1-10, find the slopes of the tangent lines (there may be more than one) at the indicated points (assume $a > 0$).

1. The circle $r = a$ at $(a, \pi/3)$.

2. The circle $r = 2a \sin \theta$ at $(a, \pi/6)$.

3. The cardioid $r = a(1 + \cos \theta)$ at $(0, \pi)$.

4. The cardioid $r = a(1 - \sin \theta)$ at $(a, 0)$.

5. The petal curve $r = a \cos 2\theta$ at the origin.

6. The petal curve $r = a \sin 2\theta$ at the origin.

7. The lemniscate $r^2 = a \sin 2\theta$ at the origin.

8. The lemniscate $r^2 = a \cos 2\theta$ at the origin.

9. The spiral $r\theta = 1$ at $(1/a, a)$; $\theta > 0$.

10. The spiral $r = \theta$ at (a, a); $\theta \geq 0$.

11. Find the points on the cardioid $r = a(1 + \cos \theta)$ where the tangent lines are (a) horizontal (b) vertical.

12. For the spiral $r\theta = 1$, find the limit of the slopes of the tangent lines at $(1/a, a)$ as $a \to 0^+$. Show that the line $y = 1$ is an asymptote of the spiral.

In Exercises 13-22, find the area of the region bounded by the given curves (assume $a > 0$).

13. $r = a \cos \theta$, $-\pi/2 \leq \theta \leq \pi/2$

14. $r^2 = a^2 \sin 2\theta$

15. $r^2 = a^2 \cos^2\theta$

16. $r = \tan 2\theta$, $0 \leq \theta \leq \pi/8$

17. $r = a(1 + \cos \theta)$

18. $r^2 = 2a^2 \cos 2\theta$

19. $r = 2 \cos \theta, r = \cos \theta, \theta = 0$, and $\theta = \pi/4$

20. $r = 2 \sin \theta, r = \sin \theta, \theta = 0$, and $\theta = \pi/2$

21. $r = \sin 2\theta$

22. $r = 4 + \sin \theta$

In Exercises 23-28, find the areas of the regions ($a > 0$).

23. Inside $r^2 = 2a^2 \cos 2\theta$, but outside $r = a$.

24. Inside $r = 3a \cos \theta$, but outside $r = a(1 + \cos \theta)$.

25. Inside $r^2 = 8 \sin 2\theta$, but outside $r = 2$.

26. Inside $r = 3$ but outside $r = 2(1 + \cos \theta)$.

27. Inside both of the curves $r = 2a \cos \theta$ and $r = 2a \sin \theta$.

28. Inside both of the curves $r = \sin \theta$ and $r = \sqrt{3} \cos \theta$.

29. The center of gravity of a triangle is located 2/3 of the way along any median. A circular sector is *almost* a triangle if the central angle is small. Therefore, the center of gravity of the circular sector in Figure 13-34 has *rectangular coordinates* approximately equal to $(\frac{2}{3}r \cos \theta, \frac{2}{3}r \sin \theta)$. Thus, the moments of the sector about the x-axis and y-axis are approximately

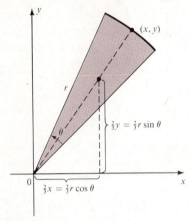

FIGURE 13–34. Exercise 29.

$$\left[\frac{2}{3}r\sin\theta\right]\left[\frac{1}{2}r^2\theta\right] \quad \text{and} \quad \left[\frac{2}{3}r\cos\theta\right]\left[\frac{1}{2}r^2\theta\right]$$

Now deduce that the center of gravity of a region bounded by the curve $r = f(\theta)$ and the lines $\theta = a$ and $\theta = b$ have rectangular coordinates

$$\bar{x} = \frac{\displaystyle\int_a^b \frac{1}{3}r^3\cos\theta\,d\theta}{\displaystyle\int_a^b \frac{1}{2}r^2\theta\,d\theta} \quad \text{and} \quad \bar{y} = \frac{\displaystyle\int_a^b \frac{1}{3}r^3\sin\theta\,d\theta}{\displaystyle\int_a^b \frac{1}{2}r^2\theta\,d\theta}$$

Find the center of gravity of the cardioid $r = a(1 + \cos\theta)$.

30. The region bounded by the lemniscate $r^2 = \cos 2\theta$ is revolved about the x-axis. Find the volume generated.

13–4 PARAMETRIC EQUATIONS

Graphs of rectangular and polar equations are curves in the plane. But it is often more practical to think of a curve as the graph of a *pair* of equations. Suppose that x and y are continuous functions of a variable t on some interval $[a, b]$; that is,

(13.22) $\quad x = x(t) \quad \text{and} \quad y = y(t) \quad t \text{ in } [a, b]^*$

Then the set of all points with rectangular coordinates $(x(t), y(t))$ for t in $[a, b]$ is a curve in the plane. The equations (13.22) are called **parametric equations** of the curve, and the variable t is called the **parameter.** The curve is the graph of the parametric equations.

It is of interest to note that a polar equation $r = f(\theta)$ gives rise to parametric equations. Since $x = r\cos\theta$ and $y = r\sin\theta$, we find that

(13.23) $\quad x(\theta) = f(\theta)\cos\theta \quad \text{and} \quad y(\theta) = f(\theta)\sin\theta$

are parametric equations with the same graph as $r = f(\theta)$. Here, the parameter is θ.

Graphs of Parametric Equations

Figure 13–35 illustrates that graphs of parametric equations can intersect themselves and have one, more than one, or no tangent lines at a point. We will return to the idea of tangent lines in a moment. Right now, let us graph some simple

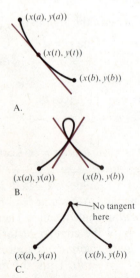

FIGURE 13–35. Graphs of parametric equations. There can be one, more than one, or no tangent lines at a point.

*We could have named these functions with the usual letters f and g; that is $x = f(t)$ and $y = g(t)$. But in this situation, it is much more suggestive to use the letters x and y for *both* the dependent variable *and* the name of the function.

parametric equations. The methods are similar to the ones used to sketch graphs of polar equations; plot points, change to rectangular coordinates (if convenient), and so on.

EXAMPLE 1   Sketch the graph of the parametric equations

$$x = t + 1 \quad \text{and} \quad y = 2t - 1$$

for (a) t in $[0, 1]$ and (b) t in $(-\infty, \infty)$.

Solution   We make a table of values for t in $[0, 1]$.

t	0	1/2	1
$x = t + 1$	1	3/2	2
$y = 2t - 1$	−1	0	1

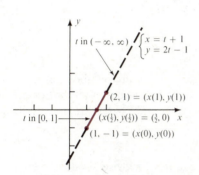

FIGURE 13–36. Example 1.

The points $(1, -1)$, $(3/2, 0)$, and $(2, 1)$ are plotted in Figure 13–36. They lie on a straight line; indeed, the graph for t in $[0, 1]$ is the line segment from $(1, -1)$ to $(2, 1)$. In part (b), t is allowed to be any real number, and the graph in this case is the entire line through those two points. ■

Instead of plotting points in Example 1, we could *eliminate the parameter;* that is, rewrite the equations without the parameter. For instance, in Example 1,

$$x = t + 1 \quad \text{and} \quad y = 2t - 1$$

We could solve for t in the first equation, $t = x - 1$, and substitute for t in the second equation

$$y = 2t - 1 = 2(x - 1) - 1 = 2x - 3$$

The graph of $y = 2x - 3$ is the entire line through $(1, -1)$ and $(2, 1)$. To obtain only the line segment for t in $[0, 1]$, we have to restrict x to the interval $[1, 2]$ because $x = t + 1$.

Solving for t in terms of x or y, as we did above, is not the only way to eliminate a parameter. Here is an example of another way.

EXAMPLE 2   Show that

$$x = [a \cos \theta] + h \quad \text{and} \quad y = [b \sin \theta] + k \quad \theta \text{ in } [0, 2\pi]$$

with $0 < b < a$ are parametric equations of an ellipse.

Solution   The parameter here is θ. We rewrite the equations as

$$\frac{x - h}{a} = \cos \theta \quad \text{and} \quad \frac{y - k}{b} = \sin \theta$$

But rather than solve one of these for θ to eliminate the parameter, we take advantage of our knowledge of the trigonometric functions and add the squares of both equations.

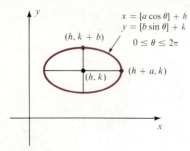

$x = [a \cos \theta] + h$
$y = [b \sin \theta] + k$

$0 \le \theta \le 2\pi$

FIGURE 13–37. Example 2.

$$\frac{(x - h)^2}{a^2} + \frac{(y - k)^2}{b^2} = \cos^2\theta + \sin^2\theta = 1$$

This is a standard form of an ellipse with its center at (h, k). See Figure 13–37. ∎

It is important to note that after eliminating the parameter, the resulting equation may have a different graph. Take, for example, the parametric equations

$$x = \cos t \quad \text{and} \quad y = \cos^2 t \quad t \text{ in } (-\infty, \infty)$$

The parameter can be eliminated by writing

$$y = x^2$$

Without further restrictions, the graph of $y = x^2$ is an entire parabola. But, because $|\cos t| \le 1$, the graph of the parametric equations is only that part of the parabola with $-1 \le x \le 1$. Therefore, to have the graphs agree, we must write

$$y = x^2 \quad -1 \le x \le 1$$

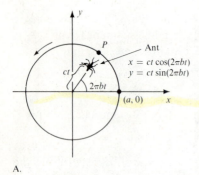

A.

$x = ct \cos(2\pi bt)$
$y = ct \sin(2\pi bt)$

EXAMPLE 3 A disc of radius a cm revolves counterclockwise about its center at the rate of b revolutions per second. An ant starts from the center and moves toward a fixed point P on the circumference at the rate of c cm/sec. Find parametric equations that describe the path of the ant relative to the xy-plane. Sketch the path.

Solution Place the disc with its center at the origin and P on the positive x-axis. At any time t seconds, the disc will have revolved through an angle of $2\pi bt$ radians and the ant will have moved a distance of ct cm from the center (Figure 13–38A). As you can see from the figure, the position of the ant at time t is

$$x = ct \cos 2\pi bt \quad y = ct \sin 2\pi bt \quad 0 \le t \le a/c$$

The path is the spiral sketched in Figure 13–38B. ∎

The curve in the next example is very interesting. Imagine a bright chalk-mark on a bicycle tire. As the bicycle moves along a level road, the chalk-mark not only moves vertically up and down, but horizontally along with the bicycle. The curve it traces out is called a *cycloid;* this curve plays an important role in the solution of two classical problems discussed in the optional reading section at the end of this chapter.

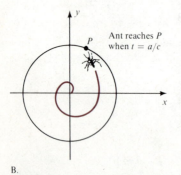

B.

FIGURE 13–38. Example 3. An ant moves from the center of a rotating disc toward a point P on the circumference. (A) Its position at time t. (B) Its path relative to the plane.

EXAMPLE 4 *(Cycloid).* Find parametric equations of the curve traced out by a point P on the circumference of a circle of radius a as the circle rolls along the x-axis. This curve is called a **cycloid.**

Solution Let the action begin with P at the origin. Now the circle starts rolling to the right; Figure 13–39A shows the path traced out by P after the radius connecting P to the center of the circle has turned through an angle of t radians.

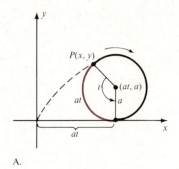

A.

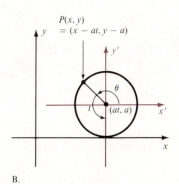

B.

FIGURE 13–39. Example 4. (A) As the circle rolls along, the point P on its circumference traces out a curve called a *cycloid*. (B) Relating the coordinates of P to t.

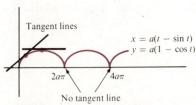

FIGURE 13–40. A cycloid. See Examples 4 and 5.

The center of the circle has moved horizontally a distance equal to the length of the arc colored in red; that is, a distance of at units.

What we want to do now is relate the coordinates (x, y) of P to the parameter t. To that end, we set up an $x'y'$-coordinate system with the center of the circle as the origin. In this system, the point P has coordinates $(x', y') = (x - at, y - a)$. But (x', y') can also be expressed in terms of the angle $\theta = 3\pi/2 - t$ (Figure 13–39B),

$$x' = a \cos \theta = a \cos\left(\frac{3\pi}{2} - t\right) = -a \sin t$$

$$y' = a \sin \theta = a \sin\left(\frac{3\pi}{2} - t\right) = -a \cos t$$

Since $x' = x - at$ and $y' = y - a$, it follows that $x - at = -a \sin t$ and $y - a = -a \cos t$ and, therefore,

$$x = a(t - \sin t) \quad \text{and} \quad y = a(1 - \cos t)$$

are parametric equations of the cycloid. The graph is sketched in Figure 13–40. Variations of the cycloid are discussed in the exercises. ■

Tangent Lines

We start with parametric equations

$$x = x(t) \quad \text{and} \quad y = y(t) \quad t \text{ in } [a, b]$$

To discuss tangent lines, we assume that x and y are differentiable functions of t on the open interval (a, b) and continuous at the endpoints. To avoid undue complications, we also assume that $x'(t)$ and $y'(t)$ are not both 0 simultaneously. With these assumptions, it follows, just as in the case of a line tangent to the graph of a polar equation, that

(13.24)

> If the parametric equations $x = x(t)$ and $y = y(t)$ satisfy the assumptions stated above, then the slope of a line tangent to the graph at $(x(t), y(t))$ is
>
> $$\text{Slope} = \frac{dy}{dx} = \frac{y'(t)}{x'(t)} \quad x'(t) \neq 0$$
>
> If $x'(t) = 0$, the tangent line is vertical at that point.

In the case of a cycloid, $x = a(t - \sin t)$ and $y = a(1 - \cos t)$; the derivatives

(13.25) $x'(t) = a(1 - \cos t) \quad y'(t) = a \sin t$

are both 0 when $t = 2n\pi$ for any integer n; that is, when $x = a(2n\pi - \sin 2n\pi) = a2n\pi$ and $y = a(1 - \cos 2n\pi) = 0$. These are the points at which P touches the x-axis. As Figure 13–40 suggests, there are no tangent lines at these points. But, according to (13.24), in most cases there is at least one tangent line.

EXAMPLE 5   Find the slope of the line tangent to the cycloid of Example 4 when $t = \pi/2$ and $t = \pi$.

Solution   It follows from (13.25) that

$$x'(\pi/2) = a \quad \text{and} \quad y'(\pi/2) = a$$

Therefore, by (13.24), the slope of the tangent line at $(x(\pi/2), y(\pi/2))$ is $a/a = 1$. When $t = \pi$, $x'(\pi) = 2a$ and $y'(\pi) = 0$, so the tangent line at $(x(\pi), y(\pi))$ is 0; that is, the tangent is horizontal (Figure 13–40).   ∎

EXAMPLE 6   The graph of the parametric equations

$$x = t^2 \quad \text{and} \quad y = t^3 - 4t \quad \text{for all } t$$

passes through the point $(4, 0)$ for $t = 2$ *and* for $t = -2$ (*Check*). Find the slopes of the tangent lines at $(4, 0)$.

Solution

$$x'(t) = 2t \quad \text{and} \quad y'(t) = 3t^2 - 4$$

are not both 0 for any t. Therefore, the slope of the tangent when $t = 2$ is

$$\frac{y'(2)}{x'(2)} = \frac{8}{4} = 2$$

The slope when $t = -2$ is

$$\frac{y'(-2)}{x'(-2)} = \frac{8}{-4} = -2$$

Thus, there are two tangent lines at $(4, 0)$. This is similar to the situation pictured in Figure 13–35B.   ∎

Suppose now that

$$x = x(t) \quad \text{and} \quad y = y(t) \quad t \text{ in } [a, b]$$

If the parameter t can be eliminated, then the higher order derivatives of y with respect to x are computed in the usual way. If not, then we use the chain rule to write

$$\frac{dy}{dx} = \frac{dy/dt}{dx/dt}$$

The derivative is usually a function of t, and if it is differentiable, then we can apply the chain rule once again to find the second derivative

$$(13.26) \quad \frac{d^2y}{dx^2} = \frac{d}{dx}\left(\frac{dy}{dx}\right) = \frac{\dfrac{d}{dt}\left(\dfrac{dy}{dx}\right)}{dx/dt}$$

EXAMPLE 7 Let

$$x = t^2 + 2t - 3 \quad \text{and} \quad y = t^3 + 1 \quad t \text{ in } [0, 1]$$

Find d^2y/dx^2 and describe the concavity of the graph.

Solution Using the method described above, we have

$$\frac{dy}{dx} = \frac{dy/dt}{dx/dt} = \frac{3t^2}{2t + 2}$$

This is differentiable for t in $[0, 1]$, and by (13.26)

$$\frac{d^2y}{dx^2} = \frac{\frac{d}{dt}\left(\frac{dy}{dx}\right)}{dx/dt} = \frac{\dfrac{(2t + 2)(6t) - 3t^2(2)}{(2t + 2)^2}}{2t + 2} \qquad \text{(Quotient rule)}$$

$$= \frac{6t^2 + 12t}{(2t + 2)^3}$$

Since the second derivative is positive for all t in $[0, 1]$, the curve is always concave up. ■

EXERCISES

In Exercises 1–16, sketch the graphs of the parametric equations. If you eliminate the parameter, make sure that the new equation has the same graph as the parametric equations.

1. $x = 2t + 3, y = t^2$; all t
2. $x = t^2, y = t/2$; all t
3. $x = t - 2, y = 2t + 3; 0 \le t \le 1$
4. $x = 4t, y = t - 1; 0 \le t \le 1$
5. $x = e^t, y = e^{2t}$; all t 6. $x = e^{2t}, y = e^t$; all t
7. $x = t, y = \sqrt{t}; t \ge 0$ 8. $x = \sqrt[3]{t}, y = t$; all t
9. $x = 4 \sin t, y = 4 \cos t$; all t
10. $x = 9 \cos t, y = 9 \sin t$; all t
11. $x = 4 \cos t, y = 9 \sin t$; all t
12. $x = 4 \cos t, y = \sin t$; all t
13. $x = \sec t, y = \tan t; -\pi/2 < t < \pi/2$
14. $x = \tan t, y = \sec t; -\pi/2 < t < \pi/2$
15. $x = \sinh t, y = \cosh t$; all t
16. $x = 2 \cosh t, y = 3 \sinh t$; all t

In Exercises 17–22, find the slope(s) of the tangent(s) at the given point on the graph.

17. $x = t, y = t^3 - 1; (1, 0)$
18. $x = e^t, y = e^{2t}; (1, 1)$
19. $x = 4 \cos t, y = 9 \sin t; (4, 0)$
20. $x = \sinh t, y = \cosh t; (0, 1)$
21. $x = t^2 - 2t + 1, y = t^4 - 4t^2 + 4; (1, 4)$
22. $x = t^2, y = t^4 + t^2; (1, 2)$
23. The parametric equations in Example 6 are $x = t^2$, $y = t^3 - 4t$ for all t. There are two tangent lines at $(4, 0)$ with slopes ± 2. Find the points where the tangents are horizontal and vertical and describe the concavity. Then use this information to sketch the graph.
24. Consider the parametric equations
$$x = \cos^3 \theta \quad \text{and} \quad y = \sin^3 \theta \quad 0 < \theta < \pi/2$$
   (a) Find the slope of the tangent line when $\theta = \pi/6$, $\pi/4$, and $\pi/3$. Also examine the slopes as $\theta \to 0^+$ and $\theta \to \pi/2^-$. Then sketch the graph.
   (b) Use symmetry to sketch the graph of $x = \cos^3 \theta$ and $y = \sin^3 \theta$, for all θ.

In Exercises 25–28, use the techniques of Exercises 23 and 24 above to sketch the graphs.

25. $x = 3t - t^3, y = t + 1$; all t

26. $x = \sin 2t, y = \sin t$; all t

27. $x = \cos 2t, y = \sin t$; all t

28. $x = \cos^4 t, y = \sin^4 t$; all t

---

29. Parametric equations are particularly useful in describing motion along curves. Suppose that the parametric equations in Exercises 1–16 represent the position of a particle at time t. Then the graphs are the paths of the particles. Go back to each exercise and mark the graph with arrows to indicate the direction of motion as t increases.

30. The position of a particle at any time t seconds is given by the parametric equations

$$x = 2 \cos t \quad \text{and} \quad y = \sin t \quad t \geq 0$$

(a) What is the path of the particle?   (b) What is its starting point?   (c) Does it move clockwise or counterclockwise?   (d) When $t = \pi/6$, at what rate are the x- and y-coordinates changing?

31. (Continuation of Exercise 30). Another particle moves with harmonic motion described by the equations

$$x = -2 \cos t \quad \text{and} \quad y = -2 \cos t \quad t \geq 0$$

(a) Find its path.   (b) At what point(s) does this path intersect the one described in Exercise 30? (c) Will the particles ever collide?

Optional Exercises

32. *(Trochoids)*. If the point P in Example 4 (about cycloids) is b units from the center with $b \neq a$, then the curve it traces out is called a *trochoid*. Suppose $0 < b < a$, and show that parametric equations for a trochoid are

$$x = at - b \sin t \quad y = a - b \cos t$$

Sketch the graph. Notice that there is a tangent line at each point.

33. (Continuation of Exercise 32). The wheels of a train have a lip on the outside to hold them on the track. As a wheel (of radius a) rolls along the track (the x-axis), a point P on the circumference of the lip (b units from the center and $b > a$) traces out a trochoid with the parametric equations derived in Exercise 32. Sketch the graph of this trochoid. What about tangent lines?

13–5   ARC LENGTH AND SURFACES OF REVOLUTION

In Section 6–5, we demonstrated that if $y = f(x)$ or $x = g(y)$ have continuous derivatives, then the lengths of their graphs are given by integrals of the form

$$(13.27) \quad \int_a^b \sqrt{1 + [f'(x)]^2}\, dx \quad \text{or} \quad \int_c^d \sqrt{1 + [g'(y)]^2}\, dy$$

What we want to do now is develop similar formulas for graphs of parametric and polar equations. After that, we discuss the surface area generated by revolving a graph about a line.

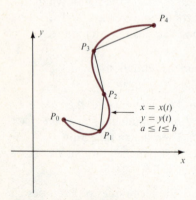

Arc Length

Suppose that $x = x(t)$ and $y = y(t)$ have continuous derivatives on an interval containing $[a, b]$. To compute the length of the graph from $t = a$ to $t = b$, we could approximate the curve by straight line segments (Figure 13–41) and then take the limit of the sums of their lengths as the distance between the points P_i approaches zero. But it seems wasteful not to use the formulas (13.27) already known.

FIGURE 13–41. Estimating the length of a curve using line segments.

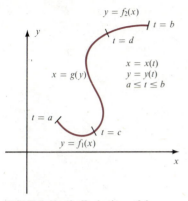

$y = f_2(x)$

$t = b$

$t = d$

$x = x(t)$
$y = y(t)$
$a \le t \le b$

$x = g(y)$

$t = a$

$t = c$

$y = f_1(x)$

FIGURE 13–42. Each piece of the curve is the graph of a function $y = f(x)$ or $x = g(y)$.

Figure 13–42 illustrates how the curve can be broken up into pieces that are graphs of functions $y = f(x)$ or $x = g(y)$. This can be done with any curve that we encounter in this book. Suppose that for $t = a$ to $t = c$, the curve is actually the graph of a function $y = f(x)$. Its length is given by the first integral in (13.27). If we make the change of variable $x = x(t)$, then $dx = x'(t)\, dt$ and the integrand becomes

$$\sqrt{1 + [f'(x)]^2}\, dx = \sqrt{1 + [f'(x(t))]^2}\, x'(t)\, dt$$
$$= \sqrt{[x'(t)]^2 + [f'(x(t))x'(t)]^2}\, dt$$

Since $y = f(x(t))$, we know that $y'(t) = f'(x(t))x'(t)$, and it follows that

$$\int_{x(a)}^{x(c)} \sqrt{1 + [f'(x)]^2}\, dx = \int_a^c \sqrt{[x'(t)]^2 + [y'(t)]^2}\, dt$$

This is the length of the curve from $t = a$ to $t = c$. From $t = c$ to $t = d$, the curve is the graph of a function $x = g(y)$. In this case, we make a change of variable $y = y(t)$ in the second integral of (13.27). We leave it for you (Exercise 31) to verify that exactly the same integrand is obtained. It follows that

(13.28)

> For parametric equations $x = x(t)$ and $y = y(t)$, with x' and y' continuous, the length of the graph from $t = a$ to $t = b$ is
> $$s = \int_a^b \sqrt{[x'(t)]^2 + [y'(t)]^2}\, dt$$

EXAMPLE 1 Find the length of one arch, from $t = 0$ to $t = 2\pi$, of the cycloid $x = a(t - \sin t)$, $y = a(1 - \cos t)$.

Solution First find the derivatives

$$x'(t) = a(1 - \cos t) \quad \text{and} \quad y'(t) = a \sin t$$

Then, according to (13.28), the length is

$$s = a \int_0^{2\pi} \sqrt{(1 - \cos t)^2 + \sin^2 t}\, dt$$

The expression under the radical simplifies to $2(1 - \cos t)$ which equals $4\sin^2(t/2)$ by one of the half-angle formulas. Since $\sin(t/2) \ge 0$ for $0 \le t \le 2\pi$, we have

$$s = 2a \int_0^{2\pi} \sin \frac{t}{2}\, dt$$

$$= 4a \left[-\cos \frac{t}{2} \right]_0^{2\pi} = 8a \text{ units} \qquad \blacksquare$$

If the curve is the graph of a polar equation $r = f(\theta)$, where f' is continuous, then (13.28) can be used by writing the parametric equations

$$x = f(\theta) \cos \theta \quad y = f(\theta) \sin \theta \qquad \left(\begin{aligned} x &= r \cos \theta \\ y &= r \sin \theta \end{aligned} \right)$$

The derivatives are

$$x'(\theta) = -f(\theta) \sin \theta + f'(\theta) \cos \theta \text{ and } y'(\theta) = f(\theta) \cos \theta + f'(\theta) \sin \theta$$

After squaring and simplifying, we find that

$$[x'(\theta)]^2 + [y'(\theta)]^2 = [f(\theta)]^2 + [f'(\theta)]^2.$$

It follows that

(13.29)

> For a polar equation $r = f(\theta)$, with f' continuous, the length of the graph from $\theta = a$ to $\theta = b$ is
>
> $$s = \int_a^b \sqrt{[f(\theta)]^2 + [f'(\theta)]^2} \, d\theta = \int_a^b \sqrt{r^2 + [r'(\theta)]^2} \, d\theta$$

EXAMPLE 2 Find the length of the spiral $r = \theta$ from $\theta = 0$ to $\theta = 2\pi$.

Solution Here, $r'(\theta) = 1$ so, by (13.29), we have

$$s = \int_0^{2\pi} \sqrt{\theta^2 + 1} \, d\theta$$

Use trigonometric substitution or Formula **21** in the Table of Integrals to obtain

$$= \left[\frac{\theta}{2} \sqrt{1 + \theta^2} + \frac{1}{2} \ln|\theta + \sqrt{1 + \theta^2}| \right]_0^{2\pi}$$

$$= \pi \sqrt{1 + 4\pi^2} + \frac{1}{2} \ln(2\pi + \sqrt{1 + 4\pi^2}) \text{ units} \qquad ■$$

Surface Area Not Responsible

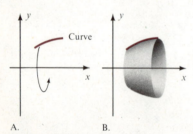

A.   B.

FIGURE 13–43. Revolving the curve in (A) about the *x*-axis results in the surface of revolution (B).

When the region under a graph is revolved about a line, the result is a solid of revolution. When only the graph itself is revolved about a line, the result is a **surface of revolution** (Figure 13–43). We want to derive a formula for computing the *surface area*, but to present a rigorous development of such a formula is beyond the scope of this book. We can, however, give an intuitive argument using the fact from geometry that the surface area S of the frustrum of a cone of slant height s and base radii r_1 and r_2 is $S = \pi s(r_1 + r_2)$; see Figure 13–44. Observe that

$$S = \pi s(r_1 + r_2)$$

(13.30)

$$= 2\pi s \left(\frac{r_1 + r_2}{2} \right)$$

$$= (2\pi r)s$$

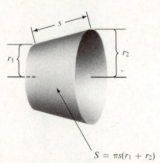

$$S = \pi s(r_1 + r_2)$$

FIGURE 13–44. The surface area of the frustrum of a cone.

where r is the *average* of the radii. Thus, the surface area is the slant height s times the circumference $2\pi r$ of a circle. Now let us move to the graph of parametric equations $x = x(t), y = y(t)$ that is revolved about the x-axis; we assume that $y(t) \geq 0$. Let ds represent the length of a small increment of the curve and think of ds as the slant height of the frustrum of a cone. From our arc length formula, we know that $ds = \sqrt{[x'(t)]^2 + [y'(t)]^2}$; therefore, by (13.30), it follows that the surface area of the strip in Figure 13–45 is approximately

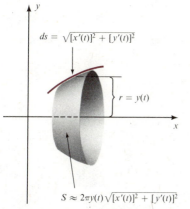

$ds = \sqrt{[x'(t)]^2 + [y'(t)]^2}$

$r = y(t)$

$S \approx 2\pi y(t)\sqrt{[x'(t)]^2 + [y'(t)]^2}$

FIGURE 13–45. The surface area of part of a surface of revolution.

(13.31) $2\pi y(t) \sqrt{[x'(t)]^2 + [y'(t)]^2}$

Thus, we can approximate the surface area by summing areas of the form (13.31). The limit of such sums, as the widths of the strips approach zero, is an integral that represents the surface area. If the graph is revolved about the y-axis, then (13.31) becomes $2\pi x(t)\sqrt{[x'(t)]^2 + [y'(t)]^2}$.

This informal discussion can be summarized as follows:

(13.32)

> If the graph of $x = x(t)$, $y = y(t)$ for $a \leq t \leq b$ is revolved about the x-axis, then the surface area generated is
>
> $$S = \int_a^b 2\pi y(t) \sqrt{[x'(t)]^2 + [y'(t)]^2} \, dt$$
>
> If it is revolved about the y-axis, the surface area is
>
> $$S = \int_a^b 2\pi x(t) \sqrt{[x'(t)]^2 + [y'(t)]^2} \, dt$$

EXAMPLE 3   Rotate one arch of the cycloid $x = a(t - \sin t), y = a(1 - \cos t)$ about the x-axis and compute the surface area.

Solution   By symmetry, we need only compute the area from $t = 0$ to $t = \pi$ and double the result. Now, as in Example 1, we have that

$$\sqrt{[x'(t)]^2 + [y'(t)]^2} = 2a \sin \frac{t}{2}$$

Since $y(t) = a(1 - \cos t) = 2a \sin^2(t/2)$, the surface area is

$$S = 2 \int_0^\pi 2\pi y(t) \sqrt{[x'(t)]^2 + [y'(t)]^2} \, dt \qquad \text{(13.32)}$$

$$= 16\pi a^2 \int_0^\pi \sin^3 \frac{t}{2} \, dt$$

$$= 16\pi a^2 \left[-2 \cos \frac{t}{2} + \frac{2}{3} \cos^3 \frac{t}{2} \right]_0^\pi \qquad \text{(Odd power of sine)}$$

$$= 16\pi a^2 \left[0 - \left(-2 + \frac{2}{3} \right) \right] = \frac{64\pi a^2}{3} \text{ square units} \qquad ■$$

If the curve that is rotated is the graph of a function $y = f(x)$, we use (13.32) and the parametric equations

$$x = t \quad \text{and} \quad y = f(t)$$

to compute the surface area. In this case $x'(t) = 1$, $y'(t) = f'(t) = f'(x)$, and $dt = dx$. Therefore,

---

(13.33)

If the graph of $y = f(x)$, $a \le x \le b$, is revolved about the x-axis, then the surface area generated is

$$S = \int_a^b 2\pi y \sqrt{1 + [f'(x)]^2} \, dx$$

If the graph of $x = g(y)$, $c \le y \le d$ is revolved about the y-axis, then the surface area is

$$S = \int_c^d 2\pi x \sqrt{1 + [g'(y)]^2} \, dy$$

EXAMPLE 4 *(Surface area of a searchlight).* The reflector of a searchlight is a parabolic mirror. If it is 2 ft across at the opening and 1 foot deep, what is its surface area?

Solution A model of the reflector is obtained by revolving the parabola $y^2 = x$, $0 \le x \le 1$, about the x-axis. To find the surface area S, we set $y = f(x) = \sqrt{x}$; then $f'(x) = 1/2\sqrt{x}$ and, by (13.33),

$$S = \int_0^1 2\pi \sqrt{x} \, \sqrt{1 + (1/4x)} \, dx$$

$$= 2\pi \int_0^1 \sqrt{x + \frac{1}{4}} \, dx$$

$$= \frac{4}{3}\pi \left[\left(x + \frac{1}{4} \right)^{3/2} \right]_0^1$$

$$= \frac{4}{3}\pi \left[\left(\frac{5}{4} \right)^{3/2} - \left(\frac{1}{4} \right)^{3/2} \right] \text{sq ft} \quad \blacksquare$$

The formula for surface area when the revolved curve is the graph of a polar equation is as follows (we leave the justification as an exercise).

(13.34)

If the graph of $r = f(\theta)$, $a \le \theta \le b$, is revolved about the x-axis, then the surface area generated is

$$S = \int_a^b 2\pi f(\theta) \sin \theta \sqrt{[f(\theta)]^2 + [f'(\theta)]^2} \, d\theta$$

EXERCISES

In Exercises 1–14, find the arc length of the given curve.

1. $x = t^2, y = t^3; 0 \leq t \leq 1$

2. $x = 2t, y = t^2 - 1; 0 \leq t \leq 1$

3. $x = \sin t, y = \cos t; 0 \leq t \leq 2\pi$

4. $x = \sin^2 t, y = \cos^2 t, 0 \leq t \leq \pi/2$

5. $r = 2 \cos \theta; 0 \leq \theta \leq \pi$

6. $r = 2 \sin \theta; 0 \leq \theta \leq 2\pi$

7. $r = 1 - \cos \theta; 0 \leq \theta \leq 2\pi$

8. $r = 1 + \sin \theta; -\pi/2 \leq \theta \leq \pi/2$

9. $x = e^t \cos t, y = e^t \sin t; 0 \leq t \leq \pi/2$

10. $x = t, y = t^2; 0 \leq t \leq 1$

11. $r\theta = 1; 1 \leq \theta \leq 2$ **12.** $r = e^\theta; 0 \leq \theta \leq 1$

13. $r = \theta; 0 \leq \theta \leq 1$

14. $r = \cos^2(\theta/2); -\pi \leq \theta \leq \pi$

In Exercises 15–22, find the surface area generated by revolving the curve about the x-axis.

15. $x = 2t^2, y = t; 0 \leq t \leq 1$

16. $x = 3t, y = 2t + 1; 0 \leq t \leq 1$

17. $y = x^2; 0 \leq x \leq 1$ **18.** $y = x^3; 0 \leq x \leq 1$

19. $x = t^2, y = 2t; 0 \leq t \leq 4$

20. $x = t^2, y = t - (t^3/3); 0 \leq t \leq 1$

21. $r = 2 \sin \theta; 0 \leq \theta \leq \pi$

22. $r = 2 \cos \theta; 0 \leq \theta \leq \pi$

In Exercises 23–26, find the surface area generated by revolving the curve about the y-axis.

23. $x = e^t \sin t, y = e^t \cos t; 0 \leq t \leq \pi/2$

24. $x = \sqrt{y}; 0 \leq y \leq 1$

25. $y = \ln x; 1 \leq x \leq 2$

26. $x = \cosh y; 0 \leq y \leq 1$

27. *(Circumference of an ellipse).* Parametric equations for an ellipse are of the form

$$x = a \cos t \quad \text{and} \quad y = b \sin t \quad 0 \leq t \leq 2\pi$$

(a) Assume $a > b$. Show that the length of one-fourth of the circumference is given by the integral

$$a \int_0^{\pi/2} \sqrt{1 - e^2 \cos^2 t}\, dt$$

where e is *eccentricity*, $\sqrt{a^2 - b^2}/a$.

(b) The integral in (a) is called an *elliptic integral;* there is no elementary method for evaluating it. Use Simpson's rule with $n = 6$ to estimate the integral above for $a = 3$ and $b = 2$.

(c) It can be shown that the entire circumference of an ellipse with semiaxes a and b is $\pi(a + b)$ (similar to $\pi(r + r) = 2\pi r$ for a circle). Compare this formula with your answer to part (b).

28. Find the surface area of the parabolic reflector of an automobile headlamp that is 22 cm at the opening and 10 cm deep.

29. *(Distance).* The wheel of an automobile, including the tire, has a radius of 18 inches. As the auto moves along a level road, what is the total distance traveled by the head of a tack stuck in the tire each time the wheel makes one revolution?

30. *(Speed).* The position of a particle at time t is given by the parametric equations

$$x = t^2 \quad \text{and} \quad y = t^3 \quad t \geq 0$$

(a) Find the distance s that it moves in the first second (s is the length of its path).

(b) Express the distance s it moves in the first t seconds as a function of t.

(c) We define the *speed* of the particle to be ds/dt. Find the speed when $t = 2$.

31. Verify that if $x = g(y)$ and a change of variable $y = y(t)$ is made in the second integral of (13.27), the result is

$$\int_{y(c)}^{y(d)} \sqrt{1 + [g'(y)]^2}\, dy = \int_c^d \sqrt{[x'(t)]^2 + [y'(t)]^2}\, dt$$

32. Verify that if $x = f(\theta) \cos \theta$ and $y = f(\theta) \sin \theta$, then $[x'(\theta)]^2 + [y'(\theta)]^2 = [f(\theta)]^2 + [f'(\theta)]^2$.

33. In the formula for surface area in polar coordinates, how does the factor $f(\theta) \sin \theta$ arise? How does the factor $\sqrt{[f(\theta)]^2 + [f'(\theta)]^2}$ arise?

Optional Exercise

34. Find the arc length of the curve $x = e^t$, $y = e^{2t}$ for $0 \le t \le 1$.

13–6 BRACHISTOCHRONES AND TAUTOCHRONES (Optional reading)*

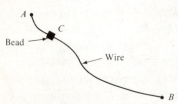

FIGURE 13–46. A thin wire joins A and B, and a bead slides without friction from A to B.

Suppose that two points A and B are joined by a thin wire, and a bead is allowed to slide without friction down the wire from A to B (Figure 13–46). The wire can be any shape, and we consider each shape as representing a *path* from A to B. For each path, the time required for the bead to slide from A to B is called the *time of descent*. There are two classical questions connected with this situation.

Question 1. What path gives the shortest possible time of descent?

Question 2. Is there a path with the property that the time of descent from A to B is *exactly the same* as the time of descent from C to B for *any* intermediate point C?

It appears, at first, that a straight line from A to B would be the path of quickest descent. But, perhaps, a more vertical descent at the beginning would provide the bead with enough velocity to get it to B in less time. Indeed, that is the case. And it seems to this author that the second question asks too much: a path on which the bead arrives at B in the same amount of time, no matter where it starts! But both questions do have answers, and it is a remarkable fact that both answers involve the same curve; namely, a *cycloid* (Example 4, Section 13–4).

The first question is called the *brachistochrone* problem (from the Greek *brachistos,* shortest, + *chronos,* time); the second is called the *tautochrone* problem (also from the Greek *tauto,* same, + *chronos,* time). The first was posed, in 1696, by Johann Bernoulli (L'Hôpital's teacher; see the footnote on page 461) as a challenge to the mathematicians of the day. It aroused great interest and was solved by Newton and Leibniz as well as Johann Bernoulli and his brother Jakob. The second question was actually posed earlier and solved in 1673 by C. Huygens (1629–1695) who applied the solution to the construction of pendulum clocks.

Our solution of the brachistochrone problem is based on the original solution by Johann Bernoulli. He began by considering an apparently unrelated problem in optics: the path that a ray of light travels from one medium into another. According to Snell's law (Example 4, Section 4–2),

$$\frac{\sin \alpha_1}{v_1} = \frac{\sin \alpha_2}{v_2}$$

where v_1 and v_2 are the velocities of light in the two mediums and α_1 and α_2 are the angles in Figure 13–47A. In Figure 13–47B, we have stratified the medium.

*Part of this discussion is adapted from George F. Simmons, *Differential Equations,* International Series in Pure and Applied Mathematics (New York: McGraw-Hill, 1972).

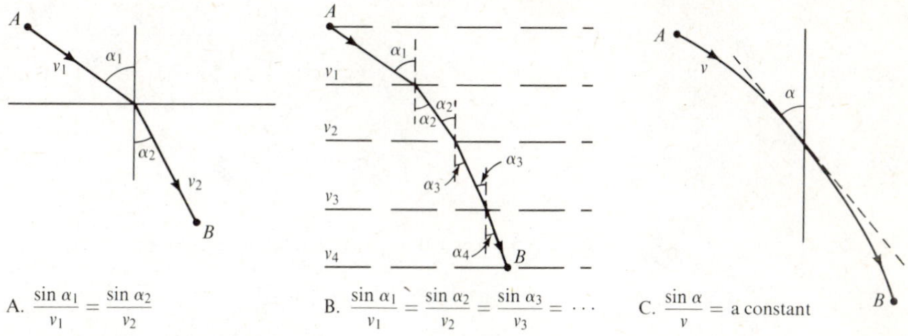

FIGURE 13–47. A ray of light seeks the path requiring the shortest time in passing from A to B.

A. $\dfrac{\sin \alpha_1}{v_1} = \dfrac{\sin \alpha_2}{v_2}$ B. $\dfrac{\sin \alpha_1}{v_1} = \dfrac{\sin \alpha_2}{v_2} = \dfrac{\sin \alpha_3}{v_3} = \cdots$ C. $\dfrac{\sin \alpha}{v} =$ a constant

In each layer, the velocity is constant, but it decreases as the medium becomes denser. As the descending ray of light passes from layer to layer, it is refracted (bent) more and more. When Snell's law is applied at the boundaries, we obtain

$$\frac{\sin \alpha_1}{v_1} = \frac{\sin \alpha_2}{v_2} = \frac{\sin \alpha_3}{v_3} = \cdots$$

That is, *these ratios are all equal to the same constant.* If we now allow the widths of these layers to approach zero, the velocity v of light decreases continuously and, at any point on the path,

(13.35) $\dfrac{\sin \alpha}{v} = c$, a constant

where α is the angle shown in Figure 13–47C. This is approximately the path of a ray of sunlight to the earth as it descends through the atmosphere of increasing density. By Fermat's *principal of least time,* we know that on the path that requires the least time for light to travel from A to B, equation (13.35) is true.

Returning now to the brachistochrone problem, we introduce a coordinate system where A is at the origin and the positive y-direction is down (Figure 13–48). We assume that the bead (like the ray of light) is capable of selecting the path down which it will slide from A to B in the shortest possible time. The discussion above shows that (13.35) holds. It is also true that the velocity of the bead at any point of its descent is completely independent of the path. In Example 4 (and the accompanying remark) in Section 6–4, we showed that the velocity depends only on the vertical distance the object moves; in the present situation,

(13.36) $v = \sqrt{2gy}$

From Figure 13–48, we also see that

(13.37) $\sin \alpha = \cos \beta = \dfrac{1}{\sec \beta} = \dfrac{1}{\sqrt{1 + \tan^2 \beta}} = \dfrac{1}{\sqrt{1 + (y')^2}}$ $(\tan \beta = -y')$

Combining optics (13.35), physics (13.36), and calculus (13.37), we have

$$c = \frac{\sin \alpha}{v} = \frac{\sin \alpha}{\sqrt{2gy}}$$

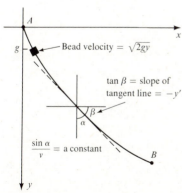

FIGURE 13–48. We assume that the bead is capable of selecting the path of shortest time of descent.

$$= \frac{1}{\sqrt{1 + (y')^2}} \cdot \frac{1}{\sqrt{2g} \sqrt{y}}$$

or

$$\sqrt{y} \sqrt{1 + (y')^2} = \frac{1}{c\sqrt{2g}}$$

We now square both sides and set $C = 1/2gc^2$ to obtain the differential equation

(13.38) $y[1 + (y')^2] = C$

If we can find a solution to (13.38) whose graph passes through A and B, then this graph is the solution to the brachistochrone problem.

If y' is replaced by dy/dx and we separate the variables, then (13.38) becomes

(13.39) $\sqrt{\dfrac{y}{C - y}} \, dy = dx$

To solve this equation we introduce a variable θ with

(13.40) $\tan \theta = \sqrt{\dfrac{y}{C - y}}$

from which it follows that

(13.41) $y = C \sin^2\theta$ (Solve (13.40) for y)

(13.42) $dy = 2C \sin \theta \cos \theta \, d\theta$

and

$$dx = \tan \theta \, dy \qquad \text{(13.39) and (13.40)}$$
$$= 2C \sin^2\theta d\theta \qquad \text{(13.42)}$$
$$= C(1 - \cos 2\theta) \, d\theta \qquad \text{(Half-angle formula)}$$

Integration now yields

$$x = \frac{C}{2}(2\theta - \sin 2\theta) + C_1$$

Since our curve is to pass through the origin, it follows from (13.40) that $x = y = 0$ when $\theta = 0$. Thus, $C_1 = 0$ and

(13.43) $x = \dfrac{C}{2}(2\theta - \sin 2\theta)$

Moreover, from (13.41), it follows that

(13.44) $y = C \sin^2\theta = \dfrac{C}{2}(1 - \cos 2\theta)$ (Half-angle)

Now set $a = C/2$ and $t = 2\theta$; then (13.43) and (13.44) become

(13.45) $x = a(t - \sin t) \quad y = a(1 - \cos t)$

the parametric equations of a *cycloid* (Example 4, Section 13–4). The value of a that makes the cycloid pass through B is the path of shortest descent time.

We now turn our attention to the second question, the tautochrone problem. It turns out that the solution is again a cycloid. To see this, let us suppose that B is at the bottom of the arch; that is, the coordinates of B in Figure 13–49 are $(\pi a, 2a)$. We will show that *no matter where it starts, the time it takes for the bead to reach B is $\pi\sqrt{a/g}$.* Suppose it starts from rest at the point (x_1, y_1). Then its velocity at any point $P(x, y)$ below (x_1, y_1) is

$$v = \sqrt{2g(y - y_1)}$$

Since $v = ds/dt$ and, by our discussion of arc length in Section 13–5, $ds = \sqrt{dx^2 + dy^2}$, we have that

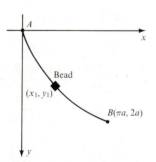

A

x

Bead

(x_1, y_1)

$B(\pi a, 2a)$

y

FIGURE 13–49. The tautochrone problem. The time it takes the bead to reach B is the same no matter what its initial position is.

$$(13.46) \quad dt = \frac{ds}{\sqrt{2g(y - y_1)}} = \sqrt{\frac{dx^2 + dy^2}{2g(y - y_1)}}$$

In order not to confuse time with the parameter t, we rewrite the parametric equations for the cycloid as

$$x = a(\theta - \sin\theta) \quad \text{and} \quad y = a(1 - \cos\theta)$$

Then, as in Example 1 of Section 13–5,

$$dx^2 + dy^2 = 4a^2\sin^2(\theta/2)d\theta^2$$

It follows from this and (13.46) that the time T of descent from (x_1, y_1) $= (a(\theta_1 - \sin\theta_1), a(1 - \cos\theta_1))$ to B is

$$T = \int_{\theta_1}^{\pi} \sqrt{\frac{4a^2\sin^2(\theta/2)}{2ga[(1 - \cos\theta) - (1 - \cos\theta_1)]}}\, d\theta$$

$$= \sqrt{\frac{2a}{g}} \int_{\theta_1}^{\pi} \frac{\sin(\theta/2)}{\sqrt{\cos\theta_1 - \cos\theta}}\, d\theta$$

$$= 2\sqrt{\frac{a}{g}}\left[-\sin^{-1}\left(\frac{\cos(\theta/2)}{\cos(\theta_1/2)}\right)\right]_{\theta_1}^{\pi} \quad \text{(Check)}$$

$$= \pi\sqrt{a/g}$$

This is what we set out to prove; the time is independent of the initial position.

REVIEW EXERCISES

In Exercises 1–14, sketch the graphs and label all important parts.

1. $\dfrac{x^2}{4} + \dfrac{y^2}{9} = 1$

2. $x^2 = 1 - \dfrac{y^2}{4}$

3. $y^2 - 4x = 4$

4. $x^2 = y - 4x$

5. $2x^2 + 2y^2 = 8$

6. $2x^2 + y^2 + 4x + 4y = 0$

7. $r = \dfrac{2}{1 - 3\cos\theta}$

8. $r = \dfrac{1}{1 + \cos\theta}$

9. $x = \cos t, y = 2\sin t$

10. $x = [3\cos t] + 1, y = [2\sin t] - 1$

11. $4x^2 - 9y^2 - 16x + 18y = 0$

12. $x^2 + 4x + 2y + y^2 - 1 = 0$

13. $x^2 = -1 + \dfrac{y^2}{4}$

14. $2x^2 + y^2 - 2y = 3$

In Exercises 15–20, find an equation of the described curve.

15. The parabola with vertex $(1, 2)$ and focus $(3, 2)$.

16. The parabola with vertex $(-2, 4)$ and directrix $y = 2$.

17. The ellipse with vertices $(5, 4)$, $(-3, 4)$, $(1, 6)$, and $(1, 2)$.

18. The ellipse with center $(-2, 3)$ and three vertices $(1, 3)$, $(-5, 3)$, and $(-2, 4)$.

19. The hyperbola with foci $(\pm 5, 0)$ and vertices $(\pm 2, 0)$.

20. The hyperbola with a vertex $(-1, 2)$ and asymptotes $y = \pm 3x + 2$.

In Exercises 21–24, change to polar coordinates.

21. $(x - 2)^2 + (y - 1)^2 = 9$ **22.** $5x^2 + 2xy = 1$

23. $\dfrac{(x - 1)^2}{3} + \dfrac{(y + 1)^2}{5} = 1$ **24.** $y^2 = 4(x - 2)$

In Exercises 25–28, change to rectangular coordinates.

25. $r^2 = 2 \sin \theta$ **26.** $x = \tan \theta$

27. $r = \dfrac{2}{1 - \cos \theta}$ **28.** $r^2 \cos 2\theta = -1$

In Exercises 29–32, sketch the graphs.

29. $r = 2(1 - \sin \theta)$

30. $3 - 2r = -r \sin \theta$

31. $r = 2 \sin 2\theta \cos 2\theta$

32. $2^{\theta} r = 1; 0 \le \theta \le 2\pi$

In Exercises 33–36, find the slope of the tangent line at the indicated points.

33. $r = 2a$ at $\left(2a, \dfrac{\pi}{6}\right)$

34. $r = a(1 - \cos \theta)$ at $(0, 0)$

35. $r = a^2 \sin 2\theta$ at $\left(a^2, \dfrac{\pi}{4}\right)$ **36.** $r\theta^2 = 1$ at $\left(\dfrac{4}{\pi^2}, \dfrac{\pi}{2}\right)$

In Exercises 37–40, find the area of the region bounded by given curves.

37. $r\theta = \dfrac{1}{4}, \dfrac{\pi}{4} \le \theta \le \dfrac{3\pi}{4}$ **38.** $r = 3 \cos \theta$

$\qquad\qquad\qquad\qquad\qquad r = 2 \sin \theta$

$\qquad\qquad\qquad\qquad\qquad 0 \le \theta \le \dfrac{\pi}{4}$

39. Inside $r^2 = 8a^2 \sin 2\theta$ **40.** $r = 2a$

$\qquad$ Outside $r = 2a$ $\qquad\qquad r = 2a \sin \theta$

$\qquad a > \dfrac{1}{4}, \; 0 \le \theta \le \dfrac{\pi}{2}$ $\qquad 0 \quad \theta \le \dfrac{\pi}{2}$

In Exercises 41–44, sketch the graphs.

41. $x = 2t - 3; y = \sqrt{t}$; for all $t \ge 0$

42. $x = \ln t, y = \ln t^2$; for all $t \ge 1$

43. $x = \sec t + 2, y = 2 \tan t - 1$; for $0 \le t < \dfrac{\pi}{2}$

44. $x = e^{\sqrt{t}}, y = e^t$; for all $t \ge 0$

In Exercises 45–48, find the slope(s) of the tangent(s) at the given point on the graph.

45. $x = t^3 - t, y = e^{2t}$; $(0, e^2)$

46. $x = \ln(t^2 - 2t), y = e^{t/3}$; $(\ln 3, e)$

47. $x = \sin t, y = \tan t$; $(0, 0)$

48. $x = t^3 - t + 1, y = t^2 - 3t$; $(1, 0)$

In Exercises 49–52, find the arc length of the given curve.

49. $x = \dfrac{t^2}{2}, y = \dfrac{t^3}{3}$; $0 \le t \le 3$

50. $r = e^{2\theta}$; $0 \le \theta \le 2$

51. $r = \theta^2$; $0 \le \theta \le 1$

52. $x = e^t \sinh t, y = e^t \cosh t$; $0 \le t \le 1$

In Exercises 53–56, find the surface area generated by revolving the curve about the x-axis.

53. $x = t, y = \dfrac{t^2}{2}$; $0 \le t \le 3$

54. $r = e^{2\theta}$; $0 \le \theta \le \dfrac{\pi}{2}$

55. $r = \sin \theta$; $0 \le \theta \le \dfrac{\pi}{2}$

56. $x = e^t \sinh t, y = e^t \cosh t$; $0 \le t \le 1$

14

VECTORS, CURVES, AND SURFACES IN SPACE

We are now at a turning point in our study of calculus. What came before is called the calculus of functions of one variable; most of the action takes place in a plane. What comes after is called the calculus of functions of several variables; most of the action takes place in space. This chapter is a transition from the old to the new. It bridges the gap in that it still treats functions of one variable, but most of the action takes place in space.

Section 14–1 discusses the notion of a three-dimensional coordinate system, otherwise known as space. Vectors, the single most important topic in this chapter, are defined in Section 14–2. Vectors are used to study lines (Section 14–4), planes (Section 14–5), space curves (Section 14–6), and motion along space curves (Section 14–7). The remaining sections discuss dot products (Section 14–3), cross products (Section 14–8), other aspects of motion such as curvature and components of acceleration (Section 14–9), and some special surfaces (Section 14–10) that will be referred to often in the chapters that follow.

14–1 COORDINATES IN THREE DIMENSIONS

A three-dimensional coordinate system is obtained by adding a third axis to the xy-plane. A z-axis is drawn through the origin perpendicular to both the x-axis and the y-axis. The xyz-system is usually pictured as in Figure 14–1; the y- and z-axes lie in the plane of the paper, and the positive x-axis projects out from the paper. This is called a *right-hand system* because if the fingers of the right hand are curled in the direction from the positive x-axis to the positive

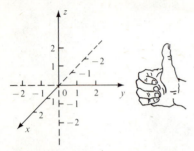

FIGURE 14–1. A right-handed three-dimensional coordinate system.

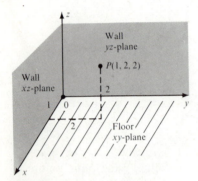

FIGURE 14–2. Using two walls and the floor of a room to visualize a three-dimensional coordinate system.

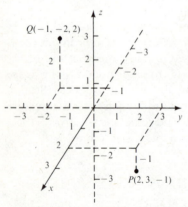

FIGURE 14–3. Locating points in space.

y-axis, then the thumb points in the direction of the positive *z*-axis. The point where the three axes meet is called the *origin*. The points on each of the axes correspond to real numbers as before. The points are labeled and each axis is called a *coordinate axis*. Each pair of coordinate axes determines a *coordinate plane*. This system is called a *three-dimensional coordinate system,* or simply *space.*

It helps to visualize space by using two walls and the floor of a room (Figure 14–2). The origin is where all three meet, the *z*-axis is where the two walls meet, and so on. The coordinate planes divide space into eight parts, and each is called an *octant*. The octant determined by the three positive axes (the room in Figure 14–2) is called the *first octant;* the other octants are not usually given names.

Points in space are located by specifying an ordered triple of numbers (a, b, c). These are, respectively, the *x-*, *y-*, *and z-coordinates* of the point. For example, the point $P(1, 2, 2)$ is displayed in Figure 14–2. Start from origin and mark off 1 unit in the positive *x*-direction, 2 units in the positive *y*-direction, and 2 units in the positive *z*-direction.

EXAMPLE 1 The points $P(2, 3, -1)$ and $Q(-1, -2, 2)$ are displayed in Figure 14–3. Neither point is in the first octant because the coordinates are not all positive. ∎

If one coordinate of a point is 0, then that point lies in one of the coordinate planes. If two coordinates of a point are 0, then that point lies on one of the coordinate axis. This is illustrated in Figure 14–4.

Distance Formula

The distance $d(P, Q)$ between two points in space is found by applying the Pythagorean theorem twice. Let $P(x_1, y_1, z_1)$ and $Q(x_2, y_2, z_2)$ be any two points. We choose points $R(x_1, y_2, z_1)$ and $S(x_2, y_2, z_1)$ so that two right triangles PRS and PQS are formed; this construction is always possible (Figure 14–5). In triangle PRS, we have

$$[d(P, S)]^2 = [d(R, S)]^2 + [d(P, R)]^2$$
$$= |x_2 - x_1|^2 + |y_2 - y_1|^2$$

In triangle PQS, we have

$$[d(P, Q)]^2 = [d(P, S)]^2 + [d(S, Q)]^2$$
$$= [|x_2 - x_1|^2 + |y_2 - y_1|^2] + |z_2 - z_1|^2$$

It follows that

(14.1)

The distance between $P(x_1, y_1, z_1)$ and $Q(x_2, y_2, z_2)$ is
$d(P, Q) = \sqrt{(x_2 - x_1)^2 + (y_2 - y_1)^2 + (z_2 - z_1)^2}$

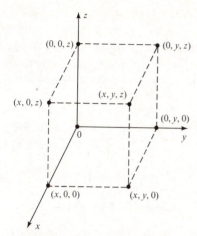

FIGURE 14–4. The coordinates of the eight vertices of a parallelepiped.

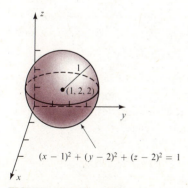

$$d(P, Q) =$$
$$\sqrt{(x_2 - x_1)^2 + (y_2 - y_1)^2 + (z_2 - z_1)^2}$$

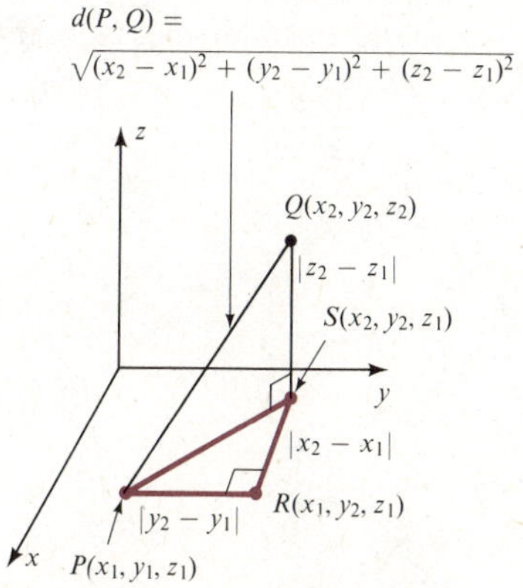

FIGURE 14–5. The distance formula.

EXAMPLE 2 Find the distance between the points $P(2, 3, -1)$ and $Q(-1, -2, 2)$.

Solution According to (14.1),

$$d(P, Q) = \sqrt{(-1 - 2)^2 + (-2 - 3)^2 + (2 + 1)^2}$$
$$= \sqrt{43} \qquad \blacksquare$$

The distance from a point P to the origin is of special interest.

(14.2)

> The distance from $P(x, y, z)$ to the origin is
> $$d(P, O) = \sqrt{x^2 + y^2 + z^2}$$

Graphs

The *graph* of an equation in three variables is the set of all points in space whose coordinates satisfy the equation. Graphs in the plane are usually curves, but because of the added dimension, graphs in space are usually surfaces. For instance, in the plane, the set of all points that are at a distance r from (h, k) is a circle $(x - h)^2 + (y - k)^2 = r^2$. But in space, the set of all points that are at a distance r from (h, k, m) is a *sphere*. The standard equation of a sphere with center (h, k, m) and radius r is

$$(x - h)^2 + (y - k)^2 + (z - m)^2 = r^2$$

EXAMPLE 3 Find the center and radius, and sketch the graph of the sphere $x^2 + y^2 + z^2 - 2x - 4y - 4z = -8$.

Solution First, we complete the squares

$$(x^2 - 2x +) + (y^2 - 4y +) + (z^2 - 4z +) = -8$$
$$(x^2 - 2x + 1) + (y^2 - 4y + 4) + (z^2 - 4z + 4) = -8 + 1 + 4 + 4$$
$$(x - 1)^2 + (y - 2)^2 + (z - 2)^2 = 1$$

It follows that the center is $C(1, 2, 2,)$ and the radius is 1; the graph is sketched in Figure 14–6. $\blacksquare$

The graph of a linear equation is another example. The plane graph of an equation $ax + by = c$ is a line. But, as we will show in a later section (after a discussion of vectors), the space graph of a linear equation in three variables

$$ax + by + cz = d$$

is a *plane*. To graph a plane, simply find any three noncolinear (not on a straight line) points whose coordinates satisfy the equation and draw the plane through them. The easiest points to find are the *intercepts* (where the plane crosses the axes).

FIGURE 14–6. Example 3.

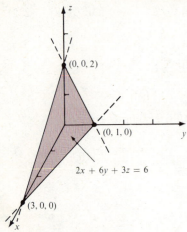

FIGURE 14–7. Example 4.

EXAMPLE 4 Sketch the plane $2x + 6y + 3z = 6$.

Solution To find the intercepts, set two of the variables equal to 0, and solve for the third. Thus, (3, 0, 0), (0, 1, 0), and (0, 0, 2) are the intercepts. Now draw the plane through these points; Figure 14–7 shows the portion of the plane that is in the first octant. ■

If one or more of the variables is missing from a linear equation, then the plane is parallel to the corresponding axes.

EXAMPLE 5 Sketch the space graphs of (a) $x = 2$ and (b) $x + 2y = 2$.

Solution These are linear equations, so their space graphs are planes.

(a) The only intercept of the plane $x = 2$ is (2, 0, 0); it can never cross the y-axis or z-axis because x is never 0. Therefore, $x = 2$ is the plane through (2, 0, 0) that is parallel to the yz-plane (Figure 14–8A).

(b) The only intercepts of the plane $x + 2y = 2$ are (2, 0, 0) and (0, 1, 0); it never crosses the z-axis (why?). Therefore, it is the plane through these points that is parallel to the z-axis (Figure 14–8B). ■

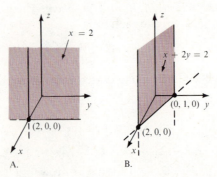

FIGURE 14–8. Example 5.   A.   B.

EXERCISES

In Exercises 1–8, plot the given points.

1. (0, 1, 2)   **2.** (−2, 0, 1)
3. (−1, 2, −2)   **4.** (0, 0, 2)
5. (−1, −2, −3)   **6.** (−1, 0, 1)
7. (1, 0, 0)   **8.** (−1, −1, −1)

In Exercises 9–12, find the distance between the points.

9. (4, −1, 2); (−1, 0, 2)   **10.** (−1, 3, 1); (0, 1, 2)
11. (−2, 3, 6); (0, 1, 5)   **12.** (8, −1, 4); (2, −1, 0)

In Exercises 13–14, use the Pythagorean theorem to show that the given points are vertices of a right triangle.

13. (1, 1, 1); (2, 1, 0); (4, 6, 2)
14. (−3, 1, 0); (2, −1, 1); (2, 4, 11)

To derive the distance formula (14.1), we found points R and S that formed two right triangles. In Exercises 15–16, find the points R and S and verify that triangles PRS and PQS are right triangles.

15. $P(−1, 2, 1)$; $Q(3, 4, −1)$
16. $P(0, −2, 3)$; $Q(−2, 4, 1)$

In Exercises 17–24, find an equation of the sphere with the given properties.

17. Center $(-1, 2, 0)$; radius 2.

18. Center $(3, -1, -4)$; radius 3.

19. Center, the origin; radius $\sqrt{3}$.

20. Center, the origin; radius $\sqrt{5}$.

21. Center $(7, -1, 4)$; tangent to the xy-plane.

22. Center $(3, -1, 2)$; tangent to the yz-plane.

23. Endpoints of a diameter are $(1, 1, 3)$ and $(3, -1, 7)$.

24. Endpoints of a diameter are $(-1, 0, 2)$ and $(-1, 4, -2)$.

In Exercises 25–30, find the center and radius of the sphere with the given equation.

25. $x^2 + y^2 + z^2 = 4$

26. $x^2 + y^2 + z^2 - 2x = 8$

27. $x^2 + y^2 + z^2 + 2x - 2y + 4z = 0$

28. $x^2 + y^2 + z^2 - 6x + 4y = 2$

29. $2x^2 + 2y^2 + 2z^2 - 2x + 4y + 6z = 0$

30. $3x^2 + 3y^2 + 3z^2 = 27$

In Exercises 31–42, sketch the plane with the given equation.

31. $y = 3$

32. $z = 2$

33. $z = -1$

34. $x = -2$

35. $x - z = 1$

36. $y - z = 1$

37. $2y + z = 2$

38. $2x - y = 2$

39. $3x - y + z = 3$

40. $-x + y - z = 1$

41. $x + y + z = 0$

42. $2x - y + 2z = 0$

Optional Exercises

In Exercises 43–50, describe the set of all points in space whose coordinates satisfy the given condition.

43. $xyz = 0$

44. $x^2 + y^2 + z^2 < 1$

45. $x^2 + y^2 = 1$

46. $|x| = 1, |y| = 1, |z| = 1$

47. $\dfrac{x^2}{4} + \dfrac{y^2}{9} = 1$

48. $yz = 0$

49. $x = y$

50. $1 < x^2 + y^2 + z^2 \leq 4$

The Vector Poem

14-2 VECTORS IN THE PLANE; VECTORS IN SPACE

; vectors up your ass; vectors in your face,

This section introduces the notion of a *vector;* it contains the basic definitions along with some of the elementary properties of vectors. Further properties are discussed in succeeding sections. We begin with an example of a vector quantity, and then use the example as a basis for the formal definition that follows.

When a particle moves from point P to point Q, we call the change in position a *displacement* and denote it by the symbol $\overrightarrow{PQ}$. This displacement is indicated geometrically by drawing an arrow with its tail at the original position P and its head, or tip, at the terminal position Q (Figure 14–9A). Displacement arrows have both *magnitude* (length or size) and *direction*. If the particle now moves from Q to R, then the net displacement from P to R is called the *resultant*, or *sum,* of the two displacements $\overrightarrow{PQ}$ and $\overrightarrow{QR}$. The resultant $\overrightarrow{PR}$ is indicated algebraically by writing

(14.3) $\overrightarrow{PR} = \overrightarrow{PQ} + \overrightarrow{QR}$

Geometrically, the resultant is the arrow from P to R. Notice that it is a diagonal of the parallelogram determined by $\overrightarrow{PQ}$ and $\overrightarrow{QR}$ (Figure 14–9B).

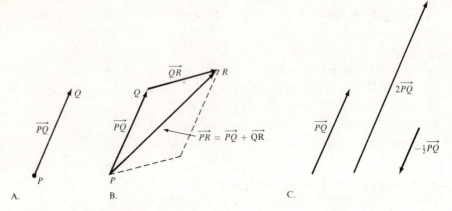

FIGURE 14–9. Displacements are vector quantities.

A. B. C.

Given a displacement $\overrightarrow{PQ}$ and a real number p, we can also form a new displacement

$$(14.4) \quad p\overrightarrow{PQ}$$

Multiplying a displacement by p changes its magnitude by a factor of $|p|$; the direction is unchanged if $p > 0$, but the direction is reversed if $p < 0$ (Figure 14–9C).

Quantities that have magnitude and direction and can be added or multiplied by real numbers according to (14.3) and (14.4) are called *vector quantities*. Displacements are just one example of a vector quantity. The arrows drawn in Figure 14–9 could also represent other vector quantities such as force, velocity, acceleration, and momentum. In contrast, quantities such as length, temperature, and mass that have only magnitude are called *scalar quantities*.

Vectors in the Plane

Vector quantities can be described mathematically if a coordinate system is introduced. We begin by considering a coordinate plane; later we consider a three dimensional system. In the plane, a displacement from $P(x_1, y_1)$ to $Q(x_2, y_2)$ can be described as an ordered pair

$$\overrightarrow{PQ} = (x_2 - x_1, y_2 - y_1)$$

because if you start at P and move $x_2 - x_1$ units in the x-direction and $y_2 - y_1$ units in the y-direction, you will end up at Q. Moreover, any ordered pair, say (a_1, a_2), can be thought of as a displacement from an arbitrary point $P(x, y)$ to the point $Q(x + a_1, y + a_2)$, as illustrated in Figure 14–10A. If another ordered pair (b_1, b_2) represents the displacement from Q to R, then the resultant displacement from P to R is the sum of the ordered pairs

$$\overrightarrow{PR} = (a_1, a_2) + (b_1, b_2) = (a_1 + a_2, b_1 + b_2) \qquad \text{(Figure 14–10B)}$$

Furthermore, multiplication of $\overrightarrow{PQ} = (a_1, a_2)$ by any real number p takes the form

$$p\overrightarrow{PQ} = p(a_1, a_2) = (pa_1, pa_2)$$

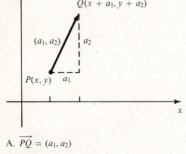

A. $\overrightarrow{PQ} = (a_1, a_2)$

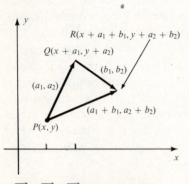

B. $\overrightarrow{PR} = \overrightarrow{PQ} + \overrightarrow{QR} = (a_1 + b_1, a_2 + b_2)$

FIGURE 14–10. Vector quantities described by ordered pairs of real numbers.

These observations motivate the following definitions:

scalar - what you do to a female fish (handwritten margin note)

(14.5)

(1) Each ordered pair (a_1, a_2) of real numbers is called a (plane) **vector.**

(2) The sum of two vectors (a_1, a_2) and (b_1, b_2) is defined as

$$(a_1, a_2) + (b_1, b_2) = (a_1 + b_1, a_2 + b_2)$$

The sum is called the **resultant** of the two vectors.

(3) Real numbers are called **scalars** and the product

$$p(a_1, a_2) = (pa_1, pa_2)$$

of a scalar and a vector is called **scalar multiplication.**

(4) The numbers a_1 and a_2 are called the **components** of the vector (a_1, a_2). That two vectors (a_1, a_2) and (b_1, b_2) are **equal** means that their corresponding components are equal; that is,

$$a_1 = b_1 \quad \text{and} \quad a_2 = b_2$$

We visualize vectors as arrows. Although technically incorrect, we will use the terms *vector* and *arrow* interchangeably. The vector (a_1, a_2) is an arrow with its tail at *any* initial point $P(x, y)$ and its head at the point $Q(x + a_1, y + a_2)$. Notice, however, that this arrow is free to move (parallel to itself) to any initial point and still be the *same vector* because the components remain the same (Figure 14–11). When the initial point happens to be the origin, the vector is called a **position vector.**

We ordinarily use boldface type to denote vectors and only occasionally resort to symbols such as $\overrightarrow{PQ}$. If $\mathbf{a} = (a_1, a_2)$ and $\mathbf{b} = (b_1, b_2)$ then

$$\mathbf{a} + \mathbf{b} = (a_1 + b_1, a_2 + b_2)$$

and, for any scalar p,

$$p\mathbf{a} = (pa_1, pa_2)$$

It follows directly from the definitions that

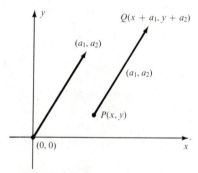

FIGURE 14–11. Both arrows represent the *same* vector because their corresponding components are equal.

(14.6)

For any vectors $\mathbf{a}, \mathbf{b}, \mathbf{c}$ and scalars p, q, we have

(1) $\mathbf{a} + \mathbf{b} = \mathbf{b} + \mathbf{a}$

(2) $(\mathbf{a} + \mathbf{b}) + \mathbf{c} = \mathbf{a} + (\mathbf{b} + \mathbf{c})$

(3) $p(\mathbf{a} + \mathbf{b}) = p\mathbf{a} + p\mathbf{b}$

(4) $(p + q)\mathbf{a} = p\mathbf{a} + q\mathbf{a}$

The **zero vector** $\mathbf{0} = (0, 0)$ cannot be depicted by an arrow because it has zero length. Clearly, $0\mathbf{a} = \mathbf{0}$ and $\mathbf{a} + \mathbf{0} = \mathbf{a}$ for all vectors $\mathbf{a}$, and $p\mathbf{0} = \mathbf{0}$ for all scalars p. The vector $-\mathbf{b}$ is interpreted to mean $(-1)\mathbf{b}$ so that

$$\mathbf{a} - \mathbf{b} = (a_1 - b_1, a_2 - b_2)$$

EXAMPLE 1   Let $\mathbf{a} = (1, 0)$ and $\mathbf{b} = (-4, 6)$. Then

(a) $\mathbf{a} + \mathbf{b} = (1, 0) + (-4, 6) = (-3, 6)$

(b) $3\mathbf{a} + 2\mathbf{b} = 3(1, 0) + 2(-4, 6)$
$$= (3, 0) + (-8, 12) = (-5, 12)$$

(c) $2\mathbf{a} - 5\mathbf{b} = 2(1, 0) - 5(-4, 6)$
$$= (2, 0) + (20, -30) = (22, -30) \quad\blacksquare$$

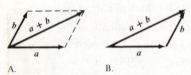

A.   B.

FIGURE 14–12. Two equivalent ways to visualize addition in terms of arrows.

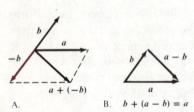

A.   B. $b + (a - b) = a$

FIGURE 14–13. Two equivalent ways to visualize a difference

Addition, in terms of arrows, can be visualized in two equivalent ways. The tails of $\mathbf{a}$ and $\mathbf{b}$ are at the same point and $\mathbf{a} + \mathbf{b}$ is a diagonal of the parallelogram determined by $\mathbf{a}$ and $\mathbf{b}$ (Figure 14–12A). Or, the tail of $\mathbf{b}$ is at the head of $\mathbf{a}$ and $\mathbf{a} + \mathbf{b}$ is the arrow from the tail of $\mathbf{a}$ to the head of $\mathbf{b}$ (Figure 14–12B). The sum $\mathbf{a} + \mathbf{b}$ is the same in both cases.

It is often necessary to represent a difference $\mathbf{a} - \mathbf{b}$ in terms of arrows. This can also be done in two equivalent ways; draw $-\mathbf{b}$ and add it to $\mathbf{a}$ (Figure 14–13A), or think of $\mathbf{a} - \mathbf{b}$ as the arrow that must be added to $\mathbf{b}$ to obtain $\mathbf{a}$ (Figure 14–13B). Note that any two vectors $\mathbf{a}$ and $\mathbf{b}$ determine a parallelogram; *one diagonal is $\mathbf{a} + \mathbf{b}$ and the other is $\mathbf{a} - \mathbf{b}$.*

The **magnitude** (or **length** or **size**) of a vector $\mathbf{a} = (a_1, a_2)$ is denoted by the symbol $|\mathbf{a}|$ and defined by

(14.7) $$|\mathbf{a}| = \sqrt{a_1{}^2 + a_2{}^2}$$

You will recognize this as the length of the line segment (or arrow) from the origin to the point (a_1, a_2) or, equivalently, from $P(x, y)$ to $Q(x + a_1, y + a_2)$. Also, note that the symbol used for magnitude is the same as that used for absolute value. Indeed, many properties are shared by magnitude and absolute value.

(14.8)

> For any vectors $\mathbf{a}$, $\mathbf{b}$ and any scalar p, we have
>
> (1) $|\mathbf{a}| \geq 0$; $|\mathbf{a}| = 0$ if and only if $\mathbf{a} = \mathbf{0}$
>
> (2) $|p\mathbf{a}| = |p||\mathbf{a}|$
>
> (3) $|\mathbf{a} + \mathbf{b}| \leq |\mathbf{a}| + |\mathbf{b}|$ (Triangle inequality)

Properties (1) and (2) are verified directly from the definition of magnitude. The proof of (3) is more complicated, but you can easily convince yourself of its validity using Figure 14–12B and the well known fact that the length of one side of a triangle is never larger than the sum of the lengths of the other two sides. (A correct proof is outlined in an exercise in the next section.)

Scalar multiplication, in terms of arrows, has the following interpretation: multiplying a vector by a scalar p changes its magnitude by a factor of $|p|$ (by 14.8.2); the direction is unchanged if $p > 0$, but the direction is reversed if $p < 0$ (Figure 14–9C).

EXAMPLE 2   Let $\mathbf{a} = (-2, 1)$ and $\mathbf{b} = (1, 3)$. Then

(a) $|\mathbf{a}| = \sqrt{(-2)^2 + 1^2} = \sqrt{5}$ and $|\mathbf{b}| = \sqrt{1^2 + 3^2} = \sqrt{10}$

(b) $|2\mathbf{a}| = |(-4, 2)| = \sqrt{(-4)^2 + 2^2} = \sqrt{20} = 2\sqrt{5}$. Notice that $|2\mathbf{a}| = 2|\mathbf{a}|$ as predicted by (14.8.2).

(c) $-\mathbf{b} = (-1, -3)$; its direction is opposite to $\mathbf{b}$.   ■

Here is an application of vector methods to navigation.

EXAMPLE 3   A jet airplane is heading due north with an airspeed of 500 mph and the wind is blowing to the southeast at 50 mph. What is the ground speed and actual flight path of the jet?

Solution   We can assume that all of the action is taking place in the xy-plane with the positive y-axis pointing north. The arrow marked $\mathbf{a} = (0, 500)$ in Figure 14–14 represents the air velocity of the jet; it is 500 units long and points due north. The arrow marked $\mathbf{w} = (50/\sqrt{2}, -50/\sqrt{2})$ represents the velocity of the wind; it is 50 units long and points southeast. To find the resultant velocity $\mathbf{v}$ of the jet, we add $\mathbf{a}$ and $\mathbf{w}$ according to (14.5).

$$\mathbf{v} = \mathbf{a} + \mathbf{w} = (0, 500) + (50/\sqrt{2}, -50/\sqrt{2})$$
$$= \left(\frac{50}{\sqrt{2}}, \frac{500\sqrt{2} - 50}{\sqrt{2}} \right)$$

Notice that this is a diagonal of the parallelogram determined by $\mathbf{a}$ and $\mathbf{w}$ (Figure 14–14). Now, the magnitude of $\mathbf{v}$ is

$$|\mathbf{v}| = \sqrt{\left(\frac{50}{\sqrt{2}} \right)^2 + \left(\frac{500\sqrt{2} - 50}{\sqrt{2}} \right)^2} \approx 466$$

Thus, the ground speed of the jet is 466 mph; the arrow $\mathbf{v}$ indicates the actual direction of flight.   ■

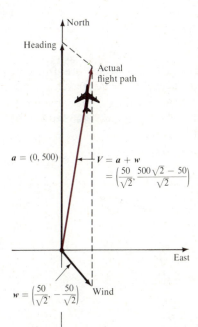

FIGURE 14–14. Example 3.

Vectors in Space

The entire discussion of vectors in the plane can be transferred over to space simply by adding a third component. The definitions analogous to (14.5) are as follows:

(14.9)

(1) Each ordered triple (a_1, a_2, a_3) of real numbers is called a (space) **vector.**

(2) The sum of two vectors $\mathbf{a} = (a_1, a_2, a_3)$ and $\mathbf{b} = (b_1, b_2, b_3)$ is defined as

$$\mathbf{a} + \mathbf{b} = (a_1 + b_1, a_2 + b_2, a_3 + b_3)$$

The sum is called the **resultant** of the two vectors.

(3) The product of a scalar p and a vector $\mathbf{a} = (a_1, a_2, a_3)$ is defined as

$$p\mathbf{a} = (pa_1, pa_2, pa_3)$$

and is called **scalar multiplication.**

(4) The numbers a_1, a_2, and a_3 are called the **components** of $\mathbf{a} = (a_1, a_2, a_3)$. If $\mathbf{b} = (b_1, b_2, b_3)$, then $\mathbf{a}$ and $\mathbf{b}$ are **equal** means that

$$a_1 = b_1, a_2 = b_2, \text{ and } a_3 = b_3$$

Vectors in space are also represented by arrows, but now they are drawn in a three dimensional coordinate system (Figure 14–15). In terms of such arrows, the sum, difference, and scalar multiplication of vectors all have the same interpretations as they do in the plane, Figures 14–9C, 14–12, and 14–13 are valid illustrations for space vectors as well as for plane vectors.

All of the addition properties (14.6) carry over to space vectors.

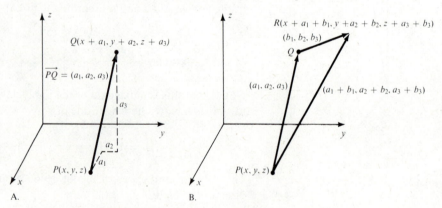

FIGURE 14–15. Vectors in space.   A.   B.

EXAMPLE 4   Let $\mathbf{a} = (1, 0, 3)$ and $\mathbf{b} = (-4, 6, 1)$. Then

(a) $\mathbf{a} + \mathbf{b} = (1, 0, 3) + (-4, 6, 1) = (-3, 6, 4)$

(b) $2\mathbf{a} - 4\mathbf{b} = 2(1, 0, 3) - 4(-4, 6, 1)$

$$= (2, 0, 6) + (16, -24, -4)$$

$$= (18, -24, 2) \quad \blacksquare$$

The zero vector in space is $\mathbf{0} = (0, 0, 0)$. The **magnitude** of a vector $\mathbf{a} = (a_1, a_2, a_3)$ is defined to be

$$|\mathbf{a}| = \sqrt{a_1^2 + a_2^2 + a_3^2}$$

This is the length of the line segment (or arrow) from the origin to the point (a_1, a_2, a_3) or, equivalently, from the point $P(x, y, z)$ to the point $Q(x + a_1, y + a_2, z + a_3)$. With this definition, all properties listed in (14.8) carry over to space.

EXAMPLE 5 Let $\mathbf{a} = (-2, 1, 3)$. Then

(a) $|\mathbf{a}| = \sqrt{(-2)^2 + 1^2 + 3^2} = \sqrt{14}$

(b) $|-3\mathbf{a}| = |(6, -3, -9)| = \sqrt{6^2 + (-3)^2 + (-9)^2}$
$$= \sqrt{126} = 3\sqrt{14}$$

Notice that $|-3\mathbf{a}| = |-3||\mathbf{a}|$ as predicted by (14.8.2). ■

Remark: As you can see, vectors in the plane and vectors in space share the same properties. Although both are important, our primary interest is in space vectors. Therefore, from now on, *the term* vector *will be taken to mean a vector in space.* On those occasions where it is important to do so, we will specify that the vectors being considered are plane vectors.

Unit Vectors

Any vector of length 1 is called a **unit vector,** such vectors are important in our work. Any nonzero vector $\mathbf{a}$ can be made into a unit vector by multiplying it by the scalar $1/|\mathbf{a}|$. This new vector is written

$$\frac{1}{|\mathbf{a}|}\mathbf{a} \quad \text{or} \quad \frac{\mathbf{a}}{|\mathbf{a}|}$$

It has length 1 because of property (2) in (14.8). It also has the same direction as $\mathbf{a}$ because $1/|\mathbf{a}|$ is positive.

EXAMPLE 6 To find a unit vector in the same direction as $\mathbf{a} = (4, -3, 2)$, we first compute $|\mathbf{a}|$.

$$|\mathbf{a}| = \sqrt{4^2 + (-3)^2 + 2^2} = \sqrt{29}$$

Then

$$\mathbf{u} = \frac{\mathbf{a}}{\sqrt{29}} = \left(\frac{4}{\sqrt{29}}, \frac{-3}{\sqrt{29}}, \frac{2}{\sqrt{29}} \right)$$

is the unit vector in the same direction as $\mathbf{a}$. ■

Of all the unit vectors, we single out the three that have the same directions as the positive coordinate axes and assign them special names and symbols.

The three vectors

(14.10) $\quad$ $\mathbf{i} = (1, 0, 0)$ $\quad$ $\mathbf{j} = (0, 1, 0)$ $\quad$ $\mathbf{k} = (0, 0, 1)$

are called the **coordinate unit vectors.**

They are drawn in Figure 14–16A. *Note:* In the plane, there are only two coordinate unit vectors: $\mathbf{i} = (1, 0)$ and $\mathbf{j} = (0, 1)$. The reason they are so important is that any vector can be written as a sum of these vectors. Given any $\mathbf{a} = (a_1, a_2, a_3)$, we can write

$$(a_1, a_2, a_3) = a_1(1, 0, 0) + a_2(0, 1, 0) + a_3(0, 0, 1)$$
$$= a_1\mathbf{i} + a_2\mathbf{j} + a_3\mathbf{k}$$

See Figure 14–16B. Thus,

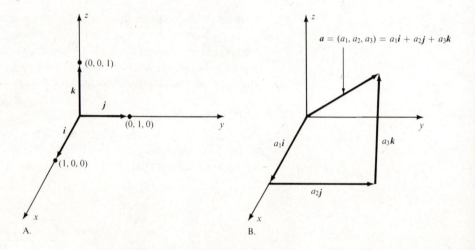

FIGURE 14–16. (A) The three coordinate unit vectors. (B) Any vector **a** can be written as the sum of its components times the coordinate unit vectors.

Given any vector $\mathbf{a} = (a_1, a_2, a_3)$, then

(14.11) $$\mathbf{a} = a_1\mathbf{i} + a_2\mathbf{j} + a_3\mathbf{k}$$

The vectors $a_1\mathbf{i}$, $a_2\mathbf{j}$, and $a_3\mathbf{k}$ are called the **vector components** of **a**.

EXAMPLE 7 $\quad$ Let $\mathbf{a} = 2\mathbf{i} - 3\mathbf{j} + \mathbf{k}$ and $\mathbf{b} = 4\mathbf{i} + 2\mathbf{k}$. Then

(a) $|\mathbf{a}| = \sqrt{2^2 + (-3)^2 + 1^2} = \sqrt{14}$ and $|\mathbf{b}| = \sqrt{4^2 + 0^2 + 2^2} = \sqrt{20} = 2\sqrt{5}$

(b) $2\mathbf{a} = 2(2\mathbf{i} - 3\mathbf{j} + \mathbf{k}) = 4\mathbf{i} - 6\mathbf{j} + 2\mathbf{k}$

(c) $3\mathbf{a} - 2\mathbf{b} = 3(2\mathbf{i} - 3\mathbf{j} + \mathbf{k}) - 2(4\mathbf{i} + 2\mathbf{k})$
$$= 6\mathbf{i} - 9\mathbf{j} + 3\mathbf{k} - 8\mathbf{i} - 4\mathbf{k}$$
$$= -2\mathbf{i} - 9\mathbf{j} - \mathbf{k} \quad \blacksquare$$

Summary

It is important to be thoroughly familiar with the dual nature of vectors. Algebraically, they are ordered triples (or ordered pairs) of real numbers; geometrically, they are arrows in space (or in the plane). The corresponding properties are listed in the following table:

TERM	ALGEBRAIC DEFINITION	GEOMETRIC INTERPRETATION
Vector	Plane: Ordered pair $\mathbf{a} = (a_1, a_2)$ Space: Ordered triple $\mathbf{a} = (a_1, a_2, a_3)$	
Addition	$\mathbf{a} + \mathbf{b} = (a_1, a_2, a_3) + (b_1, b_2, b_3)$ $= (a_1 + b_1, a_2 + b_2, a_3 + b_3)$	
Scalar Multiplication	$p\mathbf{a} = p(a_1, a_2, a_3)$ $= (pa_1, pa_2, pa_3)$	
Magnitude	$\|\mathbf{a}\| = \|(a_1, a_2, a_3)\|$ $= \sqrt{a_1{}^2 + a_2{}^2 + a_3{}^2}$	
Difference	$\mathbf{a} - \mathbf{b} = (a_1, a_2, a_3) - (b_1, b_2, b_3)$ $= (a_1 - b_1, a_2 - b_2, a_3 - b_3)$	
Unit Vector	$\dfrac{\mathbf{a}}{\|\mathbf{a}\|}$	
Coordinate Unit Vectors	$\mathbf{i} = (1, 0, 0) \quad \mathbf{j} = (0, 1, 0)$ $\mathbf{k} = (0, 0, 1)$ (In the plane: $\mathbf{i} = (1, 0)$ and $\mathbf{j} = (0, 1)$	
Vector Components	$\mathbf{a} = (a_1, a_2, a_3)$ $= a_1\mathbf{i} + a_2\mathbf{j} + a_3\mathbf{k}$	

EXERCISES

In Exercises 1–8, find $2\mathbf{a} - 3\mathbf{b}$.

1. $\mathbf{a} = (2, 0, -1), \mathbf{b} = (3, 2, 1)$
2. $\mathbf{a} = (4, -3, 6), \mathbf{b} = (-3, 2, 0)$
3. $\mathbf{a} = (0, -4, 2), \mathbf{b} = (0, 5, -4)$
4. $\mathbf{a} = (1, 0, 0), \mathbf{b} = (8, 2, 0)$
5. $\mathbf{a} = 3\mathbf{i} - 2\mathbf{k}, \mathbf{b} = 4\mathbf{j} + \mathbf{k}$
6. $\mathbf{a} = \mathbf{i} + \mathbf{j}, \mathbf{b} = 2\mathbf{i} - 3\mathbf{k}$
7. $\mathbf{a} = -\mathbf{i} + \mathbf{j} + \mathbf{k}, \mathbf{b} = 3\mathbf{i} - 2\mathbf{k}$
8. $\mathbf{a} = 4\mathbf{i} + 7\mathbf{j} - \mathbf{k}, \mathbf{b} = \mathbf{i}$

In Exercises 9–12, find $|\mathbf{a}|$, $|\mathbf{b}|$, and $|\mathbf{a} + \mathbf{b}|$; verify that $|\mathbf{a} + \mathbf{b}| \le |\mathbf{a}| + |\mathbf{b}|$.

9. $\mathbf{a} = (-1, 0, 2), \mathbf{b} = (0, 2, 0)$
10. $\mathbf{a} = (2, 1, -1), \mathbf{b} = (0, 1, 2)$
11. $\mathbf{a} = 2\mathbf{i} - 3\mathbf{j} + \mathbf{k}, \mathbf{b} = -\mathbf{i} + 2\mathbf{j}$
12. $\mathbf{a} = \mathbf{i} + \mathbf{k}, \mathbf{b} = \mathbf{j} - \mathbf{k}$

In Exercises 13–16, verify that $|p\mathbf{a}| = |p||\mathbf{a}|$.

13. $\mathbf{a} = 3\mathbf{i} - 2\mathbf{j} + \mathbf{k}, p = -3$
14. $\mathbf{a} = -\mathbf{i} + \mathbf{j} + 3\mathbf{k}; p = 2$
15. $\mathbf{a} = 4\mathbf{i} - 3\mathbf{j}; p = 1/2$
16. $\mathbf{a} = 2\mathbf{i} - 3\mathbf{j} + \mathbf{k}; p = -2$

Let $\mathbf{a} = \mathbf{i} + 2\mathbf{j} + \mathbf{k}$ and $\mathbf{b} = 2\mathbf{i} + \mathbf{j} + \mathbf{k}$. In Exercises 17–24, sketch the indicated vector.

17. $2\mathbf{a}$
18. $-\mathbf{a}$
19. $-2\mathbf{b}$
20. $\mathbf{b}$
21. $\mathbf{a} + \mathbf{b}$
22. $2\mathbf{a} + \mathbf{b}$
23. $\mathbf{a} - \mathbf{b}$
24. $\mathbf{a} - 2\mathbf{b}$

In Exercises 25–30, find a unit vector with the same direction as $\mathbf{a}$.

25. $\mathbf{a} = \mathbf{i} + \mathbf{j}$
26. $\mathbf{a} = 2\mathbf{i} - 2\mathbf{j} + \mathbf{k}$
27. $\mathbf{a} = -2(\mathbf{i} + \mathbf{j} + \mathbf{k})$
28. $\mathbf{a} = 3(\mathbf{i} - \mathbf{j} + \mathbf{k})$
29. $\mathbf{a} = -4\mathbf{j}$
30. $\mathbf{a} = 5\mathbf{k}$

31. Given $\mathbf{a} = (1, -1, 2)$, $\mathbf{b} = (0, 2, 7)$, $\mathbf{c} = (2, 0, 1)$, and $\mathbf{d} = (4, -2, 5)$.
   (a) Express $\mathbf{a} - 3\mathbf{b} + 2\mathbf{c}$ in terms of $\mathbf{i}$, $\mathbf{j}$, and $\mathbf{k}$.
   (b) Find scalars p, q, and r, such that $\mathbf{d} = p\mathbf{a} + q\mathbf{b} + r\mathbf{c}$.

32. Same as Exercise 31 for the vectors $\mathbf{a} = (0, 2, -1)$, $\mathbf{b} = (-3, 1, 0)$, $\mathbf{c} = (0, 0, 2)$, and $\mathbf{d} = (-3, -1, 5)$.

33. An airplane is heading due north with an airspeed of 200 mph and the wind is blowing from west to east at 30 mph. Make a sketch of the velocities and find the ground speed of the airplane.

34. A motor boat is heading directly across a river at 10 mph and the current flows downstream at 3 mph. Make a sketch and compute the magnitude of the resultant velocity vector of the boat.

35. Forces of 20 and 50 dynes act simultaneously on an object. The angle between the forces is $40°$. Make a sketch and find the magnitude of the resultant force on the object. *Hint:* Use the Law of Cosines.

36. A particle moves in the xy-plane so that at any time t, its position is at the tip of the position vector $\cos t\mathbf{i} + \sin t\mathbf{j}$. Make a sketch indicating the particle's position at $t = 0, \pi/2, \pi, 3\pi/2$, and 2π. Describe the path.

37. A particle moves in space so that at any time t, its position is at the tip of the position vector $\cos t\mathbf{i} + \sin t\mathbf{j} + \mathbf{k}$. Make a sketch indicating the particle's position at $t = 0, \pi/2, \pi, 3\pi/2, 2\pi$. Describe the path.

38. Same as Exercise 37 only now the position is indicated by the position vector $\cos t\mathbf{i} + \sin t\mathbf{j} + t\mathbf{k}$. (This path is called a circular helix.)

39. Suppose that Chicago is at the origin of an xy-plane. New York is (approximately) 1,000 miles from Chicago on a line making an angle of $10°$ with the positive x-axis. If the velocity vector of the prevailing wind is $25\sqrt{3}\mathbf{i} - 25\mathbf{j}$, at what constant ground speed and in what direction should a jetliner fly in order to make the Chicago-New York run in the scheduled time of 2 hours? (Assume Chicago and New York lie in the xy-plane and the velocity vector of the wind is constant.)

14–3 DOT PRODUCT

We continue our study of vectors giving special emphasis to their dual algebraic-geometric nature. The topic of this section is the *dot product* of two vectors; this notion has a variety of physical and mathematical applications.

(14.13)

If $\mathbf{a} = (a_1, a_2, a_3)$ and $\mathbf{b} = (b_1, b_2, b_3)$, then

$$\mathbf{a} \cdot \mathbf{b} = a_1 b_1 + a_2 b_2 + a_3 b_3$$

is called the **dot product** of $\mathbf{a}$ and $\mathbf{b}$. The symbol $\mathbf{a} \cdot \mathbf{b}$ is read "**a dot b.**"

EXAMPLE 1

(a) $(4, -3, 6) \cdot (-1, 5, -2) = (4)(-1) + (-3)(5) + (6)(-2) = -31$

(b) If $\mathbf{a} = 3\mathbf{i} - \mathbf{j}$ and $\mathbf{b} = 2\mathbf{i} + \mathbf{k}$, then $\mathbf{a} = (3, -1, 0)$, $\mathbf{b} = (2, 0, 1)$, and therefore,

$$\mathbf{a} \cdot \mathbf{b} = (3\mathbf{i} - \mathbf{j}) \cdot (2\mathbf{i} + \mathbf{k}) = (3)(2) + (-1)(0) + (0)(1) = 6 \quad \blacksquare$$

Observe that $\mathbf{a} \cdot \mathbf{b}$ is always a scalar, and for that reason, it is sometimes called the *scalar product* of $\mathbf{a}$ and $\mathbf{b}$. The following properties of the dot product are immediate consequences of the definition.

(14.14)

For any vectors $\mathbf{a}, \mathbf{b}, \mathbf{c}$ and scalars p, q, we have

(1) $\mathbf{a} \cdot \mathbf{a} = |\mathbf{a}|^2$

(2) $\mathbf{a} \cdot \mathbf{b} = \mathbf{b} \cdot \mathbf{a}$

(3) $\mathbf{a} \cdot (\mathbf{b} + \mathbf{c}) = \mathbf{a} \cdot \mathbf{b} + \mathbf{a} \cdot \mathbf{c}$

(4) $(p\mathbf{a}) \cdot (q\mathbf{b}) = (pq)\,\mathbf{a} \cdot \mathbf{b}$

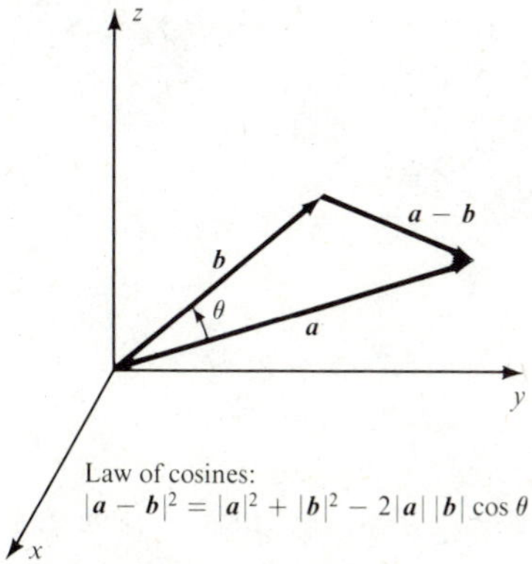

The first property above is very important, so we will prove it just for emphasis. If $\mathbf{a} = (a_1, a_2, a_3)$, then

$$\mathbf{a} \cdot \mathbf{a} = a_1 a_1 + a_2 a_2 + a_3 a_3 = a_1^2 + a_2^2 + a_3^2 = |\mathbf{a}|^2$$

The proofs of the other properties are equally straightforward.

We come now to the geometric interpretation of the dot product. Let $\mathbf{a} = (a_1, a_2, a_3)$ and $\mathbf{b} = (b_1, b_2, b_3)$ be position vectors; that is, the arrows representing them have their tails at the origin. The arrows determine two angles; the one from $\mathbf{a}$ to $\mathbf{b}$ and the other from $\mathbf{b}$ around to $\mathbf{a}$. The angle *between* $\mathbf{a}$ and $\mathbf{b}$ will always mean the one, call it θ, with $0 \leq \theta \leq \pi$.

Let us suppose for the moment that $\mathbf{a}$ and $\mathbf{b}$ are nonzero vectors and that θ satisfies the inequality $0 < \theta < \pi$. Then the vectors $\mathbf{a}, \mathbf{b}$, and $\mathbf{a} - \mathbf{b}$ form a triangle (Figure 14–17). For this triangle, the Law of Cosines yields

$$|\mathbf{a} - \mathbf{b}|^2 = |\mathbf{a}|^2 + |\mathbf{b}|^2 - 2|\mathbf{a}||\mathbf{b}|\cos\theta$$

Law of cosines:
$|a - b|^2 = |a|^2 + |b|^2 - 2|a|\,|b| \cos\theta$

FIGURE 14–17. Geometric interpretation of $\mathbf{a} \cdot \mathbf{b}$. It follows from the Law of Cosines that $\mathbf{a} \cdot \mathbf{b} = |\mathbf{a}||\mathbf{b}| \cos\theta$.

It follows that

$$2|\mathbf{a}||\mathbf{b}|\cos\theta = |\mathbf{a}|^2 + |\mathbf{b}|^2 - |\mathbf{a} - \mathbf{b}|^2 \qquad \text{(Rearrange)}$$
$$= \mathbf{a}\cdot\mathbf{a} + \mathbf{b}\cdot\mathbf{b} - [(\mathbf{a} - \mathbf{b})\cdot(\mathbf{a} - \mathbf{b})] \qquad (14.14.1)$$
$$= \mathbf{a}\cdot\mathbf{a} + \mathbf{b}\cdot\mathbf{b} - \mathbf{a}\cdot\mathbf{a} + 2\mathbf{a}\cdot\mathbf{b} - \mathbf{b}\cdot\mathbf{b} \qquad (14.14.2,3)$$
$$= 2\mathbf{a}\cdot\mathbf{b}$$

Consequently,

$$(14.15) \qquad \boxed{\mathbf{a}\cdot\mathbf{b} = |\mathbf{a}||\mathbf{b}|\cos\theta}$$

This formula was derived under the assumptions that **a** and **b** are nonzero and $0 < \theta < \pi$, but it actually holds in general. If $\mathbf{a} = \mathbf{0}$ or $\mathbf{b} = \mathbf{0}$, then both sides of (14.15) are 0. If $\theta = 0$ or $\theta = \pi$, then **a** and **b** have the same or opposite directions. In either case, we conclude that **b** is some scalar multiple of **a**; $\mathbf{b} = p\mathbf{a}$. Thus,

$$\mathbf{a}\cdot\mathbf{b} = \mathbf{a}\cdot p\mathbf{a} = p\mathbf{a}\cdot\mathbf{a} \qquad (14.14.4)$$
$$= p|\mathbf{a}||\mathbf{a}| \qquad (14.14.1)$$
$$= \pm|\mathbf{a}||\mathbf{b}| \qquad (|p||\mathbf{a}| = |\mathbf{b}|)$$

The sign depends on whether $\theta = 0$ or $\theta = \pi$, so we can replace $\pm$ by $\cos\theta$. Therefore, (14.15) *holds for all vectors* **a** *and* **b** *and all* θ *with* $0 \le \theta \le \pi$. An immediate corollary is the following useful information.

If θ is the angle between the nonzero vectors **a** and **b**, then

$$(14.16) \qquad \cos\theta = \frac{\mathbf{a}\cdot\mathbf{b}}{|\mathbf{a}||\mathbf{b}|} \quad \text{or} \quad \theta = \cos^{-1}\frac{\mathbf{a}\cdot\mathbf{b}}{|\mathbf{a}||\mathbf{b}|}$$

EXAMPLE 2 Find the angle between $\mathbf{a} = \mathbf{i} - 2\mathbf{j} + \mathbf{k}$ and $\mathbf{b} = 4\mathbf{i} + \mathbf{j} - 3\mathbf{k}$.

Solution By (14.16),

$$\cos\theta = \frac{\mathbf{a}\cdot\mathbf{b}}{|\mathbf{a}||\mathbf{b}|} = \frac{(1)(4) + (-2)(1) + (1)(-3)}{\sqrt{1^2 + (-2)^2 + 1^2}\ \sqrt{4^2 + 1^2 + (-3)^2}}$$
$$= \frac{-1}{\sqrt{6}\ \sqrt{26}} \approx -.08$$

So $\theta \approx 95°$ or 1.7 radians. ∎

EXAMPLE 3 If the edges of a cube are 1 unit long, find the angle between a diagonal and an edge.

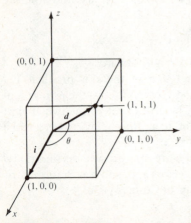

FIGURE 14–18. Example 3.

Solution If the cube is drawn as shown in Figure 14–18, then the desired angle is the angle θ between the vector **i**, representing an edge, and the vector

$\mathbf{d} = \mathbf{i} + \mathbf{j} + \mathbf{k}$ representing a diagonal. Therefore,

$$\theta = \cos^{-1} \frac{\mathbf{i} \cdot \mathbf{d}}{|\mathbf{i}||\mathbf{d}|} = \cos^{-1} \frac{1}{\sqrt{3}} \approx 55° \text{ or } 0.95 \text{ radians} \qquad \blacksquare$$

In (14.14), we listed four properties of the dot product. There is a fifth property that has many useful applications.

(14.17)

> **SCHWARZ INEQUALITY**
> For any vectors $\mathbf{a}$ and $\mathbf{b}$,
>
> $$|\mathbf{a} \cdot \mathbf{b}| \leq |\mathbf{a}||\mathbf{b}|$$

The Schwarz inequality follows from the observation that

$$\mathbf{a} \cdot \mathbf{b} = |\mathbf{a}||\mathbf{b}|\cos \theta$$

and that $|\cos \theta| \leq 1$.

Two nonzero vectors are **parallel** if one is a scalar multiple of the other; that is, they have the same or the opposite direction. Thus, $\mathbf{a}$, $2\mathbf{a}$, and $-3\mathbf{a}$ are all parallel. Two nonzero vectors are **orthogonal** or **perpendicular** if the angle between them is $\pi/2$. If $\mathbf{a}$ and $\mathbf{b}$ are perpendicular, then $\mathbf{a} \cdot \mathbf{b} = |\mathbf{a}||\mathbf{b}| \cos(\pi/2) = 0$. Conversely if $\mathbf{a}$ and $\mathbf{b}$ are nonzero vectors that form the angle θ, then $\mathbf{a} \cdot \mathbf{b} = 0$ can only mean that $\cos \theta = 0$ so $\theta = \pi/2$. It follows that

(14.18) | Two vectors $\mathbf{a}$ and $\mathbf{b}$ are orthogonal if and only if $\mathbf{a} \cdot \mathbf{b} = 0$.

By convention, we say that the zero vector $\mathbf{0}$ is parallel *and* perpendicular to every vector $\mathbf{a}$.

EXAMPLE 4 Show that the points $P(1, 3, -6)$, $Q(0, 2, 1)$, and $R(7, 2, 2)$ are the vertices of a right triangle.

Solution One way to do this is to show that the lengths of the sides (that is, the distances between the points) satisfy an equation of the form $a^2 + b^2 = c^2$. But we can also consider the sides as vectors and use (14.18). The vector $\overrightarrow{PQ}$ from $P(1, 3, -6)$ to $Q(0, 2, 1)$ is

$$\overrightarrow{PQ} = (0 - 1, 2 - 3, 1 - (-6)) = (-1, -1, 7)$$

Similarly, the vector $\overrightarrow{QR}$ from $Q(0, 2, 1)$ to $R(7, 2, 2)$ is

$$\overrightarrow{QR} = (7 - 0, 2 - 2, 2 - 1) = (7, 0, 1)$$

Since $\overrightarrow{PQ} \cdot \overrightarrow{QR} = (-1)(7) + (-1)(0) + (7)(1) = 0$, we know that they are perpendicular and the triangle is a right triangle. $\blacksquare$

Projections and Components

If $\mathbf{b} \neq \mathbf{0}$, the equation $\mathbf{a} \cdot \mathbf{b} = |\mathbf{a}||\mathbf{b}| \cos \theta$ can be rewritten as

$$|\mathbf{a}|\cos \theta = \frac{\mathbf{a} \cdot \mathbf{b}}{|\mathbf{b}|}$$

The left side is the projection of $\mathbf{a}$ onto $\mathbf{b}$ (Figure 14–19). The scalar $|\mathbf{a}|\cos \theta$ is called the **component of a on b** and is denoted by $\text{comp}_b\mathbf{a}$.

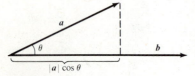

FIGURE 14–19. The projection of a on b.

(14.19)

> If $\mathbf{b} \neq \mathbf{0}$, then
>
> $$\text{comp}_b\mathbf{a} = |\mathbf{a}|\cos \theta = \frac{\mathbf{a} \cdot \mathbf{b}}{|\mathbf{b}|}$$

Figure 14–20 illustrates that $\text{comp}_b\mathbf{a}$ is positive if $0 \leq \theta < \pi/2$, is 0 if $\theta = \pi/2$, and is negative if $\pi/2 < \theta \leq \pi$.

EXAMPLE 5 If $\mathbf{a} = \mathbf{i} + \mathbf{j} - \mathbf{k}$ and $\mathbf{b} = 2\mathbf{i} - \mathbf{j} + 2\mathbf{k}$, find $\text{comp}_b\mathbf{a}$ and $\text{comp}_a\mathbf{b}$.

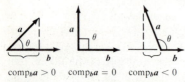

$\text{comp}_b a > 0 \quad \text{comp}_b a = 0 \quad \text{comp}_b a < 0$

FIGURE 14–20. The component of a on b can be positive, zero, or negative.

Solution By (14.19),

$$\text{comp}_b\mathbf{a} = \frac{\mathbf{a} \cdot \mathbf{b}}{|\mathbf{b}|} = \frac{2 - 1 - 2}{\sqrt{2^2 + 1^2 + 2^2}} = \frac{-1}{3}$$

$$\text{comp}_a\mathbf{b} = \frac{\mathbf{a} \cdot \mathbf{b}}{|\mathbf{a}|} = \frac{-1}{\sqrt{3}} \qquad \blacksquare$$

Work

If an object is moved a distance d in a straight line by a force $\mathbf{F}$ applied *along the line of motion*, then the work done is the magnitude of the force times the distance. Work is positive or negative depending on whether the force is applied in the direction of motion or opposite to it. That you already know. But if the force $\mathbf{F}$ is applied in any other direction, then some of the force is wasted. *Only the component of $\mathbf{F}$ along the line of motion does work* (Figure 14–21). Suppose a force $\mathbf{F}$ causes a displacement $\mathbf{D}$. Then the work W done by $\mathbf{F}$ is defined as

$$W = (\text{comp}_D\mathbf{F})|\mathbf{D}|$$
$$= (|\mathbf{F}|\cos \theta)|\mathbf{D}|$$
$$= \mathbf{F} \cdot \mathbf{D}$$

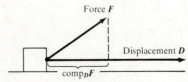

Force $\mathbf{F}$

Displacement $\mathbf{D}$

$\text{comp}_D\mathbf{F}$

FIGURE 14–21. Only the component of F along the line of motion does work.

(14.20)

> The work done by a force $\mathbf{F}$ over a displacement $\mathbf{D}$ is
>
> $$W = \mathbf{F} \cdot \mathbf{D}$$

EXAMPLE 6   Find the work done by $\mathbf{F} = 2\mathbf{i} + 2\mathbf{j} + \mathbf{k}$ in moving an object from the origin to $P(1, 1, 1)$.

Solution   The vector $\mathbf{D} = (1, 1, 1)$ represents the displacement. Thus, according to (14.20), the work done by $\mathbf{F}$ is

$$W = \mathbf{F} \cdot \mathbf{D} = (2)(1) + (2)(1) + (1)(1) = 5$$

If the magnitude of $\mathbf{F}$ is measured in newtons (N) and distance in meters (m), then the work done is 5 N-m or, equivalently, 5 joules. If the units of measurement are lbs and ft, then the work done is 5 ft-lbs. ∎

EXAMPLE 7   A weight of 50 lbs is on a 4 foot incline that makes an angle of 30° with the horizontal. How much work is done by the force of gravity as the weight slides down the incline?

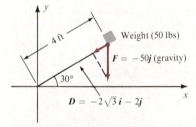

Solution   We assume that the action takes place in the xy-plane ($z = 0$). The force $\mathbf{F}$ of gravity in this case is 50 lbs straight down; that is, $\mathbf{F} = -50\mathbf{j}$. In Figure 14–22, the coordinates of the weight are $(4 \cos 30°, 4 \sin 30°) = (2\sqrt{3}, 2)$ so $\mathbf{D} = -2\sqrt{3}\mathbf{i} - 2\mathbf{j}$ is the displacement as the weight slides down the incline. Therefore, the work done by gravity is

$$W = \mathbf{F} \cdot \mathbf{D} = (-50\mathbf{j}) \cdot (-2\sqrt{3}\mathbf{i} - 2\mathbf{j})$$
$$= 100 \text{ ft-lbs} \quad ∎$$

FIGURE 14–22. Example 7.

EXERCISES

In Exercises 1–20, let $\mathbf{a} = (-1, 3, 0)$, $\mathbf{b} = (2, -1, 2)$, $\mathbf{c} = (1, 4, 1)$, and $\mathbf{d} = (3, 1, -1)$. Find the indicated scalars.

1. $\mathbf{a} \cdot \mathbf{b}$ 2. $\mathbf{c} \cdot \mathbf{d}$
3. $\mathbf{b} \cdot \mathbf{c}$ 4. $\mathbf{a} \cdot \mathbf{d}$
5. $\mathbf{a} \cdot (\mathbf{b} + \mathbf{c})$ 6. $\mathbf{d} \cdot (\mathbf{a} + \mathbf{c})$
7. $(\mathbf{c} + \mathbf{d}) \cdot 2\mathbf{a}$ 8. $(\mathbf{b} + \mathbf{c}) \cdot 4\mathbf{d}$
9. $\mathbf{a} \cdot \mathbf{a}$ 10. $\mathbf{b} \cdot \mathbf{b}$
11. $\mathbf{c} \cdot \mathbf{c}$ 12. $\mathbf{d} \cdot \mathbf{d}$
13. $4\mathbf{a} \cdot 3\mathbf{b}$ 14. $-2\mathbf{c} \cdot 8\mathbf{d}$
15. $\text{comp}_a\mathbf{b}$ 16. $\text{comp}_c\mathbf{d}$
17. $\text{comp}_{(a+b)}\mathbf{c}$ 18. $\text{comp}_{(c+d)}\mathbf{a}$
19. $\text{comp}_d(\mathbf{a} + \mathbf{b})$ 20. $\text{comp}_a(\mathbf{c} + \mathbf{d})$

In Exercises 21–24, find the angle θ (in degrees) between $\mathbf{a}$ and $\mathbf{b}$.

21. $\mathbf{a} = (1, 1, 1)$, $\mathbf{b} = (0, 1, 2)$
22. $\mathbf{a} = (-2, 1, 0)$, $\mathbf{b} = (4, 0, 1)$
23. $\mathbf{a} = \mathbf{i} + \mathbf{k}$, $\mathbf{b} = \mathbf{i} + \mathbf{j} - 2\mathbf{k}$
24. $\mathbf{a} = -\mathbf{i} + \mathbf{j}$, $\mathbf{b} = 2\mathbf{j} + 3\mathbf{k}$

In Exercises 25–26, the coordinates of three points P, Q, and R are given. Find the angle PQR (in degrees).

25. $P(1, 2, 3)$, $Q(-2, 1, 0)$, $R(4, 1, 5)$
26. $P(-1, 0, 2)$, $Q(0, 2, 3)$, $R(-1, 0, 1)$
27. Find a value of c that makes $\mathbf{a} = (2, c, 2)$ orthogonal to $\mathbf{b} = (1, 1, 1)$.
28. Same as Exercise 27 with $\mathbf{a} = (0, 3, c)$ and $\mathbf{b} = (-1, 2, 1)$.
29. Find a unit vector $\mathbf{a} = (x, y, z)$ that is orthogonal to both $\mathbf{b} = (1, 0, 1)$ and $\mathbf{c} = (0, 1, 1)$.
30. Same as Exercise 29 with $\mathbf{b} = (0, 2, 1)$ and $\mathbf{c} = (1, -1, 0)$.
31. An algebraic description of the set of all vectors orthogonal to $\mathbf{a} = \mathbf{i}$ is "the set of all vectors of the form $\mathbf{b} = x\mathbf{i} + y\mathbf{j} + z\mathbf{k}$ with $x = 0$." What is an

algebraic description of the set of all vectors orthogonal to $\mathbf{j}$? To $\mathbf{k}$? What are the geometric descriptions?

32. (Continuation of Exercise 31). An algebraic description of the set of all vectors orthogonal to $\mathbf{a} = 2\mathbf{i} - \mathbf{j}$ is "the set of all vectors of the form $\mathbf{b} = x\mathbf{i} + y\mathbf{j} + z\mathbf{k}$ with $2x - y = 0$." Check this out. What is an algebraic description of the set of vectors orthogonal to $\mathbf{a} = \mathbf{i} + \mathbf{k}$? To $\mathbf{a} = \mathbf{j} + 3\mathbf{k}$? How would you describe these sets geometrically?

33. (Continuation of Exercise 32). Find an algebraic description of the set of all vectors orthogonal to $\mathbf{a} = 2\mathbf{i} - 3\mathbf{j} + \mathbf{k}$. What is a geometric description of this set?

34. Find an algebraic description of the set of all vectors $\mathbf{c}$ that are orthogonal to both (a) $\mathbf{a} = \mathbf{i}$ and $\mathbf{b} = \mathbf{j}$, (b) $\mathbf{a} = \mathbf{i} + \mathbf{j}$ and $\mathbf{b} = \mathbf{j} + \mathbf{k}$. How would you describe these sets geometrically?

---

35. Find the work done by $\mathbf{F} = 3\mathbf{i} + 4\mathbf{j}$ newtons in moving an object along the line segment from $P(1, 2, 3)$ to $Q(2, 8, 4)$. Distance is measured in meters.

36. A carton is pulled along the (level) floor with a rope that makes an angle of $60°$ with the floor. If the force of friction (which acts opposite to the motion) is 50 lbs, what minimum magnitude of force must be applied to the rope to move the carton?

37. A weight of 150 lbs rests on an incline of $20°$. What is the minimum magnitude of the force of friction?

Power

In Section 6–4, we defined power as the *rate* at which work is done. Work $= \mathbf{F} \cdot \mathbf{D}$, and if $\mathbf{F}$ is constant, then the rate at which it is doing work is the rate at which it is moving an object along the line of $\mathbf{D}$. But the rate at which an object moves is its velocity, which is now represented by a vector $\mathbf{v}$ because it has magnitude (*speed*) and direction. It follows that the power P provided by a constant force $\mathbf{F}$ in moving an object with velocity $\mathbf{v}$ is given by the formula

$$P = \mathbf{F} \cdot \mathbf{v}$$

The units are joules/sec (watts) or ft-lbs/sec and 1 horsepower $= 550$ ft-lbs/sec $= 746$ watts.

38. A 2,000 lb automobile travels 60 mph ($=88$ ft/sec) up an incline that makes an angle of $1.08°$ with the horizontal. What horsepower is expended by the motor?

39. A force of $\mathbf{F} = 2\mathbf{i} + 2\mathbf{j} + \mathbf{k}$ newtons moves an object in the direction $\mathbf{D} = \mathbf{i} + 3\mathbf{j} + \mathbf{k}$ with a constant velocity $\mathbf{v}$ m/sec. The force does work at the rate of 5 watts. (a) What is the velocity? (b) How much work is done in 3 seconds?

---

40. A *rhombus* is a parallelogram whose sides have equal length. Show that the diagonals of any rhombus are orthogonal.

41. Use the Schwarz inequality (14.17) to prove the triangle inequality $|\mathbf{a} + \mathbf{b}| \leq |\mathbf{a}| + |\mathbf{b}|$. *Hint:* Write $|\mathbf{a} + \mathbf{b}|^2 = (\mathbf{a} + \mathbf{b}) \cdot (\mathbf{a} + \mathbf{b})$ and expand; then use (14.17) on the middle term.

42. (*Parallelogram law*). In any parallelogram, the sum of the squares of the diagonals equals the sum of the squares of the four sides. In terms of vectors, this says that $|\mathbf{a} + \mathbf{b}|^2 + |\mathbf{a} - \mathbf{b}|^2 = 2|\mathbf{a}|^2 + 2|\mathbf{b}|^2$ where $\mathbf{a}$ and $\mathbf{b}$ determine the parallelogram (draw a picture). Prove this equality.

14–4 LINES

Lines in space can be represented mathematically by three types of equations: *parametric* equations, *vector* equations, or *symmetric* equations. We begin with the parametric equations.

Let $P(x_0, y_0, z_0)$ be any point in space and let $\mathbf{d} = (a, b, c)$ be any nonzero position vector. We want to describe the line L through P that is parallel to $\mathbf{d}$. Figure 14–23 indicates that if $Q(x, y, z)$ is any other point on L, then the vector $\overrightarrow{PQ} = (x - x_0, y - y_0, z - z_0)$ must be parallel to (that is, a scalar multiple of) $\mathbf{d}$.

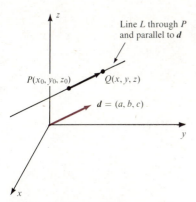

FIGURE 14–23. A point Q is on L if and only if the vector $\overrightarrow{PQ}$ is parallel to (that is, a scalar multiple of) **d**.

Therefore,

(14.21) $\quad (x - x_0, y - y_0, z - z_0) = t(a, b, c) = (ta, tb, tc)$

The converse is also true. If any vector $\overrightarrow{PQ}$ is a scalar multiple of **d**, then Q must be on L. Thus, as t ranges from $-\infty$ to ∞, all of L is traced out. Since equal vectors must have equal components, it follows from (14.21) that $x - x_0 = at$, $y - y_0 = bt$, and $z - z_0 = ct$. Therefore,

(14.22)

> The line L through $P(x_0, y_0, z_0)$ and parallel to $\mathbf{d} = (a, b, c)$ is the set of all points $Q(x, y, z)$ with
>
> $$x = x_0 + at, \quad y = y_0 + bt, \quad z = z_0 + ct \quad -\infty < t < \infty$$

The equations in (14.22) are called **parametric equations** of L; any vector parallel to L is called a **direction vector** of L.

EXAMPLE 1 Find parametric equations of the line L through $(-10, 2, 1)$ and parallel to $\mathbf{d} = (5, -4, 8)$. Where does L pierce the yz-plane?

Solution Using (14.22) with the given values, we have

$$x = -10 + 5t, \quad y = 2 - 4t, \quad z = 1 + 8t \quad -\infty < t < \infty$$

The line will pierce the yz-plane when $x = 0$; thus, we set $0 = -10 + 5t$ and find $t = 2$. For this value of t,

$$y = 2 - 4(2) = -6 \quad \text{and} \quad z = 1 + 8(2) = 17$$

Therefore, L intersects the yz-plane at $(0, -6, 17)$. ∎

EXAMPLE 2 *(Line through two points).* Find parametric equations of the line L through the points $P(2, 1, 3)$ and $Q(0, 1, 5)$.

Solution The vector

$$\overrightarrow{PQ} = (0 - 2, 1 - 1, 5 - 3) = (-2, 0, 2)$$

is a direction vector of L. To use (14.22), we choose either of the points P or Q (we choose P), and then

$$x = 2 - 2t, \quad y = 1, \quad z = 3 + 2t \quad -\infty < t < \infty$$

Remark: Notice that y is constant because the y-component of any direction vector is 0; this means that L is parallel to the xz-plane. ∎

We now take up the vector equation of a line. If Equation (14.21) is rewritten as

$$(x, y, z) - (x_0, y_0, z_0) = t\mathbf{d}$$

and we set $\mathbf{r} = (x, y, z)$ and $\mathbf{r}_0 = (x_0, y_0, z_0)$, then

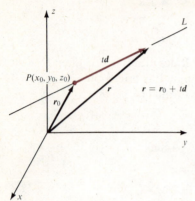

FIGURE 14–24. A vector equation of the line L.

(14.23) $\boxed{\mathbf{r} = \mathbf{r}_0 + t\mathbf{d} \quad -\infty < t < \infty}$

is a **vector equation** of the line L. The interpretation is that as t varies from $-\infty$ to ∞, the tip of $\mathbf{r}$ traces out the line L (Figure 14–24).

EXAMPLE 3   Find a vector equation of the line through $P(-2, 5, 4)$ that has a direction vector $\mathbf{d} = \mathbf{i} - 2\mathbf{j} + 3\mathbf{k}$.

Solution   Since $\mathbf{r}_0$ is the vector from the origin to P, we have

$$\mathbf{r}_0 = -2\mathbf{i} + 5\mathbf{j} + 4\mathbf{k}$$

and, by (14.23),

$$\begin{aligned} \mathbf{r} &= \mathbf{r}_0 + t\mathbf{d} \\ &= (-2\mathbf{i} + 5\mathbf{j} + 4\mathbf{k}) + t(\mathbf{i} - 2\mathbf{j} + 3\mathbf{k}) \\ &= (-2 + t)\mathbf{i} + (5 - 2t)\mathbf{j} + (4 + 3t)\mathbf{k} \end{aligned}$$

is a vector equation of the line.   ∎

And now, the third form; the symmetric equations. For parametric equations of a line L

$$x = x_0 + at, \quad y = y_0 + bt, \quad z = z_0 + ct$$

if a, b, or c is 0, then the corresponding variable x, y, or z is fixed at x_0, y_0, or z_0. If they are not 0, then we can eliminate the parameter by solving each equation for t. This process yields a set of equalities

(14.24) $\boxed{\dfrac{x - x_0}{a} = \dfrac{y - y_0}{b} = \dfrac{z - z_0}{c}}$

called **symetric equations** of the line L. As you will see in the next section, the symmetric form represents L as the intersection of three planes.

EXAMPLE 4   Find symmetric equations of the line through $P(3, -2, 1)$ and $Q(1, 0, 1)$.

Solution   We take $\overrightarrow{PQ} = (-2, 2, 0)$ as a direction vector so $a = -2$, $b = 2$, and $c = 0$. Notice that the z-component is zero; that means that $z = z_0$ and the third equation in (14.24) will not appear. Choose P as the initial point so that $x_0 = 3$, $y_0 = -2$, and $z_0 = 1$. Thus,

$$\frac{x - 3}{-2} = \frac{y + 2}{2}; \quad z = 1$$

are symmetric equations of the line.   ∎

EXAMPLE 5 *(Intersecting lines).* Find the point of intersection of the two lines

$$L_1: \quad x = 1 + 3t, y = 7 + 4t, z = -5 - 2t$$
$$L_2: \quad x = 3 - u, \ y = 3 + 2u, z = 1 - 3u$$

Solution First, notice that we use different parameters for each line because the lines may meet for different values of t and u. This situation is similar to the one encountered in graphing with polar coordinates in Chapter 13. To find the intersection, we set

(1) $1 + 3t = 3 - u$

(2) $7 + 4t = 3 + 2u$

(3) $-5 - 2t = 1 - 3u$

We need only two of these equations to find t and u; choose any two. In this example, it is easy to solve (1) for u and substitute in (2). Thus, $u = 2 - 3t$ and

$$7 + 4t = 3 + 2(2 - 3t)$$
$$= 7 - 6t$$

It follows that $t = 0$ and $u = 2 - 3(0) = 2$. You can now easily verify that both values yield the same point $(1, 7, -5)$. This is the point of intersection. ∎

Given intersecting lines L_1, with direction vector $\mathbf{d}_1$, and L_2, with direction vector $\mathbf{d}_2$, we define the angle between L_1 and L_2 to be the angle between $\mathbf{d}_1$ and $\mathbf{d}_2$. This definition needs clarification because if $\mathbf{d}$ is a direction vector of a line, so is $-\mathbf{d}$. Therefore, depending on our choice of a direction vector, the angle between the lines could be θ or $\pi - \theta$. *Let us agree to always choose the smaller angle; that is $0 \le \theta \le \pi/2$.*

EXAMPLE 6 Find the angle between the lines

$$\frac{x - 1}{-3} = \frac{y - 7}{-4} = \frac{z + 5}{2} \quad \text{and} \quad \frac{x - 1}{-1} = \frac{y - 7}{2}; \quad z = -5$$

Solution You can check that both lines contain the point $(1, 7, -5)$, so they intersect. They are both represented by symmetric equations, so the denominators are the components of their direction vectors. Thus, $\mathbf{d}_1 = (-3, -4, 2)$; the fact that z is constant for the second line means that the z-component of $\mathbf{d}_2$ is 0, and $\mathbf{d}_2 = (-1, 2, 0)$. The angle between $\mathbf{d}_1$ and $\mathbf{d}_2$ is

$$\theta = \cos^{-1} \frac{\mathbf{d}_1 \cdot \mathbf{d}_2}{|\mathbf{d}_1||\mathbf{d}_2|} = \cos^{-1} \frac{-5}{\sqrt{29} \sqrt{5}} \approx 115°$$

By our convention, $180° - \theta = 65°$ is the angle between the lines. ∎

Direction Cosines

The angles α, β, and γ that a nonzero position vector $\mathbf{a}$ makes with the coordinate vectors are called the **direction angles** of $\mathbf{a}$ (Figure 14–25). The cosines of these

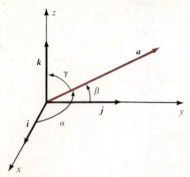

FIGURE 14–25. The direction angles of a position vector.

angles are called the **direction cosines** of **a**. If $\mathbf{a} = (a_1, a_2, a_3)$, then

$$\cos \alpha = \frac{\mathbf{a} \cdot \mathbf{i}}{|\mathbf{a}||\mathbf{i}|} = \frac{a_1}{|\mathbf{a}|} \quad \text{or} \quad a_1 = |\mathbf{a}| \cos \alpha$$

Similarly, $a_2 = |\mathbf{a}| \cos \beta$ and $a_3 = |\mathbf{a}| \cos \gamma$, and it follows that

(14.25) $\qquad \boxed{\mathbf{a} = |\mathbf{a}|(\cos \alpha \mathbf{i} + \cos \beta \mathbf{j} + \cos \gamma \mathbf{k})}$

This says that any nonzero vector can be written in terms of its direction cosines. Moreover, since the magnitude of the vector on the right must equal $|\mathbf{a}|$, it follows that

(14.26) $\qquad \boxed{\cos^2 \alpha + \cos^2 \beta + \cos^2 \gamma = 1}$

EXAMPLE 7 Find the direction cosines of $\mathbf{a} = 2\mathbf{i} + 2\mathbf{j} - \mathbf{k}$ and verify that (14.26) holds.

Solution Write the vector as in (14.25), and then the coefficients of **i**, **j**, and **k** are the direction cosines. Since $|\mathbf{a}| = 3$,

$$\mathbf{a} = 3\left(\frac{2}{3}\mathbf{i} + \frac{2}{3}\mathbf{j} - \frac{1}{3}\mathbf{k}\right)$$

Thus, $\cos \alpha = 2/3$, $\cos \beta = 2/3$, $\cos \gamma = -1/3$, and

$$\cos^2 \alpha + \cos^2 \beta + \cos^2 \gamma = \frac{4}{9} + \frac{4}{9} + \frac{1}{9} = 1 \qquad \blacksquare$$

EXAMPLE 8 A direction vector **d** of a line L has direction cosines $\sqrt{2}/2, -1/2$, and $1/2$. Find a vector equation for L given that it passes through $(4, -8, 3)$.

Solution We can use the direction cosines of **d** to formulate a parallel unit vector

$$\mathbf{u} = \frac{\sqrt{2}}{2}\mathbf{i} - \frac{1}{2}\mathbf{j} + \frac{1}{2}\mathbf{k}$$

It follows that **u** is also a direction vector of L. Since L passes through $(4, -8, 3)$, we apply (14.23) and obtain

$$\mathbf{r} = (4\mathbf{i} - 8\mathbf{j} + 3\mathbf{k}) + t\left(\frac{\sqrt{2}}{2}\mathbf{i} - \frac{1}{2}\mathbf{j} + \frac{1}{2}\mathbf{k}\right) \quad -\infty < t < \infty$$

as a vector equation of L. $\qquad \blacksquare$

EXERCISES

In Exercises 1–8, find parametric equations of the described line.

1. Passes through $P(2, -1, 3)$; direction vector $\mathbf{d} = \mathbf{i} - \mathbf{k}$.

2. Passes through $P(-1, 0, 4)$; direction vector $\mathbf{d} = 2\mathbf{i} + 3\mathbf{j}$.

3. Passes through $P(1, 3, 5)$ and $Q(5, 0, 1)$.

4. Passes through $P(8, 1, 0)$ and $Q(2, 4, 1)$.

5. Passes through $P(2, 5, 3)$ and parallel to the line in Exercise 3.

6. Passes through $P(-3, 4, 1)$ and parallel to the line in Exercise 4.

7. Passes through $P(4, 0, 1)$; direction vector has direction angles $\alpha = 60°, \beta = 120°, \gamma = 45°$.

8. Passes through $P(0, 1, -2)$; direction vector has direction angles $\alpha = 45°, \beta = 60°, \gamma = 120°$.

In Exercises 9–12, find a vector equation of the described line.

9. Passes through $P(0, 1, 2)$; direction vector $\mathbf{d} = \mathbf{i} + 2\mathbf{j} - \mathbf{k}$.

10. Passes through $P(3, -2, 1)$ and $Q(4, 1, 0)$.

11. Passes through the origin and parallel to the line in Exercise 9.

12. Passes through $P(0, -1, 4)$; direction vector has direction cosines $1/2, 1/4$, and $\sqrt{11/16}$.

In Exercises 13–16, find symmetric equations of the described line.

13. Passes through $P(8, 5, 1)$ and $Q(8, 4, 3)$.

14. Passes through $P(0, 0, 4)$; direction vector $\mathbf{d} = \mathbf{i} + \mathbf{k}$.

15. Passes through $P(-3, 2, 5)$; direction vector has direction cosines $-1/2, \sqrt{11}/6, 2/3$.

16. Passes through $P(6, 1, 0)$ and parallel to the line in Exercise 14.

17. Find a vector equation of the line through $P(-5, 6, 0)$ and parallel to the line

$$\frac{x - 1}{4} = \frac{z + 3}{3}; \quad y = 2$$

18. Find parametric equations of the line through $P(4, -1, 3)$ and parallel to the line $\mathbf{r} = (\mathbf{i} + \mathbf{k}) + t(3\mathbf{i} - \mathbf{j})$.

In Exercises 19–22, find the point of intersection and the angle between the lines.

19. $x = t, y = -t, z = -6 + 2t$ and $x = 1 - u, y = 1 + 3u, z = 2u$

20. $\mathbf{r} = (2 + 3t)\mathbf{i} - (4 + 2t)\mathbf{j} - (1 - 4t)\mathbf{k}$ and $\mathbf{r} = (6 + 4u)\mathbf{i} - (2 - 2u)\mathbf{j} - (3 + 2u)\mathbf{k}$

21. $\mathbf{r} = (1 + t)\mathbf{i} + (3 + 2t)\mathbf{j} + (2 - 5t)\mathbf{k}$ and $\mathbf{r} = (2 + 2u)\mathbf{i} + (5 + u)\mathbf{j} - (3 - u)\mathbf{k}$

22. $\dfrac{x + 1}{-2} = \dfrac{y - 3}{3} = \dfrac{z}{5}$ and $\dfrac{x - 3}{2} = \dfrac{y - 1}{-1} = \dfrac{z - 6}{3}$

23. Where does the line

$$\frac{x - x_0}{a} = \frac{y - y_0}{b} = \frac{z - z_0}{c}$$

intersect each of the coordinate planes?

24. Show that the lines $x = 3t, y = 2 + t, z = 2 - t$ and $x = 2 - u, y = u, z = 3 + u$ do not intersect.

25. If the direction angles of $\mathbf{d}$ are α, β, and γ, what are the direction angles of $-\mathbf{d}$? What is the relation between the direction cosines of $\mathbf{d}$ and $-\mathbf{d}$?

26. Find the direction cosines of a direction vector of $x = 1 + t, y = -2 + 3t, z = -t$.

27. *(Distance from a point to a line).* Figure 14–26 shows a line L and a point P not on L. For any two distinct points Q and R on L, let $\mathbf{a} = \overrightarrow{QP}$ and $\mathbf{b} = \overrightarrow{QR}$. Use the figure to find a formula for the (perpendicular) distance d from P to L in terms of $\mathbf{a}$ and $\mathbf{b}$.

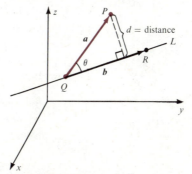

FIGURE 14–26. Exercise 27. The (perpendicular) distance from a point P to a line L.

28. (Continuation of Exercise 27). Find the distance from P to the line L.

(a) $P(0, 0, 0)$; $L: x = 2t, y = 3 - t, z = 1 + 2t$

(b) $P(-1, 3, 6)$; $L: \mathbf{r} = (\mathbf{i} - \mathbf{j}) + t(2\mathbf{i} + \mathbf{j} - \mathbf{k})$

(c) $P(0, 1, 3)$; $L: \dfrac{x - 2}{3} = \dfrac{y + 1}{4}; z = 3$

Exercises 29–33 concern a particle whose position at time t seconds is the tip of the position vector $\mathbf{r} = (1 + 2t)\mathbf{i} - (3 - t)\mathbf{j} + 4t\mathbf{k}$.

29. What is the initial position of the particle? Where is it after 1 second? What is its path?

30. When and where will it strike the xz-plane?

31. How much work is done by the force $\mathbf{F} = 2\mathbf{i} - \mathbf{j} + 2\mathbf{k}$ newtons in moving the particle through a displacement from $P(1, -3, 0)$ to $Q(7, 0, 12)$ meters?

32. If the velocity vector of the particle is $\mathbf{v} = 2\mathbf{i} + \mathbf{j} + 4\mathbf{k}$ m/sec, what is the power input of the force $\mathbf{F}$ defined in Exercise 31?

33. How close does the particle come to the origin?

34. Do the lines

$$\frac{x - x_0}{a} = \frac{y - y_0}{b} = \frac{z - z_0}{c}$$

and

$$\frac{x - x_0}{A} = \frac{y - y_0}{B} = \frac{z - z_0}{C}$$

intersect? What can you conclude about the lines, given that $aA + bB + cC = 0$? What can you conclude about the lines, given that $a/A = b/B = c/C$?

14–5 PLANES, TRAINS, & AUTOMOBILES

In this section, we discuss planes in space. A plane is defined to be the set of all vectors in space that are orthogonal (perpendicular) to some fixed nonzero vector $\mathbf{N}$. This agrees with our intuitive notion of a plane as a flat surface. It follows from this definition that every plane is the graph in space of some linear equation in x, y, and z. Conversely, the graph of every linear equation is a plane. Let us see why these statements are true.

Let $P(x_0, y_0, z_0)$ be a point in space and let $\mathbf{N} = a\mathbf{i} + b\mathbf{j} + c\mathbf{k}$ be a nonzero vector. We want to find an equation of the plane through P that is orthogonal to $\mathbf{N}$. The vector is called a **normal vector** of this plane. If we picture $\mathbf{N}$ as having its tail at P (Figure 14–27), then the point $Q(x, y, z)$ is on the plane if and only if the vector

$$\overrightarrow{PQ} = (x - x_0)\mathbf{i} + (y - y_0)\mathbf{j} + (z - z_0)\mathbf{k}$$

is orthogonal to $\mathbf{N}$; that is, Q is on the plane if and only if its coordinates satisfy the linear equation

$$\mathbf{N} \cdot \overrightarrow{PQ} = a(x - x_0) + b(y - y_0) + c(z - z_0) = 0$$

It follows that

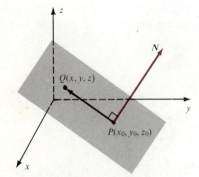

FIGURE 14–27. The point Q lies on the plane if and only if the vector $\overrightarrow{PQ}$ is orthogonal to $\mathbf{N}$.

(14.27)

An equation of the plane through $P(x_0, y_0, z_0)$ with a normal vector $\mathbf{N} = a\mathbf{i} + b\mathbf{j} + c\mathbf{k}$ is

$$a(x - x_0) + b(y - y_0) + c(z - z_0) = 0$$

Notice that the components of $\mathbf{N}$ become the coefficients of x, y, and z.

EXAMPLE 1 Find an equation of the plane through $P(2, 0, 1)$ with normal vector $\mathbf{N} = 2\mathbf{i} + 3\mathbf{j} + 4\mathbf{k}$. Sketch the plane.

Solution Applying (14.27) yields

$$2(x - 2) + 3(y - 0) + 4(z - 1) = 0$$

which reduces to

$$2x + 3y + 4z = 8$$

This is an equation of the plane. The easiest way to sketch a plane is to find its *intercepts;* that is, where it crosses the coordinate axes. Setting $y = 0$ and $z = 0$, we see that $(4, 0, 0)$ is the x-intercept; the other intercepts $(0, 8/3, 0)$ and $(0, 0, 2)$ are obtained in similar fashion. Now sketch the plane through these three points (Figure 14–28). ■

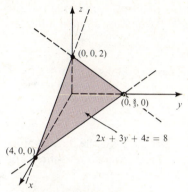

FIGURE 14–28. Example 1.

So far, we have shown every plane is the graph of a linear equation in x, y, and z. To establish the converse, we note that any linear equation in three variables x, y, and z can be written in the form

(14.28) $ax + by + cz + d = 0$

If we start with such an equation where a, b, and c are not all zero, and x_0, y_0, z_0 is any solution, then

$$ax_0 + by_0 + cz_0 + d = 0$$

Consequently, $d = -ax_0 - by_0 - cz_0$, and (14.28) becomes

$$\begin{aligned} 0 &= ax + by + cz - ax_0 - by_0 - cz_0 \\ &= a(x - x_0) + b(y - y_0) + c(z - z_0) \end{aligned}$$

But, by (14.27), this is an equation of the plane through $P(x_0, y_0, z_0)$ with normal vector $\mathbf{N} = a\mathbf{i} + b\mathbf{j} + c\mathbf{k}$. It follows that

(14.29)
> The graph of every linear equation
> $$ax + by + cz + d = 0$$
> is a plane with normal vector
> $$\mathbf{N} = a\mathbf{i} + b\mathbf{j} + c\mathbf{k}$$

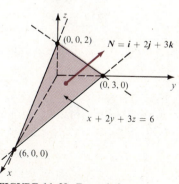

FIGURE 14–29. Example 2.

EXAMPLE 2 Find a normal vector and sketch the graph of $x + 2y + 3z = 6$.

Solution By (14.29), the coefficients of x, y, and z are the components of a normal vector. Therefore, $\mathbf{N} = \mathbf{i} + 2\mathbf{j} + 3\mathbf{k}$ is normal to this plane. The intercepts are at $x = 6, y = 3$, and $z = 2$, and the graph is sketched in Figure 14–29. ■

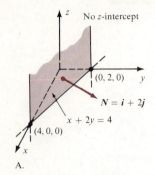

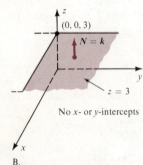

FIGURE 14–30. Example 3.

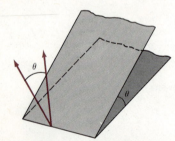

FIGURE 14–31. The angle between intersecting planes is defined to be the acute angle between the normal vectors.

CAUTION: In the two-dimensional plane, the graph of a linear equation, say $x + 2y = 4$ is a straight line. But in three-dimensional space, the graph of the linear equation $x + 2y = 4$ is a *plane;* $x + 2y = 4$ is simply a linear equation in three variables where the coefficient of z is 0.

EXAMPLE 3 Find normal vectors and sketch the (space) graphs of (a) $x + 2y = 4$ and (b) $z = 3$.

Solution

(a) A normal vector is $\mathbf{N} = \mathbf{i} + 2\mathbf{j}$; since the coefficient of z is 0, the third component of $\mathbf{N}$ is 0. The intercepts are $(4, 0, 0)$ and $(0, 2, 0)$; there is no z-intercept. This plane is parallel to the z-axis (Figure 14–30A).

(b) The coefficients of x and y are 0, and it follows that $\mathbf{N} = \mathbf{k}$ is a normal vector. This plane is parallel to the xy-plane and the only intercept is at $z = 3$ (Figure 14–30B). ■

Two planes are *parallel* if their normal vectors are parallel. In this case, the planes are identical or never intersect. When two nonparallel planes do intersect, we define the *angle* between them to be the angle between their normal vectors (Figure 14–31). Just as in the case of lines, we will always choose the angle θ with $0 < \theta \leq \pi/2$.

EXAMPLE 4 Find the angle between the planes $2x - 3y + 4z = 1$ and $-x + y = 4$.

Solution First find the normal vectors,

$$\mathbf{N}_1 = 2\mathbf{i} - 3\mathbf{j} + 4\mathbf{k} \quad \text{and} \quad \mathbf{N}_2 = -\mathbf{i} + \mathbf{j}$$

The angle between them is

$$\theta = \cos^{-1} \frac{\mathbf{N}_1 \cdot \mathbf{N}_2}{|\mathbf{N}_1||\mathbf{N}_2|} = \cos^{-1} \frac{-5}{\sqrt{29}\sqrt{2}} \approx 131°$$

By our convention, the angle between the planes is $180° - 131° = 49°$. ■

EXAMPLE 5 Find an equation of the plane through $P(1, 4, -2)$ that is parallel to the plane $-2x + y - 3z = 0$.

Solution A normal of the plane $-2x + y - 3z = 0$ is $\mathbf{N} = -2\mathbf{i} + \mathbf{j} - 3\mathbf{k}$. Since the planes are to be parallel, $\mathbf{N}$ is also normal to the plane through $P(1, 4, -2)$. Therefore, by (14.27), an equation of this plane is

$$-2(x - 1) + 1(y - 4) - 3(z + 2) = 0 \quad \text{or} \quad -2x + y - 3z = 8 \blacksquare$$

EXAMPLE 6 *(Particle striking a plane).* If the position vector of a particle at time t seconds is $\mathbf{r} = (2 + t)\mathbf{i} + (3 - 2t)\mathbf{j} + t\mathbf{k}$, at what time will the particle strike the plane $2x + y - 3z = 1$?

Solution At any time t, the particle is at the tip of $\mathbf{r}$, so its coordinates are $(2 + t, 3 - 2t, t)$. The particle strikes the plane when these coordinates satisfy the given equation of the plane; that is, when

$$2(2 + t) + (3 - 2t) - 3t = 1$$

Solving for t yields $t = 2$ seconds as the answer. ■

EXAMPLE 7 *(A plane through three points).* Find an equation of the plane that contains $P(0, 1, 2)$, $Q(1, 0, 0)$, and $R(2, 1, 3)$.

Solution Since the plane contains the points P, Q, and R, the vectors

$$\overrightarrow{PQ} = (1, -1, -2) \quad \text{and} \quad \overrightarrow{PR} = (2, 0, 1)$$

are parallel to the plane. What we want to do now is find a vector $\mathbf{N} = (a, b, c)$ that is perpendicular to *both* of them. Using dot products, we see that this happens if

$$a - b - 2c = 0 \qquad\qquad\qquad (\mathbf{N} \cdot \overrightarrow{PQ} = 0)$$
$$2a \quad\quad + c = 0 \qquad\qquad\qquad (\mathbf{N} \cdot \overrightarrow{PR} = 0)$$

Since there are two equations with three unknowns, there will, in general, be many solutions. Adding 2 times the second equation to the first equation yields $b = 5a$. The second equation alone yields $c = -2a$. The value of a is arbitrary. If we let $a = 1$, then $b = 5$ and $c = -2$, and you can easily verify that $\mathbf{N} = (1, 5, -2)$ is perpendicular to both $\overrightarrow{PQ}$ and $\overrightarrow{PR}$. Using $\mathbf{N} = (1, 5, -2)$ as a normal vector and $P(0, 1, 2)$ as a point in the plane, it follows that

$$1(x - 0) + 5(y - 1) - 2(z - 2) = 0 \quad \text{or} \quad x + 5y - 2z = 1$$

is an equation of the plane. You can check that the coordinates of both $Q(1, 0, 0)$ and $R(2, 1, 3)$ also satisfy this equation.

Remark: In Section 14–8, we discuss an easier method of finding a vector $\mathbf{N}$ perpendicular to two given vectors. ■

Next, we derive a formula for computing the (perpendicular) distance from a point $P(x_1, y_1, z_1)$ to a plane $ax + by + cz + d = 0$. The idea is to choose any point Q in the plane, and then find the component of $\overrightarrow{QP}$ on the normal vector $\mathbf{N} = (a, b, c)$. As Figure 14–32 indicates, the absolute value of this component is the distance from the point to the plane. Since any point in the plane can be chosen, we might as well take an intercept, say $Q(0, 0, -d/c)$; because not all of the coefficients are zero, it is always possible to do this. Now,

$$\mathbf{a} = \overrightarrow{QP} = \left(x_1, y_1, z_1 + \frac{d}{c}\right) \quad \text{and} \quad \mathbf{N} = (a, b, c)$$

so that the distance is

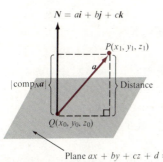

FIGURE 14–32. The (perpendicular) distance from a point to a plane.

$$|\text{comp}_N\mathbf{a}| = \frac{|\mathbf{a} \cdot \mathbf{N}|}{|\mathbf{N}|} = \frac{|ax_1 + by_1 + cz_1 + d|}{\sqrt{a^2 + b^2 + c^2}}$$

(14.30)

> The distance from the point $P(x_1, y_1, z_1)$ to the plane $ax + by + cz + d = 0$ is
>
> $$\frac{|ax_1 + by_1 + cz_1 + d|}{\sqrt{a^2 + b^2 + c^2}}$$

EXAMPLE 8 Find the distance from $P(2, 1, -1)$ to the plane $x - 2y + 4z = 9$.

Solution To apply (14.30), write the equation of the plane with 0 on the right

$$x - 2y + 4z - 9 = 0$$

and substitute the coordinates of P for x, y, and z. Thus, by (14.30),

$$\frac{|2 - 2(1) + 4(-1) - 9|}{\sqrt{1^2 + (-2)^2 + 4^2}} = \frac{13}{\sqrt{21}}$$

is the distance. ∎

Our final observation in this section relates to the symmetric equations of a line discussed in the previous section. We note that equations such as

$$\frac{x - x_0}{a} = \frac{y - y_0}{b}, \quad \frac{y - y_0}{b} = \frac{z - z_0}{c}, \text{ and } \frac{x - x_0}{a} = \frac{z - z_0}{c}$$

can easily be rewritten as linear equations in x, y, and z; therefore, each represents a plane. These three planes intersect in the line whose symmetric equations are

$$\frac{x - x_0}{a} = \frac{y - y_0}{b} = \frac{z - z_0}{c}$$

EXERCISES

In Exercises 1–10, sketch the planes.

1. $x = 1$ 2. $y = -2$

3. $z = -3$ 4. $x = 0$

5. $2x + y = 4$ 6. $y + z = 1$

7. $x + 2y + z = 3$ 8. $2x + 2y + z = 2$

9. $2x + y + 3z - 2 = 0$

10. $4x + 2y + 3z - 6 = 0$

In Exercises 11–14, find an equation of the plane through P with normal vector $\mathbf{N}$.

11. $P(3, -4, 5)$; $\mathbf{N} = (8, 2, -1)$

12. $P(0, 1, 2)$; $\mathbf{N} = (1, 1, 1)$

13. $P(1, 0, -1)$; $\mathbf{N} = (-3, 2, 0)$

14. $P(-4, 3, 0)$: $\mathbf{N} = \mathbf{i}$

In Exercises 15–22, find an equation of the described plane.

15. Through $P(6, -3, 4)$, parallel to the xy-plane.

16. Through $P(-1, 0, 4)$, parallel to the yz-plane.

17. Through $P(1, 2, 3)$, parallel to the plane $3x - 2y = 0$.

18. Through $P(-2, 0, 1)$, parallel to the plane $x + y + z = 0$.

19. Through $P(-2, 1, 0)$, $Q(4, 6, 3)$, and $R(0, 0, 5)$.

20. Through $P(1, 1, 1)$, $Q(4, -2, 1)$, and $R(-3, 8, 7)$.

21. Through $P(0, 4, 5)$, perpendicular to the planes $x - 2y - 2z = 1$ and $2x - y + z = 2$.

22. Through $P(5, -1, 0)$, perpendicular to the planes $y + z = 4$ and $x + 2y - z = 1$.

In Exercises 23–26, find the angle between the planes.

23. $2x - 3y + z + 2 = 0$ and $x + 4y - 5z - 6 = 0$

24. $x - y = 4$ and $-4x + 3y - 5z = 0$

25. $x = 3 + y$ and $y = x - z$

26. $x = 4$ and $y = 2$

In Exercises 27–30, find the distance from P to the plane.

27. $P(3, -2, 1); x - 3y + z = 2$

28. $P(0, 1, 1); x + y = -2$

29. $P(0, 0, 0); 4x + y - z = 8$

30. $P(0, 0, 0); 3x + y + z = 4$

31. Find an equation of the plane through $P(3, 0, 1)$ and perpendicular to the line $x = 2t, y = 1 - t, z = 4 + 3t$.

32. Find an equation of the plane that contains the point $P(-2, 1, 1)$ and the line $\mathbf{r} = t\mathbf{i} - (1 + t)\mathbf{j} - 3t\mathbf{k}$.

33. Find parametric equations of the line of intersection of the planes $3x - 2y + z = 0$ and $8x + 2y = 11$. *Hint:* Find two distinct points on the line.

34. Find an equation of the plane that contains the origin and is orthogonal to the line of intersection of the planes $x + y = 0$ and $2x - 3z = 0$.

35. The set of all points in space equidistant from two distinct fixed points $P(x_0, y_0, z_0)$ and $Q(x_1, y_1, z_1)$ is a plane. Find an equation of this plane.

36. The position vector of a particle at time t seconds is $\mathbf{r} = (4 - 3t)\mathbf{i} + (1 + 2t)\mathbf{j} + 5t\mathbf{k}$ feet, for $t \geq 0$. How long does it take for the particle to strike the plane $3x + 4y - z = 4$?

37. (Continuation of Exercise 36). A force of $\mathbf{F} = 2\mathbf{i} + \mathbf{j} + 6\mathbf{k}$ pounds is used to move the particle from its starting point to the plane. How much work is done?

14–6 VECTOR FUNCTIONS

In this section, we introduce the notion of a function whose values are vectors rather than numbers. These functions are used to describe curves in space. Moreover, our discussion leads quite naturally to definitions of limits, derivatives, and integrals of such functions. All of this lays the foundation for analyzing motion along curves in space, the topic of the following section.

Definitions and Graphs

We begin with some basic definitions. Let x, y, and z be three functions defined on an interval I of real numbers. For each t in I, the triple $(x(t), y(t), z(t))$ is a point in space with position vector

(14.31) $\mathbf{r}(t) = x(t)\mathbf{i} + y(t)\mathbf{j} + z(t)\mathbf{k}$

Thus, $\mathbf{r}$ is a rule that assigns to each t in I, a unique position vector $\mathbf{r}(t)$; we call this rule a **vector-valued function,** or simply, a **vector function.** The functions $x = x(t), y = y(t)$, and $z = z(t)$ are called the **component functions** of $\mathbf{r}$. If the domain of $\mathbf{r}$ is not specified, it is automatically taken to be the largest domain common to all three component functions.

EXAMPLE 1   The vector function $\mathbf{r}(t) = \ln t\mathbf{i} + 1/(t - 1)\mathbf{j} + (t + 1)\mathbf{k}$ has component functions

$$x = \ln t, \quad y = \frac{1}{t - 1}, \text{ and } z = t + 1$$

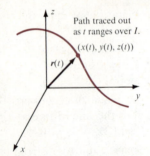

FIGURE 14–33. The functions $x = x(t)$, $y = y(t)$, and $z = z(t)$ are parametric equations of the curve.

Thus, the domain of **r** is all $t > 0$, with $t \neq 1$. The values of **r** are computed in the obvious way. For instance, at $t = 1/2$ and $t = 2$, we have

$$\mathbf{r}\left(\frac{1}{2}\right) = \left(\ln \frac{1}{2}\right)\mathbf{i} - 2\mathbf{j} + \frac{3}{2}\mathbf{k} \quad \text{and} \quad \mathbf{r}(2) = (\ln 2)\mathbf{i} + \mathbf{j} + 3\mathbf{k} \quad \blacksquare$$

As t ranges over its domain I, the tip of $\mathbf{r}(t)$ traces out a **path**, or **space curve** (Figure 14–33). The component functions

$$x = x(t) \quad y = y(t) \quad z = z(t)$$

of **r** are called *parametric equations* of the space curve.

EXAMPLE 2 Sketch the curve traced out by $\mathbf{r}(t) = (1 + 2t)\mathbf{i} + t\mathbf{j} + \mathbf{k}$, for all t.

Solution The component functions

$$x = 1 + 2t \quad y = t \quad z = 1$$

are parametric equations of a straight line. Two points on this line, corresponding to $t = 0$ and $t = 1$, are $(1, 0, 1)$ and $(3, 1, 1)$. The "curve" traced out by $\mathbf{r}(t)$ is the line through these points (Figure 14–34). $\blacksquare$

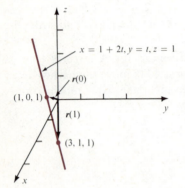

FIGURE 14–34. Example 2.

EXAMPLE 3 *(Circular helix).* Sketch the curve traced out by $\mathbf{r}(t) = a \cos t\mathbf{i} + a \sin t\mathbf{j} + bt\mathbf{k}$, for $t \geq 0$ and a, b positive.

Solution From the parametric equations

$$x = a \cos t \quad y = a \sin t \quad z = bt$$

we see that $x^2 + y^2 = a^2$. If the z-coordinate were always 0, the curve would simply wind around the circle $x^2 + y^2 = a^2$. But $z = bt$ is steadily increasing, and this makes the path resemble the coils of a spring (Figure 14–35). As an aid in sketching the graph, we make a table of values and plot a few points.

t	0	$\pi/2$	$3\pi/2$	2π
$x = a \cos t$	a	0	0	a
$y = a \sin t$	0	a	$-a$	0
$z = bt$	0	$b\pi/2$	$b3\pi/2$	$b2\pi$

Note: This curve (Figure 14–35) is called a *circular helix*. If the vector function were $\mathbf{r}(t) = a \cos t\mathbf{i} + b \sin t\mathbf{j} + ct\mathbf{k}$ with $a \neq b$, then the curve is an *elliptical helix*. $\blacksquare$

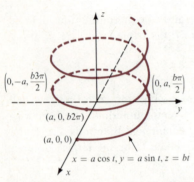

FIGURE 14–35. Example 3. A circular helix.

The formula for the arc length of a plane curve developed in Section 13–5 is valid for space curves except that one more term must be added because of the third coordinate. The proof is omitted.

(14.32)

> If the parametric equations of a space curve define functions x, y, and z with continuous derivatives, then the length of the curve from $t = a$ to $t = b$ is
>
> $$s = \int_a^b \sqrt{[x'(t)]^2 + [y'(t)]^2 + [z'(t)]^2}\, dt$$

EXAMPLE 4 Find the length of the circular helix of Example 3 from $t = 0$ to $t = 2\pi$.

Solution Using (14.32) and the parametric equations

$$x = a \cos t \quad y = a \sin t \quad z = bt$$

yields

$$s = \int_0^{2\pi} \sqrt{(-a \sin t)^2 + (a \cos t)^2 + b^2}\, dt$$

$$= \int_0^{2\pi} \sqrt{a^2 + b^2}\, dt = 2\pi \sqrt{a^2 + b^2} \text{ units} \quad \blacksquare$$

Calculus of Vector Functions

Limits, derivatives, and integrals of vector functions are defined entirely in terms of the component functions. Thus, if the limit, or the derivative, or the integral of each component function is known, then the corresponding concept of the vector function is also known. The only thing new here is that the results are expressed in terms of vector functions rather than ordinary functions.

(14.33)

> **DEFINITION OF LIMIT**
> If $\mathbf{r}(t) = x(t)\mathbf{i} + y(t)\mathbf{j} + z(t)\mathbf{k}$, then
>
> $$\lim_{t \to p}\mathbf{r}(t) = \left[\lim_{t \to p} x(t)\right]\mathbf{i} + \left[\lim_{t \to p} y(t)\right]\mathbf{j} + \left[\lim_{t \to p} z(t)\right]\mathbf{k}$$
>
> provided the limits on the right exist. This definition can be extended to one-sided limits and to limits as $t \to \pm\infty$.

Intuitively, $\mathbf{r}(t) \to \mathbf{L}$ as $t \to p$ means that $\mathbf{r}(t)$ gets closer and closer to the vector $\mathbf{L}$ as t gets closer and closer to p. This idea is expressed precisely in terms of ϵ's and δ's at the end of this section.

(14.34)

> **DEFINITION OF CONTINUITY**
> The vector function $\mathbf{r}$ is **continuous** at p if
>
> $$\lim_{t \to p} \mathbf{r}(t) = \mathbf{r}(p)$$

It follows from this definition that $\mathbf{r}$ is continuous at p if and only if all three component functions are continuous at p.

DEFINITION OF DERIVATIVE
If $\mathbf{r}(t) = x(t)\mathbf{i} + y(t)\mathbf{j} + z(t)\mathbf{k}$, then

(14.35)
$$\mathbf{r}'(t) = x'(t)\mathbf{i} + y'(t)\mathbf{j} + z'(t)\mathbf{k}$$

provided the derivatives on the right exist. Second and higher order derivatives are defined similarly.

The "dee" notation can also be used

$$\frac{d\mathbf{r}}{dt} = \frac{dx}{dt}\mathbf{i} + \frac{dy}{dt}\mathbf{j} + \frac{dz}{dt}\mathbf{k}$$

It follows from the definition that $\mathbf{r}$ is differentiable at a point p if and only if each of its component functions is differentiable at p.

EXAMPLE 5 Let $\mathbf{r}(t) = \cos t\mathbf{i} + e^{3t}\mathbf{j} + \sqrt{t}\mathbf{k}$. Then

(a) Domain of $\mathbf{r}$ is all $t \geq 0$.

(b) $\displaystyle\lim_{t \to 0^+} \mathbf{r}(t) = \left[\lim_{t \to 0^+} \cos t\right]\mathbf{i} + \left[\lim_{t \to 0^+} e^{3t}\right]\mathbf{j} + \left[\lim_{t \to 0^+} \sqrt{t}\right]\mathbf{k}$

$\qquad = \mathbf{i} + \mathbf{j}$

(c) $\mathbf{r}$ is continuous on its domain.

(d) $\mathbf{r}'(t) = [\cos t]'\mathbf{i} + [e^{3t}]'\mathbf{j} + [\sqrt{t}]'\mathbf{k}$

$\qquad = -\sin t\mathbf{i} + 3e^{3t}\mathbf{j} + (1/2\sqrt{t})\mathbf{k} \quad t > 0$

(e) $\mathbf{r}''(t) = -\cos t\mathbf{i} + 9e^{3t}\mathbf{j} - (1/4\sqrt{t^3})\mathbf{k} \quad t > 0$ ∎

The derivative can also be thought of as a limit of *difference quotients*. That is, by (14.35),

$$\mathbf{r}'(t) = x'(t)\mathbf{i} + y'(t)\mathbf{j} + z'(t)\mathbf{k}$$

$$= \left[\lim_{\Delta t \to 0} \frac{x(t + \Delta t) - x(t)}{\Delta t}\right]\mathbf{i} + \left[\lim_{\Delta t \to 0} \frac{y(t + \Delta t) - y(t)}{\Delta t}\right]\mathbf{j}$$

$$+ \left[\lim_{\Delta t \to 0} \frac{z(t + \Delta t) - z(t)}{\Delta t}\right]\mathbf{k}$$

$$= \lim_{\Delta t \to 0} \frac{\mathbf{r}(t + \Delta t) - \mathbf{r}(t)}{\Delta t}$$

The last line follows from (14.33) and the definitions of vector addition and scalar multiplication.

$$(14.36) \quad \mathbf{r}'(t) = \lim_{\Delta t \to 0} \frac{\mathbf{r}(t + \Delta t) - \mathbf{r}(t)}{\Delta t}$$

provided this limit exists.

We will use this version of the derivative in our discussion of motion in the next section.

Many of the familiar derivative formulas for ordinary functions also hold for vector functions. For example, the formulas $(\mathbf{r} + \mathbf{s})' = \mathbf{r}' + \mathbf{s}'$ and $(p\mathbf{r})' = p\mathbf{r}'$ are true and easy to verify. Even the usual product rule holds for the *dot product* of two vector functions. Let

$$\mathbf{r}(t) = x_1(t)\mathbf{i} + y_1(t)\mathbf{j} + z_1(t)\mathbf{k} \quad \text{and} \quad \mathbf{s}(t) = x_2(t)\mathbf{i} + y_2(t)\mathbf{j} + z_2(t)\mathbf{k}$$

be differentiable vector functions. Then

$$\frac{d}{dt}[\mathbf{r}(t) \cdot \mathbf{s}(t)] = \frac{d}{dt}[x_1(t)x_2(t) + y_1(t)y_2(t) + z_1(t)z_2(t)]$$

Now use the ordinary product rule on each product on the right. After some regrouping, we continue with

$$\frac{d}{dt}[\mathbf{r}(t) \cdot \mathbf{s}(t)] = [x_1(t)x_2'(t) + y_1(t)y_2'(t) + z_1(t)z_2'(t)]$$

$$+ [x_1'(t)x_2(t) + y_1'(t)y_2(t) + z_1'(t)z_2(t)]$$
$$= \mathbf{r}(t) \cdot \mathbf{s}'(t) + \mathbf{r}'(t) \cdot \mathbf{s}(t)$$

Summarizing the above discussion, we have

(14.37)

If $\mathbf{r}$ and $\mathbf{s}$ are differentiable vector functions, then

(1) $(\mathbf{r} + \mathbf{s})'(t) = \mathbf{r}'(t) + \mathbf{s}'(t)$

(2) $(p\mathbf{r})'(t) = p\mathbf{r}'(t)$, p a scalar

(3) $(\mathbf{r} \cdot \mathbf{s})'(t) = \mathbf{r}(t) \cdot \mathbf{s}'(t) + \mathbf{r}'(t) \cdot \mathbf{s}(t)$

EXAMPLE 6 Let $\mathbf{r}(t) = t\mathbf{i} - t^2\mathbf{j} + t^3\mathbf{k}$ and $\mathbf{s}(t) = \ln t\mathbf{i} + \cos t\mathbf{j} - \mathbf{k}$. Then

$$(\mathbf{r} \cdot \mathbf{s})'(t) = \mathbf{r}(t) \cdot \mathbf{s}'(t) + \mathbf{r}'(t) \cdot \mathbf{s}(t)$$

$$= [t\mathbf{i} - t^2\mathbf{j} + t^3\mathbf{k}] \cdot \left[\frac{1}{t}\mathbf{i} - \sin t\mathbf{j}\right]$$

$$+ [\mathbf{i} - 2t\mathbf{j} + 3t^2\mathbf{k}] \cdot [\ln t\mathbf{i} + \cos t\mathbf{j} - \mathbf{k}]$$

$$= 1 + t^2 \sin t + \ln t - 2t \cos t - 3t^2$$

Note: The derivative of a vector function is a vector function, but the derivative of a dot product is a scalar, or ordinary, function. ■

Integrals are also defined in terms of components.

(14.38)

> **DEFINITION OF INTEGRAL**
> If $\mathbf{r}(t) = x(t)\mathbf{i} + y(t)\mathbf{j} + z(t)\mathbf{k}$ is continuous on the interval $[a, b]$,
> then $\mathbf{r}$ is integrable and
>
> $$\int_a^b \mathbf{r}(t)\, dt = \left[\int_a^b x(t)\, dt\right]\mathbf{i} + \left[\int_a^b y(t)\, dt\right]\mathbf{j} + \left[\int_a^b z(t)\, dt\right]\mathbf{k}$$

The terminology and theory from real-valued functions carries over, including the Fundamental Theorem of Calculus.

(14.39)

> If $\mathbf{r}(t)$ is continuous and $\mathbf{R}$ is an antiderivative of $\mathbf{r}$ on $[a, b]$, then
>
> $$\int_a^b \mathbf{r}(t)\, dt = \mathbf{R}(b) - \mathbf{R}(a)$$

EXAMPLE 7   Evaluate $\int_0^1 (t\mathbf{i} - \sin t\mathbf{j} + e^t\mathbf{k})\, dt$.

Solution   The vector

$$\mathbf{R}(t) = \frac{t^2}{2}\mathbf{i} + \cos t\mathbf{j} + e^t\mathbf{k}$$

is an antiderivative of the integrand. Therefore,

$$\int_0^1 (t\mathbf{i} - \sin t\mathbf{j} + e^t\mathbf{k})\, dt = \mathbf{R}(1) - \mathbf{R}(0)$$

$$= \left[\frac{1}{2}\mathbf{i} + \cos 1\mathbf{j} + e\mathbf{k}\right] - [\mathbf{j} + \mathbf{k}]$$

$$= \frac{1}{2}\mathbf{i} + (-1 + \cos 1)\mathbf{j} + (e - 1)\mathbf{k}   \blacksquare$$

There is also the notion of the indefinite integral of a vector function, and again the same theory carries over. If $\mathbf{R}$ is any antiderivative of $\mathbf{r}$, then *every* antiderivative is of the form $\mathbf{R} + \mathbf{C}$ where $\mathbf{C}$ is a constant (vector), and we write

(14.40) $$\int \mathbf{r}(t)\, dt = \mathbf{R}(t) + \mathbf{C}$$

Alternate Definition of Limit (*Optional*)

The $\epsilon - \delta$ definition of the limit is as follows:

If $\mathbf{r} = \mathbf{r}(t)$ is defined on open intervals of the form (a, p) and (p, b), then

(14.41)

$$\lim_{t \to p} \mathbf{r}(t) = \mathbf{L}$$

means that for every $\epsilon > 0$, there is a $\delta > 0$ such that

$$|\mathbf{r}(t) - \mathbf{L}| < \epsilon \quad \text{whenever} \quad 0 < |t - p| < \delta$$

The two definitions (14.33) and (14.41) are equivalent; see Exercise 50.

EXERCISES

In Exercises 1–12, sketch the curve traced out by $\mathbf{r}(t)$.

1. $\mathbf{r}(t) = t\mathbf{i} + (t + 1)\mathbf{j}$

2. $\mathbf{r}(t) = 3t\mathbf{i} + (2t + 1)\mathbf{k}$

3. $\mathbf{r}(t) = (t - 1)\mathbf{i} + 2t\mathbf{j} - t\mathbf{k}$

4. $\mathbf{r}(t) = (t/2)\mathbf{i} - \mathbf{j} + (t - 1)\mathbf{k}$

5. $\mathbf{r}(t) = \cos t\mathbf{i} + \mathbf{j} + \sin t\mathbf{k}$

6. $\mathbf{r}(t) = 2\mathbf{i} + \sin t\mathbf{j} + \cos t\mathbf{k}$

7. $\mathbf{r}(t) = \cos t\mathbf{i} + t\mathbf{j} + \sin t\mathbf{k}$

8. $\mathbf{r}(t) = 2t\mathbf{i} + \sin t\mathbf{j} + \cos t\mathbf{k}$

9. $\mathbf{r}(t) = 2\cos t\mathbf{i} + 3\sin t\mathbf{j} + t\mathbf{k}$

10. $\mathbf{r}(t) = 2\cos t\mathbf{i} + t\mathbf{j} + 3\sin t\mathbf{k}$

11. $\mathbf{r}(t) = t\mathbf{i} + t^2\mathbf{j} + \mathbf{k}$

12. $\mathbf{r}(t) = \cosh t\mathbf{i} + \sinh t\mathbf{j} + \mathbf{k}$

In Exercises 13–16, find the length of the curve in the indicated exercise.

13. Exercise 3; $0 \le t \le 2$

14. Exercise 7; $0 \le t \le 2\pi$

15. Exercise 9; $0 \le t \le 2\pi$ (Use Simpson's rule with $n = 4$.)

16. Exercise 11; $-1 \le t \le 1$

Exercises 17–30 refer to the following four vector functions

$\mathbf{a}(t) = t^2\mathbf{i} + (t + 2)\mathbf{j} + e^t\mathbf{k}$ $\mathbf{b}(t) = \sqrt{t}\mathbf{i} + (1/t)\mathbf{j} + 2\mathbf{k}$

$\mathbf{c}(t) = \cos t\mathbf{i} + \sin t\mathbf{j} + t\mathbf{k}$ $\mathbf{d}(t) = \ln|t|\mathbf{i} - \sqrt[3]{t}\mathbf{j} + e^{4t}\mathbf{k}$

17. Find the domain of $\mathbf{a}, \mathbf{b}, \mathbf{c},$ and $\mathbf{d}$.

18. Determine where $\mathbf{a}, \mathbf{b}, \mathbf{c},$ and $\mathbf{d}$ are continuous.

19. Find $\lim_{t \to -2} \mathbf{a}(t)$.

20. Find $\lim_{t \to 1} \mathbf{b}(t)$.

21. Find $\mathbf{c}'(2\pi)$.

22. Find $\mathbf{d}'(2)$.

23. Find $(\mathbf{b} + \mathbf{d})'(t)$.

24. Find $(\mathbf{a} + \mathbf{c})'(t)$.

25. Find $(\mathbf{a} \cdot \mathbf{c})'(t)$.

26. Find $(\mathbf{b} \cdot \mathbf{d})'(t)$.

27. Find $(\mathbf{b} \cdot \mathbf{c})'(t)$.

28. Find $(\mathbf{a} \cdot \mathbf{d})'(t)$.

29. Find $\mathbf{a}''(t)$ and $\mathbf{b}''(t)$.

30. Find $\mathbf{c}''(t)$ and $\mathbf{d}''(t)$.

Evaluate the integrals in Exercises 31–34.

31. $\displaystyle\int_1^2 (t\mathbf{i} + e^t\mathbf{j} + (1/t)\mathbf{k}) \, dt$

32. $\displaystyle\int_0^\pi (\cos t\mathbf{i} + \sin t\mathbf{j} + \mathbf{k}) \, dt$

33. $\displaystyle\int_0^1 (e^t \sin t\mathbf{j} + te^t\mathbf{k}) \, dt$

34. $\displaystyle\int_0^1 (5t^2\mathbf{i} - \sqrt{t}\mathbf{j} + t^2\mathbf{k}) \, dt$

In Exercises 35–38, use the indefinite integral (14.40).

35. Find $\mathbf{r}(t)$ if $\mathbf{r}'(t) = 2t\mathbf{i} + (1/t)\mathbf{j} + \mathbf{k}$ and $\mathbf{r}(1) = \mathbf{i} + \mathbf{j} + \mathbf{k}$.

36. Find $\mathbf{r}(t)$ if $\mathbf{r}'(t) = \sin t\mathbf{i} + \cos t\mathbf{j}$ and $\mathbf{r}(0) = \mathbf{k}$.

37. Find $\mathbf{r}(t)$ if $\mathbf{r}''(t) = \mathbf{i}, \mathbf{r}'(0) = \mathbf{i} + \mathbf{j}$, and $\mathbf{r}(0) = -\mathbf{k}$.

38. Find $\mathbf{r}(t)$ if $\mathbf{r}''(t) = 6t\mathbf{i} + e^t\mathbf{j}, \mathbf{r}'(0) = \mathbf{j}$, and $\mathbf{r}(0) = \mathbf{0}$.

39. (a) If $\mathbf{r}(t) = \cos t\mathbf{i} + \sin t\mathbf{j} + \mathbf{k}$, show that $\mathbf{r}(t) \cdot \mathbf{r}'(t) = 0$.

(b) If $|\mathbf{r}(t)|$ is constant for all t, show that $\mathbf{r}(t) \cdot \mathbf{r}'(t) = 0$.

40. If $\mathbf{r}(t) = \cos t\mathbf{i} + \sin t\mathbf{j}$, show that $\mathbf{r}(t) = -\mathbf{r}''(t)$ for all t.

In Exercises 41–44, carefully draw the path traced out by $\mathbf{r}(t)$. For the indicated value t_0, compute $\mathbf{r}'(t_0)$ and draw this vector with its tail at the head of $\mathbf{r}(t_0)$. If your sketch and your computations are accurate, the *derivative vector should be tangent to the path*.

41. $\mathbf{r}(t) = t\mathbf{i} + 2t\mathbf{j} + 3t\mathbf{k}$; $t_0 = 1$ (The path is a straight line, so "tangent" in this case means lying on the line.)

42. $\mathbf{r}(t) = \cos t\mathbf{i} + \sin t\mathbf{j} + \mathbf{k}$; $t_0 = 0$ and $t_0 = \pi/2$.

43. $\mathbf{r}(t) = \cos t\mathbf{i} + \sin t\mathbf{j} + t\mathbf{k}$; $t_0 = \pi/2$ and $t_0 = 2\pi$.

44. $\mathbf{r}(t) = t\mathbf{i} + t^2\mathbf{j}$; $t_0 = 0$ and $t_0 = 1$.

45. If $\mathbf{r}$ and $\mathbf{s}$ have limits at p, show that
(a) $\lim_{t \to p}(\mathbf{r} + \mathbf{s})(t) = \lim_{t \to p}\mathbf{r}(t) + \lim_{t \to p}\mathbf{s}(t)$
(b) $\lim_{t \to p}(\mathbf{r} \cdot \mathbf{s})(t) = [\lim_{t \to p}\mathbf{r}(t)] \cdot [\lim_{t \to p}\mathbf{s}(t)]$

46. Prove derivative formulas (1) and (2) in (14.37).

Optional Exercises

50. Show that definitions (14.33) and (14.41) are equivalent; that is, each implies the other.

51. Use (14.41) to prove that
$$\lim_{t \to 0}[t\mathbf{i} + (2t - 1)\mathbf{j} + t^2\mathbf{k}] = -\mathbf{j}$$

47. Show that if f and $\mathbf{r}$ are differentiable, then
$$\frac{d}{dt}(f\mathbf{r})(t) = f(t)\mathbf{r}'(t) + f'(t)\mathbf{r}(t)$$

48. *(Chain rule).* Show that if f and $\mathbf{r}$ are differentiable and the range of f is contained in the domain of $\mathbf{r}$, then
$$\frac{d}{dt}(\mathbf{r} \circ f)(t) = \frac{d}{dt}(\mathbf{r}(f(t))) = f'(t)\mathbf{r}'(f(t))$$

49. Show that if $\mathbf{r}$ and $\mathbf{s}$ are integrable and p is a scalar, then
(a) $\displaystyle\int_a^b (\mathbf{r}(t) + \mathbf{s}(t))\, dt = \int_a^b \mathbf{r}(t)\, dt + \int_a^b \mathbf{s}(t)\, dt$
(b) $\displaystyle\int_a^b p\mathbf{r}(t)\, dt = p\int_a^b \mathbf{r}(t)\, dt$
(c) If $\mathbf{C}$ is any fixed vector, then $\int_a^b \mathbf{C} \cdot \mathbf{r}(t)\, dt = \mathbf{C} \cdot \int_a^b \mathbf{r}(t)\, dt$.

52. Use (14.41) to show that
$$\lim_{t \to p}\mathbf{r}(t) = \mathbf{L} \quad \text{if and only if} \quad \lim_{t \to p}|\mathbf{r}(t) - \mathbf{L}| = 0$$

14–7 MOTION ALONG CURVES

In this section, we study the motion of an object, or particle, moving in space. The object may be a car speeding around a racetrack, an electron being propelled through a linear accelerator, or a satellite in orbit. We assume that the motion takes place in a fixed coordinate system and that the object can be located by specifying a single point, its center of gravity.

Position, Velocity, and Acceleration

The three basic notions for analyzing motion are position, velocity, and acceleration. As a particle moves along a path, we suppose that the coordinates (x, y, z) of its position are twice differentiable functions of time
$$x = x(t) \quad y = y(t) \quad z = z(t)$$

The vector function
$$\mathbf{r}(t) = x(t)\mathbf{i} + y(t)\mathbf{j} + z(t)\mathbf{k}$$

from the origin to the particle is called the **position function** of the particle. Figure 14–36 shows the position of a particle at time t and at another time $t + \Delta t$.

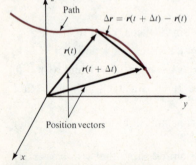

FIGURE 14–36. Δr is the change in position from time t to $t + \Delta t$.

The displacement vector $\Delta\mathbf{r} = \mathbf{r}(t + \Delta t) - \mathbf{r}(t)$ represents the *change in position*. The scalar multiple $\Delta\mathbf{r}/\Delta t$ represents the *average* change in position from time t to $t + \Delta t$, and the average change in position is called the *average velocity* over the time period Δt. Now, just as in the case of motion along a line, we define the (*instantaneous*) **velocity** to be the limit of the average velocities as Δt approaches 0; that is

$$\text{velocity} = \lim_{\Delta t \to 0} \frac{\Delta\mathbf{r}}{\Delta t} = \lim_{\Delta t \to 0} \frac{\mathbf{r}(t + \Delta t) - \mathbf{r}(t)}{\Delta t}$$

According to (14.36) in the previous section, the limit on the right is the *vector* $\mathbf{r}'(t)$. If $\mathbf{v}(t)$ denotes the velocity at time t, it follows that

$$\mathbf{v}(t) = \mathbf{r}'(t)$$

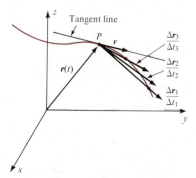

FIGURE 14–37. The vectors $\Delta\mathbf{r}/\Delta t \to \mathbf{v}$ as $\Delta t \to 0$; $\mathbf{v}$ is a direction vector for the line tangent to the path at P.

Thus, velocity is the rate of change of position with respect to time. Furthermore, the rate of change of velocity with respect to time is called the **acceleration** and is denoted by $\mathbf{a}$; that is,

$$\mathbf{a}(t) = \mathbf{v}'(t) = \mathbf{r}''(t)$$

The discussion above is completely consistent with our earlier discussion of position, velocity, and acceleration. Notice, however, that all three are now considered to be vectors and can be represented by arrows. The position vector always has its tail at the origin, but the velocity and acceleration vectors are considered to have their tails at the location of the particle. Moreover, *the arrow representing the velocity is always tangent to the path*. To see why this is so, we suppose that the particle is at point P at time t. Figure 14–37 indicates that as $\Delta t \to 0$, the vectors $\Delta\mathbf{r}/\Delta t$ approach a direction vector of the tangent line through P. It follows that this direction vector is the velocity vector $\mathbf{v}$. Figure 14–38 shows some typical velocity and acceleration vectors at various points on the path. Notice that acceleration vectors usually point toward the concave side of the path.

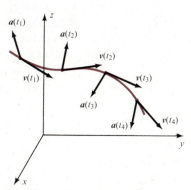

FIGURE 14–38. Typical velocity and acceleration vectors on the path of motion.

The **speed** v of a particle is defined to be the rate of change of distance (along the path) with respect to time. Speed has magnitude only and is, therefore, a scalar. If the particle starts at time t_0, then the distance s it travels along the path from t_0 to time t is given by the arc length formula (14.32)

$$s(t) = \int_{t_0}^{t} \sqrt{[x'(u)]^2 + [y'(u)]^2 + [z'(u)]^2} \, du$$

It follows from the Fundamental Theorem that

$$v(t) = s'(t) = \sqrt{[x'(t)]^2 + [y'(t)]^2 + [z'(t)]^2}$$

But the expression on the right is the length of $\mathbf{r}'(t) = \mathbf{v}(t)$ and, therefore,

$$\text{speed} = v(t) = |\mathbf{v}(t)|$$

The entire discussion above can be summarized as follows:

For a particle traveling through space, we have

(14.42)

(1) $\mathbf{r}(t)$ is the position vector; its tail is at the origin and its tip traces out the path.

(2) $\mathbf{v}(t) = \mathbf{r}'(t)$ is the velocity vector; it is tangent to the path.

(3) $\mathbf{a}(t) = \mathbf{v}'(t) = \mathbf{r}''(t)$ is the acceleration vector; it usually points toward the concave side of the path.

(4) $v(t) = s'(t) = |\mathbf{v}(t)|$ is the speed.

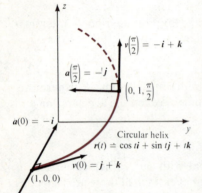

$v\left(\frac{\pi}{2}\right) = -i + k$

$a\left(\frac{\pi}{2}\right) = -j$

$\left(0, 1, \frac{\pi}{2}\right)$

$a(0) = -i$

Circular helix
$r(t) = \cos ti + \sin tj + tk$

$v(0) = j + k$

$(1, 0, 0)$

FIGURE 14–39. Example 1.

EXAMPLE 1 The position vector of a particle is $\mathbf{r}(t) = \cos t\mathbf{i} + \sin t\mathbf{j} + t\mathbf{k}$. Find its velocity, speed, and acceleration at any time t.

Solution The path of the particle is a circular helix.

$$\mathbf{v}(t) = \mathbf{r}'(t) = -\sin t\mathbf{i} + \cos t\mathbf{j} + \mathbf{k}$$
$$\text{speed} = |\mathbf{v}(t)| = \sqrt{(-\sin t)^2 + (\cos t)^2 + 1^2} = \sqrt{2}$$
$$\mathbf{a}(t) = \mathbf{v}'(t) = -\cos t\mathbf{i} - \sin t\mathbf{j}$$

Note: The acceleration is not **0** even though the speed is constant. The reason is that the velocity is constantly changing direction. Also notice that $\mathbf{v}(t) \cdot \mathbf{a}(t) = 0$, which means that **v** and **a** are always orthogonal (Figure 14–39). ∎

In this case,

Circular Motion with Constant Speed *Will Get You Nowhere!*

If a particle moves with constant speed in a circular path, then the acceleration vector always points toward the center of the circle; this is called *centripetal acceleration*. To see why this is so, we observe that a circular path lies in a single plane, and we might as well consider the circle to be in the xy-plane with radius r and center at the origin (Figure 14–40). Since the speed is constant, the angle θ

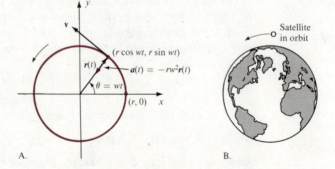

$(r \cos wt, r \sin wt)$

$r(t)$

$a(t) = -rw^2r(t)$

$\theta = wt$

$(r, 0)$

Satellite in orbit

FIGURE 14–40. (A) Circular motion with constant speed; centripetal acceleration points toward the center. (B) Example 2. Satellite in orbit around the earth.

A.

B.

from the positive x-axis to the position vector $\mathbf{r}(t)$ is changing at a constant rate ω (the Greek letter *omega*); that is, $\theta = \omega t$. It follows that the position vector is

$$\mathbf{r}(t) = r \cos \omega t\mathbf{i} + r \sin \omega t\mathbf{j}$$

Therefore,

$$\mathbf{v}(t) = -r\omega \sin \omega t\mathbf{i} + r\omega \cos \omega t\mathbf{j}$$
$$\mathbf{a}(t) = -r\omega^2 \cos \omega t\mathbf{i} - r\omega^2 \sin \omega t\mathbf{j}$$

Thus, $\mathbf{a}(t) = -\omega^2\mathbf{r}(t)$, and this shows that $\mathbf{a}$ always points toward the center of the circle; its magnitude is $|\mathbf{a}(t)| = r\omega^2$. Since the constant speed is $v = |\mathbf{v}(t)| = r\omega$, we also have the important relationship $v^2 = r|\mathbf{a}(t)|$ or

$$(14.43) \qquad \boxed{|\mathbf{a}(t)| = \frac{v^2}{r}}$$

This holds for all circular motion with constant speed.

EXAMPLE 2 *(Satellite in orbit).* Suppose that a satellite is in circular orbit 200 miles above the earth. What is its speed and period? (Assume the radius of the earth is 4,000 mi and the acceleration due to gravity is 32 ft/sec².)

Solution The radius of the circular path is $r = 4,200$ mi; the acceleration vector points toward the center of the earth and its magnitude is 32 ft/sec². To find the speed, we use (14.43) making sure the units of measurement are compatible.

$$v^2 = r|\mathbf{a}(t)| = (4,200 \text{ mi})(32 \text{ ft/sec}^2)(5,280 \text{ ft/mi})$$
$$\approx 7.1 \times 10^8 \text{ ft}^2/\text{sec}^2$$

Taking square roots, we have

$$v \approx 2.7 \times 10^4 \text{ ft/sec (about 18,409 mph)}$$

The period is the time it takes for one revolution.

$$\text{period} = \frac{2\pi r}{\text{speed}} = \frac{2\pi(4,200 \text{ mi})(5,280 \text{ ft/mi})}{2.7 \times 10^4 \text{ ft/sec}}$$
$$\approx 5,161 \text{ seconds (about 86 minutes)} \qquad \blacksquare$$

Force and Motion

Suppose a particle has constant mass m. Then Newton's second law of motion states that the product of m and the acceleration $\mathbf{a}$ of the particle equals the total external force acting on the particle.

(14.44) $\mathbf{F} = m\mathbf{a}$ (force = mass × acceleration)

If the force is a given function of time, and the initial velocity and initial position are known, then it is possible to obtain the path of the particle by integration.

EXAMPLE 3 The force acting on a particle at time t is $\mathbf{F}(t) = 6t\mathbf{i} + \mathbf{j}$. If the particle starts from the point $(3, -1, 2)$ with a velocity $\mathbf{v}(0) = 4\mathbf{k}$, find parametric equations of its path.

Solution The path is obtained by finding the position vector function $\mathbf{r}(t)$. Since $\mathbf{F} = m\mathbf{a} = m\mathbf{r}''$, we start this problem with the equation

$$\mathbf{r}''(t) = \frac{1}{m}\mathbf{F} = \frac{1}{m}(6t\mathbf{i} + \mathbf{j})$$

Integration of both sides yields

$$(14.45) \quad \mathbf{v}(t) = \mathbf{r}'(t) = \frac{1}{m}(3t^2\mathbf{i} + t\mathbf{j}) + \mathbf{C}$$

We are given that $\mathbf{v}(0) = 4\mathbf{k}$; thus, $\mathbf{C} = 4\mathbf{k}$ and (14.45) becomes

$$\mathbf{r}'(t) = \frac{1}{m}(3t^2\mathbf{i} + t\mathbf{j}) + 4\mathbf{k}$$

Integrate once again;

$$\mathbf{r}(t) = \frac{1}{m}\left(t^3\mathbf{i} + \frac{t^2}{2}\mathbf{j}\right) + 4t\mathbf{k} + \mathbf{C}$$

The starting point $(3, -1, 2)$ yields $\mathbf{r}(0) = 3\mathbf{i} - \mathbf{j} + 2\mathbf{k} = \mathbf{C}$. Therefore,

$$\mathbf{r}(t) = \left[\frac{1}{m}\left(t^3\mathbf{i} + \frac{t^2}{2}\mathbf{j}\right) + 4t\mathbf{k}\right] + [3\mathbf{i} - \mathbf{j} + 2\mathbf{k}]$$

$$= \left(\frac{t^3}{m} + 3\right)\mathbf{i} + \left(\frac{t^2}{2m} - 1\right)\mathbf{j} + (4t + 2)\mathbf{k}$$

It follows that parametric equations of the path are

$$x = \frac{t^3}{m} + 3 \quad y = \frac{t^2}{2m} - 1 \quad z = 4t + 2 \quad t \geq 0 \quad \blacksquare$$

The method of Example 3 can be applied to objects in motion near the surface of the earth. If air resistance is neglected, such objects are subject only to the force of gravity $\mathbf{F} = m\mathbf{g}$, which is constant. In this case, the action takes place in the plane determined by $\mathbf{g}$ and the initial velocity vector $\mathbf{v}_0$. This is the situation for the motion of a projectile; that is, an object launched into the air and allowed to move freely. The plane of motion is taken to be the xy-plane and the acceleration due to gravity is

$$\mathbf{g} = -32\mathbf{j}$$

which points straight down.

EXAMPLE 4 *(Path of a projectile).* A projectile is launched from the origin with an initial speed of v_0 ft/sec at an angle α from the horizontal (Figure 14–41); that is, the initial velocity vector is $\mathbf{v}_0 = v_0 \cos \alpha \mathbf{i} + v_0 \sin \alpha \mathbf{j}$. Show that the path is part of a parabola.

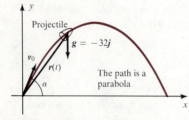

FIGURE 14–41. Example 4.

Solution   The acceleration $\mathbf{g}$ in this case is known; it points straight down and its magnitude is always 32 ft/sec². Therefore,

$$\mathbf{r}''(t) = \mathbf{g} = -32\mathbf{j}$$

Integration of both sides yields

$$(14.46) \quad \mathbf{v}(t) = \mathbf{r}'(t) = -32t\mathbf{j} + \mathbf{C}$$

Since $\mathbf{v}(0) = \mathbf{v}_0 = \mathbf{C}$, we have

$$\mathbf{C} = v_0 \cos \alpha\mathbf{i} + v_0 \sin \alpha\mathbf{j}$$

and (14.46) becomes

$$\mathbf{r}'(t) = v_0 \cos \alpha\mathbf{i} + (v_0 \sin \alpha - 32t)\mathbf{j}$$

We integrate again to obtain

$$\mathbf{r}(t) = (v_0 \cos \alpha)t\mathbf{i} + [(v_0 \sin \alpha)t - 16t^2]\mathbf{j} + \mathbf{C}$$

Since the projectile starts from the origin, we have $\mathbf{r}(0) = \mathbf{0} = \mathbf{C}$; therefore,

$$\mathbf{r}(t) = (v_0 \cos \alpha)t\mathbf{i} + [(v_0 \sin \alpha)t - 16t^2]\mathbf{j}$$

Parametric equations for the path are

$$x = (v_0 \cos \alpha)t \quad \text{and} \quad y = (v_0 \sin \alpha)t - 16t^2$$

To show that the path is a parabola, we solve the first equation for t, eliminate the parameter, and obtain

$$y = (\tan \alpha)x - \frac{16}{(v_0 \cos \alpha)^2}x^2$$

which is an equation of a parabola.   ■

EXERCISES

In Exercises 1–6, the position function of a particle is given. Find the velocity, speed, and acceleration.

1. $\mathbf{r}(t) = \mathbf{i} - 2t\mathbf{j} + (t + 1)\mathbf{k}$

2. $\mathbf{r}(t) = -3t\mathbf{i} + t\mathbf{j} + \mathbf{k}$

3. $\mathbf{r}(t) = \sqrt{t}\mathbf{i} + e^t\mathbf{j} + 4t\mathbf{k}$

4. $\mathbf{r}(t) = \ln t\mathbf{i} + 5t^2\mathbf{j} + 8t\mathbf{k}$

5. $\mathbf{r}(t) = \sin t\mathbf{i} + \cos t\mathbf{j} + t\mathbf{k}$

6. $\mathbf{r}(t) = \cosh t\mathbf{i} + \sinh t\mathbf{j} + t\mathbf{k}$

In Exercises 7–14, the position function of a particle is given. Sketch the path and the velocity and acceleration vectors at the indicated time t.

7. $\mathbf{r}(t) = t\mathbf{i} + t^2\mathbf{j}; t = 1$

8. $\mathbf{r}(t) = \cos t\mathbf{i} + \sin t\mathbf{j}; t = 0$

9. $\mathbf{r}(t) = \cosh t\mathbf{i} + \sinh t\mathbf{j}; t = 0$

10. $\mathbf{r}(t) = t\mathbf{i} + e^t\mathbf{j}; t = 0$

11. $\mathbf{r}(t) = t\mathbf{i} + (t + 1)\mathbf{j} + 2t\mathbf{k}; t = 0$

12. $\mathbf{r}(t) = 2t\mathbf{i} - \mathbf{k}; t = 2$

13. $\mathbf{r}(t) = \cos t\mathbf{i} + \sin t\mathbf{j} + \mathbf{k}; t = 2\pi$

14. $\mathbf{r}(t) = t\mathbf{i} + \mathbf{j} + t^2\mathbf{k}; t = 1$

In Exercises 15–20, the acceleration and initial position and velocity of a particle are given. Find the position functions.

15. $\mathbf{a}(t) = t\mathbf{i} - 6t\mathbf{j} + \mathbf{k}; \mathbf{r}(0) = \mathbf{0}, \mathbf{v}(0) = \mathbf{i}$

16. $\mathbf{a}(t) = 2t\mathbf{i} + t\mathbf{j} - 3\mathbf{k}; \mathbf{r}(0) = \mathbf{i} + \mathbf{j}, \mathbf{v}(0) = \mathbf{0}$

17. $\mathbf{a}(t) = \cos t\mathbf{i} + \sin t\mathbf{j}; \mathbf{r}(0) = \mathbf{k}, \mathbf{v}(0) = \mathbf{i} - \mathbf{j}$

18. $\mathbf{a}(t) = \mathbf{k}; \mathbf{r}(0) = \mathbf{i} + \mathbf{j} + \mathbf{k}, \mathbf{v}(0) = 2\mathbf{i} - 3\mathbf{j} + \mathbf{k}$

19. $\mathbf{a}(t) = e^t \mathbf{i} + \dfrac{1}{1+t^2}\mathbf{j}; \; \mathbf{r}(0) = \mathbf{v}(0) = \mathbf{0}$

20. $\mathbf{a}(t) = \cosh t \mathbf{i}; \; \mathbf{r}(0) = \mathbf{v}(0) = \mathbf{0}$

21. A particle of mass m moves along the hyperbola with parametric equations $x = \cosh \omega t, y = \sinh \omega t$, $z = 0$. Show that the force acting on the particle is directed *away* from the origin and that its magnitude is proportional to the distance from the origin.

22. A particle of mass m moves along the ellipse with parametric equations $x = a \cos \omega t, y = b \sin \omega t$, $z = 0$. Show that the force acting on the particle is directed *toward* the origin and that its magnitude is proportional to the distance from the origin.

23. A projectile is fired with an initial speed of 1,000 ft/sec at an angle of 30° with the horizontal. (Assume its path lies in the xy-plane.) (a) What is its velocity at time t? (b) What is its maximum height? (c) *What is the velocity at the maximum height?* (d) What is the range; that is, at what distance from the launch site will it strike the ground?

24. Work Exercise 23 if the angle is 60°.

25. (a) At what angle should a projectile be fired so that the maximum height is equal to the range? (b) At what angle should a projectile be fired to achieve the maximum range?

26. A batter hits a baseball to left field. As the ball leaves the bat it is 3 ft off the ground and moving with a speed of 115 ft/sec at an angle of 45° with the horizontal. If the left field fence is 12 ft high and 390 ft from home plate, did the batter hit a home run; that is, will the ball go over the fence?

27. An outfielder on a baseball team can throw a ball with an initial speed of 90 ft/sec. (a) He fields a ball and wants to throw it to home plate, a distance of 200 ft. At what angle of elevation should he throw the ball? (b) What is the maximum distance he can throw a ball without a bounce?

28. A satellite is moving at a constant speed in a circular orbit 300 miles above the earth. Find its speed and period.

29. A bullet is fired horizontally from a gun 2 meters above the ground with an initial speed of 300 meters/sec. When and where will it strike the ground? ($\mathbf{g} = -9.8\mathbf{j}$ meters/sec^2.)

30. A satellite moves in a circular orbit about the earth with a constant speed of 18,000 mph. What is the radius of the orbit?

31. A racing car is moving with a constant speed of of 90 mph around a circular track with radius of 200 ft. What is the magnitude of the centripetal acceleration?

32. About 400 years ago, Galileo showed if a projectile is fired at an angle of $45° + \theta$ or $45° - \theta$, where $0° < \theta < 45°$, the range is the same. Can you prove this?

14–8 CROSS PRODUCT

It is often necessary to find a vector that is perpendicular to two known vectors. For example, a particle moving on a circular path about the origin not only has (linear) velocity $\mathbf{v}$, but also *angular velocity* $\boldsymbol{\omega}$ about the axis of rotation. Angular velocity is a vector that is perpendicular to both the position vector $\mathbf{r}$ and the velocity vector $\mathbf{v}$ (Figure 14–42). For another example, consider a wheel spinning on an axle that is pivoted at one end, but free to turn in any direction (Figure 14–43). This is a simplified gyroscope. If the spin rate of the wheel is large enough, the axle and the wheel will actually move horizontally into the page and revolve about the pivot point O. The force that causes this motion is called *torque*, denoted by τ, the Greek letter *tau*. The torque is perpendicular to both the force of gravity $m\mathbf{g}$ and the angular velocity $\boldsymbol{\omega}$. In order to conveniently describe vectors such as $\boldsymbol{\omega}$ and τ that are orthogonal to two given vectors, the notion of a *cross product* is developed in this section. We will return to angular velocity and torque later.

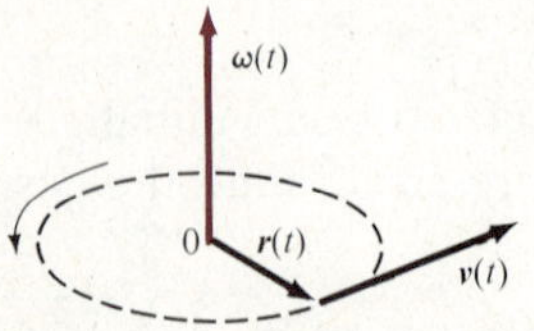

FIGURE 14–42. Angular velocity $\boldsymbol{\omega}$ is perpendicular to both $\mathbf{r}$ and $\mathbf{v}$.

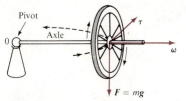

Axle will rotate about the pivot.

Pivot

Axle

τ

0

ω

$F = mg$

FIGURE 14–43. The torque τ is perpendicular to both **F** and ω.

We begin with an algebraic definition of the cross product that will quickly lead to the geometric interpretation described above.

(14.47)

The **cross product** $\mathbf{a} \times \mathbf{b}$ of two vectors

$$\mathbf{a} = a_1\mathbf{i} + a_2\mathbf{j} + a_3\mathbf{k} \quad \text{and} \quad \mathbf{b} = b_1\mathbf{i} + b_2\mathbf{j} + b_3\mathbf{k}$$

is the vector

$$\mathbf{a} \times \mathbf{b} = (a_2b_3 - a_3b_2)\mathbf{i} + (a_3b_1 - a_1b_3)\mathbf{j} + (a_1b_2 - a_2b_1)\mathbf{k}$$

Note that the cross product is a *vector* and, for that reason, it is sometimes called the **vector product.**

There is an easy way to remember this complex formula using determinants.* A *two by two determinant* is evaluated according to the formula

(14.48) $\begin{vmatrix} a_1 & a_2 \\ b_1 & b_2 \end{vmatrix} = a_1b_2 - a_2b_1$

For example,

$$\begin{vmatrix} 2 & -3 \\ 5 & 8 \end{vmatrix} = (2)(8) - (-3)(5) = 31$$

A *three by three determinant* is evaluated according to the formula

$$\begin{vmatrix} c_1 & c_2 & c_3 \\ a_1 & a_2 & a_3 \\ b_1 & b_2 & b_3 \end{vmatrix} = \begin{vmatrix} a_2 & a_3 \\ b_2 & b_3 \end{vmatrix} c_1 - \begin{vmatrix} a_1 & a_3 \\ b_1 & b_3 \end{vmatrix} c_2 + \begin{vmatrix} a_1 & a_2 \\ b_1 & b_2 \end{vmatrix} c_3$$

Note the minus sign

which, by (14.48), can be expanded as

(14.49) $= (a_2b_3 - a_3b_2)c_1 + (a_3b_1 - a_1b_3)c_2 + (a_1b_2 - a_2b_1)c_3$

Now compare the last line above with the definition of $\mathbf{a} \times \mathbf{b}$. You can see that they agree if the c's are replaced by $\mathbf{i}$, $\mathbf{j}$, and $\mathbf{k}$. Therefore, in the determinant notation, we have

(14.50) $\mathbf{a} \times \mathbf{b} = \begin{vmatrix} \mathbf{i} & \mathbf{j} & \mathbf{k} \\ a_1 & a_2 & a_3 \\ b_1 & b_2 & b_3 \end{vmatrix} = \begin{vmatrix} a_2 & a_3 \\ b_2 & b_3 \end{vmatrix} \mathbf{i} - \begin{vmatrix} a_1 & a_3 \\ b_1 & b_3 \end{vmatrix} \mathbf{j} + \begin{vmatrix} a_1 & a_2 \\ b_1 & b_2 \end{vmatrix} \mathbf{k}$

EXAMPLE 1 Let $\mathbf{a} = 2\mathbf{i} - 3\mathbf{j} + \mathbf{k}$ and $\mathbf{b} = \mathbf{i} + 4\mathbf{j} - 8\mathbf{k}$. Find $\mathbf{a} \times \mathbf{b}$ and show that it is perpendicular to both $\mathbf{a}$ and $\mathbf{b}$.

*We introduce determinants here only as a notational device. A complete discussion of determinants and their properties can be found in any book on linear algebra.

Solution

$$\mathbf{a} \times \mathbf{b} = \begin{vmatrix} \mathbf{i} & \mathbf{j} & \mathbf{k} \\ 2 & -3 & 1 \\ 1 & 4 & -8 \end{vmatrix} = \begin{vmatrix} -3 & 1 \\ 4 & -8 \end{vmatrix} \mathbf{i} - \begin{vmatrix} 2 & 1 \\ 1 & -8 \end{vmatrix} \mathbf{j} + \begin{vmatrix} 2 & -3 \\ 1 & 4 \end{vmatrix} \mathbf{k} \qquad (14.50)$$

$$= (24 - 4)\mathbf{i} - (-16 - 1)\mathbf{j} + (8 + 3)\mathbf{k} \qquad (14.48)$$

$$= 20\mathbf{i} + 17\mathbf{j} + 11\mathbf{k}$$

To prove that $\mathbf{a} \times \mathbf{b}$ is perpendicular to both $\mathbf{a}$ and $\mathbf{b}$, we show that the dot products are 0.

$$(\mathbf{a} \times \mathbf{b}) \cdot \mathbf{a} = (20)(2) + (17)(-3) + (11)(1) = 0$$
$$(\mathbf{a} \times \mathbf{b}) \cdot \mathbf{b} = (20)(1) + (17)(4) + (11)(-8) = 0 \qquad \blacksquare$$

Here are some basic properties of the cross product:

(14.51)

> (1) The cross product is *anti*commutative; $\mathbf{a} \times \mathbf{b} = -\mathbf{b} \times \mathbf{a}$.
>
> (2) Scalars can be factored out; $(p\mathbf{a}) \times (q\mathbf{b}) = (pq)\mathbf{a} \times \mathbf{b}$.
>
> (3) $\times$ distributes over $+$; $\mathbf{a} \times (\mathbf{b} + \mathbf{c}) = \mathbf{a} \times \mathbf{b} + \mathbf{a} \times \mathbf{c}$.
>
> (4) $\mathbf{a} \times \mathbf{b}$ is orthogonal to both $\mathbf{a}$ and $\mathbf{b}$.
>
> (5) $|\mathbf{a} \times \mathbf{b}|^2 = |\mathbf{a}|^2 |\mathbf{b}|^2 - (\mathbf{a} \cdot \mathbf{b})^2$
>
> (6) If $\mathbf{a}(t)$ and $\mathbf{b}(t)$ are differentiable vector functions, then
>
> $$(\mathbf{a} \times \mathbf{b})'(t) = \mathbf{a}(t) \times \mathbf{b}'(t) + \mathbf{a}'(t) \times \mathbf{b}(t)$$
>
> (Note the *order* in which the derivatives appear.)

Properties (4) and (5) are important in obtaining a geometric interpretation of the cross product; (4) was illustrated in Example 1 and (5) is proved as follows:

$$|\mathbf{a} \times \mathbf{b}|^2 = (a_2 b_3 - a_3 b_2)^2 + (a_1 b_3 - a_3 b_1)^2 + (a_1 b_2 - a_2 b_1)^2$$

After each term is squared, the result can be written as

$$|\mathbf{a} \times \mathbf{b}|^2 = (a_1^2 + a_2^2 + a_3^2)(b_1^2 + b_2^2 + b_3^2)$$
$$- (a_1 b_1 + a_2 b_2 + a_3 b_3)^2 \qquad \text{(Verify)}$$
$$= |\mathbf{a}|^2 |\mathbf{b}|^2 - (\mathbf{a} \cdot \mathbf{b})^2$$

As you can see, the calculations are lengthy, but straightforward; the proofs of the other properties are left as exercises.

If neither $\mathbf{a}$ nor $\mathbf{b}$ is $\mathbf{0}$, let θ be the angle between them. Then, using the fact that $\mathbf{a} \cdot \mathbf{b} = |\mathbf{a}||\mathbf{b}| \cos \theta$, it follows from property (5) that

$$|\mathbf{a} \times \mathbf{b}|^2 = |\mathbf{a}|^2 |\mathbf{b}|^2 - (\mathbf{a} \cdot \mathbf{b})^2$$
$$= |\mathbf{a}|^2 |\mathbf{b}|^2 - |\mathbf{a}|^2 |\mathbf{b}|^2 \cos^2 \theta$$
$$= |\mathbf{a}|^2 |\mathbf{b}|^2 (1 - \cos^2 \theta)$$
$$= |\mathbf{a}|^2 |\mathbf{b}|^2 \sin^2 \theta$$

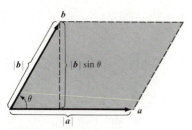

FIGURE 14–44. $|\mathbf{a} \times \mathbf{b}| = |\mathbf{a}||\mathbf{b}| \sin \theta$
= area of the parallelogram.

Noting that $\sin \theta \geq 0$ (why?), we take square roots and obtain

(14.52)

> If θ is the angle between nonzero vectors **a** and **b**, then
>
> $$|\mathbf{a} \times \mathbf{b}| = |\mathbf{a}||\mathbf{b}|\sin \theta$$

This means that *the length of* **a** $\times$ **b** *is the area of the parallelogram determined by* **a** *and* **b** (Figure 14–44).

EXAMPLE 2   Find the area of the triangle whose vertices are $P(1, 0, 0)$, $Q(0, 1, 1)$, and $R(1, 3, 2)$.

Solution   The area of the triangle is half of the area of the parallelogram determined by $\overrightarrow{PQ} = -\mathbf{i} + \mathbf{j} + \mathbf{k}$ and $\overrightarrow{PR} = 3\mathbf{j} + 2\mathbf{k}$. See Figure 14–45. Therefore,

$$\text{area of } \Delta PQR = \frac{1}{2}|\overrightarrow{PQ} \times \overrightarrow{PR}|$$

Now let us compute the cross product

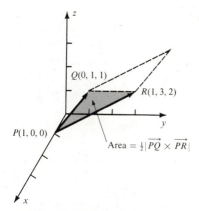

FIGURE 14–45. Example 2.

$$\overrightarrow{PQ} \times \overrightarrow{PR} = \begin{vmatrix} \mathbf{i} & \mathbf{j} & \mathbf{k} \\ -1 & 1 & 1 \\ 0 & 3 & 2 \end{vmatrix} = \begin{vmatrix} 1 & 1 \\ 3 & 2 \end{vmatrix}\mathbf{i} - \begin{vmatrix} -1 & 1 \\ 0 & 2 \end{vmatrix}\mathbf{j} + \begin{vmatrix} -1 & 1 \\ 0 & 3 \end{vmatrix}\mathbf{k}$$

$$= -\mathbf{i} + 2\mathbf{j} - 3\mathbf{k}*$$

The length of the product is $\sqrt{1^2 + 2^2 + 3^2} = \sqrt{14}$, so the area of the triangle is $\sqrt{14}/2$ square units.   ■

The cross product can also be used to derive an equation of the plane containing three noncolinear points.

EXAMPLE 3   Find an equation of the plane containing the points P, Q, and R in Example 2.

Solution   From Example 2, we know that

$$\overrightarrow{PQ} \times \overrightarrow{PR} = -\mathbf{i} + 2\mathbf{j} - 3\mathbf{k}$$

and that this vector is orthogonal to both $\overrightarrow{PQ}$ and $\overrightarrow{PR}$. It follows that this is a normal vector for the plane in question. Using $P = (1, 0, 0)$ as a point in the plane, we know from (14.27) in Section 14–5 that

$$(-1)(x - 1) + (2)(y - 0) + (-3)(z - 0) = 0 \quad \text{or} \quad -x + 2y - 3z = -1$$

is an equation of the plane.   ■

---

*To avoid errors in computing cross products, it is a good idea to quickly check that the product is orthogonal to both vectors; in this case, $(-\mathbf{i} + 2\mathbf{j} - 3\mathbf{k}) \cdot (-\mathbf{i} + \mathbf{j} + \mathbf{k}) = 0$ and $(-\mathbf{i} + 2\mathbf{j} - 3\mathbf{k}) \cdot (3\mathbf{j} + 2\mathbf{k}) = 0$.

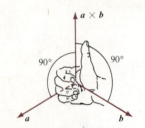

FIGURE 14–46. The vectors **a**, **b**, and **a** × **b** form a right-hand system.

From (14.52), we see that the only way for nonzero vectors to have a **0** cross product is to have $\theta = 0$ or π; that is, **a** and **b** are parallel. Thus,

$$(14.53) \quad \boxed{\mathbf{a} \times \mathbf{b} = \mathbf{0} \text{ if and only if } \mathbf{a} \text{ and } \mathbf{b} \text{ are parallel}}$$

We know that **a** × **b** is orthogonal to both **a** and **b**, and that its length is $|\mathbf{a}||\mathbf{b}|\sin\theta$. But there are *two* such vectors; one is the negative of the other. It can be shown, however, that our definition of the cross product always picks out the one that makes **a**, **b**, and **a** × **b** into a right-hand system (Figure 14–46). With this in mind, it is easy to verify that

FIGURE 14–47. Three important cross products.

$$(14.54) \quad \boxed{\mathbf{i} \times \mathbf{j} = \mathbf{k} \quad \mathbf{j} \times \mathbf{k} = \mathbf{i} \quad \mathbf{k} \times \mathbf{i} = \mathbf{j}} \qquad \text{(Figure 14–47)}$$

Referring back to Figures 14–42, we see that **r**, **v**, and the angular velocity ω form a right-hand system (think of **v** as starting from 0). In fact, ω is defined to be **r** × **v**. In Figure 14–43, the vectors ω, **F**, and torque τ form a right-hand system, and, like angular velocity, torque is also defined in terms of a cross product. We will discuss this further in the exercises.

We conclude with one more geometric application of the cross product.

EXAMPLE 4 Compute the volume of the parallelepiped with adjacent edges **a**, **b**, and **c** shown in Figure 14–48.

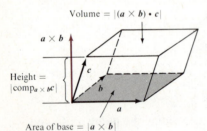

FIGURE 14–48. Example 4. Computing the volume of a parallelepiped.

Solution The base is the parallelogram determined by **a** and **b**; its area is $|\mathbf{a} \times \mathbf{b}|$. Since **a** × **b** is perpendicular to the base, the height of the parallelepiped is the absolute value of the component of **c** on **a** × **b**. By (14.19) in Section 14–3,

$$|\text{comp}_{a \times b}\mathbf{c}| = \frac{|(\mathbf{a} \times \mathbf{b}) \cdot \mathbf{c}|}{|\mathbf{a} \times \mathbf{b}|}$$

It now follows that

$$\text{Volume} = (\text{area of base})(\text{height})$$
$$= |\mathbf{a} \times \mathbf{b}| \frac{|(\mathbf{a} \times \mathbf{b}) \cdot \mathbf{c}|}{|\mathbf{a} \times \mathbf{b}|}$$
$$= |(\mathbf{a} \times \mathbf{b}) \cdot \mathbf{c}|$$

The number $(\mathbf{a} \times \mathbf{b}) \cdot \mathbf{c}$ derived above is called the *triple scalar product* of **a**, **b**, and **c**. Some of its properties are discussed in the exercises. ∎

EXERCISES

In Exercises 1–10, compute the cross products.

1. $(\mathbf{i} + \mathbf{j}) \times \mathbf{k}$ **2.** $(\mathbf{i} + \mathbf{k}) \times \mathbf{j}$

3. $\mathbf{i} \times 3\mathbf{i}$ **4.** $\mathbf{j} \times (-2\mathbf{j})$

5. $\mathbf{j} \times (3\mathbf{i} + 2\mathbf{j} - \mathbf{k})$ **6.** $\mathbf{k} \times (2\mathbf{i} - 3\mathbf{j} + 5\mathbf{k})$

7. $(\mathbf{i} - 3\mathbf{j} + 4\mathbf{k}) \times (2\mathbf{i} + \mathbf{j} - 6\mathbf{k})$

8. $(5\mathbf{i} + 3\mathbf{j} + \mathbf{k}) \times (4\mathbf{i} - \mathbf{j} - 2\mathbf{k})$

9. $(-8\mathbf{i} + \mathbf{j} + 3\mathbf{k}) \times (\mathbf{i} - 2\mathbf{j} + \mathbf{k})$

10. $(3\mathbf{i} + \mathbf{j} + \mathbf{k}) \times (-\mathbf{i} - 3\mathbf{j} + 9\mathbf{k})$

Exercises 11–18, refer to the vectors

$\mathbf{a} = 3\mathbf{i} - 2\mathbf{j} + 5\mathbf{k} \quad \mathbf{b} = \mathbf{i} + 8\mathbf{j} - 9\mathbf{k} \quad \mathbf{c} = -\mathbf{i} + \mathbf{j} - \mathbf{k}$

11. Compute $(\mathbf{a} \times \mathbf{b}) \times \mathbf{c}$ and $\mathbf{a} \times (\mathbf{b} \times \mathbf{c})$; is the cross product associative? *NO*

12. Compute $\mathbf{a} \times (\mathbf{b} + \mathbf{c})$ and $\mathbf{a} \times \mathbf{b} + \mathbf{a} \times \mathbf{c}$.

13. Compute $2\mathbf{a} \times 3\mathbf{b}$ and $6(\mathbf{a} \times \mathbf{b})$. *same*

14. Compute $\mathbf{b} \times \mathbf{c}$ and $\mathbf{c} \times \mathbf{b}$.

15. Compute $\mathbf{a} \times \mathbf{c}$; is it orthogonal to both $\mathbf{a}$ and $\mathbf{c}$? *yep*

16. Compute $|\mathbf{a} \times \mathbf{c}|^2$ and compare it with $|\mathbf{a}|^2|\mathbf{c}|^2 - (\mathbf{a} \cdot \mathbf{c})^2$.

17. Compute and compare $(\mathbf{a} \times \mathbf{b}) \cdot \mathbf{c}$ and $\mathbf{a} \cdot (\mathbf{b} \times \mathbf{c})$.

18. Compute and compare $(\mathbf{b} \times \mathbf{c}) \cdot \mathbf{a}$ and $\mathbf{b} \cdot (\mathbf{c} \times \mathbf{a})$.

19. Find $d/dt[(\mathbf{a} + t\mathbf{b}) \times (t^2\mathbf{a} + \mathbf{b})]$.

20. If $\mathbf{r}(t)$ is twice differentiable, find $d/dt[\mathbf{r}(t) \times \mathbf{r}'(t)]$.

In Exercises 21–24, find (a) the area of the triangle determined by P, Q, and R (b) an equation of the plane through P, Q, and R.

21. $P(0, 1, 2)$, $Q(-2, 1, 4)$, $R(-8, 0, 1)$

22. $P(-3, 0, 1)$, $Q(1, 0, 0)$, $R(1, 1, 0)$

23. $P(0, -1, 1)$, $Q(1, 1, 1)$, $R(0, 0, 0)$

24. $P(4, 1, 0)$, $Q(4, 0, 1)$, $R(4, 1, 3)$

In Exercises 25–28, find the volume of the described parallelepiped.

25. Adjacent sides $\mathbf{a} = \mathbf{i} + \mathbf{j}$, $\mathbf{b} = 2\mathbf{i} - \mathbf{j}$, and $\mathbf{c} = \mathbf{i} + \mathbf{j} + \mathbf{k}$.

26. Adjacent sides $\mathbf{a} = \mathbf{k}$, $\mathbf{b} = 3\mathbf{i} + \mathbf{j}$, and $\mathbf{c} = 2\mathbf{i} + \mathbf{j} + \mathbf{k}$.

27. Three vertices $P(1, 0, 0)$, $Q(3, 2, 1)$, and $R(-2, 0, 1)$ are adjacent to $S(3, 1, 2)$.

28. Three vertices $P(4, 1, 1)$, $Q(0, 0, 3)$, and $R(2, 1, 1)$ are adjacent to the origin.

---

29. *(Angular velocity)*. A particle is moving in space. Its **angular velocity** $\omega(t)$ **about the origin** is defined as $\omega(t) = \mathbf{r}(t) \times \mathbf{v}(t)$; the arrow representing ω is usually drawn with its tail at the origin. Find the angular velocity of a particle whose position function is $\mathbf{r}(t) = \cos t\mathbf{i} + \sin t\mathbf{j}$. Sketch the path and draw all three vectors $\mathbf{r}(0)$, $\mathbf{v}(0)$, and $\omega(0)$.

30. Same as Exercise 29 for a particle with position function $\mathbf{r}(t) = 3 \cos t\mathbf{i} + 2 \sin t\mathbf{j}$.

31. Can a particle moving along a straight line have a nonzero angular velocity about the origin? Test with $\mathbf{r}(t) = 2t\mathbf{i} - 4t\mathbf{j} + t\mathbf{k}$ and then test with $\mathbf{r}(t) = (2t + 1)\mathbf{i} - 3t\mathbf{j} + (t - 1)\mathbf{k}$. What is the difference between these two cases?

32. A particle of mass $m = 1$ starts from rest at the origin and moves under the influence of the force $\mathbf{F} = 6t\mathbf{i} + 4\mathbf{j} - 2\mathbf{k}$. What is its angular velocity about the origin at time t?

33. *(Torque)*. A force $\mathbf{F}(t)$ is applied to a particle. The **torque** $\tau(t)$ **exerted by** $\mathbf{F}(t)$ **about the origin** is defined to be $\tau(t) = \mathbf{r}(t) \times \mathbf{F}(t)$; the arrow representing torque is usually drawn with its tail at the location of the particle. If a particle of mass m has a position function $\mathbf{r}(t) = t\mathbf{i} + t^2\mathbf{j}$, find its torque and angular velocity at any time t (use $\mathbf{F} = m\mathbf{a}$). Sketch the path and draw in $\mathbf{r}(1)$, $\tau(1)$, and $\omega(1)$.

34. *(Central forces)*. A **central force** is one that is always parallel to the position vector. Show that *a central force exerts zero torque*.

35. *(Angular velocity and torque)*. You may already suspect that there is a connection between angular velocity and torque and, indeed, there is. Show that if a particle has constant mass m, then torque is the rate of change of $m\omega$ with respect to time; that is,

$$\tau(t) = \frac{d}{dt} m\omega(t)$$

36. *(Kepler's second law of planetary motion)*. After years of analyzing his observations, astronomer Johannes Kepler (1571–1630) formulated three laws of planetary motion. The first law states that the orbit of each planet is an ellipse. The third law relates the period with the length of the major axis of the orbit. The second law is that *the position vector of a planet* (with the sun at the origin) *sweeps out area at a constant rate*. The preceding exercises provide the basis of a beautifully simple proof of this statement.

Figure 14–49 shows a planet of mass m in orbit about the sun (assumed to be at the origin). The force (the gravitational pull of the sun) on the planet is directed towards the sun; it is a *central* force (all

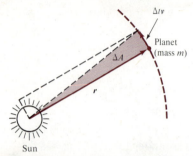

FIGURE 14–49. Exercise 36. Kepler's second law of planetary motion states that **r** sweeps out area at a constant rate.

other forces can be neglected). Over a very small positive time period Δt, the position vector **r** sweeps out an area ΔA about equal to half the area of the parallelogram determined by **r** and $\Delta t\mathbf{v}$ (Figure 14–49). Thus,

$$\Delta A = \frac{1}{2}\left|\mathbf{r} \times \Delta t\mathbf{v}\right| = \frac{1}{2m}\left|\mathbf{r} \times m\mathbf{v}\right|\Delta t$$

Use this equation and the results of Exercises 34 and 35 to prove Kepler's second law; that is, dA/dt is constant.

---

37. Prove properties (1), (2), (3), and (6) of (14.51).

38. Show that $\mathbf{a} \times \mathbf{b} = \mathbf{a} \times \mathbf{c}$ does not imply that $\mathbf{b} = \mathbf{c}$ even if $\mathbf{a} \neq \mathbf{0}$. Under what conditions will $\mathbf{a} \times \mathbf{b} = \mathbf{a} \times \mathbf{c}$ for $\mathbf{b} \neq \mathbf{c}$ and $\mathbf{a} \neq \mathbf{0}$?

39. Write $(\mathbf{a} + \mathbf{b}) \times (\mathbf{a} - \mathbf{b})$ as a multiple of $\mathbf{b} \times \mathbf{a}$.

40. *(Triple scalar product).* The **triple scalar product** of three vectors **a**, **b**, **c** is the number $(\mathbf{a} \times \mathbf{b}) \cdot \mathbf{c}$. Show that

$$(\mathbf{a} \times \mathbf{b}) \cdot \mathbf{c} = \begin{vmatrix} a_1 & a_2 & a_3 \\ b_1 & b_2 & b_3 \\ c_1 & c_2 & c_3 \end{vmatrix}$$

where the rows represent the components of **a**, **b**, and **c**.

41. Show that $(\mathbf{a} \times \mathbf{b}) \cdot \mathbf{c} = \mathbf{a} \cdot (\mathbf{b} \times \mathbf{c})$.

Optional Exercises

42. Show that $(\mathbf{a} \times \mathbf{b}) \cdot \mathbf{c} = 0$ if and only if **a**, **b**, and **c** are coplanar; that is, they lie in the same plane.

43. *(Triple vector product).* The **triple vector product** of three vectors **a**, **b**, **c** is the vector $\mathbf{a} \times (\mathbf{b} \times \mathbf{c})$. Show that $\mathbf{a} \times (\mathbf{b} \times \mathbf{c}) = (\mathbf{a} \cdot \mathbf{c})\mathbf{b} - (\mathbf{a} \cdot \mathbf{b})\mathbf{c}$.

44. Show that $\mathbf{a} \times (\mathbf{b} \times \mathbf{c}) + \mathbf{b} \times (\mathbf{c} \times \mathbf{a}) + \mathbf{c} \times (\mathbf{a} \times \mathbf{b}) = \mathbf{0}$ for all vectors **a**, **b**, and **c**.

14–9   CURVATURE AND COMPONENTS OF ACCELERATION

The acceleration of a particle traversing a path in space is caused by changes in velocity. Changes in velocity occur with a change in direction or a change in speed (Figure 14–50). In this section, we discuss each type of change and some practical applications.

FIGURE 14–50. Changes in velocity produce acceleration. The acceleration is affected by (A) changes in direction or (B) changes in speed.

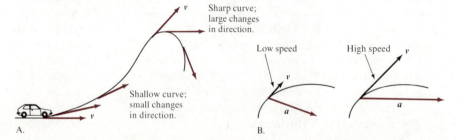

Curvature *of the spine*

The *curvature* of a path measures its change of direction. To define curvature, we use arc length s as a parameter; that is, the tip of $\mathbf{r}(s)$ is the point on the path that is at arc length s from the initial point $\mathbf{r}(0)$. See Figure 14–51. The reason we use arc length as a parameter is that $d\mathbf{r}/ds$ is always a unit vector tangent to the curve (Figure 14–52). To see why this is so, we observe that

$$\frac{ds}{dt}\frac{d\mathbf{r}}{ds} = \frac{d\mathbf{r}}{dt}$$

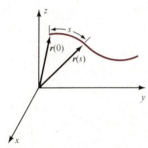

FIGURE 14–51. A curve with the arc length s as a parameter. The tip of $\mathbf{r}(s)$ is the point on the curve that is at arc length s from the initial point $\mathbf{r}(0)$.

Substituting $\mathbf{v}$ for $d\mathbf{r}/dt$ (which we know is tangent to the curve) and its length v for ds/dt, we have

$$v\frac{d\mathbf{r}}{ds} = \mathbf{v} \quad \text{or} \quad \frac{d\mathbf{r}}{ds} = \frac{\mathbf{v}}{v}$$

The vector $\mathbf{v}/v$ is tangent and of length 1.

(14.55)
> If arc length s is the parameter of a path, then
>
> $$\mathbf{T} = \frac{d\mathbf{r}}{ds}(s)$$
>
> is called the **unit tangent vector** at $\mathbf{r}(s)$.

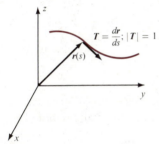

FIGURE 14–52. If the parameter is arc length s, then $\mathbf{T} = d\mathbf{r}/ds$ is always a unit tangent vector.

Although the length of $\mathbf{T}$ remains constant, its direction may change as s changes, and its derivative $d\mathbf{T}/ds = d^2\mathbf{r}/ds^2$ indicates the change in direction per unit of arc length. The magnitude of this vector is what we call curvature.

(14.56)
> If arc length s is the parameter for a path, then
>
> $$k = \left|\frac{d\mathbf{T}}{ds}(s)\right| = \left|\frac{d^2\mathbf{r}}{ds^2}(s)\right|$$
>
> is called the **curvature** of the path at $\mathbf{r}(s)$.

But we are mainly interested in motion problems where the parameter of the curve is time t rather than arc length. Therefore, as a practical matter, we will use another formula to compute the curvature.

(14.57)
> If time t is the parameter of a path and $\mathbf{r}'(t) \neq \mathbf{0}$, then
>
> $$k = \frac{|\mathbf{r}'(t) \times \mathbf{r}''(t)|}{|\mathbf{r}'(t)|^3} = \frac{|\mathbf{v}(t) \times \mathbf{a}(t)|}{|\mathbf{v}(t)|^3}$$
>
> is the curvature at $\mathbf{r}(t)$.

Let us postpone, for a moment, the derivation of this formula (the details are presented at the end of this section) and look at some interesting examples.

EXAMPLE 1 Show that the curvature of a straight line $\mathbf{r}(t) = (a_1 t + b_1)\mathbf{i} + (a_2 t + b_2)\mathbf{j} + (a_3 t + b_3)\mathbf{k}$ is 0.

Solution Since $\mathbf{r}''(t) = \mathbf{0}$ (check this out), it follows from (14.57) that $k = 0$ for all t. This is not surprising since the unit tangent vector of a straight path does not change direction. ■

EXAMPLE 2 Show that the curvature of a circle of radius r is $1/r$.

Solution We can consider the circle to lie in the xy-plane with its center at the origin. Then the position vector can be taken as $\mathbf{r}(t) = r \cos t\mathbf{i} + r \sin t\mathbf{j}$. So

$$\mathbf{r}'(t) \times \mathbf{r}''(t) = \begin{vmatrix} \mathbf{i} & \mathbf{j} & \mathbf{k} \\ -r\sin t & r\cos t & 0 \\ -r\cos t & -r\sin t & 0 \end{vmatrix} = r^2\mathbf{k}$$

Therefore, by (14.57),

$$k = \frac{|\mathbf{r}'(t) \times \mathbf{r}''(t)|}{|\mathbf{r}'(t)|^3} = \frac{r^2}{r^3} = \frac{1}{r} \qquad ■$$

Taking a cue from the last example, we define the *radius of curvature* of a path to be the reciprocal of the curvature. The radius is usually denoted by ρ, the Greek letter *rho*. Thus,

(14.58) $\quad \boxed{\rho = \dfrac{1}{k} \text{ is the \textbf{radius of curvature}.}}$

The radius of curvature of a line is taken to be ∞; for a circle of radius r, it is r.

EXAMPLE 3 Find the radius of curvature of the exponential spiral $\mathbf{r}(t) = e^{-t} \cos t\mathbf{i} + e^{-t} \sin t\mathbf{j}$, assuming the curve lies in the xy-plane.

Solution The spiral path is sketched in Figure 14–53. As $t \to \infty$, the path winds more tightly about the origin. We suspect that the curvature approaches infinity and, therefore, the radius of curvature approaches zero. Let's see if this is so.

Using the product rule for differentiation, we find that

$$\mathbf{r}'(t) = (-e^{-t}\sin t - e^{-t}\cos t)\mathbf{i} + (e^{-t}\cos t - e^{-t}\sin t)\mathbf{j}$$
$$\mathbf{r}''(t) = 2e^{-t}\sin t\mathbf{i} + 2e^{-t}\cos t\mathbf{j}$$
$$|\mathbf{r}'(t)|^3 = ([e^{-t}(\cos t + \sin t)]^2 + [e^{-t}(\cos t - \sin t)]^2)^{3/2}$$
$$= e^{-3t}2^{3/2}$$

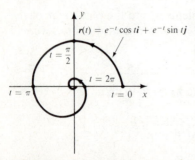

FIGURE 14–53. Example 3.

For the figure labels: $r(t) = e^{-t} \cos t\mathbf{i} + e^{-t} \sin t\mathbf{j}$, $t = \frac{\pi}{2}$, $t = 2\pi$, $t = \pi$, $t = 0$

Taking the cross product $\mathbf{r}'(t) \times \mathbf{r}''(t)$ in the usual way (we omit the details), it follows that

$$\rho = \frac{1}{k} = \frac{|\mathbf{r}'(t)|^3}{|\mathbf{r}'(t) \times \mathbf{r}''(t)|} = \sqrt{2}e^{-t}$$

The radius of curvature does, indeed, approach 0 as $t \to \infty$. ∎

Components of Acceleration

To discuss acceleration, we need to first talk about unit vectors that are *normal* to the path; that is, perpendicular to the unit tangent vector $\mathbf{T}(s)$. Since $|\mathbf{T}|^2$ is the constant 1 for all s, its derivative is 0; since $|\mathbf{T}|^2 = \mathbf{T} \cdot \mathbf{T}$, we can write

$$0 = \frac{d}{ds}(\mathbf{T} \cdot \mathbf{T}) = 2\mathbf{T} \cdot \frac{d\mathbf{T}}{ds} \tag{14.37.3}$$

It follows that $d\mathbf{T}/ds$ is perpendicular to $\mathbf{T}$. Of all the unit vectors perpendicular to $\mathbf{T}$ (there are infinitely many of them), we pick the one with the same direction as $d\mathbf{T}/ds$ and designate it by $\mathbf{N}$. See Figure 14–54.

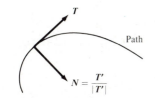

FIGURE 14–54. The vector $\mathbf{N}$ is called the *principal normal unit vector.*

(14.59)

If arc length s is the parameter of a path, then

$$\mathbf{N} = \frac{\mathbf{T}'(s)}{|\mathbf{T}'(s)|}$$

is called the **principal unit normal vector** at $\mathbf{r}(s)$.

Now we are prepared to take up the topic of acceleration. If a particle moves about a circular path of radius r (constant curvature $1/r$) with constant speed v, then, by (14.43) in Section 14–7, the acceleration vector is normal to the path with a magnitude of v^2/r. But if the curvature of the path or the speed of the particle is not constant, then the acceleration is not exactly normal to the path; it will have a nonzero tangential component.

(14.60)

The acceleration of a particle is the vector sum

$$\mathbf{a} = \frac{dv}{dt}\mathbf{T} + \frac{v^2}{\rho}\mathbf{N}$$

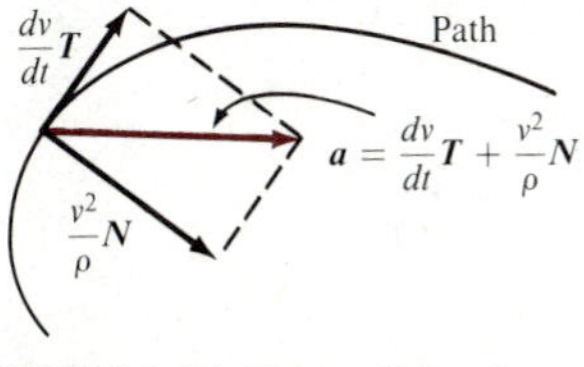

FIGURE 14–55. Tangential and normal components of acceleration.

The *tangential component* $a_T = dv/dt$ is the rate of change of speed. The *normal component* $a_N = v^2/\rho$ replaces the value v^2/r that occurs in circular motion with constant speed (Figure 14–55). The derivation of (14.60) is presented at the end of this section. Right now, let us look at some examples.

EXAMPLE 4   Find the tangential and normal components of the acceleration of a particle with position function $\mathbf{r}(t) = (t - 2)\mathbf{i} + (t - 2)^2\mathbf{j}$ when $t = 3$, assuming the path lies in the xy-plane.

Solution   The path is the parabola sketched in Figure 14–56. To find the tangential component $a_T = dv/dt$ we simply find $\mathbf{v}$, $v = |\mathbf{v}|$, and then differentiate.

$$\mathbf{v} = \mathbf{r}'(t) = \mathbf{i} + 2(t - 2)\mathbf{j}$$

$$v = |\mathbf{v}(t)| = \sqrt{1 + 4(t - 2)^2}$$

$$a_T = \frac{dv}{dt} = \frac{4(t - 2)}{\sqrt{1 + 4(t - 2)^2}}$$

When $t = 3$, $a_T = 4/\sqrt{5}$. To find the normal component, we find $[v(t)]^2$ and $\rho(t)$. In this case,

$$[v(t)]^2 = 1 + 4(t - 2)^2$$

and

$$\rho(t) = \frac{|\mathbf{v}(t)|^3}{|\mathbf{v}(t) \times \mathbf{a}(t)|} = \frac{(1 + 4(t - 2)^2)^{3/2}}{2} \qquad \text{(Check)}$$

Therefore,

$$\frac{[v(t)]^2}{\rho(t)} = \frac{2}{\sqrt{1 + 4(t - 2)^2}}$$

When $t = 3$, a_N has the value $2/\sqrt{5}$. See Figure 14–56.   ∎

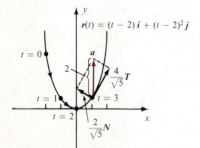

FIGURE 14–56. Example 4.

EXAMPLE 5 *(Force on a pilot).*   A pilot dives a plane along the path described in Example 4 (Figure 14–56). At the point when $t = 3$, he experiences an upward acceleration of 6 g's. If he weighs 150 lbs, what is the magnitude of the force exerted by the airplane seat on him at that moment?

Solution   The force exerted by the airplane seat is in the direction of the principal unit normal $\mathbf{N}$; its magnitude must compensate for the acceleration and the weight of the pilot. A force diagram is drawn in Figure 14–57. The magnitude of the force on the pilot caused by an acceleration of 6 g's is 6 mg, and since mg is his weight, 150 lbs, the magnitude of $\mathbf{a}$ is $6(150) = 900$ lbs. It follows from the results obtained in Example 4, that the projection of $\mathbf{a}$ onto $\mathbf{N}$ has length

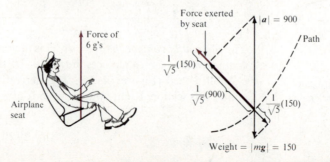

FIGURE 14–57. Example 5. The force on a pilot coming out of a dive.

$900/\sqrt{5}$. Similarly, the component of his weight that is normal to the path has length $150/\sqrt{5}$ (Figure 14–57). It follows that the magnitude of the force on the pilot is

$$\frac{900}{\sqrt{5}} + \frac{150}{\sqrt{5}} = \frac{1{,}050}{\sqrt{5}} \approx 470 \text{ lbs} \qquad \blacksquare$$

Derivation of Formulas (14.57) and (14.60)

Let t be the parameter of a curve C in such a way that arc length s, measured from $\mathbf{r}(0)$, is an increasing function of t. We will first derive formula (14.60) for the components of acceleration.

We begin with the equation

(14.61) $\quad \mathbf{v} = \dfrac{d\mathbf{r}}{dt} = \dfrac{ds}{dt}\dfrac{d\mathbf{r}}{ds} = v\mathbf{T}$

and differentiate both sides with respect to t (keeping in mind that $\mathbf{T}$ is a function of s)

$$\mathbf{a} = \frac{d^2\mathbf{r}}{dt^2} = \frac{dv}{dt}\mathbf{T} + v\frac{d\mathbf{T}}{dt}$$

$$= \frac{dv}{dt}\mathbf{T} + v^2\frac{d\mathbf{T}}{ds}$$

The last line follows from the equation $d\mathbf{T}/dt = (ds/dt)(d\mathbf{T}/ds) = v(d\mathbf{T}/ds)$. Moreover, it follows from the definition of $\mathbf{N}$ (14.59) and the definition of curvature (14.56) that $d\mathbf{T}/ds = k\mathbf{N} = \mathbf{N}/\rho$. Therefore, we have

$$\mathbf{a} = \frac{dv}{dt}\mathbf{T} + \frac{v^2}{\rho}\mathbf{N}$$

which is formula (14.60).

To derive formula (14.57) for curvature, we take the cross product of $\mathbf{v} = v\mathbf{T}$ (14.61) and $\mathbf{a}$:

$$\mathbf{v} \times \mathbf{a} = v\mathbf{T} \times \left(\frac{dv}{dt}\mathbf{T} + \frac{v^2}{\rho}\mathbf{N}\right) \qquad\qquad (14.60)$$

$$= v\frac{dv}{dt}(\mathbf{T} \times \mathbf{T}) + \frac{v^3}{\rho}(\mathbf{T} \times \mathbf{N}) \qquad\qquad (14.51.3)$$

The product $\mathbf{T} \times \mathbf{T} = \mathbf{0}$ by (14.53), and $|\mathbf{T} \times \mathbf{N}| = 1$ by (14.52). Therefore,

$$|\mathbf{v} \times \mathbf{a}| = \frac{v^3}{\rho} = kv^3$$

from which it follows that

$$k = \frac{|\mathbf{v} \times \mathbf{a}|}{v^3} = \frac{|\mathbf{v} \times \mathbf{a}|}{|\mathbf{v}|^3}$$

This is formula (14.57) and completes our derivations.

EXERCISES

In Exercises 1–10, find the curvature of the path traced out by $\mathbf{r}(t)$.

1. $\mathbf{r}(t) = 3t\mathbf{i} + t^3\mathbf{j}$ **2.** $\mathbf{r}(t) = t^2\mathbf{i} - t\mathbf{j}$

3. $\mathbf{r}(t) = t\mathbf{i} + \sin t\mathbf{j}$ **4.** $\mathbf{r}(t) = t\mathbf{i} + t^4\mathbf{j}$

5. $\mathbf{r}(t) = 2\cos t\mathbf{i} + 3\sin t\mathbf{j}$

6. $\mathbf{r}(t) = 3\cos t\mathbf{i} + 2\sin t\mathbf{j}$

7. $\mathbf{r}(t) = t\mathbf{i} + t^2\mathbf{j} + t^3\mathbf{k}$

8. $\mathbf{r}(t) = t\mathbf{i} + \ln t\mathbf{j} + t^2\mathbf{k}$

9. $\mathbf{r}(t) = \cos t\mathbf{i} + \sin t\mathbf{j} + t\mathbf{k}$

10. $\mathbf{r}(t) = t\mathbf{i} + 2\cos t\mathbf{j} + \sin t\mathbf{k}$

Exercises 11–20. For each of the curves in Exercises 1–10, suppose that $\mathbf{r}(t)$ is the position function of a particle. Find the tangential and normal components of the acceleration.

21. The graph of $y = f(x)$ in the xy-plane is traced out by a position vector of the form $\mathbf{r}(t) = t\mathbf{i} + f(t)\mathbf{j}$ (with $x = t$). Show that in this case the curvature k can be written in terms of f as follows:

(14.62)

> If $y = f(x)$ and f is twice differentiable, then
> $$k = \frac{|f''(x)|}{(1 + [f'(x)]^2)^{3/2}}$$

22. Use (14.62) to find the curvature of   (a) $y = (x/3)^3$ and   (b) $y = \sin x$. Compare your answers here with the ones for Exercises 1 and 3 above.

---

23. The pilot in Example 5 experiences an upward acceleration of 10 g's at the bottom of his dive $(t = 2)$. What is the magnitude of the force exerted by the airplane seat on him at that moment?

24. A gnat flies along a path with position function $\mathbf{r}(t) = e^{-t}\cos t\mathbf{i} + e^{-t}\sin t\mathbf{j}$; this is the exponential spiral discussed in Example 3. Find the tangential and normal components of acceleration at any time t.

25. A 2,000 lb automobile travels around a circular curve of radius 300 ft. If the road is level and the frictional force of the road on the car is 1,000 lbs, what is the maximum speed the car can travel without skidding? (Use $g \approx 32$ ft/sec².)

26. *(Banking a curve in the road).* A circular curve of radius r is banked at an angle θ so that a car can traverse the curve at v ft/sec even under icy conditions (no friction). The idea is that the normal force **F** exerted by the road on the car will have a large enough horizontal component to supply the centripetal force necessary to keep the car from skidding (Figure 14–58). The magnitude of the centripetal force must equal the normal component of the acceleration of the car.   (a) Show that the maximum possible speed v without skidding is

$$v = \sqrt{rg\tan\theta}$$

(b) If a curve of radius 300 ft is banked at an angle of 10°, at what maximum speed can it be traversed under icy conditions?

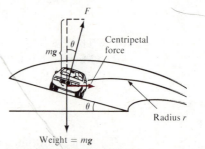

FIGURE 14–58. Exercise 26. A car traversing a banked curve under icy conditions.

27. *(Weighing yourself at the equator).* A person at the equator rotates in a circle of radius 4,000 mi once every day. What effect does this have on a person's true weight?

28. *(Swinging a pail of water).* A pail of water is swung in a vertical circle of radius 3 ft. What is the minimum speed necessary at the top of the circle for the water to remain in the pail?

in RPM   Test question

$\sqrt{96} \approx 9.8$ ft/s

14-10   QUADRIC SURFACES

In the next chapter, we begin the study of functions of more than one variable. In general, the graphs of such functions are difficult, or even impossible, to sketch. There is, however, an important class of functions that are defined by quadratic equations in three variables; their graphs, which are surfaces in space, are known as **quadric surfaces.** Because they are relatively easy to draw, quadric surfaces are used often in the illustrations and exercises of the succeeding chapters. Therefore, it is important that you become familiar with them.

Graphing in space is usually more difficult than graphing in a plane because of the extra dimension. But there are the usual graphing aids such as locating the intercepts, taking advantage of symmetry, and so on. Another helpful device is to visualize, or actually sketch, some *traces* of the surface. The intersection of a surface and a plane is called the **trace** of the surface in that plane. The most useful traces for graphing are those in planes that are parallel to the coordinate planes. You will see how they are used in the examples below.

There are nine distinct categories of quadric surfaces, and we will give an example of each. The constants a, b, and c are assumed to be positive.

EXAMPLE 1 *(Ellipsoid).*   A standard equation of an ellipsoid is

$$\frac{x^2}{a^2} + \frac{y^2}{b^2} + \frac{z^2}{c^2} = 1 \qquad \text{(Figure 14-59)}$$

The intercepts are $(\pm a, 0, 0)$, $(0, \pm b, 0)$, and $(0, 0, \pm c)$. The surface is *perfectly symmetric*; that is, it is symmetric about the axes, the coordinate planes, and the origin. This is true because each variable can be replaced by its negative. Each trace parallel to a coordinate axis is an ellipse. For example, in the plane $z = d$ (with $d < c$), the equation becomes

$$\frac{x^2}{a^2} + \frac{y^2}{b^2} = 1 - \frac{d^2}{c^2} \qquad (z = d)$$

which is an ellipse. The trace in the yz-plane ($x = 0$) is the ellipse

$$\frac{y^2}{b^2} + \frac{z^2}{c^2} = 1 \qquad (x = 0)$$

The traces in all three coordinate planes are indicated in the figure.

The numbers a, b, and c are the lengths of the semiaxes. If two of them are equal, we have a surface of revolution; if all three are equal, we have a sphere. ■

EXAMPLE 2 *(Hyperboloid of one sheet).*   A standard equation is

$$\frac{x^2}{a^2} + \frac{y^2}{b^2} - \frac{z^2}{c^2} = 1 \qquad \text{(Figure 14-60)}$$

The intercepts are $(\pm a, 0, 0)$ and $(0, \pm b, 0)$; there are no z-intercepts (why?). The surface is perfectly symmetric and is *unbounded* in the z direction; that is,

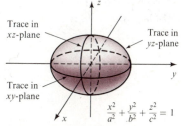

Trace in
xz-plane

Trace in
yz-plane

Trace in
xy-plane

$$\frac{x^2}{a^2} + \frac{y^2}{b^2} + \frac{z^2}{c^2} = 1$$

FIGURE 14-59. Example 1. An ellipsoid.

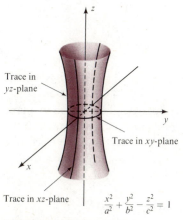

Trace in
yz-plane

Trace in xy-plane

Trace in xz-plane

$$\frac{x^2}{a^2} + \frac{y^2}{b^2} - \frac{z^2}{c^2} = 1$$

FIGURE 14-60. Example 2. A hyperboloid of one sheet.

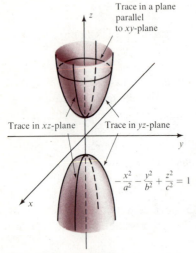

Trace in a plane parallel to xy-plane

Trace in xz-plane

Trace in yz-plane

$$-\frac{x^2}{a^2} - \frac{y^2}{b^2} + \frac{z^2}{c^2} = 1$$

FIGURE 14–61. Example 3. A hyperboloid of two sheets.

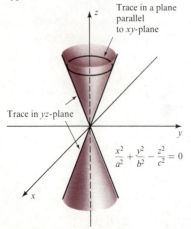

Trace in a plane parallel to xy-plane

Trace in yz-plane

$$\frac{x^2}{a^2} + \frac{y^2}{b^2} - \frac{z^2}{c^2} = 0$$

FIGURE 14–62. Example 4. An elliptic cone.

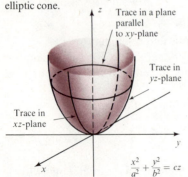

Trace in a plane parallel to xy-plane

Trace in yz-plane

Trace in xz-plane

$$\frac{x^2}{a^2} + \frac{y^2}{b^2} = cz$$

FIGURE 14–63. Example 5. An elliptic paraboloid.

z can have arbitrarily large positive or negative values. Traces in planes parallel to the xy-plane are ellipses

$$\frac{x^2}{a^2} + \frac{y^2}{b^2} = 1 + \frac{d^2}{c^2} \qquad (z = d)$$

Traces in planes parallel to the other coordinate planes are hyperbolas. For example, in the xz-plane,

$$\frac{x^2}{a^2} - \frac{z^2}{c^2} = 1 \qquad (y = 0)$$

If $a = b$, it is a surface of revolution. ∎

EXAMPLE 3 *(Hyperboloid of two sheets).* A standard equation is

$$-\frac{x^2}{a^2} - \frac{y^2}{b^2} + \frac{z^2}{c^2} = 1 \qquad \text{(Figure 14–61)}$$

The intercepts are $(0, 0, \pm c)$; there are no x- or y-intercepts. The surface is perfectly symmetrical, and although it is unbounded in the z direction, $|z|$ can never be less than c (why?). Therefore, there are two parts to the surface; the part for $z \geq c$ and the part for $z \leq -c$. Traces parallel to the xy-plane are ellipses; traces parallel to the other coordinate planes are hyperbolas. If $a = b$, it is a surface of revolution. ∎

EXAMPLE 4 *(Elliptic cone).* A standard equation is

$$\frac{x^2}{a^2} + \frac{y^2}{b^2} - \frac{z^2}{c^2} = 0 \qquad \text{(Figure 14–62)}$$

The only intercept is the origin and, again, the surface is perfectly symmetric and unbounded in the z-direction. Traces in planes parallel to the xy-plane are ellipses; traces in the yz- and xz-plane are straight lines intersecting at the origin. If $a = b$, it is a circular cone. ∎

EXAMPLE 5 *(Elliptic paraboloid).* A standard equation is

$$\frac{x^2}{a^2} + \frac{y^2}{b^2} = cz \quad c > 0 \qquad \text{(Figure 14–63)}$$

The only intercept is the origin. The surface does not extend below the xy-plane since $z \geq 0$. It is, however, symmetric with the z-axis, the xz-plane, and the yz-plane. The surface is unbounded in the positive z direction. The traces are ellipses (parallel to the xy-plane) and parabolas (parallel to the other coordinate planes). If $a = b$, it is a surface of revolution. ∎

EXAMPLE 6 *(Hyperbolic paraboloid).* A standard equation is

$$\frac{y^2}{b^2} - \frac{x^2}{a^2} = cz \quad c > 0 \qquad \text{(Figure 14–64)}$$

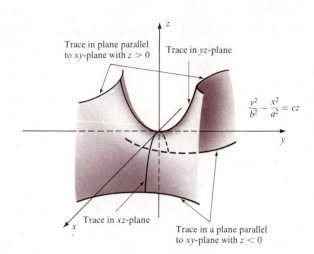

Trace in plane parallel to xy-plane with $z > 0$

Trace in yz-plane

$$\frac{y^2}{b^2} - \frac{x^2}{a^2} = cz$$

Trace in xz-plane

Trace in a plane parallel to xy-plane with $z < 0$

FIGURE 14–64. Example 6. A hyperbolic paraboloid.

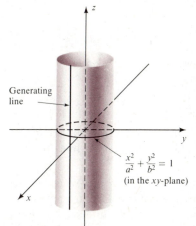

Generating line

$$\frac{x^2}{a^2} + \frac{y^2}{b^2} = 1$$
(in the xy-plane)

FIGURE 14–65. Example 7. An elliptic cylinder.

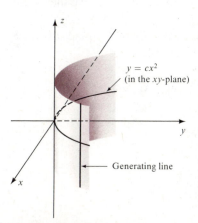

$y = cx^2$
(in the xy-plane)

Generating line

FIGURE 14–66. Example 8. A parabolic cylinder.

The origin is the only intercept. The surface is symmetric with the xz- and yz-planes and the z-axis. Traces parallel to the xy-plane are hyperbolas with their shape depending on whether $z > 0$ or $z < 0$ (why?). Traces parallel to the other coordinate planes are parabolas. Observe that the origin is a minimum point for the trace in the yz-plane, but it is a maximum point for the trace in the xz-plane. Such a point is called a *saddle point* of the surface; the figure shows why this name is appropriate. ■

If one of the variables, say z, does not appear in the equation, then the resulting quadric surface is *generated* by a line parallel to the z-axis. Such surfaces are called *cylinders*.

EXAMPLE 7 *(Elliptic cylinder).*   A standard equation is

$$\frac{x^2}{a^2} + \frac{y^2}{b^2} = 1 \qquad \text{(Figure 14–65)}$$

The only intercepts are $(\pm a, 0, 0)$ and $(0, \pm b, 0)$. The surface is unbounded in the z-direction and is perfectly symmetric. The surface is generated by a vertical line that moves along the ellipse

$$\frac{x^2}{a^2} + \frac{y^2}{b^2} = 1$$

in the xy-plane. If $a = b$, it is a circular cylinder.   ■

EXAMPLE 8 *(Parabolic cylinder).*   A standard equation is

$$y = cx^2 \qquad \text{(Figure 14–66)}$$

This surface is generated by a vertical line moving along the parabola $y = cx^2$ in the xy-plane.   ■

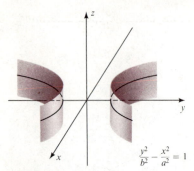

FIGURE 14–67. Example 6. A
hyperbolic cylinder.

EXAMPLE 9 *(Hyperbolic cylinder).* A standard equation is

$$\frac{y^2}{b^2} - \frac{x^2}{a^2} = 1 \qquad \text{(Figure 14–67)}$$

This surface is generated by a vertical line moving along the corresponding hyperbola in the xy-plane. ■

EXERCISES

In Exercises 1–20, the equation represents a surface in space. Identify and sketch the surface.

1. $2x^2 + 3y^2 - z^2 = 6$

2. $x^2 + y^2 + z^2 = 4$

3. $100x^2 + 25y^2 + 16z^2 = 400$

4. $x^2 + 9y^2 - 16z^2 = -16$

5. $9y^2 - 4x^2 = 36$

6. $36x^2 + 9y^2 - 4z^2 = 36$

7. $y^2 - 4x^2 = 16z$ **8.** $y^2 + 4x^2 = 16z$

9. $144x^2 + 144y^2 - 9z^2 = 0$ **10.** $x^2 - 9y^2 = 9$

11. $25x^2 + 225y^2 - 9z^2 = -225$

12. $y = x^2$

13. $9x^2 + 16z^2 = 144$

14. $4x^2 + 4y^2 - z^2 = 0$

15. $9z^2 + 16y^2 = 144x$ **16.** $x^2 + 4y^2 = 8$

17. $y = 4z^2$ **18.** $4y^2 - x^2 = 12z$

19. $x^2 + y^2 + z^2 - 4x + 6y = 3$

20. $x^2 - y^2 = 1$

In Exercises 21–26, sketch the graph of the given equation in the xy-plane. Then identify and find an equation of the quadric surface generated by revolving that curve about the indicated axis.

21. $x^2 + y^2 = 1$; x-axis

22. $x^2 + 4y^2 = 4$; y-axis

23. $y = x^2$; y-axis   **24.** $x = y^2$; x-axis

25. $x^2 - 4y^2 = 4$; x-axis

26. $y^2 - x^2 = 1$; y-axis

27. Find an equation of the ellipsoid with center at the origin and x, y, and z semiaxes of lengths 1, 2, and 3.

28. Find an equation of the circular cylinder about the z-axis generated by the vertical line through $(3, 4, 0)$.

29. Find an equation of the elliptic cone whose traces in the yz-plane are the intersecting lines $y = \pm z$ and in the xz-plane are the intersecting lines $x = \pm 2z$.

30. Identify and find an equation of the quadric surface whose trace in each plane $z = c$ ($c \geq 0$) is the ellipse $x^2 + 4y^2 = c$.

31. Identify and find an equation of the quadric surface whose traces are as follows:

in planes $z = c$; $\begin{cases} x^2 + y^2 = c^2 - 1 & \text{if } |c| \geq 1 \\ \text{empty} & \text{if } |c| < 1 \end{cases}$

in planes $x = c$; $z^2 - y^2 = 1 + c^2$

in planes $y = c$; $z^2 - x^2 = 1 + c^2$

REVIEW EXERCISES

In Exercises 1–8, sketch the graphs (in space) of the given equations.

1. $4x^2 + 9y^2 + 36z^2 = 36$ **2.** $x = 1$

3. $x + y + z = 1$ **4.** $2x + 3y + z = 6$

5. $x^2 + y^2 + z^2 - 2x - 4y = 0$

6. $x^2 + y^2 + z^2 = 1$

7. $4x^2 + 9y^2 = 36$

8. $4x^2 - 9y^2 = z$; $z \geq 0$

In Exercises 9–14, sketch the curves (in the plane or in space) traced out by the given position function.

9. $\mathbf{r}(t) = 3t\mathbf{i} + (2 - t)\mathbf{j}$

10. $\mathbf{r}(t) = \mathbf{i} + t\mathbf{j} + (2t + 1)\mathbf{k}$

11. $\mathbf{r}(t) = \cos t\,\mathbf{i} + \sin t\,\mathbf{j} + t\mathbf{k}$

12. $\mathbf{r}(t) = 2t\mathbf{i} + \cos t\,\mathbf{j} + \sin t\,\mathbf{k}$

13. $\mathbf{r}(t) = \mathbf{i} + t\mathbf{j} + e^t\mathbf{k}$

14. $\mathbf{r}(t) = t\mathbf{i} + t^3\mathbf{j} + \mathbf{k}$

15. Find the length of the curve in Exercise 11 for $0 \le t \le 2\pi$.

16. Find the velocity, speed, and acceleration of a particle with the position function in Exercise 12.

17. Find the curvature of the path traced out by the position function in Exercise 13.

18. Find the normal and tangential components of acceleration of the particle in Exercise 16.

Exercises 19–43, refer to the following four vectors

$$\mathbf{a} = 2\mathbf{i} - 3\mathbf{j} + \mathbf{k} \quad \mathbf{b} = \mathbf{i} + \mathbf{j} - 2\mathbf{k}$$
$$\mathbf{c} = 4\mathbf{i} + \mathbf{j} + 3\mathbf{k} \quad \mathbf{d} = 6\mathbf{i} - 7\mathbf{j} + 9\mathbf{k}$$

19. $\mathbf{a} \cdot 2\mathbf{b} =$

20. $3\mathbf{a} - \mathbf{b} =$

21. $\text{comp}_c\mathbf{a} =$

22. $\text{comp}_{-b}\mathbf{d} =$

23. $\mathbf{a} \times \mathbf{d} =$

24. $\mathbf{c} \times \mathbf{b} =$

25. Find a unit vector in the same direction as $\mathbf{a}$.

26. Find a unit vector opposite to $\mathbf{c}$.

27. Find the angle between $\mathbf{a}$ and $\mathbf{b}$.

28. Find the angle between $\mathbf{c}$ and $\mathbf{d}$.

29. Find a unit vector orthogonal to $\mathbf{a}$ and $\mathbf{b}$.

30. Find parametric equations of the line through $(1, -2, 1)$ with direction vector $\mathbf{c}$.

31. Find a vector equation of the line through $(5, 1, -2)$ and parallel to $\mathbf{b}$.

32. Find an equation of the plane through $(0, 1, 2)$ and perpendicular to the line in Exercise 30.

33. Find an equation of the plane through $(5, 1, -2)$ that is parallel to $\mathbf{a}$ and $\mathbf{c}$.

34. Find the point where the line of Exercise 30 meets the plane of Exercise 32.

35. $\mathbf{a} \times \mathbf{b} \times \mathbf{c} =$

36. $\mathbf{c} \times \mathbf{b} \cdot \mathbf{d} =$

37. Find the area of the parallelogram determined by $\mathbf{a}$ and $\mathbf{b}$.

38. Same as Exercise 37 for the vectors $\mathbf{b}$ and $\mathbf{c}$.

39. Find the volume of the parallelepiped determined by $\mathbf{a}$, $\mathbf{b}$, and $\mathbf{c}$.

40. Same as Exercise 39 for the vectors $\mathbf{b}$, $\mathbf{c}$, and $\mathbf{d}$.

41. Find the direction angles of $\mathbf{a}$.

42. Are any of the vectors $\mathbf{a}$, $\mathbf{b}$, $\mathbf{c}$, or $\mathbf{d}$ orthogonal to $\mathbf{i} + 2\mathbf{j} - 2\mathbf{k}$?

43. Find scalars p, q, and r such that $\mathbf{d} = p\mathbf{a} + q\mathbf{b} + r\mathbf{c}$.

Exercises 44–56 refer to the following three points

$$P(1, 1, 1) \quad Q(-2, 1, 0) \quad R(4, -2, 3)$$

44. Find the midpoint of the line segment from P to Q.

45. Find the area of the triangle PQR.

46. Find an equation of the plane through P, Q, and R.

47. Find parametric equations of the line through P and Q.

48. Find parametric equations of the line through Q and R.

49. Find an equation of the plane through P that is perpendicular to $\overrightarrow{PR}$.

50. Find an equation of the plane through Q that is perpendicular to $\overrightarrow{QR}$.

51. Find the distance from Q to the plane in Exercise 49.

52. Find the distance from P to the plane in Exercise 50.

53. Find $2\overrightarrow{PQ} + 3\overrightarrow{QR}$.

54. Find the angle between $\overrightarrow{PQ}$ and $\overrightarrow{PR}$.

55. Find the distance from P to the line through Q and R.

56. Find the angle between the plane in Exercise 50 and the line in Exercise 48.

Exercises 57–61 refer to a particle whose position at time t seconds is the tip of the position vector $\mathbf{r}(t) = (2t - 1)\mathbf{i} + 3t\mathbf{j} + (t + 4)\mathbf{k}; t \ge 0$.

57. What is the initial position of the particle? Where is it after 1 second? Describe its path.

58. When and where will it strike the yz-plane?

59. How much work is done by the force $\mathbf{F} = 2\mathbf{i} + 3\mathbf{j} - \mathbf{k}$ newtons in moving the particle through a displacement from $(-1, 0, 4)$ to $(1, 3, 5)$ meters?

60. What is the power input of the force $\mathbf{F}$ defined in Exercise 59?

61. How close does the particle come to the origin?

62. Find symmetric equations of the line of intersection of the planes $x + y = 1$ and $y + z = 1$.

63. The acceleration vector of a particle at time t is $\mathbf{a}(t) = 3t^2\mathbf{i} + (2t + 1)\mathbf{j} - \mathbf{k}$. If it starts from rest at the origin, where is it after 2 seconds?

64. Let $\mathbf{v}(t)$ be the velocity of a particle at time t. Show that $d/dt|\mathbf{v}(t)|^2 = 2\mathbf{v}(t) \cdot \mathbf{v}'(t)$.

65. A projectile is fired with an initial velocity of 500 m/sec at an angle of 30° with the horizontal (the acceleration due to gravity is $-9.8\mathbf{j}$ m/sec²). (a) What is its velocity at time t? (b) What is the maximum height reached? (c) What is the velocity at the maximum height? (d) What is the range?

66. A train is moving with a constant speed of 60 mph around a circular curve with a radius of 300 ft. What is the magnitude of the centripetal acceleration?

67. The position function of a particle is $\mathbf{r}(t) = \cos t\mathbf{i} + \sin t\mathbf{j} + t\mathbf{k}$. Find the particle's *angular velocity* about the origin.

68. (Continuation of Exercise 67). If the particle has mass $m = 10$, find the torque about the origin.

69. A highway has a circular curve of radius 200 ft that is banked at an angle of 8°. What is the maximum possible speed under icy conditions (no friction)?

70. The position function of a particle is $\mathbf{r}(t) = (1 + \cos 2t)\mathbf{i} + (\sin 2t)\mathbf{j}$. Find (a) the normal and tangential components of acceleration at time t, (b) the curvature of the path at time t.

15

PARTIAL
DIFFERENTIATION

The derivative is an extremely useful tool for studying the behavior of a function of one variable. In this chapter, we use an extension of this tool to study the behavior of a function of several variables.

The first two sections serve as an introduction to functions of several variables; they discuss graphs, limits, and continuity. Section 15–3 defines the concept of a partial derivative, which is the extension of the derivative mentioned above. In the rest of the chapter, partial derivatives are applied to problems such as approximation, rates of change, related rates, and maximum-minimum.

Three items can be singled out for special attention. The first, of course, is the partial derivative. The second is the Approximation Theorem (Section 15–4); it is the basis of the theory. The third is the gradient (Section 15–6); not only is it the basis of several applications in this chapter, but it also plays an important role in later chapters.

An Application to Economics, the optional reading section, illustrates the use of calculus in verifying an economic theory concerning marginal productivity.

15–1 FUNCTIONS OF SEVERAL VARIABLES

Until now, functions have been defined on subsets of the real line; we call them functions of one variable. In this section, the notion of a function of several variables is introduced. We begin with the definition of a function of two variables.

(15.1)

> **DEFINITION**
> Let D be a set of ordered pairs of real numbers. A rule f that assigns to each pair (x, y) in D a unique real number $f(x, y)$ is called a **function of two variables.** The number $f(x, y)$ is called the **value** of f at (x, y). The set D is called the **domain** of f and the set of all values is called the **range** of f.

The domain D is usually pictured as a subset of the xy-plane. If the domain is not specified, then it is automatically taken to be the largest set for which the expression defining f is meaningful. Here are some examples.

EXAMPLE 1   Let $f(x, y) = 4/\sqrt{y - x^2}$. Find the domain and evaluate f at $(0, 1), (1, 3)$, and $(-1, 2)$. Also find the range.

Solution   The domain D is the set of all pairs (x, y) with $y - x^2 > 0$, or $y > x^2$. This is the subset of the xy-plane above the parabola $y = x^2$ (Figure 15–1). The values of f are obtained by straight substitution.

$$f(0, 1) = \frac{4}{\sqrt{1 - 0^2}} = 4$$

$$f(1, 3) = \frac{4}{\sqrt{3 - 1^2}} = \frac{4}{\sqrt{2}}$$

$$f(-1, 2) = \frac{4}{\sqrt{2 - (-1)^2}} = 4$$

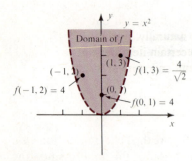

FIGURE 15–1. Example 1. Domain of $f(x, y) = 4/\sqrt{y - x^2}$.

Since the values of f are all positive and the denominator can be arbitrarily small, it follows that the range of f is the set of all positive numbers.   ∎

A function f of three variables is defined just as in (15.1), except that the domain is a set of ordered triples (x, y, z) and the values of f are denoted by $f(x, y, z)$. Now the domain is pictured as a subset of space.

EXAMPLE 2   Let $f(x, y, z) = \sin^{-1}z/(1 + \sqrt{1 - x^2 - y^2})$. Find the domain and evaluate f at $(0, 0, 0), (1, 0, 1)$, and $(0, 1, -1/2)$. Also find the range.

Solution   The domain consists of all triples (x, y, z) with $x^2 + y^2 \leq 1$ and $|z| \leq 1$ (why?). This is the *solid* cylinder pictured in Figure 15–2. The values of f are

$$f(0, 0, 0) = \frac{\sin^{-1}0}{1 + \sqrt{1 - 0^2 - 0^2}} = 0$$

$$f(1, 0, 1) = \frac{\sin^{-1}1}{1 + \sqrt{1 - 1^2 - 0^2}} = \frac{\pi}{2}$$

$$f\left(0, 1, -\frac{1}{2}\right) = \frac{\sin^{-1}(-1/2)}{1 + \sqrt{1 - 0^2 - 1^2}} = -\frac{\pi}{6}$$

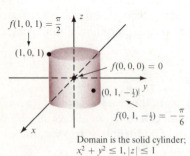

Domain is the solid cylinder; $x^2 + y^2 \leq 1, |z| \leq 1$

FIGURE 15–2. Example 2. Domain of $f(x, y, z) = \sin^{-1}z/(1 + \sqrt{1 - x^2 - y^2})$.

Since $|\sin^{-1}z| \leq \pi/2$ and the denominator is 1 or larger, the range of f is the interval $[-\pi/2, \pi/2]$.   ∎

Functions of two and three variables arise quite naturally in applications. For instance, the formula

$$A = \frac{1}{2}bh$$

for the area of a triangle defines A as a function of two variables; the base b and the height h. Here, b and h are referred to as the **independent variables,** whereas A is called the **dependent variable.** The domain of this function is understood to be all pairs (b, h) with $b \geq 0$ and $h \geq 0$. A useful function of three variables is

$$f(x, y, z) = \sqrt{x^2 + y^2 + z^2}$$

which is the length of the vector $x\mathbf{i} + y\mathbf{j} + z\mathbf{k}$. The domain here is all of space; the range is the set of nonnegative numbers. Although we shall concentrate on functions of two and three variables, it should be pointed out that functions of more than three variables also come about naturally. For example, a manufacturer may find that the cost of producing a certain item depends on four independent variables; material, labor, equipment, and storage costs.

The usual *algebra* of functions applies here. That is, we define the *sum* $f + g$ by

$$(f + g)(x, y) = f(x, y) + g(x, y)$$

the *product fg* by

$$fg(x, y) = f(x, y)g(x, y)$$

and so forth. If g is a function of one variable whose domain contains the range of f, then the *composition* $g \circ f$ is defined by

$$g \circ f(x, y) = g(f(x, y))$$

A function of several variables is a *polynomial function* if it is the sum of products of nonnegative integer powers of its variables; a *rational function* is a quotient of polynomials. Examples are

Polynomial: $f(x, y, z) = 3x^2y - 4xz^3 + 5x^2y^2z^2$

Rational: $g(x, y) = \dfrac{\sqrt{2}x^2y^2}{2xy - 9y^2}$

Graphs

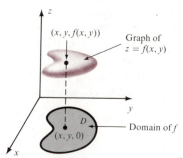

Let f be a function of two variables with domain D, a subset of the xy-plane. The **graph** of f is defined to be the graph, in space, of the equation $z = f(x, y)$. That is, the graph of f is a surface in space consisting of points $(x, y, z) = (x, y, f(x, y))$. Each point $(x, y, f(x, y))$ on the surface lies directly over or under the corresponding point $(x, y, 0)$ in D (Figure 15–3).

FIGURE 15–3. The graph of the function f is a surface in space.

EXAMPLE 3 Sketch the graph of the polynomial $f(x, y) = 4x^2 + y^2$.

Solution The domain is the entire xy-plane. The graph of f is the graph of the equation

$$z = 4x^2 + y^2$$

which is an elliptic paraboloid (Example 5, Section 14–10). The graph, sketched in Figure 15–4, shows the traces in the planes $z = c$ for $c = 1, 2, 3,$ and 4. We will make use of these traces shortly. ∎

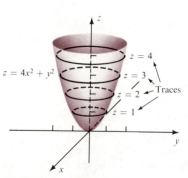

FIGURE 15–4. Example 3.

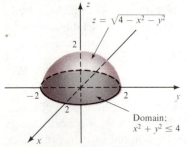

FIGURE 15–5. Example 4.

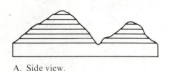

A. Side view.

B. Top view.

FIGURE 15–6. The lines joining points with the same elevation (called contour lines) provide a clear picture of the terrain.

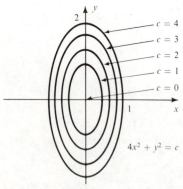

FIGURE 15–7. Level curves of $f(x, y) = 4x^2 + y^2$.

EXAMPLE 4 Sketch the graph of $f(x, y) = \sqrt{4 - x^2 - y^2}$.

Solution The domain here is the set of pairs (x, y) with $x^2 + y^2 \leq 4$; these points lie on or within the circle $x^2 + y^2 = 4$. The graph of f is the graph of

$$z = \sqrt{4 - x^2 - y^2}$$

or, equivalently,

$$x^2 + y^2 + z^2 = 4 \quad z \geq 0$$

This is the hemisphere sketched in Figure 15–5. ∎

Level Curves and Surfaces

Graphs of functions of two variables are usually too difficult to draw by hand. Functions of the type discussed in Examples 3 and 4 above are rare exceptions. (With modern technology, however, graphs can be drawn by a computer, and we will exhibit some examples in a moment.) But there is a method, discovered by map makers long ago, that helps to describe three-dimensional graphs with two-dimensional drawings. Map makers typically sketch curves that join all points with the same elevation. These curves provide a clear picture of the terrain (Figure 15–6).

We can do the same thing with the graph of a function of two variables. If c is a particular value of f, we sketch the curve *in the domain* of f that joins all points (x, y) at which $f(x, y) = c$. This curve is called a **level curve** of f. The level curves provide a clear picture of the graph of f. For instance, Figure 15–7 shows some level curves for $f(x, y) = 4x^2 + y^2$, the function in Example 3. Each is an ellipse $4x^2 + y^2 = c$; as the point (x, y) moves along the ellipse, the values of f remain constant. Notice the connection between level curves and traces of the graph; each level curve is the projection on the xy-plane of the trace in the plane $z = c$.

EXAMPLE 5 Sketch some level curves of $f(x, y) = xy$.

Solution For each number $c \neq 0$, the corresponding level curve is the hyperbola

$$xy = c$$

As a point (x, y) moves along each hyperbola, the values of f remain constant. When $c = 0$, the "curve" is the coordinate axes. Figure 15–8 shows some of the level curves and the graph of $f(x, y) = xy$; they were drawn by a computer. ∎

Figures 15–9 and 15–10 are other examples of computer generated level curves and graphs.

Drawing graphs of a function of two variables is usually difficult, but drawing graphs of a function of three variables is virtually impossible. To do so would require a four-dimensional drawing. Nevertheless, we can obtain useful information by knowing the *level surfaces* of such functions. These are graphs of

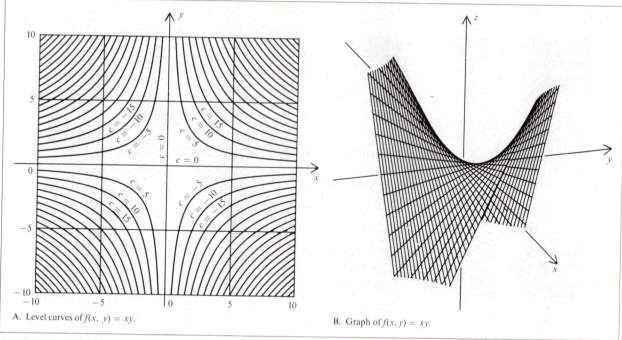

A. Level curves of $f(x, y) = xy$.

B. Graph of $f(x, y) = xy$.

FIGURE 15-8. Example 5.
Computer-generated level curves and
the graph of $f(x, y) = xy$.

equations such as $f(x, y, z) = c$. As the point (x, y, z) moves over a level surface, the values of f remain constant.

EXAMPLE 6

(a) Level surfaces of the function $f(x, y, z) = x - 3y + 4z$ are *planes* of the form $x - 3y + 4z = c$.

(b) Level surfaces of the function $f(x, y, z) = x^2 + 2y^2 + 3z^2$ are *ellipsoids* of the form $x^2 + 2y^2 + 3z^2 = c$. ■

We close with an important observation. If f is a function of two variables and we define F, a function of three variables, by $F(x, y, z) = f(x, y) - z$ then the *graph* of $z = f(x, y)$ is the *level surface* $F(x, y, z) = 0$ of F.

(15.2)
> The graph of $z = f(x, y)$ is the level surface of $F(x, y, z) = f(x, y) - z$ corresponding to $c = 0$.

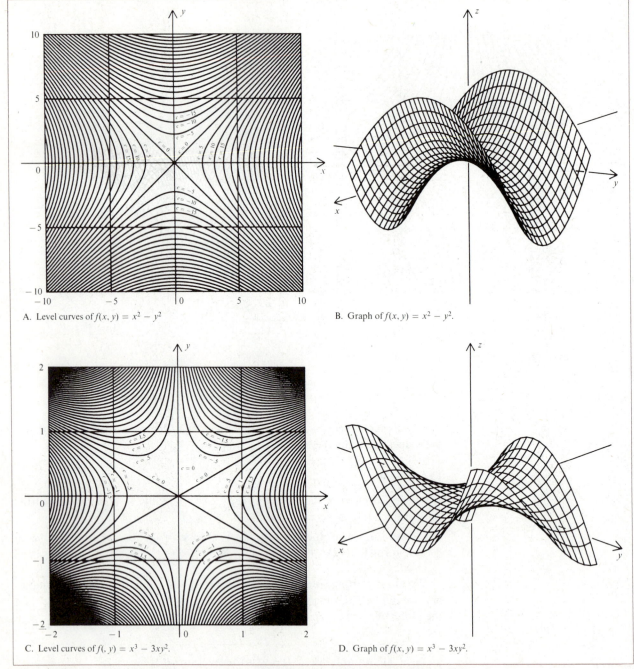

A. Level curves of $f(x, y) = x^2 - y^2$

B. Graph of $f(x, y) = x^2 - y^2$.

C. Level curves of $f(, y) = x^3 - 3xy^2$.

D. Graph of $f(x, y) = x^3 - 3xy^2$.

FIGURE 15.9 Examples of computer-generated level curves and graphs. The graph in part (B) is called a *saddle*, the graph in part (D) is called a *monkey saddle* because there are places for 2 legs and a tail. The graph of $f(x, y) = 4x^3y - 4xy^3$ (not shown) is called a *dog saddle* because there are places for all four legs.

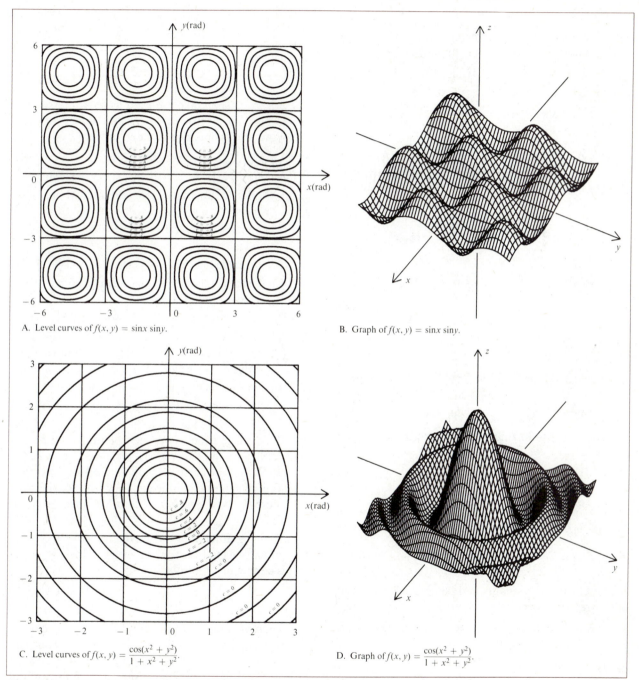

A. Level curves of $f(x, y) = \sin x \, \sin y$.

B. Graph of $f(x, y) = \sin x \, \sin y$.

C. Level curves of $f(x, y) = \dfrac{\cos(x^2 + y^2)}{1 + x^2 + y^2}$.

D. Graph of $f(x, y) = \dfrac{\cos(x^2 + y^2)}{1 + x^2 + y^2}$.

FIGURE 15–10. Examples of computer-generated level curves and graphs.

EXERCISES

In Exercises 1–8, find the domain of f. Evaluate f at the indicated points and, if possible, find the range of f.

1. $f(x, y) = 2x - 5y$; $(-2, 4), (1, 0), (2, -3)$
2. $f(x, y) = x + 3y^2$; $(0, 1), (-1, -1), (1, \sqrt{2})$
3. $f(x, y) = \dfrac{\sqrt{x + y}}{x^2}$; $(2, 1), (3, 0), (-1, 4)$
4. $f(x, y) = \dfrac{1}{\sqrt{x^2 + y^2}}$; $(1, 1), (\sqrt{2}, \sqrt{3}), (0, 1)$
5. $f(x, y, z) = xy \tan^{-1} z$; $(0, 0, 0), (1, 1, 1), (5, 0, 8)$
6. $f(x, y, z) = x^2 + yz$; $(0, 1, 2), (-2, 4, 1), (7, -1, 3)$
7. $f(x, y, z) = \sqrt{y} e^{xz}$; $(0, 1, 4), (1, 4, 0), (1, 1, 1)$
8. $f(x, y, z) = \sqrt{4 - x^2 - y^2 - z^2}$; $(0, 0, 0), (0, 2, 0), (1, 1, \sqrt{2})$

In Exercises 9–20, identify and sketch the graph of f.

9. $f(x, y) = 1$
10. $f(x, y) = -3$
11. $f(x, y) = 6 - 2x - 3y$
12. $f(x, y) = x + y$
13. $f(x, y) = \sqrt{1 - x^2 - y^2}$
14. $f(x, y) = \sqrt{9 - x^2 - y^2}$
15. $f(x, y) = x^2 + y^2$
16. $f(x, y) = 2x^2 + y^2$
17. $f(x, y) = x^2$
18. $f(x, y) = y^2$
19. $f(x, y) = y^2 - x^2$
20. $f(x, y) = 4y^2 - x^2$

In Exercises 21–30, identify and sketch some level curves of f.

21. $f(x, y) = 3x - 4y$
22. $f(x, y) = 2x$
23. $f(x, y) = y - x^2$
24. $f(x, y) = x^2 - y^2$
25. $f(x, y) = y - \sin x$
26. $f(x, y) = y - e^x$
27. $f(x, y) = ye^x$
28. $f(x, y) = y \sin x$
29. $f(x, y) = \dfrac{x^2 + y^2}{x}$
30. $f(x, y) = \dfrac{x}{x + y}$

In Exercises 31–36, identify and sketch the level surface of f corresponding to the given value of c.

31. $f(x, y, z) = 4x - y + z$; $c = 4$
32. $f(x, y, z) = z$; $c = 1$
33. $f(x, y, z) = x^2 + 3y^2 + 2z^2$; $c = 1$
34. $f(x, y, z) = x^2 + 3y^2 - 2z^2$; $c = 1$
35. $f(x, y, z) = z^2 - x^2 - y^2$; $c = 0$ and $c = 1$
36. $f(x, y, z) = y^2 - x^2 - z$; $c = 0$

In Exercises 37–40, write out expressions for

$$\frac{f(x + h, y) - f(x, y)}{h} \quad \text{and} \quad \frac{f(x, y + h) - f(x, y)}{h}$$

where $h \neq 0$. Then take the limit as $h \to 0$ (assuming the x and y are fixed). Try to make a connection between the results obtained and derivatives.

37. $f(x, y) = 3x - 2y$
38. $f(x, y) = x^2 + 5y$
39. $f(x, y) = xy$
40. $f(x, y) = x \sin y$

41. Formulate a function of two variables whose value at (x, y) is
   (a) the area of a rectangle with sides of length x and y,
   (b) the angle between the vector $\mathbf{i}$ and $x\mathbf{i} + y\mathbf{j}$,
   (c) the area of the parallelogram determined by the vectors $\mathbf{i}$ and $x\mathbf{i} + y\mathbf{j}$.

42. Formulate a function of three variables whose value at (x, y, z) is
   (a) the surface area of a rectangular parallelepiped with sides of length x, y, and z,
   (b) the angle between the vectors $\mathbf{i}$ and $x\mathbf{i} + y\mathbf{j} + z\mathbf{k}$,
   (c) the volume of the parallelepiped determined by $\mathbf{i}, \mathbf{j}$, and $x\mathbf{i} + y\mathbf{j} + z\mathbf{k}$.

43. The voltage drop V between points a and b on a wire varies directly as the distance d between a and b, and the current I, and inversely as the cross-sectional area A and the conductivity c of the wire. Write V as a function of four variables; d, I, A, and c.

44. The magnitude F of the gravitational force exerted on an object at (x, y, z) by an object at the origin is a function of five variables

$$F(x, y, z, m, M) = \frac{GmM}{x^2 + y^2 + z^2}$$

where m and M are the masses of the objects and G is the (positive) universal gravitation constant. This is Newton's *law of gravitation*. If m and M are constant, F can be considered as a function of the three variables x, y, and z. Describe the level surfaces of this function. What is their physical significance?

15–2   LIMITS AND CONTINUITY

It is time now to bring in some calculus. We begin with a discussion of limits and continuity for functions of several variables. As before, these topics are first treated informally, with the precise definitions appearing at the end of the section.

Limits

Suppose that f is a function of two variables. If the values $f(x, y)$ get closer and closer to a fixed number L as the points (x, y) get closer and closer to a fixed point (p, q), then we say that *the limit of f as (x, y) approaches (p, q) equals L*, and write

$$\lim_{(x, y) \to (p, q)} f(x, y) = L \qquad \text{(Figure 15–11)}$$

It is not required that f actually be defined *at* (p, q); only that it be defined *near* (p, q). The whole idea can be rephrased as follows:

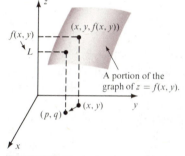

$f(x, y)$

L

$(x, y, f(x, y))$

A portion of the graph of $z = f(x, y)$.

(p, q) (x, y) y

x

FIGURE 15–11. As (x, y) gets closer and closer to (p, q), $f(x, y)$ gets closer and closer to L.

(15.3)

> If $f(x, y)$ can be made arbitrarily close to L by taking (x, y) sufficiently close but not equal to (p, q), then
> $$\lim_{(x, y) \to (p, q)} f(x, y) = L$$

Although there is one important difference (which we will talk about shortly), this is much the same concept as the limit of a function on one variable. Theorem (2.5) in Section 2–1 concerning the limits of sums, products, and quotients can be extended using the same arguments. For example, if $f(x, y)$ is close to L and $g(x, y)$ is close to M whenever (x, y) is close to (p, q), then $(f + g)(x, y) = f(x, y) + g(x, y)$ must be close to $L + M$ whenever (x, y) is close to (p, q); that is, *the limit of the sum is the sum of the limits*. The other parts are argued in a similar way.

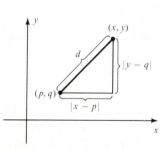

y

(x, y)

d

$|y - q|$

(p, q)

$|x - p|$

x

FIGURE 15–12. Saying that (x, y) is close to (p, q) means that the distance d between them is small. In the diagram, $|x - p| < d$ and $|y - q| < d$ (why?). Therefore, if (x, y) is close to (p, q), then x is close to p and y is close to q.

(15.4)

> If $\lim_{(x, y) \to (p, q)} f(x, y) = L$ and $\lim_{(x, y) \to (p, q)} g(x, y) = M$, then
>
> (1) $\lim_{(x, y) \to (p, q)} cf(x, y) = cL$, c any constant
>
> (2) $\lim_{(x, y) \to (p, q)} (f + g)(x, y) = L + M$
>
> (3) $\lim_{(x, y) \to (p, q)} (fg)(x, y) = LM$
>
> (4) $\lim_{(x, y) \to (p, q)} (f/g)(x, y) = L/M$, provided $M \neq 0$

This result can easily be extended to any finite number of functions.

The next step is to establish a theorem concerning the limits of the special functions $f(x, y) = x$ and $g(x, y) = y$. If (x, y) is close to (p, q), it must be true that x is close to p and that y is close to q (Figure 15–12). It follows from this observation that

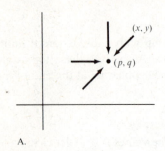

A.

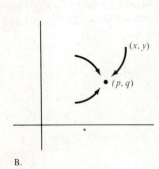

B.

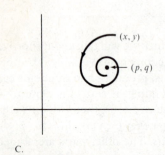

C.

FIGURE 15–13. (x, y) can approach (p, q) in many ways.

(15.5)
$$\lim_{(x, y) \to (p, q)} x = p \quad \text{and} \quad \lim_{(x, y) \to (p, q)} y = q$$

Combining (15.4) and (15.5), we know that limits of polynomial and rational functions can be found by direct substitution (provided the denominator is nonzero).

EXAMPLE 1

(a) $\displaystyle \lim_{(x, y) \to (1, 2)} 3x + y = 3(1) + 2 = 5$

(b) $\displaystyle \lim_{(x, y) \to (-1, 4)} x^2 y + 3xy^2 = (-1)^2(4) + 3(-1)(4)^2 = -44$

(c) $\displaystyle \lim_{(x, y) \to (0, 1)} \frac{y}{x^2 + y^2} = \frac{1}{0^2 + 1^2} = 1$ ∎

Analogs of theorems (2.7) and (2.8), concerning the limit of an nth root and the limit of an absolute value, are also valid for functions of two variables. Even the Squeeze Theorem (2.9) is true.

We said earlier that there is one important difference between the limits

$$\lim_{x \to p} f(x) \quad \text{and} \quad \lim_{(x, y) \to (p, q)} f(x, y)$$

The difference is that in the one variable case, x can approach p in essentially two ways, from the right or the left. In the two variable case, however, (x, y) can approach (p, q) in infinitely many ways. Figure 15–13 shows but a few of the paths along which (x, y) can appraoch (p, q). If the limit is to exist, then $f(x, y)$ must get close to the same number L no matter what path is taken by (x, y) in approaching (p, q). The next example illustrates a typical situation where this does not happen.

EXAMPLE 2 Show that $\lim_{(x, y) \to (0, 0)} xy/(x^2 + y^2)$ does not exist.

Solution The function $f(x, y) = xy/(x^2 + y^2)$ is defined everywhere except at $(0, 0)$. If (x, y) approaches $(0, 0)$ along the x-axis, then $y = 0$ and

$$f(x, 0) = \frac{x \cdot 0}{x^2 + 0^2} = 0 \quad x \neq 0$$

Similarly, if (x, y) approaches $(0, 0)$ along the y-axis, then $x = 0$ and $f(0, y) = 0$. So far, it looks like $f(x, y) \to 0$ as $(x, y) \to (0, 0)$. But notice what happens if (x, y) approaches $(0, 0)$ along the line $y = x$. In that case,

$$f(x, y) = f(x, x) = \frac{x^2}{x^2 + x^2} = \frac{1}{2} \quad x \neq 0$$

Therefore, there are points very close to $(0, 0)$ at which the value of f is 0 and other points equally close at which the value of f is $1/2$. The limit cannot exist. ∎

Continuity

Before defining continuity, we have to discuss the notions of *interior* and *boundary* points. Given a point (a, b), the set of points (x, y) whose distance from (a, b) is less than some positive number r is called an **open disc about** (a, b); this is the set of points inside the circle with center (a, b) and radius r (Figure 15–14A). Now suppose that R is some region in the xy-plane. A point (a, b) is an **interior point** of R if R contains an open disc about (a, b). See Figure 15–14B. If every point of R is an interior point, then R is said to be an **open set** (analogous to an open interval). A point (a, b) is a **boundary point** of R if every open disc about (a, b) contains points in R and also points not in R (Figure 15–14C). If R contains all of its boundary points, then R is said to be a **closed set** (analogous to a closed interval).

Returning momentarily to limits, suppose that f is defined at all points (x, y) in a region R except, possibly, at (p, q). If (p, q) is an interior point of R, then (15.3) is used to investigate the limit of f as (x, y) approaches (p, q). But, if (p, q) is a boundary point of R, then (15.3) must be amended to say that (x, y) is not only close to (p, q) but also an element of R. In this way, limits can be extended to boundary points (analogous to taking the limit at an endpoint of an interval).

We can now define continuity. Suppose f is a function of two variables with domain D. Then f is **continuous** at (p, q) if

$$\lim_{(x, y) \to (p, q)} f(x, y) = f(p, q)$$

If (p, q) happens to be a boundary point of D, the limit is taken in the sense mentioned above. If f is continuous at each point of D, then we say that f is **continuous.** Observe that if f is continuous at (p, q), then f must be defined at (p, q), and the limit must exist and equal $f(p, q)$. In geometric terms, this means that there are no holes or gaps in the graph of a continuous function.

EXAMPLE 3 Discuss the continuity of $f(x, y) = x$ and $g(x, y) = y$.

Solution Applying (15.5) to any point (p, q), we have

$$\lim_{(x, y) \to (p, q)} f(x, y) = \lim_{(x, y) \to (p, q)} x = p = f(p, q)$$

and

$$\lim_{(x, y) \to (p, q)} g(x, y) = \lim_{(x, y) \to (p, q)} y = q = g(p, q)$$

Therefore, by definition, f and g are continuous everywhere. ∎

All of the continuity theorems for functions of one variable also hold for functions of two variables. In particular, it follows from Example 3 that all polynomials are continuous as are all rational functions except at points where the denominator is zero. The composition theorem is also valid. If f is continuous at (p, q) and g is a function of one variable that is continuous at $f(p, q)$, then the composition $g \circ f(x, y) = g(f(x, y))$ is also continuous at (p, q).

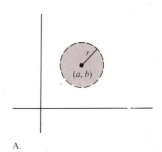

A.

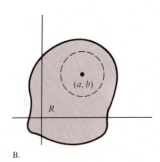

B.

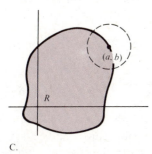

C.

FIGURE 15–14. (A) An open disc about (a, b). (B) (a, b) is an interior point of R. (C) (a, b) is a boundary point of R.

EXAMPLE 4 Discuss the continuity of $h(x, y) = \sin^{-1}xy$.

Solution The domain of h is the set of points (x, y) with $|xy| \leq 1$. To discuss continuity, we write h as a composition. Let $f(x, y) = xy$ and $g(t) = \sin^{-1}t$; then $h(x, y) = g \circ f(x, y) = g(xy) = \sin^{-1}xy$. Since f is continuous everywhere and g is continuous for $|t| \leq 1$, it follows that h is continuous for $|xy| \leq 1$. ∎

If a function is not a pure composition then a slightly different technique is used.

EXAMPLE 5 Discuss the continuity of $k(x, y) = x^2 e^{xy}$.

Solution The domain of k is the whole plane. We can view k as a product of the continuous function $h(x, y) = x^2$ and the continuous composition $g \circ f$ where $f(x, y) = xy$ and $g(t) = e^t$. Thus, k is continuous. ∎

Functions of Three Variables

The notions of limits and continuity and the results concerning them can be extended to functions of three (or any number of) variables. The notation for limits is

(15.6) $$\lim_{(x, y, z) \to (p, q, r)} f(x, y, z) = L$$

Here, (x, y, z) can approach (p, q, r) along any curve (path) in space. In space, we still talk about interior and boundary points of sets, but *open spheres* replace open discs in the definitions.

Definition of Limit *(Optional)*

(15.7)

> If a function f of two variables is defined everywhere on an open disc about (p, q), except possibly at (p, q) itself, then
>
> $$\lim_{(x, y) \to (p, q)} f(x, y) = L$$
>
> means that for any given $\epsilon > 0$, there is a $\delta > 0$ such that
>
> $|f(x, y) - L| < \epsilon$ whenever $0 < \sqrt{(x - p)^2 + (y - q)^2} < \delta$

If (p, q) is a boundary point of the domain of f, then the words "and (x, y) is in the domain of f" must be appended to the last line. In the case of a function of three variables, the appropriate notation (15.6) is used and the last line reads

$$|f(x, y, z) - L| < \epsilon \quad \text{whenever} \quad 0 < \sqrt{(x - p)^2 + (y - q)^2 + (z - r)^2} < \delta$$

EXERCISES

In Exercises 1–6, find the indicated limits.

1. $$\lim_{(x, y) \to (1, -3)} \frac{x - 4y}{8x + y}$$

2. $\lim_{(x,y)\to(0,0)} 2x^2 + 4xy + 5$

3. $\lim_{(x,y)\to(4,1)} \sqrt{3x^2y + 2xy^2}$ **4.** $\lim_{(x,y)\to(6,2)} y\sin(\pi/x)$

5. $\lim_{(x,y,z)\to(1,0,-1)} (x^2+y^2)\cos^{-1}z$

6. $\lim_{(x,y,z)\to(-2,4,0)} 2xy + 2yz + 2xz$

In Exercises 7–8, the indicated limits do not exist. Demonstrate this by finding two paths on which (x, y) approaches (p, q), and $f(x, y)$ approaches different numbers (see Example 2).

7. $\lim_{(x,y)\to(0,0)} \dfrac{x^2 - y^2}{x^2 + y^2}$

8. $\lim_{(x,y)\to(0,0)} \dfrac{x^2y}{x^4 + y^2}$

(*Hint:* Try the parabolic path $y = x^2$.)

In Exercises 9–24, find the indicated limits, if they exist.

9. $\lim_{(x,y)\to(0,0)} \dfrac{1}{\sqrt{4 - x^2 - y^2}}$ **10.** $\lim_{(x,y)\to(2,2)} \dfrac{xy}{x^2 + y^2}$

11. $\lim_{(x,y)\to(0,0)} \dfrac{\sin xy}{xy}$

12. $\lim_{(x,y)\to(0,0)} \dfrac{1 - \cos xy}{xy}$

13. $\lim_{(x,y)\to(0,0)} \dfrac{1/xy}{e^{1/|xy|}}$ **14.** $\lim_{(x,y)\to(5,0)} \dfrac{\sin xy}{xy}$

15. $\lim_{(x,y)\to(0,0)} \dfrac{3xy}{2x^2 + 5y^2}$ **16.** $\lim_{(x,y)\to(0,0)} \dfrac{5 + xy}{x^2 + y^2}$

17. $\lim_{(x,y)\to(0,0)} \dfrac{x^2 - y^2}{x - y}$

18. $\lim_{(x,y)\to(0,0)} \dfrac{x^2 - 3xy + 2y^2}{x - y}$

19. $\lim_{(x,y)\to(1,0)} \dfrac{1 - e^y}{1 - x}$

20. $\lim_{(x,y)\to(0,0)} \dfrac{x^3 - x^2y + xy^2 - y^3}{x^2 - 2xy + y^2}$

21. $\lim_{(x,y,z)\to(-2,1,3)} \dfrac{xy + yz + xz}{x^2 + y^2 + z^2}$

22. $\lim_{(x,y,z)\to(0,1,2)} \dfrac{\sqrt{x^2 + y^2 + z^2}}{e^{xyz}}$

23. $\lim_{(x,y,z)\to(0,0,0)} \dfrac{xy + yz + xz}{x^2 + y^2 + z^2}$

24. $\lim_{(x,y,z)\to(0,0,0)} \dfrac{1 + e^{xy}}{\cos xyz}$

For each function in Exercises 25–32, find the interior and boundary points of its domain, and discuss its continuity.

25. $h(x, y) = \ln(x^2 + y^2)$

26. $h(x, y) = \cos^{-1}(x + y)$

27. $h(x, y, z) = \dfrac{\sqrt{z}}{x^2 + y^2}$ **28.** $h(x, y, z) = e^{xyz}$

29. $k(x, y) = x^2 + y|x|$ **30.** $k(x, y) = \sqrt{x}e^y$

31. $k(x, y, z) = \sqrt{xy}\tan^{-1}z$

32. $k(x, y, z) = x^2 + y^2\sin z$

Optional Exercises

Use Definition (15.7) to prove the statements in Exercises 33–36.

33. $\lim_{(x,y)\to(p,q)} x = p$ and $\lim_{(x,y)\to(p,q)} y = q$

34. $\lim_{(x,y)\to(p,q)} (f + g)(x, y) = \lim_{(x,y)\to(p,q)} f(x, y) + \lim_{(x,y)\to(p,q)} g(x, y)$

35. Limits are unique; that is if $f(x, y)$ approaches L and M as $(x, y) \to (p, q)$, then $L = M$.

36. If f is continuous at (p, q) and $f(p, q) > 0$, then there is an open disc about (p, q) such that $f(x, y) > 0$ for all (x, y) that are in both the disc and the domain of f.

15-3 PARTIAL DERIVATIVES

In this section, we define the notion of a *partial derivative* and examine some of its properties. The idea is not at all complicated. For example, let

$$f(x, y) = x^2 y + x \sin y$$

The partial derivative of f with respect to x is again a function of two variables, denoted by f_x. To find this function, simply treat y as a constant and differentiate f with respect to x in the usual way. Thus,

$$f_x(x, y) = 2xy + \sin y \qquad \text{(} y \text{ treated as a constant)}$$

To find the partial derivative of f with respect to y, denoted by f_y, treat x as a constant and differentiate f with respect to y. Thus,

$$f_y(x, y) = x^2 + x \cos y \qquad \text{(} x \text{ treated as a constant)}$$

The definitions given below express these notions in formal language.

(15.8)

> If f is a function of two variables, then the **first partial derivatives of f with respect to x and y** are the functions f_x and f_y defined by
>
> $$f_x(x, y) = \lim_{\Delta x \to 0} \frac{f(x + \Delta x, y) - f(x, y)}{\Delta x}$$
>
> $$f_y(x, y) = \lim_{\Delta y \to 0} \frac{f(x, y + \Delta y) - f(x, y)}{\Delta y}$$
>
> provided these limits exist.

In the definition of f_x, y is held fixed; only x is allowed to vary. This effectively reduces f to a function of one variable x, and f_x is the usual derivative with respect to x. If x is fixed and only y is allowed to vary, then f_y is the usual derivative with respect to y.

EXAMPLE 1 If $f(x, y) = x^3 y^2 - e^{xy}$, then

$$f_x(x, y) = 3x^2 y^2 - y e^{xy} \qquad \text{(} y \text{ treated as a constant)}$$
$$f_y(x, y) = 2x^3 y - x e^{xy} \qquad \text{(} x \text{ treated as a constant)} \quad \blacksquare$$

Notation: Recall the Leibniz notation df/dx for the derivative of a function of one variable. For partial derivatives, the d becomes a script delta, ∂. Thus,

$$f_x \text{ can also be written as } \frac{\partial f}{\partial x}, \text{ and}$$

$$f_y \text{ can also be written as } \frac{\partial f}{\partial y}.$$

EXAMPLE 2 Let $z = x\sqrt{x^2 + y^2}$ and find $\partial z / \partial x$ and $\partial z / \partial y$.

Solution For $\partial z / \partial x$, treat y as a constant, and use the product rule on x.

$$\frac{\partial z}{\partial x}(x, y) = x\left[\frac{1}{2}(x^2 + y^2)^{-1/2}(2x)\right] + (x^2 + y^2)^{1/2}$$

$$= \frac{2x^2 + y^2}{\sqrt{x^2 + y^2}} \qquad\qquad \text{(Simplify)}$$

For $\partial z / \partial y$, the product rule is not necessary because x is treated as a constant.

$$\frac{\partial z}{\partial y}(x, y) = x\left[\frac{1}{2}(x^2 + y^2)^{-1/2}(2y)\right] = \frac{xy}{\sqrt{x^2 + y^2}} \qquad \blacksquare$$

Interpretation of Partial Derivatives

Suppose that f_x and f_y exist at a point (x_0, y_0). Figure 15–15A shows the plane $x = x_0$ through $(x_0, y_0, 0)$ parallel to the yz-plane. The trace of the surface $z = f(x, y)$ in this plane is a curve C_1 representing the equation $z = f(x_0, y)$. It follows from the definition of f_y and our previous theory about derivatives, that the slope of the tangent line L_1 (shown in the figure) is $f_y(x_0, y_0)$.

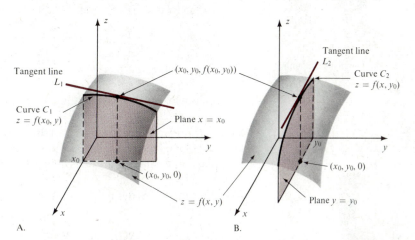

FIGURE 15–15. Interpretations of partial derivatives. (A) $f_y(x_0, y_0)$ is the slope of the tangent line L_1. (B) $f_x(x_0, y_0)$ is the slope of the tangent line L_2.

But more information is made available by putting this into the context of vectors. Since every point on the curve C_1 has coordinates of the form $(x_0, y, f(x_0, y))$, it follows that the tip of the position vector

$$\mathbf{r}(y) = x_0\mathbf{i} + y\mathbf{j} + f(x_0, y)\mathbf{k}$$

traces out C_1 as y ranges over its domain. Therefore, the derivative of $\mathbf{r}$ with respect to y

$$\mathbf{r}'(y) = \mathbf{j} + f_y(x_0, y)\mathbf{k}$$

represents *tangent vectors* to the curve C_1 (it may help to think of y as time t). It follows now that

$$(15.9) \qquad \mathbf{d}_1 = \mathbf{r}'(y_0) = \mathbf{j} + f_y(x_0, y_0)\mathbf{k}$$

is a *direction vector* for L_1. Figure 15–15B shows that a similar treatment is valid for f_x. Here, the plane $y = y_0$ through $(x_0, y_0, 0)$ is parallel to the xz-plane. The trace of $z = f(x, y)$ is a curve C_2, and the slope of the tangent line L_2 is $f_x(x_0, y_0)$.

In this case, the position vector function for C_2 is $\mathbf{r}(x) = x\mathbf{i} + y_0\mathbf{j} + f(x, y_0)\mathbf{k}$, and differentiating with respect to x, we have that

(15.10) $\quad \mathbf{d}_2 = \mathbf{r}'(x_0) = \mathbf{i} + f_x(x_0, y_0)\mathbf{k}$

is a direction vector for L_2.

Using the direction vectors $\mathbf{d}_1$ and $\mathbf{d}_2$, we can find equations of the lines L_1 and L_2. But more important (for use in the next section), we can find an equation of the plane containing L_1 and L_2. We know that

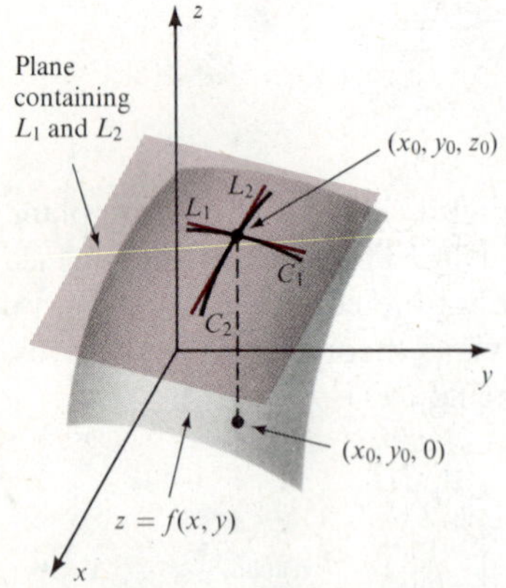

(15.11) $\quad \mathbf{N} = \mathbf{d}_1 \times \mathbf{d}_2 = \begin{vmatrix} \mathbf{i} & \mathbf{j} & \mathbf{k} \\ 0 & 1 & f_y(x_0, y_0) \\ 1 & 0 & f_x(x_0, y_0) \end{vmatrix}$

$$= f_x(x_0, y_0)\mathbf{i} + f_y(x_0, y_0)\mathbf{j} - \mathbf{k}$$

is normal to the plane and $(x_0, y_0, f(x_0, y_0))$ is on the plane. If we set $z_0 = f(x_0, y_0)$, it follows from (14.27) in Section 14–5 that

(15.12) $\quad \boxed{f_x(x_0, y_0)(x - x_0) + f_y(x_0, y_0)(y - y_0) - (z - z_0) = 0}$

is an equation of the plane containing the tangent lines L_1 and L_2 (Figure 15–16).

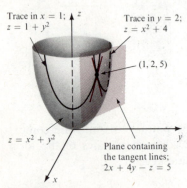

FIGURE 15–16. The plane containing the tangent lines L_1 and L_2; an equation is $f_x(x_0, y_0)(x - x_0) + f_y(x_0, y_0)(y - y_0) - (z - z_0) = 0$.

EXAMPLE 3 Consider the circular paraboloid $z = x^2 + y^2$. The traces in the planes $x = 1$ and $y = 2$ are parts of parabolas (Figure 15–17). Find an equation of the plane that contains the lines tangent to both parabolas at the point $(1, 2, 5)$.

Solution First we define $f(x, y) = x^2 + y^2$ and evaluate f_x and f_y at $(x_0, y_0) = (1, 2)$.

$$f_x(x, y) = 2x, \text{ so } f_x(1, 2) = 2$$
$$f_y(x, y) = 2y, \text{ so } f_y(1, 2) = 4$$

Now, using (15.12), we have

$$2(x - 1) + 4(y - 2) - (z - 5) = 0 \quad \text{or} \quad 2x + 4y - z = 5$$

as an equation of the plane. ■

Trace in $x = 1$; $z = 1 + y^2$

Trace in $y = 2$; $z = x^2 + 4$

$(1, 2, 5)$

$z = x^2 + y^2$

Plane containing the tangent lines; $2x + 4y - z = 5$

FIGURE 15–17. Example 3.

Another interpretation, which again exploits our previous experience with derivatives, is that the first partial derivatives signify the *rate of change* of f in directions parallel to the x- and y-axes. Moreover, if $f_x(x_0, y_0) > 0$, then the values of f will increase in the positive x-direction and decrease in the negative x-direction. If $f_x(x_0, y_0) < 0$, the opposite is true. Similar statements hold for f_y.

EXAMPLE 4 The temperature T in degrees Celsius at any point (x, y) on a circular hot plate is $T = 50 \sin xy$. If an ant is at the point $(1, 2)$, in which direction, north, south, east, or west, should it move in order to cool off the fastest (Figure 15–18)?

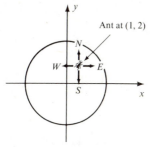

y

Ant at (1, 2)

N

W ← *→ E

S

x

FIGURE 15–18. Example 4. An ant on a hot plate.

Solution   The partial derivatives of T evaluated at (1, 2) represent the rate of change of temperature, T_x in the east-west direction and T_y in the north-south direction.

$$T_x(x, y) = 50y \cos xy, \text{ so } T_x(1, 2) = 100 \cos 2 \approx -42$$
$$T_y(x, y) = 50x \cos xy, \text{ so } T_y(1, 2) = 50 \cos 2 \approx -21$$

It follows that to obtain the greatest decrease in temperature, the ant should move toward the east (positive x-direction).

Remark: In Section 15–6, we will discover that if the ant were free to choose any direction, it would move slightly north of east rather than directly east. ■

Three Variables

For functions f of three variables, there are three first partial derivatives

$$f_x \text{ or } \frac{\partial f}{\partial x} \qquad f_y \text{ or } \frac{\partial f}{\partial y} \qquad f_z \text{ or } \frac{\partial f}{\partial z}$$

The definitions are similar; for example,

$$f_x(x, y, z) = \lim_{\Delta x \to 0} \frac{f(x + \Delta x, y, z) - f(x, y, z)}{\Delta x}$$

and so on. In practice, they are found by differentiating with respect to one variable, treating the other two as constants.

EXAMPLE 5

(a) If $f(x, y, z) = xz + x^2y + y^2z$, then

$$f_x(x, y, z) = z + 2xy, f_y(x, y, z) = x^2 + 2yz, \text{ and } f_z(x, y, z) = x + y^2$$

(b) If $w = xy^2z^3$, then

$$\frac{\partial w}{\partial x} = y^2z^3, \frac{\partial w}{\partial y} = 2xyz^3, \text{ and } \frac{\partial w}{\partial z} = 3xy^2z^2 \qquad ■$$

The interpretation of the partial derivatives as the rate of change in the x, y, and z directions holds for functions of three variables. The geometric interpretations, however, are impossible to draw, and we will not discuss them in this book.

Higher Order Partials

The first partial derivatives are again functions of two or three variables, as the case may be, and we can find their partial derivatives. These are called the **second partial derivatives** of f. For functions of two variables, there are four second partial derivatives.

$$(f_x)_x \text{ is denoted by } f_{xx} \text{ or } \frac{\partial^2 f}{\partial x^2}$$

$(f_x)_y$ is denoted by f_{xy} or $\dfrac{\partial^2 f}{\partial y \partial x}$

$(f_y)_x$ is denoted by f_{yx} or $\dfrac{\partial^2 f}{\partial x \partial y}$

$(f_y)_y$ is denoted by f_{yy} or $\dfrac{\partial^2 f}{\partial y^2}$

Functions of three variables have nine second partial derivatives; f_{xx}, f_{xy}, f_{xz}, and so forth. Of course, we can continue with third, fourth, and higher order partial derivatives.

CAUTION Observe that the order in which x and y appear in f_{xy} and $\partial^2 f / \partial y \partial x$ are reversed. The reason is that

$$f_{xy} = (f_x)_y = \frac{\partial}{\partial y}\left(\frac{\partial f}{\partial x}\right)$$

and the symbol on the right simplifies to $\partial^2 f / \partial y \partial x$.

Terminology: We usually refer to first partial derivatives as *first partials* or simply *partials* and to second partial derivatives as *second partials*. Second partials such as f_{xy} or f_{yz} are sometimes called *mixed partials*.

EXAMPLE 6 Let $f(x, y) = x^2 y + \sin xy$ and find all second partials of f.

Solution The first partials are

$$f_x(x, y) = 2xy + y \cos xy \quad \text{and} \quad f_y(x, y) = x^2 + x \cos xy$$

The second partials are

$$f_{xx}(x, y) = 2y - y^2 \sin xy$$
$$f_{xy}(x, y) = 2x + \underbrace{y(-x \sin xy) + \cos xy}_{\text{product rule}} = 2x - xy \sin xy + \cos xy$$
$$f_{yx}(x, y) = 2x + \underbrace{x(-y \sin xy) + \cos xy}_{\text{product rule}} = 2x - xy \sin xy + \cos xy$$
$$f_{yy}(x, y) = -x^2 \sin xy \quad \blacksquare$$

Notice that in the example above, the mixed partials are equal, $f_{xy} = f_{yx}$. This is always the case *provided the second partials are continuous.*

(15.13)

Let f be a function of two variables such that its second partials are continuous at (x_0, y_0). Then

$$f_{xy}(x_0, y_0) = f_{yx}(x_0, y_0)$$

A similar statement holds for functions of three variables.

We will not prove this result, but Exercise 55 is an example of a function whose mixed partials are not equal (the mixed partials are not continuous).

Before closing out this section, we want to show you that a function need not be continuous at a point even though its partials exist at that point.

EXAMPLE 7   Let f be the function defined by

$$f(x, y) = \begin{cases} \dfrac{xy}{x^2 + y^2} & \text{if } (x, y) \neq (0, 0) \\ 0 & \text{if } (x, y) = (0, 0) \end{cases}$$

In Example 2 of the preceding section, we showed that this function does not have a limit as $(x, y) \to (0, 0)$. Therefore, it cannot be continuous at $(0, 0)$. Nevertheless, using the definitions in (15.8), we can show that the partials exist at $(0, 0)$.

$$f_x(0, 0) = \lim_{\Delta x \to 0} \frac{f(0 + \Delta x, 0) - f(0, 0)}{\Delta x} \tag{15.8}$$

$$= \lim_{\Delta x \to 0} \frac{\dfrac{(0 + \Delta x)0}{\Delta x^2 + 0^2} - 0}{\Delta x}$$

$$= \lim_{\Delta x \to 0} \frac{0}{\Delta x} = 0$$

A similar calculation shows that $f_y(0, 0) = 0$. ∎

EXERCISES

In Exercises 1–20, find all first partials of the given function.

1. $f(x, y) = 3x^3y + 2xy^2$

2. $f(x, y) = 2xy^4 - 3x^2y^2$

3. $f(x, y) = \sqrt{x^2 + y^2}$ **4.** $f(x, y) = e^{xy}$

5. $g(u, v) = u \sin uv$ **6.** $g(u, v) = \ln uv$

7. $V = \dfrac{1}{3}\pi r^2 h$ **8.** $S = 2\pi rh$

9. $S = 2xy + 2yz + 2xz$ **10.** $V = xyz$

11. $f(x, y, z) = x \cos\dfrac{y}{z}$

12. $f(x, y, z) = \sin^{-1}xyz$

13. $F = \dfrac{3}{x^2 + y^2 + z^2}$

14. $G = \sqrt{x^2 + y^2 + z^2}$

15. $g(r, s, t) = rs \tan^{-1}t$ **16.** $g(r, s, t) = rs \sec t$

17. $f(x, y) = \ln\sqrt{x^2 + y^2}$

18. $f(x, y) = \tan(x - y)$

19. $f(x, y) = \dfrac{e^{xy}}{y \sin x}$ **20.** $f(x, y) = xe^{x\cos y}$

In Exercises 20–30, find f_{xy} and f_{yx}.

21. $f(x, y) = 5xy^3 - 3x^2y$

22. $f(x, y) = x^2 + 3xy + y^2$

23. $f(x, y) = x^2y \sin x$ **24.** $f(x, y) = xy \tan xy$

25. $f(x, y) = \ln(x^2 + y^2)$ **26.** $f(x, y) = x^2e^{xy}$

27. $f(x, y, z) = xe^y + ye^z + ze^x$

28. $f(x, y, z) = \ln xyz$

29. $f(x, y, z) = 3x^2yz + e^{xyz}$

30. $f(x, y, z) = x \tan^{-1}yz$

In Exercises 31–34, evaluate the partial at the given point.

31. $w = x^2 + y^2 - 2xy \cos z; \dfrac{\partial^2 w}{\partial x \partial z}(0, 1, \pi/6)$

32. $u = \dfrac{x^2 + y^2}{xz}; \dfrac{\partial^2 u}{\partial y \partial x}(1, 3, 1)$

33. $f(r, s, t) = e^{rs} \sin t; \dfrac{\partial^3 f}{\partial r \partial t \partial s}(3, 1, 0)$

34. $f(u, v, w) = e^u \ln vw; \dfrac{\partial^3 f}{\partial w \partial u \partial v}(\ln 2, 4, 1/2)$

---

35. The relationship between rectangular and polar coordinates is given by $x = r \cos \theta$ and $y = r \sin \theta$. Find x_r, x_θ, y_r, and y_θ.

36. The relationship between rectangular and polar coordinates can also be expressed as $r = \sqrt{x^2 + y^2}$ and $\theta = \tan^{-1}(y/x)$. Find r_x, r_y, θ_x, and θ_y.

37. Let C_1 and C_2 be the traces of the elliptic paraboloid $z = 2x^2 + 3y^2$ in the planes $x = 2$ and $y = 5$. Find an equation of the plane that contains the lines tangent to C_1 and C_2 at $(2, 5, 83)$.

38. Same as Exercise 37 for the hyperbolic paraboloid $z = 4y^2 - x^2$ at the point $(1, 3, 35)$.

A function f is said to be *harmonic* if it satisfies *Laplace's equation*

$$\frac{\partial^2 f}{\partial x^2} + \frac{\partial^2 f}{\partial y^2} = 0$$

In Exercises 39–46, determine which of the functions are harmonic.

39. $f(x, y) = e^x \cos y$ **40.** $f(x, y) = 3x^2 - 2y^3$

41. $f(x, y) = \ln(x^2 + y^2)$

42. $f(x, y) = \ln \sqrt{x^2 + y^2}$

43. $f(x, y) = \sin xy$

44. $f(x, y) = \tan^{-1}(y/x)$

45. $f(x, y) = \cos x \sinh y + \sin x \cosh y$

46. $f(x, y) = (x + y)^3$

A pair of functions f and g are said to satisfy the *Cauchy-Riemann equations* if

$$\frac{\partial f}{\partial x} = \frac{\partial g}{\partial y} \quad \text{and} \quad \frac{\partial f}{\partial y} = -\frac{\partial g}{\partial x}$$

In Exercises 47–52, determine which pairs of functions satisfy the Cauchy-Riemann equations.

47. $f(x, y) = x \cos y$ and $g(x, y) = x \sin y$

48. $f(x, y) = x$ and $g(x, y) = y$

49. $f(x, y) = e^x \cos y$ and $g(x, y) = e^x \sin y$

50. $f(x, y) = e^{x+y}$ and $g(x, y) = e^{x-y}$

51. $f(x, y) = \dfrac{1}{2}\ln(x^2 + y^2)$ and $g(x, y) = \tan^{-1}\dfrac{y}{x}$

52. $f(x, y) = x^2 - y^2$ and $g(x, y) = 2xy$

53. Suppose that a, b, and c are sides of a triangle, and θ is the angle between sides a and b. According to the law of cosines

$$c = \sqrt{a^2 + b^2 - 2ab \cos \theta}$$

If $a = b = 2$ and $\theta = \pi/4$, then   (a) find the rate of change c with respect to θ and   (b) find the rate of change of the area $A = \frac{1}{2}ab \sin \theta$ with respect to θ.

54. Three resistances R_1, R_2, R_3 connected in parallel produce a resistance R with

$$R = \left[\frac{1}{R_1} + \frac{1}{R_2} + \frac{1}{R_3}\right]^{-1} = \frac{R_1 R_2 R_3}{R_1 R_2 + R_2 R_3 + R_1 R_3}$$

Find $\partial R/\partial R_1$.   ↙ R_2 is a part that varies

---

55. Let f be a function defined by

$$f(x, y) = \begin{cases} \dfrac{x^3 y - xy^3}{x^2 + y^2} & \text{if } (x, y) \neq (0, 0) \\ 0 & \text{if } (x, y) = (0, 0) \end{cases}$$

Use definition (15.8), as we did in Example 7, to show that $f_x(0, y) = -y$ for all y and $f_y(x, 0) = x$ for all x. Then deduce that $f_{xy}(0, 0) \neq f_{yx}(0, 0)$.

15–4   DIFFERENTIALS AND DIFFERENTIABILITY

In this section, we investigate the change in the values of a function $z = f(x, y)$ produced by small changes in x and y. Let us first review the one variable case. In Section 4–3, we showed that for a function $y = f(x)$, the change $\Delta y = f(x_0 + \Delta x) - f(x_0)$ can be approximated by the differential dy,

$$\Delta y \approx dy = f'(x_0)\Delta x$$

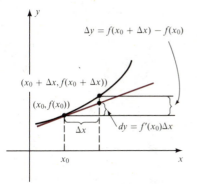

FIGURE 15–19. $\Delta y \approx dy$ for small Δx.

Figure 15–19 illustrates that dy is the difference between $f(x_0)$ and a corresponding point on the tangent line. What we want to show now is that, under certain conditions, the same sort of approximation is valid in the two variable case with the plane

(15.14) $z - z_0 = f_x(x_0, y_0)(x - x_0) + f_y(x_0, y_0)(y - y_0)$

(developed in the last section) taking the place of the tangent line. That is, if we substitute $\Delta x = x - x_0$ and $\Delta y = y - y_0$, and call $dz = z - z_0$ the *differential*, then

(15.15) $dz = f_x(x_0, y_0)\Delta x + f_y(x_0, y_0)\Delta y$

can be used to approximate the actual change

(15.16) $\Delta z = f(x_0 + \Delta x, y_0 + \Delta y) - f(x_0, y_0)$

Figure 15–20 illustrates this approximation.

Before stating the conditions and proving that Δz is approximated by dz, let us look at an example.

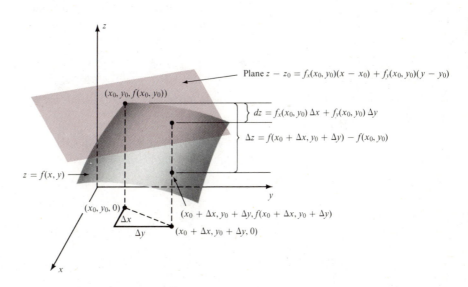

FIGURE 15–20. $\Delta z \approx dz$ for small Δx and Δy.

EXAMPLE 1 Let $z = f(x, y) = 4x^3 + 2x^2y^2 - y^3$. Compute Δz and dz as (x, y) changes from $(2, 1)$ to $(1.98, 1.03)$.

Solution In this example, $(x_0, y_0) = (2, 1)$, $\Delta x = -.02$, and $\Delta y = .03$. Therefore, by (15.16),

$$\Delta z = f(1.98, 1.03) - f(2, 1)$$
$$= [4(1.98)^3 + 2(1.98)^2(1.03)^2 - (1.03)^3] - [4(2)^3 + 2(2)^2(1)^2 - 1^3]$$
$$= -.72 \quad \text{(correct to 2 places)}$$

Now we compute dz. First,

$$f_x(x, y) = 12x^2 + 4xy^2, \text{so} \quad f_x(2, 1) = 56$$
$$f_y(x, y) = 4x^2y - 3y^2, \text{so} \quad f_y(2, 1) = 13$$

and, according to (15.15),

$$dz = f_x(2, 1)(-.02) + f_y(2, 1)(.03)$$
$$= (56)(-.02) + (13)(.03) = -.73$$

As you see, this is a close approximation to Δz. ∎

Now we will state and prove the main result of this section.

(15.17)
> **APPROXIMATION THEOREM**
> Suppose that
>
> (1) $z = f(x, y)$ is defined on a disc D about (x_0, y_0).
> (2) f_x and f_y are defined on D and are continuous at (x_0, y_0).
> (3) Δz is the change in values, $f(x_0 + \Delta x, y_0 + \Delta y) - f(x_0, y_0)$.
>
> Then
>
> $$\Delta z = f_x(x_0, y_0)\Delta x + f_y(x_0, y_0)\Delta y + \epsilon_1\Delta x + \epsilon_2\Delta y$$
>
> where $\epsilon_1 \to 0$ and $\epsilon_2 \to 0$ as Δx and $\Delta y \to 0$.

The key to the proof is two applications of the Mean Value Theorem. First we observe that Δz can be rewritten as

$$(15.18) \quad \Delta z = [f(x_0 + \Delta x, y_0 + \Delta y) - f(x_0, y_0 + \Delta y)]$$
$$+ [f(x_0, y_0 + \Delta y) - f(x_0, y_0)]$$

To use the Mean Value Theorem, we introduce two new functions of *one* variable

$$g(x) = f(x, y_0 + \Delta y) \quad \text{and} \quad h(y) = f(x_0, y)$$

Both g and h are differentiable on their domains because

$$g'(x) = f_x(x, y_0 + \Delta y) \quad \text{and} \quad h'(y) = f_y(x_0, y)$$

exist on D. Applying the Mean Value Theorem to g on the interval $[x_0, x_0 + \Delta x]$ (Figure 15–21), we obtain

$$g(x_0 + \Delta x) - g(x_0) = g'(u)\Delta x$$
$$= f_x(u, y_0 + \Delta y)\Delta x$$

for some u between x_0 and $x_0 + \Delta x$. Using the definition of g, the last equation becomes

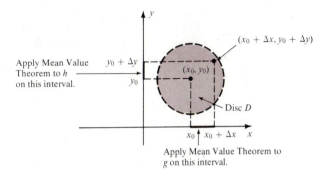

FIGURE 15-21. Diagram for the proof of (15.17).

$$(15.19) \quad f(x_0 + \Delta x, y_0 + \Delta y) - f(x_0, y_0 + \Delta y) = f_x(u, y_0 + \Delta y)\Delta x$$

A similar treatment is valid for h. Here,

$$h(y_0 + \Delta y) - h(y_0) = h'(v)\Delta y$$
$$= f_y(x_0, v)\Delta y$$

for some v between y_0 and $y_0 + \Delta y$. This equation can be rewritten as

$$(15.20) \quad f(x_0, y_0 + \Delta y) - f(x_0, y_0) = f_y(x_0, v)\Delta y$$

Substitution of (15.19) and (15.20) into (15.18) yields

$$(15.21) \quad \Delta z = f_x(u, y_0 + \Delta y)\Delta x + f_y(x_0, v)\Delta y$$

Now we use the hypothesis that f_x and f_y are continuous at (x_0, y_0) to assert that

$$f_x(u, y_0 + \Delta y) \to f_x(x_0, y_0) \quad \text{and} \quad f_y(x_0, v) \to f_y(x_0, y_0)$$

as Δx and Δy approach zero. Therefore, we can set

$$(15.22) \quad f_x(u, y_0 + \Delta y) = f_x(x_0, y_0) + \epsilon_1 \quad \text{and} \quad f_y(x_0, v) = f_y(x_0, y_0) + \epsilon_2$$

and be sure that ϵ_1 and ϵ_2 both approach zero as Δx and Δy approach zero. Substituting (15.22) into (15.21), we have

$$\Delta z = [f_x(x_0, y_0) + \epsilon_1]\Delta x + [f_y(x_0, y_0) + \epsilon_2]\Delta y$$

which is equivalent to the desired result and completes the proof.

Using the definition of dz given in (15.15), we can rewrite the conclusion of (15.17) as

$$\Delta z = dz + \epsilon_1 \Delta x + \epsilon_2 \Delta y$$

where ϵ_1 and $\epsilon_2 \to 0$ as Δx and $\Delta y \to 0$. Therefore, if Δx and Δy are small, then $\Delta z \approx dz$. That is why we call (15.17) the Approximation Theorem.

Recall that in the one variable case, the change Δx in the independent variable is usually written as dx resulting in the equation $dy = f'(x)\, dx$. Now we will do the same for functions of two variables.

(15.23)

Let $z = f(x, y)$. For the independent variables x and y, set $dx = \Delta x$ and $dy = \Delta y$. Then the **differential** dz of the dependent variable z is defined by

$$dz = f_x(x, y)\, dx + f_y(x, y)\, dy$$

It can also be written as

$$dz = \frac{\partial z}{\partial x}\, dx + \frac{\partial z}{\partial y}\, dy$$

EXAMPLE 2   The radius and height of a right circular cone are measured to be 12 and 36 cm, respectively. If the measurement is accurate to within $1/2\%$ ($= .005$), approximate the maximum possible error in calculating the volume.

Solution   The volume of a cone is $V = \frac{1}{3}\pi r^2 h$. The differential of V is

$$dV = \frac{\partial V}{\partial r}\, dr + \frac{\partial V}{\partial h}\, dh = \frac{2}{3}\pi rh\, dr + \frac{1}{3}\pi r^2\, dh$$

The possible error in the radius measurement is $\pm(12)(.005) = \pm.06$; for the height, it is $\pm(36)(.005) = \pm.18$. Therefore, the maximum error in computing the volume is approximately

$$dV = \frac{2}{3}\pi(12)(36)(\pm.06) + \frac{1}{3}\pi(12)^2(\pm.18) = \pm 81.4 \text{ cm}^3$$

Remark: In this example,

$$\frac{\partial V}{\partial r}(r, h) = \frac{2}{3}\pi rh, \text{ so } \frac{\partial V}{\partial r}(12, 36) = 288\pi$$

$$\frac{\partial V}{\partial h}(r, h) = \frac{1}{3}\pi r^2, \text{ so } \frac{\partial V}{\partial r}(12, 36) = 48\pi$$

These are the coefficients of dr and dh in the expression for dV. Since $\partial V/\partial r$ is six times the size of $\partial V/\partial h$, it follows that a small change in the radius produces about six times the effect on the volume as an equally small change in the height. We say that in this case the volume is more *sensitive* to a change in the radius than in the height. If the measurements were reversed, say $r = 36$ and $h = 12$, then, as you can easily check, the volume would be more sensitive to a change in height than a change in radius. These considerations are important in engineering problems (see Exercise 41).   ■

Functions of Three or More Variables

The discussion above can be extended to functions of three or more variables. Let $w = f(x, y, z)$ be defined on a sphere S with center (x_0, y_0, z_0). If the partials of f exist on S and are continuous at (x_0, y_0, z_0), then

$$\Delta w = f_x(x_0, y_0, z_0)\Delta x + f_y(x_0, y_0, z_0)\Delta y + f_z(x_0, y_0, z_0)\Delta z + \epsilon_1\Delta x + \epsilon_2\Delta y + \epsilon_3\Delta z$$

where ϵ_1, ϵ_2, and $\epsilon_3 \to 0$ as Δx, Δy, and $\Delta z \to 0$. The differentials of the independent variables are $dx = \Delta x$, $dy = \Delta y$, and $dz = \Delta z$, and the differential of the dependent variable w is

$$dw = \frac{\partial w}{\partial x} dx + \frac{\partial w}{\partial y} dy + \frac{\partial w}{\partial z} dz$$

The extension to functions of four or more variables is similar. In any case, the differential can be used to approximate the actual change in values. Here is an example with five variables.

EXAMPLE 3   Newton's law of gravitation states that the magnitude of the force exerted by an object of mass M at the origin on an object of mass m located at $(x, y, z) \neq (0, 0, 0)$ is,

$$F = \frac{GmM}{x^2 + y^2 + z^2}$$

where G is the universal gravitational constant, $6.672 \times 10^{-11} N\text{-}m^2/kg^2$. The differential of F is

$$dF = \frac{\partial F}{\partial m} dm + \frac{\partial F}{\partial M} dM + \frac{\partial F}{\partial x} dx + \frac{\partial F}{\partial y} dy + \frac{\partial F}{\partial z} dz$$

and it can be used to approximate the change in F due to a small change in any or all of the five variables.   ■

Differentiability and Tangent Planes

To say that a function of one variable is differentiable at a point x_0 means that $f'(x_0)$ exists. But functions of two or more variables require a stronger definition.

(15.24)

> A function $z = f(x, y)$ is **differentiable** at (x_0, y_0) if it is defined in a disc about (x_0, y_0) and Δz can be expressed in the form
>
> $$\Delta z = f_x(x_0, y_0)\Delta x + f_y(x_0, y_0)\Delta y + \epsilon_1 \Delta x + \epsilon_2 \Delta y$$
>
> where ϵ_1 and $\epsilon_2 \to 0$ as Δx and $\Delta y \to 0$.

The same definition can be extended to include functions of three or more variables. According to the Approximation Theorem, if the partials of f are defined and continuous on a disc D (or a sphere or a similar set of proper dimension), then f is differentiable on D.

Although the existence of partial derivatives does not insure continuity (Example 7 in the last section), differentiability does.

(15.25)

> If a function of several variables is differentiable at a point, then it is continuous at that point.

The proof for a function f of two variables is as follows: By definition (15.24),

$$f(x_0 + \Delta x, y_0 + \Delta y) - f(x_0, y_0) = \Delta z = f_x(x_0, y_0)\Delta x$$
$$+ f_y(x_0, y_0)\Delta y + \epsilon_1\Delta x + \epsilon_2\Delta y$$

The limit of the right side of this equation as $(\Delta x, \Delta y) \to (0, 0)$ is 0 because ϵ_1 and $\epsilon_2 \to 0$ as Δx and $\Delta y \to 0$. Therefore, the limit of the left side is also 0; that is,

$$\lim_{(\Delta x, \Delta y) \to (0, 0)} f(x_0 + \Delta x, y_0 + \Delta y) = f(x_0, y_0)$$

which is equivalent to saying that f is continuous at (x_0, y_0).

We conclude this section with a remark about tangent planes. It was mentioned earlier that for $z = f(x, y)$, the differential dz is the difference between $f(x_0, y_0)$ and the z-component of the corresponding point on the plane

$$z - z_0 = f_x(x_0, y_0)\Delta x + f_y(x_0, y_0)\Delta y \qquad \text{(Figure 15–20)}$$

If C_1 and C_2 are the traces of the surface $z = f(x, y)$ in the planes $x = x_0$ and $y = y_0$, then the plane above contains the lines tangent to C_1 and C_2 at the point (x_0, y_0, z_0), where $z_0 = f(x_0, y_0)$. This is statement (15.12) in the preceding section. It turns out that if f is differentiable at (x_0, y_0), then this plane contains the tangent lines of all smooth curves on the surface $z = f(x, y)$ that pass through (x_0, y_0, z_0). For this reason, we call it the *tangent plane* of $z = f(x, y)$ at (x_0, y_0, z_0). A tangent plane is the analog of a tangent line for functions of one variable. We will discuss tangent planes in more detail in Section 15–7, after introducing the notion of a *gradient*.

EXERCISES

In Exercises 1–20, find the differentials of the given functions.

1. $z = x^2 - 3xy + y^2$
2. $z = 4x^3 - 3x^2y^2 + y^3$
3. $z = x^2 \sin y + y^2$
4. $z = \tan^{-1}x^2y$
5. $z = xe^{xy}$
6. $z = x \ln xy$
7. $w = xy^2 \ln z$
8. $w = x \cos y + y \sin z$
9. $w = \tan^{-1}xy^2z^3$
10. $w = x^2e^{yz}$
11. $A = \dfrac{1}{2}bh$
12. $V = LWH$
13. $S = 2xy + 2yz + 2xz$ (Surface area of a rectangular parallelepiped)
14. $A = \dfrac{1}{2}r^2\theta$ (Area of a circular sector, θ measured in radians)
15. $S = \pi r \sqrt{r^2 + h^2}$ (Surface area of a cone)
16. $A = \dfrac{1}{2}(a + b)h$ (Area of a trapezoid)
17. $A = \dfrac{1}{2}ab \sin C$ (Area of a triangle)

18. $C = \cos^{-1}\left[\dfrac{a^2 + b^2 - c^2}{2ab}\right]$ (Law of Cosines)
19. $R = \left[\dfrac{1}{R_1} + \dfrac{1}{R_2} + \dfrac{1}{R_3}\right]^{-1}$ (Three resistances in parallel)
20. $D = \sqrt{x^2 + y^2 + z^2}$ (Distance to the origin)

Exercises 21–30 refer to the functions defined above. In each case, use (a) a calculator to compute the actual change in values and (b) differentials to compute the approximate change.

21. In Exercises 1, (x, y) changes from $(2, 3)$ to $(1.98, 3.01)$.
22. In Exercise 3, (x, y) changes from $(4, 0)$ to $(4.1, -.2)$.
23. In Exercises 5, (x, y) changes from $(1, 0)$ to $(.98, .03)$.
24. In Exercise 7, (x, y, z) changes from $(1, 1, 1)$ to $(1.1, 0.9, 1.1)$.
25. In Exercise 9, (x, y, z) changes from $(1, 1, 1)$ to $(0.98, 1.01, 1)$.
26. In Exercise 11, (b, h) changes from $(9, 4)$ to $(8.6, 4.2)$.

27. In Exercise 13, (x, y, z) changes from $(1, 2, 3)$ to $(0.98, 2.03, 2.95)$.

28. In Exercise 15, (r, h) changes from $(12, 6)$ to $(11.7, 6.2)$.

29. In Exercise 17, (a, b, C) changes from $(4, 8, \pi/6)$ to $(3.9, 8.2, 29\pi/180)$.

30. In Exercise 19, (R_1, R_2, R_3) changes from $(4, 8, 2)$ to $(4, 7.8, 2.3)$.

In Exercises 31–42, use differentials to work the problems.

31. Estimate the value of $\sin(89\pi/180)\cos^2(29\pi/180)$ (check your answer with a calculator).

32. Estimate the value of $\sqrt{101.2}/\sqrt[3]{26.3}$ (check your answer with a calculator).

33. The dimensions of a rectangular parallelepiped are measured as 6, 2, and 5 inches with a possible error in measurement of 1/2%. Approximate the maximum possible error in computing the surface area.

34. Same as Exercise 33 only now find the possible error in computing the volume.

35. A cylindrical steel rod has length 2 feet and radius 1 foot. About how much nickel (cubic feet) is needed to coat the rod with a thickness of 1/10 inch?

36. A rectangular box with a top has dimensions 12, 14, and 7 cm plus a coating of paint 1/10 cm thick. About how much paint (cubic cm) is on the box?

37. About how much lumber (cu ft) does it take to make a rectangular box (with top) having outside dimensions 5, 3, and 7 ft, if the lumber is 1/2 inch thick?

38. What is the approximate change in area of the right triangle in Figure 15–22A if θ changes by $\Delta\theta$ and

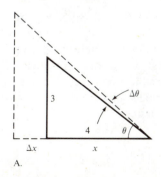

A.

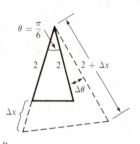

B.

FIGURE 15–22. (A) Exercise 38. (B) Exercise 39.

x changes by Δx? Is the area more sensitive to changes in θ or x?

39. Same as Exercise 38 for the isosceles triangle in Figure 15–22B. (Use the formula in Exercise 17 above.) Is the area more sensitive to changes to θ or x?

40. The position of a particle is
$$\mathbf{r}(t, p, T, g) = \sqrt{3t + p^2}\mathbf{i} + g^{4/5}\mathbf{j} + e^{tT}\mathbf{k}$$
where t is time, p is pressure, T is temperature, and g is the force of gravity. Approximately, where is the particle when $t = 0.01$, $p = 0.02$, $T = -0.1$, and $g = 32.2$? Hint: treat each component separately.

41. (Deflection of a beam). A rectangular beam with length $L = 5$m, height $h = 0.2$m, and width $w = 0.1$m is supported at both ends and subjected to a uniform load of $p = 100$kg/m. The amount of sag S at the middle of the beam, called the deflection, is given by the formula
$$S = C\frac{pL^4}{wh^3} \text{ meters}$$
where C depends on the material of the beam; for this particular beam, $C = 10^{-8}$. Assuming that a small change in any one of the variables is feasible, explain how an engineer can use the differential dS to decide which variable to change in order to reduce the deflection by the greatest amount.

42. Tip the beam in Exercise 41 over on its side so that now $h = 0.1$m and $w = 0.2$m. Will this alter the decision of the engineer?

43. Define f by
$$z = f(x, y) = \begin{cases} \dfrac{xy}{x^2 + y^2} & \text{if } (x, y) \neq (0, 0) \\ 0 & \text{if } (x, y) = (0, 0) \end{cases}$$

In Example 7 of Section 15–3, we showed that f_x and f_y are 0 at $(0, 0)$ and, surely, they exist at all other points. And yet, near the origin, dz cannot be used to approximate the change in f. For instance, $\Delta z = f(1/100, 1/100) - f(0, 0) = \frac{1}{2}$, but $dz = f_x(0, 0)(1/100) + f_y(0, 0)(1/100) = 0$. Explain.

15–5 THE CHAIN RULE

If $y = f(u)$ and $u = g(x)$, then the chain rule, developed in Section 3–4 allows us to write

$$(15.26) \quad \frac{dy}{dx} = \frac{dy}{du}\frac{du}{dx}$$

This chain rule is the most powerful tool we have for computing derivatives of functions of one variable. Our purpose in this section is to develop an equally powerful chain rule for functions of several variables.

Let $z = f(x, y)$ be differentiable on some region R in the xy-plane. Suppose that C is some curve in R, and that we need to investigate the behavior of f on C. For example, how the temperature $f(x, y)$ varies as the point (x, y) moves along a contour curve C on a map. To carry out this investigation, we suppose that C has parametric equations

$$x = x(t) \quad \text{and} \quad y = y(t)$$

where x and y are differentiable functions of t. Then z is a (composite) function of t, and dz/dt is found using the formula

$$(15.27) \quad \frac{dz}{dt} = \frac{\partial z}{\partial x}\frac{dx}{dt} + \frac{\partial z}{\partial y}\frac{dy}{dt}$$

This is one version of the *chain rule* for functions of two variables; it is the counterpart of equation (15.26) for one variable.

The proof of (15.27) is not difficult. Let t_0 and $t_0 + \Delta t$ be distinct points (so $\Delta t \neq 0$) and let Δx, Δy, and Δz be the corresponding changes in x, y, and z. Then, according to the Approximation Theorem (15.17), we can write

$$(15.28) \quad \frac{\Delta z}{\Delta t} = \frac{\partial z}{\partial x}(x(t_0), y(t_0))\frac{\Delta x}{\Delta t} + \frac{\partial z}{\partial y}(x(t_0), y(t_0))\frac{\Delta y}{\Delta t} + \epsilon_1\frac{\Delta x}{\Delta t} + \epsilon_2\frac{\Delta y}{\Delta t}$$

where ϵ_1 and $\epsilon_2 \to 0$ as Δx and $\Delta y \to 0$. Now we let $\Delta t \to 0$. Since x and y are differentiable functions of t, it follows that Δx and $\Delta y \to 0$, so the terms involving the ϵ's will approach zero. What is left is equation (15.27) where the derivatives are evaluated at t_0 and the partials are evaluated at $(x(t_0), y(t_0))$. A similar treatment is valid for functions of three or more variables. It is assumed here and throughout this section that the domains and ranges of the various functions are arranged so that the composite functions are defined.

CHAIN RULE FOR SEVERAL VARIABLES
If $z = f(x, y)$, $x = x(t)$, and $y = y(t)$ are all differentiable functions, then

$$\frac{dz}{dt} = \frac{\partial z}{\partial x}\frac{dx}{dt} + \frac{\partial z}{\partial y}\frac{dy}{dt}$$

(15.29)

If $w = f(x, y, z)$, $x = x(t)$, $y = y(t)$, and $z = z(t)$ are all differentiable functions, then

$$\frac{dw}{dt} = \frac{\partial w}{\partial x}\frac{dx}{dt} + \frac{\partial w}{\partial y}\frac{dy}{dt} + \frac{\partial w}{\partial z}\frac{dz}{dt}$$

Similar statements hold for functions of more than three variables.

EXAMPLE 1 Suppose $z = x^2y + y^3$, $x = 2t^2$, and $y = 4t^3$. Find dz/dt.

Solution We could simply substitute for x and y, expressing z as a function of t, and differentiate. But, for practice, let us use (15.29) instead.

$$\frac{dz}{dt} = \frac{\partial z}{\partial x}\frac{dx}{dt} + \frac{\partial z}{\partial y}\frac{dy}{dt}$$
$$= (2xy)(4t) + (x^2 + 3y^2)(12t^2)$$
$$= (16t^5)(4t) + (4t^4 + 48t^6)(12t^2)$$
$$= 112t^6 + 576\,t^8 \quad \blacksquare$$

Now we extend the chain rule even further. Again, let $z = f(x, y)$, but this time let x and y be functions of two variables, say $x = x(u, v)$ and $y = y(u, v)$. Thus, z can be considered as a (composite) function of u and v, and we can inquire about the partials $\partial z/\partial u$ and $\partial z/\partial v$. For $\partial z/\partial u$, we consider v to be a constant v_0 and write an equation similar to (15.28). There are only minor changes; Δt is replaced by Δu, $x(t_0)$ is replaced by $x(u_0, v_0)$, and so on. If the limit is taken as $\Delta u \to 0$, the result can be written as follows:

If $z = f(x, y)$, $x = x(u, v)$, and $y = y(u, v)$ are all differentiable functions, then

$$\frac{\partial z}{\partial u} = \frac{\partial z}{\partial x}\frac{\partial x}{\partial u} + \frac{\partial z}{\partial y}\frac{\partial y}{\partial u}$$

(15.30)

and

$$\frac{\partial z}{\partial v} = \frac{\partial z}{\partial x}\frac{\partial x}{\partial v} + \frac{\partial z}{\partial y}\frac{\partial y}{\partial v}$$

Once again, similar statements hold for functions of three or more variables. In fact, if $w = f(x, y, z, \ldots, t)$ is a differentiable function of any number of variables, and each variable, in turn, is a differentiable function of any number of variables $x = x(u, v, \ldots, s), y = y(u, v, \ldots, s)$ and so on, then statements similar to (15.30) hold. For example,

(15.31)
$$\frac{\partial w}{\partial u} = \frac{\partial w}{\partial x}\frac{\partial x}{\partial u} + \frac{\partial w}{\partial y}\frac{\partial y}{\partial u} + \frac{\partial w}{\partial z}\frac{\partial z}{\partial u} + \cdots + \frac{\partial w}{\partial t}\frac{\partial t}{\partial u}$$

This is the most general statement of the chain rule.

EXAMPLE 2 Suppose $w = x + y^2 + z^3, x = u \cos v, y = u \sin v$, and $z = uv$. Find $\partial w/\partial u$ and $\partial w/\partial v$.

Solution By (15.31), we have

$$\frac{\partial w}{\partial u} = \frac{\partial w}{\partial x}\frac{\partial x}{\partial u} + \frac{\partial w}{\partial y}\frac{\partial y}{\partial u} + \frac{\partial w}{\partial z}\frac{\partial z}{\partial u}$$
$$= (1)\cos v + 2y \sin v + 3z^2 v$$
$$= \cos v + 2u \sin^2 v + 3u^2 v^3$$

Similarly,

$$\frac{\partial w}{\partial v} = \frac{\partial w}{\partial x}\frac{\partial x}{\partial v} + \frac{\partial w}{\partial y}\frac{\partial y}{\partial v} + \frac{\partial w}{\partial z}\frac{\partial z}{\partial v}$$
$$= (1)(-u \sin v) + 2y(u \cos v) + 3z^2 u$$
$$= -u \sin v + 2u^2 \cos v \sin v + 3u^3 v^2$$

Remark: Since w is being considered as a function of u and v, its partials should be in terms of u and v. ■

Implicit Differentiation

In Section 4–4, we developed a method for finding the derivative of a function defined implicitly by an equation. That method often involved cumbersome calculations. Now the chain rule (15.29) can be used to derive a simpler method of implicit differentiation.

Suppose that an equation in x and y implicitly defines y as a differentiable function of x. Rewriting the equation with only 0 on the right, the left side can be thought of as a function $F(x, y)$ and we have

(15.32) $F(x, y) = 0$

To use the chain rule, we introduce a "dummy" variable t by setting $x = t$. Differentiating both sides of (15.32) with respect to t yields,

$$0 = \frac{dF}{dt} = \frac{\partial F}{\partial x}\frac{dx}{dt} + \frac{\partial F}{\partial y}\frac{dy}{dt}$$

Since $x = t$, it follows that $dx/dt = 1$ and $dy/dt = dy/dx$; solving the equation above yields,

$$\frac{dy}{dx} = -\frac{\partial F/\partial x}{\partial F/\partial y}$$

Here is a concise statement:

(15.33)

> If $F(x, y) = 0$ defines y as a differentiable function of x, then
>
> $$\frac{dy}{dx} = -\frac{F_x(x, y)}{F_y(x, y)}$$

EXAMPLE 3   If $y^3 - 3xy = 5x^2y^2$ defines y as a differentiable function of x, find dy/dx.

Solution   First, we rewrite the equation with only 0 on the right

$$y^3 - 3xy - 5x^2y^2 = 0$$

and then let $F(x, y)$ be the expression on the left. By (15.33), we know that

$$\frac{dy}{dx} = -\frac{F_x(x, y)}{F_y(x, y)} = -\frac{-3y - 10xy^2}{3y^2 - 3x - 10x^2y}$$

You can see that this is much simpler than our earlier method of implicit differentiation.   ■

If an equation in three variables x, y, and z defines z as a differentiable function of x and y, a similar procedure provides the partials of z. The only difference is that we must now introduce two "dummy" variables by setting $x = u$ and $y = v$. The net result is as follows:

(15.34)

> If $F(x, y, z) = 0$ defines z as a differentiable function of x and y, then
>
> $$\frac{\partial z}{\partial x} = -\frac{F_x(x, y, z)}{F_z(x, y, z)} \quad \text{and} \quad \frac{\partial z}{\partial y} = -\frac{F_y(x, y, z)}{F_z(x, y, z)}$$

EXAMPLE 4   If $4xy - x^2z^2 + yz^3 = 8 - z^4$ defines z as a differentiable function of x and y, find $\partial z/\partial x$ and $\partial z/\partial y$.

Solution   Rewrite the equation as

$$4xy - x^2z^2 + yz^3 - 8 + z^4 = 0$$

and let $F(x, y, z)$ be the expression on the left. Then, by (15.34),

$$\frac{\partial z}{\partial x} = -\frac{F_x(x, y, z)}{F_z(x, y, z)} = -\frac{4y - 2xz^2}{-2x^2z + 3yz^2 + 4z^3}$$

$$\frac{\partial z}{\partial y} = -\frac{F_y(x, y, z)}{F_z(x, y, z)} = -\frac{4x + z^3}{-2x^2z + 3yz^2 + 4z^3} \qquad \blacksquare$$

Related Rates

The chain rule can also be used to solve related rate problems.

EXAMPLE 5   The area of a triangle with sides a and b and the included angle C is given by the formula

$$A = \frac{1}{2}ab \sin C$$

If A, a, and b are changing at given rates, find the rate of change of C in terms of dA/dt, da/dt, and db/dt.

Solution   According to the chain rule (15.29), we have

$$\frac{dC}{dt} = \frac{\partial C}{\partial A}\frac{dA}{dt} + \frac{\partial C}{\partial a}\frac{da}{dt} + \frac{\partial C}{\partial b}\frac{db}{dt}$$

There are two ways to proceed from here. One is to solve for C

$$C = \sin^{-1}\left[\frac{2A}{ab}\right]$$

and compute the partials of C; in this example, that would involve a great deal of computation. It is much easier to set $F(A, a, b, C) = A - (1/2)ab \sin C = 0$ and use the implicit differentiation formulas (15.34) to compute the partials. Thus,

$$\frac{dC}{dt} = -\frac{F_A}{F_C}\frac{dA}{dt} - \frac{F_a}{F_C}\frac{da}{dt} - \frac{F_b}{F_C}\frac{db}{dt}$$

$$= -\frac{1}{-\frac{1}{2}ab \cos C}\frac{dA}{dt} - \frac{-\frac{1}{2}b \sin C}{-\frac{1}{2}ab \cos C}\frac{da}{dt} - \frac{-\frac{1}{2}a \sin C}{-\frac{1}{2}ab \cos C}\frac{db}{dt}$$

$$= \frac{2}{ab \cos C}\frac{dA}{dt} - \frac{\tan C}{a}\frac{da}{dt} - \frac{\tan C}{b}\frac{db}{dt} \qquad \blacksquare$$

$sngs$ 1, 3, 5, 7, 11, 13, 17, 19, 23, 31, 33, 34

EXERCISES

In Exercises 1–6, find dz/dt.

1. $z = x^2y + 3y^2$, $x = 3t + 1$, $y = t^2$

2. $z = 3xy^3 - 4x^2$, $x = t^2 + t$, $y = 6t$

3. $z = e^{xy}$, $x = \ln t$, $y = 3t^2$

4. $z = x \sin y$, $x = e^t$, $y = 1/t$

5. $z = \tan^{-1}xy$, $x = \tan t$, $y = e^t$

6. $z = \sqrt{x^2 + y^2}$, $x = \cos t$, $y = \sin t$

In Exercises 7–10, find dw/dt.

7. $w = x^2y + z$, $x = t^2$, $y = t^3 + 1$, $z = 2t - 5$

8. $w = \sin xyz$, $x = 1/t$, $y = \ln t$, $z = t$

9. $w = \sqrt{x^2 + y^2 + z^2}$, $x = \cos t$, $y = \sin t$, $z = t$

10. $w = \ln xyz$, $x = e^t$, $y = e^{2t}$, $z = t$

In Exercises 11–14, find $\partial z/\partial u$ and $\partial z/\partial v$.

11. $z = x^2y^3 + x \sin y$, $x = u^2$, $y = uv$

12. $z = 3x^2 + 2xy - 4y^2$, $x = 2u + 3v$, $y = u^2 - v^2$

13. $z = x^2 \ln y$, $x = uv$, $y = 3u + v$

14. $z = e^x \ln y$, $x = u^2 - 2v$, $y = v^2 - 2u$

In Exercises 15–16, find $\partial w/\partial r$, $\partial w/\partial s$, and $\partial w/\partial t$.

15. $w = x^2y^3z$, $x = 3 rst$, $y = r^2t$, $z = s^3$

16. $w = e^{xyz}$, $x = r + s + t$, $y = rst$, $z = r^2 + s^2 + t^2$

In Exercises 17–22, assume that the equation defines y as a differentiable function of x and find dy/dx.

17. $x^2 + 6xy = 5y^2 - 3$

18. $x \sin y + y \sin x = 0$

19. $e^{xy} = \tan y$

20. $3x^2y^2 - 4xy^3 = 5x^4y^4$

21. $x \sin y - y \cos x = 0$

22. $x + y = \tan y$

In Exercises 23–26, assume that the equation defines z as a differentiable function of x and y. Find $\partial z/\partial x$ and $\partial z/\partial y$.

23. $2xy^2 - 3z^2y = 4z^4$ **24.** $xyz - xz^2 + yz^3 = 0$

25. $xe^{yz} + ye^{xz} = z$ **26.** $x^2y + \sin xyz = 6$

27. If $z = f(x, y)$, $x = r \cos \theta$, and $y = r \sin \theta$, show that

$$\left[\frac{\partial z}{\partial r}\right]^2 + \frac{1}{r^2}\left[\frac{\partial z}{\partial \theta}\right]^2 = \left[\frac{\partial z}{\partial x}\right]^2 + \left[\frac{\partial z}{\partial y}\right]^2$$

28. For the same functions in Exercise 27, show that

$$\frac{\partial^2 z}{\partial x^2} + \frac{\partial^2 z}{\partial y^2} = \frac{\partial^2 z}{\partial r^2} + \frac{1}{r^2}\frac{\partial^2 z}{\partial \theta^2} + \frac{1}{r}\frac{\partial z}{\partial r}$$

29. (a) Show that $z = e^{(x + 2y)} + \ln(x^2 + 4xy + 4y^2)$ satisfies the equation

$$\frac{\partial z}{\partial y} = 2\frac{\partial z}{\partial x}$$

(b) Show that *any* differentiable function $z = f(x + 2y)$ will satisfy the equation!

30. *(Wave equation).* A function $z = f(x,t)$ where x is position and t is time, is said to satisfy a *wave equation* if

$$a^2\frac{\partial^2 z}{\partial x^2} = \frac{\partial^2 z}{\partial t^2} \quad a \text{ constant}$$

(a) Show that $z = \sin(x - t) + \cos(x + t)$ satisfies a wave equation.

(b) Show that $z = (x - 2t)^2 + (x + 2t)^3$ satisfies a wave equation.

(c) Show that $z = f(x - at) + g(x + at)$ where f and g are any functions having second partials satisfies a wave equation.

---

31. A triangle has sides a and b with the included angle C. If the area A is held constant and a and b are each increasing at the rate of 2 cm/sec, how fast is the angle C changing when $a = 3$ cm, $b = 4$ cm, and $C = \pi/4$?

32. For the same triangle in Exercise 31, suppose that C is held constant at $\pi/4$ and the area varies. Under the same conditions described above, how fast is the area changing?

33. Suppose that a particle moves along the circular helix $x = \cos t$, $y = \sin t$, $z = t$, where t is time in seconds. Suppose that its temperature T in degrees Celsius at position (x, y, z) is $T(x, y, z) = 90 \sin xyz$. When $t = \pi/4$ seconds, is it cooling off or heating up? At what rate?

34. For the particle in Exercise 33, how fast is it moving away from the origin when $t = \pi$?

35. The trunk of a tree is in the shape of a circular cylinder. If the radius and height increase (on average) about 2 cm and 3 m, respectively, per year, how fast is the amount of lumber (cubic meters) increasing when the radius is 15 cm and the height is 10 m?

36. The pressure p, volume V, and temperature T of a confined gas are related by the equation $pV = cT$, where c is a constant depending on the units of measurement. Find the time rate of change of p in terms of dV/dt and dT/dt.

37. *(Poiseuille's formula).* The rate at which the volume V of fluid flows past a cross-section of a circular tube of radius r and length L (in centimeters) is given by

$$V = \frac{\pi r^4 p}{8 \eta L}$$

where p is the difference in pressure at the two ends of the tube in dynes/cm², and η (the Greek letter *eta*) is the viscosity coefficient of the fluid. This is called *Poiseuille's formula,* and it has many applications. If the radius, pressure, and length are allowed to vary with time, how fast is V changing in terms of dr/dt, dp/dt, and dL/dt when $r = 1$ mm (1 millimeter $= 1/10$ cm), $L = 10$ cm, and $p = 2 \times 10^6$ dynes/cm²? Is V more sensitive to small changes in r or L or p?

38. The magnitude F of the gravitational force exerted on a rocket by the earth is

$$F = \frac{GmM}{r^2}$$

where G is the universal gravitational constant, m the mass of the rocket, M the mass of the earth, and r the distance from the center of the earth to the rocket. When the rocket is 4060 miles from the center of the earth (about 60 miles above the surface), it weighs 8,000 lbs, it is rising at the rate of 60 miles/sec, and it is using fuel at the rate of 100 lbs/sec. How fast is the force F decreasing at that moment? (G and M are constant.)

15–6 DIRECTIONAL DERIVATIVES AND THE GRADIENT

Example 4 in Section 15–3 is about an ant on a hot plate; we found that if the ant can move only parallel to the axes, then it should move in the positive x direction to cool off the fastest. But is this the best direction for the ant to move? Is there, perhaps, a direction in which the temperature decreases most rapidly? To answer this question, you have to know about *directional derivatives* and *gradients*. First we treat functions of two variables and afterwards functions of more variables.

Directional Derivatives

We begin with a function $z = f(x, y)$ and a point (x_0, y_0) in its domain. We already know that $f_x(x_0, y_0)$ and $f_y(x_0, y_0)$ are the rates of change of f in directions parallel to the axes. Now suppose we want to know the rates of change along lines through (x_0, y_0) that are in the xy-plane but not parallel to the axes. The first step is to pick any line L through (x_0, y_0) and indicate its direction by a direction vector parallel to it; specifically a *unit* vector $\mathbf{u}$ ($|\mathbf{u}| = 1$) as illustrated in Figure 15–23. If $\mathbf{u} = a\mathbf{i} + b\mathbf{j}$ (a vector in the xy-plane), it follows from (14.22) in Section 14–4 that

$$x = x_0 + at \quad y = y_0 + bt \quad z = 0$$

are parametric equations of L. As t ranges over its domain, the point $(x_0 + at, y_0 + bt)$ moves along L. The quotient

$$\frac{f(x_0 + at, y_0 + bt) - f(x_0, y_0)}{t}$$

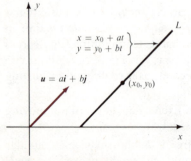

FIGURE 15–23. The unit vector $\mathbf{u}$ ($|\mathbf{u}| = 1$) is a direction vector of L.

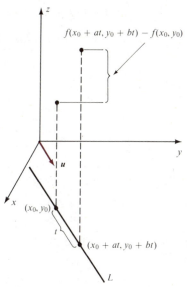

$f(x_0 + at, y_0 + bt) - f(x_0, y_0)$

(x_0, y_0)

u

t

$(x_0 + at, y_0 + bt)$

L

FIGURE 15–24. The (signed) distance between (x_0, y_0) and $(x_0 + at, y_0 + bt)$ is t.

represents the average change in f with respect to a change in t. But, since $\mathbf{u}$ is a unit vector, t is also the (signed) distance along L (Figure 15–24). Therefore, this quotient also represents the average change in f with respect to a change along L. Taking the limit as $t \to 0$ will give the (instantaneous) rate of change of f at (x_0, y_0) along L; that is, the rate of change in the direction indicated by $\mathbf{u}$. This motivates the following definition:

(15.35)

> Let f be defined on a disc about (x_0, y_0), and let $\mathbf{u} = a\mathbf{i} + b\mathbf{j}$ be a unit vector. Then the **directional derivative of f in the direction $\mathbf{u}$,** denoted by $D_{\mathbf{u}}f$, is defined as
>
> $$D_{\mathbf{u}}f(x_0, y_0) = \lim_{t \to 0} \frac{f(x_0 + at, y_0 + bt) - f(x_0, y_0)}{t}$$
>
> provided this limit exists.

Remarks:

(1) If you start at (x_0, y_0) and move in the direction indicated by $\mathbf{u}$, then the rate of change of f is given by $D_{\mathbf{u}}f(x_0, y_0)$.

(2) If $\mathbf{u} = \mathbf{i}$ then $a = 1$ and $b = 0$, so (15.35) yields

$$D_{\mathbf{i}}f(x_0, y_0) = \lim_{t \to 0} \frac{f(x_0 + t, y_0) - f(x_0, y_0)}{t} = f_x(x_0, y_0)$$

Similarly, $D_{\mathbf{j}}f(x_0, y_0) = f_y(x_0, y_0)$. In other words, the partials are directional derivatives in the positive x and y directions.

Rather than present an example computing a directional derivative from the definition, let us prove a theorem that makes it easy to find such derivatives.

(15.36)

> **DIRECTIONAL DERIVATIVES THEOREM**
> If f is differentiable on a disc about (x_0, y_0), and $\mathbf{u} = a\mathbf{i} + b\mathbf{j}$ is any unit vector, then $D_{\mathbf{u}}f(x_0, y_0)$ exists and
>
> $$D_{\mathbf{u}}f(x_0, y_0) = af_x(x_0, y_0) + bf_y(x_0, y_0)$$

To prove this result, we write

$$f(x_0 + at, y_0 + bt) - f(x_0, y_0) = f_x(x_0, y_0)at + f_y(x_0, y_0)bt + \epsilon_1 at + \epsilon_2 bt$$

where ϵ_1 and $\epsilon_2 \to 0$ as $t \to 0$. This is simply the definition (15.24) of a differentiable function with Δx and Δy replaced by at and bt. Dividing both sides of this equation by t and taking the limit as $t \to 0$ yields the desired result.

Now we present an example.

EXAMPLE 1 If $f(x, y) = 3x^2y - 4xy^3$ and $\mathbf{u} = \frac{1}{2}\mathbf{i} + \sqrt{3}/2\mathbf{j}$, find $D_\mathbf{u}f(1, -2)$.

Solution After making sure that $\mathbf{u}$ is a unit vector, we find the partials of f,

$$f_x(x, y) = 6xy - 4y^3, \text{ so } \quad f_x(1, -2) = 20$$
$$f_y(x, y) = 3x^2 - 12xy^2, \text{ so } \quad f_y(1, -2) = -45$$

Now, according to (15.36),

$$D_\mathbf{u}f(1, -2) = 20\left(\frac{1}{2}\right) + (-45)\left(\frac{\sqrt{3}}{2}\right) = \frac{20 - 45\sqrt{3}}{2} \qquad \blacksquare$$

Suppose that a nonzero direction vector $\mathbf{a}$ with $|\mathbf{a}| \neq 1$ is given. In that case, we define the rate of change of f in the direction indicated by $\mathbf{a}$ as $D_\mathbf{u}f$ where $\mathbf{u} = \mathbf{a}/|\mathbf{a}|$.

EXAMPLE 2 Let $f(x, y) = 3x^2y - 4xy^3$ as in Example 1, and find the rate of change of f at $(1, -2)$ in the direction $\mathbf{a} = 3\mathbf{i} + 4\mathbf{j}$.

Solution The unit vector having the same direction as $\mathbf{a}$ is

$$\mathbf{u} = \frac{\mathbf{a}}{|\mathbf{a}|} = \frac{3}{5}\mathbf{i} + \frac{4}{5}\mathbf{j}$$

Using the calculations derived in Example 1, we obtain

$$D_\mathbf{u}f(1, -2) = 20\left(\frac{3}{5}\right) + (-45)\left(\frac{4}{5}\right) = -24$$

as the rate of change of f in the direction indicated by $\mathbf{a}$. $\qquad \blacksquare$

The Gradient

Returning to the Directional Derivatives Theorem (15.36), we observe that

(15.37) $\quad D_\mathbf{u}f(x_0, y_0) = af_x(x_0, y_0) + bf_y(x_0, y_0)$

can be viewed as the *dot product* of the vector $\mathbf{u} = a\mathbf{i} + b\mathbf{j}$ and the vector $f_x(x_0, y_0)\mathbf{i} + f_y(x_0, y_0)\mathbf{j}$. This second vector is called the *gradient* of f.

(15.38)

> **DEFINITION**
> If the partials of $z = f(x, y)$ exist, then the **gradient of** f, denoted by ∇f, is defined as
> $$\nabla f(x, y) = f_x(x, y)\mathbf{i} + f_y(x, y)\mathbf{j}$$
> The symbol ∇f is read "del f".

The gradient of f is sometimes written *grad f*. Gradients are central to the calculus of functions of several variables and are also extremely useful in physical applications.

It follows from (15.37) and (15.38) that

(15.39) $\boxed{D_{\mathbf{u}}f(x, y) = \nabla f(x, y) \cdot \mathbf{u}}$

at *any* point (x, y) where f is differentiable and in *any* direction $\mathbf{u}$. This tells us a great deal.

GRADIENT THEOREM

If f is differentiable at (x_0, y_0) and $\nabla f(x_0, y_0) \neq \mathbf{0}$, then

(15.40)

(1) the greatest rate of increase of f at (x_0, y_0) is $|\nabla f(x_0, y_0)|$ and it occurs in the direction of $\nabla f(x_0, y_0)$,

(2) the greatest rate of decrease of f at (x_0, y_0) is $-|\nabla f(x_0, y_0)|$ and it occurs in the direction of $-\nabla f(x_0, y_0)$,

(3) the rate of change of f at (x_0, y_0) is 0 in any direction orthogonal to $\nabla f(x_0, y_0)$.

All three parts of the theorem follow from the equation

$$D_{\mathbf{u}}f(x_0, y_0) = \nabla f(x_0, y_0) \cdot \mathbf{u}$$
$$= |\nabla f(x_0, y_0)| \cos \theta \qquad (|\mathbf{u}| = 1)$$

where θ is the angle between $\nabla f(x_0, y_0)$ and $\mathbf{u}$. The greatest positive value of $D_{\mathbf{u}}f(x_0, y_0)$ is $|\nabla f(x_0, y_0)|$, and it occurs when $\theta = 0$; that is, when $\mathbf{u}$ points in the direction of $\nabla f(x_0, y_0)$. This proves part (1). The greatest negative value of $D_{\mathbf{u}}f(x_0, y_0)$ is $-|\nabla f(x_0, y_0)|$ and it occurs when $\theta = \pi$; that is, when $\mathbf{u}$ points in the direction $-\nabla f(x_0, y_0)$. This proves part (2). Part (3) is left for you to prove.

Remember the ant on a hot plate? Now, with the Gradient Theorem, we can easily help the old boy out.

EXAMPLE 3   The temperature T in degrees Celsius at any point (x, y) on a circular hot plate is $T = 50 \sin xy$, and an ant is at the point $(1, 2)$. In which direction should it move to cool off the fastest? How fast is it cooling off? In which directions would it experience no change in temperature?

Solution   The partials are

$$T_x(x, y) = 50y \cos xy, \text{ so }\quad T_x(1, 2) = 100 \cos 2 \approx -42$$
$$T_y(x, y) = 50x \cos xy, \text{ so }\quad T_y(1, 2) = 50 \cos 2 \approx -21$$

and $\nabla T(1, 2) = 100 \cos 2\mathbf{i} + 50 \cos 2\mathbf{j}$. By (15.40.2), the ant should move in the direction of the vector

$$-\nabla T(1, 2) = -100 \cos 2\mathbf{i} - 50 \cos 2\mathbf{j}$$
$$\approx 42\mathbf{i} + 21\mathbf{j}$$

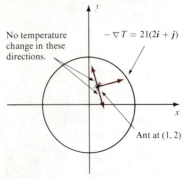

No temperature change in these directions.

$-\nabla T = 21(2i + j)$

Ant at (1, 2)

FIGURE 15–25. Example 3. Ant on a hot plate.

In this direction, it is cooling off at the rate of

$$|\nabla T| \approx \sqrt{(42)^2 + (21)^2} \approx 47° \text{ per unit distance}$$

Since $\pm (i - 2j)$ are both orthogonal to ∇T, the rate of change of T in these directions is zero (Figure 15–25). ■

Functions of More Variables

Virtually everything that was said above carries over to functions of any number of variables. For example, if $w = f(x, y, z)$ is differentiable at (x_0, y_0, z_0) and $\mathbf{u} = a\mathbf{i} + b\mathbf{j} + c\mathbf{k}$ is a unit vector, then

$$D_{\mathbf{u}}f(x_0, y_0, z_0) = \lim_{t \to 0} \frac{f(x_0 + at, y_0 + bt, z_0 + ct) - f(x_0, y_0, z_0)}{t}$$

is called the *directional derivative of f in the direction* $\mathbf{u}$. The *gradient of f* is

$$\nabla f(x, y, z) = f_x(x, y, z)\mathbf{i} + f_y(x, y, z)\mathbf{j} + f_z(x, y, z)\mathbf{k}$$

and

$$D_{\mathbf{u}} f(x, y, z) = \nabla f(x, y, z) \cdot \mathbf{u}$$

Moreover, the Gradient Theorem is valid.

EXAMPLE 4 Let $f(x, y, z) = x + y^2z$, $\mathbf{u} = 1/\sqrt{3}\mathbf{i} + 1/\sqrt{3}\mathbf{j} + 1/\sqrt{3}\mathbf{k}$, and find $D_{\mathbf{u}}f(0, 1, 3)$. In what directions will the rate of change of f be zero?

Solution We first find the partials

$$\frac{\partial f}{\partial x} = 1 \quad \frac{\partial f}{\partial y} = 2yz \quad \frac{\partial f}{\partial z} = y^2$$

and evaluate them at (0, 1, 3) to obtain

$$\nabla f(0, 1, 3) = \mathbf{i} + 6\mathbf{j} + \mathbf{k}$$

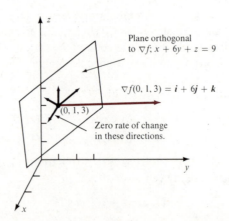

Plane orthogonal to ∇f: $x + 6y + z = 9$

$\nabla f(0, 1, 3) = i + 6j + k$

(0, 1, 3)

Zero rate of change in these directions.

FIGURE 15–26. Example 4.

Since the given **u** is a unit vector, we have

$$D_u f(0, 1, 3) = \nabla f(0, 1, 3) \cdot \mathbf{u} = \frac{1}{\sqrt{3}} + \frac{6}{\sqrt{3}} + \frac{1}{\sqrt{3}} = \frac{8}{\sqrt{3}}$$

According to part (3) of the Gradient Theorem, the rate of change of f is 0 in any direction orthogonal to $\nabla f(0, 1, 3) = \mathbf{i} + 6\mathbf{j} + \mathbf{k}$. This includes all vectors lying in the plane passing through $(0, 1, 3)$ and orthogonal to ∇f; an equation of the plane is $x + 6y + z = 9$ (Figure 15–26). ■

EXAMPLE 5   Suppose that the voltage E at each point in space is given by $E(x, y, z) = x^2 - xyz$. A particle with a positive charge tends to move in the direction of greatest voltage drop. In which direction will such a particle at $(1, 1, 1)$ move? What if it is at $(-1, -1, -1)$?

Solution   Since

$$\frac{\partial E}{\partial x} = 2x - yz \quad \frac{\partial E}{\partial y} = -xz \quad \frac{\partial E}{\partial z} = -xy$$

we have

$$\nabla E(1, 1, 1) = \mathbf{i} - \mathbf{j} - \mathbf{k}$$

The greatest voltage drop will occur in the direction of $-\nabla E$. Thus, the particle will move from $(1, 1, 1)$ in the direction of $-\mathbf{i} + \mathbf{j} + \mathbf{k}$. If it starts from $(-1, -1, -1)$, then it will move in the direction of

$$-\nabla E(-1, -1, -1) = 3\mathbf{i} + \mathbf{j} + \mathbf{k} \quad ■$$

sugg 1, 3, 5, 43, 44, 45, 46

EXERCISES

In Exercises 1–20, find the directional derivatives at point P in the indicated directions. (Save your calculations of the gradients for use in Exercises 21–40 below.)

1. $f(x, y) = x^2 + y^2$; $P(1, 1)$; $\mathbf{a} = \dfrac{1}{\sqrt{2}}\mathbf{i} + \dfrac{1}{\sqrt{2}}\mathbf{j}$

2. $f(x, y) = 2xy^2$; $P(-1, 3)$; $\mathbf{a} = \dfrac{\sqrt{3}}{2}\mathbf{i} + \dfrac{1}{2}\mathbf{j}$

3. $f(x, y) = e^{xy}$; $P(-2, 1)$; from P to $Q(0, 2)$

4. $f(x, y) = x \ln y$; $P(4, 5)$; from P to $Q(3, 6)$

5. $f(x, y) = x^2 \sin y$; $P(3, \pi/3)$; $\mathbf{a} = -3\mathbf{i} + 4\mathbf{j}$

6. $f(x, y) = xy + \tan y$; $P(-1, \pi/4)$; $\mathbf{a} = \mathbf{i} - 3\mathbf{j}$

7. $f(x, y, z) = xy^2 z^2$; $P(1, -1, 2)$; $\mathbf{a} = \dfrac{1}{2}\mathbf{i} + \dfrac{1}{2}\mathbf{j} - \dfrac{1}{\sqrt{2}}\mathbf{k}$

8. $f(x, y, z) = xy + z^2$; $P(-2, 1, 3)$;

$$\mathbf{a} = \frac{1}{3}\mathbf{i} - \frac{2}{3}\mathbf{j} + \frac{\sqrt{5}}{3}\mathbf{k}$$

9. $f(x, y, z) = 2xy^2 - 3z^4$; $P(0, 2, 1)$; from P to $Q(-3, 1, 4)$

10. $f(x, y, z) = 40 - xyz$; $P(3, 0, 1)$; from P to $Q(1, 1, 1)$

11. $f(x, y, z) = x \tan^{-1}(y + z)$; $P(1, 0, 1)$; $\mathbf{a} = \mathbf{i} - \mathbf{j} + \mathbf{k}$

12. $f(x, y, z) = x \cos yz$; $P(1, 0, 1)$; $\mathbf{a} = -\mathbf{i} + 2\mathbf{j} - 3\mathbf{k}$

13. $g(r, s) = re^{(r+s)}$; $P(0, 0)$; $\mathbf{a} = \mathbf{i}$

14. $g(r, s, t) = t \ln(r + s)$; $P(1, 0, 1)$; $\mathbf{a} = \mathbf{k}$

15. $T(x, y, z) = \dfrac{9}{x^2 + y^2 + z^2}$; $P(1, 1, 1)$; from P to $Q(0, 2, 0)$

16. $T(x, y) = \ln(x^2 + y^2)$; $P(0, 1)$; from P to $Q(8, 2)$

17. $A(a, b, C) = \dfrac{1}{2}ab \sin C$; $P(1, 1, \pi/6)$; $\mathbf{a} = \mathbf{i} + \mathbf{j} + \mathbf{k}$

18. $S(x, y, z) = 2xy + 2yz + 2xz$; $P(1, 1, 1)$; $\mathbf{a} = \mathbf{j} + \mathbf{k}$

19. $S(r, h) = \pi r \sqrt{r^2 + h^2}$; $P(3, 4)$; $\mathbf{a} = 2\mathbf{i} - \mathbf{j}$

20. $V(r, h) = \pi r^2 h$; $P(4, 1)$; $\mathbf{a} = -\mathbf{i} + 3\mathbf{j}$

21–40. For each of the functions and points P in Exercises 1–20 above, find the direction of maximum rate of increase and the directions in which the rate of change is zero.

---

41. An object traveling through space is observed to follow the path of least resistance. If the magnitude of the resistance to its motion is $R(x, y, z) = x^2 + y^2 + 4z$, in which direction will it move from the point $(3, 1, 2)$?

42. If the voltage at each point in the solid sphere $x^2 + y^2 + z^2 \le 1$ is $E(x, y, z) = 2 - x^2 - y^2 - z^2$, in which direction will a positively charged particle move from the point $(1/2, 1/3, 1/4)$?

43. Let $T(x, y) = x^2 - 2y^2$ be the temperature at (x, y). From the point $(8, 1)$, in which direction would you move to (a) heat up the fastest, (b) cool off the fastest, and (c) experience the least temperature change?

44. A circular metal plate of radius 1 sits on the xy-plane with its center at $(3, 4)$. It is heated by a heat source at the origin so that the temperature at P is inversely proportional to the distance from P to the origin. The temperature at its center is $75°F$. What is the rate of change of temperature at its center in the direction $2\mathbf{i} - \mathbf{j}$? In which directions from the center are the rates of change equal to zero?

45. A geological map shows that the altitude at point (x, y) is $A(x, y) = 100 - x^2 - y^2$ feet. If water is spilled at $(3, 4)$, in which direction will it run off?

46. A road is planned for the region of the map in Exercise 45 and it is desired to have the road perfectly level near the point $(3, 4)$. Is this possible? Explain.

47. *(Vector fields).* A function that assigns to each point in the plane a vector in the plane is called a *vector field*. For example, if $f(x, y) = 100 - x^2 - y^2$, then ∇f is a vector field because it assigns to each point (x, y), the vector

$$\nabla f(x, y) = -2x\mathbf{i} - 2y\mathbf{j}$$

Sketching a vector field means picking out a few points in the domain and drawing in the corresponding vector (or a scaled down version of it) with its tail at that point. Sketch the vector field given above using the 25 points (x, y) with $x = 0$, ± 1, ± 2 and $y = 0$, ± 1, ± 2 and scaling down each vector by a factor of $1/4$. Interpret this sketch with regard to the run off of a heavy rainfall in the region of the map of Exercise 45.

---

48. *(Gradients and the chain rule).* Let $z = f(x, y)$, and let $\mathbf{r}(t) = x(t)\mathbf{i} + y(t)\mathbf{j}$ trace out a curve contained in the domain of f. Assuming differentiability, show that

$$\frac{dz}{dt} = \nabla f(x, y) \cdot \mathbf{r}'(t)$$

49. Prove the following statements (assume differentiability):

(a) $\nabla(cf) = c\nabla f$ (b) $\nabla(f + g) = \nabla f + \nabla g$

(c) $\nabla(fg) = f\nabla g + g\nabla f$

(d) $\nabla\left(\dfrac{f}{g}\right) = \dfrac{g\nabla f - f\nabla g}{g^2}$

open book/notes

Test 3 14-8 — 15-6 except 14-10

15–7 TANGENT PLANES AND NORMAL VECTORS

Tangent lines play an important role in the study of functions of one variable, and tangent *planes* play an equally important role in the study of functions of several variables. We begin our discussion with a definition of a tangent plane. Let S be a surface in space containing the point P_0, and let C be any curve lying in S passing through P_0. Any vector tangent to C at P_0 is called a **tangent vector of S at P_0**.

DEFINITION

(15.41)

Let S be a surface in space containing the point P_0. If there is a plane T containing every tangent vector of S at P_0, then T is called the **tangent plane of S at P_0.**

Contrary to our usual procedure, we start off considering a function $w = F(x, y, z)$ of three variables; you will understand why in a moment. Let us recall that for each constant c, the equation $F(x, y, z) = c$ defines a level surface S of F. Now we can state the main result of this section.

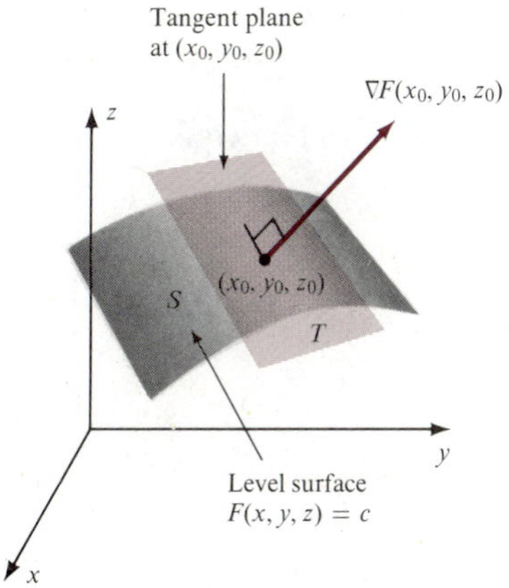

TANGENT PLANE THEOREM

(15.42)

Let S be a level surface of $w = F(x, y, z)$ containing the point $P_0(x_0, y_0, z_0)$. If F is differentiable at P_0, then there is a tangent plane T of S at P_0. Moreover, if $\nabla F(x_0, y_0, z_0) \neq \mathbf{0}$, then it is a normal vector of T.

The theorem is illustrated in Figure 15–27. Before we prove this theorem, let us look at some of its consequences. First off, an equation of the tangent plane is easily obtained because we know that ∇F is a normal vector. It follows from our previous work with planes that

---

(15.43)

If F is differentiable at (x_0, y_0, z_0) and $\nabla F(x_0, y_0, z_0) \neq \mathbf{0}$, then

$$F_x(x_0, y_0, z_0)(x - x_0) + F_y(x_0, y_0, z_0)(y - y_0)$$
$$+ F_z(x_0, y_0, z_0)(z - z_0) = 0$$

is an equation of the tangent plane.

EXAMPLE 1   Find an equation of the plane tangent to the ellipsoid $4x^2 + y^2 + 2z^2 = 4$ at the point $(1/2, 1, 1)$.

Solution   If we let $F(x, y, z) = 4x^2 + y^2 + 2z^2$, then the given ellipsoid is a level surface of F. The next step is to evaluate the partials at $(1/2, 1, 1)$,

$$F_x(1/2, 1, 1) = 4 \quad F_y(1/2, 1, 1) = 2 \quad F_z(1/2, 1, 1) = 4$$

It follows now from (15.43) that

$$4(x - 1/2) + 2(y - 1) + 4(z - 1) = 0 \quad \text{or} \quad 4x + 2y + 4z = 8$$

is an equation of the tangent plane (Figure 15–28).   ∎

The discussion now shifts to functions of two variables and planes tangent to their graphs. Given $z = f(x, y)$, we define $F(x, y, z) = f(x, y) - z$ and note

Tangent plane at (x_0, y_0, z_0)

$\nabla F(x_0, y_0, z_0)$

S (x_0, y_0, z_0)

T

y

Level surface $F(x, y, z) = c$

x

FIGURE 15–27. Plane T is tangent to the level surface S, and ∇F is a normal vector of T. This is an illustration of the Tangent Plane Theorem.

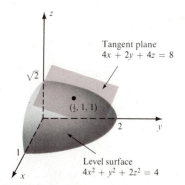

Tangent plane $4x + 2y + 4z = 8$

$\sqrt{2}$

$(\frac{1}{2}, 1, 1)$

2 y

1

Level surface $4x^2 + y^2 + 2z^2 = 4$

x

FIGURE 15–28. Example 1.

that *the level surface* $F(x, y, z) = 0$ *is the graph of* f. (This is the reason we began with functions of three variables.) Using this device, we can apply the Tangent Plane Theorem to the graph of f. Since $F(x, y, z) = f(x, y) - z$, it follows that

$$\nabla F(x, y, z) = f_x(x, y)\mathbf{i} + f_y(x, y)\mathbf{j} - \mathbf{k}$$

and, if $z_0 = f(x_0, y_0)$, then

(15.44)

> If $z = f(x, y)$ is differentiable at (x_0, y_0), then an equation of the plane tangent to the graph of f at the point (x_0, y_0, z_0) is
>
> $$f_x(x_0, y_0)(x - x_0) + f_y(x_0, y_0)(y - y_0) - (z - z_0) = 0$$

This is the same plane developed in Section 15–4 where it is used to define differentials; if x and y are close to x_0 and y_0, then z is close to z_0.

EXAMPLE 2   Let $z = x^2 + 2y^2$. Sketch the graph and find an equation of the tangent plane at $(1, 1, 3)$.

Solution   If we define $F(x, y, z) = x^2 + 2y^2 - z$, then *the level surface* $F(x, y, z) = 0$ *is the graph of* $z = x^2 + 2y^2$. At the point $(1, 1, 3)$, we have

$$\frac{\partial z}{\partial x}(1, 1) = 2 \quad \text{and} \quad \frac{\partial z}{\partial y}(1, 1) = 4$$

and it follows from (15.44) that

$$2(x - 1) + 4(y - 1) - (z - 3) = 0 \quad \text{or} \quad 2x + 4y - z = 3$$

is an equation of the tangent plane (Figure 15–29).   ∎

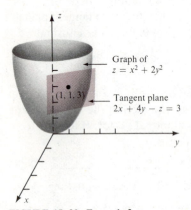

Graph of $z = x^2 + 2y^2$

$(1, 1, 3)$

Tangent plane $2x + 4y - z = 3$

FIGURE 15–29. Example 2.

Now let us prove the Tangent Plane Theorem. Let S be a level surface of $w = F(x, y, z)$. We will show that ∇F is orthogonal to every tangent vector of S at P_0. This means that all such vectors are contained in one plane which is what we want to prove.

Let C be any curve lying in the level surface S that passes through P_0 and has a tangent vector there. If

$$\mathbf{r}(t) = x(t)\mathbf{i} + y(t)\mathbf{j} + z(t)\mathbf{k}$$

is a position function for C with $\mathbf{r}(t_0) = P_0(x_0, y_0, z_0)$, then

$$\mathbf{r}'(t_0) = \frac{dx}{dt}(t_0)\mathbf{i} + \frac{dy}{dt}(t_0)\mathbf{j} + \frac{dz}{dt}(t_0)\mathbf{k}$$

is tangent to C and, therefore, it is a tangent vector of S at P_0. Since C lies in S, it follows that F is constant on C; that is

$$w(t) = F(x(t), y(t), z(t)) = c \quad \text{for all } t$$

Therefore, $w'(t_0) = 0$ and, using the chain rule on F, we have

$$(15.45) \quad 0 = F_x(x_0, y_0, z_0)\frac{dx}{dt}(t_0) + F_y(x_0, y_0, z_0)\frac{dy}{dt}(t_0) + F_z(x_0, y_0, z_0)\frac{dz}{dt}(t_0)$$

We now make the important observation that (15.45) can be written with the right side expressed as a dot product

$$0 = \nabla F(x_0, y_0, z_0) \cdot \mathbf{r}'(t_0)$$

This equation says it all; every tangent vector at P_0 is perpendicular to the gradient at P_0, so they all lie in one plane T, and ∇F is a normal vector of T. This completes the proof of the Tangent Plane Theorem.

If the gradient $\nabla F(x_0, y_0, z_0) \neq \mathbf{0}$, then it is called a **normal vector** of the level surface S at (x_0, y_0, z_0). The proof above shows that a nonzero gradient is always normal to a level surface of a function of three variables. For a function f of two variables, a similar proof would show that a nonzero gradient of f is normal to the *level curves* of f.

EXAMPLE 3   Let $f(x, y) = x^2 + 4y^2$. Sketch the level curves for $f(x, y) = 4$, 16, and 36, and the normal vectors at $(2, 0)$, $(-2, \sqrt{3})$, and $(4, \sqrt{5})$.

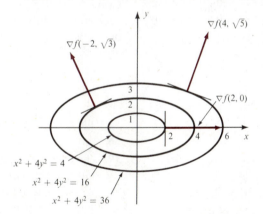

FIGURE 15-30. Example 3.

Solution   Each level curve is an ellipse (Figure 15-30). The gradient of f at (x, y) is

$$\nabla f(x, y) = 2x\mathbf{i} + 8y\mathbf{j}$$

so $\nabla f(2, 0) = 4\mathbf{i}$, $\nabla f(-2, \sqrt{3}) = -4\mathbf{i} + 8\sqrt{3}\mathbf{j}$, and $\nabla f(4, \sqrt{5}) = 8\mathbf{i} + 8\sqrt{5}\mathbf{j}$. These vectors are normal to the level curves as illustrated in the figure.   ■

EXERCISES

In Exercises 1-8, find an equation of the plane tangent to the given level surface at the indicated point. Sketch the surface and the tangent plane.

1. Sphere $x^2 + y^2 + z^2 = 4$ at $(1, 1, \sqrt{2})$.

2. Hyperboloid of one sheet $x^2 + y^2 - z^2 = 1$ at $(1, 1, 1)$.

3. Hyperboloid of two sheets $x^2 + y^2 - z^2 = -1$ at $(0, 0, -1)$.

4. Circular cone $x^2 + y^2 - z^2 = 0$ at $(1, 1, \sqrt{2})$.

5. Paraboloid of revolution $x^2 + y^2 = z$ at $(1, 1, 2)$.

6. Hyperbolic paraboloid $y^2 - x^2 = z$ at $(1, 1, 0)$.

7. Elliptic cylinder $4x^2 + y^2 = 4$ at $(1, 0, 2)$.

8. Parabolic cylinder $y = x^2$ at $(2, 4, 0)$.

In Exercises 9–14, find a vector normal to the given level *curves* at the indicated points. Sketch the curve and the vector.

9. $3x^2 + y^2 = 4$; $(1, 1)$

10. $x^2 + 3y^2 = 4$; $(1, 1)$

11. $x^2 - y = 0$; $(2, 4)$

12. $x^2 + y = 0$; $(2, -4)$

13. $x^2 - y^2 = 3$; $(-2, 1)$

14. $x^2 - y^2 = -4$; $(0, 2)$

In Exercises 15–22, find an equation of the plane tangent to the given level surfaces at the indicated points.

15. $2xy + 2yz + 2xz = 8$; $(1, 1, 3/2)$

16. $xy - z^2 = 0$; $(2, 2, 2)$

17. $z^2 + \ln|xy| = 9$; $(-1, 1, 3)$

18. $xe^y \sin z = 0$; $(1, 0, 0)$

19. $8x^2 - 3y^2 = z^2$; $(1, 1, \sqrt{5})$

20. $x^3y + \sqrt{z} = 2$; $(1, 2, 0)$

21. $\sqrt{x^2 + y^2 + z^2} = 5$; $(3, 4, 0)$

22. $xyz - 4y^3 = -30$; $(1, 2, 1)$

In Exercises 23–32, find an equation of the plane tangent to the graph of the given functions at the indicated points.

23. $z = y^2 - x^2$; $(2, 1, -3)$ **24.** $z = x^2$; $(2, 3, 4)$

25. $z = e^{-x^2 - y^2}$; $(2, -1, e^{-5})$

26. $z = \sin xy$; $(1, \pi/2, 1)$

27. $z = \frac{1}{3}\pi x^2 y$; $(2, 1, 4\pi/3)$

28. $z = \pi x \sqrt{x^2 + y^2}$; $(2, 1, 2\sqrt{5}\pi)$

29. $z = \frac{1}{2}x^2 \sin y$; $(4, \pi/6, 4)$ **30.** $z = e^{xy}$; $(1, 1, e)$

31. $z = \ln(1 + x^2 + y^2)$; $(3, 4, \ln 26)$

32. $z = e^{\sin xy}$; $(1, 0, 1)$

33. Find the point(s) on the ellipsoid $4x^2 + y^2 + 9z^2 = 36$ where the tangent plane is parallel to the plane $16x - 2y + 36z = 9$.

34. Same as Exercise 33 for the plane $8x + 2y = 5$.

35. Find the point(s) on the surface $e^x + yz^2 = 1$ where its tangent plane is perpendicular to the tangent plane of $x^2 + y^2 + z^2 = 1$ at $(0, 0, 1)$.

36. Same as Exercise 35 for the surface $xyz = 1$.

Curves and surfaces have normal vectors and tangent vectors. At points of intersection, we say that the curves or surfaces are *orthogonal* if the appropriate vectors are orthogonal. The meaning should be clear.

37. The point $(-2, -2, 2\sqrt{2})$ is contained in the intersection of the sphere $x^2 + y^2 + z^2 = 16$ and the cone $z^2 = x^2 + y^2$. Are the surfaces orthogonal at that point?

38. Same as Exercise 37 for the point $(2, -2, -2\sqrt{2})$.

39. The curve with parametric equations $x = t^2$, $y = t$, and $z = \frac{1}{2}\ln t$ (for $t > 0$) meets the surface $2x^2 + y^2 + z = 3$ at the point $(1, 1, 0)$. Is the curve orthogonal to the surface at that point?

40. Show that any radius of a sphere is perpendicular to the sphere.

Optional Exercise

41. The plane $x + y + z = 1$ and the sphere $x^2 + y^2 + z^2 = 1$ intersect in a circle containing the point $(0, 0, 1)$. Find a vector (in the plane) that is tangent to the circle at that point.

15–8   MAXIMA AND MINIMA

Maxima and minima for functions of one variable are discussed in Chapter 4. We found that if f is defined on a closed and bounded interval $[a, b]$, then the extreme values of f can occur only at

(1) critical points in the open interval (a, b), or

(2) the endpoints a and b.

The basic ideas are much the same for functions of two (or more) variables, but the definitions need to be adjusted. For example, a function f of two variables is defined on a set R in the xy-plane instead of an interval $[a, b]$. The set of *interior points* of R replaces the open interval (a, b), and the set of *boundary points* of R replaces the endpoints a and b. (The notions of interior point and boundary point were defined in Section 15–2.) The set R is *closed* if it contains its boundary points, and it is *bounded* if it can be contained in some disc of finite radius.

Suppose that a function f of two variables is defined on a set R. Then $f(p, q)$ is a **local maximum** of f if there is some disc D about (p, q) such that $f(x, y) \le f(p, q)$ whenever (x, y) is in R and in D. The value $f(p, q)$ is a **local minimum** of f if $f(x, y) \ge f(p, q)$ whenever (x, y) is in R and D. If

$$f(a, b) \le f(x, y) \le f(c, d)$$

for all (x, y) in R, then $f(a, b)$ is called the **absolute minimum** and $f(c, d)$ is called the **absolute maximum of f on** R. Any local or absolute maximum or minimum is called an **extreme value** of f. Extreme values occur at the peaks and valleys of the graph of f (Figure 15–31). The extreme value theorem for several variables is stated below without proof.

(15.46)

> **EXTREME VALUE THEOREM**
> A continuous function of two variables defined on a closed and bounded subset of the plane must attain its absolute maximum and minimum values.
> Similar statements hold for functions of three or more variables.

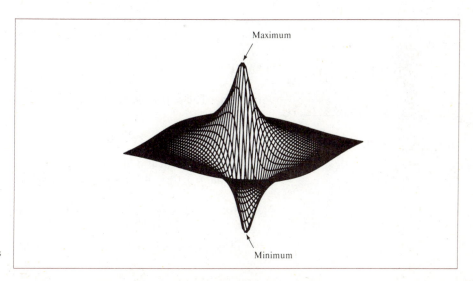

FIGURE 15–31. Local maxima and minima occur at the peaks and valleys of the graph.

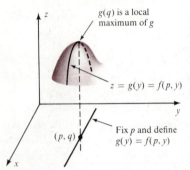

FIGURE 15–32. If $f(p, q)$ is a local maximum of f and $g(y) = f(p, y)$, then $g(q)$ is a local maximum of g.

Let g be a function of one variable. If an extreme value of g occurs at an interior point q, then q must be a critical point; that is, $g'(q) = 0$ or $g'(q)$ does not exist. We can apply this information to a function f of two variables. Suppose that an extreme value of f occurs at an interior point (p, q). We fix p and define $g(y) = f(p, y)$. It follows that g will have an extreme value at q (Figure 15–32), and either $g'(q) = 0$ or it doesn't exist. Since $g'(q) = f_y(p, q)$, we conclude that either $f_y(p, q) = 0$ or it doesn't exist. A similar argument, using $h(x) = f(x, q)$ would show the same conclusion about $f_x(p, q)$. This proves the following result:

(15.47)

> Let f have an extreme value at (p, q), an interior point of the domain of f. Then
> $$f_x(p, q) = f_y(p, q) = 0$$
> or one (or both) of the partials does not exist.

Taking our cue from the one variable case, we say that (p, q) is a **critical point** of f if $f_x(p, q) = f_y(p, q) = 0$ or if either partial does not exist at (p, q). It follows from (15.47), that

(15.48)

> Let f be a function of two variables defined on a set R. Then the extreme values of f can occur only at
>
> (1) critical points that are interior points of R, or
> (2) boundary points of R.
>
> Similar statements hold for functions of three or more variables.

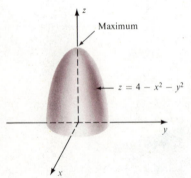

FIGURE 15–33. Example 1.

In this section, we concentrate on finding extreme values at interior points; in the next section we discuss extreme values at boundary points.

EXAMPLE 1   Let $f(x, y) = 4 - x^2 - y^2$, and find the extreme values of f.

Solution   The partial derivatives exist everywhere, so the extreme values occur only where
$$f_x(x, y) = -2x = 0 \quad and \quad f_y(x, y) = -2y = 0$$
Thus, the only critical point is $(0, 0)$. The graph of f (Figure 15–33) clearly indicates that $f(0, 0) = 4$ is an absolute maximum.   ■

EXAMPLE 2   Let $f(x, y) = \sqrt{x^2 + y^2}$ and find the extreme values.

Solution   The partials
$$f_x(x, y) = \frac{x}{\sqrt{x^2 + y^2}} \quad and \quad f_y(x, y) = \frac{y}{\sqrt{x^2 + y^2}}$$

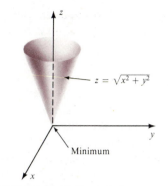

$z = \sqrt{x^2 + y^2}$

Minimum

FIGURE 15–34. Example 2.

do not exist at the origin. They do exist at all other points, but are not both zero. Therefore, the only critical point is $(0, 0)$ and the graph (Figure 15–34) shows that $f(0, 0) = 0$ is an absoltue minimum. ∎

The next example illustrates that a function of two variables, just as a function of one variable, need not have an extreme value at a critical point.

EXAMPLE 3 Let $f(x, y) = y^2 - x^2$. Show that the origin is the only critical point, but $f(0, 0)$ is not an extreme value.

Solution The partials, $f_x(x, y) = -2x$ and $f_y(x, y) = 2y$ exist everywhere, and are simultaneously zero only at the origin. Therefore, it is the only critical point. But $f(0, 0) = 0$ is not an extreme value because every disc about $(0, 0)$ contains points of the form $(x, 0)$ and $(0, y)$ with x and y nonzero. At these points

$$f(x, 0) = -x^2 < 0 \quad \text{and} \quad f(0, y) = y^2 > 0$$

Consequently, 0 is not an extreme value. The graph of this function has a *saddle point* at the origin (Figure 15–35). ∎

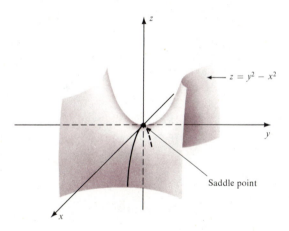

$z = y^2 - x^2$

Saddle point

FIGURE 15–35. Example 3.

In the examples so far, it has been relatively easy to determine whether or not a critical point yields an extreme value. The next example illustrates a more difficult situation.

EXAMPLE 4 Let $f(x, y) = x^3 - 3xy - \frac{3}{2}y^2$. Find the extreme values of f.

Solution The partials

$$f_x(x, y) = 3x^2 - 3y \quad \text{and} \quad f_y(x, y) = -3x - 3y$$

exist everywhere. Therefore, the only critical points are where

$$3x^2 - 3y = 0 \quad \text{and} \quad -3x - 3y = 0$$

To find the simultaneous solution, we note that the second equation yields $y = -x$ and substitute this into the first equation. Thus, $3x^2 - 3(-x) = 3x^2 + 3x = 0$ and $x = 0$ or -1. It follows that

$$(0, 0) \quad \text{and} \quad (-1, 1)$$

are the only critical points of f. But are $f(0, 0) = 0$ and $f(-1, 1) = 1/2$ extreme values? We cannot use the graph of f to answer this question because the graph is too complicated to draw. We could perform some laborious computations with points close to $(0, 0)$ and $(-1, 1)$, but fortunately there is another way to test the critical points. Let us pause here to describe the test, and return to this example afterwards. ■

The test referred to in the last example is similar to the Second Derivative Test (4.6) and we call it the *Second Partials Test*. Its proof can be found in texts on advanced calculus.

SECOND PARTIALS TEST

Suppose that f has continuous second partials on a disc about (p, q), and that $f_x(p, q) = f_y(p, q) = 0$. Define a new function

$$F(x, y) = f_{xx}(x, y)f_{yy}(x, y) - [f_{xy}(x, y)]^2$$

(15.49)

(1) If $F(p, q) > 0$ and $f_{xx}(p, q) < 0$, then $f(p, q)$ is a local maximum.
(2) If $F(p, q) > 0$ and $f_{xx}(p, q) > 0$, then $f(p, q)$ is a local minimum.
(3) If $F(p, q) < 0$, then f has a saddle point at (p, q).
(4) If $F(p, q) = 0$, then the test fails.

Remarks:

(a) Since f has continuous second partials, $f_{xy} = f_{yx}$, so either may be used in defining F.

(b) If $F(p, q) > 0$, then f_{xx} and f_{yy} must have the same sign at (p, q). Therefore, f_{yy} can be used in place of f_{xx} in parts (1) and (2) of the test.

(c) Similar tests are available for functions of three or more variables, but we will not discuss them here.

We return now to complete Example 4.

EXAMPLE 4 *(continued)* We left off with

$$f_x(x, y) = 3x^2 - 3y \quad \text{and} \quad f_y(x, y) = -3x - 3y$$

which give $(0, 0)$ and $(-1, 1)$ as the critical points. To use the Second Partials Test, we find

$$f_{xx} = 6x \quad f_{yy} = -3 \quad f_{xy} = -3$$

They are all continuous, so we form the function F

$$F(x, y) = f_{xx}(x, y)f_{yy}(x, y) - [f_{xy}(x, y)]^2$$
$$= (6x)(-3) - (-3)^2$$
$$= -18x - 9$$

Since $F(0, 0) = -9 < 0$, we know by (15.49.3) that $(0, 0)$ is a saddle point; $f(0, 0) = 0$ is not an extreme value. But $F(-1, 1) = 18 - 9 = 9 > 0$ and $f_{xx}(-1, 1) = 6(-1) = -6 < 0$; so, by (15.49.1), $f(-1, 1) = 1/2$ is a local maximum. Figure 15–36 is a computer generated sketch of the graph.   ■

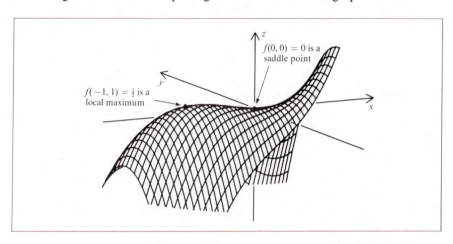

$f(0, 0) = 0$ is a saddle point

$f(-1, 1) = \frac{1}{2}$ is a local maximum

FIGURE 15–36. Example 4. Computer-generated graph of $f(x, y) = x^3 - 3xy - \frac{3}{2}y^2$.

EXAMPLE 5   A rectangular box without a top is to have a volume of 12m³. What dimensions require the least amount of material?

Solution   Let x and y be the dimensions of the base and z be the height (Figure 15–37). Minimizing the material is the same as minimizing the surface area $S = xy + 2yz + 2xz$. Using the side condition, volume $= xyz = 12$, we can substitute $z = 12/xy$ and reduce S to a function of two variables

$$S = xy + \frac{24}{x} + \frac{24}{y}$$

To find the critical points, we set the partials equal to zero.

$$S_x = y - \frac{24}{x^2} = 0 \quad \text{and} \quad S_y = x - \frac{24}{y^2} = 0$$

The first equation yields $y = 24/x^2$, and substitution into the second equation yields

$$x - \frac{x^4}{24} = 0, \text{ so } x = 0 \quad \text{or} \quad x = \sqrt[3]{24}$$

We can disregard $x = 0$ because in that case the volume would be 0. Therefore, $x = \sqrt[3]{24}$ and, solving for y, we find that $(\sqrt[3]{24}, \sqrt[3]{24})$ is the only viable critical

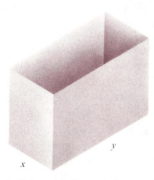

z

y

x

FIGURE 15–37. Example 5.

point in this problem. From the nature of the problem, we suspect that this point will yield a minimum value for S. But just to make sure, we use the Second Partials Test.

$$S_{xx}(x, y) = \frac{48}{x^3}, \text{ so } S_{xx}(\sqrt[3]{24}, \sqrt[3]{24}) = 2$$

$$S_{yy}(x, y) = \frac{48}{y^3}, \text{ so } S_{yy}(\sqrt[3]{24}, \sqrt[3]{24}) = 2$$

and $S_{xy}(x, y) = 1$. It follows from (15.49.2) that we have found a minimum value. Since $z = 12/xy$, the final answer is that $x = y = \sqrt[3]{24}$ and $z = 12/\sqrt[3]{(24)^2}$ are the dimensions that require the least material. ∎

EXERCISES

In Exercises 1–20, find the extreme values (if any) of f.

1. $f(x, y) = x^2 + xy + 2y^2$
2. $f(x, y) = x^2 + 2xy - y^2$
3. $f(x, y) = x^2 - 3xy - y^2$
4. $f(x, y) = 2x^2 - xy + y^2$
5. $f(x, y) = x^3 - 12xy + y^3$
6. $f(x, y) = x^3 + xy - 2y^3$
7. $f(x, y) = x^4 - 3y^3 + 2xy$
8. $f(x, y) = x^4 + 25x + y^2$
9. $f(x, y) = e^x \sin y$ 10. $f(x, y) = e^{xy}$
11. $f(x, y) = \sin x + \sin y; 0 < x, y < 2\pi$
12. $f(x, y) = \sin x + \sin y + \sin(x + y); 0 < x, y < 2\pi$
13. $f(x, y) = |x| + |y|$ 14. $f(x, y) = \frac{1}{x} + \frac{1}{y} + xy$

15. $f(x, y) = xy(x^2 + y^2 - 4)$
16. $f(x, y) = xy(x^2 - y^2)$
17. $f(x, y) = x^2 + xy^2 + y^4$
18. $f(x, y) = 2x^2 - 2xy + y^2 - 2x$
19. $f(x, y) = (x^2 + y^2)e^{x^2 - y^2}$
20. $f(x, y) = \ln(4 - \sin xy)$

21. In Example 5, the material used to construct the box costs $2/m^2$ for the sides and $3/m^2$ for the bottom. The volume is still to be 12m³. What dimensions yield the minimum cost?

22. If the box in Exercise 21 has a top made out of the same material as the bottom, what dimensions yield the minimum cost?

23. Find three positive numbers whose sum is S and whose product is as large as possible.

24. Find three positive numbers whose product is S and whose sum is as small as possible.

25. Find a vector in space whose length is 10 units and whose components have the largest possible sum.

26. Find the point in space the sum of whose coordinates is 64 and whose distance from the origin is minimum.

27. A rectangular parallelepiped is inscribed in a sphere of radius r. What dimensions yield the largest volume?

28. Same as Exercise 27 for a parallelepiped with sides parallel to the coordinate axes and inscribed in the ellipsoid $16x^2 + 4y^2 + 9z^2 = 144$.

Method of least squares. It often happens that in theory, the data collected in an experiment should lie on a straight line $y = mx + b$. But the actual data seldom do. For instance, in any given locality, we expect temperature to vary linearly with altitude. But if we measure the temperature T at various altitudes A, the points (A_1, T_1), (A_2, T_2), and so on, would probably not all lie on one line. The problem in such a case is to find a line that *best fits the data*. Let (x_1, y_1), (x_2, y_2), ..., (x_n, y_n) represent the data collected in an experiment. For any line $y = mx + b$, each point (x_i, y_i) will have a vertical deviation from the line of $d_i = y_i - (mx_i - b)$ as shown in Figure 15–38. The idea is to choose m and b so as to *minimize* the effect of the deviations. Because some deviations are positive and others are negative, we

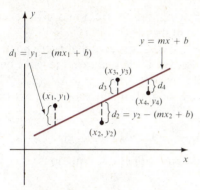

FIGURE 15–38. The *method of least squares* for finding the line that best fits the points (x_i, y_i).

square them so as to treat them equally. Thus, we are led to the problem of finding the m and b that minimize the function

$$f(m, b) = d_1{}^2 + d_2{}^2 + \cdots + d_n{}^2$$

(15.50)
$$= \sum_{i=1}^{n} (y_i - mx_i - b)^2$$

This method of finding the straight line that best fits the data is called the *method of least squares*.

In Exercises 29–30, use the method of least squares to determine the line that best fits the given points. Sketch the points and the line.

29. (1, 1), (2, 3), (4, 3)

30. (0, 1), (1, 3), (2, 2), (3, 4), (4, 5)

31. The solubility of sodium nitrate ($NaNO_3$) is linearly related to water temperature (degrees Celsius). The data collected in an experiment by Mendeléjev are (0, 66.7), (4, 71.0), (10, 76.3), (15, 80.6), (21, 85.7), (29, 92.9), (36, 99.4), (51, 113.6), and (68, 125.1) where the first coordinate is the temperature and the second is the number of parts of sodium nitrate that dissolve in 100 parts of water. Use the method of least squares to find the line that best fits the data.

32. The temperature (in degrees Fahrenheit) was taken at various altitudes (in feet) near Donner Pass in California, and the data collected was (4,000, 40), (5,000, 38), (6,000, 32), (7,000, 33), and (8,000, 30). Use least squares to find the line that best fits the data.

33. For points $(x_1, y_1), (x_2, y_2), \ldots, (x_n, y_n)$, use (15.50) and show that if $y = mx + b$ is the best fit, then

$$m\left(\sum x_i\right) + nb = \sum y_i, \text{ and}$$
$$m\left(\sum x_i{}^2\right) + b\left(\sum x_i\right) = \sum x_i y_i$$

where the sums are taken from $i = 1$ to n.

34. Use the results of Exercise 33 to show that if $y = mx + b$ is the best fit, then

$$\sum_{i=1}^{n} d_i = \sum_{i=1}^{n} (y_i - mx_i - b) = 0$$

15–9 CONSTRAINTS AND LAGRANGE MULTIPLIERS

The last section was devoted to finding extreme values at interior points. We now turn our attention to finding extreme values at boundary points. The method we shall use is called the *method of Lagrange multipliers*.

LAGRANGE'S THEOREM

Let f and g be functions of two variables with continuous partials. Let C be a level curve $g(x, y) = c$ of g, and suppose that f has an extreme value at (x_0, y_0) when its domain is restricted to C. If $\nabla g(x_0, y_0) \neq \mathbf{0}$, then there is a number λ (Greek letter *lambda*) such that

$$\nabla f(x_0, y_0) = \lambda \nabla g(x_0, y_0)$$

(15.51)

The same holds for functions f and g of three variables if "level curve" is changed to "level surface."

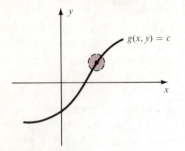

FIGURE 15–39. Every point on a level curve is a boundary point.

We will first discuss the technique of applying this theorem and give several examples, and return to its proof at the end of this section.

The domain of f is restricted to the level curve C; observe that every point of a level curve is a boundary point (Figure 15–39). If f has an extreme value on C at (x_0, y_0), then Lagrange's Theorem guarantees that $\nabla f(x_0, y_0) = \lambda \nabla g(x_0, y_0)$ for some number λ, which is called a *Lagrange multiplier*. Therefore, the extreme values of f on C can occur only at points (x, y) for which

$$f_x(x, y)\mathbf{i} + f_y(x, y)\mathbf{j} = \lambda g_x(x, y)\mathbf{i} + \lambda g_y(x, y)\mathbf{j} \quad and \quad g(x, y) = c$$

or, put another way, those points (x, y) that satisfy the three equations

(15.52)
$$
\begin{aligned}
f_x(x, y) &= \lambda g_x(x, y) \\
f_y(x, y) &= \lambda g_y(x, y) \\
g(x, y) &= c
\end{aligned}
$$

Finding extreme points by solving these simultaneous equations is called the *method of Lagrange multipliers*.

EXAMPLE 1 Let $f(x, y) = x^2 + y^2$ be defined on the ellipse $4x^2 + 9y^2 = 36$. Find the extreme values of f.

Solution If we define $g(x, y) = 4x^2 + 9y^2$, then the given ellipse is a level curve of g. Thus, the extreme values of f can occur only at points that satisfy equations (15.52); that is, for some number λ,

$$
\begin{array}{ll}
f_x(x, y) = \lambda g_x(x, y) & 2x = \lambda 8x \\
f_y(x, y) = \lambda g_y(x, y) & 2y = \lambda 18y \\
g(x, y) = 36 & 4x^2 + 9y^2 = 36
\end{array}
$$

Suppose $x \neq 0$. Then the first equation yields $\lambda = 1/4$; but if $\lambda = 1/4$, then the second equation yields $y = 0$. When $y = 0$, the third equation yields $x = \pm 3$. Therefore, under the supposition that $x \neq 0$, we know that $x = \pm 3, y = 0$, and $\lambda = 1/4$ are simultaneous solutions. Now let us suppose that $x = 0$. The third equation yields $y = \pm 2$; but if $y = \pm 2$, then the second equation yields $\lambda = 1/9$. Therefore, $x = 0, y = \pm 2$, and $\lambda = 1/9$ are also solutions. It follows that f can assume extreme values only at the points $(\pm 3, 0)$ and $(0, \pm 2)$*. We simply examine the values

$$f(3, 0) = 9, \quad f(-3, 0) = 9, \quad f(0, 2) = 4, \text{ and } \quad f(0, -2) = 4$$

and conclude that the extreme values of f are 9 and 4 (Figure 15–40). ∎

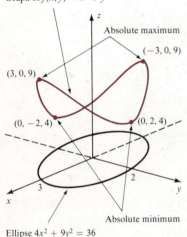

Graph of $f(x, y) = x^2 + y^2$

Ellipse $4x^2 + 9y^2 = 36$

FIGURE 15–40. Example 1.

---

*Notice that λ plays no part in the answer. It is merely a dummy variable introduced to help us find the possible extreme values.

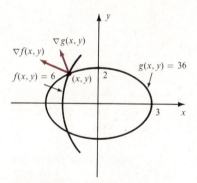

FIGURE 15–41. Lagrange's Theorem says that if $\nabla f(x, y)$ and $\nabla g(x, y)$ are not parallel, then f cannot have an extreme value at (x, y).

Let us pause a moment to give a geometric interpretation of Lagrange's Theorem using the functions f and g of Example 1. Figure 15–41 shows the ellipse $g(x, y) = 36$ and part of the level curve $f(x, y) = 6$; the latter is part of the circle $x^2 + y^2 = 6$. The curves intersect at (x, y). The gradient of f is normal to its level curve and points in the direction of greatest increase for f. The figure shows that $\nabla f(x, y)$ is not normal to the ellipse; it leans to the left. This indicates that the values of f will be larger than 6 at points on the ellipse to the left of (x, y) and smaller than 6 at points to the right. Thus, $f(x, y) = 6$ is not an extreme value. Moreover, *this is true at every point where the gradient of f is not normal to the ellipse*. It follows that extreme values can occur only at points where ∇f is normal to the ellipse. But the ellipse is a level curve of g, so ∇g is normal to it at every point. Therefore, extreme values of f occur only at points where ∇f and ∇g are both normal, hence parallel, and $\nabla f = \lambda \nabla g$ for some number λ.

The method of Lagrange multipliers works for functions of three variables, but then there are four simultaneous equations to solve.

EXAMPLE 2 Find the point on the plane $x + 5y - z = -14$ that is nearest the point $(2, 3, 4)$.

Solution The distance from (x, y, z) to $(2, 3, 4)$ is

$$\sqrt{(x - 2)^2 + (y - 3)^2 + (z - 4)^2};$$

this function is to be minimized subject to the *constraint* (or *side condition*) that $x + 5y - z = -14$. We can save some work by discarding the square root*. Thus, we define

$$f(x, y, z) = (x - 2)^2 + (y - 3)^2 + (z - 4)^2$$

and

$$g(x, y, z) = x + 5y - z;$$

and use the method of Lagrange multipliers to minimize f on the level surface $g(x, y, z) = -14$. There are four equations

$$f_x(x, y, z) = \lambda g_x(x, y, z) \qquad 2x - 4 = \lambda$$
$$f_y(x, y, z) = \lambda g_y(x, y, z) \qquad 2y - 6 = 5\lambda$$
$$f_z(x, y, z) = \lambda g_z(x, y, z) \qquad 2z - 8 = -\lambda$$
$$g(x, y, z) = -14 \qquad x + 5y - z = -14$$

We solve each of the first three equations for x, y, and z in terms of λ

$$x = \frac{\lambda + 4}{2} \qquad y = \frac{5\lambda + 6}{2} \qquad z = \frac{-\lambda + 8}{2}$$

---

*Clearly the distance function and its square will both attain a minimum value at the same point.

and substitute in the fourth

$$\frac{\lambda + 4}{2} + 5\frac{5\lambda + 6}{2} - \frac{-\lambda + 8}{2} = -14$$

Solving for λ yields $\lambda = -2$. Once λ is known, we see that $x = 1, y = -2$, and $z = 5$. Since we know from geometry that there is a point on the plane that is nearest to $(2, 3, 4)$, and since $(1, -2, 5)$ is the only point at which f can have an extreme value, we conclude that $(1, -2, 5)$ is the answer. ■

Lagrange multipliers can also be used to work some of the problems in the last section. As an illustration, we will rework Example 5.

EXAMPLE 3   A rectangular box without a top is to have volume of $12m^3$. What dimensions require the least amount of material?

Solution   We still want to minimize the surface area

$$S = xy + 2yz + 2xz$$

but instead of using the volume $xyz = 12$ to eliminate z, we will use $V = xyz$ as a *constraint*. The resulting equations are

$$S_x = \lambda V_x \qquad y + 2z = yz\lambda$$
$$S_y = \lambda V_y \qquad x + 2z = xz\lambda$$
$$S_z = \lambda V_z \qquad 2y + 2x = xy\lambda$$
$$V = 12 \qquad\qquad xyz = 12$$

Subtracting the second equation from the first yields

$$y - x = (y - x)z\lambda$$

from which it follows that $y = x$ or $z\lambda = 1$. But if $z\lambda = 1$, then the first equation yields $z = 0$ and z cannot be 0(why?). Therefore, $y = x$ and the third equation yields $\lambda = 4/x$. Substituting this into equation two yields $z = x/2$. Now we have $y = x$ and $z = x/2$, so the last equation yields

$$\frac{x^3}{2} = 12 \quad \text{or} \quad x = \sqrt[3]{24}$$

which is the same answer obtained in the last section. ■

The method also works if there is more than one constraint, but each constraint must have its own Lagrange multiplier. For instance, if there are two constraining functions g and h, then f can have extreme values only at points (x, y, z) that satisfy the equation

(15.53) $\boxed{\nabla f(x, y, z) = \lambda \nabla g(x, y, z) + \mu \nabla h(x, y, z)}$

The method is illustrated in the next example.

EXAMPLE 4   Let C be the trace of the ellipsoid $x^2 + y^2 + 2z^2 = 1$ in the plane $3x + 3y + z = 1$. Find the points on C that are nearest to, and farthest from, the origin.

Solution   If we define

$$f(x, y, z) = x^2 + y^2 + z^2 \qquad \text{(Square of the distance function)}$$
$$g(x, y, z) = x^2 + y^2 + 2z^2 \qquad \text{(Ellipsoid is a level surface)}$$
$$h(x, y, z) = 3x + 3y + z \qquad \text{(Plane is a level surface)}$$

then the problem is to find the extrema of f subject to the constraints $g(x, y, z) = 1$ and $h(x, y, z) = 1$. This gives rise to five simultaneous equations; three from (15.53) plus $g = 1$ and $h = 1$. The equations are

$$f_x = \lambda g_x + \mu h_x \qquad\qquad 2x = 2x\lambda + 3\mu$$
$$f_y = \lambda g_y + \mu h_y \qquad\qquad 2y = 2y\lambda + 3\mu$$
$$f_z = \lambda g_z + \mu h_z \qquad\qquad 2z = 4z\lambda + \mu$$
$$g = 1 \qquad\qquad x^2 + y^2 + 2z^2 = 1$$
$$h = 1 \qquad\qquad 3x + 3y + z = 1$$

Subtracting the second equation from the first, we have

$$2x - 2y = (2x - 2y)\lambda$$

from which it follows that either $x = y$ or $\lambda = 1$. If $x = y$, then from the last two equations, we have

$$2y^2 + 2z^2 = 1 \quad \text{and} \quad 6y + z = 1$$

The solution of these two equations is obtained by substituting $z = 1 - 6y$ in the first equation. This yields

$$y = \frac{12 \pm \sqrt{70}}{74} \quad \text{and} \quad z = \frac{1 \mp 3\sqrt{70}}{37}$$

so possible extreme values of f occur at

$$\left(\frac{12 + \sqrt{70}}{74}, \frac{12 + \sqrt{70}}{74}, \frac{1 - 3\sqrt{70}}{37} \right)$$

(15.54)   and

$$\left(\frac{12 - \sqrt{70}}{74}, \frac{12 - \sqrt{70}}{74}, \frac{1 + 3\sqrt{70}}{37} \right)$$

If $\lambda = 1$, it follows from the first equation above that $\mu = 0$; with $\mu = 0$ and $\lambda = 1$, the third equation yields $z = 0$. Then the last two equations become

$$x^2 + y^2 = 1 \quad \text{and} \quad 3x + 3y = 1$$

and the solution of these two equations is

$$x = \frac{1 \pm \sqrt{17}}{6} \quad \text{and} \quad y = \frac{1 \mp \sqrt{17}}{6}$$

This leads to two other possible extrema of f at the points

(15.55) $\left(\dfrac{1 + \sqrt{17}}{6}, \dfrac{1 - \sqrt{17}}{6}, 0\right)$ and $\left(\dfrac{1 - \sqrt{17}}{6}, \dfrac{1 + \sqrt{17}}{6}, 0\right)$

The way to decide between the four possible points is to evaluate f at each of them. It turns out (we spare you the details) that both points in (15.55) are 1 unit from the origin; that is the maximum distance. The second point in (15.54) is at a distance of

$$\dfrac{2952 - 24\sqrt{70}}{(74)^2} \approx 0.5 \text{ units}$$

from the origin, and it is the closest. ∎

Success at finding extreme values using Lagrange multipliers clearly depends on your ability to solve simultaneous equations. There is no single method that works in every case. Each problem has its own peculiarities; some require a great deal of ingenuity to solve.

Proof of Lagrange's Theorem (Optional)

Let $\mathbf{r}(t) = x(t)\mathbf{i} + y(t)\mathbf{j}$ be a differentiable vector function that traces out the curve C with $\mathbf{r}(t_0) = (x_0, y_0)$. Since f has an extreme value at (x_0, y_0), so does the function $h(t) = f(x(t), y(t))$, and $h'(t_0) = 0$. By the chain rule,

$$0 = h'(t_0) = f_x(x_0, y_0)\dfrac{dx}{dt}(t_0) + f_y(x_0, y_0)\dfrac{dy}{dt}(t_0)$$

$$= \nabla f(x_0, y_0) \cdot \mathbf{r}'(t_0)$$

Now $\mathbf{r}'(t_0)$ is tangent to C, the level curve of g, so $\nabla f(x_0, y_0)$ is normal to C. But so is $\nabla g(x_0, y_0)$ normal to its level curve C. Since we are in a plane, it follows that $\nabla f(x_0, y_0)$ and $\nabla g(x_0, y_0)$ are parallel, and that completes the proof.

EXERCISES

In Exercises 1–12, use the method of Lagrange multipliers to find the absolute extreme values subject to the given constraints.

1. $f(x, y) = 3x^2 + 3y^2 - 4y;\ x^2 + y^2 = 1$
2. $f(x, y) = 3x + y^2;\ 4x^2 + y^2 = 4$
3. $f(x, y) = 4x^2 - 4xy + y^2;\ x^2 + y^2 = 1$
4. $f(x, y) = x^2 + 2xy + 2y^2;\ x^2 + y^2 = 1$
5. $f(x, y) = 12y - 6xy - 4y^2 - 9x^2;\ 9x^2 + 4y^2 = 36$
6. $f(x, y) = 3x^2 - 2y^2; y = 5x^2$
7. $f(x, y, z) = xyz;\ x + 2y + 4z = 8$
8. $f(x, y, z) = x^2 + y^2 + z^2;\ 3x - 2y + 4z = 1$
9. $f(x, y, z) = x^2 + y^2 + z^2;\ x - y + z = 1$

10. $f(x, y, z) = x + y + z;\ x^2 + y^2 + z^2 = 1$
11. $f(x, y, z) = x^2 + y^2 + z^2;\ x^2 + y^2 - z^2 = 0$ and $z - x - y = 1$
12. $f(x, y, z) = x + y + z;\ x + y = 1$ and $y + z = 1$

If the domain of a function has an interior and a boundary, you can find the extreme values in the interior by the methods of the last section (partials, Second Partials Test, and so on). The extreme values on the boundary can be found by the methods of this section. BUT, extreme values for the boundary may not be extreme values on the whole domain; each one must be checked. EXAMPLE: $f(x, y) = x^2 + y^2 - 2y$ with domain $x^2 + y^2 \le 4$. You can easily check that $f(0, 2) = 0$ is an absolute minimum *on the boundary* $x^2 + y^2 = 4$, but it is *not* a minimum on the whole domain because every disc about (0, 2) contains points $(0, y)$ with $0 < y < 2$ and then $f(0, y) < 0$. In Exercises 13–16 find the absolute extreme values on the given domains.

13. The function in Exercise 1 with domain $x^2 + y^2 \leq 1$.

14. The function in Exercise 3 with domain $x^2 + y^2 \leq 1$.

15. The function in Exercise 5 with domain $9x^2 + 4y^2 \leq 36$.

16. $f(x, y) = x^2 + y^2 + 2x$ with domain $x^2 + y^2 \leq 1$.

In Exercises 17–20, find the points, if any, on the given curve that are closest to, and farthest from, the origin.

17. The circle $(x - 2)^2 + (y + 1)^2 = 1$.

18. The ellipse $(x - 8)^2 + 4(y + 5)^2 = 36$.

19. The rotated ellipse $17x^2 + 12xy + 8y^2 = 100$.

20. The rotated hyperbola $2x^2 - y^2 + 4xy - 2x + 3y = 6$.

---

21. A rectangular box without a top is to have a surface area of 128 in². What dimensions yield the greatest volume?

22. Suppose that the material of the box in Exercise 21 costs $3 per sq ft for the sides and $8 per sq ft for the bottom. What dimensions yield the least cost?

23. Find an equation of the plane with positive intercepts that passes through (2, 1, 1) and cuts off the least volume in the first octant.

24. Same as Exercise 23 with the plane passing through (2, 1, −1).

25. A triangle has sides x, y, and z and perimeter p. Setting $s = p/2$, a formula from geometry says that the area is $A = [s(s - x)(s - y)(s - z)]^{1/2}$. Find the dimensions of the triangle with perimeter p that has the largest area.

26. A tin can (with top and bottom) has a surface area of S units. Find the dimensions that yield the greatest volume.

27. (*Cobb-Douglas production function*). If $Q(C, L)$ is the quantity of a commodity produced by C units of capital input and L units of labor input, then Q is called a production function. If $Q(C, L) = C^a L^b$ where $0 < a, b < 1$, then Q is called a *Cobb-Douglas production function*. Suppose that $Q(C, L) = C^{1/3} L^{2/3}$, that each unit of capital costs F, each unit of labor costs G, and the total to be spent is H. What capital and labor input will maximize the output subject to the constraint $FC + GL = H$?

15–10   AN APPLICATION TO ECONOMICS
(Optional reading)

Economics is normally concerned with discrete quantities (that is, quantities measured in whole units) such as dollars, people, or automobiles. To apply the methods of calculus, we make the basic assumption that the magnitudes of the quantities under investigation are very large compared to the finest subdivision of the unit involved. Then the functions describing these quantities can be effectively approximated by differentiable functions. The basic assumption holds, for example, in problems involving the general economy of a nation or of a large corporation like General Motors. But even then, the methods of calculus rarely play an essential role in answering specific questions such as, "What level of production will yield General Motors a maximum profit?" Why is this so? Because the task of finding realistic revenue and cost functions is so overwhelming that, by comparison, maximizing their difference is a trivial operation. Nevertheless, calculus does play an important role in formulating and verifying general economic theory. As an illustration, we will discuss the use of Lagrange multipliers in connection with marginal productivity. This and two other examples are contained in an interesting article by Nevison.*

Economists think of the derivative (or partial derivative) as representing the change in a function that results from a change of one unit in a variable. In this connection, they use the term *marginal*. For example, if production X depends on hours of labor L, then dX/dL is the *marginal productivity* of labor and represents the change in production resulting from a change of one hour of labor. The different components that go into the production of a product are called *factors of production*. In this section, we will verify (using a simplified model) the following result from economic theory: *When the cost of a desired level of production is minimized, the marginal productivity of a dollar's worth of a factor of production is the same for all factors of production.*

Our model is a large automobile manufacturing company that has established output goals for a certain period and has minimized its costs. To keep things simple, we assume that labor L, capital K, and steel S are the factors of production. We assume further that the cost of each factor is given; labor costs w dollars per hour, capital costs r dollars per dollar, and steel costs p dollars per ton. Then the total cost C is given by

$$(15.56) \quad C(L, K, S) = wL + rK + pS$$

The amount X of automobiles produced also depends on the factors of production, and we assume that

$$(15.57) \quad X = X(L, K, S)$$

is a continuously differentiable function of the three variables. If the desired output is X_0 automobiles, then the cost C must be minimized subject to the constraint

$$(15.58) \quad X(L, K, S) = X_0$$

We now introduce a Lagrange multipler λ and use the method developed in Section 15–9 to derive the four equations

$$\frac{\partial C}{\partial L} = \lambda \frac{\partial X}{\partial L} \qquad w = \lambda \frac{\partial X}{\partial L}$$

$$\frac{\partial C}{\partial K} = \lambda \frac{\partial X}{\partial K} \qquad r = \lambda \frac{\partial X}{\partial K}$$

$$\frac{\partial C}{\partial S} = \lambda \frac{\partial X}{\partial S} \qquad p = \lambda \frac{\partial X}{\partial S}$$

$$X = X_0 \qquad\qquad X = X_0$$

If the first three equations are solved for $1/\lambda$, we have

$$(15.59) \quad \frac{1}{w}\frac{\partial X}{\partial L} = \frac{1}{r}\frac{\partial X}{\partial K} = \frac{1}{p}\frac{\partial X}{\partial S}$$

*The discussion in this section is adapted from Christopher H. Nevison, *Lagrange Multipliers: Applications to Economics*, UMAP (Unit 270), Newton, Mass., 1979.

The quantity $1/w$ represents hours of work per dollar and $\partial X/\partial L$ represents the marginal productivity of an hour's work. Therefore, $(1/w)(\partial X/\partial L)$ represents the marginal productivity of a dollar's worth of labor. Similarly, $(1/r)(\partial X/\partial K)$ is the marginal productivity of a dollar's worth of capital and $(1/p)(\partial X/\partial S)$ is the marginal productivity of a dollar's worth of steel. Therefore, equations (15.59), which must be satisfied if C is a minimum, translate into the result that *the marginal productivity of a dollar's worth of a factor of production is the same for all factors.* This is what we wanted to show and completes our discussion.

REVIEW EXERCISES

In Exercises 1–6, identify and sketch the graph (in space) of the given equation.

1. $4x^2 + 9y^2 + 36z^2 = 36$
2. $x^2 + 3y^2 - 2z^2 = 6$
3. $x^2 + 4y^2 = 4$ 4. $z^2 = x^2 + y^2$
5. $z = x^2$
6. $x^2 + y^2 + z^2 - 4x - 2z = 0$

7. Find an equation of the circular cylinder generated by the vertical line through $(2, 0, 0)$.

8. Find an equation of the quadric surface generated by revolving the parabola $y = x^2$ (in the xy-plane) about the y-axis.

In Exercises 9–14, find the domain, the range, and sketch the graph of the given functions.

9. $f(x, y) = \sqrt{1 - x^2 - y^2}$
10. $f(x, y) = 1 - x - y$
11. $f(x, y) = x^2 + 4y^2$ 12. $f(x, y) = x^2 - y^2$
13. $f(x, y) = 4 - 2x - y$ 14. $f(x, y) = x^2$

15. Find the gradient of the function in Exercise 9.
16. Find the gradient of the function in Exercise 10.
17. Sketch several level curves of the function in Exercise 11.
18. Sketch several level curves of the function in Exercise 12.
19. In which direction from the point $(1, 1)$ will the function in Exercise 13 increase most rapidly?
20. In which direction from the point $(3, 2)$ will the function in Exercise 14 decrease most rapidly?

In Exercises 21–24, identify and sketch the level surface of f corresponding to the given value of c.

21. $f(x, y, z) = 4x^2 + y^2 + 9z^2$; $c = 36$
22. $f(x, y, z) = x^2 + y^2 - z$; $c = 0$
23. $f(x, y, z) = z - x^2 - y^2$; $c = 1$
24. $f(x, y, z) = z - xy$; $c = 1$
25. Find an equation of the tangent plane of the level surface in Exercise 21 at the point $(1, 2, 2\sqrt{7}/3)$.
26. Find an equation of the tangent plane to the level surface in Exercise 22 at the point $(1, 1, 2)$.
27. Find the highest and lowest points (if there are any) of the level surface in Exercise 23.
28. Same as Exercise 27 for the surface in Exercise 24.

In Exercises 29–36, find all first partials of the given functions.

29. $f(x, y) = x^2y - xy^2$
30. $f(x, y) = x^2y + 2x^2y^2$
31. $f(x, y) = x \sin y$ 32. $f(x, y) = e^{xy}$
33. $f(x, y, z) = xy^z$; $(y > 0)$
34. $f(x, y, z) = \ln \sqrt{x^2 + y^2 + z^2}$
35. $f(x, y, z) = e^{xy} \cos xz$ 36. $f(x, y, z) = \dfrac{xy}{z}$

37. Find the local extreme values (if any) for the function in Exercise 29.
38. Same as Exercise 37 for the function in Exercise 30.
39. For the function in Exercise 31, let $x = t^2$ and $y = t^3 + t^2$, and find df/dt.
40. For the function in Exercise 32, let $x = \cos t$ and $y = \sin t$, and find df/dt.
41. For the function in Exercise 33, find $\partial^3 f/\partial y \partial x \partial z$.
42. For the function in Exercise 34, find $\partial^2 f/\partial z^2$.
43. For the function in Exercise 35, let $x = u - v$, $y = u + v$, and $z = uv$, and find $\partial f/\partial u$.

44. For the function in Exercise 36, let x, y, and z be as in Exercise 43, and find $\partial f/\partial v$.

In Exercises 45–46, assume that y is defined implicitly as a differentiable function of x, and find dy/dx.

45. $x^2y^3 - 3x \sin y + xe^y = 0$

46. $x^y + x \ln y = 0$

In Exercises 47–48, assume that z is defined implicitly as a differentiable function of x and y, and find $\partial z/\partial x$ and $\partial z/\partial y$.

47. $xe^{yz} - 2x^2z^3 + y^2z = 0$

48. $x \cos yz + y \sin xz = z$

In Exercises 49–52, find the differentials of the given functions.

49. $z = 2x^2 - 3xy + 5y^2$ **50.** $z = x^2y + \sin xy$

51. $w = xe^{yz}$ **52.** $w = \tan^{-1}xyz$

In Exercises 53–56, find the directional derivatives of f at the given point P in the indicated directions. Also find the direction in which the values of f decrease most rapidly.

53. $f(x, y) = 4x^2y$; $P(2, 1)$; $\mathbf{a} = 3\mathbf{i} + 4\mathbf{j}$

54. $f(x, y) = 3x^2 + y^2$; $P(-1, -2)$; $\mathbf{a} = \mathbf{i} - 2\mathbf{j}$

55. $f(x, y, z) = xy \tan^{-1}z$; $P(1, 1, 1)$; $\mathbf{a} = \mathbf{i} + \mathbf{j} + \mathbf{k}$

56. $f(x, y, z) = 5 - xyz$; $P(4, 1, -2)$; $\mathbf{a} = 2\mathbf{i} - \mathbf{j} + 2\mathbf{k}$

---

57. Find symmetric equations for all lines through the origin that are normal to the surface $xy + z = 1$.

58. Use the method of least squares to find the line that best fits the points $(-2, 1)$, $(0, 0)$, $(1, 1)$, $(2, 3)$, and $(3, 3)$.

59. Find the points on the intersection of $z^2 = x^2 + y^2$ and $x + y - z = -1$ that are nearest and farthest from the origin.

60. The bottom and top of a rectangular box cost twice as much per square foot as do the sides. What dimensions minimize the cost if the volume is to be 8 cu ft?

61. Let S be the sphere $x^2 + y^2 + z^2 = 1$ and let T be the sphere $(x - a)^2 + (y - b)^2 + (z - c)^2 = 1$. For what values of a, b, and c will T intersect S orthogonally?

62. The temperature at any point of the disc $x^2 + y^2 \leq 1$ is $T(x, y) = x^2 + 2y^2 - x$. In which direction from the point $(1/\sqrt{2}, 1/\sqrt{2})$ should you move to heat up the fastest?

63. What are the maximum and minimum temperatures on the disc in Exercise 62?

64. Find an equation of the plane that passes through $(1, 1, 1)$ and cuts off the least volume in the first octant.

65. Let $z = f(x, y)$ have continuous partials. Suppose that $P(x_1, y_1, z_1)$ is the point on the graph of f that is nearest to the point $Q(x_0, y_0, z_0)$ not on the graph. Show that $\overrightarrow{PQ}$ is normal to the graph at P.

16

MULTIPLE INTEGRALS

In this chapter, we extend the notion of an integral to functions of two and three variables. Section 16–1 contains a detailed development of the double integral for functions of two variables. Then, in Section 16–5, the extension to triple integrals for functions of three variables is not at all difficult. Both types, double and triple integrals, are evaluated by means of iterated integrals, which are developed in Section 16–2. The remaining sections deal mainly with applications, which include area, volume, mass, moments, moments of inertia, and surface area.

16–1 DOUBLE INTEGRALS

Our goal is to define an integral for a function of two variables. With one exception, right at the beginning, the definition is strikingly similar to that of the integral defined in Chapter 5, complete with partitions, Riemann sums, and so on.

The exception is this: on the line, integrals are defined over intervals; $\int_a^b f(x)\,dx$. The obvious extension to the plane is to define integrals over rectangles. But it turns out that the regions in the plane that are most useful for integration are not rectangles. Rather, they are regions with, perhaps, irregular shapes.

Figure 16–1 shows two regions. The first is the set of points (x, y) with $a \le x \le b$ and $h_1(x) \le y \le h_2(x)$ where h_1 and h_2 are continuous functions of x.

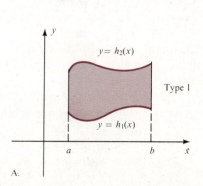

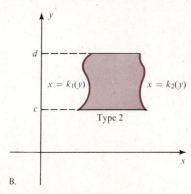

FIGURE 16–1. Regions of (*A*) type 1 and (*B*) type 2.

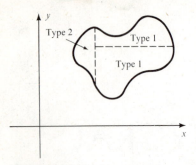

FIGURE 16–2. Breaking up a region into a finite number of regions of types 1 and 2.

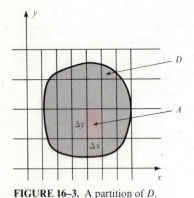

FIGURE 16–3. A partition of D.

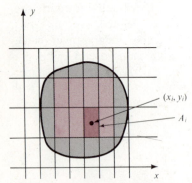

FIGURE 16–4. An inner partion of D.

We call this a region of **type 1.** The second is the set of points (x, y) with $k_1(y) \leq x \leq k_2(y)$ and $c \leq y \leq d$ where k_1 and k_2 are continuous functions of y. We call this a region of **type 2.** Any region that can be broken up into a finite number of type 1 or type 2 regions, we call an **elementary region** (Figure 16–2). For the rest of this chapter, we will consider only those functions whose domains are elementary regions.

The discussion from here on parallels the one in Chapter 5. Let D be the domain of $z = f(x, y)$. We *partition* D by forming a grid with horizontal lines, Δy apart, and vertical lines, Δx apart (Figure 16–3). Thus, D is partitioned into small rectangles A whose area is denoted by ΔA; that is,

$$\Delta A = \Delta x \, \Delta y$$

Those rectangles that are not completely contained in D are discarded, leaving what we call a (**regular***) **inner partition** P of D (Figure 16–4). Now, we number the rectangles of P in some order $A_1, A_2, \ldots, A_n$, choose one point (x_i, y_i) in each A_i, and form a **Riemann sum** corresponding to P

$$(16.1) \qquad R_P = \sum_{i=1}^{n} f(x_i, y_i) \Delta A$$

where ΔA is the area of A_i (all the rectangles have the same area). Except for f being a function of two variables instead of one and ΔA being an area instead of the length of a subinterval, this is exactly the same definition of a Riemann sum given in Section 5–2. We continue the parallel by letting $\|P\|$ denote the length of the diagonal of the rectangles of P. If the limit of the Riemann sums as $\|P\| \to 0$ exists, we call the limit the *double integral of f over D.*

DEFINITION

Let $z = f(x, y)$ be defined on D. Then the **double integral of f over** D, denoted by $\iint_D f(x, y) \, dA$, is defined by

$$(16.2) \qquad \int \int_D f(x, y) \, dA = \lim_{\|P\| \to 0} \sum_{i=1}^{n} f(x_i, y_i) \Delta A$$

provided this limit exists.

If the integral in (16.2) exists, then f is said to be **integrable over** D. It can be shown that if D is an elementary region and f is continuous on D, then f is integrable over D.

If $f(x, y) \geq 0$ on D, then each term $f(x_i, y_i) \Delta A$ of a Riemann sum is the volume of a parallelepiped with base A and height $f(x_i, y_i)$ as shown in Figure 16–5.

---

*An inner partition is *regular* if all rectangles have the same length and the same width. This is not at all necessary, but it makes computations a little bit easier.

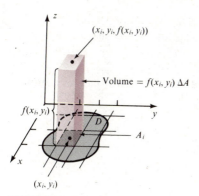

FIGURE 16–5. Each term in a Riemann sum is the volume of a parallelepiped.

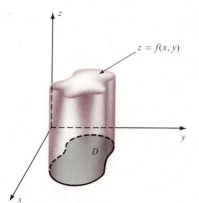

FIGURE 16–6. The volume of the solid is defined to be $V = \iint_D f(x, y) \, dA$.

The Riemann sum

$$R_P = \sum_{i=1}^{n} f(x_i, y_i) \Delta A$$

is an approximation to the volume of the solid under the graph of f; that is, the solid between the surface $z = f(x, y)$ and the domain D. If f is continuous, then as $\|P\| \to 0$, the approximations tend to a limit which we define as the volume of the solid (Figure 16–6).

(16.3) $$\text{Volume} = \iint_D f(x, y) \, dA$$

EXAMPLE 1   Let $f(x, y) = 2x + y$ be defined on the set of points (x, y) with $0 \le x \le 2$ and $0 \le y \le \sqrt{4 - x^2}$ (Figure 16–7A). Let P be the regular inner partition of squares $1/2$ units on a side (Figure 16–7B). Approximate the volume under the graph of f by computing the Riemann sum R_P taking (x_i, y_i) to be the center of the ith square.

Solution   There are eight squares, each with area $1/4$. The center of A_1 is $(1/4, 5/4)$, the center of A_2 is $(3/4, 5/4)$, and so on. Therefore,

$$R_P = \sum_{i=1}^{8} f(x_1, y_1) \Delta A$$

$$= \left[2\left(\frac{1}{4}\right) + \frac{5}{4} \right]\left(\frac{1}{4}\right) + \left[2\left(\frac{3}{4}\right) + \frac{5}{4} \right]\left(\frac{1}{4}\right) + \cdots + \left[2\left(\frac{5}{4}\right) + \frac{1}{4} \right]\left(\frac{1}{4}\right)$$

$$= \frac{66}{16} = \frac{33}{8}$$

This is an approximation to the volume under the graph of f.   ∎

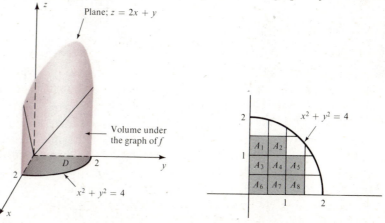

FIGURE 16–7. Example 1.

The following statements for double integrals are similar to those for the integral defined in Chapter 5. The proofs are also similar and are omitted here.

(16.4)   Continuous functions are integrable.

(16.5)   If f is integrable on D consisting of two (or more) nonoverlapping elementary regions D_1 and D_2, then

$$\int\int_D f(x, y) \, dA = \int\int_{D_1} f(x, y) \, dA + \int\int_{D_2} f(x, y) \, dA$$

(16.6) $\int\int_D cf(x, y) \, dA = c \int\int_D f(x, y) \, dA$, c a constant

(16.7) $\int\int_D \left[f(x, y) + g(x, y) \right] dA = \int\int_D f(x, y) \, dA + \int\int_D g(x, y) \, dA$

(16.8)   If $f(x, y) \geq 0$ on D, then $\int\int_D f(x, y) \, dA \geq 0$.

Just as with integrals for functions of one variable, the use of double integrals would be severely limited if they had to be evaluated by finding the limit of Riemann sums. Fortunately, there is another method called the method of *iterated integrals;* it is discussed in the next section.

EXERCISES

In Exercises 1–6, the region D is the square with vertices $(0, 0), (0, 1), (1, 1)$ and $(1, 0)$. The partition P is the set of squares $1/2$ units on a side. Compute the Riemann sum R_P for the given function and the indicated choice of points $(x_i \, y_i)$.

1. $f(x, y) = 3x + 2y$; (x_i, y_i) is the center of A_i.

2. $f(x, y) = 3x + 2y$; (x_i, y_i) is the lower left corner of A_i.

3. $f(x, y) = x^2 y$; (x_i, y_i) is the lower right corner of A_i.

4. $f(x, y) = x^2 y$; (x_i, y_i) is the center of A_i.

5. $f(x, y) = 1$; (x_i, y_i) is the center of A_i.

6. $f(x, y) = x$; (x_i, y_i) is the center of A_i.

7. In Exercise 5, $f(x, y) = 1$. What is the interpretation of $\int\int_D f(x, y) \, dA = \int\int_D dA$ in this case? Does this agree with your answer to Exercise 5?

8. In Exercise 6, $f(x, y) = x$. What is the exact value of $\int\int_D f(x, y) \, dA = \int\int_D x \, dA$? (Draw a picture.) Compare this with your answer to Exercise 6.

9. Let D be the rectangle with vertices $(0, 0), (0, 2), (1, 2)$, and $(1, 0)$, and let $f(x, y) = y$. What is the exact value of $\int\int_D f(x, y) \, dA = \int\int_D y \, dA$? (Draw a picture.)

In Exercises 10–14, let D and P be as in Example 1 (Figure 16–7). Compute R_P for the given function and the indicated choice of points (x_i, y_i).

10. $f(x, y) = y$; (x_i, y_i) is the center of A_i.

11. $f(x, y) = x$; (x_i, y_i) is the center of A_i.

12. $f(x, y) = x + 2y$; (x_i, y_i) is the upper right corner of A_i.

13. $f(x, y) = 1$; (x_i, y_i) is the upper right corner of A_i.

14. $f(x, y) = xy^2$; (x_i, y_i) is the center of A_i.

15. In Exercise 13, $f(x, y) = 1$. What is the interpretation of $\int\int_D f(x, y) \, dA = \int\int_D dA$? Why doesn't your answer to Exercise 13 compare too favorably with this interpretation? What can you do to make the comparison better and better?

16. If $f(x, y) \leq 0$ on D, what interpretation would you put on $\int\int_D f(x, y) \, dA$?

In Exercises 17–20, let D be the unit disc $x^2 + y^2 \le 1$ and compute the (exact) value of $\iint_D f(x, y)\, dA$. (Draw pictures and compute the indicated volumes.)

17. $f(x, y) = \sqrt{x^2 + y^2}$

18. $f(x, y) = \sqrt{1 - x^2 - y^2}$

19. $f(x, y) = 1 - \sqrt{1 - x^2 - y^2}$

20. $f(x, y) = x$

16-2 ITERATED INTEGRALS

Except for the most elementary cases, it is impractical to evaluate a double integral using definition (16.2). This section describes a method by which the evaluation of a double integral reduces to the evaluation of two ordinary integrals. The central idea is borrowed from our experience with partial derivatives.

Suppose $z = f(x, y)$ is continuous on the rectangle R shown in Figure 16–8; then x varies from a to b and y varies from c to d. The integral

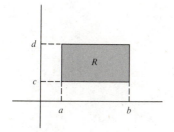

FIGURE 16–8. The domain of f.

$$(16.9) \qquad \int_c^d f(x, y)\, dy$$

is called a *partial definite integral of f with respect to y*. The symbol dy indicates that y is the variable, and x is to be treated as a constant. If x is treated as a constant, then the integral (16.9) can be evaluated in the usual way using the Fundamental Theorem of Calculus.

EXAMPLE 1 Let $f(x, y) = 3xy^2 + x^2 + 2y$, $c = 0$, and $d = 1$. Then

$$\int_c^d f(x, y)\, dy = \int_0^1 (3xy^2 + x^2 + 2y)\, dy$$

$$= \Big[xy^3 + x^2y + y^2 \Big]_0^1 \qquad (x \text{ is constant})$$

$$= x + x^2 + 1 \qquad \blacksquare$$

As this example illustrates, evaluation of (16.9) yields a function of x alone, which we designate by $A(x)$. Thus,

$$(16.10) \quad A(x) = \int_c^d f(x, y)\, dy$$

It can be shown that if f is continuous, then A is continuous, so now we can integrate A with respect to x. Since x varies from a to b, we have

$$\int_a^b A(x)\, dx = \int_a^b \left[\int_c^d f(x, y)\, dy \right] dx$$

The integral on the right is called an **iterated integral.** It is usually written without the brackets:

$$(16.11) \qquad \int_a^b \int_c^d f(x, y)\, dy\, dx$$

EXAMPLE 2   Evaluate $\int_0^3 \int_1^2 (2xy + 3y^2)\, dy\, dx$

Solution   By definition,

$$\int_0^3 \int_1^2 (2xy + 3y^2)\, dy\, dx = \int_0^3 \left[\int_1^2 (2xy + 3y^2)\, dy \right] dx$$

$$= \int_0^3 \left[xy^2 + y^3 \right]_1^2 dx \qquad (x \text{ is constant})$$

$$= \int_0^3 [(4x + 8) - (x + 1)]\, dx$$

$$= \int_0^3 (3x + 7)\, dx$$

$$= \left[\frac{3x^2}{2} + 7x \right]_0^3 = \frac{69}{2} \quad \blacksquare$$

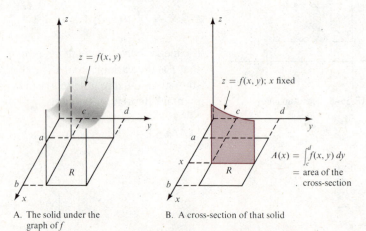

A. The solid under the graph of f

B. A cross-section of that solid

FIGURE 16–9 Geometric interpretation of the iterated integral; $\int_a^b \int_c^d f(x, y)\, dy\, dx = \int_a^b A(x)\, dx =$ volume.

Figure 16–9 illustrates a geometric interpretation of an iterated integral with f continuous and nonnegative. For each fixed x, $a \le x \le b$,

$$A(x) = \int_c^d f(x, y)\, dy$$

is the *area* of a cross-section of the solid under the graph of f. It follows from the methods employed in Section 6–2 (Volume by Slicing) that $\int_a^b A(x)\, dx$ is the volume of the solid under the graph of f. But, for a continuous nonnegative function, the double integral also yields the volume of this solid. Therefore, in this case, we have that

$$(16.12) \qquad \boxed{\int\int_R f(x, y)\, dA = \int_a^b \int_c^d f(x, y)\, dy\, dx}$$

This equality provides a relatively painless method of evaluating double integrals.

Now, let's start over again. We could have just as easily defined a partial definite integral of f with respect to x

$$\int_a^b f(x, y)\, dx$$

The symbol dx indicates that x is the variable and y is to be treated as a constant. The value will be a continuous function of y alone and, integrating that with respect to y yields the iterated integral

(16.13) $\qquad \displaystyle\int_c^d \left[\int_a^b f(x, y)\, dx \right] dy \qquad$ usually written as $\qquad \displaystyle\int_c^d \int_a^b f(x, y)\, dx\, dy$

The geometric interpretation is the same except that the cross-sections are now parallel to the xz-plane. As before, the value of the iterated integral (16.13) is the volume of the solid under the graph of f. It follows that if f is a nonnegative continuous function defined on a rectangle, then both iterated integrals are equal to the double integral, hence, equal to each other

(16.14) $\qquad \displaystyle\iint_R f(x, y)\, dA = \int_a^b \int_c^d f(x, y)\, dy\, dx = \int_c^d \int_a^b f(x, y)\, dx\, dy$

Changing from one of the iterated integrals displayed above to the other is called **reversing the order of integration.**

EXAMPLE 3   Reverse the order of integration in Example 2 and evaluate.

Solution   The result of reversing the order of integration is $\int_1^2 \int_0^3 (2xy + 3y^2)\, dx\, dy$ which, by definition, is

$$\int_1^2 \left[\int_0^3 (2xy + 3y^2)\, dx \right] dy = \int_1^2 \left[x^2 y + 3y^2 x \right]_0^3 dy \qquad (y \text{ is constant})$$

$$= \int_1^2 (9y + 9y^2)\, dy$$

$$= \left[\frac{9}{2} y^2 + 3y^3 \right]_1^2$$

$$= (18 + 24) - \left(\frac{9}{2} + 3 \right) = \frac{69}{2}$$

This agrees with the answer in Example 2.   ∎

So far, the domains of the integrands have been rectangles. But iterated integrals can also be defined over elementary regions. Suppose that $z = f(x, y)$ is continuous on a region of type 1 or type 2 (Figure 16–1 in the last section).

For type 1, $a \le x \le b$ and $h_1(x) \le y \le h_2(x)$. If we fix x and integrate first with respect to y, the result is a continuous function of x, which is then integrated with respect to x. That is,

$$(16.15) \quad \int_a^b \int_{h_1(x)}^{h_2(x)} f(x, y)\, dy\, dx = \int_a^b \left[\int_{h_1(x)}^{h_2(x)} f(x, y)\, dy \right] dx$$

For a domain of type 2, $c \le y \le d$ and $k_1(y) \le x \le k_2(y)$. Now we fix y and integrate first with respect to x:

$$(16.16) \quad \int_c^d \int_{k_1(y)}^{k_2(y)} f(x, y)\, dx\, dy = \int_c^d \left[\int_{k_1(y)}^{k_2(y)} f(x, y)\, dx \right] dy$$

EXAMPLE 4 Let $f(x, y) = 3x + 2y$ be defined on the region bounded by the line $y = x$ and the curve $y = x^2$ (Figure 16–10). This is a region of type 1 and also of type 2. Evaluate both iterated integrals (16.15) and (16.16).

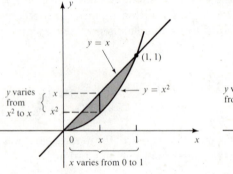

FIGURE 16–10. Example 4. Finding the limits of integration. (A) For each x with $0 \le x \le 1$, y varies from x^2 to x, so the integral is $\int_0^1 \int_{x^2}^x f(x, y)\, dy\, dx$. (B) For each y with $0 \le y \le 1$, x varies from y to $\sqrt{y}$, so the integral is $\int_0^1 \int_y^{\sqrt{y}} f(x, y)\, dx\, dy$.

A. Type 1

B. Type 2

Solution As a type 1 region (Figure 16–10A), $a = 0$ and $b = 1$. For each x with $0 \le x \le 1$, y ranges from the lower boundary $y = x^2$ to the upper boundary $y = x$. Therefore, by (16.15),

$$\int_0^1 \int_{x^2}^x (3x + 2y)\, dy\, dx = \int_0^1 \left[3xy + y^2 \right]_{x^2}^x dx \qquad (x \text{ is constant})$$

$$= \int_0^1 [(3x^2 + x^2) - (3x^3 + x^4)]\, dx$$

$$= \int_0^1 (4x^2 - 3x^3 - x^4)\, dx$$

$$= \left[\frac{4}{3}x^3 - \frac{3}{4}x^4 - \frac{1}{5}x^5 \right]_0^1 = \frac{23}{60}$$

As a type 2 region (Figure 16–10B), $c = 0$ and $d = 1$. For each y with $0 \le y \le 1$, x ranges from the left boundary $x = y$ to the right boundary $x = \sqrt{y}$. Therefore, by (16.16),

$$\int_0^1 \int_y^{\sqrt{y}} (3x + 2y)\, dx\, dy = \int_0^1 \left[\frac{3}{2}x^2 + 2yx\right]_y^{\sqrt{y}} dy \qquad (y \text{ is constant})$$

$$= \int_0^1 \left[\left(\frac{3}{2}y + 2y\sqrt{y}\right) - \left(\frac{3}{2}y^2 + 2y^2\right)\right] dy$$

$$= \int_0^1 \left(\frac{3}{2}y + 2y^{3/2} - \frac{7}{2}y^2\right) dy$$

$$= \left[\frac{3}{4}y^2 + \frac{4}{5}y^{5/2} - \frac{7}{6}y^3\right]_0^1 = \frac{23}{60}$$

The two answers agree. ■

Although the values of both iterated integrals are not always the same (see Exercises 40 and 41), the preceding discussion and examples provide convincing evidence that in most cases the values of the double integral and both iterated integrals are all equal. The formal proof of the statement below can be found in advanced texts.

(16.17) | If $z = f(x, y)$ is continuous on an elementary region, then the double integral and both iterated integrals exist and all three have the same value.

Given an iterated integral, it is often the case that reversing the order of integration transforms a difficult, or even impossible, integration problem into a relatively easy one. And, by (16.17), both integrals have the same value, provided f is continuous and its domain is elementary.

Reversing the order of integration is a two-step process.

(1) If the domain is not explicitly given, then use the limits of integration to sketch the intended domain of the integrand.

(2) Use the sketch to determine the new limits of integration in the reverse order.

All of this is illustrated in the next example.

EXAMPLE 5 Evaluate $\int_0^2 \int_y^2 2ye^{x^3}\, dx\, dy$.

Solution There is no elementary method for integrating e^{x^3} with respect to x, so we will try to reverse the order of integration. The given order $dx\, dy$ indicates that the domain is of type 2. The limits of integration indicate that y ranges from 0 to 2 and for each fixed y, x ranges from $x = y$ on the left to $x = 2$ on the right. Therefore, the domain of the integrand is the triangular region shown in Figure 16–11A. This region is also of type 1, so we can reverse the order of integration to $dy\, dx$. Now x will range from 0 to 2 and for each fixed x, y ranges from the

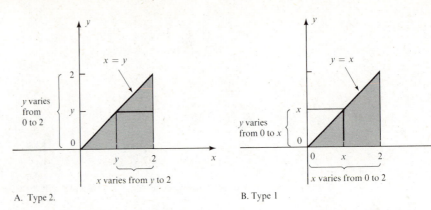

FIGURE 16–11. Example 5.

A. Type 2.

B. Type 1

lower boundary $y = 0$ to the upper boundary $y = x$ (Figure 16–11B). Thus, by (16.17),

$$\int_0^2 \int_y^2 2ye^{x^3} \, dx \, dy = \int_0^2 \int_0^x 2ye^{x^3} \, dy \, dx$$

$$= \int_0^2 \left[y^2 e^{x^3} \right]_0^x dx \qquad \text{(x is constant)}$$

$$= \int_0^2 x^2 e^{x^3} \, dx \qquad \text{(Basic form $\int e^u \, du$)}$$

$$= \frac{1}{3} e^{x^3} \Big|_0^2 = \frac{1}{3}(e^8 - 1) \qquad \blacksquare$$

Sometimes the shape of the region dictates what order of integration to use.

EXAMPLE 6 Let D be the region in the first quadrant bounded by the curves $y = x^2$ and $y = (x^2 + 1)/2$. Evaluate the double integral $\iint_D x \, dA$.

Solution By (16.17), we can use either iterated integral to evaluate the double integral. The first step is to sketch the domain; this is done in Figure 16–12. As the figure indicates, D consists of two regions of type 2, but only one of type 1. Therefore, it is easier to integrate in the order $dy \, dx$. For each fixed x with $0 \le x \le 1$, y ranges from the lower boundary $y = x^2$ to the upper boundary $y = (x^2 + 1)/2$. Therefore,

$$\iint_D x \, dA = \int_0^1 \int_{x^2}^{(x^2+1)/2} x \, dy \, dx$$

$$= \int_0^1 \left[xy \right]_{x^2}^{(x^2+1)/2} dx$$

$$= \int_0^1 \left(\frac{x^3 + x}{2} - x^3 \right) dx$$

$$= \left[-\frac{x^4}{8} + \frac{x^2}{4} \right]_0^1 = \frac{1}{8} \qquad \blacksquare$$

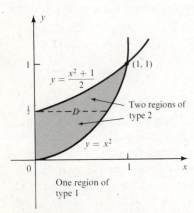

FIGURE 16–12. Example 6.

EXERCISES

To evaluate the integrals in the exercises below, you can use integration by parts, change of variables, trigonometric substitution, or any other method previously discussed.

In Exercises 1–20, evaluate the iterated integrals *in the given order*.

1. $\displaystyle\int_0^1 \int_0^3 2y^2 x \, dx \, dy$

2. $\displaystyle\int_{-1}^0 \int_0^2 (x + y) \, dy \, dx$

3. $\displaystyle\int_{-2}^1 \int_0^2 e^x \, dy \, dx$

4. $\displaystyle\int_3^4 \int_0^1 \cos y \, dx \, dy$

5. $\displaystyle\int_{-1}^1 \int_y^3 y^2 x \, dx \, dy$

6. $\displaystyle\int_{-2}^1 \int_0^x 2xy \, dy \, dx$

7. $\displaystyle\int_2^4 \int_0^x e^x \, dy \, dx$

8. $\displaystyle\int_0^3 \int_y^2 3x^2 y \, dx \, dy$

9. $\displaystyle\int_0^1 \int_{-x}^x x \, dy \, dx$

10. $\displaystyle\int_1^2 \int_{-x}^x x \sin y \, dy \, dx$

11. $\displaystyle\int_0^\pi \int_0^y \sin x \cos y \, dx \, dy$

12. $\displaystyle\int_0^2 \int_0^x e^x \, dy \, dx$

13. $\displaystyle\int_0^{\pi/2} \int_0^{\sin y} \frac{1}{\sqrt{1 - x^2}} dx \, dy$

14. $\displaystyle\int_0^1 \int_0^{\arctan y} \sec^2 x \, dx \, dy$

15. $\displaystyle\int_1^2 \int_x^{x^2} xy \, dy \, dx$

16. $\displaystyle\int_0^1 \int_{x^2}^x xy \, dy \, dx$

17. $\displaystyle\int_0^1 \int_0^x x^2 e^{xy} \, dy \, dx$

18. $\displaystyle\int_1^2 \int_0^x e^{x^2} \, dy \, dx$

19. $\displaystyle\int_1^e \int_0^x \ln x \, dy \, dx$

20. $\displaystyle\int_0^1 \int_y^1 \frac{1}{1 + y^2} dx \, dy$

In Exercises 21–28, go back to the indicated exercise above, sketch the intended domain, reverse the order of integration, and evaluate.

21. Exercise 1

22. Exercise 3

23. Exercise 5

24. Exercise 7

25. Exercise 9

26. Exercise 11

27. Exercise 13

28. Exercise 15

In Exercises 29–36, evaluate the iterated integrals (either order may be used; choose the one you think is easier).

29. $\displaystyle\int_1^2 \int_1^3 x e^{xy} \, dx \, dy$

30. $\displaystyle\int_0^1 \int_1^2 y \sin xy \, dy \, dx$

31. $\displaystyle\int_{-1}^1 \int_0^2 y \sin x^2 \, dx \, dy$

32. $\displaystyle\int_0^1 \int_0^{\arccos x} e^{\sin y} \, dy \, dx$

33. $\displaystyle\int_0^1 \int_x^1 e^{y^2} \, dy \, dx$

34. $\displaystyle\int_0^9 \int_{\sqrt{x}}^3 \sin(y^3) \, dy \, dx$

35. $\displaystyle\int_1^e \int_0^{\ln x} y \, dy \, dx$

36. $\displaystyle\int_1^2 \int_0^{\ln x} dy \, dx$

In Exercises 37–38, let $z = f(x, y)$ be continuous on D. In each case, break D into two regions of type 1, and write the double integral of f over D as the sum of two iterated integrals.

37. D is the region in the first quadrant bounded by the curves $y = \sqrt{x}$ and $y = \sqrt{2x - 16}$.

38. D is the region in the first quadrant bounded by curves $y = x^2$ and $y = 8 - x^2$, the x-axis, and the line $x = 5/2$.

39. *(Area).* If $f(x, y) \geq 0$, then $\iint_D f(x, y) \, dA$ is the volume of the solid under the graph of f. The volume of a solid of constant height 1 is just the area of the base. Therefore, $\iint_D dA$ is the area of D. Using iterated integrals find the area of the regions described in Exercises 37 and 38.

Statement (16.17) says that if f is continuous on an elementary region, then the double and iterated integrals exist and are equal. The next two exercises illustrate what can happen if f is not continuous. In both exercises, the domain D is the square with vertices $(0, 0)$, $(0, 1)$, $(1, 1)$, and $(1, 0)$.

40. Let

$$f(x, y) = \begin{cases} \dfrac{x - y}{(x + y)^3} & \text{if } (x, y) \neq (0, 0) \\ 0 & \text{if } (x, y) = (0, 0) \end{cases}$$

Show that

$$\int_0^1 \int_0^1 f(x, y) \, dy \, dx = \frac{1}{2}$$

and

$$\int_0^1 \int_0^1 f(x, y) \, dx \, dy = -\frac{1}{2}$$

Hint: Write

$$\frac{x - y}{(x + y)^3} \text{ as } \frac{x + y}{(x + y)^3} - \frac{2y}{(x + y)^3}$$

or

$$\frac{2x}{(x + y)^3} - \frac{x + y}{(x + y)^3}$$

depending on the order of integration.

41. Let $f(x, y) = 0$ if x is irrational, $f(x, y) = 1$ if x is rational and $0 \leq y \leq 1/2$, and $f(x, y) = -1$ if x is rational and $1/2 < y \leq 1$. Show that $\int_0^1 \int_0^1 f(x, y) \, dy \, dx = 0$, but that $\int_0^1 \int_0^1 f(x, y) \, dx \, dy$ does not exist!

16–3   APPLICATIONS OF DOUBLE INTEGRALS

If $z = f(x, y)$ is continuous and nonnegative on an elementary region D, then the volume V of the solid under the graph of f is defined to be

(16.18) $$V = \int \int_D f(x, y) \, dA$$ (Section 16–1)

EXAMPLE 1   Find the volume in the first octant bounded by the elliptic paraboloid $z = x^2 + 3y^2$ and the plane $x + y = 1$.

Solution   Figure 16–13 shows the solid; it is bounded above by the surface $z = x^2 + 3y^2$ and its base is the triangular region D determined by the plane $x + y = 1$. The volume is given by the double integral over D of $f(x, y) = x^2 + 3y^2$

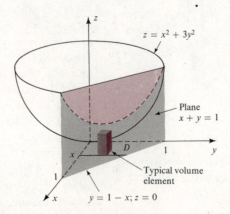

FIGURE 16–13. Example 1.

which, in turn, is evaluated by an iterated integral. To determine the limits of integration, we note that for any x with $0 \le x \le 1$, y will vary from $y = 0$ to $y = 1 - x$. Thus,

$$V = \int \int_D (x^2 + 3y^2)\, dA = \int_0^1 \int_0^{1-x} (x^2 + 3y^2)\, dy\, dx$$

$$= \int_0^1 \left[x^2 y + y^3 \right]_0^{1-x} dx$$

$$= \int_0^1 (x^2 - x^3 + 1 - 3x + 3x^2 - x^3)\, dx$$

$$= \int_0^1 (1 - 3x + 4x^2 - 2x^3)\, dx$$

$$= \left[x - \frac{3}{2}x^2 + \frac{4}{3}x^3 - \frac{1}{2}x^4 \right]_0^1 = \frac{1}{3} \quad \blacksquare$$

Let us quickly review how the definition of volume (16.18) comes about. The region D is covered by a grid, forming very small rectangles A of length Δx and width Δy. The rectangles not completely contained in D are discarded; those remaining form an inner partition P. The rectangles of P are numbered A_1, $A_2, \ldots, A_n$, and a point (x_i, y_i) is picked from each A_i. If ΔA_i represents the area of A_i, then the product $f(x_i, y_i)\Delta A_i$ is the volume of a parallelepiped with base A_i and height $f(x_i, y_i)$; see Figure 16–5 in Section 16–1. This is an approximation to the volume between the graph of f and the rectangle A_i. Therefore, the Riemann sum

$$R_P = \sum_{i=1}^{n} f(x_i, y_i)\Delta A_i$$

is an approximation to the volume V between the graph of f and the region D. The definition of volume as a double integral is the result of taking the limit of R_P as $\|P\| \to 0$.

The pattern described above is repeated so often in applications that from now on we will use a shortened version that is best demonstrated by example.

Area

Let D be an elementary region in the plane. We pick a typical small rectangle contained in D and denote its length by dx, its width by dy, and its area by dA (Figure 16–14). Thus,

$$dA = dx\, dy$$

If the areas of all such rectangles are added, the sum is an approximation to the area of D. Furthermore, the accuracy of the approximations increases as the rectangles get smaller. Therefore, we define the area A of D by

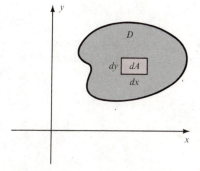

FIGURE 16–14. A typical rectangle.

$$(16.19) \quad \boxed{A = \int\int_D dA}$$

EXAMPLE 2   Use double integration to find the area of the region D bounded by the curves $y = x^2$ and $y = (x^2 + 1)/2$.

Solution   The region is sketched in Figure 16–15. Using symmetry, we can find the area in the first quadrant and multiply by 2. By (16.19) the area is a double integral which, in turn, is evaluated by an iterated integral. For any x, with $0 \le x \le 1$, y varies from $y = x^2$ to $y = (x^2 + 1)/2$. Therefore,

$$A = 2 \int_0^1 \int_{x^2}^{(x^2+1)/2} dy\, dx = 2 \int_0^1 \left(\frac{x^2 + 1}{2} - x^2 \right) dx$$

$$= \frac{2}{3} \quad \blacksquare$$

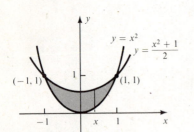

FIGURE 16–15. Example 2.

Mass, Moments, and Center of Mass

Let D be a thin plate, or lamina, in the shape of an elementary region in the plane. We suppose that the mass is continuously distributed over D and that the mass of a typical rectangle A contained in D with area dA is approximated by the product

$$\delta(x, y)\, dA$$

where $\delta(x, y)$ is the density at a typical point (x, y) in A. The sum of all such products approximates the mass of D and the accuracy of the approximations increases as the rectangles get smaller. Therefore, we define the mass M of D by

$$(16.20) \quad \boxed{M = \int\int_D \delta(x, y)\, dA}$$

The *(first) moment* of a point mass about a line is the mass times the distance to the line. Since the distance from (x, y) to the x-axis is y, the moment of a typical rectangle about the x-axis is approximately $y\, \delta(x, y)\, dA$. Therefore, we define the **(first) moment of D about the x-axis** by

$$(16.21) \quad \boxed{M_x = \int\int_D y\, \delta(x, y)\, dA}$$

Similarly, the **(first) moment of D about the y-axis** is

(16.22)
$$M_y = \int \int_D x \, \delta(x, y) \, dA$$

It follows that the coordinates $(\bar{x}, \bar{y})$ of the **center of mass** are

(16.23)
$$\bar{x} = \frac{M_y}{M} = \frac{\displaystyle\int \int_D x \, \delta(x, y) \, dA}{\displaystyle\int \int_D \delta(x, y) \, dA}$$

$$\bar{y} = \frac{M_x}{M} = \frac{\displaystyle\int \int_D y \, \delta(x, y) \, dA}{\displaystyle\int \int_D \delta(x, y) \, dA}$$

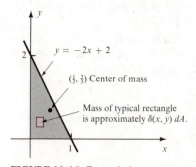

$y = -2x + 2$

$(\frac{1}{2}, \frac{2}{3})$ Center of mass

Mass of typical rectangle is approximately $\delta(x, y) \, dA$.

FIGURE 16-16. Example 3.

EXAMPLE 3   A lamina D has the triangular shape shown in Fgure 16–16 and its density at (x, y) is $\delta(x, y) = x$. Find the center of mass of D.

Solution   In general, to find the center of mass, we have to compute M, M_x, and M_y. Each is given by a double integral that is evaluated by an iterated integral. But in this example, we can do some preliminary reasoning and, per-haps, eliminate some work. The density depends only on x; as far as y is con-cerned, the density is constant. Therefore, the y-coordinate of the center of mass lies $1/3$ of the distance from a base to the opposite vertex (Example 4, Section 6–6); in this example, $\bar{y} = 2/3$. Now all we need is M and M_y to find $\bar{x}$. We ob-tain the limits of integration from Figure 16–16, and compute

$$M = \int_0^1 \int_0^{-2x+2} x \, dy \, dx = \int_0^1 \Big[xy \Big]_0^{-2x+2} dx \qquad (\delta(x, y) = x)$$

$$= \int_0^1 (-2x^2 + 2x) \, dx = \frac{1}{3}$$

$$M_y = \int_0^1 \int_0^{-2x+2} x^2 dy \, dx = \int_0^1 \Big[x^2 y \Big]_0^{-2x+2} dx \qquad (16.22)$$

$$= \int_0^1 (-2x^3 + 2x^2) \, dx = \frac{1}{6}$$

It follows now that $\bar{x} = 1/2$ and $(1/2, 2/3)$ is the center of mass.   ■

Moment of Inertia

The kinetic energy of a particle of mass m moving on a straight line with speed v is

(16.24) $\text{K. E.} = \dfrac{1}{2}mv^2$

Now consider a particle revolving about an axis at a distance r and with angular speed ω; then $v = r\omega$ and its kinetic energy is $(1/2)mv^2 = (1/2)mr^2\omega^2$. If a system of masses $m_1, m_2, \ldots, m_n$ are revolving about the same axis at distances $r_1, r_2, \ldots, r_m$ all with the same angular speed ω, then the kinetic energy of the system is

$$\text{K. E.} = \sum_{i=1}^{n} \frac{1}{2}m_i r_i^2 \omega^2 = \frac{1}{2}\omega^2 \sum_{i=1}^{n} m_i r_i^2$$

The sum on the right is a property of the system called its **moment of inertia about the axis,** and is denoted by I. Thus,

(16.25) $\text{K. E.} = \dfrac{1}{2}I\omega^2$

Comparing (16.24) and (16.25), we see that I plays the same role in rotational motion that mass does in linear motion.

Because the distance r to the axis is squared, the moment of inertia is sometimes called the *second* moment. For thin plates D with density δ, the second moments or moments of inertia I_x, I_y, and I_0 about the x-axis, y-axis, and origin are defined as follows:

(16.26) $I_x = \displaystyle\int\int_D y^2\,\delta(x, y)\,dA$

(16.27) $I_y = \displaystyle\int\int_D x^2\,\delta(x, y)\,dA$

(16.28) $I_0 = \displaystyle\int\int_D (x^2 + y^2)\,\delta(x, y)\,dA$

The second moment I_0 can also be interpreted as the moment of inertia about the z-axis.

EXAMPLE 4   Compute the moment of inertia I_y about the y-axis of the lamina D in Example 3 (Figure 16–16).

Solution   The density was given in Example 3 as $\delta(x, y) = x$. We now apply (16.27) using the same limits of integration as before.

$$I_y = \int_0^1 \int_0^{-2x+2} x^3\,dy\,dx = \int_0^1 \left[x^3 y \right]_0^{-2x+2} dx$$

$$= \int_0^1 (-2x^4 + 2x^3)\,dx$$

$$= \frac{1}{10} \quad \blacksquare$$

EXERCISES

In Exercises 1–8, use double integration to find the volume of the indicated solid.

1. The solid between the graph of $f(x, y) = xy^2$ and the rectangle with vertices $(0, -1)$, $(0, 2)$, $(1, 2)$ and $(1, -1)$.

2. The solid between the graph of $f(x, y) = x + 2y$ and the rectangle with vertices $(-1, 0)$, $(-1, 1)$, $(1, 1)$ and $(1, 0)$.

3. The solid bounded by the coordinate planes and the plane $6x + 4y + 3z = 12$.

4. The solid bounded by the coordinate planes and the plane $x + y + z = 1$.

5. The part of the solid circular cylinder $x^2 + y^2 \le 4$ between the xy-plane and the plane $y + z = 1$.

6. Same as Exercise 5, but the second plane is $x + z = 9$.

7. The intersection (at right angles) of two solid circular cylinders of radius 3. (An example is the volume bounded by the graphs of $x^2 + y^2 = 9$ and $y^2 + z^2 = 9$.)

8. The solid sphere $x^2 + y^2 + z^2 \le 4$.

In Exercises 9–14, use double integration to find the area of the described region D.

9. Bounded by $x = y$ and $x = 4y - y^2$.

10. Bounded by $y = 4x - 3$ and $y = x^2$.

11. Bounded by $y = 5 - x$ and $y = 6/x$.

12. Bounded by $y = x + 4$ and $y = e^x$.

13. Bounded by $y = \sin x$, $y = \cos x$, $x = 0$, $x = \pi/2$.

14. Bounded by $x = y^2$ and $x - y = 3$.

In Exercises 15–22, find the center of mass of a thin plate D with the indicated shape and density.

15. D is the region in Exercise 9, $\delta(x, y) = 2$.

16. D is the region in Exercise 11, $\delta(x, y) = y$.

17. D is the left half of the region in Exercise 13, $\delta(x, y) = 1$.

18. D is the region in Exercise 14, $\delta(x, y) = 1$.

19. D is the triangular region bounded by the axes and the line $2x - y = 2$, $\delta(x, y) = |y|$.

20. D is the disc $x^2 + y^2 \le 1$, $\delta(x, y) = |x|$.

21. D is the triangular region bounded by the axes and the line $x + y = 2$, $\delta(x, y) = x^2 + y^2$ (the square of the distance to the origin).

22. D is the same as in Exercise 21, $\delta(x, y) = xy$.

In Exercises 23–24, compute the indicated moment of inertia.

23. D is the semi-disc $x^2 + y^2 \le 1$, $y \ge 0$ and $\delta(x, y) = y$. Find I_x.

24. D is bounded by $x = y^3$, $x = 8$ and the x-axis; $\delta(x, y) = y^2$. Find I_y.

25. D is a thin homogeneous (that is, the density is a constant, say δ) plate in the shape of a square of side a. Find the moment of inertia with respect to (a) any side (b) either diagonal (c) a line perpendicular to the square and passing through its center (d) a line perpendicular to the square and passing through a corner (e) a line perpendicular to the square and passing through the midpoint of a side.

26. (Parallel-axis theorem). Let D be a thin homogeneous plate. If I_{cm} is the moment of inertia of D about an axis through the center of mass and I is the moment of inertia of D about a parallel axis a distance h from the center of mass, then $I = I_{cm} + Mh^2$. This is called the parallel-axis theorem. Verify this result for parts (c), (d), and (e) of Exercise 25.

27. Let D be a thin homogeneous circular disc $x^2 + y^2 \leq r^2$ with density δ. Find the moment of inertia about the z-axis. Use the parallel-axis theorem (Exercise 26) to find the moment of inertia about the line (in space) $x = r/2, y = 0$.

28. Let D be a thin homogeneous triangular plate with vertices at $(-1, -1), (0, 2)$ and $(1, -1)$ and density δ. Find the moment of inertia about the z-axis and about the line (in space) $x = -1, y = 0$.

16–4 POLAR COORDINATES

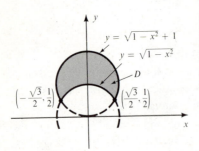

$$y = \sqrt{1 - x^2} + 1$$
$$y = \sqrt{1 - x^2}$$

$$\left(-\frac{\sqrt{3}}{2}, \frac{1}{2}\right) \qquad \left(\frac{\sqrt{3}}{2}, \frac{1}{2}\right)$$

FIGURE 16–17. If $\delta(x, y) = 1/\sqrt{x^2 + y^2}$, find the mass.

A thin plate has the shape of the region D shown in Figure 16–17, and its density at (x, y) is the reciprocal of the distance to the origin; $\delta(x, y) = 1/\sqrt{x^2 + y^2}$. The integral needed to compute the mass of D is

$$(16.29) \qquad \iint_D \delta(x, y)\, dA = \int_{-\sqrt{3}/2}^{\sqrt{3}/2} \int_{\sqrt{1-x^2}}^{\sqrt{1-x^2}+1} \frac{1}{\sqrt{x^2 + y^2}}\, dy\, dx$$

This is an example of an integration that is difficult to handle in rectangular coordinates, but is simple if polar coordinates are introduced by a change of variable.

The situation with polar coordinates is this: A region D consists of points (r, θ) with $h_1(\theta) \leq r \leq h_2(\theta)$ and $\alpha \leq \theta \leq \beta$. We assume that h_1 and h_2 are continuous and that these restrictions give each point in D a unique pair of polar coordinates (just as each point has a unique pair of rectangular coordinates). If D is the domain of a function $z = f(r, \theta)$, we want to define

$$\iint_D f(r, \theta)\, dA$$

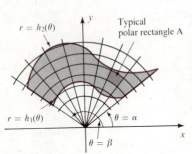

Typical polar rectangle A

$$r = h_2(\theta)$$
$$r = h_1(\theta)$$
$$\theta = \alpha$$
$$\theta = \beta$$

A.

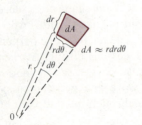

$$dA \approx r\, dr\, d\theta$$

B.

FIGURE 16–18. Partition of a region into polar rectangles.

Instead of partitioning the region into rectangles by means of horizontal and vertical lines, we form a grid using circles with centers at the origin and straight lines through the origin (Figure 16–18A). This partitions the region into what we will call *polar rectangles*. To form a Riemann sum for f, we discard all polar rectangles that are not completely contained in D, leaving an inner partition P. The polar rectangles in P are numbered $A_1, A_2, \ldots, A_n$, a point (r_i, θ_i) is chosen in each A_i, and we form the sum

$$R_P = \sum_{i=1}^{n} f(r_i, \theta_i)\Delta A_i$$

where ΔA_i is the area of A_i. If $\|P\|$ denotes the length of longest diagonal of the polar rectangles in P, then we define the double integral by

$$(16.30) \qquad \boxed{\iint_D f(r, \theta)\, dA = \lim_{\|P\| \to 0} \sum_{i=1}^{n} f(r_i, \theta_i)\Delta A_i}$$

provided this limit exists. As in previous cases, if f is continuous, then the double integral exists.

Our next task is to evaluate the double integral with an iterated integral. For rectangular coordinates, the area dA of a typical rectangle is replaced by $dy\,dx$ (or $dx\,dy$) in the iterated integral. But the situation is different for polar rectangles. Figure 16–18B shows a typical polar rectangle. The straight sides have length dr, the change in r. The curved side closest to the origin is that part of the circumference of a circle of radius r determined by a central angle of $d\theta$ radians. Therefore, the length of this side is $r\,d\theta$. It follows that dA, the area of A, is approximately $r\,dr\,d\theta$. Accordingly, in integrals involving polar coordinates,

(16.31) | dA is replaced by $r\,dr\,d\theta$

It can be shown that if dA is replaced by $r\,dr\,d\theta$, then the integral in (16.30) can be evaluated by an iterated integral

$$(16.32)\quad \int\!\!\int_D f(r,\theta)\,dA = \int_\alpha^\beta \int_{h_1(\theta)}^{h_2(\theta)} f(r,\theta)r\,dr\,d\theta$$

EXAMPLE 1   Evaluate $\int\!\!\int_D \cos\theta\,dA$ where D is the region bounded by the curve $r=1$ and the lines $\theta=0$ and $\theta=\pi/4$.

Solution   Figure 16–19 shows the region D; for each θ between 0 and $\pi/4$, r ranges from 0 to 1. Therefore, we apply (16.32) to obtain

$$\int\!\!\int_D \cos\theta\,dA = \int_0^{\pi/4}\int_0^1 (\cos\theta)r\,dr\,d\theta$$

$$= \int_0^{\pi/4}\left[(\cos\theta)\frac{r^2}{2}\right]_0^1 d\theta \qquad (\theta\ \text{constant})$$

$$= \frac{1}{2}\int_0^{\pi/4}\cos\theta\,d\theta = \frac{\sqrt{2}}{4} \qquad\blacksquare$$

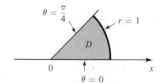

FIGURE 16–19. Example 1.

Let us return now to the problem stated in the opening paragraph.

EXAMPLE 2   A thin plate has the shape of the region D shown in Figure 16–17. Its density at (x,y) is $\delta(x,y) = 1/\sqrt{x^2+y^2}$. Find its mass.

Solution   Evaluating the integral in (16.29) is very difficult. Therefore, we will convert the whole problem to polar coordinates. To do so, we need to express the boundaries and the integrand in polar form. Referring to Figure 16–20, we see that the upper boundary is part of the circle $x^2 + (y-1)^2 = 1$, which converts to $r = 2\sin\theta$; the lower boundary is part of $x^2 + y^2 = 1$, which converts to $r = 1$. The intersection points are $(1, \pi/6)$ and $(1, 5\pi/6)$. It follows that θ varies from $\pi/6$ to $5\pi/6$ and for each fixed θ, r varies from 1 to $2\sin\theta$. Finally,

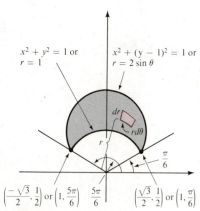

FIGURE 16–20. Example 2.

since $r^2 = x^2 + y^2$, we replace the integrand $1/\sqrt{x^2 + y^2}$ by $1/r$. Therefore, the difficult integral (16.29) becomes the simple integral

$$M = \int \int_D \delta(r, \theta)\, dA = \int_{\pi/6}^{5\pi/6} \int_0^{2\sin\theta} \left(\frac{1}{r}\right) r\, dr\, d\theta$$

$$= \int_{\pi/6}^{5\pi/6} \int_0^{2\sin\theta} dr\, d\theta$$

$$= \int_{\pi/6}^{5\pi/6} \left[r\right]_1^{2\sin\theta} d\theta$$

$$= \int_{\pi/6}^{5\pi/6} (2 \sin\theta - 1)\, d\theta$$

$$= \left[-2\cos\theta - \theta\right]_{\pi/6}^{5\pi/6} = 2\sqrt{3} - \frac{2\pi}{3} \qquad \blacksquare$$

Example 2 illustrates that an iterated integral in rectangular coordinates can be transformed into an iterated integral in polar coordinates. Three steps are involved:

(1) Rewrite the limits of integration in polar form.
(2) Replace x by $r \cos\theta$ and y by $r \sin\theta$ in the integrand; thus, $f(x, y)$ becomes $f(r \cos\theta, r \sin\theta)$.
(3) Replace $dy\, dx$ (or $dx\, dy$) by $r\, dr\, d\theta$.

This is a *change of variables* from rectangular to polar coordinates.

EXAMPLE 3   Find the moment of inertia about the x-axis of a thin homogeneous disc $x^2 + y^2 \le a^2$ with constant density δ (Figure 16–21).

Solution   In rectangular coordinates, we have the troublesome integral

$$I_x = \int_{-a}^{a} \int_{-\sqrt{a^2-x^2}}^{\sqrt{a^2-x^2}} \delta y^2\, dy\, dx$$

Let us change to polar coordinates.

(1) Rewrite the limits of integration; for each θ with $0 \le \theta \le 2\pi$, r ranges from 0 to a.
(2) The integrand δy^2 becomes $\delta r^2 \sin^2\theta$.
(3) Replace $dy\, dx$ by $r\, dr\, d\theta$.

Therefore,

$$I_x = \int_0^{2\pi} \int_0^a \delta r^2 \sin^2\theta\, r\, dr\, d\theta = \delta \int_0^{2\pi} \left[\frac{r^4}{4} \sin^2\theta\right]_0^a d\theta$$

$$= \delta \frac{a^4}{4} \int_0^{2\pi} \sin^2\theta\, d\theta$$

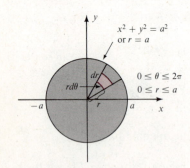

$x^2 + y^2 = a^2$
or $r = a$

$0 \le \theta \le 2\pi$
$0 \le r \le a$

FIGURE 16–21. Example 3.

$$= \delta \frac{a^4}{4} \int_0^{2\pi} \left(\frac{1}{2} - \frac{\cos 2\theta}{2} \right) d\theta$$

$$= \delta \frac{a^4}{4} \left[\frac{\theta}{2} - \frac{\sin 2\theta}{4} \right]_0^{2\pi} = \delta \pi \frac{a^4}{4} \quad \blacksquare$$

EXAMPLE 4   Find the volume inside the paraboloid $z = 4 - x^2 - y^2$ that lies above the *xy*-plane (Figure 16–22).

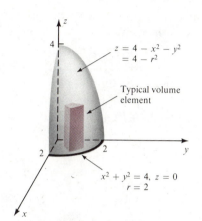

$z = 4 - x^2 - y^2$
$= 4 - r^2$

Typical volume element

$x^2 + y^2 = 4, \ z = 0$
$r = 2$

FIGURE 16–22. Example 4.

Solution   In rectangular coordinates, we have the troublesome integral

$$V = 4 \int_0^2 \int_0^{\sqrt{4-x^2}} (4 - x^2 - y^2) \, dy \, dx$$

In polar coordinates, we have the simple integral

$$V = 4 \int_0^{\pi/2} \int_0^2 (4 - r^2) r \, dr \, d\theta$$

$$= 4 \int_0^{\pi/2} \int_0^2 (4r - r^3) \, dr \, d\theta$$

$$= 4 \int_0^{\pi/2} \left[2r^2 - \frac{r^4}{4} \right]_0^2 d\theta$$

$$= 16 \int_0^{\pi/2} d\theta = 8\pi \quad \blacksquare$$

We close with the observation that if it is more convenient to integrate first with respect to θ, then you can reverse the order of integration from $r \, dr \, d\theta$ to $r \, d\theta \, dr$ by finding the new limits of integration in the usual way.

EXERCISES

In the following exercises, use either rectangular or polar coordinates (choose the one you think is easier).

In Exercises 1–6, evaluate the integrals.

1. $\int_{-1}^1 \int_0^{\sqrt{1-x^2}} e^{-(x^2 + y^2)} \, dy \, dx$

2. $\int_{-1}^1 \int_0^{\sqrt{1-x^2}} (x^2 + y^2) \, dy \, dx$

3. $\int_1^2 \int_0^x \frac{1}{\sqrt{x^2 + y^2}} \, dy \, dx$

4. $\int_{3/\sqrt{2}}^3 \int_0^{\sqrt{9-x^2}} \frac{1}{\sqrt{x^2 + y^2}} \, dy \, dx$

5. $\int_0^1 \int_0^{\sqrt{1-x^2}} \sin(x^2 + y^2) \, dy \, dx$

6. $\int_0^1 \int_{-\sqrt{4-x^2}}^{\sqrt{4-x^2}} x^2 \, dy \, dx$

7. D is the region in the first and second quadrant between the circles $x^2 + y^2 = 1$ and $x^2 + y^2 = 4$. Find $\int\int_D x^2 y \, dA$.

8. D is the rectangle with vertices $(0, 0)$, $(0, 1)$, $(1, 1)$, and $(1, 0)$. Find $\int\int_D \sin x \cos x \, dA$.

9. D is the triangle with vertices $(0, 0)$, $(0, 2)$, and $(2, 0)$. Find $\int\int_D e^y \, dA$.

10. D is the disc $x^2 + y^2 \le 1$. Find $\int\int_D (x^2 + y^2) \, dA$.

11. D is the disc $x^2 + y^2 \le 1$. Find $\int\int_D e^{\sqrt{x^2 + y^2}} \, dA$.

12. D is the region bounded by the parabola $y = x^2$ and the line $y = 1$. Find $\int\int_D x \, dA$.

In Exercises 13–16, use double integrals to find the area of the described region.

13. One loop of $r^2 = 9 \sin 2\theta$.

14. Inside the circle $x^2 + (y - 1)^2 = 1$, but outside the circle $x^2 + y^2 = 1$.

15. Inside the cardioid $r = 1 - \cos \theta$.

16. Inside the cardioid $r = 2(1 - \cos \theta)$, but outside the circle $r = 2$.

In Exercises 17–20, use double integration to find the volume of the described solid.

17. A sphere of radius a.

18. The solid bounded by the paraboloid $z = 1 - x^2 - y^2$ and the xy-plane.

19. The intersection of the solid sphere $x^2 + y^2 + z^2 \leq 4$ and the solid cylinder $x^2 + y^2 \leq 1$.

20. The solid bounded by the xy-plane and the graphs of $z = 4 - x^2 - y^2$ and the cylinder $x^2 + y^2 = 1$.

21. Let D be a thin homogeneous plate in the shape of the region between the (polar) spirals $r = \theta$ and $r = 2\theta$ for $0 \leq \theta \leq 3\pi$. If the density is $\delta = 1$, find I_0 the moment of inertia about the origin.

22. Find the center of mass of a thin homogeneous plate ($\delta = 1$) in the shape of the cardioid $r = 1 - \cos \theta$.

23. Find the moment of inertia about the z-axis of a thin homogeneous ($\delta = 1$) disc $x^2 + y^2 \leq a^2$.

24. Find the moment of inertia about the z-axis of a thin homogeneous ($\delta = 1$) plate in the shape of a triangle with verticles $(-1, 0)$, $(0, 1)$, and $(1, 0)$.

16–5   TRIPLE INTEGRALS

The extension of double integrals for functions of two variables to triple integrals for functions of three variables is easily accomplished. Some of the geometric interpretations are lost because it is impossible to draw graphs of functions of three variables, but everything else carries over.

First we define a solid of **type 1.** Suppose that D is an elementary region in the xy-plane and that g_1 and g_2 are continuous functions defined on D. Then the solid S consisting of points (x, y, z) where (x, y) is in D and $g_1(x, y) \leq z \leq g_2(x, y)$ is a solid of type 1 (Figure 16–23). A grid consists of planes parallel to the coordinate planes. A grid partitions S into *boxes* (Figure 16–24) from which we obtain an *inner partition*. Each box has volume $dV = dx\, dy\, dz$. If f is a function defined on S, we form Riemann sums and take the limit in the usual way. The limit, if it exists, is defined to be the **triple integral of f over S,**

$$\int \int \int_S f(x, y, z)\, dV$$

It can be shown that if f is continuous on S, then the triple integral exists and can be evaluated by an iterated integral

$$(16.34) \quad \boxed{\int \int \int_S f(x, y, z)\, dV = \int \int_D \left[\int_{g_1(x, y)}^{g_2(x, y)} f(x, y, z)\, dz \right] dA}$$

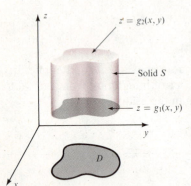

FIGURE 16–23. A solid of type 1.

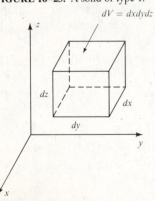

FIGURE 16–24. A typical box in a partition of S.

If D is of type 1 in the xy-plane, then the iterated integral on the right becomes

(16.35) $$\int_a^b \int_{h_1(x)}^{h_2(x)} \int_{g_1(x,\,y)}^{g_2(x,\,y)} f(x, y, z)\, dz\, dy\, dx$$

If D is of type 2 in the xy-plane, then we have

(16.36) $$\int_c^d \int_{k_1(y)}^{k_2(y)} \int_{g_1(x,\,y)}^{g_2(x,\,y)} f(x, y, z)\, dz\, dx\, dy$$

EXAMPLE 1   Let S be the parallelepiped consisting of all points (x, y, z) with $0 \le x \le 1$, $1 \le y \le 2$, and $1 \le z \le 3$. Let $f(x, y, z) = 3xz^2$ and evaluate the triple integral of f over S.

Solution   Here D is the rectangle $0 \le x \le 1$, $1 \le y \le 2$ in the xy-plane and S is the solid over D between the planes $z = 1$ and $z = 3$ (Figure 16–25). Using (16.35), we have

$$\iiint_S 3xz^2\, dV = \int_0^1 \int_1^2 \int_1^3 3xz^2\, dz\, dy\, dx$$

$$= \int_0^1 \int_1^2 \left[xz^3 \right]_1^3 dy\, dx$$

$$= 26 \int_0^1 \int_1^2 x\, dy\, dx = 13 \qquad \blacksquare$$

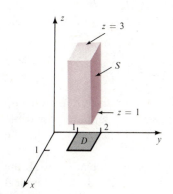

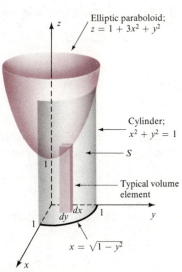

FIGURE 16–25. Example 1.

Just as the double integral over D of the constant function $f(x, y) = 1$ gives the area of D, the triple integral over S of the constant 1 gives the volume of S.

EXAMPLE 2   Use triple integration to find the volume of the solid S bounded above by the elliptic paraboloid $z = 1 + 3x^2 + y^2$, below by the xy-plane, and on the sides by the cylinder $x^2 + y^2 = 1$.

Solution   Figure 16–26 shows the part of S in the first octant; we will compute this volume and, by symmetry, multiply by 4 to get the volume of S. The figure shows that D is the disc $x^2 + y^2 \le 1$. For each fixed (x, y) in D, z varies from the lower boundary $z = 0$ to the upper boundary $z = 1 + 3x^2 + y^2$. Therefore,

$$\text{Volume} = \iiint_S dV = \iint_D \int_0^{1+3x^2+y^2} dz\, dA$$

We can integrate over D in either order, $dy\, dx$ or $dx\, dy$; we will use the latter. The figure shows a typical volume element, and the total volume is given by

$$\text{Volume} = 4 \int_0^1 \int_0^{\sqrt{1-y^2}} \int_0^{1+3x^2+y^2} dz\, dx\, dy$$

$$= 4 \int_0^1 \int_0^{\sqrt{1-y^2}} (1 + 3x^2 + y^2)\, dx\, dy$$

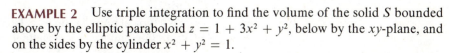

$V = \frac{4}{3}\pi r^3$

FIGURE 16–26. Example 2.

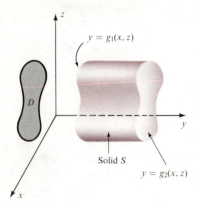

FIGURE 16-27. A solid of type 2.

$$= 4 \int_0^1 2\sqrt{1 - y^2}\, dy \qquad \text{(Now use trigonometric substitution or Formula 28)}$$

$$= 4 \left[y\sqrt{1 - y^2} + \sin^{-1} y \right]_0^1 = 2\pi \qquad \blacksquare$$

There are two other types of solids we will define. If D is an elementary region in the xz-plane and S consists of all points (x, y, z) with (x, z) in D and $g_1(x, z) \le y \le g_2(x, z)$, then S is of **type 2** (Figure 16–27). The resulting triple and iterated integrals are

$$(16.37) \qquad \int\int\int_S f(x, y, z)\, dV = \int\int_D \left[\int_{g_1(x, z)}^{g_2(x, z)} f(x, y, z)\, dy \right] dA$$

The double integral over D is evaluated in the order $dz\, dx$ or $dx\, dz$, depending on the region D. A **type 3** solid is one in which the region D is in the yz-plane and then

$$(16.38) \qquad \int\int\int_S f(x, y, z)\, dV = \int\int_D \left[\int_{g_1(y, z)}^{g_2(y, z)} f(x, y, z)\, dx \right] dA$$

The double integral over D is evaluated in the order $dy\, dz$ or $dz\, dy$, depending on the region D. Altogether, there are six possible orders of integration for triple integrals. If the integrand is continuous, then any order of integration yields the same value.

All of the applications for double integrals carry over to triple integrals. If the density of a solid S is given by a continuous density function δ, then a typical box in any partition will have (approximate) mass

$$\delta(x, y, z)\, dV$$

where (x, y, z) is a typical point in the box. Then the Riemann sums approximate the mass of S and the total mass M is defined by

$$(16.39) \qquad \boxed{ M = \int\int\int_S \delta(x, y, z)\, dV }$$

The first moments of a solid S are taken with respect to planes. For example, if (x, y, z) is a typical point in a typical box in S, then the distance to the yz-plane is x. If δ is the density function, then

$$x\delta(x, y, x)\, dV$$

approximates the moment with respect to the yz-plane. Therefore, we define M_{yz}, the moment of S with respect to the yz-plane, to be

$$(16.40) \qquad \boxed{ M_{yz} = \int\int\int_S x\delta(x, y, z)\, dV }$$

Similar definitions are made for M_{xz} and M_{xy}. The coordinates of the center of mass are

(16.41) $$\bar{x} = \frac{M_{yz}}{M} \qquad \bar{y} = \frac{M_{xz}}{M} \qquad \bar{z} = \frac{M_{xy}}{M}$$

Moments of inertia are still taken with respect to lines or points. Since the square of the distance from (x, y, z) to the z-axis is $x^2 + y^2$, we define

(16.42) $$I_z = \int \int \int_S (x^2 + y^2)\, \delta(x, y, z)\, dV$$

to be the moment of inertia of S about the z-axis. Similar definitions are made for I_x, I_y, and I_0.

EXAMPLE 3   The tetrahedron T bounded by the coordinate planes and the plane $x + y + z = 1$ has a density function $\delta(x, y, z) = xyz$. Express the mass of T as an iterated triple integral, first in the order $dy\, dz\, dx$ and then in the order $dx\, dz\, dy$.

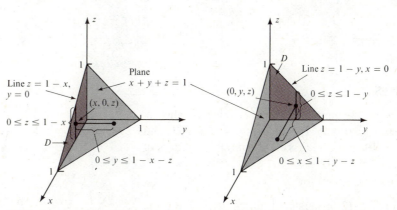

FIGURE 16–28.   Example 3.

A. Tetrahedron considered as type 2.

B. Tetrahedron considered as type 3.

Solution   For the order $dy\, dz\, dx$, we consider T as a solid of type 2; this is pictured in Figure 16–28A. The region D is the triangle in the xz-plane with vertices $(0, 0, 0)$, $(1, 0, 0)$, and $(0, 0, 1)$. For each x with $0 \le x \le 1$, z varies from $z = 0$ to (the line) $z = 1 - x$. For each $(x, 0, z)$ in D, y varies from $y = 0$ to (the plane) $y = 1 - x - z$. Therefore, the mass is

$$M = \int_0^1 \int_0^{1-x} \int_0^{1-x-z} xyz\, dy\, dz\, dx$$

For the order $dx\,dz\,dy$, we consider T as a solid of type 3; this is pictured in Figure 16–28B. The region D is now a triangle in the yz-plane. For each y with $0 \le y \le 1$, z varies from $z = 0$ to (the line) $z = 1 - y$. For each $(0, y, z)$ in D, x varies from $x = 0$ to (the plane) $x = 1 - y - z$. In this case, the mass is expressed as

$$M = \int_0^1 \int_0^{1-y} \int_0^{1-y-z} xyz\,dx\,dz\,dy$$

We leave it for you to check that both integrals yield the same mass. ∎

EXAMPLE 4 The solid S enclosed by the elliptic paraboloids $z = 3x^2 + y^2$ and $z = 4 - x^2 - y^2$ has constant density 1. Express M_{xy}, its moment with respect to the xy-plane, as an iterated triple integral.

Solution The two paraboloids intersect in the set of points (x, y, z) where

$$3x^2 + y^2 = 4 - x^2 - y^2 \quad \text{or} \quad 2x^2 + y^2 = 2$$

Therefore, with D as the elliptic region $2x^2 + y^2 \le 2$ in the xy-plane, S is a solid of type 1 (Figure 16–29). For each x with $-1 \le x \le 1$, y varies from the curve $y = -\sqrt{2 - 2x^2}$ to the curve $y = \sqrt{2 - 2x^2}$. For each $(x, y, 0)$ in D, z varies from the lower surface $z = 3x^2 + y^2$ to the upper surface $z = 4 - x^2 - y^2$. The integrand for computing M_{xy} is $z\,\delta(x, y, z)$; since $\delta = 1$ in this example, the integrand is z. Therefore, the moment is

$$M_{xy} = \int_{-1}^1 \int_{-\sqrt{2-2x^2}}^{\sqrt{2-2x^2}} \int_{3x^2+y^2}^{4-x^2-y^2} z\,dz\,dy\,dx \qquad ∎$$

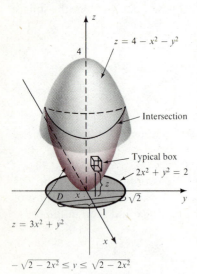

$-\sqrt{2 - 2x^2} \le y \le \sqrt{2 - 2x^2}$

FIGURE 16–29. Example 4.

EXERCISES

1. Let S be the parallelepiped bounded by the six planes $x = 0$, $x = 1$, $y = 1$, $y = 2$, $z = -1$, and $z = 3$. Evaluate $\iiint_S 8xyz\,dV$ in all six orders of integration.

2. Same as Exercise 1 for $\iiint_S (x + y + z)\,dV$.

In Exercises 3–12, evaluate the integrals *in the given order*.

3. $\displaystyle\int_0^1 \int_0^3 \int_{-1}^2 dz\,dx\,dy$

4. $\displaystyle\int_1^2 \int_{-1}^1 \int_0^4 dy\,dz\,dx$

5. $\displaystyle\int_0^1 \int_1^{2y} \int_0^x (x + 2z)\,dz\,dx\,dy$

6. $\displaystyle\int_0^2 \int_0^x \int_0^x dz\,dy\,dx$

7. $\displaystyle\int_0^2 \int_0^{2-z} \int_0^{2-z-y} dx\,dy\,dz$

8. $\displaystyle\int_0^1 \int_0^{2-2z} \int_0^{2-2z-y} dx\,dy\,dz$

9. $\displaystyle\int_0^2 \int_0^{x^2} \int_0^x 2x^2y\,dz\,dy\,dx$

10. $\displaystyle\int_0^1 \int_0^{\sqrt{y}} \int_0^y 3z^2\,dz\,dx\,dy$

11. $\displaystyle\int_0^1 \int_0^{\sqrt{1-z^2}} \int_0^{1-z} y\,dx\,dy\,dz$

12. $\displaystyle\int_1^2 \int_0^{z^3} \int_0^{y/z} dx\,dy\,dz$

In Exercises 13–16, refer back to the indicated exercise, reverse the order of integration to $dy\,dz\,dx$, and evaluate.

13. Exercise 7   **14.** Exercise 5
15. Exercise 11   **16.** Exercise 9

In Exercises 17–24, use triple integration to find the volumes.

17. The tetrahedron bounded by the coordinate planes and the plane $x/a + y/b + z/c = 1$ where a, b, and c are positive.

18. The intersection of the two solid cylinders $x^2 + z^2 \leq a^2$ and $x^2 + y^2 \leq a^2$.

19. The solid enclosed by the parabolic cylinder $z = 4 - x^2$ and the planes $y + z = 4, y = 0$, and $z = 0$.

20. The solid enclosed by the cylinder $x^2 + y^2 = 1$, the plane $x + y + z = 1$, and the coordinate planes.

21. The solid bounded by the xy-plane, the plane $z - x = 2$, and the elliptic cylinder $x^2 + 4y^2 = 4$.

22. The solid bounded by the parabolic cylinder $z = 4 - x^2$, the coordinate planes, and the plane $y = 1$.

23. The solid bounded by the elliptic paraboloids $z = x^2 + 3y^2$ and $z = 8 - x^2 - y^2$.

24. The solid bounded by the elliptic paraboloids $z = 6 - 2x^2 - y^2$ and $z = x^2 + y^2$.

25. S is a cube of side a with the origin at one corner. If the mass at any point is directly proportional to the square of its distance from the origin, find the center of mass.

26. Same as Exercise 25 for the parallelepiped bounded by the coordinate planes and the planes $x = 1$, $y = 2, z = 3$.

27. Let S be the solid consisting of the points (x, y, z) with $0 \leq x \leq 1, 0 \leq y \leq x$, and $-y^2 \leq z \leq x^2$. If the density function is $\delta(x, y, z) = 1 + x$, find the mass of S.

28. Let S be the solid enclosed by the elliptic paraboloids $z = 3 - x^2 - y^2$ and $z = x^2 + y^2 - 5$. Find M_{xz}, the moment of S with respect to the xz-plane.

16–6 CYLINDRICAL AND SPHERICAL COORDINATES

In Section 16–4, we saw that a change of variables from rectangular to polar coordinates can often convert a difficult double integral problem into a simple one. For triple integrals, there are two sets of coordinate systems that can be used for the same purpose.

Cylindrical Coordinates

Cylindrical coordinates are an extension of polar coordinates from the plane to space. A point P has **cylindrical coordinates** (r, θ, z) if z is the usual (third) rectangular coordinate, and (r, θ) are the polar coordinates of the projection P' of P onto the xy-plane (Figure 16–30). If the rectangular coordinates of P are (x, y, z), then the formulas

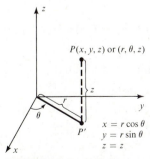

$P(x, y, z)$ or (r, θ, z)

$x = r \cos \theta$
$y = r \sin \theta$
$z = z$

FIGURE 16–30. Cylindrical coordinates.

(16.43) $$x = r \cos \theta \quad y = r \sin \theta \quad z = z$$

can be used to change from rectangular to cylindrical coordinates. The equation $r = c$, where c is a positive constant, is equivalent to $x^2 + y^2 = c^2$. Therefore, the graph of $r = c$ is a cylinder (hence the name cylindrical coordinates); the graph

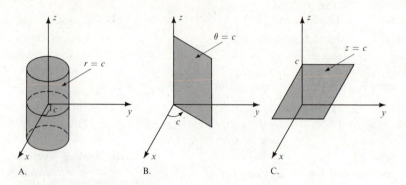

FIGURE 16–31. Graphs in cylindrical coordinates.

A.   B.   C.

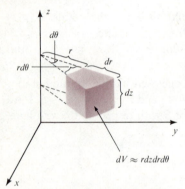

$$dV \approx r\,dz\,dr\,d\theta$$

FIGURE 16–32. A typical cylindrical box.

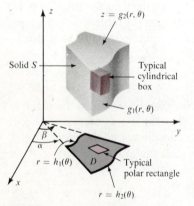

FIGURE 16–33. S consists of the points (r, θ, z) with $\alpha \leq \theta \leq \beta$, $h_1(\theta) \leq r \leq h_2(\theta)$, and $g_1(r, \theta) \leq z \leq g_2(r, \theta)$.

of $\theta = c$ is a plane containing the z-axis; the graph of $z = c$ is a plane parallel to the xy-plane (Figure 16–31).

The procedure for using cylindrical coordinates in integration is as follows: Let S be a solid in space. The grid that covers S consists of cylinders $r = c$ whose radii differ by dr, planes $\theta = c$ with the angle $d\theta$ between them, and planes $z = c$ with distance dz between them. This divides S into what we will call *cylindrical boxes;* a typical cylindrical box is shown in Figure 16–32. If f is continuous on S, we form inner partitions, Riemann sums, and take the limit in the usual way to obtain the triple integral

$$(16.44) \quad \iiint_S f(r, \theta, z)\, dV$$

Now suppose that S consists of points (r, θ, z) with $\alpha \leq \theta \leq \beta, h_1(\theta) \leq r \leq h_2(\theta)$, and $g_1(r, \theta) \leq z \leq g_2(r, \theta)$; see Figure 16–33. It can be shown that if all of the functions involved are continuous, then the integral (16.44) can be evaluated by an iterated integral. As Figure 16–32 indicates, the volume element dV must be replaced by $r\,dz\,dr\,d\theta$. Thus,

$$(16.45) \quad \boxed{\iiint_S f(r, \theta, z)\, dV = \int_{\alpha}^{\beta} \int_{h_1(\theta)}^{h_2(\theta)} \int_{g_1(r,\theta)}^{g_2(r,\theta)} f(r, \theta, z)\, r\, dz\, dr\, d\theta}$$

To make a change of variable in an integral from rectangular to cylindrical coordinates,

(1) Sketch the solid and change the limits of integration accordingly.

(2) Replace x and y by $r \cos \theta$ and $r \sin \theta$ in the integrand.

(4) Replace $dx\, dy\, dz$ (or any of the five other orders) with $r\, dz\, dr\, d\theta$ (the order is determined by the new limits of integration).

EXAMPLE 1 Find the volume of the solid S bounded by the xy-plane, the cylinder $x^2 + y^2 = a^2$, and the cone $z^2 = x^2 + y^2, z \geq 0$ (Figure 16–34).

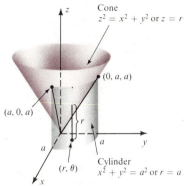

FIGURE 16–34. Example 1.

Solution It is probably easier to work this problem with double integrals, but let us see how it goes with triple integrals. In rectangular coordinates, the volume is

$$V = 4 \int_0^a \int_0^{\sqrt{a^2 - x^2}} \int_0^{\sqrt{x^2 + y^2}} dz\, dy\, dx$$

We now change to cylindrical coordinates. The figure shows that $0 \le \theta \le \pi/2$, $0 \le r \le a$, and $0 \le z \le r$. Replacing $dz\, dy\, dx$ with $r\, dz\, dr\, d\theta$, we have

$$V = 4 \int_0^{\pi/2} \int_0^a \int_0^r r\, dz\, dr\, d\theta$$

$$= 4 \int_0^{\pi/2} \int_0^a r^2\, dr\, d\theta = \frac{2\pi}{3} a^3 \qquad \blacksquare$$

Spherical Coordinates

The **spherical coordinates** of a point P are (ρ, ϕ, θ) where ρ (the Greek letter *rho*) is the distance to the origin, ϕ (the Greek letter *phi*) is the angle $0 \le \phi \le \pi$ between the positive z-axis and a line joining P to the origin, and θ is the second

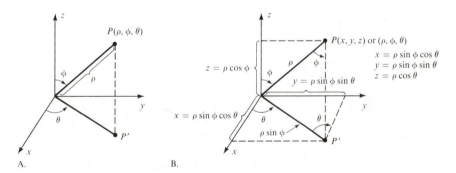

FIGURE 16–35. Spherical coordinates.

A.

B.

polar coordinate of the projection P' of P onto the xy-plane (Figure 16–35A). If the rectangular coordinates of P are (x, y, z), then the formulas

(16.46) $\qquad x = \rho \sin \phi \cos \theta \quad y = \rho \sin \phi \sin \theta \quad z = \rho \cos \phi \quad \rho^2 = x^2 + y^2 + z^2$

can be used to change from rectangular to spherical coordinates (Figure 16–35B). The equation $\rho = c$, where c is a positive constant is equivalent to $x^2 + y^2 + z^2 = c^2$. Therefore, the graph of $\rho = c$ is a sphere (hence the name spherical coordinates), the graph of $\phi = c$ is a part of a cone (unless $c = \pi/2$), and the graph of $\theta = c$ is a plane containing the z-axis (Figure 16–36).

To use spherical coordinates in integration, we partition with spheres, cones, and planes that divide a solid into what we will call *spherical boxes*. Then we

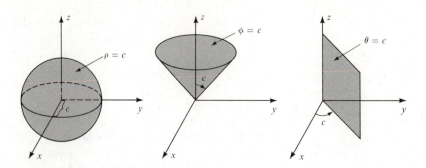

FIGURE 16–36. Graphs in spherical coordinates.

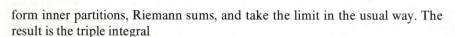

$dV \approx \rho^2 \sin \phi d\rho d\phi d\theta$

$d\phi$ $d\rho$

$\rho d\phi$

$\rho \sin \phi \, d\theta$

$d\theta$

$rd\theta = \sqrt{x^2 + y^2} \, d\theta = \rho \sin \phi d\theta$

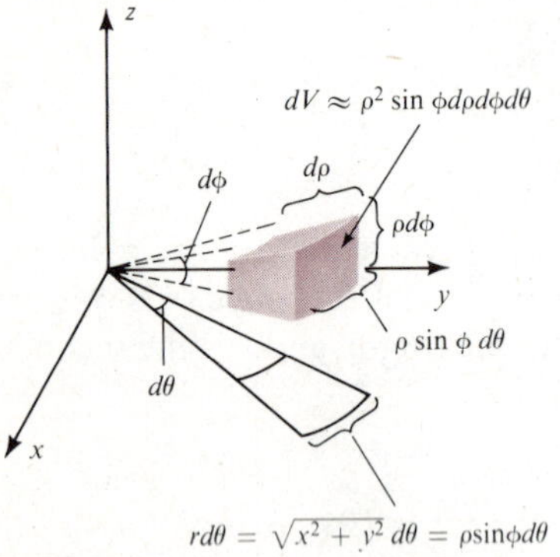

FIGURE 16–37. A typical spherical box.

Typical volume element

$\rho = a$ ϕ

θ

$0 \leq \rho \leq a$
$0 \leq \phi \leq \pi$
$0 \leq \theta \leq 2\pi$

FIGURE 16–38. Example 2.

form inner partitions, Riemann sums, and take the limit in the usual way. The result is the triple integral

$$(16.47) \qquad \iiint_S f(\rho, \phi, \theta) \, dV$$

To find an equivalent iterated integral, we look at a typical spherical box (Figure 16–37) and determine that dV must be replaced by $\rho^2 \sin \phi \, d\rho \, d\phi \, d\theta$. Therefore,

$$(16.48) \qquad \boxed{\iiint_S f(\rho, \phi, \theta) \, dV = \iiint f(\rho, \phi, \theta) \, \rho^2 \sin \phi \, d\rho \, d\phi \, d\theta}$$

where the integral on the right is an iterated integral whose limits of integration are determined by the solid S. To make a change of variable from rectangular to spherical coordinates, we follow the three steps given earlier for cylindrical coordinates making the obvious adjustments.

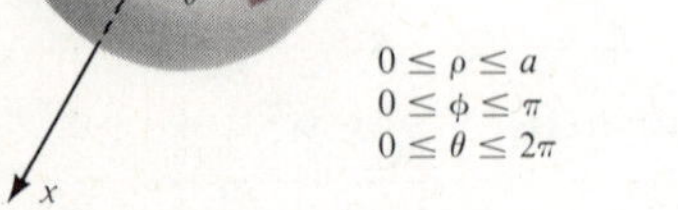

EXAMPLE 2   Find the volume of a sphere of radius a (Figure 16–38). If the density is constant $\delta = 1$, find I_z, its moment of inertia about the z-axis.

Solution   In rectangular coordinates,

$$V = 8 \int_0^a \int_0^{\sqrt{a^2 - x^2}} \int_0^{\sqrt{a^2 - x^2 - y^2}} dz \, dy \, dx$$

To change over to spherical coordinates, we determine from the figure that $0 \leq \rho \leq a, 0 \leq \phi \leq \pi$, and $0 \leq \theta \leq 2\pi$. Therefore,

$$V = \int_0^{2\pi} \int_0^{\pi} \int_0^a \rho^2 \sin \phi \, d\rho \, d\phi \, d\theta$$

$$= \frac{a^3}{3} \int_0^{2\pi} \int_0^{\pi} \sin \phi \, d\phi \, d\theta$$

$$= \frac{2a^3}{3} \int_0^{2\pi} d\theta = \frac{4}{3}\pi a^3$$

Now we find I_z. The square of the distance from (x, y, z) to the z-axis is $x^2 + y^2$, which converts to

$$x^2 + y^2 = \rho^2 \sin^2 \phi \cos^2 \theta + \rho^2 \sin^2 \phi \sin^2 \theta \qquad (16.46)$$
$$= \rho^2 \sin^2 \phi$$

Therefore, since the density is $\delta = 1$, we have

$$I_z = \int_0^{2\pi} \int_0^\pi \int_0^a \rho^4 \sin^3 \phi \, d\rho \, d\phi \, d\theta$$

$$= \frac{a^5}{5} \int_0^{2\pi} \int_0^\pi \sin^3 \phi \, d\phi \, d\theta \qquad \begin{array}{l}\text{(Write } \sin^3 \phi = \sin \phi(1 - \cos^2 \phi) \text{ or} \\ \text{use Formula \textbf{53})}\end{array}$$

$$= \frac{4a^5}{15} \int_0^{2\pi} d\theta = \frac{8}{15}\pi a^5 \qquad \blacksquare$$

EXERCISES

In Exercises 1–4, sketch the solid, change to cylindrical coordinates, and evaluate the integral.

1. $\displaystyle\int_0^1 \int_0^{\sqrt{1-x^2}} \int_0^{\sqrt{1-x^2-y^2}} z \, dz \, dy \, dx$

2. $\displaystyle\int_0^1 \int_0^{\sqrt{1-y^2}} \int_0^1 (x^2 + y^2) \, dz \, dx \, dy$

3. $\displaystyle\int_0^1 \int_0^{\sqrt{1-x^2}} \int_0^{x^2+y^2} (x^2 + y^2) \, dz \, dy \, dx$

4. $\displaystyle\int_0^2 \int_0^{\sqrt{4-y^2}} \int_0^{x^2+y^2} dz \, dx \, dy$

In Exercises 5–8, sketch the solid, change to spherical coordinates, and evaluate the integral.

5. $\displaystyle\int_0^1 \int_0^{\sqrt{1-x^2}} \int_0^{\sqrt{1-x^2-y^2}} (x^2 + y^2 + z^2) \, dz \, dy \, dx$

6. $\displaystyle\int_0^{\sqrt{2}} \int_0^{\sqrt{2-x^2}} \int_0^{\sqrt{4-x^2-y^2}} dz \, dy \, dx$

7. $\displaystyle\int_0^1 \int_{\sqrt{1-x^2}}^{\sqrt{4-x^2}} \int_0^{\sqrt{4-x^2-y^2}} dz \, dy \, dx$
$\displaystyle + \int_1^2 \int_0^{\sqrt{4-x^2}} \int_0^{\sqrt{4-x^2-y^2}} dz \, dy \, dx$

8. $\displaystyle\int_0^1 \int_0^{\sqrt{1-x^2}} \int_0^{1+x^2+y^2} dz \, dy \, dx$

In Exercises 9–14, use cylindrical or spherical coordinates to evaluate the integral (choose the coordinate system you think is more appropriate).

9. $\iiint_S (x^2 + y^2) \, dV$ where S is the cylinder $r \le 2$ between the planes $z = 0$ and $z = 2$.

10. $\iiint_S (x^2 + y^2 + z^2) \, dV$ where S is the sphere $\rho \le 2$.

11. $\iiint_S (x^2 + y^2 + z^2)^2 \, dV$ where S is the sphere $\rho \le 1$.

12. $\iiint_S x^2 y \, dV$ where S is the solid cone $r \le z \le 2$.

13. $\iiint_S dV$ where S is the solid bounded by the cylinder $x^2 + y^2 = 4$, the cone $z^2 = x^2 + y^2$ (with $z \ge 0$), and the xy-plane.

14. $\iiint_S dV$ where S is the set of points inside the sphere $\rho = 2$, but outside the sphere $\rho = 1$.

The integrals in Exercises 15–18 cannot be evaluated (as they stand) by elementary methods. Make a change of variables and evaluate them.

15. $\displaystyle\int_0^1 \int_0^{\sqrt{1-x^2}} \int_0^2 \sin (x^2 + y^2) \, dz \, dy \, dx$

16. $\displaystyle\int_0^1 \int_{-1}^1 \int_0^1 y \cos (x^2 + z^2) \, dz \, dy \, dx$

17. $\displaystyle\int_0^1 \int_0^{\sqrt{1-y^2}} \int_0^{\sqrt{1-y^2-z^2}} \sin \sqrt{(x^2 + y^2 + z^2)^3} \, dx \, dz \, dy$

18. $\displaystyle\int_0^1 \int_0^1 \int_0^1 e^{\sqrt{(x^2 + y^2 + z^2)^3}} \, dz \, dy \, dx$

In Exercises 19–24, use cylindrical coordinates.

19. Find the center of mass of the homogeneous ($\delta = 1$) solid bounded by the xy-plane, the cone $z = \sqrt{x^2 + y^2}$, and the cylinder $x^2 + y^2 = 4$.

20. Find M_{xy}, the moment about the xy-plane, of the solid in Exercise 19.

21. Find the moment of inertia of a homogeneous ($\delta = 1$) solid cylinder of radius a and height h about its axis.

22. Find the moment of inertia of the cylinder in Exercise 21 about a diameter of its base.

23. Find the moment of inertia of the cylinder in Exercise 21 about a diameter through its center.

24. Find the volume of a cone with base radius a and height h.

In Exercises 25–30, use spherical coordinates.

25. Find the volume of an "ice cream cone" bounded by the sphere $x^2 + y^2 + z^2 = 1$ and the cone $z = \sqrt{x^2 + y^2}$.

26. Find the volume of the solid bounded by the graphs of $x^2 + y^2 + z^2 = 1$, $x = y$, and $x = 2y$; x and y positive.

27. Find the moment of inertia I_z of a solid hemisphere $x^2 + y^2 + z^2 \le 4$, $z \ge 0$ if the density is $\delta(x, y, z) = \sqrt{x^2 + y^2 + z^2}$.

28. Find the moment of inertia I_0 of the hemisphere in Exercise 27.

29. Find the mass of the solid between the spheres $\rho = 1$ and $\rho = 2$ if the density is $\delta(x, y, z) = 1/(x^2 + y^2 + z^2)$.

30. Find the moment of inertia of the solid in Exercise 29 about the origin.

16–7   SURFACE AREA

In Section 12–10, we derived a formula for computing the area of a surface of revolution. Now, using double integrals, we will define what is meant by the area of the graph of a function of two variables with continuous partials, and give examples for three different functions.

Suppose that D is an elementary region in the xy-plane and that f has continuous partials on D. Let S be the surface $z = f(x, y)$. We partition D into rectangles with horizontal and vertical lines. A typical rectangle A with area $dA = dy\, dx$ is projected onto the surface S; we will call the projection an S-rectangle (Figure 16–39). The object now is to approximate the area of the S-rectangle. We choose one corner (x_0, y_0) of the rectangle A; to be consistent, we will always choose the corner nearest the origin. Let C_1 and C_2 be the traces of S in the planes $y = y_0$ and $x = x_0$ (Figure 16–39). We now project our typical rectangle A through S and onto the *tangent plane* of S at $(x_0, y_0, f(x_0, y_0))$. The projection may not be a rectangle, but it will be a parallelogram P determined by two tangent vectors $\mathbf{a}$ and $\mathbf{b}$ with their initial points at $(x_0, y_0 f(x_0, y_0))$; see Figure 16–40. We reason that the area of P is a good approximation to the area of the S-rectangle, so the next step is to compute the area of P.

The slope of $\mathbf{a}$ in the tangent plane is $f_x(x_0, y_0)$. Taking a side view through the xz-plane (Figure 16–41), we see that $\mathbf{a} = dx\, \mathbf{i} + f_x(x_0, y_0)\, dx\, \mathbf{k}$. In similar fashion, a side view through the yz-plane would indicate that $\mathbf{b} = dy\, \mathbf{j} + f_y(x_0, y_0)\, dy\, \mathbf{k}$. Since the area of a parallelogram determined by two vectors $\mathbf{a}$ and $\mathbf{b}$ is $|\mathbf{a} \times \mathbf{b}|$, we have

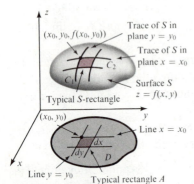

FIGURE 16–39. Approximating the area of a surface.

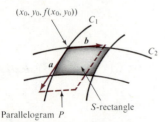

FIGURE 16–40. The vectors $\mathbf{a}$ and $\mathbf{b}$ are tangent vectors that determine a parallelogram P. The area of P approximates the area of the S-rectangle.

$$\mathbf{a} \times \mathbf{b} = \begin{vmatrix} \mathbf{i} & \mathbf{j} & \mathbf{k} \\ dx & 0 & f_x(x_0, y_0)\, dx \\ 0 & dy & f_y(x_0, y_0)\, dy \end{vmatrix}$$

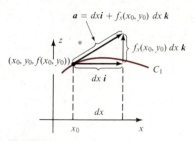

FIGURE 16–41. Side view through the xz-plane.

$$= -f_x(x_0, y_0)\, dy\, dx\, \mathbf{i} - f_y(x_0, y_0)\, dy\, dx\, \mathbf{j} + dy\, dx\, \mathbf{k}$$
$$= [-f_x(x_0, y_0)\mathbf{i} - f_y(x_0, y_0)\mathbf{j} + \mathbf{k}]\, dA \qquad (dA = dy\, dx)$$

and

$$\text{area of } P = |\mathbf{a} \times \mathbf{b}| = \sqrt{[f_x(x_0, y_0)]^2 + [f_y(x_0, y_0)]^2 + 1}\, dA$$

Because this approximates the area of a typical S-rectangle, we define the area of the surface S to be

(16.49)
$$\text{Area of } S = \int\!\!\int_D \sqrt{[f_x(x, y)]^2 + [f_y(x, y)]^2 + 1}\, dA$$

Since the integrand is continuous, the double integral can be evaluated by an iterated integral. Notice the similarity with the formula for arc length derived in Section 6–5.

The presence of the square root in formula (16.49) makes it difficult to evaluate surface area integrals in all but the most elementary cases. Changing to polar coordinates is sometimes useful, and, in certain cases, we can use Simpson's rule to estimate the surface area. Here are three examples.

EXAMPLE 1 Let D be the rectangle with vertices $(0, 0)$, $(0, 2)$, $(1, 2)$, and $(1, 0)$. Let $f(x, y) = x^{3/2}$ and compute the area of the graph of f over D (Figure 16–42).

Solution The first step is to find the integrand for the double integral in (16.49). In this example,

$$f_x(x, y) = \frac{3}{2}x^{1/2}, \text{ so } \left[f_x(x, y)\right]^2 = \frac{9}{4}x$$

Since $f_y(x, y) = 0$, the integrand is $\sqrt{(9/4)x + 1}$. The limits of integration are obtained from the figure. Therefore,

$$\text{Area} = \int\!\!\int_D \sqrt{(9/4)x + 1}\, dA = \int_0^1 \int_0^2 \sqrt{(9/4)x + 1}\, dy\, dx$$
$$= 2 \int_0^1 \sqrt{(9/4)x + 1}\, dx$$
$$= \frac{16}{27}\left[\left(\frac{13}{4}\right)^{3/2} - 1\right]$$

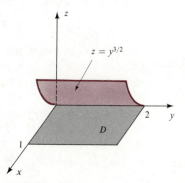

FIGURE 16–42. Example 1.

Remark: The figure and the second to last line above both indicate that the area is 2 times the length of the curve $z = x^{3/2}$ from $x = 0$ to $x = 1$. ∎

EXAMPLE 2 Find the surface area of that part of the hemisphere $z = \sqrt{4 - x^2 - y^2}$ inside the cylinder $x^2 + y^2 = a^2$ with $0 < a \le 2$ (Figure 16–43).

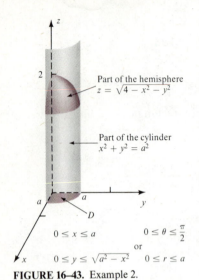

Part of the hemisphere
$z = \sqrt{4 - x^2 - y^2}$

Part of the cylinder
$x^2 + y^2 = a^2$

$0 \le x \le a$

$0 \le \theta \le \dfrac{\pi}{2}$

or

$0 \le y \le \sqrt{a^2 - x^2}$ $0 \le r \le a$

FIGURE 16–43. Example 2.

Solution First find the integrand; $z = f(x, y)$, so

$$f_x(x, y) = \frac{-x}{\sqrt{4 - x^2 - y^2}} \quad f_y(x, y) = \frac{-y}{\sqrt{4 - x^2 - y^2}}$$

Thus,

$$\left[f_x(x, y) \right]^2 + \left[f_y(x, y) \right]^2 + 1 = \frac{4}{4 - x^2 - y^2}$$

Using the figure, we let D be the region $0 \le x \le a$ and $0 \le y \le \sqrt{a^2 - x^2}$; using symmetry and (16.49), we have

$$\text{Area} = 4 \int_0^a \int_0^{\sqrt{a^2 - x^2}} \frac{2}{\sqrt{4 - x^2 - y^2}} \, dy \, dx$$

This is the type of integral that lends itself to a change of variable to polar coordinates. Notice, however, that we *first find the integrand in rectangular coordinates and then change over*. The reason is that definition (16.49) is derived using rectangular coordinates. With that point understood, we make the change to polar coordinates

$$\text{Area} = 4 \int_0^{\pi/2} \int_0^a \frac{2r}{\sqrt{4 - r^2}} \, dr \, d\theta$$

$$= 4\pi \left(2 - \sqrt{4 - a^2} \right)$$

As a check on the answer, we note that if $a = 2$, the result is 8π, the surface area of a hemisphere of radius 2. ∎

EXAMPLE 3 Use Simpson's rule with $n = 6$ to estimate the surface area of the graph of $f(x, y) = \sin y$ over the triangle D with $0 \le y \le \pi$ and $0 \le x \le \pi - y$ (Figure 16–44).

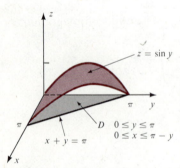

$z = \sin y$

D $0 \le y \le \pi$
$0 \le x \le \pi - y$

$x + y = \pi$

FIGURE 16–44. Example 3.

Solution The integrand is $\sqrt{\cos^2 y + 1}$ and the limits of integration are given. Therefore, the area is

$$\int_0^\pi \int_0^{\pi - y} \sqrt{\cos^2 y + 1} \, dx \, dy = \int_0^\pi (\pi - y) \sqrt{\cos^2 y + 1} \, dy$$

We will use Simpson's rule with $n = 6$ to estimate the integral on the right.

(1) Partition $[0, \pi]$ with equally spaced points

$$x_0 = 0, x_1 = \pi/6, x_2 = \pi/3, \ldots, x_6 = \pi$$

(2) In this example, $f(y) = (\pi - y) \sqrt{\cos^2 y + 1}$. Therefore, we compute

$$f(0) \approx 4.4428829$$
$$4f(\pi/6) \approx 13.8531210$$
$$2f(\pi/3) \approx 4.6832098$$

$$4f(\pi/2) \approx 6.2831853$$
$$2f(2\pi/3) \approx 2.3416049$$
$$4f(5\pi/6) \approx 2.7706243$$
$$f(\pi) = 0$$

(3) Add and multiply by $(\pi - 0)/18$. The sum is 34.3746282; therefore

$$\int_0^\pi (\pi - y) \sqrt{\cos^2 y + 1} \; dy \approx (34.3746282)\left(\frac{\pi}{18}\right)$$

$$\approx 5.9995045$$

Remark: If the integrand contains both variables, then Simpson's rule cannot be used. For example, in the integral

$$\int_a^b \int_c^d \sqrt{1 + \cos^2 x + \cos^2 y} \; dy \, dx$$

we cannot estimate the inner integral with Simpson's rule because x can be any value between a and b.   ∎

EXERCISES

Find the area of the described surfaces.

1. The graph of $z = x^2 + y$ over the triangular region with vertices $(0, 0)$, $(1, 1)$, and $(1, 0)$.

2. The graph of $z = x + 2y^2$ over the triangular region with vertices $(0, 0)$, $(2, 1)$, and $(0, 1)$.

3. That part of the hemisphere $z = \sqrt{1 - x^2 - y^2}$ inside the elliptic cylinder $x^2 + 4y^2 = 1$. *Hint.* $\int 1/\sqrt{1 - x^2 - y^2} \; dy = \sin^{-1}(y/\sqrt{1 - x^2}) + C$.

4. Same as Exercise 3 for any elliptic cylinder $a^2x^2 + y^2 = a^2$ with $0 < a \le 1$.

5. That part of the paraboloid $z = 4 - x^2 - y^2$ above the xy-plane.

6. That part of the paraboloid $z = 4 - x^2 - y^2$ inside the cylinder $x^2 + y^2 = 1$.

7. That part of the hemisphere $z = \sqrt{16 - x^2 - y^2}$ inside the cylinder $x^2 - 4x + y^2 = 0$.

8. Same as Exercise 7 for the cylinder $x^2 - 2x + y^2 = 0$.

9. That part of the plane $x + y + z = 1$ cut out by the cylinder $x^2 + y^2 = 1$.

10. Same as Exercise 9 for the plane $(x/a) + (y/b) + (z/c) = 1$ where a, b, and c are positive.

11. That part of the graph of $z = xy$ inside the cylinder $x^2 + y^2 = 4$.

12. That part of the graph of $z = x^2$ that lies over the rectangular region with vertices $(0, 0)$, $(0, 1)$, $(6, 1)$, and $(6, 0)$.

13. Use Simpson's rule with $n = 6$ to approximate the area of that part of the graph of $z = x^3 + y$ over the rectangle in Exercise 12.

14. Same as Exercise 13 for $z = y + \cos x$.

REVIEW EXERCISES

Evaluate the integrals in Exercises 1–20. For some, you may want to reverse the order of integration or make a change of variables.

1. $\displaystyle\int_0^2 \int_0^1 3xy^2 \; dy \, dx$

2. $\displaystyle\int_{-2}^0 \int_1^2 (x^2 + y^2) \; dy \, dx$

3. $\displaystyle\int_0^2 \int_0^3 \int_0^4 (x + y + z)\,dx\,dy\,dz$

4. $\displaystyle\int_{-1}^2 \int_0^4 \int_0^1 xyz\,dx\,dy\,dz$

5. $\displaystyle\int_0^1 \int_0^{\sqrt{1-x^2}} (x^2 + y^2)\,dy\,dx$

6. $\displaystyle\int_0^1 \int_0^x x^2\, e^{xy}\,dy\,dx$

7. $\displaystyle\int_0^1 \int_0^{1-x} y \sin(y + x)\,dy\,dx$

8. $\displaystyle\int_0^1 \int_0^{\sqrt{4-y^2}} e^{(x^2+y^2)}\,dx\,dy$

9. $\displaystyle\int_0^1 \int_0^{x^3} \int_0^y 2xy^2 z\,dz\,dy\,dx$

10. $\displaystyle\int_0^1 \int_0^{\arcsin x} \cos y\,dy\,dx$

11. $\displaystyle\int_0^1 \int_0^{\sqrt{1-y^2}} \int_0^{\sqrt{1-y^2-z^2}} 2z\,dx\,dz\,dy$

12. $\displaystyle\int_0^1 \int_0^{\sqrt{1-y^2}} \int_0^{\sqrt{1-y^2-z^2}} (x^2 + y^2 + z^2)\,dx\,dz\,dy$

13. $\displaystyle\int_{-1}^1 \int_{-\sqrt{1-x^2}}^{\sqrt{1-x^2}} \int_0^{1+x^2+y^2} dz\,dy\,dx$

14. $\displaystyle\int_0^1 \int_1^{\sqrt{1-x^2}} \sqrt{x^2 + y^2}\,dy\,dx$

15. $\displaystyle\int_0^1 \int_{-\sqrt{1-x^2}}^{\sqrt{1-x^2}} \sin(x^2 + y^2)\,dy\,dx$

16. $\displaystyle\int_0^1 \int_0^{\sqrt{1-x^2}} \int_0^{\sqrt{1-x^2-y^2}} (x^2 + y^2)\,dz\,dy\,dx$

17. $\displaystyle\int_{\pi/2}^{\pi} \int_0^{\tan x} \frac{1}{1 + y^2}\,dy\,dx$

18. $\displaystyle\int_0^1 \int_0^{\arctan x} \sec^2 y\,dy\,dx$

19. $\displaystyle\int_0^1 \int_0^{\arccos x} e^{\sin y}\,dy\,dx$

20. $\displaystyle\int_0^1 \int_0^1 \int_0^1 y \sin(x + z)\,dz\,dy\,dx$

In Exercises 21–24, write the double integral of a general function $z = f(x, y)$ over D as an iterated integral. For some, you may have to break D into two or more pieces.

21. D is the region bounded by the curve $x = y^2$ and the line $x + y = 2$.

22. D is the region bounded by the curves $y = \sqrt{1 - x^2}$ and $y = x^2$.

23. D is the region inside the parallelogram with vertices $(0, 0)$, $(1, 1)$, $(3, 2)$, and $(2, 1)$.

24. D is the region bounded by the curves $y = x^2/2$, $y = -x^2 + 2$, and the line $y = x$.

In Exercises 25–28, write the triple integral of a general function $z = f(x, y, z)$ over S as an iterated integral. For some, you may have to break S into two or more solids.

25. S is the solid bounded by the xy-plane and the surface $z = 4 - x^2 - y^2$.

26. S is the solid bound by the yz-plane and the surface $x = y^2 + z^2 - 1$.

27. S is the solid in the first octant bounded by the cylinder $x^2 + y^2 = 1$ and the planes $x + y + z = 1$, $z = 0$, and $z = 1$.

28. S is the solid inside the sphere $x^2 + y^2 + z^2 = 1$.

29. Use a double integral to find the area of the region in Exercise 21.

30. Use a double integral to find the area of the region in Exercise 22.

31. Use a triple integral to find the volume of the solid in Exercise 25.

32. Use a triple integral to find the volume of the solid in Exercise 26.

In Exercise 33–38, use either double or triple integrals to find the volume of the described solid.

33. The solid between the graph of $f(x, y) = 3x^2 y^2$ and the rectangle with vertices $(0, 0)$, $(1, 0)$, $(1, 2)$, and $(0, 2)$.

34. The solid bounded by the coordinate planes and the plane $x + 2y + 3z = 6$.

35. The intersection of the solid cylinders $x^2 + y^2 \le 9$ and $x^2 + z^2 \le 9$.

36. The intersection of the solid cylinder $x^2 + y^2 \le 4$ and the solid sphere $x^2 + y^2 + z^2 \le 9$.

37. The solid bounded by the surfaces $z = x^2 + y^2$ and $z = 4 - x^2 - y^2$.

38. The solid in the first octant bounded by $z = x^2 + y^2$ and the plane $x + y + z = 1$.

In Exercises 39–42, find the area of the indicated surfaces.

39. That part of the paraboloid $z = x^2 + y^2 - 4$ below the xy-plane.

40. That part of the surface $z = x^2 + y^2 - 4$ inside the cylinder $x^2 + y^2 = 4$.

41. The graph of $f(x, y) = 3x + y^2$ over the rectangle with vertices $(0, 0), (1, 0), (1, 1)$, and $(0, 1)$.

42. The graph of $f(x, y) = x^3 + y^3$ over the triangle with vertices $(0, 0), (1, 1)$, and $(1, 0)$. Use Simpson's rule with $n = 4$.

43. Find the center of mass of a hemisphere $x^2 + y^2 + z^2 = 1, z \geq 0$ if the density is $\delta(x, y, z) = x^2 + y^2 + z^2$.

44. Find the moment I_y of the hemisphere in Exercise 43.

45. An "ice cream cone" is bounded by the sphere $x^2 + y^2 + z^2 = 1$ and the cone $z = \sqrt{x^2 + y^2}$. Find its moment I_0 if the density of the cone part is the constant 1, but the density of the "ice cream" part is the constant 3.

46. Find the moment M_{xy} of the homogeneous solid with $\delta = 1$ bounded by $z = x^2 + y^2, x^2 + y^2 = 4$, and the xy-plane.

End of
2nd year
1st semester

17

ELEMENTS OF VECTOR CALCULUS

Vector calculus is a mathematical tool for describing physical phenomena in the plane and in space.* In this chapter, we define line and surface integrals. The former arise in the study of work done by a force applied along a curved path. The latter arise in the study of fluid flows, electricity, and magnetism.

Section 17–1 discusses the concepts of smooth curves and vector fields; both are basic to the material in the remainder of the chapter. Section 17–2 introduces the notion of a line integral and describes its relationship to an ordinary integral. In Section 17–3, we illustrate how the *principle of the conservation of energy* is derived mathematically. Sections 17–4 and 17–5 contain the celebrated theorems of Green (relating line integrals to double integrals), Gauss (relating surface integrals to triple integrals), and Stokes (relating line integrals to surface integrals). These theorems form the basis for most applications of vector calculus.

17–1 SMOOTH CURVES AND VECTOR FIELDS

Both topics mentioned in the title are fundamental to the material of this chapter. The idea is to introduce them and discuss their properties in a separate section now, rather than interrupt the flow of our presentation later.

Smooth Curves

Let C be a curve, or path, in the plane with parametric equations

$$x = x(t) \quad y = y(t) \quad a \leq t \leq b$$

Then C is traced out by the tip of a position (vector) function

$$\mathbf{r}(t) = x(t)\mathbf{i} + y(t)\mathbf{j}$$
$$= (x(t), y(t)) \quad a \leq t \leq b$$

---

*The American physicist J. Willard Gibbs (1839–1903) was principally responsible for consolidating the material discussed in this chapter into a cohesive field of study known as *vector analysis*.

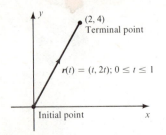

FIGURE 17–1. Orientation of a curve in the plane.

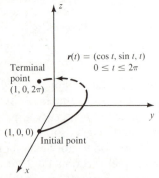

FIGURE 17–2. Orientation of a curve in space.

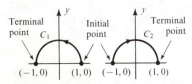

FIGURE 17–3. C_1 and C_2 are the same set of points but with opposite orientation; we write $C_1 = -C_2$.

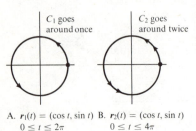

A. $r_1(t) = (\cos t, \sin t)$ B. $r_2(t) = (\cos t, \sin t)$
   $0 \le t \le 2\pi$ $0 \le t \le 4\pi$

FIGURE 17–4. $C_1 \ne C_2$.

Note: Everything that follows also holds for curves in space; simply append a third parametric equation $z = z(t)$ for the curve and a third component $z(t)\mathbf{k}$ to **r**.

(17.1)

> **DEFINITION**
> A position function **r** defined on an interval $[a, b]$ is **smooth** if
>
> (1) **r** is continuous on $[a, b]$.
> (2) **r'** is continuous on $[a, b]$.
> (3) $\mathbf{r}'(t)$ is never **0**.
>
> A curve C is **smooth** if it has at least one smooth position function.

A curve C has a natural direction, or *orientation,* induced by **r**. As t increases from a to b, the curve is mapped out from the *initial point* $\mathbf{r}(a)$ to the *terminal point* $\mathbf{r}(b)$; the orientation is indicated by arrows drawn on the curve (Figures 17–1, 17–2). Sometimes two curves consist of the same set of points, but have opposite orientations (Figure 17–3). Or, they have the same set of points, but the position function of one of them traces out the path more than once (Figure 17–4). We will need to distinguish among these curves.

Suppose that C_1 and C_2 have the same set of points and that each has a position function that traces out the path the same number of times. If C_1 and C_2 have the same orientation, then we say they are *equivalent* and write

(17.2) $C_1 = C_2$

If they have opposite orientations, we say they are *opposite* and write

$$C_1 = -C_2$$

Given a curve C with position function **r** defined on $[a, b]$, it is often necessary to find a position function for the opposite curve $-C$. One way to do that is to define a new function $\mathbf{r}_1$ by

$$\mathbf{r}_1(t) = \mathbf{r}(-t) \quad t \text{ in } [-b, -a]$$

Then $\mathbf{r}_1$ is the position function of a curve, say C_1. The initial point of C_1 is $\mathbf{r}_1(-b) = \mathbf{r}(b)$ the terminal point of C, and the terminal point of C_1 is $\mathbf{r}_1(-a) = \mathbf{r}(a)$ the initial point of C. Moreover, as t ranges over $[-b, -a]$, $\mathbf{r}_1(t)$ traces out the same set of points as **r** but in the opposite direction. In other words, $C_1 = -C$.

(17.3)

> If **r**, defined on $[a, b]$, is a position function for the curve C, then
>
> $$\mathbf{r}_1(t) = \mathbf{r}(-t) \quad t \text{ in } [-b, -a]$$
>
> is a position function for the opposite curve $-C$.

EXAMPLE 1 Let C be the helix $\mathbf{r}(t) = (\cos t, \sin t, t)$ for t in $[0, 2\pi]$ (Figure 17–2). Find a position function for $-C$.

Solution Define a new position function $\mathbf{r}_1$ on $[-2\pi, 0]$ by

$$\mathbf{r}_1 = \mathbf{r}(-t) = (\cos(-t), \sin(-t), -t)$$
$$= (\cos t, -\sin t, -t) \quad -2\pi \le t \le 0$$

The initial point is $\mathbf{r}_1(-2\pi) = (1, 0, 2\pi)$, the terminal point is $\mathbf{r}_1(0) = (1, 0, 0)$ and $\mathbf{r}_1$ traces out the helix in the opposite direction from C. ■

There is a way of *adding* certain curves. If the terminal point of C_1 coincides with the initial point of C_2, then the curve C consisting of the points in both curves along with the inherited orientation is called the *sum* of C_1 and C_2 and is denoted by

$$C = C_1 + C_2$$

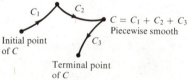

This definition can be extended to any finite number of curves (Figure 17–5).

Observe that there may be sharp corners in the sum of two or more curves; at these points, the derivative of any position function will not exist. Therefore, the sum of smooth curves may not be smooth; but it will be what we call *piecewise smooth.*.

FIGURE 17–5. The sum of smooth curves is a piecewise smooth curve.

(17.4)

> **DEFINITION**
>
> A curve C is **piecewise smooth** if it is the sum of a finite number of smooth curves.

EXAMPLE 2 Express the (unit) square with vertices $(0, 0), (0, 1), (1, 1)$ and $(1, 0)$ as the sum of four paths so that the square has a counterclockwise direction (Figure 17–6).

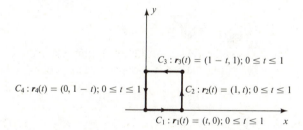

FIGURE 17–6. Example 2. The unit square is piecewise smooth.

Solution

(a) The first path, C_1, is the straight line from $(0, 0)$ to $(1, 0)$; its position function is $\mathbf{r}_1(t) = (t, 0)$ for $0 \le t \le 1$.

(b) The next path, C_2, joins $(1, 0)$ to $(1, 1)$; its position function is $\mathbf{r}_2(t) = (1, t)$ for $0 \le t \le 1$.

(c) The third path, C_3, joins $(1, 1)$ to $(0, 1)$; to make the orientation from right to left, we use the position function $\mathbf{r}_3(t) = (1 - t, 1)$ for $0 \le t \le 1$.

(d) The last path, C_4, joins $(0, 1)$ to $(0, 0)$. The position function $\mathbf{r}_4(t) = (0, 1 - t)$ for $0 \le t \le 1$ induces the correct orientation.

Figure 17–6 shows the piecewise smooth sum $C_1 + C_2 + C_3 + C_4$.   ■

A curve is called *simple* if it does not cross itself; it is called a *closed curve* if the initial and terminal points coincide. A circle or the unit square are examples of simple closed curves. If two curves C_1 and C_2 have the same initial and terminal points but no other points in common, then $C_1 + (-C_2)$, usually written as $C_1 - C_2$, is a simple closed curve.

EXAMPLE 3   Let

$$C_1 : \mathbf{r}_1(t) = (t, 2t) \quad 0 \le t \le 2$$
$$C_2 : \mathbf{r}_2(t) = (t, t^2) \quad 0 \le t \le 2$$

Then both curves have the same initial point $(0, 0)$, the same terminal point $(2, 4)$, and no other points in common. Therefore, $C_1 - C_2$ is a simple closed curve (Figure 17–7).   ■

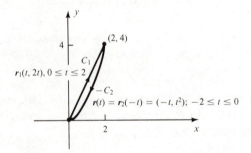

FIGURE 17–7. Example 3.
$C = C_1 - C_2$ is a simple closed curve.

Vector Fields

A **vector field** is a function $\mathbf{F}$ that assigns to each point of the plane (or space) a vector in the plane (or space). The general notation is

$$\mathbf{F}(x, y) = M(x, y)\mathbf{i} + N(x, y)\mathbf{j}$$

or

$$\mathbf{F}(x, y, z) = M(x, y, z)\mathbf{i} + N(x, y, z)\mathbf{j} + P(x, y, z)\mathbf{k}$$

Vector fields are used to describe such phenomena as gravitational attraction, electromagnetic force, and fluid flow. For example, a vector field that assigns to each point in space the vector that represents the gravitational force of the earth on a unit mass at that point, is called a *force (vector) field*. Figure 17–8 illustrates this vector field. A vector field that assigns to each particle of water in a river its velocity vector is called a *velocity (vector) field* (Figure 17–9). The figures show only a few of the vectors in each field; it should be remembered that each point is assigned a vector.

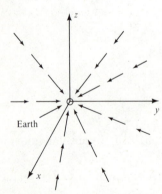

FIGURE 17–8. Force field about the earth.

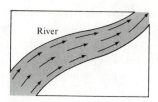

FIGURE 17–9. Velocity field.

EXAMPLE 4 Sketch the velocity field of a disc rotating at a constant rate about its center in a counterclockwise direction.

Solution At each point (x, y) on the disc, the velocity vector is tangent to the circle of radius $\sqrt{x^2 + y^2}$. The magnitude of the velocity increases as the point moves away from the center (why?). A sketch is drawn in Figure 17–10. ∎

EXAMPLE 5 *(Inverse square vector fields).*

(a) According to Newton's law of gravitation, the gravitational force **F** exerted by a particle of mass M at the origin on a unit mass at (x, y, z) is

$$\mathbf{F} = \frac{-GM}{(x^2 + y^2 + z^2)^{3/2}} (x\mathbf{i} + y\mathbf{j} + z\mathbf{k}) = \frac{-GM}{r^2} \mathbf{u} \qquad (G \text{ constant})$$

where $\mathbf{r} = x\mathbf{i} + y\mathbf{j} + z\mathbf{k}$, $r = |\mathbf{r}|$, and $\mathbf{u}$ is the unit vector $\mathbf{r}/r$.

(b) According to Coulomb's law, the electrostatic force **E** exerted by a point charge of Q coulombs at the origin on a unit charge of like sign at (x, y, z) is

$$\mathbf{E} = \frac{kQ}{(x^2 + y^2 + z^2)^{3/2}} (x\mathbf{i} + y\mathbf{j} + z\mathbf{k}) = \frac{kQ}{r^2} \mathbf{u} \qquad (k > 0)$$

where r and $\mathbf{u}$ are defined as above. Both fields are examples of what we call *inverse square* vector fields. The earth's force field (Figure 17–8) is an inverse square vector field. ∎

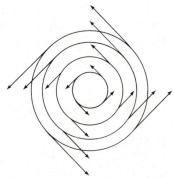

FIGURE 17–10. Example 4. Velocity field of a rotating disc.

EXAMPLE 6 *(Gradient).* The gradient of a function of two or three variables is a vector field that will be important to us. If $z = f(x, y)$, then the vector field defined by the gradient is denoted by ∇f; that is,

$$\nabla f(x, y) = f_x(x, y)\mathbf{i} + f_y(x, y)\mathbf{j}$$

If $w = f(x, y, z)$, then we use the symbol ∇f in the same way. The gradient of a function f is called a *conservative vector field* and f is called a *potential function* (this terminology is explained in Section 17–3). ∎

EXAMPLE 7 Let

$$\mathbf{F}(x, y, z) = \frac{c}{(x^2 + y^2 + z^2)^{3/2}} (x\mathbf{i} + y\mathbf{j} + z\mathbf{k})$$

where c is a constant, be an inverse square vector field. If

$$f(x, y, z) = \frac{-c}{\sqrt{x^2 + y^2 + z^2}}$$

then

$$\nabla f(x, y, z) = \frac{c}{(x^2 + y^2 + z^2)^{3/2}} (x\mathbf{i} + y\mathbf{j} + z\mathbf{k}) = \mathbf{F}(x, y, z) \qquad \text{(Check)}$$

Thus, **F** is a conservative force field with potential function f. ∎

We conclude with the observation that the concepts of limit, continuity, differentiation, and integration can be defined for vector fields by treating each component function separately as a function of two (or three) variables. For example, suppose, that $\mathbf{F}(x, y) = M(x, y)\mathbf{i} + N(x, y)\mathbf{j}$. Then $\mathbf{F}$ is continuous if M and N are continuous, $\mathbf{F}$ has continuous partials if M and N have continuous partials, and so on.

EXERCISES

In Exercises 1–6, sketch the curves with the given position function. Indicate the direction.

1. $\mathbf{r}(t) = (t + 1, 3t), 0 \le t \le 1$

2. $\mathbf{r}(t) = (2t - 1, t + 1), 0 \le t \le 1$

3. $\mathbf{r}(t) = (t - 1, t^2), 0 \le t \le 1$

4. $\mathbf{r}(t) = (t, t^2 + 1), 0 \le t \le 1$

5. $\mathbf{r}(t) = (\sin t, \cos t, t), 0 \le t \le \pi/2$

6. $\mathbf{r}(t) = (\cos t, 3 \sin t, t), 0 \le t \le \pi/2$

7–12. For each of the curves C in Exercises 1–6, write a position function for $-C$.

In Exercises 13–18, find a position function for the described curve, or write it as the sum of two or more curves and find a position function for each.

13. The semicircle $x^2 + y^2 = 1, y \le 0$ with counterclockwise orientation.

14. The circle $x^2 + y^2 = 1$ with clockwise orientation.

15. The circle $(x - 1)^2 + (y - 1)^2 = 1$ with counterclockwise orientation.

16. The ellipse $4x^2 + 9y^2 = 36$ with counterclockwise orientation.

17. The triangular path from $(0, 0, 0)$ to $(1, 1, 1)$ to $(1, 0, 0)$ and back to the origin.

18. The triangular path from $(0, 0, 0)$ to $(1, 0, 0)$ to $(2, 1, 1)$ and back to the origin.

In Exercises 19–26, sketch enough of the vector field to obtain a general pattern.

19. $\mathbf{F}(x, y) = y\mathbf{i} - x\mathbf{j}$

20. $\mathbf{F}(x, y) = x\mathbf{i} + y\mathbf{j}$

21. $\mathbf{F}(x, y) = (x^2 + y^2)^{-1/2} (x\mathbf{i} + y\mathbf{j})$

22. $\mathbf{F}(x, y) = \mathbf{i} + \mathbf{j}$

23. $\mathbf{F}(x, y, z) = -x\mathbf{i} - y\mathbf{j} - z\mathbf{k}$

24. $\mathbf{F}(x, y, z) = x\mathbf{i} + \mathbf{j} + \mathbf{k}$

25. $\mathbf{F}(x, y, z) = x\mathbf{i}$

26. $\mathbf{F}(x, y, z) = \mathbf{i} + \mathbf{j} + \mathbf{k}$

In Exercises 27–31, sketch the conservative vector field that is the gradient of the given function.

27. $f(x, y) = xy$

28. $f(x, y) = x + y$

29. $f(x, y) = x^2 + y^2$

30. $f(x, y) = x^2 + 2y^2$

31. $f(x, y, z) = 1/\sqrt{x^2 + y^2 + z^2}$

17–2   LINE INTEGRALS

A *line integral* is an integral taken over a curve C in a plane or in space. The integrand may involve a vector field $\mathbf{F}$ or a scalar (that is, real valued) function f. Although they are quite similar in nature, each type has its own notation

$$\int_C \mathbf{F} \cdot d\mathbf{r} \quad \text{and} \quad \int_C f(x, y) \, ds$$

and its own interpretations and applications.

Line Integrals of Vector Fields

Let C be a smooth curve in the plane with position function

$$\mathbf{r}(t) = x(t)\mathbf{i} + y(t)\mathbf{j}$$
$$= (x(t), y(t)) \quad a \leq t \leq b$$

(What follows is also true for curves in space; simply append a third component $z(t)\mathbf{k}$ to $\mathbf{r}(t)$.) Let $\mathbf{F}$ be a continuous vector field defined on a region containing the curve C. The definition of the line integral of $\mathbf{F}$ over C is motivated by the following question: If $\mathbf{F}$ is a force field, how much work is done by $\mathbf{F}$ in moving a particle along C?

To answer that question, we partition the interval $[a, b]$ into small subintervals. Each small subinterval will correspond to a small subarc on C (Figure 17–11). Now we approximate the work done by $\mathbf{F}$ over a typical subarc, say from $\mathbf{r}(t_{i-1})$ to $\mathbf{r}(t_i)$.

Recall from our discussion of work in Section 14–3 that only the component of the force along the line of motion does any work (Figure 17–12). For motion on a curved path, only the component of force along the line of a *tangent vector* does any work. Now, we choose any point in our subarc and compute the component of $\mathbf{F}$ on a tangent vector at that point. Figure 17–13 illustrates this computation if our choice is the left endpoint of the subarc; $\mathbf{F}(\mathbf{r}(t_{i-1}))$ is the force and $\mathbf{r}'(t_{i-1})$ is a tangent vector. The component of $\mathbf{F}$ on $\mathbf{r}'$ is

$$(17.5) \qquad \text{comp}_{\mathbf{r}'} \mathbf{F} = \frac{\mathbf{F} \cdot \mathbf{r}'}{|\mathbf{r}'|} \qquad\qquad ((14.19) \text{ in Section } 14\text{–}3)$$

If the subarc is very small and $\mathbf{F}$ is continuous, we can assume that the component (17.5) is almost constant; that is, we assume

$$(17.6) \qquad \frac{\mathbf{F}(\mathbf{r}(t)) \cdot \mathbf{r}'(t)}{|\mathbf{r}'(t)|} \approx c_i \quad \text{for } t_{i-1} \leq t \leq t_i$$

The distance that this constant force moves the particle is the arc length s_i where

$$s_i = \int_{t_{i-1}}^{t_i} \sqrt{[x'(t)]^2 + [y'(t)]^2} \, dt \qquad ((13.28) \text{ in Section } 13\text{–}5)$$

$$(17.7) \qquad = \int_{t_{i-1}}^{t_i} |\mathbf{r}'(t)| \, dt$$

Let W_i be the work done by $\mathbf{F}$ on the typical subarc. It follows from (17.6) and (17.7) that

$$W_i \approx c_i \int_{t_{i-1}}^{t_i} |\mathbf{r}'(t)| \, dt = \int_{t_{i-1}}^{t_i} c_i |\mathbf{r}'(t)| \, dt$$

$$\approx \int_{t_{i-1}}^{t_i} \frac{\mathbf{F}(\mathbf{r}(t)) \cdot \mathbf{r}'(t)}{|\mathbf{r}'(t)|} |\mathbf{r}'(t)| \, dt$$

$$= \int_{t_{i-1}}^{t_i} \mathbf{F}(\mathbf{r}(t)) \cdot \mathbf{r}'(t) \, dt$$

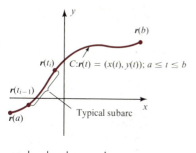

FIGURE 17–11. A partition of $[a, b]$ into subintervals induces a partition of C into subarcs.

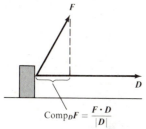

$$\text{Comp}_{\mathbf{D}}\mathbf{F} = \frac{\mathbf{F} \cdot \mathbf{D}}{|\mathbf{D}|}$$

FIGURE 17–12. Only the component of $\mathbf{F}$ on $\mathbf{D}$ does work in moving the object.

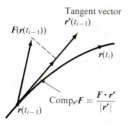

$$\text{Comp}_{\mathbf{r}'}\mathbf{F} = \frac{\mathbf{F} \cdot \mathbf{r}'}{|\mathbf{r}'|}$$

FIGURE 17–13. $\text{Comp}_{\mathbf{r}'} \mathbf{F}$ does the work.

The sum over all subarcs approximates the total work done. Since **F** is continuous and C is smooth, the integral of $\mathbf{F} \cdot \mathbf{r}'$ exists, and we define the work W done by **F** in moving a particle along C as

(17.8)
$$W = \int_a^b \mathbf{F}(\mathbf{r}(t)) \cdot \mathbf{r}'(t)\, dt$$

EXAMPLE 1 Find the work done by the force field $\mathbf{F}(x, y) = xy\mathbf{i} + 3\mathbf{j}$ in moving a particle from $(1, 0)$ to $(0, 1)$ along the quarter circle $\mathbf{r}(t) = (\cos t, \sin t)$, $0 \le t \le \pi/2$ (Figure 17–14). Force is measured in newtons, distance in meters.

Solution To use (17.8), we write out $\mathbf{F}(\mathbf{r}(t))$ and $\mathbf{r}'(t)$, compute the dot product, and integrate in the usual way.

(1) $\mathbf{F}(\mathbf{r}(t)) = \mathbf{F}(\cos t, \sin t) = (\cos t)(\sin t)\mathbf{i} + 3\mathbf{j}$

$\qquad = (\cos t \sin t, 3)$

(2) $\mathbf{r}'(t) = (-\sin t, \cos t)$

(3) $\mathbf{F}(\mathbf{r}(t)) \cdot \mathbf{r}'(t) = -\sin^2 t \cos t + 3 \cos t$

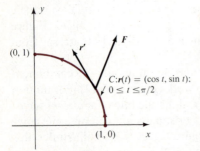

FIGURE 17–14. Example 1.

The limits of integration in this problem are $a = 0$ and $b = \pi/2$. Therefore,

$$W = \int_0^{\pi/2} (-\sin^2 t \cos t + 3 \cos t)\, dt$$

$$= -\frac{1}{3}\sin^3 t + 3 \sin t \Big|_0^{\pi/2} = \frac{8}{3} \text{joules} \qquad \blacksquare$$

The integral in (17.8) is called a *line integral* and it can be applied to any vector field. Moreover, if we write $d\mathbf{r} = \mathbf{r}'(t)\, dt$, then the integrand takes the form $\mathbf{F} \cdot d\mathbf{r}$. This is the notation commonly used.

DEFINITION

Let C be a curve with position function **r** defined on $[a, b]$. Let **F** be a vector field defined on a region D containing C. The **line integral of F over** C, denoted by $\int_C \mathbf{F} \cdot d\mathbf{r}$, is defined by

(17.9)

$$\int_C \mathbf{F} \cdot d\mathbf{r} = \int_a^b \mathbf{F}(\mathbf{r}(t)) \cdot \mathbf{r}'(t)\, dt$$

whenever the integral on the right exists.

If **F** is continuous and C is smooth, then the line integral exists. If **F** is a force field, then the work done by **F** is computed using a line integral.

Although the notation may be a bit confusing at first, evaluating a line integral is not nearly as complicated as it looks. The three-step process is as follows:

(1) Find a position function **r** for C (if one is not given), and compute **r**′.

(2) Write $\mathbf{F}(x, y)$ in terms of t; that is, write out $\mathbf{F}(\mathbf{r}(t))$.

(3) Compute the dot product $\mathbf{F}(\mathbf{r}(t)) \cdot \mathbf{r}'(t)$, and integrate from a to b (the domain of **r**) in the usual way.

EXAMPLE 2   Let $\mathbf{F}(x, y) = x^2 y\mathbf{i} + \mathbf{j}$ and evaluate $\int_C \mathbf{F} \cdot d\mathbf{r}$ where C is the straight line from $(0, 0)$ to $(2, 4)$.

Solution   We follow the three steps outlined above.

(1) $\mathbf{r}(t) = (t, 2t), 0 \leq t \leq 2$ is a position function for C (Figure 17–15); $\mathbf{r}'(t) = (1, 2)$

(2) $\mathbf{F}(\mathbf{r}(t)) = \mathbf{F}(t, 2t) = 2t^3\mathbf{i} + \mathbf{j} = (2t^3, 1)$

(3) $\mathbf{F}(\mathbf{r}(t)) \cdot \mathbf{r}'(t) = (2t^3, 1) \cdot (1, 2) = 2t^3 + 2$. Therefore,

$$\int_C \mathbf{F} \cdot d\mathbf{r} = \int_0^2 (2t^3 + 2)\, dt$$

$$= \frac{t^4}{2} + 2t \Big|_0^2 = 12 \quad \blacksquare$$

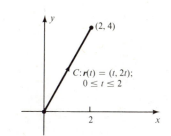

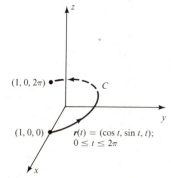

FIGURE 17–15. Example 2.

The next example is in three dimensions.

EXAMPLE 3   Let $\mathbf{F}(x, y, z) = xy\mathbf{i} + \mathbf{j} + 2z\mathbf{k}$. Let C be the helix traced out by $\mathbf{r}(t) = (\cos t, \sin t, t), 0 \leq t \leq 2\pi$ (Figure 17–16). Find the line integral of **F** over C.

Solution

(1) **r** is given, and $\mathbf{r}'(t) = (-\sin t, \cos t, 1)$

(2) $\mathbf{F}(\mathbf{r}(t)) = \mathbf{F}(\cos t, \sin t, t)$

$$= (\cos t)(\sin t)\mathbf{i} + \mathbf{j} + 2t\,\mathbf{k}$$

(3) $\mathbf{F}(\mathbf{r}(t)) \cdot \mathbf{r}'(t) = -\sin^2 t \cos t + \cos t + 2t$, so

$$\int_C \mathbf{F} \cdot d\mathbf{r} = \int_0^{2\pi} \mathbf{F}(\mathbf{r}(t)) \cdot \mathbf{r}'(t)\, dt$$

$$= \int_0^{2\pi} (-\sin^2 t \cos t + \cos t + 2t)\, dt$$

$$= -\frac{1}{3}\sin^3 t + \sin t + t^2 \Big|_0^{2\pi} = 4\pi^2 \quad \blacksquare$$

FIGURE 17–16. Example 3.

As we saw in the preceding section, a curve can have many different position functions, and two different orientations. The next example illustrates that the choice of position function does not affect the line integral, but the orientation does.

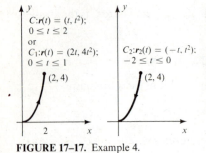

FIGURE 17–17. Example 4.

EXAMPLE 4 Let $F(x, y) = -y\mathbf{i} + x\mathbf{j}$. Let

(a) C be the curve $\mathbf{r}(t) = (t, t^2), 0 \le t \le 2$.

(b) C_1 be the curve $\mathbf{r}_1(t) = (2t, 4t^2), 0 \le t \le 1$.

(c) C_2 be the curve $\mathbf{r}_2(t) = (-t, t^2), -2 \le t \le 0$.

Then $C = C_1$ and $C = -C_2$ (Figure 17–17). Find the line integral of $\mathbf{F}$ over each curve.

Solution

(a) $\mathbf{r}'(t) = (1, 2t)$ and $\mathbf{F}(\mathbf{r}(t)) = -t^2\mathbf{i} + t\mathbf{j} = (-t^2, t)$

Therefore,

$$\int_C \mathbf{F} \cdot d\mathbf{r} = \int_0^2 \mathbf{F}(\mathbf{r}(t)) \cdot \mathbf{r}'(t)\, dt$$

$$= \int_0^2 (-t^2 + 2t^2)\, dt = \frac{1}{3}t^3 \Big|_0^2 = \frac{8}{3}$$

(b) $\mathbf{r}_1'(t) = (2, 8t)$ and $\mathbf{F}(\mathbf{r}_1(t)) = (-4t^2, 2t)$

Therefore,

$$\int_{C_1} \mathbf{F} \cdot d\mathbf{r}_1 = \int_0^1 (-8t^2 + 16t^2)\, dt = \frac{8}{3}t^3 \Big|_0^1 = \frac{8}{3}$$

(c) $\mathbf{r}_2'(t) = (-1, 2t)$ and $\mathbf{F}(\mathbf{r}_2(t)) = (-t^2, -t)$

Therefore,

$$\int_{C_2} \mathbf{F} \cdot d\mathbf{r}_2 = \int_{-2}^0 (t^2 - 2t^2)\, dt = -\frac{1}{3}t^3 \Big|_{-2}^0 = -\frac{8}{3} \qquad \blacksquare$$

It can be shown that the illustrations in Example 4 always hold.

(17.10)

> If the line integral of $\mathbf{F}$ over C exists, then
>
> (1) $\displaystyle\int_C \mathbf{F} \cdot d\mathbf{r} = \int_{C_1} \mathbf{F} \cdot d\mathbf{r}$ whenever $C = C_1$
>
> (2) $\displaystyle\int_C \mathbf{F} \cdot d\mathbf{r} = -\int_{-C} \mathbf{F} \cdot d\mathbf{r}$

Now suppose that C is a sum of curves; that is, $C = C_1 + C_2 + \cdots + C_n$. Then we define the line integral over C as the sum of the line integrals over the curves C_i.

> **DEFINITION**
> If $C = C_1 + \cdots + C_n$ and **F** is a vector field defined on a region containing C, then
>
> $$\int_C \mathbf{F} \cdot d\mathbf{r} = \sum_{i=1}^{n} \int_{C_i} \mathbf{F} \cdot d\mathbf{r}_i$$
>
> whenever the integrals on the right exist.

(17.11)

If **F** is continuous and C is piecewise smooth (that is, each C_i is smooth), then the sum on the right exists.

EXAMPLE 5   Let $\mathbf{F}(x, y) = 4x^2\mathbf{i} + 3y\mathbf{j}$ be a force field and compute the work done by **F** in moving a particle around the unit square (Figure 17–18).

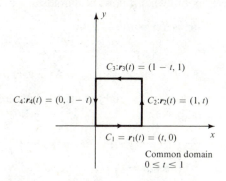

FIGURE 17–18. Example 5.

Solution   The square is written as the sum of four paths; the work done on each path is computed, and the total work is the sum. The figure indicates the position functions for each part. For C_1, $\mathbf{r}_1'(t) = (1, 0)$ and $\mathbf{F}(\mathbf{r}_1(t)) = (4t^2, 0)$; therefore,

$$\int_{C_1} \mathbf{F} \cdot d\mathbf{r}_1 = \int_0^1 4t^2 \, dt = \frac{4}{3}t^3 \Big]_0^1 = \frac{4}{3}$$

For C_2, $\mathbf{r}_2'(t) = (0, 1)$ and $\mathbf{F}(\mathbf{r}_2(t)) = (4, 3t)$; therefore,

$$\int_{C_2} \mathbf{F} \cdot d\mathbf{r}_2 = \int_0^1 3t \, dt = \frac{3}{2}$$

For C_3, $\mathbf{r}_3'(t) = (-1, 0)$ and $\mathbf{F}(\mathbf{r}_3(t)) = (4(1 - t)^2, 3)$; therefore,

$$\int_{C_3} \mathbf{F} \cdot d\mathbf{r}_3 = \int_0^1 -4(1 - t)^2 \, dt = \frac{4}{3}(1 - t)^3 \Big]_0^1 = -\frac{4}{3}$$

(Negative work means that the force is applied *against* the motion.) For C_4, $\mathbf{r}_4'(t) = (0, -1)$ and $\mathbf{F}(\mathbf{r}_4(t)) = (0, 3(1 - t))$; therefore,

$$\int_{C_4} \mathbf{F} \cdot d\mathbf{r}_4 = \int_0^1 -3(1 - t)\, dt = \frac{3}{2}(1 - t)^2 \Big|_0^1 = -\frac{3}{2}$$

The total work done is

$$W = \int_C \mathbf{F} \cdot d\mathbf{r} = \sum_{i=1}^{4} \int_{C_i} \mathbf{F} \cdot d\mathbf{r}_i = 0$$

This force field did no work at all! That is, the *net* work is zero. (We will have more to say about this phenomenon in the next section.) ■

Line integrals of vector fields can also be written in another form that will be useful in later work. Let

$$\mathbf{F}(x, y, z) = M(x, y, z)\mathbf{i} + N(x, y, z)\mathbf{j} + P(x, y, z)\mathbf{k}$$

be a vector field and let C be a curve in space with position function $\mathbf{r}(t) = x(t)\mathbf{i} + y(t)\mathbf{j} + z(t)\mathbf{k}$. Just as a change of pace, we are using space here as an illustration, but what follows also holds in a plane. If we write

$$d\mathbf{r} = dx\mathbf{i} + dy\mathbf{j} + dz\mathbf{k}$$

then

$$\mathbf{F} \cdot d\mathbf{r} = M(x, y, z)\, dx + N(x, y, z)\, dy + P(x, y, z)\, dz$$

The expression on the right is called a **differential form.** Using this form, the line integral becomes

(17.12)
$$\int_C \mathbf{F} \cdot d\mathbf{r} = \int_C M(x, y, z)\, dx + N(x, y, z)\, dy + P(x, y, z)\, dz$$

The integral on the right is evaluated as before; that is, translate everything into terms of the parameter and integrate as usual.

EXAMPLE 6 Evaluate $\int_C xy\, dx + z\, dy + yz\, dz$ where C is the line segment from $(0, 0, 0)$ to $(1, 3, 5)$.

Solution A position function is $\mathbf{r}(t) = (t, 3t, 5t)$, $0 \le t \le 1$; thus, $x = t$, $y = 3t$, and $z = 5t$. It follows that $dx = dt, dy = 3\, dt, dz = 5\, dt$,

$$xy\, dx = 3t^2\, dt, \quad z\, dy = 5t(3\, dt), \quad yz\, dz = 15t^2(5\, dt)$$

and the sum is $(78t^2 + 15t)\, dt$. Therefore,

$$\int_C xy\, dx + z\, dy + yz\, dz = \int_0^1 (78t^2 + 15t)\, dt = \frac{67}{2}$$ ■

Line Integrals of Scalar Functions

Line integrals of scalar functions are motivated by the following question: If a thin wire has the shape of a piecewise smooth curve C and a continuous density function δ, what is its mass?

To answer this question, let C be a curve in space (or in a plane) with parametric equations

$$x = x(t) \quad y = y(t) \quad z = z(t) \qquad a \le t \le b$$

We partition $[a, b]$ into small subintervals that, in turn, partition C into small subarcs. The mass of a typical subarc is approximately

$$\delta(x, y, z) \, \Delta s$$

where Δs is the length of the subarc and (x, y, z) is a typical point in it. The sum over all subarcs approximates the mass of the wire, and the approximations get closer and closer to a single number as the subarcs get smaller and smaller. We denote that number by the symbol $\int_C \delta(x, y, z) \, ds$ and define the mass M as

(17.13)
$$M = \int_C \delta(x, y, z) \, ds$$

The integral in (17.13) is also called a *line integral* and it can be defined for any scalar function f. To evaluate the integral, we convert everything into terms of the parameter. Since the arc length formula yields

$$ds = \sqrt{[x'(t)]^2 + [y'(t)]^2 + [z'(t)]^2} \, dt$$

we have

(17.14)
$$\int_C f(x, y, z) \, ds =$$
$$\int_a^b f(x(t), y(t), z(t)) \sqrt{[x'(t)]^2 + [y'(t)]^2 + [z'(t)]^2} \, dt$$

EXAMPLE 7   A wire is in the shape of the helix $\mathbf{r}(t) = (\cos t, \sin t, t)$, $0 \le t \le \pi/2$. Its density at any point is $\delta(x, y, z) = xy$. Find its mass.

Solution   The mass is given by the line integral $\int_C \delta(x, y, z) \, ds$. Here, $x = \cos t$, $y = \sin t$, and $z = t$; so $\delta(x, y, z) = (\cos t)(\sin t)$. Therefore, by (17.14),

$$M = \int_C \delta(x, y, z) \, ds = \int_0^{\pi/2} \cos t \sin t \sqrt{(-\sin t)^2 + (\cos t)^2 + 1} \, dt$$

$$= \sqrt{2} \int_0^{\pi/2} \cos t \sin t \, dt$$

$$= \sqrt{2} \left[\frac{1}{2} \sin^2 t \right]_0^{\pi/2} = \frac{\sqrt{2}}{2} \qquad \blacksquare$$

EXERCISES

In Exercises 1–14, evaluate $\int_C \mathbf{F} \cdot d\mathbf{r}$ for the given $\mathbf{F}$ and $\mathbf{r}$.

1. $\mathbf{F}(x, y) = 3x\mathbf{i} + 2y\mathbf{j}; \mathbf{r}(t) = (1 - t, 3t), 0 \le t \le 1$

2. $\mathbf{F}(x, y) = -y\mathbf{i} + x\mathbf{j}; \mathbf{r}(t) = (4t + 1, t), 0 \le t \le 2$

3. $\mathbf{F}(x,y) = 4x^2\mathbf{i} - 3y\mathbf{j}; \mathbf{r}(t) = (t, \sqrt{1 - t^2}), -1 \le t \le 1$

4. $\mathbf{F}(x, y) = x^2y\mathbf{i} + \mathbf{j}; \mathbf{r}(t) = (t, t^2), 0 \le t \le 2$

5. $\mathbf{F}(x, y, z) = x\mathbf{i} - xy\mathbf{j} + 2z\mathbf{k}; \mathbf{r}(t) = (\cos t, \sin t, t), 0 \le t \le 2\pi$

6. $\mathbf{F}(x, y, z) = x^2y\mathbf{i} + \mathbf{j} - 4z\mathbf{k}; \mathbf{r}(t) = (t, t^2, t^3), 0 \le t \le 1$

7. $\mathbf{F}(x, y, z) = -zy\mathbf{i} + zx\mathbf{j} + xy\mathbf{k}; \mathbf{r}(t) = (\cos t, \sin t, t), 0 \le t \le 2\pi$

8. $\mathbf{F}(x, y, z) = z\mathbf{i} + x\mathbf{j} - y\mathbf{k}; \mathbf{r}(t) = (t, t^2, t), 0 \le t \le 1$

9. $\mathbf{F}(x, y, z) = x^2\mathbf{i} + y^2\mathbf{j} + z^2\mathbf{k}; \mathbf{r}(t) = (\cos t, \sin t, 1), 0 \le t \le 2\pi$

10. $\mathbf{F}(x, y, z) = 2x\mathbf{i} - 3z\mathbf{j} + 4y\mathbf{k}; \mathbf{r}(t) = (t, 2t, 3t), 0 \le t \le 1$

11. $\mathbf{F}(x, y) = e^x\mathbf{i} + e^y\mathbf{j}; \mathbf{r}(t) = (2t, 1 - t), 0 \le t \le 1$

12. $\mathbf{F}(x, y) = \sin x\mathbf{i} + \mathbf{j}; \mathbf{r}(t) = (t, \sin t), 0 \le t \le \pi/2$

13. $\mathbf{F}(x, y) = \sec^2 x\mathbf{i} + y\mathbf{j}; \mathbf{r}(t) = (t, \sin t), 0 \le t \le \pi/4$

14. $\mathbf{F}(x, y) = (1/x)\mathbf{i} + (1/y)\mathbf{j}; \mathbf{r}(t) = (1 + 2t, 1 + 3t), 0 \le t \le 1$

In Exercises 15–18, evaluate the line integrals given in differential form.

15. $\int_C x\, dx + y\, dy; \mathbf{r}(t) = (t - 1, t^2), 0 \le t \le 1$

16. $\int_C x^2\, dx + y^2\, dy; \mathbf{r}(t) = (\sin t, \cos t), 0 \le t \le 2\pi$

17. $\int_C x\, dx + y\, dy + z\, dz; \mathbf{r}(t) = (\cos t, \sin t, t), 0 \le t \le \pi/2$

18. $\int_C 2xy\, dx + dy + y\, dz; \mathbf{r}(t) = (t, t^2, t^3), 0 \le t \le 1$

In Exercises 19–20, evaluate the line integrals with respect to arc length.

19. $\int_C x^3\, ds; \mathbf{r}(t) = (3t, t^3), 0 \le t \le 1$

20. $\int_C (x + y)\, ds; \mathbf{r}(t) = (\cos t, \sin t), 0 \le t \le \pi$

21. Find the work done by the force field $\mathbf{F}(x, y) = xy\mathbf{i} + 3y\mathbf{j}$ in moving a particle from $(1, 1)$ to $(2, 10)$ along the following paths (force in newtons, distance in meters):

   (a) C is the line segment from $(1, 1)$ to $(2, 10)$.

   (b) C_1 is the sum of the line segments from $(1, 1)$ to $(2, 1)$ and from $(2, 1)$ to $(2, 10)$.

   (c) C_2 is the part of the parabola $y = 3x^2 - 2$ joining $(1, 1)$ and $(2, 10)$.

   (d) $C_3 = C - C_2$.

22. Same as Exercise 21 for the force field $\mathbf{F}(x, y) = x^2\mathbf{i} + y\mathbf{j}$.

23. Same as Exercise 21 for the force field $\mathbf{F}(x, y) = y\mathbf{i} + x\mathbf{j}$.

24. Same as Exercise 21 for the force field $\mathbf{F}(x, y) = x^2y\mathbf{i} + \mathbf{j}$.

25. Find the work done by the force field $\mathbf{F}(x, y, z) = yz\mathbf{i} + xz\mathbf{j} + xy\mathbf{k}$ in moving an object from $(1, 0, 0)$ to $(0, 1, \pi/2)$ along the following paths (force in pounds, distance in feet):

   (a) C is the line segment from $(1, 0, 0)$ to $(0, 1, \pi/2)$.

   (b) C_1 is the sum of the line segments from $(1, 0, 0)$ to $(0, 0, 0)$ and from $(0, 0, 0)$ to $(0, 1, \pi/2)$.

   (c) C_2 is part of a circular helix of radius 1 that joins the two points.

   (d) $C_3 = C - C_2$.

26. Same as Exercise 25 for the force field $\mathbf{F}(x, y, z) = x\mathbf{i} + y\mathbf{j} + z\mathbf{k}$.

27. Same as Exercise 25 for the force field $\mathbf{F}(x, y, z) = xy\mathbf{i} + \mathbf{j} + \mathbf{k}$.

28. Same as Exercise 25 for the force field $\mathbf{F}(x, y, z) = y\mathbf{i} + \mathbf{j} + z\mathbf{k}$.

29. A positive charge of 1 C is located at the origin. Coulomb's law says that the force exerted on a negative charge of q C located at (x, y, z) is

$$\mathbf{F} = \frac{-kq}{(x^2 + y^2 + z^2)^{3/2}} (x\mathbf{i} + y\mathbf{j} + z\mathbf{k}) \quad \text{newtons}$$

where k is a constant. How much work (joules) is done by this force in moving a particle with a negative charge of 10 C once around the circular path $x^2 + y^2 = 1, z = 0$ (distance in meters)?

30. How much work is done by the force in Exercise 29 in moving the particle on a straight path from $(2, 2, 2)$ to $(1, 1, 1)$?

31. A wire is bent in the shape of the helix $\mathbf{r}(t) = (\cos t, \sin t, t), 0 \le t \le \pi$. If the density is constant, $\delta = 1$, find the mass of the wire.

32. Find the center of mass of the wire in Exercise 31. (The moments are defined as usual; that is, $M_{yz} = \int_C x \, \delta(x, y, z) \, ds$, and so on.)

33. Suppose $\mathbf{r}(t) = (\cos t, 0, t), 0 \le t \le \pi$. Let $f(x, y, z) = 3x^2 + 2yz$, and show that

$$\int_C \nabla f \cdot d\mathbf{r} = f(\mathbf{r}(\pi)) - f(\mathbf{r}(0))$$

17-3   CONSERVATIVE VECTOR FIELDS

Some vector fields $\mathbf{F}$ are *path independent;* that is, given two points A and B, the value of the line integral of $\mathbf{F}$ is the same over any piecewise smooth path from A to B. Such vector fields are also called *conservative,* for reasons that will be explained shortly. If $\mathbf{F}$ is path independent, then we use the notation

$$\int_A^B \mathbf{F} \cdot d\mathbf{r}$$

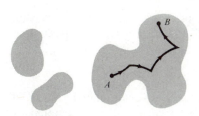

A. Not connected.   B. Open and connected.
FIGURE 17-19. Regions in the plane.

to denote the line integral of $\mathbf{F}$ over any piecewise smooth path from A to B.

There is a theorem concerning path independent vector fields that is reminiscent of the Fundamental Theorem of Calculus. To state the theorem, we need to define an open connected region. A region R in the plane (or space) is *open* if every point in R is the center of some disc (or sphere) that is contained in R. It is *connected* if any two points in R can be joined by a piecewise smooth curve contained in R (Figure 17-19). Now we can state the theorem.

FUNDAMENTAL THEOREM

Let $\mathbf{F}$ be a continuous vector field on an open connected region R in the plane.

(1) Suppose $\mathbf{F}$ is path independent. Fix (x_0, y_0) in R and define a function f on R by

(17.15)
$$f(x, y) = \int_{(x_0, y_0)}^{(x, y)} \mathbf{F} \cdot d\mathbf{r}$$

Then f has continuous partials and $\nabla f = \mathbf{F}$.

(2) If g is any function such that $\nabla g = \mathbf{F}$, then $\mathbf{F}$ is path independent and

$$\int_{(x_0, y_0)}^{(x_1, y_1)} \mathbf{F} \cdot d\mathbf{r} = g(x_1, y_1) - g(x_0, y_0)$$

The proof is presented at the end of this section. We state here that the theorem also holds in three dimensions. At each step in the proof, you can easily see what adjustments should be made for the three variable case.

The theorem says that if you know a function f with $\nabla f = \mathbf{F}$, then f can be used to evaluate the line integral of $\mathbf{F}$.

EXAMPLE 1 Let $\mathbf{F} = (8xy + 3y^3)\mathbf{i} + (4x^2 + 9xy^2)\mathbf{j}$ and find $\int_C \mathbf{F} \cdot d\mathbf{r}$ where C is the path consisting of four line segments from $(0, 0)$ to $(1, 3)$ to $(3, 1)$ to $(3, 0)$ to $(1, 1)$; see Figure 17–20.

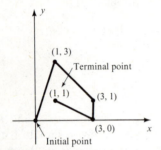

FIGURE 17–20. The curve C in Example 1.

Solution Without the Fundamental Theorem, you would have to compute the sum of the line integrals over each segment. But you can easily verify that $\mathbf{F}$ is the gradient of

$$f(x, y) = 4x^2y + 3xy^3$$

Therefore, by part (2) of the theorem, $\mathbf{F}$ is path independent and

$$\int_C \mathbf{F} \cdot d\mathbf{r} = \int_{(0, 0)}^{(1, 1)} \mathbf{F} \cdot d\mathbf{r} = f(1, 1) - f(0, 0) = 7 \qquad \blacksquare$$

The Fundamental Theorem has applications in mechanics. It is here that the name *conservative* arises. Suppose a particle of mass m moves along a smooth curve C with position function $\mathbf{r}$ defined on $[a, b]$. Then

$$(17.16) \quad \mathbf{v} = \mathbf{r}'(t) \quad \mathbf{a} = \mathbf{v}'(t) \quad v = |\mathbf{v}(t)|$$

are, respectively, the velocity vector, the acceleration vector, and the speed of the particle at time t. The *kinetic energy* of the particle at time t, denoted by $K(\mathbf{r}(t))$, is defined by

$$(17.17) \quad K(\mathbf{r}(t)) = \frac{1}{2}m[v(t)]^2$$

According to Newton's second law of motion, the force $\mathbf{F}$ acting on the particle at time t is

$$(17.18) \quad \mathbf{F}(\mathbf{r}(t)) = m\mathbf{a}(t) = m\mathbf{v}'(t)$$

The work done by $\mathbf{F}$ in moving the particle along C is given by the line integral

$$W = \int_C \mathbf{F} \cdot d\mathbf{r} = \int_a^b \mathbf{F}(\mathbf{r}(t)) \cdot \mathbf{r}'(t)\, dt$$

$$= m \int_a^b \mathbf{v}'(t) \cdot \mathbf{v}(t)\, dt \qquad (17.16), (17.18)$$

Since $\mathbf{v}' \cdot \mathbf{v} = (1/2)[\mathbf{v} \cdot \mathbf{v}]' = (1/2)[v^2]'$ (check this out), we continue with

$$= \frac{m}{2} \int_a^b [v^2]'(t)\, dt$$

$$= \frac{1}{2}m[v(b)]^2 - \frac{1}{2}m[v(a)]^2$$

$$= K(\mathbf{r}(b)) - K(\mathbf{r}(a)) \qquad (17.17)$$

This shows that *the work done by* **F** *equals the change in kinetic energy* as the particle moves from $\mathbf{r}(a)$ to $\mathbf{r}(b)$.

Now suppose that **F** is path independent. By part (1) of the Fundamental Theorem, there is a function f such that $\nabla f = \mathbf{F}$. The function f is called a *potential function* of **F** and $-f(\mathbf{r}(t))$ is called the *potential energy* of the particle at time t. Since $\nabla f = \mathbf{F}$, it follows from part (2) of the theorem that the work integral above also has the value

$$W = \int_C \mathbf{F} \cdot d\mathbf{r} = f(\mathbf{r}(b)) - f(\mathbf{r}(a))$$

It follows that

$$f(\mathbf{r}(b)) - f(\mathbf{r}(a)) = K(\mathbf{r}(b)) - K(\mathbf{r}(a))$$

or

$$(17.19) \quad K(\mathbf{r}(a)) - f(\mathbf{r}(a)) = K(\mathbf{r}(b)) - f(\mathbf{r}(b))$$

Equation (17.19) represents the *principle of the conservation of energy;* it says that *if a particle is acted upon by a path independent force, then the sum of the potential and kinetic energies is conserved.* That is why a path independent vector field is called **conservative.** Here is another interesting observation about conservative vector fields.

(17.20) A continuous vector field **F** defined on an open connected set R is conservative if and only if its integral around any piecewise smooth closed curve in R is 0.

The proof of (17.20) is as follows: If **F** is conservative, then the integral depends only on the endpoints of the path. But the endpoints of a closed curve are identical; therefore, the integral is 0. Now suppose that the integral over any closed curve is 0 and that (x_0, y_0) and (x_1, y_1) are in R. Given any two smooth curves C_1 and C_2 from (x_0, y_0) to (x_1, y_1), we know that $C = C_1 - C_2$ is a piecewise smooth closed curve (Figure 17–21) so that

$$0 = \int_C \mathbf{F} \cdot d\mathbf{r} = \int_{C_1} \mathbf{F} \cdot d\mathbf{r} - \int_{C_2} \mathbf{F} \cdot d\mathbf{r}$$

This shows that the integral over C_1 equals the one over C_2; that is, **F** is path independent, hence conservative.

Statement (17.20) says that a conservative force field does zero work in moving an object around a closed path! Since the earth's gravitational force field is conservative (see Exercise 23), it does zero work in moving, for example, a satellite around a closed orbit.

Given a vector field **F**, it is natural to ask whether or not **F** is conservative and, if so, how is a potential function for **F** determined? The answer to the second question is illustrated in Examples 3 and 4 below. To answer the first question,

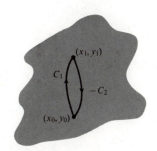

FIGURE 17–21. $C_1 - C_2$ is a closed curve.

we suppose that $F(x, y) = M(x, y)\mathbf{i} + N(x, y)\mathbf{j}$ where M and N have continuous partials throughout the whole plane. If F is conservative, then there is some function f such that $\nabla f = F$; that is,

$$f_x(x, y)\mathbf{i} + f_y(x, y)\mathbf{j} = M(x, y)\mathbf{i} + N(x, y)\mathbf{j}$$

Thus, $f_x = M$ and $f_y = N$. If M and N have continuous partials, then f has continuous second partials, in which case,

$$M_y = f_{xy} = f_{yx} = N_x$$

It can be shown that the converse is also true. That is,

(17.21)

> Suppose $F = M(x, y)\mathbf{i} + N(x, y)\mathbf{j}$ where M and N have continuous partials throughout the plane. Then F is conservative if and only if
> $$M_y = N_x$$

EXAMPLE 2   Determine whether or not $F(x, y) = 4x^2 y\mathbf{i} + 3y^2\mathbf{j}$ is conservative. If so, find a potential function.

Solution   Since

$$\frac{\partial}{\partial y}(4x^2 y) = 4x^2 \quad \text{and} \quad \frac{\partial}{\partial x}(3y^2) = 0$$

it follows from (17.21) that F is not conservative, so there is no potential function.   ∎

EXAMPLE 3   Determine whether or not $F(x, y) = (6x + y)\mathbf{i} + (x + 2y)\mathbf{j}$ is conservative. If so, find a potential function.

Solution   The component functions of F have continuous partials everywhere and

$$\frac{\partial}{\partial y}(6x + y) = 1 = \frac{\partial}{\partial x}(x + 2y)$$

So F is conservative. Now we want to find a function f with

(17.22) $f_x = 6x + y \quad \text{and} \quad f_y = x + 2y$

Integrating the first equation with respect to x (treating y as a constant) yields

(17.23) $f(x, y) = 3x^2 + xy + g(y)$

where $g(y)$ is a function of y alone; that is the "constant of integration" when we integrate with respect to x. By (17.22), we want $f_y = x + 2y$; so we differentiate (17.23) with respect to y,

$$f_y = x + g'(y)$$

set the result equal to $x + 2y$,

$$x + g'(y) = x + 2y$$

and solve for g'. It follows that $g'(y) = 2y$, so $g(y) = y^2 + C$. Therefore, by (17.23),

$$f(x, y) = 3x^2 + xy + y^2 + C$$

Check: $f_x = 6x + y$ and $f_y = x + 2y$; so $\nabla f = (6x + y)\mathbf{i} + (x + 2y)\mathbf{j} = \mathbf{F}.$ ∎

The counterpart of (17.21) in space is as follows:

(17.24) Suppose $\mathbf{F}(x, y, z) = M(x, y, z)\mathbf{i} + N(x, y, z)\mathbf{j} + P(x, y, z)\mathbf{k}$ where M, N, and P have continuous partials throughout space. Then $\mathbf{F}$ is conservative if and only if

$$M_y = N_x \quad M_z = P_x \quad N_z = P_y$$

EXAMPLE 4 Determine whether or not

$$\mathbf{F}(x, y, z) = yz\mathbf{i} + (xz + e^z \cos y)\mathbf{j} + (xy + e^z \sin y)\mathbf{k}$$

is conservative. If so, find a potential function.

Solution We leave it to you to verify that the partials are continuous throughout space and satisfy the equations in (17.24). Now we want to find a function f with $\nabla f = \mathbf{F}$; that is, a function $w = f(x, y, z)$ with

(17.25) $f_x = yz \quad f_y = xz + e^z \cos y \quad f_z = xy + e^z \sin y$

Integrating the first equation with respect to x, holding y and z constant, yields,

(17.26) $f(x, y, z) = xyz + g(y, z)$

where the "constant of integration" is a function of y and z. Now we differentiate (17.26) with respect to y, set the result equal to the second equation in (17.25) and solve for g.

$$f_y = xz + g_y(y, z) = xz + e^z \cos y$$

The conclusion is that $g_y(y, z) = e^z \cos y$. Integrating g_y with respect to y yields

$$g(y, z) = e^z \sin y + h(z) \qquad\qquad (h \text{ is a function of } z)$$

Now (17.26) becomes

(17.27) $f(x, y, z) = xyz + e^z \sin y + h(z)$

It remains only to find the function h. Differentiating (17.27) with respect to z, and setting the result equal to the third equation in (17.25), we have

$$f_z = xy + e^z \sin y + h'(z) = xy + e^z \sin y$$

Therefore, $h'(z) = 0$, so h is a constant function and the final answer is

$$f(x, y, z) = xyz + e^z \sin y + C$$

Check: $f_x = yz, f_y = xz + e^z \cos y$, and $f_z = xy + e^z \sin y$; so $\nabla f = \mathbf{F}$. ∎

CAUTION The hypotheses of both (17.21) and (17.24) state that the component functions M, N, and P must have continuous partials throughout the plane or space. This condition can be relaxed somewhat to include any region without "holes"; that is, any region R with the property that if C is a closed curve in R, then all points inside the curve are also in R. If R does have a hole (even a single missing point), then $\mathbf{F}$ may fail to be conservative even though M, N, and P have continuous partials on R and satisfy the equations relating their partials (see Exercise 27).

Proof of the Fundamental Theorem

For part (1), (x_0, y_0) is fixed and (x, y) is any other point in R. First we pick a point (x_1, y) in R with $x_1 \neq x$ and join (x_0, y_0) to (x_1, y) with any smooth curve in R. Then we join (x_1, y) to (x, y) with a *horizontal line segment*. This can always be done because R is open and connected (Figure 17–22A). Since $\mathbf{F}$ is path independent, we can write

$$f(x, y) = \int_{(x_0, y_0)}^{(x_1, y)} \mathbf{F} \cdot d\mathbf{r} + \int_{(x_1, y)}^{(x, y)} \mathbf{F} \cdot d\mathbf{r}$$

In the first integral on the right, both limits have fixed first coordinates so its value does not depend on x. Therefore,

$$\frac{\partial f}{\partial x}(x, y) = 0 + \frac{\partial}{\partial x} \int_{(x_1, y)}^{(x, y)} \mathbf{F} \cdot d\mathbf{r}$$

Writing $\mathbf{F} \cdot d\mathbf{r}$ in differential form $M(x, y)\, dx + N(x, y)\, dy$ and observing that $dy = 0$ on the horizontal line segment, we continue with

$$\frac{\partial f}{\partial x}(x, y) = \frac{\partial}{\partial x} \int_{(x_1, y)}^{(x, y)} M(x, y)\, dx$$

$$= M(x, y)$$

The last line follows from the Fundamental Theorem of Calculus because y is fixed so that the integrand may be considered as a continuous function of x alone. Similarly, if we choose a path with a vertical line segment (Figure 17–22B) and differentiate with respect to y, then a similar argument shows that

$$\frac{\partial f}{\partial y}(x, y) = N(x, y)$$

This proves that f has continuous partials and $\nabla f = \mathbf{F}$.

For part (2), suppose that $\nabla g = \mathbf{F}$; we want to show that $\mathbf{F}$ is path independent and the line integral can be evaluated using the values of g at the endpoints. Both objectives can be proved at once by fixing two points (x_0, y_0) and (x_1, y_1)

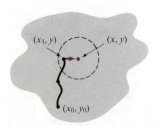

A.

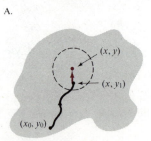

B.

FIGURE 17–22. Proving the Fundamental Theorem.

in R and letting C be *any* smooth path from the first point to the second. (If C is only piecewise smooth, then it can be broken up into smooth parts.) Let $\mathbf{r}$, defined on $[a, b]$, be a position function for C. The composition $g \circ \mathbf{r}$ is a scalar function defined on $[a, b]$ and, by the chain rule,

$$(17.28) \quad (g \circ \mathbf{r})'(t) = \frac{\partial g}{\partial x}(\mathbf{r}(t))\frac{dx}{dt}(t) + \frac{\partial g}{\partial y}(\mathbf{r}(t))\frac{dy}{dt}(t)$$

$$= \nabla g(\mathbf{r}(t)) \cdot \mathbf{r}'(t)$$

Since $\nabla g = \mathbf{F}$, we have

$$\int_C \mathbf{F} \cdot d\mathbf{r} = \int_a^b \nabla g(\mathbf{r}(t)) \cdot \mathbf{r}'(t)\, dt$$

$$= \int_a^b (g \circ \mathbf{r})'(t)\, dt \qquad\qquad (17.28)$$

$$= g \circ \mathbf{r}(b) - g \circ \mathbf{r}(a) \qquad \text{(Fund. Th. of Calculus)}$$

$$= g(\mathbf{r}(b)) - g(\mathbf{r}(a))$$

$$= g(x_1, y_1) - g(x_0, y_0)$$

This completes the proof.

EXERCISES

In Exercises 1–14, determine whether or not the given vector field is conservative. If so, find a potential function.

1. $\mathbf{F}(x, y) = \cos x\mathbf{i} + \sin y\mathbf{j}$

2. $\mathbf{F}(x, y) = 3y\mathbf{i} + 2\sqrt{y}\mathbf{j}$

3. $\mathbf{F}(x, y) = 2xy^3\mathbf{i} + 6x^2\mathbf{j}$

4. $\mathbf{F}(x, y) = y\mathbf{i} + x\mathbf{j}$

5. $\mathbf{F}(x, y) = 6xy^3\mathbf{i} + 9x^2y^2\mathbf{j}$

6. $\mathbf{F}(x, y) = \sqrt{y}\mathbf{i} + (x/2\sqrt{y})\mathbf{j}$

7. $\mathbf{F}(x, y, z) = yz\mathbf{i} + xz\mathbf{j} + xy\mathbf{k}$

8. $\mathbf{F}(x, y, z) = e^{yz}(\mathbf{i} + xz\mathbf{j} + xy\mathbf{k})$

9. $\mathbf{F}(x, y, z) = \cos y\mathbf{j}$

10. $\mathbf{F}(x, y, z) = 3xy\mathbf{i} + x\mathbf{j} - 6z^2\mathbf{k}$

11. $\mathbf{F}(x, y, z) = 2xy^3z\mathbf{i} + x^2z\mathbf{j} + x^2\mathbf{k}$

12. $\mathbf{F}(x, y, z) = e^z\mathbf{k}$

13. $\mathbf{F}(x, y, z) = e^z(y\mathbf{i} + x\mathbf{j} + xy\mathbf{k})$

14. $\mathbf{F}(x, y, z) = 3x\mathbf{i} + y^2\mathbf{j} + \mathbf{k}$

In Exercises 15–20, refer back to the given exercise and evaluate the given integral.

15. Exercise 1; $\int_C \mathbf{F} \cdot d\mathbf{r}$ where $\mathbf{r}(t) = (\pi t/2, \pi t)$, $0 \le t \le 1$.

16. Exercise 3; $\int_C \mathbf{F} \cdot d\mathbf{r}$ where $\mathbf{r}(t) = (1 - 2t, 3t)$, $0 \le t \le 1$.

17. Exercise 5; $\int_C \mathbf{F} \cdot d\mathbf{r}$ where C is the sum of three line segments from $(0, 1)$ to $(-3, 1)$ to $(4, 8)$, to $(2, 2)$.

18. Exercise 7; $\int_C \mathbf{F} \cdot d\mathbf{r}$ where $\mathbf{r}(t) = (t, 2t, 3t)$, $0 \le t \le 1$.

19. Exercise 9; $\int_C \mathbf{F} \cdot d\mathbf{r}$ where $\mathbf{r}(t) = (t, 2t, t^2)$, $0 \le t \le \pi$.

20. Exercise 11; $\int_C \mathbf{F} \cdot d\mathbf{r}$ where $\mathbf{r}(t) = (t, t^2, t^3)$, $0 \le t \le 1$.

21. A conservative force field $\mathbf{F}$, measured in newtons, moves an object with mass m so that its position at any time t is $\mathbf{r}(t) = (\cos t, \sin t, t^2)$; distance is measured in meters. (a) How much work is done by the force during the time interval from $t = 0$ to $t = 1$? (b) What is the increase (or decrease) in potential energy of the particle during that time? (c) Find the force $\mathbf{F}$ when $t = \pi/2$.

22. Same as Exercise 21 for the position function $\mathbf{r}(t) = (3t^2, t^3 + 4t^2 - 3, t^4)$ from $t = 0$ to $t = 1$.

23. Gravitational force fields are of the form

$$F = \frac{C}{(x^2 + y^2 + z^2)^{3/2}}(x\mathbf{i} + y\mathbf{j} + z\mathbf{k}) \quad C \text{ a constant}$$

(a) Show that **F** is conservative by finding a potential function for it. (b) Show that if $a^2 + b^2 + c^2 = p^2 + q^2 + r^2$, then zero work is done by this force in moving an object from (a, b, c) to (p, q, r).

24. A rocket weighing 2 tons falls to earth from a height of 1,000 miles. How much work is done by gravity? (Let the center of the earth be the origin and take the radius of the earth to be 4,000 miles. The weight of an object near the surface of the earth is the magnitude of the force of gravity on that object.)

25. An elevator carries 10 people to the 50th floor of an office building and returns to ground level empty. If the elevator weighs 5,000 lbs, the people weigh 1,500 lbs, and the 50th floor is 500 ft above ground level, how much work is done by the elevator against gravity for the round trip? (See the remarks in Exercise 24.)

Exercises 26 and 27 illustrate the delicate nature of the results stated in (17.20) and (17.21).

26. Let $\mathbf{F}(x, y) = x^2\mathbf{i} + yx^4\mathbf{j}$. Show that (a) $\int_C \mathbf{F} \cdot d\mathbf{r} = 0$ if C is any closed curve of the form $x^2/a^2 + y^2/b^2 = 1$, and (b) $\int_C \mathbf{F} \cdot d\mathbf{r} \neq 0$ if C is the boundary of the unit square. Is **F** conservative?

27. Let

$$\mathbf{F}(x, y) = \frac{-y}{x^2 + y^2}\mathbf{i} + \frac{x}{x^2 + y^2}\mathbf{j}$$

Show that (a) $M_y = N_x$ for all (x, y) except the origin, and (b) $\int_C \mathbf{F} \cdot d\mathbf{r} \neq 0$ if C is the circle $x^2 + y^2 = 1$. Is **F** conservative?

17–4 GREEN'S THEOREM*

Green's Theorem establishes a relationship between double integrals and line integrals. It is a singularly beautiful and important part of mathematics and has far-reaching applications in physics and engineering.

Here is a rough idea of Green's Theorem. Given a region D, its boundary C (Figure 17–23), and a vector field $\mathbf{F}(x, y) = M(x, y)\mathbf{i} + N(x, y)\mathbf{j}$, then finding the line integral of **F** around C is the same as finding the double integral of $\partial N/\partial x - \partial M/\partial y$ over D. In symbols,

$$(17.29) \quad \int_C M\, dx + N\, dy = \iint_D \left(\frac{\partial N}{\partial x} - \frac{\partial M}{\partial y}\right) dA$$

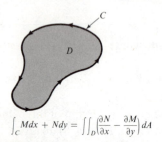

$$\int_C M dx + N dy = \iint_D \left(\frac{\partial N}{\partial x} - \frac{\partial M}{\partial y}\right) dA$$

FIGURE 17–23. Illustration of Green's Theorem.

EXAMPLE 1 Let $\mathbf{F}(x, y) = x^2\mathbf{i} + x\mathbf{j}$ and let D be the unit disc $x^2 + y^2 \leq 1$; the boundary of D is the unit circle C (Figure 17–24). Verify that equation (17.29) holds.

Solution First we evaluate the line integral. A position function for C is $\mathbf{r}(t) = (\cos t, \sin t), 0 \leq t \leq 2\pi$. Therefore,

$$x = \cos t \quad dx = -\sin t\, dt$$
$$y = \sin t \quad dy = \cos t\, dt$$
$$M(x, y) = x^2 = \cos^2 t \quad \text{and} \quad N(x, y) = x = \cos t$$

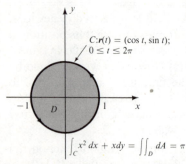

$$\int_C x^2 dx + x dy = \iint_D dA = \pi$$

FIGURE 17–24. Example 1.

*George Green (1793–1841) was a famous English mathematician and physicist who did extensive work in electricity and magnetism.

The resulting line integral is

$$\int_C M \, dx + N \, dy = \int_0^{2\pi} (\cos^2 t)(-\sin t \, dt) + (\cos t)(\cos t \, dt)$$

$$= \int_0^{2\pi} (-\cos^2 t \sin t + \cos^2 t) \, dt$$

$$= \left[\frac{\cos^3 t}{3} + \frac{t}{2} + \frac{\sin 2t}{4} \right]_0^{2\pi} = \pi$$

Now we evaluate the double integral. Since

$$\frac{\partial N}{\partial x} - \frac{\partial M}{\partial y} = \frac{\partial}{\partial x}(x) - \frac{\partial}{\partial y}(x^2) = 1 - 0$$

it follows that

$$\int\int_D \left(\frac{\partial N}{\partial x} - \frac{\partial M}{\partial y} \right) dA = \int\int_D dA = \text{area of } D = \pi$$

The two values agree, so (17.29) holds. ∎

EXAMPLE 2   Let $\mathbf{F}(x, y) = x^2 y \mathbf{i} + y^3 \mathbf{j}$, let D be the unit square, and let C be its boundary (Figure 17-25). Verify that equation (17.29) holds.

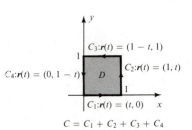

$C_3 : r(t) = (1 - t, 1)$

$C_2 : r(t) = (1, t)$

$C_4 : r(t) = (0, 1 - t)$ D

$C_1 : r(t) = (t, 0)$

$C = C_1 + C_2 + C_3 + C_4$

FIGURE 17-25. Example 2.

Solution   First the line integral. The boundary C is broken up into four line segments whose position functions are indicated in the figure. In this example, $M(x, y) = x^2 y$ and $N(x, y) = y^3$. Therefore,

$$\int_{C_1} M \, dx + N \, dy = \int_0^1 0 \, dt = 0 \qquad \text{(Because } y = 0\text{)}$$

$$\int_{C_2} M \, dx + N \, dy = \int_0^1 t^3 \, dt = \frac{1}{4} \qquad \text{(Because } x = 1, y = t\text{)}$$

$$\int_{C_3} M \, dx + N \, dy = \int_0^1 -(1 - t)^2 \, dt = -\frac{1}{3} \quad \text{(Because } x = 1 - t, y = 1\text{)}$$

$$\int_{C_4} M \, dx + N \, dy = \int_0^1 -(1 - t)^3 \, dt = -\frac{1}{4} \quad \text{(Because } x = 0, y = 1 - t\text{)}$$

Summing these values yields,

$$\int_C M \, dx + N \, dy = -\frac{1}{3}$$

Now the double integral. Since

$$\frac{\partial N}{\partial x} - \frac{\partial M}{\partial y} = \frac{\partial}{\partial x}(y^3) - \frac{\partial}{\partial y}(x^2 y) = 0 - x^2$$

it follows that

$$\iint_D \left(\frac{\partial N}{\partial x} - \frac{\partial M}{\partial y}\right) dA = \int_0^1 \int_0^1 -x^2 \, dy \, dx = -\frac{1}{3}$$

The two values agree and, again, (17.29) holds. ∎

The examples above show that, at least in some cases, Green's Theorem allows you to evaluate tedious line integrals using simple double integrals. In each example, the region D was both type 1 and type 2. For such regions, equation (17.29) is easy to verify. Figure 17–26A shows such a region with its boundary $C = C_1 - C_2$ written in terms of functions of x. If $\mathbf{F}(x, y) = M(x, y)\mathbf{i} + N(x, y)\mathbf{j}$, then

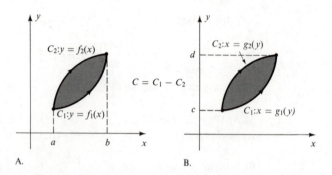

FIGURE 17–26. Green's Theorem for a region that is type 1 and type 2.

$$\int_C M(x, y) \, dx = \int_{C_1} M(x, y) dx - \int_{C_2} M(x, y) \, dx$$

$$= \int_a^b M(x, f_1(x)) dx - \int_a^b M(x, f_2(x)) \, dx$$

$$= \int_a^b [M(x, f_1(x)) - M(x, f_2(x))] \, dx$$

$$= \int_a^b \left[-M(x, y)\right]_{f_1(x)}^{f_2(x)} dx$$

$$= \int_a^b \left[\int_{f_1(x)}^{f_2(x)} -\frac{\partial M}{\partial y}(x, y) \, dy\right] dx$$

$$= \iint_D -\frac{\partial M}{\partial y}(x, y) \, dA$$

Figure 17–26B shows the same region with its boundary written in terms of functions of y. Applying an argument similar to the one above to $N(x, y)$ would show that

$$\int_C N(x, y) \, dy = \iint_D \frac{\partial N}{\partial x}(x, y) \, dA$$

We leave the details to you. Adding the two equations yields (17.29).

It can be shown that Green's Theorem also applies to regions that are of type 1 or of type 2. In fact, the theorem applies to any region you will encounter in this book. For simplicity, we will state the theorem formally for regions of types 1 or 2, and then indicate how it can be extended to more general regions.

(17.30)

> **GREEN'S THEOREM**
> Let D be a region of type 1 or 2 and let C be its boundary oriented counterclockwise. If M and N are functions with continuous partials on an open region R containing both D and C, then
>
> $$\int_C M\,dx + N\,dy = \int\int_D \left(\frac{\partial N}{\partial x} - \frac{\partial M}{\partial y}\right) dA$$

In the following examples it is assumed that the boundary has the proper orientation.

EXAMPLE 3   Find $\int_C (x + y)\,dx + (x + xy)\,dy$ where C is the boundary of the region bounded by the curves $x = y^2$ and $x = y + 2$.

Solution   Figure 17–27 shows the region (of type 2) and the boundary C. According to Green's Theorem,

$$\int_C (x + y)\,dx + (x + xy)\,dy = \int\int_D \left[\frac{\partial}{\partial x}(x + xy) - \frac{\partial}{\partial y}(x + y)\right] dA$$

$$= \int_{-1}^{2}\int_{y^2}^{y+2} y\,dx\,dy$$

$$= \int_{-1}^{2} (y^2 + 2y - y^3)\,dy = \frac{27}{12} \quad \blacksquare$$

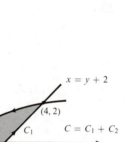

FIGURE 17–27. Example 3.

Green's Theorem can be extended to more general regions. Figure 17–28A shows the region D between two quarter circles and its boundary C. The dashed line divides D into two subregions D_1 and D_2 of type 1, each with its own boundary. The theorem applies to each part. The sum of the double integrals over

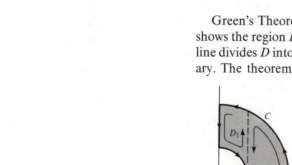

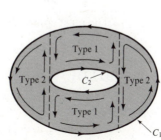

FIGURE 17–28. Green's Theorem also applies to these regions.

A.

B.

D_1 and D_2 equals the double integral over D. We claim that the sum of the line integrals over the boundaries of D_1 and D_2 equals the line integral over C. The reason is that although both boundaries include the dashed line, it is traversed once in each direction; so the line integrals over that part cancel each other, and what is left is the line integral over C. Figure 17–28B shows a region D with a hole; the boundary consists of two curves C_1 and C_2. The curves are oriented according to the **left-hand rule.** That is, as you traverse any part of the boundary, the region is always on your left. We say that this is a **positive orientation** of the boundary. The figure illustrates that D can be subdivided into regions of type 1 or 2. Observe that each of the dashed partition lines is traversed once in each direction. It can be shown that the double integral over D equals the sum of the line integrals over C_1 and C_2. That is, Green's Theorem applies to such regions.

EXAMPLE 4   Evaluate $\int_C y^2\, dx + 3\, dy$ where C is the boundary of the region between the two semicircles shown in Figure 17–29.

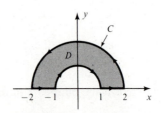

FIGURE 17–29. Example 4.

Solution   Green's Theorem applies, so

$$\int_C y^2\, dx + 3\, dy = \int\int_D \left[\frac{\partial}{\partial x}(3) - \frac{\partial}{\partial y}(y^2) \right] dA$$

$$= \int\int_D -2y\, dA$$

$$= \int_0^\pi \int_1^2 (-2r\sin\theta)\, r\, dr\, d\theta \qquad \text{(Change variables)}$$

$$= -\frac{14}{3} \int_0^\pi \sin\theta\, d\theta = -\frac{28}{3} \qquad \blacksquare$$

Green's Theorem also provides an elegant method for computing areas. Observe that if $\partial N/\partial x - \partial M/\partial y = 1$, then

$$\int_C M\, dx + N\, dy = \int\int_D dA = \text{area of } D$$

Any M and N with continuous partials can be used. For example, if D is a region and its boundary C is a simple closed curve, then setting $M = 0$ and $N = x$ yields

$$\int_C x\, dy = \int\int_D dA = \text{area of } D$$

The most commonly used formulas are as follows:

(17.31)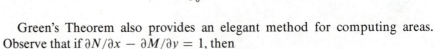
$$\text{Area} = \int_C x\, dy \quad \text{or} \quad \int_C -y\, dx \quad \text{or} \quad \frac{1}{2}\int_C -y\, dx + x\, dy$$

EXAMPLE 5   Find the area of the ellipse

$$\frac{x^2}{a^2} + \frac{y^2}{b^2} = 1$$

Solution   We will use the third formula in (17.31) because, in this case, it is the easiest. A position function for the boundary is $\mathbf{r}(t) = (a\cos t, b\sin t), 0 \le t \le 2\pi$ (Figure 17–30). Therefore,

$$\text{Area} = \frac{1}{2}\int_C -y\, dx + x\, dy$$

$$= \frac{1}{2}\int_0^{2\pi} [(-b\sin t)(-a\sin t) + (a\cos t)(b\cos t)]\, dt$$

$$= \frac{ab}{2}\int_0^{2\pi} (\sin^2 t + \cos^2 t)\, dt$$

$$= \pi ab \qquad \blacksquare$$

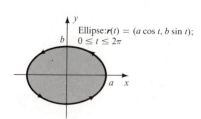

Ellipse: $\mathbf{r}(t) = (a\cos t, b\sin t)$; $0 \le t \le 2\pi$

FIGURE 17–30. Example 5.

Interpretations of Green's Theorem

Let $\mathbf{F}(x, y) = M(x, y)\mathbf{i} + N(x, y)\mathbf{j}$ be a vector field. The **divergence** of $\mathbf{F}$, denoted by div $\mathbf{F}$, is the scalar function

$$(17.32) \quad \text{div } \mathbf{F}(x, y) = \frac{\partial M}{\partial x}(x, y) + \frac{\partial N}{\partial y}(x, y)$$

The **scalar curl** of $\mathbf{F}$ is also a scalar function:

$$(17.33) \quad \text{scalar curl } \mathbf{F}(x, y) = \frac{\partial N}{\partial x}(x, y) - \frac{\partial M}{\partial y}(x, y)$$

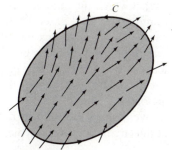

FIGURE 17–31. The velocity field of a fluid through a region bounded by C.

Both div $\mathbf{F}$ and scalar curl $\mathbf{F}$ arise in the study of fluid flow, electricity, and magnetism. Let us think of $\mathbf{F}$ as representing the velocity vector field of a fluid. Suppose that a curve C encloses a region D through which fluid flows; C has positive orientation (Figure 17–31). The *flux* of $\mathbf{F}$ across C is defined as the rate at which the fluid crosses C in a direction perpendicular to C. Figure 17–32 indicates that the flux across a typical arc length element ds is

$$\mathbf{F} \cdot \mathbf{n}\, ds$$

where $\mathbf{n}$ is a unit normal to the curve C pointing outward as you traverse C. It follows that the flux of $\mathbf{F}$ across all of C is given by the line integral

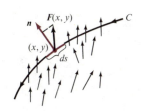

FIGURE 17–32. Computing the flux of $\mathbf{F}$ across C.

$$(17.34) \quad \text{flux} = \int_C \mathbf{F} \cdot \mathbf{n}\, ds$$

If $\mathbf{r}(t) = x(t)\mathbf{i} + y(t)\mathbf{j}$ is a position function of C, then the vector $y'(t)\mathbf{i} - x'(t)\mathbf{j}$ is perpendicular to C (why?) and points outward. Thus, the unit normal is

$$\mathbf{n} = \frac{1}{|\mathbf{r}'|}(y'\mathbf{i} - x'\mathbf{j})$$

and

$$\mathbf{F} \cdot \mathbf{n} \, ds = \underbrace{\frac{My' - Nx'}{|\mathbf{r}'|}}_{\mathbf{F} \cdot \mathbf{n}} \underbrace{|\mathbf{r}'| \, dt}_{ds}$$

$$= My' \, dt - Nx' \, dt$$

$$= M \, dy - N \, dx$$

Therefore, (17.34) becomes

$$(17.35) \quad \text{flux} = \int_C -N \, dx + M \, dy$$

Applying Green's Theorem (with the roles of M and N reversed), the flux of the fluid through the boundary is

$$\text{flux} = \int_C -N \, dx + M \, dy = \int \int_D \left(\frac{\partial M}{\partial x} + \frac{\partial N}{\partial y} \right) dA$$

$$= \int \int_D \text{div} \, \mathbf{F} \, dA \qquad (17.32)$$

This measures the rate (the amount per unit of time) at which the fluid *diverges* across the boundary. A zero flux means that the amount of fluid entering the region is offset by the same amount leaving the region. A negative flux means that more fluid is entering the region than is leaving it; either the fluid is being compressed or there is a *sink* (a point where fluid leaves without passing through the boundary) within the region. A positive flux means the opposite; either the fluid is expanding or there is a *source* within the region.

The *circulation* of $\mathbf{F}$ around C measures the movement of the fluid in the direction tangent to C. Figure 17–33 indicates that the circulation over a typical arc length element ds is

$$\mathbf{F} \cdot \mathbf{T} \, ds$$

where $\mathbf{T}$ is the unit tangent vector

$$\mathbf{T} = \frac{\mathbf{r}'}{|\mathbf{r}'|} = \frac{1}{|\mathbf{r}'|} (x' \mathbf{i} + y' \mathbf{j})$$

Therefore,

$$\mathbf{F} \cdot \mathbf{T} \, ds = \frac{Mx' + Ny'}{|\mathbf{r}'|} |\mathbf{r}'| \, dt$$

$$= M \, dx + N \, dy$$

Again, by Green's Theorem, it follows that

$$\text{circulation} = \int_C M \, dx + N \, dy = \int \int_D \left(\frac{\partial N}{\partial x} - \frac{\partial M}{\partial y} \right) dA$$

$$(17.37) \qquad = \int \int_D \text{scalar curl} \, \mathbf{F} \, dA \qquad (17.33)$$

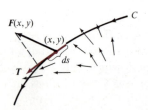

$F(x, y)$

(x, y)

ds

T

C

FIGURE 17–33. Computing the circulation of $\mathbf{F}$ around C.

This measures the way in which the fluid *curls* around the boundary. A positive circulation means a net movement in the same direction as C; a negative circulation means a net movement opposite to C; and a zero circulation means a net movement perpendicular to C.

Similar interpretations can be made if **F** represents an electric or magnetic field.

EXERCISES

Unless otherwise stated, assume the boundary curves have positive orientation. In Exercises 1–6, evaluate the line integrals using Green's Theorem.

1. $\int_C xy\, dx + x^2\, dy$ where C is the unit square.

2. $\int_C x^2\, dx - y^2\, dy$ where C is the unit circle.

3. $\int_C (3x^2 + y)\, dx + (2x + y^3)\, dy$ where C is the unit circle.

4. $\int_C (e^x + 2y)\, dx + (x - e^y)\, dy$ where C is the unit square.

5. $\int_C e^x\, dx + \sin y\, dy$ where C is the cardioid $r = 1 - \cos \theta$.

6. $\int_C \sin x\, dx - \cos y\, dy$ where C is the circle $r = 2 \sin \theta$.

In Exercises, 7–18, evaluate the line integrals with or without Green's Theorem (choose the method you think is easier.)

7. $\int_C (2y + \tan x)\, dx - (3x + e^y)\, dy$ where C is the rectangle with vertices $(1, 1), (2, 1), (2, 2)$, and $(1, 2)$.

8. $\int_C y^2\, dx + x^2\, dy$ where C is the unit circle.

9. $\int_C 2xy^3\, dx + 3x^2\, y^2\, dy$ where C is the unit circle.

10. $\int_C 4y^4\, dx + 16xy^3\, dy$ where C is the ellipse $x^2 + 2y^2 = 1$.

11. $\int_C y\, dx - x\, dy$ where C is the ellipse $9x^2 + 4y^2 = 36$.

12. $\int_C x^2\, y\, dx + dy$ where C is the unit square.

13. $\int_C (xe^y + \tan^{-1} x)\, dx + \ln y\, dy$ where C is the boundary of the region bounded by the curves $y = x^2$ and $y = x$.

14. $\int_C (y^3 + \sin x)\, dx + (x^2 - \ln y)\, dy$ where C is the boundary of the region bounded by the curves $y = \sqrt{x}$ and $y = x$.

15. $\int_C 3y^2\, dx + 4x\, dy$ where C is the boundary of the region between the two semicircles shown in Figure 17–29.

16. $\int_C 4y\, dx + x^2\, dy$ where C is the same as in Exercise 15.

17. $\int_{C_1} \mathbf{F} \cdot d\mathbf{r} - \int_{C_2} \mathbf{F} \cdot d\mathbf{r}$ where $\mathbf{F}(x, y) = (x - y^3)\mathbf{i} + (y + x^3)\mathbf{j}$, C_1 is the circle $x^2 + y^2 = 4$, and C_2 is the circle $x^2 + y^2 = 1$ both oriented counterclockwise (draw a picture).

18. $\int_{C_1} y^2\, dx - \int_{C_2} y^2\, dx$ where C_1 and C_2 are the same as in Exercise 17.

In Exercises 19–22, find the area of the regions enclosed by the given curves.

19. C has a position function $\mathbf{r}(t) = (t^2, -\sin t)$, $-\pi \le t \le \pi$.

20. C is composed of the graphs of $y = x^3$ and $y = x^2$ between $x = 0$ and $x = 1$.

21. C has a position function $\mathbf{r}(t) = (t^2 - 3, \frac{1}{3}t^3 - t)$, $-\sqrt{3} \le t \le \sqrt{3}$.

22. C is the graph of $x^{2/3} + y^{2/3} = 1$.

23. Show that Green's Theorem applies to the region in Figure 17–34A. (Subdivide it into regions of type 1 or 2 and show that each subdivision line is traversed once in each direction.)

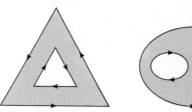

A. Exercise 23.   B. Exercise 24.

FIGURE 17–34.

24. Same as Exercise 23 for the region in Figure 17–34B.

25. A thin homogeneous plate (with $\delta = 1$) is in the shape of a region D with boundary C. Use Green's

Theorem to show that the moments M_y and I_y are given by

$$M_y = \int_C \frac{1}{2}x^2 \, dy \quad \text{and} \quad I_y = \int_C \frac{1}{3}x^3 \, dy$$

26. Derive formulas similar to those above for $M_x, \bar{x}, \bar{y}$, I_x, and I_0.

27. A water drain is located at the origin. In the northern hemisphere, water near the drain will flow in a clockwise direction. The velocity field of the water is approximately

$$\mathbf{F} = \frac{y}{x^2 + y^2}\mathbf{i} - \frac{x}{x^2 + y^2}\mathbf{j}$$

Find the circulation of $\mathbf{F}$ around the unit circle. (*Note:* You cannot use Green's Theorem because $\mathbf{F}$ is not defined at the origin.) Does the sign of the answer agree with the interpretation of circulation given in the text?

28. Let $\mathbf{F}(x, y) = (x^2 + y^2)\mathbf{i} + 2xy\mathbf{j}$. Compute the flux of $\mathbf{F}$ across the unit circle and the circulation of $\mathbf{F}$ around the unit square.

29. Show that a conservative vector field has zero circulation around any piecewise smooth closed curve.

17–5 GAUSS' THEOREM; STOKES' THEOREM*

Green's Theorem relates line integrals to double integrals. The theorems mentioned in the title are extensions of Green's Theorem to space. Stokes' Theorem relates line integrals to surface integrals, and Gauss' Theorem relates surface integrals to triple integrals. Such relationships are important in the study of three dimensional fluid flows, electric fields, and magnetic fields.

Surface Integrals

A surface integral, as you might expect, is a double integral over a surface in space. The definition of a surface integral is motivated by the following question: Given a surface S and a vector field, $\mathbf{F}$, what is the flux of $\mathbf{F}$ across S in a direction perpendicular to S?

To answer this question, we set up the following situation. Let D be a region in the xy-plane to which Green's Theorem applies and let C be its boundary. Let f be a function that has continuous partials throughout a region containing both D and C, and let S be the surface $z = f(x, y)$. Now suppose that $\mathbf{F}$ is a continuous vector field defined on a region (in space) containing S. To find the flux of $\mathbf{F}$ across S, we project a typical rectangle in D, with area dA, up to S forming what we call a typical S-rectangle (Figure 17–35). Recalling our experience with surface area (Section 16–7), the element dS that approximates the area of the S-rectangle is

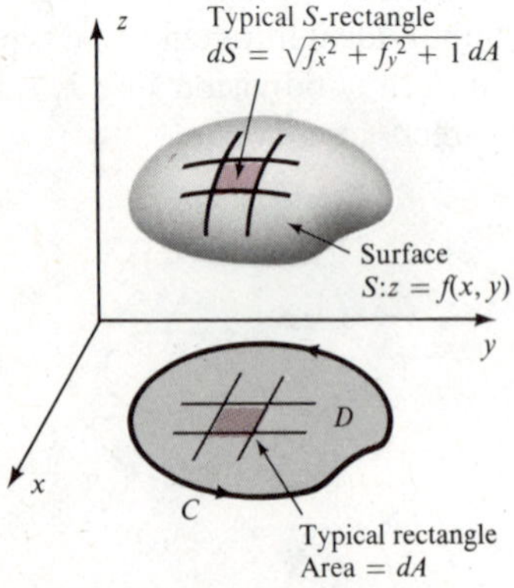

$$(17.38) \quad dS = \sqrt{f_x^2 + f_y^2 + 1} \, dA \quad\quad (16.49)$$

Furthermore,

$$(17.39) \quad \mathbf{n} = \frac{-f_x\mathbf{i} - f_y\mathbf{j} + \mathbf{k}}{\sqrt{f_x^2 + f_y^2 + 1}}$$

Typical S-rectangle
$dS = \sqrt{f_x^2 + f_y^2 + 1} \, dA$

Surface
$S: z = f(x, y)$

D

C

Typical rectangle
Area $= dA$

FIGURE 17–35. The graph of $z = f(x, y)$ is a surface.

---

*Karl Friedrich Gauss (1777–1855), whom many consider to be the greatest mathematician of all time, made important contributions to virtually every branch of mathematics. George G. Stokes (1819–1903) was an English mathematician.

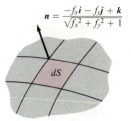

$$n = \frac{-f_x\mathbf{i} - f_y\mathbf{j} + \mathbf{k}}{\sqrt{f_x^2 + f_y^2 + 1}}$$

Part of the surface S

FIGURE 17–36. A unit outward normal vector.

is a unit vector pointing outward from S. See Figure 17–36 (also see Figure 16–40 in Section 16–7; $\mathbf{n} = (\mathbf{a} \times \mathbf{b})/|\mathbf{a} \times \mathbf{b}|$). The component of $\mathbf{F}$ normal to S is $\mathbf{F} \cdot \mathbf{n}$, and the flux of $\mathbf{F}$ through the typical S-rectangle is $\mathbf{F} \cdot \mathbf{n}\, dS$. Now we sum over all S-rectangles, take the limit as the rectangles get smaller, and define the total flux to be this limit. It is denoted by the symbol

$$\int\int_S \mathbf{F} \cdot \mathbf{n}\, dS$$

which is called the **surface integral of F over** S. To evaluate this integral, let $\mathbf{F}(x, y, z) = M(x, y, z)\mathbf{i} + N(x, y, z)\mathbf{j} + P(x, y, z)\mathbf{k}$, which we abbreviate as

$$\mathbf{F} = M\mathbf{i} + N\mathbf{j} + P\mathbf{k}$$

Then it follows from (17.38) and (17.39) that

(17.40)
$$\int\int_S \mathbf{F} \cdot \mathbf{n}\, dS = \int\int_S \frac{-Mf_x - Nf_y + P}{\sqrt{f_x^2 + f_y^2 + 1}}\, dS$$
$$= \int\int_D (-Mf_x - Nf_y + P)\, dA$$

The definition of a surface integral can be extended to any surface you will encounter in this book, including the boundary surface of a solid. We have used the surface $z = f(x, y)$ for simplicity.

EXAMPLE 1   Let $\mathbf{F}(x, y, z) = x\mathbf{i} + y\mathbf{j} + z\mathbf{k}$ and let S be the part of the paraboloid $z = 4 - x^2 - y^2$ shown in Figure 17–37. Find the flux of $\mathbf{F}$ across S.

Solution   The flux is given by a surface integral. To apply (17.40), we note that

$$M = x \quad N = y \quad P = z = 4 - x^2 - y^2$$

that

$$f_x = \frac{\partial z}{\partial x} = -2x \quad \text{and} \quad f_y = \frac{\partial z}{\partial y} = -2y$$

and that D is the quarter circle $x^2 + y^2 \le 4$, x and $y \ge 0$. Therefore,

$$\int\int_S \mathbf{F} \cdot \mathbf{n}\, dS = \int\int_D (2x^2 + 2y^2 + 4 - x^2 - y^2)\, dA$$
$$= \int\int_D (x^2 + y^2 + 4)\, dA$$
$$= \int_0^{\pi/2} \int_0^2 (r^2 + 4)r\, dr\, d\theta \qquad \text{(Change variables)}$$
$$= 6\pi \quad \blacksquare$$

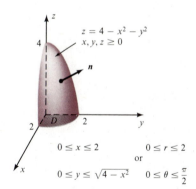

$z = 4 - x^2 - y^2$
$x, y, z \ge 0$

$0 \le x \le 2$ $0 \le r \le 2$
or
$0 \le y \le \sqrt{4 - x^2}$ $0 \le \theta \le \frac{\pi}{2}$

FIGURE 17–37. Example 1.

EXAMPLE 2   According to Coulomb's law, the electric field **E** due to a point charge of Q coulombs located at the origin is the inverse square field.

$$\mathbf{E}(x, y, z) = \frac{kQ}{(x^2 + y^2 + z^2)^{3/2}} \mathbf{r}(x, y, z) = \frac{kQ}{r^2} \mathbf{u}(x, y, z)$$

where k is a universal constant ($k \approx 9 \times 10^9$ $N\text{-}m^2/C^2$), $\mathbf{r}$ is the position vector, $r = |\mathbf{r}|$ and $\mathbf{u}$ is the unit vector $\mathbf{r}/r$. Find the flux of **E** through the surface of a sphere of radius a (Figure 17–38).

Solution   As the figure indicates, $r = a$ and the unit outward normal is $\mathbf{n} = \mathbf{r}/a = \mathbf{u}$. Therefore, we can compute $\mathbf{E} \cdot \mathbf{n}$ directly

$$\mathbf{E} \cdot \mathbf{n} = \mathbf{E} \cdot \mathbf{u} = \left[\frac{kQ}{a^2} \mathbf{u} \right] \cdot \mathbf{u} = \frac{kQ}{a^2}$$

and, since the surface area of the sphere is $4\pi a^2$,

$$\text{flux} = \int \int_S \mathbf{E} \cdot \mathbf{n} \, dS = \frac{kQ}{a^2} \int \int_S dS = 4\pi kQ$$

Notice that the flux does not depend on the radius of the sphere because a is not in the final answer. In fact, as we will show in Example 4, it does not even depend on the surface being a sphere!   ∎

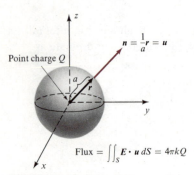

Point charge Q

$\mathbf{n} = \dfrac{1}{a}\mathbf{r} = \mathbf{u}$

$\text{Flux} = \int\!\!\int_S \mathbf{E} \cdot \mathbf{u} \, dS = 4\pi kQ$

FIGURE 17–38. Example 2.

Gauss' Theorem

If **F** is a three dimensional vector field, we define the **divergence** of **F** by

(17.41) $\text{div } \mathbf{F} = \dfrac{\partial M}{\partial x} + \dfrac{\partial N}{\partial y} + \dfrac{\partial P}{\partial z}$

The divergence is often denoted by the symbol

$\nabla \cdot \mathbf{F}$

If we think of ∇ as $(\partial/\partial x, \partial/\partial y, \partial/\partial z)$ and $\partial/\partial x$ "times" M as $\partial M/\partial x$, then

$$\nabla \cdot \mathbf{F} = \left(\frac{\partial}{\partial x}, \frac{\partial}{\partial y}, \frac{\partial}{\partial z} \right) \cdot (M, N, P)$$

(17.42)

$$= \frac{\partial M}{\partial x} + \frac{\partial N}{\partial y} + \frac{\partial P}{\partial z}$$

$$= \text{div } \mathbf{F}$$

In the plane, the flux across the boundary curve equals the double integral of the divergence over the region; see (17.36). In space, the flux across the boundary surface equals the triple integral of the divergence over the solid. This is Gauss' (Divergence) Theorem.

> **GAUSS' (DIVERGENCE) THEOREM**
> Let W be a solid of type 1, 2, or 3 and let S be the boundary
> surface of W. If $\mathbf{F}$ is a vector field whose component functions
> have continuous partials in a region containing W and S, then
>
> $$\int\int_S \mathbf{F} \cdot \mathbf{n}\, dS = \int\int\int_W \nabla \cdot \mathbf{F}\, dV$$

(17.43)

Gauss' Theorem applies to a much larger class of solids, but we have used those of types 1, 2, and 3 for simplicity. We will not prove the theorem.

EXAMPLE 3   Let W be the solid cylinder $x^2 + y^2 \le 4$ between the planes $z = 0$ and $z = 3$. Let S be its surface boundary, and let $\mathbf{n}$ be the unit outward normal to S (Figure 17–39). If $\mathbf{F}(x, y, z) = (x^2 + e^{yz})\mathbf{i} + (y - z\tan^{-1} x)\mathbf{j} + x^2y^3\mathbf{k}$, find $\int\int_S \mathbf{F} \cdot \mathbf{n}\, dS$.

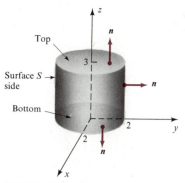

FIGURE 17–39. Example 3.

Solution   Direct evaluation of this integral would require separate evaluations over the top, bottom, and side surfaces (Figure 17–39); each one involves a difficult computation. But using Gauss' Theorem, we have

$$\int\int_S \mathbf{F} \cdot \mathbf{n}\, dS = \int\int\int_W \nabla \cdot \mathbf{F}\, dV$$

$$= \int\int\int_W (2x + 1)\, dV \qquad (17.42)$$

$$= \int_{-2}^{2} \int_{-\sqrt{4-x^2}}^{\sqrt{4-x^2}} \int_0^3 (2x + 1)\, dz\, dy\, dx$$

We can evaluate this integral as it stands, or change to cylindrical coordinates,

$$= \int_0^{2\pi} \int_0^2 \int_0^3 (2r\cos\theta + 1)\, r\, dz\, dr\, d\theta$$

In either case, the answer is 12π.   ∎

EXAMPLE 4   Let S be the surface boundary of a solid containing the origin. We assume that Gauss' Theorem applies to this solid. Show that the flux through S of the electric field $\mathbf{E}$ defined in Example 2 is $4\pi kQ$.

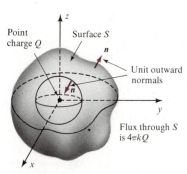

FIGURE 17–40. Example 4.

Solution   We cannot apply Gauss' Theorem directly because $\mathbf{E}$ is not defined at the origin. If, however, we draw a sphere S_1 about the origin that is contained inside of the solid bounded by S (see Figure 17–40), and let W be the solid inside of S but outside of S_1, then the theorem does apply to W and $\mathbf{E}$. The boundary of W consists of the two surfaces, S and S_1. It follows that

$$\int\int_S \mathbf{E} \cdot \mathbf{n}\, dS + \int\int_{S_1} \mathbf{E} \cdot \mathbf{n}\, dS = \int\int\int_W \nabla \cdot \mathbf{E}\, dV$$

You can easily check that $\nabla \cdot \mathbf{E} = 0$, so that

$$\text{flux} = \int\int_S \mathbf{E} \cdot \mathbf{n} \, dS = -\int\int_{S_1} \mathbf{E} \cdot \mathbf{n} \, dS = 4\pi kQ$$

The last equality follows from the result in Example 2 once you realize that the unit outward normals for S_1 must point away from W; that is, they point *toward* the origin rather than away from it (Figure 17–40).   ■

Stokes' Theorem

If $\mathbf{F}$ is a three dimensional vector field, we define the **curl** of $\mathbf{F}$ by

(17.43) $\text{curl } \mathbf{F} = (P_y - N_z)\mathbf{i} + (M_z - P_x)\mathbf{j} + (N_x - M_y)\mathbf{k}$

The curl is often denoted by the symbol

$$\nabla \times \mathbf{F}$$

because the determinant

$$(17.44) \quad \nabla \times \mathbf{F} = \begin{vmatrix} \mathbf{i} & \mathbf{j} & \mathbf{k} \\ \dfrac{\partial}{\partial x} & \dfrac{\partial}{\partial y} & \dfrac{\partial}{\partial z} \\ M & N & P \end{vmatrix}$$

represents the vector defined above.

The circulation around the boundary of a region in the plane equals the double integral of the *scalar* curl over the region; see (17.37). In space, the circulation around the boundary of a surface is the surface integral of the curl over the surface. This is Stokes' Theorem.

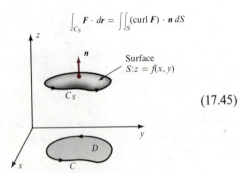

$$\int_{C_S} \mathbf{F} \cdot d\mathbf{r} = \int\int_S (\text{curl } \mathbf{F}) \cdot \mathbf{n} \, dS$$

FIGURE 17–41. Stokes' Theorem.

(17.45)

STOKES' THEOREM
Let D be a region in the plane to which Green's Theorem applies and let S be the surface $z = f(x, y)$ where f has continuous partials. Let C be the boundary of D and C_S be the corresponding boundary of S. Orient C_S according to the positive orientation of C (Figure 17–41). If $\mathbf{F}$ has continuous partials in a region containing S and C_S, and $\mathbf{n}$ is the unit outward normal, then

$$\int_{C_S} \mathbf{F} \cdot d\mathbf{r} = \int\int_S (\nabla \times \mathbf{F}) \cdot \mathbf{n} \, dS$$

Stokes' Theorem applies to a much larger class of surfaces, but we have used $z = f(x, y)$ for simplicity. We will not prove Stokes' Theorem.

EXAMPLE 5   Let S be that portion of the paraboloid $z = x^2 + y^2$ between the plane $z = 0$ and $z = 1$ (Figure 17–42). Let $\mathbf{F}(x, y, z) = 2z\mathbf{i} - 3x\mathbf{j} + 4y\mathbf{k}$ and verify Stokes' Theorem in this case.

Solution   We want to show that the two integrals in (17.45) have the same value. First, the line integral. The boundary C_S has a position function $\mathbf{r}(t) = (\cos t, \sin t, 1), 0 \le t \le 2\pi$. Therefore,

$$\int_{C_S} \mathbf{F} \cdot d\mathbf{r} = \int_0^{2\pi} \mathbf{F}(\mathbf{r}(t)) \cdot \mathbf{r}'(t) \, dt$$

$$= \int_0^{2\pi} (2, -3 \cos t, 4 \sin t) \cdot (-\sin t, \cos t, 0) \, dt$$

$$= \int_0^{2\pi} (-2 \sin t - 3 \cos^2 t) \, dt$$

$$= \left[2 \cos t - \frac{3}{2}t - \frac{3}{4} \sin 2t \right]_0^{2\pi}$$

$$= -3\pi$$

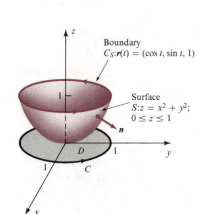

Boundary
$C_S : r(t) = (\cos t, \sin t, 1)$

Surface
$S : z = x^2 + y^2$;
$0 \le z \le 1$

n

D

C

FIGURE 17-42. Example 5.

Now the surface integral

$$\nabla \times \mathbf{F} = \begin{vmatrix} \mathbf{i} & \mathbf{j} & \mathbf{k} \\ \dfrac{\partial}{\partial x} & \dfrac{\partial}{\partial y} & \dfrac{\partial}{\partial z} \\ 2z & -3x & 4y \end{vmatrix} = 4\mathbf{i} + 2\mathbf{j} - 3\mathbf{k}$$

According to formula (17.40),

$$\iint_S (\nabla \times \mathbf{F}) \cdot \mathbf{n} \, dS = \iint_D (-Mf_x - Nf_y + P) \, dA$$

where $M = 4$, $N = 2$, $P = -3$, $f_x = 2x$, $f_y = 2y$, and D is the circular region $x^2 + y^2 \le 1$ (Figure 17-42). Changing to polar coordinates, we have

$$\iint_S (\nabla \times \mathbf{F}) \cdot \mathbf{n} \, dS = \iint_D (-8x - 4y - 3) \, dA$$

$$= \int_0^{2\pi} \int_0^1 (-8r \cos \theta - 4r \sin \theta - 3) \, r \, dr \, d\theta$$

$$= -3\pi$$

which agrees with our first answer.   ∎

EXERCISES

In Exercises 1–6, evaluate the surface integrals $\iint_S \mathbf{F} \cdot \mathbf{n} \, dS$ using formula (17.40).

1. $\mathbf{F} = 2x\mathbf{i} + 2y\mathbf{j} + z\mathbf{k}$; S is the surface $z = x^2 + y^2$ between the planes $z = 0$ and $z = 4$.

2. $\mathbf{F} = 3y\mathbf{i} - x\mathbf{j} + 5\mathbf{k}$; S is the surface $z = 4 - x^2 - y^2$ above the xy-plane.

3. $\mathbf{F} = \cos x\mathbf{i} + \sin y\mathbf{j} + z\mathbf{k}$; S is the part of the plane $x + y + z = 1$ in the first octant.

4. $\mathbf{F} = x^2\mathbf{i} + y^2\mathbf{j} + z^2\mathbf{k}$; S is the part of the plane $2x + 3y + 4z = 12$ in the first octant.

5. $\mathbf{F} = x\mathbf{i} + y\mathbf{j} + z^2\mathbf{k}$; S is the hemisphere $x^2 + y^2 + z^2 = 1, z \ge 0$.

6. $\mathbf{F} = 3\mathbf{i} + 4\mathbf{j} + 5\mathbf{k}$; S is the hemisphere $x^2 + y^2 + z^2 = 4, z \geq 0$.

In Exercises 7–12, use Gauss' Theorem to evaluate the surface integrals $\int \int_S \mathbf{F} \cdot \mathbf{n} \, dS$.

7. $\mathbf{F} = x^3\mathbf{i} + x^2y\mathbf{j} + x^2z\mathbf{k}$; S is the boundary of the region bounded by the cylinder $x^2 + y^2 = 4$, the xy-plane, and the plane $z = 2$.

8. $\mathbf{F} = x^3\mathbf{i} + y^3\mathbf{j} + z^3\mathbf{k}$; S is the same as in Exercise 7.

9. $\mathbf{F} = xy\mathbf{i} + yz\mathbf{j} + xz\mathbf{k}$; S is the boundary of the region bounded by the coordinate planes and the plane $x + y + z = 1$.

10. $\mathbf{F} = \sin y\mathbf{i} + \cos x\mathbf{j} + z^2\mathbf{k}$; S is the same as in Exercise 9.

11. $\mathbf{F} = (\tan^{-1}x - \sin^{-1}yz)\mathbf{i} + (x^5 + e^{xz})\mathbf{j} + (\sinh z - \cosh y)\mathbf{k}$; S is the boundary of the solid bounded by the six planes $x = \pm 1, y = \pm 2, z = \pm 3$.

12. $\mathbf{F} = (x + y^2)\mathbf{i} + (y - \tan^{-1}xz)\mathbf{j} + (z + e^{xy})\mathbf{k}$; S is the same as in Exercise 11.

In Exercises 13–16, verify Stokes' Theorem (see Example 5).

13. $\mathbf{F} = z^2\mathbf{i} + x^2\mathbf{j} + y^2\mathbf{k}$; S is the paraboloid $z = x^2 + y^2$ between the planes $z = 0$ and $z = 1$.

14. $\mathbf{F} = x^2\mathbf{i} + z^2\mathbf{j} + y^2\mathbf{k}$, same S as in Exercise 13.

15. $\mathbf{F} = x\mathbf{i} + 4z\mathbf{j} + y\mathbf{k}$, S is the part of $z = 4 - x^2 - y^2$ above the xy-plane.

16. $\mathbf{F} = 5x\mathbf{i} - 4\mathbf{j} + 3z\mathbf{k}$; same S as in Exercise 15.

In Exercise 17–22, evaluate the given integrals. You can evaluate directly, use Gauss' Theorem, or Stokes' Theorem (choose the method you think is easiest.)

17. $\int_C (4z + e^x) \, dx + (x^3 - \tan^{-1}y) \, dy + (y + z^3) \, dz$ where C has a position function $\mathbf{r}(t) = (\cos t, \sin t, 1), 0 \leq t \leq 2\pi$.

18. $\int \int_S \mathbf{F} \cdot \mathbf{n} \, dS$ where $\mathbf{F} = x^2\mathbf{i} + y^2\mathbf{j} + z^2\mathbf{k}$ and S is the boundary of the cube bounded by the coordinate planes and the planes $x = 1, y = 1, z = 1$.

19. $\int \int_S (\nabla \times F) \cdot \mathbf{n} \, dS$ where $\mathbf{F} = 2y\mathbf{i} + z^2\mathbf{j} + x^2z\mathbf{k}$ and S is the the upper part of the ellipsoid $9x^2 + 4y^2 + z^2 = 36, z \geq 0$.

20. $\int_C yz \, dx + xz \, dy + xy \, dz$ where C is the triangle with vertices $(1, 1, 1), (1, 2, 1)$ and $(1, 2, 0)$.

21. $\int \int_S \mathbf{F} \cdot \mathbf{n} \, dS$ when $\mathbf{F} = x^3\mathbf{i} + y^3\mathbf{j} + z^3\mathbf{k}$ and S is the sphere $x^2 + y^2 + z^2 = 1$.

22. $\int \int_S (\nabla \times F) \cdot \mathbf{n} \, dS$ where $\mathbf{F} = x\mathbf{i} + y\mathbf{j} + z\mathbf{k}$ and S is the hemisphere $x^2 + y^2 + z^2 = 1, z \geq 0$.

For the exercises below, let f and g be scalar functions with continuous second partials. The symbol $\nabla^2 f$ denotes the scalar function

$$\nabla^2 f = \frac{\partial^2 f}{\partial x^2} + \frac{\partial^2 f}{\partial y^2}$$

The symbols C_S, S, and W have the same meaning as in the text. Assuming that the theorems of Gauss and Stokes can be applied, show that the following equations are true.

23. $\int \int_S \nabla f \cdot \mathbf{n} \, dS = \int \int \int_W \nabla^2 f \, dV$

24. $\int \int_S (f\nabla g) \cdot \mathbf{n} \, dS = \int \int \int_W (f\nabla^2 g + \nabla f \cdot \nabla g) \, dV$

25. $\int \int_S (f\nabla g - g\nabla f) \cdot \mathbf{n} \, dS$
$= \int \int \int_W (f\nabla^2 g - g\nabla^2 f) \, dV$

26. $\int \int_{C_S} f\nabla g \cdot d\mathbf{r} = \int \int_S (\nabla f \times \nabla g) \cdot \mathbf{n} \, dS$

REVIEW EXERCISES

In Exercises 1–8, let C be that part of the parabola $y = 2x^2 - 3x + 1$ joining $(0, 1)$, to $(2, 3)$, and let C_1 be the line segment from $(0, 1)$ to $(2, 3)$.

1. Find a position function for C and for $-C$.

2. Find a position function for C_1 and for $-C_1$.

3. Let $\mathbf{F} = x^2y\mathbf{i} - (x^3/2)\mathbf{j}$ newtons. Find the work done by $\mathbf{F}$ in moving a particle from $(0, 1)$ to $(2, 3)$ along the path C (distance measured in meters).

4. Find the work done by the force $\mathbf{F}$ in Exercise 3 in moving a particle from $(0, 1)$ to $(2, 3)$ along C_1.

5. Let $F(x, y) = (e^x + \sin y)i + (e^y + x \cos y)j$. Compute the circulation of F around the closed curve $C - C_1$. Interpret the answer in terms of a fluid flow.

6. Same as Exercise 5 for the field $F = xyi + j$.

7. Let $F(x, y) = (x^2 + \sin y)i + \cos xj$. Compute the flux of F across the closed curve $C - C_1$. Interpret the answer in terms of a fluid flow.

8. Same as Exercise 7 for the field $F(x, y) = yi + xj$.

In Exercises 9–12, let S be the surface $z = x^2 + y^2$ between the planes $z = 0$ and $z = 4$.

9. Find a position function for the boundary curve C_S.

10. Find the flux of $F(x, y, z) = xi + yj + zk$ across S.

11. Evaluate $\int \int_S F \cdot n \, dS$ where $F(x, y, z) = (x^3 + yz)i + (y^3 - \sin xz)j + (z^2 + \ln|xy|)k$.

12. Let $F(x, y, z) = z^2i + x^2j + y^2k$. Verify that $\int_{C_S} F \cdot dr = \int \int_S (\nabla \times F) \cdot n \, dS$.

In Exercises 13–18, evaluate the line integral directly or use Stokes' Theorem.

13. $\int_C F \cdot dr$ where $F = xyi + yzj + xzk$ and $r(t) = (t, t^2, t^3), 0 \le t \le 1$.

14. $\int_C F \cdot dr$ where $F = x^2i + y^2j + z^2k$ and $r(t) = (t, t^2, t), 0 \le t \le 1$.

15. $\int_C (z + \sin x) \, dx + (x^2 + \sin y) \, dy + (3y + \ln|z|) \, dz$ and C is the square with vertices $(0, 0, 1)$, $(0, 1, 1)$, $(1, 1, 1,)$, and $(1, 0, 1)$.

16. $\int_C yz \, dx + xz \, dy + xy \, dz$ and C is as in Exercise 15.

17. $\int_C \tan^{-1} x \, dx + \sin^{-1} y \, dy + \cos^{-1} z \, dz$ and $r(t) = (\cos t, \sin t, 1), 0 \le t \le 2\pi$.

18. $\int_C xy \, dx + xz \, dy + yz \, dz$ and C is as in Exercise 17.

In Exercises 19–22, evaluate the surface integral $\int \int_S F \cdot n \, dS$ directly or use Gauss' Theorem.

19. $F = xi + yj + zk$ and S is the surface of the tetrahedron bounded by the coordinate planes and the plane $x + y + z = 1$.

20. $F = yi - 3xj + zk$ and S is that part of $z = x^2 + y^2 - 4$ below the xy-plane along with the circular region $x^2 + y^2 \le 4$ in the xy-plane.

21. $F = x^3i + y^3j + z^3k$ and S is that part of $z = 1 - x^2 - y^2$ above the xy-plane along with the circular region $x^2 + y^2 \le 1$ in the xy-plane.

22. $F = xyi + yzj + xzk$ and S is the same as in Exercise 19.

In Exercises 23–28, determine whether or not the given field is conservative. If it is, find a potential function.

23. $F = 3x^2 \sin y \cos zi + x^3 \cos y \cos zj - x^3 \sin y \sin zk$

24. $F = e^{xy}(x^2y + 2x)i + x^3e^{yx}j$

25. $F = (y/x)i + (1 + \ln|xy|)j$

26. $F = xyzi + x^2yj + zk$

27. $F = y^zi + xzy^{z-1}j + xy^z \ln yk$

28. $F = \dfrac{1}{\sqrt{x^2 + y^2 + z^2}}(xi + yj + zk)$

29. A thin wire in the shape of $x^2 + y^2 = 1$ has a density function $\delta(x, y) = 1 + x^2$. Find its mass.

30. A thin wire with $\delta = 1$ is in the shape of the helix $x = \cos 3t, y = \sin 3t, z = t^2, 0 \le t \le 2\pi$. Find its mass.

31. Use Green's Theorem to find the area between the x-axis and one arch of the cycloid $x = t - \sin t$, $y = 1 - \cos t, 0 \le t \le 2\pi$.

32. (a) Let F be a vector field with continuous partials. Show that $\text{div}(\text{curl } F) = 0$. (b) Let f be a scalar function with continuous partials. Show that curl $(\nabla f) = 0$.

33. The electric field E produced by a charge of Q coulombs at the origin is an inverse square field.

$$E = \frac{kQ}{(x^2 + y^2 + z^2)^{3/2}}(xi + yj + zk)$$

(Example 2, Section 17–5)

The force F it exerts on a charge q of like sign is $F = qE$. Find the work done by an electric field produced by 10 C at origin in moving a charge of 3 C around the helix $r(t) = (\cos t, \sin t, t), 0 \le t \le 2\pi$.

34. (Continuation of Exercise 33). How much total work is done in moving the particle along the four line segments from $(1, 0, 0)$ to $(1, 1, 1)$ to $(-3, 1, 4)$ to $(0, 0, 2\pi)$ to $(1, 0, 2\pi)$?

35. (Continuation of Exercise 33). Let S be the surface of the cylindrical can $x^2 + y^2 = 1, -1 \le z \le 1$. (a) What is the flux of E across the top of the can, $x^2 + y^2 \le 1, z = 1$? (b) What is the flux of E

across the lateral surface of the can? *Hint:* See Example 4 in Section 17–5.

36. *(Heat flow).* Let W be a solid in space and let $T(x, y, z)$ be the temperature at (x, y, z). Then $\mathbf{F} = -k\nabla T$, where k is a positive constant depending on the heat conductivity of the material, represents the flow of heat through W. If W is the solid sphere $x^2 + y^2 + z^2 \leq 1$ and $T(x, y, z) = x^4 + y^4 + z^4$, compute the heat flow (the flux of T) across the boundary surface of W.

DIFFERENTIAL EQUATIONS

The study of differential equations is the most important part of mathematics for understanding the sciences. The reason is that many relationships found in nature are most easily expressed in the language of differential equations. We have already seen many examples of this in earlier chapters. In Sections 4–7 and 4–8, we used differential equations to enrich our discussion of velocity and acceleration; in Section 5–7, we used them to compute the escape velocity of a rocket; in Chapter 7 they are used to motivate the definition of the natural logarithm and to solve problems about pollution, population growth, and radioactive decay. The list below contains some of the new applications presented in this chapter.

(1) *Free fall with air resistance;* computing terminal velocity.

(2) *Population growth in a limited environment;* the logistic equation.

(3) *Chemical reactions;* the concentration of iodine in the reaction

$$CH_3COCH_3 + I_2 \rightarrow CH_3COCH_2I + HI$$

(4) *Path of a missile fighting a crosswind.*

(5) *Motion with variable mass;* the burnout velocity of a rocket.

(6) *Forced vibrations and resonance;* the violent motion of a bridge under the impact of men marching across it.

Section 1 contains the basic terms and definitions used throughout the chapter. Each of the next four sections introduces a new class of first order differential equation—separable, homogeneous, exact, and first order linear—along with a method for solving it and a description of some of its applications. The last three sections are devoted to second order linear equations and their applications.

18–1 INTRODUCTION

An equation involving a dependent variable and its derivatives with respect to one or more independent variables is called a **differential equation.** If there is more than one independent variable, then partial derivatives are involved, and the equation is called a *partial differential equation.* If there is only one independent variable, then only ordinary derivatives are involved, and the equation is called an *ordinary differential equation.* In this book, we will concentrate on ordinary equations.

Some examples of ordinary equations are

(1) $y' = 2x$

(2) $\dfrac{ds}{dt} + s = t$

(3) $u'' + u = 0$

In equation (1), y is assumed to be a function of x; in (2), s is a function of t; in equation (3), u is assumed to be a function of some one variable, say v, even though it does not appear in the equation. The **order** of a differential equation is the highest order derivative appearing in it. Thus, equations (1) and (2) are **first order,** and (3) is a **second order** equation. A **solution** of a differential equation is any function $y = f(x)$ that satisfies the equation. For instance, the function $y = x^2 + 1$ is a solution of $y' = 2x$ because $y' = (x^2 + 1)' = 2x$.

EXAMPLE 1 Show that $s = t - 1 + 2e^{-t}$ is a solution of

$$\frac{ds}{dt} + s = t$$

Solution This is equation (2) above. With $s = t - 1 + 2e^{-t}$, we have that

$$\frac{ds}{dt} + s = (1 - 2e^{-t}) + (t - 1 + 2e^{-t}) = t$$

which is the original differential equation. ■

EXAMPLE 2 Show that $u = 2 \cos v - 3 \sin v$ is a solution of

$$u'' + u = 0$$

Solution This is equation (3) above; it is a second order equation, so we have to differentiate twice. With $u = 2 \cos v - 3 \sin v$, it follows that

$$u' = -2 \sin v - 3 \cos v$$

and that

$$u'' + u = (-2 \cos v + 3 \sin v) + (2 \cos v - 3 \sin v) = 0$$

which is the original equation. ■

The solutions of equations (1), (2), and (3) noted above are called *particular* solutions. For equation (1), $y' = 2x$, it is easy to see that

(4) $y = x^2 + C$

is a solution for any choice of the constant C. Conversely, every solution of (1) is of this form. Therefore, we call (4) the **general solution** of equation (1). It is more difficult to show (but we will do so in a later section) that

(5) $s = t - 1 + Ce^{-t}$ and $u = C_1\cos v + C_2\sin v$

where the C's are arbitrary constants, are general solutions to equations (2) and (3), respectively. General solutions to ordinary equations contain the same number of arbitrary constants, called **parameters,** as the order of the equation; first order equations have one parameter, second order have two parameters, and so on. To *solve* a differential equation means to find the general solution.

In applications, there may be conditions, explicitly stated or implied by the problem, that pick out a particular solution from the general solution. For example, the value of the solution or its derivatives at one or more points may be specified. Such conditions are called **initial** conditions.

EXAMPLE 3   Find the solution of

$$\frac{ds}{dt} + s = t$$

that satisfies the initial condition $s(0) = 3$.

Solution   According to (5) above, the general solution is

$$s = t - 1 + Ce^{-t}$$

Now we set $t = 0$, $s(0) = 3$, and solve for C

$$3 = s(0) = -1 + C$$

so $C = 4$, and $s = t - 1 + 4e^{-t}$ is the desired particular solution.   ∎

A solution of a differential equation is often an equation that implicitly defines the dependent variable as a function of the independent variable.

EXAMPLE 4   Show that $y = \sin^{-1}xy$ is an (implicit) solution of

$$y'\sqrt{1 - x^2y^2} = xy' + y$$

Solution   We are assuming that $y = \sin^{-1}xy$ implicitly defines a function $y = f(x)$. To show that this function satisfies the equation, we use implicit differentiation

$$y' = [\sin^{-1}xy]' = \frac{1}{\sqrt{1 - x^2y^2}}[xy]'$$

$$= \frac{xy' + y}{\sqrt{1 - x^2y^2}} \qquad \text{(Product rule)}$$

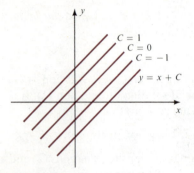

FIGURE 18–1. Family of curves for $y' = 1$.

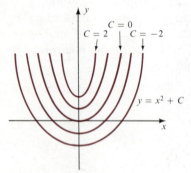

FIGURE 18–2. Family of curves for $y' = 2x$.

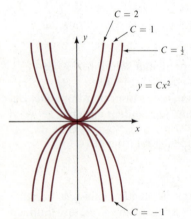

FIGURE 18–3. Example 5. Family of curves for $xy' = 2y$.

Multiplying through by $\sqrt{1 - x^2 y^2}$ yields the given differential equation; this shows that $y = \sin^{-1} xy$ is an (implicit) solution. ∎

It is often desirable to think of the general solution in terms of a *family of curves*. For instance, the general solution to $y' = 1$ is $y = x + C$, which represents a family of parallel lines with slope 1 (Figure 18–1). The general solution of $y' = 2x$ is $y = x^2 + C$, which represents a family of parabolas (Figure 18–2). These are examples of *one parameter* families. Conversely, a one parameter family of curves may represent a first order differential equation.

EXAMPLE 5 Given the family of parabolas $y = Cx^2$ (Figure 18–3), find the differential equation they represent.

Solution We are looking for a differential equation for which $y = Cx^2$ is the general solution. To find the equation, we differentiate with respect to x, $y' = 2Cx$, and then eliminate C from the two equations. This can be done by solving either equation for C and substituting in the other. The resulting equation in this example is

$$ y' = 2 \left(\frac{y}{x^2} \right) x \quad \text{or} \quad xy' = 2y \qquad ∎ $$

In applications, it is usually necessary to express given information as a differential equation. Here are two examples.

EXAMPLE 6 The rate of change of a function is inversely proportional to its value. Express this information as a differential equation.

Solution Let $y = f(x)$ be the function; then y is the value and y' is the rate of change. The differential equation that expresses the given information is

$$ y' = \frac{k}{y} \qquad ∎ $$

EXAMPLE 7 A particle moves along a coordinate line so that its acceleration a at time t is proportional to the square of its position s. Express this information as a (second order) differential equation.

Solution The acceleration a is the second derivative d^2s/dt^2 of position. Therefore, the differential equation is

$$ \frac{d^2 s}{dt^2} = ks^2 \qquad ∎ $$

Once the differential equation is set up, the next step, of course, is to solve it. Methods for solving differential equations are discussed in the succeeding sections.

EXERCISES

In Exercises 1–14, determine whether or not the given functions (some are defined implicitly) are solutions to the corresponding differential equations.

1. $y = x(C + \ln x); (x + y) = xy'$

2. $x^2 + xy - y^3 = C; (2x + y) + (x - 3y^2)\dfrac{dy}{dx} = 0$

3. $u = Ce^v; vu' = e^{2v}$

4. $u = v^2 + C; uv + u' = 0$

5. $s = C_1\sin 2t + C_2\cos 2t; s'' + 4s = 0$

6. $s = C_1\sinh 2t + C_2\cosh 2t; s'' - 4s = 0$

7. $y^2 = x^2 - Cx; x^2 + y^2 = 2xyy'$

8. $y = Ce^{y/x}; y' = y^2/(xy - x^2)$

9. $\tan^{-1}y = x + y + C; 1 + y^2 + y^2y' = 0$

10. $y + \sin y = x; \cos x + \cos y = y'$

11. $y = C_1e^{2x} + C_2e^{3x}; y'' - 5y' + 6y = 0$

12. $y = C_1e^x + C_2e^{-x}; y'' + 5y' - 6y = 0$

13. $xy = \ln y + C; y' = y^2/(1 - xy)$

14. $x^2 + y^2 = C; yy' + x = 0$

Exercises 15–20 refer to the exercises above. Find the particular solution that satisfies the given condition(s).

15. Exercise 1; $y(1) = 3$

16. Exercise 5; $s(0) = 2, s'(0) = 3$

17. Exercise 7; $y(1) = 4$ 18. Exercise 9; $y(0) = 1$

19. Exercise 11; $y(0) = 5, y'(0) = 14$

20. Exercise 13; $y(0) = 1$

In Exercises 21–30, find a differential equation with the given family of curves.

21. $y = 2x + C$ 22. $y = 3x^2 + C$

23. $y = Ce^x$ 24. $y = C\sin x$

25. $y = e^{Cx}$ 26. $y = C\ln x$

27. All lines through the origin.

28. All lines with slope -2.

29. All circles with centers at the origin.

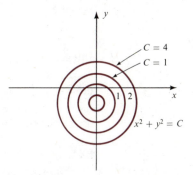

FIGURE 18–4. Exercise 29.

30. All circles through the origin with centers on the x-axis.

In Exercises 31–37, express the given information as a differential equation.

31. A particle moves along a coordinate line and its velocity v at time t is inversely proportional to its position s.

32. The velocity of the particle in Exercise 31 is proportional to the square root of its position.

33. Express Newton's second law of motion $F = ma$ as a second order differential equation.

34. A function equals the square of its derivative.

35. A function equals the cube of its derivative.

36. The rate of change of population is directly proportional to the present population.

37. (Newton's law of cooling). The rate at which a hot body cools is proportional to the difference between its temperature T and the constant temperature C of the surrounding medium.

18-2 SEPARABLE EQUATIONS

Many applications give rise to differential equations that can be written in the form

$$(18.1) \qquad g(y)y' = f(x)$$

Such equations, called **separable equations**, were introduced in Section 5–7 and used in Section 7–4 to discuss pollution, population growth, and radioactive decay. In this section, we review how to solve separable equations and then describe some new applications.

Equation (18.1) is solved by *separating the variables.* We multiply both sides by dx and apply the definition $dy = y'\, dx$ introduced in Section 4–3; then (18.1) becomes

$$g(y)\, dy = f(x)\, dx$$

Integrating both sides of this equation yields

(18.2) $$\int g(y)\, dy = \int f(x)\, dx + C$$

as the general solution of (18.1).

EXAMPLE 1   Show that

$$y' = 2xy$$

is separable and find the general solution.

Solution   The variables can be separated by writing

$$\frac{1}{y}\frac{dy}{dx} = 2x$$

and then

$$\frac{1}{y}dy = 2x\, dx$$

Applying (18.2), we integrate both sides and obtain

$$\ln y = x^2 + C$$

This is an implicit solution; it can be solved for y using exponentials

$$y = e^{x^2 + C} = Ke^{x^2} \qquad\qquad (K = e^C) \qquad\blacksquare$$

A separable (or any type of) differential equation is often written in **differential form**

(18.3) $$M(x, y)\, dx + N(x, y)\, dy = 0$$

The procedure for solving the equation is the same as before.

EXAMPLE 2   Solve the equation

$$dx - y(1 + x^2)\, dy = 0$$

Solution   This is the differential form (18.3) with $M(x, y) = 1$ and $N(x, y) = -y(1 + x^2)$. We separate the variables first, and then integrate.

$$y\, dy = \frac{1}{1 + x^2}dx$$

$$\frac{1}{2}y^2 = \tan^{-1}x + C \qquad \text{(Integrate both sides)}$$

which is the (implicit) general solution. ∎

Remark: From now on, we will not call attention to the fact that a solution is implicit or explicit. We will usually leave an implicit solution as it stands unless it is necessary, as in some applications, to find an explicit solution.

Separable equations represent a certain class of differential equation; other classes, along with their special methods of solution, are introduced in succeeding sections of this chapter. Given a differential equation, your ability to solve it will depend, in part, on your skill at recognizing the class to which it belongs.

EXAMPLE 3 Classify the equations

$$dx - (x + y)\,dy = 0 \quad \text{and} \quad -x\,dx + x^2y\,dy = 0$$

as to separable or nonseparable and solve the separable one(s).

Solution The first equation is not separable; the plus sign prevents us from separating the variables (check this out for yourself). The second equation is separable because it can be written as $x^2y\,dy = x\,dx$, and then

$$y\,dy = \frac{1}{x}\,dx$$

Integrating both sides yields

$$\frac{1}{2}y^2 = \ln x + C$$

as a general solution. ∎

Remark on rigor: In some parts of mathematics, algebra for example, it is entirely appropriate to be as mathematically precise as possible. But introductory discussions of differential equations, whose primary purpose is to help the student understand the nature and importance of these equations, tend to be less precise. For instance, you may have noticed that in the last example, we divided by x^2 without saying that $x \neq 0$, and that the integral of $1/x$ was found to be $\ln x$ instead of the more precise $\ln |x|$. This practice is maintained throughout our presentation.

Applications

You have already seen (Section 7–4) that some problems about pollution, population growth, and radioactive decay give rise to separable equations of the form

$$y' = ky \quad \text{or} \quad \frac{1}{y}\,dy = k\,dt \qquad \text{(\textit{t} is time)}$$

Let us now look at some other applications.

EXAMPLE 4 *(Free fall with air resistance).* A 320 lb object is dropped from a stationary balloon. If its velocity is $v = v(t)$ and the air resistance is $v/10$ lbs, express v as a function of t and find the terminal velocity (that is, the limiting value of v as t tends to infinity).

Solution Let us first solve the problem for any freely falling body. According to Newton's second law of motion, the force F acting on an object of mass m will produce an acceleration $a = F/m$ or $F = ma$. A body in free fall under the influence of the earth's gravity $g(\approx 32 \text{ ft/sec}^2)$ is subject to the force $F = mg$. If y represents the distance down to the body from some fixed height, then its acceleration is d^2y/dt^2, and the resulting differential equation is

$$m\frac{d^2y}{dt^2} = mg$$

If we now assume that air exerts a retarding force *(drag)* that is proportional to the velocity, then the total force acting on the body is $mg - k(dy/dt)$ and the equation that describes the motion is

(18.4) $$m\frac{d^2y}{dt^2} = mg - k\frac{dy}{dt}$$

Now we let $v = dy/dt$ be the velocity, so (18.4) becomes

$$m\frac{dv}{dt} = mg - kv$$

Dividing by m and separating the variables, we obtain

$$\frac{dv}{g - \dfrac{k}{m}v} = dt$$

Integrating both sides yields

$$-\frac{m}{k}\ln\left(g - \frac{k}{m}v\right) = t + C$$

and solving for v, we have

$$g - \frac{k}{m}v = C_1e^{-kt/m}$$

Since $v = 0$ when $t = 0$, it follows that $C_1 = g$, so

(18.5) $$v = \frac{mg}{k}(1 - e^{-kt/m}) \quad \text{ft/sec}$$

is the velocity at any time t. From (18.5), we see that the terminal velocity is

(18.6) $$\lim_{t \to \infty} v = \frac{mg}{k} \quad \text{ft/sec}$$

The graph of v is illustrated in Figure 18–5.

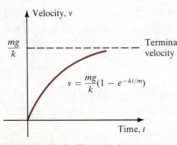

FIGURE 18–5. The velocity versus time curve of a freely falling object with air resistance proportional to the velocity (Example 4).

In our problem, $k = 1/10$, $mg = 320$ lbs, and $m = 320/32 = 10$ slugs. Thus, by (18.5), we have

$$v = 3,200(1 - e^{-t/100}) \text{ ft/sec}$$

and, by (18.6), the terminal velocity is 3,200 ft/sec. ■

Another application is in the study of *population growth in a limited environment*. Under ideal conditions, populations tend to grow exponentially. But the ideal conditions cannot last indefinitely because the food supply may diminish, and the environment may become overcrowded or polluted. The differential equation that is used as a model for population growth in a limited environment is

(18.7) $$\frac{dy}{dt} = ky(S - y)$$

where $y = y(t)$ is the population at time t, k is a constant depending on the particular population under observation, and S is the maximum population the given environment will support. Equation (18.7) is called a **logistic equation;** a detailed account of this equation, including its derivation, was presented in Section 9–7.

To solve (18.7), we separate the variables

$$\frac{1}{y(S - y)} dy = k \, dt$$

and integrate the left side using a partial fraction decomposition or Formula **49** in the Table of Integrals. After solving the result for y, we find that

(18.8) $$y = \frac{S}{1 + Ce^{-Skt}}$$

is the general solution to the logistic equation. If $y_0 = y(0)$ is the initial population, then setting $t = 0$ in (18.8) yields $C = (S - y_0)/y_0$. Thus, (18.8) becomes

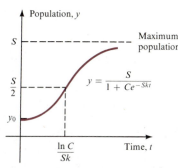

Population, y

S ---- Maximum population

$\dfrac{S}{2}$

$y = \dfrac{S}{1 + Ce^{-Skt}}$

y_0

$\dfrac{\ln C}{Sk}$ Time, t

FIGURE 18–6. The logistic model of exponential growth. At $t = (\ln C)/Sk$, the population is $S/2$; this is the point of maximum rate of growth.

(18.9) $$\boxed{y = \frac{S}{1 + Ce^{-Skt}} \quad \text{where } C = \frac{S - y_0}{y_0}}$$

The graph of y is illustrated in Figure 18–6. Observe that the most rapid rate of growth occurs at the inflection point.

EXAMPLE 5   Twenty fruit flies are put into an environment that will support about 1,000 flies. Experiments by Raymond Pearl in the early 1920's show that if t is measured in days, then the logistic equation for this situation is

$$\frac{dy}{dt} = 1.6 \times 10^{-4} y(1,000 - y)$$

How many fruit flies will there be after 25 days?

Solution   Applying (18.9) with

$$k = 1.6 \times 10^{-4} \quad S = 1{,}000 \quad C = \frac{1{,}000 - 20}{20} = 49$$

the population after 25 days will be

$$y(25) = \frac{1{,}000}{1 + 49e^{-(.16)(25)}} = 527 \text{ fruit flies} \quad \blacksquare$$

Still another application is to *chemical reaction rates*. Suppose that chemical A combines with chemical B to form one or more products P; this is usually written as

$$A + B \rightarrow P$$

The rate at which P is produced depends, in part, on how frequently the molecules of A and B collide. This, in turn, is jointly proportional to the number of molecules of A and B that are present. The number of molecules is given in *moles* and one mole is 6×10^{23} molecules. The concentration of a chemical A is measured in moles per liter and denoted by the symbol $[A]$.

EXAMPLE 6   Suppose that equal amounts of chemicals A and B combine to form P and the rate of reaction is governed by the differential equation

$$\frac{d}{dt}[A] = -0.01[A][B]$$

If, at time $t = 0$, $[A] = [B] = a$ moles per liter, find the concentration of A as a function of time.

Solution   Let y denote the concentration of A at time t. Since the concentration of A and B are the same throughout the reaction, the given differential equation can be written in terms of y as

$$\frac{dy}{dt} = -0.01y^2$$

This is a separable equation that is easily solved

$$y^{-2} \, dy = -0.01 \, dt, \text{ so } \quad y^{-1} = 0.01t + C$$

Setting $t = 0$ and $y = a$ yields $C = 1/a$; it follows that

$$y = \frac{1}{0.01t + (1/a)}$$

is the concentration of A (and B) at time t. $\blacksquare$

EXERCISES

In Exercises 1–20, the given equations are separable. Find the general solutions and check your answers *before* looking in the back of the book.

1. $\sin x \, dx + \cos y \, dy = 0$ **2.** $x \, dy - y \, dx = 0$

3. $2x(y^2 + 1) + y' = 0$

4. $\tan t - s^2 s' = 0$

5. $\dfrac{1}{\ln u} = \dfrac{u}{v^2}\dfrac{dv}{du}$

6. $y' = e^x - x$

7. $y' = \sin^{-1} x$

8. $y' = x - 1 + xy - y$

9. $e^y \sin x\, dx - \cos x\, dy = 0$

10. $x^2 y' - yx = 0$

11. $2y\, dx + (xy + 5x)\, dy = 0$

12. $x^2\, dy - y^2\, dx = 0$

13. $x \tan y\, dx - \sec x\, dy = 0$

14. $x^2\, dy - y(x^2 + 1)\, dx = 0$

15. $s' = e^{t-s}$

16. $\cos x\, dy - y\, dx = 0$

17. $y\sqrt{x^3 + 1}y' = -x^2(y^2 + 1)$

18. $e^y\, dy + y^{-1}\, dx = 0$

19. $y \ln x\, dx = x\, dy$

20. $(1 + x^2)\, dy = (1 + y^2)\, dx$

(g) $dy + 2y\, dx = e^{-x}\, dx$

(h) $u'v = (1 + v)\csc u$

(i) $(s^2 + t^2)s' = s^2$

(j) $(x \sin y - y)y' = \cos y$

(k) $\sqrt{1 - y^2} = y'\sqrt{1 - x^2}$

(l) $ts' + s = \sin t$

(m) $(2xy + y^2)\, dx = (y - x^2 - 2xy)\, dy$

(n) $y' - y = \cos x$

(o) $\sqrt{s}ts' = 3$

(p) $(xe^{y/x} + y) = xy'$

(q) $y' - 4x^2 y = x$

(r) $u^2 u' = uv - v^2$

(s) $u' + u \tan v = \sec v$

(t) $ts' = -s + \sqrt{s^2 + t^2}$

In Exercises 21–30, find the particular solutions that satisfy the given condition. Some refer to the exercises above.

21. $\sqrt{s}ts' + t = 0;\ s(1) = 4$

22. $y' = x \cos^2 y;\ y(0) = \pi/4$

23. $y' = (1 + y^2)x;\ y(2) = 0$

24. $x\, dy = \sqrt{1 - y^2}\, dx;\ y(1) = 1/2$

25. Exercise 1; $y(0) = 0$

26. Exercise 3; $y(2) = 0$

27. Exercise 7; $y(0) = 1$

28. Exercise 9; $y(0) = 0$

29. Exercise 11; $y(1) = 1$

30. Exercise 15; $s(0) = 0$

31. Determine which of the equations below are separable and solve them.

(a) $(x^2 + y)\, dx + (x + 2y)\, dy = 0$

(b) $tse^s\, ds = -(t^2 + 1)\, dt$

(c) $(x + y)y' + x - y = 0$

(d) $y'(y + x^3) = 3x^2 y$

(e) $r \cos \theta + \sin \theta \dfrac{dr}{d\theta} = 0$

(f) $1 + r \cos \theta + \sin \theta \dfrac{dr}{d\theta} = 0$

32. If y represents the amount present of a radioactive substance at time t that decays at a rate proportional to y, then

$$\frac{dy}{dt} = -ky$$

where k is a positive constant. Show that the amount present at any time t is $y(0)e^{-kt}$.

33. Radioactive beryllium is sometimes used to date deep-sea artifacts. Beryllium decays according to the equation $(dy/dt) = -1.5 \times 10^{-7}y$. What is the half-life of beryllium (in years)?

34. What function is equal to the cube of its derivative and also takes the value 1 at 0?

35. What is the most general function with the property that the square of the function plus the square of its derivative equals 1?

36. Newton's law of cooling says that the rate at which a body cools is proportional to the difference in temperature between it and its surroundings. If an object at 110°F is placed in a medium of constant temperature 10°F and after one hour the temperature of the object is 60°, what will its temperature be in another 80 minutes?

37. In Example 4, suppose that the air resistance is proportional to v^2. What form does equation (18.5) take in this case? What is the terminal velocity?

38. At a constant temperature, the rate of change of atmospheric pressure p with respect to altitude h is proportional to p. If $p = 30$ at sea level and $p = 28$ when $h = 3,000$ ft, find the pressure at 10,000 ft.

39. An object starts from 0 and travels along a coordinate line toward the point A which is 6 ft to the right of 0. If its initial velocity is 30 ft/sec and its velocity thereafter is proportional to the square of its distance from A, will it ever reach the point A? Explain.

40. The present population of a colony of ants is 10,000 and the logistic equation for this colony is

$$\frac{dy}{dt} = 3 \times 10^{-6} y(40,000 - y)$$

How long (in days) will it take for the colony to double in size?

41. The chemical reaction $CH_3COCH_3 + I_2 \rightarrow CH_3COCH_2I + HI$, which combines equal amounts of acetone and iodine, is governed by the differential equation

$$\frac{dy}{dt}[I_2] = -4 \times 10^{-4}[CH_3COCH_3][I_2]$$

where t is measured in seconds. If the beginning concentrations of acetone and iodine are both 5 moles/liter, find the concentration of iodine as a function of time.

18–3   HOMOGENEOUS EQUATIONS

A first order equation

(18.10) $$\frac{dy}{dx} = f\left(\frac{y}{x}\right)$$

where the right side is any continuous function of y/x is called **homogeneous.** For example,

$$x\,dy - (x + y)\,dx = 0$$

is homogeneous because it can be rewritten as

$$\frac{dy}{dx} = \frac{x + y}{x} = 1 + \frac{y}{x}$$

and the right side is a function of y/x. We solve (18.10) by making a change of variable

(18.11) $$v = \frac{y}{x}$$

Then $y = vx$, and the product rule yields

18.12) $$\frac{dy}{dx} = v + x\frac{dv}{dx}$$

Substituting (18.11) and (18.12) into (18.10) yields

$$v + x\frac{dv}{dx} = f(v)$$

which is equivalent to the separable equation

(18.13) $$\frac{1}{f(v) - v} dv = \frac{1}{x} dx$$

After finding the general solution of this equation, we must, of course, replace v by y/x to obtain a solution of (18.10).

EXAMPLE 1 Solve the equation $x \, dy - (x + y) \, dx = 0$.
Solution First rewrite the equation as

$$\frac{dy}{dx} = \frac{x + y}{x} = 1 + \frac{y}{x} = f\left(\frac{y}{x}\right)$$

If we let $v = y/x$, then $f(v) = 1 + v$ and (18.13) reads

$$\frac{1}{(1 + v) - v} dv = \frac{1}{x} dx$$

whose solution is $v = \ln x + C$. Replacing v by y/x yields

$$\frac{y}{x} = \ln x + C \quad \text{or} \quad y = x \ln x + Cx$$

as a general solution to the original equation.

Check. $$x \, dy = x \, d(x \ln x + Cx) = x\left[x\left(\frac{1}{x}\right) + \ln x + C\right] dx$$

$$= (x + x \ln x + Cx) \, dx$$

and

$$(x + y) \, dx = (x + x \ln x + Cx) \, dx$$

Therefore, $x \, dy - (x + y) \, dx = 0$, which is the original equation. ∎

We summarize the above procedure as follows:

(18.14)

To solve a homogeneous equation

$$\frac{dy}{dx} = f\left(\frac{y}{x}\right)$$

(1) The substitution $v = y/x$ converts the given equation into the separable equation

$$\frac{1}{f(v) - v} dv = \frac{1}{x} dx$$

(2) Solve the separable equation and replace v by y/x.

EXAMPLE 2 Solve $(x^2 + y^2)\, dx - 2xy\, dy = 0$.

Solution First rewrite the equation as

$$\frac{dy}{dx} = \frac{x^2 + y^2}{2xy} = \frac{1}{2}\left[\frac{x}{y} + \frac{y}{x}\right] = f\left(\frac{y}{x}\right)$$

If $v = y/x$, then

$$f(v) - v = \frac{1}{2}\left(\frac{1}{v} + v\right) - v$$

$$= \frac{1}{2}\left(\frac{1}{v} - v\right)$$

$$= \frac{1 - v^2}{2v}$$

Therefore, the resulting differential equation in (18.14) is

$$\frac{2v}{1 - v^2}\, dv = \frac{1}{x}\, dx$$

whose general solution is $-\ln(1 - v^2) = \ln x + C$. Using exponentials, this is equivalent to $x(1 - v^2) = K$ where $K = e^{-C}$. Replacing v by y/x yields

$$x\left(1 - \frac{y^2}{x^2}\right) = K \quad \text{or} \quad x^2 - y^2 = Kx$$

as the general solution. The family of curves is the family of hyperbolas shown in Figure 18–7. ∎

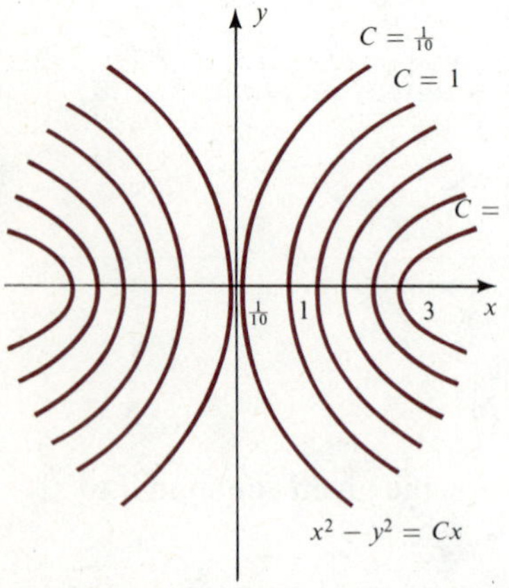

$x^2 - y^2 = Cx$

FIGURE 18–7. Example 2. The family of curves for $(x^2 + y^2)\, dx - 2xy\, dy = 0$.

The family of curves for a homogeneous equation has a simple geometric property. Let B be any curve and let k be any nonzero constant. We denote by kB the set of points (kx, ky) where (x, y) is a point of B. Now suppose B is a member of the family of curves of the homogeneous equation $y' = f(y/x)$. Then it is easy to see that replacing x by kx and y by ky does not change the equation, so that kB is again in the family. The converse can also be shown; that is, if kB is in the family whenever B is, then the differential equation must be homogeneous. For instance, in Figure 18–7, let D be curve $x^2 - y^2 = x$, so $C = 1$. If $k = 1/3$, then we replace x by $x/3$, y by $y/3$ and the resulting curve is

$$\left(\frac{x}{3}\right)^2 - \left(\frac{y}{3}\right)^2 = \frac{x}{3} \quad \text{or} \quad x^2 - y^2 = 3x$$

which is the curve with $C = 3$. If $k = 10$, the result is the curve with $C = 1/10$ shown in the figure.

We conclude this section with an application of homogeneous equations.

EXAMPLE 3 A missile fired from a point 15 miles due east of a stationary target has a homing device that keeps it always heading toward the target. If the speed of the missile is 200 mph and the wind is blowing due south at 50 mph, what is the path of the missile?

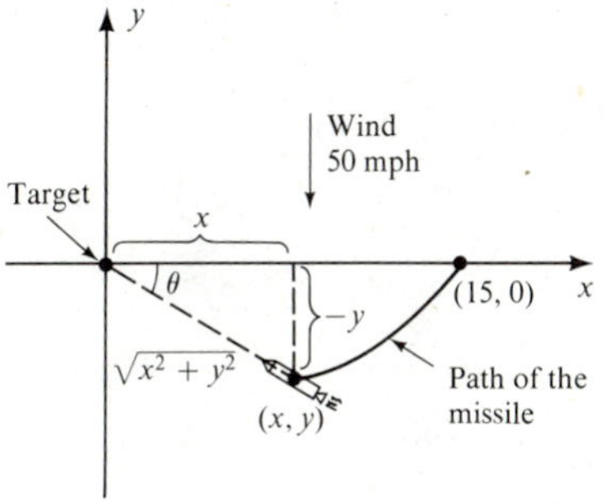

Solution   We set up a coordinate system with the target at the origin as shown in Figure 18–8. At any point (x, y) on the path, the components of the missile's velocity are

$$\frac{dx}{dt} = -200 \cos \theta \quad \text{and} \quad \frac{dy}{dt} = -50 + 200 \sin \theta$$

Therefore, dividing the second equation by the first, we have

$$\frac{dy}{dx} = \frac{-50 + 200 \sin \theta}{-200 \cos \theta} = \frac{-50 + 200(-y/\sqrt{x^2 + y^2})}{-200(x/\sqrt{x^2 + y^2})}$$

$$= \frac{-50\sqrt{x^2 + y^2} - 200y}{-200x}$$

$$= \frac{1}{4}\sqrt{1 + (y/x)^2} + (y/x)$$

This is a homogeneous equation. We let $v = y/x$ and, by (18.14), the resulting equation is

$$\frac{1}{\frac{1}{4}\sqrt{1 + v^2}}\, dv = \frac{1}{x}\, dx$$

The integral of the left side can be evaluated with trigonometric substitution or Basic Formula **17**. In either case, we have

$$4 \ln (v + \sqrt{1 + v^2}) = \ln x + C$$

We evaluate C by noting that y, and hence v, is 0 when $x = 15$. It follows that $C = -\ln 15$ and if v is replaced by y/x, then

$$4 \ln \left(\frac{y}{x} + \sqrt{1 + (y/x)^2}\right) = \ln x - \ln 15$$

This is equivalent to

$$[y + \sqrt{x^2 + y^2}]^4 = (15)^{-1} x^5$$

which is an equation of the path of the missile.   ■

FIGURE 18–8. Example 3.

EXERCISES

In Exercises 1–18, the given equations are homogeneous. Find the general solutions.

1. $(x - 2y) + x\dfrac{dy}{dx} = 0$

2. $(2x + y)\, dy - (4x + y)\, dx = 0$

3. $2y\, dx - x\, dy = 0$

4. $3x\, dy + y\, dx = 0$

5. $x^2 y' = y^2 + 2xy$

6. $x^2 y' = 4y^2 - xy$

7. $x^2 y' = x^2 + xy + y^2$

8. $y^2y' = y^2 + xy + x^2$

9. $(2x - y)\,dy = (4y - 3x)\,dx$

10. $(x + y)\,dx = (x - y)\,dy$

11. $(2x^2 - y^2) + 3xy\,y' = 0$

12. $(3x + y)\,dy = (2x - 3y)\,dx$

13. $2xy' = y - 6x$

14. $2xy' = y + 3x$

15. $xy' = y[1 + \ln(y/x)]$

16. $x^2y' + xy + 2y^2 = 0$

17. $x^2y' = x^2 + y^2$

18. $(y^2 + x^2)\,dx = xy\,dy$

19. Exercise 3 above is separable as well as homogeneous. Solve it now as a separable equation and compare your answers.

20. Do the same as Exercise 19 with the equation in Exercise 4.

21. Some of the equations below are separable, some are homogeneous, and some are neither. Find the ones that are separable and/or homogeneous and solve them.

(a) $(x^2 + y)\,dx = (x + 2y)\,dy$

(b) $(x + y)y' + (x - y) = 0$

(c) $ue^v\,du - v\,du = 0$

(d) $y' = 3x^2y/(y + x^3)$

(e) $(y^2 + x^2)y' = y^2$

(f) $1 + r\cos\theta + \sin\theta\,\dfrac{dr}{d\theta} = 0$

(g) $dy + 2y\,dx = e^{-x}dx$

(h) $(xe^{y/x} + y)\,dx = x\,dy$

(i) $r\theta\,dr + \cos\theta\,d\theta = 0$

(j) $(x\sin y - y)y' = \cos y$

(k) $(1 - x) = (1 + y^2)y'$

(l) $ts' + s = \sin t$

(m) $(2xy + y^2)\,dx = (x^2 - y)\,dy$

(n) $y' - y = \cos x$

(o) $u^2u' + (v^2 - uy) = 0$

(p) $y' + yx = 4x$

(q) $dy - 4x^2y\,dx = x\,dx$

(r) $\cos u \sin v\,du = \tan v\,dv$

(s) $u' + u\tan v = \sec v$

(t) $s' = \dfrac{-s + \sqrt{s^2 + t^2}}{t}$

22. In Example 3 , suppose that the speed of the wind is a mph and that the missile is fired from a point c miles due east of the target with a speed of b mph. Show that an equation of the path of the missile is

$$y = \frac{1}{2}\left[\frac{x^{m+1}}{c^m} - \frac{c^m}{x^{m-1}}\right]$$

where $m = a/b$.

23. A boat is launched on one side of a river and continuously heads toward the point on the opposite shore directly across from the launch site. Suppose that the river is $1/2$ mile wide, its current has a speed of 3 mph, and the boat also has a speed of 3 mph. At what point on the opposite shore will the boat land?

18–4 EXACT EQUATIONS

A first order equation

(18.15) $M(x, y)\,dx + N(x, y)\,dy = 0$

is called an **exact differential equation** if there is a function $z = f(x, y)$ such that

(18.16) $\dfrac{\partial f}{\partial x} = M$ and $\dfrac{\partial f}{\partial y} = N$

In this case, (18.15) can be written as

(18.17) $\dfrac{\partial f}{\partial x}\,dx + \dfrac{\partial f}{\partial y}\,dy = 0$

The left side of (18.17) is called the *differential of f* and is denoted by *df* (see Section 15–4). It follows that (18.15) takes the form $df = 0$ and that $f(x, y) = C$ is the general solution.

(18.18)

> If $M(x, y)\, dx + N(x, y)\, dy$ is exact and
>
> $$\frac{\partial f}{\partial x} = M \quad \text{and} \quad \frac{\partial f}{\partial y} = N$$
>
> then $f(x, y) = C$ is the general solution.

EXAMPLE 1 Solve $y\, dx + x\, dy = 0$.

Solution This equation can be solved easily by separating the variables, but let us use the method discussed above. In this example, $M(x, y) = y$ and $N(x, y) = x$. By inspection, we recognize that if $f(x, y) = xy$, then

$$\frac{\partial f}{\partial x} = y = M \quad \text{and} \quad \frac{\partial f}{\partial y} = x = N$$

Therefore, by (18.18), $xy = C$ is the general solution.

Check: Differentiating the solution yields

$$0 = C' = (xy)' = x\frac{dy}{dx} + y$$

so $y\, dx + x\, dy = 0$, which is the original equation. ■

In Example 1, the function $f(x, y) = xy$ was found by inspection. Clearly this technique is impractical in all but the simplest cases. In what follows, we will state a *test* for exactness and develop a *method* for finding the function f.

First, suppose that (18.15) is exact and that f is a function satisfying (18.16). Assuming that f has continuous partials, we know from Section 15–3 that the mixed partials of f are equal. Therefore, if (18.15) is exact, then

(18.19) $$\frac{\partial M}{\partial y} = \frac{\partial^2 f}{\partial y \partial x} = \frac{\partial^2 f}{\partial x \partial y} = \frac{\partial N}{\partial x}$$

The converse statement is also true, and it follows that

(18.20)

> The equation $M(x, y)\, dx + N(x, y)\, dy = 0$ is exact if and only if
>
> $$\frac{\partial M}{\partial y} = \frac{\partial N}{\partial x}$$

Rather than prove the converse statement in general, we will show, by example, how to construct the function f, given that $M_y = N_x$. The method will be familiar to those of you who studied Chapter 17, where we referred to $\mathbf{F}(x, y) = M(x, y)\mathbf{i} + N(x, y)\mathbf{j}$ as a *conservative force field* and to f as a *potential function*.

EXAMPLE 2   Show that $(2x + y) \, dx + (x - 3y^2) \, dy = 0$ is exact and solve it.

Solution   Notice that this equation is not separable. Nor is it homogeneous, so previously discussed methods do not apply. We test for exactness using (18.20).

$$\frac{\partial M}{\partial y} = \frac{\partial}{\partial y}(2x + y) = 1 \quad \text{and} \quad \frac{\partial N}{\partial x} = \frac{\partial}{\partial x}(x - 3y^2) = 1$$

The next step is to construct the function f that satisfies

$$(18.21) \quad \frac{\partial f}{\partial x} = 2x + y \quad \text{and} \quad \frac{\partial f}{\partial y} = x - 3y^2$$

Integrating the first equation with respect to x (treating y as a constant) yields

$$(18.22) \quad f(x, y) = x^2 + xy + g(y)$$

where $g(y)$ is a function of y alone; it is the "constant of integration" when we integrate with respect to x. By (18.21), $f_y = x - 3y^2$, so we differentiate (18.22) with respect to y, set the result equal to $x - 3y^2$, and solve for g.

$$\frac{\partial f}{\partial y} = x + g'(y) = x - 3y^2$$

It follows that $g'(y) = -3y^2$ and $g(y) = -y^3$; we usually omit the constant C here because it will appear later in the general solution. Now, by (18.22),

$$f(x, y) = x^2 + xy - y^3$$

and it follows from (18.18) that $x^2 + xy - y^3 = C$ is the general solution of the given equation.   ■

We summarize the procedure as follows:

(18.23)

> If $M(x, y) \, dx + N(x, y) \, dy = 0$ is exact, then
>
> (1) Set $f(x, y) = \int M(x, y) \, dx$. The constant of integration is a function g of y alone.
>
> (2) For the function f obtained in (1), compute f_y, set that equal to N, and solve for g.
>
> (3) Then $f_x = M, f_y = N$, and, by (18.18),
>
> $$f(x, y) = C$$
>
> is the general solution.

EXAMPLE 3 Solve $(-y \sin x)\, dx + (\cos x - \cos y)\, dy = 0$.

Solution Since

$$\frac{\partial}{\partial y}(-y \sin x) = -\sin x = \frac{\partial}{\partial x}(\cos x - \cos y)$$

the equation is exact. We now use (18.23) to find a function f with

$$\frac{\partial f}{\partial x} = -y \sin x \quad \text{and} \quad \frac{\partial f}{\partial y} = \cos x - \cos y$$

Integrating the first equation with respect to x yields

$$f(x, y) = y \cos x + g(y)$$

Next, we differentiate with respect to y, set the result equal to $\cos x - \cos y$, and solve for g.

$$\frac{\partial f}{\partial y} = \cos x + g'(y) = \cos x - \cos y$$

It follows that $g(y) = -\sin y$ and that

$$f(x, y) = y \cos x - \sin y$$

Therefore, $y \cos x - \sin y = C$ is the general solution. ∎

Integrating Factors (Optional)

It may seem from the discussion so far that exact equations are highly specialized and, therefore, limited in scope. It can be shown, however, that every first order equation that has a general solution can (at least theoretically) be transformed into an exact equation and the solution found by the methods discussed above. We will illustrate the procedure here for some special cases only.

The equation

(18.24) $2y\, dx + x\, dy = 0$

is not exact because

$$\frac{\partial}{\partial y}(2y) = 2 \quad \text{and} \quad \frac{\partial}{\partial x}(x) = 1$$

But if we multiply both sides of the equation by x, then

$$2xy\, dx + x^2\, dx = 0$$

is exact; the left side is $d(x^2 y)$, so a general solution for (18.24) is $x^2 y = C$. The multiplier x used above is called an *integrating factor*.

DEFINITION

A nonzero function μ is called an **integrating factor** of the equation
$M(x, y) \, dx + N(x, y) \, dy = 0$ if

(18.25)
$$\mu(M \, dx + N \, dy) = 0$$

is exact.

It is frequently possible to discover integrating factors by inspection. Certain combinations of differentials suggest trying a particular multiplier.

EXAMPLE 4 Solve $y \, dx + (x^2 y - x) \, dy = 0$.

Solution We rearrange the equation by writing

(18.26) $\quad x^2 y \, dy - (x \, dy - y \, dx) = 0$

The expression in parentheses reminds us of the differential

$$d\left(\frac{y}{x}\right) = \frac{x \, dy - y \, dx}{x^2}$$

Therefore, dividing both sides of (18.26) by x^2 yields

$$y \, dy - \frac{x \, dy - y \, dx}{x^2} = 0$$

or

$$y \, dy + d\left(\frac{y}{x}\right) = 0$$

Integration now shows that a general solution of (18.26) is

$$\frac{1}{2} y^2 + \frac{y}{x} = C \quad \blacksquare$$

The method of inspection usually depends on recognizing combinations of $x, y, dx,$ and dy. Some of these combinations are as follows:

(18.27) $\qquad d(xy) = x \, dy + y \, dx$

(18.28) $\qquad d\left(\frac{y}{x}\right) = \frac{x \, dy - y \, dx}{x^2}$

(18.29) $\qquad d\left(\frac{x}{y}\right) = \frac{y \, dx - x \, dy}{y^2}$

(18.30) $\quad d(x^2 + y^2) = 2(x \, dx + y \, dy)$

EXAMPLE 5 Solve $(y + x^3 y^2) \, dx + x \, dy = 0$.

Solution We rearrange to get a familiar combination

$$(y \, dx + x \, dy) + x^3 y^2 \, dx = 0$$

The expression in parentheses is $d(xy)$; if we multiply through by $(xy)^{-2}$, then

$$(xy)^{-2}(y\,dx + x\,dy) + x\,dx = 0$$

The left side is $d\,[-(xy)^{-1} + (x^2/2)]$ so

$$-\frac{1}{xy} + \frac{x^2}{2} = C$$

is a general solution. ■

Here is an application of exact equations.

EXAMPLE 6 It was pointed out in Chapter 12 that a parabolic mirror with a light source at the focus will reflect the light in rays parallel to the axis. Show that this is the only shape with this property.

Solution We revolve a curve $y = f(x)$ about the x-axis to generate a surface of revolution and consider the light source to be at the origin (Figure 18–9). We want to prove that if the light is reflected in a beam, then the curve is a parabola with the origin as its focus. Using the figure and some geometry, we have that $\phi = \beta$ and $\theta = \alpha + \phi$. By the assumed reflection property, it follows that $\alpha = \beta$, so $\theta = 2\beta$. Since $\tan \theta = y/x$, we have

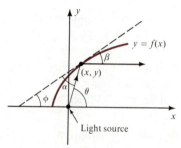

FIGURE 18–9. Example 6.

$$\frac{y}{x} = \tan \theta = \tan 2\beta = \frac{2 \tan \beta}{1 - \tan^2\beta}$$

Converting to derivatives ($\tan \beta = dy/dx$) yields

$$\frac{y}{x} = \frac{2(dy/dx)}{1 - (dy/dx)^2}$$

Solving for dy/dx gives

$$\frac{dy}{dx} = \frac{-x \pm \sqrt{x^2 + y^2}}{y} \qquad \text{(Quadratic formula)}$$

$$x\,dx + y\,dy = \pm \sqrt{x^2 + y^2}\,dx$$

Using (18.30), we can write

$$\pm \frac{d(x^2 + y^2)}{2\sqrt{x^2 + y^2}} = dx$$

Integration yields $\pm \sqrt{x^2 + y^2} = x + C$ and simplification of this equation gives

$$y^2 = 2Cx + C^2$$

as an equation of the curve. This is a parabola with its focus at the origin. ■

We conclude this section with a statement concerning integrating factors that are functions of x alone.

If, in the equation $M\,dx + N\,dy = 0$, we have

$$\frac{1}{N}\left(\frac{\partial M}{\partial y} - \frac{\partial N}{\partial x}\right) = g(x)$$

(18.31)

is a function of x alone, then

$$\mu = e^{\int g(x)dx}$$

is an integrating factor.

The proof of (18.31) is outlined in an exercise. We will give many examples of this integrating factor in the next section, where it is used to solve linear differential equations.

EXERCISES

In Exercises 1–10, test for exactness and then solve the equations.

1. $(3x^2 - y^2)\,dx + (3 - 2xy)\,dy = 0$

2. $(3x^2 + y)\,dx + (x - 2y)\,dy = 0$

3. $(3u^2 + v)\,du + u\,dv = 0$

4. $(e^u + 2uv^2)\,du + 2u^2v\,dv = 0$

5. $y\cos xy\,dx + (x\cos xy + \sin y)\,dy = 0$

6. $(\sin y - y\sin x)\,dx + (x\cos y + \cos x)\,dy = 0$

7. $y\left(e^{xy} + \dfrac{1}{x}\right)dx + (xe^{xy} + \ln x)\,dy = 0$

8. $(y - x^3)\,dx + (x + y^3)\,dy = 0$

9. $\left(\dfrac{1}{1 + x^2} + \cos y\right)dx = x\sin y\,dy$

10. $(e^y + \cos x\cos y)\,dx = (\sin x\sin y - xe^y)\,dy$

Exercises 11–20, are *optional*. If the equation is not exact, then find an integrating factor and solve.

11. $(y - x)\,dy + y\,dx = 0$

12. $(y^3 - x)\,dy + y\,dx = 0$

13. $2x\,dx + 2y\,dy = (x^2 + y^2)\,dx$

14. $x\,dx + y\,dy = \sqrt{x^2 + y^2}\,dx$

15. $(y + x)\,dy = (y - x)\,dx$

16. $(x + y)\,dx = (x - y)\,dy$

17. $y\,dx - x\,dy = y^3\,dy$

18. $y\,dx + (1 - x)\,dy = 0$

19. $x\,dy + y\,dx = x^3y^4\,dy$

20. $x\,dy + y\,dx = x^3y^2\,dx$

21. The equation $\frac{1}{3}yx^3\,dy + \frac{1}{2}y^2x^2\,dx = 0$ is both separable and exact. Solve it both ways. Do you get equivalent answers? Which method is easier?

22. The equation $(3x + y)\,dx + (x - 4y)\,dy = 0$ is both homogeneous and exact. Solve it both ways. Do you get equivalent answers? Which method is easier?

23. Classify each of the following equations as separable, homogeneous, exact, or none of the above. Then solve the equations in the first three classifications.

(a) $(x^2 + y)\,dx + (x + 2y)\,dy = 0$

(b) $uv + (e^u + 1)u' = 0$

(c) $(s + t)\,ds - (s - t)\,dt = 0$

(d) $y' = 3x^2y/(y + x^2)$

(e) $(x\cos y)\,dx + dy = 0$

(f) $(1 + r\cos\theta)\,d\theta + \sin\theta\,dr = 0$

(g) $dy + (2y - e^{-x})\,dx = 0$

(h) $(du/dv) = (1 + v)/u^2$

(i) $(x^4 + y^4)\,dy = x^4\,dx$

(j) $(x\sin y - y)y' = \cos y$

(k) $(1 + s)\,ds = t\,dt$

(l) $ts' + s = \sin t$

(m) $(2xy + y^2)\,dx = (y - x^2 - 2xy)\,dy$

(n) $y' - y = \cos x$

(o) $\sqrt{xy}\,dy = 3\,dx$

(p) $(xe^{y/x} + y)\,dx = x\,dy$

(q) $dy = (4x^2y + x)\,dx$

(r) $u^2v' + (v^2 - uv) = 0$

(s) $u' + u \tan v = \sec v$

(t) $xy' + y = \sqrt{x^2 + y^2}$

Optional Exercise

24. Prove (18.31) by following the steps below.

(1) Multiply both sides of the original equation by $\mu = e^{\int g(x)dx}$; the result is

$$e^{\int g(x)dx}M(x, y)\,dx + e^{\int g(x)dx}N(x, y)\,dy = 0$$

Then μ is an integrating factor provided we can show that this equation is exact; that is, provided we can show that

$$\frac{\partial}{\partial y}(e^{\int g(x)dx}M(x, y)) = \frac{\partial}{\partial x}(e^{\int g(x)dx}N(x, y))$$

(2) Compute the partial on the left keeping in mind that the first factor μ is a function of x alone, so $\partial\mu/\partial y = 0$.

(3) Compute the partial on the right.

(4) Now use the hypothesis

$$\frac{1}{N}\left(\frac{\partial M}{\partial y} - \frac{\partial N}{\partial x}\right) = g(x)$$

to show that the partials obtained in (2) and (3) are equal. This completes the proof.

18-5   FIRST ORDER LINEAR EQUATIONS

Among the most important differential equations are the *linear equations,* in which the derivative of highest order is a linear function of the derivatives of lower order. A **first order linear equation** can be written in the form

(18.32) $y' + g(x)y = f(x)$

where f and g are functions of x alone. In this section, we illustrate how to solve such equations and describe some of their applications. Second order linear equations are discussed in the next section.

To solve (18.32), we rewrite it as

$$[g(x)y - f(x)]\,dx + dy = 0$$

and look for an integrating factor.* This equation is in the form $M(x, y)\,dx + N(x, y)\,dy$ with $M(x, y) = g(x)y - f(x)$ and $N(x, y) = 1$, and you can easily verify that

$$\frac{1}{N}\left(\frac{\partial M}{\partial y} - \frac{\partial N}{\partial x}\right) = g(x)$$

Therefore, applying (18.31), given at the end of the preceding section, we know that $e^{\int g(x)dx}$ is an integrating factor. If both sides of (18.32) are multiplied by this factor, the result is

(18.33) $e^{\int g(x)dx}y' + yg(x)e^{\int g(x)dx} = f(x)e^{\int g(x)dx}$

---

*Those of you who did not study integrating factors in Section 18–4 can move directly to equation (18.33) where we have multiplied both sides of (18.32) by $e^{\int g(x)dx}$. This function is called an *integrating factor.*

The left side is the derivative of the product $e^{\int g(x)dx}y$, so we have

$$\frac{d}{dx}(e^{\int g(x)dx}y) = f(x)e^{\int g(x)dx}$$

The general solution is now obtained by multiplying through by dx and integrating both sides. The examples illustrate how the method works.

EXAMPLE 1 Solve $y' + 2xy = e^{-x^2}$.

Solution Here, $g(x) = 2x$ and $f(x) = e^{-x^2}$. The integrating factor is

$$e^{\int g(x)dx} = e^{\int 2xdx} = e^{x^2}$$

We multiply through by this factor to obtain

$$e^{x^2}y' + y(2xe^{x^2}) = 1$$

The left side is the derivative of the product $e^{x^2}y$, so

$$\frac{d}{dx}(e^{x^2}y) = 1 \quad \text{or} \quad d(e^{x^2}y) = dx$$

Integration of both sides yields the general solution

$$e^{x^2}y = x + C \qquad \blacksquare$$

This method for solving first order linear equations can be summed up very simply.

(18.34)

> To solve $y' + g(x)y = f(x)$,
>
> (1) Multiply through by $e^{\int g(x)dx}$.
> (2) Write the result as $d(e^{\int g(x)dx}y) = f(x)e^{\int g(x)dx}\,dx$.
> (3) Integrate.

EXAMPLE 2 Solve $y' - y = 4$.

Solution This equation can be solved as a separable equation

$$\frac{1}{y+4}dy = dx$$

but let us instead solve it as a linear equation to further illustrate the method. The equation $y' - y = 4$ is linear with $g(x) = -1$ and $f(x) = 4$. The integrating factor is

$$e^{\int g(x)dx} = e^{\int(-1)dx} = e^{-x}$$

Multiplying through by e^{-x} yields

$$e^{-x}y' - e^{-x}y = 4e^{-x}$$

Notice that the left side is the derivative of the product $e^{-x}y$; thus,

$$\frac{d}{dx}(e^{-x}y) = 4e^{-x} \quad \text{or} \quad d(e^{-x}y) = 4e^{-x}\,dx$$

and integration gives

$$e^{-x}y = -4e^{-x} + C \quad \text{or} \quad y = -4 + Ce^{x}$$

as the general solution.

Check. $y' - y = Ce^{x} - (-4 + Ce^{x}) = 4$, which is the original equation. ■

EXAMPLE 3 Show that $xy' + 2y = e^{x}$ is linear and solve it.

Solution Rewrite the equation as

$$y' + \frac{2}{x}y = \frac{1}{x}e^{x}$$

which is linear with $g(x) = 2/x$ and $f(x) = e^{x}/x$. In this case, the integrating factor is

$$e^{\int g(x)dx} = e^{\int (2/x)dx} = e^{2\ln x} = x^{2}$$

Applying (18.34), we have

$$x^{2}y' + 2xy = xe^{x}$$

Notice that the left side is the derivative of $x^{2}y$ and that the right side must be integrated by parts. The result of integration is that

$$x^{2}y = xe^{x} - e^{x} + C$$

is the general solution. ■

Applications

First order linear equations can be applied to a wide variety of practical problems some of which we have already studied. For example, the equation for population growth

$$y' = ky$$

can be written and solved as a linear equation, $y' - ky = 0$. Another example is the equation

$$\frac{dv}{dt} = g - \frac{k}{m}v$$

that describes free-falling motion with air resistance; it can be solved as a linear equation, $v' + (k/m)v = g$. Let us now look at some other examples.

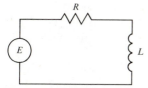

FIGURE 18–10. Example 4. A simple RL-circuit.

EXAMPLE 4 *(RL-circuits).* The simple electric circuit shown in Figure 18–10 contains a source of electromotive force (emf) labeled E, a resistance R, and an inductance L. The emf produces a current I that, according to Kirchhoff's law, satisfies the differential equation

$$(18.35) \quad L\frac{dI}{dt} + RI = E$$

The resistance R and inductance L are constant; E can be constant or vary with time depending on the source. Solve (18.35) under the condition that E is a constant E_0 that produces an initial current of I_0.

Solution Rewrite the equation as

$$\frac{dI}{dt} + \frac{R}{L}I = \frac{E_0}{L}$$

The integrating factor is

$$e^{\int (R/L)\,dt} = e^{Rt/L}$$

Multiplying through by this factor yields

$$\frac{d}{dt}(Ie^{Rt/L}) = \frac{E_0}{L}e^{Rt/L}$$

and integration gives

$$Ie^{Rt/L} = \frac{E_0}{R}e^{Rt/L} + C$$

Since $I = I_0$ when $t = 0$, it follows that $C = I_0 - (E_0/R)$ and that the last equation simplifies to

$$(18.36) \quad I = \frac{E_0}{R} + \left(I_0 - \frac{E_0}{R}\right)e^{-Rt/L}$$

This solution of (18.35) describes the current in the circuit at any time t. ■

Each of the applications mentioned above, including the one in Example 4, can also be worked by separating the variables. Here is an example that needs the linear equation method.

EXAMPLE 5 *(Motion with variable mass).* A rocket is fired straight up burning fuel at the constant rate of a slugs/sec. We let v be the velocity of the rocket, Bv be the drag due to air resistance, and we suppose that the velocity u of the exhaust gas is constant (Figure 18–11). Because the mass M of the rocket is steadily decreasing as the fuel burns and the motion of the exhaust is opposite to that of the rocket, it can be shown that Newton's second law of motion leads to the equation

$$(18.37) \quad F = M\frac{dv}{dt} - ua$$

The force F in this case is the weight gM plus the drag Bv so that (18.37) becomes

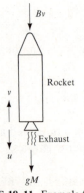

EXAMPLE 18–11. Example 5. Motion with variable mass.

(18.38) $-gM - Bv = M\dfrac{dv}{dt} - ua$

The minus signs indicate the negative direction of F. Let M_1 be the mass of the rocket and payload, let M_2 be the initial mass of the fuel, and let $M_0 = M_1 + M_2$.

It follows that until the fuel runs out (at time $t = M_2/a$), the total mass M at time t is

$$M = M_0 - at$$

We substitute this into (18.38) and rearrange to obtain

(18.39) $\dfrac{dv}{dt} + \dfrac{B}{M_0 - at}v = \dfrac{ua}{M_0 - at} - g$

This is the linear differential equation that describes the motion of the rocket.
    To solve (18.39), we find an integrating factor. Since

$$\int \frac{B}{M_0 - at}\, dt = -\frac{B}{a}\ln(M_0 - at)$$

we know that e to that power, which yields $(M_0 - at)^{-B/a}$, is an integrating factor. Multiplying through, we have

$$\frac{d}{dt}[v(M_0 - at)^{-B/a}] = ua(M_0 - at)^{(-B/a)-1} - g(M_0 - at)^{-B/a}$$

Integration yields

$$v(M_0 - at)^{-B/a} = \frac{ua}{B}(M_0 - at)^{-B/a} + \frac{g}{a - B}(M_0 - at)^{-(B/a)+1} + C$$

and simplification gives

$$v = \frac{ua}{B} + \frac{g}{a - B}(M_0 - at) + C(M_0 - at)^{B/a}$$

Since $v(0) = 0$, we can solve for C and obtain

(18.40) $v = \dfrac{ua}{B} + \dfrac{g}{a - B}(M_0 - at) - \left[\dfrac{ua}{B} + \dfrac{gM_0}{a - B}\right]\left[\dfrac{M_0 - at}{M_0}\right]^{B/a}$

as the velocity of the rocket.
    Of particular interest to an aeronautical engineer is the *burnout velocity;* that is, the velocity at the moment the fuel is exhausted. At that time $t = M_2/a$, and the mass $M_0 - at$ is simply M_1, the mass of the rocket and payload. From (18.40), we see that the burnout velocity is

(18.41) $v = \dfrac{ua}{B} + \dfrac{g}{a - B}M_1 - \left[\dfrac{ua}{B} + \dfrac{gM_0}{a - B}\right]\left[\dfrac{M_1}{M_0}\right]^{B/a}$

In Exercise 39, you are asked to use this formula to compute the burnout velocity of the V-2 rockets used by Germany in World War II.   ■

EXERCISES

In Exercises 1–10, use the method discussed in this section to solve the equations.

1. $y' + y = x$

2. $y' - y = x$

3. $xy' - 3y = x^4$

4. $y' + y = 2xe^{-x}$

5. $y' + y = (1 + e^{2x})^{-1}$

6. $y' \sin x + y \cos x = 1$

7. $(2y - x^3) \, dx = x \, dy$

8. $x \, dy = (y + x^2 \sin x) \, dx$

9. $dy = (e^{2x} + 3y) \, dx$ **10.** $x^3 y' + 3x^2 y = 1$

In Exercises 11–30, classify and solve the equations.

11. $y' - y = 5$

12. $(x^2 + y) \, dx + (x + 2y) \, dy = 0$

13. $(x^2 + y^2) \, dx = 2xy \, dy$

14. $(x + y)y' + x - y = 0$

15. $(x + y^2) \, dy = (x^2 - y) \, dx$

16. $r \cos \theta + (\sin \theta)r' = 0$

17. $y\sqrt{1 - x^2}y' = x$ **18.** $dy + 2y \, dx = e^{-x} \, dx$

19. $xy' = y(1 + \ln(y/x))$

20. $\sqrt{1 - y^2} = y'\sqrt{1 + x^2}$

21. $y' + y = e^{-x}$ **22.** $xy' = 2(1 - y)$

23. $(1 + e^x)y' + e^x y = 1$ **24.** $y' + y = (1 + e^x)^{-1}$

25. $s' = t(1 + s)^{-1}$ **26.** $s' = (s + t)/(s - t)$

27. $(1 + x)y' + 2y = 4(1 + x)$

28. $t \, ds = (s + 1) \, dt$

29. $y' - xy = x$ **30.** $y' + (y/x) = x^{-2}$

31. Verify that $(y - 2x) \, dx + x \, dy = 0$ is homogeneous, exact, and linear. Solve it by all three methods. Are the answers equivalent?

Exercises 32–37 refer to Example 4 and the circuit in Figure 18–10.

32. Suppose the initial current equals E/R. Show that in this case the inductance does not effect the circuit. What is $I(t)$?

33. Find the current at 2 seconds if $L = 2$, $R = 5$, $E = 15$, and $I(0) = 4$.

34. Describe the behavior of $I(t)$ if   (a) $I(0) > E/R$ and   (b) $I(0) < E/R$. Is I increasing or decreasing? What happens as t increases?

35. The *time constant* of the circuit is $t = L/R$. If $I(0) = 0$, what percentage of the maximal current is achieved at $t = L/R$?

36. (a) Show that Ohm's law $(E = IR)$ holds whenever I is at a maximum or minimum value.   (b) If I is at a minimum, is E increasing or decreasing? What if I is at a maximum?

37. (a) Solve equation (18.35) if the initial current is I_0 and $E = E_0 e^{-t}$; this is the case when E is a battery whose power diminishes with time.   (b) Solve (18.35) if the initial current is I_0 and $E = E_0 \sin(120\pi t)$; this is the case when E is a generator producing alternating current at 60 cycles/sec.

Exercises 38–39 refer to Example 5 and the rocket in Figure 18–11.

38. If the drag force is neglected (that is, $B = 0$), use equation (18.39) to show that the burnout velocity in this case is

$$u \ln\left(\frac{M_0}{M_1}\right) - \frac{gM_2}{a}$$

39. The famous V-2 rocket used by Germany in World War II and later by the United States for rocket research has the following specifications:

Weight without fuel: 8,900 lbs
Fuel weight: 19,400 lbs
Fuel consumption: 275 lbs/sec
Exhaust gas velocity: 6,560 ft/sec
Drag constant: 0.1 slug/sec

Use these data to compute the burnout velocity of a V-2 rocket.

18–6   HOMOGENEOUS SECOND ORDER LINEAR EQUATIONS

We take up now the topic of second order linear differential equations. These equations are of great significance in science, especially in the theories of vibrations and electricity. In this section and the next, we develop a method of solving such equations, and in Section 18–8, we describe some applications.

A differential equation of the form

$$y^{(n)} + g_1(x)y^{(n-1)} + \cdots + g_n(x)y = f(x)$$

where f and the g's are functions of x alone, is called an **n th order linear equation.** Our discussion of such equations is limited to second order equations with constant coefficients; that is, the g's are constant functions. If $f(x) = 0$ for all x, then the equation is said to be *homogeneous*. (The meaning here of the term homogeneous is different from that in Section 18–3.) If $f(x) \neq 0$ for some x, then the equation is called *nonhomogeneous*. We treat the homogeneous case in this section and the nonhomogeneous case in the next.

The general second order homogeneous linear equation with constant coefficients has the form

(18.42) $y'' + by' + cy = 0$

Much of the theory of these linear differential equations is contained in the following result:

(18.43)

> If the functions y_1 and y_2 are any two solutions of (18.42) with the property that neither is a constant multiple of the other, then
>
> $$y = C_1y_1 + C_2y_2$$
>
> is the general solution of (18.42).

Suppose that y_1 and y_2 are solutions of (18.42). Then, for any constants C_1 and C_2, we have

$$(C_1y_1 + C_2y_2)'' + b(C_1y_1 + C_2y_2)' + c(C_1y_1 + C_2y_2)$$
$$= C_1(y_1'' + by_1' + cy_1) + C_2(y_2'' + by_2' + cy_2) = 0 + 0 = 0$$

proving that $C_1y_1 + C_2y_2$ is, indeed, a solution of (18.42). We are not prepared, however, to prove the important part of (18.43); that is, that *every* solution is of this form. The proof that $C_1y_1 + C_2y_2$ is the general solution can be found in texts devoted to differential equations.

Our starting point in solving (18.42) is to consider $y = e^{mx}$ as a possible solution. The reason for choosing this function is that its derivatives are constant multiples of itself; $y' = me^{mx}$ and $y'' = m^2e^{mx}$. Substitution into (18.42) yields

$$m^2e^{mx} + bme^{mx} + ce^{mx} = 0$$

or

(18.44) $(m^2 + bm + c)e^{mx} = 0$

and, since e^{mx} is never 0, we see that (18.44) holds if and only if

(18.45) $m^2 + bm + c = 0$

Equation (18.45) is called the **auxiliary** (or **characteristic**) equation of (18.42). The value of the auxiliary equation is this:

(18.46)
> If r is a root of the auxiliary equation $m^2 + bm + c = 0$, then $y = e^{rx}$ is a solution of the differential equation $y'' + by' + cy = 0$.

EXAMPLE 1   The auxiliary equation of $y'' - 2y' - 3y = 0$ is $m^2 - 2m - 3 = 0$. Since $m^2 - 2m - 3 = (m - 3)(m + 1)$, we know that $m = 3$ is a root. It follows from (18.46) that $y = e^{3x}$ is a solution of $y'' - 2y' - 3y = 0$. We can easily verify this.

$$(e^{3x})'' - 2(e^{3x})' - 3e^{3x} = 9e^{3x} - 6e^{3x} - 3e^{3x} = 0 \qquad \blacksquare$$

Moreover, with the aid of (18.43), we can use the auxiliary equation to find the general solution.

EXAMPLE 2   Solve $y'' - 2y' - 3y = 0$.

Solution   The auxiliary equation $m^2 - 2m - 3 = (m - 3)(m + 1)$ has roots $r = 3$ and $r = -1$. It follows from (18.46) that $y = e^{3x}$ and $y = e^{-x}$ are solutions. Furthermore, the quotient

$$\frac{e^{3x}}{e^{-x}} = e^{4x} \quad \text{(not a constant)}$$

shows that neither is a constant multiple of the other. Now (18.43) yields

$$y = C_1 e^{3x} + C_2 e^{-x}$$

as the general solution. $\blacksquare$

In many cases, the roots of the auxiliary equation $m^2 + bm + c = 0$ can be found by factoring. But if the factors are not evident, you can always use the quadratic formula, which, in the present case, takes the form

$$m = \frac{-b \pm \sqrt{b^2 - 4c}}{2}$$

There are two distinct real roots, one real root, or two distinct complex roots depending on whether the discriminant $b^2 - 4c$ is positive, zero, or negative. Each possibility leads to different forms of general solutions of the differential equation. We will discuss each case separately, give an example, and then present a summary of the methods.

Case I: Distinct real roots ($b^2 - 4c > 0$)

There are two distinct real roots m_1 and m_2 of the auxiliary equation (as was the case in Example 2 above). By (18.46), $y_1 = e^{m_1 x}$ and $y_2 = e^{m_2 x}$ are solutions, and neither is a multiple of the other because their quotient is the nonconstant function $e^{(m_1 - m_2)x}$. Therefore, by (18.43),

(18.47) $y = C_1 e^{m_1 x} + C_2 e^{m_2 x}$

is the general solution.

EXAMPLE 3   Solve $y'' + (3/2)y' - y = 0$.

Solution   The auxiliary equation is $m^2 + (3/2)m - 1 = 0$ and the quadratic formula yields the distinct roots

$$\frac{-\dfrac{3}{2} \pm \sqrt{\dfrac{9}{4} + 4}}{2} = \frac{-\dfrac{3}{2} \pm \dfrac{5}{2}}{2} = \frac{1}{2} \text{ or } -2$$

Therefore, $y = C_1 e^{x/2} + C_2 e^{-2x}$ is the general solution.   ∎

Case II: One real root ($b^2 - 4c = 0$)
   This case happens when $c = b^2/4$ and the auxiliary equation is a perfect square

$$m^2 + bm + \frac{b^2}{4} = \left(m + \frac{b}{2}\right)^2 = 0$$

Therefore, $m = -b/2$ and consequently we have only one solution $y_1 = e^{-bx/2}$ of the differential equation. But (18.43) requires that we have two solutions, so we set out to find another. It turns out that $y_2 = xe^{-bx/2}$ is also a solution. To verify this, some straightforward derivative computations yield

$$y_2' = \left(-\frac{b}{2}x + 1\right)e^{-bx/2}$$

$$y_2'' = \left(-b + \frac{b^2}{4}x\right)e^{-bx/2}$$

and, since $c = b^2/4$, we have $y_2'' + by_2' + cy_2 = 0$ (check out these computations). Moreover, the quotient of the two solutions y_1 and y_2 is the nonconstant function x. Now (18.43) gives

(18.48) $y = C_1 e^{-bx/2} + C_2 xe^{-bx/2}$

as the general solution.

EXAMPLE 4   Solve $y'' - 4y' + 4y = 0$,

Solution   The only root of the auxiliary equation $m^2 - 4m + 4 = 0$ is $m = 2$. It follows from (18.48) that the general solution is

$$y = C_1 e^{2x} + C_2 xe^{2x}$$ ∎

Case III: Distinct complex roots ($b^2 - 4c < 0$)

Only a few important facts from the theory of complex numbers are needed to treat this case.

A **complex number** is a number of the form $u + iv$ where u and v are real numbers and

$$i^2 = -1 \quad \text{or} \quad i = \sqrt{-1}$$

All of the basic arithmetic properties of real numbers carry over to complex numbers with the important added capability of obtaining square roots of negative numbers. For example,

$$\sqrt{-4} = i2 \quad \text{because } (i2)^2 = i^2 2^2 = -4$$

Similarly, $(-i2)^2 = -4$, so the equation $x^2 + 4 = 0$ has two complex roots $x = \pm i2$. Any quadratic equation $m^2 + bm + c = 0$ with $b^2 - 4c < 0$ will also have two complex roots. If $m^2 - 2m + 5 = 0$, then the quadratic formula yields

$$m = \frac{2 \pm \sqrt{4 - 20}}{2} = \frac{2 \pm \sqrt{-16}}{2} = \frac{2 \pm i4}{2} = 1 \pm i2$$

The complex number $u - iv$ is called the **conjugate** of $u + iv$. Thus, for the equation above, the root $1 - i2$ is the conjugate of the root $1 + i2$. As you can see, the quadratic formula *always* yields conjugate roots.

Our discussion of complex numbers will be complete once we have established the following important result due to Euler.

(18.49)

> **EULER'S FORMULA**
>
> $e^{it} = \cos t + i \sin t$

We prove Euler's formula by recalling the Maclaurin series for e^x, $\cos x$, and $\sin x$ presented in the table at the end of Section 11–8:

$$e^x = 1 + x + \frac{x^2}{2!} + \frac{x^3}{3!} + \cdots$$

$$\cos x = 1 - \frac{x^2}{2!} + \frac{x^4}{4!} - \cdots$$

$$\sin x = x - \frac{x^3}{3!} + \frac{x^5}{5!}$$

We *define* e^{it} to be the series above for e^x with x replaced by it; that is,

$$e^{it} = 1 + it + \frac{(it)^2}{2!} + \frac{(it)^3}{3!} + \cdots$$

Since $i^2 = -1$, $i^3 = -i$, $i^4 = 1$, and so on, we see that

$$e^{it} = 1 + it + \frac{(it)^2}{2!} + \frac{(it)^3}{3!} + \cdots$$

$$= 1 + it - \frac{t^2}{2!} - i\frac{t^3}{3!} + \cdots$$

$$= \left(1 - \frac{t^2}{2!} + \cdots\right) + i\left(t - \frac{t^3}{3!} + \cdots\right)$$

$$= \cos t + i \sin t$$

which is Euler's formula.

Let us now return to the problem of solving an equation $y'' + by' + cy = 0$ with $b^2 - 4c < 0$. The roots of the auxiliary equation $m^2 + bm + c = 0$ will be conjugate complex numbers, say $u \pm iv$. If the pattern established for the earlier cases holds true (and it does), then

(18.50) $y_1 = e^{(u+iv)x} = e^{ux}e^{ivx}$ and $y_2 = e^{(u-iv)x} = e^{ux}e^{-ivx}$

are solutions of the differential equation. In light of Euler's formula, the solutions in (18.50) become

$$y_1 = e^{ux}e^{ivx} = e^{ux}(\cos vx + i \sin vx)$$
$$y_2 = e^{ux}e^{-ivx} = e^{ux}(\cos vx - i \sin vx)$$

But we are interested in real-valued solutions, and we can obtain such solutions as follows:

$$\frac{y_1 + y_2}{2} = e^{ux} \cos vx \quad \text{and} \quad \frac{y_1 - y_2}{2i} = e^{ux} \sin vx$$

Since neither of these solutions is a constant multiple of the other, (18.43) gives

(18.51) $y = e^{ux}(C_1 \cos vx + C_2 \sin vx)$

as the general solution.

EXAMPLE 5   Solve $y'' - 2y' + 5y = 0$.

Solution   The auxiliary equation is $m^2 - 2m + 5 = 0$ and the quadratic formula yields

$$m = \frac{2 \pm \sqrt{4 - 20}}{2} = \frac{2 \pm \sqrt{-16}}{2} = 1 \pm i2$$

Applying (18.51) with $u = 1$ and $v = 2$, the general solution is

$$y = e^x(C_1 \cos 2x + C_2 \sin 2x) \qquad \blacksquare$$

EXAMPLE 6   Find the solution of $y'' = -9y$ that satisfies the initial conditions $y(0) = 0$ and $y'(0) = 3$.

Solution With the equation written as $y'' + 9y = 0$, the auxiliary equation is $m^2 + 9 = 0$ whose roots are $\pm i3$. In this case, $u = 0$, $v = 3$, and it follows that the general solution is

$$y = C_1 \cos 3x + C_2 \sin 3x$$

Since $y(0) = 0$, we see that $C_1 = 0$, so $y = C_2 \sin 3x$; since $y'(0) = 3$, we see that $C_2 = 1$. Therefore, the solution asked for is

$$y = \sin 3x \quad \blacksquare$$

Remark: In Section 8–3, the second order equation

$$s'' = -b^2 s$$

was found to describe *simple harmonic motion*. The general solution was given in that section as

$$\boxed{s = A \cos (bt + C)}$$

which does not look like the general solution that would be obtained by the methods discussed here. But if we write $A \cos (bt + C) = A \cos bt \cos C - A \sin bt \sin C$, let $C_1 = A \cos C$, and $C_2 = -A \sin C$, then the general solution above becomes

$$s = C_1 \cos bt + C_2 \sin bt$$

which is of the same form that is obtained in Example 6. We will return to a discussion of simple harmonic motion in Section 18–8.

Summary

Given the equation $y'' + by' + cy = 0$ with auxiliary equation $m^2 + bm + c = 0$, there are three cases:

(1) If $b^2 - 4c > 0$, there are distinct real roots m_1 and m_2, and $y = C_1 e^{m_1 x} + C_2 e^{m_2 x}$ is the general solution.

(2) If $b^2 - 4c = 0$, there is one (real) root m, and $y = C_1 e^{mx} + C_2 x e^{mx}$ is the general solution.

(3) If $b^2 - 4c < 0$, there are (conjugate) complex roots $u \pm iv$, and $y = e^{ux}(C_1 \cos vx + C_2 \sin vx)$ is the general solution.

EXERCISES

In Exercises 1–20, solve the second order equations.

1. $y'' + 6y' + 9y = 0$
2. $y'' - 10y' + 25y = 0$
3. $y'' - y' - 12y = 0$
4. $y'' + 6y' + 5y = 0$
5. $y'' + 4y = 0$
6. $y'' - 9y = 0$
7. $y'' + 4y' = 0$

8. $y'' - 2y' = 0$

9. $y'' - 2y' + 2y = 0$

10. $y'' - 4y' + 5y = 0$

11. $y'' - 2y' + y = 0$

12. $y'' + 4y' + 4y = 0$

13. $y'' + y' - 6y = 0$

14. $y'' - 2y' - 3y = 0$

15. $y'' + 8y = 0$

16. $y'' + 5y = 0$

17. $y'' - 2y' + 4y = 0$

18. $y'' - y' + y = 0$

19. $2y'' - 4y' + y = 0$ (Divide through by 2)

20. $4y'' + y' - 5y = 0$

In Exercises 21–26, find the particular solution that satisfies the given initial conditions. Some refer to the exercises above.

21. $y'' - 8y' + 15y = 0$; $y(0) = 1$ and $y'(0) = 5$

22. $y'' + 3y' - 4y = 0$; $y(0) = 2$ and $y'(0) = -3$

23. $y'' - 8y' + 16y = 0$; $y(0) = 3$ and $y'(0) = 13$

24. $y'' - 4y' + 4y = 0$; $y(0) = 2$ and $y'(0) = 3$

25. Exercise 5; $y(0) = 4$ and $y'(\pi) = 6$

26. Exercise 15; $y(0) = 0$ and $y'(0) = 1$

18–7 NONHOMOGENEOUS SECOND ORDER LINEAR EQUATIONS

We continue our study of second order linear equations with constant coefficients by taking up the nonhomogeneous case

$$(18.52) \quad y'' + by' + cy = f(x)$$

where f is a nonzero function. Part of the task of solving such an equation involves finding a general solution of the **associated homogeneous equation**

$$(18.53) \quad y'' + by' + cy = 0$$

To see why this is so, we suppose that y is the general solution and y_P is a particular solution of (18.52). Then

$$y'' + by' + cy = f(x) \quad \text{and} \quad y_P'' + by_P' + cy_P = f(x)$$

and it follows that

$$(y - y_P)'' + b(y - y_P)' + c(y - y_P) = f(x) - f(x) = 0$$

Therefore, $y - y_P = y_H$ is the general solution of the homogeneous equation (18.53). In other words,

(18.54)

> If y_P is a particular solution of $y'' + by' + cy = f(x)$, then the general solution is
>
> $$y = y_P + y_H$$
>
> where y_H is the general solution of the associated homogeneous equation $y'' + by' + cy = 0$.

According to (18.54), the problem of finding the general solution of $y'' + by' + cy = f(x)$ reduces to two subproblems: (1) finding some particular solution of the equation, and (2) finding the general solution of the associated homogeneous equation. The second subproblem is solved using the methods discussed in the preceding section. Let us now discuss how to find a particular solution.

Undetermined Coefficients

The *method of undetermined coefficients* is a method for finding a particular solution y_P of $y'' + by' + cy = f(x)$ for certain types of functions $f(x)$; we will treat the cases in which $f(x)$ is an exponential, a sine or cosine, or a combination of such functions. These are the cases that occur most frequently in applications.

To find a particular solution of

(18.55) $y'' + by' + cy = ke^{ax}$ (k and a are constants)

we reason as follows. The only function whose second derivative plus b times its first derivative plus c times itself equals the exponential ke^{ax} is some constant multiple of e^{ax}. That is, we guess that $y_P = Ae^{ax}$ and try to find the *undetermined coefficient* A. The result of substituting Ae^{ax} into the left side of (18.55) is

$$Aa^2 e^{ax} + Abae^{ax} + Ace^{ax} = A(a^2 + ba + c)e^{ax}$$

Therefore, *if a is not a root of the auxiliary equation $m^2 + bm + c = 0$* (that is, if $a^2 + ba + c \neq 0$), then $y_P = Ae^{ax}$ is a solution of (18.55) if and only if

(18.56) $A = \dfrac{k}{a^2 + ba + c}$

EXAMPLE 1   Find a particular solution of $y'' - 3y' + 2y = 5e^{3x}$ and then find the general solution.

Solution   The discussion above leads us to guess that $y_P = Ae^{3x}$ is a particular solution for some number A. After checking that 3 is not a root of the auxiliary equation $m^2 - 3m + 2 = 0$, (18.56) yields

$$A = \frac{5}{3^2 - 3(3) + 2} = \frac{5}{2}$$

so $y_P = (5/2)e^{3x}$ is a particular solution.

Check. $\left(\dfrac{5}{2}e^{3x}\right)'' - 3\left(\dfrac{5}{2}e^{3x}\right)' + 2\left(\dfrac{5}{2}e^{3x}\right) = \dfrac{45}{2}e^{3x} - \dfrac{45}{2}e^{3x} + 5e^{3x} = 5e^{3x}$

To find the general solution of the equation, we need to find the general solution of the associated homogeneous equation $y'' - 3y' + 2y = 0$. Since $m^2 - 3m + 2 = (m - 1)(m - 2)$, we know that $y_H = C_1 e^x + C_2 e^{2x}$ is such a solution. Therefore, applying (18.54) yields

$$y = \frac{5}{2}e^{3x} + C_1 e^x + C_2 e^{2x}$$

as the general solution of $y'' - 3y' + 2y = 5e^{3x}$.   ∎

The method described above works whenever a is not a root of $m^2 + bm + c = 0$. What happens when a is a root? In this case, we guess that

(18.57) $y_P = Axe^{ax}$

is a particular solution. Substitution of this function into (18.55) yields

(18.58) $A = \dfrac{k}{2a + b}$

provided that a is not a double root of $m^2 + bm + c = 0$ (that is, provided that $2a + b \neq 0$). If a is a double root, then

(18.59) $y_P = \dfrac{k}{2}x^2 e^{ax}$

is a particular solution. We ask you to verify the last two assertions in Exercise 29.

EXAMPLE 2   Find a particular solution of $y'' - 6y' + 8y = e^{2x}$.

Solution   Here, 2 is a root (but not a double root) of $m^2 - 6m + 8 = 0$, so we use (18.58) to write

$$A = \frac{1}{2(2) - 6} = -\frac{1}{2}$$

Therefore, by (18.57), $y_P = -(1/2)xe^{2x}$ is a particular solution.

Check. $\left(-\dfrac{1}{2}xe^{2x}\right)'' - 6\left(-\dfrac{1}{2}xe^{2x}\right)' + 8\left(-\dfrac{1}{2}xe^{2x}\right)$

$$= -\frac{1}{2}[(2^2 - 6(2) + 8)xe^{2x} + (2^2 - 6)e^{2x}]$$

$$= -\frac{1}{2}(0 - 2e^{2x}) = e^{2x} \qquad \blacksquare$$

Now suppose that we want to find a particular solution of

(18.60) $y'' + by' + c = p \cos vx + q \sin vx$

Since derivatives of sines and cosines are multiples of sines and cosines, we guess that a particular solution is

(18.61) $y_P = A \cos vx + B \sin vx$

and try to find the undetermined coefficients A and B. The method of computing A and B is illustrated in the next example.

EXAMPLE 3   Find a particular solution of $y'' + y' = 6 \cos 3x + 2 \sin 3x$.

Solution   We guess that $y_P = A \cos 3x + B \sin 3x$ is a solution. Substituting this solution into the equation yields

$$(A \cos 3x + B \sin 3x)'' + (A \cos 3x + B \sin 3x)'$$
$$= (-9A \cos 3x - 9B \sin 3x) + (-3A \sin 3x + 3B \cos 3x)$$
$$= (-9A + 3B)\cos 3x + (-3A - 9B)\sin 3x$$

The last line will equal $6 \cos 3x + 2 \sin 3x$ if A and B satisfy the following two equations

$$-9A + 3B = 6$$
$$-3A - 9B = 2$$

The simultaneous solutions are $A = -2/3$ and $B = 0$, so

$$y_P = A \cos 3x + B \sin x = -\frac{2}{3}\cos 3x$$

is a particular solution.

Check. $\left(-\dfrac{2}{3}\cos 3x\right)'' + \left(-\dfrac{2}{3}\cos 3x\right)'$

$$= 9\left(\frac{2}{3}\right)\cos 3x + 3\left(\frac{2}{3}\right)\sin 3x = 6 \cos 3x + 2 \sin 3x \qquad \blacksquare$$

Remark: If $y'' + by' + c = p \cos vx + q \sin vx$ and iv is a root of $m^2 + bm + c = 0$, then the method used in Example 3 breaks down because the co-efficients sum to zero. In this case, try

$$y_P = x(A \cos vx + B \sin vx)$$

as a trial solution.

The method of undetermined coefficients is summarized in the following table:

(18.62)

---

(1) A particular solution of $y'' + by' + cy = ke^{ax}$ is

$$\frac{k}{a^2 + ba + c}e^{ax} \quad \text{or} \quad \frac{k}{2a + b}xe^{ax} \quad \text{or} \quad \frac{k}{2}x^2 e^{ax}$$

depending on whether a is not a root, a root but not a double root, or a double root of $m^2 + bm + c = 0$.

(2) A particular solution of $y'' + by' + cy = p \cos vx + q \sin vx$ is of the form

$$A \cos vx + B \sin vx \quad \text{or} \quad x(A \cos vx + B \sin vx)$$

depending on whether iv is not or is a root of $m^2 + bm + c = 0$.

(3) We combine (1) and (2) as follows: A particular solution of $y'' + by' + cy = e^{ux}(p \cos vx + q \sin vx)$ is of the form

$$e^{ux}(A \cos vx + B \sin vx)$$

provided $u + iv$ is not a root of $m^2 + bm + c = 0$.

Variation of Parameters (Optional)

Another method of determining a particular solution is called *variation of parameters*. Although we do not need variation of parameters for the applications of the next section, the method is important enough to describe here. We will present it in outline form and then give an illustrative example. Again we begin with an equation

$$y'' + by' + cy = f(x)$$

and suppose that $y_H = C_1 y_1 + C_2 y_2$ is the general solution of the associated *homogeneous* equation $y'' + by' + cy = 0$. We now *vary the parameters* C_1 and C_2 by replacing them with functions $u = u(x)$ and $v = v(x)$; our interest is to find a particular solution of the form

(18.63) $\quad y_P = uy_1 + vy_2$

We begin by computing the derivative of y_P and rearranging the terms

$$y_p' = (uy_1' + u'y_1) + (vy_2' + v'y_2)$$
$$= (uy_1' + vy_2') + (u'y_1 + v'y_2)$$

Another differentiation will involve the second derivatives of u and v. We avoid this complication by requiring that

(18.64) $\quad u'y_1 + v'y_2 = 0$

Then $y_P' = uy_1' + vy_2'$ and the second derivative is

$$y_P'' = uy_1'' + u'y_1' + vy_2'' + v'y_2'$$

Substituting y_P'' and y_P' into the original equation yields

$$u(y_1'' + by_1' + cy_1) + v(y_2'' + by_2' + cy_2) + u'y_1' + v'y_2' = f(x)$$

Since y_1 and y_2 are solutions of the homogeneous equation, the expressions in the parentheses are both 0, so the equation above reduces to $u'y_1' + v'y_2' = f(x)$. This, along with (18.64) provides two equations involving u' and v'.

(18.65) $\quad u'y_1 + v'y_2 = 0$
$\quad\quad\quad\quad u'y_1' + v'y_2' = f(x)$

Once the equations are solved for u' and v', the functions u and v can be found by integration.

EXAMPLE 4   Find a particular solution of $y'' + y = \sec x$; then find the general solution.

Solution   The associated homogeneous equation is $y'' + y = 0$. The roots of the auxiliary equation $m^2 + 1 = 0$ are $\pm i$, so

(18.66) $\quad y = C_1 \cos x + C_2 \sin x$

is a general solution. Now we replace C_1 and C_2 by functions u and v and seek a particular solution $u \cos x + v \sin x$. Applying (18.65), with $y_1 = \cos x$ and $y_2 = \sin x$, we have two equations

$$u' \cos x + v' \sin x = 0$$
$$-u' \sin x + v' \cos x = \sec x$$

We multiply the first equation by $\sin x$, the second by $\cos x$, and add; the result is $v' = 1$. We solve for u' with a similar computation; the result is $u' = -\sin x \sec x = -\tan x$. Integration now yields

$$u = \int -\tan x \, dx = \ln \cos x \quad \text{and} \quad v = \int 1 \, dx = x$$

(The constants of integration are not necessary here.) Therefore,

$$y_P = (\ln \cos x) \cos x + x \sin x$$

is a particular solution.

Check. $\quad y_P' = [(\ln \cos x)(-\sin x) + (\cos x)(-\tan x)] + [x \cos x + \sin x]$
$\qquad\quad = (\ln \cos x)(-\sin x) + x \cos x$
$\qquad y_P'' = [(\ln \cos x)(-\cos x) + (-\sin x)(-\tan x)] + [-x \sin x + \cos x]$
$\qquad\quad = -(\ln \cos x)(\cos x) + \sin^2 x \sec x - x \sin x + \cos x$

Thus, $y_P'' + y_P = \sin^2 x \sec x + \cos x = \sin^2 x \sec x + \cos^2 x \sec x = \sec x$.

Now that a particular solution is known, we combine it with (18.66) and obtain the general solution

$$y = [(\ln \cos x)(\cos x) + x \sin x] + C_1 \cos x + C_2 \sin x$$
$$= [\ln \cos x + C_1] \cos x + [x + C_2] \sin x \qquad \blacksquare$$

EXERCISES

In Exercises 1–10, find a particular solution using undetermined coefficients.

1. $y'' + 2y' - 24y = 3e^x$
2. $y'' + 4y' + 3y = 2e^x$
3. $y'' - 10y' + 41y = \sin x$
4. $y'' + y = \cos x$
5. $y'' - 4y' + 4y = e^{2x}$
6. $y'' + y' - 6y = e^{2x}$
7. $y'' + y' - 12y = e^{3x}$
8. $y'' + 2y' + y = e^{-x}$
9. $y'' + y = e^{2x} \cos 3x$
10. $y'' + y' = e^x \sin x$

Exercises 11–16 are *optional.* Find a particular solution using a variation of parameters.

11. $y'' + y = -\tan x$
12. $y'' + 4y = \sec 2x$
13. $y'' + y = \csc x$
14. $y'' + y = \cot x$
15. $y'' + 2y' + y = e^{-x}/x^2$
16. $y'' - 2y' + y = e^x/x^2$

In Exercises 17–28, find general solutions using the method you prefer. Some refer to the exercises above.

17. $y'' - y = 2e^x$ 　　**18.** $y'' + y = 3e^{-x}$
19. $y'' + y = \sin x$ 　　**20.** $y'' + y = \cot^2 x$

29. Suppose that $y'' + by' + cy = ke^{ax}$. (a) Show that if a is a root (but not a double root) of

$m^2 + bm + c = 0$, then

$$y_P = \frac{k}{2a + b}xe^{ax}$$

is a particular solution. (b) If a is a double root, show that $y_P = kx^2e^{ax}/2$ is a particular solution.

18-8   VIBRATIONS

In general, vibrations are the result of a system in equilibrium being disturbed by some outside force. In this section, we will study vibrations in terms of a mechanical system consisting of a spring that is anchored at one end and a weight that is attached to the other end (Figure 18–12). The topic is developed in three stages. First, we neglect all frictional forces; the resulting motion is an *undamped vibration*. Next, we include the frictional forces; the resulting motion is a *damped vibration*. Finally, we impose a continuous outside force on the system, and the resulting motion is a *forced vibration*.

Undamped Vibrations

The system in Figure 18–12 is in equilibrium when the spring is stretched s units; that is $Mg - ks = 0$, where k is the spring constant. If the weight is now displaced y units from equilibrium, then the total force on it is $Mg - k(s + y) = Mg - ks - ky = -ky$. That is, the effective force on the weight is $F_s = -ky$. Therefore, neglecting all other forces, Newton's second law of motion leads to the differential equation

$$M\frac{d^2y}{dt^2} = -ky \quad \text{or} \quad \frac{d^2y}{dt^2} + \frac{k}{M}y = 0$$

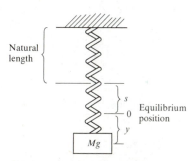

FIGURE 18–12. Mechanical system consisting of a spring and a weight of mass M.

This is the situation we described in Section 8–3, entitled *Simple Harmonic Motion*.

It is convenient to write the last equation above as

$$(18.67) \quad \frac{d^2y}{dt^2} + c^2y = 0$$

where $c = \sqrt{k/M}$. The roots of the auxiliary equation $m^2 + c^2 = 0$ are $\pm ic$, and it follows from previous work that the general solution of (18.67) is

$$(18.68) \quad y = C_1 \cos ct + C_2\sin ct$$

If the weight is initially pulled down y_0 units and released with zero velocity, then the initial conditions are

$$y(0) = y_0 \quad \text{and} \quad y'(0) = 0$$

It is easy to verify that under these conditions, $C_1 = y_0$ and $C_2 = 0$, so (18.68) becomes

(18.69) $y = y_0 \cos ct$

Equation (18.69) describes what is called *simple harmonic motion;* its graph is shown in Figure 18–13. The *amplitude* (maximum displacement) is y_0; the *period* (time required for a complete cycle) is

(18.70) $T = \dfrac{2\pi}{c} = 2\pi\sqrt{M/k}$ time per cycle

and the *frequency* (number of cycles per unit of time) is

(18.71) $f = \dfrac{1}{T} = \dfrac{1}{2\pi}\sqrt{k/M}$ cycles per unit time

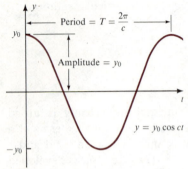

FIGURE 18–13. Undamped vibrations.

EXAMPLE 1 A 10 lb weight stretches a spring 2 inches. If the weight is now pulled down an additional 3 inches and released, what is an equation of the resulting motion if all other forces are neglected? What is the period and frequency of the vibrations?

Solution An equation of the motion is provided by (18.69); that is, $y = y_0 \cos ct$ where $c = \sqrt{k/M}$ and it remains only to find the constants y_0, M, and k. We are given that $y_0 = 3$ and that the weight is 10 lbs; so $M = 10/32$. Furthermore, the force of 10 lbs stretches the spring 2 inches, so $k = 5$. Therefore, $c = \sqrt{k/M} = 4$ and

$$y = 3 \cos 4t$$

describes the motion. Equations (18.70) and (18.71) provide the period T and frequency f

$$T = \frac{2\pi}{4} = \frac{\pi}{2} \quad \text{and} \quad f = \frac{1}{T} = \frac{2}{\pi} \qquad \blacksquare$$

Damped Vibrations

Now we consider the effect of a damping force F_d due to the friction in the spring or the viscosity of the medium (air, water, oil, and so on) through which the weight moves. We assume that this force opposes the motion and that it is proportional to the velocity; that is, $F_d = -a(dy/dt)$. The total force now is the damping force F_d plus the spring force F_s, so the equation of motion is

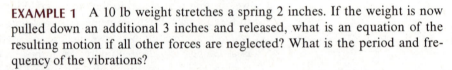

$$M\frac{d^2y}{dt^2} = F_d + F_s \quad \text{or} \quad \frac{d^2y}{dt^2} + \frac{a}{M}\frac{dy}{dt} + \frac{k}{M}y = 0$$

Again it is convenient to rewrite this equation as

(18.72) $\dfrac{d^2y}{dt^2} + 2b\dfrac{dy}{dt} + c^2y = 0$

where $b = a/2M$ and $c = \sqrt{k/M}$. The roots of the auxiliary equation $m^2 + 2bm + c^2 = 0$ are

$$m_1 \text{ and } m_2 = \frac{-2b \pm \sqrt{4b^2 - 4c^2}}{2} = -b \pm \sqrt{b^2 - c^2}$$

General solutions of (18.72) depend on these roots, and there are three cases.

Case I *(Overdamped motion).* Here we assume that the damping force F_d is large compared to the force of the spring F_s; that is, $b > c$ or $b^2 > c^2$. Then m_1 and m_2 are distinct real roots and the general solution of (18.72) is

$$y = C_1 e^{m_1 t} + C_2 e^{m_2 t}$$

With the initial conditions $y(0) = y_0$ and $y'(0) = 0$, we can evaluate C_1 and C_2 to obtain

$$(18.73) \quad y = \frac{y_0}{m_2 - m_1}(m_2 e^{m_1 t} - m_1 e^{m_2 t})$$

The shape of the graph depends, of course, on the values of m_1 and m_2, but a typical case is shown in Figure 18-14. Like a pendulum swinging in molasses, the motion is not oscillatory. In this case, we say that the system is *overdamped*.

Case II *(Critically damped).* Here we assume that the damping and spring forces are such that $b = c$. Then $b^2 = c^2$ and $m_1 = m_2 = -b = -c$ is a double root. The general solution of (18.72) is $y = C_1 e^{-ct} + C_2 t e^{-ct}$, and the initial conditions yield

$$(18.74) \quad y = y_0 e^{-ct}(1 + ct)$$

The graph is shown in Figure 18-15. Again, there is no oscillatory motion, but now if the damping force is decreased by even the smallest amount, vibrations will occur. Therefore, we say that this system is *critically damped*.

Case III *(Underdamped).* Here the damping force is overcome by the spring force and $b^2 < c^2$. The roots m_1 and m_2 are complex conjugates, which we write as $-b \pm i\alpha$ where $\alpha = \sqrt{c^2 - b^2}$. The general solution of (18.72) is

$$y = e^{-bt}(C_1 \cos \alpha t + C_2 \sin \alpha t)$$

and the initial conditions yield

$$(18.75) \quad y = \frac{y_0}{\alpha} e^{-bt} (\alpha \cos \alpha t + b \sin \alpha t)$$

If we introduce $\theta = \tan^{-1}(b/\alpha)$ and refer to the diagram in Figure 18-16, then

$$\cos(\alpha t - \theta) = \cos \alpha t \cos \theta + \sin \alpha t \sin \theta$$

$$= \frac{1}{\sqrt{\alpha^2 + b^2}} (\alpha \cos \alpha t + b \sin \alpha t)$$

$$= \frac{1}{c} (\alpha \cos \alpha t + b \sin \alpha t)$$

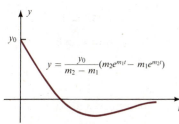

FIGURE 18-14. Overdamped motion. (The shape of this graph depends on the values of m_1 and m_2. In some cases, for example, the graph resembles the one in Figure 18-15 below.)

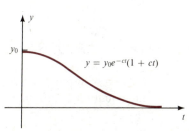

FIGURE 18-15. Critically damped motion.

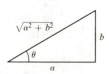

FIGURE 18-16. $\theta = \tan^{-1}(b/\alpha)$

It follows that (18.75) can be rewritten as

(18.76) $y = \dfrac{y_0 c}{\alpha} e^{-bt}\cos(\alpha t - \theta)$

The motion is oscillatory with an amplitude that decreases exponentially (Figure 18–17). The function is not periodic in the strict sense, but it does cross the t-axis at regular intervals. Therefore, we can talk about its "period"

$$T = \dfrac{2\pi}{\alpha} = \dfrac{2\pi}{\sqrt{c^2 - b^2}}$$

and its "frequency" $f = 1/T$. This is called *underdamped* motion.

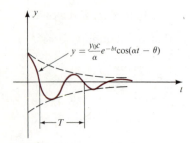

$y = \dfrac{y_0 c}{\alpha} e^{-bt}\cos(\alpha t - \theta)$

FIGURE 18–17. Underdamped motion.

EXAMPLE 2 Describe the motion of the weight in Example 1 if the damping force is $F_d = -a(dy/dt)$ where (a) $a = 5$ (b) $a = 5/2$ and (c) $a = 5/8$.

Solution From Example 1, we know that $y_0 = 3$, $c = 4$, and $M = 10/32$.

(a) If $a = 5$, then $b = a/2M = 8 > c$. Therefore, the motion is overdamped and is described by equation (18.73) with $y_0 = 3$ and

$$m_1, m_2 = -b \pm \sqrt{b^2 - c^2} = -8 \pm 4\sqrt{3}$$

(b) If $a = 5/2$, then $b = 4 = c$. This motion is critically damped and is described by equation (18.74)

$$y = 3e^{-4t}(1 + 4t)$$

(c) If $a = 5/8$, then $b = 1 < c$. This motion is underdamped and is described by equation (18.76) with $y_0 = 3$, $\alpha = \sqrt{c^2 - b^2} = \sqrt{15}$, $b = 1$, and $\theta = \tan^{-1}(1/\sqrt{15}) \approx 0.25$. ∎

Forced Vibrations

Now we extend the study of vibrations to include an external force F_e. Such a force, for example, may result from vibrations of the wall to which the spring is anchored or a magnetic field acting on the weight. The most important case is a periodic external force, say $F_e = F_0 \cos \omega t$. Then the equation of motion is

(18.77) $M\dfrac{d^2 y}{dt^2} + a\dfrac{dy}{dt} + ky = F_0 \cos \omega t$

General solutions y_H of the associated homogeneous equation have already been discussed in the first two parts of this section. Therefore, to find a general solution of (18.77), we need only find some particular solution. Using the method of undetermined coefficients, we try $A \cos \omega t + B \sin \omega t$ as a possible solution. The result (we spare you the details) is

(18.78) $y_P = \dfrac{F_0}{\sqrt{(k - \omega^2 M)^2 + \omega^2 a^2}} \cos(\omega t - \phi)$

where $\phi = \tan^{-1}[\omega a/(k - \omega^2 M)]$. Thus, the general solution of (18.77) is

$$y = y_H + y_P$$

Whenever there is damping, then there are three possibilities for y_H, but in every case, $y_H \to 0$ as $t \to \infty$. Therefore, we call y_H the *transient* part of the general solution. The particular solution y_P is called the *steady-state* part. In practice, we neglect the transient part and assert that for large t, the motion is described by (18.78); this motion is called a *forced vibration*.

The amplitude

$$(18.79) \qquad \frac{F_0}{\sqrt{(k - \omega^2 M)^2 + \omega^2 a^2}}$$

of the forced vibration described by (18.78) can be used to explain the violent motion of a bridge under the impact of strong pulsating winds or of the tramping of marching feet. If the motion is lightly damped (a is very small) and the impressed frequency of the wind or the marching feet is very close to the natural frequency of the bridge (this is the case when $\omega \approx \sqrt{k/M}$), then the denominator in (18.79) is small, so the amplitude of the vibration is very large (see Exercise 5). This phenomenon is known as *resonance* (Figure 18–18).

FIGURE 18–18A. The Tacoma Narrows Bridge. It was destroyed during a windstorm on November 7, 1940.

"The bridge collapses, people die."

Rhythm of foot-tappers suggested as the cause

KANSAS CITY, Mo. (UPI)—Foot-tapping revelers might have touched off a rhythmic vibration that caused the "sky bridges" at the Hyatt Regency Hotel to collapse in a deadly avalanche of concrete and steel, experts theorized yesterday.

Although it will be months before investigators complete their work, one engineer said his first thought when he heard of the tragedy was the possibility that harmonic vibrations had put more stress on the catwalk-like sky bridges than they could handle.

"You get something going at a certain frequency. That's what happens on bridges when you get cars going. They develop a certain frequency and vibrate until they collapse," said the engineer, who asked not to be identified.

Kansas City theoretical physicist John Gamble, who arrived at the hotel a half hour after the collapse during the weekly Tea Dance, used the example of troops crossing a bridge to explain the vibration principle.

"When crossing bridges, troops suppose to be out of step and not in step. In step, their marching could set up a sympathetic vibration.

"The energy supplied is larger the more people there are. In dancing to music, for example, you supply lots of energy and the vibration is going to get large. And if you have more energy than the structural design can take, then you have things happening like bridge collapses," Gamble said.

Dr. Stanley T. Rolfe, a professor in the engineering department at the University of Kansas, has participated in investigations of structural failures of bridges and the collapse of the roof of Kemper Arena in Kansas City in 1979. He emphasized that only a full investigation can uncover how large a role, if any, sympathic vibration played in the Hyatt collapse.

Rolfe explained that in the most famous example, the Galloping Gertie Bridge collapsed over the Tacoma Narrows in Washington on Nov. 7, 1940, when wind started the bridge vibrating up and down until it twisted itself apart.

"It's a very unusual thing for structures," Rolfe said.

FIGURE 18–18B. Newspaper article from the *San Francisco Examiner,* July 19, 1981.

EXERCISES

Exercises 1–5 refer to the following system: A weight of 16 lbs stretches a spring 8 inches. The weight is then pulled down an additional 4 inches and released.

1. Describe the motion if friction is neglected. What is the frequency of the vibrations? What is the amplitude?

2. Describe the motion if friction produces a damping force $F_d = -2(dy/dt)$.

3. Describe the motion if friction produces a damping force $F_d = -\sqrt{5}(dy/dt)$.

4. Describe the motion if friction produces a damping force $F_d = -dy/dt$. What is the period?

5. Describe the motion for large t if the damping force is $F_d = -0.1(dy/dt)$ and the system is subjected to an external force $F_e = 5 \cos 2t$. What is the amplitude of the forced vibrations?

Exercises 6–7 refer to the simple RLC electrical circuit shown in Figure 18–19. It contains a resistance of R ohms, an inductance of L henrys, a capacitance of C farads, and a generator (or battery) that produces an electromotive force of E volts. According to Kirchhoff's laws, the current I amperes at time t satisfies the differential equation

$$L\frac{d^2I}{dt^2} + R\frac{dI}{dt} + \frac{1}{C}I = \frac{dE}{dt}$$

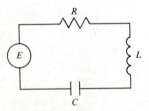

FIGURE 18–19. Exercises 6 and 7.
A simple RLC circuit.

6. Find I as a function of t if $L = 1$, $R = 0$, $C = 1/4$, and E is constant.

7. Find I as a function of t under the same conditions as Exercise 6 only now $E = 3 \sin t$.

8. One end of a stiff wire is attached to the center of flat disc of mass M, and the other end is attached to the ceiling. If the disc is initially twisted through an angle of θ_0 and then released, the resulting motion satisfies the differential equation

$$\frac{1}{2}Mr^2\frac{d^2\theta}{dt^2} = -k\theta$$

where k is a constant depending on the wire and r is the radius of the disc. Find θ as a function of t.

9. (Airplane roll). Figure 18–20A shows an airplane with the right aileron up and the left aileron down. This increases the lift on the left wing, decreases the lift on the right wing, and produces a torque (moment). This torque causes the airplane to rotate about the x-axis and is called *roll*. As the bank angle ϕ changes with time, it satisfies the differential equation

$$\phi'' + B\phi' = L$$

where B and L are constants depending on the mass distribution of the airplane, the aerodynamic properties of the airplane, and the extent to which the ailerons are deflected. Find ϕ as a function of time, given the initial conditions $\phi(0) = 0$ and $\phi'(0) = 0$.

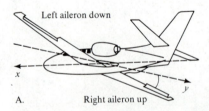

A.   Right aileron up

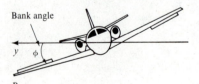

B.

FIGURE 18–20. Exercise 9.

10. Assume that

$$\frac{k}{M} - \frac{a^2}{2M^2} > 0$$

and show that in the forced vibration (18.78), the maximum amplitude occurs when the frequency $\omega/2\pi$ equals

$$\frac{1}{2\pi} \sqrt{\frac{k}{M} - \frac{a^2}{2M^2}}$$

REVIEW EXERCISES

In Exercises 1–30, solve the differential equations.

1. $y' + y \tan x = 2 \sec x$ **2.** $y'' + y' - 6y = 0$

3. $y'' - 8y' + 16y = 0$ **4.** $y' + y = e^{4x}$

5. $y'' + 2y' + y = e^x$ **6.** $xy' + y = (x - 2)^2$

7. $(x^2 + 1) dy = x(3 - 2y) dx$

8. $(x + y) dy = (y - x) dx$

9. $(xe^{y/x} + y) dx = x \, dy$ **10.** $y' = e^{x-y}$

11. $y' + y = e^x$ **12.** $xy' = x^2 + 3y$

13. $xy' = \sqrt{x^2 + y^2}$ **14.** $y' + 2xy = e^{-x^2}$

15. $y'' + y' + (3/2)y = 0$ **16.** $y'' + 4y = 0$

17. $y'' + 4y = 3 \sin x$ **18.** $2y' - y = e^{x/2}$

19. $x \, dy - y \, dx = xy^2 \, dx$ **20.** $y'' + 4y = \tan 2x$

21. $x^3 y' + 3x^2 y = \sin x$

22. $(x + y) dx + (x + y^2) dy = 0$

23. $2y \, dx + x \, dy = 0$ **24.** $y' - y = 2$

25. $y'' - 3y' + 2y = e^{5x}$

26. $y\sqrt{1 - x^2} y' = \sqrt{1 - y^2}$

27. $xy' + y = x^2 \cos x$

28. $(x^2 + y) dx - x \, dy = 0$

29. $(x^2 + x)y' = y^2 - y - 2$

30. $y'' - 4y' + 3y = 0$

---

31. As a raindrop of mass m falls, the air resistance is proportional to the velocity. Write a differential equation that describes the motion.

32. A submarine shuts down its engines and glides horizontally through the water. The only force acting on it is due to friction, which is proportional to the velocity. Write a differential equation that describes the motion. Will the submarine ever come to rest?

33. Within the earth itself a body is attracted toward the earth's center with a force that is directly proportional to its distance from the center. Suppose that a straight tunnel is drilled through the earth connecting two points on the surface. If a ball is dropped into one end and friction is neglected, then it will come to rest at the other end and return. Find the time required for the round trip. Notice that the time is independent of the two points on the earth's surface.

34. (Continuation of Exercise 33). If the tunnel passes through the earth's center, what is the velocity of the ball when it reaches the center?

35. A tank contains 200 gal of water. Brine containing 2 lbs of salt per gal runs in at the rate of 2 gal/min and the mixture leaves the tank at the same rate. Find the amount of salt in the tank at time t. When will the tank contain 300 lbs of salt?

36. A piece of iron at 90°C is immersed in water kept at a constant 50°C. If the iron is 89°C when $t = 100$, what is its temperature at time t?

37. A spherical buoy of radius R floats half submerged in the water. If it is depressed (by the wind, for instance), then a force equal to the excess displaced water will push it up. If friction is neglected, the buoy will oscillate up and down. Find the frequency of the oscillation.

38. Answer the same question in Exercise 37 if the buoy is cylindrical with a 6 inch radius and a weight of 200 lbs (water weighs 62.5 lbs/cu ft).

APPENDIX A

A SHORT TABLE OF INTEGRALS

Basic Formulas

1. $\int x^n \, dx = \begin{cases} \dfrac{x^{n+1}}{n+1} + C, n \neq -1 \\ \ln|x| + C, n = -1 \end{cases}$

2. $\int e^x \, dx = e^x + C$

3. $\int a^x \, dx = \dfrac{a^x}{\ln a} + C$

4. $\int \sin x \, dx = -\cos x + C$

5. $\int \cos x \, dx = \sin x + C$

6. $\int \sec^2 x \, dx = \tan x + C$

7. $\int \csc^2 x \, dx = -\cot x + C$

8. $\int \sec x \tan x \, dx = \sec x + C$

9. $\int \csc x \cot x \, dx = -\csc x + C$

10. $\int \tan x \, dx = -\ln|\cos x| + C$

11. $\int \cot x \, dx = \ln|\sin x| + C$

12. $\int \sec x \, dx = \ln|\sec x + \tan x| + C$

13. $\int \csc x \, dx = \ln|\csc x - \cot x| + C$

14. $\int \dfrac{dx}{\sqrt{a^2 - x^2}} = \sin^{-1} \dfrac{x}{a} + C$

15. $\int \dfrac{dx}{a^2 + x^2} = \dfrac{1}{a} \tan^{-1} \dfrac{x}{a} + C$

16. $\int \dfrac{dx}{|x|\sqrt{x^2 - a^2}} = \dfrac{1}{a} \sec^{-1} \dfrac{x}{a} + C$

17. $\int \dfrac{dx}{\sqrt{a^2 + x^2}} = \ln|x + \sqrt{a^2 + x^2}| + C$

18. $\int \dfrac{dx}{\sqrt{x^2 - a^2}} = \ln|x + \sqrt{x^2 - a^2}| + C$

19. $\int \dfrac{dx}{a^2 - x^2} = \dfrac{1}{2a} \ln\left|\dfrac{x + a}{x - a}\right| + C$

20. $\int \dfrac{dx}{x\sqrt{a^2 - x^2}} = -\dfrac{1}{a} \ln\left|\dfrac{a + \sqrt{a^2 - x^2}}{x}\right| + C$

Radicals $\sqrt{a^2 + x^2}$

21. $\int \sqrt{a^2 + x^2} \, dx = \dfrac{x}{2} \sqrt{a^2 + x^2} + \dfrac{a^2}{2} \ln|x + \sqrt{a^2 + x^2}| + C$

22. $\int x^2 \sqrt{a^2 + x^2} \, dx = \dfrac{x}{8}(a^2 + 2x^2)\sqrt{a^2 + x^2} - \dfrac{a^4}{8} \ln|x + \sqrt{a^2 + x^2}| + C$

23. $\int \dfrac{\sqrt{a^2 + x^2}}{x} \, dx = \sqrt{a^2 + x^2} - a \ln\left|\dfrac{a + \sqrt{a^2 + x^2}}{x}\right| + C$

24. $\int \dfrac{\sqrt{a^2 + x^2}}{x^2} \, dx = -\dfrac{\sqrt{a^2 + x^2}}{x} + \ln|x + \sqrt{a^2 + x^2}| + C$

25. $\int \dfrac{x^2}{\sqrt{a^2 + x^2}} \, dx = \dfrac{x}{2} \sqrt{a^2 + x^2} - \dfrac{a^2}{2} \ln|x + \sqrt{a^2 + x^2}| + C$

26. $\int \dfrac{dx}{x^2 \sqrt{a^2 + x^2}} = -\dfrac{\sqrt{a^2 + x^2}}{a^2 x} + C$

27. $\int \dfrac{dx}{(a^2 + x^2)^{3/2}} = \dfrac{x}{a^2 \sqrt{a^2 + x^2}} + C$

Radicals $\sqrt{a^2 - x^2}$

28. $\displaystyle\int \sqrt{a^2 - x^2}\, dx = \frac{x}{2}\sqrt{a^2 - x^2} + \frac{a^2}{2}\sin^{-1}\frac{x}{a} + C$

29. $\displaystyle\int x^2\sqrt{a^2 - x^2}\, dx = \frac{x}{8}(2x^2 - a^2)\sqrt{a^2 - x^2} + \frac{a^4}{8}\sin^{-1}\frac{x}{a} + C$

30. $\displaystyle\int \frac{\sqrt{a^2 - x^2}}{x}\, dx = \sqrt{a^2 - x^2} - a\ln\left|\frac{a + \sqrt{a^2 - x^2}}{x}\right| + C$

31. $\displaystyle\int \frac{\sqrt{a^2 - x^2}}{x^2}\, dx = -\sin^{-1}\frac{x}{a} - \frac{\sqrt{a^2 - x^2}}{x} + C$

32. $\displaystyle\int \frac{x^2\, dx}{\sqrt{a^2 - x^2}} = \frac{a^2}{2}\sin^{-1}\frac{x}{a} - \frac{1}{2}x\sqrt{a^2 - x^2} + C$

33. $\displaystyle\int \frac{dx}{x^2\sqrt{a^2 - x^2}} = -\frac{\sqrt{a^2 - x^2}}{a^2 x} + C$

34. $\displaystyle\int \frac{dx}{(a^2 - x^2)^{3/2}} = \frac{x}{a^2\sqrt{a^2 - x^2}} + C$

35. $\displaystyle\int (a^2 - x^2)^{3/2}\, dx = \frac{x}{8}(5a^2 - 2x^2)\sqrt{a^2 - x^2} + \frac{3a^4}{8}\sin^{-1}\frac{x}{a} + C$

Radicals $\sqrt{x^2 - a^2}$

36. $\displaystyle\int \sqrt{x^2 - a^2}\, dx = \frac{x}{2}\sqrt{x^2 - a^2} - \frac{a^2}{2}\ln|x + \sqrt{x^2 - a^2}| + C$

37. $\displaystyle\int x^2\sqrt{x^2 - a^2}\, dx = \frac{x}{8}(2x^2 - a^2)\sqrt{x^2 - a^2} - \frac{a^4}{8}\ln|x + \sqrt{x^2 - a^2}| + C$

38. $\displaystyle\int \frac{\sqrt{x^2 - a^2}}{x}\, dx = \sqrt{x^2 - a^2} - a\sec^{-1}\frac{x}{a} + C$

39. $\displaystyle\int \frac{\sqrt{x^2 - a^2}}{x^2}\, dx = -\frac{\sqrt{x^2 - a^2}}{x} + \ln|x + \sqrt{x^2 - a^2}| + C$

40. $\displaystyle\int \frac{x^2}{\sqrt{x^2 - a^2}}\, dx = \frac{x}{2}\sqrt{x^2 - a^2} + \frac{a^2}{2}\ln|x + \sqrt{x^2 - a^2}| + C$

41. $\displaystyle\int \frac{dx}{x^2\sqrt{x^2 - a^2}} = \frac{\sqrt{x^2 - a^2}}{a^2 x} + C$

42. $\displaystyle\int \frac{dx}{(x^2 - a^2)^{3/2}} = -\frac{x}{a^2\sqrt{x^2 - a^2}} + C$

Radicals $\sqrt{ax + b}$

43. $\displaystyle\int \frac{\sqrt{ax + b}}{x}\, dx = 2\sqrt{ax + b} + b\int \frac{dx}{x\sqrt{ax + b}}$

44. $\displaystyle\int \frac{\sqrt{ax + b}}{x^2}\, dx = -\frac{\sqrt{ax + b}}{x} + \frac{a}{2}\int \frac{dx}{x\sqrt{ax + b}}$

45. $\displaystyle\int \frac{dx}{x\sqrt{ax + b}} = \begin{cases} \dfrac{2}{\sqrt{-b}}\tan^{-1}\sqrt{\dfrac{ax + b}{-b}} + C & \text{if } b < 0 \\[2ex] \dfrac{1}{\sqrt{b}}\ln\left|\dfrac{\sqrt{ax + b} - \sqrt{b}}{\sqrt{ax + b} + \sqrt{b}}\right| + C & \text{if } b > 0 \end{cases}$

46. $\int \dfrac{dx}{x^2\sqrt{ax + b}} = -\dfrac{\sqrt{ax + b}}{bx} - \dfrac{a}{2b}\int \dfrac{dx}{x\sqrt{ax + b}}$

47. $\int \dfrac{x}{\sqrt{ax + b}}\, dx = \dfrac{2}{3a^2}(ax - 2b)\sqrt{ax + b} + C$

Linear $ax + b$

48. $\int x(ax + b)^n\, dx = \begin{cases} \dfrac{(ax + b)^{n+1}}{a^2}\left[\dfrac{ax + b}{n + 2} - \dfrac{b}{n + 1}\right] + C & \text{if } n \neq -1, -2 \\[3mm] \dfrac{x}{a} - \dfrac{b}{a^2}\ln|ax + b| + C & \text{if } n = -1 \\[3mm] \dfrac{1}{a^2}\left[\ln|ax + b| + \dfrac{b}{ax + b}\right] + C & \text{if } n = -2 \end{cases}$

49. $\int \dfrac{dx}{x(ax + b)} = \dfrac{1}{b}\ln\left|\dfrac{x}{ax + b}\right| + C$

50. $\int \dfrac{dx}{x^2(ax + b)} = -\dfrac{1}{bx} + \dfrac{a}{b^2}\ln\left|\dfrac{ax + b}{x}\right| + C$

Trigonometric

51. $\int \sin^2 ax\, dx = \dfrac{x}{2} - \dfrac{1}{4a}\sin 2ax + C$

52. $\int \cos^2 ax\, dx = \dfrac{x}{2} + \dfrac{1}{4a}\sin 2ax + C$

53. $\int \sin^n ax\, dx = \dfrac{-\sin^{n-1}ax \cos ax}{na} + \dfrac{n - 1}{n}\int \sin^{n-2}ax\, dx$

54. $\int \cos^n ax\, dx = \dfrac{\cos^{n-1}ax \sin ax}{na} + \dfrac{n - 1}{n}\int \cos^{n-2}ax\, dx$

55. $\int \sin^n ax \cos^m ax\, dx = \dfrac{-\sin^{n-1}ax \cos^{m+1}ax}{a(m + n)} + \dfrac{n - 1}{m + n}\int \sin^{n-2}ax \cos^m ax\, dx$

$= \dfrac{\sin^{n+1}ax \cos^{m-1}ax}{a(m + n)} + \dfrac{m - 1}{m + n}\int \sin^n ax \cos^{m-2}ax\, dx$

(If $m = -n$, use Formulas 57 or 58)

56. $\int \sec^n ax\, dx = \dfrac{\sec^{n-2}ax \tan ax}{a(n - 1)} + \dfrac{n - 2}{n - 1}\int \sec^{n-2}ax\, dx$

57. $\int \tan^n ax\, dx = \dfrac{\tan^{n-1}ax}{a(n - 1)} - \int \tan^{n-2}ax\, dx$

58. $\int \cot^n ax\, dx = -\dfrac{\cot^{n-1}ax}{a(n - 1)} - \int \cot^{n-2}ax\, dx$

59. $\int \sin ax \cos bx\, dx = -\dfrac{\cos(a + b)x}{2(a + b)} - \dfrac{\cos(a - b)x}{2(a - b)} + C$

60. $\int \sin ax \sin bx\, dx = \dfrac{\sin(a - b)x}{2(a - b)} - \dfrac{\sin(a + b)x}{2(a + b)} + C$

61. $\int \cos ax \cos bx\, dx = \dfrac{\sin(a - b)x}{2(a - b)} + \dfrac{\sin(a + b)x}{2(a + b)} + C$

62. $\int x^n \sin ax \, dx = -\frac{1}{a} x^n \cos ax + \frac{n}{a} \int x^{n-1} \cos ax \, dx$

63. $\int x^n \cos ax \, dx = \frac{1}{a} x^n \sin ax - \frac{n}{a} \int x^{n-1} \sin ax \, dx$

64. $\int \sin^{-1} ax \, dx = x \sin^{-1} ax + \frac{1}{a} \sqrt{1 - a^2 x^2} + C$

65. $\int \cos^{-1} ax \, dx = x \cos^{-1} ax - \frac{1}{a} \sqrt{1 - a^2 x^2} + C$

66. $\int \tan^{-1} ax \, dx = x \tan^{-1} ax - \frac{1}{2a} \ln(1 + a^2 x^2) + C$

67. $\int x^n \sin^{-1} ax \, dx = \frac{x^{n+1}}{n+1} \sin^{-1} ax - \frac{a}{n+1} \int \frac{x^{n+1}}{\sqrt{1 - a^2 x^2}} \, dx$

68. $\int x^n \cos^{-1} ax \, dx = \frac{x^{n+1}}{n+1} \cos^{-1} ax + \frac{a}{n+1} \int \frac{x^{n+1}}{\sqrt{1 - a^2 x^2}} \, dx$

69. $\int x^n \tan^{-1} ax \, dx = \frac{x^{n+1}}{n+1} \tan^{-1} ax - \frac{a}{n+1} \int \frac{x^{n+1}}{1 + a^2 x^2} \, dx$

Exponential and Logarithmic

70. $\int x^n e^{ax} \, dx = \frac{1}{a} x^n e^{ax} - \frac{n}{a} \int x^{n-1} e^{ax} \, dx$

71. $\int x^n b^{ax} \, dx = \frac{x^n b^{ax}}{a \ln b} - \frac{n}{a \ln b} \int x^{n-1} b^{ax} \, dx$

72. $\int e^{ax} \sin bx \, dx = \frac{e^{ax}}{a^2 + b^2} (a \sin bx - b \cos bx) + C$

73. $\int e^{ax} \cos bx \, dx = \frac{e^{ax}}{a^2 + b^2} (a \cos bx + b \sin bx) + C$

74. $\int \ln ax \, dx = x \ln ax - x + C$

75. $\int x^n \ln ax \, dx = \frac{x^{n+1}}{n+1} \ln ax - \frac{x^{n+1}}{(n+1)^2} + C$

ANSWERS TO UNDERLINED EXERCISES*

CHAPTER 1

Section 1–1 (page 7)

1.

3.

5.

7.

9.

11. $\begin{cases} > 0 \text{ on } (2, \infty) \\ = 0 \text{ at } 2 \\ < 0 \text{ on } (-\infty, 2) \end{cases}$
13. $\begin{cases} > 0 \text{ on } (-\infty, 5/6) \\ = 0 \text{ at } 5/6 \\ < 0 \text{ on } (5/6, \infty) \end{cases}$
15. $\begin{cases} > 0 \text{ on } (0, 3) \\ = 0 \text{ at } 0 \text{ and } 3 \\ < 0 \text{ on } (-\infty, 0) \text{ or } (3, \infty) \end{cases}$

17. $\begin{cases} > 0 \text{ on } (-\infty, -2) \text{ or } (-1, \infty) \\ = 0 \text{ at } -2 \text{ and } -1 \\ < 0 \text{ on } (-2, -1) \end{cases}$
19. $\begin{cases} > 0 \text{ on } (-2, 1/3) \\ = 0 \text{ at } 1/3 \text{ and } -2 \\ < 0 \text{ on } (-\infty, -2) \text{ or } (1/3, \infty) \end{cases}$

21. $\begin{cases} > 0 \text{ on } (0, 2) \text{ or } (-\infty, -1) \\ = 0 \text{ at } 0, -1, 2 \\ < 0 \text{ on } (-1, 0) \text{ or } (2, \infty) \end{cases}$
23. $\begin{cases} > 0 \text{ on } (0, \infty) \\ = 0 \text{ at } 0, -1 \\ < 0 \text{ on } (-\infty, -1) \text{ or } (-1, 0) \end{cases}$

25. $\begin{cases} > 0 \text{ on } (-\infty, -2) \text{ or } (2, \infty) \\ = 0 \text{ at } -2, 2 \\ < 0 \text{ on } (-2, 2) \end{cases}$
27. $\begin{cases} > 0 \text{ on } (-\infty, -2) \text{ or } (0, \infty) \\ = 0 \text{ at } 0 \\ < 0 \text{ on } (-2, 0) \end{cases}$

29. $\begin{cases} > 0 \text{ on } (3, 4) \\ = 0 \text{ at } 4 \\ < 0 \text{ on } (-\infty, 3) \text{ or } (4, \infty) \end{cases}$
31. $\begin{cases} > 0 \text{ on } (2, \infty) \\ = 0 \text{ at } 1 \\ < 0 \text{ on } (-\infty, 1) \text{ or } (1, 2) \end{cases}$

33. 5 **34.** 3 **35.** 4 **37.** 8 **39.** $\begin{cases} x & \text{if } x \geq 0 \\ -x & \text{if } x < 0 \end{cases}$ **40.** $\begin{cases} y & \text{if } y \geq 0 \\ -y & \text{if } y < 0 \end{cases}$

41. $|x - 4| = 1/10$ **43.** $|3u + 2| < .01$ **45.** The distance between x and -3 is less than or equal to 1. **47.** The distance between $2t$ and 1 is less than 0.01.
49. $-3.5 < x < -2.5$ **51.** $1/3 < x < 3$ **53.** $1/4 \leq x \leq 3/4$ **55.** $2.5 < x < 3.5$

57. $\begin{cases} > \text{ on } \left(-\infty, \dfrac{-3 - \sqrt{13}}{2}\right) \text{ or } \left(\dfrac{-3 + \sqrt{13}}{2}, \infty\right) \\ = 0 \text{ at } (-3 \pm \sqrt{13})/2 \\ < \text{ on } \left(\dfrac{-3 - \sqrt{13}}{2}, \dfrac{-3 + \sqrt{13}}{2}\right) \end{cases}$ **58.** $> 0 \text{ on } (-\infty, \infty)$

---

*A solutions manual, prepared by Professor Edward Keller, is available through your bookstore. It contains complete explanations of those answers with underlined numbers.

59. (a) $-\sqrt{6} < x < -\sqrt{2}$ or $\sqrt{2} < x < \sqrt{6}$ (b) $-\sqrt{10} < x < -\sqrt{8}$ or $\sqrt{8} < x < \sqrt{10}$ **60.** No x (empty set) **61.** (a) $[-9, 18]$ (b) $[-12, 6]$ (c) $[0, 36]$
62. $(1, \infty)$ **63.** (a) 4 (b) 3 **64.** 7

Section 1–2 (page 15)

1. $\sqrt{5}, (3/2, 4)$ **3.** $\sqrt{41}, (-1/2, -1)$ **5.** $5, (3/2, 2)$

7.

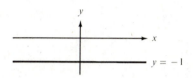

9.

11.

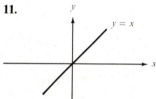

13.

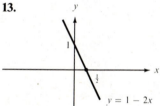

15.

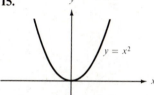

17.

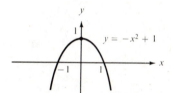

19.

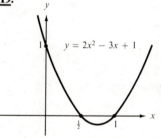

21.

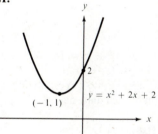

23.

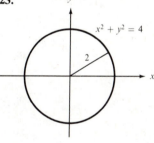

25.

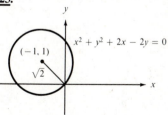

27.

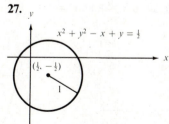

29.

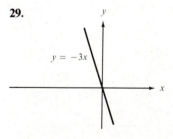

31.

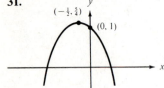

33. No graph

35.

37.

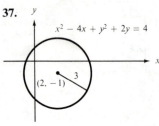

39.

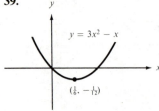

41. x-axis

43.

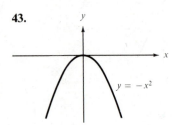

45. $x^2 + y^2 = 1$ **46.** $x^2 + y^2 = 16$ **47.** $(x + 1)^2 + (y - 3)^2 = 2$ **49.** $(x - 1)^2 + (y + 2)^2 = 37$ **51.** $(x - 5)^2 + (y - 2)^2 = 25$ **53.** $6, 27, x^2 + 2xh + h^2 - 2x - 2h + 3$ **55.** Yes   **57.** Yes; no, for the triangle in Exercise 56. **58.** The midpoint of each diagonal is $(1, 1/2)$; yes.

Section 1–3 (page 22)

1. $5, 5, 5$ **3.** $-4, 1.2, 2 + h$ **5.** $3 - 3\sqrt{2}, 25.79, 5 - 5h + h^2$ **7.** $0, \approx 0.56, 1$
9. $1.9, 1.99, 1.999$ **11.** $2.1, 2.01, 2.001$ **13.** $[3, \infty), [0, \infty), 1$ **15.** $[-\sqrt{3}, \sqrt{3}], [0, \sqrt{3}], \sqrt{2}$ **17.** all $x \neq -1$, all $x \neq 0, 1$ **19.** x in $[-2, 0)$ or $(0, 2], [0, \infty), \sqrt{2}/2$

21.

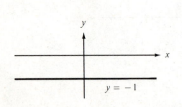

23.

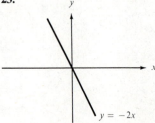

25.

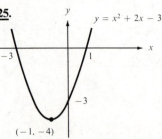

27.

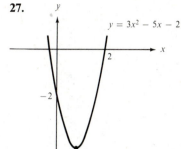

$y = 3x^2 - 5x - 2$

29.

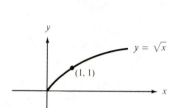

$y = \sqrt{x}$

(1, 1)

31.

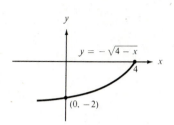

$y = -\sqrt{4-x}$

(0, -2)

33.

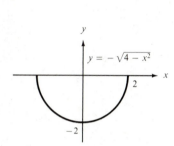

$y = -\sqrt{4-x^2}$

35.

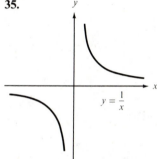

$y = \dfrac{1}{x}$

37.

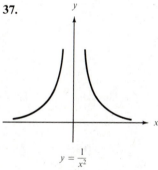

$y = \dfrac{1}{x^2}$

39.

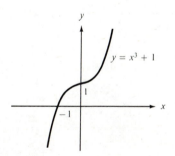

$y = x^3 + 1$

41. $\begin{cases} > 0 \text{ on } (1/3, \infty) \\ = 0 \text{ at } x = 1/3 \\ < 0 \text{ on } (-\infty, 1/3) \end{cases}$ **43.** $\begin{cases} > 0 \text{ on } (0, \infty) \text{ or } (-\infty, -2) \\ = 0 \text{ at } -2 \text{ and } 0 \\ < 0 \text{ on } (-2, 0) \end{cases}$

45. $\begin{cases} > 0 \text{ on } (2, 3) \\ = 0 \text{ at } 2 \text{ and } 3 \\ < 0 \text{ on } (-\infty, 2) \text{ or } (3, \infty) \end{cases}$ **47.** $\dfrac{x^2 - 3x + 2}{x - 3}$ $\begin{cases} > 0 \text{ on } (1, 2) \text{ or } (3, \infty) \\ = 0 \text{ at } 1 \text{ and } 2 \\ < 0 \text{ on } (-\infty, 1) \text{ or } (2, 3) \end{cases}$

49. > 0 on $(0, \infty)$; not defined for $x \le 0$ **51.** < 0.15 **52.** < 0.05 **53.** $< 1/40$
54. $< .0002$

Section 1–4 (page 30)

1. $3\Delta x$ **3.** $-5\Delta x$ **5.** $2x\Delta x - 3\Delta x + \Delta x^2$ **7.** $\Delta x - 8x\Delta x - 4\Delta x^2$

9. $3x^2\Delta x + 3x\Delta x^2 + \Delta x^3$ **11.** $\dfrac{-\Delta x}{x(x + \Delta x)}$ **12.** $\dfrac{\sqrt{x} - \sqrt{x + \Delta x}}{\sqrt{x}\,\sqrt{x + \Delta x}}$

13.

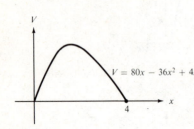

$y = |x - 1|$

15.

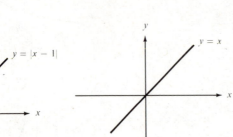

$y = x$

17.

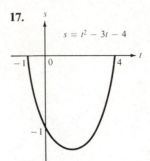

$s = t^2 - 3t - 4$

19.

$V = 80x - 36x^2 + 4x^3$

21.

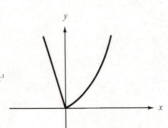

23.

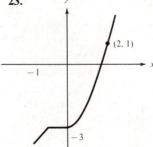

$(2, 1)$

25. ≈ 1.68, ≈ 0.07 **27.** 336, 256, [0, 10]   **28.** 8, 3, -1; 2 and 4 sec   **29.** 36π ft³;

$\dfrac{4\pi}{3}\Delta r(27 + 9\Delta r + \Delta r^2)$ **30.** $4\Delta x(20 - 18x - 9\Delta x + 3x^2 + 3x\Delta x + 4\Delta x^2)$

31. $16\Delta t(-2t - \Delta t + 10)$; 80 ft and 16 ft   **33.** $A = 3b^2/2$; $[0, \infty)$

34.

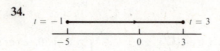

$t = -1$ $t = 3$

35.

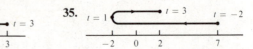

$t = 1$ $t = 3$ $t = -2$

37.

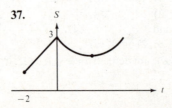

39. $P = 2W + \dfrac{200}{W}$; $(0, \infty)$ **40.** $d = 500t$ **41.** $V = \pi h^3 / 12$

43. $A = 12r - \left(\dfrac{\pi}{2} + 2\right)r^2$ **45.** $A = \sqrt{3}x^2/4$; $[0, \infty)$ **47.** $y = \sqrt{9 - (x + 1)^2} + 2$; $[-4, 2]$ **48.** $A = \dfrac{x^2}{4\pi} + \left(\dfrac{20 - x}{4}\right)^2$ **49.** $C = a\left(2\pi r^2 + \dfrac{V}{2r}\right)$

Section 1–5 (page 37)

7. 1 **9.** $-\dfrac{1}{2}$ **11.** 3 **13.** $-3/4$ **15.** 5 **17.** $y = -\dfrac{1}{5}x + \dfrac{9}{5}$ **19.** $y = \dfrac{1}{5}x + \dfrac{13}{5}$

21. $y = \dfrac{3}{4}x - \dfrac{17}{24}$ **22.** $y = -2x + 13$ **23.** 4 **25.** 11 **27.** 2.71 **29.** $-8/3$

31. 0 **33.** 3 **35.** $2x + \Delta x - 2$ **37.** $1/(\sqrt{x + \Delta x} + \sqrt{x})$ **38.** $\dfrac{-1}{x(x + \Delta x)}$

39. $3x^2 + 3x\Delta x + \Delta x^2$ **41.** $\overline{AB}$ has slope -2, $\overline{AC}$ has slope $1/2$ **43.** $x = 1$

45. (a) $y = 1$ (b) $x = 1$ (c) $y = -1$ (d) $x = -1$ **46.** $y = -\dfrac{\sqrt{3}x}{3} + \dfrac{2\sqrt{3}}{3}$

47. $S = -25,000t + 250,000$ **48.** $F = \dfrac{9}{5}C + 32$; 37 **49.** 63 ft/sec; 207 ft/sec

50. 144 ft/sec; 48 ft/sec **51.** $-16, -48, -80, -112$ ft/sec; the minus signs indicate that motion is in the negative (down) direction.

52.

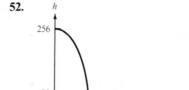

53. ≈ 19.5 in²/in **54.** ≈ 105.7 in³/in

Section 1–6 (page 45)

1. 8 **3.** 1/4 **5.** 1 **7.** 1,000 **9.** 3 **11.** -4 **13.** -5 **15.** 0 **16.** -2 **18.** 1/2
20. $2 + h$ **21.** 16 **23.** 1 **25.** 2 **26.** 10 **28.** π **30.** 7 **31.** 2 **33.** 1/9 **35.** 1/3

36.

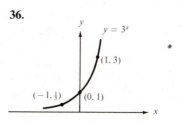

38.

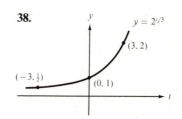

40.

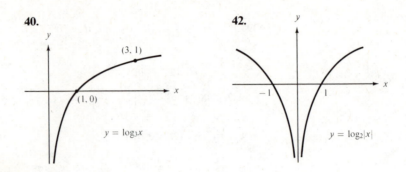

42.

43. $Q = 10(1/2)^{t/2}$, 5/8 gr, $-5/4$ gr/yr, the amount is decreasing   **44.** ≈ 73 (*Hint:* $c = 5$)   **45.** Slightly more than \$39 billion   **46.** \$2,208   **47.** \$12,191
48. 139 months or 11.6 yrs   **49.** 2, 2.5937425, 2.7048138, 2.7169236

Section 1–7 (page 53)

1. $\pi/4$; $\sqrt{2}/2$, $\sqrt{2}/2$, 1   **3.** $-270°$; 1, 0, undefined   **5.** $-5\pi/4$; $\sqrt{2}/2$, $-\sqrt{2}/2$, -1
7. $315°$; $-\sqrt{2}/2$, $\sqrt{2}/2$, -1 **9.** $\pi/2$; 1, 0, undefined   **11.** (0, 1)   **13.** ($\sqrt{2}/2$, $\sqrt{2}/2$)
15. ($-1/2$, $\sqrt{3}/2$)   **16.** (3, 0)   **18.** ($3\sqrt{2}/2$, $3\sqrt{2}/2$)   **20.** ($-3\sqrt{3}/2$, 3/2)   **21.** (0, r)
23. ($-r\sqrt{2}/2$, $-r\sqrt{2}/2$)   **25.** ($r \cos \alpha$, $r \sin \alpha$)   **26.** 4/5, 3/5, 4/3   **28.** 4/5, $-3/5$,
$-4/3$ **30.** $-\sqrt{3}/3$, $-\sqrt{2/3}$, $\sqrt{2}/2$ **32.** $-\sqrt{3/7}$, $2\sqrt{7}/7$, $-\sqrt{3}/2$ **34.** $\sqrt{2}/2$,
$-\sqrt{2}/2$, -1 **36.** $\pi/4$, $5\pi/4$ **37.** all t **38.** $\pi/4$, $5\pi/4$ **39.** $\pi/6$, $\pi/2$, $5\pi/6$, $7\pi/6$,
$3\pi/2$, $11\pi/6$ **40.** $\pi/2$, $3\pi/2$

41.

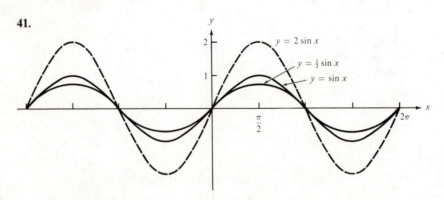

42.

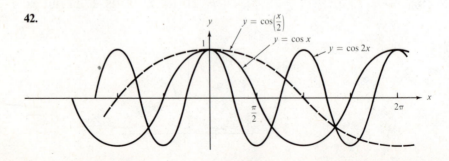

43.

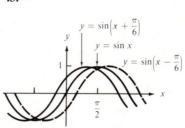

44.

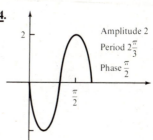

45. $L = 5 \cot \alpha, 5/\sqrt{3}$ ft, $5\sqrt{3}$ ft **46.** $L = 5x/7$ **47.** $h = 5.1 \tan \alpha, \approx 1.2$ m/sec

48. $s = \dfrac{10\sin(90 + \alpha)°}{\sin(38 - \alpha)°}, \approx 23$ miles

50.

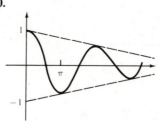

51.

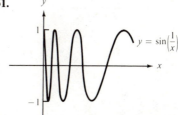

Section 1–8 (page 59)

1. 40 **3.** 0 **5.** 1/9 **7.** 1 **9.** 2 **11.** $\sin x^2, \sin^2 x$ **13.** $98x^2 - 1, 14x^2 - 7$
15. both $x^{3/2}$ **17.** both x **19.** both x **21.** Composition; first $x^2 + 1$, then square
23. Quotient; square divided by composition **25.** Composition; quotient and
square root **27.** Composition; first sine, then exponential **29.** Composition;
first exponential, then sine **31.** Composition of a composition; first $x^2 + 1$, then
cube root, then sine **33.** Composition; first $2x$ + composition $(x + 1)^3$, then fourth
power **35.** Composition of a composition of a composition; first $x + 1$, then square,
then cosine, then cube **37.** Composition; first 2^x, then 2^{2^x} **39.** Product of
compositions **43.** Even **45.** Odd **47.** Neither **49.** Odd **51.** Neither
53. fg and $f + g$ are even **54.** fg even, $f + g$ odd **55.** fg odd, $f + g$ could be neither
56. (c) $f = (g + h)/2$ **57.** (a) The graph is symmetric about the y-axis.
(b) The graph is symmetric about the origin.

Section 1–9 (page 66)

1. $f^{-1}(x) = x/3$ **3.** $f^{-1}(x) = x$ **5.** $f^{-1}(x) = 2x - 4$ **7.** $f^{-1}(x) = 2(1 - x)/3$
9. $f^{-1}(x) = x^2, x \le 0$ **11.** $f^{-1}(x) = -4x^2, x \ge 0$ **13.** $f^{-1}(x) = x^2 - 4, x \ge 0$
15. $f^{-1}(x) = (x - 2)^2, x \ge 2$ **17.** $f^{-1}(x) = \sqrt[3]{x + 2} - 1$, all x
19. $f^{-1}(x) = (x - 1)^3 + 2$, all x

21.

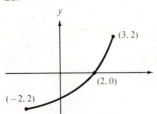
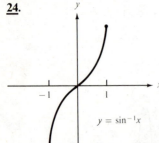

22.

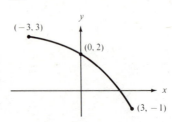
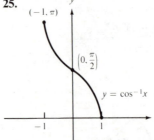

23.

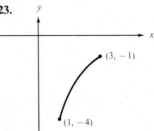
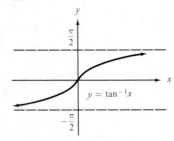

24.

25.

26.

<u>**27.**</u> 4, 1 **28.** −4 **29.** 3 **30.** 73 <u>**31.**</u> about 20.5 days

Review Exercises for Chapter 1 (page 68)

1. $(-2.01, -1.99)$ **3.** $(1.95, 2.05)$ **5.** $\begin{cases} > 0 \text{ on } (0, 3/2) \\ = 0 \text{ at } 0, 3/2 \\ < 0 \text{ on } (-\infty, 0) \text{ or } (3/2, \infty) \end{cases}$

7. $\begin{cases} > 0 \text{ on } (0, 1) \text{ or } (3, \infty) \\ = 0 \text{ at } 0, 1, 3 \\ < 0 \text{ on } (-\infty, 0) \text{ or } (1, 3) \end{cases}$ **9.** $\begin{cases} > 0 \text{ on } (5, \infty) \\ = 0 \text{ at } 5 \\ < 0 \text{ on } (0, 5) \end{cases}$

11. $\begin{cases} > 0 \text{ on } (-6, -5) \text{ or } (1, \infty) \\ = 0 \text{ at } -6, 1 \\ < 0 \text{ on } (-\infty, -6) \text{ or } (-5, 1) \end{cases}$ **13.** No inverse **15.** $f^{-1}(x) = (x + 1)/2$

17. $f^{-1}(x) = (5 + \sqrt{1 + 4x})/2, x > -1/4$ **19.** $f^{-1}(x) = \sqrt[3]{x - 1}$
21. $f^{-1}(x) = 1/x$ **23.** No inverse **25.** Circle; center $(3/2, -2)$, radius $\sqrt{13}/2$
27. Single point $(-1, 2)$ **29.** 2 **31.** $h^2 + h - 1$ **33.** 9 **35.** $y = (-x + 9)/2$
37. $y = (x + 4)/\sqrt{15}$ **39.** 0 **41.** $2x + \Delta x + 1$ **43.** $-1/x(x + \Delta x)$ **45.** 4/5,
$-3/5, -4/3$ **47.** $-\sqrt{3}/2, -1/2, \sqrt{3}$ **49.** Composition of a composition
51. Composition of a power and sum **53.** $L = [y^2 + (2y/(y - 6))^2]^{1/2}; 4\sqrt{10}$
55. $S = 2\pi r^2 + (60/r), r > 0$ **57.** $D = [(x + 1)^2 + (x^2 - 3x - 4)^2]^{1/2}$
59. $W = 110x - 4x^2, 0 \le x \le 25/2$ **61.** 228π cu in/in
63. $R = (400 - 15x)(x + 15)$ cents; $64.60, $65.10, $64.40 **65.** $A = 4 \tan \alpha$,
$0 \le \alpha < \pi/2$

67.

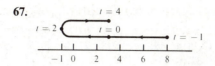

CHAPTER 2

Section 2–1 (page 81)

1. 8   **3.** 1/4   **5.** −1/9   **7.** −1/12   **9.** 1/4   **11.** $1/2\sqrt{7}$ **13.** Left and right limits do not exist   **15.** 1.0986   **16.** 1.0000   **17.** 0.4343   **18.** 0.0175   **19.** 3   **21.** 25   **23.** −1/16   **25.** −3   **27.** Does not exist   **29.** $1/2\sqrt{2}$ **31.** 12   **33.** 0   **34.** 1   **35.** 4   **37.** 0   **39.** −8   **41.** $4x - 1$ **43.** $3x^2 - 4$ **45.** $-1/(x + 1)^2$ **47.** $-1/2x^2$ **49.** $1/2\sqrt{x + 4}$ **51.** $-2/(x - 1)^{3/2}$ **53.** $(1.0986)3^x$ **55.** −2 ft/sec, 0 (at rest)   **57.** True   **59.** False; M could be 0   **61.** True; if h is close to 0, then $p + h$ is close to p, so $f(p + h)$ is close to L **63.** False (see remark after Example 4)   **65.** False (see Exercise 51 in Section 1–7)

Section 2–2 (page 89)

1. 5   **3.** −7   **5.** 4   **7.** 4/3   **9.** $\sqrt{3}$ **11.** 2   **13.** 1/6   **15.** $-1/4\sqrt{2}$ **17.** $2^{16} = 65,536$ **19.** 0   **21.** 1   **22.** 0   **23.** 0.6000   **24.** 0   **25.** −0.5   **26.** 0   **27.** 0   **29.** $1/\pi$ **31.** $-\pi/x^2$ **33.** $-7.2x$ **35.** $1/18.62\sqrt{x}$ **37.** $-9.31/2x^{3/2}$ **39.** $2x$ **40.** $(0.6931)2^x$ **41.** $v(t) = 2$ **43.** $v(t) = -8t + 6$ **45.** $v(t) = t^2 - 1$ **47.** (a) $2x, 3x^2, 4x^3$ (b) $5x^4$ (c) nx^{n-1} (d) Yes   **48.** $-|x| \le xf(x) \le |x|$ **49.** $\delta = \epsilon$ **50.** $\delta = \epsilon/3$ **51.** $\delta = \epsilon/2$ **52.** $\delta = 3\epsilon/2$ **53.** $\delta = \epsilon$ **54.** $\delta = \epsilon$

Section 2–3 (page 96)

7. all x **9.** [0, 1)   **11.** all x **13.** all $x \ne \pm n\pi/2, n = 1, 3, 5, \ldots$ **15.** all $x \ne \pm 2n\pi, n = 0, 1, 2, \ldots$ **17.** all $x < 0$ **19.** all x **21.** all $x \ne 0$ **23.** all positive $x \ne 2$

25. all x except 0.9999 and 1.0001   **27.** 0   **29.** 1   **31.** $\frac{1}{2}\sin(-1) \approx -0.4207$

33. Does not exist   **35.** −8   **37.** $a = 3, b = 4$ **39.** (a) −3   (b) 16   (c) 8   (d) 1   **40.** If h is close to 0, then $p + h$ is close to p; thus, $f(x) \to L$ as $x \to p$ is equivalent to $f(p + h) \to L$ as $h \to 0$.   **41.** f is continuous at p is equivalent to $f(x) \to f(p)$ as $x \to p$ which, in turn, is equivalent to $f(p + h) \to f(p)$ as $h \to 0$.   **42.** (a) $-|x| \le f(x) \le |x|$ so $f(x) \to 0 = f(0)$ as $x \to 0$ (b) at points $x \ne 0$, the limit does not exist.   **44.** False   **46.** False   **49.** False   **51.** True   **53.** False   **55.** False   **56.** f is continuous at p if, for every $\epsilon > 0$, there is a $\delta > 0$ such that $|f(x) - f(p)| < \epsilon$ whenever $|x - p| < \delta$.

Section 2–4 (page 104)

1. $\delta = $ any positive number   **2.** $\delta = \epsilon$ **3.** $\delta = \epsilon/3$ **4.** $\delta = \epsilon/5$ **5.** $\delta = 5\epsilon$ **6.** $\delta = \min\{2 - \sqrt{4 - \epsilon}, \sqrt{4 + \epsilon} - 2\}$ **7.** $\delta = \min\{4 - \sqrt{16 - \epsilon}, \sqrt{16 + \epsilon} - 4\}$ **8.** $\delta = \min\{7 - (\sqrt{7} - \epsilon)^2, (\sqrt{7} + \epsilon)^2 - 7\}$ **9.** $\delta = \min\{27 - (3 - \epsilon)^3, (3 + \epsilon)^3 - 27\}$ **10.** Note that $2x^2 - 3x - 9 = (x - 3)(2x + 3)$. If $|x - 3| < 1$, then $x < 4$ and $2x + 3 < 11$. Therefore let $\delta = \min\{1, \epsilon/11\}$.

Review Exercises for Chapter 2 (page 105)

1. -12 **3.** $-1/9$ **5.** $\sqrt{2}/2$ **7.** Does not exist **9.** $1/2\sqrt{10}$ **11.** 27 **13.** 8
15. 1 **17.** $3/4$ **19.** ≈ 0.9933 **21.** 6 **23.** 8 **25.** $-1/1024$ **27.** Does not exist
29. $8/5$ **31.** 0 **33.** $-10x$ **35.** $-9/x^2$ **37.** $\approx (1.386)4^x$ **39.** $-4/x\sqrt{x}$
41. $\approx (0.9999)(2.718)^x$ **43.** $x \neq n\pi$, n an integer **45.** $x \neq -2$ **47.** $x \neq n\pi/2$,
n an odd integer **49.** $0 \leq x < 1$ **51.** all x **53.** all x **55.** $-\sqrt{6}/18$ **57.** $a = 12$,
$b = 4$ **59.** $v = -3$ ft/sec **61.** $v = 2t - 4$ ft/sec **63.** True **65.** False **67.** True
69. True **71.** True **74.** (a) -5; given ϵ, let $\delta = \epsilon/4$ (b) 14; given ϵ, let $\delta = \epsilon/3$
(c) 8; given ϵ, let $\delta = \epsilon$ (d) 1; given ϵ, let $\delta = \min\{1 - \sqrt{1 - \epsilon}, \sqrt{1 + \epsilon} - 1\}$

CHAPTER 3

Section 3–1 (page 117)

1. 0 **4.** 1 **6.** -2 **9.** $-2x$ **11.** $8x - 3$ **13.** $-4x + 4$ **16.** 0 **18.** $12x - 5$
20. $3x^2$ **22.** $12x^2 - 4x$ **24.** $-1/9x^2$ **26.** $-14/x^3$ **28.** $-3/2\sqrt{1 - x}$
30. $-1/x\sqrt{6x}$ **32.** $-2, 0, 2$ **33.** $3, 0, 3$ **34.** $1/2, 1, 2$, gets larger and larger
35. $y = (-x + 4)/\sqrt{3}$ **37.** Yes; positive **38.** Not differentiable (not even
continuous) at b; not differentiable at d (continuous, but sharp corner); differentiable
at f and negative **39.** $g'(c) = $ slope $= -1$ **40.** $g'(e) = 0$ (horizontal tangent)
41. 6π cm²/cm **43.** $80 - 72x + 12x^2$ in³/in **45.** 64 ft/sec up; 32 ft/sec down;
5 sec **51.** $1/3\sqrt[3]{x^2}$ **52.** $-1/3x\sqrt[3]{x}$

Section 3–2 (page 126)

1. $27x^2 - 6x + 6$ **3.** $-t - \dfrac{4}{5}t^3$ **5.** $-2u^{-3/2} + 4.5u^{3.5}$ **7.** $(-3/2\sqrt{x})$

$+ (1/3x\sqrt[3]{x})$ **9.** $15t^4 - 16t^3 + 42t - 28$ **11.** $(u^2 - 4u - 1)/(2 - u)^2$

13. $\dfrac{3}{4}x^{-1/4} + \dfrac{7}{10}x^{-3/10} + \dfrac{7}{12}x^{-5/12} + \dfrac{8}{15}x^{-7/15}$ **15.** $(2t^{-5} - 4t^{-6}$

$+ 2t^{-7})/(t^{-3} - t^{-4})^2$ **17.** $-(1 + 2u)/(1 + u + u^2)^2$ **19.** $4x^3 - 12x^2 + 12x - 4$
21. $4t^3 - 3t^2 + 2t + 3$ **23.** $(6u^5 + 18u^4 + 12u^3 + 10u + 10)/(u + 1)^4$ **25.** 0
27. 7.2 **29.** $26 - 54x$ **31.** $3.57t^{-0.3}$ **33.** $60x^3 + 12x - 42$ **35.** $4/(x + 1)^3$

37. $192x - 18$ **39.** $\dfrac{-15}{16}x^{-7/2}$ **41.** $-2GmMr^{-3}$ **43.** $y = 11x - 36$; $y = -4x$;

$y = 16x + 24$; positive on $(-\infty, -2/3)$ or $(2, \infty)$; negative on $(-2/3, 2)$; zero at
$-2/3$ and 2 **44.** 20π in²/in **45.** T is increasing about 4% per year **47.** 0.747%/torr
48. Right for $t > 2$, left for $t < 2$, rest at $t = 2$ **51.** 2 ft/sec; 196π ft³/ft

53. $\left[\dfrac{1}{g(x + h)} - \dfrac{1}{g(x)}\right]\Big/h$; $[g(x) - g(x + h)]/hg(x + h)g(x)$; g is continuous

(because it is differentiable); the limit of the product in (3) is $-g'(x)/g^2(x)$

55. $16/90$ ft/sec **58.** $f(x) = 4$ **59.** $f(x) = 7x + 3$ **61.** $\dfrac{1}{3}x^3 + \dfrac{1}{2}x^2 + 9$

62. $\dfrac{1}{4}x^4 - \dfrac{1}{3}x^3 + \dfrac{1}{2}x^2 + 7x - \dfrac{17}{12}$

Section 3–3 (page 133)

5. $6x + e^x$ **7.** $x^4e^x + 4x^3e^x$ **9.** $e^x/7$ **11.** $e^t(t^2 + 3t + 2)$
13. $e^x(x - 1)^2/(x^2 + 1)^2$ **15.** $-(x - 1)^2/e^x$ **17.** $e^x(2x + 1)/2\sqrt{x}$
19. $(1 - 2x)/(2\sqrt{x}e^x)$ **21.** $e^x(x^2 + 4x + 2)$ **23.** $e^x(x^3 - 2x^2 + 2x)/x^4$
25. $(2 - 4x + x^2)/e^x$ **27.** $y = -x + 2$ **29.** $y = 3ex - 2e$ **31.** $P = P_0e^{(t\ln2)/10}$
32. $P = 3 \times 10^7 e^{(0.0427)t\ln2}$ **33.** $Q = 3e^{-(t\ln2)/1600}$ **34.** $x + e^x$ **36.** $(x^2/2) - 2e^x$

38. $\dfrac{1}{4}x^4 - \dfrac{1}{2}e^x$ **40.** xe^x **41.** x^3e^x **43.** e^x/x **44.** e^x/x^2

Section 3–4 (page 140)

1. $3(x^2 - 2x + 1)^2(2x - 2)$ **3.** $(6x + e^x)/2\sqrt{3x^2 + e^x}$

5. $4(x^{-1} + \sqrt{x})^3\left(-x^{-2} + \dfrac{1}{2}x^{-1/2}\right)$ **7.** $(2x + 1)e^{(x^2+x+1)}$ **9.** $-e^{1/x}/x^2$

11. $e^{\sqrt{x}}/2\sqrt{x}$ **13.** $(2x^3/\sqrt{x^4 + 1})e^{\sqrt{x^4+1}}$ **15.** $2xe^{x^2}e^{e^{x^2}}$
17. $(x^2 - 3x + 1)(2x^3 + x)^2[2(2x - 3)(2x^3 + x) + 3(6x^2 + 1)(x^2 - 3x + 1)]$
19. $[-5e^{-5x}(x^2 - 3) - 4xe^{-5x}]/(x^2 - 3)^3$ **21.** $4(x^{-1} + 2 + x)^3(-x^{-2} + 1)$
23. $[8t(t^2 + 1)^3 + 3]/3[(t^2 + 1)^4 + 3t]^{2/3}$ **25.** $50(3x - 1)/(4x + 7)^3$
27. $exe^{-1}e^{x^e}$ **29.** $e^{4t}(8t + 33)/2\sqrt{t + 4}$ **31.** $2e^{2x}(x^2 - 2x - 5)/(x^2 - 5)^3$
33. $320(2 - 4x)^3$ **35.** $6(x^2 - 3)(5x^2 - 3)$ **37.** $2e^{x^2}(2x^2 + 1)$
39. $e^{\sqrt{x}}(\sqrt{x} - 1)/4x\sqrt{x}$ **41.** $1/3$ **42.** 7 **43.** 9 **45.** $y = -189x - 351$
46. 384 lb/sec **47.** 8 cm/sec **48.** 36π ft³/min **49.** $3/\pi$ ft/min **51.** $\approx 3\%$; 23.4 yrs

53. e^{x^2} **55.** $\dfrac{1}{2}x^2 + \dfrac{1}{3}x^3 + e^{2x}$ **57.** $\dfrac{1}{5}(x^2 + 1)^5$

Section 3–5 (page 147)

1. $7\cos 7x$ **3.** $2x + 6\sec^2 2x$ **5.** $-e^{-t}(2\pi\sin 2\pi t + \cos 2\pi t)$ **7.** $24\cos(3x + 2)$
9. $(2t + 1)\sec^2(t^2 + t + 1)$ **11.** $(-\sin t)e^{\cos t}$ **13.** $6\sin^2 2x\cos 2x$
15. $[(3x + 1)\sec^2 x - 3\tan x]/(3x + 1)^2$
17. $-9\cos^2\sqrt{3x + 1}\sin\sqrt{3x + 1}/2\sqrt{3x + 1}$ **19.** $2\sin t\cos te^{\sin^2 t}$ **21.** $-\sin x$
23. $2\cos^{-3}x\sin x$ (write $y = \cos^{-2}x$) **25.** $2e^{-t}\sin t$ **27.** $2(\cos^2 x - \sin^2 x) = 2\cos 2x$
29. $-[\cos(1/x) + 2x\sin(1/x)]/x^4$ **31.** $(6\sin x + 2)\cos x$ **33.** $3e^{6x}/\sqrt{e^{6x} + 1}$

35.

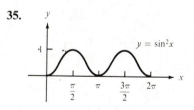

$y = \sin^2 x$

36. 5 ft/rad **37.** $-5\pi/54$ ft/sec **38.** $\pi/3 \approx 1$ mi/sec (change degrees to radians)

39. $(c^2/2)\cos x$ **41.** $V = \dfrac{\pi}{3}\left(6 - \dfrac{3x}{\pi}\right)^2\sqrt{36 - \left(6 - \dfrac{3x}{\pi}\right)^2}$

43. $f'(2/\pi) = 1, g'(2/\pi) = 4/\pi$ **45.** $e^x + \cos x$ **47.** $(\cos 3x)/3$ **49.** $\sin e^x$ **51.** $\sin^2 x$

Section 3–6 (page 155)

1. $-1/2$ **2.** $1/4$ **3.** g not defined at 1   **4.** $1 + \sqrt{2}$ **5.** u not differentiable at 0
7. $\pi/2$ **9.** $(1 + \sqrt{7})/6$

11.

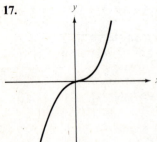

13.

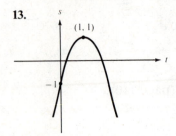

15.

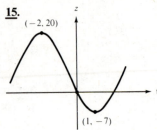

17.

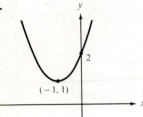

19.

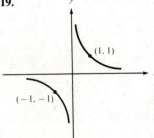

21.

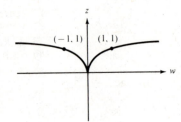

23.

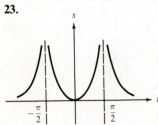

25.

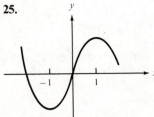

26.

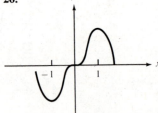

27.

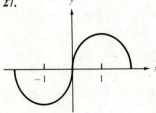

28.

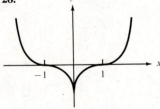

29. After 10 days   **31.** $0 \le x \le 10$ **33.** $0 \le h \le 4\sqrt{6}$ **35.** $32.3 \le p \le 50$
36. $36.7 \le p \le 50$

37.

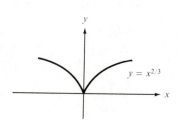

(a)   (b)   (c)   (d)

Review Exercises for Chapter 3 (page 157)

1. 0   **3.** $6x - 2$ **5.** $-9/x^2$ **7.** $1/2\sqrt{x + 1}$ **9.** $32x^3 - 18x^2 + 10x$
11. $14/3\sqrt[3]{t + 1}$ **13.** $3x^2 + 6x + 2$ **15.** $(-6x + 11)/2e^x\sqrt{3x - 4}$
17. $5(4x^3 - 3x^2)^4(12x^2 - 6x)$ **19.** $e^{\sqrt{2x+1}}/\sqrt{2x + 1}$ **21.** $-e^{-t}(2\pi \sin 2\pi t + \cos 2\pi t)$
23. $(3x - 4)^2(-12x + 13)/(5 - 4x)^3$ **25.** $-120x(3 - (6 - (1 - x^2)^3)^4)^4$
$\times (6 - (1 - x^2)^3)^3(1 - x^2)^2$ **27.** $(-6 \sin \sqrt{1 - 6x})/\sqrt{1 - 6x}$ **29.** 0
31. $(2x \cos(1/x) - \sin(1/x))/x^4$ **33.** $(x^3 + 7)(72x^4 + 126x)$

35. $e^x(e^x + 2)/4(e^x + 1)^{3/2}$ **37.** $y = c$ **39.** $y = 2x^2$ **41.** $y = 5x - \dfrac{3}{2}x^2$

43. $y = \dfrac{1}{4}x^4 - x^3 + 2x^2 - x$ **45.** $y = e^{x^2}$ **47.** $-\dfrac{1}{3}\cos 3x$ **49.** $y = e^x\sin x$

51. $8/27$ **53.** $(6 - \sqrt{156})/12$ **55.** Not defined at $\sqrt{2}$

57.

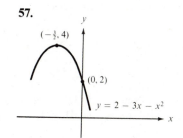

$(-\tfrac{3}{2}, 4)$ $(0, 2)$ $y = 2 - 3x - x^2$

59.

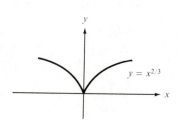

$y = x^{2/3}$

61.

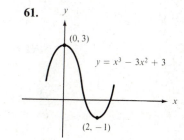

$(0, 3)$ $y = x^3 - 3x^2 + 3$ $(2, -1)$

63.

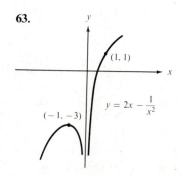

$(1, 1)$ $y = 2x - \dfrac{1}{x^2}$ $(-1, -3)$

65.

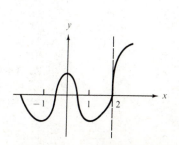

67. $y = 11x - 10$ **69.** $y = 2x + 5$ **71.** 7   **73.** $-15\sqrt{3}$ ft/sec   **75.** $0 < x < 10$
77. $4\pi/3$ ft^2/sec   **79.** About $0.005\, Q_0$ gr/day

CHAPTER 4

Section 4–1 (page 167)

1. Abs. min $(-3/2, -1/4)$ **3.** Abs. max $(1/2, 5/4)$ **5.** Abs. min $(-2, 0)$ and
$(0, 0)$; abs. max $(1, 3)$; loc. max $(-4/3, 32/27)$ **7.** Abs. max $(-1, 4)$;
abs. min $(3/2, -11/16)$; end max $(2, 1)$ **9.** Abs. min $(1, 3)$ **11.** Abs. min $(0, 0)$
13. Abs. max $(0, 0)$ **15.** Abs. min -324 **17.** Abs. max $37/2$; abs. min -126
19. Abs. max $\left(\sqrt{60} - \dfrac{3}{2}\right)\left(\dfrac{80}{\sqrt{60}} - 2\right)$; Abs. min 0   **21.** Abs. max $p/2$; abs. min $-p/2$
23. Abs. min 8   **25.** Abs. min $\sqrt{3}/2$ **27.** Abs. min $p^2/2(\pi - 4)$;
abs. max $p^2(6 - \pi)/2$ **28.** Abs. min $26/15$; abs. max $2\sqrt{10}/3$ **29.** Abs. min occurs
at $x = \sqrt{3}\pi p/(9 + \sqrt{3\pi})$; abs. max occurs at $x = p$

Section 4–2 (page 173)

1. $-18, 18$ **3.** 3,600 ft^2 **5.** About 67 ft by 53.7 ft   **7.** 5/3 in   **8.** 1,357 shirts;
$6.10/shirt   **11.** Min: let side of square be $L/(\pi + 4)$; max: use it all for a circle
12. $\pi/4$ **13.** $2\pi(1 - \sqrt{2/3})$ **14.** About 55 miles   **15.** 25 ft^2 **16.** 25 ft^2
17. $w = 4\sqrt{3}, h = 4\sqrt{6}$ **19.** $\sqrt{60}$ by $80/\sqrt{60}$ **21.** $y = -2x + 4$ **22.** No such line
23. $5\sqrt{5}$ ft   **24.** Stay underground   **25.** Head for the point $2/\sqrt{11}$ miles from C
26. $r = \dfrac{1}{2}\sqrt[3]{V/\pi}$ **27.** $r = (V/\pi[(4 + 2\sqrt{2}) + (2/3)])^{1/3}$ **28.** $(\pm\sqrt{5/2}, 5/2)$
31. No such rectangle   **32.** $r = 1/2, h = 1$ **33.** $r = \sqrt{200/3}$ **35.** About 9.9 ft
37. Isosceles, with equal sides $(p - b)/2$ **38.** Yes   **39.** Equilateral, with sides
$2\sqrt{A/\sqrt{3}}$ **40.** Equilateral, with sides $p/3$

Section 4–3 (page 180)

1. $\Delta y = 0.0302, dy = 0.03$ **3.** $\Delta y = 0.070501, dy = 0.07$ **5.** $\Delta y = 0.0011149$,
$dy = 0.0011111$ **7.** $\Delta y = -0.1, dy = -0.09615$ **9.** 2.015   **11.** 10.0677
13. 0.5151   **15.** 1.03   **20.** $\pm 6\%$ **22.** 0.96π cu in   **24.** 0.0396 cu ft   **25.** 1,037 cu cm
26. Increase by about 0.0038 cm   **27.** Leave about 0.036 in at the top and 0.018 in
on each side

Section 4–4 (page 186)

1. $12x\sqrt{y}/(2\sqrt{y} - 1)$ **3.** $4\sqrt{y^2 + y\sqrt{y}}/(2\sqrt{y} + 1)$ **5.** $-(y + 2x)/(2y + x)$
7. $-y^2/x^2$ or $(1 - y)/(x - 1)$ **9.** $y \cos xy/(1 - x \cos xy)$
11. $-\sin(x + y)/(1 + \sin(x + y))$ **13.** $-\sqrt{y/x}$ **15.** $-2x(x^3 + y^3)/y^5$
17. $2y^2(y + x)/x^4$ or $2(y - 1)/(x - 1)^2$ **19.** $4y + \sqrt{3}x = 8\sqrt{3}$ **21.** $5x - 4y = 12$
23. $bx + ay = 2ab$ **25.** $(-2/\sqrt{5}, 1/\sqrt{5})$ and $(2/\sqrt{5}, -1/\sqrt{5})$ **26.** $y' = -x/y$ is
the slope of the tangent; the radius joins $(0, 0)$ to (x, y), so its slope is y/x.
28. There are none   **29.** $-\sqrt{3}$ in/sec   **30.** $-\sqrt{3}$ in/sec

Section 4–5 (page 191)

1. 2.5 ft/sec   **3.** Distance between them is decreasing 4 mph   **4.** Apart at $3/\sqrt{13}$
units/sec; together at $4/\sqrt{10}$ units/sec   **6.** $9/\sqrt{5}$ ft/sec   **8.** $10/3\pi$ ft/min

10. Increasing at $(5Rn + 6)/10$ cubic units/min **12.** 100 ft/sec $\approx$ 68 mph
13. 50π ft/sec $\approx$ 107 mph **17.** 1/4 ft/min **16.** Moves at a constant rate
17. 5 ft/sec **18.** 60 ft/sec; 0 (because the rock hits the ground when $t = \sqrt{3}$)
19. -250 units/sec **21.** 0.04π cu cm/sec **22.** 0.0225 cm/sec **23.** The water
does not rise because ice displaces its own weight. **24.** 300 mph **25.** 60π miles/min
26. $3/\pi$ ft/min **27.** $1/27\pi$ ft/min **28.** 138/13 and 92/13 miles

Section 4–6 (page 197)

3.

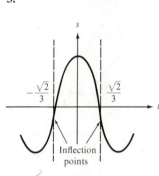

5.

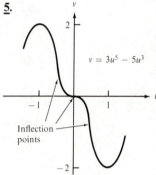

$v = 3u^5 - 5u^3$

7.

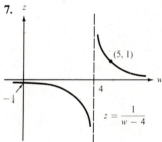

$z = \dfrac{1}{w-4}$

9.

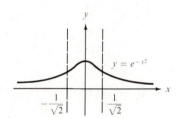

$y = e^{-x^2}$

11.

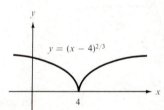

$y = (x-4)^{2/3}$

12.

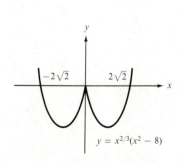

$y = x^{2/3}(x^2 - 8)$

13.

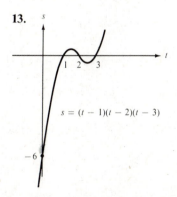

$s = (t-1)(t-2)(t-3)$

15.

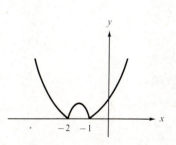

17.

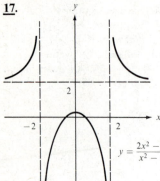

$$y = \frac{2x^2 - 1}{x^2 - 4}$$

19.

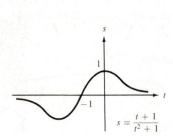

$$s = \frac{t + 1}{t^2 + 1}$$

20.

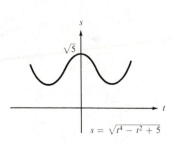

$$s = \sqrt{t^4 - t^2 + 5}$$

21. $\frac{1}{3}x^3 - x^2 + \frac{2}{3}$ **23.** $2, -1$ **24.** $s = 2 + t + \sin \pi t$

25.

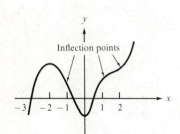

26.

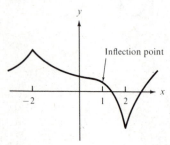

Inflection point

27.

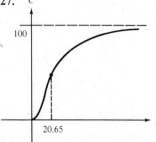

28. 1/3 dollars **30.** $A = C/x$; $A' = [xC' - C]/x^2$; $A' = 0$ when $xC' - C = 0$; that is, when $C' = C/x = A$

Section 4–7 (page 204)

1. $5x + C$ **3.** $3x^2 - 4x + C$ **5.** $2t^3 - \frac{3}{2}t^2 + t + C$ **7.** $\frac{2}{3}t^{3/2} - 2t^{1/2} + C$

9. $\frac{3}{4}x^{4/3} + \frac{1}{3}x^{-3} + C$ **11.** $3x^3 + C_1x + C_2$ **13.** $-\sin x + C_1x + C_2$

15. $e^t + \frac{1}{2}t^2 + C_1t + C_2$ **17.** $-\cos t - \sin t + C_1t + C_2$

18. $\cos t - \sin t + C_1t + C_2$ **19.** $\frac{1}{2}x^{-1} + \frac{4}{3}x^{3/2} + C_1x + C_2$ **21.** $3x - x^2 + 1$

23. $\frac{2}{3}x^{3/2} - 2x - 4$ **25.** $e^t + \tan t$ **27.** $\frac{1}{6}x^3 + \frac{1}{2}x^2 - \frac{7}{2}x + \frac{17}{6}$

28. $\frac{4}{15}x^{5/2} - \frac{16}{3}x + \frac{49}{5}$ **29.** $1 - \frac{1}{2}t - \cos t$ **30.** $\frac{1}{2}t^2 + e^t + 2t$ **31.** $e^x \sin x + C$

33. $x^3\cos x + C$ **35.** $e^x/(x + 1) + C$ **37.** $\dfrac{1}{3}e^{3t} + C$ **39.** $\dfrac{1}{10}(x^2 + 1)^5 + C$

41. $2\sqrt{x^3 + x} + C$ **43.** $\dfrac{1}{2}\sin^2 t + C$ **45.** 19 ft to right of 0; moving left

46. At 0 (or equilibrium point); moving up   **53.** $100e^{-x/10}$ **54.** $5e^6$ units left of 0

Section 4–8 (page 209)

1.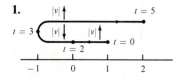

2.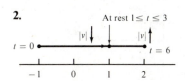

3.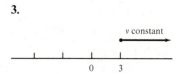

5.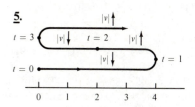

7.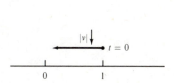

9. $7, -2, -2$ **11.** $-(1/3)e^{-3} - 1, e^{-3}, -3e^{-3}$ **13.** 206 ft/sec   **15.** 56/3 ft; 4 ft/sec²
17. 6 hrs and 180 miles; 30 sec and 1/4 miles   **19.** 1,600 ft, 20 sec   **21.** 356 ft
22. $\sqrt{10}$ sec   **23.** 2 sec   **24.** 320 ft/sec   **25.** About 0.05 hrs or 3 min
26. 90 mph; about 5.4 mi   **27.** 20 sec   **28.** 100 mph; 5/18 mi   **29.** 100/3 sec
30. 4 ft; 5 ft   **31.** 14 sec   **32.** About 98 sec after cut-off (Note: the net acceleration
for the first minute is $52 - 32 = 20$ ft/sec²)   **33.** 133 ft   **35.** $(35 - 5e^{-9})/3$; slowing
down

Section 4–9 (page 215)

1. 1.325   **3.** 1.332   **5.** -1.663 **6.** 0.755   **7.** 1.293   **8.** 1.353   **10.** 0.682
11. 0.352   **12.** -0.641 **13.** 2.709   **14.** -3.485 **15.** $(0.249, -0.621)$

Review Exercises for Chapter 4 (page 215)

1.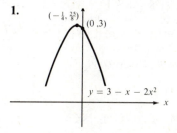

3.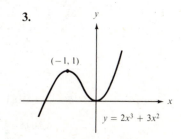

5.

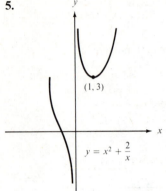

$y = x^2 + \dfrac{2}{x}$

$(1, 3)$

7.

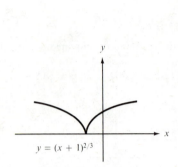

$y = (x + 1)^{2/3}$

9.

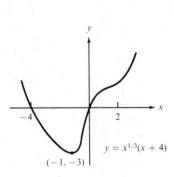

$y = x^{1/3}(x + 4)$

$(-1, -3)$

-4

11. 3.9875   **13.** -1.02 **15.** 0.9930   **17.** $6x^2 \sqrt{y}/(2\sqrt{y} + 1)$
19. $(3y - 2x)/(2y - 3x)$ **21.** $[y^2 + y\sin(x/y)]/x\sin(x/y)$ **23.** $2(y - 1)/(x - 1)^2$
25. $\sin y/\cos^3 y$ **27.** $x^3 + 2x^2 - 2x + C$ **29.** $(2/3)t^{3/2} + (1/t) + C$
31. $-(1/2\pi)\cos 2\pi t + C$ **33.** $(1/4)e^{2x} + (1/9)\cos 3x + C_1 x + C_2$ **35.** $e^x \sin x + C$
37. $(1/20)t^5 - (1/4)t^4 + (1/6)t^3 + C_1 t + C_2$ **39.** $4y = 5x - 16$ **41.** $y = -x + 2$
43. $-3/4, -5/3, 0$ **45.** $-5/4, 2, 4$ **47.** Min: use 3.77 inches for the circle; max:
use it all for the circle.   **49.** 0.65 mi   **51.** Constant rate k **53.** 274   **55.** 0
57. 0.000259 sec   **59.** Reflect town A across the river and join to town B with a
straight line; where the line crosses the river is the location of the station.   **61.** 40 mph

63.

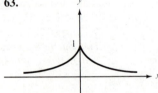

65. $(-1, 1)$

CHAPTER 5

Section 5–1 (page 225)

1. $9/2, 3, 15/4, 33/8$ **2.** $8, 10, 9, 17/2$ **3.** $\pi, 3.259, 3.206, 3.184$ **4.** $3/2, 2, 7/4, 13/8$
5. $1, 1, 1, 1$ **6.** 0.6931 (correct to four places)   **7.** $1/3$ **8.** $1/4$ **9.** 90 ft-lbs   **10.** 4 ft
11. $9/2$ ft   **12.** 8 ft   **13.** $3/2$ ft   **14.** $1/3$ ft   **15.** $1/4$ ft   **16.** 90 ft   **17.** 4 ft
18. True   **19.** True   **20.** $m(b - a) \le A$ **21.** $A \le B$ **22.** True   **23.** True
24. $B - A$ **25.** $A = C + D$

Section 5–2 (page 234)

1. $9/4$ **3.** $27/2$ **5.** $21/2$ **7.** 18   **9.** $9\pi/4$ **11.** $\pi/2$ **13.** $\dfrac{4\pi}{3} + \dfrac{\sqrt{3}}{2}$ **14.** $\dfrac{8\pi}{3} - 2\sqrt{3}$

15. $8/3$ **17.** $2/3$ **19.** 2   **21.** 84 ft-lbs   **22.** 8 ft   **23.** $9/4$ ft   **24.** $27/2$ ft
25. $21/2$ ft   **26.** 18 ft   **27.** 84 ft   **28.** 8 ft   **29.** True   **30.** True   **31.** True

32. $\int_a^b f(x)\,dx \le \int_a^b g(x)\,dx$ **33.** True   **34.** True   **35.** $A + c(b - a)$
38. (a) 1   (b) 2   (c) $2 + \pi$ (d) 2

Section 5–3 (page 241)

1. $G(x) = 3(x - 1)$ **3.** $G(x) = x^2$ **5.** $G(x) = (3x^2 - 4x + 1)/2$
7. $G(x) = (-x^2 + 2x + 15)/2$ **9.** 12   **11.** 4   **13.** 1/2   <u>**15.**</u> 11   **17.** 16   **19.** 0
<u>**21.**</u> 14   **23.** 14/3   **24.** 6   **25.** 2   <u>**27.**</u> 9/4   **29.** $2(1 - e^{-3})$ **31.** $-22/3$ **33.** 15/4
<u>**35.**</u> $8(\sqrt{2} - 1)$ **37.** Area; $\pi/4$ **38.** Fund. Th; 16/3   **39.** Not possible as yet

40. Not possible as yet   **41.** Area; $\dfrac{\pi}{3} + \dfrac{\sqrt{3}}{2}$ **42.** Not possible as yet

<u>**43.**</u> Fund. Th; $(e^3 - 1)/3$ <u>**44.**</u> Fund. Th; 0   **45.** $f(x)$ **46.** Differentiate both sides
of the equation $\int_0^x f(u)\,du = x^2 + 3x$ to obtain $f(x) = 2x + 3$.   **47.** $f(x) = 2x + 3$
and $a = -4$ or 1   **49.** $F'(x) = G'(g(x))g'(x) = f(g(x)) \cdot 2 = 2f(2x)$ **50.** (a) $2xf(x^2)$
(b) $(\cos x)f(\sin x)$

Section 5–4 (page 249)

<u>**1.**</u> 4   **3.** 0   <u>**5.**</u> 1   **7.** $(e^4 - 1)/4$ **9.** Not possible as yet   **10.** 1/2   <u>**11.**</u> $\dfrac{\pi}{3} + \dfrac{\sqrt{3}}{2}$

13. $\dfrac{2\pi}{3} - \dfrac{\sqrt{3}}{2}$ **15.** 18/5   **17.** 46/15   **19.** 38/3   **21.** 9/2   <u>**23.**</u> 2/3, 2   **24.** 0, 2

25. 0, 8/3   <u>**26.**</u> (a) 2, 2/3, 4/3 sq units   (b) 2/3 ft to the right, 2 ft   (c) 2/3 million
27. (a) 8/3, 4/3, 4/3   (b) At the starting point, 8/3 mi   (c) 0 (breaks even)
28. (a) 8   (b) 5/2, 11/2,   (c) At time b it is 3 feet to the left of where it was at
time a; it has moved a total of 8 ft   (d) The company lost $3 million
<u>**30.**</u> $1.35 million   **31.** $2 million   **32.** About $1,704,277   <u>**33.**</u> $550,000
34. 2 ft to right; 17/4 ft   <u>**35.**</u> $(-2, 0)$, 17/2 ft   **36.** No, he will lose $6,666 before
he starts making money; $\sqrt{3} \approx 1.7$ years   **37.** $\pi/2$ **38.** $1 + \pi/4$

Section 5–5 (page 256)

1. 4   **3.** 5/2   <u>**5.**</u> 4   **7.** 37/3   **9.** -36 <u>**11.**</u> 3/8   **13.** 9, -9 **14.** $2(\sqrt{5} - \sqrt{2})$,
$2(\sqrt{2} - \sqrt{5})$ **15.** -14 **16.** 9   **17.** $-38/3$ **18.** 4   <u>**19.**</u> 17   <u>**20.**</u> 0, 1/3, 0;
average value approaches 0   **21.** 1, 4/3, 2; average value approaches ∞
<u>**22.**</u> 29,800 kw; 1,000 kw; 20,200 kw   <u>**23.**</u> $\pi/2$

28. $\dfrac{1}{b - a} \int_a^b f'(x)\,dx = \dfrac{1}{b - a}\Big[f(x)\Big]_a^b = \dfrac{f(b) - f(a)}{b - a}$

29. f is not integrable (see Exercise 40 in Section 5–2); $|f(x)| = 1$ for all x, so it is
integrable.

Section 5–6 (page 262)

1. $\left(\dfrac{1}{5}\right)(x^3 + 1)^5 + C$ <u>**3.**</u> $2(x^2 + 2)^{1/2} + C$ <u>**5.**</u> $\left(\dfrac{1}{40}\right)(4x + 1)^{10} + C$

<u>**7.**</u> $\left(\dfrac{1}{18}\right)(x^3 - 3)^6 + C$ **9.** $\left(\dfrac{1}{3}\right)(x^2 + 2x - 3)^3 + C$ **11.** $2(x^2 + 3x - 2)^{1/2} + C$

13. $\left(\dfrac{1}{6}\right)(x^2 + 2x - 3)^3 + C$ **15.** $4(x^3 - 4x^2 + x)^{1/2} + C$ <u>**17.**</u> $-\left(\dfrac{1}{5}\right)\left(1 + \dfrac{1}{x}\right)^5 + C$

19. $\left(\dfrac{1}{2}\right)(1 + \sqrt{x})^4 + C$ **21.** $2(1 + \sqrt[3]{x})^{3/2} + C$ **23.** $-2(1 + \sqrt{x})^{-1} + C$

25. $\dfrac{1}{5}x^5 + \dfrac{8}{3}x^3 + 16x + C$ **26.** $\dfrac{1}{2}x^2 + 2x + \dfrac{1}{x} + C$ **27.** $\left(\dfrac{1}{12}\right)(3x + 1)^4 + C$

29. $\dfrac{1}{7}x^7 + \dfrac{3}{5}x^5 + x^3 + x + C$ **31.** $\dfrac{1}{a(r + 1)}(ax + b)^{r+1} + C$

33. $\dfrac{1}{an(r + 1)}(ax^n + b)^{r+1} + C$ **35.** 7/576   **36.** 175/2   **37.** 7/192   **38.** 19/3

41. (b)(d)(f)   **42.** $\left(\dfrac{1}{3}\right)\sin 3x + C$ **43.** $-4\cos(x/4) + C$ **44.** $-\left(\dfrac{1}{2}\right)\cos x^2 + C$

45. $\left(\dfrac{1}{3}\right)\sin x^3 + C$ **46.** $-e^{-x} + C$ **47.** $\left(\dfrac{1}{2}\right)e^{x^2} + C$ **48.** $\sin e^x + C$ **49.** $e^{\sin x} + C$

Section 5–7 (page 269)

1. $y^2 = -x^2 + C$ **3.** $y = \left(\dfrac{1}{4}x^2 + C\right)^2$ **5.** $e^y = e^x + C$ **7.** $\cos y = -\sin x + C$

9. $(y^2 + 1)^{-2} = 4x^2 + C$ **11.** Not separable   **13.** $(1 + y^2)^{1/2} = \dfrac{1}{3}(1 + x^2)^{3/2} + C$

15. Not separable   **17.** $y^{3/2} = 9x^{1/2} + C$ **19.** $(1 + y^2)^{-1/2} = x + C$

21. $s = \left(\dfrac{1}{3}\right)[(1 + t^2)^{3/2} - 1]$ **22.** $v = \left(\dfrac{1}{2}t + 1\right)^2$ **23.** 335/12 ft

24. $R = 400x - 4{,}000x^{-1/2}$ **25.** \$9,867   **26.** About 7,078 mi   **27.** $s = 2gR^2/v^2$; as t increases, v approaches 0 and s increases without bound.

Section 5–8 (page 274)

1. $11/32 \approx 0.3438$; 1/3; 1/3   **3.** 0.6153; 0.6094; 0.6095   **5.** 1.8961; 2.0045; 2
7. $84/256 \approx 0.3281$ **9.** 0.6065   **11.** 2.0523   **13.** 1.1481   **15.** 0.7854   **17.** 5.7833 mi
18. 2.8916 mi/min $\approx$ 173.5 mph   **19.** (a) 63.69   (b) 64.08   (c) 64.11
20. About 0.119 liters/sec   **21.** 89.667   **23.** (a) $6.729 \times 10^{-4}(M = 1)$
(b) $5.208 \times 10^{-4}(M = 24)$

Review Exercises for Chapter 5 (page 279)

1. $\dfrac{2}{3}(x + 1)^{3/2} + C$ **3.** Not possible as yet   **5.** $\dfrac{1}{3}(x^2 + 1)^{3/2} + C$ **7.** $4\tan x + C$

9. Not possible as yet   **11.** $-\dfrac{1}{5}\left(1 + \dfrac{1}{x}\right)^5 + C$ **13.** $-\dfrac{3}{4}(1 - \sqrt[3]{x})^4 + C$

15. $\dfrac{1}{10}(x^2 - 1)^5 + C$ **17.** $\dfrac{1}{2}x^2 - 3x - \dfrac{1}{x} + C$ **19.** Not possible as yet

21. $\dfrac{5}{6}x^{6/5} - \dfrac{20}{23}x^{23/20} + C$ **23.** $-2(1 + \sqrt{x})^{-1} + C$ **25.** 0   **27.** 1/3

29. $\dfrac{8\pi}{3} - 2\sqrt{3}$ **31.** 0.7854   **33.** 0   **35.** $-422/5$ **37.** 0.8814   **39.** $\dfrac{1}{2}[24^{2/3} - 5^{2/3}]$

41. $4y^{1/2} = x^2 + C$ **43.** $\sqrt{1 + y^2} = x + C$ **45.** $y = -1/(\tan x + C)$

47. $y^{3/2} = \dfrac{3}{2}e^x + C$ **49.** Not separable   **51.** 3   **53.** 11/6   **55.** 180 in-lbs

59. $v = t^2 - 8t + 12$; speeding up for $2 < t < 4$ and $t \ge 6$, slowing down for $0 \le t < 2$ and $4 < t < 6$ **61.** $s(6) = 0$; 64/3 ft

CHAPTER 6

Section 6–1 (page 286)

1. 1/6   **2.** 1/2   **3.** 15/16   **4.** $\sqrt{2} - 1$ **5.** 5/4   **6.** 3/4   **7.** 1   **9.** 72   **11.** 125/48

13. 9   **14.** 8/3   **15.** 9/2   **16.** $\dfrac{\pi}{2} + \dfrac{4}{3}$ (area of the semicircle plus area of the rest)

17. 1/3   **19.** $2\sqrt{2}/3$ **21.** 1/2   **23.** ≈ 0.4887 **24.** Curves intersect when $x = 0$, $(3 \pm \sqrt{13})/2$;   Area ≈ 11.718222 **25.** 9/2   **26.** 1   **27.** 0.30675

28. $\displaystyle\int_0^2 (\sqrt{4 - (x - 2)^2} - x)\, dx$ **29.** $\displaystyle\int_{(4-\sqrt{15})/2}^{(4+\sqrt{15})/2} (2\sqrt{4 - (y - 2)^2} - 1)\, dy$

30. $\displaystyle 2\int_0^4 (\sqrt{4 + y} - \sqrt{4 - y})\, dy$ **31.** 5/2

Section 6–2 (page 290)

1. $\left(\dfrac{1}{25}\right)(5x + 3)^5 + C$ **2.** $\left(\dfrac{2}{3}\right)(x^3 + 1)^{1/2} + C$ **3.** $\left(\dfrac{2}{9}\right)(x^3 + 5)^{3/2} + C$

4. $\left(\dfrac{1}{10}\right)(x^2 + 2x - 3)^5 + C$ **5.** $\left(\dfrac{1}{5}\right)\left(1 - \dfrac{1}{x}\right)^5 + C$ **6.** $\left(\dfrac{2}{5}\right)(1 + \sqrt{x})^5 + C$

7. 1/4   **8.** 112/3   **9.** 0   **10.** 81/8   **11.** $\pi r^2 h/3$ **12.** $k^2 h/3$ **13.** $\left(\dfrac{2}{3}\right)r^3 \tan \alpha$

14. 2   **16.** 8/15   **17.** $4\pi r^3/3$ **18.** $4\sqrt{3}/15$ **20.** $3\pi/20$ **21.** $\pi/2$ **22.** $9\pi/2$

23. $\displaystyle\int_0^\pi \sin^2 x\, dx$ **26.** $\displaystyle\int_a^b \pi[f(x)]^2\, dx$ **27.** $\displaystyle\int_a^b \pi([g(x)]^2 - c^2)\, dx$ **28.** About 5,480 cu ft

29. About 7,552 cu ft   **30.** $\sqrt{3}\pi r^3/2$ **31.** $16r^3/3$ **32.** $8r^3/3$ **33.** $2\sqrt{3}$

Section 6–3 (page 299)

1. $32\pi/5$ **3.** $176\pi/15$ **5.** π **7.** $\pi/6$ **8.** $5\pi/6$ **9.** $\pi/2$ **10.** $4\pi/15$ **11.** $16\sqrt{2}\pi/5$
13. $88\sqrt{2}\pi/15$ **15.** $\pi(16 - 6\sqrt{3})/3$ **17.** $8\pi/3$ **19.** $3\pi/10$ **21.** $64\pi/5$ **22.** $\pi/2$
23. $8\pi/3$ **24.** Disc; $15\pi/2$ **26.** Shell; $3\pi/2$ **28.** Shell; $32\sqrt{2}\pi/3$
29. $2\pi[r^2 d - (d^3/3)]$ **30.** About 13.1 gal   **31.** $3/2\pi$ ft/sec   **32.** $\pi k R^4/2$;
0.8π cu cm/sec   **33.** $\pi(e^2 - 3)/2$ **34.** $\pi(e - 2)$ **35.** 32π **36.** $\pi/2$
37. $\pi(4\sqrt{2} - 3)/2$ **38.** ≈ 299.18696 (exact answer is $32\pi^2$)

Section 6–4 (page 307)

1. $(1 + x^{2/3})^{3/2} + C$ **2.** $\dfrac{1}{2}e^{x^2} + C$ **3.** $\dfrac{1}{2}e^{2x} + C$ **4.** $-2\cos\sqrt{x} + C$ **5.** 7/3

6. 74/3   **7.** 6 in-lbs   **8.** 90; 270 in-lbs   **10.** 4   **11.** 24,543,693 ft-lbs;
49,087,385 ft-lbs   **12.** About 4.1; 8.3 hrs   **13.** About 15.8 hrs   **14.** About 20.2 hrs; no
16. $3k/10$ J   **17.** About 11.9 in-lbs   **19.** 900 ft-lbs   **20.** 1,150 ft-lbs
21. (a) 2,500 ft-lbs;   (b) about 18 ft/sec   **22.** 44 hp   **23.** 48.8 hp

24. 1.786×10^{-4} GMm **26.** $F = \dfrac{6}{32}s^{1/3}$; $W \approx 0.75$ ft-lbs

27. $k = 4$; natural length 3 in

Section 6–5 (page 312)

1. ≈ 15.17 **3.** $123/32$ **5.** ≈ 6.31 **6.** ≈ 1.44 **8.** ≈ 1.44 **9.** $\displaystyle\int_0^1 (1 + e^{2x})^{1/2}\, dx$

11. $\displaystyle\int_0^2 (1 + 4y^2)^{1/2}\, dy$ **12.** $\displaystyle\int_0^1 \left[1 + \left(3x + \frac{1}{2}\right)^2\right]^{1/2} dx$ **13.** 5.1087569 **14.** 1.4789778

15. $3\sqrt{101}$ ft/sec **17.** $\displaystyle\int_1^3 \left[1 + \left(5x - \frac{17}{2}\right)^2\right]^{1/2} dx$ **18.** About 8,120 ft

19. $\displaystyle 2\int_0^{150} \left[1 + \frac{1}{144}(e^{x/150} - e^{-x/150})^2\right]^{1/2} dx$ **20.** 6

Section 6–6 (page 319)

1. $(9/7, 6/7)$ **3.** $-6/13$; $-17/13$ ft/sec **5.** $(3/4, 3/10)$ **7.** $(1/2, 2/5)$
9. $(9/20, 9/20)$ **11.** $(0, 4/3\pi)$ **12.** $(4/3\pi, 4/3\pi)$ **13.** $(4\sqrt{2}/3\pi, 4(2 - \sqrt{2})/3\pi)$
15. $(4/3, 2/3)$ **16.** $(7/6, 7/6)$ **17.** $(8/9, 8/9)$ **18.** $\left(1, \dfrac{2\pi + 1}{\pi + 2}\right)$

19. $\left(\dfrac{3\pi + 43}{3\pi + 24}, \dfrac{3\pi + 23}{3\pi + 24}\right)$ **21.** ≈ 4.88 mph **22.** $\approx 483,000$ ft **23.** (a) $\pi/2$ (b) $\pi/5$
24. (a) $2\pi/15$ (b) $\pi/6$

Section 6–7 (page 324)

1. -0.37 **3.** $7/2$ **5.** 2 **7.** $28/9$ **9.** 3.6 **11.** $\$1.3225$ **12.** 2 **13.** $5.68' \approx 5'8''$
14. $+0.33$ **15.** -0.09 **16.** -0.014 **17.** $14/3$ hrs **18.** 0.18 **19.** 0.4554167
20. $\approx 5.734'$ or $5'9''$ **21.** About 4,912 hrs **22.** $\approx .7913$ **23.** 957 hrs
24. (a) $1/2$ (b) 30 days

Section 6–8 (page 328)

1. $5,208,333$ lbs **3.** 111 lbs **4.** $\displaystyle\int_1^2 100(y - 1)\sqrt{4 - y^2}\, dy$ **5.** $\displaystyle\int_0^4 125y\sqrt{4 - y}\, dy$

6. $\displaystyle\int_0^5 125y\sqrt{9 - y}\, dy$ **8.** $800/3$ lbs **10.** (a) $1,200\pi$ lbs (b) $3,600\pi$ lbs (at depth y,
the force on a thin strip of surface is $50y(2\pi r)\Delta y$)

Review Exercises for Chapter 6 (page 329)

1. $1/3$ **3.** ≈ 2.448478 **5.** 1 **7.** $-\left(\dfrac{2}{3}\right)\left[\left(\dfrac{5}{4}\right)^{3/2} - \left(\dfrac{3}{2}\right)^{3/2}\right]$ **9.** 0 **11.** $9/2$

13. $125/4$ cu in **15.** $\pi^2/2$ **17.** $\pi/2$ **19.** ≈ 10.5131 **21.** 36 in-lbs **23.** 900 lbs
25. $8\pi/3$ **27.** $(0, 4r/3\pi)$ **29.** $1,276,272$ ft-lbs **31.** ≈ 0.71 **33.** ≈ 0.21
35. $(1, 1/3)$; $2\pi/3$; 2π

CHAPTER 7

Section 7–1 (page 336)

1. $1/x$ **3.** $3/x$ **5.** $-9/(8 - 18x)$ **7.** $4x/(x^4 - 1)$
9. $(3x^2 + 6x + 1)/3(x^2 + 1)(x + 3)$ **11.** $(\sec^2 x)/\tan x$ **13.** $[2\ln(x + 1)]/(x + 1)$
15. $1 + \ln x$ **17.** $-(1 + 2\ln(1/x))/x^3$ **19.** $1/x \ln x$ **21.** $3y$
23. $-((y/x) + \ln y)/((x/y) + \ln x)$ **25.** $6xy + (y/x)$

27.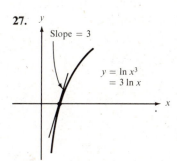

29.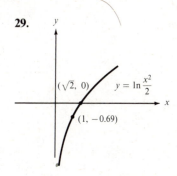

31.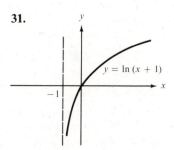

33.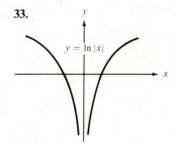

35. $y = 3x - 2$ **37.** 0.6930 **39.** Yes   **41.** $v = 2t/(t^2 + 1)$; $a = (2 - 2t^2)/(t^2 + 1)^2$

42. $\pi \ln 2$ **43.** $2\pi \ln 2$ **45.** $\ln x = \ln(x^{1/q})^q = q \ln x^{1/q}$; therefore, $\dfrac{1}{q}\ln x = \ln x^{1/q}$

47. Domain is $(-\infty, \infty)$; range is $(0, \infty)$; graph is the reflection of $y = \ln x$ through the line $y = x$.

48. Take the derivative of both sides with respect to x; the result is $\dfrac{E'(x)}{E(x)} = 1$.

Section 7–2 (page 341)

1. Yes; f' always positive   **3.** No; f not one-to-one   **5.** Yes; f' positive for $x > 2$ so f is one-to-one on $[2, \infty)$ **7.** Yes; $f' = \cos x$ is positive on $(-\pi/2, \pi/2)$ **8.** No; $\cos x$ is not one-to-one on $[-\pi/2, \pi/2]$ **9.** $f'(x) = 1/(1 + x^2)$ is always positive
11. 4   **13.** $1/12$ **15.** 3   **17.** $1/3$ **19.** $1/9$ **21.** $1/2$ **23.** 1   **25.** $1/9$ **27.** -4
29. 1   **30.** 1   **31.** $2/\sqrt{3}$ **32.** -1 **33.** Since $E'(x) = E(x) > 0$, E has an inverse L. Then $E(L(x)) = x$ for $x > 0$. Differentiate with respect to x to obtain $E(L(x))L'(x) = 1$.

Section 7–3 (page 348)

1. $-2e^{-2x}$ **3.** $10xe^{5x^2}$ **5.** $e^{\sqrt{x}}/2\sqrt{x}$ **7.** $xe^x(x+2)$ **9.** $(\cos x)e^{\sin x}$
11. $-e^{-x}(\sin x + \cos x)$ **13.** $(-e^{1/x})/x^2$ **15.** 1 **17.** $1/(e^x+1)$ **19.** $(1+\ln x)x^x$
21. $ye^{xy}/(1-xe^{xy})$ **23.** $2xy/(1-ye^y)$

25.

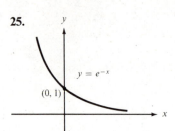

27.

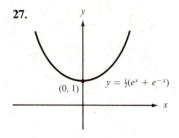

29.

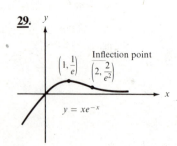

31.

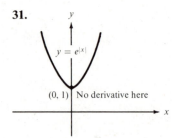

33. $y = 2e^{-7x}$ **35.** $y = e^{(5x-2)}$ **37.** $y = x+1$ **39.** $(-1/3, e)$ **41.** $y = 20e^{5x}$;
$x = (\ln 2)/5$ **43.** Increasing for $k > 0$; decreasing for $k < 0$ **44.** 1 or -5
46. (a) $f'(x) = e^x - 1 > 0$ for $x > 0$, so f is increasing; (b) $f(0) = e^0 - 1 - 0 = 0$;
(c) It follows from (a) and (b) that $f(x) = e^x - 1 - x > 0$ for $x > 0$; therefore,
$e^x > 1 + x$

Section 7–4 (page 353)

1. $(1/2)\ln(10/3) \approx .6020$ **3.** $(1/2)\ln(3/4) \approx -.1438$ **4.** $(\ln 2)/5 \approx .1386$
5. $(1/1.3)\ln(8/7) \approx .1027$ **7.** $10^4 \ln 2 \approx 6{,}932$ days ≈ 19 years **9.** About 461 min
or 7.7 hrs **10.** 1,024 million **11.** About 10 hrs **12.** About 291 million
13. About 57.6 yrs **15.** About 54% **17.** About 6,799 yrs old **18.** About 79%
19. About 18 yrs **20.** About 29°C **22.** About 1.6 min **23.** $5(e-1) \approx 8.6$ ft
24. $Dy - C \ln y = A \ln x - Bx + a$ constant

Section 7–5 (page 360)

1. $-1/x$ **3.** $(2x-5)/(x^2-5x+2)$

5. $[x(x^2-3)^2 \sqrt{x^2+1}]\left[\dfrac{1}{x} + \dfrac{4x}{x^2-3} + \dfrac{x}{x^2+1}\right]$ **7.** $\dfrac{1}{x} + \dfrac{4x}{x^2-3} + \dfrac{x}{x^2+1}$

9. $\dfrac{1}{x} + \dfrac{x}{x^2+4} - \dfrac{3}{x+2}$ **11.** $y\left[\dfrac{1}{x} + \dfrac{x}{x^2+4} - \dfrac{3}{x+2}\right]$

13. $y\left[\dfrac{1}{x} + 1 - \tan x - \dfrac{1}{x+1}\right]$ **15.** $8x[\ln(x^2 + 1)]^3/(x^2 + 1)$ **17.** $e^{7x}\left[\dfrac{1}{x} + 7\ln x\right]$

19. $\left(\dfrac{1}{2}\right)\left(\dfrac{1}{x+1} - \dfrac{1}{x-2}\right)$ **21.** $-(1/2)\ln|4 - x^2| + C$ **23.** $(1/2)(4 - x^2)^{-1} + C$

25. ≈ 0.3783266 **27.** $(1/2)\ln|x^2 + 8x + 9| + C$ **29.** $(1/4)(\ln x)^4 + C$
31. $(2/3)\ln|1 + x\sqrt{x}| + C$ **33.** ≈ 0.4999773 **35.** $x - 4e^{-x} + C$
37. $\ln(e^x + 4) + C$ **39.** $-(e^x + 4)^{-1} + C$ **41.** ≈ 0.4606 **42.** Shell method yields
$\pi(1 - e^{-1}) \approx 1.9858653$ **43.** $\ln 2 \approx .6931472$ **44.** $7 + \ln(1/2) \approx 6.3068528$ ft
45. Solid generated by revolving the region bounded by $y = e^x$, $x = 0$, and $y = 2$
about the y-axis.

Section 7–6 (page 366)

1. $6^x \ln 6$ **3.** $[4^{\sqrt{x}} \ln 4]/2\sqrt{x}$ **5.** $3^{\sin x}(\cos x) \ln 3$ **7.** $(2x - 3)/(x^2 - 3x + 1) \ln 5$
9. $-1/(x^2 - 1) \ln 7$ **11.** $x10^{\sqrt{x^2+1}} \ln 10/\sqrt{x^2 + 1}$ **13.** $1/x(\ln x)(\ln 3)$
15. $x2^x(2 + x \ln 2)$ **17.** $(x + 1)^{(x+1)}[1 + \ln(x + 1)]$ **19.** $x \cos x(\sin x)^{(x-1)}$
$+ (\sin x)^x \ln(\sin x)$ **21.** $x^{x^x}[x^{(x-1)} + x^x \ln x + x^x(\ln x)^2]$ **23.** $2^x/\ln 2 + C$
25. $3^{x^2}/(2 \ln 3) + C$ **27.** ≈ 361.37663 **29.** $2\sqrt{1 + 10^x}/\ln 10 + C$
31. ≈ 0.7403627 **33.** $(\ln 2)(\ln|\log_2 x|) + C$ **35.** Take ln of both sides;
$x = (\ln 2)/(\ln 3 - \ln 2) \approx 1.7095113$ **36.** Use (7.49) to write
$\dfrac{\ln x}{\ln 2} = 1 + \dfrac{\ln x}{\ln 3}$; then $\ln x = (\ln 2)(\ln 3)/(\ln 3 - \ln 2)$ and $x \approx 6.541$

37. $\dfrac{2}{\ln 3} - \dfrac{1}{\ln 2} \approx 0.3777834$ **38.** $\pi\left(1 - \dfrac{1}{2\ln 2}\right) \approx 0.8754126$

Section 7–7 (page 370)

1. Separable and linear; $y = Ce^{4x} - (3/4)$ **3.** Separable and linear; $y = Ce^{x^2/6}$

5. Neither   **7.** Both; $s = 1 - Ce^{-t^2/2}$ **9.** Linear: $y = \dfrac{x^2}{3} + \dfrac{C}{x}$

11. Linear; $y = (x + C)e^{-x}$ **13.** Linear; $y = (x + C)/(1 + e^x)$ **14.** Linear;
$y = e^{-x}[\ln(1 + e^x) + C]$

15. Separable; $s + (s^2/2) = (t^2/2) + C$ **17.** Linear; $\dfrac{4}{3}(1 + x) + \dfrac{C}{(1 + x)^2}$

19. Both; $y = Ce^{x^2/2} - 1$ **21.** About 40.5 days   **22.** About 100 days

23. $y = 0.1 + 0.4e^{-t/10}$ **25.** $I = \dfrac{E}{R} + Ce^{-Rt/L}$ is the general solution

(a) if $I(0) = E/R$, then $C = 0$ and $I(t) = E/R$ (b) if $I(0) > E/R$, then $C > 0$
and $I(t)$ decreases to E/R as t increases; if $I(0) < E/R$, then $C < 0$ and $I(t)$
increases to E/R as t increases   (c) E/R **26.** (a) 10 min   (b) $y = 2$
$- 0.0015(10 - t)^3$ lbs/gal, $0 \le t < 10$ **27.** 20 fans

Review Exercises for Chapter 7 (page 371)

1. $(x + 4)/3$ **3.** $(x - 2)^2$ **5.** $1/3$ **7.** 7   **9.** $1/6$ **11.** 1   **13.** $4x/(x^4 - 1)$

15. $\cot x$ **17.** $(\cos t)e^{\sin t}$ **19.** $\dfrac{1}{3}\left[\dfrac{2x}{x^2 + 1} + \dfrac{2x}{x^2 + 2}\right]$ **21.** $(2x^2 + 1)e^{x^2}$ **23.** $3^t \ln 3$

25. $x^x(1 + \ln x)$ **27.** $9^{\sqrt{x}} \ln 9/2\sqrt{x}$ **29.** $x^3 4^x(4 + x \ln 4)$ **31.** $(1/8)\ln|1 + 8x| + C$

33. $-[\ln x]^{-1} + C$ **35.** $\dfrac{x^2}{2} - \dfrac{1}{2x^2} + C$ **37.** $(1/4)\ln(x^4 + 1) + C$ **39.** $(1/2)(\ln 2)^2$

41. $-3^{-2x}/2 \ln 3 + C$ **43.** $(1/\ln 9)\ln(1 + 9^x) + C$ **49.** $y = e^{-3x}$ **51.** $y = Ce^x - 1$
53. $s = Ce^{2t} - e^t$ **55.** $(x + C)(10 - x)$ **57.** ≈ 6.67 lbs/sq in **59.** ≈ 96.5 ft
61. $3\pi/\ln 4 \approx 6.7985402$ **65.** About 11 days **67.** $y - 2 = (-2 \ln 2)(x - 2)$
69. \$12,169

CHAPTER 8

Section 8–1 (page 379)

13. $2\pi/3$ **15.** $\pi/6$ or $5\pi/6$ **17.** $\pi/2$ **19.** Sec x and csc x are never zero; cot $x = 0$
for $x = \pi/2$ or $3\pi/2$ **20.** Secant: $\pi/2$ or $3\pi/2$; cosecant and cotangent: 0 or π
21. $6x \sec^2(3x^2)$ **23.** $4 \sin(2x)\cos(2x)$ **25.** $2x \sec(x^2)\tan(x^2)$ **27.** $-2 \cot x \csc^2 x$
29. $-4x(x^2 + 1)\csc(x^2 + 1)^2\cot(x^2 + 1)^2$ **31.** $(y \sec^2 xy)/(1 - x \sec^2 xy)$
33. $-(y + \sin xy \tan xy)/x$ **35.** $-\cot y \sin^2 y$ **37.** $-(1/2)\cot x^2 + C$
39. $(1/\sqrt{3}) - (1/2)$ **41.** $-\ln|\cos e^x| + C$ **43.** $2 \ln|\sec \sqrt{x} + \tan \sqrt{x}| + C$
45. $-\csc x + C$ **47.** $-(1/3)\cot^3 x + C$ **49.** $\ln|\tan x| + C$

53. $\left(\dfrac{1}{2}\right)\left(\dfrac{\pi}{6} + \dfrac{\sqrt{3}}{3}\right)\left(\dfrac{\pi\sqrt{3}}{3} + 2\right)$ **55.** π

Section 8–2 (page 389)

1. $-\pi/6$ **3.** $5\pi/6$ **4.** $\pi/4$ **5.** 0 **7.** $4/\sqrt{17}$ **9.** 0 **11.** π **13.** $6x^2/\sqrt{1 - 4x^6}$
15. $1/2x\sqrt{x - 1}$ **17.** $e^{\sin^{-1}x}/\sqrt{1 - x^2}$ **19.** $1/|x|\sqrt{25x^2 - 1}\sec^{-1}5x$
21. $-x/\sqrt{1 - x^2}$ **23.** -1 **24.** (a) $\cos(\sin^{-1}x) = \sqrt{1 - x^2}$ (draw a triangle)

(b) $\sin^{-1}(\cos x)$ is the number whose sine is $\cos x$; that is, $\sin^{-1}(\cos x) = \dfrac{\pi}{2} - x$

25. $(1/2)\sin^{-1}2x + C$ **27.** $(1/\sqrt{5})\tan^{-1}\sqrt{5}x + C$ **29.** $\sec^{-1}\sqrt{3}x + C$ **31.** $\pi/6$

33. $-\sqrt{4 - x^2} + C$ **35.** $\dfrac{1}{3}\sec^{-1}\dfrac{x}{3} + C$ **37.** $(\tan^{-1}\sqrt{x})^2 + C$

39. $\alpha = \cos^{-1}(r/R)^4 \approx 1.5$ radians **40.** (a) $\pi/6$ (b) Use the shell method;
$\pi(2 - \sqrt{3})$ **41.** 3.1415918 **43.** $2\sqrt{7}$ ft **46.** (a) $\tan^{-1}(x + 1) + C$
(b) $\sin^{-1}(x + 2) + C$

Section 8–3 (page 396)

1. $y = (1/2)\cos\left(2t + \dfrac{\pi}{2}\right)$ **2.** $y = (x^4/12) + (x^3/6) - (x^2/2) + 5x + 1$

3. $y = 8e^{-2x}$ **4.** $y = 1 + 2e^{-x^2/2}$ **5.** $y = \sqrt{2}\cos\left(3t - \dfrac{\pi}{4}\right)$ **6.** $y = 3e^{\tan^{-1}x} - 1$

7. $y = \ln|\sin^{-1}x + 2|$ **8.** $s = \sqrt{2}\cos\left(5t + \dfrac{\pi}{4}\right)$ **9.** $s = 2\cos\left(8\pi t + \dfrac{\pi}{3}\right)$

10. $s = 5\cos\left(2t + \dfrac{\pi}{2}\right)$ **11.** $A = 2, T = 2, f = 1/2$

12. $s(0) = -2; v(0) = 0;$ max v at $t = 2n + (1/2)$; min v at $t = 2n - (1/2)$;
max a at $t = 2n$; min a at $t = 2n - 1$

13. $s = 5\cos\sqrt{8}t$ **15.** $Q = Q_0\cos\left(\dfrac{1}{\sqrt{LC}}t\right)$

16. (a) About 4.1×10^4 cycles/sec (b) $t = \pi\sqrt{LC}/2$ **17.** $s = 7\cos\left(\dfrac{4\pi}{3}t + \dfrac{\pi}{2}\right)$

18. $s = (1.7\sin 80°)\cos\left(16\pi t + \dfrac{\pi}{2}\right)$

19. Set $a = -A$ and $c = C - \dfrac{\pi}{2}$; $p = A\cos C$ and $q = -A\sin C$

Section 8–4 (page 402)

1. $4\cosh(4x)$ **3.** $(-\sin x)\sinh(\cos x)$ **5.** $2x/(1 - x^4)$ **7.** $-1/\sqrt{1 - e^{2x}}$
9. $1/2\sqrt{x^2 - x}$ **11.** $\cosh^{-1}(x/2) + C$ **13.** $(1/3)\sinh^{-1}3x + C$ **15.** $\operatorname{sech}^{-1}|2x| + C$

17. $\ln(3 + \sqrt{10})$ **19.** $(1/2)\ln(14/15)$ **21.** $\sinh 1 = (1/2)\left(e - \dfrac{1}{e}\right)$
23. $v = \sqrt{32/k}\tanh(\sqrt{32k}t)$; $\sqrt{32/k}$

Review Exercises for Chapter 8 (page 406)

1. $3\tan^2(x^3 + x)\sec^2(x^3 + x)(3x^2 + 1)$ **3.** $2x\sec^2x^2$ **5.** $4\sec^24x$
7. $(2xy\sec^2x^2y)/(1 - x^2\sec^2x^2y)$ **9.** $(3x^2 + 2x)/\sqrt{1 - (x^3 + x^2)^2}$

11. $y\left[\dfrac{x}{\sqrt{1 - x^2}} + \sin^{-1}x\right]$ **13.** $1/\sqrt{(1 - x^2)(1 - (\sin^{-1}x)^2)}$ **15.** $10x\cosh(5x^2 + 4)$

17. $(-\tanh x)(\operatorname{sech}^2\ln\operatorname{sech} x)$ **19.** $1/(1 - x^2)\sqrt{1 + (\tanh^{-1}x)^2}$ **21.** $\sin x^3 + C$

23. $\sin^2\sqrt{x} + C$ **25.** $\dfrac{1}{5}\sec^5x + C$ **27.** $\tan(x^2 + x) + C$ **29.** $\sin^{-1}(x/3) + C$

31. $\tan^{-1}(5x) + C$ **33.** $\tan^{-1}(x^2) + C$ **35.** $\dfrac{1}{4}\ln\left|\dfrac{x}{x + 4}\right| + C$

37. $2\ln\left|\dfrac{x}{2} + \sqrt{(x^2/4) - 1}\right| + C$ **39.** $-\ln\left|\dfrac{1 + \sqrt{1 - x^4}}{x^2}\right| + C$

41. $\ln|x + \sqrt{4 + x^2}| + C$ **43.** $y = A\cos(4t + C)$ **45.** $s = \sqrt{2}\cos\left(6t + \dfrac{\pi}{4}\right)$

47. $s = 3\cos\left(4t + \dfrac{\pi}{6}\right)$ **49.** (a) 1.59×10^4 cycles/sec (b) $t = 1.57 \times 10^{-5}$sec

51. $\dfrac{1}{2}\left(e - \dfrac{1}{e}\right)$ **53.** $y = -\left(\dfrac{1}{\sqrt{3}}\right)\left(x - \dfrac{\pi}{6}\right) + 1$ **55.** Concave upward **57.** $\pi^2/8$

CHAPTER 9

Section 9–1 (page 416)

1. $-x\cos x + \sin x + C$ **3.** $(1/3)x\sin 3x + (1/9)\cos 3x + C$ **5.** $(1/2)xe^{2x}$
$- (1/4)e^{2x} + C$ **7.** $x\ln 4x - x + C$ **9.** $x\sin^{-1}(x/2) + \sqrt{4 - x^2} + C$
11. $(1/13)e^{2x}(2\sin 3x - 3\cos 3x) + C$ **13.** $(1/17)e^x(\cos 4x + 4\sin 4x) + C$
15. $(1/10)\sec 5x\tan 5x + (1/10)\ln|\sec 5x + \tan 5x| + C$ **17.** $-xe^{-x} - e^{-x} + C$
19. $(1/2)x^2\ln x - (1/4)x^2 + C$ **21.** $(1/2)(x^2 - 1)\ln(x + 1) - (1/4)x^2 + (1/2)x + C$
23. $(2/3)x(x + 1)^{3/2} - (4/15)(x + 1)^{5/2} + C$ **25.** $(1 + 2e^3)/9$
27. $(\pi/4) - (1/2)\ln 2$ **29.** $-(e^{\pi/2} + 1)/5$

Section 9–2 (page 421)

1. $-x\cos x + \sin x + C$ **2.** $(1/9)\sin 3x - (1/3)x\cos 3x + C$ **3.** $1/2$

4. $(\pi/4) - (1/2)\ln 2$ **5.** $-xe^{-x} - e^{-x} + C$ **6.** $e^x(\cos x + \sin x)/2 + C$

7. $x(\sin \ln x - \cos \ln x)/2 + C$ **8.** $(x^2/2)\ln x - (x^2/4) + C$

9. $(\ln|\csc x - \cot x| - \csc x \cot x)/2 + C$ **10.** $x\tan x + \ln|\cos x| + C$ **11.** $2 - \sqrt{3}$

13. $-x^2e^{-x} - 2xe^{-x} - 2e^{-x} + C$ **15.** $(1/2)e^{x^2} + C$ **17.** $\sin^{-1}x^2 + C$

19. $-x^2\cos x + 2x\sin x + 2\cos x + C$ **21.** $-(3/4)x(1-x)^{4/3}$
$- (9/28)(1-x)^{7/3} + C$ **23.** $\ln|x-1| - \ln|x-2| + C$

25. $(1/4)(\sec 2x \tan 2x + \ln|\sec 2x + \tan 2x|) + C$ **27.** $(e^{2x}/13)(2\sin 3x$
$- 3\cos 3x) + C$ **29.** $x\tan^{-1}2x - (1/4)\ln(1 + 4x^2) + C$

35. $\dfrac{x^{n+1}}{n+1}\ln ax - \dfrac{x^{n+1}}{(n+1)^2} + C$; $4\ln 6 - (1/4)\ln 3 - (15/16)$ **39.** $2\pi^2$

41. $(\pi/2) - 1$ **42.** (a) $1 - \left(\dfrac{2}{a_0^2}r_1^2 + \dfrac{2}{a_0}r_1 + 1\right)e^{-2r_1/a_0}$ (b) $1 - $ (a positive number)

(c) 0.99 **43.** \$1,759   **44.** (a) $x - 1 + Ce^{-x}$ (b) $(1/x)(\sin x + C) - \cos x$

Section 9–3 (page 428)

1. $\sin x - (1/3)\sin^3 x + C$ **2.** $(1/4)\tan 4x - x + C$

3. $(1/5)\cos^5 x - (1/3)\cos^3 x + C$ **4.** $(1/9)\tan^3 3x + (1/3)\tan 3x + C$

5. $(1/8)\tan^8 x + (1/6)\tan^6 x + C$ **6.** $(5/16)x - (1/4)\sin 2x + (3/64)\sin 4x$
$+ (1/48)\sin^3 2x + C$ **7.** $(1/16)x - (1/64)\sin 4x + (1/48)\sin^3 2x + C$

8. $-(1/10)\cos 5x - (1/2)\cos 2x + C$ **9.** $(1/7)\sec^7 x - (2/5)\sec^5 x + (1/3)\sec^3 x + C$

10. $(1/12)\tan^4 3x - (1/6)\tan^2 3x - (1/3)\ln|\cos 3x| + C$

11. $-(1/2)\ln|1 + 2\cos x| + C$ **12.** $\tan x + C$ **13.** $(1/6)\tan^2 3x + C$

15. $-\sin x - \csc x + C$ **17.** $(1/4)x^4\ln 3x - (1/16)x^4 + C$ **19.** $(1/2)\tan^{-1}(x/2) + C$

21. $(1/4)\sec^3 x \tan x + (3/8)\sec x \tan x + (3/8)\ln|\sec x + \tan x| + C$

23. $(2/5)\sin^{5/2}x + C$ **25.** $e^{2x}(\sin 4x - 2\cos 4x)/10 + C$ **27.** $-3x^{2/3}\cos\sqrt[3]{x}$
$+ 6x^{1/3}\sin\sqrt[3]{x} + 6\cos\sqrt[3]{x} + C$ **28.** $(x+1)\ln(x+1) - (x+1) + C$ **29.** 0

30. π **31.** $\dfrac{\sin(a-b)x}{2(a-b)} - \dfrac{\sin(a+b)x}{2(a+b)} + C$; $\dfrac{\sin(a-b)x}{2(a-b)} + \dfrac{\sin(a+b)x}{2(a+b)} + C$

32. $\ln(2 + \sqrt{3})$ **33.** $a = -2\cos t \sin t = -\sin 2t$ ft/sec^2; $s = (1/2)t + (1/4)\sin 2t$
$+ C$ ft   **34.** $3/\sqrt{2}$ amps   **35.** (a) $\bar{x} = 1/(e-1), \bar{y} = (e+1)/4$

36. $(1/4)x^2 - (x/4)\sin 2x - (1/8)\cos 2x + C$ **37.** $(\pi/2, 3/8)$

Section 9–4 (page 434)

1. $(x/2)\sqrt{x^2-4} + 2\ln|x + \sqrt{x^2-4}| + C$ **2.** $(x/2)\sqrt{x^2+4} - 2\ln|x$
$+ \sqrt{x^2+4}| + C$ **3.** $(1/16)\tan^{-1}(x/2) + x/8(x^2+4) + C$

4. $\sqrt{5} + (1/2)\ln(2 + \sqrt{5})$ **5.** $\pi/6$ **6.** $\ln|x + \sqrt{x^2-3}| - (\sqrt{x^2-3})/x + C$

7. $-x/5\sqrt{x^2-5} + C$ **8.** $-(\sqrt{2+x^2}/2x) + C$ **9.** $-(\sqrt{2+x^2}/x)$
$+ \ln|x + \sqrt{2+x^2}| + C$ **10.** $(x/8)(25 - 2x^2)\sqrt{5-x^2} + (75/8)\sin^{-1}(x/\sqrt{5}) + C$

11. $(x/2)\sqrt{1-9x^2} + (1/6)\sin^{-1}3x + C$

12. $\left(\dfrac{x+1}{2}\right)\sqrt{4-(x+1)^2} + 2\sin^{-1}\left(\dfrac{x+1}{2}\right) + C$ **13.** $x\sec^{-1}\sqrt{x} - \sqrt{x-1} + C$

15. $\pi/108$ **17.** $\left(\dfrac{1}{2\sqrt{3}}\right)\ln\left|\dfrac{x+\sqrt{3}}{x-\sqrt{3}}\right| + C$ **19.** $(-\sqrt{1-x^2}/x) + C$

22. $\sqrt{5} + (1/2)\ln(2 + \sqrt{5})$ **23.** $(\pi/4)[3\sqrt{2} - \ln(1 + \sqrt{2})]$
24. $(\delta/8)[3\sqrt{2} - \ln(1 + \sqrt{2})]$ **25.** $\sqrt{2} - \ln(1 + \sqrt{2})$
26. $\bar{x} = [\ln(13/4)]/\tan^{-1}(3/2), \bar{y} = (1/16) + [3/104 \tan^{-1}(3/2)]$

Section 9–5 (page 440)

1. $5\ln|x + 2| - 5\ln|x + 3| + C$ **2.** $2\ln|x - 1| + \ln|x - 2| - 3\ln|x - 3| + C$
3. $(1/2)x^2 - 2x + (1/4)\ln|x - 1| + (27/4)\ln|x + 3| + C$ **4.** $\ln|x - 2|$
$- 2(x - 2)^{-1} + C$ **5.** $(1/4)x^4 + (4/3)x^3 + 6x^2 + 32x - 32(x - 2)^{-1}$
$+ 80\ln|x - 2| + C$ **6.** $(1/4)\ln|x^2 - 1| - (1/4)\ln|x^2 + 1| + C$
7. $\ln(x^2 + 1) - 4(x^2 + 1)^{-1} + C$ **8.** $\ln|x - 1| - (1/2)\ln(x^2 + 4)$
$- (1/2)\tan^{-1}(x/2) + C$ **9.** $\ln|x - 1| - (1/2)(x^2 + 4)^{-1} + (1/8)x(x^2 + 4)^{-1}$
$+ (1/16)\tan^{-1}(x/2) + C$ **10.** $2\ln|x - 4| + 2\ln|x + 1| - (3/2)(x + 1)^{-2} + C$
11. $\pi^2/32$ **12.** $\sin^{-1}(x/2) + C$ **13.** $x\cot^{-1}x + (1/2)\ln(1 + x^2) + C$
15. $2(\sin 3 - \sin 2)$ **17.** $(3/8)\tan^{-1}x + (1/2)x(1 + x^2)^{-1}$
$+ (1/8)x(1 - x^2)(1 + x^2)^{-2} + C$ **19.** $\ln|x + 1| - (1/2)\ln(x^2 + 1) - \tan^{-1}x + C$
21. $(\pi/12) + (\sqrt{3}/2) - 1$ **23.** $-(1/2)\ln|y - 1| + (1/2)\ln|y - 3| = x + C$
25. $y = A\cos(\sqrt{2}x + C)$ **27.** $y = S/(1 + Ce^{-8kx})$ **28.** $Ce^{(dy+bx)} = x^ay^c$
29. $y = (1/4)x^4 + (2/3)x^3 - (3/2)x^2 + C$ **31.** $y = 200/(1 + 3e^{-0.3t})$

Section 9–6 (page 445)

1. $\frac{1}{2}\tan^{-1}[(x + 1)/2] + C$ **2.** $\sin^{-1}[(x + 2)/\sqrt{5}] + C$ **3.** $\tan(x/2) + C$

4. $\frac{1}{\sqrt{3}}\ln\left|\frac{\tan(x/2) + \sqrt{3}}{\tan(x/2) - \sqrt{3}}\right| + C$ **5.** $2\sqrt{x} - 2\tan^{-1}\sqrt{x} + C$

6. $-\frac{4}{27}(2 - 3x)^{3/2} + \frac{2}{45}(2 - 3x)^{5/2} + C$ **7.** $\frac{2}{3}(x + 4)^{3/2} - 8(x + 4)^{1/2} + C$

8. $-\frac{1}{4}(2x + 3)^{-1} + \frac{5}{8}(2x + 3)^{-2} + C$ **9.** $2\sqrt{1 + e^x} + \ln\left|\frac{\sqrt{1 + e^x} - 1}{\sqrt{1 + e^x} + 1}\right| + C$

10. $-\frac{1}{3}x + \frac{5}{6}\tan^{-1}\left(2\tan\frac{x}{2}\right) + C$ **11.** $\frac{1}{5}\cos^5x - \frac{1}{3}\cos^3x + C$

13. $-2\left(1 + \tan\frac{x}{2}\right)^{-1} - \ln|1 + \sin x| + C$ **15.** $\frac{1}{3}x^3\ln x - \frac{1}{9}x^3 + C$

17. $2e^{\sqrt{x}}(\sqrt{x} - 1) + C$ **19.** $\frac{x}{8} - \frac{1}{32}\sin 4x + C$ **21.** $\sqrt{x^2 + 2x + 2} + C$

23. $\sqrt{x^2 + 2x + 2} - \ln|(x + 1) + \sqrt{x^2 + 2x + 2}| + C$

25. $\frac{3}{4}\ln|x - 3| + \frac{1}{4}\ln|x + 1| + C$ **27.** $2x\cos x - 2\sin x + x^2\sin x + C$

29. $\frac{1}{5}(1 + x^2)^{5/2} - \frac{1}{3}(1 + x^2)^{3/2} + C$

Review Exercises for Chapter 9 (page 449)

1. $\tan x(\ln \tan x - 1) + C$ **3.** $(3/2)(x\ln x - x) + C$
5. (Reduction formula)$(1/6)\cos^5x \sin x + (5/24)\cos^3x \sin x + (5/32)\sin 2x$
$+ (5x/16) + C$ **7.** $-(1/6)\cos 3x - (1/14)\cos 7x + C$ **9.** $x\sqrt{4 + x^2}$
$+ 4\ln(x + \sqrt{4 + x^2}) + C$ **11.** $(1/8)(\sin x)(1 + 2\sin^2x)\sqrt{1 + \sin^2x}$

$- (1/8)\ln|\sin x + \sqrt{1 + \sin^2 x}| + C$ **13.** $\ln|x - 3| - \ln|x + 1| + C$ **15.** $2\ln|x - 1|$
$- 2(x - 1)^{-1} + C$ **17.** $(1/2)\tan^{-1}[(x + 2)/2] + C$

19. $(1/\sqrt{3})\ln\left|\dfrac{\tan(x/2) + 2 - \sqrt{3}}{\tan(x/2) + 2 + \sqrt{3}}\right| + C$

21. $-(x/8)(8x^2 - 125)\sqrt{25 - 4x^2} + (1{,}875/16)\sin^{-1}(2x/5) + C$

23. $-(x/8)\sqrt{25 - 4x^2} + (25/16)\sin^{-1}(2x/5) + C$ **25.** $\dfrac{1}{\ln 3}x3^x - \dfrac{1}{(\ln 3)^2}3^x + C$

27. $(x^2/2)\sin 2x - (1/4)\sin 2x + (x/2)\cos 2x + C$ **29.** $(1/8)\sin 4x$
$+ (1/20)\sin 10x + C$ **31.** $(1/2)\tan 2x + (1/6)\tan^3 2x + C$
33. $x - (8/\sqrt{15})\tan^{-1}[(4\tan(x/2) + 1)/\sqrt{15}] + C$ **35.** $(1/8)\ln|4x - 3|$
$+ 11/8(4x - 3) + C$ **37.** $(2/3)\ln|x + 1| + 3\ln|x + 3| - (11/3)\ln|x + 4| + C$
39. $\ln|x - 2| - (1/2)\ln|x^2 + 1| - 2\tan^{-1}x + C$ **41.** $(2/3)x^3 + 4x$
$- (8/\sqrt{2})\tan^{-1}(x/\sqrt{2}) + C$ **43.** $(x/2)\sqrt{4x^2 - 9} - (9/4)\ln|2x + \sqrt{4x^2 - 9}| + C$
45. $(x/8)(2x^2 - 9)\sqrt{x^2 - 9} - (81/8)\ln|x + \sqrt{x^2 - 9}| + C$ **47.** $(4/9)(1 + x)^{9/4}$
$- (4/5)(1 + x)^{5/4} + C$ **49.** $-(1/2)\cos 2x + (1/2)\cos^3 2x - (3/10)\cos^5 2x$
$+ (1/14)\cos^7 2x + C$ **51.** $(x^2/2)\sin^{-1}x - (1/4)\sin^{-1}x + (x/4)\sqrt{1 - x^2} + C$

53. $(x/25\sqrt{x^2 + 25}) + C$ **55.** $2\ln|x - 1| - \ln|x| - \dfrac{x}{(x - 1)^2} + C$

57. $(1/225)[10x\sin 5x - (25x^2 - 2)\cos 5x] + C$ **59.** $(1/4)\sin^4 x - (1/6)\sin^6 x + C$
61. $\ln|\sec e^x + \tan e^x| + C$ **63.** $\tan e^x + C$ **65.** $(2/3)(1 + e^x)^{3/2} + C$
67. $2\tan^{-1}\sqrt{x} + C$ **69.** $(-1/4)x^2 e^{-4x} - (1/8)xe^{-4x} - (1/32)e^{-4x} + C$
71. $e^{\tan x} + C$ **73.** $(1/x)\sqrt{9 - 4x^2} - 2\sin^{-1}(2x/3) + C$ **75.** $(x^3/3) + 2x^2$
$- 5x + C$ **77.** $-2\sqrt{1 + \cos x} + C$ **79.** $x\tan^{-1}5x - (1/10)\ln(1 + 25x^2) + C$
81. $(1/13)e^{2x}(2\sin 3x - 3\cos 3x) + C$ **83.** $2\sqrt{x}\sin\sqrt{x} + 2\cos\sqrt{x} + C$
85. $(1/2)(x^2 - 1)e^{x^2} + C$ **87.** $-x + \sqrt{2}\tan^{-1}(\sqrt{2}\tan x) + C$ **89.** $\ln|x - 1|$
$- (x - 1)^{-1} + C$ **91.** $(-\sqrt{4 - x^2}/4x) + C$ **93.** $2\sin\sqrt{x} + C$
95. $(1/2)(\sin^{-1}x)^2 + C$ **97.** $e^x(\cos 2x + 2\sin 2x) + C$ **99.** $(x - 1)\ln\sqrt{x - 1}$
$- (x/2) + C$ **101.** $5\pi^2/32$ **103.** $y = (1/2)x^2 - (1/2)x + (1/4) + (3/4)e^{-2x}$
105. $\bar{x} = (2e^2 + 6)/(e^2 + 1), \bar{y} = (3e^4 + 1)/8(e^2 + 1)$ **107.** $100/(1 + e^{-7})$

CHAPTER 10

Section 10–1 (page 460)

1. $-\infty, \infty$ **3.** $\infty, -\infty$ **5.** $-\infty, -\infty$ **7.** $\infty, -\infty$ **9.** $\infty, -\infty$ **11.** $1/2, 1/2$
13. $0, 0$ **15.** $\infty, -\infty$ **17.** $-\infty, \infty$ **19.** $1/2, 1/2$ **21.** 0 **22.** ∞

23.

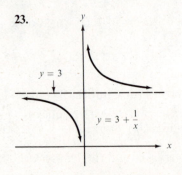

25.

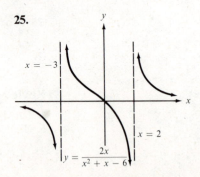

27.

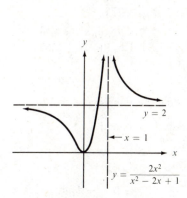

$$y = \frac{2x^2}{x^2 - 2x + 1}$$

29.

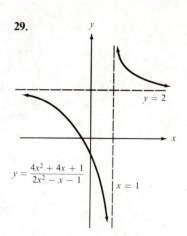

$$y = \frac{4x^2 + 4x + 1}{2x^2 - x - 1}$$

Section 10–2 (page 468)

1. 1/3   **3.** 1/2   **5.** 1   **7.** ln(2/3)   **9.** 0   **11.** 1/2   **13.** 0   **15.** e **17.** e^3 **19.** $-\infty$
21. 0   **23.** −1   **25.** 1   **27.** 1   **29.** 0   **31.** 1/2   **33.** e **35.** 1/3   **37.** ∞   **39.** ∞
41. 1/2   **43.** 1/4   **45.** 1/3   **47.** 1/3   **49.** e^{-6}

Section 10–3 (page 474)

1. Diverges   **3.** $\pi/2$ **5.** Diverges   **7.** 3   **9.** −1/4   **11.** 1/4   **13.** 1/2
15. Diverges   **17.** Diverges   **19.** Diverges   **21.** Diverges   **23.** Diverges
25. −ln(1/2)   **27.** 1/2   **29.** Converges to $1/(r - 1)$ if $r > 1$ and diverges for $r \le 1$
30. Converges to $1/(1 - r)$ if $r < 1$ and diverges for $r \ge 1$ **31.** Converges;
$1/(1 + x^3) < 1/x^2$ **33.** Diverges; $1/\ln x > 1/x$

35. Area $= \displaystyle\int_1^\infty (1/x)\,dx = \infty$; Volume $= \displaystyle\int_1^\infty \pi(1/x)^2\,dx = \pi$

36. Area $= \displaystyle\int_0^1 (1/\sqrt{x})\,dx = 2$; Volume $= \displaystyle\int_0^1 \pi(1/x)\,dx = \infty$ **37.** $(\pi/2) - 1$

38. $\pi/2$ **39.** (a) $12,500   (b) $28,125   **41.** (a) 1   (b) $e^{-0.5} \approx 0.61$
42. $3(.61) - 2(.39) = \$1.05$

43. $\displaystyle\int_1^\infty \frac{\cos x}{x^2}\,dx = \int_1^\infty \frac{f(x)}{x^2}\,dx - \int_1^\infty \frac{g(x)}{x^2}\,dx$; both integrals converge by comparing
the integrands to $1/x^2$ **44.** Let $u = 1/x$ and $dv = \sin x\, dx$

Review Exercises for Chapter 10 (page 479)

1. $-\infty, \infty$ **3.** ∞, ∞ **5.** −6, −6   **7.** 0,0   **9.** 2/3, 2/3   **11.** 0   **13.** Vertical,
$x = -7/2$; horizontal, $y = 1$ **15.** Vertical, $x = 2$; horizontal, $y = 0$ **17.** 1
19. −1   **21.** $-\infty$ **23.** 1   **25.** 2   **27.** 0   **29.** ∞   **31.** 1/6   **33.** $1/\ln 2$ **35.** 1/2
37. 1   **39.** 2   **41.** 2/3   **43.** Diverges   **45.** Diverges   **47.** Diverges   **49.** $\ln|1 + \sqrt{2}|$
51. Converges; compare $1/x^2$ **53.** Converges; compare $e^{-x/2}$

CHAPTER 11

Section 11–1 (page 487)

1. 0   **3.** 0   **5.** 0   **7.** Does not exist   **9.** 6   **11.** 1   **13.** 0   **15.** e **17.** $\pi/2$ **19.** 1
21. 0   **23.** e^{-3} **24.** 1, 3, 6, 10   **26.** 1, 3/2, 11/6, 25/12   **28.** 1/2, 2/3, 3/4, 4/5
29. $-\ln 2, -\ln 3, -\ln 4, -\ln 5$ **30.** 1/2, 3/4, 7/8, 15/16   **31.** Nondecreasing and bounded by 3

32. Nondecreasing (the integrand is positive) and bounded by $\displaystyle\int_0^\infty \frac{1}{1+x^2}\, dx = \pi/2$

33. Nondecreasing and bounded by 1   **34.** Nondecreasing and bounded by 1
35. $0.3333\ldots = 1/3$

Section 11–2 (page 494)

1. 3   **2.** 3/2   **3.** $S_n = -\ln(n+1)$; diverges (to $-\infty$)   **4.** $S_n = \sqrt{n} - 1$;
diverges (to ∞)   **5.** 3   **6.** 1/4   **7.** 2/3 $(r = -1/2)$ **8.** Diverges (to ∞)   **9.** 1/12
11. Diverges   **13.** Diverges   **15.** 1   **17.** 5/2   **19.** 5/3   **21.** 27/10
23. $\Sigma(1/10)^n = 10/9$ **24.** $(12/100)\Sigma(1/100)^n = 12/99$
26. Yes; $0.999\ldots = (9/10)\Sigma(1/10)^n = 1$ **27.** 50 meters   **31.** The ball could, at least theoretically, bounce infinitely many times, but all of the bouncing takes place in a *finite amount of time* $t = (2v/g)\Sigma(1/2)^n = 4v/g$.

32. Compare with $\displaystyle\int_1^\infty (1/x^2)\, dx$ (draw a picture)

Section 11–3 (page 501)

1. Test fails   **2.** Diverges   **3.** Diverges   **4.** Test fails   **5.** $> 1/n$; diverges
6. $< 1/n^2$; converges   **7.** $> 1/k$; diverges   **8.** $< 1/n^2$; converges   **9.** Diverges
10. Converges   **11.** Converges   **12.** Converges   **13.** Diverges; nth term
15. Converges; integral or comparison   **17.** Converges; comparison   **19.** Converges; comparison   **21.** Diverges; integral   **23.** Converges; comparison   **25.** Converges; comparison   **27.** Diverges; integral   **29.** Converges; compare with $(n^2/3^n) + (2/3)^n$
31. 0.40   **32.** 0.92   **33.** 1.08   **34.** 0.58   **35.** $(1/10)\Sigma(9/10)^{n-1} = 1$
36. $(9/10)^4 = 0.6561$; this is the probability that at least the first four bulbs will light up.

Section 11–4 (page 507)

1. Converges   **2.** Converges   **3.** Test fails   **4.** Diverges   **5.** Converges
6. Converges   **7.** Converges   **8.** Test fails   **9.** Converges   **10.** Diverges
11. Converges   **12.** Diverges   **13.** Diverges; compare $1/k$ **15.** Converges; compare $1/k^2$ **17.** Converges; root or comparison   **19.** Converges; ratio or compare $1/n^2$ **21.** Converges; ratio   **23.** Converges; limit comparison
25. Diverges; nth term   **27.** Converges; ratio   **29.** Diverges; limit comparison
31. The limit of the ratios are $(2/e) < 1$ and $(3/e) > 1$ **32.** 3/4   **33.** (a) 4   (b) ∞
34. 10   **35.** $20

Section 11–5 (page 513)

1. Abs conv   **3.** Cond conv   **5.** Cond conv   **7.** Abs conv   **9.** Cond conv
11. Diverges   **13.** Abs conv   **15.** Abs conv   **17.** Abs conv   **19.** Cond conv
21. Abs conv   **23.** Abs conv   **25.** Abs conv   **27.** Diverges   **29.** Cond conv
31. 0.63   **32.** 0.82   **33.** 0.22   **34.** 0.84

Section 11–6 (page 518)

<u>**1.**</u> $(-2, 2)$ **3.** $[-1, 1]$ <u>**5.**</u> $[17/9, 19/9)$ **7.** $(-1, 1)$ <u>**9.**</u> $(-\infty, \infty)$

<u>**11.**</u> Converges only at $x = -1$ **13.** $(-1, 1)$ <u>**15.**</u> $\left(-2 - \dfrac{1}{p}, -2 + \dfrac{1}{p}\right)$

17. $(-7, 13)$ **19.** $(-9, 7)$ <u>**21.**</u> $(-1, 3)$; use the nth term test at the endpoints
<u>**23.**</u> $(-1/2, 1/2)$ **25.** $(-\infty, \infty)$ **27.** $(-1/e, 1/e)$ **29.** r in both cases
30. Yes; $\Sigma nx^{n-1} = (1/x)\Sigma nx^n = 1/(1-x)^2$

Section 11–7 (page 528)

1. $x/(1-x)$, x in $(-1, 1)$ **2.** $1/(3-x)$, x in $(-3, 3)$ <u>**3.**</u> $1/(1 + x^2)$, x in $(-1, 1)$
4. $1/(x-2)$, x in $(2, 4)$ **5.** $-1/(1 + x^2)$, x in $(-1, 1)$ **6.** $1/(1 - x^2)$, x in $(-1, 1)$
<u>**7.**</u> $-2x/(1 + x^2)^2$, x in $(-1, 1)$ **8.** $6x/(3 - x^2)^2$, x in $(-\sqrt{3}, \sqrt{3})$ **9.** $-\ln(1 - x)$,
x in $(-1, 1)$ **10.** $x - \ln(1 + x)$, x in $(-1, 1)$ <u>**1.**</u> $\tan^{-1}x$, x in $(-1, 1)$ **12.** $\ln x$,
x in $(0, 2)$ **13.** $1/x$, x in $(0, 4)$ **15.** $1/(5 - x)$, x in $(-5, 5)$ <u>**17.**</u> $2/(5 - x)^3$,
x in $(-5, 5)$ **19.** $16/(1 - 4x)$, x in $(-1/4, 1/4)$ <u>**21.**</u> $1/2$ **23.** $16/9$
25. $\tan^{-1}(0.3) \approx 0.291$ **27.** $1/17$ <u>**29.**</u> 2   <u>**31.**</u> 0.1823 **32.** -0.1054 **33.** 0.0997

34. -0.1974 <u>**35.**</u> $\displaystyle\sum \frac{(-1)^n}{2n + 1}x^{4n+2}$ **36.** $\displaystyle\sum \frac{(-1)^n}{n + 1}x^{2n+2}$ **37.** (a) ∞ (d) $f(x) = e^x$
39. The differentiated series is Σx^{n-1} which diverges at $x = -1$.

Section 11–8 (page 537)

1. $\Sigma(-1)^n x^{2n}$ on $(-1, 1)$ <u>**3.**</u> $\Sigma - x^{n+1}/(n + 1)$ on $(-1, 1)$ <u>**5.**</u> $\Sigma x^{2n}/(2n)!$
on $(-\infty, \infty)$ **7.** $\Sigma(-1)^n x^{n+1}/n!$ on $(-\infty, \infty)$

9. $1 + \displaystyle\sum \frac{(-1)^{n-1}[1 \cdot 3 \cdot 5 \cdots (3 - 2n)]}{2^n n!}x^n$ on $(-1, 1)$

10. $(1/2) + (1/2)\displaystyle\sum \frac{(-1)^n 2^{2n}}{(2n)!}x^{2n}$ on $(-\infty, \infty)$ **11.** $\displaystyle\sum \frac{(-1)^n 5^{2n}}{(2n)!}x^{2n}$ on $(-\infty, \infty)$

13. $\Sigma(-1)^{n+1}(x - 1)^n/n$ **15.** $e^3\Sigma(-1)^n(x + 3)^n/n!$ **16.** $10\Sigma[(x - 1)\ln 10]^n/n!$
<u>**17.**</u> $\Sigma(-1)^n(x - 2)^n/2^{n+1}$ **18.** $\Sigma(-1)^n(x - \pi/4)^n/\sqrt{2}(2n)!$

<u>**19.**</u> $\dfrac{1}{2} + \dfrac{\sqrt{3}}{2}\left(x - \dfrac{\pi}{6}\right) - \dfrac{1}{4}\left(x - \dfrac{\pi}{6}\right)^2 - \dfrac{\sqrt{3}}{12}\left(x - \dfrac{\pi}{6}\right)^3$

21. $1 + 2\left(x - \dfrac{\pi}{4}\right) + 2\left(x - \dfrac{\pi}{4}\right)^2 + \dfrac{8}{3}\left(x - \dfrac{\pi}{4}\right)^3$ **23.** $-x^2/2$

<u>**25.**</u> $1 + \dfrac{1}{3}x - \dfrac{1}{9}x^2 + \dfrac{5}{81}x^3$ **27.** 1.6487 <u>**29.**</u> 0.3103

31. (a) $1 + \dfrac{1}{2}(x - 1) - \dfrac{1}{8}(x - 1)^2 + \dfrac{1}{16}(x - 1)^3$

(b) $\sqrt{2} + \dfrac{1}{2\sqrt{2}}(x - 2) - \dfrac{1}{16\sqrt{2}}(x - 2)^2 + \dfrac{1}{64\sqrt{2}}(x - 2)^3$

(c) $2 + \dfrac{1}{4}(x - 4) - \dfrac{1}{64}(x - 4)^2 + \dfrac{1}{512}(x - 4)^3$ **33.** 0.78540 **34.** 0.98279

Review Exercises for Chapter 11 (page 542)

1. 0   **3.** e^3 **5.** ∞ **7.** $7/4$ **9.** Diverges   **11.** $7/5$ **13.** Converges to 0
15. Diverges   **17.** Converges   **19.** Converges   **21.** Converges   **23.** Diverges

25. Abs conv **27.** Abs conv **29.** Cond conv **31.** $(-2, 2)$ **33.** $[1/2, 3/2)$
35. $[-21/10, -19/10)$ **37.** $x/(1 - x^2)$ **39.** $4x^3/(1 - x^4)^2$ **41.** $2/(1 - x)^2$
43. $\Sigma(-1)^{n+1}(2x)^{2n}/(2n)!$ **45.** $\Sigma(x \ln 2)^n/n!$ **47.** $\Sigma(1 - 2x)^n/en!$ **49.** $127/999$
51. 0.57 **53.** 0.7904 **55.** 0.905

CHAPTER 12

Section 12–2 (page 552)

1.

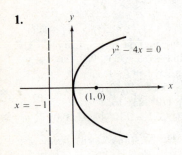

3.

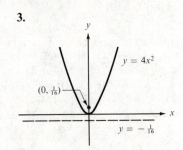

5.

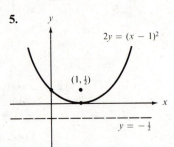

7. $(0, -2), (-1/16, -2), x = 1/16$, opens left **9.** $(1, 1), (1, 0), y = 2$, opens down
11. $(-2, -2), (-9/4, -2), x = 7/4$, opens left **13.** $x^2 = 12y$
15. $(x + 1)^2 = 16(y - 4)$ **17.** $7(x + 1)^2 = y + 3$ **19.** $(x - 6)^2 = -8(y - 1)$

21. $2(x - 1)^2 = y - 1$ **23.** $8/3, 2\pi$ **25.** $\sqrt{5} + \dfrac{1}{2}\ln(2 + \sqrt{5})$

26. $y - y_1 = -c(x - x_1)/(2ay_1 + b)$ **28.** $x^2 = 225y$, 374 meters **29.** $y = (\tan \theta)x$
$- [16x^2/(v_0\cos \theta)^2]$ **30.** $\pi/4$ **31.** $\alpha_1 = \tan^{-1}[(y_1{}^2 - 2px_1 + 2p^2)/y_1(x_1 + p)]$
34. $4.6° = 4°36'$ **36.** $y = 6x - 9, y = -2x - 1$

Section 12–3 (page 558)

1. Circle; center $(0, 0)$, radius 2 **3.** Parabola; vertex $(0, 1)$, opens down

5.

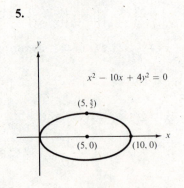

7.

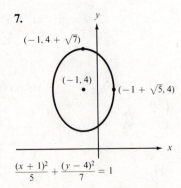

9. Circle; center $(-1, 4)$, radius $\sqrt{5}$ **11.** Ellipse; vertices $(2 \pm 5/2, 1)$ and
$(2, 1 \pm 5/3)$ **13.** Circle; center $(-5/2, 3/2)$, radius $\sqrt{34}/2$ **15.** Ellipse; vertices
$(-2 \pm 2, -1)$ and $(-2, -1 \pm \sqrt{2})$ **17.** $4(x - 2)^2 + (y - 2)^2 = 4$
19. $32(x + 1)^2 + 36(y - 2)^2 = 1{,}152$ **21.** $5(x - 2)^2 + 9(y - 3)^2 = 45$
23. $4(x + 3)^2 + 16(y + 2)^2 = 64$ **25.** $4(x - 2)^2 + 3(y - 1)^2 = 12$ **27.** πab
28. $(4/3)\pi ab^2$ **29.** $(1/3)\pi abh$ **32.** $(-2, -1.5)$ and $(2.8, 0.9)$; $(0.8, \pm 1.8)$ and
$(-0.8, \pm 1.8)$ **34.** These are the ellipses in the second part of Exercise 32. The
particles will *not* collide. **35.** The square of the distance is $(x - c)^2 + b^2(a^2 - x^2)/a^2$
for $-a \le x \le a$. The derivative is never zero on $[-a, a]$ so the extreme points occur
at the endpoints; max at $(-a, 0)$ and min at $(a, 0)$. **36.** 9.07×10^7 mi, 9.45×10^7 mi
37. 0.96 **39.** (c) $32x^2 + 36y^2 = 1{,}152$

Section 12–4 (page 564)

1.

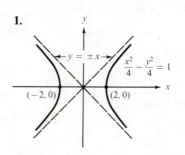

3.

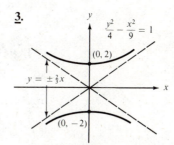

5. Ellipse **7.** Hyperbola; center $(-1, 2)$, vertices $(-1 \pm 4, 2)$ **9.** Parabola
11. Ellipse **13.** Hyperbola; center $(2, -2)$, vertices $(2 \pm \sqrt{2}, -2)$ **15.** Hyperbola;
center $(2, 1)$, vertices $(2, 1 \pm 1)$ **17.** $15x^2 - y^2 = 15$ **19.** $5y^2 - 4(x - 1)^2 = 20$
21. $36(x + 1)^2 - 4(y - 2)^2 = 144$ **23.** $5(y - 3)^2 - 20x^2 = 64$
25. $4[\sqrt{2} - \ln(\sqrt{2} + 1)]$ **26.** $8\pi(2 - \sqrt{2})/3$ **27.** Place A at $(-25, 0)$ and B at
$(25, 0)$. The source lies on the left branch of the hyperbola with center $(0, 0)$, foci at
$(\pm 25, 0)$, and vertices at $(\pm 15, 0)$. **28.** About $(-81, 106)$ **29.** $b^2(x - h)^2$
$- a^2(y - k)^2 = a^2 b^2, x \ge h + a$ **32.** (c) $32x^2 - 4y^2 = 128$

Section 12–5 (page 570)

1.

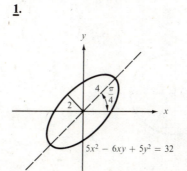

3.

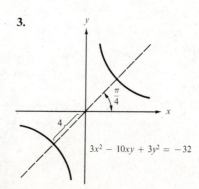

5. No graph   **7.** Parabola   **9.** Ellipse; $4x'^2 + 9y'^2 = 36$ **11.** Hyperbola;
$x'^2 - y'^2 = 10$ **13.** Two intersecting lines   **15.** Parabola   **17.** Hyperbola
19. Ellipse   **21.** Hyperbola

Review Exercises for Chapter 12 (page 571)

1. Ellipse   **3.** Parabola   **5.** Circle   **7.** Hyperbola   **9.** Hyperbola   **11.** Ellipse
13. Parabola   **15.** Parabola   **17.** Hyperbola   **19.** $(y - 2)^2 = 8(x - 1)$
21. $2(x + 1)^2 = y + 5$ **23.** $(x - 1)^2 + 4(y - 4)^2 = 16$ **25.** $(x + 2)^2$
$+ 9(y - 3)^2 = 9$ **27.** $21x^2 - 4y^2 = 84$ **29.** $9x^2 - (y - 2)^2 = 9$ **31.** Reflective
surface A is the right branch of a hyperbola with a horizontal transverse axis.
Surface B is a parabola placed with the indicated opening at the focus of the left
branch of the hyperbola and shaped so that its focus coincides with the focus of the
right branch of the hyperbola.

CHAPTER 13

Section 13–1 (page 580)

1.

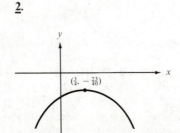

$y = x^2 + 4x$

$(-2, -4)$

2.

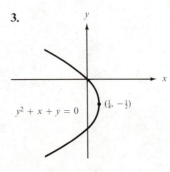

$(\frac{3}{4}, -\frac{23}{40})$

$2x^2 + 5y - 3x + 4 = 0$

3.

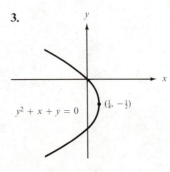

$y^2 + x + y = 0$

$(\frac{1}{4}, -\frac{1}{2})$

4.

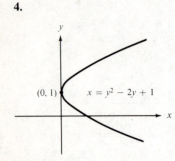

$(0, 1)$ $x = y^2 - 2y + 1$

5.

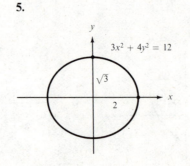

$3x^2 + 4y^2 = 12$

$\sqrt{3}$

2

6.

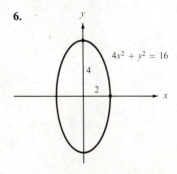

$4x^2 + y^2 = 16$

4

2

7.

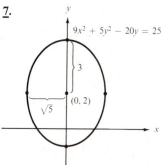

$9x^2 + 5y^2 - 20y = 25$

3

$(0, 2)$

$\sqrt{5}$

8.

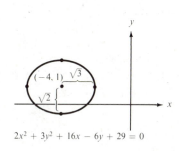

$(-4, 1)$ $\sqrt{3}$

$\sqrt{2}$

$2x^2 + 3y^2 + 16x - 6y + 29 = 0$

9.

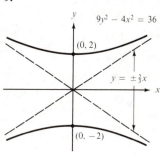

$9y^2 - 4x^2 = 36$

$(0, 2)$

$y = \pm\frac{2}{3}x$

$(0, -2)$

10.

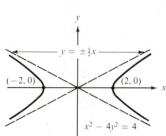

$y = \pm\frac{1}{2}x$

$(-2, 0)$ $(2, 0)$

$x^2 - 4y^2 = 4$

11.

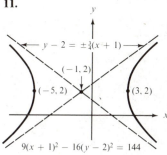

$y - 2 = \pm\frac{3}{4}(x + 1)$

$(-1, 2)$

$(-5, 2)$ $(3, 2)$

$9(x + 1)^2 - 16(y - 2)^2 = 144$

12.

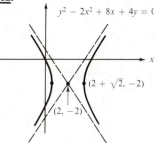

$y^2 - 2x^2 + 8x + 4y = 0$

$(2 + \sqrt{2}, -2)$

$(2, -2)$

13. Ellipse   **14.** Hyperbola   **15.** Parabola   **16.** Hyperbola   **17.** $x^2 = 12y$
18. $7(x + 1)^2 = y + 3$ **19.** $4(x - 2)^2 + (y - 2)^2 = 4$ **20.** $4(x - 2)^2$
$+ 3(y - 1)^2 = 12$ **21.** $15x^2 - y^2 = 15$ **22.** $y^2 - 7x^2 = 9$

23. $\sqrt{5} + \dfrac{1}{2}\ln(2 + \sqrt{5})$ **24.** $8\pi(2 - \sqrt{2})/3$ **25.** πab **26.** $\pi abh/3$

27. If A and B are at $(-25, 0)$ and $(25, 0)$, the source is on the left branch of the
hyperbola with foci at $(\pm 25, 0)$ and vertices at $(\pm 15, 0)$.   **28.** Ellipse;
counterclockwise   **29.** The branch of hyperbola $b^2(x - h)^2 - a^2(y - k)^2 = a^2b^2$
with $x \geq h + a$ (because $\cosh t \geq 1$)   **30.** Closest $(a, 0)$; farthest $(-a, 0)$
31. $9.07 \times 10^7, 9.45 \times 10^7$ miles   **32.** Parabola   **37.** Reflective surface A is the
right branch of a hyperbola with a horizontal transverse axis. Surface B is a parabola
placed with the indicated opening at the focus of the left branch of the hyperbola
and shaped so that its focus coincides with the focus of the right branch of the
hyperbola.

Section 13–2 (page 587)

1. $r = 3$; circle   **3.** $r^2 = \csc 2\theta$, hyperbola   **5.** $r(a \cos \theta + b \sin \theta) = c$, line
7. $x^2 + y^2 = x$; circle   **9.** $y = x \tan c$, line   **11.** $y^2 = 1 - 2x$, parabola
13. $4x^2 + 3y^2 - 4y = 4$, ellipse

15.

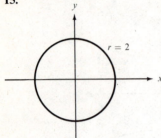

$r = 2$

17.

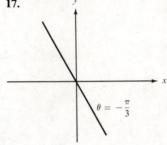

$\theta = -\dfrac{\pi}{3}$

19.

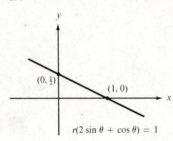

$(0, \tfrac{1}{2})$ $(1, 0)$

$r(2 \sin \theta + \cos \theta) = 1$

21.

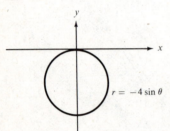

$r = -4 \sin \theta$

23.

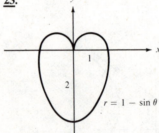

$r = 1 - \sin \theta$

25.

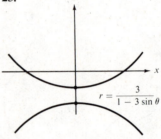

$r = \dfrac{3}{1 - 3 \sin \theta}$

27.

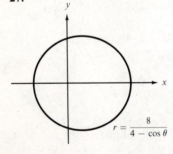

$r = \dfrac{8}{4 - \cos \theta}$

29.

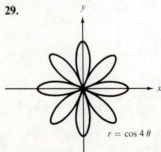

$r = \cos 4\theta$

31.

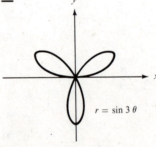

$r = \sin 3\theta$

33.

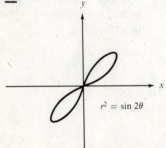

$r^2 = \sin 2\theta$

35.

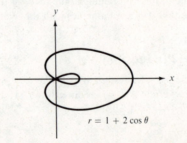

$r = 1 + 2 \cos \theta$

37.

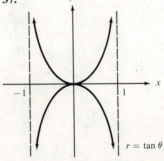

-1 1

$r = \tan \theta$

39.

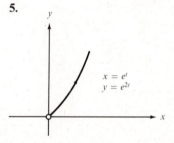

$r\theta = 1$

45. $(\sqrt{2}, \pi/4), (0, 0)$ **46.** $(0, 0), (1/2, 7\pi/6), (1/2, 11\pi/6)$ **47.** (a) Collide at $(\sqrt{2}, \pi/4)$, but not at $0, 0,$ (b) Collide at $(1/2, 7\pi/6)$ and $(1/2, 11\pi/6)$, but not at $(0, 0)$.

Section 13–3 (page 591)

1. $-\sqrt{3}/3$ **2.** $\sqrt{3}$ **3.** 0 **5.** ± 1 **7.** 0 and ∞ (vertical tangent)
9. $(a\cos a - \sin a)/(-a\sin a - \cos a)$ **11.** (a) $(0, \pi), (3a/2, \pm\pi/3)$ (b) $(2a, 0)$, $(a/2, \pm 2\pi/3)$ **12.** $0; y = r\sin\theta = (\sin\theta)/\theta \to 1$ as $\theta \to 0$ **13.** $\pi a^2/4$ **15.** $\pi a^2/2$
17. $3\pi a^2/2$ **19.** $(3\pi/16) + (3/8)$ **21.** $\pi/2$ **23.** $a^2(\sqrt{3} - (\pi/3))$ **24.** πa^2
25. $4(\sqrt{3} - (\pi/3))$ **27.** $a^2((\pi/2) - 1)$ **29.** $(5a/6, 0)$

Section 13–4 (page 597)

1.

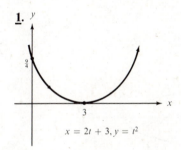

$x = 2t + 3, y = t^2$

3.

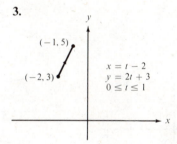

$(-1, 5)$
$(-2, 3)$
$x = t - 2$
$y = 2t + 3$
$0 \le t \le 1$

5.

$x = e^t$
$y = e^{2t}$

7.

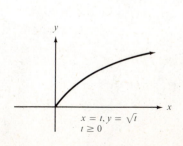

$x = t, y = \sqrt{t}$
$t \ge 0$

9.

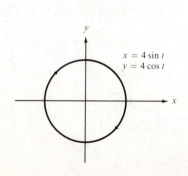

$x = 4\sin t$
$y = 4\cos t$

11.

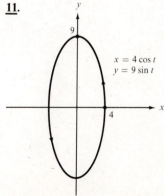

$x = 4\cos t$
$y = 9\sin t$

13.

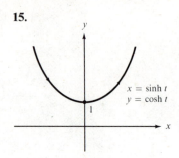

$x = \sec t$
$y = \tan t$
$-\frac{\pi}{2} < t < \frac{\pi}{2}$

15.

$x = \sinh t$
$y = \cosh t$

17. 3   **19.** No slope (vertical)   **21.** 0 when $t = 0$, 8 when $t = 2$

23.

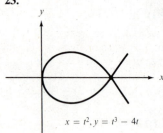

$x = t^2, y = t^3 - 4t$

24.

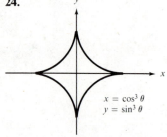

$x = \cos^3 \theta$
$y = \sin^3 \theta$

25.

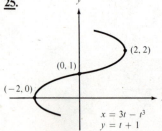

$(2, 2)$
$(0, 1)$
$(-2, 0)$
$x = 3t - t^3$
$y = t + 1$

27.

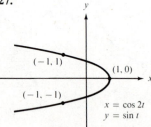

$(-1, 1)$
$(1, 0)$
$(-1, -1)$
$x = \cos 2t$
$y = \sin t$

30. Ellipse, $(2, 0)$, counterclockwise, $x'(\pi/6) = -1, y'(\pi/6) = \sqrt{3}/2$ units/sec
31. Line segment from $(-2, 2)$ to $(2, 2)$, $(\pm 2/\sqrt{5}, \pm 2/\sqrt{5})$; no

32.

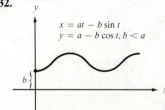

$x = at - b \sin t$
$y = a - b \cos t, b < a$

33.

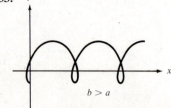

$b > a$

Section 13–5 (page 603)

1. $(13\sqrt{13} - 8)/27$ **3.** 2π **5.** 2π **7.** 8 **9.** $\sqrt{2}(e^{\pi/2} - 1)$ **11.** $[-(\sqrt{5}/2)$
$+ \ln(2 + \sqrt{5})] - [-\sqrt{2} + \ln(1 + \sqrt{2})]$ **13.** $(\sqrt{2}/2) + (1/2)\ln(1 + \sqrt{2})$
15. $\pi(17^{3/2} - 1)/24$ **17.** $\pi[(9\sqrt{5}/16) - (1/32)(\ln(1 + \sqrt{5}/4) - \ln(1/2))]$
19. $8\pi(17^{3/2} - 1)/3$ **21.** $4\pi^2$ **23.** $2\sqrt{2}\pi(2e^\pi + 1)/5$ **25.** $\pi[2\sqrt{5} - \sqrt{2}$
$+ \ln(2 + \sqrt{5}) - \ln(1 + \sqrt{2})]$ **27.** (b) 3.9663669 **29.** 12 ft (the length of one
arch of the cycloid) **30.** (a) $(13\sqrt{13} - 8)/27$ (b) $s = (1/27)[(4 + 9t^2)^{3/2} - 8]$
(c) $4\sqrt{10}$

34. $e\sqrt{(1/4) + e^2} - (\sqrt{5}/4) + (1/4)\ln(e + \sqrt{(1/4) + e^2}) - (1/4)\ln\left(1 + \dfrac{\sqrt{5}}{2}\right)$

Review Exercises for Chapter 13 (page 607)

1. Ellipse **3.** Parabola **5.** Circle **7.** Hyperbola **9.** Ellipse **11.** Hyperbola
13. Hyperbola **15.** $(y - 2)^2 = 8(x - 1)$ **17.** $(x - 1)^2 + 4(y - 4)^2 = 16$
19. $21x^2 - 4y^2 = 84$ **21.** $r^2 - 2r(2 \cos \theta + \sin \theta) = 4$ **23.** $(3 + 2 \cos^2\theta)r^2$
$+ (6 \sin \theta - 10 \cos \theta)r = 7$ **25.** $(x^2 + y^2)^{3/2} = 2y$ **27.** $y^2 = 4(x + 1)$
29. Cardioid point down **31.** Eight pedal curve $(2 \sin 2\theta \cos 2\theta = \sin 4\theta)$
33. $-\sqrt{3}$ **35.** -1 **37.** $1/12\pi$ **39.** $2a^2[\sqrt{3} - (\pi/2)]$ **41.** The part of the parabola
$y^2 = (x + 3)/2$ above the x-axis **43.** That part of the right branch of the hyperbola
$2(x - 2)^2 - (y + 1)^2 = 2$ lying above the line $y = -1$ **45.** e^2 **47.** ± 1
49. $(10^{3/2} - 1)/3$ **51.** $(5^{3/2} - 8)/3$ **53.** $(\pi/8)[57\sqrt{10} - \ln(3 + \sqrt{10})]$ **55.** $\pi^2/2$

CHAPTER 14

Section 14–1 (page 612)

9. $\sqrt{26}$ **11.** 3 **13.** Sides have lengths $\sqrt{2}$, $\sqrt{33}$, and $\sqrt{35}$ **15.** $R(-1, 4, 1)$;
$S(3, 4, 1)$ **17.** $(x + 1)^2 + (y - 2)^2 + z^2 = 4$ **19.** $x^2 + y^2 + z^2 = 3$
21. $(x - 7)^2 + (y + 1)^2 + (z - 4)^2 = 16$ **23.** $(x - 2)^2 + y^2 + (z - 5)^2 = 6$
25. Origin, 2 **27.** $(-1, 1, -2)$, $\sqrt{6}$ **29.** $(1/2, -1, -3/2)$, $\sqrt{14}/2$ **31.** Through
$y = 3$ and parallel to xz-plane **33.** Through $z = -1$ and parallel to xy-plane
35. Through line $x - z = 1$ and parallel to y-axis **37.** Through line $2y + z = 2$
and parallel to x-axis **39.** Through points $(1, 0, 0)$, $(0, -3, 0)$, and $(0, 0, 3)$
41. Through points $(0, 0, 0)$, $(1, -1, 0)$, and $(0, 1, -1)$ **43.** All three coordinate
planes **44.** Interior of the unit sphere **45.** A circular cylinder of radius 1 with
the z-axis as its center **46.** The surface of the unit cube **47.** An elliptic cylinder
with the z-axis as its center **48.** All points on either the xz-plane or the xy-plane
49. Plane through the z-axis and the line $x = y$ in the xy-plane **50.** Points outside
the unit sphere, but on or inside the sphere with center $(0, 0, 0)$ and radius 2

Section 14–2 (page 622)

1. $(-5, -6, -5)$ **3.** $(0, -23, 16)$ **5.** $6\mathbf{i} - 12\mathbf{j} - 7\mathbf{k}$ **7.** $-11\mathbf{i} + 2\mathbf{j} + 8\mathbf{k}$
9. $\sqrt{5}, 2, 3$ **11.** $\sqrt{14}, \sqrt{5}, \sqrt{3}$ **13.** $3\sqrt{14}$ **15.** $5/2$ **25.** $(\mathbf{i} + \mathbf{j})/\sqrt{2}$
27. $-(\mathbf{i} + \mathbf{j} + \mathbf{k})/\sqrt{3}$ **29.** $-\mathbf{j}$ **31.** (a) $5\mathbf{i} - 7\mathbf{j} - 17\mathbf{k}$ (b) $p = 2, q = 0, r = 1$
33. 202 mph **35.** About 67 dynes **36.** The unit circle **37.** A unit circle with
center $(0, 0, 1)$ and parallel to the xy-plane

38.

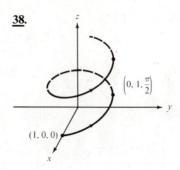

$\left(0, 1, \frac{\pi}{2}\right)$

$(1, 0, 0)$

39. About 463 mph in a direction making an angle of about 14° with the positive x-axis

Section 14–3 (page 627)

1. -5 **3.** 0   **5.** 6   **7.** 22   **9.** 10   **11.** 18   **13.** -60 **15.** $-5/\sqrt{10}$ **17.** 11/3
19. $3/\sqrt{11}$ **21.** 39°   **23.** 107°   **25.** 14°   **27.** -4 **29.** $(1, 1, -1)/\sqrt{3}$
31. All vectors $x\mathbf{i} + z\mathbf{k}$; all vectors $x\mathbf{i} + y\mathbf{j}$; coordinate planes   **32.** All vectors
$x\mathbf{i} + y\mathbf{j} + z\mathbf{k}$ with $x + z = 0$; with $y + 3z = 0$; planes in space   **33.** All vectors
$x\mathbf{i} + y\mathbf{j} + z\mathbf{k}$ with $2x - 3y + z = 0$; a plane   **34.** (a) All vectors $x\mathbf{i} + y\mathbf{j} + z\mathbf{k}$
with $x = y = 0$; with $x + y = 0$ and $y + z = 0$; lines in space   **35.** 27 joules

36. 100 lbs   **37.** A little more than 51 lbs   **38.** About 6 hp   **39.** (a) $\mathbf{v} = \dfrac{5}{9}\mathbf{D}$ m/sec

(b) 15 joules   **41.** $|\mathbf{a} + \mathbf{b}|^2 = |\mathbf{a}|^2 + 2\mathbf{a}\cdot\mathbf{b} + |\mathbf{b}|^2 \le |\mathbf{a}|^2 + 2|\mathbf{a}\cdot\mathbf{b}| + |\mathbf{b}|^2 \le |\mathbf{a}|^2$
$+ 2|\mathbf{a}||\mathbf{b}| + |\mathbf{b}|^2 = (|\mathbf{a}| + |\mathbf{b}|)^2$; therefore, $|\mathbf{a} + \mathbf{b}| \le |\mathbf{a}| + |\mathbf{b}|$

Section 14–4 (page 633)

1. $x = 2 + t, y = -1, z = 3 - t$ **3.** $x = 1 + 4t, y = 3 - 3t, z = 5 - 4t$
5. $x = 2 + 4t, y = 5 - 3t, z = 3 - 4t$ **7.** $x = 4 + t, y = -t, z = 1 + \sqrt{2}t$
9. $\mathbf{r} = t\mathbf{i} + (1 + 2t)\mathbf{j} + (2 - t)\mathbf{k}$ **11.** $\mathbf{r} = t\mathbf{i} + 2t\mathbf{j} - t\mathbf{k}$ **13.** $x = 8; \dfrac{y - 5}{-1} = \dfrac{z - 1}{2}$

15. $\dfrac{x + 3}{-1/2} = \dfrac{y - 2}{\sqrt{11}/6} = \dfrac{z - 5}{2/3}$ **17.** $\mathbf{r} = (-5 + 4t)\mathbf{i} + 6\mathbf{j} + 3t\mathbf{k}$

18. $x = 4 + 3t, y = -1 - t, z = 3$ **19.** $(2, -2, -2), 90°$ **21.** $(2, 5, -3), \approx 94°$
22. $(-1, 3, 0), \approx 70°$

23. $\left(x_0 - \dfrac{az_0}{c}, y_0 - \dfrac{bz_0}{c}, 0\right), \left(x_0 - \dfrac{ay_0}{b}, 0, z_0 - \dfrac{cy_0}{b}\right), \left(0, y_0 - \dfrac{bx_0}{a}, z_0 - \dfrac{cx_0}{a}\right)$

24. The simultaneous solution of $3t = 2 - u$ and $2 + t = u$ is $t = 0$ and $u = 2$,
but then the z's are not equal.   **25.** $\pi - \alpha, \pi - \beta, \pi - \gamma$; negatives of each other

27. $d = \sqrt{|\mathbf{a}|^2 - (\text{comp}_\mathbf{b}\mathbf{a})^2} = \dfrac{1}{|\mathbf{b}|}\sqrt{|\mathbf{a}|^2|\mathbf{b}|^2 - (\mathbf{a}\cdot\mathbf{b})^2}$ **28.** (a) $\sqrt{89}/3$ (b) $5\sqrt{2}$

(c) 14/5   **29.** $(1, -3, 0), (3, -2, 4)$, the half line $x = 1 + 2t, y = -3 + t, z = 4t$
for $t \ge 0$ **31.** 33 joules   **32.** 11 watts   **33.** $\sqrt{209/21}$ m   **34.** Yes, at (x_0, y_0, z_0);
direction vectors (and the lines) are orthogonal; direction vectors (and the lines) are
parallel

Section 14–5 (page 638)

11. $8x + 2y - z = 11$ **13.** $3x - 2y = 3$ **15.** $z = 4$ **17.** $3x - 2y + 1 = 0$
19. $7x - 6y - 4z + 20 = 0$ **21.** $4x + 5y - 3z = 5$ **23.** $\approx 52°$ **25.** $\approx 35°$
27. $8/\sqrt{11}$ **29.** $8/3\sqrt{2}$ **31.** $2x - y + 3z = 9$ **32.** $x + y + 1 = 0$
33. $x = 2 + 2t, y = 3 - 8t, z = -22t$

35. $(x_1 - x_0)\left(x - \dfrac{x_1 + x_0}{2}\right) + (y_1 - y_0)\left(y - \dfrac{y_1 + y_0}{2}\right)$

$+ (z_1 - z_0)\left(z - \dfrac{z_1 + z_0}{2}\right) = 0$ **36.** 2 sec   **37.** 5 ft-lbs

Section 14–6 (page 645)

1. Line $x = t, y = t + 1, z = 0$ **3.** Line $x = t - 1, y = 2t, z = -t$ **5.** Unit circle
with center $(0, 1, 0)$ and parallel to xz-plane

7.

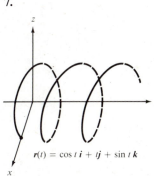

$r(t) = \cos t\, i + tj + \sin t\, k$

9.

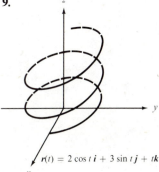

$r(t) = 2 \cos t\, i + 3 \sin t\, j + tk$

11. Parabola $y = x^2$ in the plane $z = 1$ **13.** $2\sqrt{6}$ **14.** $2\pi\sqrt{2}$
15. $2\pi\sqrt{5}(2 + \sqrt{2})/3 \approx 16$ **16.** About 2 units   **17.** All t; $t > 0$; all t; all $t \neq 0$,
19. $4\mathbf{i} + e^{-2}\mathbf{k}$ **21.** $\mathbf{j} + \mathbf{k}$ **23.** $((1/2)t^{-1/2} + t^{-1})\mathbf{i} - (t^{-2} + (1/3)t^{-2/3})\mathbf{j} + 4e^{4t}\mathbf{k}$
25. $(3t + 2)\cos t + (1 - t^2)\sin t + (1 + t)e^t$ **27.** $(t^{-1} + (1/2)t^{-1/2})\cos t$
$- (t^{1/2} + t^{-2})\sin t + 2$ **29.** $2\mathbf{i} + e^t\mathbf{k}$; $- (1/4)t^{-3/2}\mathbf{i} + 2t^{-3}\mathbf{j}$ **31.** $(3/2)\mathbf{i}$
$+ (e^2 - e)\mathbf{j} + \ln 2\mathbf{k}$ **33.** $(1/2)(e \sin 1 - e \cos 1 + 1)\mathbf{j} + \mathbf{k}$ **35.** $t^2\mathbf{i} + (1 + \ln t)\mathbf{j} + t\mathbf{k}$
36. $(1 - \cos t)\mathbf{i} + \sin t\mathbf{j} + \mathbf{k}$ **37.** $((1/2)t^2 + t)\mathbf{i} + t\mathbf{j} - \mathbf{k}$ **38.** $t^3\mathbf{i} + (e^t - 1)\mathbf{j}$
39. (b) $|\mathbf{r}(t)|^2 = \mathbf{r}(t) \cdot \mathbf{r}(t) = $ constant; therefore, $0 = [\mathbf{r}(t) \cdot \mathbf{r}(t)]' = 2\mathbf{r}(t) \cdot \mathbf{r}'(t)$

42.

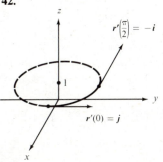

$r'\left(\dfrac{\pi}{2}\right) = -i$

$r'(0) = j$

43.

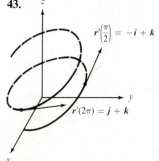

$r'\left(\dfrac{\pi}{2}\right) = -i + k$

$r'(2\pi) = j + k$

Section 14–7 (page 651)

1. $-2\mathbf{j} + \mathbf{k}$; $\sqrt{5}$; $\mathbf{0}$ **3.** $(1/2)t^{-1/2}\mathbf{i} + e^t\mathbf{j} + 4\mathbf{k}$; $\sqrt{\left(\dfrac{1}{4t}\right) + e^{2t} + 16}$; $-(1/4)t^{-3/2}\mathbf{i} + e^t\mathbf{j}$

5. $\cos t\mathbf{i} - \sin t\mathbf{j} + \mathbf{k}$; $\sqrt{2}$; $-\sin t\mathbf{i} - \cos t\mathbf{j}$

7.

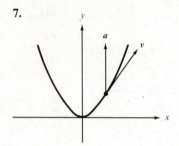

9.

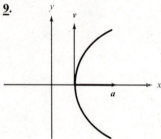

11.

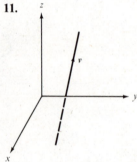

13.

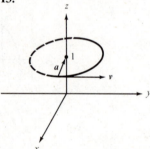

15. $(t + (1/6)t^3)\mathbf{i} - t^3\mathbf{j} + (1/2)t^2\mathbf{k}$ **17.** $(1 + t - \cos t)\mathbf{i} - \sin t\mathbf{j} + \mathbf{k}$
19. $(e^t - t - 1)\mathbf{i} + [t \tan^{-1}t - (1/2)\ln(1 + t^2)]\mathbf{j}$ **23.** (a) $500\sqrt{3}\mathbf{i} + (500 - 32t)\mathbf{j}$
ft/sec (b) 3,906.25 ft (c) $500\sqrt{3}$ ft/sec (d) $\approx 27,063$ ft   **25.** (a) $\approx 76°$
(b) 45°   **26.** Yes   **27.** (a) 26° (b) 253 ft   **29.** 191.7 meters away in 0.64 sec
30. 4,125 miles   **31.** 87.12 ft/sec²

Section 14–8 (page 656)

1. $\mathbf{i} - \mathbf{j}$ **3.** 0   **5.** $-\mathbf{i} - 3\mathbf{k}$ **7.** $14\mathbf{i} + 14\mathbf{j} + 7\mathbf{k}$ **9.** $7\mathbf{i} + 11\mathbf{j} + 15\mathbf{k}$ **11.** No
13. Both equal $-132\mathbf{i} + 192\mathbf{j} + 156\mathbf{k}$ **15.** $-3\mathbf{i} - 2\mathbf{j} + \mathbf{k}$; yes   **16.** Both equal 14
17. Both equal 28   **19.** $3t^2(\mathbf{b} \times \mathbf{a})$ **20.** $\mathbf{r}(t) \times \mathbf{r}''(t)$ **21.** (a) $\sqrt{83}$ (b) $2x - 18y$
$+ 2z + 14 = 0$ **23.** (a) $\sqrt{6}/2$ (b) $-2x + y + z = 0$ **25.** 3   **27.** 11
29. $\boldsymbol{\omega}(t) = \mathbf{k}$ **30.** $\boldsymbol{\omega}(t) = 6\mathbf{k}$ **31.** Yes, if the line does not pass through the origin.
32. $\boldsymbol{\omega}(t) = -t^4(\mathbf{j} + 2\mathbf{k})$ **33.** $\boldsymbol{\tau}(t) = 2tm\mathbf{k}$, $\boldsymbol{\omega}(t) = t^2\mathbf{k}$ **34.** If $\mathbf{r}$ and $\mathbf{F}$ are parallel,
then $\boldsymbol{\tau} = \mathbf{r} \times \mathbf{F} = \mathbf{0}$

35. $\dfrac{d}{dt} m\boldsymbol{\omega}(t) = m\dfrac{d}{dt}[\mathbf{r}(t) \times \mathbf{v}(t)] = m[\mathbf{r}(t) \times \mathbf{a}(t)] + m[\mathbf{v}(t) \times \mathbf{v}(t)] = \mathbf{r}(t) \times m\mathbf{a}(t)$

$= \mathbf{r}(t) \times \mathbf{F}(t) = \tau(t)$ **38.** $\mathbf{a} \times \mathbf{b} = \mathbf{a} \times (\mathbf{a} + \mathbf{b})$ for any $\mathbf{a}$ and $\mathbf{b}$; if $\mathbf{a}$ and $\mathbf{b} - \mathbf{c}$
are parallel   **39.** $2(\mathbf{b} \times \mathbf{a})$ **42.** $(\mathbf{a} \times \mathbf{b}) \cdot \mathbf{c}$ is the volume of the parallelepiped
determined by $\mathbf{a}, \mathbf{b}$, and $\mathbf{c}$

Section 14–9 (page 664)

1. $(2/3)t(1 + t^4)^{-3/2}$ **3.** $|\sin t|(1 + \cos^2 t)^{-3/2}$ **5.** $6(4 \sin^2 t + 9 \cos^2 t)^{-3/2}$
7. $2(9t^4 + 9t^2 + 1)^{1/2}(9t^4 + 4t^2 + 1)^{-3/2}$ **9.** $1/2$ **11.** $6t^3(1 + t^4)^{-1/2}$;
$6t(1 + t^4)^{-1/2}$ **13.** $(-\cos t \sin t)(1 + \cos^2 t)^{-1/2}$; $|\sin t|(1 + \cos^2 t)^{-1/2}$
15. $-5(\cos t \sin t)(4 \sin^2 t + 9 \cos^2 t)^{-1/2}$; $6(4 \sin^2 t + 9 \cos^2 t)^{-1/2}$
17. $2(2t + 9t^3)(1 + 4t^2 + 9t^4)^{-1/2}$; $2(9t^4 + 9t^2 + 1)^{1/2}(1 + 4t^2 + 9t^4)^{-1/2}$
19. $0, 1$ **21.** $|\mathbf{r}' \times \mathbf{r}''| = |(\mathbf{i} + f'\mathbf{j}) \times f''\mathbf{j}| = |f''|$ and $|\mathbf{r}'|^3 = (1 + [f']^2)^{3/2}$
22. (a) $162|x|(81 + x^4)^{-3/2}$, which agrees with Exercise 1 where $x = 3t$
(b) $|\sin x|(1 + \cos^2 x)^{-3/2}$ **23.** $1,650$ lbs   **24.** $-\sqrt{2}e^{-t}$; $\sqrt{2}e^{-t}$ **25.** $40\sqrt{3}$ ft/sec
≈ 47.2 mph   **26.** $|\mathbf{F}| \cos \theta = mg$ = weight of the car and $|\mathbf{F}| \sin \theta = mv^2/r$
= centripetal force; therefore, $\tan \theta = v^2/rg$ (b) 41 ft/sec ≈ 28 mph
27. Weight is reduced by centripetal acceleration mv^2/r **28.** $\sqrt{96} \approx 9.8$ ft/sec

Section 14–10 (page 668)

1. Hyperboloid of one sheet (Figure 14–60)   **3.** Ellipsoid; semiaxes of 2, 4, and 5
(Figure 14–59)   **5.** Hyperbolic cylinder (Figure 14–67)   **7.** Hyperbolic paraboloid
(Figure 14–64)   **9.** Circular cone (Figure 14–62)   **11.** Hyperboloid of two sheets
(Figure 14–61)   **13.** Elliptic cylinder generated by a line parallel to the y-axis
15. Elliptic paraboloid opening out of the page ($x \geq 0$)   **17.** Parabolic cylinder
generated by a line parallel to the x-axis   **19.** Sphere with center at $(2, -3, 0)$
and radius 4   **21.** Sphere; $x^2 + y^2 + z^2 = 1$ **23.** Circular paraboloid; $x^2 + z^2 = y$
25. Hyperboloid of two sheets $x^2 - 4y^2 - 4z^2 = 4$ **27.** $36x^2 + 9y^2 + 4z^2 = 36$
28. $x^2 + y^2 = 5$ **29.** $x^2 + 4y^2 - 4z^2 = 0$ **30.** Elliptic paraboloid; $x^2 + 4y^2 = z$
31. Hyperboloid of two sheets; $-x^2 - y^2 + z^2 = 1$

Review Exercises for Chapter 14 (page 668)

1. Ellipsoid with center $(0, 0, 0)$ and semiaxes 3, 2, and 1   **3.** Plane through $(1, 0, 0)$,
$(0, 1, 0)$, and $(0, 0, 1)$ **5.** Sphere with center at $(1, 2, 0)$ and radius $\sqrt{5}$ **7.** Elliptic
cylinder   **9.** Line through $(3, 1)$ and $(0, 2)$ **11.** Circular helix   **13.** Curve $z = e^y$ in
the plane $x = 1$ **15.** $2\sqrt{2}\pi$ **17.** $e^t(1 + e^{2t})^{-3/2}$ **19.** -6 **21.** $8/\sqrt{26}$ **23.** $-20\mathbf{i}$
$- 12\mathbf{j} + 4\mathbf{k}$ **25.** $(2\mathbf{i} - 3\mathbf{j} + \mathbf{k})/\sqrt{14}$ **27.** $\approx 109°$ **29.** $(\mathbf{i} + \mathbf{j} + \mathbf{k})/\sqrt{3}$
31. $\mathbf{r}(t) = (5 + t)\mathbf{i} + (1 + t)\mathbf{j} - 2(1 + t)\mathbf{k}$ **33.** $5x + y - 7z = 40$ **35.** $10\mathbf{i} + 5\mathbf{j}$
$- 15\mathbf{k}$ **37.** $5\sqrt{3}$ **39.** 40 **41.** About $58°, 143°, 74°$ **43.** $p = 2, q = -2, r = 1$
45. $3\sqrt{11}/2$ **47.** $x = 1 - 3t, y = 1, z = 1 - t$ **49.** $3x - 3y + 2z = 2$
51. $\sqrt{11}/2$ **53.** $12\mathbf{i} - 9\mathbf{j} + 7\mathbf{k}$ **55.** $\sqrt{11/6}$ **57.** $(-1, 0, 4), (1, 3, 5)$, straight line
59. 12 joules   **61.** $\sqrt{17}$ when $t = 0$ **63.** $(4, 14/3, -2)$ **65.** (a) $\mathbf{v}(t) = 250\sqrt{3}\mathbf{i}$
$+ (250 - 9.8t)\mathbf{j}$ m/sec   (b) $3,189$ m   (c) $250\sqrt{3}\mathbf{i}$ (d) $22,092$ m
67. $(\sin t - t \cos t)\mathbf{i} - (t \sin t + \cos t)\mathbf{j} + \mathbf{k}$ **69.** 30 ft/sec ≈ 20 mph

CHAPTER 15

Section 15–1 (page 678)

1. All x and y; $-24, 2, 19$; all z **3.** $x + y \geq 0, x \neq 0$; $\sqrt{3}/4, \sqrt{3}/9, \sqrt{3}$; all $z \geq 0$
5. All x, y, and z; $0, \pi/4, 0$; all w **7.** $y \geq 0$, all x and z; $1, 2, e$; all $w \geq 0$

9. Plane $z = 1$ **11.** Plane $2x + 3y + z = 6$ **13.** Hemisphere $x^2 + y^2 + z^2 = 1$; $z \geq 0$ **15.** Circular paraboloid; $z = x^2 + y^2$ **17.** Parabolic cylinder, $z = x^2$
19. Hyperbolic paraboloid; $z = y^2 - x^2$ **21.** Lines, $3x - 4y = c$ **23.** Parabolas; $y = x^2 + c$ **25.** Sine curves; $y = c + \sin x$ **27.** Exponential curves; $y = ce^{-x}$
29. Circles passing through the origin with centers at $x = c/2$ **31.** Plane; $4x - y + z = 4$ **33.** Ellipsoid; $x^2 + 3y^2 + 2z^2 = 1$ **35.** Cone for $c = 0$; hyperboloid of two sheets for $c = 1$

37. $\dfrac{[3(x+h) - 2y] - [3x - 2y]}{h} \to 3$ and $\dfrac{[3x - 2(y+h)] - [3x - 2y]}{h} \to -2$

as $h \to 0$; the first is the derivative of f if y is treated as a constant and the second is the derivative of f if x is treated as a constant   **38.** $2x, 5$ **39.** y, x **40.** $\sin y, x \cos y$
41. (a) $f(x, y) = xy$ (b) $f(x, y) = \cos^{-1}(x/\sqrt{x^2 + y^2})$ (c) $f(x, y) = |y|$
42. (a) $f(x, y, z) = 2(xy + xz + yz)$ (b) $f(x, y, z) = \cos^{-1}(x/\sqrt{x^2 + y^2 + z^2})$
(c) $f(x, y, z) = |z|$ **43.** $V = kdI/cA$ (k a constant)

Section 15–2 (page 682)

1. $13/5$ **3.** $2\sqrt{14}$ **5.** π **7.** Along the x-axis, $f(x, 0) \to 1$; but along the y-axis, $f(0, y) \to -1$ **8.** Along the x-axis, $f(x, 0) = 0$; but along the parabola $y = x^2$, $f(x, x^2) = 1/2$ **9.** $1/2$ **11.** 1   **13.** 0   **15.** No limit   **17.** 0   **19.** No limit (try the path $y = \ln x$)   **21.** $-5/14$ **23.** No limit   **25.** Domain: all (x, y) except $(0, 0)$; $(0, 0)$ is a boundary point, all others are interior; continuous on its domain
27. Domain: all (x, y, z) with $x^2 + y^2 \neq 0$ and $z \geq 0$; xy-plane and upper z-axis comprise the boundary; continuous on its domain   **29.** Domain: all (x, y); all points are interior, no boundary; continuous   **31.** Domain: all (x, y, z) with $xy \geq 0$; the xz- and yz-planes are boundaries, all points with $xy > 0$ are interior; continuous on its domain   **33.** Given $\epsilon > 0$, let $\delta = \epsilon$. Then $|x - p| < \epsilon$ whenever $0 < \sqrt{(x-p)^2 + (y-q)^2} < \delta = \epsilon$

Section 15–3 (page 689)

1. $9x^2y + 2y^2$; $3x^3 + 4xy$ **3.** $x(x^2 + y^2)^{-1/2}$; $y(x^2 + y^2)^{-1/2}$ **5.** $uv \cos uv + \sin uv$; $u^2\cos uv$ **7.** $2\pi rh/3$; $\pi r^2/3$ **9.** $2(y + z)$; $2(x + z)$; $2(x + y)$ **11.** $\cos(y/z)$; $(-x/z)\sin(y/z)$; $(xy/z^2)\sin(y/z)$ **13.** $-6x(x^2 + y^2 + z^2)^{-2}$; $-6y(x^2 + y^2 + z^2)^{-2}$; $-6z(x^2 + y^2 + z^2)^{-2}$ **15.** $s \tan^{-1}t$; $r \tan^{-1}t$; $rs(1 + t^2)^{-1}$ **17.** $x(x^2 + y^2)^{-1}$; $y(x^2 + y^2)^{-1}$ **19.** $e^{xy}(y \sin x - \cos x)/y \sin^2x$; $e^{xy}(xy - 1)/y^2\sin x$ **21.** $15y^2 - 6x$
23. $x^2\cos x + 2x \sin x$ **25.** $-4xy(x^2 + y^2)^{-2}$ **27.** e^y **29.** $6xz + (xyz + 1)ze^{xyz}$
31. 1   **33.** $4e^3$ **35.** $\cos \theta$; $-r \sin \theta$; $\sin \theta$; $r \cos \theta$ **37.** $8x + 30y - z = 83$
38. $-2x + 24y - z = 35$ **39.** Harmonic   **41.** Harmonic   **43.** No   **45.** Harmonic
47. No   **49.** Yes   **51.** Yes   **53.** (a) $\sqrt{2/(2 - \sqrt{2})}$ (b) $\sqrt{2}$

Section 15–4 (page 696)

1. $(2x - 3y)\,dx + (-3x + 2y)\,dy$ **3.** $2x \sin y\,dx + (x^2\cos y + 2y)\,dy$
5. $e^{xy}(xy + 1)\,dx + x^2e^{xy}\,dy$ **7.** $y^2 \ln z\,dx + 2xy \ln z\,dy + (xy^2/z)\,dz$

9. $(1 + x^2y^4z^6)^{-1}[y^2z^3\,dx + 2xyz^3\,dy + 3xy^2z^2\,dz]$ **11.** $\dfrac{1}{2}(h\,db + b\,dh)$

13. $2(y + z)\,dx + 2(x + z)\,dy + 2(x + y)\,dz$
15. $\pi(r^2 + h^2)^{-1/2}[(2r^2 + h^2)\,dr + rh\,dh]$

17. $\dfrac{1}{2}(b \sin C\,da + a \sin C\,db + ab \cos C\,dC)$

19. $(R/R_1)^2 \, dR_1 + (R/R_2)^2 \, dR_2 + (R/R_3)^2 \, dR_3$ **21.** 0.1011; 0.1 **23.** 0.0092; 0.01
25. −0.00015; 0 **27.** −0.2622; −0.26 **29.** −0.2479; −0.2418 **31.** 0.7651
33. ± 1.04 in² **34.** ± 0.9 in³ **35.** 0.157 ft³ **37.** 5.9 ft³ **38.** $3\Delta x + 12.5\Delta\theta$; more
sensitive to change in θ **39.** $\Delta x + \sqrt{3}\Delta\theta$; more sensitive to θ **41.** $dS = 0.0078 \, dp$
$+ 0.6250 \, dL - 7.8125 \, dw - 11.7185 \, dh$; the sag is most sensitive to changes in height,
so increase the height to decrease the sag. **43.** The partials are not continuous at (0, 0)

Section 15–5 (page 703)

1. $48t^3 + 18t^2 + 2t$ **3.** $(3t + 6t \ln t)e^{3t^2\ln t}$ **5.** $e^t(\sec^2 t + \tan t)(1 + e^{2t}\tan^2 t)^{-1}$
7. $7t^6 + 4t^3 + 2$ **9.** $t/\sqrt{1 + t^2}$ **11.** $7u^6v^3 + 2u \sin uv + u^2v \cos uv$;
$3u^7v^2 + u^3\cos uv$ **13.** $2uv^2 \ln(3u + v) + [3u^2v^2/(3u + v)]$; $2u^2v \ln(3u + v)$
$+ [u^2v^2/(3u + v)]$ **15.** $72r^7s^5t^5$; $45r^8s^4t^5$; $45r^8s^5t^4$ **17.** $-(x + 3y)/(3x - 5y)$
19. $-ye^{xy}/(xe^{xy} - \sec^2 y)$ **21.** $-(\sin y + y \sin x)/(x \cos y - \cos x)$
23. $y^2/(3zy + 8z^3)$; $(4xy - 3z^2)/(6zy + 16z^3)$ **25.** $-(e^{yz} + yze^{xz})/(xye^{yz}$
$+ xye^{xz} - 1)$; $-(xze^{yz} + e^{xz})/(xye^{yz} + xye^{xz} - 1)$ **29.** (b) Set $t = x + 2y$; then
$z_x = f'(t)t_x = f'(t)$ and $z_y = f'(t)t_y = 2f'(t) = 2z_x$ **30.** (c) Set $r = x - at$ and
$s = x + at$; then $z = f(r) + g(s)$, so $z_x = z_r r_x + z_s s_x = z_r + z_s$; $z_{xx} = z_{rr} + z_{ss}$;
$z_t = -az_r + az_s$; $z_{tt} = a^2(z_{rr} + z_{ss}) = a^2z_{xx}$ **31.** $-7/6$ rad/sec **32.** $7/\sqrt{2}$ cm²/sec
33. Heating up about 41.6 degrees/sec **35.** 0.4 m^3/yr **37.** $100\pi\eta^{-1}(dr/dt)$
$+ \pi 10^{-5}(8\eta)^{-1}(dp/dt) - \pi(4\eta)^{-1}(dL/dt)$; most sensitive to the radius
38. $-2.3 \times 10^{-14}GM$ lbs/sec

Section 15–6 (page 709)

1. $2\sqrt{2}$ **3.** 0 **5.** $9(2 - \sqrt{3})/5$ **7.** $-2(1 + \sqrt{2})$ **9.** $-60/\sqrt{19}$ **11.** $\pi/4\sqrt{3}$
13. 1 **15.** $2/\sqrt{3}$ **17.** $(2 + \sqrt{3})/4\sqrt{3}$ **19.** $56\pi/5\sqrt{5}$ **21.** $\mathbf{i} + \mathbf{j}$; $\pm(\mathbf{i} - \mathbf{j})$

23. $\mathbf{i} - 2\mathbf{j}$; $\pm(2\mathbf{i} + \mathbf{j})$ **25.** $\sqrt{3}\mathbf{i} + \dfrac{3}{2}\mathbf{j}$; $\pm\left(\mathbf{i} - \dfrac{2\sqrt{3}}{3}\mathbf{j}\right)$

27. $\mathbf{i} - 2\mathbf{j} + \mathbf{k}$; $x\mathbf{i} + y\mathbf{j} + z\mathbf{k}$ where $x - 2y + z = 0$ **29.** $2\mathbf{i} - 3\mathbf{k}$; $x\mathbf{i} + y\mathbf{j} + z\mathbf{k}$
where $2x - 3z = 0$ **31.** $\pi\mathbf{i} + 2\mathbf{j} + 2\mathbf{k}$; $x\mathbf{i} + y\mathbf{j} + z\mathbf{k}$ where $\pi x + 2y + 2z = 0$
33. $\mathbf{i}$, $\pm\mathbf{j}$ **35.** $\mathbf{i} + \mathbf{j} + \mathbf{k}$; $x\mathbf{i} + y\mathbf{j} + z\mathbf{k}$ where $x + y + z = 0$ **37.** $\mathbf{i} + \mathbf{j} + \sqrt{3}\mathbf{k}$;
$x\mathbf{i} + y\mathbf{j} + z\mathbf{k}$ where $x + y + \sqrt{3}z = 0$ **39.** $17\mathbf{i} + 6\mathbf{j}$; $\pm(6\mathbf{i} - 17\mathbf{j})$
41. $-3\mathbf{i} - \mathbf{j} - 2\mathbf{k}$ **43.** (a) $4\mathbf{i} - \mathbf{j}$ (b) $-4\mathbf{i} + \mathbf{j}$ (c) $\pm(\mathbf{i} + 4\mathbf{j})$
44. $-(6/\sqrt{5})°$/unit distance; $\pm(4\mathbf{i} - 3\mathbf{j})$ **45.** $6\mathbf{i} + 8\mathbf{j}$

47.

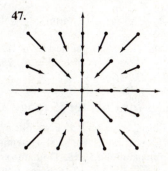

Section 15–7 (page 713)

1. $x + y + \sqrt{2}z = 4$ **3.** $z = -1$ **5.** $2x + 2y - z = 2$ **7.** $x = 1$ **9.** $6\mathbf{i} + 2\mathbf{j}$
11. $4\mathbf{i} - \mathbf{j}$ **13.** $-4\mathbf{i} - 2\mathbf{j}$ **15.** $5x + 5y + 4z = 16$ **17.** $-x + y + 6z = 20$
19. $8x - 3y - \sqrt{5}z = 0$ **21.** $3x + 4y = 25$ **23.** $4x - 2y + z = 3$
25. $4x - 2y + e^5z = 11$ **27.** $x + y - (3/4\pi)z = 2$ **29.** $2x + 4\sqrt{3}y - z = 4$
$+ (2\pi/\sqrt{3})$ **31.** $3x + 4y - 13z = 25 - 13\ln 26$ **33.** $(\pm 12/\sqrt{53}, \mp 6/\sqrt{53},$
$\pm 12/\sqrt{53})$ **34.** $(\pm 6/\sqrt{5}, \pm 6/\sqrt{5}, 0)$ **35.** $(0, 0, z)$ and $(0, y, 0)$ for any y or z
37. Yes; $(-4, -4, 4\sqrt{2}) \cdot (-4, -4, -4\sqrt{2}) = 0$

39. Yes. They meet when $t = 1$; $\mathbf{r}'(1) = 2\mathbf{i} + \mathbf{j} + \dfrac{1}{2}\mathbf{k}$ is *tangent* to the curve,

$\nabla = 4\mathbf{i} + 2\mathbf{j} + \mathbf{k}$ is *normal* to the surface, and these vectors are parallel.
41. Set $F(x, y, z) = x^2 + y^2 + z^2 - x - y - z$. We want a vector in the plane
$x + y + z = 1$ that is perpendicular to $\nabla F(0, 0, 1)$; an answer is $\mathbf{a} = \mathbf{i} - \mathbf{j}$ with
its tail at $(0, 0, 1)$.

Section 15–8 (page 720)

1. 0 min   **3.** $(0, 0)$ is a saddle point; no extreme values   **5.** $(0, 0)$ is a saddle point;
$f(4, 4)$ is a min   **7.** Saddle point at $(0, 0)$; $f(18^{-1/5}, -2(18)^{-3/5})$ is a min
9. No critical points   **11.** $f(\pi/2, \pi/2)$ is max and $f(3\pi/2, 3\pi/2)$ is a min; $(\pi/2, 3\pi/2)$
and $(3\pi/2, \pi/2)$ are saddle points. The test fails at $(0, \pm\pi)$, $(\pm\pi, \pm\pi)$, $(\pm\pi, \mp\pi)$,
and $(\pm\pi, 0)$, but these are easily seen to be saddle points   **13.** The test cannot be used,
but $f(0, 0) = 0$ is a min   **15.** $f(1, 1)$ and $f(-1, -1)$ are min; $f(1, -1)$ and $f(-1, 1)$
are max; $(0, 0)$, $(0, \pm 2)$, and $(\pm 2, 0)$ are saddle points   **17.** Test fails, but $f(0, 0) = 0$
is a min because $f(x, y) \geq 0$ (if $x \geq 0$, then $f(x, y) \geq 0$; if $x < 0$, then $x^2 + xy^2 + y^4$
$> (x + y^2)^2 \geq 0)$ **19.** $f(0, \pm 1)$ are saddle points; $f(0, 0)$ is min   **21.** $x = y = \sqrt[3]{16}$,
$z = 12/\sqrt[3]{16^2}$ **23.** All three equal $S/3$ **25.** All components are $10/\sqrt{3}$
27. A cube with sides $2r/\sqrt{3}$ **28.** $6/\sqrt{3}, 12/\sqrt{3}, 8/\sqrt{3}$ **29.** $4x - 7y + 7 = 0$
30. $9x - 10y + 12 = 0$ **31.** $y = 0.9x + 67.5$ (rounded to one decimal place)

Section 15–9 (page 726)

1. Max $f(0, -1) = 7$; min $f(0, 1) = -1$ **3.** Max $f(\pm 2/\sqrt{5}, \mp 1/\sqrt{5}) = 5$;
min $f(\pm 1/\sqrt{5}, \pm 2/\sqrt{5}) = 0$ **5.** $-36 \pm 27\sqrt{3}$ **7.** Min $64/27$ **9.** Min $1/3$
11. Min at $(-1 + (1/\sqrt{2}), -1 + (1/\sqrt{2}), -1 + \sqrt{2})$ **13.** Max $f(0, -1) = 7$;
min $f(0, 2/3) = -4/3$ **15.** Max at $(-2/3, 2)$; min at $(-1, -3\sqrt{3}/2)$

17. Nearest $\left(-\dfrac{2}{\sqrt{5}} + 2, \dfrac{1}{\sqrt{5}} - 1\right)$; farthest $\left(\dfrac{2}{\sqrt{5}} + 2, -\dfrac{1}{\sqrt{5}} - 1\right)$

19. Nearest $(2, 1)$ and $(-2, -1)$; farthest $(2, -4)$ and $(-2, 4)$ **21.** Base $8\sqrt{2/3}$ by
$8\sqrt{2/3}$; height $4\sqrt{2/3}$ **23.** $x + 2y + 2z = 6$ **25.** $x = y = z = p/3$
27. $C = H/3F, L = 2H/3G$

Review Exercises for Chapter 15 (page 729)

1. Ellipsoid   **3.** Elliptic cylinder   **5.** Parabolic cylinder   **7.** $x^2 + y^2 = 4$
9. Domain: $x^2 + y^2 \leq 1$; range: $0 \leq z \leq 1$; hemisphere   **11.** Domain; xy-plane;
range: $z \geq 0$; elliptic paraboloid   **13.** Domain: xy-plane; range: all z; plane
15. $(-x\mathbf{i} - y\mathbf{j})/\sqrt{1 - x^2 - y^2}$ **17.** Ellipses with centers at the origin   **19.** $-2\mathbf{i} - \mathbf{j}$
21. Ellipsoid   **23.** Elliptic paraboloid   **25.** $2x + y + 3\sqrt{7}z = 18$ **27.** Lowest

(0, 0, 1); no highest point   **29.** $f_x = 2xy - y^2, f_y = x^2 - 2xy$ **31.** $f_x = \sin y, f_y = x \cos y$ **33.** $f_x = y^z, f_y = xzy^{z-1}, f_z = xy^z \ln y$ **35.** $f_x = e^{xy}(-z \sin xz + y \cos xz)$, $f_y = xe^{xy}\cos xz, f_z = -xe^{xy}\sin xz$ **37.** None   **39.** $2t \sin(t^3 + t^2)$ $+ t^2(3t^2 + 2t)\cos(t^3 + t^2)$ **41.** $y^{z-1}(1 + z \ln y)$ **43.** $e^{u^2-v^2}[2u \cos(u^2v - uv^2)$ $- (2uv - v^2)\sin(u^2v - uv^2)]$ **45.** $-(2xy^3 - 3 \sin y + e^y)/(3x^2y^2 - 3x \cos y + xe^y)$ **47.** $-(e^{yz} - 4xz^3)/(xye^{yz} - 6x^2z^2 + y^2); -(xze^{yz} + 2yz)/(xye^{yz} - 6x^2z^2 + y^2)$ **49.** $(4x - 3y)\,dx + (-3x + 10y)\,dy$ **51.** $e^{yz}(dx + xz\,dy + xy\,dz)$ **53.** $112/5$; $-\mathbf{i} - \mathbf{j}$ **55.** $(\pi + 1)/2\sqrt{3}; -(\pi/4)\mathbf{i} - (\pi/4)\mathbf{j} - (1/2)\mathbf{k}$ **57.** $x = y = z$; $x = y = -z; x = y = 0$ **59.** $(-1 \pm \sqrt{1/2}, -1 \pm \sqrt{1/2}, -1 \pm \sqrt{2})$ **61.** $a^2 + b^2 + c^2 = 2$ **63.** $9/4$ and $-1/4$

CHAPTER 16

Section 16–1 (page 734)

1. $5/2$ **3.** $5/32$ **5.** 1 **6.** $1/2$ **7.** Area of D **8.** $1/2$ **9.** 2 **11.** $11/8$ **13.** 2 **15.** Area of $D = \pi$; $\|P\|$ too large   **17.** $2\pi/3$ **19.** $\pi/3$

Section 16–2 (page 741)

1. 3 **3.** $2(e - e^{-2})$ **5.** $14/5$ **7.** $3e^4 - e^2$ **9.** $2/3$ **11.** $-\pi/2$ **13.** $\pi^2/8$ **15.** $27/8$ **17.** $(e - 2)/2$ **19.** $(e^2 + 1)/4$ **21.** $\int_0^3 \int_0^1$ **23.** $\int_{-1}^1 \int_{-1}^x + \int_1^3 \int_{-1}^1$ **25.** $\int_{-1}^0 \int_{-y}^1 + \int_0^1 \int_y^1$ **27.** $\int_0^1 \int_{\sin^{-1}x}^{\pi/2}$ **29.** $\frac{1}{2}e^6 - e^3 - \frac{1}{2}e^2 + e$ **31.** 0 **33.** $(e - 1)/2$ **35.** $(e - 2)/2$ **37.** $\int_0^8 \int_0^{\sqrt{x}} + \int_8^{16} \int_{\sqrt{2x-16}}^{\sqrt{x}}$ **38.** $\int_0^2 \int_0^{x^2} + \int_2^{5/2} \int_0^{8-x^2}$ **39.** (a) $64/3$ (b) $99/24$

Section 16–3 (page 747)

1. $3/2$ **3.** 4 **5.** 4π **7.** 144 **9.** $9/2$ **11.** $(5/2) + 6 \ln(2/3)$ **13.** $2(\sqrt{2} - 1)$ **15.** $(12/5, 3/2)$ **17.** $((\pi\sqrt{2} - 4)/4(\sqrt{2} - 1), 1/4(\sqrt{2} - 1))$ **19.** $(1/4, -1)$ **21.** $(4/5, 4/5)$ **23.** $4/15$ **25.** (a) $\delta a^4/3$ (b) $\delta a^4/12$ (c) $\delta a^4/6$ (d) $2\delta a^4/3$ (e) $5\delta a^4/12$ **26.** $I = \delta a^4/6 = I_{cm}(h = 0)$; for a line through a corner, $h = a/\sqrt{2}$, and $I = (\delta a^4/6) + (\delta a^2)(a/\sqrt{2})^2 = 2\delta a^4/3$; for a line through a midpoint, $h = a/2$ and $I = (\delta a^4/6) + (\delta a^2)(a^2/4) = 5\delta a^4/12$ **27.** $\pi\delta r^4/2; 3\pi\delta r^4/4$

Section 16–4 (page 751)

1. $(\pi/2)(1 - e^{-1})$ **3.** $\ln(\sqrt{2} + 1)$ **5.** $(\pi/4)(1 - \cos 1)$ **7.** $62/15$ **9.** $e^2 - 3$ **11.** 2π **13.** $9/2$ **15.** $3\pi/2$ **17.** $4\pi a^3/3$ **19.** $4\pi(8 - 3\sqrt{3})/3$ **21.** $729\pi^5/4$ **23.** $\pi a^4/2$

Section 16–5 (page 756)

1. 24 **3.** 9 **5.** $2/3$ **7.** $4/3$ **9.** 32 **11.** $5/24$ **13.** $4/3$ **15.** $5/24$ **17.** $abc/6$ **19.** $128/5$ **21.** 4π **23.** $8\pi\sqrt{2}$ **25.** $\bar{x} = \bar{y} = \bar{z} = 7a/12$ **27.** $3/5$ **28.** 0

Section 16–6 (page 761)

1. $\pi/16$ **3.** $\pi/12$ **5.** $\pi/10$ **7.** $7\pi/6$ **9.** 16π **11.** $4\pi/7$ **13.** $16\pi/3$
15. $\pi(1 - \cos 1)/2$ **17.** $\pi(1 - \cos 1)/6$ **19.** $(0, 0, 3/4)$ **21.** $\pi a^4 h/2$
23. $\frac{1}{4}\pi a^2 h \left(a^2 + \frac{1}{3} h^2 \right)$ **25.** $\pi(2 - \sqrt{2})/3$ **27.** $128\pi/9$ **29.** 4π

Section 16–7 (page 765)

1. $(3\sqrt{6} - \sqrt{2})/6$ **3.** $2\pi/3$ **5.** $\pi(17\sqrt{17} - 1)/6$ **7.** $16(\pi - 2)$ **9.** $\sqrt{3}\pi$
11. $2\pi(5\sqrt{5} - 1)/3$ **13.** ≈ 217

Review Exercises for Chapter 16 (page 765)

1. 2 **3.** 108 **5.** $\pi/8$ **7.** $\frac{1}{2}\cos 1 + \sin 1 - 1$ **9.** $1/85$ **11.** $\pi/8$ **13.** $3\pi/2$

15. $\pi(1 - \cos 1)/2$ **17.** $\pi^2/8$ (Observe that $\pi/2 \le x \le \pi$) **19.** $e - 1$

21. $\displaystyle\int_{-2}^{1} \int_{y^2}^{2-y} f(x, y)\, dx\, dy$ **23.** $\displaystyle\int_{0}^{1} \int_{y}^{2y} + \int_{1}^{2} \int_{2y-1}^{y+1} f(x, y)\, dx\, dy$

25. $\displaystyle\int_{-2}^{2} \int_{-\sqrt{4-x^2}}^{\sqrt{4-x^2}} \int_{0}^{4-x^2-y^2} f(x, y, z)\, dz\, dy\, dx$

27. $\displaystyle\int_{0}^{1} \int_{0}^{1-x} \int_{1-x-y}^{1} + \int_{0}^{1} \int_{1-x}^{\sqrt{1-x^2}} \int_{0}^{1} f(x, y, z)\, dz\, dy\, dx$ **29.** $9/2$ **31.** 8π **33.** $8/3$

35. 144 **37.** 4π **39.** $\pi(17\sqrt{17} - 1)/6$ **41.** $\frac{1}{2}(\sqrt{14} + 5\ln(2 + \sqrt{14})/\sqrt{10}$

43. $(0, 0, 5/12)$ **45.** $2\pi(\sqrt{2} - 1)/5\sqrt{2}$

CHAPTER 17

Section 17–1 (page 774)

1.

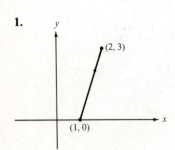

3.

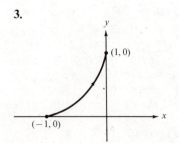

5.

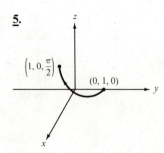

7. $\mathbf{r}(t) = (1 - t, -3t), -1 \le t \le 0$ **9.** $\mathbf{r}(t) = (-t - 1, t^2), -1 \le t \le 0$
11. $\mathbf{r}(t) = (-\sin t, \cos t, -t), -\pi/2 \le t \le 0$ **13.** $\mathbf{r}(t) = (\cos t, \sin t), \pi \le t \le 2\pi$
15. $\mathbf{r}(t) = (1 + \cos t, 1 + \sin t), 0 \le t \le 2\pi$ **17.** $C_1 + C_2 + C_3$ where
$\mathbf{r}_1(t) = (t, t, t), 0 \le t \le 1; \mathbf{r}_2(t) = (1, 1 - t, 1 - t), 0 \le t \le 1; \mathbf{r}_3(t) = (1 - t, 0, 0,),$
$0 \le t \le 1$

ANSWERS TO UNDERLINED EXERCISES **909**

19.

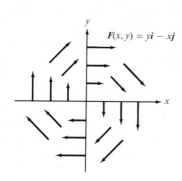

$F(x, y) = yi - xj$

21.

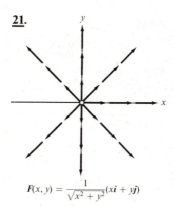

$F(x, y) = \dfrac{1}{\sqrt{x^2 + y^2}}(xi + yj)$

23.

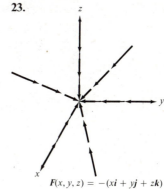

$F(x, y, z) = -(xi + yj + zk)$

25.

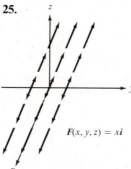

$F(x, y, z) = xi$

27.

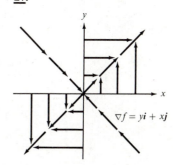

$\nabla f = yi + xj$

29.

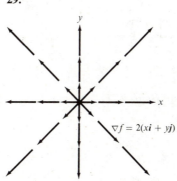

$\nabla f = 2(xi + yj)$

31.

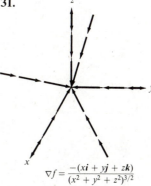

$\nabla f = \dfrac{-(xi + yj + zk)}{(x^2 + y^2 + z^2)^{3/2}}$

Section 17–2 (page 782)

1. 15/2 **3.** 8/3 **5.** $4\pi^2$ **7.** $2\pi^2$ **9.** 0 **11.** $e^2 - e$ **13.** 5/4 **15.** 0 **17.** $\pi^2/8$
19. $27(2\sqrt{2} - 1)/2$ **21.** (a) 315/2 joules (b) 150 joules (c) 627/4 joules
(d) 3/4 joules **23.** (a) (b) (c) 19 (d) 0 **25.** all 0 **27.** (a) $(5 + 3\pi)/6$
(b) $(2 + \pi)/2$ (c) $(4 + 3\pi)/6$ (d) 1/6 **29.** 0 **31.** $\sqrt{2}\pi$

Section 17–3 (page 789)

1. $\sin x - \cos y$ **3.** Not conservative   **5.** $3x^2 y^3$ **7.** xyz **9.** $\sin y$ **11.** Not conservative   **13.** xye^z **15.** 3   **17.** 96   **19.** 0   **21.** (a) 2 m joules   (b) decrease of 2 m joules   (c) $m(-\mathbf{j} + 2\mathbf{k})$ **23.** (a) $-C(x^2 + y^2 + z^2)^{-1/2}$ (b) The potential function takes the same value at both points   **25.** 750,000 ft-lbs   **26.** (b) The integral around the square is $7/6$; $\mathbf{F}$ is not conservative   **27.** (a) $N_x = M_y = (y^2 - x^2)/(x^2 + y^2)$ for $(x, y) \neq (0, 0)$ (b) The integral around the unit circle is 2π; $\mathbf{F}$ is not conservative

Section 17–4 (page 797)

1. $1/2$ **3.** π **5.** 0 ($\mathbf{F}$ is conservative)   **7.** -5 **9.** 0   **11.** -12π **13.** $(e - 3)/2$ **15.** $6\pi - 28$ **17.** $45\pi/2$ **19.** 4π **21.** $8\sqrt{3}/5$

25. (a) $\int_C \frac{1}{2}x^2\, dy = \int \int_D x\, dA = M_y$ (b) $\int \frac{1}{3}x^3\, dy = \int \int_D x^2\, dA = I_y$

26. $M_x = \int_C -\frac{1}{2}y^2\, dx$, $\bar{x} = \int_C \frac{1}{2}x^2\, dy \bigg/ \int_C x\, dy$, and so on   **27.** -2π

28. (a) 0   (b) 0

Section 17–5 (page 803)

1. -24π **3.** $\dfrac{13}{6} - \cos 1 - \sin 1$ **5.** $11\pi/6$ **7.** 40π **9.** $1/8$ **11.** $12\pi + 16\sinh 3$

12. 144   **13.** 0   **15.** 0   **16.** 0   **17.** $3\pi/4$ **19.** -12π **21.** $12\pi/5$

Review Exercises for Chapter 17 (page 804)

1. $\mathbf{r}(t) = (t, 2t^2 - 3t + 1), 0 \le t \le 2$; $\mathbf{r}(-t) = (-t, 2t^2 + 3t + 1), -2 \le t \le 0$ **3.** $-10/3$ joules   **5.** 0; net flow is perpendicular to $C - C_1$ **7.** $16/3$; there is a source inside $C - C_1$ **9.** $\mathbf{r}(t) = (2\cos t, 2\sin t, 4), 0 \le t \le 2\pi$ **11.** $160\pi/3$ **13.** $27/28$ **15.** 1   **17.** 0 (the integrand is conservative)   **19.** $1/2$ **21.** $3\pi/4$ **23.** $f(x, y, z) = x^3 \sin y \cos z$ **25.** $f(x, y) = y \ln|xy|$ **27.** $f(x, y, z) = xy^z$ **29.** 3π **31.** 3π **33.** $30k[1 - (1 + 4\pi^2)^{-1/2}]$ **35.** (a) $2kQ\pi(1 - 1/\sqrt{2})$ (b) $4kQ\pi/\sqrt{2}$

CHAPTER 18

Section 18–1 (page 811)

1. Yes   **3.** No   **5.** Yes   **7.** Yes   **9.** Yes   **11.** Yes   **13.** Yes   **15.** $C = 3$ **17.** $C = -3$ **19.** $C_1 = 1, C_2 = 4$ **21.** $y' = 2$ **23.** $y' = y$ **25.** $xy' = y \ln y$ **27.** $y' = y/x$ **29.** $yy' + x = 0$ **30.** $y' = (y^2 - x^2)/2xy$ **31.** $s' = k/s$ **33.** $s'' = F/m$ **35.** $(y')^3 = y$ or $y' = \sqrt[3]{y}$ **37.** $T' = -k(T - C), k > 0$

Section 18–2 (page 816)

1. $-\cos x + \sin y = C$ **3.** $x^2 + \tan^{-1}y = C$ **5.** $\ln(\ln u) + v^{-1} = C$ **7.** $y = x \sin^{-1}x + \sqrt{1 - x^2} + C$ **9.** $-\ln \cos x + e^{-y} = C$ **11.** $x^2 y^5 e^y = C$ **13.** $\cos x + x \sin x - \ln \sin y = C$ **15.** $e^s - e^t = C$

17. $\dfrac{2}{3}(x^3 + 1)^{-1/2} + \dfrac{1}{2}\ln(y^2 + 1) = C$ **19.** $\dfrac{1}{2}[\ln x]^2 = \ln y + C$

21. $s^{3/2} + t^{3/2} = 9$ **23.** $\tan^{-1}y = (x^2/2) - 2$ **25.** $-\cos x + \sin y = -1$

27. $y = x \sin^{-1}x + \sqrt{1 - x^2}$ **29.** $x^2y^5e^y = e$

31. (b) $(s - 1)e^s + \frac{1}{2}t^2 + \ln t = C$;   (e) $\ln r + \ln \sin \theta = C$;

(h) $v + \ln v + \cos u = C$;   (k) $\sin^{-1}x = \sin^{-1}y + C$ (o) $\frac{2}{3}s^{3/2} = 6t^{1/2} + C$

33. About 4.6 million years   **35.** $\sin(x + C)$

37. $\sqrt{mg/k}\dfrac{1 - e^C}{1 + e^C}$ where $C = -2\sqrt{gk/m}\,t$; terminal velocity is $\sqrt{mg/k}$

39. No; if s represents the distance from A, then $v = (ds/dt) = ks^2$ and $s = 6/(5t + 1)$
40. About 9 days   **41.** $[I_2] = [4 \times 10^{-4}t + (1/5)]^{-1}$

Section 18–3 (page 821)

1. $y = x + Cx^2$ **3.** $y = Cx^2$ **5.** $y = Cx^2/(1 - Cx)$ **7.** $\tan^{-1}(y/x) = \ln x + C$
9. $y = x + C(y + 3x)^5$ **11.** $(x^2 + y^2)^3 = Cx^2$ **13.** $(y + 6x)^2 = Cx$ **15.** $y = xe^{Cx}$

17. $\dfrac{2}{\sqrt{3}}\tan^{-1}\dfrac{2}{\sqrt{3}}\left(\dfrac{y}{x} - \dfrac{1}{2}\right) - \ln x = C$ **18.** $y^2 - 2x^2 \ln x = Cx^2$

21. Separable: (c), (i), (k), (p), (r); Homogeneous: (b), (e), (h), (o), (t)   **23.** At the point 1/4 mile below the intended destination

Section 18–4 (page 828)

1. $x^3 - xy^2 + 3y = C$ **3.** $u^3 + uv = C$ **5.** $\sin xy - \cos y = C$

7. $e^{xy} + y \ln x = C$ **9.** $\tan^{-1}x + x \cos y = C$ **11.** $\ln y + \dfrac{x}{y} = C$

13. $\ln(x^2 + y^2) = x + C$ **15.** $\ln\sqrt{x^2 + y^2} = \tan^{-1}(x/y) + C$
17. $(x/y) = (x^2/2) + C$ **19.** $-(xy)^{-2} = y^2 + C$

21. Separable: $\dfrac{1}{3}\ln y = -\dfrac{1}{2}\ln x + C$; exact: $y^2x^3 = C$. These answers are equivalent; the first may be rewritten as $y^{1/3} = C_1x^{-1/2}$ or $y^{1/3}x^{1/2} = C_1$ and raise both sides to the sixth power.   **23.** Separable: (b), (e), (h), (k), and (o); homogeneous: (c), (i), (p), (r), and (t); exact: (a), (d), (f), (j), and (m)

Section 18–5 (page 834)

1. $y = x - 1 + Ce^{-x}$ **3.** $y = x^4 + Cx^3$ **5.** $y = e^{-x}\tan^{-1}e^x + Ce^{-x}$
7. $y = -x^3 + Cx^2$ **9.** $y = -e^{2x} + Ce^{3x}$ **11.** Linear; $y = Ce^x - 5$
13. Homogeneous; $x^2 - y^2 = Cx$ **15.** Exact; $y^3 + 3xy - x^3 = C$

17. Separable; $\dfrac{1}{2}y^2 + \sqrt{1 - x^2} = C$ **19.** Homogeneous; $y = xe^{Cx}$

21. Linear; $y = (x + C)e^{-x}$ **23.** Linear; $y = (x + C)/(1 + e^x)$

25. Separable; $2s + s^2 = t^2 + C$ **27.** Linear; $y = \dfrac{4}{3}(1 + x) + C(1 + x)^{-2}$

29. Linear; $y = -1 + Ce^{x^2/2}$ **31.** Homogeneous: $(1 - (y/x))^{-1/2} = Cx$; exact: $yx - x^2 = C$; linear: $xy = x^2 + C$; all are equivalent   **33.** $3 + e^{-5}$ **35.** About 63%

37. (a) $I = \dfrac{E_0}{R - L}e^{-t} + \left[I_0 - \dfrac{E_0}{R - L}\right]e^{-Rt/L}$

(b) $I = \dfrac{E_0}{\sqrt{R^2 + (120\pi L)^2}} \sin(120\pi t - \alpha) + \left[I_0 + \dfrac{120\pi E_0 L}{R^2 + (120\pi L)^2} \right] e^{-Rt/L}$, where

$\alpha = \tan^{-1}(120\pi L/R)$ **39.** 5,298 ft/sec (about 1 mile/sec)

Section 18-6 (page 840)

1. $e^{-3x}(C_1 + C_2 x)$ **3.** $C_1 e^{4x} + C_2 e^{-3x}$ **5.** $C_1 \cos 2x + C_2 \sin 2x$ **7.** $C_1 + C_2 e^{-4x}$
9. $e^x(C_1 \cos x + C_2 \sin x)$ **11.** $e^x(C_1 + xC_2)$ **13.** $C_1 e^{2x} + C_2 e^{-3x}$
15. $C_1 \cos 2\sqrt{2}x + C_2 \sin 2\sqrt{2}x$ **17.** $e^x(C_1 \cos \sqrt{3}x + C_2 \sin \sqrt{3}x)$ **19.** $C_1 e^{(2+\sqrt{2})x/2}$
$+ C_2 e^{(2-\sqrt{2})x/2}$ **21.** e^{5x} **23.** $e^{4x}(3 + x)$ **25.** $4 \cos 2x + 3 \sin 3x$

Section 18-7 (page 846)

1. $-\dfrac{1}{7}e^x$ **3.** $\dfrac{1}{170}(\cos x + 4 \sin x)$ **5.** $\dfrac{1}{2}x^2 e^{2x}$ **7.** $\dfrac{1}{7}xe^{3x}$

9. $\dfrac{1}{40}e^{2x}(3 \sin 3x - \cos 3x)$ **11.** $[\ln(\sec x + \tan x)]\cos x$ **13.** $(\ln \sin x) - x$

15. $-e^{-x}\ln x$ **17.** $(C_1 + x)e^x + C_2 e^{-x}$ **19.** $C_1 \sin x + \left(C_2 - \dfrac{x}{2} \right) \cos x$

21. $-\dfrac{1}{7}e^x + C_1 e^{4x} + C_2 e^{-6x}$ **23.** $\left(\dfrac{1}{2}x^2 + C_1 x + C_2 \right) e^{2x}$

25. $\dfrac{1}{40}e^{2x}(3 \sin 3x - \cos 3x) + C_1 \cos x + C_2 \sin x$

27. $C_1 \sin x + [C_2 + \ln(\sec x + \tan x)]\cos x$

Section 18-8 (page 852)

1. $y = 4 \cos 2t$, $1/\pi$, 4 **2.** Critical damping, $y = 4e^{-2t}(1 + 2t)$ **3.** Overdamped,
$y = 2[(1 - \sqrt{5})e^{(-1-\sqrt{5})t} - (-1 - \sqrt{5})e^{(1-\sqrt{5})t}$ **4.** Underdamped,
$y = (8/\sqrt{3})e^{-t}[\cos(\sqrt{3}t - \pi/6)]$; $2\pi/\sqrt{3}$ **5.** $y = 25 \cos(2t - (\pi/2))$; 25
6. $I = C_1 \cos 2t + C_2 \sin 2t$ **7.** $I = C_1 \cos 2t + C_2 \sin 2t + \cos t$
8. $\theta = \theta_0 \cos\sqrt{2k/Mr^2}\,t$ **9.** $\phi = (Lt/B) - (L/B^2)(1 - e^{-Bt})$

Review Exercises for Chapter 18 (page 853)

1. $y = 2 \sin x + C \cos x$ **3.** $y = e^{4x}(C_1 + C_2 x)$ **5.** $\dfrac{1}{4}e^x + e^{-x}(C_1 + C_2 x)$

7. $(x^2 + 1)(3 - 2y) = C$ **9.** $\ln x + e^{-(y/x)} = C$ **11.** $y = \dfrac{1}{2}e^x + Ce^{-x}$

13. $y\sqrt{x^2 + y^2} + x^2 \ln(y + \sqrt{x^2 + y^2}) + y^2 = 3x^2 \ln x + Cx^2$
15. $y = e^{-x/2}(C_1 \cos(\sqrt{5}/2)x + C_2 \sin(\sqrt{5}/2)x)$ **17.** $y = C_1 \cos 2x$
$+ C_2 \sin 2x + \sin x$ **19.** $(x/y) + (x^2/2) = C$ **21.** $x^3 y = C - \cos x$ **23.** $x\sqrt{y} = C$

25. $y = C_1 e^x + C_2 e^{2x} + \dfrac{1}{12}e^{5x}$ **27.** $y = x \sin x + 2 \cos x - (2/x)\sin x + C/x$

29. $y = [Cx^3 + 2(x + 1)^3][(x + 1)^3 - Cx^3]^{-1}$ **31.** $y'' + \dfrac{b}{m}y' = g$

33. About 85 minutes **35.** 139 min **37.** $(1/2\pi)(\sqrt{3g/2R})$

INDEX

Production Supervision: Pam Price, Laurie Greenstein
Copy Editing: Michael Holmes
Proofreading: Victor Meyers
Text Design: Jon Fratis
Cover Design: Kenny Beck
Cover Photograph: Glen Wexler
Illustrations: Boardworks
Composition: Typothetae

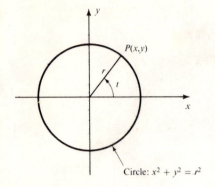

Circle: $x^2 + y^2 = r^2$

$$\sin t = \frac{y}{r}$$

$$\cos t = \frac{x}{r}$$

$$\tan t = \frac{y}{x}$$

$$\sin A = \frac{\text{side opposite}}{\text{hypotenuse}} = \frac{a}{c}$$

$$\cos A = \frac{\text{side adjacent}}{\text{hypotenuse}} = \frac{b}{c}$$

$$\tan A = \frac{\text{side opposite}}{\text{side adjacent}} = \frac{a}{b}$$

$$\cos^2 t + \sin^2 t = 1, \quad 1 + \tan^2 t = \sec^2 t, \quad \cot^2 t + 1 = \csc^2 t$$

$$\sin(-t) = -\sin t, \quad \cos(-t) = \cos t, \quad \tan(-t) = -\tan t$$

$$\cos t = \sin\left(\frac{\pi}{2} - t\right) \quad \sin t = \cos\left(\frac{\pi}{2} - t\right)$$

Addition Formulas

$$\sin(s \pm t) = \sin s \cos t \pm \cos s \sin t$$

$$\cos(s \pm t) = \cos s \cos t \mp \sin s \sin t$$

$$\tan(s \pm t) = \frac{\tan s \pm \tan t}{1 \mp \tan s \tan t}$$

Double-Angle Formulas

$$\sin 2t = 2 \sin t \cos t$$

$$\cos 2t = \cos^2 t - \sin^2 t = 2 \cos^2 t - 1 = 1 - 2 \sin^2 t$$

$$\tan 2t = \frac{2 \tan t}{1 - \tan^2 t}$$

Half-Angle Formulas

$$\sin^2 t = \frac{1 - \cos 2t}{2} \quad \cos^2 t = \frac{1 + \cos 2t}{2}$$

$$\tan t = \frac{\sin 2t}{1 + \cos 2t} = \frac{1 - \cos 2t}{\sin 2t}$$